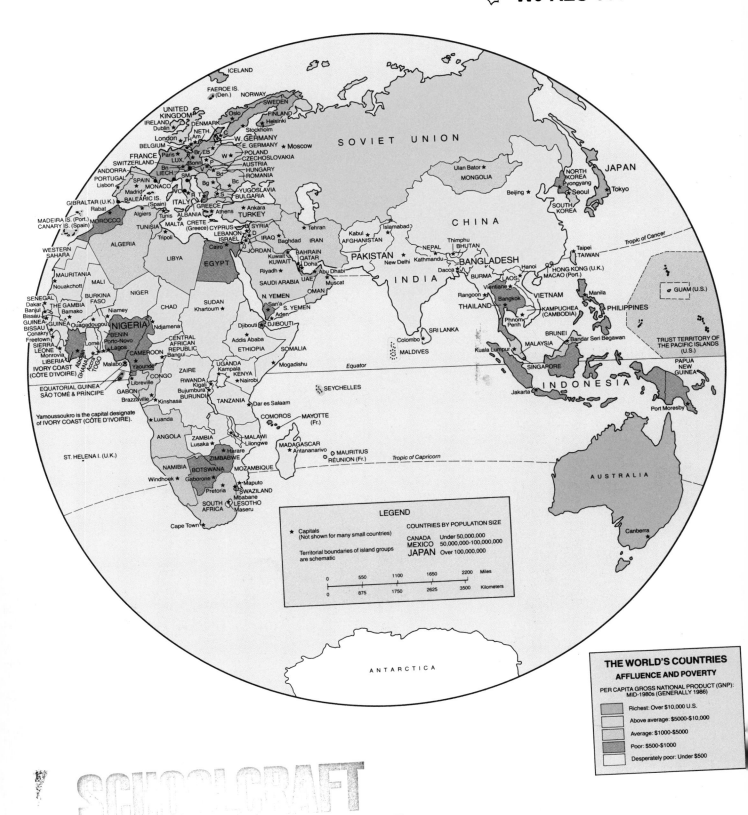

ICELAND
FAEROE IS. (Den.)
NORWAY
SWEDEN
FINLAND
Oslo
Helsinki
UNITED KINGDOM
IRELAND
Dublin
DENMARK
London
NETH.
Stockholm
W. GERMANY
Moscow
SOVIET UNION
BELGIUM
E. GERMANY
POLAND
FRANCE
Paris
LUX.
CZECHOSLOVAKIA
SWITZERLAND
Bonn
AUSTRIA
HUNGARY
ROMANIA
ANDORRA
LIECH.
Ulan Bator
MONGOLIA
NORTH KOREA
JAPAN
PORTUGAL
MONACO
YUGOSLAVIA
BULGARIA
Beijing
Pyongyang
Seoul
Tokyo
Lisbon
SPAIN
Madrid
ITALY
GREECE
SOUTH KOREA
GIBRALTAR (U.K.)
BALEARIC IS. (Spain)
ALBANIA
Athens
Ankara
TURKEY
CHINA
Rabat
Algiers
Tunis
MALTA
CRETE (Greece)
CYPRUS
SYRIA
Tehran
MADEIRA IS. (Port.)
MOROCCO
TUNISIA
LEBANON
ISRAEL
IRAQ
IRAN
Islamabad
Taipei
TAIWAN
Tropic of Cancer
CANARY IS. (Spain)
Tripoli
Baghdad
Kabul
AFGHANISTAN
NEPAL
Thimphu
BHUTAN
WESTERN SAHARA
ALGERIA
LIBYA
Cairo
EGYPT
JORDAN
Kuwait
QATAR
BAHRAIN
PAKISTAN
New Delhi
Kathmandu
BANGLADESH
HONG KONG (U.K.)
MACAO (Port.)
GUAM (U.S.)
MAURITANIA
MALI
NIGER
CHAD
SUDAN
Khartoum
Riyadh
Doha
Abu Dhabi
UAE
Muscat
INDIA
Dacca
BURMA
LAOS
Hanoi
Manila
NOUAKCHOTT
SENEGAL
Dakar
Banjul
THE GAMBIA
BURKINA FASO
Niamey
Ndjamena
SAUDI ARABIA
San'a
N. YEMEN
S. YEMEN
Aden
OMAN
Rangoon
Vientiane
Bangkok
VIETNAM
PHILIPPINES
GUINEA-BISSAU
Bamako
Ouagadougou
Djibouti
DJIBOUTI
THAILAND
Phnom Penh
KAMPUCHEA (CAMBODIA)
TRUST TERRITORY OF THE PACIFIC ISLANDS (U.S.)
Bissau
Conakry
GUINEA
BENIN
NIGERIA
CENTRAL AFRICAN REPUBLIC
Addis Ababa
SRI LANKA
BRUNEI
Freetown
SIERRA LEONE
Lome
Porto-Novo
Lagos
Bangui
ETHIOPIA
SOMALIA
Colombo
MALAYSIA
Bandar Seri Begawan
Monrovia
LIBERIA
IVORY COAST (CÔTE D'IVOIRE)
GHANA
ACCRA
TOGO
CAMEROON
Malabo
Yaounde
UGANDA
Kampala
KENYA
MALDIVES
PAPUA NEW GUINEA
EQUATORIAL GUINEA
SÃO TOMÉ & PRÍNCIPE
Libreville
GABON
CONGO
ZAIRE
RWANDA
Kigali
BURUNDI
Bujumbura
Nairobi
Mogadishu
Equator
SINGAPORE
INDONESIA
Yamoussoukro is the capital designate of IVORY COAST (CÔTE D'IVOIRE).
Brazzaville
Kinshasa
TANZANIA
Dar es Salaam
SEYCHELLES
Jakarta
Port Moresby
Luanda
COMOROS
MAYOTTE (Fr.)
ST. HELENA I. (U.K.)
ANGOLA
ZAMBIA
Lusaka
MALAWI
Lilongwe
MADAGASCAR
Antananarivo
MAURITIUS
RÉUNION (Fr.)
Tropic of Capricorn
Harare
ZIMBABWE
MOZAMBIQUE
AUSTRALIA
NAMIBIA
BOTSWANA
Windhoek
Gaborone
Maputo
Pretoria
SWAZILAND
Mbabane
LESOTHO
Maseru
Canberra
SOUTH AFRICA
Cape Town

ANTARCTICA

LEGEND

★ Capitals
(Not shown for many small countries)

Territorial boundaries of island groups are schematic

COUNTRIES BY POPULATION SIZE

CANADA — Under 50,000,000
MEXICO — 50,000,000-100,000,000
JAPAN — Over 100,000,000

| 0 | 550 | 1100 | 1650 | 2200 | Miles |
| 0 | 875 | 1750 | 2625 | 3500 | Kilometers |

THE WORLD'S COUNTRIES

AFFLUENCE AND POVERTY

PER CAPITA GROSS NATIONAL PRODUCT (GNP): MID-1980s (GENERALLY 1986)

Richest: Over $10,000 U.S.

Above average: $5000-$10,000

Average: $1000-$5000

Poor: $500-$1000

Desperately poor: Under $500

SOME BASIC PATTERNS OF WORLD ECONOMIC GEOGRAPHY

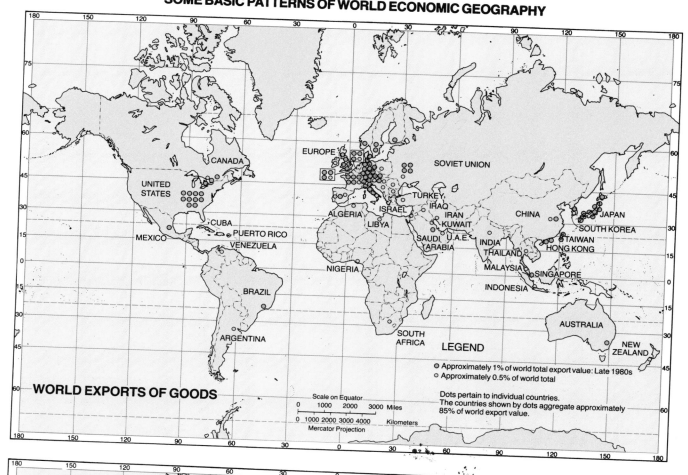

WORLD EXPORTS OF GOODS

LEGEND

⊙ Approximately 1% of world total export value: Late 1980s
○ Approximately 0.5% of world total

Scale on Equator
0   1000   2000   3000 Miles
0 1000 2000 3000 4000 Kilometers
Mercator Projection

Dots pertain to individual countries.
The countries shown by dots aggregate approximately
85% of world export value.

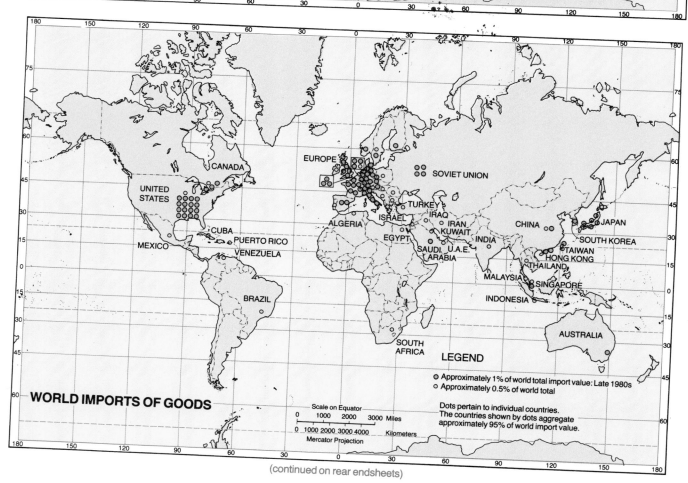

WORLD IMPORTS OF GOODS

LEGEND

⊙ Approximately 1% of world total import value: Late 1980s
○ Approximately 0.5% of world total

Scale on Equator
0   1000   2000   3000 Miles
0 1000 2000 3000 4000 Kilometers
Mercator Projection

Dots pertain to individual countries.
The countries shown by dots aggregate
approximately 95% of world import value.

(continued on rear endsheets)

WORLD REGIONAL GEOGRAPHY

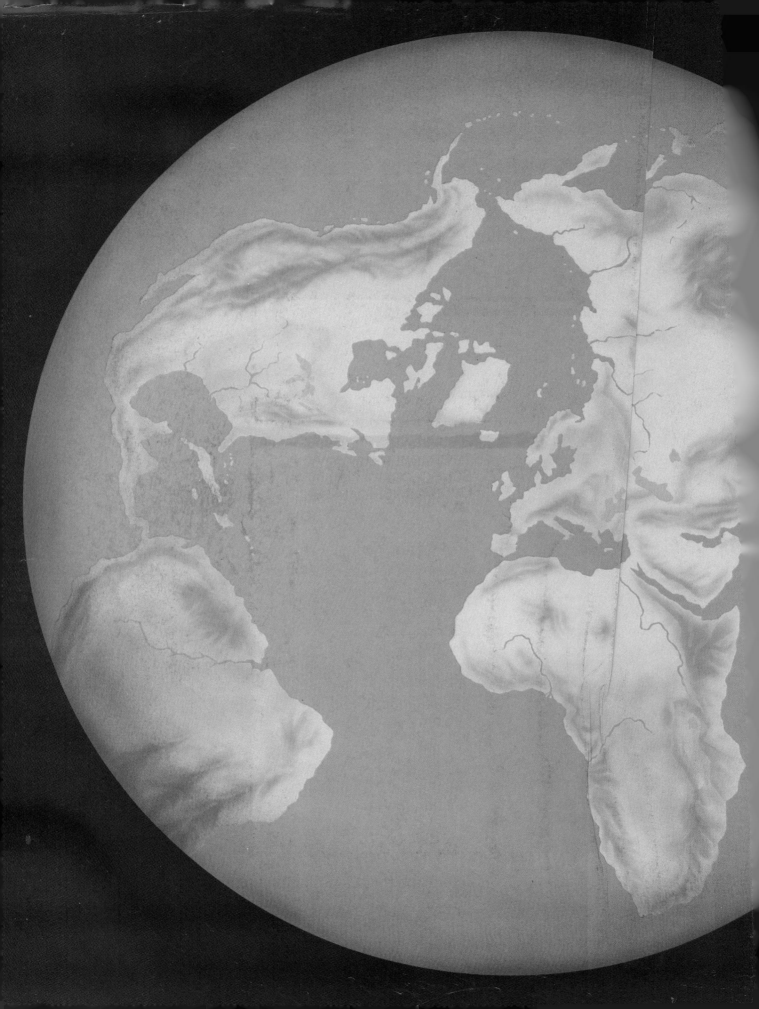

WORLD REGIONAL GEOGRAPHY

Jesse H. Wheeler, Jr.
University of Missouri-Columbia

J. Trenton Kostbade
University of Missouri-Columbia

 SAUNDERS COLLEGE PUBLISHING

Philadelphia Ft. Worth Chicago
San Francisco Montreal Toronto
London Sydney Tokyo

Text Typeface: Meridien
Compositor: Waldman Graphics, Inc.
Acquisitions Editor: John Vondeling
Developmental Editor: Maureen Iannuzzi
Managing Editor: Carol Field
Project Editor: Maureen Iannuzzi and Becca Gruliow
Copy Editor: Ann Blum
Manager of Art and Design: Carol Bleistine
Art and Design Coordinator: Doris Bruey
Text Designer: Adrianne Dudden
Cover Designer: Lawrence R. Didona
Text Artwork: Tasa Graphic Arts, Inc.
Layout Artist: Dorothy Chattin
Director of EDP: Tim Frelick
Production Manager: Bob Butler

Cover Credit: Paul Ambrose Studios/Fran Heyl Associates

Printed in the United States of America

World Regional Geography

ISBN: 0-03-005371-4

Library of Congress Catalog Card Number: 89-042922

0123   032   987654321

Preface

MAJOR OBJECTIVE OF THE BOOK

World Regional Geography is an explanatory geographic survey of eight major world regions. The book builds on the foundations of a widely used older text, *Regional Geography of the World*, which went through three editions published by Holt, Rinehart and Winston between 1955 and 1975. Massive rewriting, revamping, and updating under the direction of Saunders College Publishing has produced a volume we believe to be a far more flexible and attractive instrument for teaching and learning world regional geography on an introductory college level. The major objective continues to be the presentation of basic concepts and supporting facts about contemporary world geography for the general education of college students.

Emphasizing both human and physical geography, the book surveys each region as to location, component countries, world role, distinctive physical and cultural characteristics, relation to other world areas, and major problems. Major attention is also given to important individual countries and groups of countries within each world region. Although the principal aim is general education in world geography to help remedy the serious geographic illiteracy that many recent surveys of the American public have revealed, the material in the book is also extremely valuable preparation for students who may be attracted to become geography majors.

ALLIED OBJECTIVES

Although the book is centrally concerned with the specific geographical circumstances of each major world region and its major countries, much material of a more generalized kind is introduced and discussed as a means of achieving the following allied objectives:

- To give a general introduction to *geographical ideas*—the ways of thinking that characterize the modern discipline of geography. Some of these ideas are embodied in short quotations from prominent geographers in a prefatory section entitled "Introductory Views."

- To introduce *regional interpretation and analysis*—how one goes about the study of an area from a geographic point of view.

- To introduce *maps and their uses*—the kinds of maps, map projections, map scale, the design of maps to portray geographical ideas, and the importance of maps as part of one's intellectual equipment.

- To present certain *world patterns of distribution* (population, landforms, climate, economic activities, and so on) as context for the characterization of the major world regions and important countries that are the main focus of attention. These maps also foster global thinking on the part of students.

- To introduce and explain some of the *natural and human processes* that underlie the areal differentiation and spatial order of the world.

- To discuss *current world problems* such as rapid population increase in "developing" areas, food and water supply, employment and income, health, housing, energy supply, environmental pollution, cultural and political conflict, and runaway urbanism.

- To introduce some generalized ideas and specific examples of *current world development*.

- To give an opportunity for students to learn something of the nature and opportunities of *geography as a professional field*. This is done most explicitly in a section at the end of Chapter 1.

- To provide some acquaintance with *geographical bibliography*—where to find material of a geographical character. This is done in a section of up-to-date Readings and References.

OVERALL ORGANIZATION AND SCOPE

The book surveys eight major world regions: Europe, the Soviet Union, the Middle East (Southwest Asia and North Africa), the Orient (Indian subcontinent, Southeast Asia, and East Asia), the Pacific World (Oceania, including Australia and New Zealand), Africa (the continent and some Indian Ocean and Atlantic islands), Latin America, and Anglo-America (the United States and Canada, with Greenland as a marginal appendage). Within each major region, the individual countries are often grouped regionally for analysis, and particularly large or influential countries are treated in greater depth than others. Pertinent background material on systematic physical and cultural geography is presented at appropriate places, including sidebars entitled "Editorial Comments" (for examples, see the sidebars entitled "Culture" (page 9) and "Glaciation" (page 91)). A general introduction to geography as a field of learning is presented in Part 1, "Thinking About Geography."

LEVEL OF LANGUAGE

In general, the book is written in nontechnical language for undergraduate college and university students, many of whom have had no previous formal training in the discipline of geography. However, formal geographical terminology is introduced and explained at a level appropriate to an introductory college geography text. Readability is enhanced by short, well-written quoted selections and by varied approaches to the individual regions and countries. The book considers each region and country on its merits, and basic concepts and facts are presented in a manner appropriate to the individual area rather than under a set list of topics.

ADAPTABILITY TO VARIED COURSES

This book has been explicitly designed to be a flexible instrument for the teaching of college-level courses in world regional geography. Such courses are extremely varied in length, outlook, organization, and students. Thus this textbook offers options in content and organization to suit differing needs. Every section and subsection is designed to be intelligible on its own account. Hence teachers can mix and match materials as they choose to fit their purposes and needs.

The authors cannot overstress their attempt to make the book sufficiently flexible to fit many course structures, depths, and sequences. Some teachers may wish to follow the present order of chapters, while others will want to vary the order. A major theme woven into the various sections of the book is the worldwide impact of Westernization, beginning in Europe and spreading from there to other regions. In conformity with this theme, Europe comes first in the series of regional chapters, but the book has been organized in such a way that it is perfectly feasible to begin a course with any of the other regions. The structuring of the individual regional chapters is such that any given chapter or major subsection of a chapter is intelligible without excessive dependence on preceding material.

If time does not permit the entire book to be covered, selected parts may easily be combined to provide material of the appropriate length. For example, teachers may wish to assign the general introductory chapters on the various regions, together with selected other chapters or parts of chapters, to provide a very general survey of the world and intensive case studies of certain countries or regions. Teachers may wish to assign certain chapters to be discussed in class, may assign other chapters to be read without class discussion, and may wish to lecture on the subject matter of some chapters without assigning them. General introductory material in Chapters 1 to 3 may or may not be assigned, according to the amount of previous training in geography that students have received.

MAPS AND PHOTOS

Nearly all of the place names mentioned in the text are shown on one or more of well over 100 full-color maps. Map colors were chosen carefully to be pedagogically effective as well as pleasing to the eye. A considerable number of maps showing the world or regional distribution of population, cultivated land, climate, major crops, minerals, industrial concentrations, and other elements of geographical significance are included. The maps have been designed, and 200 photos (all but a few in full color) have been chosen, to fit the text matter, and all maps and photos are keyed to the reading. As a result, instructors may not find it necessary to require students to purchase an atlas.

FEATURES OF THE TEXT
Chapter Titles and Subheads

We have tried to provide chapter titles and subheads that are evocative and dynamic, for example, the title of Chapter 1, "Learning to Think Geographically," or the subhead "Northern Mexico: Localized Development in a Dry and Rugged Outback." The numerous subheads form a running outline of the text matter, and they facilitate assignments.

Explanation and Summation by Numbered Points

Student comprehension of the material is facilitated in many places by the use of numbered points for explanations, summaries, and comparisons. Examples include an explanation of Europe's worldwide expansion in the colonial age (pages 71–72), some reasons for recent economic difficulties in Britain (pages 130–131), some general characteristics of the Orient as a major world region (page 353), and some shared traits of the countries of Tropical Africa (pages 498–504).

International Focus

The book contributes explicitly to international understanding by generalized discussions of world geographic distributions and problems, and the presentation of current world events in their geographic contexts. Hence there is conformity to recent nationwide emphasis on greater internationalization of college and university curricula. A subsection entitled "The Geographic Effects of Places on Each Other" (pages 55–56) stresses the interconnectedness of places and thus contributes to global thinking.

Material from Allied Disciplines

General education is fostered by the frequent use of interpretive historical or other material from allied disciplines. Use of such material is very selective, but the authors believe the total coverage contributes substantially to a general understanding of the modern world.

Use of Statistics

Recent statistics are used selectively and comparatively to underscore important concepts about places and areas. Databases on such topics as area, population, health, income, trade, production, and climate are provided in tables of data for regions and countries at relevant places. Both non-metric data and metric data are cited in text and tables, with non-metric data cited first and metric equivalents (enclosed in parentheses in text matter) following.

References and Readings

A References and Readings section at the end of the book emphasizes books and articles dating from the later 1970s or the 1980s. The titles are grouped by major world region except for an introductory general section on the discipline of geography and its systematic subfields. Some exceptionally well-written recent articles from such magazines as *Fortune* or *The Economist* are cited in addition to numerous articles and books from the professional literature of the field of geography and allied fields.

ANCILLARIES

The following ancillaries accompany this text and are available upon request from the publisher:

- An Instructor's Manual containing suggestions for alternative course structures and the use of maps as teaching and study aids.
- A Study Guide to aid students in self-evaluation on text material.
- Four-Color Transparencies of more than 100 maps from the text.
- 35-mm Slides of maps and photographs.
- A Test Bank featuring a large selection of multiple-choice questions keyed to text.
- A Computerized Test Bank with over 1000 multiple-choice questions (Macintosh or IBM compatible).
- A Map-Pak containing unlabeled maps to be used in class or as an independent-study aid.

AUTHORSHIP

The work of two authors is very intermixed. Wheeler was responsible for general editing, most cartography, the selection and captioning of photographs, and preparation of References and Readings. Wheeler wrote most of Chapters 1 and 3 and the sections on East Central Europe, the Soviet Union, the Middle East, Africa, Introduction to the Orient, Introduction to Latin America (with extensive assistance by Brian K. Long) and Caribbean America (assisted by Steven F. Fair). Kostbade wrote Chapter 2, the Europe chapters except the section on East Central Europe, and the chapters on the Indian Subcontinent, Southeast Asia, the Chinese Realm, Japan and Korea, the Pacific World, the regions of Latin America (except Caribbean America), and Anglo-America. But all the chapters represent a close collaboration of the two authors, with each contributing to the other's chapters in significant ways. Some material written originally by Richard S. Thoman has been retained here and there in Chapters 7, 24, and 25, woven into the revised chapters' structures.

ACKNOWLEDGMENTS

We would like to thank the many professors who read and critiqued the manuscript or otherwise assisted in the preparation of the book. Of the authors' own colleagues at the University of Missouri–Columbia, Joseph H. Astroth, Jr., read the chapter on the Indian Subcontinent in galley proof and gave invaluable assistance on cartographic matters; Christopher L. Salter read the chapters on the Chinese Realm and Japan and Korea in galley proof; Walter A. Schroeder, Joseph J. Hobbs, and Steven F. Fair classroom-tested extensive portions of the manuscript; Gail S. Ludwig commented on the book from the standpoint of geographic education; and Joseph H. Astroth, Jr., William A. Noble, and Gary E. Johnson supplied some fine photographs of Asia and the Pacific World. Walter A. Schroeder made the original sketches for the diagrams in Chapter 2.

Many professors from a variety of universities and colleges critiqued the entire book in manuscript. They include Peggy Alexander, Jefferson College; Thomas D. Anderson, Bowling Green State University; Marvin Baker, University of Oklahoma; C. Taylor Barnes, United States Air Force Academy; Stephen S. Birdsall, University of North Carolina at Chapel Hill; Allen D. Bushong, University of South Carolina; David M. Eldridge, College of DuPage; Harold M. Elliott, Weber State College; James C. Hughes, Slippery Rock University; John C. Lewis, Northeast Louisiana University; Gordon R. Lewthwaite, California State University, Northridge; Ronald E. Nelson, Western Illinois University; Richard Pillsbury, Georgia State University; Milton D. Rafferty, Southwest Missouri State University; Gregory Rose, Ohio State University–Marion; Rose Sauder, University of New Orleans; and Ronald V. Shaklee, Youngstown State University. All suggestions from this varied group of reviewers were carefully considered and many were heeded. We, of course, bear full responsibility for the result. We greatly appreciate the time, care, and perceptiveness devoted to the reviewing task by the critics.

We gained valuable insights and perspectives on the book from the standpoint of classroom teaching through innumerable conversations with graduate teaching assistants in the large introductory-level "Regions and Nations of the World" courses at the University of Missouri–Columbia. Special acknowledgement is made to the perceptive comments of certain recent teaching assistants, including Ann Wright McInerny, H. Todd Stradford, Jr. (who also supplied some of the book's finest photographs), Russell Ivy, Virginia Thompson, Elizabeth Cook Gardner, and Julia Goodell Barnes, but we are indebted to a long line of other talented teachers far too numerous to cite individually.

Special tribute must be paid to Mary A. Wilson, who typed the manuscript in several versions. Her superb word-processing skills were critically important as the text moved through various stages over a period of years. She worked with enthusiasm and aplomb, often under great pressure of time. Katie D'Agostino must also be specially recognized for her work with the tables. Her painstaking work with varied statistical sources is appreciated. Jennifer Nichols assisted with cartography, particularly computerized cartography under the guidance of Joseph H. Astroth, Jr. Their joint efforts created a fine double-page panel of maps on the agricultural geography of the United States. Mary Lyon compiled and drew the base for the endpaper map of countries. Richard Egan checked place names on maps, and Norman Fry updated the maps of Soviet minerals. The original maps for the panels on London and New York were executed by Dorothy Woodson and Max Gilland, respectively. Margery Wheeler gave perceptive assistance with various writing tasks, and David Wheeler made helpful judgments on African development problems from an economist's perspective. We are grateful for their help, as well as that of many other people not named. The profoundest debt is owed to our wives and children for their extraordinary support and forbearance over the long period that the book was in preparation.

Full thanks are expressed to John Vondeling, Associate Publisher, who supported the project enthusiastically; Kate Pachuta, Associate Editor, who assisted with reviews and permissions; Maureen Iannuzzi and Becca Gruliow, Project Editors; Carol Bleistine, Manager of Art and Design; Carol Field, Managing Editor, and others who were associated with the project at Saunders College Publishing. Dennis Tasa of Tasa Graphic Arts, Inc., Tijeras, New Mexico, applied superb cartographic skills to the production of a memorable series of four-color maps from bases developed originally by Jesse H. Wheeler, Jr.

Thanks are also due the people, firms, and government agencies who permitted the reproduction of textual matter, photographs, and maps. Such permissions are separately acknowledged where they occur in the text. In editing the varied readings, mostly short, that are sprinkled through the text matter, minor changes in spelling, capitalization, and punctuation have been made to give uniformity to the book as a whole.

We hope this book can supply useful geographical perspectives for college and university students of the 1990s who must cope with the increasingly complex geography of an interdependent world.

Jesse H. Wheeler, Jr.
J. Trenton Kostbade

Columbia, Missouri
July, 1989

List of Maps

Contents

PART I THINKING ABOUT GEOGRAPHY

3 Key Topics in the Geographic Interpretation of Countries and Regions   46

PART II   EUROPE

4 Geography and Human Experience in Europe   60

5   Environmental and Historical Factors in Europe's Economic Success 80

6   Changing Fortunes in the Varied British Isles 104

9 Geographic Personalities of Europe's Outer Regions 184

PART III   THE SOVIET UNION

PART IV   THE MIDDLE EAST

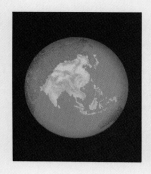

PART V THE ORIENT

PART VI THE PACIFIC WORLD

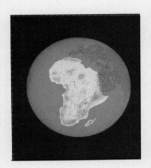

PART VII   AFRICA

PART VIII   LATIN AMERICA

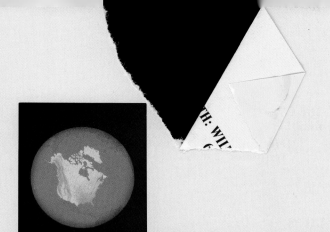

PART IX   ANGLO-AMERICA

PART I

THINKING ABOUT GEOGRAPHY

Introductory Views

"A liberal education should develop in each individual the realization that his own country, region, and ethnic, religious, or linguistic group is but one among many, each with differing characteristics, and that other countries, regions, or social groups are not necessarily queer, or irrational, or inferior. It may be argued that one cannot see his own country and culture in perspective until he has studied other lands and peoples. Only then can one realize that his own civilization is but one of a family of civilizations with common elements yet distinctive characteristics, evolving through time from common antecedents with differentiation but with much cultural borrowing, facing similar problems yet with particular combinations of attitudes, policies, technologies, climates, soils, minerals, and evolved economic systems."

Chauncy D. Harris, "The Geographic Study of Foreign Areas and Cultures in Liberal Education," in *Geography in Undergraduate Liberal Education* (Washington, D.C.: Association of American Geographers, 1965), p. 25.

"Geography is an integrative subject first and last, dealing with things in association, not merely elements taken separately or items listed in inventory form."

Robert S. Platt, "Introductory Field Study," in *Perspective in the Study of Geography* (University of Chicago, Department of Geography, 1951), p. 12.

"The two great themes of geography are the relationship between nature and society and the relationship between location and society."

B. L. Turner II, "Bring Geography Back to American Universities," *The Christian Science Monitor*, November 19, 1987, p. 15.

"In historical think...ng there has been great debate about whether history is a science or an art. It was perhaps an unnecessary debate because history, like geography, is both. Geography is a science in the sense that what facts we perceive must be examined, and perhaps measured, with care and accuracy. It is an art in that any presentation (let alone any perception) of those facts must be selective and so involve choice, and taste, and judgment."

H. C. Darby, "The Problem of Geographical Description," Institute of British Geographers, Publication No. 30, *Transactions and Papers,* 1962, p. 6.

"As a science, geography rests on observation, insists upon measurement, calls for classification, and is finally judged by the intellectual probity of its explanation."

Hugh C. Prince, "The Geographical Imagination," *Landscape,* Winter 1961–1962, p. 23.

"Good geographical writing quite definitely is an art, and the highest form of the geographer's art is writing evocative descriptions that facilitate an understanding and an appreciation of places, areas, and regions."

John Fraser Hart, "The Highest Form of the Geographer's Art," *Annals,* Association of American Geographers, 72 (1982), p. 27.

"Land and life is what geography has always been about."

D. R. Stoddart, "To Claim the High Ground: Geography for the End of the Century," Institute of British Geographers, *Transactions New Series,* 12 (1987), p. 334.

CHAPTER 1

Learning to Think Geographically

London on the River Thames in southeastern England is a major node in the worldwide system of places that geographers study. Tower Bridge is at the lower left, with new London Bridge next upstream. The financial district (City of London) is at the lower right.

THE main reason for studying geography is to gain a better understanding and appreciation of the world in which we live. Of course, many subjects other than geography contribute to this end. Geography shares a common fund of information with them, but each subject has its own ways of thinking about the world and of putting its findings to use. Geography's thinking is essentially *spatial:* The subject looks at things as they associate themselves spatially in the world and its parts, and it tries to discern *spatial order* in the heterogeneous array of things that covers the globe. No matter how the field is defined—study of areas, study of relationships between people and environment, study of spatial organization—geography is always characterized by a desire to find and understand the underlying spatial order and associations of things. *Applied geography* attempts to use geographic understandings in practical ways, for example, to help solve locational problems in the field of planning.

THE DAILY REALITY OF GEOGRAPHY

Geography is not just a body of knowledge developed by geographers and taught in schools or applied to the solution of human problems. Geography is an important part of the reality we experience daily. It lies all about us, conditions our lives, and, indeed, includes us. We see its variety when we travel, and we see its effects when we follow the news. Each of us has some conception of what is meant by the "geography" of areas; that is, we are aware of land, people, climate, vegetation, structures, and so on. But the geography we experience has so many components that we have difficulty in comprehending it in an orderly way. Unless we have learned to think geographically, the things we perceive are apt to seem a rather formless jumble, and some important elements may escape our notice. How, then, may we bring our perceptions into better order? Of what does an area's geography consist? One thing is *layout* or *configuration*—how different features are arranged in relation to each other. For example, in the prologue to his great trilogy of novels called *U.S.A.,* John Dos Passos described the layout of the United States as "the world's greatest rivervalley fringed with mountains and hills." The country might also be described as a ring of great metropolitan areas surrounding the world's most productive block of farm land. The essential layout of the entire world could be summarized as a single

expanse of ocean surrounding a few continental masses, of which by far the largest in area, population, and resources is the land mass comprising Eurasia and Africa. At the opposite extreme is the layout of the "personal space" in which our life routines are set. Maps and mental images of layout, from a world scale to an intimate local scale, help us feel at home in the world by providing a spatial framework within which to fit the daily flow of events that affect our lives.

But the geography of an area includes far more than simple layout. It also encompasses *interaction* of many kinds. Natural elements within areas interact with each other to create the distinctive natural environments that are settings for human activity. People interact with each other and with nature as they use, organize, and modify the earth. Such interactions not only are a part of geography, but they constantly create new geography. The results are often visible in the changing *look of the land*—still a third facet of geographic reality.

DIFFERENCES, SIMILARITIES, REGULARITIES, AND INTERACTIONS ON THE EARTH: THE BASES OF GEOGRAPHIC STUDY

Human attempts to portray and understand geographic phenomena are very old. In the ancient world, the field of geography first developed as a normal outgrowth of the curiosity of people regarding lands which were different from their own. As knowledge of the world increased through discovery and exploration, an almost endless variety of landscapes and modes of life was found to exist from place to place over the world. Scholars were intrigued by these differences in areas, and they sought to describe and account for them. Thus through observation, description, and analysis grew the modern field of geography. Today the impact of modern technology and modern forms of economic, social, and political organization is bringing about far-reaching changes in all parts of the world. It is tending to make various parts of the world more similar, at least outwardly. Yet striking differences among areas continue to exist. Countries and regions continue to differ widely in location, in world impact, in types of people and combinations of resources, and in ways and levels of living. They differ in physical character and political character, in economic mainstays, in internal arrangement and organization, in external connections and relationships, in landscapes, in potentialities, and in problems. The understanding of such differences continues to be a major objective of geographic study. But the study of geography is also motivated by the desire to discover and understand similarities and regularities that exist in otherwise diverse places and the interactions of peoples and places with each other. We turn now to a major intellectual tool for geographic work: the concept of spatial associations.

RECOGNIZING SPATIAL ASSOCIATIONS

To think geographically about the varied phenomena on the face of the earth is to think in terms of *spatial associations*. This is sometimes called the "spatial approach" or "spatial tradition" in the field of geography; but rightly considered, all geographic work is spatial in some sense. The geographer learns to look for the locational (spatial) and functional relationships of things and to view associated things as composites. Such composites (sets of geographic features) are spatial associations. They may be comprised of things which are located together or things that lie apart but have functional interconnections. A rock type, a vegetation type, and the soil type formed from the rock and vegetation, all occupying the same area, make up a spatial association. The same is true of an irrigated valley, a snowclad mountain range (which may be hundreds of miles away), the river that brings irrigation water from mountain snow fields to the valley, and the city whose businesses and institutions service the agricultural valley. There are countless examples of spatial associations. The important thing, however, is not to enumerate them, but to learn to recognize them and to get into the habit of thinking of the earth and its parts as being comprised of them (Figs. 1.1 and 1.2).

Figure 1.1   *The Merced River, shown here in Yosemite National Park, is one of numerous streams carrying snow meltwater from the high Sierra Nevada to the floor of California's Central Valley. The deepest snows of the United States accumulate in the Sierra. They are formed of moisture brought to the mountains by westerly winds off the nearby Pacific Ocean. The melting snows of these and other Western mountains have provided a vital flow of water for California's huge economic and urban development.*

Figure 1.2 *The California Aqueduct, shown in this view, is a relatively new conduit carrying water southward from the Sierra Nevada and other mountains to the Los Angeles region. Some of the water goes to the outlying Los Angeles satellite of Palmdale (top of view),* an important center of aerospace industries and space flight based on nearby Edwards Air Force Base. Together, Figures 1.1 and 1.2 provide a fine example of a *spatial association interconnecting humid mountains with intensively developed lowlands dependent upon mountain water.*

SPATIAL ASSOCIATIONS OF PEOPLE AND ENVIRONMENT

Suppose one flew eastward along the Tropic of Cancer from the Atlantic Ocean to India. For thousands of miles across Africa and southwestern Asia, little would be glimpsed except the barren landscape of the world's greatest deserts. Then, over India, the pattern would change to a mat of tiny crop fields speckled with innumerable villages. Explanation of the change would obviously involve spatial associations among climate, agriculture, and settlement. The study of such associations between humanity and the physical environment—often called the "man-land" or "ecological" approach—has been a major focus of interest in the field of geography from the very beginning. This way of thinking about spatial order is concerned with understanding the physical environment itself, the culturally conditioned *perception* and uses of that environment by human groups, and the spatial results of those uses. The approach is exemplified frequently in this text. Some illustrations that come easily to mind are found in discussions of the use of resources in different areas, the pressure of population on resources in accordance with varying population densities and technologies, the preferences for upland settlement in some areas and lowland settlement in others, the problems of using subarctic and tropical soil types, and questions of exploitation of mineral resources in "developing" countries for use in "developed" countries.

Early ecological investigations by geographers, including many studies in the early 20th century, were concerned less with the impact of people on the environment than with the environment's impact on people. Such an impact was assumed, and the geographer then set out to demonstrate it and elucidate it. This school of thought has been called *environmental determinism.* It is generally rejected in modern geography, which is interested both in the environmental conditions of human life and in humanity's impact on the environment.

Cultural Effects on Environmental Relationships

The interrelations of people with their environment are conditioned by (1) their culture, which determines many of their felt needs; (2) their technological capacities; and (3) their attitudes toward and perceptions of nature. Resources may be thought of as aspects of nature that people can use; which parts of nature they find useful depends upon their culture.[1] Thus, the members of a tribe of hunter-gatherers in a tropical forest do not see their environment as presenting the same difficulties and potentials that may be seen in the same area by an American mining engineer. The tribesmen and the engineer will see different resources and will have different technologies for exploiting them. So far as the inventions which make aspects of nature usable—that is, make resources of them—spring from the cultural back-

---

[1]For the guidance of students who have had no previous course work in college or high school geography, a limited number of important technical terms are explained in sidebars (Editorial Comments) such as the one concerning culture.

Editorial Comment

CULTURE

The concept of *culture* is much used in geography, as in other disciplines that study people and their works. It is a very useful concept, but so inclusive that scholars do not agree on any exact definition. A group's culture includes the values, beliefs, and aspirations, modes of behavior, social institutions, knowledge, and skills which are transmitted and learned within the group. It also includes a material aspect— the group's material possessions and products. Important for geography is the fact that people see their environment, and relate to it, in terms of their culture. Thus, two human groups with different cultures may see the same environment as offering quite different opportunities and difficulties and may relate to it in quite different ways. In so doing, each group will alter the environment in its own way, thereby creating a distinctive *cultural landscape* (the landscape as modified by people). Cultures, of course, are constantly changing. They change by the adoption of inventions (material and nonmaterial) made by their own members, and by the learning and adapting of traits from the cultures of other groups with whom they come in contact. Many of the characteristics of regions of the earth are better understood when one keeps in mind the concept of cultures as distinctive group ways of life (which include distinctive ways of looking at and using the environment), whose elements often spread to modify other cultures and cultural landscapes.

grounds of the inventors and users, culture may be said to **create** resources, whose exploitation then has very real effects on the natural environment being used.

LEARNING THE LANGUAGE OF MAPS

The map is the geographer's primary tool for investigating spatial order on the earth. By and large, the field of geography is concerned with things that can be mapped. A map is a portrayal of location: It shows where things and places on the earth are located in relation to each other. Thus, maps are indispensable tools for discovering, examining, and portraying spatial layouts and associations. They are also indispensable records of humanity's exploration, occupation, use, and organization of the earth. They provide a vital context for comprehending each day's news, as well as for understanding many works of literature and scholarship, and they are valuable aids for helping us plan the future. No agency or company concerned with planning facilities and operations which involve location can accomplish its purpose properly without the use of maps.

Maps can be used, at least in theory, to portray almost anything that varies from one place to another. Some maps show the distribution and interrelations of things which are fixed in place—structures, terrain features, and so on—while other maps portray the flow of people, goods, and ideas from place to place. A map communicates locational facts and concepts to us by means of a specialized "language" that has to be mastered if the map is to be properly understood and used. This language has several vital elements, which we may designate as *scale, projection, classification and symbolization,* and *design.* Each of these is examined below.

MAP SCALE

A map is the opposite of a microscope. Microscopes enlarge things that otherwise would be too small to be studied. A map, by contrast, is a reducer; it enables us to comprehend an extent of earth-space by reducing it to the size of a single sheet. The amount of reduction is indicated by the *scale,* that is, the actual distance on the earth that is represented by a given linear unit on the map. A common way of denoting scale is to use a *representative fraction* such as 1/10,000 or 1/10,000,000, wherein one linear unit on the map (inch, centimeter, etc.) equals 10,000 or 10,000,000 such units on the ground. A large-scale map is one with a large representative fraction (for example, 1/10,000 or even 1/1000); a small-scale map has a small representative fraction (for example, 1/1,000,000 or 1/10,000,000). A large-scale map permits more detail to be shown for a mapped area, whereas small-scale maps permit fewer details and require an increasing degree of generalization as the scale of the map decreases. On a map with a scale of 1/5000, a house can be represented in its true dimensions by a black square; on a map with a scale of 1/5,000,000, it would be impossible to represent a single house at true scale. In order to convey relative magnitudes effectively, symbols on a small-scale map (for instance, symbols denoting cities of different sizes) often occupy disproportionately large portions of the map sheet. Often it proves desirable to highlight selected items and places by using a connected series of maps at successive scales. An example in this text would be the panel of maps showing New York City in its regional setting at three different scales (see Fig. 28.8). Examination of the panel shows that as the scale grows larger, an increasing number of statements about Manhattan Island becomes possible.

MAP PROJECTIONS

The term *map projection* has a rather arcane ring, and it is true that the construction of projections is a complex matter. But the major ideas that underlie them are not difficult to grasp. A projection is simply a way to minimize distortion in one or more properties of a map (direction, distance, shape, or area). All maps introduce a certain amount of distortion, because it is not possible to represent the curved surface of the earth with complete accuracy on a flat sheet of paper. One can no more do it than one can flatten the skin of an orange without tearing it. If the area being represented is very small, the distortion may be slight enough to be disregarded, but maps that represent larger areas may introduce very serious distortions. A classic instance is the well-known *Mercator projection* (see Fig. 24.1, inset), originally designed as an aid to navigation. On the projection, every straight line has a constant bearing (direction). To achieve this purpose, both the parallels of latitude and the meridians of longitude are shown as straight lines. On a globe, the lines of longitude draw closer together from the equator to the poles, where they converge, but on a Mercator map these lines remain parallel. For this reason, a Mercator map greatly exaggerates the east-west dimension of areas near the poles. In addition, it exaggerates the north-south dimension, since the parallels are not spaced evenly between the equator and the poles, as they are on a globe, but gradually draw farther and farther apart with increasing distance from the equator. On a Mercator map, the *mathematical location* (latitude and longitude) of every place is accurate (see Editorial Comment: Latitude

and Longitude), but areas near the poles are stretched out of shape and are greatly distorted in size, with the familiar result that Greenland appears bigger than South America, although in reality it is only one eighth as large. Such a map cannot be used for visual comparison of the sizes of continents or countries; for that purpose, we must use an *equal-area projection* (Fig. 1.3). Almost endless ways have been devised to project the curved surface of the globe onto a plane, but none makes it possible to simultaneously represent true direction, true distance, true shape, and true area for a particular segment of earth-space.

CLASSIFICATION AND SYMBOLIZATION

Maps make it possible for us to extract certain items from the totality of things, to see the patterns of distribution they form, and to compare these patterns with each other. No map is a complete record of an area but represents a selection of certain details that the mapmaker records to accomplish a particular purpose. These details are shown by *symbols*. Before the symbols can be chosen, the unprocessed data on which they are based must be classified (grouped into classes) in order to provide categories that the symbols will represent. *Classification* and *symbolization* go hand in hand. The classes may be categories of physical or cultural forms (rivers, roads, settlements, and so on), aggregates such as 100,000 people or 1 million tons of coal, or averages such as an average population density of 68 persons per square mile within a defined area. Aggregates or averages frequently are ordered into ranked categories in a graded series (for example, population densities of 0–49, 50–99, 100 or over per square mile) for portrayal on the map. The ranks are not self-evident but must be selected by the mapmaker, sometimes by use of elaborate statistical procedures. To represent the varied phenomena with which the subject of geography deals, the mapmaker has available a wide range of symbols: lines, dots, circles, squares, and others. Many are illustrated by maps in this text.

The Concept of Pattern

Dots are commonly used symbols for portraying quantities, and the *dot map* is one of the commonest types of maps showing the distribution of people or things on the earth. In a dot map of population, for example, each dot represents a stated number of people, and the dot is placed as near as possible to the center of the geographic space occupied by these people. In interpreting such a map, we are not

Editorial Comment

LATITUDE AND LONGITUDE

The term *latitude* denotes position with respect to the equator and the poles. Latitude is measured in degrees, minutes, and seconds. The equator, which circles the globe midway between the poles, has a latitude of 0°. All other latitudinal lines are parallel to the equator and to each other, and therefore are called parallels. Places north of the equator are said to be in locations of north latitude; places south of the equator, in locations of south latitude. The highest latitude a place can have is 90°. Thus, the latitude of the North Pole is 90°N, and that of the South Pole is 90°S. Places near the equator are said to be in *low latitudes;* places near the poles, in *high latitudes* (Fig. 1.3). The Tropics of Cancer and Capricorn, at 23½°N and 23½°S, and the Arctic and Antarctic Circles, at 66½°N and 66½°S, form convenient and generally realistic boundaries for the low latitudes and the high latitudes, respectively (although a more symmetrical arrangement would place the boundaries at 30°N and S and 60°N and S). Places occupying an intermediate position with respect to the poles and the equator are said to be in *middle latitudes.* The middle latitudes have a marked seasonal quality in their yearly temperature regimes, which contrasts strongly with the relatively constant and monotonous heat of low-latitude lowlands and with the extremes of cold which prevail for much of the year in the high latitudes.

The *meridians of longitude* are imaginary straight lines connecting the poles. Every meridian runs due north and south. The meridians converge at the poles and are widest apart at the equator. Longitude, like latitude, is measured in degrees, minutes, and seconds. The meridian which is most often used as a base (starting point) runs through the Royal Astronomical Observatory in Greenwich, England. It is known as the *meridian of Greenwich,* or *prime meridian.* Every point on a given meridian has the same longitude. The prime meridian has a longitude of 0°. Places east of the prime meridian are in east longitude; places west of it are in west longitude. The meridian of 180°, exactly halfway around the world from the prime meridian, is the dividing line between places east and west of Greenwich. The longitude of New York City is 73°58′ West of Greenwich, or simply 73°58′W. At the equator, 1° of longitude is equivalent to 69.15 statute miles (111.29 km). Since the east-west circumference of the earth decreases toward the poles, however, a degree of longitude at the Arctic Circle is equivalent to only 27.65 miles (44.50 km).

greatly concerned with the individual dots, but with the way the dots are arranged; that is, with the *pattern* that they form. Some maps show a rather even spacing of the dots, but other maps will show the dots strongly clustered in some areas and very sparse in others. The best means of portraying almost any distribution on the earth's surface is to show it on a map. One can readily imagine what a hopeless task it would be to describe in any detail the distribution of the world's people without reference to a map. But by making a map, we are able to see the broad pattern at a glance (see Fig. 3.2).

As used by geographers, the word *pattern* generally connotes a set of locations seen as a composite. Common types of patterns are comprised of geographic elements that repeat themselves in an orderly way (for example, rectangular fields in the American Midwest, or world types of climate), are concentrated or dispersed in a definable manner (for example, the world's population), move about in characteristic paths or waves (for example, overseas migration of Europeans since the 15th century), or connect with each other in a network or system (for example, the highways of the United States). The recognition, description, analysis, and evaluation of patterns are major facets of geographic activity. The comparison of different patterns to see what correlations exist (for example, between rainfall and crop yields) underlies much of the scientific work in the field. Such work utilizes both maps and sophisticated statistical techniques.

Problems of Portraying and Explaining Spatial Distributions

It is not quite so simple as it may sound to make maps that help show and explain where things are on the earth. To do so, the mapmaker must answer various questions. The first is: How can the thing in which the investigator is interested be defined and measured? This can be difficult. An example is the world map of population (see Fig. 3.2). Here the subject is the number of people in the different parts of the world. But even in this *apparently* simple case, the investigator must define the areal units for which data are to be sought. For instance, will figures be required for the entire United States, for states, for counties, for cities, or for some other areal units? And what comparable units will be required for other countries? The mapmaker must also decide what census data are sufficiently up-to-date for use, what published estimates are adequate for countries lacking census data, and what data will have to be estimated by the investigator. Frequently, geographers are interested in investigating the spatial distribution of more elusive phenomena (such as standard of living or economic health), which present conceptual difficulties of definition as well as difficulties in measurement. Maps showing cities by different size-of-population categories (for example, Fig. 7.4) present difficulties of defining "city," because larger cities have now generally outgrown their political boundaries and have become clusters of suburbs around an original central city. Consequently, the investigator must ask how much territory properly belongs to a particular urban cluster. The maps of London (see Fig. 6.8) illustrate this question. How much of southeastern England should be defined as actually and realistically a part of London? The answer will greatly influence the population size given.

A second question that the investigator must face is, in simplest terms: Where is it? That is, what is the location or distribution of the thing under study? The answer commonly involves mapping, but there are other ways of expressing a distribution, for example, statistical tables, graphs, and verbal descriptions. In the case of world population, an answer to the question "where" has been given in this text as a dot map on which each large dot represents approximately 10 million people and each small dot represents approximately 5 million people. Some major aspects of the distribution have also been summarized in Table 3.1. The table could easily be expressed in percentage terms and/or converted into a graph.

One of the complexities of geography is that different scales and different ways of showing distributions can give somewhat different impressions concerning the same distribution. For example, other methods of showing the distribution of world population might be used and might convey rather different ideas from Figure 3.2. Some alternative methods might include the use of a single symbol in each country, with the size of the symbol proportional to the country's population, or drawing a map on which the area of each country was shown pro-

portional to its population. Or, if symbolization by dots were retained, a map using one dot per 100,000 people, or one dot per 1 million people. Change of scale would also be likely to somewhat alter impressions. Figure 3.2 is a small-scale map, that is, it shows a large area in a small page space, with relatively little detail and a relatively high degree of generalization. A larger-scale map of the world would require a larger page space, allowing more detail with less generalization. The student might note that Figure 3.2 does not give impressions of the distribution of population in Europe that are identical to those given by the population map of Europe (see Fig. 4.5). The latter is drawn at a larger areal scale (more page space to show the same area), and it also takes a somewhat different approach to mapping population, using densities instead of numbers, and using symbolization by ranked categories rather than by dots. This does not mean that either map is false. It merely illustrates the fact that alternative scales and methods can give somewhat different impressions of, or portray different aspects of, the answer to the *apparently* simple question of where something is or how it is distributed.

Once a subject is defined and measured and its distribution determined and expressed, a third question arises: Why is it located or distributed the way it is? The search for answers generally involves an attempt to explain relationships between the mapped subject and other phenomena that apparently influence it or influence people's decisions about it. In some cases, generally accepted broad theories are available about the behavior and relations of things under investigation. Some of these theories have been developed in the field of geography, and others can be borrowed from other disciplines. They suggest hypotheses (tentative explanations) concerning reasons for particular distributions. Much of the research time of geographers is devoted to developing theories and hypotheses concerning locations and distributions, and testing whether hypothesized relations really do help to explain actual distributions. The student will find that this general mode of thought is frequently used in this text (often without explicit identification), as both historical and relational ideas are introduced as to why things are found where they are.

MAP DESIGN

Proper *design* to communicate facts and ideas effectively is an important facet of map "language." The symbols used must be clear, appropriate to the task, and sharply differentiated from each other. If several different distributions are juxtaposed on the same map, it is desirable that the user be able to pick out each individual distribution independently of the others. If orders of magnitude are shown by gradations in symbol sizes and letter sizes, the gradations should be easily readable. For an example of map design embodying these principles, see the urban and industrial map of the British Isles (Fig. 6.7). How many separate distributions can you find on this map? Note that the largest (and generally most important) cities are shown by the largest symbols and their names are in the largest and heaviest type. Throughout this text, gradation of symbol and letter sizes offers a convenient way to differentiate places according to levels of importance.

REGIONS AS TOOLS FOR GEOGRAPHIC THINKING

The concept of the *region* is a very old and useful device for discerning and analyzing spatial order on the earth. It is used endlessly in geographic work. For example, as previously noted, much geographic work centers on maps showing spatial distributions, and these maps cannot be used effectively unless they are regionalized in some way. This can be illustrated by the dot map of world population (see Fig. 3.2). The map conveys an overall impression of very uneven distribution, but more precise analysis will be facilitated by even such a simple regionalization as dividing the map into areas having dense populations, areas having little or no population, and areas in between. Once this is done, we can begin to look for other phenomena associated spatially with the presence or absence of population and to generate hypotheses concerning possible causal relationships between population occurrence and other phenomena. Regions, then, are basic tools for geographic thinking.

KINDS OF REGIONS

As customarily used by geographers, the term *region* refers to an area of considerable size which has a substantial degree of internal unity or homogeneity and which differs in significant respects from adjoining areas. A region may be wholly contained within a single country, or it may include several different countries or parts of countries. Many kinds of regions are used in geographic study. They may or may not be defined and delimited with mathematical precision. Some regions are distinguished on a physical

basis; others have a cultural, economic, social, or political basis; and still others are composites. Regions may be defined on the basis of a single characteristic or of multiple characteristics. In defining and interpreting regions, geographers make use of three fundamental ideas: (1) *generalized areal homogeneity*, (2) *coherent functional organization*, and (3) *regional distinctiveness*. Each of these ideas is associated with a particular category of regions.

The *homogeneous region*, also known as the *uniform* or *formal region*, is an area having a unitary quality "because of the relative homogeneity of a certain characteristic, or characteristics, throughout. Put another way, it is an area within which a certain characteristic or combination of characteristics falls within a specified range of variation. Examples are a lowland region, containing only land below a certain elevation; a lowland forest region, containing only land below a certain elevation and forested to a specified degree; an industrial region, containing only statistical units with more than a specified percent of their labor force in manufacturing."[2]

The *functional region*, also called the *nodal region*, is a coherent structure of areal units organized into a functioning system by lines of movement or influence that converge on a central node or trunk. An example is the "milkshed" (milk supply area) of a city, exhibiting a pattern of dairy farms reached by transportation lines over which fresh milk moves to processing and distribution points, and ultimately to consumers in the city. A more complex example is the territory served by a commercial metropolis. It is comprised of a hierarchy of local and regional commercial centers (crossroads stores, hamlets, villages, small towns, small cities, and so on) which are bound to each other and to the central metropolis by the flow of people, goods, and information over an organized network of transportation and communication lines. Still a third instance is the territory served by a railway system, with its pattern of feeder lines branching from a central trunk.

Many regions, sometimes called *general regions*, are too varied and loosely knit to qualify in any strict sense as "homogeneous regions" or "functional regions" but are recognized simply on the basis of overall distinctiveness. The "major world regions" that form the basic framework of this text are best thought of as "general regions."

The concept of regions has a utility that extends beyond formal course work in geography.

---

[2]J. Trenton Kostbade, "The Regional Concept and Geographic Education," *Journal of Geography*, 67 (1968), 8. Used by permission of the *Journal of Geography*.

Regions are areal categories, and categories are essential to thought. Regionalization is either a type of classification or a process very similar to classification. Man cannot think without grouping his infinitely varying observations and experiences into categories; and he cannot orient himself with respect to the areally differentiated surface of the earth, and think about that differentiation, without grouping the infinite variety of places on the surface of the earth into regions. Thus regional concepts are not merely the tools of geographers. They are necessities for thinking man, as is indicated by their widespread use outside the field of geography and in everyday life and discourse.[3]

_____
[3]J. Trenton Kostbade, "The Regional Concept and Geographic Education," _Journal of Geography,_ 67 (1968), 7. Used by permission. Wording follows author's original manuscript.

AIM OF THE REGIONAL APPROACH

The essential aim of the regional approach to geographic study, as exemplified in this text, is to provide a description of the world in terms of the principal differences and similarities among its parts. So far as possible, the text attempts to make this description more meaningful and comprehensible by making it explanatory, that is, by explaining the differences and similarities. Broadly speaking, the general procedure is to focus on an area and attempt to identify its principal traits or characteristics as a part of the world. These traits may represent similarities to other regions, or they may represent distinctive attributes that are peculiarly characteristic of the particular region—things of which it has notably much or little compared to other regions. The latter give distinctiveness to the region in question, differentiate

Figure 1.3 Map of the major world regions that form the basic framework of this text. Some regions overlap others.

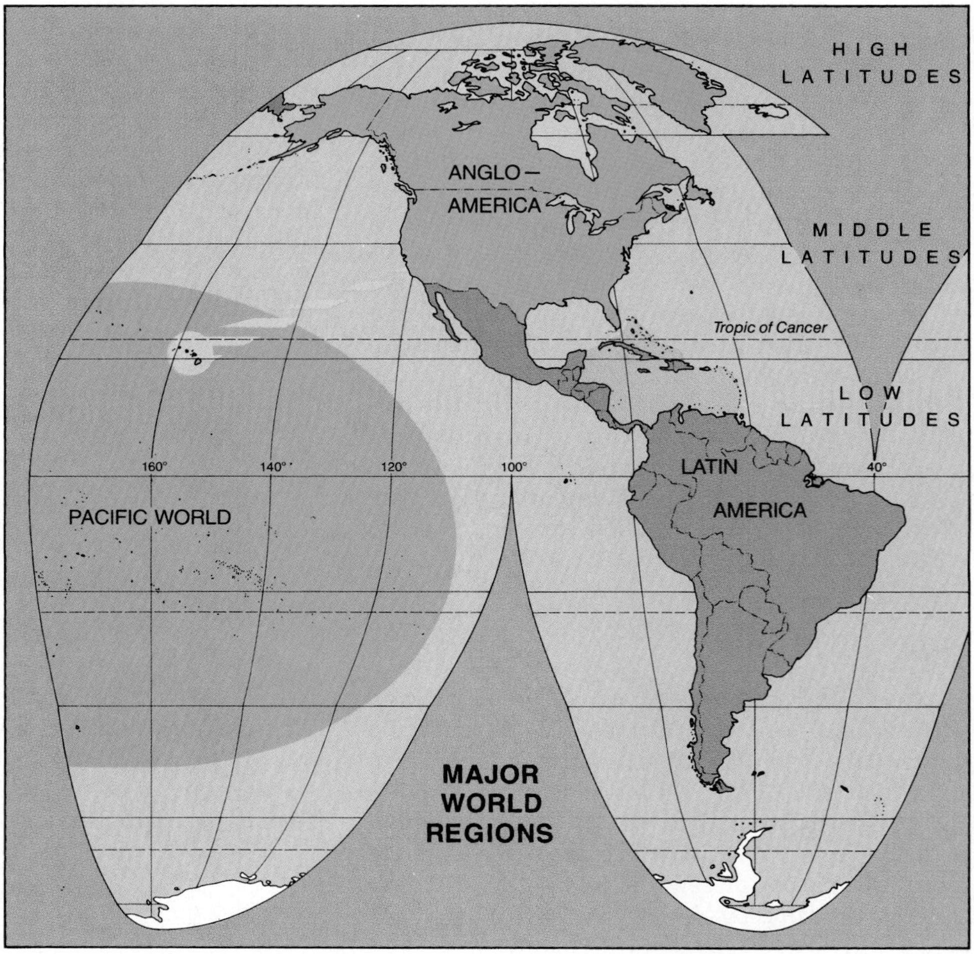

it from other regions, and have probably led us to see it as a distinguishable region in the first place. The region is then described in terms of these major characteristics or traits. The unique combination of traits in a region is sometimes said to give the region its "character" or "personality," analogous to the character or personality conferred on a person by his or her unique combination of traits.

But such a description could be a mere list of traits. It takes on more meaning when explanation of the traits is sought. Frequently, the region's distinguishing traits can be explained at least partially by their interrelations with each other or with other traits which come to light only when explanation of those originally perceived is attempted. Some traits may only be explainable, if at all, by tracing their particular historical development and attempting to grasp the conditions that originally brought them about and may have subsequently modified them. Frequently, certain traits of a region will be explain-

able partly in terms of their relation to things or places outside the region under examination, for example, the relation of iron-ore mining in Australia to the growth of the Japanese steel industry. The end result, if we are sufficiently skillful and fortunate, should be a picture of the region as a distinguishable part of the world in terms of a group of outstanding traits which characterize it and a heightened understanding of the region through the interrelations, origins, and evolution of these traits.

THE MAJOR WORLD REGIONS

The *major world regions* discussed in this text are Europe, the Soviet Union, the Middle East, the Orient, the Pacific World, Africa, Latin America, and Anglo-America (Figs. 1.3 and 1.4). Although these regions are commonly recognized as grand divisions of the world, geographers often differ as to the precise area

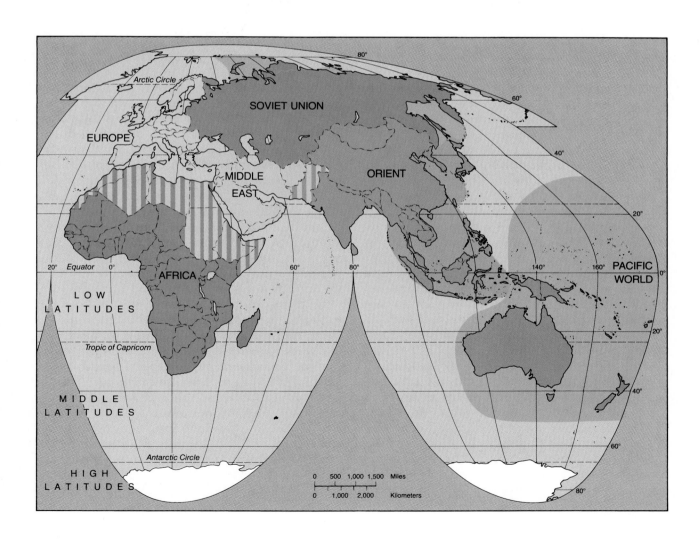

Figure 1.4 Three styles of religious architecture near the Moscow Kremlin mirror influences from three major world regions. The minarets on the building at the right suggest the Middle East, where the Islamic (Muslim) religion originated. The central building with its four columns is in a style developed in Europe, and the multicolored domes of St. Basil's Cathedral at the left are an inheritance of the Soviet Union from the Orthodox Christian culture of Old Russia. All three buildings were originally places of worship, although none is used as such today.

which should be included in each. Here the eight world regions are delimited as groups of countries for convenience in presenting regional concepts on an introductory level. However, the student should not assume that the political lines used as regional boundaries necessarily represent sharp lines of division in other respects. In physical, cultural, and economic terms, the major world regions tend to be separated by *zones of transition,* where the characteristics of one region change gradually to those of the next. This tendency to merge gradually rather than to be separated by sharp lines is also true of regions of a smaller order of size. Since regions are merely convenient devices useful in generalizing about the world, it is more important to grasp the particular set of features which characterizes the *core* of a region than it is to search for an exact *line* where the characteristics of one region end and those of another region begin. The student will note that some major world regions overlap others. For example, ''Africa'' overlaps the ''Middle East,'' since North Africa is included in both regions. This should present no difficulty so long as one conceives of regions simply as useful tools designed to further geographic study. For various reasons, it seems desirable to consider North Africa in the context of both ''Africa'' and the ''Middle East.'' This procedure appears clearly to enhance our understanding of both world regions and thus to be consistent with the aims of geographic study.

WHAT CAN I *DO* WITH GEOGRAPHY?

This text is designed for general education. It is hoped that the student will emerge with such values as a better perspective on international affairs, an enhanced appreciation of travel, an inkling of the world's true variety and complexity, and better insight into some other subjects of study. This general introduction to the subject matter of geography may also motivate some students to consider the career values inherent in geographic ways of thinking.

The ability to think habitually in terms of relative location and associations of geographic features does not come automatically in the course of an education that lacks explicit geographic training, nor does an easy familiarity with the world of maps. In a world preoccupied with such matters as resource management, spatial planning, and international security, the skills and knowledge of geographers are highly relevant. In the United States and many other countries, thousands of geographers already work professionally in education, research, government, and the business world. They gather and process geographic information, make it available to interested parties, use it to help solve human problems, and carry on research in order to create new geographic knowledge. In their daily work, geographers make use of a variety of skills, including library skills; field-observation skills; interrogation skills (the use of interview and questionnaire techniques);

statistical skills, including computer processing; cartographic and remote-sensing skills; skills in geographic analysis and synthesis; and presentation skills—conveying geographic knowledge to other persons by written, oral, and visual means. Geographers are employed in a wide spectrum of occupations. This includes teaching at all levels, cartographic work, planning, and various other kinds of work in both the public and private sectors. Representative jobs entail such activities as planning and executing maps; studying land use; writing environmental-impact statements; interpreting air photographs; gathering and processing geographic information for intelligence agencies, planning agencies, or industrial development agencies; serving as governmental planners at levels varying from the locality to the nation; working for tourist agencies; or performing a variety of tasks related to marketing and the placement of facilities such as health-care services, recreational services, detention services, and so on. Even for persons not specifically labeled "geographers," the skills of geography may prove highly valuable and salable assets in the employment market. At a professional level, these skills can be acquired only by much more advanced education and training than are relevant to this introductory and broad-scale text.

CHAPTER 2

Processes that Create Spatial Order and Differentiate the World

Canyon de Chelly in the Arizona section of the Colorado Plateau provides spectacular evidence of natural processes at work. The fainter imprint of human processes can be seen in Indian homes (summit) and fields (canyon floor) on the Navajo Reservation.

THE patterns of spatial order that geographers try to identify and elucidate result from processes through which nature and humans determine the contents of parts of the earth (see chapter-opening photo and Fig. 2.1). Natural processes endow the world and its parts with minerals and rocks, landforms, climates, water features, plant and animal communities, and soils—the interlocking assemblage of features that we call the natural environment. For purposes of study, the distribution of single elements such as amount of precipitation or degree of slope may be mapped and analyzed, but the ultimate aim is to see the individual elements of the environment in their spatial and functional relationships with each other and with humans. In this text, attention is focused especially on the ways that particular combinations of natural features and their use by groups of people confer distinctiveness on countries and regions. People, in their turn, create spatial patterns and differentiate the world through such processes as natural population change, migration, innovation and diffusion of innovations, natural-resource exploitation, site selection and development, specialization and trade, inertia, and geopolitical organization and interaction. All these natural and human processes are ongoing; they have been ceaselessly active in the past and will presumably continue into the future. They produce successive modifications in the geography of all parts of the world, although some modifications occur at a faster pace than others. At a global scale, the ongoing processes form a worldwide system of people and environment, the nature of which is under investigation by scientists, including geographers. Consideration of the system as a whole lies for the most part beyond the scope of this text, which is focused on the examination of individual parts of the world. But the idea of processes that create spatial order and differentiate the world is a vital part of the explanatory description and analysis of areas. Each area is in flux. Natural processes continue to modify it, while people continue to investigate, reevaluate, and change both the area and their own use of it in accordance with their changing needs and aspirations, their changing social, economic, and political circumstances, and their changing technological competence.

NATURAL PROCESSES AND THEIR GEOGRAPHIC RESULTS

Five major groups of natural processes are of prime importance in terms of their geographic results: (1) processes that create rocks; (2) processes that create landforms; (3) processes that create climates; (4) processes that create wild plant and animal communities; and (5) processes that create soils. We may refer to them more succinctly as rock-forming, land-shaping, climatic, biotic, and soil-forming processes. Many photographs in this chapter illustrate the results.

ROCK-FORMING PROCESSES AND THE MAJOR TYPES OF ROCKS

The study of rocks is carried on mainly by geologists, but the student of geography is concerned with rocks in various ways. Such interest centers on the fact that the rocks which are under, or in some areas are at, the surface of the earth vary widely in physical character from one place to another and that these variations are often related to such other characteristics of places as landforms, soils, and mining.

Rocks are composed of minerals, which may be chemical elements or combinations of elements. Minerals that are economically useful to people may or may not occur in rocks in sufficient concentrations to be minable at a profit. A mineral concentration which can be profitably mined is known as an *ore.*

Figure 2.1   Numerous physical and human processes have created this scene in the American West. The service area occupies a niche in Lake Powell, the reservoir of Glen Canyon Dam on the Colorado River. Rock-forming, land-shaping, climatic, hydrologic, and biotic processes contributed to the natural setting, and various technological, economic, cultural, and political processes played a role in creating the reservoir and the service area. Hence, geographic interpretation of even this small scene is a complex matter. The rock seen here is sandstone.

Sedimentary, Igneous, and Metamorphic Rocks

The rocks of the earth form three major groups according to mode of origin: sedimentary rocks, igneous rocks, and metamorphic rocks.

Sedimentary rocks have been formed from sediments (sand, gravel, clay, or lime)—deposited by running water, wind, or wave action either in bodies of water or on land—which have consolidated into rock over a span of geologic time by pressure of the accumulated deposits or the cementing action of chemicals contained in waters percolating through the rock materials. The main classes of sedimentary rocks are sandstone (see chapter-opening photo and Figs. 2.1 and 2.9), shale (formed principally of clay), and limestone (Fig. 2.2). Sedimentary rocks generally are comprised of multiple layers or *strata,* and the term *stratified rocks* often is used in reference to

them (see Fig. 2.5). In a coal-bearing area, the layers of coal commonly are interspersed (''interbedded'') with layers of sandstone and/or shale. *Igneous,* or *volcanic, rocks* are formed by the cooling and solidifying of molten materials, either by slow cooling within the earth (as in the case of granite; Fig. 2.3) or by more rapid cooling when poured out on the surface by volcanic activity (as in the case of basalt). *Metamorphic rocks* are formed from igneous or sedimentary rocks through changes occurring in the rock structure as a result of heat, pressure, or the chemical action of infiltrating waters. Marble, formed from pure limestone, is a common example of a metamorphic rock.

The major types of rocks have certain general, although not invariable, associations with other geographic characteristics and features of areas. Igneous and metamorphic rocks, for instance, being generally harder and more closely knit together than sedimentary rocks, are usually more resistant to wearing down by weathering and erosion. They tend to form uplands in areas where the sedimentary rocks have been weathered into lowlands. (But in purely sedimentary areas, the more resistant sedimentaries tend to form higher lands than the less resistant ones.) In general, sedimentary rocks weather into more fertile types of soil than igneous or metamorphic rocks. However, certain types of volcanic rock break down into extremely fertile soils, and, within the sedimentaries themselves, there are great variations in this respect, with limestones usually forming soils of greater natural fertility than do shales or sandstones. Sedimentary rocks are also significantly different from igneous and metamorphic rocks with respect

Figure 2.2   Both the English and French sides of the English Channel are lined with spectacular exposures of the pure limestone called chalk. *The "White Cliffs of Dover" are the best known. Seen here is an even higher cliff at Beachy Head on England's Channel coast near the resort of Brighton.*

Figure 2.3   One of the world's most famous exposures of granite is revealed here in California's Yosemite Valley, a part of the high mountains called the Sierra Nevada. El Capitan is at the left, Bridal Veil Falls at the right. The falls plunge over the edge of a hanging valley, formed when glacial erosion scrubbed away the lower part of the valley.

to associated minerals of economic value. Coal and petroleum, for example, are customarily found in sedimentary areas, whereas metal-bearing ores are found in igneous or metamorphic rocks except for certain deposits (particularly of iron ore) formed of materials removed from these rocks by percolating waters and redeposited in sedimentary areas.

LAND-SHAPING PROCESSES AND THEIR RESULTS

The main processes that shape the earth's surface into landforms are (1) *warping, folding,* and *faulting* of the rock layers of the earth's crust (known collectively as diastrophic processes or "diastrophism"), (2) *vulcanism,* and (3) *weathering, erosion,* and *deposition.* Warping, folding, faulting, and vulcanism are called the *tectonic processes;* their basic long-term effect is to bring about unevenness in the land surface. Weathering, erosion, and deposition are the *gradational processes;* their long-term tendency is to wear

down the higher parts of the surface and fill in the lower parts until the surface is reduced to a uniform plain.

Warping, Folding, and Faulting

Forces in the earth whose nature is not well understood subject the solid rocks of the crust to enormous stresses and slowly warp and bend, or break, them. Masses of rock may be warped upward to form domelike or arched structures, or downward to form structural basins. Such structures may be purely local, or they may be subcontinental in size. Often the uplift or depression is accomplished without much structural change in the rocks themselves. For example, in the Grand Canyon, the Colorado River has cut its way through sedimentary rocks that initially lay at a much lower elevation but have been subjected to warping processes which lifted the entire mass of rocks thousands of feet while leaving the horizontally bedded rock layers essentially undisturbed. On the other hand, intense bending of rock

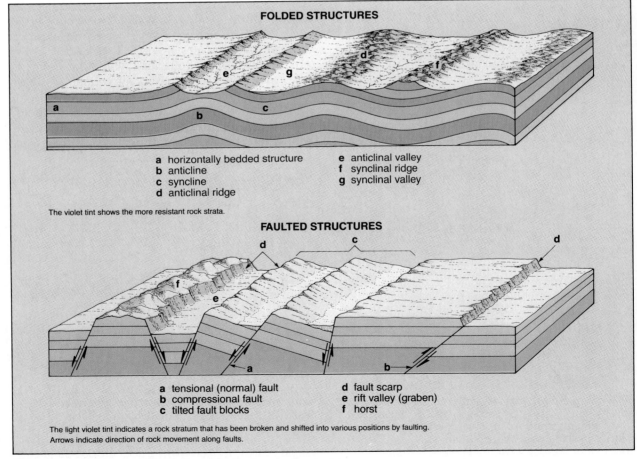

Figure 2.4   Diagrams of folded and faulted rock structures.

Figure 2.5 *Intense folding has left these strata vertical. The locale is the anthracite coal-mining region in eastern Pennsylvania.*

Figure 2.6 *This westward view from Zabriskie Point in southeastern California shows the severely eroded fault scarp that forms the western wall of the rift valley called Death Valley. The scarp rises from an elevation well below sea level on the valley floor to 11,045 feet (3367 m) at the summit of snowcapped Telescope Peak on the distant horizon.*

layers ("folding") may produce an accordion-like series of upfolds (*anticlines*) and downfolds (*synclines*) (Figs. 2.4 and 2.5). Such structures are very common in mountainous areas of the world. In some instances, the anticlines form the present ridges and the synclines form valleys, but in other instances a complex erosional history has produced synclinal ridges or anticlinal valleys.

When they are crowded together or pulled apart by irresistible forces in the crust, rocks may break and the separate portions move past each other along the line of fracture. This process is known as *faulting,* and the break in the rocks is called a *fault* (Fig. 2.4). A break due to rock masses being pulled apart is a *tensional* or "*normal*" *fault,* whereas a break due to rocks being pushed together until one mass rides up over the other is a *compressional fault.* Horizontal displacement of rocks along faults may damage fences, transportation lines, or buildings, but the most pronounced landform effects occur when rocks are displaced vertically. Such displacements of rock masses upward or downward along faults have often been great enough to create mountains of a type called *block mountains.* In the most common instance, the rock mass comprising the mountain is tilted in such a way as to create a steep face ("fault scarp") on the side where the fault occurs and a longer gentler "back slope" behind the scarp. The Sierra Nevada and the Wasatch Range of western United States (see Fig. 28.21) are "tilted block mountains" of this kind. In the Sierra Nevada, the fault scarp faces eastward, while in the Wasatch Range, it faces westward. Some block mountains have been formed between roughly parallel faults and have a scarp face on two sides. A landform of this type is called a *horst* (Fig. 2.4). A steep-sided trough known as a *rift valley* or *graben* is formed

when a segment of the crust is displaced downward between parallel tensional faults or when segments of the crust which border it ride upward along parallel compressional faults (Fig. 2.4). The Great Rift Valley of eastern Africa and southwestern Asia (see Fig. 21.2), Death Valley in California (see Figs. 2.6 and 28.22), and the Rhine Rift Valley on the border between West Germany and France are well-known examples. Faulting is very widespread in mountainous regions. In one part of the American West, a whole succession of mountain ranges separated by flat, detritus-filled basins ("basin and range topography") has been produced by faulting.

Vulcanism

In some parts of the world, the present terrain was shaped initially by outpourings of volcanic material from vents or cracks in the earth (Fig. 2.7). A volcanic cone such as Mount Fuji or Mount Vesuvius is a familiar instance. In places, enormous amounts of molten material issuing in repeated flows through large fissures in the crust have buried the previous topography to a depth of hundreds or thousands of feet. Prominent examples are found in Ethiopia, interior India, and the Columbia–Snake Plateau of

Figure 2.7 *Extrusive vulcanism produced this landscape in arid Baja California, Mexico. The volcanic peak that dominates the view is called Las Tres Vírgenes (The Three Virgins). Note the scanty cover of drought-resistant vegetation in this extremely dry area.*

northwestern United States. This type of volcanic activity is known as *extrusive vulcanism.* The cooling and solidifying of the extruded molten matter often has produced the dark-colored rock called *basalt. Intrusive vulcanism* takes place when molten material rising through the crust penetrates the existing rocks and cools and solidifies without reaching the surface. Granite and other ''intrusive'' igneous rocks are formed in this way. Uparching followed by erosion may cause the overlying rocks to be stripped away, revealing the intrusive igneous formations. These may then be sculptured into hills or mountains by erosive processes. For example, many of the present mountain ranges in the Colorado and Wyoming Rockies originated in this manner.

The Global System of High Mountains and Associated Tectonic Phenomena

Warping, folding, faulting, and vulcanism are interrelated processes, although the precise connections among them are by no means entirely clear. In recent geologic time, tectonic activity has reached its maximum intensity in a branching worldwide zone of crustal instability, a good part of which is under the sea. Such activity is believed to be associated with the movement and interplay of crustal masses known as plates (see Editorial Comment: Plate Tectonics). The land portions of the zone of instability comprise a system of high mountains which may be thought of as radiating in several arms from the Pamir Knot in Asia (see landform map, Fig. 2.10). Most of the truly high mountains of the world—for example, the Alps, Pyrenees, Atlas, Himalayas, Rockies, and Andes—lie within this system. Most

mountains in the system are very young, geologically speaking, and the tectonic processes associated with mountain building are still active. This fact often has disastrous consequences for people who live in or near the mountains, as the global system of young mountains is the main zone within which earthquakes and active volcanoes occur. The earthquakes result from rock movement along faults. Most of the devastating earthquakes of human history, including many recent ones, have occurred in or near the young-mountain belt. Active volcanoes occur in lines or clusters in some parts of the belt—for example, in Central America, in the Alaska Peninsula and Aleutian Islands, and in Java and other islands in Indonesia. These groupings of volcanoes are irregularly distributed and are, indeed, by no means typical of the young-mountain system as a whole, most parts of which show no evidence of recent volcanic activity.

Weathering, Erosion, and Deposition

Landforms that have been shaped initially by tectonic processes are then sculptured by gradational processes; indeed, the two types of processes go on simultaneously. One of the gradational processes is *weathering.* When rocks are subjected to weathering, they gradually crumble, and the fragments move downslope by gravity or are transported away by moving water, ice, or wind and deposited elsewhere. *Mechanical weathering* involves such processes as (1) the formation of cracks or joints in rocks as a result of stresses, (2) the wedging action of ice when water freezes in rock crevices, or (3) the expansive force exerted by plant roots growing in crevices. *Chemical weathering* is more complex than mechanical weathering. It involves interaction between surface or underground water and certain chemicals in rocks. Generally the water with which rocks come in contact contains weak acids acquired when falling raindrops interact with carbon dioxide in the atmosphere (forming carbonic acid) or when water comes in contact with decomposing plant life and absorbs acids in this manner. Especially when fortified by acids, water dissolves some of the minerals in rocks and carries them away, thus weakening the rocks and facilitating mechanical weathering (which in turn exposes still more rock surfaces to chemical action). In other instances, rock minerals may swell when they interact with water and thus break the bonds that hold individual mineral grains together. In still other cases, minerals may be changed in form or structure, becoming softer or more easily dissolved.

Editorial Comment

PLATE TECTONICS

Scientists who study the structure and movements of the earth's rocky crust—the *lithosphere*—have come to believe that the crust is comprised of immense segments called *plates*, which are thought to be about 60 miles (c. 100 km) thick but may be thousands of miles across. Eurasian, North American, South American, African, Australian, Antarctic, Pacific, and Nazca (southeastern Pacific) Plates have been identified. The plates are in motion, and they bear the continents and segments of the oceans about. The branching linear system of young mountains, frequent earthquakes, and active vulcanism on the globe is thought to coincide with the interfaces where different plates are in contact. In such zones the plates slowly grind past each other or the edge of one plate may submerge beneath the edge of another. In certain places, for example, the north-south oriented Mid-Atlantic Ridge, it appears that new crustal material is upwelling to replace material that disappears beneath plates. Crustal activity at the margins of plates is thought to generate earthquakes, active vulcanism, and high mountains that result from the crumpling upward of rock layers. The general name applied to crustal activity of the kind described is *plate tectonics.*

A second gradational process is *erosion,* the wearing away and shaping of the earth's surface by running water (Fig. 2.8), ice, or wind. Erosion by the cutting action of streams is extremely important in shaping the form of the land. In this process, the force of running water is augmented by fragments of rock which are carried along by the stream and act as a kind of file, abrading the stream channel and helping the stream to cut its valley. As the valley is deepened, its edges tend to crumble, erode, and slump, with the result that the valley grows wider as it is cut deeper. In arid areas deepening tends to occur more rapidly relative to widening than is the case in humid areas, and streams that have risen in more humid areas carve deep, steep-walled canyons ("inverted mountains"; see chapter-opening photo and Fig. 2.9). In humid areas, while streams are deepening valleys, weathering and erosion of the sides of the valleys occur more actively than they do in arid lands. A plateau may thus be carved into a mass of hills or mountains consisting of stream valleys separated by ridges which may be capped by remnants of the original surface. The hills and low mountains of the "Appalachian Plateau" in eastern United States (see Fig. 28.2) were formed in this way. Long-continued erosion will reduce a highland to a lowland plain, but, at any stage, the process may be interrupted by upwarping. Many land surfaces as they exist today are primarily a reflection of some stage in the destruction of a raised surface by stream erosion.

Ice and wind are active agents of the erosional process in some areas. The grinding, scouring movement of glaciers over some parts of the land surface during the Great Ice Age left many characteristic features behind when the ice melted (see Editorial Comment: Glaciation in Chapter 5), and glaciation is still active today in restricted areas (see Fig. 28.31). The wind is particularly active as an erosive force in

Figure 2.8   A symmetrical pattern of erosion in the tropical East African country of Kenya. Streams and their tributaries eroding headward will eventually destroy all the divides (interfluves) in this view.

Figure 2.9 Canyon-making is still in progress at Glen Canyon on the Colorado River. Here an enormous chunk of sandstone has fallen into Lake Powell, leaving the fresh scar at the right. The white stripe at the cliff bottom has been created by fluctuating water levels in the Lake Powell reservoir.

arid and semiarid areas where the protective cover of vegetation on the surface of the land is thin or absent.

Material transported by water, ice, or wind eventually is deposited by the agent that removed it; actually, it may be picked up and redeposited many times. Large areas of the world are plains underlain by sedimentary rocks that were formed by the deposition of vast quantities of sediments in shallow seas. Prominent examples shown on maps in this text are the Interior Plains of North America (see Fig. 26.9), the North European Plain (see Fig. 5.7), and the interior plains of South America drained by the Amazon, Orinoco, and Paraná Rivers (Figs. 2.10 and 24.1). Recently deposited sediments not yet consolidated into rocks form alluvial plains in river valleys; major areas are found along the lower Mississippi River (see Fig. 28.2), the Amazon, the Huang He (Yellow River) of North China (see Fig. 14.2), and many other rivers. Often a distinction must be made between older alluvium and newer alluvium, as in northern India. Here the main plain of the Ganges and Indus rivers (see Fig. 15.3), formed of older alluvium, is hundreds of miles wide, but with relatively narrow strips of newer alluvium forming the present flood plains of the rivers, which flow in valleys eroded into the main plain. Some of the flattest areas in the world are composed of recent alluvium in river flood plains and deltas, in the beds of vanished lakes, or in stretches of former sea bottom that have been upwarped slightly to form dry land.

The Major Classes of Landforms

The results of these land-shaping processes are commonly classified into four very general types of landforms: plains, plateaus, hill lands, and mountains (see map, Fig. 2.10). Each type is susceptible to further subdivision, and the distinctions between the types are not always clear and objective.

In general, a *plain* is a relatively level area of slight elevation (although some parts of the "High Plains" in the western United States reach elevations of over 5000 feet [over 1500 m] above sea level). Some plains are flat (Fig. 2.11), but most of them exhibit gentle to moderate slopes. However, it is the horizontal rather than the vertical dimension that predominates in plains country. Most of the world's people live on plains, so that consideration of plains and their significance becomes a highly important facet of geographic study.

To a degree, a *plateau* is simply an elevated plain (chapter-opening photo). Most areas recognized as plateaus lie at elevations of 2000 feet (610 m) or

Figure 2.10 Highly generalized map of world landforms, modified from a map by the United States Department of Agriculture.

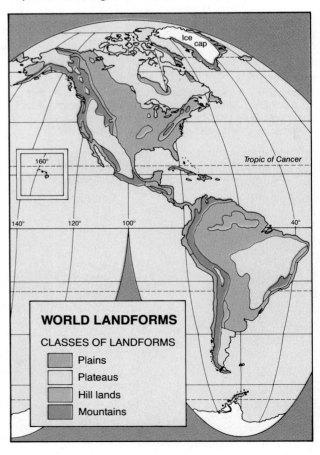

WORLD LANDFORMS

CLASSES OF LANDFORMS

Plains

Plateaus

Hill lands

Mountains

more, although some are considerably lower. To qualify as plateaus in the strictest sense, such areas should be terminated on at least one side by a steep edge, or *escarpment*, marking an abrupt transition from the plateau surface to areas at a lower elevation. However, the word *plateau* is often used loosely in referring to relatively level areas lying at considerable heights, whether terminated by a definite escarpment or not. Well-defined plateaus are often spoken of as "tablelands." To complicate matters, many plateaus in areas with considerable rainfall have been cut into predominantly hilly or even mountainous country by their many streams, yet are still commonly referred to as plateaus.

In *hill lands* and *mountains,* the vertical dimension predominates. No sure way has been devised to distinguish between these two classes of landforms (Fig. 2.12). In general, mountains are higher and more rugged than hill lands, are more of a barrier to movement, and generally offer fewer possibilities for human settlement and use. Whereas many areas of hill land support moderate to dense populations, most mountain areas are sparsely populated. Local usage of the term *mountain* varies greatly from place to place over the earth. In plains country, areas of a

Figure 2.11 *A flat plain vegetated by steppe grassland in western New South Wales, Australia, near the mining town of Broken Hill.*

few hundred feet elevation which stand out conspicuously above their surroundings may be called "mountains" by the local inhabitants. However, the term is most commonly used in reference to comparatively rugged areas lying at least 2000 to 3000 feet (*c.* 600 to 900 m) above sea level.

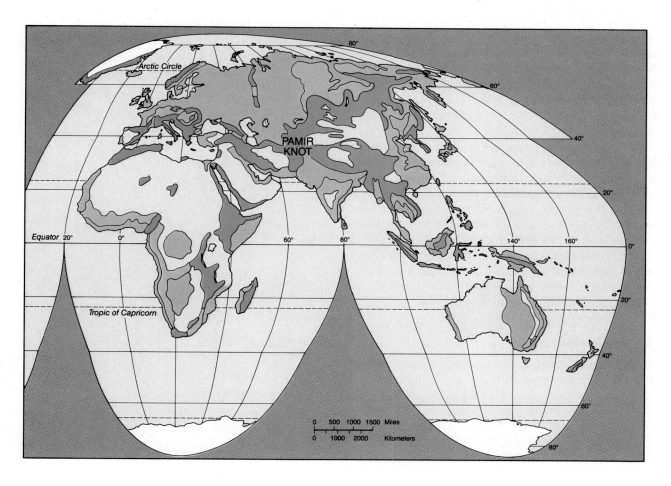

Figure 2.12   *The ridge that bounds this central Pennsylvania valley might be classified as either a high hill or a low mountain. The view features Amish ("Pennsylvania Dutch") farms in the Appalachian Ridge and Valley Section.*

CLIMATIC PROCESSES AND THEIR GEOGRAPHIC RESULTS

The great climatic variety of the earth results primarily from areal differences in the processes by which the earth's atmosphere is heated and cooled. The fundamental processes are the radiation and absorption of heat energy. Solar energy (insolation) reaches the earth in the form of short-wave radiation, some of which is visible as light. A considerable part of the total radiation never reaches the ground but is reflected back into space or absorbed by water vapor, carbon dioxide, other gases, or dust particles in the atmosphere. Most of the solar energy that reaches the land or ocean surface is absorbed, and some of it is returned to the atmosphere in the form of long-wave radiation, which generates heat and is the principal agent that warms the atmosphere. Thus the air is heated mainly from the underlying land or water surface, although ultimately and indirectly, of course, the source of the heat is the sun.

Air Temperature

Of primary geographic consequence is the fact that the earth varies greatly from place to place in the total amount of energy received annually from the sun and in the resulting air temperatures. If we omit the effects of cloud cover, we may say that the lower the latitude of a place (the nearer the equator it lies), the more solar energy it receives annually. One rea-

son is that on the average throughout the year, the sun's rays strike the earth more nearly vertically at lower latitudes, concentrating a given amount of solar energy on a smaller extent of surface than at higher latitudes (Fig. 2.13). A second reason is that when they approach the earth at a more nearly vertical angle in lower latitudes, the solar rays must pass through a smaller thickness of absorbing and reflecting atmosphere before reaching the surface than is the case at higher latitudes (Fig. 2.13). However, many areas near the equator have a relatively abundant cloud cover, which reflects or absorbs much of the incoming solar radiation and reduces the amount reaching the surface. The result is that the areas of highest annual solar energy actually received at the surface appear to be in the vicinity of 20° to 30°N and S of the equator. In these zones, the average angle of receipt of the sun's rays is still close to vertical, while the cloud cover tends to be much less.

Processes transferring heat from one part of the earth to another are essential in maintaining earth's habitability. The differences in energy received from the sun are such that places near the equator receive more energy annually than they lose by radiation to

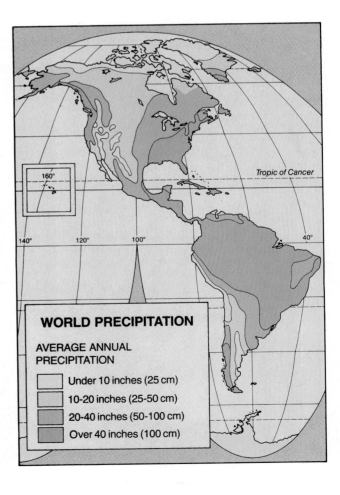

WORLD PRECIPITATION

AVERAGE ANNUAL PRECIPITATION

Under 10 inches (25 cm)

10-20 inches (25-50 cm)

20-40 inches (50-100 cm)

Over 40 inches (100 cm)

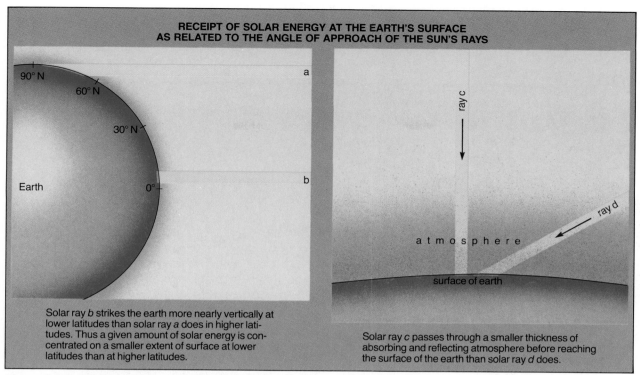

RECEIPT OF SOLAR ENERGY AT THE EARTH'S SURFACE
AS RELATED TO THE ANGLE OF APPROACH OF THE SUN'S RAYS

Solar ray *b* strikes the earth more nearly vertically at lower latitudes than solar ray *a* does in higher latitudes. Thus a given amount of solar energy is concentrated on a smaller extent of surface at lower latitudes than at higher latitudes.

Solar ray *c* passes through a smaller thickness of absorbing and reflecting atmosphere before reaching the surface of the earth than solar ray *d* does.

Figure 2.13   Diagrams showing receipt of solar energy as related to earth curvature and thickness of atmosphere.

Figure 2.14   World precipitation map, modified from a map by the United States Department of Agriculture.

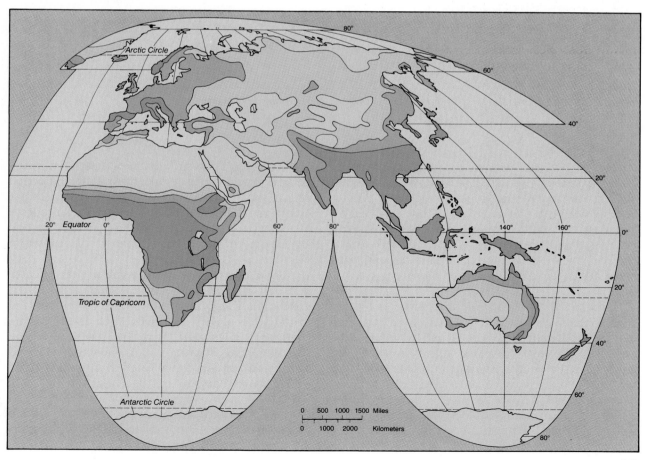

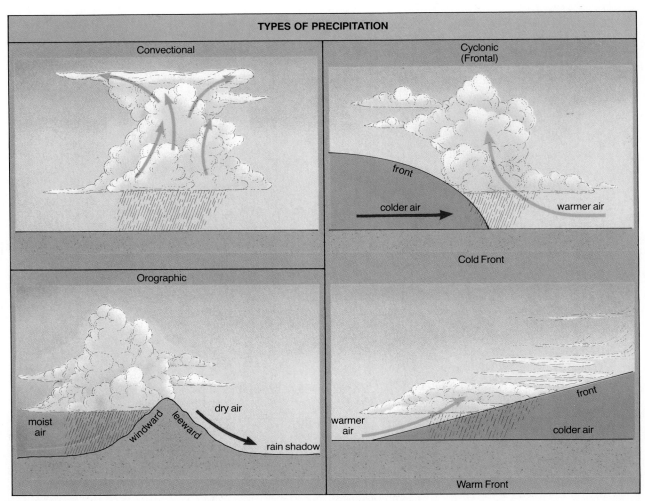

Figure 2.15 *Diagrams showing origins of convectional, orographic, and cyclonic precipitation.*

outer space, while places near the poles radiate more energy than they receive. If heat were not transferred after its original receipt, low-latitude areas would become steadily hotter and high-latitude areas steadily colder until both were uninhabitable. But warm air masses and ocean currents originating in lower latitudes transport heat to higher latitudes and thus maintain an equilibrium ("heat balance") on the earth.

Extremes of heat and cold are tempered by the equalizing effects of moving air and ocean waters, but great differences in annual and seasonal temperatures from place to place still remain. In low-lands near the equator, temperatures remain high throughout the year, whereas in areas near the poles temperatures remain low for most of the year. In the intermediate (middle) latitudes, well-marked seasonal changes of temperature occur, with warmer temperatures associated generally with the period of

high sun (sun's rays striking the surface more nearly vertically) and cooler temperatures with the period of low sun (angle of impact more oblique). In addition, variability of temperature in these latitudes is increased by periodic incursions of polar or tropical air masses, bringing unseasonably cold or warm weather.

Absorption of earth radiation by air that is denser and that contains more water vapor and dust is greater than absorption by air that is thinner and contains less of these materials. Such differences are characteristic between air at lower elevations and that at higher elevations. Consequently, temperatures at all latitudes are affected by altitude. On the average, air temperature decreases about 3.6°F (2.0°C) for each increase of 1000 feet (305 m) in altitude. As a result, a place on the equator at an elevation of several thousand feet may have an average annual temperature comparable to that of a

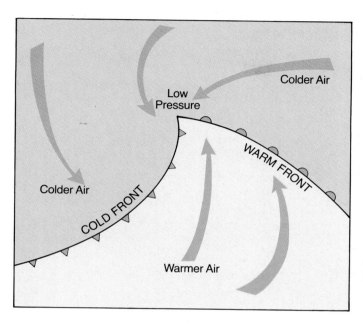

Figure 2.16   Diagram of a cyclone (Northern Hemisphere). The entire system is moving west to east in the airstream of the westerly winds.

place in a middle-latitude lowland, although the tropical place will not experience the seasonal temperature changes of the middle latitudes.

Precipitation

Processes which cool the air are responsible for precipitation (see Fig. 2.14, p. 29) and diagram, Fig. 2.15). Precipitation results when water vapor in the atmosphere is cooled to the point that it condenses and thus changes from a gaseous to a liquid or solid form. How much cooling is necessary depends on the original temperature of the air and the amount of water vapor in it, as warm air can hold more water vapor than cool air. Differences from place to place in type of air and occurrence of cooling processes sufficient to give precipitation are great enough that average annual precipitation ranges from practically none in many places to 400 inches (*c.* 1000 cm) or more annually in a few places.

For cooling and precipitation to occur, it is generally necessary for air to rise. This may be brought about in several ways. In equatorial latitudes, or in the high-sun period elsewhere, air heated by intense surface radiation may rise rapidly, cool rapidly, and produce a heavy downpour of rain. Precipitation that originates in this way is called *convectional precipitation* (Fig. 2.15). *Orographic* (mountain-associated) *precipitation* results when moving air strikes a topographic barrier and is forced upward. Some of the heaviest rainfall and snowfall in the world is produced in this way. Most of the moisture falls on the windward side of the barrier, and the lee side is apt to be excessively dry. Such dry areas are said to

be areas of *rain shadow.* In some parts of the world, the existence of arid or semiarid lands is primarily due to rain shadow. *Cyclonic* or *frontal precipitation* is generated in traveling low-pressure cells (*cyclones*), which bring different air masses into contact (Figs. 2.15 and 2.16). A cyclone normally comprises a very extensive segment of the atmosphere and may overlie hundreds of thousands of square miles of the earth's surface. In the atmosphere, air moves from high pressure to low, just as water runs downhill, and a cyclone is a portion of the atmosphere into which different air masses are drawn. One air mass will normally be cooler, drier, and more stable than the other. Such masses do not mix readily, but each tends to retain its separate identity. They are in contact within a zone which may be 3 to 50 miles (*c.* 5 to 80 km) wide, and is called a *front.* A front is named according to the air mass which is advancing. In a *cold front,* the cold air pushes in under the warm air and forces it upward and back. In a *warm front,* the warm air rides up over the cold air and gradually pushes it back. But whether the front is a warm front or a cold front, precipitation is apt to result when the warmer air mass rises and condensation takes place. The middle latitudes, especially in winter, experience a succession of traveling cyclones moving from west to east in an airstream called the *westerly winds* (Fig. 2.17). These cyclones rotate slowly and may be thought of as somewhat analogous to whirls and eddies that are carried along by a running stream. Normally, two fronts are experienced as a cyclone passes. First comes the warm front, and then the cold front. But the cold front moves faster and eventually overtakes the warm front. When this oc-

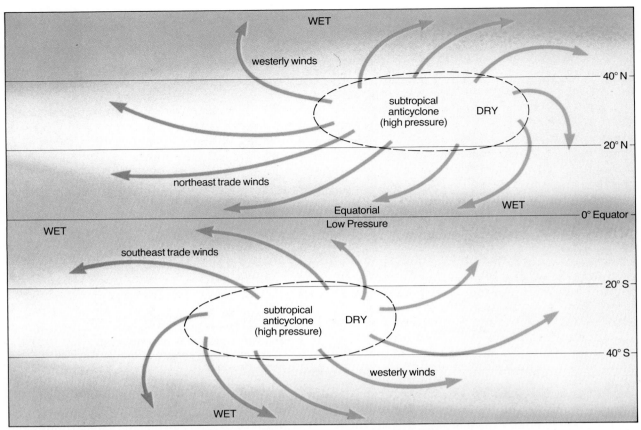

WET

westerly winds

40° N

subtropical
anticyclone
(high pressure)   DRY

20° N

northeast trade winds

WET

Equatorial
Low Pressure   0° Equator

WET

southeast trade winds

20° S

subtropical
anticyclone
(high pressure)   DRY

40° S

westerly winds

WET

*Figure 2.17   Idealized wind and pressure systems.
Irregular shading indicates wetter areas.*

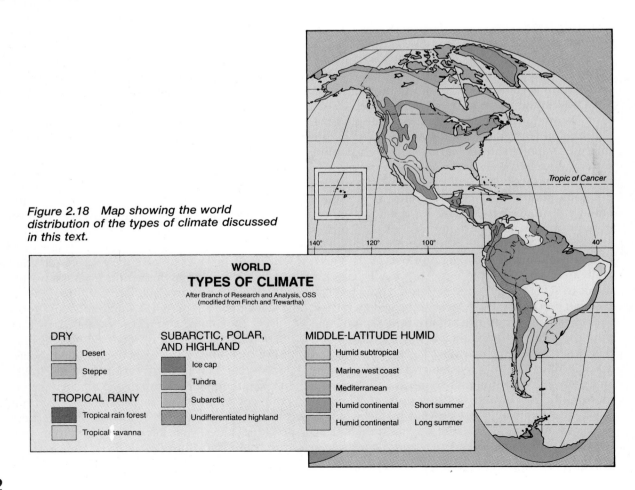

*Figure 2.18   Map showing the world
distribution of the types of climate discussed
in this text.*

Tropic of Cancer

140°   120°   100°   40°

WORLD
TYPES OF CLIMATE
After Branch of Research and Analysis, OSS
(modified from Finch and Trewartha)

DRY

Desert

Steppe

TROPICAL RAINY

Tropical rain forest

Tropical savanna

**SUBARCTIC, POLAR,
AND HIGHLAND**

Ice cap

Tundra

Subarctic

Undifferentiated highland

MIDDLE-LATITUDE HUMID

Humid subtropical

Marine west coast

Mediterranean

Humid continental   Short summer

Humid continental   Long summer

curs, the cyclone is said to be occluded, and it disappears from the atmospheric pressure map.

Large areas of the earth receive very low amounts of precipitation (see Fig. 2.14). Often such areas are primarily the result of high atmospheric pressure. Some parts of the atmosphere generally exhibit high pressures; these are known as semipermanent high-pressure cells or *anticyclones* (Fig. 2.17). In such a high, the air is descending and thus becomes warmer as it comes under the increased pressure (weight) of the air above it. As it warms, its capacity to hold water vapor increases, and the result is minimal precipitation. The streams of dry, stable air that move outward from the anticyclones are apt to bring prolonged drought to the areas that lie in their path. Such is often the effect of the *trade winds*—streams of air that originate in semipermanent anticyclones on the margins of the tropics and are attracted equatorward by a semipermanent low-pressure cell, the *equatorial low.* In some parts of the world, cold ocean waters appear to be mainly responsible for the existence of coastal deserts. Air moving from sea to land is warmed; instead of yielding precipitation, its capacity to hold water vapor is increased. But many,

perhaps most, areas of excessively low precipitation stem from a combination of the foregoing influences. The Sahara of northern Africa, for example, seems to be primarily the result of high atmospheric pressures but is also due in part to the rain-shadow effect of the Atlas Mountains and the presence of cold Atlantic waters along its western coast.

The Major Types of Climate

The foregoing processes produce an infinite variety of local climates. But it is possible to group these into a limited number of major *types of climate*, each of which occurs in more than one part of the world (Fig. 2.18). In the *ice-cap climate, tundra climate,* and *subarctic climate,* the dominant feature is the long, severely cold winter. In *desert* and *steppe climates* the dominant feature is aridity or semiaridity. Deserts and steppes occur both in low latitudes and in middle latitudes; the principal area, sometimes called the ''Dry World,'' extends in a broad band across northern Africa and southwestern and central Asia. Rainy low-latitude climates include the *tropical rain forest* type and the *tropical savanna* type—the critical dif-

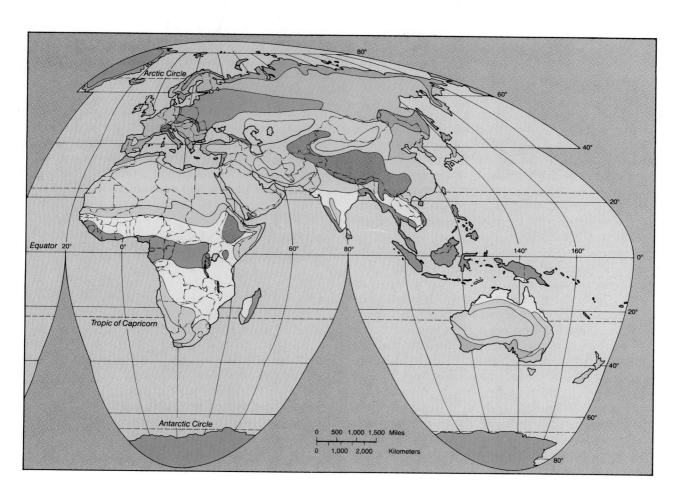

Figure 2.19   Map showing major categories of world vegetation, modified from a map by the United States Department of Agriculture.

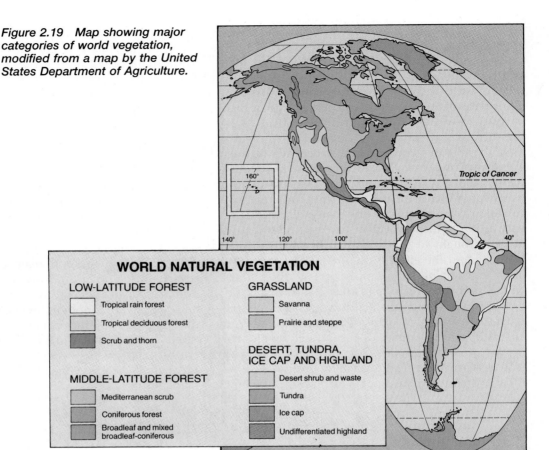

WORLD NATURAL VEGETATION

LOW-LATITUDE FOREST

☐ Tropical rain forest

☐ Tropical deciduous forest

☐ Scrub and thorn

MIDDLE-LATITUDE FOREST

☐ Mediterranean scrub

☐ Coniferous forest

☐ Broadleaf and mixed broadleaf-coniferous

GRASSLAND

☐ Savanna

☐ Prairie and steppe

DESERT, TUNDRA, ICE CAP AND HIGHLAND

☐ Desert shrub and waste

☐ Tundra

☐ Ice cap

☐ Undifferentiated highland

ference between them being that the tropical savanna type has a pronounced dry season, whereas the dry season is short or absent in the tropical rain forest climate. The humid middle-latitude climates have mild to hot summers and winters ranging from mild to cold. There are several relatively distinct types of climate in this group. The *marine west coast climate,* occupying the western sides of continents in the higher middle latitudes, has winters that are greatly moderated by the effects of warm ocean currents and summers that tend to be on the cool side, although still warm enough for a great variety of crops. The *mediterranean (dry-summer subtropical) climate* customarily occupies an intermediate location between a marine west coast climate on the poleward side and a steppe or desert climate on the equatorward side; in the high-sun period, it lies under high atmospheric pressures and is rainless; in the low-sun period, it lies in the westerly wind belt and receives precipitation of cyclonic or orographic origin. The *humid subtropical climate* occupies locations on the southeastern margins of continents and is characterized by hot summers, mild to cool winters, and ample precipitation for agriculture. The *humid*

continental climate lies poleward of the humid subtropical type; it has cold winters, warm to hot summers, and sufficient rainfall for agriculture, with the greater part of the precipitation in the summer half-year. *Highland climates* exhibit a range of conditions according to altitude and exposure to wind and sun.

BIOTIC PROCESSES AND NATURAL VEGETATION

Biotic processes are those which clothe the earth with living things, including humans. Geographic interest in these processes and their results centers on those aspects that differ from one area to another. It does not focus primarily on aspects of biology which are universal or have little effect in rendering one area distinct from another.

One of the most striking and significant features of most areas is their natural vegetation—the part of their vegetation cover that is not cultivated or domesticated. The natural vegetation of an area is striking because it contributes so much to the appearance of an area's landscapes. It is significant because it

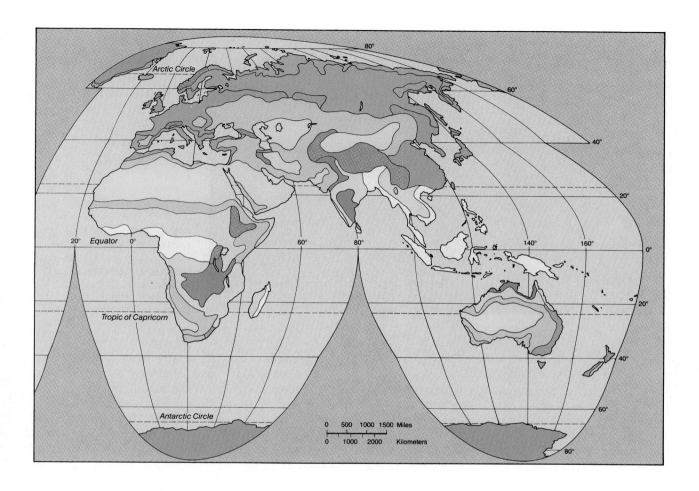

provides resources or hindrances for human activities and because it is very closely related to many other aspects of the area's natural environment. The particular type of natural vegetation in an area is influenced strongly, through various processes, by climatic conditions, light level and duration, slope and drainage, rock and soil type, and the area's wild fauna, as well as the evolutionary history of the flora itself. In addition, despite being called "natural" vegetation, it is often greatly influenced by such actions of people as burning, clearing, and the grazing of livestock. In turn, the natural vegetation exercises a powerful influence on certain other characteristics of an area, such as the quality of the soil, the speed of erosion, the nature of wild animal life, and the presence or absence of some economic potentials.

The geographic importance of different processes related to natural vegetation varies with the degree of generalization one employs. On a broadly generalized worldwide scale of observation and comparison, climatic conditions related primarily to intake of water by plants through their root systems and loss of it to the atmosphere by transpiration are the most important. Major types of natural vegetation are closely enough related to major types of climate that the vegetative results of the climatic conditions are often used to name the climate types, for example, the tropical rain forest type of climate or the tundra type of climate. However, when observation becomes more detailed, other factors assume considerable importance, and variations in natural vegetation are found to be related to various aspects of the natural environment other than climate.

Forests

When one looks at a map showing the broadly generalized world pattern of vegetation types (Fig. 2.19), it is apparent that broadleaf forest tends to occupy the areas which are climatically most favorable to vegetative growth. Where heat and moisture are continuously, or almost continuously, available, the *tropical rain forest* is found (Fig. 2.20). It is composed of broadleaf trees and is evergreen because the climate imposes no dormant season. The trees are comprised of more species per unit area than in any other forest, due to the ability of many species to compete in this optimum environment. Many trees

Figure 2.20 Clearings of settlers in a tropical rain forest along the Amazon River near Manaus, Brazil. The houses are built on stilts as a means of escaping the annual floods.

grow straight and very tall in their competition with each other for sun, although smaller species requiring less sun flourish under the shade of the forest giants. At ground level, the forest floor is so well shaded that there is little undergrowth until some circumstance, such as the occurrence of a river, breaks the dense leaf canopy overhead. In tropical areas with a noticeable dry season, but still enough moisture available for tree growth, the tropical rain forest is replaced by *tropical deciduous forest.* The broadleaf trees lose their leaves and cease to grow during the dry season and then put on foliage and resume their growth during the wet season. The tropical deciduous forest approaches the luxuriance of rain forest in wetter areas but declines to low, sparse scrub in drier areas. In middle-latitude areas with hot, wet summers, a *midlatitude deciduous forest* is found, composed of broadleaf species (oaks, maples, and the like) that are different from those in the tropical areas. Here the cold temperatures of winter freeze the water within reach of plant roots and render it seasonally unavailable. The trees shed their leaves and cease to grow, thus reducing water loss, and then resume their foliage and grow vigorously during the hot, wet summer when climatic conditions resemble those of rainy tropical lands.

Where water availability is high, needleleaf trees such as pine, spruce, and fir cannot, in general, compete successfully with the broadleaf trees. But in middle-latitude forested areas where water conditions are somewhat less favorable, the needleleaf evergreen trees tend to become dominant. These trees, which are often called coniferous because most of them bear seed cones, are adaptable to a wide range of climatic conditions. In particular, they can stand long periods of freezing temperatures with attendant lack of water. The most extensive needleleaf evergreen forests lie north of the broadleaf forests in the regions of subarctic climate which cover great areas of North America and Eurasia. They are generally called the northern coniferous or boreal forests, or, after Russian terminology, the *taiga.* The trees in these forests reach a considerable size and density in more favorable southerly locations but become progressively smaller and sparser toward the north (see Fig. 26.5). Such forests are absent in the Southern Hemisphere because land masses in that hemisphere do not reach far enough south to provide the long, cold winters and short, warm summers in which the taiga forests develop. However, needleleaf evergreen forests occupy some locations where temperature conditions are less severe than those of the taiga, for example:

1. In certain situations hot and wet enough for broadleaf trees, a permeable sandy soil allows much water to percolate downward out of reach of tree roots, thus decreasing the available moisture and giving an advantage to needleleaf over broadleaf trees. An outstanding instance of this is the pine forests of the coastal plain along the Gulf of Mexico and Atlantic coasts of the American South. Many experts believe, however, that repeated burning of this area helped to drive out the broadleaf trees and establish the dominance of pine. (Needleleaf trees are generally less vulnerable to fire than are broadleaf trees.)

2. Needleleaf conifers are also dominant in some areas along the west coasts of continents which have the marine west coast type of cli-

mate (see Fig. 27.5). These areas are very wet, with mild winters and summers that are warm but seldom hot. Although they appear to offer good conditions for broadleaf trees and stands of these are found, the forest is predominantly needleleaf. The reasons for this are not fully known, but the circumstance is particularly fortunate for the United States and Canada, where the most extensive stands occur. Here the redwoods of northern California and southern Oregon and the Douglas fir of Washington, Oregon, and southern British Columbia grow to unusual size in the favorably moist conditions and are highly valuable commercially.

3. Finally, needleleaf forests are also found in certain tropical areas, principally in uplands or on very permeable soils.

As a rule, broad types of vegetation do not change abruptly along a line of division, although such a change may be represented symbolically by a line on a map. Rather, one type tends to grade into another through a transition zone in which broadleaf trees and needleleaf trees compete with each other and are intermingled. Such transitional areas are designated as areas of *mixed forest.* They occur both north and south of the broadleaf forest zone of eastern United States, cover sizable areas in Europe and the Soviet Union, and occur in various other parts of the world. Within them, stands of broadleaf trees often occupy places which are more favorable to tree growth, and needleleaf stands occupy less favorable spots.

Grasslands

Where still less moisture is available, trees cannot flourish and grass becomes the dominant vegetation. If there were no trees, grass would be dominant over most of the world, as it is extremely adaptable to a very wide range of climatic conditions. But in areas wet enough to support forests, it tends to be shaded out; thus, it is ordinarily dominant only in areas with environments hostile to trees. One reason for its hardiness is that so much of the plant is underground in an extensive root system.

Grasslands that have short grass are called *steppes* (chapter-opening photo of Chapter 20). They are the semiarid transition zones between humid areas and truly arid deserts. They occur in areas with widely differing temperature conditions, both in the tropics and in middle latitudes. In general, their grasses grow higher and denser toward their wetter edges, and lower and sparser toward their drier edges. The amount of precipitation under which a steppe develops depends largely on temperature. Where temperatures are higher, evaporation is faster and an

Figure 2.21   *A view of vegetation and land use in the tropical savanna climate of southeastern Zaire. In the foreground, patches of manioc (the darkish area) and corn are bordered by coarse savanna grass. Beyond are the huts of a village lining both sides of a dirt road.*

area can have a considerable amount of precipitation and still be too dry for forest; while areas with cooler temperatures have less evaporation and thus require a lower amount of precipitation to rule out forest and leave the land to grass. The major area of steppe in the United States occupies a broad band of country in the Great Plains along the eastern foot of the Rocky Mountains..

However, very large areas of grassland occur in some parts of the world where conditions do not appear to rule out a forest cover. These are the *prairies* of the middle latitudes and the *tropical savannas* of the low latitudes. The world's greatest expanse of prairie, now almost completely destroyed in favor of agriculture, was a triangular area in North America with its corners in Alberta, Texas, and Indiana (see Fig. 28.18). Before the area was settled, tall grasses ranging from knee-high to horse-high were dominant, although heat and moisture are sufficient for forest growth. The reasons for this anomaly are in dispute. There is a strong presumption that one important factor was repeated burning, which helped grass to replace forest, the latter being more vulnerable to fire.

Tropical savannas occur, like tropical deciduous forests, in low-latitude areas with marked wet and dry seasons (Fig. 2.21). The savannas have scattered trees, and the deciduous forests are likely to have some grass on the forest floor. The boundary between them is often indistinct and debatable, with forest gradually giving way to savanna as the dry season becomes longer and grass predominates more

Figure 2.22 World map showing major soil categories, modified from a map by the United States Department of Agriculture. A systematic classification scheme more recently developed is known as the Seventh Approximation. It utilizes terms such as aridisols (for most soils of deserts and semideserts) or mollisols (for most mid-latitude grassland soils). A world map embodying this classification can be found in Goode's World Atlas (Chicago: Rand McNally & Company).

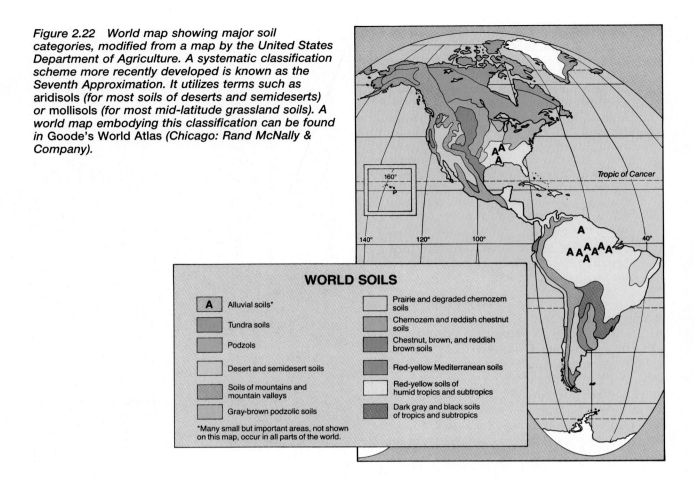

WORLD SOILS

A	Alluvial soils*		Prairie and degraded chernozem soils
	Tundra soils		Chernozem and reddish chestnut soils
	Podzols		Chestnut, brown, and reddish brown soils
	Desert and semidesert soils		Red-yellow Mediterranean soils
	Soils of mountains and mountain valleys		Red-yellow soils of humid tropics and subtropics
	Gray-brown podzolic soils		Dark gray and black soils of tropics and subtropics

*Many small but important areas, not shown on this map, occur in all parts of the world.

and more over trees. As in the case of prairies, the origin of the vast savannas of the tropics is a matter of scientific debate, with the presumption that repeated burning has probably been at least one important factor favoring grasses over trees. According to local conditions, savanna grasses vary between about 1 foot and 12 feet (0.3 to 3.66 m) in height.

Vegetation of Deserts, "Mediterranean" Climate, and Tundra

Three extreme climatic environments in which aridity plays a large role—desert, "mediterranean" climate, and tundra—are characterized by distinctive types of vegetation. The *deserts* of middle and low latitudes are too dry to support a cover of either trees or grass (see Figs. 2.7 and 26.11). For the most part, they have scattered *desert shrub* vegetation, although exceptional areas may have no vegetation at all. Desert shrub consists of bushy plants which are *xerophytic;* that is, they possess special adaptations for withstanding drought. Among these adaptations are (1) small leaves, providing a minimal leaf surface for transpiration (loss of water to the atmosphere), (2) special qualities of leaf and bark to reduce transpi-

ration, and (3) extensive or deep root systems to gather moisture.

A rather similar, although more luxuriant, vegetation characterizes areas of the "mediterranean" type of climate. This climate exhibits an extreme quality in the desert-like dryness of its hot summers, which are combined with mild, wet winters. Due to the summers, the vegetation must be xerophytic. It consists primarily of many species and sizes of shrubs. These are predominantly evergreen and broadleaf, with adaptations such as thick bark and waxy or leathery leaves to retard transpiration and thus conserve moisture during the dry summers. Such vegetation is called by a variety of names, including *maquis* and *chaparral* (see Fig. 28.23).

Still another extreme environment is the *tundra* (see Fig. 26.4). Here vegetation must be adapted to survive a long, cold winter when moisture is unobtainable because it is frozen, and a very short and very cool summer. Tundra vegetation includes mosses, lichens, shrubs dwarfed trees, and some grass in favored spots. Large areas have practically no vegetation. As in deserts after a rare summer rainfall, ephemeral wild flowers will bloom during the summer.

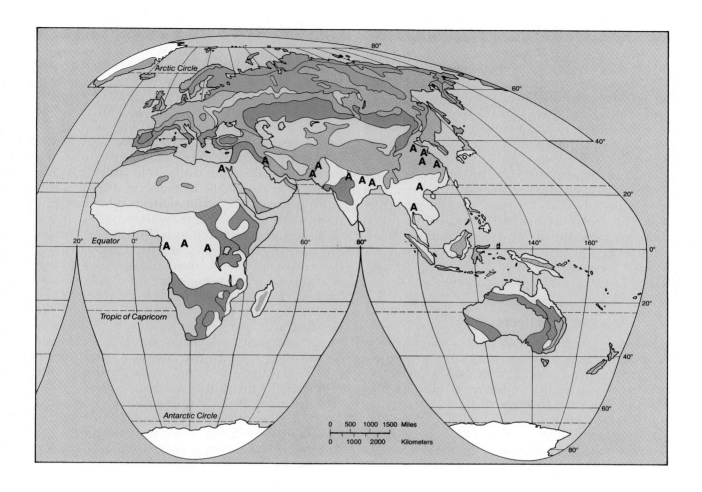

Summary

As we have seen, a process of competition among plants for moisture appears to explain in broad outline the distribution of major types of natural vegetation in the world. Broadleaf forests generally require more moisture than other types of natural vegetation and, where they can flourish, tend to exclude other types. Needleleaf forests can exist within a wider range of moisture conditions. They can and do exist in very moist conditions but are usually excluded from such areas by the competition of broadleaf trees. However, they can also flourish in areas that have a smaller total of available moisture and in areas where a long dry season must be endured each year. Hence, they often, but not always, occupy areas less favorable in moisture supply than those occupied by the broadleaf type of forest. Grass can grow in a still wider range of conditions, from areas wet enough for rain forest to areas of near-desert conditions. It tends to be dominated or excluded by forest in wetter areas, however, and to be left as the dominant natural vegetation only in areas too dry for forest growth. Where even grass is largely excluded, shrubs and tundra plants are dominant.

SOIL-FORMING PROCESSES AND SOIL QUALITY

Variations from place to place in the thin earth mantle of decomposed rock and decayed organic material called *soil* are among the most significant of geographic patterns (Fig. 2.22). The quantity and quality of foodstuffs and other vegetable products which humanity can wrest from various parts of the earth are intimately related to the characteristics of the soil and to the ability of farmers to take advantage of these characteristics or to overcome them. And most other development of areas is severely handicapped if the soil cannot be made to yield an adequate agricultural production.

Soils vary enormously and in extremely complex ways in response to a wide variety of factors. These are often quite intricate in their operation. Some soils, even with little care, will yield abundant and varied agricultural products for many years. Some will yield very small quantities or only a small variety of useful products or will deteriorate rapidly with use. Almost every intermediate point on the spectrum of soil quality between these extremes occurs in various places on the earth, presenting humanity with innumerable opportunities and problems in the exploitation of this vital resource.

Despite this complexity, however, it is possible to call attention to a few major circumstances and processes which help to make the geography of soils comprehensible in broad outline. This possibility is due to the fact that the characteristics of soils are closely related to other major aspects of nature—notably to climate, vegetation, and rock types. Many of the important characteristics of a soil are often due to the type of rock whose weathering has provided the initial soil material. Other soil characteristics are due to the effects of temperature and moisture, while still others are due to the effects of the natural vegetation whose decomposed remains, called *humus*, are incorporated into the soil.

Leaching and Its Effects

Leaching is a very important process contributing to broad and significant differences in the soil from one area to another. Plants are nourished by taking into their roots, in solution, a variety of minerals from the soil. The more adequate the mineral content of the soil, the more fertile the soil will be, other things being equal. Leaching is the process by which mineral plant foods are removed from the soil in solution. Since plants cannot utilize minerals unless the latter are dissolved in water, adequate water content within the soil is absolutely essential. Excess water, however, will dissolve the important minerals and carry them by underground percolation into strata beyond the reach of plant roots (eventually to rivers and the sea). Thus, leaching is a process of impoverishment of the soil, lowering its useful mineral content. As this process depends upon moisture supply, there is a strong general tendency for areas which receive more precipitation to have soils that are more leached. Because they are more lacking in minerals required by plants, these soils are less fertile than many soils in areas of lower precipitation. Such areas are generally wet enough to have a natural vegetation of forest. Thus, as a broad generalization, we can say that humid, naturally forested areas tend to have soils impoverished by leaching. On the other hand, the soils of drier areas in which the natural vegetation is grassland or desert vegetation tend to be leached relatively little or not at all. Unfortunately, the high content of mineral nutrients in such soils may be of limited utility or no utility because of inadequate moisture for most plants. In addition, too high a concentration of mineral salts, as in many dry-area soils, may itself be toxic to useful plants.

The amount of leaching a soil is subjected to is not entirely a function of the amount of precipitation in the area. The type of vegetation and temperature conditions can also influence it strongly. Areas with a natural vegetation of needleleaf forest tend to have heavily leached soils, partly because the mat of needles which accumulates on top of the soil imparts an acid character to water seeping through it. The resulting acidity of the water in the soil tends to speed the dissolution of mineral nutrients. Acidity of soil water under broadleaf forests tends to be less strong, with the result that soils in such areas, particularly in the middle latitudes, tend to be somewhat better supplied with nutrients. Warm temperatures also speed leaching. Where winters are not cold enough for the soil to be frozen, the leaching process can proceed the year around instead of only part of each year. In addition, high temperatures bring about a general speeding up of chemical reactions, such as the dissolution of mineral plant foods. For the foregoing reasons, very heavily leached soils are characteristic both of subarctic areas, with their vegetation of needleleaf forest, and of many tropical areas that have heavy precipitation and high temperatures. Such areas stand at the opposite end of the scale of soil leaching from deserts and dry grasslands. Heavily leached soils in the tropics, like those of subarctic areas, tend to be highly acid in reaction, due to the fact that bases are leached out. A highly acid chemical environment is unfavorable for most crop plants.

Decay and Absorption of Organic Material

A second important process contributing to broad differentials of soil fertility in different parts of the world is the decay and absorption into the soil of organic material. Such material is derived principally from the vegetation which grows on the land: natural vegetation before use of the soil or in intervals of nonuse, and crop remains in soils being farmed. This decomposed organic soil material, called *humus*, is a very important element in soil productivity. It adds to the store of mineral plant foods in the soil, especially the very critical element nitrogen; contributes to a physical structure of the soil which favors air circulation and root development; adds to the ability of a soil to store water, and retards leaching. Rock particles without organic material do not form a true soil.

Humus accumulation in the soil under natural conditions varies greatly from one part of the earth to another. It is generally highest in grassland soils because dryness slows the action of microorganisms which destroy organic material. It is somewhat lower in midlatitude broadleaf forest areas, which are wetter. It is very low in three types of areas: (1) deserts, which have little plant cover to provide humus; (2) wet tropical areas, where heat, moisture, and the activity of microorganisms are so intense as to destroy and remove most dead organic material before it can be incorporated in the soil; and (3) areas with a needleleaf forest vegetation, where the mat of needles on the surface of the soil is resistant to decomposition—an effect especially marked in cold environments where the action of microorganisms is greatly retarded.

Other Factors Affecting Soil Quality

The amount of leaching and of humus are important determinants of the quality of soils, but there are, in addition, a number of other factors which display less regular patterns of relationship with climate and vegetation.

1. Some areas have stony soils where the rock has not been completely weathered. Such soils are likely to be difficult to work.

2. The soils of poorly drained areas tend to be waterlogged to such an extent that air in the soil, necessary for most plants, is deficient. However, they have a very high organic content and often become quite fertile when drained.

3. Soils which are too sandy are usually poor for several reasons. Sand is composed of particles of relatively large size compared to those that compose silt or clay. It offers a relatively small particle-surface area within a given volume compared with soils made up of finer particles. This handicaps it in supporting plant life, since plant roots feed only from the surface area of particles. In addition, sand holds water less well than soils composed of finer particles and therefore often provides drought conditions for plants in areas where moisture might otherwise be adequate.

4. Claypan or hardpan soils often develop in very level areas in both middle and low latitudes. These have a relatively impermeable layer not far below the surface, due to a high concentration of small particles at certain depths as a result of water movements within the soil. Sometimes these particles are actually cemented. The result is retardation of downward and upward movement of water within the soil. After a rain, the soil tends to waterlog, as water cannot permeate downward through the pan; in a dry period, the soil becomes excessively dry, as water from below cannot move upward through the pan by capillary action to replace that lost by evaporation and transpiration. The pan may also physically inhibit the downward penetration of roots. However, it may be valuable for some kinds

of agriculture, for instance in ponding water in a rice paddy.

Certain widespread conditions and processes often result in soils superior to those of surrounding areas with similar climatic and vegetative conditions.

1. In many places, soils resulting from the weathering of underlying limestone have superior character. A high lime content in a soil contributes to giving it an internal structure with abundant pore space for circulation of water and air, and it counteracts acidity. In humid areas, lime is often applied to cultivated land for these purposes.

2. Soils formed from alluvium deposited on the flood plains of rivers are often, although not always, unusually productive (Fig. 2.23). They are level and easily cultivated and may be subject to natural renewal of fertility by flooding. They are likely to have a rich humus content derived from swamp and marsh vegetation; they are also likely, of course, to require extensive flood control and drainage operations.

3. A third type of soil which is often unusually productive is soil derived from loess (see Fig. 28.19). Loess is material which has been picked up, transported, and deposited in its present location by wind. This means that it consists of very fine particles. Such particles, as mentioned previously, afford a large amount of feeding surface for plants, along with adequate pore space for the necessary air and water in the soil.

4. Finally, certain volcanic materials weather into exceptionally fertile soil. This is especially noteworthy in a number of tropical areas where generally poor soils may be interspersed with much better ones derived from volcanic deposits. In some locations, soil fertility is renewed intermittently by volcanic ash settling out of the air or by new lavas being eroded, transported by streams, and deposited as alluvium.

The Human Role in Determining Soil Quality

People themselves are agents of the greatest importance in determining the quality of soils. In cultivating the land, or grazing animals on it, they can either increase or decrease soil fertility. Their activities in decreasing fertility generally receive more notice. They include (1) destruction of the plant cover by overgrazing or clean-tilled cultivation, leading to excessive erosion; (2) repeated burning of the vegetation cover, thus destroying humus content; and

Figure 2.23 *The Nile Valley and Delta in Egypt is a famous exhibit of thick alluvium. This has been formed mainly of fertile particles eroded from rainy volcanic highlands in Ethiopia and then brought to Egypt by the annual Nile floods. Since most of this material now settles in Lake Nasser, the reservoir of Egypt's High Aswan Dam, the fertility of the land is no longer renewed to the same degree by the annual flooding. This view shows land in late autumn after the summer flood is over. The crop is sugarcane. The scene is in Middle Egypt across the river from the Karnak Temple at Luxor.*

(3) cropping the land too continuously, thus depleting reserves of mineral nutrients. However, people also can do many things—given proper knowledge, incentive, and means—to improve the quality of soils. Bogs can be drained; mineral content built up by fertilizers; nitrogen added by growing legumes, whose roots serve as hosts to nitrogen-fixing bacteria; acidity corrected by liming; erosion checked by proper cultivation and crop selection; and organic content increased by manuring. Thus, the activities of humans must be added to such natural factors as leaching, humus supply, and material of origin as a major reason why soils exhibit great differences from one region to another.

HUMAN GEOGRAPHIC PROCESSES

Defining natural processes that create spatial order and help differentiate the world into distinctive areas is less difficult than defining human processes that accomplish these ends. This is due to the fact that humans make decisions on the basis of motives which may be capricious and may, in fact, be unknown to the investigator. Human decision-making itself can be viewed as a process determining where people live, how they support themselves, what structures they create, what organizations they develop, and what relationships they evolve with each other and with nature. General examination of the decision process in the abstract is beyond the purview of this text. We can, however, profitably consider certain processes that spring from human de-

cision and have definable geographic results. We look first at the human geographic process of *natural population change* and the associated process of *migration.* This is conveniently followed by a discussion of the process of *innovation and diffusion of innovations.* Some illustrations of the geographic effects of migration and diffusion are then drawn from European expansion overseas. This is succeeded by a discussion of the interrelated processes of *natural-resource exploitation, site selection and development,* and *specialization and trade.* Finally, the geographic effects of cultural, political, and economic *inertia* and *geopolitical organization and interaction* are considered.

NATURAL POPULATION CHANGE

''Natural population change'' may be defined as the net increase of a population from an excess of births over deaths or the net decrease where deaths exceed births. The process is affected by such factors as individual decisions to have or not have children; the age at marriage; biological levels of fertility; the availability of contraceptives; the incidence of diseases and other hazards of life; the availability of medical services; religious or other cultural sanctions; and government policies concerning population increase and control. These factors are themselves a product of complex physical, biological, economic, and cultural circumstances. In fact, natural population change is intertwined with the entire life experience of a population.

The process of natural population change and the closely related process of migration have created the very uneven pattern of world population distribution (see Fig. 3.2) and are continuing to alter it. Today the rates of natural population change exhibit great variety from one country to another. For example, most African, Asian, and Latin American countries are increasing in population at rapid rates, whereas the countries of Europe have slow rates of increase and a few are actually decreasing in population. Scholars concerned with population have noted a *demographic transition*, whereby a country initially has a high birth rate, a high death rate, and a low rate of natural increase, then moves through a middle stage of high birth rate, low death rate (mainly a consequence of improved medical services and improved food supply), and high rate of natural increase, and ultimately reaches a third stage of low birth rate, low or medium death rate, and low or negative rate of increase. The majority of countries are now in the middle stage. Most of them are characterized by rates of natural increase ranging from 2.0 to 3.5 percent a year. By contrast, the European countries, the United States, Canada, Australia, New Zealand, the Soviet Union, Japan, Uruguay, and possibly China have reached the final stage. Nearly all have increase rates of 1.0 percent a year or less, for example, United States, 0.7; Soviet Union, 1.0; United Kingdom, 0.2; West Germany, −0.1 (1988 estimates). Differences in rates of natural population increase from country to country and from one major world region to another are expected to greatly change the map of world population distribution in the next few decades. Related changes in many other geographic patterns (land use, means of livelihood, urbanization, and so on) will also occur.

MIGRATION

Human migration—from country to country, continent to continent, old farm lands to new farm lands, countryside to city, city core to city periphery—is a major facet of human history. It has repeatedly changed the world's geography and is doing so in many places today. Migration may be temporary, periodic, or permanent; it brings about the settlement of empty lands, changes the population mix of lands already settled, may result in the displacement of one people by another, and often causes massive social problems. People migrate to escape disease or oppressors, to avoid starvation, to seek a better life, or simply because they are forced to move and are unable to help themselves. The origins, destinations, routes, and ultimate impacts of migration are topics of high geographic importance.

Innumerable examples of migration as a geographic process are cited in this text, for example,

the incursions of Europeans into all parts of the world since the 15th century; the prehistoric incursions of Asians and Australoids into the Pacific islands; the vast movements of people from countryside to city that have characterized most countries in the 20th century; the enslavement and transportation of Africans to the New World in the colonial age; and the flight of many millions of refugees from warfare, genocide, drought, flood, and famine in the 20th century.

INNOVATION AND DIFFUSION OF INNOVATIONS

The history of humanity's occupation, use, and organization of the earth has been punctuated by *innovation*—the discovery of new ways of thinking and new ways to do things. Examples are limitless—for example, the domestication of particular plants and animals, methods of tillage, industrial processes and equipment, the engineering of structures, the art of writing, improved methods of transportation and communication, languages, religions, value systems, and ideologies. The *diffusion* of innovations, generally to some areas and not to others, has brought about widespread geographic change and has been responsible for much of the earth's differentiation into the distinctive areas surveyed in this text. The collection of innovations that we call the Industrial Revolution diffused outward from western Europe to other parts of the globe, resulting in great geographic changes, many of which are still in progress. In this text we will not often examine the process of innovation and diffusion *per se*. However, we will frequently assess the geographic results of the fact that some areas discover or adopt new things and others do not, or do so more slowly. In the following section, we look at some global results of the diffusion of innovations from Europe.

Geographic Results of European Overseas Expansion

Since about 1500 A.D., the most far-reaching human influence on the geography of the world has been exerted by people of European culture who initially fanned out over the globe by sea from the European homeland. This movement, the "Expansion of Europe," is described in some detail in Chapter 4. Throughout this text, reference is continually made to the impact of Europeans on the geography of areas they have colonized, controlled, or influenced. The activities of Europeans revolutionized the geography of large sections of the earth. Most areas outside of Europe have been controlled politically by a European colonial power at one time or another, and most of the international boundaries that appear on the world political map were established by Europeans. The present division of the world into nationalistic sovereign states originated mainly in Europe, and European ideas about government and politics greatly influence the political life of most countries. Europeans have migrated in large numbers to overseas areas and have carried with them the languages and cultures of their homelands. Today most people in the Americas, Australia, and New Zealand, as well as minorities in South Africa and various other countries, speak European tongues as "first" or "home" languages. Even in postcolonial countries of Asia and Africa where there are few people of European descent, educated persons are generally conversant with the language of the European colonial power that was formerly in control. In many developing countries where regional languages are used, such a European language (for example, English in India) affords a common means of communication for educated persons from all sections of the country. Many young independent countries actually employ European tongues as official languages. It may be noted that the widespread use of these languages greatly simplifies the work of the United Nations and other international organizations. The Industrial Revolution which began in Europe has transformed large sections of the earth and is still spreading. Europeans organized, and in large measure still operate, the worldwide system of transportation, communication, and trade. Even the world's commercial agriculture has been developed mainly by Europeans or their descendants.

NATURAL-RESOURCE EXPLOITATION

The success of Europeans in developing a new industrial civilization and then diffusing portions of it to other lands has been closely related to Europe's possession or acquisition (often by conquest) of important natural resources that European culture developed the means to exploit. The general process of exploiting soils, minerals, and other natural resources has, of course, made possible the survival and spread of the human race, but some peoples and countries have surpassed others in gaining prosperity and power from resource exploitation. The United States—a country originally endowed with superlative natural resources, which it exploited with a technology partly borrowed from Europe and partly developed on its own—is a prominent example. The process of natural-resource exploitation creates

landscape patterns both directly and indirectly, and it contributes greatly to the disparities among areas in economic and social well-being and in political and economic power. Hence, it is an important factor in the analysis of spatial order and area differentiation in this text.

SITE SELECTION AND DEVELOPMENT

The exploitation of natural resources and the general organization of the earth's surface by humans necessitate the selection and development of particular sites, each characterized by a pre-existing association of locational traits. In early California, for example, Franciscan missionaries sited their missions in places where there was cultivable land accessible to irrigation water and near concentrations of Indians who might be Christianized and used as labor. Many of the earliest American farmers in the Midwest chose sites accessible to timber, running water, and small prairie grasslands providing good soil. Any number of cities in the world illustrate the choice of sites for particular locational reasons; for example, the choice of a site at the southern tip of Africa by the Dutch East India Company to establish the maritime way-station that became the modern city of Cape Town. Present-day facilities such as new mines, factories, shopping centers, planned "New Towns," or superhighways are not sited haphazardly, but only after careful study of environmental and locational factors. Knowing why a particular site was chosen for a particular enterprise, why the development of the site proceeded in a certain way, and what the consequences have been is vital in geographic understanding.

SPECIALIZATION AND TRADE

In the beginning, the sparse population of the prehistoric world may have been unspecialized and self-sufficient. But very early in humanity's prehistory, it proved advantageous for people to specialize and to trade the products and services generated. Specializations developed by area, and the results have increasingly differentiated the world through the development of trading cities, of particular industries in particular places, and of agricultural specialties in areas offering advantageous conditions. Attempts to minimize costs and maximize returns in specialized production and in trade require choices in regard to labor, markets, materials, transportation, and political security for investments. Such choices have long changed the map and are still changing it drastically today. They create definable patterns of spatial distribution, and they contribute to the differentiation of the world into distinctive areas. It is difficult to overstress specialization and trade as factors contributing to the geographic patterns of the present world. Again and again this text will examine the specialized character of areas and some of the reasons that underlie particular specialties.

INERTIA

One human process that tends to perpetuate certain spatial patterns and maintain regional distinctiveness is cultural, political, and economic inertia. People are often very averse to changing their traditional habits, beliefs, and value systems in favor of adopting innovations. A handy example of such continuity and its geographic results is the "Pennsylvania Dutch" (German) country of southeastern Pennsylvania, described in Chapter 28. Here the Amish, a conservative branch of the Mennonite religious group, adhere stoutly to the horse and buggy for transportation, cling to traditional religious practices, family relationships, systems of inheritance, farm management, and vernacular architecture, and maintain a distinctive "Old World" type of agricultural landscape. In another instance, currently important in the United States, fixed investment costs for economic enterprises such as steel mills may lead to their persistence long after they have been surpassed in efficiency by innovative newer types of equipment. The industrial world is sprinkled with obsolescent structures and equipment that have persisted long after their maximum usefulness has passed. Even in relatively dynamic and "progress-oriented" societies, many people—such as workers in outmoded plants—have a stake in resisting change. Such resistance appears glaringly in certain "developing" countries where a privileged elite is averse to any change which might threaten the group's political and economic control.

GEOPOLITICAL ORGANIZATION AND INTERACTION

Throughout human history, the search for security, order, and dominance has led to the formation of political units; to the institution of state policies to achieve political, economic, and social goals; to colonialism; and to war. The results are evident in the worldwide system of states and in the innumerable geographic features that have resulted from decisions by governments. An obvious example is the network of national political boundaries. Such lines separate one sovereignty from another, and they articulate a patchwork of political territories that have generally originated in warfare, colonialism, or other aggressive behavior, not infrequently by groups seeking security. As a consequence of government policies, any number of geographic features have been instituted deliberately or have occurred inadvertently. The Soviet Union, with its highly centralized system, affords many prime examples. But so does the United States, in which, for instance, the decision of the federal government to build the Interstate Highway System has wrought vast geographic changes. The network of new thruways has great effects on the increase or decline of jobs in particular areas, the places people choose to live, and the places where people prosper or where they suffer unemployment. The geographic effects of such government decisions spread in ever-widening ripples, and they add to the increasingly dynamic quality of world geography and of present-day studies by geographers.

CHAPTER 3

Key Topics in the Geographic Interpretation of Countries and Regions

Extreme poverty on display in Calcutta, India. Like these houseless people, the majority of earth's inhabitants are poor, non-Western, and culturally diverse. The poverty shown in the photo is exceptionally severe, but it is closer to the world's economic norm than is the relative affluence of most Westerners.

THIS chapter is intended as a specific aid to the study of the regional chapters that follow. It introduces a set of "key topics" that will be repeatedly touched on as different countries and regions are studied. The authors hope that an introductory acquaintance with these recurring topics will make the subsequent regional discussions more easily comprehensible. The topics include (1) location, (2) population, (3) political status, (4) natural environment, (5) type of economy, (6) internal arrangement and organization, (7) external connections and relationships, (8) characteristic landscapes and their origins, and (9) problems.

LOCATION

One of the reasons for studying geography is to learn the location of important features on the earth's surface—in other words, to learn where things are in the world. One tries to acquire a mental map on which countries, important cities, rivers, mountain ranges, climatic zones, agricultural and industrial areas, and other features are plotted in their correct relation to each other. No person can claim to be truly educated who does not carry this kind of map in his or her head. However, it is not enough simply to know the *facts* of location; one must also develop an understanding and appreciation of the *significance* of location. To a considerable degree, the geographic characteristics of any area are due, directly or indirectly, to its location. Various models (idealized conceptions) have been created by scholars to convey general ideas about the effects of location within specified sets of circumstances. However, these useful formulations cannot take account of the full complexities of the real world, where each area has a different location and hence a unique set of locational factors to be examined.

Perhaps an illustration will point up these remarks. Let us compare the location of Great Britain with that of New Zealand (Fig. 3.1). Each is an island area, Great Britain being a single large island, while New Zealand has two main islands about equal in

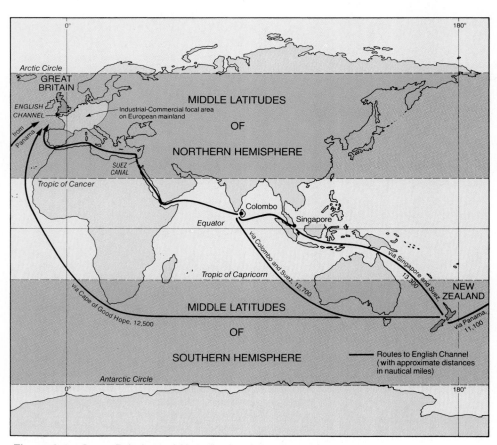

Figure 3.1   Great Britain and New Zealand compared in location.

size. Both come within the influence of westerly winds, which, blowing off the surrounding seas, bring abundant rain and moderate temperatures throughout the year. The climates of Great Britain and New Zealand are remarkably similar, in spite of the fact that these areas are located in opposite hemispheres and are about as far from each other as it is possible for two places on the earth to be.

Thus, in certain ways, the locations of Great Britain and New Zealand are similar. In other respects, however, their locations are vastly different. Great Britain is located in the Northern Hemisphere, which contains the bulk of the world's land and most of the principal centers of population and industry; New Zealand is on the other side of the equator, in the Southern Hemisphere. Great Britain is located near the center of the world's land masses (see Fig. 6.5) and is separated by only a narrow channel from the densely populated industrial areas of western continental Europe; New Zealand is surrounded by vast expanses of ocean. Great Britain is located in the western seaboard area of Europe where many major ocean routes of the world converge; New Zealand is far away from the centers of world commerce. For more than four centuries, Great Britain has shared in the development of northwestern Europe as a great organizing center for the world's economic and political life; New Zealand, meanwhile, has lived in comparative isolation. Great Britain, in other words, has had a *central* location within the existing frame of human activity on the earth, whereas New Zealand has had a *peripheral* location. These differences of location help to explain why Great Britain has become a densely populated industrial area and an important center of political and economic power, while New Zealand has remained a sparsely populated pastoral country of much less significance in world affairs. Centrality of location is a highly important factor to consider in assessing the geography of any country, region, or other place.

The student should realize that factors of location are not constant but are relative to the circumstances of a particular time. In other words, *the significance of location changes as circumstances change*. During the early centuries when the borderlands of the Mediterranean Sea contained the principal centers of European culture and political power, Great Britain was an unimportant area situated on the very edge of the known world, its location being scarcely more advantageous than that of New Zealand. Not until much later, when the European center of gravity had shifted from the Mediterranean region to the shores of the North Sea and the age of oceanic expansion had commenced, did the location of Great Britain become highly advantageous.

POPULATION

Population is a topic of major importance in geographic study. More than any other element, it is human life that gives character and geographic significance to areas. The numbers, density, distribution, and qualities of the population, together with population changes or trends, supply an essential background or focus in studying the geography of any area. These aspects of population will be illustrated in an introductory way through a brief consideration of world population as a whole. (See also the discussion of natural population change and migration in Chapter 2.)

POPULATION NUMBERS

In mid-1988, the total population of the world was estimated at 5.1 billion, with an annual rate of increase of 1.7 percent. No very accurate figure for the world's present population is yet possible, due to a lack of census data for some countries as well as to great variations in the accuracy of census-taking from one country to the next. Estimated populations of the eight major world regions are given in Table 3.1. The following important points should be noted: (1) the huge total population of the Orient, which has several times as many people as Africa, Europe, the Middle East, or Latin America; (2) the remarkably close similarity in numbers between the Soviet Union and Anglo-America, two regions about the same size and to some degree analogous in their middle-latitude or high-latitude environments, types of mineral wealth, and histories of settlement; and (3) the extremely small number of people in the Pacific World as compared with other world regions—partly a consequence of aridity in the island continent of Australia.

POPULATION DENSITY

The total area of the earth's surface is approximately 197 million square miles (510 million sq km). The land surface, however, including inland waters, comprises only about 57.4 million square miles (148.7 million sq km), or slightly more than 29 percent of the total. A population total of 5.1 billion gives a figure of 89 people per square mile (34 per sq km) as the average density of population for the entire world. This figure is somewhat greater than the estimated average density for the United States in 1988 (68 per square mile, or 26 per sq km).

Figures on population density for the world as a whole are practically meaningless, due to the

TABLE 3.1   THE MAJOR WORLD REGIONS: BASIC DATA

Region	Area (million sq mi)	Area (million sq km)	Estimated Population (millions, mid-1988)	Estimated Rate of Natural Increase (%, 1988)	Population Density (per sq mi)	Population Density (per sq km)	Infant Mortality Rate[c]	Urban Population (%)	Cultivated Land (% of total area)	Per Capita GNP ($US: 1986)
Europe	1.9	4.9	497	0.3	264	102	13	75	29	8170
Soviet Union	8.7	22.4	286	1.0	33	13	25	65	10	7400
Middle East[b]	7.0	18.0	492	2.9	71	27	105	39	8	1500
Orient[b]	8.0	20.8	2805	1.8	350	135	71	26	18	970
Pacific World[b]	3.5	9.0	27	1.2	8	3	36	70	6	9000
Africa[b]	11.7	30.3	623	2.9	53	21	110	30	6	620
Latin America	7.9	20.5	429	2.2	54	21	57	68	9	1720
Anglo-America[b]	8.3	21.5	272	0.7	33	13	10	74	12	17170
World	57.4	148.7	5128	1.7	89	34	77	41	11	3010
United States	3.6	9.4	246	0.7	68	26	10	74	20	17500

[a]In general, tables of statistical data in this text are based on standard sources such as those listed under Part I in the References and Readings at the end of the book. The primary sources for Table 3.1 and comparable tables for individual major world regions are the Population Reference Bureau, *World Population Data Sheet* (annual) and the *Britannica Book of the Year*, 1988. Many figures in the tables are rough approximations based on data of varying accuracy reported by governments. Some parts of the world have not been accurately surveyed, and many countries do not have adequate census data on which to base population estimates. Hence, wide discrepancies exist in estimates cited by various sources.
[b]Certain units are included in two different regional totals; for example, Egypt is included under both Africa and the Middle East.
[c]Annual number of deaths to infants less than age 1 year per 1000 births.

extremely uneven distribution of population, resources, and productive facilities over the earth. However, such figures do furnish a rough yardstick with which to measure the densities of various countries and regions.

POPULATION DISTRIBUTION

The distribution of population over the earth is extraordinarily uneven (Figs. 3.2 and 3.3). The majority of the world's people are concentrated in three major clusters. Two clusters are in the Orient, and together they comprise about one half of humankind. The larger of the two, or East Asian cluster, lies in China, Japan, and Korea, with a southward extension from China into Vietnam. The second, or South Asian cluster, lies mainly in India, but also in Bangladesh and Pakistan. A third major cluster occupies most of Europe and portions of the Soviet Union. Elsewhere in the world smaller concentrations of population are scattered about here and there, separated from each other by stretches of lightly settled or unoccupied land.

Areas in which population is extremely sparse or absent make up more than three fourths of the earth's land surface. Such areas fall into the following four principal categories: (1) arctic and subarctic areas, where settlement is hampered by excessive cold; (2) areas of desert and dry grassland, which are handicapped by lack of moisture; (3) areas of rugged highland, where settlement is restricted by steep slopes and high altitudes; and (4) areas of tropical rain forest and tropical grassland or scrub, where excessive heat and moisture, dense forest growth or rank grasses, and infertile soils discourage fixed settlements. Of the four main types of "negative areas," the tropical forests and grasslands seem to offer the best possibilities for large future increases in population. Indeed, some areas of this type, particularly in the Orient, already support large and dense populations. However, areas as large or larger in Latin America, Africa, and the East Indies are still quite sparsely populated.

POPULATION TRENDS

Various periods of history have seen great changes in the numbers, density, and distribution of people over the earth and in different countries and regions. Some of the most significant changes have taken place during the past three or four centuries. This period has witnessed an unparalleled increase in total population numbers. It is thought that since 1650 the world's population has increased ten times over, from approximately 500 million to about 5 billion. At the present time (late 1980s), this increase is con-

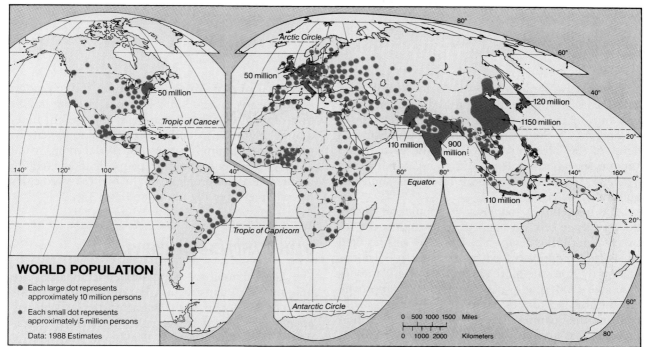

Figure 3.2   A French geographer, the late Jean Brunhes, maintained that the two most significant maps were the map of population and the map of rainfall. Much geography can be read or inferred from the map of population shown above. Among other things the map shows that the areas where large numbers of people have found it desirable to live are rather limited in extent. There are striking concentrations in the intensively cultivated farming regions of southern and eastern Asia and in the industrialized regions of Europe, the western Soviet Union, the eastern United States, and Japan. Note the tendency of people to congregate on the margins of continents, near the sea. Geographers are students of regions, but they are also students of distributions—not only of people, but of crops, livestock, rainfall, minerals, industries, and other features on the earth. They are particularly interested in the various ways that different distributions are related to each other. Compare the map of population with the map of cultivated land (Fig. 3.3) for similarities and differences in the pattern of distribution.

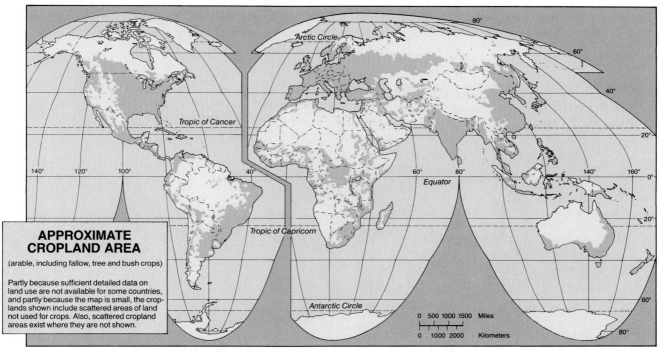

Figure 3.3   World map of cultivated land. Source: U.S. Department of Agriculture.

tinuing at an estimated rate of 85 to 90 million a year. Fears are frequently expressed that humanity will outrun its food supply unless the increase in numbers is checked. However, some optimistic students of this problem believe that the earth could support several times its present population if all the available resources were more intensively and scientifically managed.

The tremendous increases in population during the past three centuries appear to be mainly the result of a declining death rate. The latter has resulted partly from improved medical and sanitary facilities and partly from the more abundant, varied, and dependable food supply made possible by improved agricultural techniques, better transportation facilities, and the opening of new lands for cultivation in the Americas and elsewhere.

At the present time, population is increasing in nearly all the world's countries, but the rate of increase is much greater in some countries than in others. Sharp upward trends of population in such countries as Kenya (an estimated natural increase of 4.1 percent per year), Mexico (2.4 percent), or India (2.0 percent) contrast strikingly with the slow population growth of France (0.4 percent) or West Germany (−0.1 percent). Percentage rates of population growth seem to be slowing in most countries, but the existing potential for increase in numbers is so great that there may be more than 6 billion people on the planet by the year 2000 and possibly 10 billion people ultimately.

During the past three or four centuries, the world pattern of population distribution has been altered considerably. This was brought about principally by two great movements of population: the migration of Europeans to new lands overseas, and the migration of rural dwellers to cities. At the present time, a huge migration from the countryside to the city in many "developing" countries is depriving rural areas of energetic and able people and is creating intractable social problems in the overcrowded cities (see chapter-opening photo and Fig. 3.4).

OTHER ASPECTS OF POPULATION

The general topic of population encompasses not only numbers, density, distribution, and movements or trends but also such characteristics as physical appearance, language, religion, food habits, educational levels, adequacy of livelihood, and general cultural heritage. The peoples of various areas are vastly different in these respects. Such differences are of fundamental concern and interest to the student of geography.

POLITICAL STATUS

We live in a world which has been divided into a multitude of political units—sovereign states, subdivisions of these states (provinces, states, departments, counties, and so on), and the various odds and ends of dependent political units and their subdivisions that still remain as legacies from a colonial age that is almost over. Separated by boundaries drawn on maps and in many cases shown by mark-

Figure 3.4   School children in a poor quarter of Cairo, Egypt, symbolize the explosive population growth that impedes economic advance in most of the world's "less-developed" countries.

ers on the ground, these political divisions form a complicated patchwork enclosing most of the land surface of the earth.

The political status of any area is an important feature of its geography. In a broad sense, political status includes not only the political organization of the area in terms of functional political units but also the distinctive role of the area in the political world. Although the latter is sometimes difficult to define precisely, it should not be ignored in geographic study.

The key unit in the worldwide political structure is the sovereign state. Each state occupies territory and administers the people and resources of that territory. In most matters, the state, whether large or small, is its own master within its own boundaries. Despite the many attempts at supranational organization, the system of sovereign states shows a remarkable persistence and seems likely to be the basic international frame within which citizens and their nations will operate for a long period of time. This fact is a major reason for the emphasis on the geography of individual sovereign states in this text.

An important geographic circumstance in any state is the degree to which it has attained political cohesion throughout its extent. Some states, such as Denmark or New Zealand, are very cohesive politically, but other states exhibit separatist tendencies in certain areas. Indonesia, Belgium, and Sudan are notable examples of states in which strong *centrifugal tendencies* exist.

NATURAL ENVIRONMENT

Natural environment refers to the total complex of natural conditions and resources occurring in an area. The principal elements of the natural environment are landforms, climate, natural vegetation, soils, native animal life, underground and surface waters, and mineral resources. These elements vary considerably in relative significance from one area to another and from one period of history to another. To a considerable degree, the significance of a particular environmental feature depends on the type of culture prevailing at a given time in the area where the feature is found. For example, the rich coal deposits of western Pennsylvania were of little or no use to the American Indians who originally inhabited that area, but the present inhabitants, armed with a different technology and living within the framework of an industrial civilization, find these same coal deposits to be of great significance as sources both of wealth and of problems.

One should beware of easy generalizations about the "influence" of the natural environment on human life. The connections or relationships which exist between people and their environment are often extremely complicated. Any particular set of environmental features offers various possibilities for human use. Social, cultural, economic, technological, psychological, and historical factors will condition the actual employment to which such features will be put by a human group.

An American geographer, the late Robert S. Platt, summed up these ideas effectively in the section quoted below:[1]

> The importance of our natural environment is obvious. People have been, and still are, conscious of the significance to them in everyday life of the weather, the soil, minerals, mountains, and plains. Travelers long ago observed, and still do, that differences from place to place in natural environment are associated with differences in the lives of people. . . .
>
> [However] it is misleading to advance the hypothesis of an active influence of natural environment tending to shape human life in the natural and proper way and to look for coincidences between environment and life as evidence confirming this hypothesis. Increasing evidence shows that the hypothesis of a simple and direct relationship is not thus confirmed, that there is no proper natural way of shaping life but innumerable ways, not sorted out by nature but reduced by man's choices past and present. *People live differently in similar environments and differently at different times in the same environment,* without feeling any environmental pressure to lessen these differences.
>
> . . . . *In any given type of regional environment, people have alternative ways of living, apparently many in some places and few in others.* Probably everywhere the conceivable number of possibilities is far greater than people can imagine and far beyond the range of choice now open to them.
>
> Actually, people are limited by things other than natural environment, though set within the confines of that environment. Particularly are they limited by habits they have learned and facilities available to them, accumulated through an unbroken series of choices and rejections in the entire course of their history—in other words, by the cultural heritage of the group to which they belong, by their culture defined in the broadest sense. The choices made in the past which now limit people in their activities have been impelled not by natural environment but by the play of history thereafter embodied in their culture.

---

[1]Robert S. Platt, "Environmentalism Versus Geography," *American Journal of Sociology,* 53 (1948), 351–352. Reprinted by permission of The University of Chicago Press. (Some portions italicized in editing.)

For example, in the Great Lakes region farmers have been limited agriculturally to a certain range of possibilities, mainly involving cereals and livestock. But if some of our ancestors at the dawn of civilization had not chosen to domesticate certain grasses as cereals and certain four-footed animals as livestock but, instead, had chosen to develop fungus growths or edible insects as a basis of productive culture, our mode of life might be now utterly different in ways which the natural environment might support as well as it does our present agriculture, or conceivably better.

TYPE OF ECONOMY

One of the most significant characteristics of an area is its type of economy—in other words, the kind of mechanism which humans have developed in the area as a means of satisfying their needs and wants for goods and services. An important aspect of an area's economy is its *occupational structure,* or the numbers and proportion of people employed in each of the principal means of livelihood (see Figs. 3.5–3.7). In about one half of all independent countries, more people are still employed in *agriculture* than any other occupation. Heavy dependence on agriculture is especially characteristic of less-developed countries. In countries with more advanced technologies, a large share of the people are engaged in *manufacturing, transportation, trade,* and *personal*

Figure 3.6   *Bargeloads of coal in America's Ohio Basin awaiting movement to industrial markets (mainly thermal-electric power plants) symbolize several occupations: mining, manufacturing, transportation, and trade.*

and professional services. But even in highly industrialized countries, agriculture is a significant means of livelihood. However, the proportion of the population engaged in this occupation is generally much smaller than it is in developing countries, although agricultural production per worker and per unit of land is often much higher.

The more sparsely settled regions of the world are characterized by such occupations as *grazing* (*nomadic herding* or *livestock ranching*; see chapter-opening photo of Chapter 20), *hunting, fishing, trapping,* or the *forest industries* (see Fig. 27.5). In scattered areas, *mining* serves as a source of livelihood. Mining areas are distributed irregularly over the earth wherever commercial deposits of minerals are found.

Besides the occupational stucture, it is also important to consider the characteristic *units of economic organization* in an area (farms, factories, stores, and so on), together with the lines of transportation and communication which connect these units. Other pertinent questions regarding the economy of the area include the following:

1. What is the general status of technology—advanced, moderately developed, primitive, or a mixture of these?
2. What are the principal commodities produced, and in what quantities?
3. Is production primarily of a commercial or a subsistence character?
4. Is the economy largely self-contained, or is there a significant dependence upon trade with other areas?
5. What are the principal imports and exports?

Figure 3.5   *The world's agriculture covers a wide spectrum, from subsistence hoe-culture to large-scale mechanized operations such as the wheat harvest seen here on a collectivized farm in the Soviet Union's black-earth belt. The locale is the Bashkir Autonomous Soviet Socialist Republic between the Volga River and the Ural Mountains.*

Figure 3.7 One of the great trends in the present world is an expansive development of services, which *are increasingly international in character. The photo shows Disney World in Tokyo, Japan, with some American visitors.*

6. Have the productive facilities of the area been largely financed by domestic capital, or by outside capital?

7. What is the adequacy of livelihood provided by the economy?

8. What are the major economic trends in the area?

INTERNAL ARRANGEMENT AND ORGANIZATION

In studying an area geographically, one is constantly concerned with the ways in which the features that give it character are arranged and organized spatially—that is, how they are distributed within the area, how different distributions are related to each other, and how the geographic features of the area are linked together by natural processes or human enterprise into coordinated structures and functioning systems. To think geographically about the internal make-up of an area is to think in terms of such spatial patterns, associations, and interconnections. Students of geography should make a conscious effort at all times to relate the geographic features and distributions of an area to each other and to think in terms of overall spatial patterns into which details can be fitted. This inevitably means to think in terms of maps, for maps portray spatial patterns and associations with a clarity that is generally beyond the reach of words. Students should frequently refer to maps as they read geographic ma-

terial, should use outline maps for note taking, should make sketch maps to show important spatial relationships, and should try to visualize map relationships when these are stated in words in text material.

EXTERNAL CONNECTIONS AND RELATIONSHIPS

The overall pattern of an area's economic, cultural, social, and political interaction with other parts of the world is an important component of its total geography. Such connections and relationships involve trade flows, monetary flows, and population movements; cultural, technical, and informational exchanges; political, strategic, and military influences and alignments; and the day-to-day flow of personal and group decisions that affect human use and organization of the earth. This exceedingly broad topic cuts across all the other topics considered in this chapter, including the topic of internal arrangement and organization. A little reflection will show how intimately the internal make-up of most areas is related to their interaction with the outside world. For example, coffee is an important crop in the tropics of southern Brazil, and the landscape gives abundant evidence of this fact, but the geographic pattern of this part of Brazil would be very different were it not for the extremely large market for imported coffee afforded by the United States.

THE GEOGRAPHIC EFFECTS OF PLACES ON EACH OTHER

The Brazilian example just given is a local illustration of an important general truth: places have effects on each other's geographies. In this respect, no place is truly an island, although some are more insular than others; nor are places the same in the magnitude, intensity, frequency, and reach of the geographic effects they engender. Such effects may be physical, economic, political, cultural, or social, and they are brought about by trade, investment, migration, military action (or the threat of it), and other mechanisms of spatial interaction. Especially large and complex effects are generated by such large and powerful nations as the United States, the Soviet Union, or Japan, but many smaller and less powerful places have an impact disproportionate to their size. Examples are tiny Kuwait on the Persian/Arabian Gulf, which has a worldwide reach because of its vast oil and gas resources; and the small island country of Singapore, which affects other places through its dynamic economic development and its strategic position on a strait connecting two oceans.

The geographic impact of places on each other is a very useful consideration in studying geography. Awareness of this factor leads to enhanced appreciation of the world as an extremely complicated, interconnected system in which conditions, activities, events, and changes in one part affect various other parts in diverse ways. The same mode of thought is applicable to the internal geography of countries. Within a given country, certain areas will exert greater geographic effects than others. Prominent examples include: (1) major metropolitan areas, which provide unusually abundant and diversified opportunities for employment, and exert much control over the national economy; (2) major manufacturing areas, especially those that specialize in products that are of basic importance to the country's economy as a whole; (3) major seaports and other centers of commerce; (4) unusually productive or distinctive agricultural areas, particularly those that produce an important surplus for export; (5) areas that produce minerals on which the country's economy is very dependent; (6) areas that differ markedly in culture from the rest of the country, especially if this results in political disaffection which threatens the country's cohesion and stability; (7) areas with a pronounced strategic importance in wartime; and (8) areas of unusual poverty that consume a disproportionate share of the nation's revenues. In this text, certain areas are often treated at greater length than other areas that have a larger extent or population but are seen as less critical in their impacts on the country or the world as a whole.

CHARACTERISTIC LANDSCAPES AND THEIR ORIGINS

One of the factors that lends color and interest to geography as a field of study is the opportunity to observe and interpret the extraordinary variety of landscapes that the world affords. The study of landscapes carries intellectual as well as esthetic rewards. A great deal of the human record, and nature's works also, are inscribed in the earth's surface and call for careful mapping and rigorous analysis if one is to truly understand and appreciate what is written there. From the earliest times, the landscape has been a basic source of data in geographic work, and the marked differences in landscapes over the earth have stimulated much geographic inquiry. Some landscapes are intricate, irregular, and hard to characterize, while others are regular, sharply defined, and conducive to precise geographic description and characterization. One may, for instance, contrast the complex land-use pattern of the "Pennsylvania Dutch" country, adjusted to the contours of a rolling surface (Fig. 3.8), with the rectangular pattern of large fields and straight roads in the Interior Plains of the United States or Canada. In the study of landscapes, the geographic mind is concerned with the features that are typical, in other words, with land-

Figure 3.8   *The care with which "Pennsylvania Dutch" (German) farmers manage and conserve their rolling Piedmont land is strikingly evident in this irregular mosaic of contouring and strip-cropping in Lancaster County, Pennsylvania. Erosion is kept in check by level furrows that follow contours and by the planting of crops in strips that alternate close-growing crops such as hay or wheat with clean-tilled crops such as corn or tobacco. This complex landscape gives no hint of the rectangularity that characterizes aerial views of the American Midwest.*

scape units that repeat themselves and are subject to mapping, classification, and orderly analysis. In general, the interest of the geographer in a landscape is different from that of the tourist. The tourist tends to be attracted by features that are quaint, unique, exotic, or bizarre; the geographer is excited by the things that are *representative;* his or her concern is to discover what the landscapes of an area are like *on the average.*

PROBLEMS

The people of almost any area are confronted with serious problems which must be solved if the potentialities of the area for affording a good life are to be realized. Most of these problems fall under the heading of economic problems, political problems, social problems, resource problems, or a mixture of these. Space will not permit an extended discussion of the endless variety they assume, although this text incorporates discussions of them in many places. Some broad categories of problems that are important on a world basis follow: (1) problems of resource sufficiency and regional inequalities, (2) problems of regional self-determination, (3) problems of ethnic relations, (4) problems of ideologic confrontation, and (5) problems of security against violence. Each is discussed briefly in the following paragraphs.

PROBLEMS OF RESOURCE SUFFICIENCY AND REGIONAL INEQUALITIES

In the late 20th century, no country or major world region has all the resources it needs to meet the expectations of its people for security and a good life. Anglo-America would come the closest, but even this resource-rich region does not have a sufficiency of many critical items (oil, for example, or bauxite from which to secure aluminum needed for aircraft production). For many ''less-developed'' countries, it is much simpler to enumerate the few resources present than the appalling number that are insufficient. Actually, the concept of resource sufficiency is rather elusive. Skilled application of technology can often overcome resource inadequacies, but the cost may be so high that many areas cannot afford to pay it.

What do we mean by *resources*? The *Random House Dictionary of the English Language* defines *resources* as ''the collective wealth of a country or its means of producing wealth''—a view encompassing the natural environment, labor, capital, and technology. The world exhibits a highly uneven distribution of resources and fruits of resource use. Such inequalities are a problem within most countries, but they are the most apparent and highly publicized when entire countries and major world regions are compared on a world scale. Some areas, known as the ''developed'' areas, are comparatively well provided with resources and have a high level of living. The areas characterized as ''less developed,'' ''developing,'' or ''emerging,'' by contrast, generally have abundant labor but have serious gaps in many other resources and a relatively low level of living. Countries generally classified as ''less developed'' comprise roughly three fourths of the world's people but, by one estimate, control only one fifth of the world's wealth (1980s). Their plight is the subject of much international study and concern.

PROBLEMS OF REGIONAL SELF-DETERMINATION

In most countries, there are political stresses caused by the desire of people in certain parts of the country to have a greater voice in managing their affairs. The basis for such desires (and demands) may be economic, or cultural, or both. In many cases, the historical memories of the people look back to a time when they were, in fact, a separate political entity. In extreme cases, there may be attempts at secession. It is not uncommon for proponents of secession or greater self-determination to resort to terrorism. Separatist sentiments may or may not focus on a language different from that of the country as a whole. Examples of the quest for greater self-determination will be met with frequently in this text: in French Canada, the Basque Country of Spain, the Flemish-speaking and Walloon (French-speaking) parts of Belgium, the Sikh area in northwestern India, and many other places. Several nations have been created since World War II on the basis of separatist aspirations—Pakistan and Bangladesh being prominent examples. In Nigeria, the attempted secession of the southern Ibo-speaking area (Biafra) failed after a bloody civil war. In Canada, there has been a tug-of-war between the central government in Ottawa and various provinces (notably French-speaking Quebec and oil-rich Alberta) that desire to maintain or increase their already large powers of self-government.

PROBLEMS OF ETHNIC RELATIONS

Ethnic stresses often lie at the heart of political separatism, but they are potent sources of trouble even where separatism is not involved. A fine example of the latter is the stormy history of race relations in the United States during the 1950s and 1960s, growing out of the long previous exploitation of the black population. Anti-Semitism in Tsarist Russia and the Soviet Union offers another instance. Many grave international problems have an ethnic basis. *Ethnocentrism*—regarding one's own group as superior and as setting proper standards for others—is common and too often shades into an intolerant pride in race, religion, language, or historical memories on the part of an in-group. Such feelings may lead to prejudicial acts against other groups for whom the in-group feels a particular hostility, or concerning whom it harbors particular fears. Ethnocentrism has plagued the human race throughout its history. It has been a factor in many savage episodes during the 20th century, and today it continues to present a highly visible face in many parts of the world.

PROBLEMS OF IDEOLOGIC CONFRONTATION

Ideologic confrontation is not the same as ethnic confrontation, although ethnocentric feelings may become an ideology in a sense (for example, Naziism in Germany). Most commonly, the term *ideology* refers to a system of political and/or economic beliefs such as communism, capitalism, autocracy, or democracy. Many aspects of world geography reflect the clash of ideologies—for example, the existence of Taiwan as a non-Communist political entity separate from Communist China, or the division of Germany into West Germany (non-Communist) and East Germany (Communist).

PROBLEMS OF SECURITY AGAINST VIOLENCE

In recent decades, a wave of armed conflicts, terrorism and other crimes, arms races, and belligerent rhetoric has made security against violence a major concern of political units and organizations at all levels from localities to the United Nations. The existence of violence on a large scale has become a hallmark of such unfortunate places as Lebanon, Iran, Kampuchea (Cambodia), and Northern Ireland. Military alliances such as the North Atlantic Treaty Organization or the Warsaw Pact have become an important part of the political map. Huge expenditures to support the armed forces of nations and dissident groups have robbed peoples of vital financial resources needed for economic and social betterment. Such expenditures, characteristic of both the "developed" and the "less-developed" parts of the world, are officially justified as necessary for defense or, in some cases, for the redress of perceived grievances. The resulting financial burdens are augmented by the costs of supporting internal police forces to combat terrorism and general lawlessness or, in many dictatorships, to discourage political dissent of any kind. Such situations and problems will be encountered and discussed at appropriate places in this text, although they have many ramifications that must be left to academic disciplines other than geography.

All the foregoing types of problems are interlocked, and no sizable area in the world is free from any of them. These problems form a necessary part of any consideration of world regional geography. Solutions will be slow in coming, but anyone who attempts to follow world events in these times will be well served by some knowledge of their various geographic contexts.

DIFFICULTY OF EXPLAINING AN AREA

To account for the complex web of features and relationships that make an area geographically distinctive is a challenging task. One must accept the fact that to explain fully the geography of an area may be very difficult or impossible due to the presence of historical or other factors whose effects cannot be known with certainty. However, this should not deter a search for such understanding as the available data will permit. Much that is useful and enlightening can be learned about the geography of areas, even if all the facts cannot be fully explained.

PART II

EUROPE

CHAPTER 4

Geography and Human Experience in Europe

Roman Catholicism has made a deep imprint on Europe's cultural geography. Seen here are Bishops assembled for a Papal Mass at the Vatican.

MANY areas have made important contributions to shaping the present world but none more impressively than Europe. Such European events as modern nation-building, the development of science and technology, the rise of advanced economies, and expansion overseas are central aspects of world history. Today, the region continues to play a vital role in the world's economic, political, and intellectual life. There is, of course, a darker side to European dynamism. Warfare initiated by Europeans and backed by advanced European technology and military technique has often brought disaster to the Europeans themselves and to other peoples. The world wars of 1914–1918 and 1939–1945, fought principally in Europe, were especially shattering. Since World War II, the region has endured an armed confrontation between Communist and non-Communist powers, with the threat of nuclear destruction ever present.

Neither the positive achievements of Europe nor its episodes of military aggression, destructive warfare, and fear can be properly evaluated outside the context of the region's geography. For example, one cannot comprehend Europe's Industrial Revolution without some knowledge of the nature and distribution of European mineral and waterpower resources. Nor can one understand military strategy without an appreciation of terrain, routeways, and population concentrations. From prehistoric times the events of both peace and war have been conditioned by geographic circumstances and, in turn, have left their mark on Europe's increasingly complex geography. To provide geographic context for the region's dramatic history and present dilemmas is a major aim of this chapter and the one that follows. At the same time, the flow of history is used to interpret the spatial attributes and geographic distinctiveness of present-day Europe and its parts. The analysis begins with a definition of Europe and then proceeds to an overview of the region's varied countries and cultures. Next, the impact of Europe on the rest of the world is evaluated. Finally, in Chapter 5, the region's diversified environment and complex economy are examined, especially with regard to interrelations between environmental features and economic life.

DEFINITION AND BASIC MAGNITUDES

In this book *Europe* is defined as the countries of Eurasia lying west of the Soviet Union and Turkey. This definition follows a more modern rather than a more traditional usage. Traditionally Europe has been regarded as a continent, separated from Asia by the Ural Mountains, Ural River, Caspian Sea, Caucasus Mountains, Black Sea, Turkish Straits (Bosporus, Sea of Marmara, Dardanelles), and Aegean Sea (see maps, Figs. 5.1 and 10.2). Thus, it has included parts of the present-day Soviet Union and Turkey. But increasingly during the 20th century, the term *Europe* has been used to refer to the remainder of the continent after the culturally different Soviet and Turkish parts are excluded, and that usage is followed here. Thus defined, Europe is a great peninsula of Eurasia, fringed by lesser peninsulas and islands, and bounded by the Arctic and Atlantic oceans, the Mediterranean Sea, Turkey, the Black Sea, and the USSR (Figs. 4.1 and 5.1).

Understanding the geographic reality of Europe requires a grasp of some basic magnitudes that are very different from those of the United States. For example, all of Europe has an area only about two thirds as great as that of the 48 conterminous states of the United States (excluding Alaska and Hawaii). This means that Europe's many countries tend to be rough equivalents of American states in terms of area. The larger European countries are on the same general scale as the larger states in the western part of the conterminous United States, although even the largest country, France, is not as big as Texas. The small ones are on the same areal scale as the smaller states along the Atlantic seaboard of the United States, and the most common sizes for European countries are close to those of middle-sized American states such as Ohio or Tennessee. One of the striking features about Europe is the large impact on the world that societies nourished on some of these small pieces of land have had. Some countries are actually smaller in area than any American state. These include Luxembourg, Malta in the Mediterranean Sea south of Italy, and the "microstates" of Andorra, in the Pyrenees Mountains between France and Spain; Monaco, an enclave along France's Mediterranean coast; Liechtenstein, between Switzerland and Austria; San Marino, surrounded by Italy, and the Vatican State (Vatican City), surrounded by Rome (Fig. 4.1). The microstates, tiny in population as well as in area, are separate political entities as a result of various oddities of Europe's complex history and politics.

A different set of comparative magnitudes emerges when European countries are compared in population with American states. The largest state, California, would be a rather populous country if it were European, but even California is greatly exceeded in population by six European countries. Most European countries are comparable in population to relatively populous American states. Cer-

EUROPE

INTRODUCTORY LOCATION MAP

POPULATION
OF COUNTRY
(MILLIONS)

Over 50
20-50
10-20
Under 10

⊕ Capital of Country
● Largest city of Country
✪ Capital and largest city

*Figure 4.1 Europe, showing political units as
of 1989. Note the fact that in most European
countries, the political capital is also the
largest city (metropolitan area).*

tain patterns in Europe with respect to total populations of countries are both apparent and important (Table 4.1). One is that there are four "giant" countries—West Germany, Italy, the United Kingdom, and France—which far outweigh any others in population. Their respective populations range from about 61 to 56 million, and together the four countries represent a little less than half of Europe's population, or almost as many people as there are in the United States. Another pattern is that many of the smaller countries occupy territories dominated by northerly climates and/or mountainous terrain. A third pattern is that the countries of the Communist east tend to be middle-sized in population. All told, Europe's population of half a billion divides rather neatly into approximately one half in the four "giant" states, a little more than one quarter in the Communist east, and about one quarter in the smaller countries of the non-Communist west.

A third useful comparison between European countries and American states is by density of population. Even excluding great disparities among the microstates, Europe exhibits an extreme range—from over 1000 persons per square mile (over 386 per sq km) in the Netherlands (excluding inland waters) to about 6 per square mile (2 per sq km) in Iceland (see Table 4.1). Comparisons with American states reveal about the same range, although in general the European countries are more densely populated. In mid-1988, only 21 of the 50 American states had estimated densities over 100 per square mile (over 40 per sq km), as against 23 of 27 European countries, excluding the microstates.

REGIONAL GROUPS OF COUNTRIES IN EUROPE

Most European countries differ from their neighbors in language and all have distinct "personalities" of their own. But regional groups of countries can be distinguished, and this procedure is helpful in gaining an introductory general picture of such a complex part of the world. One way of grouping the countries for geographic analysis follows:

1. At the northwest, separated from the mainland by the English Channel and the North Sea, are the *countries of the British Isles* (Fig. 4.1). They include the United Kingdom of Great Britain and Northern Ireland, and the Republic of Ireland. Some factors that contribute to their geographic distinctiveness are their insular locations, their use of English as their main language, close trade and immigration ties with each other, and a long history of unhappy and sometimes bloody political rela-

tionships. But the United Kingdom is far more industrialized, urbanized, and populous than is the Irish Republic.

2. The *countries of West Central Europe* are France, West Germany, East Germany, Belgium, the Netherlands, Luxembourg, Switzerland, Austria, and the microstates of Liechtenstein, Andorra, and Monaco. German, French, and Dutch are the main languages of this regional group. Highly industrialized and urbanized societies with great population densities and high levels of living are characteristic, and the countries carry on an intense foreign trade with each other and with the wider world. Together with the United Kingdom and northern Italy, they form one of the world's greatest concentrations of manufacturing and are the greatest single focus of world trade. The agriculture of the countries of West Central Europe is also highly developed and exceptionally productive. Although these countries have warred against each other on numerous occasions, their relations in the period since World War II have generally been cooperative. This has been least true of relations involving East Germany, which is a Communist-ruled Soviet satellite obliged to harmonize its policies with those of the USSR. East Germany's close ties to the "East bloc" of Communist nations would possibly justify its inclusion in the countries of East Central Europe, but it stands apart from those nations in its different language and culture, higher technological level, and historical development within Germany.

3. The *countries of Northern Europe* include Denmark, Norway, Sweden, Finland, and Iceland. They are sometimes called the Fennoscandic countries or, except for Finland, the Scandinavian countries. Small populations, productive economies, and high levels of education and income are characteristic. Related Germanic languages are spoken, aside from Finnish, which belongs to a different linguistic family. Most environments are distinctly northern, and all of these countries except Denmark have very sparsely populated northern and/or mountainous sections occupying most of their area.

4. The *countries of Southern Europe* include Portugal and Spain on the Iberian Peninsula; Italy; Greece; the microstates of San Marino and Vatican City, each an enclave within Italy; Malta; and the British colony of Gibraltar, commanding the famous strait of the same name. Because of proximity to the Mediterranean Sea and the fact that they share a "mediterranean" climate and ecology, these are often called the Mediterranean countries. The usage is not very exact, since Albania and parts of France and Yugoslavia also have med-

TABLE 4.1 EUROPE: BASIC DATA

Political Unit[a]	Area (thousand sq mi)	Area (thousand sq km)	Estimated Population (millions, mid-1988)	Estimated Annual Rate of Natural Increase (%, 1988)	Estimated Population Density, 1988 (per sq mi)	Estimated Population Density, 1988 (per sq km)	Infant Mortality Rate	Urban Population (%)	Cultivated Land (% of total area)	Per Capita GNP ($US: 1986)
British Isles										
United Kingdom	94.5	244.8	57.1	0.2	604	233	9.5	91	29	8920
Ireland	27.1	70.3	3.5	0.8	129	50	8.7	56	14	5080
Totals	121.7	315.1	60.6	0.2	498	192	9.5	89	25	8700
West Central Europe										
France	211.2	547.0	55.9	0.4	265	102	8.0	73	34	10740
West Germany	96.0	248.6	61.2	−0.1	638	246	9.5	94	30	12080
East Germany	41.8	108.3	16.6	0.0	397	153	9.6	77	46	5600 [b]
Belgium	11.7	30.4	9.9	0.1	843	325	9.4	95	25	9230
Netherlands	14.4	37.3	14.7	0.4	1021	394	8.0	89	23	10050
Luxembourg	1.0	2.6	0.4	0.1	368	142	9.0	78	44	15920
Switzerland	15.9	41.3	6.6	0.3	414	160	6.9	61	10	17840
Austria	32.4	83.9	7.6	0.0	235	91	11.2	55	19	10000
Totals	424.5	1099.4	172.9	0.1	407	157	8.9	82	32	10830
Northern Europe										
Denmark	16.6	43.1	5.1	−0.1	307	119	8.4	84	62	12640
Norway	125.2	324.2	4.2	0.2	34	13	8.5	71	3	15480
Sweden	173.7	450.0	8.4	0.1	48	19	5.9	83	7	13170
Finland	130.1	337.0	4.9	0.3	38	15	5.8	62	7	12180
Iceland	39.8	103.0	0.2	0.9	6	2	5.4	90	under 1	13370
Totals	485.4	1257.3	22.8	0.1	47	18	6.9	77	7	13270
Southern Europe										
Italy	116.3	301.2	57.3	0.0	493	190	10.1	72	41	8570
Spain	194.9	504.8	39.0	0.4	200	77	9.0	91	41	4840
Portugal	35.6	92.1	10.3	0.3	290	112	15.8	30	39	2230
Greece	50.9	131.9	10.1	0.2	198	77	12.3	58	30	3680
Malta	0.12	0.3	0.4	0.7	2850	1100	13.0	85	44	3470
Totals	397.8	1030.4	117.1	0.3	294	114	10.4	73	39	6330
East Central Europe										
Poland	120.7	312.7	38.1	0.7	316	122	17.5	61	47	3890 [b]
Czechoslovakia	49.4	127.9	15.6	0.2	316	122	13.9	74	40	5400 [b]
Hungary	35.9	93.0	10.6	−0.2	295	114	19.0	58	57	1710 [b]
Romania	91.7	237.5	23.1	0.5	252	97	25.6	49	44	2020 [b]
Bulgaria	42.8	110.9	9.0	0.2	210	81	14.5	65	37	2800 [b]
Albania	11.1	28.7	3.1	2.1	279	108	43.0	34	25	930 [b]
Yugoslavia	98.8	255.8	23.6	0.6	239	92	27.1	47	31	2120 [b]
Totals	450.4	1166.5	123.1	0.6	273	105	21.0	57	41	3050
Grand Totals	1879.8	4868.7	496.5	0.3	264	102	13.0	75	29	8170

[a]Microstates and other units omitted from the table include *Andorra* (181 sq mi/468 sq km; population 50,000; per capita GNP $9000 in 1983); the *Faeroe Islands* (540 sq mi/1399 sq km; population 47,000; per capita GNP $11,200 in 1985); *Gibraltar* (2.25 sq mi/5.8 sq km; population 31,000; per capita GNP $4550 in 1985); *Greenland* (840,000 sq mi/2,175,000 sq km; population 55,000; per capita GNP $7330 in 1985); *Liechtenstein* (62 sq mi/160 sq km; population 28,000; per capita GNP $20,960 in 1980); *Monaco* (0.73 sq mi/1.9 sq km; population 29,000); *San Marino* (24 sq mi/61 sq km; population 22,000; per capita GNP $8250 in 1980); *Svalbard* (24,208 sq mi/62,698 sq km; population 3500); *Vatican City State* (0.17 sq mi/0.44 sq km; population 1000).

[b]1984 or earlier years in the 1980s.

iterranean attributes, but in this text it is convenient to place those countries in other regional groups. The natural environment of Southern Europe—featuring mountains, subtropical temperatures, dry summers, and the nearness of most places and people to the sea—would remind many Americans of southern California. Despite remarkable cultural achievements in earlier times, this is today a relatively poor part of Europe, with much less industrial development than Europe's main industrial focus in West Central Europe and Great Britain. It becomes increasingly nonindustrial and poverty-stricken toward the south.

5. Along or near the borders of the Soviet Union are the *countries of East Central Europe*—Poland, Czechoslovakia, Hungary, Romania, Bulgaria, Albania, and Yugoslavia. These countries, along with East Germany and part of Austria, came under Communist control in the aftermath of World War II. In most of them, this was the result of occupation by the victorious Soviet armies that pushed German forces out. Soon after the war, the occupation resulted in the installation of national Communist governments subservient to the USSR. In Yugoslavia and Albania, national Communist resistance forces, rather than the Soviet army, seized power, and the result has been Communist regimes independent of the USSR.

Soviet hegemony in most of East Central Europe since World War II is only the latest episode in a long history of control over parts of this region by outside powers—notably Austria, Germany, Russia, and Ottoman Turkey. Before World War II, the region was a poor and rather agricultural part of Europe. It is still relatively poor and agricultural, although a number of massive industrial developments have been achieved under the Communists. Culturally this is, with some exceptions, the Slavic part of Europe. Slavic languages are spoken by approximately two thirds of the people of the region, as they are by most of the people of the USSR. The main non-Slavic exceptions are Hungarian, which is not closely related to any other European national language; Albanian, of which the same can be said; and Romanian, which is basically a Romance language, but with many Slavic words and expressions.

THE KALEIDOSCOPE OF LANGUAGES AND RELIGIONS

Europe's diverse languages and religions are powerful indicators of culture, badges of nationality, and repositories of achievement. They stand very high among the factors that differentiate Europe geographically. During the long course of European history, they have often played an important role in political conflict, including warfare, repression, discrimination, and terrorism. Language and religious patterns are interrelated, but for clarity they will be discussed separately, commencing with language.

Europe emerged from prehistory as the homeland of many different tribal peoples. In ancient and medieval times, certain peoples experienced periods of vigorous expansion, and their languages and cultures became widely diffused. First came the expansion of the Greeks and the Celtic peoples; later that of the Romans, the Germanic (Teutonic) peoples, and the Slavic peoples. As each expansion occurred, traditional languages persisted in some areas but were displaced in others. Each of the important languages eventually developed many local dialects. With the rise of centralized national states in early modern times, particular dialects became the bases for standard national languages. Thus, the dialect of Paris evolved as standard French, and the dialect of London as standard English. But people in many localities continued to use their accustomed dialects and do so today. Europe's language pattern, then, is complex: With certain major exceptions, every country has its own national language, and many local dialects and minor languages persist in isolated or deeply rural areas or in zones along international frontiers. We turn now to a consideration of the successive language expansions and contractions that occurred, each leaving its mark on the territorial pattern of languages.

EXPANSION AND DECLINE OF THE GREEK AND CELTIC LANGUAGES

The first millennium B.C. witnessed a great expansion of the Greek and Celtic peoples. In peninsulas and islands bordering the Aegean and Ionian seas, the early Greeks evolved a civilization that reached unsurpassed heights of philosophical inquiry and literary and artistic expression. Greek adventurers, traders, and colonists used the Mediterranean Sea as their highway to spread classical Greek civilization and its language along much of the Mediterranean shoreline. Evidence of the early geographic range and subsequent influence of the Greek language and culture is apparent in the many Greek elements in modern European languages. But over the course of centuries, the use of Greek in most areas disappeared as new peoples and languages expanded. Today it remains the spoken language of Greece itself, but the dialects in use are very different from the classical Greek in which works were written that form one of the major foundations of Western civilization (see

Editorial Comment

WESTERN CIVILIZATION

"Western civilization" and related terms such as *the West* and *Westernization* are indispensable to a study of world geography, since they connote innumerable traits that give geographic distinctiveness to places. In general, Western civilization refers to the sum of values, practices, and achievements that had roots in ancient Greece, Rome, and Palestine, subsequently flowered in Europe, and are still being developed and modified there and in other areas to which they have been diffused. The term incorporates a set of languages, religious practices associated with "Western" Christian churches (the Roman Catholic church and the Protestant churches, but not the Orthodox Eastern churches), systems of law, and systems of social, economic, and political organization. Among the core conceptions of modern Western civilization are strong commitments (not always successful or effective) to education, experimental science, technological progress, economic development, democratic representative government, and explicit protection of individual rights and liberties. A strong reliance on private capitalism developed prior to and during the Industrial Revolution (see Chapter 5), although this has been modified by widespread socialistic tendencies during the 20th century. *Westernization* has reference to the process whereby non-Western societies acquire Western traits, which are adopted with varying degrees of thoroughness. As used today, *the West* refers most centrally to western Europe, Anglo-America, Australia, and New Zealand, although the concept embraces other countries—such as the Latin American countries, Israel, and South Africa—in greater or less degree.

Editorial Comment). However, the "pure" form of modern Greek taught in schools diverges less from the ancient language.

Europe's Celtic languages expanded at roughly the same time as Greek and, like Greek, are represented today only by remnants. The expansion of preliterate Celtic-speaking tribes took place from nuclear areas in what is now southern Germany and Austria. Much of continental Europe and even the British Isles were occupied. Conquest and cultural influence by later arrivals, mainly Romans and Germans, eventually eliminated the Celtic languages except for a few pockets that survive today in relatively remote peninsulas and islands: Breton in Brittany, Welsh in Wales, Irish Gaelic in western Ireland, and Scottish Gaelic in northwestern Scotland. Few people speak a Celtic language exclusively; in nearly all cases English or (in Brittany) French is also spoken. Recently the Celtic tongues have tended to become symbols for nationalist aspirations, sometimes backed by sporadic terrorism. A case in point is the agitation by some Bretons for greater autonomy or even separation from France.

EUROPE'S MAJOR LANGUAGES TODAY: ROMANCE, GERMANIC, AND SLAVIC

In present-day Europe the overwhelming majority of the people speak Romance, Germanic, or Slavic languages. The Romance languages evolved from Latin, originally the language of ancient Rome and a small district around it. In a few centuries before and just after the birth of Christ the Romans subdued territories extending from Great Britain to northern Africa and western Asia, and the use of Latin was spread to this large empire (Fig. 4.2). Latin did not necessarily displace the languages of conquered peoples (although often it did), but the upper classes in each Roman province were encouraged to learn it, and it became increasingly common as the language of administration, commerce, and education within the Roman Empire. In the eastern parts of the empire, both Latin and Greek were widely used, but Latin had by far the greater impact in the west. Over a long period, extending well beyond the collapse of the western part of the empire in the 5th century, regional dialects of Latin survived and gradually evolved into Italian, French, Spanish, Portuguese, and other Romance languages of today. Each Romance language is now centered in a particular national state, but often the language frontier does not coincide with the political frontier. For example, Italian is the principal language not only of Italy itself, but also of Corsica and the city of Nice in France, and of some parts of southern Switzerland. The French language extends beyond the borders of France to western Switzerland and also to southern Belgium, where it is known as Walloon. Among lesser Romance languages are the Catalan of northeastern Spain, long a basis for separatist aspirations in the province of Catalonia; the language called

Figure 4.2   Roman expansion in Europe, North Africa, and southwestern Asia is still evidenced by the remains of numerous stone structures. The photo shows a well-preserved Roman theater at the ancient city of Gerasa (Jerash) near Amman, Jordan.

Provençal, similar to Catalan and spoken in some rural parts of southern France; and minor dialects in some Swiss mountain valleys. The Romanian language originated as a Romance language but has acquired an extensive Slavic vocabulary.

In the middle centuries of the first millennium A.D., the power of Rome declined and a prolonged expansion by Germanic and Slavic peoples began. Germanic peoples first appear in history as a group of tribes inhabiting the coasts of Germany and much of Scandinavia. Subsequently, they expanded southward into Celtic lands east of the Rhine. Roman attempts at conquest were repelled, and the Latin language had little impact in Germany. In the 5th and 6th centuries A.D., Germanic incursions overran the Western Roman Empire, but in many areas the conquerors eventually were absorbed into the culture and language of their Latinized subjects. However, German expanded into, and has remained, the language of Austria, Luxembourg, Liechtenstein, the greater part of Switzerland, the previously Latinized part of Germany west of the Rhine, and parts of easternmost France (Alsace and part of Lorraine). It remains the language of present-day West and East Germany.

The Germanic languages of Europe include many languages other than German itself. In the Netherlands, Dutch developed as a language closely related to dialects of northern Germany, while Flemish, almost identical to Dutch, became the language of northern Belgium. Except for Finnish, which belongs to another language family, the present languages of Northern Europe are descended from the same ancient Germanic tongue. Modern Icelandic has diverged least from the parent language and would be largely unintelligible to a speaker of Danish, Norwegian, or Swedish. The latter languages are, in general, mutually intelligible.

English is basically a Germanic language, although it has many words and expressions derived from French, Latin, Greek, and other languages. Originally it was the language of the German tribes known as Angles and Saxons who invaded England in the 5th and 6th centuries A.D. The Norman conquest of England in the 11th century established French for a time as the language of the court and the upper classes. Modern English retains the Anglo-Saxon grammatical structure, but it has borrowed great numbers of words from French and other languages (Fig. 4.3). English is now the principal language in most parts of the British Isles, having been imposed by conquest or spread by cultural diffusion to areas outside England.

Slavic languages are dominant in most of East Central Europe. The main ones today include the Russian and Ukrainian of the Soviet Union; Polish; Czech and Slovak in Czechoslovakia; Serbian and Croatian in Yugoslavia; and Bulgarian. They apparently originated in eastern Europe and Russia and were spread and differentiated from each other during the Middle Ages as a consequence of migrations and cultural contacts involving various peoples. Over a period of many centuries, the Slavic languages lost much ground to an expansion of German power and culture eastward, but this long trend was abruptly reversed by Germany's defeats in the two

Figure 4.3 *The English language of today has roots in both Anglo-Saxon and French. The French component is a heritage from invaders under Duke William of Normandy who took control of England in 1066 as a result of victory over the native English in the Battle of Hastings. The battle, fought a few miles inland from the English Channel, takes its name from the seaside town of Hastings, shown in the photo. Today, Hastings is a popular Channel resort.*

world wars of the 20th century. After World War II the Slavic frontier moved westward as a result of the flight, expulsion, or transfer of millions of German-speaking people from parts of East Central Europe and the Soviet Union which they had previously inhabited.

A few languages in present-day Europe are very distinct in that they are not related to any of the groups just discussed. Some are survivals of very ancient languages that have persisted from prehistoric times in isolated (usually mountainous) locales. The two outstanding examples are the Albanian language in the Balkan Peninsula and the Basque language spoken in or near the western Pyrenees Mountains of Spain and France. Some languages that are not related to others in Europe do have relatives in the Soviet Union. They reached their present locales through migrations of peoples westward. The prime examples are the Finnish language and Hungarian, also called Magyar. History records many invasions of Europe from the east like that of the Magyars in the 9th century. Most invaders were eventually defeated, sometimes by new invaders, but the Magyars were able to establish themselves in the plains of Hungary. Surrounded by Slavs and Germans (Austrians), they have maintained their distinctive language there.

RELIGIOUS DIVISIONS

In the Roman Empire, Christianity arose, spread, survived persecution, and became a dominant institution in the late empire. It survived the empire's fall

and continued to spread, first within Europe and then to other parts of the world. Today it remains the principal faith of a Europe that is very secularized. In total number of adherents the Roman Catholic Church is Europe's largest religious division, as it has been since the Christian church was first established. From the days of the Roman Empire onward, the authority of the Popes at Rome, exercised through an international hierarchy of church officials, put a lasting stamp on European civilization (see chapter-opening photo). Today the areas that remain predominantly Roman Catholic include Italy, Spain, Portugal, France, Belgium, the Republic of Ireland, large parts of the Netherlands and West Germany, Austria, Poland, Hungary, and probably Czechoslovakia. Important Catholic minorities are also present in the United Kingdom, Yugoslavia, and other countries.

During the Middle Ages a center of Christianity existed in Constantinople (now Istanbul, Turkey) as a rival to Rome. From it, the Orthodox Eastern Church spread to become and remain the dominant form of Christianity in Greece, Bulgaria, Romania, and the Serbian part of Yugoslavia, as well as in Russia to the east. Various national churches such as the Greek Church and the Russian Church exist within the Orthodox communion.

In the 16th century the Protestant Reformation took root in various parts of Europe (Fig. 4.4), and a subsequent series of bloody religious wars, persecutions, and counterpersecutions left Protestantism dominant in Great Britain, northern Germany, the Netherlands, and the five countries of Northern Europe. Except for the Netherlands, where a higher

Figure 4.4 *In Geneva, Switzerland, the Protestant Reformation is commemorated by a lengthy stone wall, a portion of which is shown here. Known as the Reformation Monument, the wall is dominated by large statues of Protestant religious leaders, including John Calvin (second from left). Geneva, where Calvin lived and preached, was an important center of the Reformation. This religious movement drastically changed the cultural and political order in Europe.*

Catholic birth rate has reversed the balance, these areas are still mainly Protestant. In addition, important Protestant minorities are found in various other countries.

A non-Christian element in Europe is represented by the Islamic (Muslim) faith, which is the principal religion of Albania and is the faith of minorities in Yugoslavia and Bulgaria. Islam was once more widespread in the Balkan Peninsula, where it was established by the Ottoman Turks during their period of rule from the 14th century to the 19th century. Under the Roman Empire, Jewish minorities spread from Palestine into Europe, where they have since persisted in many countries despite recurrent persecution. Today the number of European Jews is much smaller than it was before the anti-Semitic Nazi regime in Germany killed an estimated 6 million Jews and caused the flight of hundreds of thousands more, mainly to Israel.

EUROPE'S GLOBAL IMPORTANCE

In recent centuries, the peoples of Europe have exerted a profound influence on the rest of humanity. Far more than any other region, Europe has shaped the human geography of the modern world. From tentative beginnings in the 15th century, European seamen, traders, soldiers, colonists, and administrators burst upon the world in succeeding centuries, and by the 19th century had created a world that was mostly dominated by Europeans politically and was organized into an economic system in which world trade and finance centered on Europe. In their fairly recent histories most countries outside Europe have had the experience of being European colonies—in most cases formally and in some informally. Almost invariably, the attempt to understand the geography of other world regions requires an assessment of the influence of Europe on their development. The latter process was both complex and reciprocal. Europe's own societies, economies, and governments were changing and developing during the centuries that the region was reaching out to dominate the world. Such transformations were in part a response to overseas contacts. Europe, in turn, radiated its own changes to the rest of the world. The entire process is modern history's most important example of spatial interaction and cultural diffusion. From it emerged the political and economic world that we know today.

HOW EUROPE SHAPED THE WORLD IN THE COLONIAL AGE

The process of exploration and discovery by which Europeans filled in the world map began with 15th-century Portuguese expeditions down the west coast of Africa, and then expanded with dazzling successes that soon put Europe in sea contact with most of the world. In 1488 a Portuguese captain, Bartholomew Diaz, sailing far south of the equator along Africa's Atlantic coast, rounded the Cape of Good Hope and opened the way for subsequent European voyages into the Indian Ocean. Then, in 1492, America was discovered when a Spanish expedition commanded by a Genoese (Italian), Christopher Columbus, crossed the Atlantic and touched land in the Bahama Islands and the islands of Cuba and Hispaniola. In 1497–1499, the seaway to the coveted spices of tropical Asia was inaugurated by a Portuguese, Vasco da Gama, who voyaged by the Cape route from Portugal to India and back. A quarter century later, these early feats of exploration culminated in the first circumnavigation of the globe. This tremendous voyage was the work of a Spanish expedition led initially by yet another Portuguese, Ferdinand Magellan. After crossing the Atlantic, the expedition rounded South America through the Straits of Magellan and then sailed across the broad Pacific to the

Philippine Islands, where Magellan was killed by tribesmen. Subsequent misadventures left a tiny remnant of survivors who finally reached Spain by way of the Indian Ocean, the Cape of Good Hope, and the long northward journey along the west coast of Africa. Worldwide exploration, both coastal and inland, continued apace for centuries, with leadership being wrested from the Spanish and Portuguese by other Europeans—notably the Dutch, French, and English. By the 19th century, attention was focusing mainly on the exploration of the remaining unknown areas in the polar regions, interior Africa, and western North America.

Missionaries, soldiers, and traders were seldom far behind the explorers. It was not uncommon for the same man to play two or more of these roles. Trading posts, Christian missions, and military garrisons to protect them were soon established along many coasts, or, in some cases, inland. Inland posts were especially notable in Latin America, where they were often a response to the highland location of Indians who could be Christianized, robbed of any accumulated store of precious metals, and put to work in mines or on the land. Many of the European outposts came to dominate the areas where they were established, and eventually many became the bases from which complete colonial control was established over non-European lands and peoples.

Settlement by European colonists, on differing scales and at different times, took place in many non-European areas. Spanish colonists founded the city of Santo Domingo, in the present Dominican Republic, before the year 1500; and even in the early 20th century, English colonization in parts of Africa was continuing. In previously well-populated or environmentally difficult areas European populations remained small minorities, but in other circumstances local versions of European societies took such firm root that they became numerically dominant. Today, the descendants of transplanted Europeans comprise most of the population in the United States, Canada, parts of Latin America, Australia, and New Zealand.

In establishing their colonies and carrying on their various overseas enterprises, Europeans not only migrated to non-European areas, but they often transferred non-Europeans from place to place. The most notable instance was the large-scale movement of slaves from Africa to the New World. Such transfers greatly influenced the ultimate racial and cultural make-up of sizable areas, and they affected Europe itself to a minor degree. Also of great importance in shaping the world's subsequent geography was the transfer of plants and animals from one place to another. A few major examples are the

diffusion of horses, hogs, and cattle to North America from Europe; of tobacco and corn (maize) from the Americas to Europe and other parts of the world; of rubber from South America to Asia; of the potato from the Americas to Europe, where it became a major food; and of coffee from the Middle East to Latin America, where it became the principal basis of a number of regional economies.

A worldwide system of trade focused on Europe was a major outcome of European expansion. The system principally involved the movement of raw materials and food products from the rest of the world to Europe, and the return sale of European manufactured goods to non-European areas. In general this exchange favored Europe, and wealth from the exchange mainly accumulated there. The infusions of money enabled certain European cities, particularly London, to become the centers of world finance. These cities continued to be aggrandized by an expanding worldwide flow of loans and investments out of Europe and a return flow of interest, dividends, and profits into the region. To a large extent the trade and financial patterns described in this paragraph continue today, although other centers of commerce and finance (notably in the United States and Japan) have risen to challenge Europe's dominance.

INTERPRETATIONS OF EUROPEAN DOMINANCE

What factors account for the rather sudden rise of Europe to world dominance? How was this small region able to transform immensely larger portions of the world and to maintain its commanding position for so long? Many interpretations are possible, with various scholars emphasizing such factors as the following:

1. Parts of Europe had already developed capitalist institutions by the end of the Middle Ages, and the colonial age saw a further rapid development of capitalism in the region. Profit became a highly acceptable motive, and the relative freedom of action afforded by the capitalistic system provided opportunities for gaining wealth by taking risks. Energetic entrepreneurs found it possible to mobilize capital into large companies and to exploit both European and overseas labor. Such ventures often involved unscrupulous, illegal, or even barbaric practices, and these led to increasing regulation by governments as time went on. But the sheer energy and ingenuity exhibited by Europeans in pursuit of profits under the system of capitalism is seen by some as a major

explanation for the world dominance that Europe achieved.

2. By the end of the Middle Ages, the Europeans had reached a level of technology generally superior to that of non-Europeans with whom they came in contact during the early Age of Discovery. In particular, their achievements in shipbuilding, navigation, and the manufacture and handling of artillery gave them decided advantages in seeking out new lands to exploit and establishing themselves there. With the passage of time, European technology continued to develop at an increasing pace, and the technological gap between Europe and most other areas widened.

3. Europe's scientific prowess must not be overlooked as an explanation for the region's rise to world dominance. Technology is partly applied science, and the foundations of modern science were being constructed almost entirely in Europe during the centuries when European influence was becoming paramount. Until recently, the great names of science have been overwhelmingly the names of Europeans; as late as World War II, the development of atomic fission in the United States was carried out by a team comprised largely of prominent refugee European scientists.

4. The aggressive confidence of Europeans in the superiority of their own civilization is another explanation that has been offered for European success. Abundant proof exists that such feelings were very prevalent and were often coupled with a strong sense of a mission to bring European culture to peoples and lands beyond the seas. Europeans often displayed great zeal in propagating their own ideas and institutions among alien peoples.

Space does not permit further examination of the arguments and speculations that have been put forward to explain Europe's rise to dominance in the outer world. The foregoing explanations are undoubtedly significant, and they indicate the complexity of this important historical problem. We now turn to a related matter of critical significance—the loss of Europe's primacy in the present century.

DECLINE OF EUROPE'S PRIMACY

In the 20th century, Europe's preeminence in the world has been lost. European colonial empires have disappeared except for a few small remnants. In many fields Europe remains prominent, but in most fields it has lost the commanding position it once held. What has caused the relative decline in its for-

tunes? Some important reasons and circumstances are as follows:

1. Europe suffered enormous casualties and damage in World Wars I and II, which were initiated and mainly fought in Europe. Recovery was eventually achieved with American aid, but the region's altered position could not be reversed. The wars destroyed Europe's ability to maintain its predominance in the face of such trends as anticolonialism, the rise of the United States and the Soviet Union to world power, and the spread of modern industry to many areas outside Europe.

2. Rising nationalism in the colonial world during the 20th century resulted in the virtual end of the European colonial empires within a surprisingly short time after World War II. Taking advantage of a weakened Europe and a mounting disapproval of colonialism in the world at large, one European colony after another gained independence. In most colonies European control was thrown off with little or no bloodshed, although in some colonies there was much fighting. Colonies such as French Indochina and Portuguese Angola witnessed civil wars between rival factions as well as armed struggles between native insurgents and the troops of the European power. Opposition to continued European control was often spearheaded by leaders who had been educated in Europe and had absorbed nationalistic ideas there. Nationalism—the claim of a nationality to have its own political unit—is a sentiment that seems to have originated in Europe. By propagating it, the Europeans helped bring about their own downfall in the colonial world.

3. Europe's predominance was seriously eroded by the rising stature of the United States and the Soviet Union. These enormous countries, each far larger than any European country, have outstripped Europe in military power and world influence. The enhanced positions they have achieved in the 20th century are based on their large populations, areas, resources, and economies; their ability to develop and mass-produce advanced weapons; and political integration enabling them to harness national energies to achieve national goals.

4. Europe once enjoyed a near-monopoly in exports of manufactured goods, but, in the 20th century, manufacturing has developed in many countries outside Europe. Japan is a prime example. Since the middle of the 19th century, this Asian country has emerged from an isolated medievalism to become an industrial giant competing with Europe in markets

all over the world. The United States began its industrialization earlier than Japan, but it, too, reached industrial maturity in the 20th century and became a vigorous competitor of Europe. Actually the United States, like Europe, has now become an "older" industrial society faced with severe competition not only from Japan but from units such as South Korea or Taiwan, which have developed ultramodern industries very recently. Cheap labor is a factor drawing industry to poor countries outside Europe and the United States, but such countries also are advantaged by rising levels of education and technical competence. Both Europe, which began the Industrial Revolution, and the United States, which inherited the Revolution very early, are seeing their industrial leadership eroded as industrialism continues to spread. Unemployment caused by foreign competition is requiring the older societies to rethink their situations in order to determine the industries and other activities in which they can hope to excel.

5. Europe's ability to assert itself in world affairs has been weakened by the region's new dependence on outside sources of energy. European industry and transportation once depended very largely on Europe's own coal resources to supply power, but coal is now outdated for many uses, and the region's coal fields have become excessively costly to exploit. Despite the recent development of North Sea oil and gas resources (which primarily benefit Great Britain and Norway), Europe is now very dependent upon Middle Eastern oil and other imported sources of power.

POPULATION AS AN ELEMENT IN EUROPE'S CONTINUING IMPORTANCE

Despite the relative decline in its fortunes, Europe continues to be a very important part of the world. A major reason lies in the region's large and technically skilled population. In sheer numbers, Europeans represent one of the great masses of world population. Among major world regions considered in this text, Europe still ranked third in population in mid-1988, after the Orient and Africa, although it would soon be surpassed by rapid growth in the Middle East (see Table 3.1). Its estimated total of 497 million people was nearly as large as that of the Soviet Union and the United States combined. In a space only half the size of the United States are concentrated one of every ten people in the world. Birth rates are low in this region of relative affluence and high urbanization—so low that a few countries, in-

cluding West Germany, are actually decreasing in population, while the whole region is increasing much more slowly than most of the world. But a population of this magnitude, well educated and highly productive in many fields, will long continue to be a very significant factor in world affairs.

Major Population Axes and Their Significance

Except for some northern, rugged, or infertile areas, European population density is everywhere greater than the world average. However, the greatest congestion is found along two axes of exceptional industrialization and urbanization near coal or hydroelectric power sources. One axis extends north-south from Great Britain to Italy (Fig. 4.5). In addition to large parts of Great Britain, it includes extreme northeastern France, most of Belgium and the Netherlands, West Germany's Rhineland, northern Switzerland, and—across the Alps—northern Italy. The second axis of densest population extends west-east from Great Britain to southern Poland, and continues into the Soviet Union. It corresponds to the first axis as far east as the Rhineland, from whence it forms a relatively narrow strip eastward across West and East Germany, western Czechoslovakia, and Poland. Although the two belts represent a minor part of Europe's total extent, they contain more large cities and have a greater value of industrial output than the rest of Europe combined. Only in eastern North America and in Japan are there urban-industrial strips of the same order of importance.

The European belts coincide with major routeways that were in use very early for migration, trade, and military movement. Many cities along these routes are quite old; in the Middle Ages some were already famous for trade fairs, where merchants gathered to buy and sell goods. With the rise of modern industry, coal fields along the west-east axis became important; and in the southern part of the north-south axis, the age of electricity saw the development of many industries based on the hydroelectric resources of the Alps. Both belts have benefited from the fact that they traverse areas of relatively good soil.

PRODUCTIVITY AND AFFLUENCE

Economic productivity is one of the most significant reasons for the continuing importance of Europe. This region continues to form a major part of the world economy, providing an immense output of

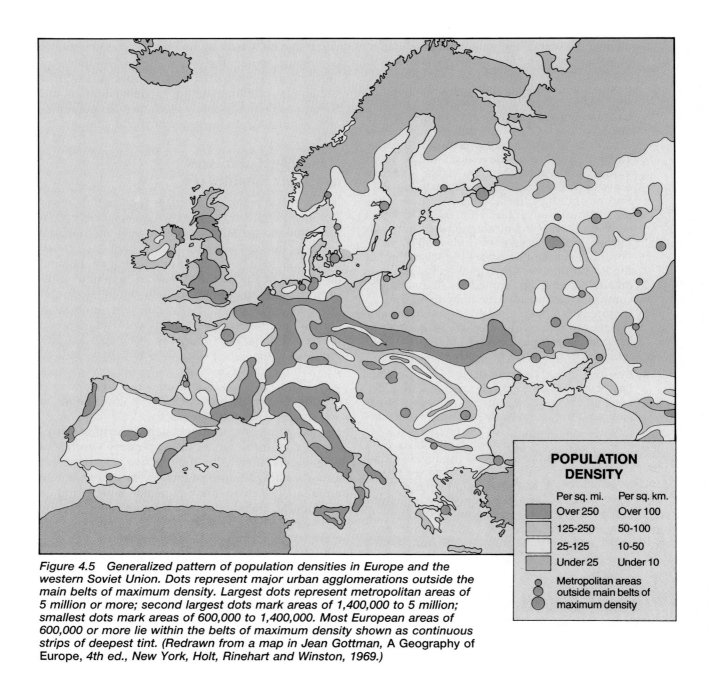

Figure 4.5 Generalized pattern of population densities in Europe and the western Soviet Union. Dots represent major urban agglomerations outside the main belts of maximum density. Largest dots represent metropolitan areas of 5 million or more; second largest dots mark areas of 1,400,000 to 5 million; smallest dots mark areas of 600,000 to 1,400,000. Most European areas of 600,000 or more lie within the belts of maximum density shown as continuous strips of deepest tint. (Redrawn from a map in Jean Gottman, A Geography of Europe, *4th ed., New York, Holt, Rinehart and Winston, 1969.)*

POPULATION DENSITY

Per sq. mi.	Per sq. km.
Over 250	Over 100
125-250	50-100
25-125	10-50
Under 25	Under 10

Metropolitan areas outside main belts of maximum density

goods and services, and serving as a major market for other parts of the world. Europe is still an important supplier of manufactured goods, despite increasing competition from other lands, and its worldwide network of services such as ocean shipping, air transportation, communications, and financial services still plays a large role in transacting the world's business. Not so well recognized is the fact that Europe is also a major agricultural area, where farmers extract some of the world's highest yields from farms that are generally rather small. Despite their high population densities, most European countries have some agricultural exports, often of a very specialized kind, and a number of countries export a greater value of agricultural produce than they import. The high level of output is partly a result of government tariff protection and subsidies. European governments foster agriculture in order to keep in the good graces of farmers politically and somewhat lessen strategic vulnerability from excessive dependence on imported food.

Levels of output, and therefore of affluence, vary greatly within Europe, although the region as a whole is one of the "developed" and relatively rich parts of the world. In general, Southern Europe and East Central Europe are the least affluent of Europe's major areas as measured by per capita GNP (gross national product: roughly equivalent to income; see Table 4.1). But an average country in either Southern Europe or East Central Europe would be a relatively rich country in Asia, Africa, or Latin America. Ireland is the poorest European country outside of Southern and East Central Europe, but even its per capita GNP of about $5000 in 1986 was much higher than a world average of $3000. Other indicators of level of living, such as infant mortality, average life expectancy, or rate of literacy, confirm the fortunate situation of most Europeans as compared with the poverty-stricken majority of the world's population (see Table 3.1).

WORLD STRATEGY: EUROPE HOLDS THE ECONOMIC BALANCE

When such high levels of productivity are attained by a group of countries totaling about half a billion persons, it is obvious that the aggregate output is huge. Europe's share of the world economy is, in fact, so large that the region holds the economic balance in the confrontation between superpowers which has dominated world politics and strategy for several decades. This is especially true in manufacturing. For example, take the case of steel—a vital material in the manufacture of armaments. In 1986, the combined output of the steel industries of non-Communist Europe amounted to 19 percent of world production. Soviet satellite countries in eastern Europe accounted for another 9 percent; the USSR itself for 22 percent; Japan for 14 percent; and the United States for 10 percent. Hence the power of the United States and non-Communist Europe was supported by about 29 percent of world steel output, and Soviet power by about 31 percent. With the output of non-Communist Europe shifted to the other side, the balance would have read United States 10 percent; Soviet Union and Europe 50 percent. In many other industries, too, and the skills that support them, non-Communist Europe represents a similar crucial balance from a strategic point of view. If the economic (and military) weight of non-Communist Europe was added to that of the Soviet bloc or even neutralized, the results might well be catastrophic for the United States. Hence, Europe remains a critical arena of contested ground in the long confrontation between the superpowers. Needless to say, the Europeans derive no comfort from their position as a potential battlefield of two giant countries bristling with both non-nuclear and nuclear weapons, although fears and tensions of all

the parties have recently been allayed a bit by some initial moves of the superpowers to scale down their armaments.

THE SEARCH FOR DEVELOPMENT AND SECURITY THROUGH INTERNATIONAL COOPERATION

A significant movement toward greater economic, military, and political cooperation among European countries has helped Europe recover from World War II and has enhanced the region's stature and distinctiveness in the postwar world. Searching for development and security by means of common action, various countries have banded together in a series of organizations designed to foster unity. Unfortunately, this effort has largely been split into cooperation among non-Communist ("Western") countries on the one hand and Communist ("Eastern") countries on the other. And within each group of cooperating countries, the old self-centered nationalisms are far from dead. Thus, one must view the new internationalism as a hopeful portent, while recognizing that nothing resembling a true "United States of Europe" has emerged and possibly never will.

REASONS FOR EUROPE'S NEW AGE OF COOPERATION

What is responsible for the new age of cooperation in a region long notorious for national quarrels and strife? Important among the many motivations are a long-standing ideal of European unity, unfortunate experience in disastrous wars, perceived military and political threats after World War II, and a search for economic betterment through the concerted action of nations. There has long been an ideal of unity in Europe, which probably traces back in greatest part to Roman emperors who imposed a common system of law and order over much of Europe, and medieval Popes who claimed (and often exerted) authority over all European Christians, including heads of state. For many centuries a measure of unity did, in fact, prevail in Europe, and present-day Europeans are heirs of this tradition. But the tradition was not strong enough to prevent incessant wars after the grip of the Roman Empire loosened. Lack of unity climaxed in the fearful bloodletting and physical devastation of two general wars in Europe within the space of about 30 years in the 20th century (World War I, 1914–1918; World War II, 1939–1945). These wars were separated by only 21 years of uneasy "peace," during which violent national quarrels between European states, including economic strife, were common. In these difficult years the peoples of Europe suffered from economic stagnation and depression, periods of severe inflation, class conflicts, and unemployment. After World War II economic recovery was slow, and there were fears in western Europe of possible military aggression by the Soviet Union and/or a political takeover by powerful Communist parties in certain Western countries, notably Italy and France. In the meantime, the USSR and the Communist governments of the satellite countries east of the Iron Curtain saw American military power in Europe as a major threat, which might be countered by economic and military cooperation under the leadership of the Soviet Union. The desire for economic recovery and military security in both the Western and Eastern nations set the stage for the remarkable period of supranational organization described in the following paragraphs.

COOPERATION IN THE NON-COMMUNIST WEST

As they considered Europe's plight after World War II, many influential people in western Europe took the view that new catastrophes could best be avoided and more peaceful, secure, and prosperous societies achieved through a close alignment with the United States and an evolutionary development of cooperation among the non-Communist European states. Some idealists hoped that unification of a "United States of Europe" might eventually result. But the crisis atmosphere that enveloped non-Communist Europe led statesmen to focus their immediate attention on economic and military cooperation within organizations of sovereign states. In 1948 a major step was taken when the Organization for European Economic Cooperation (OEEC) was formed among many European states to coordinate the use of American aid under the Marshall Plan. The United States had proffered this aid to bolster Europe against Communism. The Marshall Plan was terminated in 1952, but OEEC continued to function. In 1960 the name of the organization was changed to Organization for Economic Cooperation and Development (OECD), and the membership has expanded to include not only non-Communist Europe but also the United States, Canada, Japan, Australia, New Zealand, and Turkey (Fig. 4.6).

But the most significant economic organization among Europe's "Western" nations has come to be the European Economic Community (EEC), often called the European Common Market (Fig. 4.6). It was established by six countries—France, West Germany, Italy, Belgium, Luxembourg, and the Netherlands—under the Treaty of Rome in 1957. They had already been experimenting with the elimination of political barriers to the unified operation of their coal and steel industries by organizing themselves into the European Coal and Steel Community (1951), and they now looked toward the removal of all economic barriers among them. In 1973 the United Kingdom, Denmark, and Ireland also became member nations; then in 1981 Greece, and in 1986 Portugal and Spain. Other European nations will possibly join in the future. Cyprus, Malta, and Turkey are associate members.

The EEC was designed to secure the benefits of large-scale production by pooling the resources—natural, human, and financial—and markets of its members. Hence tariffs have been eliminated on goods moving from one member state to another, and restrictions on the movement of labor and capital between member states have been greatly reduced. Trusts and cartels formerly restricted competition within and between these countries (often with government encouragement), but such monopolistic organizations are now discouraged. Meanwhile, a common set of tariffs is maintained for the Community's entire area to regulate imports from the outside world, and a common system of price supports for agriculture has replaced the individual systems of member states. The founders of the EEC anticipated that free trade within such a populous and highly developed bloc of countries would (1) stimulate investment in mass-production enterprises, which could now, wherever they were located, sell freely into all EEC countries; and (2) encourage a productive geographic specialization, with each part of the Community expanding lines of production for which it was best suited. It was hoped that participation in the Community would enable each country to achieve greater production, larger exports, lower costs to consumers, higher wages, and a higher level of living than it could achieve on its own.

The Common Market has made impressive progress toward its ambitious goals. The Community as a whole has prospered, and its economy has grown mightily during the years since the Treaty of Rome, although the initial boom was succeeded by more difficult times during the 1970s and early 1980s. It is possible that this success would have been attained without the formation of the Market. Switzerland, a close neighbor of the Common Market but not a member, had a higher income per capita than any nation in the world in 1986. Norway

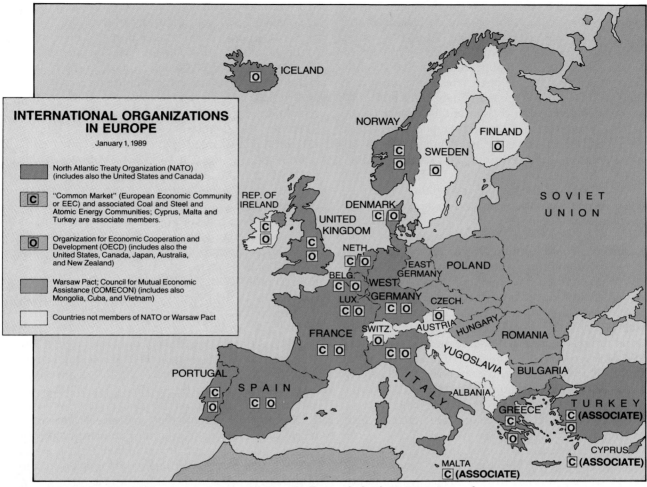

Figure 4.6 *Map showing some key organizations in Europe's intricate structure of political, strategic, and economic relationships. Members of the European Economic Community (EEC or "Common Market") also form the European Coal and Steel Community and the European Atomic Energy Community. The three organizations are so closely knit that they are often thought of collectively as the "European Community" or "European Communities" (EC).*

and Sweden, two other European states that do not belong to the Market, also rank near the top of the world scale in affluence. But Switzerland, Norway, and Sweden depend heavily on trade and other dealings with the Common Market; thus their prosperity is closely tied to that of the Market. The same is true of other European states that are not EEC members. Six nonmembers—Iceland, Norway, Sweden, Switzerland, Austria, and Finland—actually enjoy the same free-trade privileges for trade in nonagricultural products as those of member states. (The six countries belong to an organization called the European Free Trade Association, or EFTA, which maintains free trade among the members but allows each to set its own tariffs in trading with the outside world. Portugal was a member until its entry into

the Common Market in 1986.) The Common Market has also made preferential trade and economic aid treaties with many other countries—mainly ex-colonies of EEC members—in areas outside Europe.

The EEC has elaborate political and administrative machinery. Its policies are set by meetings of representatives of the member governments (ministers or heads of government). Research, monitoring, advice, and execution of policies are provided by a European Commission appointed by the members and supported by a large and very highly paid executive bureaucracy located at Brussels, Belgium. Thus, that city has a good claim to being the capital of the Common Market. But the Community also has a Court of Justice, which sits at Luxembourg City, to decide disputes over Community regulations and affairs, and a parliament—the European Parliament—which meets at Strasbourg, in France on the border with West Germany. The latter body long consisted of members appointed by governments of the member states and had practically no power. But during the 1970s the member states made important changes by (1) giving the Parliament some authority over the budget of the Community, which is financed by contributions from member states and a percentage of the revenue from import tariffs on goods entering the Community; and (2) introducing, in 1979, the election of members of the Parliament by direct vote of the people in member countries. These moves were intended to foster political unity by giving the populace of the member countries a more direct voice in matters of common concern. Evolution toward a multinational European state is one of the Community's goals, but it has made much less progress than economic unification.

The institutions that govern the EEC also govern two other "Communities"—the European Coal and Steel Community, and the European Atomic Energy Community—that have the same membership as the EEC but were founded under separate treaties in the 1950s. Increasingly the three are known collectively as the European Community, European Communities, or "EC." For most purposes they have become one body, although they retain their separate legal personalities.

The complete unification of non-Communist Europe, or even of the Common Market countries, still appears a long way off. This is especially true of political unification, but economic unification is also very incomplete. Even within the "free-trade" area, there are some non-tariff trade barriers, such as "quality standards" (refusal of individual countries to admit goods that do not meet specified criteria as to quality). Such economic measures as a uniform currency, financial policy, taxation system, welfare system, development policy, and corporate law are apparently harder to achieve than the elimination of tariffs. Europe is still very much a region of individual sovereign countries, despite its impressive advances in economic cooperation and its continuing moves toward political coordination and effective international institutions.

In the sphere of military defense, the key international body in non-Communist Europe is the North Atlantic Treaty Organization (NATO). This military alliance was formed in 1949 and includes the United States, Canada, a majority of the non-Communist European nations, and Turkey (see Fig. 4.6). The member countries are pledged to settle disputes among themselves peacefully, to keep their in-

dividual and joint defense capacities in good order, to consider an attack upon any of them in Europe or North America as an attack upon all, and to come to the aid of the country or countries being attacked. The NATO countries provide support to the organization through an international bureaucracy located in Brussels. It deals with such matters as political coordination, planning, finances, and research. The Supreme Headquarters Allied Powers Europe (SHAPE), commanded by the Supreme Allied Commander Europe (SACEUR) is near Mons, Belgium, some 20 miles (c. 30 km) from Brussels. The commander and his staff control the joint military plans, operations, assignments, and exercises of the member countries except for France, which maintains an independent command structure but has liaison with the NATO structure through a military mission. International naval forces within NATO (again with the exception of France) are controlled by two naval commands that have responsibilities, respectively, for the North Atlantic and the English Channel–southern North Sea.

The task of SACEUR is not only military but also diplomatic, as NATO's members are sovereign countries that negotiate within NATO to advance their own interests according to their own political circumstances and subordinate their military affairs to NATO only according to their own choice. However, many of the frequent internal conflicts within the organization are negotiated at higher control levels, involving heads of government and foreign ministers of the NATO countries. Controversy is often generated by the pressure of public opinion concerning western Europe's defense needs and the region's military relationships with the United States. More specifically, it is generated by such questions as (1) the superiority or inferiority of NATO forces (both conventional and nuclear) to Soviet forces, (2) the proper force levels of American nuclear-warfare capacity in Europe, (3) the level of economic effort in support of military power that is to be expected of each NATO member (there is often resistance to calls for economic sacrifice to promote defense), (4) the actual willingness of the United States to expose itself to nuclear attack in support of Europe, (5) the validity of American perceptions of danger from the Soviet Union (many west Europeans downplay the danger), and (6) the general reliability of American leadership (many west Europeans question it). The effect of these questionings on the future of NATO remained uncertain in the late 1980s. But toward the end of the organization's first four decades of existence, during which war in Europe had been avoided, the governments of member states were giving no indication that NATO was about to collapse or have its forces seriously weakened. It remained a vital part of Europe's political geography.

COOPERATION IN THE COMMUNIST EAST

The answer of the Communist east to the Common Market and NATO has been the creation of an international economic organization called the Council for Mutual Economic Assistance (COMECON) and a military alliance called the Warsaw Pact. The Soviet Union and the European countries of Poland, East Germany, Czechoslovakia, Hungary, Romania, and Bulgaria belong to both organizations, which have their respective headquarters in Moscow. Cuba, Mongolia, and Vietnam are additional members of COMECON. The European Communist state of Yugoslavia belongs to neither organization, although it cooperates in certain ways with COMECON. Another European Communist country, Albania, was a founding member of both COMECON and the Warsaw Pact but has ceased participation. COMECON has been instrumental in increasing trade and other economic interactions among its members, which formerly did much less business with each other. The armed forces of all members of the Warsaw Pact are under a unified command headed by a Soviet marshal. These forces are formidable, but the Soviet Union must reckon with the possibility of disloyalty among non-Soviet elements if conflict with the West should occur.

CHAPTER 5

Environmental and Historical Factors in Europe's Economic Success

Close proximity of industry and agriculture is common in Europe. Shown here are crop fields and a huge thermal-electric power station in West Germany.

E now turn our attention to some key features of Europe's diverse environment and the use of that environment to create a highly productive economy. Success has been so marked that most Europeans enjoy a prosperity still out of reach for most of the world. Problems such as resource inadequacy and noncompetitive industries are looming as the region moves into a "post-industrial" era, but Europe has repeatedly shown an ability to surmount difficulties and it possesses great assets provided by nature or developed by resourceful people over thousands of years. The objective of this chapter is to introduce the region's physical and economic geography in a historical context.

THE ENVIRONMENTAL SETTING OF EUROPEAN ACHIEVEMENT

The application of human creativity to a varied and useful environment has been an important feature of Europe's long record of accomplishment. The roots of this development within Europe far antedate the Age of Discovery. Later technological advances and economic growth were funded in a major way by profits from exploitation of a worldwide colonial realm. But it should not be forgotten that Europeans also exploited each other: The unbelievably primitive working conditions of early English coal mines are only one of many possible illustrations. Having made these points, we still must underscore the ingenuity displayed by Europe's scientists, farmers, engineers, and entrepreneurs in the study and development of the region's rocks, minerals, landforms, climates, waters, soils, and biotic resources. Writers and artists also have been inspired by Europe's restless seas, intricate shorelines, scenic mountains and rivers, and changeable climates. On the whole, the region has provided a good home for people of many cultures. Today, no other major world region contains such a small proportion of unproductive land. In the following paragraphs, certain significant features of Europe's diverse environment are briefly discussed.

IRREGULAR OUTLINE

One of the most noticeable characteristics of Europe as seen on a map is its extremely irregular outline. The main peninsula of Europe is fringed by numerous smaller peninsulas, including the Scandinavian, Iberian, Italian, and Balkan peninsulas of the second

order of size, the still smaller peninsulas of Jutland, Brittany, and Cornwall, and many others (Fig. 5.1). Offshore are a multitude of islands, including such large and well-known islands as Great Britain, Ireland, Iceland, Sicily, Sardinia, Corsica, and Crete. Around the indented shores of Europe, arms of the sea penetrate the land and countless harbors offer a protection for shipping. The complex mingling of land and water has created an environment which provides many opportunities for maritime activity, and much of Europe's history from the earliest times has focused on maritime trade, sea fisheries, and sea power. Today, the region remains the leading focus of the world's sea routes.

NORTHERLY LOCATION

Another striking environmental characteristic of Europe is its northerly location. Much of Europe, including some of the most densely populated areas, lies north of the conterminous 48 states (Fig. 5.2). Despite their moderate climate, the British Isles are in the latitude of Canada, and even Athens, Greece, is only slightly farther south than St. Louis, Missouri. One effect of northerly latitude that visitors to such cities as Berlin or Stockholm will notice is the long duration of daylight on summer days and the brevity of daylight in the winter.

TEMPERATE CLIMATE

As might be inferred from the presence of some of the world's most densely populated areas in latitudes corresponding to those of Canada, the climate of Europe is more temperate than its northerly location would suggest. Winter temperatures, in particular, are very mild for the latitude (Fig. 5.3). For example, London, England, has approximately the same average temperature in January as Richmond, Virginia, which is 950 miles (c. 1500 km) farther south; Reykjavik, Iceland, is nearly as warm in January as St. Louis, Missouri, 1750 miles (c. 2800 km) farther south; and Tromsö, on the coast of Norway 3° north of the Arctic Circle, is slightly warmer in January than Chicago, Illinois, about 1900 miles (c. 3000 km) to the south.

Such anomalies of temperature are largely due to the influence of relatively warm currents of water which wash the western shores of Europe during the winter. These currents, originating in tropical parts of the Atlantic Ocean, drift to the north and east and make the waters around Europe in winter much warmer than the latitude would warrant. The entire movement of water is known as the *North Atlantic*

Figure 5.1   Reference map of the main islands, peninsulas, seas, straits, and rivers of Europe. A few lakes are also shown.

Figure 5.2   *Europe and the Mediterranean Sea compared in latitude and area with the United States and Canada. Most islands have been omitted.*

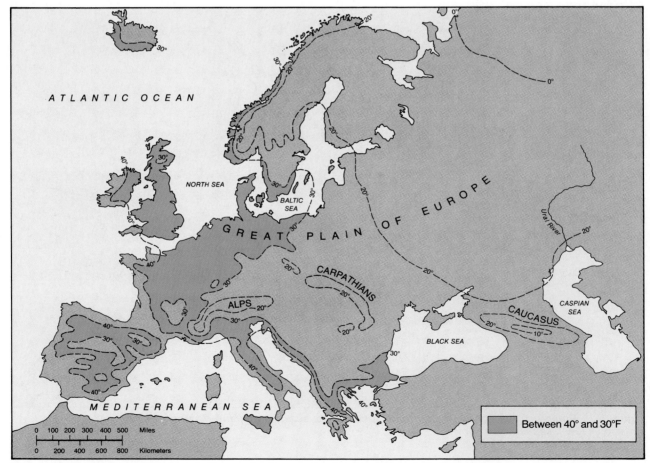

Figure 5.3   *Actual temperatures in Europe in January in degrees Fahrenheit. The dashed lines, known as isotherms, connect places having the same average January temperatures. Metric values: 10°F/−12.2°C; 20°F/−6.7°C; 30°F/−1.1°C; 40°F/4.4°C.*

Drift, and the portion that skirts the eastern shore of the United States before curving off toward Europe is called the *Gulf Stream*. Winds blow predominantly from the west (from sea to land) along the Atlantic coast of Europe. In winter, the moving air absorbs heat from the ocean and transports it to the land, making temperatures abnormally high. In the summer, the climatic roles of water and land are reversed. Instead of being warmer than the land, the ocean becomes cooler. This is due to the fact that the surface layers of the land are heated more rapidly by the more intense solar radiation of summer whereas in the fluid sea, with its horizontal and vertical currents, each unit of the sun's radiation becomes distributed through a far greater mass of material and its heating power is reduced. Hence, the air brought to the land by westerlies in the summer has a cooling effect. When coupled with the effect of distance from the equator in Europe's relatively high latitudes, the result is that in most of the region, summers are not extremely hot. Only the southern

fringes of Europe have hot-month temperatures averaging over 70°F (21°C) (Fig. 5.4 and Table 5.1).

The same winds which bring warmth in winter and coolness in summer also bring abundant moisture to the land. Most of this falls in the form of rain, although in the higher mountains and more northerly areas, there is considerable snow. Ample, well-distributed, and relatively dependable moisture has always been one of Europe's major assets. Actually, the absolute amount of precipitation is not great. Although a few highlands receive 100 inches (*c.* 250 cm) or more, the general average in European lowlands is only 20 to 40 inches (*c.* 50 to 100 cm) (Fig. 5.5). In some parts of the world nearer the equator, this amount would be distinctly marginal for agriculture. In most of Europe, however, it is sufficient for a wide range of crops due to the mild temperatures and high atmospheric humidity, which lessen the rate of evaporation and thus make the precipitation more effective for crop growth. These advantages are not enjoyed by Mediterranean Europe,

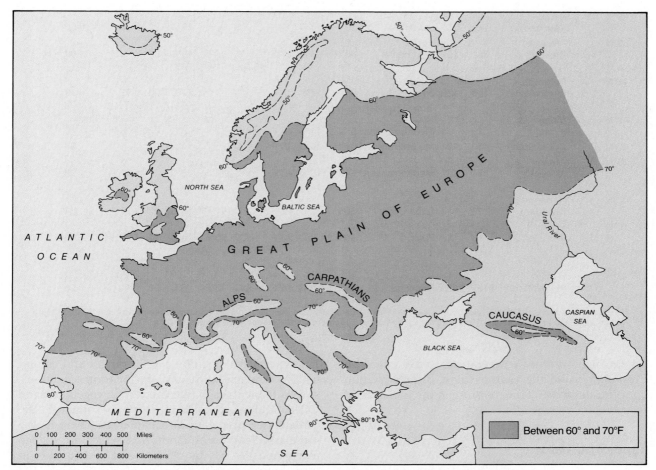

Figure 5.4 *Actual temperatures in Europe in July. Metric values: 50°F/10.0°C; 60°F/15.6°C; 70°F/21.1°C; 80°F/26.7°C. (Figs. 5.3 and 5.4 redrawn from Jean Gottman; A Geography of Europe, 4th ed., New York, Holt, Rinehart and Winston, 1969.)*

TABLE 5.1 CLIMATIC DATA FOR SELECTED EUROPEAN STATIONS[1]

Latitude (to Nearest Whole Degree)	Station	Elevation Above Sea Level (ft)	(m)	Type of Climate	Average Temperature Annual		January (or Coolest Month)		July (or Warmest Month)		Precipitation Annual Average (in)	(cm)	% Occurring April–September
53°N	Dublin (Ireland)	266	81	Marine west coast	50°F	10°C	41°F	5°C	59°F	15°C	29	75	49
56°N	Glasgow (Scotland)	29	9	Marine west coast	48°F	9°C	39°F	4°C	59°F	15°C	41	104	44
52°N	London (England)	16	5	Marine west coast	50°F	10°C	39°F	4°C	64°F	18°C	23	58	50
49°N	Paris (France)	174	53	Marine west coast	52°F	11°C	37°F	3°C	66°F	19°C	23	59	53
51°N	Brussels (Belgium)	328	100	Marine west coast	50°F	10°C	36°F	2°C	64°F	18°C	31	79	50
56°N	Copenhagen (Denmark)	72	22	Marine west coast	46°F	8°C	32°F	0°C	64°F	18°C	24	60	54
60°N	Bergen (Norway)	144	44	Marine west coast	46°F	8°C	36°F (Feb.)	2°C	59°F	15°C	77	196	45
53°N	Berlin (Germany)	180	55	Humid continental	50°F	10°C	32°F	0°C	66°F	19°C	22	56	58
52°N	Warsaw (Poland)	351	107	Humid continental	46°F	8°C	28°F (Feb.)	−2°C	66°F	19°C	19	47	63
45°N	Belgrade (Yugoslavia)	456	139	Humid continental	54°F	12°C	32°F	0°C	73°F	23°C	28	70	56
43°N	Marseilles (France)	10	3	Mediterranean	57°F	14°C	43°F	6°C	73°F	23°C	22	55	40
42°N	Rome (Italy)	430	131	Mediterranean	61°F	16°C	46°F	8°C	75°F	24°C	29	75	28
38°N	Athens (Greece)	351	107	Mediterranean	64°F	18°C	48°F	9°C	82°F	28°C	16	40	22
66°N	Haparanda (Sweden)	23	7	Subarctic	36°F	2°C	12°F	−11°C	61°F	16°C	22	55	53
70°N	Vardö (Norway)	49	15	Tundra	36°F	2°C	23°F (Feb.)	−5°C	50°F (Aug.)	10°C	17	43	49
47°N	Santis (Switzerland)	8187	2465	Highland	28°F	−2°C	16°F	−9°C	43°F	6°C	98	249	56

[1]In general, the climate statistics used in this text are rounded data based on standard sources that report temperatures in °C and precipitation in millimeters. Conversions and roundings may produce apparent small discrepancies in figures.

where high summer temperatures cause excessive evaporation and the yearly regime of precipitation concentrates most of the moisture in the winter half-year when crops need it least.

Types of Climate

In most parts of the world, climate tends to be fairly uniform over wide areas. This has allowed classification of the earth's surface into climatic regions, each characterized by a particular type of climate.

The types are defined by certain ranges and combinations of temperature and precipitation conditions. Associated with each type are certain vegetation and soil conditions, these being strongly influenced by climate (see discussion in Chapter 2). There is also some association with landforms, which are shaped in varying degrees by climate and which may, in turn, contribute to the shaping of climate types. The climatic effects of high mountains are a familiar instance of the latter. Due to the many associations between climate and other geographic features,

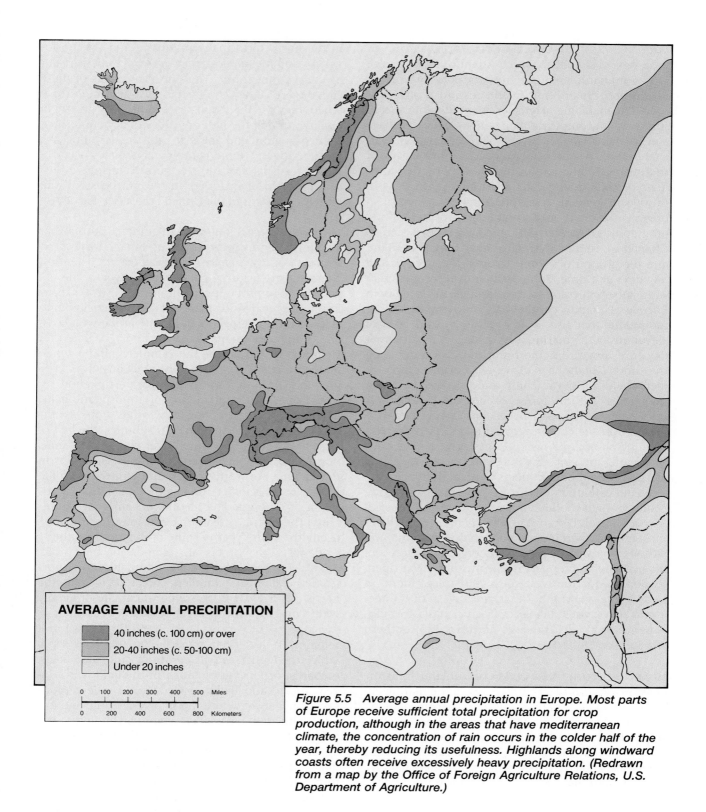

AVERAGE ANNUAL PRECIPITATION

- 40 inches (c. 100 cm) or over
- 20-40 inches (c. 50-100 cm)
- Under 20 inches

Figure 5.5   Average annual precipitation in Europe. Most parts of Europe receive sufficient total precipitation for crop production, although in the areas that have mediterranean climate, the concentration of rain occurs in the colder half of the year, thereby reducing its usefulness. Highlands along windward coasts often receive excessively heavy precipitation. (Redrawn from a map by the Office of Foreign Agriculture Relations, U.S. Department of Agriculture.)

maps showing climatic types and regions are highly useful in geographic study. Such maps embody various classifications, one of which is shown in this text on the map of world types of climate (see Fig. 2.18). A glance at this map reveals that Europe, despite its modest dimensions, has every type of climate except desert and the tropical rainy types. We now survey this remarkable diversity, type by type, with special reference to associated natural features and Europe's productive agriculture.

MARINE WEST COAST CLIMATE. This is the European type of climate in which Atlantic influences are most dominant. It extends from the coast of Norway to northern Spain, and inland toward central Europe. The main characteristics are mild winters, cool summers, and ample rainfall, with many drizzly days and frequent cloud and fog. Throughout the year, changes of weather follow each other in rapid succession as different air masses temporarily dominate or collide with each other along fronts. As in Europe in general, most precipitation is frontal in origin or results from a combination of frontal and orographic (highland) influences (see discussion in Chapter 2). In the lowlands, winter snowfall is light, and the ground is seldom covered for more than a few days at a time. Some highlands receive enough snow to support ski resorts, although this activity in Europe is mainly confined to the Alps and other mountains that rise far above the general level of elevations in the marine west coast climate. Summer days are longer, brighter, and more pleasant than the short, cloudy days of winter, but even in summer, there are many chilly and overcast days. An impressive natural forest of broadleaf deciduous trees such as oak and beech has largely been replaced over the centuries by crop land (see chapter-opening photo) and lush pasture land (see Fig. 6.3). Most soils are not first rate, due in part to being leached of important mineral plant foods by the wetness; but careful, intensive, and scientific agriculture results in high production in many areas. The frost-free season of 175 to 250 days is long enough for most crops grown in the middle latitudes to mature, although most areas have summers that are too cool for heat-loving crops such as corn (maize) to ripen. Regions of marine west coast climate are found in parts of the world other than northwestern Europe, a notable example being the Pacific coast of Anglo-America from San Francisco northward to southern Alaska (see Fig. 2.18).

HUMID CONTINENTAL CLIMATES. Inland from the coast in western and central Europe, the marine climate gradually changes. Winters become colder, summers hotter; the annual precipitation becomes somewhat less, with more precipitation in summer than in winter; the percentage of cloudiness and fog decreases; and snow accumulates for longer periods. Influences of maritime air masses from the Atlantic are still strong, but these effects are modified increasingly by continental air masses from inner Asia as one moves eastward toward Russia. At a considerable distance inland, conditions become sufficiently different for two new climate types to be distinguished: the *humid continental short-summer climate* in the north (principally in Poland and Czechoslovakia), and the *humid continental long-summer climate* in the warmer south. The European areas of these climates are largely confined to East Central Europe, where they are the major types. Soils vary greatly in quality; among the best are those that were formed from alluvium and loess in grasslands along the Danube River. Humid continental climates occur in widespread parts of the Northern Hemisphere outside Europe, for example, in the northeastern and midwestern parts of the United States (see Fig. 2.18).

MEDITERRANEAN CLIMATE. Southern Europe has an unusually distinctive type of climate—the *mediterranean* or *dry-summer subtropical climate*. The most characteristic feature of this climate type is the occurrence of a pronounced autumn and winter maximum of precipitation. The total yearly precipitation is less, on the average, than in the marine west coast and humid continental climates, and very little precipitation occurs during the summer months, when temperatures are most advantageous for crop growth. The distinctive pattern of precipitation results from a seasonal shifting of atmospheric belts. In winter, the belt of westerly winds shifts southward, bringing precipitation. In summer, the belt of high atmospheric pressures over the Sahara shifts northward, bringing desert conditions. Summers in this climate are warm to hot, and winters are mild. In lowlands, snow is rare, although it may accumulate to considerable depths on adjacent mountains. The frost-free season is very long, lasting practically the entire year in some of the more southerly lowlands. The winters are famous for their mild, bright, sunny weather—a great attraction to visitors from the damper, cloudier, cooler regions to the north. Most areas in this type of climate originally were covered with drought-resistant trees such as the cork oak or various pines, but little of this forest remains. It has been replaced by the wild scrub which the French call *maquis*, or by cultivated fields, orchards, or vineyards. Much of the land is in rugged, rocky, and badly eroded slopes where thousands of years of deforestation, overgrazing, and unwise cultivation have taken their toll. Landforms and soils form a varied patchwork, amid which there are agricultural districts of many sizes. The most productive tend to occupy patches of flat alluvium or, in the case of some Italian districts, soils formed of volcanic material. The subtropical temperatures make possible a great variety of crops, but irrigation to counteract the dry summers is a necessity for most of them. In the United States, a mediterranean type of climate occurs in parts of central and southern California.

SUBARCTIC AND TUNDRA CLIMATES. Not all the climates of Europe can be considered temperate. Some northerly sections experience the harsh conditions associated with subarctic and tundra climates. The *subarctic climate*, characterized by long, severe winters and short, rather cool summers, covers most of Finland, the greater part of Sweden, and some of Norway. Agriculture is handicapped by the brief and undependable frost-free season, coupled with thin, highly leached, acidic soils. Human settlement is scanty, and most of the land is covered by a forest of needleleaved conifers such as spruce and fir. The trees are watered by melting snow and a moderate rainfall. In the *tundra climate*, which occupies northernmost Norway and much of Iceland, cold winters combine with brief, cool summers and strong winds to create conditions hostile to tree growth. An open windswept landscape results, covered with lichens, mosses, grass, low bushes, dwarf trees, and wild flowers. These arctic wilds often teem with small mammals, fish, and migratory birds, but human inhabitants are few. Agriculture in a normal sense is not feasible, although Norway's Lapps use the tundra to graze reindeer, and Icelanders graze sheep there. Sea fishing is carried on by small communities that dot the coasts. Europe's subarctic and tundra climates are duplicated on an immensely larger scale in the Soviet Union and North America.

HIGHLAND AND ICE-CAP CLIMATES. The higher mountains of Europe, like high mountains in other parts of the world, have a *highland climate* varying in character according to altitude and differential exposure to sun, wind, and precipitation. Given enough height, the variety can be startling. The Italian slope of the Alps, for instance, ascends from subtropical conditions at the base of the mountains to tundra and ice-cap climates at the highest elevations. The *ice-cap climate* experiences temperatures that average below freezing every month in the year, enabling ice fields and glaciers to be preserved. It is characteristically a climate of polar latitudes, observable on a massive scale in Greenland. The latter island, a self-governing unit affiliated with Denmark, is covered by a huge ice sheet except for a fringe of tundra around the coasts. Smaller ice fields exist in Iceland and in the Arctic islands of Svalbard, controlled by Norway.

ADVANTAGES AND LIMITATIONS OF A VARIED TOPOGRAPHY

Not only is Europe varied climatically, but its topographic features are also extremely diversified. The nature, intricate layout, and interrelations of these features must be understood if one is to make sense of Europe's history, culture, and current activities. Each class of features—plains (Fig. 5.6), plateaus, hill lands, mountains, and water bodies—has made its contribution to European enterprise, conferring advantages in many cases but setting limitations in others. The whole physical assemblage (Fig. 5.7), overspread with a varied plant life and enriched by

Figure 5.6 Plains are widespread in Europe, and most population, industry, and agriculture are found on them. Seen here is a potato harvest on a plain in Germany. Potatoes are a major European crop.

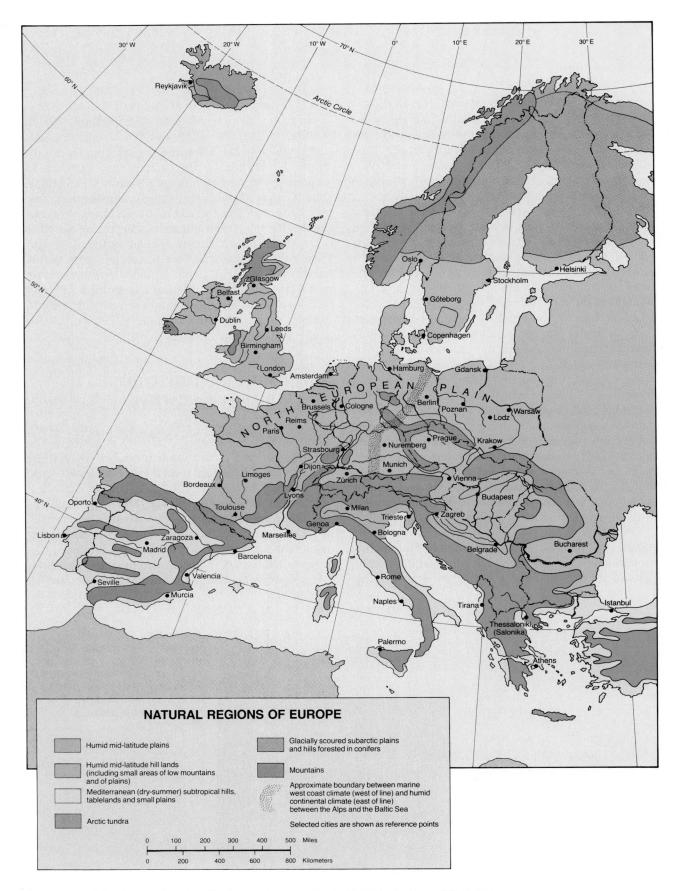

NATURAL REGIONS OF EUROPE

Humid mid-latitude plains

Humid mid-latitude hill lands (including small areas of low mountains and of plains)

Mediterranean (dry-summer) subtropical hills, tablelands and small plains

Arctic tundra

Glacially scoured subarctic plains and hills forested in conifers

Mountains

Approximate boundary between marine west coast climate (west of line) and humid continental climate (east of line) between the Alps and the Baltic Sea

Selected cities are shown as reference points

| 0 | 100 | 200 | 300 | 400 | 500 | Miles |

| 0 | 200 | 400 | 600 | 800 | Kilometers |

Figure 5.7 The "natural regions" of the map are attempts to show natural features as composite habitats. Each color-symbol represents a distinctive association of landforms and climate, with natural vegetation also stated or inferred. The result is a broad-scale view of the natural setting for human activity in Europe.

GLACIATION

During the Great Ice Age massive continental ice sheets formed over the Scandinavian Peninsula and Scotland. They moved outward to cover Finland; much of Russia; the North European Plain in Denmark, Poland, Germany, and the eastern Netherlands; and the British Isles except for a strip of southern England. For perhaps a million years, ice sheets alternately advanced and melted back as glacial and interglacial periods succeeded each other. The latest retreat of the ice possibly began about 35,000 years ago. Such "continental" ice sheets still cover approximately 10 percent of the world's land in Greenland and Antarctica, but they covered an estimated 28 percent at their maximum extent, when much of Europe, Asia, and North America lay under them.

In many landscapes the effects of glaciation are still very evident. In some areas the ice sheets scoured the preceding surface deeply, removing most of the soil and gouging hollows which became lakes and swamps when the ice melted. Today landscapes of *glacial scouring*, with thin soil, much bare rock, and many lakes, characterize most of Norway and Finland, much of Sweden, parts of the British Isles, and Iceland (which had a local ice sheet of its own). In such areas the ice often rearranged the preexisting drainage pattern, leaving rivers which now follow very irregular courses and have many rapids and waterfalls. Thus were created many favorable sites for hydroelectric installations. Today such sites are utilized extensively by Norway, Sweden, Finland, and Iceland—nations that were poor in mineral fuels and found it profitable to harness falling water for at least a good share of their electric-power needs. Even Norway, with its recently developed oil and gas fields under the North Sea, continues to generate practically all its electricity by use of hydropower.

In other glaciated areas, it was *glacial deposition* that had the predominant effect on the present landscape. As it ground along, the ice accumulated earth material on its underside and subsequently deposited it, either as *outwash*—carried by sheets of meltwater from underneath the ice—or in the form of *morainal material*—dropped in place as the ice melted. Today, the landscapes formed of outwash are likely to be flat, whereas moraines have a more irregular surface. Moraines may, in fact, form low ridges such as the *terminal moraines* that were formed by long-continued deposition at the front of an ice sheet during times when melting balanced the ice movement to such an extent that the ice front remained stationary. Glacial deposits of varying thickness were left behind on most of the North European Plain. Unfortunately, most of the deposits were so sandy that present soils formed from them are not very fertile, although they have often been made productive by careful handling and large applications of fertilizer. However, as mentioned earlier in the chapter, a belt of windblown material called *loess* was deposited at the southern edge of the plain and provides exceptionally fertile soils there. This material appears to have been picked up by winds from barren surfaces after the glacial retreat and deposited where wind velocities were checked.

---

the human associations and constructions of an eventful history, presents a series of distinctive and often highly scenic landscapes.

One of the most prominent surface features of Europe is a plain which extends from the Pyrenees Mountains across western and northern France, central and northern Belgium, the Netherlands, northern Germany, Denmark, and Poland, and stretches without a break far into the Soviet Union (Fig. 5.7). Known as the North European Plain, it has outliers in Great Britain, the southern part of the Scandinavian Peninsula, and southern Finland. For the most part the plain is not flat but is undulating or rolling in surface, with many gentle hills and low escarpments. Here and there are flat stretches, generally on alluvial land. The North European Plain contains the greater part of Europe's cultivated land, and it is underlain in some places by deposits of coal, iron ore, natural gas, potash, and other minerals which have been important in Europe's industrial development. Many of the largest European cities—including such major cities as London, Paris, and Berlin—have developed on the plain. From northeast France and Belgium eastward, a band of especially dense population extends along the southern edge of the plain; this has apparently been true since prehistoric times. It coincides with (1) extraordinarily fertile soils formed from deposits of windblown dust in glacial times—a material known by the German word *loess;* (2) an important natural routeway skirting the highlands to the south; and (3) a large share of the mineral deposits just mentioned, on and around which many cities and city clusters have grown.

South of the northern plain, Europe is predominantly mountainous or hilly, although the hills and mountains enclose many relatively level plains, valleys, and plateaus of quite varied size. The hill lands tend to be relatively old geologically; the same is true of some areas high enough to be classed as low mountains (for example, Germany's famous Black Forest: see map, Fig. 7.7). Having been exposed to weathering and erosion for such a long time, these uplands are often rather smoothed and rounded in outline. But many mountains in southern Europe are geologically young. They are often high and ruggedly spectacular, with jagged peaks and snow-capped summits that stand steeply above densely populated lowlands or the sea. They reach their peak of height and grandeur in the Alps (see Fig. 7.1). This chain, occupying the southeastern edge of France, southern Switzerland, the northern edge of Italy, a narrow fringe of southern Germany, most of Austria, and part of northwestern Yugoslavia (see Fig. 9.11), has many peaks above 12,000 feet ($c.$ 3660 m). Its highest summit, Mont Blanc (see Fig. 8.4), rises to 15,771 feet (4807 m)—more than 1000 feet higher than any peak in the High Sierra of California or the Colorado Rockies. The high mountains of southern Europe are a considerable barrier to traffic, although on the whole a less formidable barrier than their appearance on a map would suggest, since they are cut through by many passes, river valleys, and tunnels. In some areas, particularly the Balkan Peninsula, they have tended to isolate small groups of people and thus have aided in the development of many small, distinctive cultural regions.

To the north, the North European Plain is fringed by glaciated lowlands and hill lands in Finland and in eastern Sweden and by rugged, ice-scoured mountains in western Sweden and in most of Norway. The British Isles also have considerable areas of glacially scoured hill country and low mountains (see chapter-opening photo of Chapter 6), along with lowlands where glacial deposition occurred.

HOW RIVERS HAVE AIDED EUROPEAN DEVELOPMENT

As one might expect in such a humid area, Europe has numerous river systems (see Editorial Comment on River Terms and map, Fig. 5.1). They generally have their sources in the region's mountains and hill lands. The larger systems drain basins of considerable size, although none is very impressive in world terms. The Danube Basin, Europe's largest, ranks no better than 31st in the world (Victor Showers, *World Facts and Figures* [New York: John Wiley & Sons, 1979], p. 106). Nor do European rivers compare favorably in length. The Danube, at 1770 miles (2850 km), is the only one in the world's top 50 (it ranks 27th [Showers, p. 10]). But despite their modest dimensions, Europe's rivers are very important economically as carriers of goods, suppliers of water, and generators of hydroelectricity. Most of them serve at least one of these functions, and some are important for all three.

Rivers were very important parts of Europe's transport system at least as far back as Roman times, and probably before; they still are, transporting very large quantities of bulk cargo at low cost, mostly in motorized (self-propelled) barges. The more important rivers with respect to transportation—such as the Thames, Seine, Scheldt, Rhine, and Elbe—are largely those of highly industrialized areas in northwestern Europe. Farther east on the North European Plain, the Oder and the Vistula have some importance. Except in Great Britain, the general pattern of these rivers is to have their sources in the hills and mountains south of the northern plain, and then to flow generally northward across the plain to mouths on the English Channel, North Sea, or Baltic Sea. Although in their natural state they carried shallow-draft commercial vessels or even war fleets such as those of the Vikings, many rivers have required improvement in order to accommodate the barges of modern inland-waterway commerce. This has involved dredging, course straightening, bank control, and, in some cases, dams on main streams or their tributaries to control water levels. In addition, the rivers are interconnected and supplemented by a system of canals so extensive that one could cross Europe on inland waterways from the Soviet Union into France.

Important seaports have developed along the lower courses of many rivers, and some have become major cities. London on the Thames, Antwerp on the Scheldt, Rotterdam in the delta of the Rhine, and Hamburg on the Elbe are outstanding examples (see chapter-opening photos of Chapters 1 and 8). Often, the river mouths are exceptionally wide and deep, allowing ocean ships to come a considerable distance inland. This is true even of short rivers such as the Thames and Scheldt. Such enlarged river mouths are *estuaries*, which have been formed by submergence of the lower ends of river valleys. The main port city on each estuary has tended to develop at the head (upstream end) of the estuary, which may be as much as 50 to 100 miles ($c.$ 80 to 160 km) from the open sea (see discussion of London,

RIVER TERMS

Geographic discussion of rivers requires that certain terms be understood. A few of the simplest are introduced here. A *river system* is a river together with its tributaries. A *river basin* is the whole area drained by a river system. The *source* of a river is its place of origin; the *mouth* is the place where it empties into another body of water. As they near the sea, many rivers become sluggish, depositing great quantities of sediments to form *deltas*, and often dividing into a number of separate channels, known as *distributaries*. The Rhine River divides in this manner—its two principal distributaries being the Lek and the Waal (see Fig. 8.1).

including maps, in Chapter 6). As shipping has increased rapidly in size in modern times—requiring deeper channels—fewer ships have been able to come as far upstream, and port facilities have had to be extended seaward from the original port city. Some ships, especially the enormous oil tankers built in recent decades, have become so large that entirely new ports have had to be developed in locations with deeper water and more room to maneuver. Often the new facilities are in the general vicinity of the older ports, but some are sited on previously undeveloped sections of coast away from major river mouths.

The Rhine and the Danube are European rivers of exceptional fame and importance. Both are major international rivers, touching or crossing the territory of many countries. In the case of the Rhine, these countries include Switzerland—where the river rises in the Alps—Liechtenstein; Austria; France; West Germany; and the Netherlands, where the delta and mouths of the Rhine are located (see Figs. 7.7, 8.1, and 8.4). Highly scenic for much of its course, the river is Europe's most important inland waterway. Along or near it, a striking axis of intense urban-industrial development and extreme population density has developed. This is the economic heart of Europe, unmatched in significance by any other area of equal size. At its North Sea end, the Rhine axis is linked to world commerce by Europe's leading seaport, Rotterdam, located in the Netherlands on a Rhine distributary, the Lek (see Map 8.1). The port's enormous traffic, largest in the world, symbolizes the importance of the urban-industrial strip that Rotterdam chiefly serves.

The Danube (German: Donau; see Fig. 9.13) touches or crosses more countries—eight in all—than does any other river in the world. Within Europe, the Danube is also unusual in its southeasterly direction of flow—from its source in the Black Forest of southwestern West Germany to its delta in Romania and the Soviet Union on the semienclosed Black Sea (see Figs. 7.7 and 9.11). On the way, the river also crosses or forms a border of Austria, Czechoslovakia, Hungary, Yugoslavia, and Bulgaria. Thus it flows through much less intensely developed parts of Europe than those served by the Rhine, and its destination is a sea that is far more lacking in centrality and commerce than the busy North Sea into which the Rhine empties. For these reasons, the Danube is less important commercially than the Rhine, although it does carry a useful traffic of intense interest to the countries along its route.

Less useful commercially than the rivers discussed above are those of Northern Europe (Scandinavia, Finland, Iceland) and Mediterranean Europe (Iberian Peninsula, southeastern France, Italy, and southern parts of the Balkan Peninsula, including Greece). In both regions rivers tend to be short, rock-strewn, and swift, although the sluggish Po River crossing the plain of northern Italy is a conspicuous exception. In parts of Northern Europe where forest resources are considerable (Sweden, Finland, Norway), rivers are used to float logs (see Fig. 9.4). Swift-flowing rivers are used to generate hydroelectricity on a major scale in both the glacially scoured terrain of Northern Europe and the high mountains of Mediterranean Europe. But the barge transportation that facilitates commerce and industry on the North European Plain, or to a lesser degree along the Danube, is notably absent from both the extreme north and the extreme south of Europe.

Climatic circumstances often limit the usefulness of both the southern and the northern rivers. In the south, the mediterranean climate causes great variations in water flow from season to season. Rivers are often choked with water and are liable to flood during wet autumn and winter seasons, but then shrink to mere trickles in the desert-like summers.

Some have been dammed to control flooding and create reservoirs needed for irrigation. Spain, in particular, has developed a number of large irrigation schemes in recent years. Of all the southern rivers, the most useful is undoubtedly the Rhone, not only because of the river itself but also because of the transportation corridor its valley provides. This river rises in Switzerland, flows into France to the important city of Lyons, and then turns south to enter the Mediterranean near another major French city, Marseilles. From Lyons to the Mediterranean, the Rhone follows a trough between the Alps to the east and the highland called the Massif Central to the west. North of Lyons, the trough is followed by a navigable Rhone tributary called the Saône; and the entire passageway is commonly called the Rhone-Saône Trough or Valley (see Figs. 7.1 and 7.4). It has been since prehistoric times a major connection between the Mediterranean coast and the plains of northern France, despite the handicap to water traffic caused by the strong current and various rapids in the Rhone. Today the valley is the leading traffic artery of France, followed by highways and railways connecting Paris, Lyons, and Marseilles. The latter cities, both very ancient, rank next in size after Paris among French metropolitan areas. Not only is Marseilles France's third metropolis in population, but it is the second-ranking seaport in all of Europe, after Rotterdam, as measured by tonnage of seaborne freight handled. Much of its traffic consists of imported oil serving French and West German needs. France has built dams and other control works along the Rhone and has been constructing a chain of nuclear power plants along it. Such plants need very large quantities of cooling water.

The usefulness of rivers emptying into the Baltic Sea is somewhat lessened by the fact that they are ordinarily frozen for a considerable period in midwinter. Parts of the Baltic Sea itself freeze in winter; indeed, the Gulfs of Bothnia and Finland are normally frozen for 3 months or more. The Baltic and its gulfs, being almost completely enclosed by land and continuously supplied with fresh water by a large number of rivers, are less saline than the ocean and so are especially susceptible to freezing. However, extensive use of icebreakers lessens the handicap to transportation. In most other parts of Europe sea ice is uncommon.

USING THE MAP OF NATURAL REGIONS

In reviewing the preceding topical discussions of Europe's environmental features, it may be helpful to study closely the map of natural regions (see Fig. 5.7). This map attempts to combine landforms, climate, and, to some degree, natural vegetation, into a set of environmental associations, each of which occupies considerable land and is found in more than one area. The colors on the map—for example, "humid mid-latitude plains" or "glacially scoured subarctic plains and hills forested in conifers"—represent *types* of natural regions, and the individual areas of occurrence are *specific* natural regions. Only one of the latter—the North European Plain—is actually named on the map. The intent is not to provide an index map of environmental place names, but to portray the *pattern of distribution* of all the major environmental associations within Europe as a whole. Place names of specific natural regions can be found on index maps in the subsequent regional chapters on Europe. Some students may find the "natural regions" map a useful device for reviewing their mental maps of the locations of environmental features and areas mentioned in the text but not named on this map—for example, mountain ranges such as the Alps, or major seas, islands, peninsulas, and rivers.

EUROPE'S HIGHLY DEVELOPED, DIVERSIFIED ECONOMY: OLD ACHIEVEMENTS AND NEW DIRECTIONS

Diversity is a keynote of the European economy, as it is of the European environment. Numerous forms of economic activity are highly developed, and much variety exists from one country and region to another, both in the relative emphasis given to each of the major pursuits and in the relative intensity of development. The net result gives most European countries a high standing in productivity and affluence compared to other countries of the world, but at the same time sizable differentials persist within Europe itself. Not only is the latter true when entire countries are compared, but it is also true of regions within individual countries. Most countries of Europe contain regions that fall well below countrywide averages of productivity and income. These may be agricultural regions handicapped by adverse environments and/or overpopulation; regions relatively remote from major market areas; old urban-industrial regions afflicted with failing resources and industries which have become noncompetitive; or, in the case of West Germany, regions near the economic barrier presented by the boundary with the Communist countries. In some cases, notably in the United Kingdom, such areas with lagging economies also include slums in large central cities. Here unemployment is chronic, physical facilities are outdated and shabby, and the quality of life suffers. It is the policy of many European governments, as well as of the European Economic Community, to subsidize economic development in "lagging regions," but such efforts have usually had questionable success. Apparently the disabilities of these areas are often too deep-seated to respond readily to government intervention.

Historically, it has been the addition of large manufacturing, commercial, financial, and associated service sectors to an economy originally dominated by agriculture that has raised Europe to its relatively high economic position. This was the first region of the world to witness the evolution of an agricultural society into a "developed" industrial one, and industry remains vitally important to European prosperity today. The key event called the *Industrial Revolution* was preceded by a slow rise in Europe's power and capabilities, which laid the foundations for it; and the Revolution has been followed by vast advances in productive (and destructive) knowledge and technology. Even the devastation of the two world wars of the 20th century did not stop the economic dynamism of Europe, which seized upon American aid after World War II to rapidly reorganize, rebuild, and attain new heights of production and affluence. We will now examine European industrialization: its world importance, geographic distribution, historical evolution, resources, problems, major products, and spatial interties.

EUROPEAN INDUSTRIALIZATION

One of the most significant facts of global geography is the high proportion of the world's manufacturing capacity and output that is European. This proportion is decreasing, but between one quarter and one third of world output of most major industrial products still originates in Europe. Not even the United States, which is the world's foremost industrial country, can match the collective output of the European countries. Europe still contains, as it has for a long time, the leading single concentration of manufacturing on the globe. Industrialization goes far toward explaining the highly urban character of most European countries, as well as their high standing with respect to overall productivity and average income. These relationships are most apparent in areas along the major population axes described in Chapter 4, especially the north-south axis centering on the Rhine. But industrial development is widely

distributed in other parts of Europe, including peripheral areas in Southern and East Central Europe, where industry has traditionally been scanty or virtually nonexistent, but has increased markedly during recent decades.

This concentration of manufacturing, and the industrialized and urbanized societies associated with it, reflect the fact that Europe was the place where large-scale manufacturing industry, using machines driven by inanimate power, first arose. The main impact, which we have come to call the Industrial Revolution, began to be felt about the middle of the 18th century. Generally the Industrial Revolution is thought of as occurring first in Great Britain and then spreading to continental Europe, New England in the United States, and other parts of the world. It is still convulsing many countries today, although some of the world's "less-developed" countries remain relatively untouched.

Before its 18th-century Revolution, Europe had been witnessing for centuries a series of developments that paved the way for modern industry organized in factories and using machines driven by inanimate power. But the onset of the Industrial Revolution was marked by a new surge of inventiveness in various fields. Industrial inventions and innovations, many of which were developed by British inventors, made waterpower on an increased scale and then steam power available to turn machines in the new factories. Invention of a practical steam engine in Great Britain made coal a major resource (partly through making it possible to solve drainage problems in coal mines by more efficient pumps) and greatly increased the amount of power available. New processes and equipment made iron relatively abundant and cheap. Very important in raising the output of iron was the development of a way to use coke instead of the charcoal previously used for smelting iron ore in blast furnaces. (Coke made from coal could be provided more cheaply and in larger quantities than the charcoal made from wood. Furthermore, coke was harder than charcoal and would continue to burn under a charge of ore heavy enough to crush charcoal and smother a charcoal fire. Hence, larger blast furnaces were now possible—and steam engines provided the stronger air blast that the new furnaces required.) The invention of new industrial machinery, soon made of iron and driven by steam engines, multiplied the output of manufactures, principally textiles at first. Then, in the 19th century, British inventors developed processes that allowed steel, a metal superior in strength and versatility, to be made on a large scale and cheaply for the first time. Such developments led the world into the modern age of large-scale,

COAL

Coal is the residue from organic matter that has been compressed for a long period of time under overlying layers of the earth. Different original materials and different degrees and lengths of compression have resulted in a variety of coals rather than a single uniform substance. The common ranks of coal as classified on the basis of hardness and heating value (caloric content) are anthracite, bituminous, and lignite (sometimes called brown coal). Peat, which is coal in the earliest stages of formation, may be added at the lower end of the scale, since it can be burned if dried. Lignite is mostly used in areas where adequate deposits of bituminous and anthracite coal are lacking—being principally used as household fuel (in the form of briquettes) and for raising steam, which may be used in generating electricity (see chapter-opening photo). Bituminous coals are used in larger quantities than any other rank. Gas is manufactured from them, and they are used for raising steam. Some bituminous coals can be used for making coke, which is coal with the volatile elements baked out, leaving a high content of carbon. Coke is used in great quantities to manufacture iron on a large scale in blast furnaces, where it burns with intense heat. Anthracite coal burns with a hot flame and almost no smoke. It is excellent for domestic heating and can be used to produce gas and raise steam. It is not ordinarily used for coking. Often coals are referred to according to the uses for which their special properties fit them, as flame coals, gas coals, or coking coals. Not only the amount of coal, but the *kind of coal* available is very important to a country's industries. Concern over atmospheric pollution by the sulfur in coal smoke has recently increased the demand for low-sulfur coals.

mechanized industrial production. They were accompanied by sweeping changes of a social and economic character. Cities based primarily on industry grew rapidly, and the conditions of life for urban dwellers and industrial workers became an important issue. Governments gradually turned their attention to the regulation of entrepreneurs and sanctioned the formation of labor unions in order to prevent abuses growing out of callousness, greed, and the sheer bigness and impersonality of the new industrial enterprises.

CONTINUITY AND CHANGE: FROM INDUSTRIAL REVOLUTION TO POST-INDUSTRIAL SOCIETIES

The developments of the Industrial Revolution were built upon earlier foundations and have been continued and elaborated to the present time. Scholars have traced a slowly rising trend of European technical inventiveness since at least the late Middle Ages, involving such things as clocks, gears, waterwheels, the invention of printing in 15th-century Germany, improvements in shipbuilding, drainage of mines by pumping to permit deeper working, improvements in artillery and other armaments, and many others. In time, the practical improvements leading to greater production were furthered by mathematical and other scientific discoveries and by the development of a scientific viewpoint toward nature and its possibilities. At the same time—still previous to the 18th century—capitalist values, practices, and institutions were evolving and becoming prominent—for example, unfettered private ownership of property; favorable attitudes toward profit-making, including interest from loans; commercial banking and credit institutions; double-entry bookkeeping; and corporate forms of business organization. Rising commercial and industrial interests were protected by the increased law and order imposed by newly centralized states—especially Great Britain, the Netherlands, and France—and were aided in making vast profits in overseas trade by these governments' successful imperialism and colonialism.

With the passage of time, industrial technology has advanced from the crude steam engines and textile machines of 18th-century Britain to modern electronics, computer-controlled factory robots, and the invasion of space. While it has advanced, it has spread geographically. Early coal-based industry first spread from Britain to Belgium, and then more widely in Europe and the eastern United States. It began to spread to some non-European areas in the 19th century, and today's more modern technology is continuing to spread with a rising tempo. In Europe itself, an industrial boom after World War II was marked by intensified industrial development in western Europe and by massive investment in new state-owned industrial capacity in the less-developed Communist east.

A number of problems are associated with Europe's intensive industrialization. One of these is the current inadequacy of European energy and raw material resources. When the region's industrial rise began, the resource situation was quite favorable. There were fewer people, the scale of production was much smaller, and resources were generally sufficient for the simpler requirements of the time. A relatively primitive technology was served by extensive forests, numerous small-scale sources of waterpower, and a large variety of minerals in widespread locations. When coal became the main source of power and heat, Europe was favored greatly by its many usable coal fields. These fields proved to be the main localizing factor for much of Europe's industrial and urban development. During the early period of industrialization, electricity was still in the future and coal was expensive to transport; consequently, the new factories were generally built on or near the coal fields. The working population, in turn, clustered near the factories and mines (Fig. 5.8). Thus developed a group of industrial districts which incorporated mines, factories, and urban areas. These older districts are still very prominent on the map of Europe today. Notable examples include

Figure 5.8 *The mining of coal has been an important European occupation for centuries. Seen here are miners coming off shift at a British colliery.*

most major cities in Great Britain, although not London; the east-west line of cities strung across Belgium south of Brussels; the huge urban agglomeration in West Germany called the Ruhr; and the Polish industrial cities near the Czech border (see Fig. 7.1). Economic depression and urban blight are common in these places, which had the advantage of an early start but now suffer from changed industrial circumstances. The changes include a drastic shift in the resource situation; increasing competition from non-European industries; competition from newer products made elsewhere (for example, new synthetic fibers competing with the cottons and woolens of old textile centers); and decreasing requirements for labor due to automation of factories.

Europe's Industrial Resources Today

The European resource picture, once so favorable, has changed for the worse, although the region still possesses substantial resources. Only a few countries—notably Sweden, Finland, Norway, and Austria—still have surpluses of wood. Waterpower also falls far short of need, despite an intensive development of hydroelectricity in such places as the Alps, the Scandinavian Peninsula, and along the Rhone and Danube rivers. Most deposits of metallic ores are too depleted or too small to be important today. But the most far-reaching change is the altered situation of coal. Worked long and intensively, most European coal fields have become expensive to mine and have declined in production. Many were phased out as Europe shifted to cheaper imported oil for power after World War II. Coal is still widely used, but it has been outdistanced by oil, and some of its former market has been taken over by natural gas or atomic energy. Because domestic oil resources were so deficient (and largely remain so today despite new production from North Sea oil), most of the oil used in Europe since World War II has had to be imported, primarily from the Middle East or, in East Central Europe and East Germany, from the Soviet Union. Thus Europe, which was once relatively self-sufficient in industrial resources, now must import (and pay for) massive and growing amounts of many metals and other raw materials, oil, and even some coal from the United States. The energy deficiency has thus far been the most serious, not only because of the financial drain on the economies of European countries but also because of the political vulnerability of many countries to Middle Eastern oil suppliers. The main European responses to the energy problem have been (1) programs of nuclear power development, which often are opposed by considerable bodies of public opinion; (2) imposition of high energy prices to encourage conservation and raise revenue; and (3) helping the Soviet Union finance and build a huge pipeline that now brings natural gas from Soviet Siberia to west European markets.

To keep matters in perspective, it must be noted that a number of European countries do have mineral resources of considerable importance. The United Kingdom became a large producer and net exporter of oil, and also a large producer of natural gas, when its offshore fields in the North Sea were brought in during the 1970s. Norway's oil and gas fields in the North Sea likewise have yielded a large production, which meets the country's internal needs and generates large exports of both fuels. For several decades, the Netherlands has been a major producer and exporter of natural gas. Major iron-ore fields are worked in eastern France, northern Sweden, and northern Spain. Coal from southern Poland is a mainstay of that country's economy; the United Kingdom and West Germany are still sizable coal producers; and both East and West Germany produce considerable quantities of potash (used almost entirely for fertilizer). Yugoslavia, Hungary, Greece, and southern France all produce bauxite, an ore that yields aluminum. There are many other minerals and mining areas scattered about Europe. But even in the case of resources with which individual countries are well supplied, there is a tendency for Europe as a whole to need large imports from the outside world.

Moving Toward "Post-Industrial" Societies

Europe's industrial situation has not only been worsened by resource inadequacies, but also by increasing competition from overseas manufacturers. In both instances, the European situation resembles that of the United States. Among the developments that have created new competitive pressures are (1) the rise of a giant industrial economy in a rebuilt Japan after World War II; (2) the continued spread of manufacturing, often of a high-technology character, to low-wage countries such as Taiwan or Brazil in the "developing" parts of the world; and (3) a drastic lowering of ocean transport costs by huge new ships that can carry great stacks of standardized containers (trailers without their wheels; Fig. 5.9) or shiploads of raw material in bulk. Such carriers have benefited Europe and the United States, but they have been even more crucial to new overseas industries attempting to force an entry into world markets. In order to survive, com-

panies in high-wage countries adopt newer and more efficient machinery embodying the latest advances in technology. This procedure lowers costs, but deprives many workers of jobs. Hence, the new international competition indirectly generates still another major problem: *technological unemployment* (the supplanting of workers by technology). This problem actually goes back to the beginnings of technology itself and was acute during the early stages of the Industrial Revolution when many handicraft workers were being supplanted by the new machines. In the long run, the advance of technology and rising levels of affluence tend to create new and expanded employment opportunities, but the short-run effects of technological unemployment are often severe. Increasingly, labor disputes and political issues arise in both western Europe and the United States over the actions of companies in scaling down their work forces, or even closing plants and abandoning work forces entirely.

There is much evidence that the older and more advanced industrial societies are in the process of becoming "post-industrial"; that is, they will eventually be societies in which industrial workers are as uncommon as farmers are now in most "industrial" societies. The introduction and improvement of computer-controlled robots (machines programmed to do specific jobs without further direction) on industrial production lines is moving the possibility of such societies closer. Serious unemployment in western Europe during the 1980s is in part related to this long-term trend, although short-term economic difficulties are also involved, and perhaps more heavily. As the trend toward automation continues, many monotonous but rather well-paid jobs in industrial production seem likely to be eliminated. Hence, the more advanced and prosperous European countries are beset by such concerns as (1) where new jobs are to come from (presumably from the paperwork and other "service" sectors of the economy); (2) how the transition is to be made without too much suffering; and (3) how income is to be allocated between a few highly productive industrial workers who tend robots, and a predominant mass of service employees whose output per worker will often be less.

Major Products and Spatial Interties of European Industries Today

As an intensively developed, highly diversified industrial complex, Europe produces the full range of modern industrial commodities, and its industries are served by a close network of efficient transpor-

Figure 5.9 Container ships such as the one in this view have revolutionized ocean transportation in recent decades by providing a new and more convenient way to carry varied types of cargo.

tation and communication facilities. There are three main types of commodities: metal goods, chemicals, and textiles. The most important of the metalworking industries is the iron and steel industry, which is concentrated mainly in older industrial districts, such as West Germany's Ruhr district, or in new coastal plants accessible to large ships bringing iron ore or coal from overseas. Many metals other than iron and steel are produced in Europe, generally from imported ores. The most valuable products manufactured from metals are machinery and transportation equipment. Such items form the largest single category of exports from each of the four largest west European producers of crude steel (West Germany, Italy, France, and the United Kingdom), and they are very important in the export trade of many other European countries.

Huge quantities of chemicals are manufactured, used, and traded by European nations. Many industrial processes require chemicals of some type, and chemicals are used to make agricultural fertilizers and innumerable consumer products. Europe is able to supply many chemical raw materials in quantity—especially potash, salt, limestone, coal, and natural gas—but petroleum, the material needed in greatest quantity, must be mainly imported. Not only does the refining of petroleum provide indispensable fuels and lubricants, but petroleum products and natural gas are used to make a great variety of "petrochemicals" such as synthetic fibers. In recent decades Europe has seen a great upsurge in petroleum refining and petrochemical manufacture, found principally in seaports such as Rotterdam, London, Marseilles, and

Hamburg, or in major inland industrial cities or districts, particularly along the Rhine River.

The manufacture of textiles is a very widespread branch of European industry. Until recently, cotton and woolen textiles were the most important. Today they suffer from non-European competition and are often associated with relatively depressed industrial areas such as Lancashire and Yorkshire in Great Britain (see Chapter 6). By contrast, newer synthetic-fiber industries using wood, coal, and oil as raw materials are much more dynamic and now have a larger production than the older textile industries.

Finally, we must note that Europe continues to be world famous for many types of goods requiring painstaking care and a high degree of skill in manufacture. Swiss watches, German optics, Scotch whiskies, and Paris fashions are cases in point. Even goods of this type, however, have come under increasing competition in world markets, particularly from Japan. Figure 7.1 shows locations of the main districts and cities where most of Europe's manufacturing is carried on.

Europe's industries are linked to their suppliers, markets, and control centers by an elaborate system of transportation and communication. The water transport net has been described previously, and many of its features are shown by Figure 7.1. Rail transportation in Europe is fast and efficient. The French, along with the Japanese, have pioneered ultrafast passenger trains, and West Germany has been experimenting with even faster trains that will ride on a cushion of air. A sizable and spreading net of motor thruways—still little developed in East Central Europe, Greece, Iberia, Ireland, and Northern Europe—had its origins in the divided highways called *autobahns* that Germany began developing before World War II. The internal airnet is not so highly developed as that of the United States, being, in fact, less needed in a region of such short distances and efficient surface transport. But air service is quite adequate to the need, and about a dozen airports with worldwide connections (led by London's Heathrow) are among the world's 75 largest in number of passengers handled. Since World War II there has been a huge development of oil and gas pipelines, many of which cross international frontiers (see Fig. 7.10).

Another instance of the very international character of Europe's spatial interties lies in electric-power grids. The power systems of different countries often are linked to each other. Peaks of electricity supply or consumer demand in France, for example, may not coincide in time with peaks in Switzerland, and both countries can benefit from the larger total capacity of their linked systems. France's large thermal capacity may be able to transmit supplemental power to Switzerland in the winter when hydropower output from Swiss mountain streams is low; the Swiss, in turn, can return the power in the spring when their streams are running full and hydropower production is high.

THE IMPORTANT CONTRIBUTION OF AGRICULTURE AND FISHERIES

Agriculture was the original foundation of Europe's economy, and it is still a very important component. Since the prehistoric time when farming spread from the adjacent Middle East, Europe has been one of the world's outstanding agricultural areas. Food provided by the region's agriculture allowed Europe to become a relatively well-populated area at an early time. After about 1500, an ascending curve of agricultural improvement began. Introduction of important new crops such as the potato played a part, but so did practical improvements such as new crop rotations and scientific advances that produced a better knowledge of the chemistry of fertilizers. The rise of industrial cities provided growing markets for European farmers, and the latter were given some protection by tariffs or direct subsidies to encourage production and support incomes. Both of these governmental favors are now enjoyed by farmers in the Common Market.

Today, the result of Europe's long agricultural evolution is a farming sector that stands very high in overall volume of production and in yields per unit of farm land. This relatively small part of the world produces more than 70 percent of the world's wine and olive oil (largely from the Mediterranean region, although some wine is produced outside that region); more than 50 percent of the world's rye (concentrated on poor soils of the North European Plain) and beet sugar; more than 30 percent of its oats and barley; more than 20 percent of its wheat, flax, and pigs; and more than 10 percent of its citrus fruit (from Southern Europe), cattle, and sheep. Most European countries have export surpluses of certain agricultural specialties. The flowers of the Netherlands are a particularly famous case, although the same country's exports of dairy products are more lucrative. A few countries such as the Netherlands and Denmark actually manage an overall net export of agricultural commodities. But as a region, Europe is by no means agriculturally self-sufficient. Sizable imports of supplemental agricultural commodities are required, despite the huge output from Europe's own farms, the bulk of which are efficient high-yield producers. The main imports are tropical products such as coffee and tea that Europe cannot grow; feed grains (for example, corn) and soybeans for livestock production; and wheat.

The agriculture of Europe is as diverse as other aspects of the region. An enormous number of small rural regions with distinctive landscapes exist, and such landscapes have been formed largely by the detailed interaction of distinctive agricultural practices with distinctive physical environments over decades, centuries, or millennia. Picturesque agricultural countrysides are often the objective of European tourism. Hikers carefully exploring some intriguing geographic corner are a common sight.

Thus the agricultural geography of Europe could be detailed into hundreds of regions. But its major emphases can be summarized in broad generalizations about a few large and very generalized regions. For this purpose, regional groups of countries are used in the following discussion.

The Agricultural Core in Northwestern and Central Europe

Europe's most productive area agriculturally is a cluster of countries in northwestern and central Europe: the United Kingdom, Ireland, the Benelux countries (Belgium, Netherlands, and Luxembourg), West and East Germany, Switzerland, Austria, and Denmark. This may be considered the agricultural core, with a total value of output and average yield per unit of land greater than any other European area of comparable size. In these countries, farming is directed primarily toward milk, beef, and pork production. Some farms are specialized on either dairy or meat-animal production, but farms which combine the two are probably more common. Amounts of land in pasture tend to be high, and crop land is devoted largely to fodder crops—that is, to growing feed for livestock (see Fig. 6.14). This area is too far north for good results with corn grown for grain (which is the main feed crop in the United States), so the area relies on a variety of other crops. For example, barley is important in many areas, and there is much more use of root crops and tubers for feed than is true of the United States. A prime instance is potatoes (see Fig. 5.6), which are used in the United States almost exclusively for human food but in Europe are used extensively to feed livestock as well as people. In the densely populated European core, root crops and tubers serve the need for a high output of calories per unit of land cultivated, but they do so at the cost of labor intensity—that is, of high input of labor per unit of land and output. Such intensity is generally characteristic of these countries, despite a considerable amount of mechanization. Farms tend to be smaller than American farms, and maximum effort is applied to secure high yields from the limited land available (see Fig. 5.10). This results not only in high labor intensity but also in high capital intensity—for instance, in very heavy applications of fertilizer to soils that are generally mediocre in natural fertility. The effort is repaid in yields (outputs per unit of land) of crops and animal products that are among the highest in the world. The whole system is encouraged by the huge urban consuming markets in this part of Europe and by high price supports for farm products—supports set by the European Economic Community (EEC) in the Common Market countries and by national governments in others. Exceptionally fertile soils, such as those formed of loess, tend to show greater than average concentrations of wheat (grown primarily for human food) and sugar beets. Neither of these crops is well adapted to poor soils. But even where wheat and sugar beets are of major importance, there is still a large output of livestock and dairy products. The animals are fed partly on sugar-beet pulp left as a residue when the juice is pressed from the beets to make sugar. In East and West Germany, considerable land is used for rye, a food grain that grows relatively well in poor sandy soils. The world's main "rye belt" extends along the sandy North European

Plain from Germany across Poland and into the Soviet Union.

Agriculture in the Dry-Summer Subtropics of Southern Europe

A second major agricultural region comprises the four mountainous countries of Southern Europe—Greece, Italy, Spain, and Portugal—with their mediterranean or dry-summer subtropical type of climate (see Fig. 9.8). Compared with the agricultural core discussed previously, Southern Europe is distinguished by a greater emphasis on wheat, although supplementary imports are generally necessary; far greater amounts of land in tree crops and other perennial crops, particularly grapevines; and a unique emphasis on fruits and vegetables grown for export to European countries farther north. The subtropical mediterranean climate, with its rainy winters, allows winter wheat, planted in the fall and harvested in the spring, to be grown. Olive trees and other drought-resistant trees and vines can survive the hot dry summers. When irrigation is provided to counteract the summer dryness, an all-year farm production of remarkable variety and intensity becomes possible. Some irrigation has been used in parts of Southern Europe for thousands of years, and new projects have been pushed vigorously, notably in Spain, in recent years. But most of the land is still not irrigated, while much of it is too mountainous

Figure 5.10   Terraced vineyards on a steep slope in West Germany's Rhine Gorge indicate the highly intensive character of agriculture in Europe's agricultural core.

for any agricultural use except grazing, most often by sheep and goats. Close browsing by these animals has bared the soil on many steep slopes to erosion, with consequent flooding, siltation, and waterlogging of farm land in adjoining lowlands. These countries of Southern Europe are less industrialized and urbanized than are those of northwestern Europe. Consequently, they are much more characterized by overcrowded countrysides and rural poverty, and their inhabitants find it more difficult to secure an adequate level of nutrition. The hard-pressed population can maximize calories by eating vegetable foods rather than feeding them to livestock (which entails a great net loss of calories) in order to convert them to meat and milk. The result is much less emphasis on livestock in the agriculture of Southern Europe than in the more prosperous agricultural core. Protein in the Southern European diet is more apt to come from peas and beans than from expensive meat and milk.

The Grain-and-Livestock Agriculture of East Central Europe: Relative Poverty in a Communist Setting

East Central Europe is another large agricultural area that is relatively poor. As in Southern Europe, there is a strong emphasis on food grains, especially wheat. Rye complements wheat in Poland, as does corn in the Balkan Peninsula, where summers are hotter and the crop is grown largely for human food rather than for livestock feed as in the United States. Although many areas are mountainous, nature has provided larger areas of cultivable land in East Central Europe than in Southern Europe, and there is somewhat more emphasis here on the growing of crops to feed livestock, especially pigs. As might be expected in this less urbanized area, farming is less productive per unit of land and per person than in Europe's agricultural core, and farm incomes are lower. Another regional characteristic stems from politics. These countries—Poland, Czechoslovakia, Hungary, Romania, Bulgaria, Yugoslavia, and Albania—are all under Communist governments, as is somewhat more prosperous and productive East Germany in the agricultural core. Except in Poland and Yugoslavia, most agriculture has been collectivized, although elements of private enterprise are still found. As in the Soviet Union, the latter include private garden plots that workers on collective or state-owned farms are allowed to till as private enterprises. In Poland and Yugoslavia, resistance to collectivization was so great that the collectives were allowed to disband in the 1950s, although

some large state-owned farms still exist and many farmers belong to cooperatives. (For a fuller discussion, see Chapter 9.)

The Scanty Agriculture of the Northern Periphery

Four countries of Northern Europe—Iceland, Norway, Sweden, and Finland—form a fourth region with respect to agriculture. These countries on Europe's northern periphery are notable mainly for their very low proportions of agricultural land. Iceland is largely a cold and rocky, although scenically spectacular, wasteland. Most of Norway is mountainous, with much wasteland and considerable forest; and most of Sweden and Finland is forested. Dairy farming is the dominant form of agriculture in all four countries, although some areas have other specialties such as the sheep that are grazed on the sparse pastures of Iceland's tundra. Adequate markets and government subsidies and social programs enable the relatively few farmers to have a level of living well above that of farmers in Southern or East Central Europe.

The Special Circumstances of France

France is a special case, and the country may be considered a fifth agricultural "region." France is the largest country of Europe in area, three quarters the size of Texas. Its large areas of plains, along with favorable climatic conditions, give it more arable land (although not a larger *proportion* of arable land) than any other country in Europe. Coupled with these circumstances is a population density that is relatively low by European standards. In northeastern France, much of the land is covered by highly fertile loess soils; climatically, the country divides between the mild, wet marine west coast climate of northwestern Europe and the dry-summer subtropical climate of Mediterranean Europe. This environmental diversity is reflected in French agriculture, which exhibits a mixture of the milk and meat production characteristic of Europe's agricultural core, the large acreages of wheat found on the good lands of the core, and the wine, fruit, and vegetable production characteristic of Southern Europe. Relative abundance of land, at least by European standards, is associated with an agriculture somewhat less intensive and productive per unit of land than in most of the agricultural core but able to regularly export considerable quantities of basic foodstuffs, especially wheat. These exports go mainly to other Common Market countries.

The Role of Fisheries in Europe's Food Economy

Throughout history fishing has been an important part of the European food economy, and the coasts are still thickly dotted with fishing ports. At times control of fishing grounds has been a major commercial and political objective of nations, even resulting in warfare. Fisheries are particularly important in shallow seas that are rich in the small organisms known as *plankton*. These organisms are the principal food for schools of herring, cod, and other fish of commercial value. The Dogger Bank in the North Sea and waters off Norway and Iceland are major fishing areas, and two nearby countries—Norway and Denmark—are Europe's leading nations in total catch. A third Scandinavian country, Iceland, which has few other resources and few people, depends mainly on exports of fish and fish products (77% of all exports by value in 1986).

CHAPTER 6

Changing Fortunes in the Varied British Isles

England's heather-carpeted Lake District, associated with great poets, is seen in late autumn. It features 16 "finger lakes" of glacial origin, radiating from a cluster of water-soaked mountains. The treeless moors with their stone fences are representative of British highlands. The lake is Wast Water, seen from the west.

T HE British Isles lie off the northwest coast of Europe. There are approximately 5500 islands in the group, but most of them are small, and only two islands, Great Britain and Ireland, are of major consequence. The largest island, Great Britain, lies only 21 miles (34 km) across the Strait of Dover from France, and the whole island group is generally considered part of Europe.

Two countries occupy the islands: the Republic of Ireland, with its capital at Dublin, and the United Kingdom of Great Britain and Northern Ireland, with its capital at London. The latter country is usually referred to by shorter names such as "United Kingdom," "UK," "Britain," or "Great Britain." It incorporates the island of Great Britain, plus the northeastern corner of Ireland and most of the smaller islands, including the Isle of Wight, Isle of Man, Hebrides, Orkneys, Shetlands, and Channel Islands (Figs. 6.1 and 6.2). Altogether the United Kingdom comprises about four fifths of the area and 94 percent of the population of the British Isles (Table 6.1).

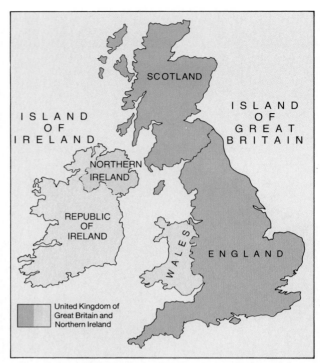

Figure 6.1 *The two main islands and the major political divisions of the British Isles.*

POLITICAL SUBDIVISIONS OF THE UNITED KINGDOM

The United Kingdom is sometimes incorrectly referred to as "England." However, England is merely the largest of four main subdivisions of the country, the others being Scotland, Wales, and Northern Ireland (Fig. 6.1). These were originally independent territories. Wales was conquered by England in the Middle Ages but preserves some cultural distinctiveness associated with the Welsh language, still spoken

	Area		Population	Approximate Density	
Political Unit	Thousand sq mi	Thousand sq km	(millions: 1987 estimates)	per sq mi	per sq km
United Kingdom	94.5	244.8	56.8	600	230
England	50.3	130.3	47.3	940	365
Wales	8.0	20.8	2.8	350	135
Scotland	30.4	78.7	5.1	165	65
Northern Ireland	5.5	14.2	1.6	290	115
Isle of Man[a]	0.225	0.583	0.068	300	120
Channel Islands[a]	0.075	0.194	0.135	1800	695
Republic of Ireland	27.1	70.2	3.5	130	50
British Isles	121.7	315.2	60.3	495	190

TABLE 6.1   BRITISH ISLES: AREA AND POPULATION DATA

[a]In a strict sense, the Isle of Man and Channel Islands are not included in the United Kingdom, both being dependencies of the British crown. But in practical effect they are part of the United Kingdom and are so regarded in this chapter.

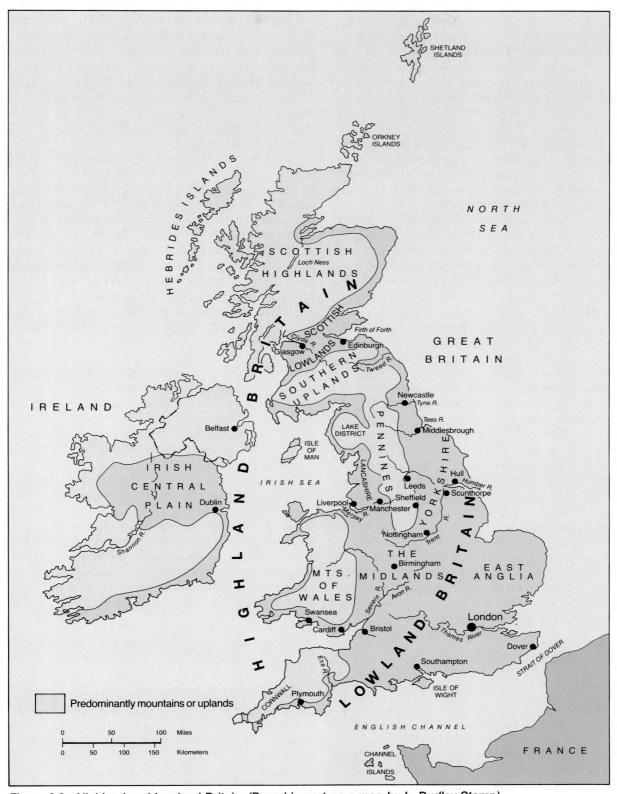

Figure 6.2 Highland and Lowland Britain. (Based in part on a map by L. Dudley Stamp)

by about one fifth of its population. Northern Ireland, together with the rest of Ireland, was twice conquered by England. The earlier conquest in the Middle Ages was followed by a lapse of English control during the Wars of the Roses in the second half of the 15th century. A second conquest was completed in the 17th century. One of its outcomes was the settlement of Scottish and English Protestants in the North, where they became numerically and politically dominant. In 1921, when the Irish Free State (later the Republic of Ireland) was established in the Catholic part of the island, the predominantly Protestant North, for economic as well as religious reasons, elected to remain with the United Kingdom. The North was given a large degree of autonomy, including its own parliament at Belfast to legislate in matters of local concern. However, these arrangements were suspended and direct British rule was interposed as a result of the grim and prolonged violence that broke out between the region's Protestants and Catholics toward the end of the 1960s. Despite many attempts at a settlement, this situation continued without basic change into the 1980s. But Northern Ireland, like the other subdivisions of the United Kingdom, has continued to send representatives to the Parliament at London.

Scotland was first joined to England when a Scottish king inherited the English throne in 1603, and the two became one country under the Act of Union passed in 1707. Scotland has no separate parliament, but it does have special administrative agencies in Edinburgh which deal with Scottish affairs. It also has its own system of courts and law. Wales has even less autonomy than does Scotland, although its largest city, Cardiff, functions in some measure as an administrative capital. But the Welsh, like the Scots, feel ethnically distinct from the English, and the separate concerns of these peoples—and of the Northern Irish as well—are recognized by the inclusion of a Secretary of State for Scotland, one for Wales, and one for Northern Ireland in the United Kingdom's government.

WORLD IMPORTANCE OF THE UNITED KINGDOM

The international dominance once attained by a country as small as the United Kingdom and the strong imprint its influence has left on the world are among the more remarkable aspects of modern history. When British power was at its peak in the century between Britain's defeat of Napoleonic France and the outbreak of World War I (1815 to 1914),

the United Kingdom was generally considered the world's greatest power. Its overseas empire eventually covered a quarter of the earth, and the influence of its language and culture spread still farther. Until the late 19th century, it was the world's foremost manufacturing nation. Its navy dominated the seas, and its merchant marine carried half or more of the world's ocean trade. It was the world's greatest trading nation, and the invested profits from its industries and commerce helped finance the development of much of the rest of the world.

From the late 19th century onward, the rate of economic growth in the United Kingdom tended to be slower than that of many other industrialized or industrializing countries. A relative decline in the country's economic importance set in, with an acceleration after World War II. Britain was still the richest country in Europe as late as 1940, but it has now lost its lead to West Germany and France in gross national product (GNP), and it has dropped below most countries of West Central and Northern Europe in per capita output and income (see Table 4.1). Relative decline economically has been accompanied by a drop in political stature. The large countries once included in the British Empire have all gained independence, and Britain's remaining overseas possessions are scattered and small.

But the United Kingdom is still a country of much consequence, although no longer the leader, in manufacturing, trade, finance, and international politics. With France, it is one of only two European countries possessing nuclear weapons, and its forces are an important part of the North Atlantic Treaty Organization (NATO). It plays a major role in the European Communities and it is associated with many of its former colonies in a worldwide Commonwealth of independent countries. The cultural impact exerted on many parts of the world during Britain's period of ascendancy continues to foster awareness of Britain and various cultural relationships between the British and other peoples.

REGIONAL VARIETY ON THE ISLAND OF GREAT BRITAIN

Aside from minor islands and a small part of Ireland, the physical base of the United Kingdom is Great Britain, an island about 600 miles (somewhat under 1000 km) long by 50 to 300 miles (80 to nearly 500 km) wide. For its small size, this island provides a homeland with a notable variety of landscapes and modes of life. Some of this variety can be attributed to a broad fundamental division of the island be-

Figure 6.3 *Fields in the English Lowland normally are fenced by hedgerows, seen here in southwestern England near the mouth of the River Severn. Hedgerow trees blend in the distance to give a misleading impression of forest. This landscape was created centuries ago by the Enclosure Movement, in which large landowners appropriated unfenced land tilled by villagers. Hedges were planted so that sheep could be raised under a system of controlled grazing.*

tween two contrasting parts of the British Isles, *Highland Britain* and *Lowland Britain* (Fig. 6.2).

HIGHLAND BRITAIN

Highland Britain embraces Scotland, Wales, Ireland, and parts of northern and southwestern England. It is predominantly an area of treeless hills, uplands, and low mountains, formed of rocks that are generally older and harder than those of Lowland Britain (see chapter-opening photo). All three of the major classes of rocks—igneous, metamorphic, and sedimentary—are well represented. Broad areas are moors, tenanted by sheep farmers or uninhabited. (In its most common usage, the term *moor* means a rainy, deforested upland, covered with grass or heather and often underlain by water-soaked peat. Moors are important water sources for densely populated British lowlands.)

The principal highlands include the Scottish Highlands, the Southern Uplands of Scotland, the Pennines and the Lake District of northern England, the mountains which occupy most of Wales, the uplands of the Cornwall Peninsula in southwestern England, and the mountainous and hilly rim of Ireland. The mountains are low—Ben Nevis in the Scottish Highlands is the highest at 4406 feet (1343 m)—but many slopes are quite steep, and the mountains often look high because they rise precipitously from a base at or near sea level.

Highland Britain includes two important lowlands: (1) the boggy, agricultural Central Plain of Ireland; and (2) the Scottish Lowlands, a densely populated industrialized valley separating the rugged Scottish Highlands from the gentler and more fertile Southern Uplands of Scotland. The Scottish Lowlands comprise only about one fifth of Scotland's area, but they incorporate more than four fifths of its population and its two main cities, Glasgow (1.8 million in the metropolitan area, including Glasgow city plus suburban and satellite districts near by)[1] on the west, and Edinburgh (650,000), Scotland's political capital, on the east.

LOWLAND BRITAIN

Practically all of Lowland Britain lies in England; it is often called the English Lowland. In contrast to the old hard rocks of Highland Britain, the Lowland section is mainly an area of younger, softer sedimentary rocks (including unconsolidated sediments of glacial, riverine, or marine origin) that have produced better soils than those of the highlands and a gentler topography (Fig. 6.3). Most of Lowland Brit-

---

[1]Unless otherwise noted, city populations in this text are rounded approximations for metropolitan areas as of the middle 1980s. They are only crudely comparable from one country to another, due to variations in the accuracy of statistics and the actual definition of *metropolitan area.* Such figures, based on standard sources such as the *Rand McNally Commercial Atlas* and the *Britannica Book of the Year,* do enable students to broadly group urban areas by orders of magnitude—an activity far more meaningful than attempts to memorize populations city by city.

Figure 6.4   Early Sunday morning view of the High Street of Dorchester, located in Lowland Britain near the English Channel. Dorchester, a market town in the county of Dorset, is the "Casterbridge" of Thomas Hardy's Wessex Novels. Later in the day, the street will be choked with midsummer traffic to and from the nearby resort of Weymouth on the Channel. British towns such as Dorchester often lack bypasses for through traffic, resulting in serious congestion along narrow streets such as the High Street shown here. Dorchester was founded in Roman times, and the town center exhibits varied building styles dating from different historical periods.

ain is overspread with a continuous mosaic of well-kept pastures, meadows, and crop fields, fenced by hedgerows and punctuated by closely spaced villages, market towns (Fig. 6.4), and industrial cities. Within Lowland Britain, the largest industrial and urban districts, aside from London, lie in the midlands and the north, around the margins of the Pennines (Fig. 6.7). Here clusters of manufacturing cities have risen in the midst of the coal fields that provided most of the power for English factories during the great surge of industrialization before World War I and continued to be overwhelmingly dominant until the rise of oil, natural gas, and nuclear energy after World War II.

INSULARITY AS A FACTOR IN THE EARLY RISE OF BRITAIN

The rise of Great Britain to world importance began after Europe established contact by sea with the Americas and the Orient. Following the voyages of

Columbus and Vasco da Gama at the end of the 15th century, the nations along the western seaboard of Europe launched an intense competition for trade and colonies in overseas areas. By the 18th century, Britain had outdistanced her continental rivals in this struggle. In doing so, the island was able to exploit certain geographic advantages as against its continental competitors.

Perhaps the greatest of these advantages was the fact that Britain had no land frontiers to defend. A strong navy was sufficient to protect the island and thus to allow Britain to choose the time and place in which to engage in land fighting. In contrast, the continental powers had to be constantly on the alert against possible military moves by their neighbors. Of necessity, their attention was strongly divided between Europe itself and the colonial areas, whereas Britain's island security made possible a greater concentration of effort in the colonial field.

The Age of Discovery changed immensely the significance of Britain's location. Previously, Great Britain had been an island outpost on the edge of the known world. During the Middle Ages, the is-

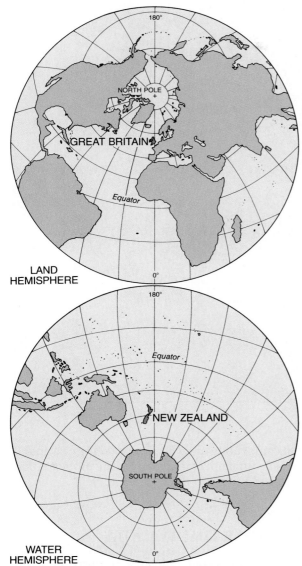

Figure 6.5 *In the top map, note how the major land masses are grouped around the margins of the Atlantic and Arctic oceans. The British Isles and the northwestern coast of Europe lie in the center of the "land hemisphere," which comprises 80 percent of the world's total land area and has approximately 91 percent of the world's population. New Zealand lies near the center of the opposite hemisphere, or "water hemisphere," which has only 20 percent of the land and 9 percent of the population.*

land was mainly a pastoral area supplying raw and semifinished wool to continental industries. Its peripheral location and type of economy have caused Britain to be described as the Australia of the time. Following the great discoveries in the 15th and 16th centuries, however, Great Britain found itself at the

front door of Europe opening toward the new lands overseas (Fig. 6.5). Britain's position between the coast of northwestern Europe and the open Atlantic meant that continental shipping would have to pass by at close range, especially since most ships preferred the narrow English Channel to the stormy ocean outlet of the North Sea between Scotland and Norway. Thus, Britain lay across or near the lines of communication connecting the continent with overseas areas, whereas the continental countries had no similar advantage with respect to Britain. This advantage of position could be exploited either peaceably to expand trade or militarily to intercept and harry the overseas shipping of continental rivals.

CHANGING FORTUNES OVERSEAS: FROM BRITISH EMPIRE TO COMMONWEALTH OF NATIONS

A major result of the United Kingdom's rise to dominance of the world's oceans was its acquisition of a large part of the world as a colonial empire. The major possessions were:

1. The Indian subcontinent, spasmodically conquered from native rulers and Britain's European rivals in the 17th, 18th, and 19th centuries.

2. Many Caribbean islands, acquired and settled in the 17th century.

3. Canada and the eastern part of the United States, taken over and settled in the 17th and 18th centuries.

4. Australia and New Zealand, settled in the late 18th and the 19th centuries.

5. Parts of Southeast Asia and China, appropriated mainly in the 19th century.

6. Much of Africa, where slaving outposts were established very early, but conquest and occupation took place mainly in the 19th and early 20th centuries.

Control over some territories, such as Palestine in the Middle East, came as late as World War I, while one major possession, the United States, was lost as early as 1783. Since World War II, almost all of the British Empire has gained independence. But the former colonies around the world retain many British imprints, and many units are still associated with the United Kingdom as equals in an international organization known as the Commonwealth of Nations (Fig. 6.6).

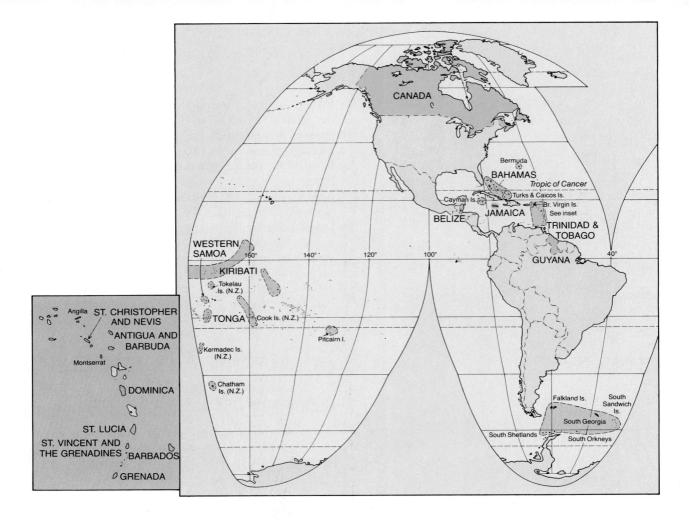

THE GEOGRAPHIC IMPACT OF BRITAIN'S INDUSTRIAL REVOLUTION

In the 18th and 19th centuries, the United Kingdom was transformed geographically by the Industrial Revolution (see discussion in Chapter 5). The rapid development of new industrial machines and techniques, together with the availability of power, materials, capital, labor, markets, and entrepreneurial skills, made possible the long period of industrial expansion that began accelerating around 1750. Population increased rapidly and concentrated on the coal fields, where mining and industrial towns and adjacent seaports experienced a mushroom growth (Fig. 6.7). Britain became the world's first essentially industrial and urban nation.

Five industries were of outstanding prominence in the nation's industrial rise: coal mining, iron and steel, cotton textiles, woolen textiles, and shipbuilding. The early start of these industries compared with those of other countries gave the United Kingdom industrial preeminence through most of the 19th century. Their exported surpluses paid for massive imports of food and raw materials and made Britain the hub of world trade. They made Britain wealthy. But their difficulties in the face of growing interna-

tional competition and changing patterns of demand have been fundamental to many of the problems that have faced Britain in the 20th century.

THE BRITISH COAL INDUSTRY

Coal became a major industrial resource between 1700 and 1800, when it first became fuel for the blast furnace and the steam engine. It was coked for use in blast furnaces, and coke replaced the charcoal previously used for smelting iron ore. With the development of an economically practical steam engine in Britain, coal supplanted human muscles and running water as the principal source of industrial energy. Since these 18th century developments, coal has been a key factor in the economy of the United Kingdom. Its use has changed the face of Britain. Electricity was still in the future during the early period of industrialization, and coal was expensive to transport; consequently, the new factories were generally built on or near the coal fields. The working population, in turn, clustered near the factories and mines (see Fig. 5.8). Thus developed a group of industrial districts which incorporated mines, factories, and urban areas. These districts contained a sizable share of the British population, and do so today.

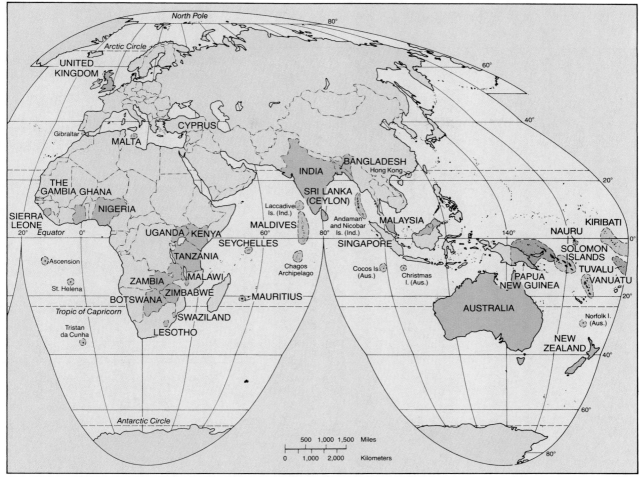

Figure 6.6 *The Commonwealth of Nations as of 1988. Names in capitals indicate sovereign states. Dependent units are indicated by capital and small letters; those not otherwise identified as to ownership are administered or protected by the United Kingdom. Small island dependencies are encircled by dashes. Some minor islands, often uninhabited, are not indicated on the map, and Antarctic claims are not shown.*

Coal production rose until World War I: 6 million tons (later estimate) in 1770; 65 million in 1854; 287 million, the all-time high, in 1913. Coal powered domestic industry and also became a major export used by railways, industries, and shipping over much of the world. However, the British coal industry has followed an erratic up and down course since 1913, with the net result being a drastic long-term decline in production and employment. Between 1913 and the 1980s, the annual coal output fell by more than one half, the number of jobs in coal mining was cut by more than four fifths, and practically all the coal export trade disappeared. Major factors in this decline were (1) increasing competition from foreign coal in world markets, (2) depletion of Britain's own coal that was easiest and cheapest to mine (although large reserves of less easily mined coal still remain), and (3) increasing substitution of other energy sources for coal—both in overseas countries and in Britain itself.

The Place of Coal in Britain's New Age of Diversified Energy

In Britain's own energy consumption, coal overshadowed all other sources as late as 1950. But this dominance was eroded in the 1950s and 1960s by a growing dependence on imported oil, natural gas imported in liquid form, and electricity from British nuclear plants. In this period, British port areas—especially the lower Thames Estuary downstream from London, the Mersey Estuary near Liverpool, Milford Haven in Wales, and Fawley near Southampton—became the sites of one of the world's largest oil refining industries.

In the late 1960s and the 1970s, Britain's energy situation was again revolutionized by large-scale development of newly discovered oil and gas fields underneath the North Sea. Norway and the Netherlands shared in this development, but their output was far surpassed by production from the British sector. How long Britain can sustain major production

113

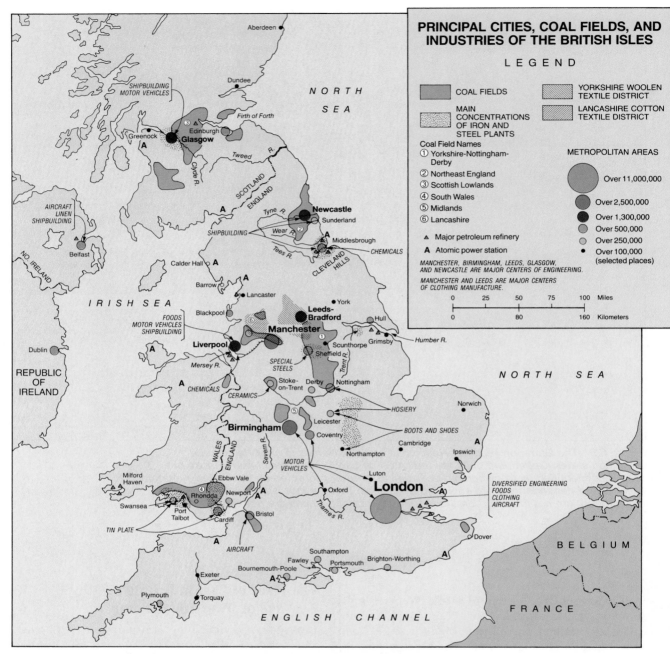

Figure 6.7 *Industrial and urban map of the British Isles. "Coal Fields" mark concentrations of coal mines but do not portray the full extent of known coal deposits. Some of the more important industrial emphases of different cities and areas are indicated by italicized words and arrows. The canal symbol from the Mersey estuary to Manchester marks the Manchester Ship Canal.*

of oil and gas is uncertain, but the North Sea finds have already given the British economy a much needed boost.

By the 1980s, coal's principal remaining role in Britain was to supply the major part of the fuel for generating electricity. Whether it can hold this position in the face of competition from gas, oil, and nuclear energy is an important question in British economic life, not least because of employment dif-

ficulties in coal-mining areas. Oil and gas major competitors for some time, and nuclear plants have been looming larger. In fact, by 1986 more than a dozen nuclear stations in widely dispersed coastal locations (Fig. 6.7) were producing 20 percent of all British electricity. Hydropower plays little part in energy supply. A few hydrostations in the Scottish Highlands provide approximately 1 percent of the national output of electricity.

Despite the decline of coal, this mineral's mark remains indelibly printed on Britain's geography. Its earlier predominance goes far toward explaining the development and location of such populous urban-industrial areas as those in the Scottish Lowlands, Northeast England, Yorkshire, Lancashire, the Midlands, and South Wales (Figs. 6.7 and 7.1). And its decline, which has been even more pronounced in terms of employment than in terms of production (due to mechanization), presents the United Kingdom with continuing problems of employment and regional redevelopment in depressed mining areas.

The Major Coal Fields

The United Kingdom's major coal fields lie mostly around the edges of the highlands on the island of Great Britain (Fig. 6.7). In England, the Pennines are ringed on three sides by a series of fields. The *Lancashire fields* to the west and the smaller, scattered *Midlands fields* to the south of the Pennines provided fuel for the development of great industrial districts centering on Manchester and Birmingham. But more important than these fields today is the *Yorkshire-Nottingham-Derby field* in the lowland along the eastern flank of the Pennines. This field is by far the leading British field in coal production, has the largest and most cheaply mined reserves of any field in the United Kingdom, and is the field best able to hold its own in the increasing competition of coal with other fuels. Another English field, the Northumberland-Durham or *Northeast England field,* lies on the North Sea coast between the Tees River and the Scottish border. Outside of England, the *South Wales field* occupies valley floors on the south flank of the Welsh mountains, and the *Scottish Lowlands fields* stretch from sea to sea across the narrow waist of Scotland.

IRON AND STEEL

Steel was a scarce and expensive metal before the Industrial Revolution. Iron, none too cheap or plentiful itself, was more commonly used. It was made in small blast furnaces by heating iron ore over a charcoal fire, with temperatures raised by an air blast from a primitive bellows. Supplies of iron ore in Great Britain were ample for the needs of the time, but the island was largely deforested by the 18th century and charcoal was in short supply. By 1740 Britain was importing nearly two thirds of its iron from countries with better charcoal supplies, notably Sweden and Russia.

This heavy import dependence then changed when 18th-century British ironmasters made revolutionary changes in iron production: replacement of charcoal by coke; improvement of the blast mechanism; invention and use of a refining process, called *puddling,* to make the iron more malleable; and adoption of the rolling mill in place of the hammer for final processing. Thus, the fuel shortage was solved, the scale and pace of production were expanded, and the island's general changeover to machine industry was promoted. The output of iron in the United Kingdom expanded by something like 190 times between 1720 and 1855; by the latter date, the country was producing half the world's iron, much of which was exported.

Further advances in production resulted when British inventors developed the Bessemer converter and the open-hearth furnace in the 1850s. These allowed steel, a metal superior in strength and versatility, to be made on a large scale and cheaply for the first time. In the Bessemer converter, steel is made rapidly from molten iron by passing a blast of air through it to burn out impurities. In the open-hearth furnace it is made by heating iron with gas flames over a period of many hours and drawing off the impurities. In both processes quality is controlled by the addition of carbon and alloys, but it is controlled better in the slower open-hearth process. After these inventions iron became primarily a raw material for steel, although great quantities of iron are still consumed directly by fabricating industries. The Bessemer converter and the open-hearth furnace still play a role in the world's steel industry, but a newer oxygen converter, in which an oxygen blast is used to process molten pig iron, has been replacing them in new and modernized plants since the 1950s. The oxygen process is by far the main way that steel is now made in the United Kingdom.

The early iron industry in Great Britain relied entirely on domestic ores. A major problem in iron making is to bring together economically the industry's bulky raw materials, but suitable ores were often present in British coal fields. For a century after 1750, nearly all the ore used was "coal-measures" ore. Depletion of this ore led in 1850 to the start of large-scale iron mining outside the coal fields and in the 1870s to the first large imports of foreign ore. Since then the industry has depended in part on low-grade domestic ores from the English Lowland (see Fig. 6.7) but increasingly on high-grade imported ores that now supply all but a small fraction of the iron content of all ore used.

Although British inventors and entrepreneurs led the world into the age of cheap, mass-produced, and extensively used iron and steel, Britain soon lost its leadership in steel production. The United States and Germany surpassed the British output before 1900. The Soviet Union surpassed Britain in the 1930s, and Japan did so in the 1960s. By 1987 the United Kingdom ranked only eighth in the world.

Iron and Steel Districts

In harmony with the widespread distribution of coal and ore deposits, British iron and steel plants developed in widely scattered locations. All the coal-field industrial areas shared in the growth of the industry, as did the main areas of domestic iron-ore production just south of the Humber Estuary and in Northamptonshire to the east of Birmingham. The industry is still widely scattered in Britain, but three main districts now account for most of the output: (1) The leading district is *Eastern England,* with plants at Scunthorpe near the Humber, in and around Sheffield, and in the Northamptonshire ore fields. (2) *South Wales* and (3) the *lower Tees River area* are the other districts of greatest consequence today. Both are located near major coal fields, both used local iron ore initially, and both now use foreign ore exclusively. A minor share of British steel is produced in other districts, particularly the Scottish Lowlands around Glasgow. Employment in the steel industry has suffered from increased competition and mechanization—the same factors that have undercut employment in coal mining. Since many steel districts also mine coal, they are subject to the combined effects of decline in two major sources of employment. Rebuilding based on newer types of industry has not yet been highly successful, although local progress has been made.

Two historically famous places associated with the early development of the iron and steel industry are Birmingham (2.7 million) and Sheffield (700,000). The district around Birmingham, especially the "Black Country" to the west and northwest of the city, was the leading center for the British iron industry until the middle of the 19th century, and Birmingham was the first city to become world famous for large-scale iron production. But the city lost its major importance in this industry long ago and turned to the manufacture of diverse metal products such as brassware and, later, automobiles. Birmingham also has important service functions associated with its position as the metropolis of the densely populated and highly industrialized West Midlands region. Sheffield developed a tradition of high-grade steel manufacture very early, so that "Sheffield steel" came to mean steel of extraordinary quality, generally for use in tools and cutlery. Something of this specialty has continued, although the Sheffield area is an important producer of more ordinary grades of steel.

COTTON MILLING IN LANCASHIRE

During the period of Britain's industrial and commercial ascendancy, cotton textiles were its leading export by a wide margin, amounting to almost a quarter of all exports by value in 1913. Although for a time it was rather scattered, the enormous cotton industry eventually became concentrated in Lancashire, in northwestern England, with Manchester (2.8 million) the principal commercial center (Fig. 6.7). Raw cotton was imported, and manufactured textiles were exported, through the Mersey River port of Liverpool (1.5 million), although Manchester itself became a supplementary port after completion of a ship canal to the Mersey Estuary in 1894. Manchester is still the business center of the industry and a center of finishing industries (primarily bleaching, dyeing, and printing) and clothing manufacture, but most of the spinning of cotton yarns and weaving of cloth is now carried on in other cities and towns of Lancashire. However, the industry has declined to a mere shadow of its former size and importance.

A hand-labor textile industry in Lancashire dated from the Middle Ages and was transformed by the Industrial Revolution of the 18th century. By the early 18th century the industry not only used local wool but also was importing flax from Ireland, had originated new fabrics using linen in combination with small amounts of imported cotton, and was selling part of the product in tropical markets in connection with the flourishing slave trade of Liverpool. As it affected this early Lancashire textile industry, the Industrial Revolution involved several lines of development. First a series of inventors, in good part Lancashire men, mechanized and thus speeded the spinning and weaving processes. Then power was applied to drive enlarged and improved machines— first falling water from Pennine streams, then steam raised by coal from underlying Lancashire fields. Finally, in 1793, the American invention of the cotton gin, which separated the seeds from the raw cotton fibers economically, made cotton a relatively cheap material from which inexpensive cloth could be manufactured.

The stage was now set for the spectacular rise of the Lancashire cotton industry and, incidentally, for the spread of cotton farming in the American South, which supplied most of the raw material. Liverpool merchants financed much of the industrial development of Lancashire, but the cotton factories were built inland on the streams and the coal. In addition to sources of power, the inland locations gave access to supplies of soft water from the Pennines for use in the finishing processes and in boilers. For over a century, until World War I, Lancashire dominated world trade in cotton and cotton textiles.

But 1913 saw the peak of British production; after that, the decline of the industry was about as spectacular as its former rise. The root of the trouble was increasing foreign competition. The cotton industry is peculiarly adaptable to areas that are commencing to industrialize and has often been the first large-scale industry to be established. Among the reasons for this are (1) the comparatively low level of skill required of the labor force in cotton mills, especially those manufacturing the cheaper grades of cloth; (2) the almost universal importance of the product; and (3) the transportability of the raw material. Once the industry has been started in such an area, it often has had the advantages of (1) cheap labor, connected with the area's relatively low level of living; (2) newer equipment; and (3) a protective tariff to keep out competition from older producers, such as Britain.

In the 20th century, country after country has surpassed the United Kingdom in cotton textile production and exports, with a resulting slide in employment and great economic distress in Lancashire. About nine tenths of the British cotton industry is still concentrated in Lancashire, but the number of workers in Britain has dropped to about 50,000, from a peak of 600,000 just before World War I. Production is now more concentrated on quality fabrics, which competitors find more difficult to duplicate, and is more geared to the home market and the use of synthetic fibers such as nylon and rayon. However, Lancashire's future as an industrial area lies mainly with other industries such as engineering (a general name for a wide range of industries that make machinery, vehicles, and tools), chemicals, and clothing.

THE WOOLEN TEXTILE INDUSTRY OF YORKSHIRE

The United Kingdom has one of the largest woolen textile industries in the world and is a major exporter of woolens. Most of the British woolen industry is concentrated in western Yorkshire (the "West Riding") in Leeds-Bradford (1.5 million) and a large number of smaller textile towns and cities (see Fig. 6.7). This area on the east slope of the Pennines contains the largest single concentration of woolen and worsted textile plants in the world. Bradford is the largest producing center; Leeds, the main city in size, is a diversified manufacturing center in which

engineering industries and clothing manufacture are more important than the woolen industry.

In the Middle Ages, Great Britain, an island well supplied with grazing land, exported raw wool and then woolen cloth to continental Europe. Thus, the Industrial Revolution, as it applied to woolen manufacture, involved the mechanization of an industry that had long been important, rather than, as in the case of cotton milling, the development of a new industry based on a fiber that had never been of prime importance. With the mechanization of the woolen industry, production expanded rapidly and soon outran Britain's output of raw wool. The country became not only the leading exporter of woolen goods but also the leading importer of raw wool, positions it held for many decades but has now lost to various other countries.

The west Yorkshire area was the leader among several important woolen-manufacturing districts in Great Britain before the Industrial Revolution. It was well suited for the industry through its resources of water from the Pennine streams, which supplied both power and abundant soft water for cleaning wool and finishing cloth. With the coming of the Industrial Revolution, this was the quickest of the woolen-manufacturing areas to adopt the new machinery, first used in adjacent Lancashire, and had the advantage of being located on a coal field. The industry largely died out in competing areas—except for some upland sections of Scotland, such as the Tweed Valley, producing distinctive tweeds and tartans—and became concentrated more and more in west Yorkshire.

The British woolen industry was less spectacular in its rise and has undergone a less drastic decline than the cotton industry. Although it did gain a position of world leadership, its growth was somewhat checked by the rapid rise of the cotton industry after the invention of the cotton gin and by the fact that many countries already had woolen industries by the time of the Industrial Revolution. It neither exported as much nor was as dependent on the export market as the cotton industry, since Britain's home consumption of woolens has been large. But it, too, reached an export peak before World War I and has since declined. However, the decline has been greatly checked by the importance of the home market as compared with the market for cotton goods; by the greater labor skills required in its processes, which makes it more difficult to initiate in new industrial areas; and by its greater emphasis on quality, for which the British industry has long had a world-wide reputation. Nevertheless, Yorkshire, like Lancashire, now relies mainly—in both employment and value of production—on industries such as machinery and clothing that formerly were of secondary importance.

SHIPBUILDING ON THE TYNE AND CLYDE

When wood was the principal material in building ships, Great Britain, as a major seafaring and shipbuilding nation, was presented with increasingly serious difficulties due to deforestation. Large timber imports were brought from the Baltic area and from America; and the importation of completed ships from the American colonies, principally New England, was so great that an estimated one third of the British merchant marine was American-built by the time of the American Revolution. During the first half of the 19th century, American wooden ships posed a serious threat to Britain's commercial dominance on the seas.

But in the later 19th century, ocean transportation was transformed by iron and steel ships propelled by steam. Britain led the way and was building four fifths of the world's seagoing tonnage by the 1890s. Two districts developed as the world's greatest centers of shipbuilding. These were (1) the Northeast England district, with a great concentration of shipyards along the lower Tyne River between Newcastle (1.3 million) and the North Sea, and with other yards along the rivers Wear and Tees; and (2) the western part of the Scottish Lowlands, with shipyards lining the River Clyde for miles downstream from Glasgow (see Fig. 6.7). Both districts had a seafaring tradition—Newcastle in the early coal trade and in fishing, and Glasgow in trade with America; both were immediately adjacent to centers of the iron and steel industry; and both had suitable waterways for location of the yards. Their ships were built mainly for the United Kingdom's own merchant marine—as they still are—but even the minority built for foreign shipowners were an important item in Britain's exports.

During the 20th century, Britain's share in world shipbuilding has declined steadily except for short-term recoveries after both world wars. Its output fell from about three fifths of the world total in 1913 to one third just before World War II and is not more than 1 or 2 percent today. New shipyards in rising competitors such as Japan and South Korea have tended to have advantages in cheaper skilled labor, as well as the ability to start with the latest equipment and techniques. But the actual tonnage of ships launched by British shipbuilders has fallen much less than has the British share in an expanded world industry. Many shipyards have had to close, with attendant employment loss, but both the Clyde and Northeast England continue to be active centers of the industry.

A LAND OF CITIES

The Industrial Revolution gave a powerful impetus to the growth of cities at the same time that Britain's total population was increasing rapidly. In the 19th century, the United Kingdom became the world's first predominantly urban nation, and it remains one of the most highly urbanized in the world. Most of the large cities, with the notable exception of London, are on or near the coal fields which supplied the power for early industrial growth. The most striking cluster is comprised of cities on or near the coal fields that flank the Pennines to the east, south, and west, that is, in the industrial districts of Yorkshire, the Midlands, and Lancashire, respectively (see Fig. 6.7). In this area, measuring only about 80 miles (c. 130 km) from east to west and 100 miles (c. 160 km) from north to south, about two dozen metropolitan areas have a combined population of approximately 14 million people. The largest, Manchester and Birmingham, have more than 2½ million each, and two others, Leeds-Bradford and Liverpool, have 1½ million each. Lesser urban clusters occur on or near the other major coal fields in the Scottish Lowlands, Northeast England, and South Wales. In striking contrast, the largest metropolitan areas that lie at a considerable distance from the coal fields—aside from London—are generally smaller and more widely spaced. Most of them are in southern England. For many years, they have tended to grow more rapidly and be more prosperous than the coal-field industrial cities. The latter tend to be physically unattractive and too dependent on old and failing industries. The transportability of electricity, natural gas, and oil has made it unnecessary for industries to locate near coal, and recent development has tended to favor the relatively pleasant environments of southern England.

THE UNIQUENESS OF LONDON

Britain's largest urban agglomeration differs from the other major cities in location, in historical background, and in the nature of its economy. The London metropolitan area, with about 12 million people, sprawls over a considerable part of southeastern

England (Fig. 6.8). Thus, it differs locationally in being relatively far from the island's major coal fields and their industrial centers. Its historical background is also different in that London was already a major city before the Industrial Revolution. The city already had 700,000 people in the late 17th century, when most of the present cities were little more than hamlets. And finally, London differs in that its economy is not based so much on industry. Although the metropolitan area does contain a great deal of manufacturing, this sector does not dominate. Instead, it shares leadership with several other major elements, some of which developed before the Industrial Revolution and have been greatly expanded since. The latter include commerce, government, finance and insurance, and corporate administration. Basic elements more recently developed include communications media (publishing, television, films) and tourism.

THE FUNDAMENTAL IMPORTANCE OF COMMERCE

Outstanding importance as a seaport and commercial center was the original foundation for London's development. Trade generated population growth, profits and financial operations, including insurance (the insurance business began with marine insurance covering risks to ships and cargoes). Population, trade, and wealth attracted government. Finance, trade, and government attracted corporate offices. The large, growing, and increasingly prosperous population provided both a labor force and a large market for manufacturing.

London became an important seaport as early as the Roman occupation of England in the 1st century A.D. and has continued so to the present. Certain aspects of London's location and evolution as a port not only help account for the city's own development but also throw some light on port cities in general. This is not to say that other ports in the world have evolved precisely like London, but they often do exhibit circumstances and tendencies comparable to London's. Such factors include:

1. *Location relatively near major trading partners.* London is located in the corner of Britain nearest the continent, which was the only important area with which England traded before the Age of Discovery and is still of major importance to British trade today.

2. *Location in a productive area.* London is located in the English Lowland, which was the most productive and populous part of Great Britain in the preindustrial period, supplying most of the wool which was Britain's main export, and providing most of the British market for imports.

3. *Location in a physical situation facilitating transport connections with the surrounding region.* From the site of London, a major route to the interior, with branches, was provided by the Thames River and its tributaries. Before the age of the railroad and the motortruck, water transport had greater advantages over land transport than it has today, and these natural inland waterways were extensively used for transporting goods to and from London. During the Industrial Revolution, their utility was enhanced by canals (little used today) which connected the Thames system with other British rivers. Furthermore, London became a major junction of roads as early as Roman times, and, again, the city's location provided advantages. The lower Thames was originally fringed by tidal marshes forming a major barrier to transportation, but the original site of London provided firm ground that penetrated these marshes and approached the river bank on both sides. Highways could avoid the marshes by converging on London and crossing the river there.

4. *Location on a river permitting relatively deep penetration of ocean ships into the land.* A seaport set inland in such a manner can serve more trading territory in the surrounding countryside, within a given radius, than can a port on a comparatively straight stretch of coast or on a promontory. This principle can be easily illustrated by drawing an arc of 100 miles (161 km) radius from the Channel port of Dover, which is closer to the continent, and another arc of 100 miles radius from London and comparing the respective sizes of the areas falling within the two arcs (Fig. 6.9). Inland location of a port tends to maximize the distance goods can be carried toward many destinations by relatively cheap water transportation (ocean vessels) and to minimize the more expensive land transport.

But while London, like many other ports, originally developed well inland, later growth of the port has been downstream toward the ocean. This has been necessary to accommodate the growth in size (and hence the efficiency) of ships, since large ships cannot go as far up the river. Although ships have been growing larger for centuries, in the industrial age they have been growing at an accelerating pace, especially with the advent of oil supertankers and other very large ships in the last few decades. In addition, more efficient loading and unloading

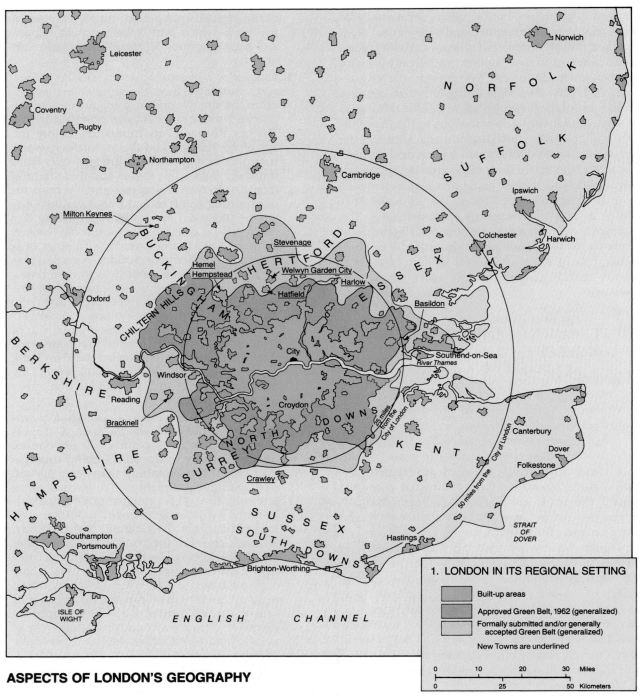

1. LONDON IN ITS REGIONAL SETTING

Built-up areas

Approved Green Belt, 1962 (generalized)

Formally submitted and/or generally accepted Green Belt (generalized)

New Towns are underlined

ASPECTS OF LONDON'S GEOGRAPHY

Figure 6.8   On the map "London in Its Regional Setting," small areas of parks and other open space within the main built-up area of London appear in the tint for "Approved Green Belt" but are not, strictly speaking, part of the Green Belt. The historic area of Westminster as outlined on the map "The Lower Thames" has been enlarged northward to incorporate more of the West End. Built-up areas shown on the maps are generalized from the Atlas of Britain and Northern Ireland (1963), by permission of Clarendon Press, Oxford. Green Belt after Ministry of Housing and Local Government, shown on Ordnance Survey base, Crown copyright. Various details of maps 2, 3, and 4 after J. T. Coppock and Hugh C. Prince (eds.), Greater London (Faber and Faber, 1964). All the Docks except the Tilbury Docks have been closed and are being redeveloped for other uses.

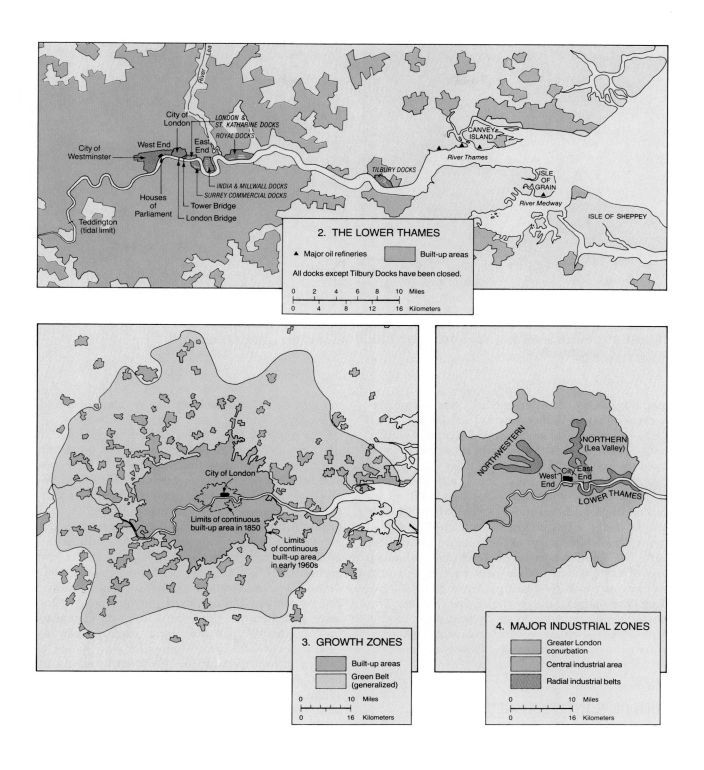

City of Westminster

West End

City of London

East End

LONDON & ST. KATHARINE DOCKS

ROYAL DOCKS

River Lea

Houses of Parliament

Tower Bridge

London Bridge

INDIA & MILLWALL DOCKS

SURREY COMMERCIAL DOCKS

Teddington (tidal limit)

TILBURY DOCKS

CANVEY ISLAND

River Thames

ISLE OF GRAIN

River Medway

ISLE OF SHEPPEY

2. THE LOWER THAMES

▲ Major oil refineries   Built-up areas

All docks except Tilbury Docks have been closed.

0 2 4 6 8 10 Miles
0   4   8   12   16 Kilometers

City of London

Limits of continuous built-up area in 1850

Limits of continuous built-up area in early 1960s

3. GROWTH ZONES

Built-up areas

Green Belt (generalized)

0   10 Miles
0   16 Kilometers

NORTHWESTERN

NORTHERN (Lea Valley)

West End

City

East End

LOWER THAMES

4. MAJOR INDUSTRIAL ZONES

Greater London conurbation

Central industrial area

Radial industrial belts

0   10 Miles
0   16 Kilometers

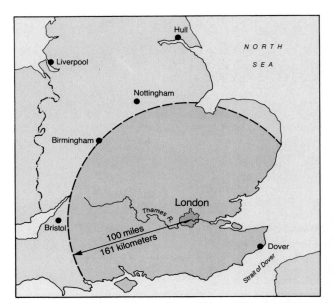
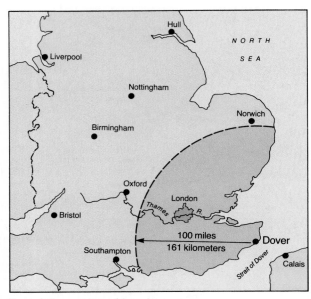

Figure 6.9   Colored areas lie within a 100-mile (161-km) radius of the center of London and Dover, respectively.

methods, such as *containerization* (loading large pre-packed metal containers instead of the individual items or cartons), have developed recently (see Fig. 5.9). These methods require large areas of relatively open ground along the waterfront for stacking and moving containers, and such areas cannot generally be found in the congested older port built up for earlier ships and methods. Instead, they must be sought in the areas downstream from the original port. Thus, a large share of the facilities in the older part of the London port have been closed, with much loss of employment, while new port areas, using modern methods that require fewer people, have been developing far out on the Thames Estuary. This story is being repeated in many port areas around the world. The extreme instance is supertanker terminals that are located well offshore and are connected to the land by pipelines.

LONDON'S INTERNAL GEOGRAPHY

The Internal Regionalization of Cities

It is well known and easily observed that cities possess distinctive internal geographies. As a city grows, various activities (functions) become more or less segregated into particular areas. The city comes to consist of a collection of recognizable "regions," such as commercial sections, industrial sections, and residential sections of different types. In actuality, each section of any size is generally distinguished,

not by a complete monopoly of any one function in it, but by a distinctive combination of functions.

The primary force behind this differentiation of the city into sections appears to be the economic competition of different functions—that is, of people and institutions desiring to carry on different activities—for locations within the city. This is a very complicated competition, because both the kinds of locations prized and the amount which can be paid for a given location vary from one function to another. For example, the kind of location desired by an industrial firm may not be the same as the kind desired by a householder looking for a place to live or by a company proposing to operate a department store, and these possible competitors will vary as to the amounts they feel able to pay for any given location. Theoretically, the competitor who stands to gain most from occupation of a given location can bid highest for it and is likely to occupy it. The location's value for various potential users is likely to be affected by such factors as topography, government policies, the types of development already present in the area and in adjacent areas, and, especially, accessibility. The factor of accessibility is generally important enough that the parts of the city most easily accessible to the greatest number of people—the central business district and to a lesser extent areas along major thoroughfares—tend to have the highest land and rental values. Since these various factors are subject to change (even topography can be changed extensively with modern equipment), the character of sections of the city may also change. In fact, the kind and rate of change going on within it is often one of the most significant characteristics of a section of the city.

Several generalizations concerning the typical arrangement of sections that cities develop have been proposed by scholars. Two of these, the *concentric zone generalization* and the *sector generalization,* are used in the present description of London.[2] The concentric zone generalization holds that cities become articulated into contrasting zones arranged as concentric rings around the central business district (CBD). Thus as one progresses outward in any direction from the center of a city, he or she should pass in sequence from (1) the central business district to (2) a zone of poor housing and light industry to (3) a zone of working people's housing to (4) a zone of middle-class housing to (5) a suburban zone of commuters' towns. The sector generalization, on the other hand, holds that the contrasting sections of a city tend to be arranged like wedge-shaped slices of a pie, with the inner points of the slices converging on the central business district. London's internal arrangement does not fit either generalization with any great exactness, and it may be that no city does so. Nevertheless, it is apparent that the metropolis is arranged in some respects and to some degree concentrically and in some respects and to some degree in sectors. The two generalizations offer useful viewpoints from which to survey London's historical development and present structure.

London's Inner Core: The City and the West End

However the structure of London is seen, that structure focuses, as in most cities, on an inner core which is the city's central business district. In the case of London, as not uncommonly in other cities, the inner core is not only the main business center of the present city but also the historical nucleus around which the present metropolis has developed.

The core consisted initially of two small cities on the north bank of the Thames. One was the walled port and commercial City of London, developed originally by the Romans. The other, a bit upstream to the west, was Westminster, which became the seat of the British monarchs during the Middle Ages (see chapter-opening photo of Chapter 1 and Figs. 6.8, 6.10, and 6.11.)

The City of London and Westminster grew together physically during the latter 17th century, and then—slowly at first, more rapidly with the advent of railroads in the 19th century, and still more rapidly in the 20th century—the present huge metropolis grew around them. As this occurred, they themselves developed as the heart of the metropolis and of the whole country. Together they occupy an area which extends for about 5 miles (8 km) from east to west along the Thames and is much narrower from north to south. One geographer assigns them, along with some immediately adjacent sections which have become part of the core, an area of about 11 square miles (28 sq km) and says of their development:

> Concentrated in these few square miles are the functions of the Crown and Government, Church, Law, Press, Finance, Banking and Insurance, the major commodity and wholesale markets, most of the institutions of the University of London, the teaching hospitals, the headquarters of the majority of Britain's leading business enterprises, nationalized industries and corporations, and the professional organizations and trade unions. Central London is also the nation's main cultural, entertainment, shopping and tourist centre and the focus of communication by road, rail, cable, and wireless.[3]

Although they function jointly as the central core of London, the two original cities have maintained separate and well-recognized identities. The resident population of the ancient City of London has declined almost to the vanishing point, while several hundred thousand people flow into "The City" each day to work. Primarily housing corporate headquarters and financial institutions, this part of London is possibly the leading center of private international financial transactions in the world. Modest building heights were long maintained by building regulations, thus allowing the 17th-century dome of St. Paul's Cathedral to be the visual focus (Fig. 6.10). But restrictions on height have been relaxed since

---

[2]For a discussion of these generalizations, see Raymond E. Murphy, *The American City: An Urban Geography* (2nd ed.; New York: McGraw-Hill Book Company, 1974). The sector generalization has been developed especially with reference to investigations of the expansion and migration of urban residential areas, but growth and arrangement of other functions in a linear or sectoral pattern have often been observed. Another well-known generalization of urban growth and structure, not used here, is called the *multiple nuclei generalization.* It was first proposed in Chauncy D. Harris and Edward L. Ullman, "The Nature of Cities," *Annals of the American Academy of Political and Social Science,* 242 (November 1945), 7–17.

[3]D. F. Stevens, "The Central Area," Chap. 7 in J. T. Coppock and Hugh C. Prince (eds.), *Greater London* (London: Faber and Faber Limited, 1964), p. 167.

Figure 6.10   The great dome of St. Paul's Cathedral, seen here from the Thames River, is a famous landmark in the City of London. Very evident in the view is a hodgepodge of building styles from different architectural eras. Many new buildings were thrown up hastily from the rubble left by German bombing in World War II. Critics complain that undistinguished designs have wrecked the skyline and sullied the magnificence of the cathedral.

World War II, and new office towers now dominate the skyline (Fig. 6.11). Among the many well-known institutions of The City are the Bank of England and the association of insurance underwriters and brokers called Lloyd's.

The part of the central core which has developed in and around Westminster continues to have, as it had from the beginning, a somewhat different character. This section, generally called "The West End," continues to be the seat of the British government, containing the Houses of Parliament, the government ministries, and the principal residence of the monarch at Buckingham Palace. It also shares some of the office development of The City. But it is much more diversified than The City, accounting for most of the other functions of the core mentioned in the previous quotation, plus sizable although declining residential areas, some of which are very fashionable. A tremendous variety of activities, architectural styles, and even ethnic groups is the keynote of the West End. Numerous public and private parks provide greenery and open space to an extent seldom found in the world's great cities.

Concentric Aspects of London

Seen from the point of view of the concentric zone generalization, London is comprised of a series of rings around the inner core. As one progresses outward from the core, each ring is progressively newer

as a part of the London agglomeration. In addition, population density reaches a peak in the ring immediately around the core and successive rings outward are less densely populated. As shown in the "Growth Zones" section of Figure 6.8, these rings may be summarized as follows:

1. *An area built up before 1850 and, along with the core, comprising the city as of that date.* Although the first railway reached London in 1836, this part of the city developed essentially before the railway age. Buildings and tracts of many periods are now intermingled, with large modern urban renewal projects set amidst housing first built in the 19th century and earlier. Population density is the highest in London, although it is decreasing because of the migration of residents to more outlying zones. Some of this zone was originally developed with mansions for the well-to-do and some of it with row housing for workers. One-time mansions have been transmuted into apartment buildings, hotels, shops, or offices. Mixed with this dense residential development are many warehousing and wholesaling activities, as well as railroad lines, which are rather closely spaced here as they converge toward a ring of terminals in and around the edge of the inner core.

The part of this zone lying immediately east of The City is often referred to as the East End. Its character contrasts very greatly with that of both The City

Figure 6.11   Since World War II, a forest of new skyscrapers has sprouted in the ancient City of London. A comparable cluster has emerged in a district of central Paris. Just as in Manhattan's Wall Street financial district, the demand for office space at the center of affairs has propelled buildings upward. The congested area of "The City" shown in the photo is a major world center of private financial and commercial decision-making.

and the West End. The East End developed very early as a densely populated industrial quarter. It was heavily bombed in World War II and has witnessed very extensive urban redevelopment. Sharp reductions in resident population and population densities have occurred as workers have moved toward London's outskirts.

This inner zone, especially the East End, was the locale of much of London's manufacturing until quite recently. Small factory and shop industries in such lines as clothing, furniture, and precision instruments were strongly represented. But as a means of lowering urban densities, it became the policy of successive British governments after World War II to promote and subsidize the relocation of both manufacturing and people from the metropolitan center to the suburbs and other regions of the country. This policy was launched at a time when noncentral locations were offering increasing advantages for industry with regard to such factors as cheap space for plants and unimpeded access for trucks. The result was an all-too-successful decentralization of manufacturing and its jobs, to the point where the government is now trying to relieve central-area unemployment with policies that it hopes will favor the reindustrialization of central London and other inner cities in Britain.

2. *A zone built up in the past century, as railroads and then, to some extent, buses and automobiles extended commuting ranges, and as outlying industrial and business centers grew.* The map shows that the continuously built-up area of the city, roughly circular in form, was approximately 10 miles (16 km) in diameter in 1850 but is now approximately 25 miles (40 km) across. Thus the greater part of the present continuously built-up area, called the Greater London Conurbation, is a product of the past century; in fact, about half of it is a product of the period since World War I, as the built-up area approximately doubled during the interwar period.

This increase in the built-up area was not synchronized with the increase in the city's population. Population growth in the conurbation during the past century took place mainly in the earlier part of the period. From 1851 to 1911 the population of the conurbation increased from 2.7 million to 7.3 million. From 1911 to 1939 it increased only from 7.3 million to 8.7 million, while the built-up area doubled in size; and since 1939 the conurbation's population has actually decreased to around 6.8 million. Thus continued expansion of the periphery of the city was accompanied by progressive depopulation at the center. The new building on the periphery was done to standards of much lower density than in older residential sections, with fewer dwellings per

unit of area and more open spaces retained. Population decline began in The City during the 1850s and has spread outward until, after 1951, it characterized most of the conurbation. The zone of increasing population has moved outward ahead of the zone of decrease and now does not lie within the conurbation at all but in surrounding suburbs and satellites which are part of a London Region but are not part of the continuously built-up conurbation.

3. *The Metropolitan Green Belt.* This ring lies just outside the conurbation, that is, just outside the continuously built-up area. The ring varies in width but is generally on the order of 15 to 25 miles (24 to 40 km) wide. Within it, strong restrictions are enforced against new development in order to check the further spread of the conurbation and to preserve large areas of open countryside adjacent to it. Green belt proposals began to be made in Britain in the later 19th century, and implementation by London authorities began in the 1930s. After World War II the idea of green belts around British cities became part of national planning policy. Local authorities are encouraged to submit plans for green belt areas; if these plans are approved by the government, the local authorities may then control land uses and development and prevent further building in the designated areas. There is much urban development within London's Green Belt, as many suburbs and satellite towns already existed when planning controls were first applied. There has been a strong tendency to expand the belt but not to be completely rigid about shutting off further development. Local governments within the Green Belt take a prominent part in such decisions, within the frame of national guidelines.

The Metropolitan Green Belt has effectively checked the spread of the built-up area of Greater London, which has expanded little since 1939, and has apparently maintained more open space than would be present without the controls. But population within the belt has continued to increase fairly rapidly. Some of this increase was planned in order to accommodate the desired thinning of Greater London's population. To further the latter objective, the British government acquired land after World War II and built nine new satellite cities in or near the Green Belt. These were designated New Towns (Fig. 6.8), and now have a combined population of around 625,000. Costly commuting to London has been lessened by the success of these planned cities in attracting industry and other employment. Some of the other large British cities also have New Towns in their environs.

4. *The "Outer Southeast," or territory surrounding London outside the Green Belt.* This outer fringe of the

Figure 6.12 London has close relationships with a host of important service facilities spread over southeastern England. Among them are Oxford University and Cambridge University ("Oxbridge"), where a large share of Britain's leaders have been trained. Pictured here are King's College Chapel and adjacent buildings at Cambridge. From the chapel, a well-known Christmas Eve service is annually broadcast and telecast to the world.

metropolitan complex has become the main area of rapid growth in all of Britain. Large numbers of its residents commute to work in London, although a great deal of employment is provided by urban places (such as Oxford and Cambridge; Fig. 6.12) within the outer zone itself. Two important circumstances that have fostered the growth of the Outer Southeast and its commuter traffic to London are (1) the continued growth of office employment in London, accompanied by a decline in London's population and work force; and (2) the restrictions on development closer to London in the Green Belt. The British government, desirous of lessening urban congestion by encouraging the outward shift of some of London's population, has subsidized the development of selected places in the Outer Southeast that are called Expanded Towns. These places have contractual arrangements with London to absorb people willing to relocate from the conurbation. To provide jobs in order to lessen long-distance commuting, the government also encouraged industries to shift from London to the Expanded Towns. Recently, however, the further suburbanization of London industry has been seen as hurtful to London itself.

Sectoral Aspects of London

To visualize London and its region as a structure of concentric rings points up some aspects of the metropolis but obscures others. This is because some things are arranged more in a pattern of sectors radiating outward from near the metropolitan center than in a pattern of rings. Particularly prominent among these sectors are three belts of industrial development.

1. *The lower Thames industrial belt* is intertwined physically with the port of London and has close functional relations with it. Both the port and the associated industrial activities have been shifting downstream, away from central London and out toward deeper water and greater space. Until very recently a large share of London's shipping was accommodated in and near central London in a series of docks: immense enclosed basins dug in the alluvium along the river. But most of the docks have now been closed, and new container-ship and other facilities have been built downstream. Many of the industries in the lower Thames belt are those which might be expected in a major port—notably the processing of imported commodities such as oil or sugar. But the belt also includes engineering industries which have grown out of former shipbuilding activity along the river, and nuisance industries, such as certain chemical plants, which foul the air and are typically located downstream in relatively open and isolated areas.

2. *The northwestern industrial belt* has also developed in a strategic position with respect to transportation. It extends outward through the side of London which lies closest to the industrial Midlands, thus combining immediate access to London's market and labor with superior access to Britain's other economic center of gravity around the Pennines. Clusters of factories are strung along the numerous railroads and major highways which connect London with the Midlands and industrial areas beyond, and residential development is interspersed between the factory clusters. Small and medium-sized plants producing consumer goods are the most common, although these plants vary widely in scale of operations and type of product.

3. *The northern industrial belt* follows the valley of the Lea River, which enters London from the north and flows into the Thames about 2 miles (3 km) east of The City. Poorly drained, marshy, and subject to flooding, this valley was avoided by housing development as London grew, leaving it available for industrial sites with little competition. It has canal transportation as well as roads and railroads. Industrial development is especially intense in the lower end, but extends far up the valley to the north, where it is interspersed with large reservoirs and recreational areas such as parks and playing fields. Working people's housing lines both sides of the valley on slightly higher, better-drained land.

The industrial sectors just described are Greater London's main concentrations of manufacturing, but various smaller industrial areas are scattered through the conurbation. However, the sectors that lie between the three main industrial belts, and most of the area south of the Thames, tend to be predominantly residential. They comprise a sea of housing, thinning and of generally newer construction toward its outer edges, dotted with shopping areas and cut through by commercial streets. This housing tends to be poorer in quality, more crowded, and more densely occupied near the industrial sectors. In contrast, exceptionally prized residential areas occupied by more well-to-do people often occur on heights where the Thames Valley, in which the Greater London Conurbation lies, gives way to somewhat higher ground to the north and south. Two interruptions in this residential pattern are noteworthy. One is the sector or belt along the Thames upstream from central London in the direction of the royal residence at Windsor. This area is marked by unusual expanses of park land, numerous large reservoirs that store Thames water for the city, and intensive recreational use of the river. The other is the borough of Croydon, in the southern section of the conurbation, where the first considerable center of office employment outside of central London has been developing.

FOOD AND AGRICULTURE IN BRITAIN

One of the traditional characteristics of the United Kingdom has been its large dependence on imported food. In fact, it was the largest net importer of food in the world until the 1970s, when a few countries such as West Germany and Japan surpassed it. Countries all over the world are increasing their dependence on food purchased in international markets, due to such factors as rising affluence, population increase, failing agriculture, and cheaper shipping rates. Consequently, Britain's role as a market for the world's agricultural surpluses has become less unique.

The necessity for importing food into Britain stems in part from unfavorable natural conditions for agriculture. In most of the country, crop production is unrewarding or impossible (Fig. 6.13). Poor soils, steep slopes, cool summers, and excessive cloudiness and moisture are the principal handicaps. Highland Britain suffers the most on all these counts, but various parts of Lowland Britain are also affected. Comparatively slight rises in elevation bring sharp decreases in temperature and increases in precipitation.

In Highland Britain, the warmest month generally averages below 60°F (16°C), and precipitation is nearly always over 40 inches (c. 100 cm) and in many places more than 60 inches (c. 150 cm) a year. Moorland, often waterlogged and boggy, is estimated to cover more than a quarter of the land in the island of Great Britain (see chapter-opening photo). In terms of economic use, such land is mainly unimproved natural pasture ("rough grazing"), generally with a low density of both animals and people. Moorland is used primarily to graze sheep, although small numbers of beef cattle also are grazed.

Land and climate are more favorable to agriculture in Lowland Britain (Figs. 6.3 and 6.14), especially in the sunnier, drier, more level, and more fertile east. Even in Lowland Britain, however, much of the land is in pastures and meadows, although these are usually sown and tended and have a high carrying capacity. Somewhat under one third of the land in the United Kingdom is regularly plowed and planted in tilled crops (29% by one count). Eastern England, the area least affected by the wetness and cloudiness induced by winds off the Atlantic, is the largest area where arable land predominates over grass.

Britain's dependence on imported food arose with the growth of its population during the 19th century, from about 11 million to about 38 million. In earlier times the land had been adequate to feed the smaller population and it had afforded an export of wool and sometimes of grain. During the early part of the Industrial Revolution, improvements in agriculture amounted to an accompanying Agricultural Revolution. Among the advances were (1) new implements, (2) new crops such as potatoes and turnips, (3) new crop rotations which eliminated the frequent fallow periods of previous times, (4) increased application of fertilizers, (5) scientific stock breeding, and (6) the extension of crop production into marginal areas. Food production increased so much that it almost kept pace with population growth for many decades. But through the 19th century, food imports increased steadily, especially imports of wheat, which was basic in the diet and relatively easy to transport. A decision to rely on the cheapest wheat available, imported or British, was embodied in the repeal of the Corn Laws in 1846 ("corn" means wheat or grain in Britain). The Corn Laws had forbidden the importation of wheat until the price of British wheat rose to a specified point, thus protecting the British producer.

Although imports continued to rise, the full effect of the free importation of grain was not felt until the 1870s and subsequent decades. By that time trans-

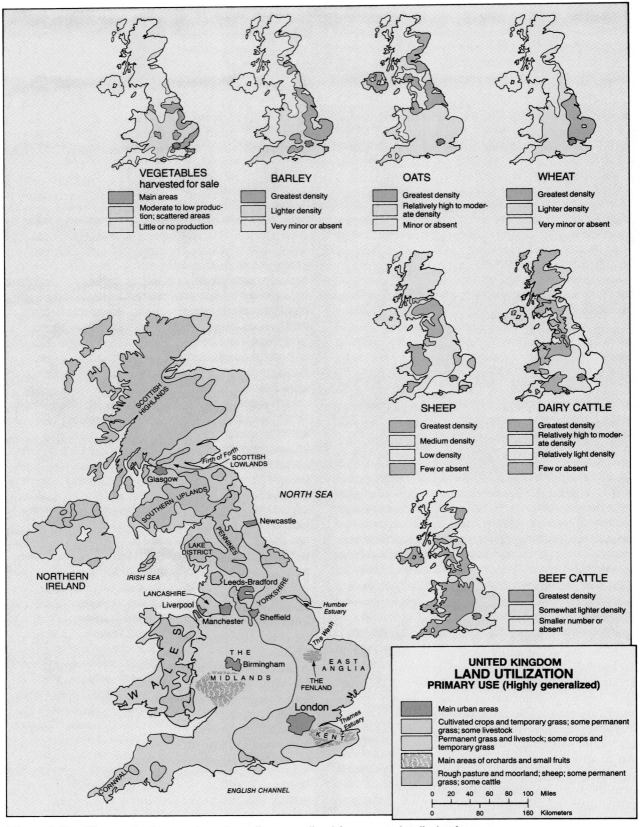

Figure 6.13 *All maps in the panel are broadly generalized from very detailed reference maps in* The Atlas of Britain and Northern Ireland *(Oxford, Clarendon Press, 1963, by permission). Many small tracts of forest and woodland are not shown. The area shown on the map as the Fenland, bordering the broad but shallow east-coast bay called The Wash, is a tract of low, flat land that once was marsh or sea bottom but has been reclaimed for agriculture by diking and artificial drainage. The resulting farm land is the most fertile and productive area of its size in the British Isles. The Fenland, devoted largely to cultivated crops rather than to grass and livestock, is especially notable for vegetables, potatoes, fruits, wheat, and sugar beets.*

Figure 6.14 *Intensive animal husbandry, seen in the section of eastern England known as East Anglia. Although England's crop-growing is heavily concentrated in East Anglia and other areas along the eastern margins of England, livestock are also important. The cattle herd in the foreground is composed of mixed breeds raised in other areas but brought here for final fattening. In the background is a barley field; beyond that, trees along hedgerows give the countryside a wooded appearance.*

portation facilities had become adequate to pour floods of grain into Britain cheaply from new producing areas in the Americas and Australia. The price of wheat fell to levels at which many British farmers, handicapped by relatively high land and production costs, could not survive. Widespread depopulation of rural areas and a drastic decline in British wheat acreage and production followed. Most other crops were also affected, although to a lesser degree.

This disaster to British cultivators accentuated a major characteristic of the United Kingdom's agriculture: its high concentration on grass farming for milk and meat production. Much of Britain had always been primarily pastoral, but now much of the land on which other crops were no longer profitable reverted to grass, for which the climate was relatively favorable. Meat and, to an even greater degree, fresh milk are more difficult and expensive to transport than grains and many other vegetable products; consequently, they afford local producers a degree of advantage over competitors located a greater distance from the market. Thus, grasslands, dairy and beef cattle, and sheep have become the dominant elements in British agriculture. Land in crops is used primarily to produce barley or other supplementary animal feeds. Less than one tenth of the United Kingdom is used for producing vegetable products directly consumed for human food—mainly wheat, potatoes, sugar beets, vegetables, and fruits—and substantial portions of some of these are fed to ani-

mals. Even so, the United Kingdom imports large quantities of meat and dairy products as well as other foods.

This dependence on imported food is not due to lack of efficiency in British agriculture but to the high density of population and the relatively high plane of living. Crop yields and the output of animal products per unit of land are among the highest in the world. Output per farm worker in Britain's highly mechanized agriculture may well be the highest in Europe. Total food production has increased sharply since World War II, mainly an outcome of various government policies aimed at reducing imports. Results have been achieved principally by inducing farmers to increase the acreage in feed crops, mainly barley, at the expense of grassland, since a given acreage of such crops can generally feed more animals than the same acreage of grass, although the feed crops may be more expensive to produce. There has also been a large increase in wheat production. These expansions have required various subsidies to farmers and have been expensive. However, despite the high cost in subsidies and food prices, Britain's membership in the Common Market (since 1973) probably will continue to foster the tendency to expand British agricultural production. Up to now, the agricultural policy of the Common Market has emphasized subsidies to agriculture, including high import tariffs on agricultural products entering the Market from outside. As long as this policy holds, British farmers will continue to receive incentives in the form of high and protected prices, and the British people, like the other peoples in the Common Market, will eat more expensive food, produced to a greater degree within the home country or, at least, within the EEC.

TWENTIETH-CENTURY DIFFICULTIES: THE DRAG OF OLD INDUSTRIAL AREAS

The United Kingdom has experienced an almost constant succession of economic difficulties since World War I, punctuated by military and political struggles which have tended to accelerate the country's decline in power and influence. The economic difficulties have largely been associated with the declining fortunes of the very industries upon which former growth was based, especially coal mining, cotton textiles, shipbuilding, and iron and steel production. All major industrial districts in Britain except the London area have recently been classed, both by the British government and the Common Market, as areas of chronic economic distress. A number of these areas have been chronically depressed most of the time since World War I. Overall, the economic downturn in Britain has been relative rather than absolute: There has been quite a substantial increase in standard of living over the long term, with a few sharp reversals interspersed; but except for a few promising recent years in the 1980s, gains have been much slower than in most industrial countries. Britain has been left in a middle position among European countries with regard to per capita income, rather than in the top position it still enjoyed at the outbreak of World War II. At various times, the slow growth in overall wealth has been accompanied by a great deal of unemployment in the industrial and mining districts. The British problem of slow growth and faltering old industrial areas may have relevance for the United States, since some scholars believe that the latter nation may be commencing to follow the painful British path.

Essentially, the economies of Britain's older industrial districts have become shaky because their basic industries have been increasingly unable to compete in world markets. Britain's dependence on trade requires that British industries be competitive internationally if the country is to prosper. But for the past 70 years or so, the older industries—especially coal, cotton textiles, shipbuilding, and iron and steel—have had much difficulty in marketing their products. Consequently their sales, output, employment, and relative wage levels have tended to decline drastically, although the patterns and extent of decline have differed somewhat from industry to industry. Whole communities in the coal-field industrial areas have experienced chronic distress because they were so specialized in one or two faltering industries. When this occurs, continuing unemployment can be avoided if the community can develop or attract new industries rapidly enough or if it loses workers rapidly enough by migration to areas where employment is growing. Britain's old industrial areas have witnessed both the development of new employment sources and migration to growing areas—mainly in southeast England around London or in such overseas countries as Australia or Canada. But most of the time neither process has been rapid enough to avert unemployment and its depressing effect on wages. To complicate the problem, even some of Britain's industries developed in the 20th century—such as the automobile manufacturing of the Midlands and southeast England, and the aircraft industry of southern England—have been experiencing the same competitive difficulties.

Industrial difficulties in older industrial areas and countries, of which Britain is the most outstanding example, are a product of many circumstances, but the most vital circumstance—at least for the United Kingdom—is the increasing degree of international industrial competition that has developed in the 20th century, especially since World War II. Intensified competition is partly the logical outcome of the continued spread of industrialization around the world. It has been accentuated in recent decades by the rapid improvement of communications and transportation. This has brought a precipitous drop in the costs of maintaining contact with distant production facilities and markets and of shipping materials, subassemblies, and finished products for long distances. Distance from competitors no longer offers much protection, and distance from markets is no longer much of a handicap. Competition has also been accentuated by the systematic and scientific development of new products having the potential to compete with and replace older ones. Probably the outstanding 20th-century example is the replacement of older means of transport by the automobile and truck and the attendant replacement of coal by oil. Another is the development of synthetic fibers, made from wood or petroleum, to compete with cotton and wool.

In the strenuous competition for international markets, older industrial areas tend to lose to newer ones. Some of the major reasons follow:

1. *Loss of a resource advantage.* A common instance is a change for the worse in the significance of local coal. Location on a coal field was the main basis for the development of many older districts, which eventually lost the advantage because of (a) depletion of the local coal, (b) the rise of lower-cost coal fields elsewhere, or (c) replacement of the coal by oil or natural gas. Great Britain was fortunate in the discovery and development of major oil and natural gas reserves in the British portion of the North Sea during the period from the 1960s onward. But although this new resource wealth helps the country as a whole, it does not solve the problem of unemployed coal miners.

2. *Cheaper labor in newer areas.* In older districts there has been time for unionization to occur and to increase wages, whereas a new area is likely to have an unorganized work force to whom even low wages seem a great improvement. With lower wages, prices can be lower.

3. *Technological advantages of newer areas.* Such areas can be developed with the newest and most efficient technology. In older areas there is a large investment in older plant and equipment, which is difficult to write off, scrap, and replace with the latest technology. In some cases strong unions may oppose the new technology because it provides fewer jobs.

For example, in the British iron and steel industry much new equipment has been unused or "over-manned" with more workers than necessary because of union resistance to the job loss involved. At times, British governments have had to subsidize losses by this industry (at the taxpayers' expense) in amounts of hundreds of millions of pounds.

4. *Technological unemployment resulting from the efforts of older industries to survive.* Even if an industry of an older area survives the competitive struggle, unemployment and hardship are apt to be visited upon the community, because the only way the industry is likely to survive is by increased mechanization, automation, and the use of robots. Production may be maintained or even expanded by this tactic, but workers will be fewer in number.

5. *Conservatism of management.* After years or decades of success, there is often a tendency for the management as well as the workers in an older industrial area to be somewhat resistant to necessary changes. There is a natural tendency to be complacent in the belief that procedures which have worked in the past will continue to succeed in the future. Many charges of this nature have been leveled against British management.

The expanding employment sector in the United Kingdom, as in other long-industrialized countries, has become the service sector—a term embracing all workers not directly involved in producing material goods. Apparently this trend must continue if an increase in employment is to be enjoyed: Agriculture has fallen to low levels of employment, and industrial jobs are being lost to competition and to automation. Increasingly, Britain—along with other "industrial" nations—is a country of white-collar workers. The expanding service sector which employs them is highly varied in its make-up, some of its major parts being government, finance, transportation, insurance, health services, management, sales, education, research, personal services, the professions, tourism, and communications media.

There are many unanswered questions about this service-sector-oriented, "post-industrial" society toward which the industrialized Western nations are evolving. Some of these questions follow:

1. Will these societies be able to maintain themselves industrially with lightly manned, highly automated manufacturing industries?

2. Can the service industries provide an adequate supply of exports needed in international trade? The United Kingdom has long supported itself in part by "invisible exports" such as capital, financial services, insurance services, ocean shipping and airline services, and

tourism; and many other Western nations are now heavily involved in this kind of activity. But industrial exports are still important to them.

3. Can service jobs be expanded rapidly enough to supply high employment in the face of declining industrial employment?

4. Can service jobs be made productive enough to provide high incomes to those holding them? There is reason to doubt the ability of many parts of the service sector to increase productivity at rates comparable to those of large-scale and highly mechanized agricultural or manufacturing industries. If such doubts prove justified, the countries with a high dependence on services are likely to experience slow economic growth and chronic problems.

5. Will new service employment locate in the afflicted old industrial areas where it is especially needed, or will it mainly grow elsewhere? In Britain the main growth of such employment, as with employment in newer forms of manufacturing, has taken place in London and surrounding parts of southeast England, rather than in the depressed coalfield industrial areas. This has accentuated the sharp contrast between the relatively prosperous London region and the rest of the country. The British government has been attempting to push the development of new office complexes in needy cities away from London, but such efforts have not significantly changed the overall situation.

The answers to the foregoing questions will go far toward determining the future of Western countries both in and outside of Europe. The United Kingdom will be a key country to watch because the process of change is relatively far along there and is proceeding rapidly.

IRELAND

Ireland is a land of hills and lakes, marshes and peat bogs, cool dampness and verdant grassland. The island consists of a central plain surrounded on the north, south, and west by hills and low, rounded mountains. The climate is marine to an extreme, with average temperatures ranging from a little above 40°F (4°C) in January to about 60°F (16°C) in July, and rainfall from over 100 inches (c. 250 cm) in western uplands to about 30 inches (76 cm) in eastern lowlands. Glaciation blocked drainage lines and has left many lakes connected by sluggish rivers, of which the most famous is the River Shannon. Marshes and waterlogged moorland are common, and peat bogs cover one seventh of the island. Although nearly all the natural forest is gone, the cool dampness fosters luxuriant grass, giving the countryside its prevailing and proverbial green. The island's open, varied, and verdant rural landscapes are a major asset to a growing tourist industry.

THE IMPACT OF NATURE AND HISTORY ON IRISH INDUSTRIALIZATION

The Republic of Ireland, which occupies a little more than four fifths of the island, is in the midst of a transition from an agricultural to an industrial economy. Industrialization was long delayed and only began to build up rapidly in the 1950s and 1960s. Manufacturing has become a larger employer than agriculture and supplies more exports by value. But the transition is far from complete. Agriculture is still more important as an employment sector here than is true of most countries in western Europe, and the Republic is a considerable *net* exporter of agricultural commodities (exports more than it imports) and a *net* importer of manufactures. Another indicator of the transitional nature of this still-industrializing economy is the relatively low income level of the population. In per capita income Ireland ranks near the bottom of Europe's countries west of the Iron Curtain and is exceeded by East Germany and Czechoslovakia east of the Curtain. Switzerland's per capita income is well over three times that of Ireland.

Both nature and history contributed to the long delay in Irish industrialization. Some of the delay was undoubtedly due to an almost total lack of the mineral resources which facilitated early industrial development next door on the island of Great Britain. The island of Ireland has practically no coal, aside from an abundance of the very low-grade coal called peat. Nor were other important minerals found in quantity until the discovery of lead, zinc, and a few others after 1960. But Ireland's delayed industrialization is also a consequence of its long subordinate relationship to England. After centuries of intermittent intrusion and strife, England completed an effective conquest of Ireland in the 17th century and subsequently made the island a formal part of the United Kingdom. But Ireland's status was actually colonial, and the island was governed for about two centuries under probably the harshest rule imposed on any British colonial territory. Most of the land was expropriated and divided among large estates held by English landlords, who were often absentees living in Britain. The Irish peasants became tenants on these estates and were reduced to

extremes of poverty, with the proceeds of their labor drained off—and often out of Ireland—in extortionate rents and taxes. Irish trade and industry were penalized, restricted, and nearly destroyed at times by government measures designed to favor British competitors. Ireland, especially the part now constituting the Republic, became a land of deep poverty, sullen hostility, and periodic violence.

The Protestant North of Ireland—now known as Northern Ireland and still a part of the United Kingdom—was somewhat less handicapped by British policies than was the Catholic South. There was a strong element of religious discrimination in British rule, Catholics being regarded as potentially disloyal to the Protestant crown. But Northern Ireland had a Protestant majority, whereas the rest of the island was almost universally Catholic. The Protestants were the descendants of Scottish Presbyterians and other British Protestants who were brought to the north of Ireland in the 17th century under the auspices of the British crown. They were intended to form a nucleus of loyal population in a hostile, conquered country. Protestant entrepreneurs were able to develop shipbuilding and linen textile industries in and near the North's main city, Belfast (700,000). These industries have declined recently, but Northern Ireland, which has the lowest per capita income of any of the main sections of the United Kingdom, has been attracting new industries seeking cheap labor. Thus industrialization continues, despite the North's recent history of political violence.

In the Republic of Ireland, the recent surge of industrial development has been accomplished under a program designed to attract foreign-owned plants by a combination of cheap labor, tax concessions, and help in financing plant construction. Hundreds of plants have been attracted, although most of them are rather small. They are owned mainly by American, British, and West German companies, and their major products are processed foods, textiles, office machinery, organic chemicals, and clothing. Most of the new plants have been built in and around the two main cities—Dublin (1.1 million) and Cork (200,000)—or in the vicinity of the Shannon international airport in western Ireland. Power to support this development is provided by imported oil, small hydroelectric installations, and electric generating stations that burn peat.

AGRICULTURAL CHARACTERISTICS OF THE IRISH REPUBLIC

With 13 percent of its labor force on farms (1985), the Republic of Ireland still depends on agriculture to a marked degree. The same physical and climatic handicaps to cultivation that affect Great Britain are present but to an even greater extent. Only about 14 percent of the Republic is classed as arable land, 70 percent is pasture, and most of the rest is wasteland. Grazing and feeding of cattle is the main element in the agricultural economy, as it is in Northern Ireland. Much of the arable land is devoted to fodder crops, which are supplemented by imported feeds. Meat, dairy products, and live feeder cattle for fattening elsewhere are major Irish exports, although agricultural exports are now surpassed in value by exported manufactures. In addition to supplemental grain imported to feed animals, there are imports of wheat and many other human foods produced in inadequate quantities on Ireland's farms or not at all.

Although the large estates of the period of English rule are gone and the Irish farmer now tends to be a landowner, there is marked rural poverty. Farms tend to be both small and undermechanized. This is especially true in western Ireland, where a particularly rugged and wet environment is inhabited by an unusually dense farm population. The high density originally stemmed from flight to the region to escape the full weight of English exactions. In addition to rural poverty, another consequence of Ireland's high dependence on agriculture is the continued close economic relationship of the Republic with the United Kingdom. Britain's need for large agricultural imports, together with its physical proximity, has long made it a logical market for Irish agricultural products. About one third of Ireland's exports are sold to Britain, and about two fifths of its imports, mainly manufactures, come from there (middle 1980s).

REPOPULATION

Ireland's population history is almost unique. Few of the world's countries have smaller populations today than they had a century and a half ago, but the island of Ireland has only about 5 million people now (3.5 million in the Republic; 1.6 million in Northern Ireland) compared with 8.2 million in 1841. In the 18th and early 19th centuries, its population increased very rapidly on the precarious basis of one newly introduced subsistence crop, the po-

tato. In the late 1840s, this crop failed for several years in a row, and the "potato famine" which resulted claimed the lives of nearly a million Irish people in 5 years, while another 1½ million fled the island, mainly to the United States, in the decade following 1846. A pattern of heavy emigration was established which continued for more than a century. The United States continued to receive emigrants, and England became a major destination as well. Between 1841 and 1961, the population declined in 10 out of 12 intercensal decades, as emigration outweighed natural increase. But this pattern was reversed in the 1960s. A fairly rapid population increase set in and has continued, although high unemployment and considerable emigration are still characteristic.

PARTITION AND RELIGIOUS VIOLENCE

Armed resistance regained Ireland its independence in 1921. But the Protestant majority in the North did not want to become a minority in a heavily Catholic united Ireland. The result was an agreement whereby the present Northern Ireland remained part of the United Kingdom. This, however, left about one third of Northern Ireland's people as a Catholic minority in a unit where the Protestant majority continued to practice a repressive discrimination in many areas of life. For half a century, the result was sporadic terrorism—mainly by the underground organization called the Irish Republican Army (IRA)—in an attempt to impel Britain to give up Northern Ireland. In 1969, the tensions in Northern Ireland spawned a continuing series of mass Catholic protests against discrimination, accompanied by repressive government actions and campaigns of terror by one wing of the IRA (the Provisional Wing or "Provos") and by extremist Protestant groups. British troops were sent to occupy the area. Through the 1970s and into the late 1980s, the situation was little changed as the punishing sequence of riots, assassinations, bombings, sporadic firefights, imprisonments, and prison hunger strikes continued. The violence has occasionally extended to Dublin, where the Irish government has not supported the IRA campaign, and to London. A poll some years ago indicated that about one fifth of the people of the Republic were in sympathy with the IRA.

CHAPTER 7

France and Germany in the Intricate Geography of West Central Europe

The Palace of Versailles near Paris is a reminder of the centuries when Royalist France was Europe's strongest power. The photo shows an expensively decorated section of the huge building constructed in the 17th century by King Louis XIV.

T HE west central portions of the European mainland across the English Channel and North Sea from the British Isles are occupied by a group of highly developed industrial nations that have long had close relations with each other. Germany[1] and France are the largest nations in the group, and the only ones which have ranked as great powers in the 20th century. Four of the remaining countries—the Netherlands, Belgium, Luxembourg, and Switzerland—are found in a historic buffer zone separating the two major nations. Austria, although it is not in the buffer zone, may be included in the group because of its close proximity to both West and East Germany, its extensive trade with West Germany, and its close relationships with Germany, both culturally and historically. Monaco and Andorra, adjoining France, and Liechtenstein, between Switzerland and Austria, are relatively insignificant microstates with a quasi-independent status; proximity alone entitles them to be numbered with the countries of this group.

Even for complex Europe, the geography of West Central Europe is unusually intricate. Spatial variations are great, layers of history are encountered at every turn, and functional linkages are extraordinarily complicated. The individual countries differ sharply in their human geographies and national perspectives, and the whole area is differentiated into a mosaic of regions and localities exhibiting great physical and cultural variety. Interwoven with the whole is a tight mesh of transport and communication links, together with international political relationships reflecting the experience of many centuries. In this chapter the geographic intricacy of West Central Europe will be examined country by country, but certain attributes that characterize the countries as a whole will be noted at the outset. Among the most striking are topographic variety, economic interdependence, historical antagonism, and a burgeoning cooperation.

The physical area occupied by the countries of West Central Europe may be thought of as three concentric and widely different arcs of land: (1) an outer arc of lowland plains bordering the Bay of Biscay, English Channel, and North and Baltic seas; (2) a central arc of hills, low mountains, and small plains; and (3) an inner arc consisting of the high Alps mountains, with another high range, the Pyrenees, offset to the west (maps, Figs. 7.1, 5.7). France and Germany include portions of all three arcs. The Netherlands and northern Belgium lie within the outer arc of plains, southern Belgium and Luxembourg are in the central hilly zone, and Switzerland and Austria lie principally within the inner arc of Alpine mountains. However, the main populated areas of Switzerland and Austria are in the southern fringes of the central hilly arc.

In the 19th and early 20th centuries, the largest coal and iron ore deposits in Europe (Fig. 7.2) provided bases for large-scale industrial development. Consequently, a series of great industrial agglomerations ("major industrial concentrations") now stretches from France to Poland along the main axis of mineral deposits (Fig. 7.1). Many individual industrial cities lie outside the main clusters, and some of these—such as Paris—are sufficiently important to be considered "major industrial concentrations" in their own right. In the typical agglomeration, coal mining (iron mining in Lorraine) and iron and steel production have been economic foundations, with chemicals, textiles, and heavy engineering also common. Interchange of resources and products has been fundamental to this development, although often hindered in the past by political antagonisms among the countries involved. After World War II, the need for interchange to promote economic recovery and progress led to the elimination of trade barriers within the framework of the European Coal and Steel Community and the more extensive European Economic Community (Common Market), both formed initially by France, West Germany, the Benelux countries, and Italy (see Chapter 4).

Although the industrial agglomerations based on coal and iron ore are still of utmost importance to the economies of their countries, they have become problem areas in recent decades. Coal production has declined in the face of competition from imported oil, and, to some extent, lower-cost coal imported from the United States. Lorraine iron ore has been replaced, in markets other than the Lorraine-Saar district itself, by higher-grade ores transported with increasing cheapness from distant sources. Steel mills have been automated to use significantly less labor, and new mills have been built in coastal locations to use imported materials. Many thousands of jobs have been lost, and the attraction of new industries is a major concern of all these areas.

---

[1]In this chapter, the name "Germany" will be applied to the German nation prior to the end of World War II, or to the two German republics of today taken together, whereas "West Germany" (the German Federal Republic) and "East Germany" (the German Democratic Republic) will be applied to the political units into which Germany was separated after World War II.

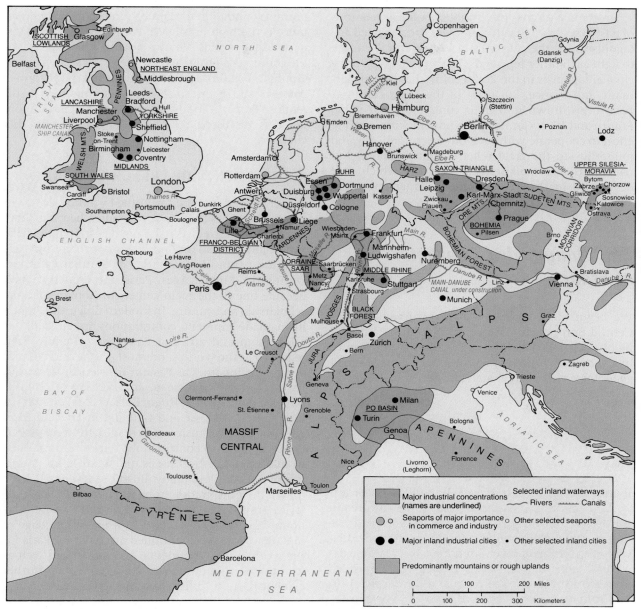

Figure 7.1 Industrial concentrations and cities, seaports, internal waterways, and highlands in West Central Europe. Older industries such as coal mining, heavy metallurgy, heavy chemicals, and textiles cluster around highland margins in the congested districts shown as "major industrial concentrations." Local coal deposits provided fuel for the Industrial Revolution in most of these districts, which have shifted increasingly to newer forms of industry as older industries have declined. But the new and more diversified industries have also proliferated in metropolitan London, Paris, and smaller industrial cities away from the old "major industrial concentrations." Relative sizes of dots and circles indicate rough groupings of industrial cities and seaports according to importance. Rivers and canals shown on the map are navigable by barges and, in some instances, by ships.

In the meantime, the troubled political relations of past times have greatly improved. Deep-seated antagonisms led the countries of West Central Europe to fight each other repeatedly in past wars. The principal antagonists in the 20th century have been Germany and France, but most of the other countries have been drawn into their contention and strife. France and West Germany are now making a strong effort to cooperate with each other economically and militarily—to the great relief of many nations, but particularly the small Benelux countries, which were devastated by the fighting among larger powers in World War II.

The Many-Sided Character of France

We begin our country-by-country survey of West Central Europe with France, an old, important, and highly developed nation where the play of history in a varied environment has produced a many-sided geography rich in regional nuances and historical associations.

FRANCE'S INTERNATIONAL IMPORTANCE

France has long been one of the world's most important countries. It is the largest country by area in Europe (see Table 4.1) and is Europe's leading nation in value of agricultural output. In the preindustrial age, the manpower supported by a large agricultural area helped make France the dominant power on the European mainland during the 17th and 18th centuries and part of the 19th. Present-day France remains one of the "giants" of Europe in population, although it is rather sparsely populated by the crowded standards of West Central Europe (see Table 4.1). The country is one of the world's major industrial and trading nations, and it is an atomic power by virtue of manufacturing its own atomic weapons. France's geographic position in Europe has proved extremely important in the strategy of the two 20th-century world wars, and it still adds to France's importance in political-military affairs. France has lost its former significance as a major colonial power, but it still maintains special relations with, and often exerts a strong influence on, a large group of former French colonies that have gained independence since World War II. Still another important foundation for France's international importance is the high prestige of French culture in many parts of the world. We turn now to a brief survey of the country's cultural achievements, its strategic geographic position, and its relationships with former colonies and a few small overseas units that it still retains. France's industrial and commercial importance is discussed at greater length later in this chapter.

ACHIEVEMENTS OF FRENCH CULTURE

France was conquered for Rome by Julius Caesar in the first century B.C. and thereafter became thoroughly Romanized. It has thus had a literate civilization longer than any other European country except the peninsular countries of Southern Europe. During the Middle Ages, Paris, the French capital, became one of the main centers of Roman Catholic scholarship and culture. In modern times, the French have made eminent contributions in all fields of scholarship and art, and Paris has generally been regarded as the world's foremost artistic and cultural center. Such names as Pasteur in science; Descartes, Pascal, Voltaire, Rousseau, and Bergson in philosophy; Molière, Racine, Hugo, Balzac, Zola, and Proust in literature; Matisse and Gauguin in art; and Bizet and Debussy in music come readily to mind as evidence of France's contributions to civilization. Even to make such a short list is almost to falsify by selection. In addition, many illustrious foreigners have drawn their inspiration from and pursued their studies in France. The position of French as an international language, particularly as a major language of diplomacy, and the continued attraction of Paris for the world's artists, students, and tourists, as well as for important international conferences, are further evidences of the cultural prestige of France.

STRATEGIC IMPORTANCE OF THE FRENCH BEACHHEAD

During the 20th century, France's strategic position has been that of a beachhead. A major path to the conquest of Germany in two world wars has lain through France. Location at the western end of Europe (except for Spain and Portugal, which have been militarily weak and relatively isolated behind the Pyrenees), sea frontage on three sides, proximity to Great Britain, and good communications with the continent to the east have combined to make France a critical area in the wars and hence a critical area in the politics of the 20th century.

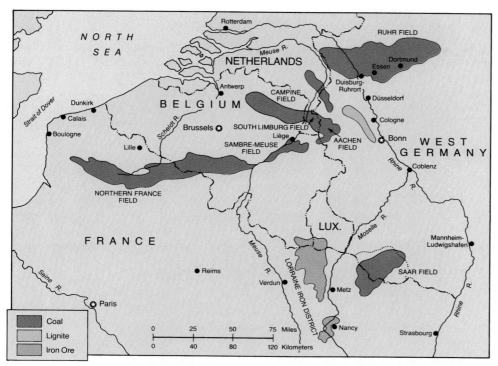

Figure 7.2 *Major coal fields of West Central Europe. Note the line of fields stretching across western Germany, the Low Countries, and northeastern France. Stars show national capitals. (After a map by the U.S. Geological Survey)*

SPECIAL OVERSEAS RELATIONSHIPS: THE LEGACY OF EMPIRE

The world significance of France has been heightened by its former importance as an imperial and colonial power. Before it began to dissolve in 1954, the French overseas empire totaled about 4½ million square miles (11½ million sq km—21 times the size of France itself) and had a population of more than 70 million—a colonial aggregate second only to that of Great Britain in population and second to none in area. Most of this empire was in Africa, although there were some important possessions elsewhere (Fig. 7.3). Actual settlement overseas by Frenchmen was seldom on a large scale—the principal exception being Algeria—but political control was often accompanied by a strong penetration of French culture. Many of the leading figures in France's former possessions speak French, have received the same type of education that Frenchmen receive, often including higher education in French universities, and have had political or military training and experience in the service of the French government. In the Americas, areas of French-descended population and/or culture exist as survivals from an earlier empire that was lost before most of the more recent possessions were acquired.

Most of France's colonial empire gained independence between 1946 and 1962. This process involved two major armed struggles, one in Vietnam (1946–1954) and the other in Algeria (1954–1962). In both places, large French military efforts proved unable to suppress revolts carried on by guerrilla warfare, and France was eventually driven to evacuate the territories and grant independence. The Algerian struggle split France's population into violently opposed groups and nearly plunged the country into a military coup and/or civil war. It caused the writing and adoption of a new French constitution, setting up the current Fifth Republic. Meanwhile, as the Algerian struggle went on, France initiated a series of steps which eventually bestowed national independence on most of its other colonial possessions.

Despite the loss of political jurisdiction, however, France continues to maintain special relationships, closer and more formalized in some cases than in others, with most of its former possessions. They receive economic, technical, and military aid and have special trading privileges in the French market.

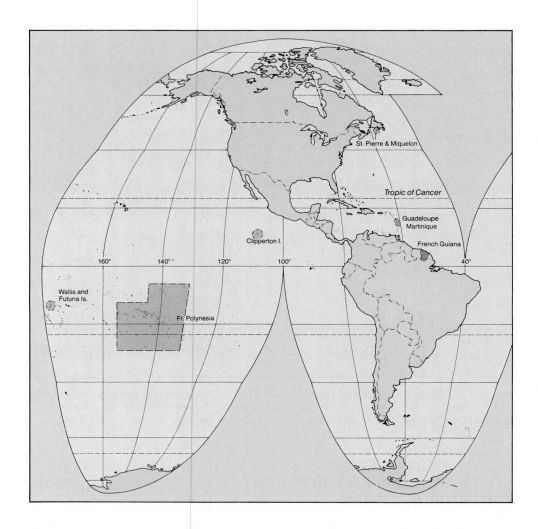

Many have been accorded favored treatment by the European Common Market, as have many former colonies of other Common Market members. Meanwhile, France continues its affiliation with a variety of small overseas departments and overseas territories in widely separated parts of the world.

TOPOGRAPHY AND FRONTIERS

France was once a very great power (chapter-opening photo). In the 17th and 18th centuries, its power was a primary factor in European politics. In the early 19th century, under Napoleon I, it came near to subjugating the entire continent before it was checked by a continental coalition in alliance with Great Britain. France's power was based on the fact that the French kings unified a large and populous national territory at a relatively early date. Before Germany and Italy achieved political unification in

the later 19th century, France was a larger country relative to its European neighbors than it is today.

PHYSICAL AND CULTURAL UNITY OF THE FRENCH HEXAGON

Modern France lies mostly within an irregular hexagon framed on five sides by seas and mountains (Fig. 7.4). In the south, the Mediterranean Sea and Pyrenees Mountains form two sides of the hexagon; in the west and southwest, the Bay of Biscay and the English Channel form two sides; in the southeast and east, a fifth side is formed by the Alps and Jura mountains and, farther north, the Vosges Mountains with the Rhine River a short distance to the east. Part of the sixth or northeastern side of the hexagon is formed by the low Ardennes Upland, which lies mostly in Belgium and Luxembourg; but broad lowland passageways lead into France from West Ger-

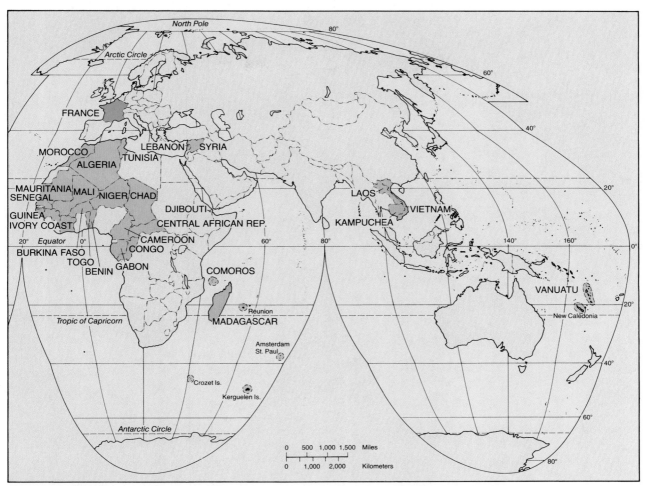

Figure 7.3 *France and its overseas departments and territories (small letters) as of 1988, and former French possessions that have received independence since 1940 (capital letters). Guadeloupe, Martinique, French Guiana, and Réunion are overseas departments. St. Pierre and Miquelon, New Caledonia, French Polynesia, Wallis and Futuna Islands, and Mayotte (in the Comoros) are overseas territories. "French Southern and Antarctic Lands" (not named on map) is an overseas territory composed of the islands between the Tropic of Capricorn and Antarctica, plus France's claims in Antarctica. A few tiny coastal areas that were ceded to India by France are not shown.*

many both north and south of the Ardennes. Aside from the large Massif Central in south central France, all the country's major highlands are peripheral. Even the Massif Central, a hilly upland with low mountains in the east, is skirted by lowland corridors which connect the extensive plains of the north and west with the Mediterranean littoral. These corridors include the Rhone-Saône Valley between the Massif to the west and the Alps and Jura to the east, and the Carcassonne Gap between the Massif and the Pyrenees. Thus, France is a country whose frontiers coincide, except in the northeast, with seas, mountains, or the Rhine, and which has no serious internal barriers to movement.

The French hexagon began to form a recognizable political unit at a very early time. It entered history as Roman Gaul, conquered mostly by Julius Caesar in the 1st century B.C. The Romans found a degree of cultural unity already established within it among the Gallic tribes. The Rhine was the frontier between these tribes and the Germans to the east, and, failing to conquer the Germans, the Romans established the Rhine as the frontier of the Empire. Five hundred years of Roman occupation and Latin-

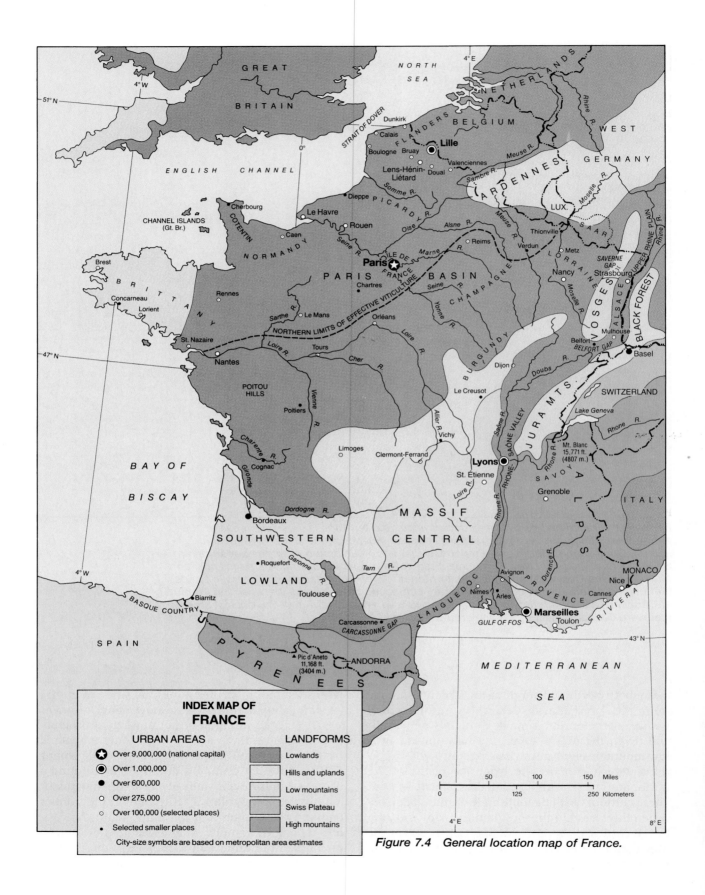

GREAT
BRITAIN

NORTH
SEA

NETHERLANDS

BELGIUM

WEST
GERMANY

ENGLISH CHANNEL

STRAIT OF DOVER

Dunkirk

Calais

Boulogne   Bruay   **Lille**

Lens-Hénin-
Liétard   Douai

Valenciennes

FLANDERS

Sambre R.

ARDENNES

Meuse R.

Rhine R.

LUX.

Moselle R.

SAAR

CHANNEL ISLANDS
(Gt. Br.)

Cherbourg

Le Havre

Dieppe

PICARDY

Somme R.

Oise R.

Alsne R.

Reims

CHAMPAGNE

Marne

Thionville

Verdun

Metz

LORRAINE

SAVERNE
GAP

Strasbourg

UPPER RHINE PLAIN

Rhine R.

Caen

Rouen

Seine R.

Paris

ILE DE
FRANCE

Chartres

Seine R.

Nancy

Moselle R.

ALSACE

VOSGES MTS.

BLACK FOREST

COTENTIN

NORMANDY

PARIS   BASIN

NORTHERN LIMITS OF EFFECTIVE VITICULTURE

Yonne R.

Mulhouse

Belfort

BELFORT GAP

Basel

Brest

BRITTANY

Rennes

Le Mans

Sarthe R.

Orléans

Loire R.

BURGUNDY

Dijon

Doubs R.

JURA MTS.

SWITZERLAND

Concarneau

Lorient

St. Nazaire

Nantes

POITOU
HILLS

Tours

Cher R.

Loire R.

Vienne R.

Le Creusot

Saône R.

RHONE-SAÔNE VALLEY

Lake Geneva

Rhone R.

Mt. Blanc
15,771 ft.
(4807 m.)

SAVOY

Poitiers

Limoges

Clermont-Ferrand

Allier R.

Vichy

Lyons

St. Étienne

Loire R.

Grenoble

ALPS

ITALY

BAY OF

BISCAY

Charente R.

Cognac

Gironde

Bordeaux

SOUTHWESTERN

Dordogne R.

MASSIF

CENTRAL

Roquefort

Garonne R.

Tarn R.

LOWLAND

Toulouse

Rhone R.

Durance R.

Avignon

PROVENCE

Nîmes   Arles

Nice

Cannes

MONACO

RIVIERA

Biarritz

BASQUE COUNTRY

Carcassonne

CARCASSONNE GAP

LANGUEDOC

Marseilles

GULF OF FOS   Toulon

SPAIN

PYRENEES

Pic d'Aneto
11,168 ft.
(3404 m.)

ANDORRA

MEDITERRANEAN

SEA

INDEX MAP OF
FRANCE

URBAN AREAS

✪ Over 9,000,000 (national capital)

◉ Over 1,000,000

● Over 600,000

◎ Over 275,000

○ Over 100,000 (selected places)

• Selected smaller places

City-size symbols are based on metropolitan area estimates

LANDFORMS

Lowlands

Hills and uplands

Low mountains

Swiss Plateau

High mountains

0   50   100   150   Miles

0   125   250   Kilometers

Figure 7.4   General location map of France.

ization increased the unity of Gaul and also increased the distinction between it and unoccupied Germany. During the Middle Ages the land within the hexagon was politically fragmented, like most of Europe, but Germanic invaders were absorbed and a considerable degree of unity was maintained in a culture now becoming French. An exception was the area nearest the Rhine, where Germanic languages and culture became established for some distance west of the river. In the later Middle Ages, the physical character and underlying cultural unity of the territory within the hexagon were important factors aiding the French kings in their successful attempt to build a unified and centralized state extending over a broad area. This state emerged in the 15th, 16th, and 17th centuries and attained, in general, the boundaries of ancient Gaul, except near the Rhine.

MOUNTAIN FRONTIERS OF THE SOUTH AND EAST

The state which emerged as modern France was to a considerable degree a natural fortress, as two of its three land frontiers lay in rugged mountain areas. The Pyrenees, along the border with Spain, are a formidable barrier with two peaks over 11,000 feet (c. 3350 m), and only one pass lower than 5000 feet (1524 m). In general, the lowest elevations in the Pyrenees are found in the west, where the small linguistic group called the Basques—neither French nor Spanish in language—occupies an area extending into both France and Spain.

In southeastern France, the Alps and Jura mountains are followed by France's boundaries with Italy and Switzerland. The sparsely populated Alpine frontier between France and Italy is even higher than the Pyrenees. Mont Blanc, the highest mountain in Europe, reaches 15,771 feet (4807 m), and several other summits exceed 12,000 feet (3658 m). The mountains are cut by several deep passes which have long been important as routes; and a highway tunnel beneath the northern shoulder of Mont Blanc has provided an important new connection between France and Italy since it was opened in 1963. But the number of important routes through the mountains is limited, and the Alps have been an important defensive rampart throughout France's history. The Jura Mountains, on the French-Swiss frontier, are a very different kind of range but are also difficult to cross. Their rounded crests generally reach only 3000 to 5500 feet (approximately 900–1700 m) but are arranged in a series of long ridges separated by deep valleys. Easy passes through the ridges are comparatively few.

TERRITORIAL COMPONENTS OF FRANCE'S NORTHEASTERN FRONTIER

France's northeastern frontier, extending from Switzerland to the North Sea, has been less defensible than the mountain frontiers in the south and, unlike them, extends for long distances through well-populated areas where different cultures—the French and the Germanic—meet and interlock. It was France's misfortune that a stronger industrial and military power coalesced across the frontier during the 19th century. Although France's eventual decline from its position as a great power was probably inevitable when the United States and the Soviet Union began to emerge as giant, effectively organized powers, it was most closely and directly associated with the rise of Germany.

Belgium and Luxembourg, although they lie between France and Germany, have not effectively separated them in times of conflict. Hence, the French-German frontier may be thought of as a zone reaching to the North Sea and including these smaller countries. The border zone comprises six well-defined physical sections: the Belfort Gap, the Upper Rhine Plain, the Vosges Mountains, Lorraine, the Ardennes Upland, and the northern plains.

THE BELFORT GAP OR GATE. Also known as the Burgundian Gate, this is a lowland corridor between the northern end of the Jura Mountains and the southern end of the Vosges Mountains. It forms an easy, although narrow, passageway between the Rhone-Saône Valley and interior France to the west and the Rhine Valley and West Germany to the east.

THE UPPER RHINE PLAIN. From the Rhine port of Basel, in Switzerland, to a point north of the French Rhine port of Strasbourg, the boundary between France and West Germany follows the Rhine River. Here the river flows through a level-floored alluvial valley formed originally by the foundering of a section of the earth's crust. The valley, known technically as a rift valley or *graben*, lies between ranges of low mountains which are rather similar in physical character: the Vosges Mountains in France and the Black Forest in West Germany. These mountains are remnants of a continuous structure which existed before the central section slipped downward to form the rift valley. The floor of the valley is often called the Upper Rhine Plain. The French section of the plain, about 10 to 20 miles (c. 15 to 30 km) wide, lies in the old province of Alsace.[2] Located outside France's strongest defense lines, the plain has been an important pawn, and prize, in French-German struggles.

THE VOSGES MOUNTAINS. For about 70 miles (a little more than 100 km), this low, wooded range overlooks the flat Upper Rhine Plain. Lower uplands continue northward into West Germany beyond the Saverne Gap at the north end of the mountains proper. Although the highest summit is below 4700 feet (1433 m), the Vosges range presents a steep face to the east and in the past has constituted a military barrier of considerable value.

LORRAINE. Between the Vosges Mountains and the Ardennes Upland, a rolling lowland connects France and West Germany across the old French province of Lorraine. It is sometimes called the Lorraine Gate. To the west, it opens into France's largest lowland, which is centered on Paris and is often referred to as the Paris Basin. Between the West German border and Paris, the terrain is interrupted at intervals by long escarpments facing eastward. These escarpments, formed by the outcropping of resistant layers among the sedimentary rocks which underlie the Paris Basin, have proved to be useful military barriers when properly fortified in past wars. But the lowland gateway of Lorraine has nonetheless been an easier military route into France than the upland to the north or the mountains to the south.

THE ARDENNES UPLAND. This low plateau centers in southern Belgium and extends into adjacent Luxembourg and West Germany and, for a very short distance, into France. It is one of many remnants in western and southern Europe of the ancient Hercynian mountains. These remnants appear today as comparatively low, discontinuous uplands. Europe's principal coal deposits are found around the margins of these old highlands.

Stream valleys deeply entrenched below a rolling, forested surface make the Ardennes a considerable barrier to transportation, especially to east-west transportation, as the major valleys trend north

and south. This area has generally been considered unfavorable terrain for military movements, although Germany launched two major offensives across it in World War II.

THE NORTHERN PLAINS. For about 80 miles (c. 130 km), from the northern edge of the Ardennes to the North Sea, the French-Belgian frontier runs through low plains, which are a segment of the North European Plain and thus are continuous across Belgium and the Netherlands into northern Germany. In both world wars, the plains of Belgium, forming a broad corridor equipped with excellent transportation facilities, were an important avenue of invasion from Germany into France. The seaward portion of these plains is the region of Flanders, famous for the bitterly contested battles fought there between the German armies and the British and French in World War I.

MAJOR FRENCH-GERMAN CONFLICTS AND THEIR GEOGRAPHIC CONSEQUENCES

As we have seen, the Franco-German border zone consists of three lowland passageways—the northern plains, Lorraine, and the Belfort Gap—separated by two upland barriers—the Ardennes and the Vosges—plus Alsace lying outside the Vosges barrier. This border zone, along with some nearby areas, has been the arena for much of modern Europe's disastrous military history.

The Franco-Prussian War was fought here in 1870–1871. German forces, led by the German state of Prussia, broke into France through Lorraine. French armies were destroyed around Metz and against the edge of the Ardennes, Paris was invested, and France was compelled to surrender. As a result, Germany was unified into one country under Prussian leadership, and it succeeded France as the leading power of continental Europe. At the end of the war, France was compelled to cede to Germany Alsace and northeastern Lorraine, where a frontier was established which left Nancy and Verdun in France but placed Metz and Strasbourg in Germany. The areas acquired by Germany were (and are) predominantly German in language. They also contain major mineral resources: potash in southern Alsace, and salt and iron ore in Lorraine. The Lorraine iron-ore deposits are Europe's largest, and most of them went to Germany. Minerals from Alsace-Lorraine supplied much raw material for the subsequent spectacular development of chemical and iron-and-steel industries in Germany, while even the minor share

---

[2]At the time of the French Revolution, France's ancient provinces were abolished as administrative territorial units, and the present structure of small, less autonomous units called "departments" was instituted. But many of the provincial names from pre-Revolutionary France (Alsace, Lorraine, Burgundy, Champagne, Brittany, Normandy, and so on) are still well-recognized, significant, and useful regional designations.

of the iron-ore deposits left to France in the French-speaking part of Lorraine was very important in France's more modest development of heavy industry.

Between 1914 and 1918, the "Western Front" of World War I centered in the Franco-German border zone and adjacent northeastern France. Germany's 1914 offensive against France followed the northern plains route, engulfing most of Belgium. Stopped about 15 miles (*c.* 25 km) from Paris in the Battle of the Marne, the Germans fell back a few miles and entrenched. Four years of trench warfare along various lines in northeastern France followed, with the French supported from the beginning by British forces and eventually by American armies. The devastation and loss of life were enormous. France bore the brunt of the Allied effort and lost 1.3 million men, more than 3 percent of her total population. Close to half a million French perished in one battle alone—the 6-month defense of Verdun in the Lorraine passageway in 1916. The enormity of such losses comes into perspective when one realizes that the United States lost about 100,000 men out of a much larger population in World War I and about 60,000 in the more recent and traumatic Vietnam War. As a result of her "victory," France regained the areas lost in 1871 and also gained control of the Saar, a coal-bearing heavy industrial area just east of Lorraine. The Saar was not ceded to France but was placed under an international political administration and attached to France economically. It was eventually returned to Germany as a result of an overwhelming pro-German vote in a plebiscite in 1935.

A third French-German conflict arose as part of World War II (1939–1945). France was again invaded, in 1940, by the northern plains route. This time the Netherlands as well as Belgium was overrun by the German assault. But the movement through Belgium proved to be only a powerful feint. When large French and British forces had been drawn into northern Belgium to meet it, the main German thrust was delivered across the Ardennes Upland. This section had always been considered too difficult for major military operations and was only lightly defended. But it proved to be an insufficient barrier in a new age of mechanized and motorized warfare. German armies broke across and out of the Ardennes and then turned north to reach the English Channel, thus trapping the Allied armies in the north in a huge pocket against the sea. These trapped armies were destroyed or captured, except for some 338,000 men who were evacuated to Britain through the Channel port of Dunkirk. The Germans

were then able to brush aside the remaining French forces as they rapidly overran much of the country, sending it down to stunning defeat.

Four years later, in 1944, liberation from German occupation again made France a battlefield. After intensive preparatory bombing, which reached far into France, Allied armies landed in Normandy and swept the German forces back across northern France. Their advance was stubbornly contested, especially in Normandy, and many French communities were reduced to ruins, although Paris was spared. A secondary invasion force landed on the Mediterranean coast and moved into Germany by way of the Rhone-Saône Valley and the Belfort Gap. Liberated France emerged from the war with most of its transportation equipment and one eighth of its housing destroyed, with wheat production at well under half of the prewar level, with most of its machine tools removed to Germany, and industrial production at about one fifth of the prewar output. About 600,000 men had died in service, compared to about 400,000 for the United States with a population more than three times that of France.

The New Era of Franco-German Cooperation

Since World War II France and West Germany have achieved a drastic reversal of previous relations. The border zone has become a scene of cooperative efforts rather than antagonism and warfare. Progress has come about mainly within the framework of the developing European "communities"—the Coal and Steel Community, the Economic Community, and the Atomic Energy Community (see Chapter 4). Results of the new cooperation are to be seen in the amicable settlement of the Saar question and the canalization of the Moselle River. Following the collapse of Germany in World War II, control over the Saar's economy passed once more to France, and the area was given a quasi-independent political status. The resulting dispute between France and West Germany was ended, however, in 1957 by return of the Saar to West Germany following a plebiscite in which the Saar's population voted, as it had in 1935, to be reunited with the German state. Agreements concerning the Saar guaranteed France's right to purchase a substantial part of the Saar's annual coal production, and West Germany agreed to assist France in canalizing the Moselle River. This stream flows from the Lorraine heavy-industrial region through highlands to the Rhine (see Figs. 7.1 and 7.4). Although the Moselle had long been canalized for barge traffic in Lorraine itself, the section from Lorraine to the Rhine was practically useless for through-navigation by large barges due to frequent rapids and stretches of shallow water. Canalization of the lower river, from Metz to Coblenz, was completed in 1964, providing Lorraine with good water access to the Rhine. The necessary engineering involved the construction of 13 dams, which not only deepened the Moselle but also made possible the generation of large amounts of power at ten hydroelectric stations attached to the dams.

FRANCE'S FLOURISHING AGRICULTURE

Agriculture is a more important part of the economy in France than it is in most highly developed countries. As late as the early 20th century, about one half of the French were still peasants in a rural economy with a large subsistence component. Since then, and especially since World War II, rapid economic change has been accompanied by a massive migration from farms to cities. The number of farmers has fallen to the point that only about 7 percent of the country's labor force is now in agriculture. But this is still higher than in most developed countries, and, although their numbers have fallen, France's farmers have become more efficient. Hence, agricultural output has increased greatly, which is a common experience in industrializing countries.

Not only is agriculture important in France, but France is important in agriculture. Despite its relatively small size on a world scale, the country is both a major producer and a major exporter of agricultural products. Two products—wheat and wine—are of primary importance in this connection, but France also has large exports of barley, cereal products, corn, dairy products, fruits, and vegetables. Its sizable agricultural imports are mainly those appropriate to a country having a generally high standard of living, with meat, certain fruits and vegetables, and tropical products such as coffee leading the import list.

For a country with four fifths the area of Texas to attain such large farm exports, some human and physical advantages for agriculture must obviously be present. One advantage often cited is the relatively small population compared with France's area. But one must remember that a density of 265 persons per square mile is low only by west European standards; it is almost four times the density of the United States. France's membership in the Common Market is a major advantage. This gives French farmers free access to a huge and affluent consuming market in countries too crowded to produce enough agricultural output for their own needs and somewhat protected by tariffs from non-Market producers. In addition, the Common Market budget provides price supports for many agricultural products, with French farmers being the leading beneficiaries.

France's physical advantages for agriculture include topographic, climatic, and soil conditions. The main topographic advantage is the high proportion of plains. Large areas, especially in the north and west, are level enough for cultivation. Climatically, most of the country has the moderate temperatures and year-round moisture of the marine west coast climate but without the extreme wetness that handicaps many parts of the British Isles. The wettest lowland areas are in the northern part of the country, and it is here that crop and dairy production is especially intense. Southern France experiences notably warmer temperatures, allowing a sizable production of corn in the southwestern plains. An extraordinary development of vineyards is concentrated particularly in the south and southwest, although important vineyard areas are found as far north as Champagne, east of Paris. Finally, northeastern France from the frontier to beyond Paris has fine soils developed from loess. As in much of Europe, these exceptionally fertile soils are used principally to produce wheat and sugar beets. This wheat belt in northeastern France is the country's chief breadbasket; the beet production from the same area provides not only sugar but also livestock feed (beet residues from processing) for meat and milk production. Both wheat and beets tend to occupy the best soils because both are crops unusually sensitive to soil fertility.

FRENCH CITIES AND INDUSTRIES
THE PRIMACY OF PARIS

Paris is the greatest urban and industrial center of France, completely overshadowing all other cities in both population and manufacturing development. With an estimated metropolitan population of about 10 million, it is by far the largest city on the mainland of Europe (excluding Moscow in the Soviet Union). Paris is located at a strategic point relative to natural lines of transportation, but it is especially the product of the growth and centralization of the French government and of the transportation system created by that government. Like London, it has no major natural resources for industry in its immediate vicinity yet is the greatest industrial center of its country.

The city began on an island in the Seine River, which offered a defensible site and facilitated crossing of the river. The early growth of Paris, which dates from Roman times if not before, must have been furthered by its location in a highly productive agricultural area and at a focus of navigable streams as well as of land routes crossing the Seine. The rivers which come together in the vicinity of Paris include the Seine itself, which comes from the southeast and flows northwestward from Paris to the English Channel; the Marne, which comes from the east to join the Seine; and the Oise, which joins the Seine from the northeast (see Fig. 7.4). In recent centuries, these rivers have been improved and canals have been provided where necessary to give Paris waterway connections with seaports on the lower Seine, with the coal field of northeastern France, with Lorraine, and with the Saône and Loire rivers (see Fig. 7.1).

In the Middle Ages, Paris became the capital of the kings who gradually extended their effective control over all of France. As the rule of the French monarchs became progressively more absolute and centralized, their capital, housing the administrative bureaucracy and the court, grew in size and came to dominate the cultural as well as the political life of France. When, in relatively recent times, national road and rail systems were built, their trunk lines were laid out to connect Paris with the various outlying sections of the country. The result was a radial pattern, with Paris at the hub. As the city grew in population and wealth, it became increasingly a large and rich market for goods.

The local market, plus transportation advantages and proximity to the government, provided the foundations upon which a huge industrial complex developed. Speaking broadly, this development comprises two major classes of industries. On the one hand, Paris is the principal producer of the high-quality luxury items—fashions, perfumes, cosmetics, jewelry, and so on—for which France has long been famous. The trades that produce these items are very old. Their growth before the Revolution of 1789 was based in considerable part on the market provided by the royal court; since the Revolution, it has been favored by the continued concentration of wealth in the city. On the other hand, and more important now, Paris has become the country's leading center of engineering industries and secondary metal manufacturing. These industries are concentrated in a ring of industrial suburbs that has sprung up in the 19th and 20th centuries. Automobile manufacturing is the most important single industry in this group, but a great variety of metal goods is produced. Other industries produce chemicals, rubber goods, printed and published materials, foods and beverages, leather, glass, and many other goods. In addition, the city's highly diversified economic base includes an endless variety of services for a local, regional, national, and worldwide clientele. In recent decades, there has been a tendency, encouraged by French national and regional planning, for routine production facilities of manufacturing companies to be located, or relocated, away from Paris, while the Paris agglomeration retains the company headquarters, research and development efforts, and types of production unusually dependent on labor skills.

The Seine Ports

Two ports near the mouth of the Seine handle much of the overseas trade of Paris. Rouen (400,000), the medieval capital of Normandy, is located at the head of navigation for smaller ocean vessels using the Seine. However, with the increasing size of ships in recent centuries, more and more of Rouen's port functions have been taken over by Le Havre (275,000), located at the entrance to the wide estuary of the Seine. Both cities have varied industries, and the lower Seine area in general is a major district for refining the oil which France imports in large quantities.

URBAN AND INDUSTRIAL DISTRICTS ADJOINING BELGIUM

France's second-ranking urban and industrial area is composed of cities clustered on or near the country's principal coal field, located in the north near the Belgian border. A bit north of the coal field, several cities, of which Lille is the largest, form a metropolitan agglomeration of about 1 million people. On the coal field itself a number of mining and industrial centers in the vicinity of Lens and Douai aggregate more than 500,000 people, and smaller centers are scattered over the field.

This region, generally called the Nord, is both a major contributor to, and a major problem in, the economy of France. In the 19th and early 20th centuries, it developed a rather typical concentration of the coal-based industries of that period, with close juxtaposition of coal mining, steel production, cotton and woolen textile plants, coal-based chemicals, and heavy engineering. It still is of major importance in these lines, but they have not been growth industries in the second half of the 20th century. The region is now burdened with extraordinary employment problems and has become an area of out-migration. Many thousands of jobs have been lost as its industries have declined and/or rationalized production to use less labor in order to meet competitive challenges. At the same time, the region, with its amenity-poor environment, has not been able to attract sufficient industry of newer types to solve the problem, despite various subsidies and other encouragements for new development provided by the French government.

URBAN AND INDUSTRIAL DEVELOPMENT IN SOUTHERN FRANCE

France's third largest urban-industrial cluster centers on the metropolis of Lyons (French: Lyon; metropolitan population about 1.2 million), in southeastern France. Here the valleys of the Rhone and Saône rivers join, providing routes through the mountains to the Mediterranean, northern France, and Switzerland. Lyons, located at this junction, has been an important city since pre-Roman times. For several centuries, silk production was the basic manufacturing industry of the city and for many smaller towns in the vicinity; the district is still the leading silk-manufacturing area of Europe. In the 19th and 20th centuries, however, nylon and rayon textiles, chemical products, and engineering industries have eclipsed silk in the district's industrial structure. The synthetic fiber and chemical industries are largely outgrowths of the silk industry.

The Rhone-Saône Valley, which connects Lyons with the Mediterranean and leads northward from Lyons toward Paris, is of major importance to France, both as a routeway and as an energy producer. Since prehistoric times, the valley has been a major connection between the North European Plain and the Mediterranean. It now forms the leading transportation artery in France, providing links between Paris and the coast of the Mediterranean via superhighways, rails that carry some of the fastest trains in the world, and the barge waterway formed by the Rhone, Saône, and connecting rivers and canals to the north. A major oil pipeline from Marseilles to West Germany also follows the Valley. In addition, this corridor between the Alps and Jura on the east and the Massif Central on the west plays a prominent role in French energy production. Formerly, the prime energy source for the Lyons area was coal mined in small fields around the heavy-industrial center of St. Étienne (325,000) in the eastern edge of the Massif Central. More recently, France has stressed hydroelectric and atomic energy to com-

pensate for the inadequacy of its fossil fuel supplies, and the Rhone Valley plays a major role in both. A series of massive dams on the Rhone and its tributaries produce about one tenth of France's electricity, and further power is provided by several large nuclear plants in the Valley that use uranium from the nearby Massif Central and foreign sources.

France's principal Mediterranean cities—Marseilles, Toulon, and Nice—are strung along the indented coast east of the Rhone delta. Marseilles (French: Marseille; 1.1 million) ranks in a class with Lyons and Lille in metropolitan population and is France's leading seaport. It is located on a natural harbor far enough from the mouth of the Rhone to be free of the silt deposits that have clogged and closed the harbors of ports closer to the river mouth which were once more important than Marseilles. Although it is a very old city, perhaps the oldest in France, Marseilles' modern growth was closely associated with the acquisition of France's trans-Mediterranean empire in North Africa and with the building of the Suez Canal, both in the 19th century. The French imperial system emphasized trade with the empire, especially with Algeria, and the Suez Canal made the Mediterranean a major world thoroughfare, while the Rhone-Saône trench provided an easy route between Marseilles and northern France. Rapid growth of port traffic and manufacturing in the Marseilles area during recent years has been particularly associated with increasingly large oil imports and related development of the petrochemical industry.

Toulon (400,000), southeast of Marseilles, is France's main Mediterranean naval base, while Nice (450,000), near the Italian border, is the principal city of the French Riviera, probably the most famous resort district in Europe. Along this easternmost section of the French coast, the Alps come down to the sea to provide a spectacular shoreline dotted with beaches. The mountains also provide a slightly drier and warmer variety of mediterranean climate than is usual in these latitudes by affording some protection from northerly and westerly winds. A string of resort towns and cities along the coast includes, near Nice, the tiny principality of Monaco, which is nominally independent but is closely related to France economically and administratively.

Deposits of bauxite scattered along the Mediterranean littoral are among France's more important mineral resources. To manufacture aluminum economically from bauxite requires large quantities of low-cost electric power, and France's aluminum works are mostly located in Alpine valleys near hydroelectric installations. Surplus ore is exported through Toulon.

MINING AND HEAVY INDUSTRY IN LORRAINE

Lorraine, in eastern France, is crucially important in the French economy, but the area has an exceptionally severe and chronic problem of unemployment. Lorraine's importance stems from the fact that it contains the principal concentration of heavy industry in the country and has sizable resources related to this type of development. The largest iron-ore production in Europe comes from ore bodies extending northward from the vicinity of Nancy (275,000), past Metz (175,000), and into southern Luxembourg and Belgium (Fig. 7.2). The main output of iron and steel in France derives from plants scattered along the ore belt for about 60 miles (*c.* 100 km). Coal fields extend from West Germany's Saar across the border into adjacent Lorraine. In addition, large deposits of salt near Nancy, along with the coal, form the basis for important chemical industries.

Contraction of employment in Lorraine's mills and mines has given this region a problem typical of that experienced by many old heavy-industrial and mining areas in western Europe and the United States. The problem in Lorraine is due in part to competition from new and more efficient plants and superior resources elsewhere. Lorraine iron ore is large in quantity but low in quality. It has been losing markets outside Lorraine to richer ores from overseas; and the region's coal industry has long been afflicted by the competition of imported oil and gas. New iron and steel plants in western Europe tend to be located on the seaboard—as at Dunkirk in northern France and on the Bay of Fos adjacent to Marseilles in Mediterranean France—for convenient access to ores and fuels brought in from overseas by increasingly large and efficient ships. Production costs of European coal are so high that American and other foreign coal fields often find it possible to compete in European markets. Under such circumstances, older iron and steel plants in the interior, and mines as well, may become so unprofitable that they are closed, or they may cut their work forces drastically by operating at lower levels and/or mechanizing and automating production to a greater degree. Like other European governments with similar problems, the French government is attempting to aid distressed areas by attracting new and more diversified development there. In Lorraine, this has had limited success thus far. One problem in such areas is lack of amenities, the qualities of natural and cultural environments that people find agreeable. Belching chimneys, grimy buildings, and industrial waste piles, to say nothing of outmoded tenements and general congestion left over from a bygone age,

give negative impressions to newer industries seeking a place to locate. Even with the most valiant efforts, it is difficult to make an old heavy-industrial district an attractive place to live, and new firms are likely to choose less cluttered and more livable surroundings.

ALSACE

German-speaking Alsace, adjoining Lorraine in eastern France, includes the steep eastern front of the Vosges Mountains and the long, narrow plain between the foot of the mountains and the Rhine River boundary with Germany. This French sector of the Upper Rhine Plain is densely populated and productive. Large deposits of potash near Mulhouse (225,000) in southern Alsace provide the basis for a considerable chemical industry. Streams descending from the Vosges furnish waterpower, and a major hydroelectric power and navigation project along the Rhine has been built during recent decades, involving five low dams on the Rhine, several diversion canals parallel to the river, and eight hydropower stations on the canals. Barges leave the Rhine at intervals to traverse the canals. Strasbourg (400,000), France's major Rhine River port, has thus acquired improved water connections southward to Basel, Switzerland, to complement its connections to Germany and the Low Countries in one direction and to Lorraine and Paris in another. The city is the seat of the European Parliament elected by citizens of the Common Market countries to further the cause of west European coordination and unity. Although Strasbourg is in France, its population speaks German; this fact, together with location on the Rhine boundary between two old and powerful antagonists, gives the city symbolic significance as a center of European amity.

THE DISPERSED CITIES AND INDUSTRIES OF WESTERN FRANCE AND THE MASSIF CENTRAL

Western France and the Massif Central are considerably less populous, urban, and industrial than the parts of France to the northeast and east. The largest urban center, Bordeaux, has only slightly more than 600,000 people in its metropolitan area, and the next two cities, Toulouse and Nantes, have only about 525,000 and 475,000, respectively. These cities are very widely spaced. Bordeaux is a seaport at the head of the broad estuary of the Garonne River, which is called the Gironde. The city has given its name to the wines from the surrounding region, exported through Bordeaux for centuries. Nantes is the corresponding seaport where the other main river of western France, the Loire, comes to the Atlantic. Toulouse, a very old and historic city with modern machinery, chemical, and aircraft plants, developed on the Garonne well upstream from Bordeaux, at the point where the river comes closest to the Carcassonne Gap. Lying between the Pyrenees and the Massif Central, the Gap provides a lowland route from France's Southwestern Lowland to the Mediterranean. Only two other cities in western France and the Massif—an area as large as West Germany—have metropolitan populations of 250,000 or more. The larger of the two, St. Étienne, has been mentioned previously in connection with the nearby Rhone-Saône Valley. The smaller city, Clermont-Ferrand (250,000), is the home of the Michelin tire company, which developed radial automobile tires. Some industrial development, often of a very specialized character (for example, Cognac brandy and Limoges chinaware), is scattered throughout western France and the Massif away from the main cities. But the open, vineyard-dotted countrysides of the western lowlands and the livestock pastures that dominate the broken topography of the Massif are a far cry from the industrial clutter of the Belgian border or the huge cityscape of metropolitan Paris.

SOME GENERAL CHARACTERISTICS OF FRENCH URBAN AND INDUSTRIAL DEVELOPMENT

France is a country which shows very striking contrasts in the nature of its development before and since World War II. From the Industrial Revolution until recent decades, the country tended to fall behind its neighbors, except for Spain and Italy, in industrialization and modernization. It changed enough to become a considerable industrial power, but more slowly than most of its neighbors, and it remained more rural and less urban and industrial. Differences in this respect between France and the United Kingdom or Germany were particularly marked. Then, after the shock of World War II, France acquired a new dynamism. New development of the most modern types has taken place at an impressive pace and scale. The old patterns often continue alongside the new, and the result is a situation of striking contrasts. There is, for instance, the contrast between the many massive and modern—although not always esthetically pleasing—apartment-type suburban developments around Paris and the larger provincial cities and the fact that many of the French still live in housing built before World War I, or even before the 20th century. Or the contrast between the small artisan's shop manufacturing luxury goods by hand and the large factories and power plants of a new industrial age. Despite the rapid development in recent years, the effects of the past are still manifest in many ways. For example, they are apparent in the relatively small size of French cities. Paris is a giant, but it is the only city with a metropolitan population of much more than 1 million, and only three other cities—Lyons, Marseilles, and Lille—reach the 1 million mark.

Analysts have blamed various factors for the slow growth of the French economy and the country's retarded industrialization between about 1800 and 1950. No definitive explanation can be given, but the following factors have been suggested and are worthy of note:

1. Slow population growth for more than a century before World War II restricted the size of the domestic market and labor force.

2. France's transport system was centered on Paris to such a degree that areas away from Paris often lacked adequate connections with each other.

3. France's coal resources were smaller and of poorer quality than those of major competitors during the period of coal-fueled industrialization in the 19th and early 20th centuries.

4. Capital to fund new industries was less available than in major competitor countries due to inadequacies in the French banking system and a national tendency to hoard precious metals as a protection against inflation and political instability.

5. There was a tendency before World War II for French governments to protect French enterprises from the spur of competition by tariffs against foreign goods and by supporting cartel-type monopoly arrangements within the country.

6. France had no pressing need to capture foreign markets, due to the country's near self-sufficiency in agriculture and possession of a large colonial empire managed as a privileged trading bloc for France.

7. Certain cultural attitudes of the French may have slowed economic growth, for example, (a) the tendency to emphasize individualism as against the regimentation of an industrialized society, (b) the tendency to assign greater status to professional careers than to business careers, and (c) the tendency to place a greater value on a modest security with adequate lei-

Figure 7.5 Present-day France is a very advanced country in science and technology. One facet of this is experimentation with solar power. The photo shows a solar-electric generating station in the foothills of the Pyrenees Mountains southeast of Toulouse. The scrubby vegetation is characteristic of France's mediterranean climatic region.

sure than on the maximization of material output and consumption.

In the decades following World War II, France's old patterns were rapidly eroded by relatively dynamic new development (Fig. 7.5). Population growth, encouraged by government subsidies for children, speeded up. Inadequate coal resources became less of a handicap as other energy sources increased in importance. Protection of inefficient producers from competition was weakened or abolished by France's membership in the Common Market. The French government promoted combinations of companies into larger units capable of higher efficiency and international competitiveness. And France was well situated to move into newly developing industries and technologies. It had a large pool of labor available for transfer from an overmanned agriculture, a relatively modest share of its manpower and capital tied up in older industries, and an excellent educational system. Not the least of its advantages was the adoption of a future-oriented national planning system.

NATIONAL AND REGIONAL PLANNING

The new dynamism of France has been achieved under a system of deliberate planning. Since 1947, the country has engaged in both national and regional planning with the stated objectives of improving na-

tional output and living standards, reducing the considerable differences in prosperity of different sections of the country, and making France a pleasant place to live in a technological age. Unlike the all-encompassing state plans of Communist countries, this type of planning is envisaged as complementary to the activities of private enterprise, with the state using various means to influence the decisions of private companies while at the same time cooperating with them. A few key industries in France are government-owned, but government efforts are directed principally toward the encouragement of efficiency on the part of private enterprise and providing an adequate "infrastructure" (for example, good transportation facilities, water supply and sewage facilities, cheap electrical power, a skilled and educated populace) to support productive activities.

In addition to fostering overall national growth and improvement, French planning is directed toward reducing imbalances of long standing among the regions of France. In the mid-1960s, the government called particular attention to the contrasting levels of development among four major regions—the Paris region, Northeast, Southeast, and West (Fig. 7.6). The most significant contrasts were (1) between the Paris region and the rest of the country, and (2) between the less developed and poorer West and the rest of the country. Since modern economic growth centers strongly in cities, the French planners have been attempting to divert development from Paris to the main cities of other regions. Their tools have been restrictions and penalties on various types

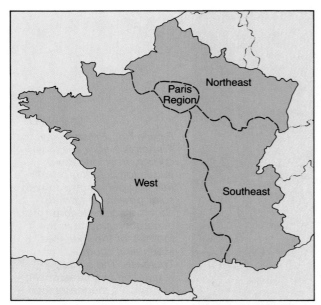

Figure 7.6 *France, with its four regions of contrasting levels of development. In order, highest to lowest: (1) Paris Region, (2) Northeast, (3) Southeast, and (4) West. Why? (Source:* France Town and Country Environment Planning, *p. 10.)*

of new development in the Paris region, a varied array of financial rewards for firms undertaking to locate facilities in regions of greater need, and direct location or relocation of government-owned facilities in the regional centers. An outstanding success in promoting regional centers has been the rapid growth of Toulouse in southwestern France as a major European center of aircraft and aerospace industries. Within the Paris region itself, a similar policy of economic and population decentralization has been followed, with functions such as wholesale markets and universities removed from the old city to the suburbs and several large new planned cities constructed in the suburbs.

While there has been a real decentralization of growth in France during the planning period, important questions are being asked about the results:

1. Paris and its region have lost hundreds of thousands of manufacturing jobs. Although transfer of these jobs was sought by the planners, there is now an increasing question as to whether the process has gone so far that it is hurting the working-class people of Paris and even threatening the whole Paris economy. Has the process gone far enough? Or too far?

2. Most of the decentralized industries have not dispersed to the far corners of France, but to a ring of smaller cities (such as Rouen and

Tours) not too far from Paris, and to lesser places within the same distance zone. Is this not at least a partial failure, considering that the objectives have been more focused on needier areas, especially in the West? Apparently businesses have very often chosen lesser subsidies and locations not too far from Paris instead of the larger subsidies they could have had for more distant locations in more depressed areas.

3. Some analysts believe that decentralization might have occurred without any planning policy. They maintain that after manufacturing industries reach a certain stage of development, the work in them becomes so standardized and routine that little skill is required from the work force, and that such businesses then tend to relocate from industrial cities to previously nonindustrial places where lower wages can be paid to unskilled and nonunionized labor. This point of view sees French economic decentralization as a normal result of the evolution of industries, and it questions much of the planning process as being an expensive irrelevancy.

However valid such questions and criticisms may be, there is no doubt that some decentralization of jobs and prosperity in France is essential for political stability and national cohesion. Although economic underprivilege is a potent cause of regional resentment and disaffection practically anywhere, in France and many other countries it is given urgency by the fact that parts of the poorer regions are ethnically different from the rest of the country. In such places, the relative poverty often aggravates political tensions that already exist on ethnic grounds. France has two main areas of this kind: (1) the Brittany peninsula, where feelings of ethnic separateness derive from the ancient Celtic cultural heritage; and (2) the island of Corsica, which is ethnically Italian. In both places, economic grievances and ethnic nationalism have fused into movements to achieve separation from France or at least a greater degree of autonomy.

The Fateful Role of Germany

As a result of the post–World War II political struggle between the Soviet Union and the Western bloc of nations, there have been two Germanies since 1949. The Federal Republic of Germany, commonly called West Germany, was formed from territories occupied by Western nations after the war and is allied with the West. The German Democratic Republic, or

East Germany, was formed from Soviet-occupied territory and is allied with the Soviet Union and its bloc. West Germany's capital is Bonn on the Rhine River; East Germany's is East Berlin. West Berlin is part of West Germany, although it is surrounded by East Germany on all sides.

GERMANY'S HIGH IMPORTANCE TODAY

Both Germanies are highly important in today's Europe, with West Germany much more so. Both are medium-sized countries in area even by European standards (see Table 4.1), but they are very densely populated—which is mainly a reflection of their many large cities—and they rank very high in total population. West Germany's 61.2 million people (1988 estimate) make it the leading European country in this regard; East Germany, with 16.6 million, ranks ninth. In both countries, population growth ceased in the 1970s. East Germany's population is stable; West Germany's is declining slightly as the death rate has come to exceed the birth rate.

If the two Germanies were reunited, their combined population would exceed that of any other European country by a wide margin. This is a factor of no mean weight in world affairs, but economic considerations weigh even more. Collectively, the two countries now have a preeminent position within Europe, and each is highly important within its European bloc. West Germany is Europe's leading industrial and trading state, and its economic health is critically important to most of non-Communist Europe. Country after country depends on West Germany as its leading trading partner—its main foreign market, main source of imports, or both. But West Germany's economic impact is also very strong east of the Iron Curtain. In the 1970s and 1980s, as Japanese manufacturers offered increasing competition on world markets, the West Germans sought expanded markets in the East bloc, and the Federal Republic became the leading Western country in trade with Communist Europe and the Soviet Union. East Germany, although much smaller as an industrial power, is technologically advanced. Within the Soviet-controlled bloc, it is second in total industrial importance only to the gigantic USSR itself. The potential strength of a reunited Germany has to be viewed uneasily by peoples and statesmen (not least the Soviets) who remember how Germany, aided by a few weak allies, overran large territories and then stood off a good part of the world for several years before ultimate defeat in two world wars.

THE DIVERSIFIED GERMAN ENVIRONMENT: ASSETS AND LIABILITIES

Germany arrived at its favorable economic position by skillfully exploiting advantages of centrality within Europe, together with certain key features of a diverse environment. In the following paragraphs, the general character of the environment is sketched, and critical features such as the Rhine, the loess belt, and the Ruhr coal field are then singled out for special attention in subsequent discussions of Germany's human geography.

THE NORTH GERMAN PLAIN

Broadly conceived, the terrain of Germany can be divided into a low-lying, undulating plain in the north and higher country to the south (Fig. 7.7). The lowland of northern Germany is a part of the much larger North European Plain (see Figs. 5.7 and 10.2). The German segment of the plain seldom rises much above 500 feet (c. 150 m) in elevation and is mostly below 300 feet (c. 90 m). It is predominantly an area of former glaciation, and the present topography bears much evidence of glacial action. Among the more prominent glacial features are the low ridges or *terminal moraines*, formed by deposition at the front of an advancing ice sheet during a long period when melting counterbalanced forward movement and kept the ice front stationary. These moraines, which are especially numerous east of the Elbe River, rise intermittently above a surface that is fairly level but seldom flat and is principally composed of infertile sandy or gravelly land punctuated here and there by stretches of heavy clay, a considerable number of bogs, and, in the northeast, many lakes. Along the coasts and the river estuaries bordering the North Sea lie expanses of flat, somewhat more fertile land that in most instances had to be drained artificially before use and resembles the diked lands of the adjoining Netherlands and Belgium.

THE VARIED TERRAIN OF CENTRAL AND SOUTHERN GERMANY

The central and southern countryside is much less uniform than the northern plain (Fig. 7.8). To the extreme south are the moderately high German (Bavarian) Alps (maximum elevation 9721 feet [2963 m]), fringed by a flat to rolling piedmont or foreland that slopes gradually toward the east-flowing upper Danube River. Between the Danube and the north-

Figure 7.7   General location map of Germany.

Figure 7.8   A scene in the uplands of southern West Germany. The long, narrow, unfenced strips are a survival from the pattern of landholdings in medieval times. Villages were surrounded by "open fields" in which each family was allocated a set of separated strips to till. Such fragmented farms are rarities in today's Europe but are still common in some other parts of the world, particularly the Middle East.

ern plain is a complex series of uplands, highlands, and depressions. The higher lands are predominantly composed of rounded, forested hills or low mountains. Interspersed with these are agricultural lowlands draining to the Rhine, Danube, Weser, or Elbe.

CLIMATE, SOILS AND VEGETATION

The climate of Germany is maritime in the northwest and becomes increasingly continental toward the east and south. The annual precipitation in both West and East Germany is adequate but not excessive, with most places in the lowlands receiving an average of 20 to 30 inches (*c.* 50 to 75 cm) per year. Farmers, however, must contend with soils that generally are rather infertile. This is especially true of the uplands and of most of the sandy North German Plain. The following areas, however, are major exceptions:

1. The southern margin of the North German Plain constitutes a belt with soils of great fertility formed from deposits of loess (windblown silt). This belt continues beyond Germany to both east and west.
2. The long, nearly level, alluvial Upper Rhine Plain, extending from Mainz to the Swiss border, and mostly enclosed by highlands on both sides, is also an area unusually favored by its soils.

Archaeologic evidence indicates that these two areas have been exceptionally productive agriculturally since the prehistoric beginnings of agriculture in Germany.

The dense forest that covered most of Germany at the beginning of German history (1st century B.C.) has now been reduced to scattered tracts. Conifers predominate, although deciduous species are abundant. Forests are found today on well-drained areas of poor soil and on rough highland areas. Some highlands are actually called "forest," the most famous being the Black Forest, the low mountains along the southeast side of the Upper Rhine Plain.

Prewar Germany was famous for its state-administered program of forest management; both German states still carry on this tradition. It is not uncommon when traversing a forested area to suddenly realize that the trees are arranged in neat rows, having been planted. Despite a large output of wood products, however, German demand is such that supplementary imports are needed, especially in West Germany.

LOCATIONAL RELATIONSHIPS OF GERMAN CITIES

Within the environmental framework just described, the locations of Germany's major cities tend to indicate the most economically advantageous areas and those most strategic with respect to transportation. The most outstanding aspect observable in this respect is the string of metropolitan areas in the western part of West Germany, along and near the Rhine River (see Fig. 7.7). Four of the eight largest areas—Düsseldorf, Cologne, Wiesbaden-Mainz, and Mannheim-Ludwigshafen-Heidelberg—are on the Rhine itself. The other four are on important Rhine tributaries near the Rhine—Essen on the Ruhr, Wuppertal on the Wupper, Frankfurt-on-the-Main, and Stuttgart on the Neckar. This north-south belt of cities includes several other large metropolitan centers such as Dortmund, Duisburg, and Aachen, plus numerous smaller places. The concentration of seven of the aforenamed cities at or just north of the boundary between the Rhine Uplands and the North German Plain reflects conditions other than the historic function of the Rhine and its valley as a very major transport artery. Historically, some of the seven cities profited from their location on the agriculturally excellent loess soil just north of the Uplands. Others developed as early industrial centers before the Industrial Revolution, drawing on the Uplands for ores, waterpower, wood for charcoal, and surplus labor. Then, in the 19th and early 20th centuries, all the cities grew in connection with the de-

155

velopment of the Ruhr coal field, located at the south edge of the Plain and east of the Rhine, and the smaller Aachen field. Except for Aachen and Cologne, they became parts of "the Ruhr," Europe's greatest concentration of coal mines and heavy industries. Still other important German cities lie apart from the Rhine and Ruhr; these include (1) Hanover, Leipzig, and Dresden, spread along the loess belt across Germany; (2) the two major North Sea ports of Hamburg, located well inland on the Elbe estuary, and Bremen up the Weser estuary; (3) two cities in the southern uplands—Munich, north of a pass route across the Alps, and Nuremberg; and (4) Berlin, whose large size belies its unlikely location in a sandy and infertile area of the northern plain. Berlin is mainly an artificial product of the central governments, first of Prussia and then of Germany, which developed it as their capital.

Of the cities just named, Essen, Dortmund, Duisburg, and Berlin are in the largest metropolitan complexes. Essen, Dortmund, and Duisburg have city-proper populations in the general range of 500,000 to 600,000, but the whole Inner Ruhr, in which the three cities lie (see Fig. 7.11), can be considered to have about 5 million people. Berlin's metropolitan area contains approximately 4 million people, of whom more than two thirds are on the West Berlin side. Hamburg's metropolitan population is more than 2 million, and Munich, Stuttgart, Frankfurt, and Cologne approach 2 million. The Mannheim and Düsseldorf areas have substantially more than 1 million people each, and the remaining cities mentioned in the preceding paragraph have 500,000 to 1 million. All the cities with more than 1 million people but East Berlin are in West Germany, which is Europe's most urbanized nation (94% urban by one estimate) except for Belgium (95%; see Table 4.1). Aside from East Berlin, the only East German cities in the entire list of large cities just cited are Leipzig and Dresden. Nevertheless, East Germany is highly urbanized, with a population that is 77 percent urban (United States, 74%).

HOW GERMANY EVOLVED: PEOPLE, MIGRATIONS, AND TERRITORIAL CHANGES

EARLY MIGRATIONS AND CONQUESTS

The present location of Germany and the Germans is only the latest phase in a series of large fluctuations on the map of Europe. Prehistoric German tribes are thought to have spread from a region of origin around the Baltic Sea. By the 1st century B.C.,

they occupied a wide expanse from the Rhine River in the west, where they confronted the Romans, to the vicinity of the Vistula in what is now Poland. Later on, some German tribes occupied southern Russia north of the Black Sea. In the 5th and 6th centuries A.D., the Germans broke the Rhine and Danube frontiers of the Western Roman Empire and spread across western Europe as a generally thin stratum of conquerors. They eventually reached as far as present Tunisia in North Africa and divided the conquered territories into a fluctuating collection of tribal kingdoms. In the course of time, most Germans were absorbed into the peoples of the occupied areas, but Germanic speech was permanently extended into some areas near the German homeland: the Netherlands and northern Belgium, where German evolved into Dutch and Flemish; Luxembourg; modern Germany west of the Rhine; Alsace and part of Lorraine in France; most of Switzerland; and Austria. England also became Germanic in speech as a result of these post-Roman conquests, but the English language was strongly modified by later French-speaking conquerors.

The early medieval expansion of "Germany" westward was accompanied by contraction on the east. The Germans who moved west not only were attracted to opportunity in the disintegrating Roman Empire, but some were fleeing from ferocious new invaders who entered Europe by way of the steppes of Asia and southern Russia. A particularly devastating invasion was that of the Huns who penetrated France in the 5th century before being turned back by an alliance of Germans and Romans. As the German tribes drifted west, much of the territory they abandoned was occupied by a new element, the Slavs. By about 800 A.D., Slavic peoples occupied what is now Germany to the line of the Elbe River and the Saale, as well as western Czechoslovakia to the south. This line marking the western limit of Slavic advance is approximately the modern line of division between East and West Germany, and between West Germany and Czechoslovakia. Thus, it is again the line between Slavic (Russian) and German control, although the population of East Germany remains German.

THE GERMAN PUSH EASTWARD

Following the Slavic advance, a new phase of German expansionism set in. For about 12 centuries, from 800 A.D. to the 20th century, German control, influence, and population pushed eastward into territories occupied by Slavs and some other peoples in eastern Europe. Along the Baltic coast, medieval German crusaders, traders, soldiers, and settlers conquered or dominated territories as far north as the

Gulf of Finland. When Tsarist Russia later absorbed most of these areas, the German elements remained important economically and in the politics and administration of the Tsarist state. Farther south on the North European Plain, centuries of German conquest and colonization eastward were climaxed in the late 1700s when the German state of Prussia, with its capital at Berlin, joined Russia and the German state of Austria in "partitioning" Poland. The latter country disappeared from the map, and German control reached almost to Warsaw. Farther south, Austria built an empire between the 16th and 19th centuries that included modern Czechoslovakia, parts of southern Poland and the adjacent Soviet Union, Hungary, and much of Romania and Yugoslavia. Enclaves of German settlement extended much farther east than the areas actually under Prussian or Austrian political control. For example, there were sizable German settlements along the middle Volga River in Russia. In the 19th century, groups of these Volga Germans migrated to central Kansas, bringing with them strains of wheat superior to those known in the American steppes at that time.

TWENTIETH-CENTURY CATACLYSMS AND THEIR EFFECTS

The eastward push of Germanism was checked by Germany's defeat in World War I. German and Austrian political control was rolled back in eastern Europe, although German populations remained in place and Germanic commercial interests and cultural influence continued to be strong. Poland was reconstituted, resulting in a considerable loss of territory by Germany and Austria, although Germany retained East Prussia. Germany was obliged to cede to Poland: (1) the "Polish Corridor" to the Baltic Sea between East Prussia and the rest of Germany, (2) part of the Upper Silesian mining and industrial district just north of Czechoslovakia, and (3) territories between the Corridor and Upper Silesia in what became western (now central) Poland (Figs. 7.9 and 9.11). In addition, the German city of Danzig (now Gdansk, Poland) on the Baltic was separated from Germany as a "Free State." To the south, the Austrian Empire, allied with Germany in the war, disintegrated, and Austrian control over all territo-

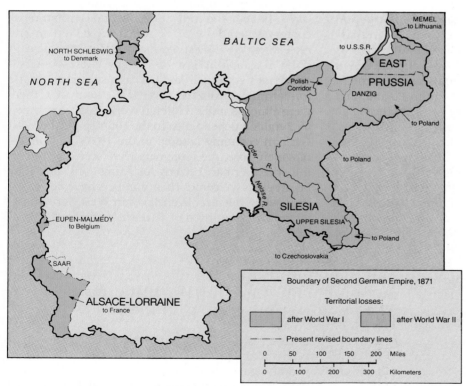

Figure 7.9 *Territorial losses of Germany resulting from World War I and World War II. Minor frontier cessions of territory to the Netherlands, Belgium, and Luxembourg following World War II are not shown. (See Lewis M. Alexander, "Recent Changes in the Benelux-German Boundary," Geographical Review, v. 43 [1953], pp. 69–76; (map compiled from Isaiah Bowman, The New World: Problems in Political Geography, 4th ed., New York: Harcourt, Brace & World, 1928, and other sources). On the map, Danzig is enclosed by dots.*

ries was lost except for the small German-speaking state of Austria itself.

After World War I, German minorities of varying size remained in various states in eastern Europe. They played major roles in the renewed German expansionism of the 1930s that led to World War II. Germany's dictator, Adolf Hitler, claimed that all Germans properly belonged within the German state, and pro-German Nazi parties were formed within the German minorities to agitate for Hitler's demands. Such pressures, backed by renewed German military might, were used by Germany to absorb Austria, Czechoslovakia, and Danzig in the late 1930s. Similar pressures on Poland were rebuffed, and this led to the German attack on Poland in 1939, which opened World War II in Europe. For the second time in a quarter of a century, a conflict between Germans and Slavs in eastern Europe had led to a general European war that broadened into a World War. (World War I had begun with an Austrian attack on the country of Serbia, now a part of Yugoslavia.)

Defeat in World War II drastically redefined Germany's eastern boundary, split the country's remaining territory into Eastern and Western segments, and brought about a massive redistribution of Europe's German population. Germany lost sizable territories in the east when: (1) East Prussia was taken away and split between Poland and the USSR, (2) the part of Silesia that had remained in German hands after World War I was ceded to Poland, and (3) a very large slice of territory between the Baltic Sea and Czechoslovakia also went to Poland (Figs. 7.9 and 9.11). Then, in 1949, the two German states of today were created as a result of quarrels between the Soviet Union on the one hand and their Western allies of World War II on the other.

The post–World War II territorial changes were associated with the relocation of millions of Germans. Some of these fled westward in the spring of 1945 to escape the advancing Russian armies. The merciless behavior of German forces in the Soviet Union and eastern Europe during the war gave good reason for ethnic Germans to fear reprisals. Many others who did not flee were forcibly expelled from East Central European countries to what was left of Germany. As a consequence, there are very few ethnic Germans left today in East Central Europe. The two cataclysms of the early 20th century have reversed 1200 years of German eastward expansion.

Even within Germany itself, large westward movements of population continued until 1961, due to the flight of East Germans to West Germany. At first the boundary between East and West could be crossed with relative ease. Even after it was closed by wires, mines, and patrols, it was still possible to cross from East Berlin to West Berlin and thence to the main part of West Germany. By 1961, more than 13 million West Germans—almost one fourth of the country's population—were migrants from the east or the children of such migrants. The main influx had come shortly after the war, with the pace then slowing. Only about 1½ million of the newcomers derived from East Germany, but this flow was such a drain on both the prestige and the economic strength of East Germany (and hence of the USSR) that the "Berlin Wall" was built in 1961. This physically sealed off the last open boundary between East and West Germany. Nearly all of the flow from the east has been halted, although a trickle of bold spirits have continued to escape and others have died in the attempt.

The influx of expellees and refugees probably helps to account for the rapid rebuilding and expansion of West Germany's economy during the 1950s and 1960s. Such a mass of destitute newcomers was initially a burden. But history has shown that refugees can also constitute an abundant supply of cheap labor to further economic growth if other conditions are right. The creation of the American economy is a case in point. If the immigrant labor force is educated, skilled, and industrious, as the German newcomers generally were, so much the better. At any rate, after 1961 West Germany felt it expedient to bring in a rapid inflow of foreign "guest workers," sometimes with their families, as a substitute for the former inflow from the east. By the late 1970s there were almost 2 million foreign workers, accompanied by 2 million dependents, in the country. As the West German economy faltered in the 1970s and unemployment rose, the foreigners were seen as competing with German citizens for scarce jobs, and they became less welcome. Their numbers have decreased somewhat, but sizable enclaves in West German cities are still comprised of Turkish, Yugoslav, or other non-German groups.

GERMANY'S HISTORIC PROBLEM OF DISUNITY

Since 1945, political disunity has been a major characteristic and problem of Germany. But this situation is nothing new in German history. A "unified" Germany existed only between 1871 and 1945 and fell far short of including all Europeans of German language and culture. But the country which emerged in 1871 was Europe's most populous, and its military power grew to the point that some observers believed it had the power not only to dominate Europe and Russia but to use this dominance as a springboard for domination of the world. The latent possibilities are still such that unity or disunity for Germany remains a major question for other countries, both European and non-European.

Although at times there has been a widespread and powerful German passion for unity, the roots of disunity run deep. The Germans who crossed the Rhine and Danube frontiers to overrun the Western Roman Empire in the 5th century were organized into large tribal units which were often in conflict with each other. When these units ceased their wanderings and took up a settled life in their old homelands east of the Rhine and in territories previously Roman, their disunity was expressed in a variety of kingdoms which were often in conflict with each other and had little internal unity and order.

During the latter Middle Ages and early modern times, relatively unified states were built in England, France, Spain, and some other parts of Europe, but disunity actually increased in Germany. Political maps of Germany in these periods show a patchwork of political units so varied and numerous as to be almost incomprehensible. Increasingly, these units acted as sovereign powers. The nominal rulers of Germany, whether styled kings of Germany, "Holy Roman Emperors," or both, lost authority within Germany as they concentrated again and again on unsuccessful attempts to revive a "Roman" empire by the conquest of Italy. Such attempts took them and their armies southward across the Alps and out of Germany for long periods of time and brought them into conflicts with the Popes of the Catholic Church, who were politically dominant in central Italy. The Papacy struck back by giving its moral sanction to rebellion and political opposition in Germany itself.

A further element of disunity was induced by the success of the religious teachings of Martin Luther in the 16th century and the consequent rise of Protestantism in Germany. Thus, religious antagonism between Catholics and Protestants was added to the competition of many small German states with each other, and there was intervention by outside powers with various political and religious motives. Germany became the main theater of action in the Thirty Years War of 1618–1648, wherein the country was ravaged not only by Germans pitted against each other but by the armies of intervening foreign powers—most notably France, Austria, and Sweden. The slaughter and famine were so devastating that many regions of Germany lost more than half of their populations. The peace that was finally arranged marked the failure of Austria, a German state with a non-German empire, as a candidate to dominate and unify Germany. It also fixed a religious pattern with Lutheranism dominant on the North

German Plain, Catholicism dominant in the southern uplands, and the areas along the Rhine more mixed. Furthermore, it left Germany with hundreds of political units exercising sovereign powers, for the most part under the absolute rule of some princely family.

THE RISE OF PRUSSIA: A CASE STUDY IN GEOPOLITICAL EXPANSION

The state that was to end the condition of near-chaos in Germany was Brandenburg, later to be called Prussia. Originally, the state was established, before 1000 A.D., as one of a series of "Marks" along Germany's eastern frontier. A Mark was a frontier principality which had the function of protecting Germany from non-German peoples—in this case, Slavs—and at the same time advancing German control and settlement. The original North Mark, which became the Mark of Brandenburg, was a block of territory between the Elbe and Oder rivers on the North European Plain. Although Berlin eventually became the main city of the area, Brandenburg was the leading center during the early centuries and gave its name to the Mark.

The environment of Brandenburg did not favor economic abundance or seem to provide a likely core for the origin of a great state. It featured belts of hilly terminal moraines and wide shallow valleys originally cut by glacial runoff and subsequently followed by small rivers. Marshes, swamps, and many lakes dotted the landscape. Soils were sandy and poor, supporting mixed forests consisting largely of conifers. The easterly location of Brandenburg within Germany gave the area more freezing weather than areas farther west. No minerals of importance have been found.

Within somewhat fluctuating boundaries, the Mark of Brandenburg existed as a reasonably important but unremarkable German state until the early 17th century. By gaining territory both east of the Oder and west of the Elbe, it gradually took the form of a relatively narrow east-west belt on the North European Plain. Berlin, originally a Slavic fishing village on the small Spree River, was founded as a German town in 1240 and replaced Brandenburg as the capital in the 15th century. Then, in the two centuries between 1600 and 1800, this unlikely candidate expanded and developed into one of the major powers of Europe. Dynastic inheritance in an age of absolute monarchies played a part in this process, as did warfare and rising military and diplomatic power. The major acquisitions in the 17th century were along the Baltic coast. Here, East Prussia was acquired from Poland and from the German military-religious order called the Teutonic Knights. The ruler of Brandenburg later began to style himself King of Prussia (1701), and the whole state became known as Prussia rather than Brandenburg. By this time, Brandenburg had also acquired, in the settlement of the Thirty Years War (1648), the coastal region known as East Pomerania, stretching from the mouth of the Oder almost to Danzig (now Gdansk in Poland). Lesser acquisitions in the 1600s also gave Brandenburg some detached enclaves of territory among the welter of states in northwestern Germany. The major acquisitions to the eastward gave the state access to the sea, which it had lacked, and a much-expanded base of territory, manpower, and resources. Resources that were especially important included better agricultural land in Pomerania and also timber, which was a valuable export to economically more developed western Europe.

In the 18th century, Prussia developed a centralized state bureaucracy of high repute, as well as a renowned professional army, and the kingdom expanded spectacularly. At one point in the century, the Prussian army, directed by the military genius of King Frederick the Great (ruled 1740–1786), fought to a standstill the combined armies of Austria, France, and Russia. Acquisitions during the century included (1) additional detached enclaves in western Germany; (2) Baltic coastal lands, including Danzig, which linked by land the previous acquisitions in Pomerania and East Prussia; (3) Silesia, comprising the plains along the upper Oder River and parts of the Sudeten Mountains to the south; and (4) late in the century, large parts of what are now central and eastern Poland. Silesia, which was conquered from Austria in 1741, contained major mineral resources—notably coal, copper, lead, and zinc. Their further development and use were skillfully fostered to provide the Prussian state with a major base for metallurgy. Upper Silesia, the coal-field area marked today by industrial and mining cities clustered around Katowice, is still the industrial heart of a European country, but today that country is Poland.

The rise of Prussia was interrupted temporarily by France's domination of Europe under Napoleon, but the end of the Napoleonic period (1815) saw Prussia larger and stronger than ever. The Prussian response to conquest by Napoleonic France was a vigorous restructuring which included universal military service for males and the founding of the University of Berlin, many of whose studies were allied to state objectives. In 1815, the new Prussian army played a major role in Napoleon's final defeat at Waterloo (near Brussels, Belgium). The European settlement that followed gave Prussia so much territory in northwestern Germany that it now occupied most of the North German Plain and most of the Rhine Uplands as well. The new acquisitions included iron ores in the Rhine Uplands, metal and textile industries, and the Ruhr coal field. The Ruhr was not yet important, but it was soon to become Germany's main source of energy and Europe's greatest center of iron and steel production.

Between 1815 and 1871, Prussia brought about the unification of Germany and gained dominance over it. By 1834, it had achieved the formation of a German "Zollverein" or customs union, bringing almost all the German states except Austria into a free-trade area marked by rapid economic development and integration. At the same time that it was pursuing economic unification, Prussia also moved toward political unification by making agreements with other German states under which they abrogated some, but not yet all, of their sovereignty. In 1866, a short, sharp war with Austria was won by Prussia, and Austria ceased to be a viable rival for the leadership of Germany. Then, in 1870, Prussia's aggressive diplomacy prompted a military attack by France. Most of Germany united with Prussia, and a devastating German counterattack swept through Lorraine and on to Paris. With France decisively defeated, Prussian prestige and power induced the other German states, except for Austria, to give up their sovereignty to a united German Empire.

The new Empire which was proclaimed in 1871 was the leading power on the continent of Europe until it dissolved in 1918 and 1919 after Germany's defeat in World War I. Officially, it was the "Second Reich" (Second Empire) of Germany, the first having been the medieval "Holy Roman Empire." Within it, the King of Prussia was also the Emperor of Germany, and Berlin was the capital of both Prussia and Germany. The Empire was federal in form, thus acknowledging the long-standing ties of Germans to their individual states such as Bavaria and Saxony, but Prussia was larger and incomparably more important than any of the other federated states. Democratic elections and parliamentary government were instituted, but electoral laws gave disproportionate voting power to the wealthy and to rural areas, thus favoring the power and position of the landed nobility (Junkers), especially of Prussia. The greatest of these was the King/Emperor (Kaiser) himself. The Chancellor (prime minister) could not be changed by the parliament, and therefore by elections, but only by the Kaiser. Thus, despite democratic forms and procedures, the Second Reich was rather autocratic and was largely built under Prussian control and to a Prussian model.

Economically, the new "Imperial Germany" was vastly successful. Conditions improved so markedly that the large outmigration which had long characterized Germany, and had brought the ancestors of many Americans to the United States, dried to a trickle. Rapid population growth soon gave Germany a clear lead over other European countries in total population. The development of a scientific, protected, and subsidized agriculture made the country nearly self-sufficient in food despite the generally poor soils and dense population. Coal, chemical, metallurgical, and engineering industries boomed, and Germany became second only to the much larger United States in heavy industry (Fig. 7.10). A world-famous educational system and scientific establishment supported these achievements. Germany's railway system, owned and operated by the state and supplemented by river improvements and canals, tied the country together and allowed a relatively wide areal spread of development and prosperity. Manipulation of railway freight rates helped bring about types and locations of development that were desired by the government. The German ports, especially Hamburg, became major centers of world trade, as did two foreign ports which largely served Germany: Rotterdam in the Netherlands, and Antwerp in Belgium. Between 1884 and 1900, a colonial empire was acquired, mainly in Africa. Germany's colonial tenure was brief, however, as the victorious Allies took over control of the colonies at the end of World War I. In the social field, the Second Reich, despite its general conservatism, pioneered many of the social insurance and welfare policies which later became identified with the "welfare states" of post–World War II Europe and North America.

Now the military and political catastrophes of the 20th century have again swung the pendulum toward German disunity. In terms of territorial disunity, the effects of World War I were relatively minor: Poland was revived and was given the "Polish Corridor" to the sea along the lower Vistula River, thus separating East Prussia from the main body of Germany. Austria, shorn of its empire, was forbidden "Anschluss" (union with Germany) by the victorious Allies. But within Germany itself, there was profound social and political disunity under the democratic Weimar Republic of 1919–1933, and this eventually led to dominance by the Nazi Party and totalitarian rule in Adolf Hitler's "Third Reich" of 1933–1945. Under Hitler, a fiercely nationalistic drive for German unity was fostered, its aim being the inclusion of all ethnic Germans as citizens of the Reich. German Jews were unacceptable for citizenship and were consigned to death camps, along with other Jews who fell into Nazi hands in Poland, France, Czechoslovakia, and elsewhere.

The result of the Third Reich's aggressions was a truncated Germany. At the end of World War II, all territories east of the line formed by the Oder River and its tributary, the Neisse, were lost to Poland except for northern East Prussia, which went to the USSR. The occupying Allied powers soon split into two antagonistic camps, each of which sponsored a German regime of its own persuasion. In 1949, the Federal Republic of Germany (West Germany) was formed from the American, British, and French zones of occupation, and the German Democratic Republic (East Germany) was formed from the Soviet zone of occupation.

The depth of division and antagonism between today's two Germanies is well symbolized by the fortified Iron Curtain boundary between them, including the Berlin Wall between East and West Berlin. Nevertheless, both republics are German, have developed as integrated parts of one Germany, and contain divided segments of many German families. Thus, even though reunification seems politically impossible in the foreseeable future, underlying forces for more cooperative relations exist. During the era of "détente"—the thaw in relations between the Soviet and Western blocs which characterized the 1970s—these forces bore some fruit. Trade between East and West Germany increased greatly, the few access routes between the main body of West Germany and West Berlin could be used without the former harassments and interruptions, and West Germans were allowed limited visits to families and friends in East Germany. This momentum continued into the 1980s, even while massive armies and missile forces remained in place along both sides of the boundary.

In view of Germany's industrial and military capacity, the reunification of the country under either Soviet or Western political auspices and influence would be a very serious development. The old question of German unity and disunity is now caught up in larger issues of world power as never before. Hence, the present politico-geographic division seems likely to persist even if the two Germanies manage to maintain a cooperative posture.

GEOGRAPHIC EVOLUTION AND PROBLEMS OF GERMANY'S INDUSTRIAL ECONOMY

Germany's development into one of the world's greatest industrial areas has roots in the distant past. Certain areas, such as the Harz Mountains, were exceptionally important centers of European mining in the Middle Ages, and many urban handicraft industries were also well established in medieval times. Between the end of the Thirty Years War (1648) and

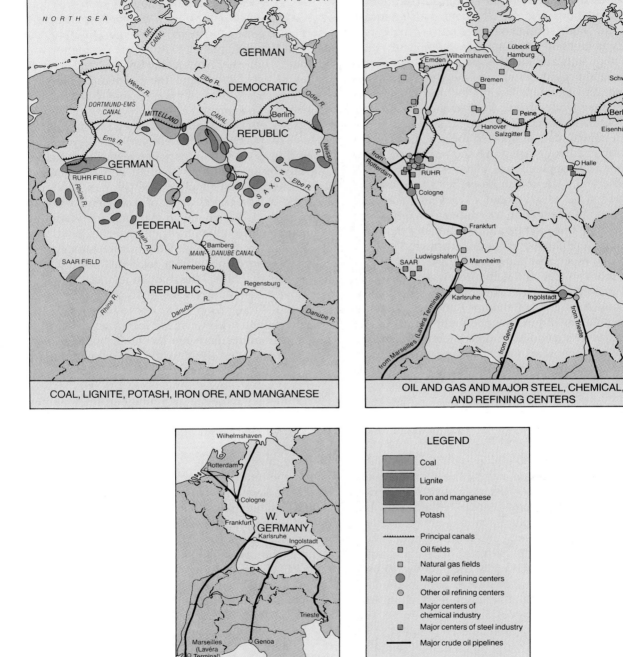

COAL, LIGNITE, POTASH, IRON ORE, AND MANGANESE

OIL AND GAS AND MAJOR STEEL, CHEMICAL, AND REFINING CENTERS

CRUDE OIL PIPELINES FROM OCEAN PORTS TO WEST GERMANY

LEGEND

Coal

Lignite

Iron and manganese

Potash

⋯⋯⋯ Principal canals

▫ Oil fields

▫ Natural gas fields

● Major oil refining centers

○ Other oil refining centers

▪ Major centers of chemical industry

▪ Major centers of steel industry

━━ Major crude oil pipelines

0   50   100   Miles

0   80   160   Kilometers

(for both main maps)

Figure 7.10 Major minerals and iron-and-steel manufacturing centers, chemical-manufacturing centers, and oil-refining centers in Germany. The part of the Main-Danube canal from Nuremberg to the Danube is still under construction (1989).

1800, handicraft manufacturing also spread widely among the rural peasantry, becoming particularly common in many poor agricultural areas, such as the Black Forest, where a second source of income was badly needed.

By about 1800, there were three German regions with unusually high concentrations of small-scale industries. One was the Rhine Uplands, in which industry spread along both sides of the river north of Wiesbaden and also away from the river in both directions, but especially toward the east. A second was Saxony, north of, and including, the Erzgebirge (Ore Mountains) on the German-Czech border. The third was Silesia, consisting of the plains along the upper Oder River—containing the major mineral resources of Upper (southeastern) Silesia—and the Sudeten Mountains adjoining the plains. None of the three regions had coal-based industrial economies by 1800. They all mined metallic ores, smelted the ores with charcoal, used the waterpower of mountain streams, and produced textiles as well as metals and metal goods. In addition to land transport, all three areas had waterway connections that made more

economical the acquisition of materials and the distribution of products. The Rhine and its tributaries served the Rhine Uplands; Saxony was served by the Elbe and its major tributary the Saale; and the Oder served Silesia. The three areas lay at or near the contact line between the North European Plain and the highlands that edge it to the south. Prussia controlled part of the Rhine Uplands industrial region and all of Silesia.

In the 19th and early 20th centuries, each of the foregoing regions developed into a major European center of coal-based factory industry. As the Industrial Revolution spread from Britain and Belgium into Germany, each area was found to have coal resources that were conveniently located and, in the case of the Rhine area and Upper Silesia, extremely large. The Ruhr coal field, located at the north edge of the Rhine Uplands and extending northward under the adjoining plain on the east side of the Rhine (Fig. 7.11), is the greatest coal field in Europe, and the Upper Silesia field is also of major importance. In both areas, the abundance of coal favored the growth of large iron and steel industries. As these

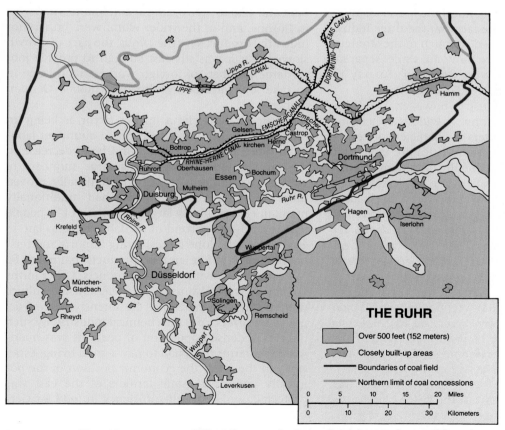

Figure 7.11 **The urban pattern of West Germany's Ruhr district, a major world center of heavy industry.**

grew in scale, they were increasingly dependent on imported ores, although the Rhine Uplands continued to produce some ore. Secondary development—such as metal goods, machinery, and coal-based chemicals—was evolved much more in the Ruhr than in Silesia, perhaps because of the Ruhr area's much better market position. Today, "the Ruhr" denotes an area of somewhat indefinite extent beyond the coal field proper. Often the term *Inner Ruhr* is used for the more concentrated urban-industrial zone incorporating Essen, Duisburg, and Dortmund, whereas the *Outer Ruhr* includes Düsseldorf and other more widely spaced cities.

In Saxony, coal resources were much poorer. Brown coal and lignite fields were eventually developed. Heat produced by brown coal and lignite per unit of weight is only about one quarter of that produced by bituminous coal, and only bituminous coal will yield coke for the iron and steel industry. Hence, an iron and steel complex did not arise in Saxony, although factory-type industries grew rapidly and promoted the growth of Leipzig, Dresden, and other cities. Machinery, precision instruments, textiles, and chemicals were (and are) the main emphases. Saxony's industrial area is roughly triangular and is often called the Saxon Triangle (see Fig. 7.1). Pertinent to the theme of political unity and disunity discussed previously is the fact that the three industrial regions which supplied so much of the strength of Imperial Germany are now in three separate countries: the Ruhr in West Germany, Saxony in East Germany, and Silesia in Poland.

Much of German industry, however, is not located in the Ruhr or Saxony. A wide scatter of industry and of urban centers is characteristic, especially in West Germany. Many cities are large, but none is so gigantic and nationally dominant as London or Paris. The absence of a single dominant center and the large number and wide scatter of fairly large industrial cities apparently resulted in good part from two circumstances: (1) *Germany was divided for a long time into petty states, a number of which eventually became internal states within German federal structures.* These units often promoted the growth of their own capitals, and many of the large cities of today were once the capitals of independent German states. (2) *German cities were able to secure power from a distance during the age of industrialization.* In the sequence of German development, railways generally reached major cities before coal became practically an industrial necessity. Thus, when coal was needed, the cities away from the fields could generally get it and continue to industrialize and grow. Waterways also helped many cities get coal; electricity then began to provide power from a distance.

Today, there is a major tendency in West Germany, as in many other advanced industrial countries, for factory industry to shift from the cities to more suburban locations and to rural areas. The attractions appear to be lower wages and lower site costs. The shift is allowed by the wide availability of power in the form of electricity, access given by superhighways, and the ability and willingness of labor to commute long distances by car or rail.

West Germany's economy recovered rapidly from its post–World War II state of depression after a new currency was provided in 1948 to replace the worthless Third Reich currency and the cigarettes which had become common tender, and after a government was formed in 1949 empowered to act. Former production levels were regained and surpassed so rapidly that the term "economic miracle" became common. Various circumstances facilitated the process, not least the rather free-wheeling free-enterprise approach followed by the government and massive aid received from the United States. Much of the physical plant was still intact and operable, despite wartime destruction; the skills of the population were still there and were supplemented by those of millions of refugees from the east; and labor was cheap for a time, as workers were abundant and glad to be working again. Markets at home, in Europe, and in the wider world were hungry for industrial products. By the 1960s, the pace of growth slowed, but it was still fast enough to provide jobs for large numbers of foreign workers who came in to take jobs vacated by upwardly mobile German workers.

Some analysts cite a further factor as being an important contributor of cheap and abundant labor for a time. This was a rapid and large decrease in the farm population, releasing large numbers of workers to nonfarm work, which usually proved more remunerative. Germany had traditionally maintained a relatively large proportion of its people on the land—23 percent of the labor force as late as 1948. This was done by restricting imports and raising the prices of foreign food products that might be too competitive with German products. Other forms of subsidy were also used, and scientific agriculture was encouraged, especially the intensive use of artificial fertilizers from the chemical industry. By such policies, a dense population of peasants was maintained on farms too small to pay a good living, especially in the west; the economic position of the politically important estate farmers of the east was protected; and near self-sufficiency in food for Germany was achieved despite the dense population and much poor soil. There was a strong emphasis on the production and consumption of agricultural products especially suited to German conditions and needs. For example, crops were emphasized that would do well on the country's poor soils: rye for bread, cabbages, and potatoes. Hogs, which would consume more farm wastes, were favored over cattle grown for meat; and sausages were emphasized to stretch the meat supply and give variety in taste. The situation then changed when the European Common Market deprived the West German farmer of protection against European competitors at the same time that industrial expansion opened new jobs. The resulting exodus of farmers from the land was so great that the West German farm labor force declined from 5.2 million to 1.7 million in the period 1950–1976, thus freeing millions of workers for industrial and service employment. Even this decline, with the accompanying consolidation of farms, still leaves most farms too small to provide what modern West Germans would consider a good material standard of living for their operators.

The pace of development slowed during the 1970s and 1980s, although West Germany continued to be a very wealthy and productive country and the average citizen probably continued to make economic gains. West German coal, already losing ground in the 1960s, became increasingly unable to compete with the imported petroleum which has now replaced it as the main energy source. West Germany has no advantage over other countries in buying expensive petroleum, and the cost and supply of energy have become important and divisive matters. New plants to generate electricity with atomic energy have sparked violent anti-nuclear protests, and the need for cheap energy led West Germany to quarrel with the United States in the early 1980s in order to complete a deal for Soviet natural gas from a new Russia-to-Europe pipeline. The export of German industrial products to world markets became less easy as Japanese and other foreign competition stiffened. To a considerable degree, West Germany turned eastward to Communist Europe and the USSR as a replacement for markets lost elsewhere. Soviet orders to German factories for metal and machinery were one of the attractions of the gas pipeline deal.

Expansion of industrial jobs in West Germany turned into contraction in all major industrial lines in 1970. A steady growth in unemployment followed, as trade and service occupations could not be expanded rapidly enough to take up the slack. The country lost 12 percent of its industrial jobs between 1970 and 1976 alone, and the mid-1980s still saw no solution. The decline of production in the face of competition, the placement of German-owned factories abroad to secure cheaper labor or market advantages, and the automation of West German factories all played a part in the decline. Some observers think that West Germany may have an especially hard time in shifting to new high-technology industries and service-based industries due to its unusual success and heavy investments in more traditional industries such as steel, machinery, and automobiles. One unfortunate result of the difficulties has been an increasing animosity on the part of some Germans toward the numerous foreign workers who still inhabit sections of most West German cities. However, despite the various difficulties of West Germany's "economic miracle," that nation ranks with the United States and Japan as one of the world's three largest exporters of goods.

East Germany, also established in 1949, is Communist ruled and Soviet dominated, but it is still German. Even Soviet exploitation and the inefficiencies of Communist bureaucrats attempting to plan and control an economy seem unable to destroy a certain German capacity to produce. In East Germany's early years much of its industrial plant was dismantled and sent to the Soviet Union. Then for still more years it was drained of wealth to help rebuild the USSR from the ravages inflicted by German armies in World War II. Before the Berlin Wall sealed the last loophole, much of the population fled to the West, thus draining East Germany of needed labor, but the country has done much better economically since then. In per capita output and income it is surpassed by all but a few European nations outside the Iron Curtain, but it stands at or very near the top among the world's Communist nations (recent data are insufficient for exact comparisons). Under Soviet direction, money was poured into a large heavy-industrial complex at Eisenhüttenstadt ("Steel Mill City"), on the Oder River to the southeast of Berlin, although neither coal nor iron ore was cheaply available. Part of the plant operates, but the rest was never finished. Consumer goods production was restricted in order to expand investment in more traditional East German industries such as chemicals and machinery. Strong emphasis has been put on intensive development of the country's brown coal deposits, including attempts to derive a process for using brown coal to make coke for the iron and steel industry. More recently, the economy has been stressing such high-technology lines as electronics, machine tools, instruments, automation, and chemicals. Somewhat greater success and prosperity have apparently attended this attempt to make East Germany the leading high-technology focus of the Soviet satellite empire, but the Wall is still considered necessary to keep large numbers of East Germans from fleeing to West Germany.

CHAPTER 8

The Vitality of West Central Europe's Smaller Countries

The commercial and industrial importance of the Benelux countries is shown vividly in this view of Rotterdam in the Netherlands. Botlek Harbor (nearest the camera) features berths for large ships, oil refineries, petrochemical plants, and other installations of the world's largest seaport. In the distance the New Waterway leads from the North Sea to the older port area at Rotterdam city.

ALTHOUGH France and the two Germanies occupy most of West Central Europe and dominate its affairs, the region also contains a group of smaller countries that have great vitality and are far from unimportant. Three of them—the Netherlands, Belgium, and Austria—have held large empires in past times, while a fourth, Switzerland, has long been prominent on the world stage as a neutral go-between in world conflicts and an international banking haven for owners of capital seeking security and anonymity for their funds. Still another country, Luxembourg, is far smaller but has some international importance connected with its sizable steel industry, close economic relations with Belgium and the Netherlands, and membership in European organizations such as NATO and the Common Market. The vitality and resourcefulness of all these countries in handling their affairs will be the major theme of this chapter. We look first at Belgium, the Netherlands, and Luxembourg—the "Benelux" countries—and then at Switzerland and Austria.

Overachievement in the Benelux Countries: Prosperity from Limited Resources and Trading Advantages

The small countries of Belgium, the Netherlands, and Luxembourg have been closely associated throughout their long histories. During some periods they have been included in a single political unit, although they have been politically separated during the greater part of the past four centuries. For several decades they have been attempting to strengthen their mutual relations and to weld themselves into an economic union, while maintaining their respective political sovereignties. The union has been designated "Benelux" from the first syllable of each country's name, and the three nations are often referred to collectively as "the Benelux countries." Although their absorption into the European Economic Community (EEC) has de-emphasized the initial concept of a Benelux customs union, the name Benelux is still common. An older name often applied to the three is "the Low Countries." In its strictest sense, however, this term is properly applied only to the two larger countries, Belgium and the Netherlands, and will be so used here.

LANDS OF LOW RELIEF

"Low Countries" is very descriptive of Belgium and the Netherlands, since approximately the northern two thirds of the former and practically all of the latter consist of a very low plain facing the North Sea (Fig. 8.1). This plain is the narrowest section of the great plain of northern Europe, which extends from France into the Soviet Union. In the Low Countries the plain seldom reaches as much as 300 feet (91 m) above sea level. Large sections near the coast, especially in the Netherlands, are actually below sea level. They are protected from flooding only by a coastal belt of sand dunes, by man-made dikes, and by constant artificial drainage.

The only land in the Low Countries with even a moderate elevation lies mainly in Belgium, south of the line formed by the Sambre and Meuse rivers. Here the Ardennes Upland rises in some places above 2000 feet (610 m), although its surface is more commonly about 1200 feet (366 m) above sea level. Much of the Ardennes is an area of little relief, although in places intrenched rivers produce a more rugged terrain. The edges of the Ardennes overlap Belgium's frontiers, extending without a break into the Rhine Uplands of the German Federal Republic on the east and for a short distance into France on the west. To the south, the Ardennes includes the northern half of the tiny country of Luxembourg. Southern Luxembourg and a small adjoining tip of Belgium have a lower, rolling terrain similar to that of neighboring French Lorraine.

A CROWDED BUT AFFLUENT CORNER OF EUROPE

High population densities, high productivity, and good incomes are outstanding characteristics of the Benelux countries. Excluding small Malta and a number of tiny microstates, the Netherlands and Belgium are the most densely occupied countries in Europe (see Table 4.1). Their smaller neighbor, Luxembourg, is less crowded, but even it has one of the higher European densities. Despite their small areas, the three countries aggregated 25 million people in mid-1988—a figure nearly as large as the population of Canada.

In many countries of the world, an unusually high density of population is associated with a low standard of living. In the Benelux countries, however, the reverse is true. By effective utilization of such opportunities and resources as their small and crowded territories afford, the peoples of these coun-

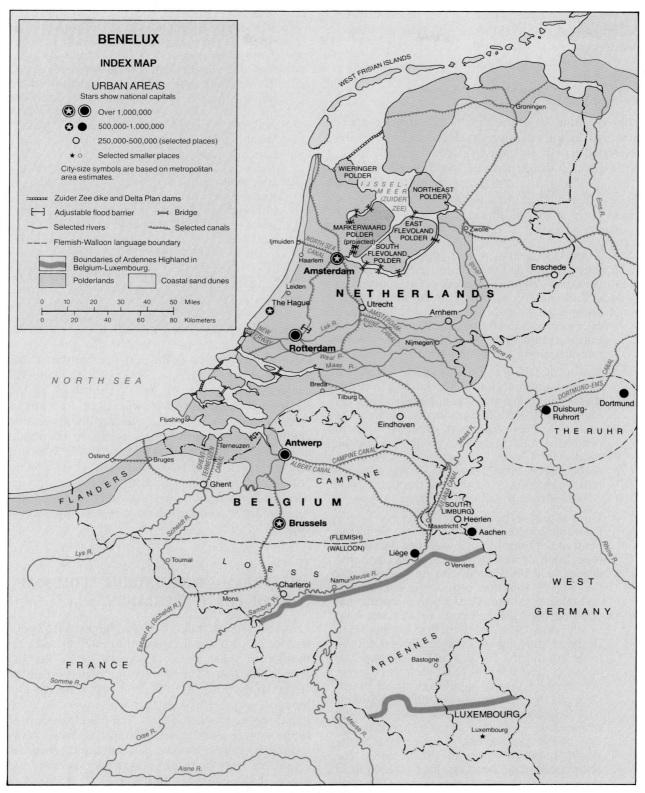

Figure 8.1   General location map of the Benelux countries: Belgium, the Netherlands, and Luxembourg. Note the extensive area of polder land reclaimed from the sea.

tries are able to maintain standards of living which are very high on a world basis. In Europe, these three countries—led by Luxembourg—all rank in the top half in per capita GNP (gross national product).

THE VITAL ROLE OF TRADE

The economic life of the Benelux countries is characterized by an intensive development of three interrelated activities—industry, agriculture, and trade. An especially distinctive and significant feature of their economies is a remarkably high development of, and dependence on, international trade. Several factors help to explain this unusual development of foreign commerce:

1. Trade is essential to such small countries if they are to make maximum use of a limited variety of internal resources. Such resources as Luxembourg's iron ore or the Netherlands' natural gas must be either left in the ground or exported for comparatively small returns unless complementary materials needed for manufacturing can be imported.

2. The need of these countries to trade is matched by their ability to trade. Specializing in activities which offer the greatest possibilities for effective use of limited resources, the Benelux nations are able to export large surpluses of certain manufactures and, in the case of the Netherlands, livestock products, vegetables, and flowers. Entry into the European Common Market has removed the tariff and associated barriers to larger markets in Europe.

3. The position of the Benelux countries is highly favorable for trade. These countries lie in the heart of the most highly developed part of Europe. Their nearest neighbors, West Germany, France, and the United Kingdom, are among the world's foremost producing, consuming, and trading nations. The resulting commercial opportunities are reflected in the fact that nearly three fifths of Benelux foreign trade (imports plus exports) is accounted for by trade among the three Benelux countries or between these countries and their three larger neighbors. A second significant aspect of position lies in the location of the Netherlands and Belgium at or near the mouth of the Rhine. This river is the greatest inland waterway of Europe, and one of the greatest in the world. Location where the Rhine meets the sea permits the Low Countries to handle in transit much of the foreign trade of Switzerland, eastern France, and, especially, West Germany.

4. The trade of the Low Countries has profited somewhat from the fact that both the Netherlands and Belgium were able to exploit large colonial empires in the tropics. Before they became independent, the Netherlands East Indies and the Belgian Congo offered assured markets for goods and capital, and allowed the home countries to act as European entrepôts for tropical agricultural products and certain minerals, particularly nonferrous metals. Wealth from this trade was important in financing industrialization, especially in the Netherlands. Today, industrial and trading specialties connected originally with this colonial trade still function in the Low Countries, although only a tiny fraction of their foreign trade is with their former colonies.

IMPORTANCE OF THE DUTCH AND BELGIAN PORTS

Exploitation of their commercial opportunities by the Low Countries is reflected in the presence of three of the world's major port cities within a distance of about 80 miles (*c.* 130 km): Rotterdam (see chapter-opening photo) and Amsterdam in the Netherlands, and Antwerp in Belgium. In terms of tonnage of goods handled, Rotterdam has been one of the two or three leading seaports in the world during the 20th century, and since World War II it has become incontestably the world's top-ranking port. Antwerp is far behind Rotterdam in tonnage handled but is still one of the world's leading ports. Amsterdam is a distant third, but even its trade is large.

AMSTERDAM—THE HISTORIC "COLONIAL" PORT OF THE NETHERLANDS

Amsterdam (1.9 million) is the largest metropolis and the constitutional capital of the Netherlands, although the government is actually located at The Hague (800,000). The city was the main port of the Netherlands during the 17th and 18th centuries and for most of the 19th century, while the Dutch colonial empire was being built. The principal element in this empire was the Netherlands East Indies (now Indonesia), in Southeast Asia, although territories elsewhere were held for various periods. Until the loss of Indonesia after World War II, Amsterdam built a growing trade in imported tropical specialties from the colonies. Many of these imports were re-exported, often after processing, to other European countries, and Amsterdam became one of the major European entrepôts for tropical products. It also became, and has remained, an important manufacturing center. Profits accumulated during these centuries supplied funds for the large overseas investments of the Netherlands and made Amsterdam an important financial center, which it continues to be.

Amsterdam also remains an important port, but it has lost much of its relative position in this respect during the past century. The original sea approach via the Zuider Zee proved too shallow for the larger ships of the 19th century. This problem was solved for a time by the opening in 1876 of the North Sea Canal. But even successive expansions of the canal have not given access to the sea comparable to that of Rotterdam. Location away from the main mouths and channel of the Rhine has proved a further disadvantage, despite improved canal connections to the river. And the loss of Indonesia has greatly decreased the port's importance as an entrepôt for tropical commodities, although continuing ties with its former colony have given Amsterdam a large Indonesian community which imparts a distinctive ethnic and cultural flavor to the city.

ROTTERDAM—THE MAJOR PORT FOR RHINE TRAFFIC

Rotterdam (1.1 million) is better situated with respect to Rhine shipping than either Amsterdam or Antwerp, being located directly on one of the navigable distributaries of the river rather than to one side (see Figs. 7.1 and 8.1). Accordingly, Rotterdam controls and profits from the major portion of the river's transit trade, receiving goods by sea and dispatching them upstream by barge, and receiving goods downstream by barge and dispatching them by sea. Two developments have mainly been responsible for the port's tremendous expansion. First was the opening in 1872 of the New Waterway, an artificial channel to the sea superior to the shallow and treacherous natural mouths of the Rhine and the sea connections of either Amsterdam or Antwerp. Second, and more fundamental, has been the increasing industrialization of areas near the Rhine, particularly the Ruhr district, for which Rotterdam has become the main sea outlet. A large addition to Rotterdam's facilities known as Europoort has been developed since 1958. It lies along the New Waterway at the North Sea entrance to the Rotterdam port area and is specially designed to handle supertankers and other large carriers of bulk cargoes.

THE SCHELDT PORT OF ANTWERP

Antwerp (1.1 million), located about 50 miles (80 km) up the Scheldt River, is primarily a port for Belgium itself but also accounts for an important share of the Rhine transit trade. Belgium's coast is straight and its rivers shallow, so that the deep estuary of the Scheldt gives Antwerp the best harbor in the country, even though it must be reached through the Netherlands. The city is a major focus in Belgium's dense railway net and has river and canal connections to Ghent, Brussels, and Liège. Transit trade is facilitated by Antwerp's position in relation to the major inland industrial areas of northwestern continental Europe. It is slightly closer than either Rotterdam or Amsterdam to these areas, and it has somewhat shorter road and rail connections.

MANUFACTURING IN BENELUX

All the Benelux countries are highly industrialized. Before major discoveries of natural gas were made in the northeastern Netherlands at the end of the 1950s, Belgium and Luxembourg were clearly better provided with domestic mineral resources. This asset helped them become somewhat more industrialized than the Netherlands. The differential is decreasing, although Belgium continues to have a far greater development of heavier types of industry and Luxembourg continues to exhibit far greater dependence upon such industries than does the Netherlands. The Netherlands, however, with its advantageous trade position and natural gas, has now become the leading country within Benelux in total value of industrial output.

HEAVY INDUSTRY IN THE SAMBRE-MEUSE DISTRICT

One of Europe's major coal fields crosses Belgium in a narrow east-west belt about 100 miles (161 km) long (see Fig. 7.2). It follows roughly the valleys of the Sambre and Meuse rivers and extends into France on the west and West Germany on the east. Liège (750,000) and smaller industrial cities strung along this Sambre-Meuse field account for most of Belgium's metallurgical, heavy chemical, and other heavy industrial production. A sizable iron and steel industry developed here, originally using local resources. These included small iron-ore deposits, with the ore being smelted first with local charcoal and later with coke made from local coal. Now, however, the industry must import its iron ore and even much of its coking coal and coke, which the Belgian fields can no longer supply in adequate quantities at low cost. Liège, the largest manufacturing center on the coal field, has metallurgical industries dating back to handicraft days and was the first city in continental Europe to develop modern, large-scale iron and steel manufacture following the Industrial Revolution. The Sambre-Meuse region has also developed a remarkable specialty in smelting nonferrous ores. Zinc ore came originally from the Ardennes, but now the ores—principally zinc, tin, and lead—are imported, and much of the metal is then exported.

The Sambre-Meuse district displays the customary problems of old, coal-based heavy industrial districts. Coal production has declined drastically in the face of competition from imported oil and even a certain amount of coal imported from better and less depleted fields in America and elsewhere. International competition has also adversely affected the iron and steel industry, which is now in a depressed state. Newer industrial plants with more modern equipment have generally located in northern Belgium, where unions are weaker and wages lower. Such plants even include an iron and steel plant built relatively near the coast at the seaport of Ghent (475,000). Hence Belgium's serious national problem of unemployment is localized particularly in the Sambre-Meuse Valley.

Throughout Belgium, the cities and even the smaller towns are generally characterized by some industrial development. Away from the coal fields, the favored types of industry tend to be those with smaller power requirements and a great need for relatively cheap labor. Belgium's capital and largest city, Brussels (nearly 2½ million), is the leading center of such industries. Some of these lighter industries are modern developments, but one, the textile industry, is very old. As far back as the 12th century, there was a major development of commercial cloth production and associated international trade in Flanders, roughly the area north of the Scheldt River and west of Antwerp.

MANUFACTURING IN LUXEMBOURG

Luxembourg's industrial structure shows a dependence on foreign trade proportionately equal to, or greater than, that of Belgium. The country's most important exports come from the iron and steel industry, carried on in several small centers near the southern border where the Lorraine iron-ore deposits of France overlap into Luxembourg. Production

is on a large scale, and the very small home market both necessitates and permits export of most of the metal. In return, Luxembourg buys fuel, alloy metals, and many other products. Luxembourg has been diversifying its manufacturing industries, and the iron and steel industry now provides a considerably smaller share of total exports than formerly.

INDUSTRIES IN THE NETHERLANDS

Until after World War II, the Netherlands had a smaller development of industry than might have been expected of a country in the heart of western Europe. It had no internal power resources except one small coal field (now closed) in the southernmost province called Limburg. The country's dense population provided cheap labor, and its location allowed cheap import of fuel and cheap export of products. So the Netherlands developed no heavy industrial region, but had a rather scattered development of light industries such as textiles and electrical equipment.

This situation changed after World War II as imported oil increasingly supplanted coal in the industries of western Europe. More oil imports enter Europe via Rotterdam-Europoort than through any other port. Part of it is forwarded as crude oil, primarily by pipeline to West German refineries along or near the Rhine (see Fig. 7.10), but part is refined at or near Rotterdam before forwarding. This activity has generated one of the world's largest concentrations of refineries. Rotterdam's oil industry reflects in part the Netherlands' important financial and managerial interest in the worldwide operations of one of the world's biggest corporations, the Royal Dutch Shell Company. Petrochemical plants located near the refineries represent the main branch of the sizable Dutch chemical industry, although some chemical plants make use of natural gas, blast furnace wastes, or by-products of the large slaughtering and meat-packing industry. Petroleum products and chemicals are important components of the Netherlands' varied exports. Amsterdam participates in these activities (including oil refining) along with Rotterdam, although its share is comparatively small.

Port locations and/or dependence on port functions also characterize many other important Dutch industries and activities. Shipbuilding was once a major industry but is relatively minor today. The repair and servicing of ships continues on a large scale, however, as does the operation of a good-sized merchant fleet. These activities reflect not only the country's present emphasis on trade but also a long maritime tradition. The Dutch played a prominent part in European discoveries and colonization overseas and for a brief time in the 17th century were probably the world's greatest maritime power. Another port-located industry is iron and steel. A large plant at Ijmuiden on the North Sea Canal uses imported coal and ore.

Certain other industries of consequence are found to some extent in the major ports but more often are located in smaller Dutch cities. Among them are food-processing, electrical-machinery, textile, and apparel industries. Some food industries process imported foods for European distribution, but more of them handle products of the remarkably productive Dutch agriculture. Certain industries tend toward interior locations having relatively cheap, productive labor and good transport connections. Among them, the electrical and electronics industry is outstanding. A Dutch company, Philips, with headquarters at Eindhoven, is one of the world's larger companies in this business and has plants scattered throughout the Netherlands.

THE INTENSIVE AGRICULTURE OF THE LOW COUNTRIES

Both the Netherlands and Belgium are outstanding agriculturally. They are characterized particularly by extremely high yields per unit of land. In fact, yields are so high that the total output of farm products is surprisingly great for such small countries. Such productivity results mainly from the following circumstances:

1. Parts of each country have very fertile soils, the largest such areas being the polders (diked and drained lands) in the Netherlands and the loess lands of Belgium.

2. Despite rapid decreases of farm population in recent times, there is still an intensive use of labor on the small farms that characterize these countries.

3. There is an intensive application of capital in such forms as water control, agricultural machines, knowledge, and, especially, fertilizer. Both countries are export producers of nitrogen fertilizers made from oil, natural gas, or coal.

4. Agriculture in the Low Countries has profited from Common Market tariff protection and national subsidies for agricultural producers. These have stimulated spectacular increases in output, to the extent that overproduction of certain products such as milk and butter has

become a more worrisome problem than food supply, even in such crowded countries as these.

5. Belgium and the Netherlands profit agriculturally from trading relationships that enable the two countries to specialize in farm commodities particularly suited to their conditions. The dense and affluent populations in and near the Low Countries provide a large market for such specialties. Requirements for other commodities such as grains are met in good part by imports.

The Low Countries exhibit three main types of agricultural land: the *polder lands, loess lands,* and *infertile lands.* Each is characterized by distinctive natural conditions and agricultural emphases.

DAIRY FARMING IN THE POLDER LANDS

Polder lands are those which have been surrounded by dikes and artificially drained (see Fig. 8.1). The process of turning former swamps, lakes and shallow seas into agricultural land has been going on for more than seven centuries. An individual polder, of which there are a great many of various sizes, is an area enclosed within dikes and kept dry by constant pumping into the drainage canals which surround it. About 50 percent of the Netherlands now consists of an intricate patchwork of polders and canals, while Belgium has a narrow strip of such lands behind its coastal sand dunes. The polder lands of the Netherlands extend for about 180 miles (290 km) between the Belgian and German borders in an irregularly shaped belt that is roughly parallel to the seacoast and is separated from the sea by dunes, swamps, or dikes.

The polders are the best lands in the Netherlands, and their production is the heart of Dutch agriculture. The reclaimed soil is very rich, water supply is subject to considerable control, and the canals form a complete transport network. The drier polders are often used for crop farming and produce huge harvests. Most polders, however, are kept in grass, and dairy farming is the main type of agriculture (Figs. 8.2 and 8.3). It thrives so well that large quantities of dairy products are exported, largely to other countries in the EEC. The production of vegetables, fruits, and horticultural specialties—such as cut flowers, potted plants, and flower bulbs—is also of considerable importance in the polder zone. Dairying and other specialized types of agriculture supply about one fifth of the total exports of the Netherlands and more than compensate, in value at least, for imports of cereals and some other agricultural commodities.

Figure 8.2 Transporting dairy cattle by boat to a pasture in the Dutch polders. The flat landscape is laced with drainage canals.

The main categories of exports provided by the Netherlands' agriculture include (1) meat, (2) evaporated, condensed, and dry milk, and cheese and butter, (3) vegetables, (4) flowers and bulbs, and (5) eggs.

THE FERTILE LOESS LANDS

A second section of outstanding fertility and agricultural production is found in Belgium. In central Belgium, a gently rolling topography was mantled in glacial times by a blanket of loess, a fine dust picked up by the wind from glacial debris. This material has formed soils of exceptional fertility in Belgium, as in many other places. These soils are found in a belt across the country between the Sambre-Meuse Valley on the south and a line just beyond Brussels on the north.

Three characteristics distinguish the agriculture of the loess belt: (1) an emphasis on wheat, which has been able to survive foreign competition on these good lands; (2) production of sugar beets, a crop requiring an exceptionally fertile soil; and (3) production of fodder crops, which give very high yields and permit the loess area to be unusually productive in a combination of dairying and output of beef and pork. This combination of food, feed, and livestock production supports the largest and most prosperous farms in Belgium.

Figure 8.3 Cheeses stacked in a marketplace at Alkmaar, Netherlands, symbolize the importance of dairy farming as one element in the rich economic diversity of the Benelux countries, Switzerland, and Austria.

THE INFERTILE LANDS

The remaining lands of the Low Countries, and most of Luxembourg as well, are relatively infertile. Sandy soils typical of the North European Plain are dominant in northern Belgium and the parts of the Netherlands that lie outside the polder belt. The Ardennes Upland in southern Belgium and Luxembourg also forms a part of the infertile lands. Livestock production for meat and milk is the characteristic form of agriculture in the infertile lands. Crops are grown mainly for feed. Yields are lower than in the polders and the loess lands, and much of the land is not used for agriculture at all.

However, one naturally infertile area—the part of Belgium called Flanders—is exceptional. Here yields are almost as high as on the naturally better lands, and rural population density is very high. In Flanders, intensive agriculture and systematic improvement of the land have a continuous history dating back at least to the 11th century. The precocious urban, commercial, and industrial development of the area stimulated Flemish agriculture during these early times. A garden type of agriculture developed, centered today on pork, dairy, vegetable, and root-crop production, but making some use of almost every crop that can be grown in the Low Countries and using soils almost completely transformed by centuries of improvement.

MAJOR PROBLEMS OF THE BENELUX NATIONS

The Benelux countries share the general problems of the modern world. Like many other countries, they are concerned with achieving and maintaining international peace and cooperation, national security, and internal unity, freedom, and economic well-being. However, due to peculiarities of position, resources, population, and economic development, the Benelux countries are faced with certain types of problems which present themselves in distinctive form or with particular urgency.

THE BELGIAN PROBLEM OF NATIONAL UNITY

In recent decades an intensification of internal regional antagonisms and ethnic separatism has troubled a number of west European countries. Belgium presents an extreme case of such difficulties. An east-west line just south of Brussels approximates a line of sharp division between the country's two major linguistic groups. South of the line live the French-speaking Belgians known as Walloons. To the north, Flemish, a close relative of Dutch, is the dominant language. An exception is Brussels, which is north of the line but is officially bilingual, with most of its people actually speaking French. When Belgium first became an independent country, by revolt and separation from the Netherlands in 1830, the Walloons were the more numerous group and were politically and economically dominant in the country. During the 19th century the French-speaking Sambre-Meuse area of coal mining and heavy industry was the most rapidly developing area economically, and it continued to be Belgium's economic focus until after World War II. Flanders, the homeland of the Flemish-speaking Belgians (Flemings), remained a less industrial and poorer region. A small French-speaking minority formed its upper class and dominated its economic life. Only in the latter part of the 19th century was Flemish allowed as a language for Flemish legal and administrative matters and in grammar schools, and only in the 1920s did it become a language of instruction in Flemish universities. Beginning in the 1930s a series of laws have equalized the status of the two linguistic groups but have introduced social and economic rigidities without resolving antagonisms between the groups. In each region it is now required that business, education, and administration be carried on exclusively in the regional language, and Brussels is required to function as a bilingual city. But rivalry and antagonism between the groups continues, and it has been spurred since World War II by differing regional fortunes. Through a higher birth rate the former Flemish minority has now become an increasing majority. The coal-based industries of the Walloon Sambre-Meuse area have expanded only slowly or have actually contracted in employment, while formerly backward Flanders has, along with Brussels, attracted large Belgian and foreign industrial investments to take advantage of its coastal location and labor that is still somewhat cheaper. Economic competition has thus reinforced ethnic antagonism, and violent incidents have occurred. Consequently, it is increasingly advocated that Belgium be split into a loose federation. Many fear that this would, and indeed will, lead to complete dissolution of the country.

POPULATION AND PLANNING PROBLEMS

Countries as intensively developed as the Netherlands and Belgium, with relatively high standards of living and high aspirations, must consider how development can be guided to provide an environment offering pleasant living conditions for their crowded populations. In the Netherlands such planning problems are especially acute, as the country has a greater density of population than Belgium and also has a more rapidly growing population. In this situation the Dutch are trying to limit urban sprawl, provide more outdoor and waterfront recreational areas, and in general maintain a pleasant and varied environment. They are especially concerned with maintaining the open space that is semi-encircled by Amsterdam, Utrecht, Rotterdam, the Hague, and other cities. These urban areas are tending to coalesce into what is referred to as Randstad Holland—the ''Ring City'' of Holland.

RECLAMATION OF NEW LAND

Land reclamation has been one response to population pressure in the Low Countries. In the past the objective has been new farm land. The past century and a half has seen much artificial improvement and expanded cultivation of such infertile areas as the peat bogs scattered in the sandy eastern Netherlands, the sandy Campine (Flemish: Kempen) southeast of Antwerp, and the Ardennes. However, the agricultural area is now tending to stabilize or contract, and current reclamation, which is most active in the Netherlands, is directed more toward sites for city extensions, new cities, industries, and recreation.

Reclamation of the Zuider Zee

One major reclamation work in the Netherlands has been the formation of new polders from the former Zuider Zee (see Fig. 8.1). A massive 18-mile-long dike was completed across the entrance of the Zuider Zee in the early 1930s, and a sizable area of sea bottom has now been reclaimed. By 1989 four of the five large polders planned were under cultivation. Meanwhile the remainder of the Zuider Zee had become a freshwater lake, the Ijsselmeer. The completed project would add about 7 percent—all excellent farm land—to the land area of the Netherlands. However, the two Flevoland polders near Amsterdam are now being developed mainly for urban and recreational use, and the same type of development is projected for Markerwaard Polder if a decision is ever made to proceed with the costly work of reclamation.

The Delta Plan

Another massive and much-publicized project of the Netherlands, called the Delta Plan, is aimed primarily at flood control and only incidentally at land reclamation. Its purpose is to prevent a recurrence of the devastating flood of February 1953, when the sea broke through the dikes in the southwestern part of the country during a storm, taking 1800 lives and causing great property damage. It is also intended to combat salinification, especially in lower sections of rivers, and hence to increase agricultural output; to connect islands to the mainland by roads across the dikes; and to stimulate the tourist industry. The heart of the Delta Plan is the construction of dikes connecting the islands in the triple delta of the Rhine, Maas (Meuse), and Scheldt with each other and with the adjoining mainland. The New Waterway, giving Rotterdam access to the sea, and the Scheldt River outlet for Antwerp will remain unobstructed, but the other channels through the delta are being closed by huge dikes. The scheme involves various engineering works besides the major dikes. By 1989, construction was far advanced.

The Contrasting Development of Switzerland and Austria

The two small countries of Switzerland and Austria, located in the heart of Europe, have often been contrasted and otherwise compared. Despite certain environmental and cultural similarities, these neighbors have been remarkably different in their historical development. Switzerland represents perhaps the world's foremost example of the economic and political success of a small nation, whereas Austria has experienced great economic and political difficulties, although these have lessened markedly in recent years. The contrasting development of the two countries is a major thread in the following discussion.

PHYSICAL SIMILARITIES OF SWITZERLAND AND AUSTRIA

With respect to physical environment, Switzerland and Austria have much in common. More than half of each country is occupied by the high and rugged Alps (Fig. 8.4). North of the Alps both countries include part of the rolling morainal foreland of the mountains. The Swiss section of the Alpine Foreland is often called the Swiss Plateau. It lies mostly between 1500 and 3000 feet (about 450–900 m) in elevation and extends between Lake Geneva, on the French border, to the southwest, and Lake

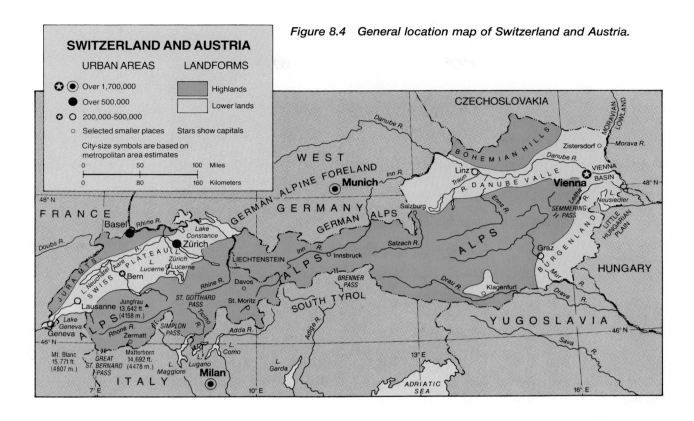

Figure 8.4 General location map of Switzerland and Austria.

Constance (German: Bodensee), on the German border, to the northeast. The Austrian section of the foreland, slightly lower in elevation, lies between the Alps and the Danube River from Salzburg on the west to Vienna on the east (Fig. 8.5). The Swiss and Austrian sections are separated from each other by a third portion of the foreland in southern Germany. North of the foreland both Switzerland and Austria include mountains or hills which are much lower in elevation and smaller in areal extent than the Alps. These highlands differ in character. The Jura Mountains, on the border between Switzerland and France, consist largely of parallel ridges formed of sedimentary rocks, while the Bohemian Hills of Austria, on the border with Czechoslovakia, represent the irregular, eroded southern edge of the old Bohemian massif and are formed of igneous and metamorphic rocks. North of the Jura Mountains in Switzerland the area around Basel opens onto the Upper Rhine Plain of Germany and France, while in eastern Austria the Vienna Basin and a strip of lowland to the south (Burgenland) adjoin the Little Hungarian Plain on the east and the Moravian Lowland of Czechoslovakia on the north. The Alps extend from Switzerland southward into Italy and southwestward into France, and from Austria southward into Italy and Yugoslavia.

THE DISPARITY BETWEEN RESOURCES AND ECONOMIC SUCCESS

Based on natural resources alone, one might expect Austria to be the more successful country economically. It has more arable land than Switzerland, and more per person, due primarily to Austria's greater proportion of non-mountainous terrain. It also has more forested land than Switzerland, and more per person. This results mainly from the fact that the Austrian Alps have somewhat lower elevations than the Swiss Alps and hence a smaller proportion of land above the tree line. The same contrast holds true for minerals. Austria has no mineral resources of truly major size, but it does have a varied output of minerals which are valuable collectively. Switzerland, on the other hand, is almost devoid of significant mineral resources. Both countries depend heavily on abundant waterpower, but Austria's potential is greater.

But in economic success the advantage lies with Switzerland, which in 1986 was the world's richest nation as measured by per capita GNP. Austria, by contrast, had a per capita GNP less than three fifths that of Switzerland in 1986. It lags behind most countries of northwestern Europe in this respect (see

Figure 8.5   A view of the Austrian section of the Alpine Foreland (Danube Valley). This photo of the deeply intrenched Danube and the rolling surface of the Foreland was taken about midway between Linz and the West German border.

Table 4.1). But this must be seen in proper perspective, as Austria is a relatively well-off country today—having, in fact, a per capita output greater than that of such countries as the United Kingdom or Italy. This is due largely to Austria's great economic growth since World War II.

The advantages of Austria with regard to natural resources have been far overweighed by factors in historical development that have favored Switzerland. The primary factor seems to have been the long period of peace enjoyed by Switzerland.

ROLE OF SWITZERLAND AS A NEUTRAL BUFFER STATE

Except for some minor internal disturbances in the 19th century, Switzerland has been at peace inside stable boundaries since 1815. The basic factors underlying this long period of peace seem to have been (1) Switzerland's position as a buffer between larger powers, (2) the comparative defensibility of much of the country's terrain, (3) the relatively small value of Swiss economic production to an aggressive state,

(4) the country's value as an intermediary between belligerents in wartime, and (5) Switzerland's own policy of strict and heavily armed neutrality. The difficulties which a great power might encounter in attempting to conquer Switzerland have often been popularly exaggerated, since the Swiss Plateau, the heart of the country, lies open to Germany and France, and even the Alps have frequently been traversed by strong military forces in past times. On the other hand, resistance in the mountains might be difficult to thoroughly extinguish. In World War II, Switzerland was able to hold a club over the head of Germany by mining the tunnels through which Swiss rail lines avoid the crests of the Alpine passes. Destruction of these tunnels would have been very costly to Germany, as well as to its military partner, Italy, since the Swiss railways were depended on to carry much traffic between them.

THE PRODUCTIVE SWISS ECONOMY

During well over a century and a half of peace, the Swiss have had the opportunity to develop an economy finely adjusted to the country's potentials and opportunities. In particular, they have skillfully exploited the fields of banking, tourism, manufacturing, and agriculture.

INTERNATIONAL BANKING

Favored by Swiss law, the country's banks have earned a worldwide reputation for security, discretion, and service. The result has been a massive inflow of capital—including some of dubious origins—to be reinvested by Swiss banks. Switzerland's largest city, Zürich (790,000), has become one of the major centers of international finance, and other Swiss cities are also heavily involved.

TOURISM

Switzerland has scenic resources in abundance, and the country has long been a major tourist destination. Development of winter sports has made snow an important resource, and Alpine ski resorts such as Zermatt, Davos, and St. Moritz have become world famous. The tourist business is well organized and regulated, with special training programs for personnel and enforced standards of service.

MANUFACTURING AND URBAN DEVELOPMENT

Less publicized than banking and tourism, but very important, is Switzerland's high development of manufacturing. This is focused on lines for which the country is particularly suited. They are based primarily on hydroelectricity from mountain streams and the skills of Swiss workers and management. Since most raw materials and supplementary fuels must be imported, industries are specialized along lines that minimize the importance of bulky raw materials and derive much of their value from skilled design and workmanship. The major products are (1) metal goods and machinery—often designed to order; (2) watches; (3) chemicals, especially pharmaceuticals; (4) textiles, generally of very high quality and with a large component of synthetic materials; and (5) aluminum, made in plants attracted to Switzerland by relatively cheap hydropower.

Reliance on hydroelectricity has facilitated the development of many small industrial centers. Most of these are in the Swiss Plateau, but some are in the Jura or the Alps. Five of the six largest cities—Zürich, Geneva, Bern (the capital), Lausanne, and Lucerne—are strung along the Plateau. Except for Zürich, these cities are between 150,000 and 450,000 in metropolitan population. The remaining

city, Basel (590,000), lies beyond the Jura at the point where the Rhine River turns north between Germany and France. Basel is the head of navigation for Rhine barges that connect land-bound Switzerland with ocean ports and other destinations via Europe's dense river and canal network.

SWITZERLAND'S DAIRY-ORIENTED AGRICULTURE

Swiss agriculture is highly specialized on dairy farming, which represents an adjustment to lands that are generally better suited to pasture and hay than to cultivated crops. Dairy production is centered in small, intensively worked farms on the rolling to hilly surface of the Swiss Plateau, although mountain pastures in the Alps are also used. Switzerland's high pastures are called "alps," and they have given their name to the mountains. *Transhumance* is an ancient practice here, as it is in many other mountainous regions of the world. The term refers to the seasonal migration of farm people and their livestock, utilizing high mountain pastures in summer and valley floors in winter. In Switzerland today there is a tendency for farmers to keep only the young stock on the high pastures in summer and to keep most animals at lower levels the year round. Here they are stall-fed, partly on imported feeds. Swiss dairies provide for the country's own needs and produce enough of a surplus to allow a modest export of cheeses, chocolate, and condensed milk. The dairy specialization is only possible because of Switzerland's ability to pay for the imports that cover a large share of its food needs.

THE NATIONAL UNITY OF SWITZERLAND

In their successful pursuit of economic goals, the Swiss have been aided by an effective national unity expressed in a stable, democratic, and competent government. The unity of the Swiss is the more remarkable in that it embraces a population divided in both language and religion. Approximately two thirds of the Swiss speak German as a native tongue, nearly one fifth speak French, and about one tenth speak Italian. Some recent immigrants speak such languages as Spanish or Turkish, and a little less than 1 percent of the Swiss speak Romansch, an almost extinct descendant of Latin which has been preserved in the mountains of southeastern Switzerland. A religious division also exists, since approximately 48 percent of the Swiss are listed as Roman Catholics and about 44 percent as Protestants.

Figure 8.6 The great achievement of the Swiss government in creating transportation routes through the rugged Alps is graphically illustrated by this view of bridges and a tunnel in precipitous terrain.

The internal political organization of Switzerland expresses and makes allowance for the ethnic diversity of the population. Originally the country was a loose alliance of small sovereign units known as cantons. When a stronger central authority became desirable in the 19th century, not only were the customary civil rights of a democracy guaranteed, but governmental autonomy was retained by the cantons except for limited functions specifically assigned to the central government. Although the functions allotted the central government have tended to increase with the passing of time, each of the local units (now 26 in number) has preserved a large measure of authority. Local autonomy is supplemented by the extremely democratic nature of the central government. In no country are the initiative and the referendum more widely used. Through these devices most important legislation is submitted directly to the people for their decision. Thus guarantees of fundamental rights, local autonomy, and close governmental responsiveness to the will of the people have successfully been used to foster national unity despite the potential handicaps of ethnic diversity and local particularism.

The central government, in turn, has pushed economic development vigorously. Two outstanding accomplishments have been the construction and operation of Switzerland's railroads (Fig. 8.6) and the development of the hydroelectric-power generating system. Despite the difficult terrain the Swiss government has built a railway network of great density, which carries not only Switzerland's own traffic but a volume of international transit traffic sufficiently large to be an important source of revenue. The highly developed hydroelectric power system, utilizing the many torrential streams of the mountains, yields about three fifths of the country's electric output and places Switzerland very high among the world's nations in hydroelectricity produced per capita. More recently, there has been a development of privately owned nuclear plants that now generate nearly all electricity other than hydro.

Thus internal unity and effective government have contributed greatly to Switzerland's economic success. On the other hand, economic success has undoubtedly reinforced the internal unity and political stability of the country, and the long period of peace has provided highly favorable conditions for both political and economic adjustment. Few modern nations have been so fortunate.

AUSTRIA: PROBLEMS AND PROSPECTS

In contrast to Switzerland's happy circumstances, Austria has been the victim during the 20th century of a series of military and political disasters which have required large and difficult readjustments of its economic life. These difficulties have been occasioned by the two world wars and the political and economic arrangements following them. Austria emerged in its present form as a defeated remnant of the Austro-Hungarian Empire when that empire disintegrated in 1918 under the stress of war and internal difficulties. Austria's population is essentially German in language and cultural background, and there is evidence that a majority wished to unite with Germany following World War I. This was forbidden by the victors, and Austria became an independent national state. Torn by internal strife and with an economy seriously disoriented by the loss of its empire, the country limped through the interwar period until absorbed by Nazi Germany, against the will of a majority, in 1938. In 1945, it was reconstituted a separate state, but, as in the case of Germany, was divided into four occupation zones administered, respectively, by the United States, the United Kingdom, France, and the Soviet Union. Vienna, like Berlin, was placed under joint occupation by the four powers. During 10 years of occupation the Soviet Zone, already badly damaged in the war, was subjected to extensive removal of industrial equipment for reparations. In 1955, the occupation was ended by agreement among the four powers. At the request of Austria, and the insistence of the Soviet Union, it was stipulated that Austria would become a permanently neutral, unaligned country on the Swiss model.

As the core area of an empire of 50 million people, Austria developed a diversified industrial economy during the years prior to 1914. Iron and steel were manufactured in a number of small centers, of which Graz was the most important. Production of a variety of secondary metal goods was centered largely in Vienna. Textiles were manufactured in the Alps and at Vienna. Wood industries made use of the Alpine forests. These industries developed within, and were dependent on, an empire which was an extraordinarily self-contained economic unit. Austrian ore was smelted mainly with coal drawn from Bohemia and Moravia, now in Czechoslovakia.

The Austrian textile industry specialized to a considerable extent on spinning, leaving much of the weaving to be done in Bohemia. In general, Austrian industry was not outstandingly efficient, but it had the benefit of a protected market in the agricultural parts of the empire to the east—areas now included in Hungary, Romania, Czechoslovakia, Yugoslavia, Poland, and the Soviet Union. In turn, Austria drew foodstuffs and some industrial raw materials from these areas, while Austrian agriculture was relatively neglected.

When the empire disintegrated at the end of World War I, the areas which had formed Austria's protected markets were incorporated in the independent states referred to in the preceding paragraph (except for the Soviet Union, which later gained a share of these lands only as a result of World War II). These states, motivated by a desire to develop industries of their own, began to erect tariff barriers. Other industrialized nations began to compete with Austria in their markets. The resultant decrease in Austria's ability to export to its former markets made it more difficult for the country to secure the imports of food and raw materials which its unbalanced economy required. Such difficulties were further increased after World War II by the absorption of East Central Europe into the Communist sphere. The necessary reorientation of Austrian industries toward new markets, at present found principally in West Germany, Italy, Switzerland, and other countries of free Europe, has not been easy. In these markets Austrian products must compete with the products of domestic industries (which may enjoy superior resources and/or tariff protection) and with the products of other industrialized exporting nations. As a consequence, Austrian exports of goods have been consistently inadequate to pay for necessary imports, which include foodstuffs, fuels, many raw materials, and certain types of manufactured goods. However, the country has been able to compensate for this deficit by revenues from a growing tourist industry, and the former large infusions of foreign aid have become unnecessary.

AUSTRIAN ECONOMIC READJUSTMENT

Austria made notable progress after World War I, and has made especially rapid progress since World War II, in building a successful economy suited to its status as a small independent state. This has involved a closer adaptation of the economy to domestic resources and a more intensive exploitation of them. Thus an economic structure of surprising diversity, considering the country's size, has developed. It consists of a blend of newer industries, usually closely related to the country's natural resources, with older industries which have persisted since imperial times and are often not so closely connected with domestic natural resources.

THE PRINCIPAL AUSTRIAN INDUSTRIES

Forest Industries and Tourism

From the viewpoint of export production, Austria's forest resources have been of fundamental importance. Approximately two fifths of Austria consists of forested areas (a higher proportion than in any European countries except Finland and Sweden), while most surrounding countries have a shortage of wood. In this situation large increases in the production and export of a variety of wood products have been possible. Timber, paper, and artificial fibers are the main items. In addition, the attraction of forested mountain areas has been a major factor in a large development of tourism. The drawing power of Austria's Alpine resorts and its scenery, only slightly less imposing than Switzerland's, is supplemented by carefully fostered cultural attractions, such as the world-famous music festivals at Salzburg and Vienna.

The Iron and Steel Industry

Austria's iron and steel industry supplies about one tenth of the country's exports of goods and provides material for an even greater value of exported manufactures. Although not large compared to that of many countries, the industry is larger than it probably would be except for a series of historical accidents. After it was shorn of important markets and sources of coal by the breakup of the Empire in 1918, it never regained its pre–World War I production until the period of German domination. Nazi Germany, however, for reasons of security connected with Austria's interior European location, more than doubled the country's steel capacity. Then American economic aid provided further help after the war. However, the industry has had to make progress in the face of increasingly inadequate natural resources. A considerable part of its iron-ore requirements is supplied by mines in the eastern Alps, but large supplemental imports are necessary. Austrian coal production has never been large and now consists only of lignite, which is of little use in the steel industry. Consequently, the industry must now import all of its coal and coke, must secure sizable amounts of ore from suppliers a considerable distance away, and then must sell much of its output in highly competitive foreign markets. The main centers of steelmaking are in mountain valleys to the near north of Graz, where plants are near ore production, and at Linz on the Danube.

The Rising Importance of Hydroelectric Power

Increased power development has been a basic aspect of Austria's recent progress. Hydroelectricity is the key element. Most installations have been built in the Alps, although several major dams have also been constructed on the Danube. Increasing supplies of relatively cheap electric power have been important to Austrian development in a number of ways. Among other things, the country has been converted from a sizable net importer of electric power to a small net exporter. At the same time, Austrian electricity consumption has risen rapidly, to the benefit of the standard of living and of many industries. Especially notable among the industries that have benefited from increased power supplies are the chemical industry, which also has available extensive salt deposits near Salzburg, and the aluminum industry, which requires very large amounts of cheap power.

Clothing and Crafts

Certain consumer goods industries which were natural to an imperial center such as Vienna have survived to play some role in modern Austria's economic revival. These include clothing manufacture and such arts and crafts production as fine glassware, porcelain, and jewelry. In the case of clothing, recent progress has been connected with Austria's growing reputation as a winter sports center and the related possibilities for setting styles.

Engineering and Textile Industries

Although the aforementioned industries provide Austria's main lines of net exports, certain other industries are important within the country in both employment and production. Outstanding among these are the engineering (machine-building) and textile industries. Various branches of both industries contribute quite importantly to exports, but development in these lines is not sufficient to satisfy the internal market, and Austria is a heavy net importer of machinery and textiles.

TRENDS IN AUSTRIAN AGRICULTURE

Since the end of the Empire, Austrian agriculture has moved toward greater specialization on dairy and livestock farming. This has been accompanied by a reduction in the farm population and the number of farms. Dairy and livestock production is a logical use of the large areas of Austria that are best adapted to pasture. An enlarged acreage devoted to feed crops such as barley and corn has accompanied the increased emphasis on animal products. At the same time there has been a decrease in the acreage of food crops (such as rye and potatoes, formerly grown in large part on poor land). Modest exports of dairy products and live animals have been achieved, but Austria is still far behind western Europe's leading countries in such exports and in overall agricultural efficiency. Many small and poor farms still exist, and 8 percent of the Austrian labor force is still in agriculture—a relatively high proportion by west European standards. Meanwhile large imports of food and feed remain necessary.

Overall, Austria has made very impressive progress since the end of the Empire, and especially since World War II, in building a viable and affluent economy within a small state. Development has stressed economic lines particularly adapted to Austrian resources, although mixing them with survivals from the imperial period. The country has come a long way since 1918, when Austria was forced to begin a new life as a small independent country instead of being a center of empire. The disastrous state of its economy in the early years necessitated recurrent foreign aid, which was needed in some cases to avert actual famine. Not until the 1950s did it finally overcome the need for help from outside. Austria's current affluence points to its remarkable success in overcoming major obstacles. That it is still considerably poorer than Switzerland is hardly surprising, in view of the very different histories of the two countries.

THE ROLE OF VIENNA

A significant aspect of Austrian readjustment since World War I has been the somewhat lessened importance of the famous Austrian capital, Vienna (1.9 million). Although Vienna is still by far the largest city and most important industrial center of Austria, there has been a pronounced tendency for population and industry to shift away from the capital and toward the smaller cities and mountain districts. From a position as the capital of a great empire, Vienna has regressed to a more modest status as the capital of a small country, and the resultant damage to its situation is reflected in the fact that its present metropolitan population is about the same as the population of the city proper half a century ago. Vienna's failure to match the growth of other Austrian cities also reflects the increased importance of forests and waterpower in the Austrian economy, as well as the dispersal of Nazi war industries into less vulnerable locations away from the capital.

Vienna's importance, however, has now persisted since Roman times and is based on more than purely Austrian circumstances. The city is located at the crossing of two of the European continent's major natural routes: (1) the Danube Valley route through the highlands separating Germany from the Hungarian plains and southeastern Europe, and (2) the route from Silesia and the North European Plain to the head of the Adriatic Sea (Figs. 8.4 and 9.11). The latter route follows the lowland of Moravia to the north and makes use of passes of the eastern Alps, especially the Semmering Pass, to the south. These routes have been important corridors of movement throughout European history. Whenever political conditions have permitted, Vienna has been a major focus of transportation and trade, and it is now regaining some of its old importance in this respect as restrictions imposed by the Iron Curtain are relaxed. But the city's position has also made it a major strategic objective in time of war—a fact that brought it great damage in World War II and that contributed to the long occupation of Austria after that war.

CHAPTER 9

Geographic Personalities of Europe's Outer Regions

Northern, Southern, and East Central Europe exhibit a prominent array of "primate" cities that are very dominant within their respective countries. One such city seen in this view is Sofia, Bulgaria's capital. High-rise buildings with standard architecture of the Communist world are very prominent here.

P RESENT-DAY Europe's intensely developed focal area in West Central Europe and the British Isles is ringed by three outer groups of countries—here termed Northern Europe, Southern Europe, and East Central Europe—that form rather well-defined and distinctive regions. Their geographic personalities will be explored in this chapter. Area and population data for the regions and their individual countries can be found in Table 4.1.

The Cohesive and Prosperous Countries of Northern Europe

The five countries of Denmark, Norway, Sweden, Finland, and Iceland are grouped regionally in this text as the countries of Northern Europe (Fig. 9.1). This accords with a geographic concept that is well established in the countries themselves. The peoples of these lands recognize their close relationships with each other and habitually group themselves and their countries geographically under the regional term *Norden*, or ''The North.''

A more common term referring to some or all of these countries is Scandinavia or the Scandinavian countries. But this regional name is somewhat ambiguous, being used occasionally to refer only to the two countries which occupy the Scandinavian Peninsula, Norway and Sweden; more often to include these countries plus Denmark; and sometimes being extended to include these three plus Iceland. When Finland is included in the group, the term *Fennoscandia* or Fennoscandian countries is used, thus taking account of the greater difference of the Finns in ancestry, historical relationships, and language as compared with the other nations. These differences, however, are overshadowed by many cultural, economic, environmental, and other similarities between Finland and the other countries of the group.

DOMINANT TRAITS OF NORTHERN EUROPE AS A REGION

''The North'' is a good descriptive term for these lands. No other highly developed countries have their principal populated areas so near the pole. Located in the general latitude of Alaska and occupying a geographic position in Eurasia somewhat analogous to that of Alaska in North America, the countries of Northern Europe represent the northernmost concentration of advanced industrial civilization in the world.

CLIMATIC EFFECTS OF THE ATLANTIC

West winds from the Atlantic, warmed here in winter by the North Atlantic Drift, moderate the climatic effects of northern location considerably. Temperatures average above freezing and harbors are ordinarily ice-free in winter over most of Denmark and along the coast of Norway. But away from the direct influence of the west winds, winter temperatures average below freezing and are particularly severe at elevated, interior, and northern locations. In summer the ocean tends to be a cooling rather than a warming influence, and most of Northern Europe has Fahrenheit temperatures in July that average no higher than the 50s or low 60s (10° to 18°C). Highlands have temperatures sufficiently low that a number of glaciers exist, both on the Scandinavian Peninsula and in Iceland. Despite the overall moderation of the climate as compared with what might be expected from the latitude, the populations of the various countries tend to cluster in the southern sections, and all countries except Denmark have considerable areas of sparsely populated terrain where the problems of development are largely those of overcoming a northern environment.

HISTORICAL AND CULTURAL UNITY

Close historical interconnections and cultural similarities are more important factors in the regional unity of Northern Europe than partial similarity of environmental problems, however. Historically, each country has been more closely related to others of the group than to any outside power. In the past, warlike relations often prevailed, and for considerable periods some countries ruled others of the group. But since the early 19th century relations between them have been peaceful. This was true in 1905 when Norway separated from Sweden, and again in 1918 when Iceland separated from Denmark. In this peaceful climate the feelings of relationship among the Northern European peoples have come to be expressed in close international cooperation among the five countries.

Cultural similarities among the countries of Northern Europe are many. Similarities of language, religion, and form of government are important examples. The languages of Denmark, Norway, and Sweden are descended from the same ancient tongue and are mutually intelligible, although not identical. Icelandic, although a branch of the same root, is

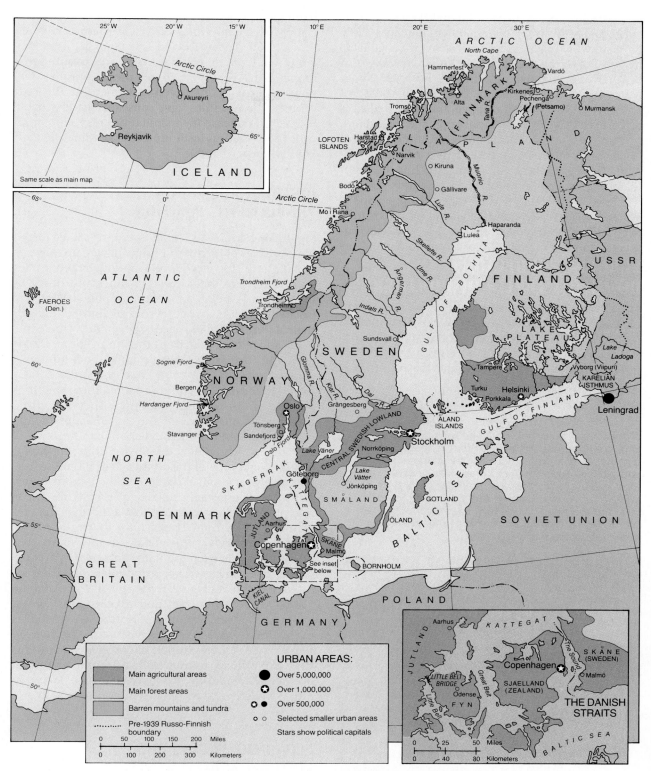

Figure 9.1   *Index map of Northern Europe. Small tracts of agricultural land are scattered through the areas of forest, mountain, and tundra, and small forest tracts occur within the agricultural areas. City size symbols are based on metropolitan-area estimates.*

more difficult for the other peoples only because it has changed less from the original Germanic language and borrowed less from other languages. Only Finnish, which belongs to a different language family, is entirely distinct from the other languages of Northern Europe. Even in Finland, however, about 6 percent of the population is of Swedish descent and speaks Swedish as a native tongue. Swedish is recognized as a second official language in Finland.

Among these countries there are no exceptions to the cultural unity embodied in a common religion. The Evangelical Lutheran Church is the dominant religious organization in each country. Eighty-eight to 97 percent of the respective populations adhere to it, at least nominally. It is a state church, supported by taxes levied by the respective governments, and is probably the most all-embracing organization outside of the state.

The countries of Northern Europe also exhibit basic similarities with respect to law and political institutions. These countries have very old traditions of individual rights, broad political participation, limited governmental powers, and democratic control. Thus, old foundations have been available for building modern democracies, and the countries of

Northern Europe are recognized as outstanding strongholds of democratic institutions. Iceland claims to have the world's oldest legislature, founded in A.D. 930. Today Iceland is a republic, as is Finland, while the other three states are constitutional monarchies (Fig. 9.2). In all these countries real power rests with an elected parliament. In the 20th century, the countries have consciously and actively worked to increase their similarities respecting legal codes and political institutions by coordinating their laws wherever feasible.

DISADVANTAGES OF AN "IN-BETWEEN" POSITION

Small populations and limited resources have forced the countries of Northern Europe to give up imperial ambitions during recent times, although their armies and fleets were the scourge of much of Europe in times past. A policy of neutrality plus a relatively isolated position in one corner of Europe allowed them a long period of peace between the Napoleonic Wars and World War II. In the 20th century, however, the increasing strategic importance of North Atlantic air and sea routes has jeopardized their safety. In the present world situation, these countries occupy an "in-between" position. They lie on the most direct routes between the United States and the western coreland of the Soviet Union. The coast of Norway, which adjoins the Soviet Union in the far north, offers some of the world's best and most strategically located naval harbors in the famous fjords. This coast is especially suitable as a base of submarine operations against North Atlantic shipping and was so used by the German navy in World War II. Denmark, Norway's neighbor to the south, lies across the outlet from the Baltic Sea, on which some of the main Soviet seaports and naval bases are located. Finland lies between the Soviet Union and the Scandinavian Peninsula, and Sweden, the largest and most powerful country of Northern Europe, lies in the midst of these various positions. The changed significance of their position has presented the countries of Northern Europe with a common problem of national security, evidenced by the fact that only Sweden escaped military occupation in World War II.

ECONOMIC, SOCIAL, AND CULTURAL ACHIEVEMENTS

Small size and resource limitations have made it necessary for each of the countries of Northern Europe to build a highly specialized economy in attempting

Figure 9.2 Northern, Southern, and East Central Europe were powerfully influenced in past times by strong kingdoms and empires that repeatedly gained and lost territories as their military fortunes waxed and waned. Even small Denmark was once a formidable power. Today a reminder of its past military prowess is seen in the daily changing of the guard at the royal palace in Copenhagen. Denmark is one of several European countries that are constitutional monarchies.

to attain a high standard of living. Success in such endeavors has been so marked that these countries are probably known as much for high living standards as for any other characteristic. In all five countries, high standards of health, education, security for the individual, and creative achievement are evidenced by impressive health statistics, long life expectancy, almost nonexistent illiteracy, and disproportionately great achievements in the arts and science.

In their attack on economic problems, the countries of Northern Europe have tended to employ a moderate socialism, consciously seeking a "middle way" between uncontrolled capitalism and communism. They have attempted to put a floor under the living standard of every member of the community, while closely limiting the accumulation of wealth. Great emphasis on conservation of resources, the exercise and general acceptance of economic control and initiative by the state, and often the development of resources cooperatively by the state and private enterprise are prominent features of economic life in these countries. At the same time private business and ownership are fostered by the state in many lines of activity, as is trade unionism. Parallel with the development of this "middle way," the countries of Northern Europe have experienced the world's greatest development of the private cooperative type of economic enterprise, reaching into almost every phase of production, distribution, and consumption.

All the Northern European countries share the general traits of the region in greater or less degree, but each country has its own mix of geographic characteristics and national experiences. We turn now to brief analyses of the individual countries, stressing the distinctive emphases in each.

AGRICULTURE'S KEY ROLE IN DENMARK

Denmark has the somewhat paradoxical distinction of possessing the largest city in Northern Europe and of being at the same time the most dependent on agriculture of any country in the region. The Danish capital of Copenhagen (nearly 1½ million) has well over a fourth of Denmark's population in its metropolitan area. Denmark has a much greater density of population than the other countries of Northern Europe, a fact accounted for by the presence of Copenhagen, the greater productivity of the land agriculturally, and the lack of any sparsely populated zone of frontier settlement.

COPENHAGEN AND THE DANISH STRAITS

Copenhagen lies on the island of Sjaelland (Zealand) at the extreme eastern margin of Denmark (see Fig. 9.1). Sweden lies only 12 miles (c. 20 km) away across The Sound, the main passage between the Baltic and the wide Kattegat and Skaggerak straits leading to the North Sea. Copenhagen grew beside a natural harbor well placed to control all traffic through The Sound, which is the most direct and most used channel for traffic in and out of the Baltic. For many years before the 17th century, Denmark controlled adjacent southern Sweden, as well as the less favored alternative channels to The Sound—the Great Belt and Little Belt. Toll was levied on all shipping passing to and from the Baltic. Although the days of levying toll are now long past, Copenhagen still benefits from its strategic location. The city does a large transit and entrepôt business in North Sea-Baltic trade and has encouraged this business by setting up a free zone in its harbor, where goods destined for redistribution may be landed without paying customs duties. Increasing trade has led to the development of industry, and the city is the principal industrial center of Denmark as well as its chief port and capital. Within a diversified industrial structure, food processing, chemicals, and engineering are outstanding branches. The food-processing industry prepares products of Denmark's specialized agriculture for export and processes the many foods that Denmark must import. Among the products of the chemical industry are many organic chemicals derived from by-products of Danish agriculture's meat production.

FACTORS IN DENMARK'S AGRICULTURAL SUCCESS

Denmark stands out sharply from the other countries of Northern Europe in the nature of its land and the place of agriculture in its economy. The topography of the country results principally from glacial deposition. The western part of the peninsula of Jutland consists mainly of sandy outwash plains (deposited by glacial meltwater) and coastal dunes. Eastern Jutland and the Danish islands exhibit a rolling topography of ground and terminal moraines (irregular sheets or ridges of material deposited directly by the glacier). The highest hill in the country is less than 600 feet (less than 183 m) in elevation. Although sandy areas of the west are not very fertile in their natural condition, they have mostly been reclaimed and are cultivated, while the clay soils which characterize the moraine areas, with greater natural fertility, support a very intensive and productive agriculture upon which the country's prosperity is largely based.

Most of Denmark is available for farming, and about three fifths of the entire country is cultivated—the largest proportion of any European country. It is fortunate that so much of Denmark is arable because the country is practically without natural resources except for soil, climate, and its strategic position for trade. Danish agriculture is so efficient that only 6 percent of the working population is employed on the land. Agriculture, however, is basic to the country's economy. Many additional workers are engaged in processing and marketing agricultural products, and many others in supplying the needs of the farms. More than one third of Denmark's exports come from agriculture.

Few countries or areas which depend so heavily on agriculture are as materially successful as Denmark has been during the 20th century. Danish agriculture is based on a highly specialized and very consciously and carefully fostered development of animal husbandry, which began to be emphasized when the competition of cheap grain from overseas brought ruin to the previous Danish system of grain farming in the latter part of the 19th century. Until recently, Danish animal husbandry was primarily dairy farming, but in recent years the production and sale of animals for meat has become dominant over dairying.

The following are some favorable factors that have contributed to the success of Danish animal husbandry:

1. The land and climate of the country are well suited to the strong emphasis on fodder crops such as barley, fodder beets, and potatoes.

2. In recent years it has been both possible and profitable for Denmark to import large quantities of supplementary feeds.

3. Nearby markets—notably in West Germany, the United Kingdom, and Sweden—have been adequate to absorb the bulk of Denmark's meat and dairy exports.

4. The Danish state has long supported agriculture by such measures as encouraging the replacement of large estates by family farms, financing agricultural research and education, and extending financial aid to private reclamation projects, especially in the sandy west.

5. Cooperative societies have developed to a point that practically all Danish farmers are members. The cooperatives process and export the farmers' products, provide farm supplies at advantageous prices, and enforce rigid standards of quality control. They make the Danish farmer his own middleman to some extent and give him the benefit of large-scale marketing and buying.

6. Probably underlying much of the success of Danish agriculture, and especially the success of the cooperative movement, is a very high level of education. The traditional school system, noted for high standards, is supplemented by various forms of adult education. An emphasis on the continuing education of adults is characteristic not only of Denmark but of all the countries of Northern Europe.

Thus Danish agriculture has become noted for its efficiency, rationality, and prosperity. To maintain these characteristics, rapid changes have been necessary in recent years. A relative shift from dairy to meat production, a rapid decline in the farm labor force, larger farms, increased mechanization, and rising output are all striking trends.

REGIONAL AND NATIONAL EMPHASES IN NORWAY

Norway stretches for well over a thousand miles (c. 1600 km) along the west side and around the north end of the Scandinavian Peninsula. The peninsula, as well as Finland, occupies part of the Fennoscandian Shield, a block of very ancient and hard rocks that is similar geologically to the Canadian or Laurentian Shield of North America. The western margins of the block in Norway are extremely rugged, being composed of mountains, uplands, and steep-sided valleys that were heavily eroded by the continental glaciation which centered in Scandinavia. Most areas have little or no soil to cover the rock surface, scraped bare by glaciation. About 70 percent of Norway is classified as wasteland. Only about 3 percent is classed as arable or pasture land, and the remainder is forest land.

The nature of the terrain hinders not only agriculture but also transportation. Glaciers deepened the valleys of streams flowing from the mountains into the sea, and when the ice melted, the sea invaded the valley floors. Thus were created the famous fjords—long, narrow, and deep extensions of the sea into the land, usually edged by steep valley walls (Fig. 9.3). Some run 100 miles (161 km) or more inland. The difficulties of building highways and railroads parallel to such a coast are obvious. Coastal steamers and planes must bear a major transport burden. Coastal shipping lanes are sheltered from Atlantic storms by a screen of small islands paralleling the coast. The fjords of Norway provide some of the finest harbors in the world, but

Figure 9.3   This large plant, located at the head of a steep-walled fjord on Norway's southwestern coast, uses hydroelectricity to produce aluminum metal from imported alumina (the intermediate-stage product from bauxite ore). Aluminum ingots on the wharf await shipment to market. The plant, called the Ardal and Sunndal Aluminum Works, is about 70 miles (113 km) southwest of Trondheim.

most of them have practically no hinterland. Much of Norway's population is scattered along them in many relatively isolated clusters. Many settlements are small and dependent on some combination of farming, fishing, and forestry. Larger towns and cities are industrial, trading, and service centers.

FRONTIER DEVELOPMENT IN NORTHERN NORWAY

The northern sections of Norway, Sweden, and Finland and adjacent parts of the Soviet Union are the homeland of the Lapps (they prefer the name Sami), a minority people whose best-known occupation is reindeer herding on a nomadic or seminomadic basis. Today some Lapps continue to gain their livelihood in this manner, but a large majority are settled in permanent homes and support themselves primarily by farming, fishing, or a combination of these. Reindeer herding was dealt a heavy blow in 1986 when radioactive fallout from the Chernobyl nuclear explosion in the Soviet Union contaminated the wild forage used by Lapps to support their herds.

The Norwegian coast as far south as Trondheim exhibits for the most part a scatter of small villages with marginal economies tied to fishing or forestry. Larger towns are few, and none is larger than about 50,000 in population. For some years after World War II, Norway followed a policy (more recently somewhat de-emphasized) of promoting industrial development along its northern coast. As a result, the larger settlements grew rapidly. Fish processing, metal industries, and textile plants were developed, along with iron-ore exports from Norwegian deposits near the port of Kirkenes. The port of Narvik continues to be the Atlantic outlet for the large iron-ore production of northern Sweden, a function which made Narvik a major German military objective in World War II.

THE IMPORTANCE OF HYDROPOWER ON NORWAY'S SOUTHWEST COAST

The southwestern coastal area of Norway, from Trondheim south, has a number of advantages over the coast farther north. Agricultural conditions are slightly more favorable, some areas of forest occur, and the location is better for utilization of the hydroelectric power which is one of Norway's primary resources (hydroelectricity represents 99.7% of all electricity generated). Along the southwest coast a considerable electrometallurgical and electrochemical industry has come into being. The raw materials are imported for the most part and the products exported. The most important of many products is aluminum. Three main urban communities—Bergen, Trondheim, and Stavanger—are spaced along this coast. All are between 125,000 and 250,000 in metropolitan population.

THE SOUTHEASTERN CORE REGION

More than half of the population of Norway lives in the southeast, which centers on the capital, Oslo (725,000). This is the core region of modern Norway. Here, where valleys are wider and the land is less rugged, are found the most extensive agricultural lands and the largest forests in the country. Streams coming down from the mountains to the west and north furnish power for sawmilling, pulp and paper production, metallurgy, electrochemical industries, and industries which produce various consumer goods for Norwegian consumption. Oslo lies at the head of Oslo Fjord where several valleys converge. It is the principal seaport and industrial, commercial, and cultural center of Norway as well as its capital and largest city.

NORWAY'S RESOURCES AND NATIONAL ECONOMY

Apart from its scanty agricultural base, Norway has rich natural resources relative to the size of its population, which is small and increasing at a very slow rate. Key resources include spectacular scenery, waterpower, fish, forests, and a variety of minerals. The country's resource position has been enhanced greatly by the recent discoveries and development of large oil and natural gas deposits located far offshore in the Norwegian sector of the North Sea floor. Like the United Kingdom, Norway has become an exception among European countries in being a net exporter of crude oil and gas. In a few years such exports grew from nothing to the country's largest exports by value, and Norway advanced economically from being a reasonably well-off country to one of Europe's very richest as measured by per capita GNP.

Meanwhile, older economic emphases remain important. Manufacturing is a large employer and provides exports of goods related to Norway's resources. Prominent examples include nonferrous metals such as aluminum (which are processed with Norway's abundant and cheap electricity), fish products, and wood products. Norway has had a major fishing industry for centuries. Today it is one of Europe's two leading nations (with Denmark) in quantity of fish caught. An ancient seafaring tradition going back to the Vikings is expressed in one of the world's largest merchant fleets. The country's agriculture is severely retarded by the environment, but about 6 percent of all Norwegians still farm (1987). Farm output falls far short of domestic needs, and large quantities of most foods are imported.

THE DIVERSIFIED GEOGRAPHY OF SWEDEN

Sweden is the largest in area and population (8.4 million) and the most diversified of the countries of Northern Europe. In the northwest it shares the mountains of the Scandinavian Peninsula with Norway; in the south it has rolling, fertile farm lands like those of Denmark; in the central area of the great lakes another relatively extensive area of good farm land occurs. To the north, between the mountains and the shores of the Baltic Sea and Gulf of Bothnia, Sweden consists mainly of ice-scoured, forested uplands similar to those which constitute the greater part of Finland. A smaller area of the latter type occurs south of Lake Vätter.

HIGH IMPORTANCE OF SWEDISH ENGINEERING AND METALLURGY

The high development of engineering and metallurgical industries is the feature which most distinguishes the Swedish economy from that of the other countries of Northern Europe. These industries have evolved out of a centuries-old tradition of metal mining and smelting. The high-grade iron ores of the Bergslagen region, just north of the Central Swedish Lowland, have been extracted and made into iron and steel since medieval times. Originally the fuel used to smelt the ore was charcoal, made from the wood of the surrounding forest. When coal and coke began to replace charcoal as fuel in Europe's iron industries, Sweden was handicapped by the absence of any major coal resources. In the age of electricity the Swedes responded to this problem by a heavy reliance on electric furnaces using current from hydropower plants; more recently this kind of smelting has been supplemented by a Swedish-invented oxygen process. The electric process is expensive but gives an extraordinary control of quality, and "Swedish steel" became practically a synonym for steel of the highest grade. The modern Swedish steel industry still centers in the Bergslagen region, but the output from that area is supplemented by the production from some large conventional plants that have been built on the Baltic and Bothnian coasts to use imported coal and coke. Although large quantities of ordinary steel are now produced in Sweden, it is still the ultra-fine steels and the products made from them that impart a special character to Swedish metallurgy. Some ore continues to come from the Bergslagen district, but this source is overshadowed today by the output of high-grade ore from Kiruna and Gällivare. These are mining settlements located in the northern wilderness of Sweden far above the Arctic Circle. The northern mines produce mainly for export, through Luleå on the Bothnian coast in summer when the Baltic is free of ice, and through Norway's ice-free Atlantic port of Narvik the year round.

The Swedish steel industry's emphasis on skill and quality carries over to the finishing and fabricating industries that use the steel. Among Swedish specialties yielding a large volume and value of exports are automobiles, industrial machinery (for example, paper-milling equipment, lifting and loading machines, drills, and cream separators), office machinery, telephone equipment, and electric-transmission equipment. The country's reputation for skill in design and quality of product also extends to items such as ball bearings, cutlery, tools, surgical instruments, and the glassware and furniture manufactured in Småland, the infertile and rather sparsely populated plateau south of the Central Lowland. Robots and many other high-technology items are increasingly important in Sweden's export trade.

Swedish manufacturing is principally carried on in numerous small industrial centers in the Central Lowland. These places, ranging in population from about 100,000 to mere villages, have grown up in an area favorably situated with respect to minerals, forests, waterpower, labor, food supplies, and trading possibilities. The Central Lowland has been the historic core of Sweden and has long maintained an important agricultural development and a relatively dense population.

FOREST INDUSTRIES

Except in the far north, where the forest cover is sparser and the trees are stunted, practically all of Sweden is naturally forested, mainly with coniferous softwoods. The wood-products industries based on these forests provide important exports (about one

Figure 9.4 Logs in central Sweden. Most of Sweden's forest industries are located further north in the zone of subarctic climate.

fourth of all exports in the middle 1980s). These industries include sawmilling, pulp milling, papermaking, the manufacture of wood chemicals and synthetic fabrics, and the production of fabricated articles such as plywood. Logging and wood industries are characteristic of most of Sweden (Fig. 9.4), but the main concentration is found in areas to the north of the Central Lowland. Here other economic opportunities are less abundant, large quantities of good timber are available, and even before the age of the truck, logs could be transported with relative ease—in winter by sled and in summer by floating them to mills on the numerous rivers. Most of the sawmills and pulp mills are located in industrial villages and towns which dot the coast of the Gulf of Bothnia at the mouths of rivers. Power for the milling operations is supplied by numerous hydroelectric stations, and much electricity is transmitted by high-tension systems to the central and southern parts of the country, where growing demands cannot be met by streams which are now almost completely developed for power. Despite widespread fears concerning atomic power, the Swedes have built a series of nuclear stations that generated 51 percent of the nation's electric output in 1986 (hydropower, 44%; imported fossil fuels, 5%). However, this may be a passing phase as the Swedish government, beset by intensified anti-nuclear protests since the Chernobyl disaster in the USSR, has announced plans to eliminate Sweden's nuclear stations by the end of the century.

SWEDEN'S SELF-SUFFICIENT AGRICULTURE

Unlike Denmark, Sweden has not developed a specialized export-oriented agriculture; and unlike Norway, Sweden commands enough good land to supply all but a minor share of its small population's food requirements. More than 90 percent of Sweden's territory is nonagricultural, but there is good

193

land in Skåne, at the southern tip of the peninsula, and in the Central Lowland; and there are patches of land in some other regions that are at least usable. Skåne, with good morainal soils like those of adjacent Denmark, and with Sweden's mildest climate, is the prime agricultural area. Its wheat-oriented agriculture is the country's main source of bread grain. In the Central Lowland, fairly fertile patches of farm land, derived largely from glacial and marine deposits, are scattered through a forested, lake-strewn terrain. Fodder crops support a livestock and dairy production that is marketed in the region's industrial towns and cities. Elsewhere in Sweden, agriculture is restricted or precluded by bare rock surfaces, infertile soils, and, in the north, a harsh climate.

STOCKHOLM AND GÖTEBORG

Stockholm (1.4 million) is the second largest city in Northern Europe (Fig. 9.5). The location of the capital reflects the role of the Central Lowland as the early core of the Swedish state and the early orientation of that state toward the Baltic and trans-Baltic lands. Stockholm is the principal administrative, financial, and cultural center of the country and shares in many of the manufacturing activities typical of the Central Lowland.

But in the past century, as Sweden has come to do more and more trading via the North Sea, Stockholm has been displaced by Göteborg (675,000), at the other end of the Central Lowland, as the leading port of the country and, in fact, of all Northern Europe. In addition to its advantage of position, the latter city has a harbor which is ice-free the year round, whereas icebreakers are needed to keep Stockholm's harbor open in midwinter. Göteborg combines its trade functions with numerous manufacturing activities.

THE SWEDISH POLICY OF NEUTRALITY AND PREPAREDNESS

In the Middle Ages and early modern times, Sweden was a powerful and imperialistic country. Finland was conquered in the 12th and 13th centuries, and in the 17th century the Baltic became almost a Swedish lake. During the 18th century, however, the rising power of Russia and to some extent of Prussia put an end to Swedish imperialism, aside from a brief campaign in 1814 through which Sweden won control of Norway from Denmark. Since 1814 Sweden has never been engaged in a war, and it has become known as one of Europe's most successful neutrals. The long period of peace has undoubtedly been partially responsible for the country's success in attaining a high level of economic welfare and a reputation for social advancement. At present, however, Sweden is carrying a heavy burden of armaments. The dangers inherent in the international situation have led Sweden's government to build up strong armed forces while maintaining the nation's traditional policy of neutrality.

FINLAND AND ITS RUSSIAN CONNECTION

Conquered and Christianized by the Swedes in the 12th and 13th centuries, Finland was ceded in 1809 by Sweden to Russia, and it was controlled by the latter nation until 1917. Under both the Swedes and

Figure 9.5 *Some of Northern Europe's distinctive urban architecture is shown in this view of Stockholm. Close-set buildings of the Old City surround the church with the prominent green spire in the center of the photo. Part of the harbor with ships anchored at wharves can be seen at the upper right.*

the Russians, the country's status was that of a semi-autonomous grand duchy, and its people developed their own culture and feelings of nationality to such an extent that the opportunity for independence provided by the collapse of Tsarist Russia toward the end of World War I was eagerly seized.

Most of Finland is a sparsely populated, glaciated, subarctic wilderness of coniferous forest, ancient igneous and metamorphic rocks, thousands upon thousands of lakes, and numerous swamps. Despite this somewhat hostile environment, Finland was primarily an agricultural country until very recently, and the majority of the population is concentrated in relatively fertile and warmer lowland districts scattered through the southern half of the country. Hay, oats, and barley are the main crops of an agriculture which is predominantly directed toward livestock production, especially dairying. With the aid of subsidies, Finnish agriculture has been able to reach a level of production that provides approximate self-sufficiency in cereals and small exports of dairy products and meat. Much woodland is present on most farms, and farmers tend to combine agriculture with forestry.

To pay for a wide variety of imports, the nation depends quite heavily on exports of forest products. About two thirds of Finland is forested, mainly in pine or spruce, and in 1986 approximately two fifths of the country's exports consisted of paper, lumber, wood pulp, and other timber products of a type largely duplicating the forest exports of Sweden. Forest production is especially concentrated in the south central part of the country, often referred to as the Lake Plateau. A poor and rocky soil discourages agriculture here, but the timber is of good quality and a multitude of lakes, connected by streams in interlocking systems, aid in transporting the logs.

Helsinki (900,000), located on the coast of the Gulf of Finland, is the national capital, largest city, main seaport, and principal commercial and cultural center. It is also the most important and diversified industrial center. Smaller industrial centers include Tampere (250,000), located in the southwestern interior, and the port of Turku (225,000), at the southwestern corner of the country. Each is important for textile production. The industries of these cities were powered originally by hydroelectricity, but this source has now been surpassed by nuclear power and almost equaled by electricity from imported fossil fuels.

PROBLEMS OF A BUFFER STATE

For centuries Finland has been a buffer between Russia, Scandinavia, and the former eastward reach of German power. Currently it has a peculiar political status and situation, due largely to events of World War II. In 1939, shortly after the outbreak of the war, Finland refused to accede to Soviet demands for the cession of certain strategic frontier areas and, despite a valiant resistance, was overwhelmed by the Soviet Union in the "Winter War" of 1939–1940. Then, in an attempt to regain what had been lost, Finland fought with Germany against the Soviet Union from 1941 to 1944, and again was defeated as German power waned.

The peace settlement following the war left Finland still in existence as an independent country, although shorn of considerable areas in the east and north which were annexed by the USSR (see Fig. 9.1). In the southeast the Karelian Isthmus between the Gulf of Finland and Lake Ladoga passed into Soviet control, and with it went the city of Viipuri, now Vyborg, which had been Finland's main timber port. More territory was lost along the central part of the eastern frontier, and in the north the small port of Petsamo (now Pechenga) and the area around it was ceded. The latter cession cut Finland off from access to the Arctic Ocean and gave the Soviet Union rich nickel mines and smelting facilities, as well as a land frontier with Norway.

Besides territorial concessions the Soviet Union insisted that Finland exhibit a reasonably friendly and cooperative attitude toward the USSR. In view of these events and circumstances, plus the growth of her huge neighbor's power, Finland has felt compelled to follow a very careful line of neutrality with respect to big-power quarrels. At the same time, however, the Finns have increasingly participated in inter-Scandinavian cooperation, both as a matter of practical interest and as an expression of their basically Western preference and orientation.

Following the war the economy of Finland was heavily burdened by (1) the necessity for rebuilding the northern third of the country, which was devastated by retreating German soldiers after Finland surrendered to the USSR; (2) the necessity for resettling one tenth of the total population of the country after they fled as refugees from the areas ceded to the Soviets; (3) the loss of the ceded areas themselves, some portions of which were of disproportionate importance in the prewar economy of the country; and (4) the necessity for making large reparations payments to the Soviet Union. In spite of these various difficulties, however, the Finns were able to make a rapid recovery from the war period. In the course of this the economy of the country was drastically changed in some respects. One of the most striking changes was the rise of metalworking industries. This was made necessary by the fact that the Soviet Union required a large part of the reparations to be paid in metal goods. Several small steel plants and other metalworking establishments were built to meet this demand, and they have continued to operate since the end of reparations. The output includes a variety of machines, other secondary metal products, and basic metals. Metal goods now have about the same value as forest products in Finland's export trade. The country has developed important export specialties in wood-processing machinery of various kinds and specialized ships such as icebreakers.

ICELAND: AFFLUENCE FROM BLEAK SURROUNDINGS

Iceland is a fairly large, mountainous island in the Atlantic Ocean just south of the Arctic Circle. Its rugged surface shows the effects of intense glaciation and vulcanism. Some upland glaciers and many active volcanoes and hot springs remain. The vegetation consists mostly of tundra, with considerable grass in some coastal areas and valleys. Trees are few, being discouraged by summer temperatures averaging about 52°F (11°C) or below as well as by the prevalence of strong winds. The cool summers are also a great handicap to agriculture. Mineral resources are almost nonexistent.

Despite the deficiencies of its environment, however, Iceland has been continuously inhabited at least since the 9th century and is now the home of a progressive and democratic republic with a population of about 250,000 in 1988. Practically all the population lives in coastal settlements (Fig. 9.6), with the largest concentration in the vicinity of the capital, Reykjavik, which itself has about 125,000 people in its metropolitan area. Due to the proximity of the relatively warm North Atlantic Drift, the coast of Iceland, and especially the southern coast where Reykjavik is located, have winter temperatures which are unusually mild for the latitude. Reykjavik has an average January temperature that is practically the same as the average for New York City.

The economy of Iceland depends basically on fishing, farming, and the processing of their products. Agriculture is centered around the raising of cattle and sheep. Farm land is used for hay and pasture and a limited production of potatoes and hardy vegetables. Although some agricultural products are exported, the real backbone of the economy is fishing. In 1986, fish products supplied about three fourths of all exports by value. Hence they are the main source of payment for the many kinds of goods which must be imported. Manufacturing is mostly, although not entirely, confined to food processing.

Figure 9.6   *A residential section in Akureyri, Iceland's second largest city. Above the spic-and-span houses of this north-coast city towers one wall of the long fjord on which Akureyri is located.*

It has been encouraged somewhat in recent times by the development of a small part of the island's considerable potential of hydroelectric power. One result has been the establishment of an aluminum mill that processes imported materials for export. Hydropower supplies 95 percent of Iceland's electricity.

For centuries Iceland was a colony of Denmark. In 1918 it became an independent country under the same king as Denmark and in 1944 declared itself a republic. It attracted much notice at the beginning of World War II because of its strategic position along major sea and air routes across the North Atlantic. The island was used by British and American forces as an air and sea base during the war. As a member of the North Atlantic Treaty Organization (NATO), Iceland is host to a small American military contingent which maintains base facilities at the Keflavik international airport near Reykjavik.

GREENLAND, THE FAEROES, AND SVALBARD

Two countries of Northern Europe, Denmark and Norway, possess outlying islands of some significance. Denmark holds Greenland, the world's largest island, off the coast of North America, and the Faeroe Islands between Norway and Iceland. Both of these areas are now considered integral, although self-governing, parts of Denmark, and their peoples have equal political rights with other Danish citizens. Although the area of Greenland, approximately

840,000 square miles (2.2 million sq km), is nearly one fourth that of the United States, about 85 percent is covered by an ice cap, and the population, mainly distributed along the southwestern and southeastern coasts, amounts to only about 55,000 persons. Nearly all of this population is of mixed Eskimo and Scandinavian descent. The principal means of livelihood are fishing, hunting, trapping, a limited amount of sheep grazing, and the mining of zinc and lead. Under the auspices of NATO, of which Denmark is a member, joint Danish-American air bases are maintained in Greenland; and, as a part of its early warning system, the United States has a giant radar installation at Thule in the remote northwest.

The Faeroes are a group of treeless islands where approximately 47,000 people of Norwegian descent make a living by fishing and grazing sheep.

Norway controls the island group of Svalbard, located in the Arctic Ocean and commonly known as Spitsbergen. Although largely covered by ice, the main island (called Spitsbergen) has the only substantial deposits of high-grade coal known to exist in Northern Europe. Mining operations are carried on by Norwegian and Soviet Russian companies, with the coal being shipped to Norway and to the Soviet port of Murmansk. Norway also holds the volcanic island of Jan Mayen in the Arctic plus two small islands in the far South Atlantic, and claims a share of Antarctica.

Ancient Splendor and Modern Problems in the Countries of Southern Europe

On the south the continent of Europe is separated from Africa by the Mediterranean Sea, into which three large peninsulas extend (Fig. 9.7). To the west, south of the Pyrenees Mountains, is the Iberian Peninsula, unequally divided between Spain and Portugal. In the center, south of the Alps, is the Italian Peninsula and its southern offshoot, the island of Sicily. To the east, between the Adriatic and Black seas, is the Balkan Peninsula, from which the Greek subpeninsula extends still farther south between the Ionian and Aegean seas. The four main countries which occupy the two western peninsulas and the Greek subpeninsula—Portugal, Spain, Italy, and Greece—along with the small island country of Malta, the mainland microstates of Vatican City (enclosed within the city of Rome) and San Marino, and the British colony of Gibraltar—may be conven-

iently grouped as the countries of Southern Europe. Three of the peninsular countries include islands in the Mediterranean, the largest of which are Sicily and Sardinia, held by Italy; Crete, held by Greece; and the Balearic Islands, held by Spain. These islands are governed as integral parts of their respective countries. Some Mediterranean islands lie outside these countries, the most notable being Corsica, which is a part of France, and Cyprus (see Chapter 13). The independent island state of Malta was a possession of the United Kingdom until 1964. Gibraltar, a peninsula on the south coast of Iberia, remains a British possession despite Spanish demands that it be returned to Spain.

The countries of Southern Europe exhibit many natural and cultural similarities as a group. Most of their natural characteristics, however, while tending to differentiate them from other parts of Europe, are shared with lands of northern Africa and southwestern Asia which front on the Mediterranean. Throughout the Mediterranean area the broad pattern of natural features tends to be much the same, despite differences in detail from place to place. There also tends to be a broad similarity in cultural practices. However, the countries of Southern Europe are distinguished as a group from their Mediterranean neighbors by important cultural differences, including differences in religion and language. Most African and Asian lands fronting on the Mediterranean are predominantly Muslim in religion, whereas Roman Catholicism is the prevailing religious faith in Spain, Portugal, and Italy, and Orthodox Christianity is dominant in Greece. The Spanish, Portuguese, Italian, and Greek languages are, of course, quite distinct from Arabic, the principal language in most of the non-European Mediterranean countries. Moreover, the countries of Southern Europe have shared in the development of Western or Occidental culture, whereas most of the African and Asian countries of the Mediterranean realm have principally been influenced by the culture of Islam.

THE DISTINCTIVE MEDITERRANEAN CLIMATE

A distinctive natural characteristic of Southern Europe is its climate, which typically combines mild, rainy winters with hot, dry summers. The Mediterranean area has given its name to this particular combination of climatic qualities. In systems of climatic classification a "mediterranean" type of climate is customarily recognized, although the designation "dry-summer subtropical," also in common

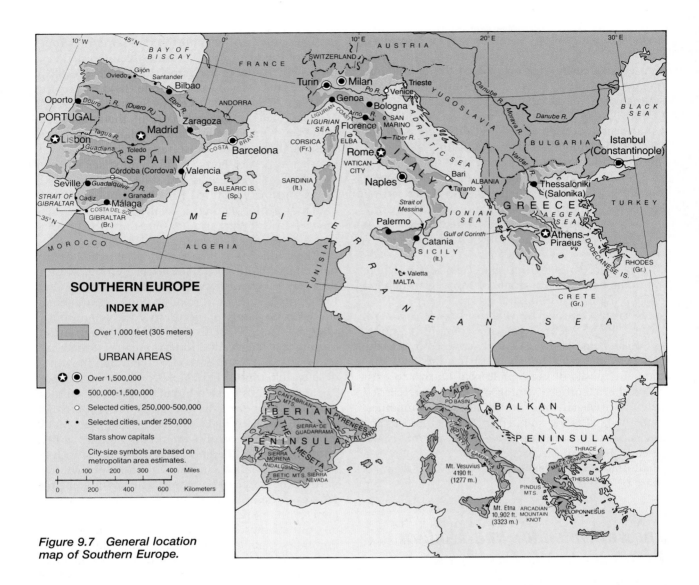

Figure 9.7 General location map of Southern Europe.

use, is perhaps more descriptive. Other areas having this type of climate occur in southern California, central Chile, southwestern South Africa, and southern Australia (see world climatic map, Fig. 2.18).

Generally speaking, the countries of Southern Europe experience temperatures averaging 40° to 50°F (4° to 10°C) in the coldest month and 70° to 80°F (21° to 27°C) in the warmest month. The total precipitation received during a year varies considerably from place to place, in response to differences in elevation and exposure to rain-bearing winds. The general average is between 15 and 35 inches (*c.* 40–90 cm) per year, with most places falling in the lower half of this range. But regardless of total precipitation, most areas experience the characteristic seasonal regime of relatively moist winters and dry summers.

In Southern Europe the characteristics associated with the mediterranean climate become increasingly pronounced toward the south. The northern extremities of both Spain and Italy have atypical climatic characteristics. Except for a strip along the Mediterranean, northern Spain has a marine climate like that of northwestern Europe, cooler and wetter in summer than the typically mediterranean areas, while the basin of the Po River in northern Italy is distinguished by cold-month temperatures in the lowlands averaging just above freezing and a relatively wet summer. Much of the high interior plateau of Spain, the Meseta, cut off from rain-bearing winds, has somewhat less precipitation and colder winters than is typical of the mediterranean climate, although the seasonal regime of precipitation is characteristically mediterranean. Table 9.1 illustrates

TABLE 9.1   CLIMATIC DATA FOR SELECTED SOUTHERN EUROPEAN STATIONS

Climatic Type or Area and Station	January Average Temperature		July Average Temperature		Average Annual Precipitation		Average Precipitation June–August	
Typically mediterranean								
Athens	48°F	9°C	82°F	28°C	16″	40 cm	1″	3 cm
Rome	46°F	8°C	75°F	24°C	29″	75 cm	2″	5 cm
Palermo	52°F	11°C	75°F	24°C	31″	78 cm	2″	4 cm
Valencia	50°F	10°C	76°F	25°C	15″	39 cm	2″	5 cm
Seville	50°F	10°C	79°F	26°C	22″	56 cm	1″	2 cm
Lisbon	52°F	11°C	72°F	22°C	28″	71 cm	1″	2 cm
Spanish Meseta								
Madrid	41°F	5°C	75°F	24°C	17″	44 cm	2″	5 cm
Po Basin								
Milan	34°F	1°C	73°F	23°C	36″	90 cm	8″	22 cm
Northern Spain (marine west coast climate)								
Oviedo	44°F	7°C	64°F	18°C	37″	94 cm	6″	16 cm

some of the climatic characteristics of Southern Europe.

THE ECOLOGY OF MEDITERRANEAN AGRICULTURE

In the areas of mediterranean climate, frosts are rare, the summers are hot and sunny, and thus the temperature regime is, in general, excellent for agriculture. But total precipitation is generally low, and a summer drought must be faced each year. Thus water is a critical factor. Where irrigation water is available during the summer months, a notable variety of crops can be produced in abundance. But in areas which are not irrigated—and these comprise the great majority—inadequate amounts or seasonal deficiencies of moisture limit the range of agricultural possibilities.

THE BASIC PATTERN OF AGRICULTURE

Agriculture in the Mediterranean Basin is principally based on crops that are naturally adapted to the prevailing climatic regime of winter rainfall and summer drought (Fig. 9.8). Winter wheat is the single most important crop. Barley, a more adaptable grain, tends to supplant wheat in some areas that are particularly dry or infertile, such as the southern part of the interior plateau of Spain. Other typical crops are

olives, grapes, and vegetables. The olive tree and the grapevine have extensive root systems and certain other adaptations which allow them to survive the summer droughts, and they yield for many years. Olive oil is a major source of fat in the typical Mediterranean diet, and virtually all of the world supply is produced in countries that touch or lie near the Mediterranean Sea. The principal use of grapes is for wine, a standard household beverage in Southern Europe and a major export product. Where irrigation water is lacking, types of vegetables are grown which will mature during the wetter winter season or in the spring. The most important are several kinds of beans and peas. These are a source of protein in an area where meat animals make only a limited contribution to the food supply. Feedstuffs are not available in sufficient quantities to fatten large numbers of animals, and parched summer pastures further inhibit the development of the meat supply. Extensive areas which are too rough for cultivation are used for grazing, but their carrying capacity is generally low. Sheep, which can survive on a sparser pasturage than cattle, are the favored animals. They are kept only partially for meat, and the total amount they supply is relatively small. In many places grazing depends on a system of transhumance utilizing lowland pastures during the wetter winter and mountain pastures during the summer. In some areas nonfood crops supplement the basic Mediterranean products. An example is tobacco, which is an export of Greece.

Figure 9.8 A classic Mediterranean landscape in Tuscany, north central Italy, near Florence. The stone town of San Gimignano on the hilltop overlooks small farm homes and vegetation characteristic of the Mediterranean world. Note the close-packed rows of grapevines in the vineyard (deep green) to the immediate left of the stone houses in the center of the view.

Areas in which the supply of available moisture is either considerably above or considerably below average tend to diverge from the normal pattern of agriculture just described. The drier areas depend more on barley than wheat, and the very driest areas depend mainly on grazing sheep and goats. Some wet and rough areas which have remained in forest are also grazed, particularly oak forests where pigs can feed on the fallen acorns or mast. The bark of one type of oak, the cork oak, supplies cork exports for Portugal and Spain, with Portugal being the world's leading supplier.

THE INTENSIVE AGRICULTURE OF IRRIGATED AREAS

Mediterranean agriculture comes to its peak of intensity and productivity in areas where the land is irrigated. In such areas relatively abundant and dependable supplies of moisture allow full exploitation of the subtropical temperatures, and the growing of fruits and vegetables, often with a large proportion

destined for export, tends to supplement and sometimes largely to displace other types of production.

Although irrigation on a small scale is found in many parts of Southern Europe, a few irrigated areas stand out from the rest in size and importance. They usually have high population densities. Among these major areas of irrigation farming are northern Portugal, the Mediterranean coast of Spain, the northern coast of Sicily, and the Italian coastal areas near the city of Naples. The largest and most important irrigated area of all, the plain of the Po River in northern Italy, uses irrigation water to supplement year-round rainfall and is discussed separately.

In northern Portugal irrigation is used mainly to intensify a type of agriculture that does not differ radically from the normal Mediterranean type. In this area, however, irrigated corn (maize) replaces wheat as the major grain, and some cattle are raised on irrigated meadows. Grapes and sheep are the other agricultural mainstays.

In the coastal regions of Spain that front on the Mediterranean Sea, irrigation has made possible the development of extensive orchards. Oranges are the most important product, and the preeminent orange-growing district of Spain, around the city of Valencia, has given its name to a type of orange. Vegetables, and in some places even tropical fruits such as dates and bananas, are grown to supplement wheat, vines, and olives.

As an agricultural area, northern Sicily is mainly differentiated from ordinary Mediterranean areas by its concentration of irrigated citrus groves. Lemons are particularly important. The district around Naples, known as Campania, is more intensively farmed, productive, and densely populated than any other agricultural area of comparable size in Southern Europe. Irrigation water applied to exceptionally fertile soils formed of volcanic debris from Mount Vesuvius supports a remarkable variety of production that includes almost every crop grown in Southern Europe. Despite the intensive cultivation and agricultural productivity of this area, however, the overcrowded rural population on its tiny farms is notably poor.

AGRICULTURE IN AREAS WITH SUMMER RAINFALL

In the northern parts of the countries of Southern Europe sizable areas are found which do not have truly mediterranean climates, and these areas exhibit corresponding differences in agriculture. In northern Spain wheat becomes subordinate to corn and rye, and the summer rainfall permits a greater development of cattle raising than is customary in Southern Europe. Another large area that diverges from the normal pattern of climate and agriculture is the Po Basin in northern Italy. This area is outstandingly important in Italian agriculture. Considerable rain falls during the summer on the level plain of the Po River, and the surrounding mountains provide superior water supplies for irrigation. In this area corn, grown both in irrigated and unirrigated fields, and irrigated rice become important cereal crops in addition to wheat. Vineyards are supplemented by orchards of peaches and other temperate fruits. The plain of the Po is also the center of Italian production for such industrial crops as sugar beets and hemp and for cattle, nourished on fodder crops and irrigated meadows. Crop yields and the general welfare of the farmers stand at considerably higher levels than in the more typically Mediterranean areas of central and southern Italy. The parts of northern Greece known as Macedonia and Thrace are also atypical. They are distinguished by a strong tendency to substitute cotton for grapes and olives to supplement grains and tobacco, which are the agricultural mainstays.

RELIEF AND POPULATION DISTRIBUTION

In terrain and population distribution as well as in climate and agriculture the countries of Southern Europe present various points of similarity. Rugged terrain predominates in all four major countries, and lowland plains occupy a relatively small part of the total land area. Individual plains tend to be small and to face the sea. They are separated from each other by the sea and by mountainous territory. Population distribution corresponds in a general way with topography, with the lowland plains being densely populated and the mountainous areas much less so, although a number of comparatively rough areas attain surprisingly high densities. Thus on the whole the picture of population distribution is one of relatively isolated areas of dense population facing the sea and separated from one another by large areas of comparatively low density.

In the Iberian Peninsula the greater part of the land consists of a plateau, the Meseta, with a surface lying at a general elevation of between 2000 and 3000 feet (c. 600–900 m). The plateau surface is interrupted at intervals by deep river valleys and ranges of mountains rising above the general level (Figs. 9.7 and 5.7). Population density is restricted,

mainly by lack of rainfall, to figures ranging generally between 25 and 100 per square mile (c. 10–40 per sq km). For the most part, the plateau edges are steep and rugged. The Pyrenees and the Cantabrian Mountains border the Meseta on the north, and the Betic Mountains, culminating in the Sierra Nevada, border it on the south. Most of the population of Spain and Portugal is distributed peripherally on discontinuous coastal lowlands which ring the peninsula.

In Italy the Alps and the Apennines are the principal mountain ranges. Northern Italy includes the greater part of the southern slopes of the Alps. The Apennines form the backbone of the peninsula, extending from their junction with the southwestern end of the Alps to the toe of the Italian boot, and appearing again across the Strait of Messina in Sicily. The Apennines vary considerably in height and appearance from place to place. East of Rome the mountains reach more than 9000 feet (c. 2750 m) in elevation. In Sicily, Mount Etna, a volcanic cone, reaches 10,902 feet (3323 m). Near Naples, Mount Vesuvius, which rises to 4190 feet (1277 m), is one of the world's most famous volcanoes. West of the Apennines, most of the land between Florence on the north and Naples on the south is occupied by a tangled mass of lower hills and mountains, often of volcanic origin. Between these highlands and the sea there is often a narrow coastal plain. Both Sicily and Sardinia, the two largest Italian islands, are predominantly mountainous or hilly.

Parts of the Italian highlands have population densities of more than 200 per square mile (c. 75 per sq km). Yet even these areas are much less densely populated than are most Italian lowlands. The largest lowland, the Po Plain, contains almost half of the entire Italian population, with non-metropolitan densities frequently reaching 200 to 500 per square mile (c. 75–200 per sq km). Other lowland areas with extremely high population densities are the narrow Ligurian Coast centering on Genoa, the plain of the Arno River as far inland as Florence, Campania around Naples, much of the eastern coastal plains and hills of the peninsula, and the northern and eastern coastal areas of Sicily.

In Greece most of the peninsula north of the Gulf of Corinth is occupied by the Pindus Range and the ranges which branch from it. Extensions of these ranges form islands in the Ionian and Aegean seas. Greece south of the Gulf of Corinth, commonly known as the Peloponnesus, is composed mainly of the Arcadian mountain knot. Along the coasts of Greece many small lowlands face the sea between mountain spurs and contain the majority of the people. A particularly famous lowland, although far from the largest, is the Attic Plain, still dominated as in ancient times by Athens and its seaport, the Piraeus. Larger lowlands are found in Thessaly and to the north in Macedonia and Thrace.

HISTORICAL CONTRASTS IN WEALTH AND POWER

The countries of Southern Europe are relatively poor in today's Europe. In each of the four main countries, current conditions offer a striking contrast to a past period of economic and political power and leadership. In Greece this period of past glory is the most remote, centering in the 5th and 4th centuries B.C., when Greek city-states were spreading the seeds of Western civilization through the Mediterranean area. To some degree there was a rebirth of Greek power and influence in the Middle Ages, when Constantinople was the capital of a Byzantine Empire which was largely Greek in population and control.

Italy's main period of former eminence was the centuries when the Roman Empire embraced the whole Mediterranean Basin and lands beyond. During the later Middle Ages, some centuries after the final collapse of the Roman Empire in the 5th century A.D., many of the Italian cities became independent centers of trade, wealth, and power. Venice became the center of a maritime empire within the Mediterranean area, as did Genoa to a lesser extent. Such inland cities as Milan, Bologna, and Florence also prospered and grew powerful on the basis of their trade with Europe north of the Alps. The growth of a hostile Turkish Empire astride routes to the East, the unification of larger states such as France and Spain, and the discovery of sea routes to the East which bypassed the Mediterranean were factors contributing to the end of this second period of Italian preeminence. Following its appearance in the 19th century, the modern unified state of Italy made an attempt to emulate ancient Rome. A colonial empire was acquired which included Libya, Ethiopia (for a brief period), and some smaller territories. All these colonies were lost as a result of World War II.

The main period of Spanish power and influence began in the Middle Ages when Spain stood as the bulwark of Christian Europe against Muslim civilization and, in a struggle lasting for centuries, eventually expelled the Moors from Europe. In the same year that this expulsion was finally accomplished, 1492, Christopher Columbus, an Italian navigator in the pay of the Spanish court, crossed the Atlantic and discovered the lands that were eventually to be called the Americas. For a century thereafter Spain stood at its greatest peak of power and prestige. It was the greatest power not only in Europe, but in the entire world, and built one of the largest empires ever known, in areas as diverse and widely separated as Italy, the Netherlands, North and South America, Africa, and the Philippine Islands. This empire shrank in size with the gradual decline in the relative power of Spain and had disappeared by the 1980s. The Spanish-held Canary Islands, off Morocco, are governed as an integral part of Spain.

Portugal also played a part in expelling the Moors from Iberia and took the lead in the 15th century in seeking a sea route around Africa to the Orient. The first Portuguese expedition to succeed in the voyage, headed by Vasco da Gama, returned from India in 1499. For the better part of a century thereafter, Portugal dominated European trade with the East and built an empire there and across the Atlantic in Brazil. However, commencing in the latter part of the 16th century there was a rapid decline in Portuguese fortunes. This was associated in part with the conquest of the small Portuguese homeland by Spain, which held it from 1580 to 1640. Meanwhile other European powers, particularly the Netherlands, offered increasingly successful competition in trade and colonization. By 1640 many of Portugal's possessions had fallen to the Dutch and could not subsequently be regained. However, the Portuguese held Brazil until the 19th century, when it gained independence, and in the 20th century Portugal became the last major European colonial nation to lose its principal overseas possessions.

FACTORS IN SOUTHERN EUROPE'S RETARDED DEVELOPMENT

Today (late 1980s), all four major countries of Southern Europe rank far behind Europe's richest countries in per capita GNP (as does Malta). The relative poverty of the region is a characteristic that has persisted stubbornly despite some rapid progress in recent times. Many of the difficulties of these countries are related to the fact that until very recently their industrial development was greatly retarded. During the 19th century and the early part of the 20th century, while northwestern Europe and the United States were forging ahead industrially, the Southern European countries remained predominantly agricultural. Even in the mid-1980s, more than one fourth of Greece's employment was still in agriculture, and Italy—the most industrialized country of Southern Europe—still had a larger proportion of its labor force in agriculture (one tenth) than was common in northwestern Europe. Despite large outmigrations from rural areas in recent years, too many Southern Europeans still depend on inadequate agricultural resources.

During the 19th and early 20th centuries Southern Europe was characterized by the following traits that have generally been inimical to industrialization.

1. The countries of the region were not major trading nations and did not develop an appreciable class of wealthy merchants—a group which has led the way in organizing and financing industrial development in many of the world's countries.

2. The people of the Southern European countries were generally poor and uneducated, offering little in the way of skilled labor or a promising market for products of industry.

3. Governments in the region were generally undemocratic, frequently unstable, often ineffective, and usually little oriented toward economic development.

4. Political, economic, and social affairs were controlled to a large extent by wealthy landowners whose interests lay in maintaining the agrarian societies that they dominated.

5. Internal transportation systems were poorly developed and could not easily be improved due to the rugged terrain.

6. Populations in the individual political units, especially before the unification of Italy in 1870, were small and hence offered only limited scope for the development of viable markets and industries.

7. In addition to the handicaps summarized in the preceding points, some of the natural resources upon which early industrialization was frequently based in more fortunate countries have not been plentiful in Southern Europe. A prominent example is wood. In some areas outside of Southern Europe wood was important in early industrial growth, either as a raw material or as a fuel for smelting metal-bearing ores. But in most of Southern Europe usable timber has been scarce due to dryness and centuries of deforestation. Many "forested" areas are covered only with scrub, a vegetation so typical as to have special names, such as *maquis* or *garrigue* in French and *macchia* in Italian. Waterpower is another resource that was important in early industrial growth in some of the world's countries but has been of limited utility in Southern Europe. Al-

though some parts of the region offered considerable scope for early industries powered by waterwheels, many streams in the areas of mediterranean climate are dry channels during part of each year. But the most significant resource deficiencies in Southern Europe probably have been shortages of coal and iron ore. In the age of the early Industrial Revolution, before hydroelectricity and oil became major sources of power, coal was the dominant power source in the development of many of the world's industrial districts. Although each of the four main countries of Southern Europe has at least some minor coal resources and production, there is only one sizable coal field in the whole region. The same generalizations can be made with respect to iron ore. In both cases the exception is northern Spain, where both coal and iron ore occur in the coastal district, including Bilbao (975,000) and smaller cities to the west. Here an industrial area developed relatively early, but it did not become a major European district. It was handicapped not only by the general political, economic, and social conditions in Spain but also by the fact that its coal includes only a limited proportion suitable for coking.

Had it not been for deficiencies with respect to these critical natural resources, some of the mineral resources which are relatively abundant in Southern Europe might have been of much greater use. Small deposits of a great many metals and other minerals do exist in the region, and some are abundant enough to make the countries of Southern Europe important world suppliers. Outstanding among these is mercury, of which Spain is the leading exporter.

MAJOR INDUSTRIAL AREAS PRIOR TO WORLD WAR II

Only three sizable industrial areas developed in Southern Europe before World War II. The most important of these by far was located in the Po Plain of northern Italy. In this area, imported coal and raw materials, hydroelectric power from stations in the surrounding mountains, and cheap Italian labor formed the basis for an industrial area specializing in textiles, but including some heavy industry and engineering. An area very similar in its foundations and specialties, but developed on a much smaller scale, came into existence in the northeastern corner of Spain, in and around Barcelona (4 million). A third area, somewhat diversified but especially focused on heavy industry, developed in the coal and iron mining section of northern coastal Spain. It is noteworthy that two of the three areas—the Po Plain and northern Spain—are not typically Mediterranean in environment, and the third area, northeastern Spain, lies near the foot of mountains with humid conditions that provide the basis for hydroelectric power development. But in contrast to the three main industrial concentrations, most of Southern Europe remained rural, agricultural, and very poor.

ECONOMIC PROGRESS SINCE WORLD WAR II

In the decades following 1950 the economies of the four main countries of Southern Europe began to show a new and unusual dynamism. They made some real progress in narrowing the economic gap between them and the countries of northwestern Europe, which were themselves continuing to develop rapidly. Industrial and service activities led the way, and all four countries showed a decline in the proportion of the population still employed in a generally poverty-stricken agriculture.

A number of factors played a part in the revitalization of the Southern European economies:

1. Improved transportation made the importing of fuels (before the oil price-rise in the 1970s) and materials for manufacturing more economical. Thus countries poor in these resources were less handicapped relative to richer countries than formerly. Ocean transport was especially important in this respect. Even today, internal land transport is not really adequate in Southern Europe except in much of Italy. In that country a tradition of road building that dates back to Roman times is expressed today in a relatively good railway system and thruway (*autostrada*) system. Due to the rugged terrain, these routes are often characterized by spectacular scenery and massive feats of engineering.

2. Coal has ceased to be the dominant source of power, its place being taken by oil or, in Italy, by natural gas and oil. All four of the main countries in Southern Europe have become major importers of oil, as have their industrial competitors in northwestern Europe. Thus Southern Europe is on more of an even footing with northwestern Europe in securing power than it was when it had to depend on large imports of coal. In Italy the power situation was also improved after World War II by the discovery and rapid development of large natural gas deposits, principally under the Po Plain. Despite this new energy source, however, expanding oil imports were so fundamental to Italy's industrial expansion and rising consumption that dependence on foreign energy has increased markedly in the postwar era.

3. Market conditions for Southern European exports improved greatly in the decades after World War II, due largely to the rising prosperity of northwestern Europe. These exports included goods such as subtropical fruits and vegetables, Italian-built automobiles, and, less obviously, labor. Streams of temporary migrants from Southern Europe found jobs in booming, labor-short industries to the north. Absence of the migrants relieved unemployment, and their remittances added purchasing power in their home areas.

4. There were other massive infusions of foreign money into the Southern European countries after World War II. Each country has been the recipient of much American economic and military aid in one form or another. Favorable economic circumstances and government policies have attracted a considerable amount of foreign investment in industrial and other facilities. And, not least, the countries of Southern Europe have been the beneficiaries of a tremendous boom in tourism. This picked up momentum in the 1950s and began to reach flood tide by the 1960s. To increasingly prosperous Europeans and Americans, Southern Europe offers a combination of spectacular scenery (Fig. 9.9), historical depth exceeding that of even the other parts of Europe, a sunny subtropical climate, the sea, and beaches. The ancient cities, often bustling and modern around their monuments of the past, were supplemented as attractions by new or greatly expanded resort areas in such places as Spain's Costa Brava and Costa del Sol, the Balearic Islands, the Italian Riviera, and various others. Tourism has focused particularly on larger and more famous attractions such as Rome, Athens, or the Costa Brava, but innumerable areas from Oporto to the Aegean islands have benefited from tourist flows sometimes outnumbering the natives.

THE PO BASIN: SOUTHERN EUROPE'S MAJOR INDUSTRIAL AREA

The Po Basin is the economic heart of modern Italy and the most highly developed and productive part of Southern Europe. Comprising just under a fifth of Italy's area, it contains about two fifths of Italy's population (more prosperous by far than the rest)

Figure 9.9 The small island of Mykonos in the Aegean islands of Greece is one of many famous tourist sites in the Mediterranean world. The light-colored architecture, the fishing boats, and the deforested slopes are characteristically Mediterranean. Vegetation on the slope in the background has turned brown in the midsummer drought. Both the island and its only town are named Mykonos.

and accounts for approximately half of its agricultural production and two thirds of its industrial output. From an economic standpoint, peninsular Italy is largely a poverty-stricken and overpopulated appendage to the Po area. These differences between north and south in Italy are manifest within the country by strong regional feelings embodying a degree of antagonism and mutual disdain, as well as by government programs to subsidize development of the poorer south.

The modern industrial development of the Po area began in the 19th century on the foundations of the region's already outstanding position in agriculture and trade. A gradual development of textile industries succeeded here, while industrial attempts further south generally failed. Imported coal and raw materials were necessary, but the northern factories had year-round streams for waterpower and were located in the richest market area in Italy. In addition, there was some accumulation of capital from trade, as some of the northern cities had been way-stations for centuries on routes leading northward from the Mediterranean across the Alpine passes.

With the 20th century came the age of hydroelectric power development in Italy. The best possibilities were in the Alps and to a lesser degree the northern Apennines—the mountain areas framing the Po Plain on three sides and not afflicted by the dryness of the Mediterranean summer. By World War II the Po area was one of the major textile manufacturing regions of Europe. It still imported coal to supplement its hydroelectric power, and it still imported raw materials. The latter were processed with relatively cheap Italian labor, and much of the finished product was exported. By this time various engineering industries had been added as a secondary element in the industrial economy of the area. Imported iron and steel were used to manufacture a growing variety of machines, particularly those types which emphasized the input of cheap skilled labor and minimized the amount of metal needed.

A very rapid further expansion of Po Basin industry occurred after World War II. Italian industrial production, most of it in the Po Basin, more than doubled from 1951 to 1961. It continued to grow rapidly through the 1960s, although the rate of growth slowed somewhat after that. Prominent components of this expansion were (1) the growth of engineering and chemical industries; (2) the development of a sizable iron and steel industry, mainly in northern Italy, but partly in peninsular Italy; and (3) a shift to domestic natural gas and imported oil as the predominant sources of power, but with hydroelectric energy continuing to play an important role. Today textile industries remain important in the Po area, but they have become secondary to engineering.

Many cities share the region's industry, but the leading ones are Milan (4 million), Turin (1.6 million), and Genoa (850,000). Milan is the region's industrial capital, the largest city of Italy, and the country's most important center of finance, business administration, and railway transportation. Milan is located between the port of Genoa and important passes, now shortened by railway tunnels, through the Alps to Switzerland and the North Sea countries. Turin is the leading center of Italy's automobile industry, although several other cities have a share. It is particularly associated with the Fiat company, which is by far the largest Italian auto manufacturer. The port of Genoa functions as a crucial part of the Po Basin industrial complex, although it is not actually in the basin. It is reached from the Milan-Turin area by passes across the narrow but rugged northwestern end of the Apennines. The city dominates the overseas trade of the Po area and is Italy's leading seaport. Genoa is now much larger and economically more important than its old medieval commercial and political rival, Venice (425,000), whose location on islands at the eastern edge of the Po Plain is relatively unfavorable for serving the main present-day industrial centers of northern Italy.

205

MAJOR URBAN CENTERS OUTSIDE THE PO BASIN

The generally low level of industrial development in Southern Europe, aside from northern Italy and the two smaller industrial areas of northeastern and northern Spain, is indicated by the basically nonindustrial character of the leading cities. The latter are mostly political capitals or ports serving especially productive agricultural areas, or both. They are, of course, not without industries appropriate to a port or to any large city, particularly one forming a reservoir of cheap labor.

In *Greece* the largest city by an overwhelming margin is Athens. With a metropolitan population of more than 3 million (including the port of Piraeus), amounting to about one third of the entire population of Greece and growing rapidly, Athens has no national rival. It is not only the national capital and the main port but also the main industrial center. This development is logical enough in a situation where imported fuels and materials are used to manufacture products that are marketed almost entirely in a domestic market whose main center is Athens itself. Greece's second city, Salonika (Thessaloniki: 725,000), is the port for a particularly productive agricultural area, some of it on reclaimed and irrigated lands on the flood plains of the Vardar and other rivers. Salonika is also the sea outlet for an important route through the mountains of southern Yugoslavia formed by the combined valleys of the Morava and Vardar rivers.

In *Italy* the largest cities outside of the northern industrial area are the capital, Rome (3.2 million), and Naples (2.8 million). Rome (Fig. 9.10) has a location that is roughly central within Italy. It lies a few miles inland about halfway down the west coast of the peninsula, in a position to some extent intermediate between the country's contrasting northern and southern regions. It is predominantly a governmental, religious, and tourist center in a region which, relative to its population size, is poorly developed agriculturally and industrially. Rome's central location and historical prestige made it the favored choice for national capital when Italy emerged as a unified state in the 19th century. Naples is the port for the populous and productive, but miserably poor, Campanian agricultural region described earlier. It is also the main urban center of one major focus of Italy's tourist industry, with attractions such as Vesuvius, the ruins of Pompeii, and the island of Capri in the vicinity.

In *Spain* the three largest cities are Madrid (4.5 million), the capital, and the two Mediterranean ports of Barcelona (4 million) and Valencia (1.3 million). Madrid was deliberately chosen as the capital

Figure 9.10   Christmas shoppers in Rome, Italy. Many famous old cities of Southern Europe not only house priceless historical relics but also contain elegant downtown sections catering to the privileged of the present day.

of Spain in the 16th century because of its location near the mathematical center of the Iberian Peninsula, approximately equidistant from the various peripheral areas of dense population and sometimes of separatist political tendencies. Located in a poor countryside, it has had little economic excuse for existence during most of its history. But its position as the capital has made it the center of the Spanish road and rail networks and thus has given it certain business advantages. In Spain's recent economic boom it has begun to attract industry on a considerable scale.

Barcelona is something of an exception in Southern Europe in that it is both its country's major port and the center of an industrial district of some consequence, the main such district in Spain. Industrial development up to the present resembles that of the Po Basin at an earlier stage when textiles were still dominant there. The main elements undergirding Barcelona's industrialization have been (1) the early formation of a commercial class which supplied financing; (2) the availability of hydroelectric power, mainly from the Pyrenees; (3) cheap and sufficiently skilled labor; and (4) imported raw materials.

Valencia, Spain's third city in population, is the business center and port for an unusually productive section of coastal Spain. An extensive development of irrigation in the area originated with the Moors in the Middle Ages. Today the density of population on irrigated land resembles that of Italy's Campania. But here much of the agricultural effort goes into producing oranges, which are Spain's largest single agricultural export.

The two large cities of *Portugal* are both seaports on the lower courses of rivers which cross the Meseta and reach the Atlantic through Portugal. Lisbon (2.3 million), on a magnificent natural harbor at the mouth of the Tagus River, is both the leading seaport of the country and the capital. The smaller city of Oporto (1.3 million), at the mouth of the Douro River, is the regional capital of northern Portugal and the commercial center for Portugal's trade in the famous port wine, which comes from terraced vineyards along the hills overlooking the Douro.

ECONOMIC STAGES AND TRENDS: GETTING AWAY FROM AGRICULTURAL ECONOMIES

Since World War II, Italy, Spain, Portugal, and Greece have all been developing from basically agricultural to basically industrial countries, although still at an early stage of industrialization. They still have unusually large proportions of people employed in agriculture by the standards of "developed" countries, and all rely on exports of agricultural products to a significant degree. But they all now depend much more heavily on exporting manufactures. "Depend" is meant literally, because all are deficient in food production and must pay for large food imports, especially grains, as well as manufactures and services in which their own production is absent or deficient. In all of these countries, large tourist industries "exporting" sun, scenery, historic sites, and beaches supplement goods exports to an important degree in paying for the needed imports, as do the remittances sent home by their many migrant workers employed in more prosperous European countries.

An early stage of industrialization is reflected in the unusual importance of certain types of manufacturing industries: textiles and clothing in all four countries; shoes in Spain and Italy. These are highly competitive industries, requiring considerable labor, in which firms are strongly driven to seek low labor costs. Although such industries have old histories in Southern Europe, there has been an upsurge of development recently in this region with its surplus agricultural population and low wages. Domestic firms have played a role in this growth, but much capital has moved in from the higher-wage countries of northwestern Europe. In the early 1980s there was a drastic slowing of development as a result of depressed markets during a long period of recession. Growth has also been retarded by increased competition from factories using still cheaper labor in "third world" countries.

Italy is somewhat exceptional in this pattern. It conforms in many ways to the generalizations just discussed, but it also has an uncharacteristically large development of "later stage" industries. Automobile production is a good example, as is the making of office machinery. These industries, along with many of the "early stage" industries, grew out of commercial and industrial development in the Po Basin dating back to the Middle Ages.

As one goes farther south in Italy, the country becomes progressively poorer in both environment and economy. Commencing shortly after World War II the Italian government began to establish institutions and policies for developing the poverty-stricken southern peninsula and the Italian islands. Agricultural development and land reform and, somewhat later, the introduction of subsidized industries, have been f atured. Controlling interests in many of Italy's largest companies are owned by two holding companies which are, in turn, controlled by the government. This leverage, along with other pressures and inducements, has been used by the state to get some major industrial expansions—most notably in iron and steel—in the south. But the differential between north and south remains and is widely perceived in Italy as a major national problem. Within Southern Europe, Spain is second to Italy in the development of "later stage" industries—such as auto production in Valencia—and has a similar north-south contrast in development and income.

GEOPOLITICAL ASPECTS OF SOUTHERN EUROPE

The fact that they have failed to attain or sustain great-power status in modern times has not made the countries of Southern Europe unimportant. Rather, they have retained a considerable political significance deriving in part from their strategic geographic positions. Their international political orientation has been a matter of grave concern to various outside nations, which have often vied with each other for power and influence in Southern Europe. This competition has both reflected and contributed to the tendency of Southern European countries toward political instability and drastic political change.

GREECE—A MEDITERRANEAN GATEKEEPER

The international importance of modern Greece stems largely from the fact that the Greek peninsula lies between the Aegean Sea and the entrance to the Adriatic, and thus commands the routes of access to the Mediterranean from the Balkan Peninsula and Black Sea. In addition, the important route of sea trade running through the Mediterranean and Red seas can be effectively threatened from Greece. The strategic implications of its position have made Greece to some extent a pawn between land power and sea power in past times. After its liberation from Turkey in the first half of the 19th century, Greece was customarily under the influence and protection of Great Britain. As a major sea power vitally dependent on the Mediterranean-Red Sea route, the latter nation was anxious to keep strong land powers, such as Russia, away from Mediterranean shores. Following World War II, Greece became the scene of a civil war between Communist and anti-Communist forces, and it seemed for a time that the country might be added to the list of Soviet satellite nations. Victory by the anti-Communist elements was underwritten by American aid, and alliance with the United States continued to be a fixed element in Greece's foreign relations until it was weakened by American acceptance of a Greek military dictatorship from 1967 to 1974, and by American failure to prevent Turkey's conquest of much Greek-inhabited territory in Cyprus in 1974. In 1981 the Greeks elected a socialist government that used anti-American rhetoric and campaigned on a platform of opposition to American bases and nuclear weapons in Greece, as well as to Greek membership in NATO. But this stance then seemed to erode somewhat due to Greek fears of Turkey and Greece's felt need for support in a continuing dispute with Turkey concerning jurisdiction within the Aegean Sea area. The dispute has recently been focusing on undersea oil deposits claimed by both countries.

THE ITALIAN ARENA

Italy's position has made it a natural sea outlet for parts of central Europe, a potential threat to the security of the sea route through the Mediterranean, and a land bridge extending most of the way across the Mediterranean toward northern Africa. In World War II Italian participation on the side of Germany forced Britain to largely abandon the strategic Mediterranean-Red Sea route and to rely on the old pre-Suez route around Africa for traffic with areas surrounding the Indian Ocean. Separated from Africa only by the narrow stretch of water between Sicily and Tunisia, the Italian causeway also permitted German and Italian land forces to wage African campaigns in relative security from Allied sea power.

The population of Italy, second largest in Europe, and the growing output of its industries add to the country's international importance. Several outside nations have been deeply involved or interested in Italian affairs during recent times. In the latter stages of World War II the country was the scene of a desperate and destructive struggle between Allied and German armies. After the war a contest for influence in Italy was carried on between the Soviet Union and the Western bloc of nations headed by the United States. This contest was mainly fought in the arena of Italian internal politics, with Soviet influence being exerted through the powerful Italian Communist Party and American support being given the anti-Communist elements of the country. The victory went to the West, but the Italian Communist Party, now less subservient to Moscow, remains the largest in the world outside of Communist countries. Nevertheless, Italy is closely linked with the Western nations in NATO, the Common Market, and other international organizations.

But since the late 1960s unrest has grown and Italy has become especially known among Western nations for violent political protest. Both extreme leftist groups, which despise even the Communist Party as being too conservative, and extreme rightist groups reborn from the Fascism of prewar Italy have carried on terrorist campaigns to disrupt a society that has long been dominated politically by the often corrupt, church-oriented Christian Democratic Party. Assassinations, bombings, kidnappings, and riots have been common occurrences. These political disturbances augment a generally high crime rate and lawlessness deriving from the great power of the Mafia criminal society based in Sicily.

GEOPOLITICAL CHANGE IN SPAIN

In Spain the Civil War of 1936–1939 reflected the general tendency of internal political struggles in Southern Europe to be directly influenced by the action of outside powers. In this war the Fascist or Insurgent faction eventually triumphed with the aid of an entire army provided by Fascist Italy and an air force and other special units largely provided by Nazi Germany. The losing republican or Loyalist side profited somewhat from aid by the Soviet Union, at that time a major antagonist of Germany and Italy on the European stage. In addition, a considerable number of volunteers from several nations fought in the Loyalist armies. The Insurgent victory was regarded as a considerable defeat for Britain and France, neither of which had intervened directly in the war, since it placed a hostile Spain on the flank of important British sea routes and left France almost surrounded by Fascist countries. As it turned out, the new Spanish government refrained from entering World War II, although it maintained a generally hostile attitude toward the Allied nations. Gradually Spain has become more closely aligned with the Western nations, particularly the United States. In 1953 an agreement was signed to permit the establishment of American air and naval bases in Spain. The fanatically anti-Communist stance of the Franco dictatorship (1939–1975), plus the physical remoteness of Spain from Communist-controlled territory and the country's sea frontage on both the Atlantic and the Mediterranean, made the Spanish bases militarily attractive to the United States. They were still occupied by American forces in early 1989 under a new eight-year agreement. However, Spain's democratic, socialist-led government was facing much internal opposition to the bases, especially their nuclear weaponry, and the Spanish government, in response to internal political pressures, was making demands that American forces in Spain be reduced.

Since Franco's death in 1975 a somewhat shaky constitutional monarchy in Spain has been able to continue the American connection despite internal opposition and has been able to draw closer to the democracies of western Europe in political and economic relations. Like Portugal, Italy, and Greece, Spain now belongs both to NATO and to the European Communities (EC). Its democratic government, like the preceding dictatorship, is constantly harassed by terrorist violence aimed at securing independence for the Basque linguistic region of northern Spain. The violence continues despite the fact that the government has granted considerable autonomy to the Basque region, as well as to the culturally distinct Catalan region (Catalonia) focusing on Barcelona, and other regions. Political stability also is endangered by dictatorial sentiments among some of the population and many army officers and units. An army group even seized control of the parliament in 1981, but this attempted coup in favor of military dictatorship was suppressed.

The Gibraltar Question

Spain has long been involved in a dispute with the United Kingdom over possession of Gibraltar. Britain took the Gibraltar peninsula from Spain in the early 18th century and has held it since that time as a fortified base guarding the eastern entrance to the Strait of Gibraltar. It is now a self-governing British colony with an area of about 2 square miles and a population of 31,000. With the dissolution of the British Empire and the decline in British power, Spain's claims to Gibraltar were revived. They have been pursued despite a referendum in 1967 in which the population voted almost unanimously to retain its ties with Britain.

THE SMALLER STRATEGIC SIGNIFICANCE OF PORTUGAL

Portugal has had a lesser degree of political and strategic significance during recent times than Greece, Italy, or Spain, and it has been less involved in international conflicts, aside from unsuccessful colonial struggles in Africa. The country's proximity to Atlantic sea routes made it an object of British support during most of its modern history, and the lack of any strong power in a position to challenge British influence gave Portugal a relatively tranquil existence in international politics. It remained neutral in World War II, but since the war it has aligned itself with the West through membership in NATO. Internally, however, its politics in recent years have been turbulent. In 1974 a revolution occasioned by discontent over the colonial wars in Africa overthrew a right-wing dictatorship that had held power for a long time. A parliamentary democracy was established, but it has been volatile and unstable, with its tranquility upset by struggles among democrats of various persuasions and Communists. Support by friendly regimes in western Europe has been important in the establishment and maintenance of this fragile democracy in Portugal. Here again we see the pervasive Southern European tendencies toward political instability and international involvement in each country's internal affairs.

POLITICAL AND ECONOMIC READJUSTMENT IN MALTA

The island state of Malta, which was granted independence by Britain and became a member of the Commonwealth of Nations in 1964, has based its livelihood in the past primarily on its strategic position. Located in the narrow waist of the Mediterranean, some 60 miles (97 km) from Sicily and 180 miles (290 km) from Africa, the two small islands of the present Maltese state, Malta and Gozo, have been a naval base for many centuries. The islands were long a stronghold of the crusading Knights of St. John. From 1800 to 1964 they were in British hands, and the income from British naval expenditures and personnel became their main source of support. The state of Malta has an area of 122 square miles (316 sq km) and an estimated population (1988) of about 340,000, for an average density of approximately 2800 people per square mile (about 1100 per sq km). Two thirds of the population lives in a multi-city urban complex that includes the small capital and port of Valletta (9000), but rural density is still extremely high. Within the country as a whole, levels of income are low, although incomes are better in Malta's relatively large colony of foreign residents, mainly British. In 1979 the British bases were closed and readjustment became necessary. Emigration and a heightened development of tourism appear to offer the principal hope for improving the economic situation. A determined effort is being made to expand tourist facilities and increase the number of visitors, to whom Malta can offer both natural and historical attractions. Meanwhile, a number of light manufacturing industries, especially clothing, have been attracted by very generous financial concessions.

Legacies of Empire in East Central Europe

In the aftermath of World War II, Communist parties took control of eight countries occupying the eastern margins of Europe between the Baltic, Black, and Adriatic seas. Since then, the national life of these countries has been restructured along Communist lines. They have exhibited the customary features of Communist states: government by one party; planning and direction of the economy by organs of the state; abolition of private ownership (with some exceptions) in the fields of manufacturing, mining, transportation, finance, commerce, and services; and measures, more thoroughgoing in some countries than in others, to convert agriculture from a private basis to a socialized basis. But the countries of Communist Europe must not be thought of as replicas of each other. Each country has implemented the Communist system in a somewhat different way (Yugoslavia differing the most widely from the rest), each one has its own distinctive relationships within the Communist world and with non-Communist nations, and each one operates within its own frame of physical, cultural, economic, and historical circumstances. One of the countries, East Germany, has been considered with West Central Europe in Chapter 7 but will be brought into the present discussion at appropriate points. The remaining seven countries—Poland, Czechoslovakia, Hungary, Romania, Bulgaria, Yugoslavia, and Albania—are grouped here as the countries of East Central Europe (Fig. 9.11). They have a combined area of 450,000 square miles (about 1.2 million sq km—the size of Texas plus California and Maine) and had an estimated population total of 123 million in 1988. Their area is only one nineteenth as great as that of their eastern neighbor, the Soviet Union, but their population amounts to 43 percent of the total for the Soviet

Union or, combined with East Germany's 16.6 million, a little more than half of the USSR total.

RECENT RELATIONSHIPS WITH THE SOVIET UNION

Comparisons between Communist Europe and the Soviet Union are very appropriate, for six of the eight countries in Communist Europe lie within the economic and strategic orbit of the USSR, and their productive capacities and markets contribute in no small measure to the USSR's position as a major world power. This situation has persisted since the closing stages of World War II, when Soviet troops pursued the retreating Germans across eastern Europe and in the process brought most of the region under Soviet control. During a transitional period from 1944 to 1948, local Communist parties were given such backing by the USSR as was necessary for them to gain control of the governments in their respective countries. Since then, the Soviet Union has made some concessions to the desire of these countries to run their own affairs, but it has adhered to a policy of insisting that the direction of affairs be in the hands of the Communist Party and that party leaders in each country be responsive to the desires and policies of the USSR. In 1956 Soviet troops crushed an uprising against the Communist government of Hungary, and in 1968 the Soviets, in company with East Germany, Poland, Hungary, and Bulgaria, effected a military occupation of Czechoslovakia after the Communist leadership of that nation had introduced political reforms regarded as dangerously liberal by the USSR. In the early 1980s threats of similar action were instrumental in keeping a restive Poland in the grip of its obedient Communist regime.

Two nations of Communist Europe—Yugoslavia and Albania—have had a very different postwar history than the others. In Yugoslavia, Marshal Tito's Communist Party led the Partisan resistance against the Nazi occupation in World War II and was able to take immediate control of the country when the war ended. Soviet troops entered Yugoslavia briefly but were withdrawn for service elsewhere as the fighting swirled on toward Germany. Tito's forces then established an independent Communist regime that soon quarreled with the Soviet Union. Yugoslavia thereupon sought and received aid from Western nations and has since continued to carry forward a program of separate Communist development. Many political and commercial links between Yugoslavia and other Communist countries have now been restored, but the Yugoslav pattern of development still diverges more from Soviet theory and

Figure 9.11   Index map of East Central Europe. The "mountainous areas" are rather broadly generalized to bring out the major outlines of the topography. The Bakony Forest, separating the Great Hungarian and Little Hungarian Plains, and the narrow highland spur extending eastward from Slovenia in northern Yugoslavia are hilly rather than truly mountainous; the same is true of certain other areas shown in the same manner. For identification of cities shown by letter in the major industrial concentrations, see Figure 7.1.

211

practice than is true of any other European Communist state, and Yugoslavia's economic relations with the non-Communist world are closer than those of any other Communist country.

In Yugoslavia's neighbor, Albania, several partisan groups carried on resistance against occupation in World War II, and at the end of the war the strongest group, devoted to the Soviet dictator, Stalin, took control of the country. After Stalin's death in 1953 the country continued to follow his policies. This led to a complete break in relations with the USSR in 1961 after the latter repudiated Stalinism. Albania has long maintained a severely isolationist and suspicious attitude in its dealings with both East and West—an attitude symbolized by the electrified fence along its borders. While this general stance continues, there have recently been some beginnings toward a more open relationship with the outer world. For example, a new railway to neighboring Yugoslavia has tied Albania to the European railnet for the first time.

Thus neither Yugoslavia nor Albania can be termed a "client" or "satellite" of the USSR, although these terms still seem appropriate to the other European Communist states, including East Germany. But even these states are much freer of outright Soviet dictation than they were in the Stalin era. The different countries have been able in some measure to reassert their separate identities and to make more decisions on the basis of national self-interest. Meanwhile, they maintain close economic and military relationships with the USSR and with each other, and all of them belong to the Soviet-sponsored Council for Mutual Economic Assistance (COMECON or CEMA) and the Warsaw Pact military alliance. For the USSR the European client states are a buffer zone providing defense in depth against a possible resurgence of German militarism, a forward strategic position vis-à-vis western Europe, and a reservoir of manpower, resources, production, and markets.

POLITICAL INSTABILITY IN THE "SHATTER BELT"

The extension of Soviet power into East Central Europe at the end of the World War II was in keeping with the general history of the region throughout modern times. In the Middle Ages several peoples in the region—the Poles, Czechs, Magyars, Bulgarians, and Serbs—enjoyed political independence for long periods and at times controlled extensive territories outside their homelands. Their situation then deteriorated as stronger powers—the Germans and Austrians on the west, and the Ottoman Turks and Rus-

sians on the east—pushed into East Central Europe and carved out empires. In the process the empires frequently collided, and the local peoples were caught in numerous wars which devastated great areas, often resulted in a change of masters, and sometimes brought about large transfers of population from one area to another. There were many revolts against ruling powers, which were often put down with great severity. Sizable sections of East Central Europe were virtually depopulated for long periods of time. Such conditions reached their zenith in the long history of wars and disorders that attended the rise and slow decline of the empire of the Ottoman Turks. In the 15th, 16th, and 17th centuries Turkish armies overran the Balkan Peninsula and then pushed northward through Hungary to the borders of Austria, until their advance was checked at Vienna in 1683. Thereafter, the Turks were gradually pushed back. The Ottoman possessions in Europe were taken over by Austria, Hungary, or Russia; or local peoples were able to achieve self-government and independence. In some areas from which the Turks were expelled, the Austro-Hungarian Empire brought in colonists of many nationalities; the result was an intricate intermixture of peoples. The principal area where this occurred was the Banat, a fertile plains region that is now divided between Yugoslavia and Romania.

In the 18th century the national independence of Poland was extinguished in a series of partitions which gave control of the east and center to Russia, the south to Austria, and the north and west to Germany (Prussia). A differential economic development then took place, with the German sections advancing at a faster rate than those held by Austria or Russia.

Thus for several centuries the map of East Central Europe remained in flux as empires rose and fell, boundaries shifted, and populations were passed back and forth from one imperial master to another. The fragmented and unstable pattern of nationalities and political units has led some students of this region to call it the "Shatter Belt" or "Crush Zone" of Europe.

ORIGINS OF THE PRESENT PATTERN OF COUNTRIES

The present pattern of countries in East Central Europe resulted from the disintegration of the Ottoman Turkish, Austro-Hungarian, German, and Tsarist Russian empires in the 19th and early 20th centuries. This process was hastened by World War I, in which Germany, Austria-Hungary, and Turkey were on the losing side. Russia, originally allied with the victors, suffered disastrous defeats and then withdrew from the war following the Bolshevik Revolution of 1917. It was not represented at the Paris Peace Conference of 1919, where the victorious Western powers rearranged the political map of East Central Europe in an attempt to satisfy the aspirations of the various nationalities that had been included in the old empires. Poland was reconstituted as an independent country. Czechoslovakia, the homeland of two closely related Slavic peoples, the Czechs and Slovaks, was carved out of the Austro-Hungarian Empire as an entirely new country. Hungary, greatly reduced in size, was severed from Austria. The Kingdom of Serbia, which had won independence from Turkey in the 19th century, was joined with the small independent state of Montenegro and several regions taken from Austria-Hungary to form the new Kingdom of the Serbs, Croats, and Slovenes, later known as Yugoslavia. Romania, which had been independent of Turkish control since the mid-19th century, was enlarged by territories taken from Austria-Hungary, Russia, and Bulgaria. Independence from Turkey had been achieved by Bulgaria in the second half of the 19th century and by Albania immediately before the outbreak of World War I. The sovereignty of these nations was confirmed by the peace conference.

BOUNDARY QUESTIONS AND IRREDENTISM

The territorial settlement in East Central Europe following World War I did not make the region politically stable. Both the internal unity and the external relations of the different countries were troubled by a complex and interrelated set of boundary issues and minority questions. No sooner had the new nations been established than quarrels began developing over frontier questions involving claims to territory on historical and/or ethnic grounds. Not only did the nations quarrel among themselves, but serious disputes also erupted between East Central European countries and neighboring countries such as Germany, the Soviet Union, and Italy. The political uproar resulted in good part from the inability of the Paris peacemakers to disentangle ethnic groups and to satisfy the historical claims of nations to territory when the new frontiers were drawn. The delimitation of political boundaries gave each nation frontier territories once ruled by others, and it left a complex pattern of ethnic minorities in country after country. Irredentism—the demand for an international transfer of territory in order to place a minority in its alleged homeland—was epidemic. Even if such demands were absent, regional minorities were prone to agitate for more self-government or even independence. In Yugoslavia, for example, the large Croatian minority agitated for more freedom from the dominant Serbians. Alleged mistreatment of ethnic German minorities by Czechs and Poles, together with Germany's claims to frontier territories surrendered after World War I, were used by Hitler to justify his seizure of Czechoslovakia in 1938–1939 and the subsequent German invasion of Poland that set off the European phase of World War II in September 1939.

Boundary Changes and Population Transfers since 1939

As a result of World War II, several important shifts of disputed territory took place (see Figs. 7.9 and 9.11). For example, Romania was forced to cede Bessarabia (a pre–World War I possession of Russia) and the northern part of adjoining Bukovina to the Soviet Union. The Soviets gained a common frontier with Hungary by annexing Ruthenia, or Carpathian Ukraine, a mountainous area at the eastern end of prewar Czechoslovakia. A number of boundary changes benefiting Yugoslavia took place along the Yugoslav-Italian frontier. In 1954 the two countries signed an agreement (confirmed by a treaty in 1976) which placed the disputed port city of Trieste under Italian administration but made it a free port. The city's environs were transferred to Yugoslavia.

But the most drastic territorial shifts occurred in Poland. In eastern Poland the Soviet Union annexed a wide strip amounting to about 46 percent of Poland's prewar area. The Poles, however, were compensated for the loss when they took over large areas from Germany—areas included in the prewar German provinces of Pomerania and Silesia immediately east of the Oder and Neisse rivers and the southern half of East Prussia. Poland also took control of Gdansk (German: Danzig). The sum of these areas amounts to only one half as much in square miles as the area taken by the Soviet Union from Poland, but the industrial development, mineral wealth, and agricultural productivity of Poland's new territories (the "Regained Territories" or "Western Territories" in Polish terminology) more than counterbalance her losses to the USSR, at least in an economic sense. The northern half of former East Prussia was taken over by the Soviet Union as a result of the war. Thus the prewar "Polish Corridor" (between Germany proper and East Prussia) no longer exists, since Germany has lost the whole of East Prussia.

In addition to boundary adjustments, and often as a consequence of them, there have been extensive population transfers in East Central Europe since the

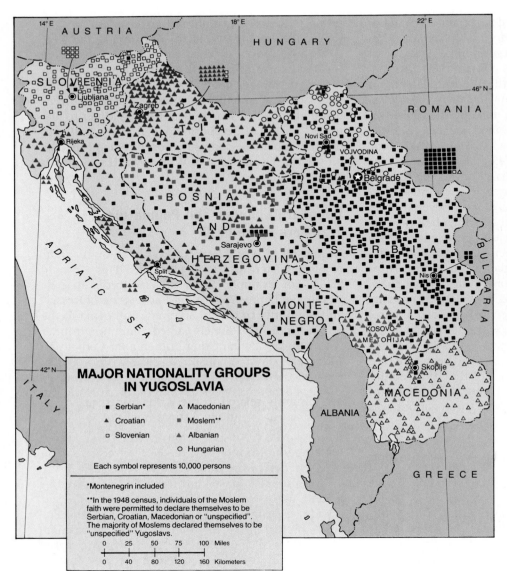

MAJOR NATIONALITY GROUPS IN YUGOSLAVIA

- ■ Serbian*
- △ Macedonian
- ▲ Croatian
- ▪ Moslem**
- □ Slovenian
- ▲ Albanian
- ○ Hungarian

Each symbol represents 10,000 persons

*Montenegrin included

**In the 1948 census, individuals of the Moslem faith were permitted to declare themselves to be Serbian, Croatian, Macedonian or "unspecified". The majority of Moslems declared themselves to be "unspecified" Yugoslavs.

| 0 | 25 | 50 | 75 | 100 | Miles |
| 0 | 40 | 80 | 120 | 160 | Kilometers |

Figure 9.12 This map has been redrawn from a map in George W. Hoffman and Fred Warner Neal, Yugoslavia and the New Communism *(New York, Twentieth Century Fund, 1962), p. 30. By permission of the authors and the Twentieth Century Fund. Names in the largest type denote the six socialist republics that constitute the Socialist Federal Republic of Yugoslavia. The Vojvodina and Kosovo-Metohija (Kosmet) are "autonomous regions" attached to Serbia. (Based on* Stanovnistvo Po Narodnosti (Population by Ethnic Nationality), *Vol. IX, Belgrade, 1954; final results of the population census of March 15, 1948.)*

beginning of World War II. Such transfers, involving millions of people—Germans, Poles, Hungarians, Italians, and others—have notably simplified the ethnic pattern, but at enormous cost to people who often were uprooted without ceremony, lost all their possessions (and not infrequently their lives), and then were dumped as refugees in a "homeland" that many had never seen. Nearly one third of the prewar populations of Poland and Czechoslovakia were composed of minorities, especially Germans, but today these minorities have largely disappeared. Millions of Jews were systematically killed during the German occupation in World War II. Ethnic minorities now constitute less than 1 percent of Poland's population and less than one tenth of Czechoslovakia's. The two countries have transferred most of their German population to Germany, including nearly all who lived in Poland's Western Territories. The only countries in East Central Europe that still have truly large minorities are Yugoslavia, a country in which the largest nationality, the Serbs, comprises only 35 percent of the population (Fig. 9.12), and Romania, whose principal minority is the Hungarians in the mountainous area called Transylvania. Irredentism and frontier questions are not very active in East Central Europe today, although some of the old disputes continue to simmer. Soviet control has put a quieting blanket over most of the region. One active quarrel is between Yugoslavia and Albania, which are Communist countries but are not Soviet satellites. Yugoslavia has an Albanian minority in the far south across the border from Albania, which alleges that its kinsmen are not well treated. Meanwhile another quarrel, between Hungary and Romania, concerns Romania's alleged mistreatment of the Hungarian minority in Transylvania. The dispute centers on the Romanian government's recent destruction of large numbers of villages, often centuries old, in order to establish new Communist-style ''agrotowns'' in Transylvania.

THE SLAVIC REALM OF EUROPE

The postwar expulsion of Germans from East Central Europe and the extermination or flight of more than 3 million Jews during the period of Nazi control intensified the Slavic character of the region. Practically all of Europe's Slavic people live here, and they form a large majority within East Central Europe as a whole. Their original homeland is thought to have lain in the area between the Vistula and Dnieper rivers. In the early Middle Ages, groups of Slavs began migrating into other parts of East Central Europe, as well as eastward into Russia. The Elbe River seems to have marked the limit of their penetration toward the west. Later on there were eastward movements by groups of Germans who established themselves among the Slavic peoples of East Central Europe and Russia. Sometimes the way was cleared for such groups by warfare, but in other situations German colonists came by invitation of local rulers, as their skills in agriculture, mining, industry, and trade were valued highly. Persistent colonization by Germans over a period of centuries resulted in the large German minorities found in several countries of East Central Europe prior to World War II.

THE MAJOR SLAVIC GROUPS

The Slavic peoples are often grouped into three large divisions: (1) *East Slavs* of the Soviet Union, including Russians, Ukrainians, Belorussians (White Russians), and Ruthenians; (2) *West Slavs*, including Poles, Czechs, and Slovaks; and (3) *South Slavs*, including Serbs, Croats, Slovenes, Bulgarians, and Macedonians. Another group, the Romanians, came under the influence of the later Roman Empire, and the Romanian language is classed with the Romance languages derived from Latin. However, the Romanians have been much affected by Slavic influences, and their language contains many Slavic words and expressions.

Although the various Slavic peoples speak related languages, such languages may not be mutually intelligible. For example, the Polish language is not easily understood by a Czech, although Poles and Czechs are customarily grouped as West Slavs. Other significant differences also exist. Serbian and Croatian, for example, are essentially one language in spoken form, but the Serbs, like the Bulgarians and most of the East Slavs, use the Cyrillic alphabet (based on the Greek), while the Croats use the Latin alphabet, as do the Slovenes and the West Slavs. A religious division exists between the West Slavs, Croats, and Slovenes, who are Roman Catholics for the most part, and the East Slavs, Romanians, Bulgarians, Serbs, and Macedonians, who adhere principally to various branches of the Orthodox Christian Church. At times, religious differences have been important factors in conflicts among the peoples of the region.

NON-SLAVIC PEOPLES

The principal non-Slavic elements in East Central Europe, excluding the Romanians, are the Hungarians and the Albanians. The Hungarians, also known as Magyars, are the descendants of nomads from the east who settled in Hungary in the 9th century. They are distantly related to the Finns and speak a language which is entirely distinct from the Slavic languages. Roman Catholicism has been the dominant religious faith in Hungary, although substantial Calvinist and Lutheran groups have long existed there. The Albanians speak an ancient language which is not related to the Slavic languages except in a very distant sense. Excluding Turkey, Albania is the only European country in which Muslims form a majority of the population, although Yugoslavia and Bulgaria have Muslim minorities forming less than one tenth of their respective populations. These Islamic elements are a legacy from the long period of Turkish rule in the Balkan Peninsula prior to World War I. It should be noted that Albania, in keeping with its strict Communist orthodoxy, is East Central Europe's only country that is officially atheistic. The actual religious beliefs of the people are little known.

A CHECKERBOARD OF HABITATS

The peoples of East Central Europe inhabit a region in which cultural and political complexity is matched by, and is in some measure a product of, great physical diversity. The mountains, forests, and swamps of the region inhibited contacts between neighboring peoples in earlier times (and in some areas still do), and thus contributed to local and regional differences in culture and political allegiance.

A glance at Figure 9.11 shows that East Central Europe forms a kind of crude checkerboard of mountains and plains. The mountains come in a wide range of elevations and degrees of ruggedness. Some are high, rugged, snowcapped, and spectacular; others are lower and more subdued, but they may compensate for their lack of scenic splendor by mineral wealth and the quiet beauty of long-inhabited landscapes. The plains exhibit varying degrees of flatness, fertility, warmth, and humidity; agriculturally they range from some of Europe's better areas to some of Europe's poorest. A handful of Europe's largest rivers carry the drainage of the region to the Black Sea, the Baltic, or the North Sea. We look now at the varied habitats of East Central Europe, commencing at the north, as a series of alternating zones of plains and mountains: the northern plain, the central mountain zone, the Danubian plains, and the southern mountain zone.

THE NORTHERN PLAIN

Most of Poland lies in the northern plain, between the Carpathian and Sudeten mountains on the south and the Baltic Sea on the north. The plain is a segment of the North European Plain, which extends westward from Poland into Germany and eastward into the Soviet Union. Central and northern Poland are comprised of land that is rather sandy and infertile, with many swamps, marshes, and lakes. The level expanses of plain are broken at intervals by terminal moraines, the low, regular, elongated hills created by the continental ice sheets that covered this area during the Ice Age. At the south the plain gradually rises to low uplands. Here, where Poland touches Europe's loess belt, are found the country's most fertile soils.

Poland's largest rivers, the Vistula (Polish: Wisla) and Oder (Polish: Odra), wind across the northern plain from Upper Silesia to the Baltic Sea. The Oder is the country's main internal waterway, carrying some barge traffic between the important Upper Silesian industrial area (with which it is connected by a canal) and the Baltic. The main seaports of Poland—Gdansk (formerly Danzig), Gdynia, and Szczecin (formerly Stettin)—are along or near the lower courses of these rivers. Gdansk and Gdynia form a metropolitan area of about 900,000 people, while metropolitan Szczecin has about 450,000.

THE CENTRAL MOUNTAIN ZONE

The central mountain zone is formed by the Carpathian Mountains and lower ranges rimming the western part of Czechoslovakia. The Carpathians ex-

tend in a giant arc for about a thousand miles (c. 1600 km) from Slovakia and southern Poland to south central Romania. Geologically, these mountains are a continuation of the Alps. They are lower than the Alps, however, and are cut by a greater number of easy passes. Elevations of more than 8000 feet (2438 m) are reached only in the Tatra Mountains of Slovakia and Poland and in the Transylvanian Alps of Romania.

West of the Carpathians, lower mountains enclose the hilly basin of Bohemia, the industrial core of Czechoslovakia. On the north, the Sudeten Mountains and Ore Mountains separate Czechoslovakia from Poland and East Germany. Between these ranges, at the Saxon Gate, the valley of the Elbe River provides a lowland connection and a navigable waterway leading from the Bohemian Basin to the highly developed industrial region of Saxony in East Germany and, farther north, to the West German seaport of Hamburg and the North Sea. However, the imposition of the Iron Curtain has largely severed traffic on the Elbe between Czechoslovakia and Hamburg, and Czechoslovakia's oceanborne trade now moves primarily through the Polish seaport of Szczecin. To the southwest the Bohemian Forest occupies the frontier zone between Czechoslovakia and West Germany. Lower highlands border Bohemia on the southeast.

Between the mountain-rimmed upland basin of Bohemia and the Carpathians of Slovakia, a convenient and historic passageway is provided by the lowland corridor of Moravia. Through this corridor run major routes of transportation connecting Vienna and the Danube Valley with the plains of Poland. To the north, near the Polish frontier, the corridor narrows at the Moravian Gate between the Sudeten Mountains and the Carpathians. Just beyond this gateway, East Central Europe's most important concentration of coal mines and iron-and-steel plants has developed in the Upper Silesian-Moravian coal field of Poland and Czechoslovakia.

THE DANUBIAN PLAINS

Two major lowlands, bordered by mountains and drained by the Danube River and its tributaries, comprise the Danubian plains. One of these, the Great Hungarian Plain, occupies two thirds of Hungary and smaller adjoining portions of Romania, Yugoslavia, and the Soviet Union. The Great Hungarian Plain is very level in most places and contains much poorly drained land in the vicinity of its rivers. To the northwest an outlier, the Little Hungarian Plain, extends into the margins of Czechoslovakia and Austria. The second major lowland drained by the Danube is comprised of the plains of Walachia and Moldavia in Romania, together with the northern fringe of Bulgaria. The Danubian plains contain the most fertile large agricultural regions in East Central Europe. Broad expanses of level land with deep, rich soils and a humid continental climate resembling that of parts of the American Midwest provide good natural conditions for growing corn and wheat, which occupy the greater part of the arable land.

The Danube River, which supplies a navigable water connection between these lowlands and the outside world, is the longest river in Eurasia west of Russia's Volga. It rises in the Black Forest of southwestern West Germany and follows a winding course of some 1750 miles (c. 2800 km) to the Black Sea, which it enters through three main channels. The Danube is customarily divided into three principal sections: Upper, Middle, and Lower. The Upper Danube, above Vienna, is fed principally by tributaries from the Alps. This section of the river is swift and hard to navigate, although large barges use it as far upstream as Regensburg, West Germany. Below Vienna the Middle Danube flows leisurely across the Hungarian plains past the Czechoslovak river port of Bratislava and the Hungarian and Yugoslav capital cities of Budapest (Fig. 9.13) and Belgrade. In the border zone between Yugoslavia and Romania, the river follows a winding series of gorges through a belt of mountains about 80 miles (c. 130 km) wide where the Carpathians reach southward to merge with the Balkan Range. The easternmost gorge is the famous Iron Gate. Beyond the Iron Gate the Lower Danube forms the boundary between the level plains of southern Romania and the low plateaus of northern Bulgaria. The river then turns northward into Romania and enters the Black Sea through a low, marshy delta. No river in the world touches so many different countries as the Danube—eight in all.

Traffic on the Danube is relatively modest compared with the enormous tonnage of bulk goods carried by barges on the Rhine or the traffic carried by road and rail in the East Central European countries that the Danube touches. The river flows mostly through agrarian areas, and heavy industries—the most important generators of barge traffic—are found in only a few places along its banks.

THE SOUTHERN MOUNTAIN ZONE

The southern mountain zone occupies most of the Balkan Peninsula. Bulgaria, Yugoslavia, and Albania, the East Central European countries that share this zone, are very mountainous, although Yugoslavia and Bulgaria contain substantial areas of lowland, and Albania has lowlands along the coast.

In Bulgaria the principal mountains are the Balkan Range, extending east-west across the center of the country, and the more extensive Rhodope Mountains in the southwest. These are rugged mountains that attain heights of more than 9000 feet (c. 2750 m) in a few places. Between the Rhodope and the Balkan Range is the productive valley of the Maritsa River, constituting, together with the adjoining Sofia Basin, the economic core region of present-day Bulgaria. North of the Balkan Range a low plateau, covered with loess and cut by deep river valleys, slopes to the Danube.

In central and southern Yugoslavia a tangled mass of hills and mountains constitutes a major barrier to travel. Through this difficult region a historic

Figure 9.13 *The Danube River at Budapest. Government buildings line the waterfront beyond the Chain Bridge. The domed structure is Hungary's national Parliament. The view was taken from Buda on the higher west bank, looking toward Pest across the river.*

lowland passage connecting the Danube Valley with the Aegean Sea follows the trough of the Morava and Vardar rivers. At the Aegean end of the passage is the Greek seaport of Salonika (Thessaloniki), which handles some traffic for Yugoslavia. The Morava-Vardar corridor is linked with the Maritsa Valley by an important east-west route which leads through the high basin in which Sofia, the capital of Bulgaria (see chapter-opening photo) is located.

Along the rugged, island-fringed Dalmatian Coast of southwestern Yugoslavia, the Dinaric Alps rise steeply from the Adriatic Sea. The principal ranges run parallel to the Adriatic Coast and impose a succession of rocky heights crossed by only a few significant passes. Much of this mountainous region of western Yugoslavia is characterized by karst or sinkhole topography caused by the solvent action of underground waters in limestone bedrock. The dry, inhospitable karst plateaus are among the more desolate parts of Europe. Railroads crossing the mountains at the north, where they are narrowest, give access to some small Yugoslav seaports and to the Italian seaport of Trieste. The spectacular Dalmatian Coast, recently made more accessible by improved roads and better rail and air service, is the main locale for Yugoslavia's flourishing tourist industry—the most important one in East Central Europe.

THE VARIED RANGE OF CLIMATES

The climates of East Central Europe are transitional between the marine climate of northwestern Europe, the extreme continental climates of Russia, and the dry-summer subtropical (mediterranean) climate of Southern Europe. In most areas winters are colder than in western Europe, although not so cold as in Russia. Typical average temperatures in January range from 28°F (−2°C) at Warsaw to 32°F (0°C) at Belgrade. Only in sheltered valleys and coastal lowlands of the extreme south do midwinter average temperatures rise significantly above freezing. An example is provided by the small seaport and resort city of Dubrovnik on the Dalmatian Coast, which has a January average of 48°F (9°C). Summer temperatures are generally higher than those in northwestern Europe, the July average being 66°F (19°C) at Warsaw and 73°F (23°C) at Belgrade.

In most lowlands of East Central Europe the average annual precipitation is between 20 and 25 inches (c. 50–65 cm), with a summer maximum except in southern areas near the Mediterranean. In the Danubian plains hot summers decrease the efficiency of the rainfall, and periods of drought make agriculture somewhat more hazardous than in the more dependable climate of northwestern Europe. But in most years moisture is sufficient for a good harvest. This humid continental long-summer climate of the Danubian plains is comparable in many respects to the climate of the corn and soybean region (Corn Belt) in the United States Midwest, but the plains of Poland have a less favorable humid continental short-summer climate that is comparable to the climate of the American Great Lakes region. The most exceptional climatic area is the Dalmatian Coast. Here temperatures are subtropical and precipitation, concentrated in the winter, reaches as much as 180 inches (c. 460 cm) per year on some windward slopes facing the Adriatic. This is Europe's greatest precipitation.

CHANGING PATTERNS OF RURAL LIFE

In most parts of East Central Europe the nature of the climate has a more decisive impact on large masses of people than is true in the highly industrialized lands of northwestern Europe. Although industry and urbanism have made impressive strides since World War II, great areas in East Central Europe still have a strongly rural and agricultural cast. Increasing numbers of workers in rural areas of East Central Europe commute to jobs in urban places, but most rural people still support themselves wholly or primarily by agriculture. On the whole, their efforts do not provide a very good livelihood, despite significant gains in recent years. Great numbers of rural homes now have electricity, and overall levels of rural income, diet, health, education, and general welfare are much improved over what they were before World War II. But even this improvement still leaves rural levels of living in the respective countries far below those of urban places, and it leaves rural dwellers in the region as a whole well below the general level of their counterparts in the countries bordering the North Sea.

LAND REFORM BEFORE THE COMMUNIST ERA

The discrepancy in rural levels of living between East Central Europe and the North Sea lands has endured throughout modern times. For centuries prior to World War I, and in some areas up to World War II, the countryside of Poland, Slovakia, Hungary, and Romania was dominated by estates worked by tenants or hired labor, and the history of estates was often one of exploitation or neglect of the laborers or tenants. Bulgaria, Serbia, Montenegro, and Albania were exceptions in that they were mainly lands of small landowning peasant farmers, but the latter were severely handicapped by ignorance, brigandage, feuds, governmental exploitation, inadequate and fragmented landholdings, poor transportation, and, in mountainous areas, the poor and scanty soil. The great majority of these peasants were among the most poverty-stricken in Europe.

In most East Central European lands held by Germany and Austria, estates were prominent, but many farmers owned the land they worked and some of them were well off. This was particularly true in certain areas where manufacturing developed, attracting surplus population off the land and providing urban markets for large amounts of farm produce. The height of rural prosperity was reached in Bohemia and Moravia, the industrial heart of the Austro-Hungarian Empire. But this area was very exceptional. In most parts of East Central Europe population accumulated on the land, the amount of land per person became smaller, strips of the fragmented small farms were subdivided and resubdivided among heirs, and the general situation of farmers deteriorated.

In the 19th and early 20th centuries some of the pressure was relieved by the emigration of millions of East Central Europeans to overseas lands, principally the United States. Those who remained clamored for land, and the result was a series of land-reform programs in various countries of the region. Such programs commenced in the 19th century but did not gain much headway until after World War I. The main intent was to break up large holdings and sell the land to small farmers or landless laborers. In many instances this took the form of expropriating and breaking up properties belonging to the aristocracy of defunct empires. After World War I, Hungarian landlords in Slovakia were dispossessed, as were Russians in eastern Poland. In Hungary, however, the politically powerful Hungarian landowners generally managed to retain their properties, as did many large native landowners in Poland and Romania.

On the whole, redistribution of land was most successful, and benefits to farmers were greatest, in Czechoslovakia. This was especially true in the more developed western provinces of Bohemia and Moravia. Elsewhere in East Central Europe, the transfer of land to peasant ownership did not generally result in notable improvement of the standard of living, although it did help to keep down discontent among the peasants, for whom ownership of land had an intense emotional significance. Many landless farm laborers were too poor to buy land, even under generous governmental credit arrangements, and farmers who did acquire land generally lacked the necessary capital, machinery, and knowledge to make effective use of it. Machinery would have been at a serious disadvantage in any case, due to the fragmented arrangement of most farms and the small size of individual strips of land.

In today's terms, East Central Europe before the Communist era was a rather representative case of a "developing" area agriculturally, with many characteristics still to be found in such places as India or the Middle East. Poverty was not so abysmal as in some developing countries today, but the general pattern of organization in the countryside was much the same and the range of problems was comparable. This was also the case in Russia before the Bolshevik Revolution of 1917.

We turn now to East Central Europe's experience with the collectivized system of agriculture that was the Communist response to the agricultural problem in Russia and then was imposed on East Central Europe after World War II.

THE UPS AND DOWNS OF AGRICULTURAL DEVELOPMENT UNDER COMMUNISM

Following World War II, the remaining large private holdings in East Central Europe were liquidated by the new Communist governments, and programs of collectivized agriculture on the Soviet model were introduced. Some farm land was placed in large state-owned farms on which workers were paid wages, but most of the land was organized into collective farms owned and worked jointly by peasant families who shared the proceeds after the operating expenses of the collective had been met. Farm families normally were allowed to retain garden plots of 1 to 3 acres (0.4–1.2 hectares) and limited numbers of livestock for their individual use, but most privately owned land and livestock became the property of the collectives. Collectivization met with strong resistance in the countryside, and in Yugoslavia and Poland it was discontinued in the 1950s and the collectives were allowed to disband. Today all but a minor share of the cultivated land in the latter countries is privately owned. The remaining East Central European countries pushed ahead with collectivization, and now have nearly all of their farm land in collective or state ownership.

For some years after collectivization was introduced, many features of the system were adverse to agricultural expansion. Severe taxation drained wealth from the farms in order to support industrialization. Efficiency suffered under excessively rigid control by government planners and collective-farm managers, who often were more politically orthodox and loyal than expert in agriculture. The lack of a personal stake in the land, plus low government-controlled prices for farm products, created apathy and indifference among the collectivized farmers, many of whom deserted the farms for work in expanding industries.

More recently, however, policies have been changed to stimulate agricultural output. Control and planning have tended to become less rigid and doctrinaire, giving collective farmers greater latitude in making production and marketing decisions for themselves. Systems of pricing and taxation have been changed to give greater financial incentives for increased output, and farmers in some countries have been given more freedom for individual enterprise on their private plots within the collectives. The level of investment in machinery, fertilizer, land improvement, and education has been increased. With this new flexibility and encouragement, agricultural output has expanded rapidly. By one estimate, food production in East Central Europe increased by an estimated 24 percent during the 1970s. This rise was slightly faster than that of western Europe and much faster than that of the USSR, but it was considerably slower than that of the United States and many of the world's "developing" countries. It was enough faster than East Central Europe's moderate population growth that the region's food output per person increased considerably. But the region continued to lag well behind the North Sea countries in productivity of agricultural land and labor, and all the East Central European countries have been requiring net grain imports to meet their food demands in the 1980s. Very wide disparities in food availability have also characterized the region's countries, from the relative abundance of Hungary, with its good land and a Communist system notably liberalized in the direction of private enterprise, to the food scarcities of Poland, where peasant farmers have had little incentive to produce and sell because a lagging industrial economy has offered them so little to buy.

Certain agricultural trends within the region as a whole tend to parallel recent trends in the USSR (see Chapter 10). Agriculture has become more mechanized, diversified, and intensive, with more attention given to protective foods (vegetables, fruits, meat, milk, eggs) for sale in urban markets, "industrial crops" such as sugar beets and sunflowers, and green forage and silage crops to feed dairy cattle. The countries where agriculture has been collectivized also resemble the USSR in that a sizable share of the vegetables, poultry, eggs, and milk that reach the market come from privately owned plots and animals of collective-farm members.

THE PATTERN OF CROPS AND LIVESTOCK

Despite the recent attempts to diversify and intensify agriculture, farming in East Central Europe from a production standpoint remains primarily a crop-growing enterprise based on corn and wheat production. Corn and wheat are the leading crops in the five southern countries of the region—Hungary, Romania, Yugoslavia, Bulgaria, and Albania. All these countries except Albania share the "Corn Belt" type of climate (humid continental long-summer), which in the United States has proved to be an excellent climate for growing not only the corn and soybeans that are the dominant crops today but also for wheat. Small and mountainous Albania, handicapped by a lack of good land and the dry Mediterranean summers, is a very minor agricultural producer but does grow some wheat with the aid of winter rainfall and some irrigated corn in the summer. In Czechoslovakia and Poland with their more northerly and cooler climates, corn is unimportant; but wheat production is a major part of agriculture as far north as southern Poland. Poland's wheat area, just north of the Sudeten and Carpathian mountains, has relatively good soils formed of loess. But most of the country is north of the loess belt and has poor sandy soils like those of northern Germany. Here rye and potatoes become the main crops.

Livestock raising is a prominent part of agriculture throughout East Central Europe, although the total value of animal products is considerably less than that of crops. Large amounts of crops, however, are fed to livestock. The region's meat supply comes primarily from pigs and only secondarily from cattle, although cattle are raised in varying numbers in all countries. The four southern countries have large areas of poor upland pasture that are not very suitable for cattle but provide grazing for millions of sheep, which are the main source of meat in Bulgaria and Albania. Milk consumption per capita is much less in East Central Europe than it is in northwestern Europe, and the lag in its availability is symptomatic of the more general lag in agricultural intensity and productivity within East Central Europe compared with the North Sea countries.

A number of agricultural products are less fundamental to East Central Europe than the major grains and livestock but are by no means unimportant. Some of these products are widespread: sugar beets in all the more fertile agricultural areas from southern Poland south; deciduous orchards throughout the region, furnishing among other things some distinctive regional beverages such as plum brandies; grapes grown for wine production from Hungary south; and sunflowers grown for vegetable oil (which is extracted from their seeds), also from Hungary south. Limited areas, especially in the warmer climates of the south, grow specialty crops not included in the previous list. A good example is Bulgaria's Maritsa Valley, which grows cotton and strawberries and provides other Communist countries with exports of tomatoes or other fresh vegetables and fruits. Being on the northern edge of Europe's mediterranean climate, the Valley has rather dry—although not completely rainless—summers, but the lack of moisture is overcome, and agriculture is intensified, by East Central Europe's largest development of irrigation.

Figure 9.14   Eger, in Hungary, was founded about ten centuries ago and is one of the many European cities that were more prominent in the past than they are today. Older low-rise buildings lining narrow streets contrast with standardized high-rise structures dating from the Communist era.

URBAN DEVELOPMENT

The strongly rural character of much of East Central Europe is mirrored in a relative shortage of cities. Particularly lacking are large urban constellations and individual cities of truly major size, although small regional centers, often old and historic, are widespread (Fig. 9.14). Only two areas—the Upper Silesian-Moravian coal field of Poland and Czechoslovakia, and the Bohemian Basin of Czechoslovakia—exhibit a closely knit web of numerous urban places. Even the agglomeration of medium-sized mining and metallurgical centers in Upper Silesia-Moravia—a Ruhr in miniature—aggregates only about 3.5 million people, compared with around 5 million for West Germany's Inner Ruhr or about 7.5 million for the total Ruhr by a broader definition. The Polish part of Upper Silesia-Moravia, with about 2.7 million people, is somewhat larger than Poland's capital, Warsaw (2.2 million). In all the other countries, the largest cities are national capitals. None of the capitals, including Warsaw, ranks in Europe's top ten cities in metropolitan population. The largest, Budapest, has about 2.6 million people, Bucharest has 2.2 million, Prague, Sofia, and Belgrade have between 1 and 1.5 million, and Albania's small capital, Tirana, has only about 200,000. In every instance the capital is its country's leading center of diversified industries, and in every instance except Poland and Yugoslavia it far overshadows other urban areas in size. Poland's multi-city metropolitan complex in Upper Silesia is bigger than Warsaw, but the third city, Lodz (1.1 million), conforms to the East Central European pattern in being only half the size of Warsaw. Lodz has long been Poland's main center of textile industries and is sometimes called the "Polish Manchester." In Yugoslavia, Zagreb, the main city in the important area called Croatia, has a population in the city proper that is about three fourths the size of that in Belgrade city. Before independence the two cities developed in different empires, with Belgrade in Serbia being under Turkish control, and Zagreb under Austrian; today the cities function as capitals of the two largest socialist republics among the six in the Yugoslav federal state. The term *primate city* is sometimes applied to a city so dominant within its country as most East Central European capitals are. Primate cities are generally (although by no means exclusively) found in "developing" countries, and their presence is often symptomatic of retarded industrial development. Certainly the lack of industry contributed powerfully to the fact that East Central Europe has so few cities of much size. The lack of big cities is very apparent if one compares the map of East Central Europe (see Fig. 9.11) with the map of Germany (see Fig. 7.7).

INDUSTRIAL DEVELOPMENT AND RESOURCES

Prior to World War II, large sections of East Central Europe had been little affected by the industrial and urban modes of life that swept over northwestern Europe after the Industrial Revolution. In most areas, life had a deeply rural quality that was matched only in the most isolated outlying sections of the countries around the North Sea. In all countries, industries grew considerably in number and size through the period between the two world wars, but they were concentrated very heavily in a handful of large urban districts or national capitals. Various other cities scattered through East Central Europe had a certain amount of industry, but it was generally of a very simple type. Such industries as existed often were financed and controlled from western Europe, and industrial life had a markedly "colonial" quality. Western Czechoslovakia was an exception to the common pattern. As the industrial heart of the Austro-Hungarian Empire, the Czech lands in Bohemia and Moravia had developed diversified, advanced, and efficient industries that were able to adapt to new conditions and continue to operate successfully when the empire disintegrated at the end of World War I. Another exception was the heavy-industrial area in Upper Silesia then held by Germany.

INDUSTRIAL TRENDS IN THE COMMUNIST ERA

The accession of Communism in East Central Europe inaugurated a new industrial and urban era. With minor exceptions the existing industries were taken out of private hands, and national economic plans modeled on the Five-Year Plans of the Soviet Union (see Chapter 10) were developed in the various countries. Under the Communists the older industrial districts have continued to grow in size and complexity, but Communist planners in all countries have placed much emphasis on the development of new industries in weakly industrialized areas. By this means they have sought (1) to raise incomes and standards of living in less-developed areas so as to make all regions more nearly uniform in socioeconomic levels; (2) to increase the contribution of less-developed areas to national economies; (3) to induce workers to remain in their home areas instead of migrating to the older, more congested industrial and urban districts; and (4) to keep down political discontent in less-developed areas by providing better economic and cultural opportunities there. There is evidence of partial success in achieving these objectives, although it should be noted that the older industrial districts continue to be the predominant centers and in most instances have suffered only a moderate decline, or have actually increased, in relative importance. Many of the newer industries in outlying areas have been criticized as uneconomic, but there have been undeniable social and political gains from their establishment. These have not, however, been sufficient to alter the imbalance among major regions in socioeconomic level. For instance, in Yugoslavia wide discrepancies continue to exist between the more developed and prosperous northern part of the country—with its larger cities and industries, gentler terrain, better soils, and closer historic relationships with the Western world—and the mountainous, less-developed south, with its history of stagnation or regression during the long period of Turkish control.

Initially the central planning agencies of the Communist governments maintained a rigid control over individual industries. Plants were directed to produce certain goods in quantities determined by the planners, and the success of a plant was judged by its ability to meet production targets rather than by its ability to sell its products at a profit in competition with other plants. Political reliability and conformity to the central plan were qualities much desired in plant managers. This system worked fairly well for relatively simple heavy-industrial products such as steel or bulk chemicals, but it resulted in shoddy consumer goods, which were often in short supply or in mountainous oversupply because production was not being determined by consumer demand.

For some time, however, the system has been changing, although with more speed in some countries than in others. There is now a strong tendency to allow the managers and labor forces of individual plants greater latitude in making their own decisions with regard to production and marketing, although such decisions must still be made within the general frame of national economic plans originated by central government bureaus. Plants are judged not on their ability to simply produce predetermined quantities of specified goods, but on their ability to actually market their products on a profitable basis. Workers in plants that make a profit receive better wages than those in plants that do not, and plants that operate at a loss are apt to be merged with other enterprises or allowed to go bankrupt. This development began first in Yugoslavia and has gone farthest there and in Hungary. It is a particularly desirable change at the present time when more stress is being placed on improving the output and quality of consumer goods. But the most remarkable East European achievement in providing a reasonable supply of consumer goods has been brought about by Hungary's reintroduction of private enterprise, to the extent of about 12,000 companies licensed to operate as of the early 1980s. However, in the late 1980s, industrial performance in East Central Europe still lagged to the point that a reformist regime in the USSR began putting increased pressure on the satellite countries for economic "liberalization."

In the early stages of industrialization, Communist planners did not aim at a balanced development at all types of industry. Instead, they stressed the types that were deemed most essential to industrial development as a whole. New investment was channeled heavily into a relatively few fields: mining, iron and steel, machinery, chemicals, construction materials, and electric power. Development of a strong base of heavy industries and sources of power was viewed as an essential prerequisite for overall industrial development, and investment in such industries was favored at the expense of consumer-type industries and agriculture. Today heavy industries are still pushed, but more attention is gradually being paid to diversified industries making consumer goods.

As a means of facilitating industrial development, great stress has been placed on the improvement of transportation. This effort has been devoted primarily to the railroads, which carry most long-distance freight in East Central Europe. Water transportation is poorly developed compared with that in northwestern Europe, although some efforts are being made to improve it. Long-distance truck transportation is not highly developed, but it is growing, and trucks are used a great deal for short hauls within cities or to connect cities and towns with their immediate hinterlands.

The planned expansion of mining and industry in East Central Europe has produced notable increases in the total output of minerals, manufactured goods, and power. The largest absolute increases have occurred in the countries that were already the most industrialized: Czechoslovakia and Poland. But the most spectacular percentage increases have occurred in the other countries, which began their postwar industrialization from a much smaller base. For example, between 1953 and 1971, Czechoslovakia increased its output of electric power by nearly 300 percent, but in Bulgaria and Romania the increase was more than 1000 percent.

THE SPOTTY DEVELOPMENT OF MINERAL WEALTH

East Central Europe has not proved unusually rich in minerals on which to base industrial development, although every country has mined a variety of useful deposits. Coal, including the low-grade coal called brown coal or lignite, has been the most valuable mineral resource to date. It fueled the early industrialization of Poland and Czechoslovakia which gave those countries an industrial lead over the rest of the region, and even in the age of oil and

gas it remains by far the leading single energy source in East Central Europe as a whole.

The region's principal coal deposits are concentrated in the Upper Silesian-Moravian field, now lying mostly in southern Poland but partly in adjacent Czechoslovakia. This coal field is the second largest in Europe, after the German Ruhr. Before World War I the Upper Silesian (now Polish) part of the field was in Germany and underwent major development as part of Germany's industrialization. After World War I, this section was split between Germany and Poland, and after World War II it became Polish. Before World War I the smaller southern (Moravian) part of the coal field was part of the Austrian Empire and became a major center of coal and steel production for that state. At the end of the war this section became part of newly created Czechoslovakia and has remained so. Thus a typical European coal-and-steel industrial district of major importance developed under Germany and Austria and has been inherited and expanded by Poland and Czechoslovakia. The usual cluster of cities in such an area is present, centering on Katowice in Poland and Ostrava in Czechoslovakia. This grimy urban belt, littered with waste piles and encrusted with soot, is the main center of coal mining and heavy industry in East Central Europe; and when the Polish economy is functioning even reasonably well the Upper Silesian output makes Poland one of the world's leading producers and exporters of coal. Steel produced here is highly important to a wide range of Polish industries, a notable share of which are concentrated in or relatively near this Upper Silesian ''black country.''

Czechoslovakia's industries are much less concentrated in the Moravian coal field area. Much of the Bohemian Basin, semi-surrounded by mountains and focusing on Prague, is a region of industrial cities with a highly diversified production. Again, the basic development occurred before World War I. In that period Bohemia became the main industrial region of the Austrian Empire. Subsequently it did rather well as part of the new democratic state of Czechoslovakia, and output in many lines has been further expanded under the post–World War II Communist government. Within the Basin, Pilsen (city proper, 165,000) is a famous center of steel, armaments, machinery, and beer production, but the Czechoslovak capital, Prague, has a much larger and more diversified overall industrial development. Throughout Bohemia's modern industrial history, not only the nearby Moravian coal resources but also Bohemia's own lignite deposits—the largest in East Central Europe—have played a major role.

Elsewhere in East Central Europe, however, mineral resources have offered fewer advantages for large-scale industrialization. A wide variety of minerals is found in the region, but most deposits are too small or too low in quality to be of major use. Exceptions include very large bauxite deposits in Hungary and Yugoslavia, and large lead and zinc deposits in Polish Upper Silesia, in Yugoslavia, and in Bulgaria. Most of the region is deficient in high-grade power resources and in iron ore. Only Romania has been a significant oil producer, from fields strung in a long arc through the Carpathian foothills, but even it has become a net importer of oil. The country's gas deposits—mainly in Transylvania—have also been East Central Europe's largest, and like the oil are commencing to run low. In the other countries oil imports are a major necessity and economic burden except in Albania, which has become a modest net exporter from small coastal fields. The USSR is the major oil supplier for the region as a whole, and some countries import large quantities of Soviet natural gas. Good coal resources are absent or small in all East Central European countries except Poland and Czechoslovakia, and thermal electricity is generated largely from lignite deposits, which are widespread. Yugoslavia, however, relies on hydropower for about one third of its electricity requirements, and about four fifths of Albania's power comes from this source. The two countries, both of which have heavy precipitation, have harnessed their mountain streams extensively to provide current.

Mineral Inadequacies in the New Age of Iron and Steel

The insufficiency of mineral resources in East Central Europe was highlighted after World War II when every one of the Communist governments undertook to build one or more large iron-and-steel plants regardless of whether national resources were sufficient. This construction program included the expansion of some older plants as well as the building of entirely new ones. It has greatly increased the region's steel capacity and output, thus conforming to the standard Communist practice of stressing iron and steel as an underpinning for industrial expansion in general. Most of the new plants are located in or near the principal industrial districts of the different countries. They are *integrated* plants, with coke ovens, blast furnaces, steel furnaces, and rolling mills which shape hot steel into bars, sheets, rods, or pipes. The largest steel mill in all of East Central Europe is the Lenin Works at Nowa Huta ("New Furnace"), a suburb of the industrial and university city of Krakow (825,000) not far to the east of Poland's Upper Silesian industrial complex.

The expansion of steel milling was carried through in defiance of resource inadequacies, since no country had really adequate iron-ore resources, and only Poland and Czechoslovakia had enough coking coal. The necessary materials are imported, with the Soviet Union being the main source in most instances. This underscores a fundamental fact about East Central Europe's industrialization in the Communist era: its heavy dependence on the Soviet Union for vital minerals. Not only iron ore and coal, but oil, gas, and other minerals are imported in great quantities and over long distances by rail or pipeline from Soviet mines and wells. In return, the USSR is a large importer of East Central Europe's industrial products.

During the 1970s, however, the trade orientation of the region toward the Soviet Union weakened. Trade and financial relations with Western countries, companies, and banks expanded rapidly. With relatively little to sell to the West, the East Central European countries borrowed massively from Western governments and banks to pay for imports from the West. This course was followed particularly by Poland and Romania. New industrial plant and equipment acquired in this way was supposed to be paid for by the goods that would be produced for export to the West. But then the Western economies began to slump, lowering the demand for imports. Combined with mismanagement by Communist bureaucrats, this led to a situation by the 1980s in which large debts to Western governments and banks needed to be repaid if countries were to maintain any credit at all, but exports with which to pay were not being produced or could not be sold. The result was a severe check to economic expansion, falling standards of living as governments squeezed out the needed money from their people, and, in Poland, social action through the formation of an independent union called Solidarity, which paralyzed the country and threatened to upset Communist dominance. This threat was only averted by violence of the security police and army against the people and the imposition for a time of martial law, followed by a harsh new set of government controls and rules after martial law was lifted. The events of the 1970s and 1980s in East Central Europe made the future course of the Communist experience very hard to predict, although ultimate Soviet control over the satellite realm had been reasserted in Poland and seemed likely to continue indefinitely. Much depended on the internal evolution of the USSR, where a recently launched program of economic and political reform faced many uncertainties. The Soviet situation is discussed in Chapter 10.

PART III

THE SOVIET UNION

Geography and Planning in the Soviet Union

Many aspects of the Soviet Union are reflected in this view of a collective farm in the Moldavian SSR. Modernism and traditional rurality coexist in a way no longer characteristic of the United States and various other countries where advanced technology has triumphed in both city and countryside.

WE live in a time when the world's geography is shaped increasingly by planning, often on a very large scale. Planned enterprises such as the Interstate Highway System of the United States not only cause direct changes in spatial organization and cultural landscapes but also generate side effects that change geography, often in unforeseen ways. In most countries such planning is done piecemeal by government or private enterprise as need arises and political and economic support is forthcoming. But in the Communist world, planning becomes more all-embracing. It is done on a centralized basis for entire nations, of which the most powerful and influential is the Union of Soviet Socialist Republics. In the Soviet Union the creation or reshaping of geography by centralized national planning has gone farther than anywhere else. Since the Russian Revolution of 1917, Soviet planners have made important changes in their country's spatial organization, have greatly altered the interaction of people and nature, and have added many new elements to the cultural landscape. They have accelerated the harnessing of resources, and their activities have imposed increasing cultural uniformity on a varied country. Backed by strong governmental authority, the planners have created new cities and transport links, enlarged older cities, carried through a massive expansion of mining and manufacturing, increased the amount of cultivated land, and reorganized the countryside by collectivized agriculture. Standardized apartment blocks have inundated the cities of the land from Arkhangelsk (Archangel) on the White Sea to Odessa on the Black Sea, and from Riga on the Baltic to Vladivostok on the Sea of Japan (see photo of Khabarovsk in the Soviet Far East, Fig. 11.13). Such changes have not obliterated the geographic patterns inherited from the older Russia of the tsars, but many of those patterns have been greatly altered. In this chapter, we will survey Soviet Russia's geographic inheritance, how it came to be, and how it has been reshaped—often at enormous human cost—under Soviet planning.

OVERVIEW OF BASIC CHARACTERISTICS

Although it is politically one country, the USSR is so huge, distinctive, and powerful that it deserves to be called a major world region. Its physical dimensions, population, resources, and internal variety are

RUSSIAN AND SOVIET AREAL NAMES

In commencing a geographic study of the Soviet Union, students should become acquainted with certain areal names for the entire country and its principal parts. Union of Soviet Socialist Republics, frequently shortened to USSR or Soviet Union, is the official name of the Soviet state. This political unit came into existence following the overthrow of the last of the Romanov tsars (Nicholas II) in 1917. Pre-Revolutionary Russia is often spoken of as Old Russia, Tsarist Russia, or Imperial Russia. Post-Revolutionary Russia is often called Soviet Russia. The name Russia is widely used to refer to the entire country either before or after the Revolution, although some

teachers and scholars prefer to restrict the present use of this name to the huge Russian Soviet Federated Socialist Republic (RSFSR), the largest of fifteen Union Republics that comprise the USSR. The part of the country west of the Ural Mountains and north of the Caucasus Mountains has been known historically as European Russia, although the name is used less commonly today due to a widespread tendency to consider the entire Soviet Union a unit separate from Europe. Similarly, the areas beyond the Urals and Caucasus have been called Asian Russia, Soviet Asia, or the eastern regions. Here Transcaucasia is the

area in and south of the Caucasus Mountains (see Fig. 11.1); Siberia has long been a general name for the area between the Urals and the Pacific; and Soviet Middle Asia is the name used in this text for the five arid republics with large Muslim populations immediately east and north of the Caspian Sea. For statistical and planning purposes the Soviets subdivide the area traditionally called Siberia into three economic regions: Western Siberia, Eastern Siberia, and the Far East. Similarly, what is called here Soviet Middle Asia is divided into two regions: Central Asia (four republics) and Kazakhstan (the Kazakh SSR; Fig. 10.1).

comparable to those of regions containing many countries. Despite serious physical handicaps, this continent-sized country has the political and economic power of a superstate.

LAND FRONTIERS

An eventful geopolitical history has given the Soviet Union a common land frontier with no less than 12 other countries in Eurasia (Fig. 10.1). Along the 8000-mile (13,000-km) boundary between the Black Sea and the Pacific Ocean, the USSR directly borders Turkey, Iran, Afghanistan, China, Mongolia, and North Korea. Pakistan and India also lie close by, although neither has an actual boundary with the USSR. In the Pacific Ocean, narrow water passages separate the Soviet islands of Sakhalin and the Kurils from Japan. If India, Pakistan, and Japan are included, the Soviet Union in Asia is a near neighbor of about half the world's people. However, too much can be made of proximity as seen on a map, since the Soviet frontiers in Asia lie mostly in mountainous (Fig. 10.2) or arid regions that form sparsely populated zones of separation between the USSR and the core regions of the bordering countries. Some territorial changes took place along the USSR's Asian frontiers in the World War II period. The southern half of Sakhalin Island and the Kuril Islands were annexed from Japan, both being areas originally ceded to Japan by Tsarist Russia. In addition, the small semi-independent region called Tannu Tuva was absorbed. Meanwhile Mongolia has continued to be a Soviet "satellite" (since 1924). In 1979 the Soviet Union set off an international furor when it invaded Afghanistan to bolster a Communist government threatened by guerrilla uprisings. This conflict was still unresolved in early June 1989, although the Soviet troops had left the country (see Afghanistan discussion in Chapter 13).

In the west the Soviet Union has common land boundaries with Romania, Hungary, Czechoslovakia, Poland, Finland, and Norway. The first four countries are satellites with Communist governments largely subservient to the Soviet Union. Finland's actions are also strongly influenced by the USSR, although Finland has a Western-type democratic government and is not a satellite. Norway is even farther outside the Soviet orbit, being a NATO member fully aligned with the Western bloc of nations. The Soviet Union's western frontiers are very different from its frontiers between the Black Sea and the Pacific. This is especially true of the 900-mile-long (1450-km) boundary zone between the Black Sea and the Baltic Sea. Here the boundary line passes mostly through well-populated lowlands that have long been disputed territory between the Russian state and its western neighbors. Many invasions of Russia have come from the west, including the French under Napoleon I in 1812 and more recently the German armies in 1941. During the World War II period (1939–1945) the Soviets expanded their national territory westward in several areas (see Fig. 9.11, inset, and Fig. 9.1). These territorial changes, which are discussed more fully in Chapters 7, 9, and 11, gave the Soviet Union frontiers with Hungary and Norway; added the small Baltic countries of Estonia, Latvia, and Lithuania to the USSR as Union Republics; and removed German East Prussia from the political map (it was divided between Russia and Poland). Many geostrategic advantages accrued to the Soviet Union from the various boundary shifts, in which a total of eight European nations yielded territory. Most of the Soviet gains were lands that had been part of Tsarist Russia, but some were acquired for the first time.[1]

AREA, POPULATION, PHYSICAL HANDICAPS, AND RESOURCES

With an area of 8.65 million square miles (22.4 million sq km), the Soviet Union is smaller than Africa but larger than any other major world region delineated in this text (see Table 3.1). Its estimated population of 286 million in mid-1988 was smaller than that of most major world regions but larger than that of Anglo-America or the Pacific World. Among individual countries the USSR ranks third in population, after China and India. The overall rate of natural population increase is relatively slow (see Table 3.1), although some non-Russian nationalities are increasing at a much faster rate. Slowing of the nationwide growth rate in recent decades has been closely linked to rapid urbanization. At the beginning of the Soviet era the USSR was largely inhabited by peasants living in agricultural villages, but by the late 1980s about two thirds of all Soviet citizens were reported to be urban dwellers. Shortage of urban living space, and the resources and life styles of families in which both parents are full-time breadwinners with relatively low incomes has caused most parents to limit their number of children to one, two, or none at all.

Stretching nearly halfway around the globe in northern Eurasia (see Fig. 10.1), the Soviet Union

---

[1]For a fuller account, see W. G. East, "The New Frontiers of the Soviet Union," *Foreign Affairs*, 29 (1951), 591–607.

Figure 10.1   *General reference map of the Soviet Union.*

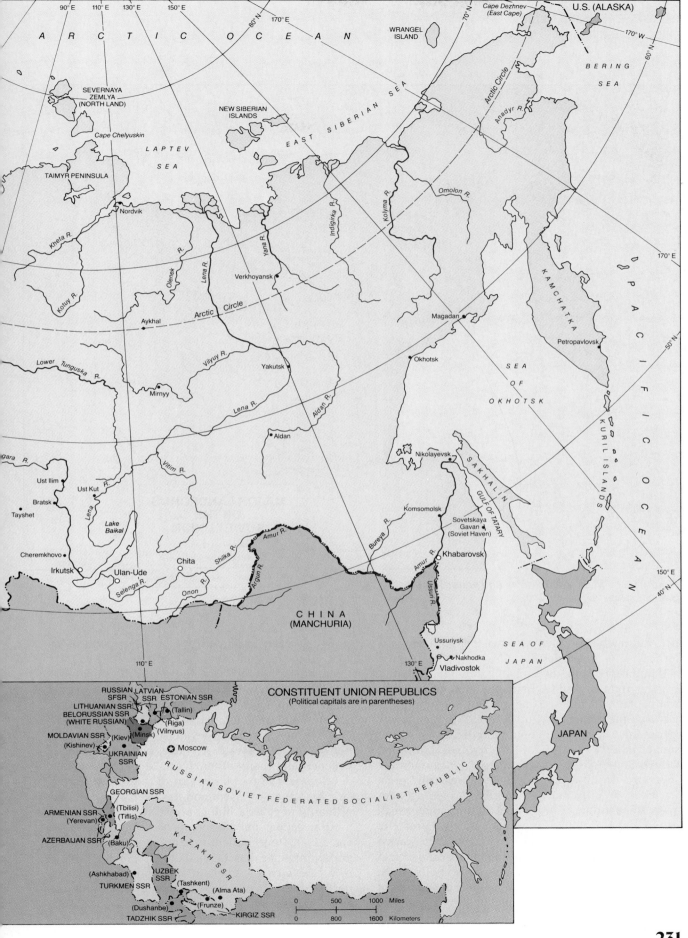

ARCTIC OCEAN

90° E 110° E 130° E 150° E 170° E 80° N 70° N Cape Dezhnev (East Cape) U.S. (ALASKA) 170° W 60° N

WRANGEL ISLAND

BERING SEA

SEVERNAYA ZEMLYA (NORTH LAND)

NEW SIBERIAN ISLANDS

EAST SIBERIAN SEA

Arctic Circle

Cape Chelyuskin

LAPTEV SEA

TAIMYR PENINSULA

Anadyr R.

Omolon R.

170° E

Nordvik

Kheta R.

Koltuy R.

Olenek R.

Lena R.

Indigirka R.

Yana R.

Kolyma R.

Verkhoyansk

KAMCHATKA

Arctic Circle

Aykhal

Magadan

Petropavlovsk

50° N

Lower Tunguska R.

Vilyuy R.

Yakutsk

Okhotsk

SEA OF OKHOTSK

Mirnyy

Lena R.

Aldan R.

Nikolayevsk

SAKHALIN

KURIL ISLANDS

PACIFIC OCEAN

Aldan

gara R.

Ust Ilim

Ust Kut

Vitim R.

Komsomolsk

GULF OF TATARY

Sovetskaya Gavan (Soviet Haven)

150° E

Bratsk

Lena R.

Bureya R.

Tayshet

Lake Baikal

Shilka R.

Amur R.

Amur R.

Khabarovsk

40° N

Cheremkhovo

Ulan-Ude

Chita

Argun R.

Selenga R.

Onon R.

Ussuri R.

Irkutsk

CHINA (MANCHURIA)

Ussuriysk

SEA OF JAPAN

110° E

130° E

Nakhodka

Vladivostok

JAPAN

CONSTITUENT UNION REPUBLICS
(Political capitals are in parentheses)

RUSSIAN SFSR LATVIAN SSR ESTONIAN SSR

LITHUANIAN SSR (Tallin)

BELORUSSIAN SSR (WHITE RUSSIAN) (Riga)

(Vilnyus)

MOLDAVIAN SSR (Kiev) (Minsk)

(Kishinev) ★ Moscow

UKRAINIAN SSR

RUSSIAN SOVIET FEDERATED SOCIALIST REPUBLIC

GEORGIAN SSR

ARMENIAN SSR (Tbilisi)

(Yerevan) (Tiflis)

KAZAKH SSR

AZERBAIJAN SSR (Baku)

(Ashkhabad) UZBEK SSR

TURKMEN SSR (Tashkent) (Alma Ata)

(Dushanbe) (Frunze)

TADZHIK SSR KIRGIZ SSR

0 500 1000 Miles

0 800 1600 Kilometers

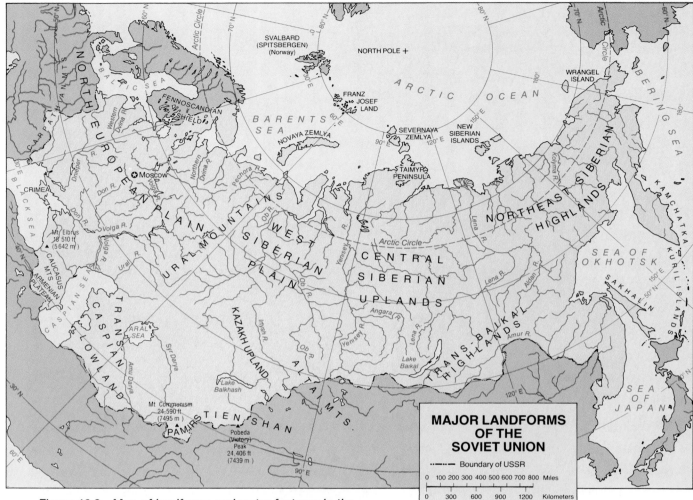

Figure 10.2   Map of landforms and water features in the Soviet Union.

has formidable problems associated with climate, terrain, and sheer distance. Most of the country is handicapped by excessive cold, infertile soils, marshy terrain, aridity, or ruggedness. Such environmental difficulties go far toward explaining the fact that the greater part of the USSR is very lightly inhabited. The average population density of 33 per square mile (13 per sq. km) is only half that of the United States. Outlying population nodes often are separated from other populated areas by great stretches of unproductive terrain. Providing adequate transportation and communication under such uneconomic conditions is a major national problem.

On the other hand, the Soviets have the varied and widespread natural resources one would expect to find in a nation of continental size. In both Tsarist Russia and the Soviet Union, economic development has relied very heavily on the country's own resources. Development in the western plains and the

Urals has been facilitated by (1) major deposits of coal, iron ore, and other minerals (see Fig. 11.4); (2) a large amount of arable land in the steppe climate of the south—where the fertile *chernozem* (black-earth) soils are dominant (see Fig. 11.1)—and in the humid continental short-summer climate of the center; (3) extensive forests in the center and north; and (4) the water resources of the Volga, Don, Dnieper, and many other rivers. Modern economic development in the Asian part of the country came largely in the Soviet era and has been concentrated primarily in Siberia, the enormous region of forest, tundra, steppe, and mountains east of the Urals (see Fig. 11.1). Development on a more modest but still important scale has taken place in Soviet Middle Asia, the area of dry lands east and immediately north of the Caspian Sea, and in mountainous Transcaucasia between the Black Sea and the Caspian. Within the Asian sector as a whole, there has been a major reliance on such resources as varied metals and fuels; scattered areas of arable land, including the eastern part of the chernozem belt; superabundant forests in Siberia; the vast hydropower potential of the Yenisey, Angara, and other rivers; and supplies of irrigation water originating in high mountains that lie adjacent to dry lands in Soviet Middle Asia and Transcaucasia.

MAJOR NATURAL AREAS

The innumerable localities that comprise the Soviet Union exhibit great environmental diversity. Even places located very near each other may have quite significant differences, although they appear on small-scale maps under some generalized category such as "plains." For example, in an area of recent glaciation the flat and swampy flood plain of a river will offer a different set of environmental conditions from the irregular terrain of a nearby moraine (area of glacial sediment). Such differences may be critically important to the local inhabitants, but they have to be subsumed under larger categories if one is to gain much overall conception of an area the size of the USSR. Here we introduce a few large areas with distinctive natural characteristics. Some are distinguished on a physiographic basis and others on the basis of natural vegetation and climate; hence they overlap each other. These selected areas— which often take the form of west-east belts—are examined later in greater detail but are outlined here as a setting for the country's cultural, political, economic, and historical development.

Physiographic Areas

One of the world's largest plains covers nearly all the USSR from the low Ural Mountains westward. It is often called the Russian Plain, although it includes sizable areas where Ukrainians or other non-Russians are in the majority. Stretching from the Arctic Ocean to the Black and Caspian seas, this plain represents a broadened extension of the North European Plain (see Fig. 10.2). About one fourth of it is drained by the Volga River. Beyond the Urals lies the flat and swampy West Siberian Plain drained by the great Ob River and bounded on the east by another huge river, the Yenisey. In central Siberia between the Yenisey and Lena rivers, the terrain is very hilly and includes some low mountains. This area is often called the Central Siberian Uplands. A belt of high and rugged mountains—the Caucasus, Pamirs, and others—frames the country in the south and east from the Black Sea to Bering Strait.

Vegetation Belts

The physiographic areas just described are cut across by a series of major west-to-east belts marked by distinctive kinds of natural vegetation. Their existence is linked closely to variations in climate. Along the Arctic Ocean the bleak *tundra* stretches through 11 time zones from the Norwegian frontier to the Pacific (see Fig. 11.1). South of it lies the world's largest forest, the Soviet *taiga* or northern coniferous forest. Still farther south, parallel belts of *mixed forest* (at the north) and *steppe* (at the south) stretch across the western Soviet Union and deep into Siberia. Together the mixed forest and steppe comprise a "Fertile Triangle" or "Slavic Coreland" (see Fig. 11.1) containing by far the greater part of the Soviet Union's population and economy. Extensive *deserts,* together with a fringe of steppes, occupy the plains and hilly uplands of Soviet Middle Asia.

NATIONALITIES AND POLITICAL ORGANIZATION

Relative physical homogeneity over large areas is thus a basic geographic characteristic of the USSR. But culturally the pattern is more intricate. The country comprises a mosaic of peoples speaking more than 100 languages, who are organized politically into many territories based on nationality. Chief among these are the 15 Soviet Socialist Republics (*Union Republics*), headed by the gigantic Russian Soviet Federated Socialist Republic or RSFSR (see Fig. 10.1, inset), which incorporates more than three fourths of the area and about half the population of the Soviet Union (see Table 11.1). The RSFSR is the republic of the Great Russians (generally referred to as Russians), the USSR's dominant nationality. Like other Union Republics except some of the smaller ones, the RSFSR is subdivided for administration into *oblasts* (or in a few instances into similar units called *krays*), which are broadly comparable in area and population to American states. An oblast or kray is basically an economic area focusing on its largest city, which serves as the political capital and for which the unit is named. The most populous (although far from the largest in area) are Moscow Oblast and Leningrad Oblast, the units centering on the USSR's two main cities. Each oblast or kray, in turn, is further subdivided into *rayons,* which are roughly comparable to American counties.

The 14 smaller Union Republics are located around the margins of the RSFSR. They fall into three regional groups: a western group of six republics between the Black and Baltic seas; a Transcaucasian group of three bordering Turkey and Iran; and a Middle Asian group of five bordering Iran, Afghanistan, and China (see Fig. 10.1). Area and population data for the republics are given in Table 11.1. The remaining nationalities of the USSR are organized into *Autonomous Soviet Socialist Republics (ASSRs)* or lesser autonomous units. Each is subordinated politically to the Union Republic within which it is located, the majority being under the RSFSR.

Thus the Soviet Union is formally a multinational state, with a federal structure prescribed by a written constitution. Most nationalities are allocated territorial units having their own governments. Each unit has an internal hierarchy of legislative councils called "soviets," and each unit sends representatives to the Supreme Soviet, which meets in Moscow. The Supreme Soviet is bicameral, being comprised of the Soviet of Nationalities, in which the nationality units are represented, and the Soviet of the Union, in which representation is proportional to population. The President of the Soviet Union is the presiding officer of the Supreme Soviet's Presidium (inner governing body). The future status of the Supreme Soviet and the President was unclear in early 1989 (see below). Administration of the government is carried on by a great number of ministries with headquarters in Moscow. The Chairman of the Council of Ministers functions as the Prime Minister or Premier of the USSR. Each ministry has an organizational structure extending down to Union Republic level and below. But until the present (early 1989), the vital decisions at every level of government have actually been made by the Communist Party, with the small Politburo of the Party's Central Committee exercising power at the highest level. Within the Politburo the chief authority is vested in the General Secretary of the Communist Party, who may or may not also be the country's President and/or Premier. At every governmental level, the legislative and administrative structures are paralleled by the Party structure, and every decision has been subject to Party approval. The Party has appointed all administrative officials and approved all candidates for legislative office. Legislative proceedings have given approval to decisions already taken by the Party. A Party Congress in 1988 approved proposals to create an additional legislative body, the Congress of People's Deputies (selected by a more open process), and to confer greater powers on the President of the USSR. The course of further governmental change under the *glasnost* (openness) and *perestroika* (restructuring) policies instituted in the regime of Mikhail Gorbachev (1985–) could not be forecast at the time of writing.

The Communist Party's monopoly of authority has been a strong force for uniformity in all spheres of Soviet life. Economic and urban uniformity is especially striking, but a trend toward increasing cultural uniformity imposed by Russians is also very evident. It is true that Soviet policy has permitted the different nationalities to retain their own languages and other elements of their traditional cultures. Alphabets have been created for nationalities that previously had no written language, and Party-approved newspapers and books are published in most of the existing languages. These procedures have helped keep down discontent, and they have facilitated mass education. But in spite of such cultural concessions to non-Russians, a definite tendency toward "Russification" has been apparent for many years. It was an important trend even before the 1917 Revolution. As the empire of the Russian tsars expanded from the original small state of Muscovy around Moscow, non-Russian peoples were not absorbed on a basis of equality with Russians but were ruled as colonials by Russian governors. Russian culture was diffused throughout the empire by Russian officials, soldiers, traders, priests, and immigrant farmers. Russification was the most thoroughgoing in sparsely populated areas where the inhabitants were too weak, undeveloped, and culturally fragmented to offer much resistance. But many peoples annexed by the tsars proved to have such cohesive cultures that the new Russian ways had relatively little impact. Such was the case with the various Muslim peoples, and the Christian Georgians and Armenians as well. Even such Slavic kinsmen as the Ukrainians insisted on retaining their cultural autonomy. Within the Russian ruling classes, the Russian Orthodox Church, and the intelligentsia, a cult of Russian superiority was very strong at times. The Russian language, especially when purified of foreign intrusions, was held to be the most beautiful and evocative of tongues; Russian Orthodoxy was regarded as the one true faith, and the very earth of Russia—especially rural Russia—was looked on as holy. Such feelings generated a sense of mission not only to protect the sacred heritage but also to spread it to others and to punish or disadvantage those who refused to accept it. Belief in the superiority of one's own culture and the sense of a mission to spread its benefits to others is, of course, a widespread phenomenon on the earth (not least in the United States), but it seems to have been embraced with special zeal by some Russians.

Favoritism toward Russians and Russianness has demonstrably carried over from Old Russia to the Soviet era. Russians hold the overwhelming majority of the top posts in the Party, the government, and the military of the USSR and RSFSR; they are very prominent in positions of responsibility even within the 14 non-Russian Union Republics (some positions can be filled *only* by Russians). Much of the ongoing process of Russification is due to the migration of large numbers of Russians to cities and industries in non-Russian republics. In some republics and many cities, Russians now outnumber the titular nationality. Such migration is often a response to labor demand in new industries that cannot meet their requirements from the titular nationality. In Estonia or Latvia, for example, a low birth rate has brought population growth and local labor supply to a standstill, while among Muslims in Soviet Middle Asia education and technical expertise are often retarded. Since Stalin's time, labor has been much freer to move about in search of work, and a pronounced southward migration of millions of Russians into the Soviet Union's "Sunbelt" has developed. The scenic and climatically favored city of Alma Ata in the far south of the Kazakh SSR, for example, is largely Russian in population, despite the fact that it is the largest city and capital of the Republic. Meanwhile, throughout the Soviet Union Russification is fostered by compulsory teaching of Russian in all schools, as well as by the general conditions for advancement within Soviet society. Non-Russians who aspire to rise above an ordinary station must learn to speak Russian fluently.

Unrest among some non-Russian ethnic groups recently led to large public demonstrations demanding greater freedom for such nationality units as the Latvian SSR and the Estonian SSR. In Transcaucasia, the Christian Armenians and the Muslim Azerbaijanis have been embroiled in a violent dispute concerning the future political status of a small territory that is mainly inhabited by Armenians but has been governed by the Azerbaijan SSR. Recent Georgian agitation for freedom was bloodily repressed by troops.

WORLD IMPORTANCE OF OLD RUSSIA AND THE USSR

This unwieldy amalgam of lands and peoples has been an important country for centuries. It has made notable contributions to the humanities and science, and its influence on world events has often been very large. Such novels as Tolstoy's *War and Peace* and Dostoevsky's *The Brothers Karamazov* are universally regarded as masterpieces, as are Tchaikovsky's symphonies, concertos, and ballets. The Russian psychologist Pavlov made famous studies of the conditioned reflex, and the Russian chemist Mendeleev helped formulate the periodic table of chemical elements. Such achievements symbolize the flowering of Russian culture and science that took place between the middle of the 19th century and World War I. In the Soviet era, creative work has often proved incompatible with pressures for political conformity, resulting in a considerable exodus of talent to the West and harassment of talented dissenters who have remained in the USSR. Artistic creativity has often been severely hampered by censorship. Scientific and technological advance has been more prominent, although political pressures have often had a dampening effect even there.

On various occasions events in Russia have had a large impact on world history. A number of these have been Russian triumphs over powerful invaders, who as a consequence suffered shattering losses in military and political power. Outstanding examples include the defeat of invading Swedish forces under a warrior-king, Charles XII, in 1709; French and associated European forces under Napoleon I in 1812; and German and associated European forces sent into the Soviet Union by Hitler in 1941–1945. In each case the Russians lost early battles and much territory but eventually inflicted crushing and decisive defeats on the invaders. The success of Old Russia and the Soviet Union in withstanding invasion by some of the most formidable armies Europe has ever produced was due in part to environmental rigors faced by invaders and the defense in depth afforded by a huge country with poor roads. But it was also due to Russian generalship and willingness to take great losses in combat and scorch the earth of the motherland in front of the invader. Primarily as a result of defeat in Russia, 18th-century Sweden lost its commanding stature as a European military power, Napoleonic France lost its temporary hegemony over most of Europe, and Nazi Germany's temporary dominance in continental Europe was demolished by the combined forces of the Soviet Union, the United States, Great Britain, France, and lesser allies.

The Soviet Union's success in withstanding the German onslaught that began in June 1941 came as a surprise to many outside observers who had predicted that the USSR would prove too weak and disunited to resist for more than a few weeks or months. But the crucially important cities of Moscow and Leningrad held out through brutal sieges, and late in 1941 the German push eastward was decisively checked in the huge Battle of Stalingrad (now Volgograd) along the Volga River. Thereafter the Germans were gradually pushed back, and by the end of the war in 1945 Soviet armies had taken Berlin and effected a junction with American forces at the Elbe River. Warsaw Pact and NATO forces still face each other today in a partitioned Germany (see Chapters 4 and 7). Soviet success in World War II was certainly due in part to assistance by the United States, United Kingdom, and other allies. These nations sent large shipments of munitions and other supplies to the USSR, and their military campaigns in North Africa, Italy, and northwestern Europe diverted sizable German forces from the Soviet front. Nevertheless, the failure of powerful Germany to conquer the USSR was a clear indication that the strength of the Soviet Union had been underrated and that Soviet power would henceforth be a major force in world affairs.

THE RUSSIAN REVOLUTION

Soviet military strength forged in World War II and subsequent years has undergirded the worldwide challenge to the Western democracies that began with yet another momentous event in Russia: the Russian Revolution of 1917. This was really two revolutions. The first began as a general uprising that overthrew the last of the Romanov tsars early in 1917 as a protest against archaic institutions and the mismanagement of a disastrous war (the Russian sector or "Eastern Front" of World War I). Later in the year came the Bolshevik Revolution, often called the October Revolution (old tsarist calendar) or November Revolution (new Soviet calendar), in which the Bolshevik faction of the Communist Party, led by Vladimir Ilyich Lenin (1870–1924), seized control of the government, made a separate peace with Germany and its allies, and then managed to survive a difficult period of civil war and foreign intervention between 1917 and 1921. Since the Revolution, Soviet economic, political, and military expansion has succeeded to the point that the Soviet Union has become one of the two superpowers of the world. Its attitudes and actions have become major determinants of world events. Hence it is highly important that the geographic character of this formidable country and the geographic bases of its power and influence be understood.

AGENTS THAT HAVE SHAPED RUSSIAN AND SOVIET GEOGRAPHY

Not only is the geography of the Soviet Union a matter of international strategic concern, but it makes an unusually instructive study in the shaping of geography by diverse agents. The latter have left their imprint in the form of geographic distributions, political boundaries, regional characteristics, transportation links, and other geographic features. The

Soviet Union's physical geography, for example, reflects the long-continued action of natural processes on a continental scale. All the major rock-forming and land-shaping processes through geologic time contributed to the country's varied physiography and minerals. In the Ukraine, the ancient iron-bearing rocks of Krivoy Rog were formed in one geologic age, the coal deposits of the Donets Basin in another. The USSR's climates, with their associated plant life and soils, derive in large measure from the interplay of continental and oceanic air masses. Polar continental air from Siberia induces the bitterest cold of winter, and moisture-laden maritime air imported from the Atlantic Ocean by westerly winds is the source of most of the country's precipitation and has a strong tempering effect on the winter cold in the western part of the USSR. The cultural landscape reflects the interaction of many peoples with a difficult environment, particularly the interaction of Slavic peoples with forest and steppe, and of Middle Asian and Transcaucasian peoples with mountains and arid environments.

The country's complicated political geography has been shaped by the interaction—often warlike—of many peoples and states over a long period. In particular, it has resulted from encounters between Russians and various neighbors from whom Russians have acquired territory, or to whom they have yielded it. This has been a story of shifting boundaries, uneasy frontiers, and transfers of populations from one sovereignty to another.

Meanwhile the country has repeatedly felt the shaping influence of strong-willed leaders such as Tsar Peter the Great (ruled 1682–1725) or Joseph Stalin (dictator, 1924–1953). These and other strong rulers not only acquired new territory, but they reshaped geography in other ways. A famous example at the beginning of the 18th century was Peter the Great's construction of the new city of St. Petersburg (now Leningrad), commencing in 1703, as a major facet of the Tsar's program to Europeanize Russia and direct its energies westward. This prodigious city-building enterprise, which brought forth a beautiful new capital and "window on the West" in the Neva River swamps at the head of the Gulf of Finland, has been well described in a biography of Peter by Robert K. Massie[2]:

> The ceaseless building operations required an appalling amount of human labor. To drive the piles into the marshes, hew and haul the timbers, drag the stones, clear the forests, level the hills, lay out the

---

[2]Robert K. Massie, *Peter the Great: His Life and World* (New York: Alfred A. Knopf, 1980), pp. 360–361. Reprinted by permission of Alfred A. Knopf, Inc.

streets, build docks and wharves, erect the fortress, houses and shipyards, dig the canals, soaked up human effort. To supply this manpower, Peter issued edicts year after year, summoning carpenters, stonecutters, masons and, above all, raw, unskilled peasant laborers to work in St. Petersburg. From all parts of his empire an unhappy stream of humanity—Cossacks, Siberians, Tatars, Finns—flowed into St. Petersburg. They were furnished with a traveling allowance and subsistence for six months, after which they were permitted, if they survived, to return home, their places to be taken by a new draft the following summer. Local officials and noblemen charged by Peter with recruiting and sending along these human levies protested to the Tsar that hundreds of villages were being ruined by the loss of their best men, but Peter would not listen.

The hardships were frightful. Workers lived on damp ground in rough, crowded, filthy huts. Scurvy, dysentery, malaria and other diseases scythed them down in droves. Wages were not paid regularly and desertion was chronic. The actual number who died building the city will never be known; in Peter's day, it was estimated at 100,000. Later figures are much lower, perhaps 25,000 or 30,000, but no one disputes the grim saying that St. Petersburg was "a city built on bones."

Figure 10.3 It is customary in the Soviet Union for brides to leave their bouquets at a memorial to the dead of World War II, called the Great Patriotic War in the USSR. The wedding party in the photograph has come to the Eternal Flame on the Tomb of the Unknown Soldier in Kiev.

In actuality, Russian and Soviet history is replete with disasters compared to which the loss of life in building St. Petersburg was very small. Human geographic patterns of many kinds have been shaped or drastically altered by recurrent wars, famines, plagues, and internal political disorders. Large areas were laid waste and the inhabitants slaughtered in the Tatar invasion from Asia in the 13th century; this was followed in the 14th century by the frightful Black Death (bubonic plague), which killed millions. More immediately relevant to the present geography are the calamities of the 20th century. Huge casualty lists in World War I and the subsequent civil war and famine were possibly surpassed after 1928 by the estimated 20 million or more who died in Stalin's purges and agricultural collectivization. The German invasion in World War II took at least another 20 million lives. Such disasters not only involved direct slaughter, or deaths from starvation and disease, but also a reduction in the number of births. Had these events not taken place, it is estimated that the Soviet Union would now have a substantially larger population. World War II had many other drastic effects on Soviet life and the country's human geography. The war caused the relocation of millions of people; enormous physical damage to settlements, factories, and livestock; and a major eastward shift in industrial development. It is no wonder that the Soviet people regard the German invasion as a national catastrophe. The memory of

Soviet resistance is kept alive by great numbers of monuments, to which newlyweds come after the marriage ceremony for the bride to leave her flowers (Fig. 10.3).

Among major shaping agents of Old Russian and Soviet geography, we note finally the successive cultures and ideologies—most notably Slavic, Viking, Orthodox Christian, Tatar, European capitalist, and European Communist—that have made major imprints. The coal mines and steel mills of the Ukraine, for example, were initially a product of west European capitalist enterprise; and innumerable features of the USSR's present collectivized economy are expressions of an ideology developed originally by west European Communist thinkers.

In the remainder of this chapter we shall look at some of the geographic effects of the various shaping agents as we consider Old Russia and the Soviet Union within the context of historical evolution and Soviet planning.

TERRITORIAL AND CULTURAL EVOLUTION

How did a single country gain control over such an immense block of land? How did it develop so many strands in its cultural fabric? Why does it incorporate

Figure 10.4 *A portion of Red Square, with the low skyline of central Moscow in the background. Lenin's Tomb of red marble is at the left, with the Kremlin Wall and two of its watchtowers farther left. A crowd has gathered to watch the ceremonial changing of the guard at the tomb. The red building with white-capped towers is the Historical Museum.*

such a medley of peoples? The answers are very important to an understanding of the present geography. The Bolshevik Revolution was a decisive cataract in the stream of Russian history, but it did not represent a complete break with the past. Continuities from pre-Revolutionary Russia are very evident in the USSR today, and many monuments from the country's past have been preserved by the Soviet state. The Moscow Kremlin, surrounded by its crenellated walls of red brick (Fig. 10.4), not only houses the central government of the USSR but also presents an array of tsarist palaces and Orthodox churches that have become museums toured by millions of Soviet and foreign visitors. Leningrad preserves the tsarist past on an even more lavish scale. Such preservation provides Soviet citizens with evidence of the tsarist extravagances and repression that helped bring on the 1917 Revolution, but it also serves a far more important political purpose by fostering a cohesive pride in the country's long history as an important power and the fount of a distinctive culture.

The political and cultural origins of the Russian state lie more than a thousand years in the past. The developing story centers about the Slavic peoples who colonized Russia from the west, interacted culturally with many other peoples, stood off or outlasted invaders, and conquered in a piecemeal fashion the giant territory their state now incorporates.

EARLY SCANDINAVIAN AND BYZANTINE INFLUENCES

Slavic peoples have inhabited Russia since the early centuries of the Christian era. During the Middle Ages Slavic tribes living in the forested regions of western Russia came under the influence of Viking adventurers from Scandinavia known as Rus or Varangians. This was one facet of the age of Viking expansion which deeply affected the North Atlantic and Mediterranean worlds. The warlike newcomers carved out trade routes and planted settlements along the rivers and portages connecting the Baltic and Black seas. In the 9th century they were instrumental in organizing a number of principalities in the upper basin of the Dnieper River and in the region extending northward from the Dnieper headwaters to the Gulf of Finland. One of the units, the principality of Kiev, ruled by a nobility of mixed Scandinavian and Slavic descent, achieved mastery over the others and became a powerful state. The culture it developed was the foundation on which the Great Russian, Ukrainian, and Belorussian cultures later arose.

Kievan Russia was much affected by its contacts with Constantinople (now Istanbul). This famous city, located on the straits connecting the Black Sea with the Mediterranean, was the capital of the Eastern Roman or Byzantine Empire, which endured for

Figure 10.5   In medieval Russia, Novgorod on the Volkhov River was a powerful city-state, although later it was added to the expanding domain of Moscow. The view shows preserved and reconstructed Orthodox Christian cathedrals and other buildings in the Novgorod Kremlin. The Volkhov River is in the foreground.

nearly a thousand years after the collapse of the Western Roman Empire in the 5th century A.D. It became an important magnet for Russian trade, and the Russians borrowed heavily from it culturally. The Orthodox Christian faith was formally accepted from the Byzantines by Grand Duke Vladimir I, ruler of Kiev, in 988. He had his subjects baptized, and Orthodox Christianity became a permanent feature of Russian life and culture.

After Constantinople fell to the Ottoman Turks, a Muslim people, in the 15th century, the Russian tsars carried on the traditions of the Eastern Empire. They proclaimed Moscow to be the "Third Rome," and the Russian Orthodox Church was made the official church of the Russian Empire. Since the 1917 Revolution, atheism has been officially fostered by the Soviet state and organized religion has been repressed and harassed in various ways. But the state has never taken the final step of closing all the churches, and public worship continues on a reduced scale. Meanwhile, the country's culture, including its cultural landscape, continues to bear the mark of the Orthodox Christian past. The multicolored onion domes of St. Basil's Cathedral (now a museum of atheism) share Red Square with Lenin's Tomb, and the picturesque Bell Tower of Ivan the Great in the Kremlin is a familiar sight to all Moscow visitors. Greater tolerance toward religion is a feature of the *glasnost* (openness) policy introduced by Mikhail Gorbachev, but in early 1989 it was not possible to predict the ultimate results.

THE TATAR INVASION

Another cultural influence reached the Russians from the heart of Asia. Since the earliest known times the steppe grasslands of southern Russia had been the habitat of nomadic horsemen of Asian origin. During the later days of the Roman Empire and in the Middle Ages these grassy plains, stretching far into Asia, provided a passageway for inroads into Europe by the Huns, Bulgars, and other nomads. In the 13th century the Tatars appeared. Many steppe and desert peoples were in their ranks but they were led by Mongols. In 1237 Batu Khan ("Batu the Splendid"), grandson of Genghis Khan, launched a devastating invasion which brought all the Russian principalities except the northern one of Novgorod (Fig. 10.5) under Tatar rule. The Khanate of Kipchak, established by Batu, endured for more than two centuries. It was one of the main divisions in a Mongol empire stretching eastward to China and southward into the Middle East. Eventually it broke apart into the separate khanates of Kazan, Astrakhan, Crimea, and Sibir. The Tatars collected taxes and tributes from the Russian principalities, but generally allowed the rulers of these units considerable latitude in matters of local government. Even the princes of Novgorod, who were not technically under Tatar control, paid tribute in order to avoid trouble. The Tatars, called by Russians the Golden Horde (after the brightly colored tents in which they lived), used the southern steppes to graze their livestock, and they established their main camps and political centers there. When their power declined in the 15th century, the rulers of the Moscow principality were able to begin a process of territorial expansion which resulted in the formation of present-day Russia. One of the most enduring legacies of the Tatar invasion is the continued presence of millions of Tatars along the Volga River. Their political unit, the Tatar Autonomous SSR, is one of the Soviet Union's most populous and important nationality units below the status of Union Republic. In fact, it is larger and economically more productive than several of the Union Republics. It contains major oil fields, the USSR's largest truck plant, and the important city of Kazan, which is the republic's capital.

TERRITORIAL EVOLUTION UNDER THE TSARS

From the 15th century until the 20th century, the Russian monarchy reached outward from Muscovy, its original domain in the Moscow region, and built the immense Russian Empire by accretion around this small nuclear core. This was imperialism by land in the era when the maritime powers of western Europe were expanding by sea, and it brought under tsarist control a host of alien peoples. The motivations were diverse. One very prominent motive was to quell raids by troublesome neighbors, particularly nomadic Muslim peoples in the southern steppes and deserts. Many Russian cities, such as Volgograd (originally Tsaritsyn and then Stalingrad) and Saratov on the Volga River, were founded as fortified outposts on the steppe frontier.

Another motivation running through several centuries of Russian history was the desire of a landlocked country for good seaports. Even today this aspiration has not entirely been met. The Soviet Union has a longer sea frontage than any other country, but to only a limited degree does it have unrestricted access to the world ocean. Most of its main ports are located on seas that have entrances controlled by other countries. The water passages that give access to the Baltic Sea from the Atlantic Ocean (the Danish Straits and the Kiel Canal) pass through Danish, Swedish, or West Germany territorial waters, while the entrance to the Black Sea is controlled by Turkey (see Fig. 10.1). The entrances to the Sea of Japan are controlled or flanked by Japan and South Korea. Acquisition of the Kuril Islands from Japan as a result of World War II gave unrestricted access to the Sea of Okhotsk, but that body of water borders remote and sparsely populated regions, is frozen for a considerable period in winter, and is used relatively little by shipping (although it is reputed to contain a major concentration of submarines carrying nuclear missiles). Most of the Arctic Ocean ports are small and except in the extreme west are closed by ice for much of the year. Nearly all of the major Soviet ports are hampered to some degree by winter freezing, although ports on the Black Sea, Baltic Sea, and Sea of Japan are now kept open all winter with icebreakers. Only in the far northwest does a stretch of coast bordering the Barents Sea give unrestricted access to the open ocean. A tongue of water from the North Atlantic Drift warms the coast in winter and makes it possible for icebreakers to keep open the harbor of Murmansk, which has become a major naval base, cargo port, and fishing port in the Soviet era.

In the vast wilderness of Siberia, the search for valuable furs and minerals (especially gold) was a powerful motive for early expansion, and the missionary impulse of Orthodox priests also played a role. Land hunger and a desire to escape serfdom and taxation led to the flight of many peasants into the fertile black-earth belt, and many landlords (such as younger sons in landed families where the land was inherited by the eldest son) also moved there with their serfs. Possibly the most important cause of expansion was simply the desire of rulers for power and status—a thread which runs through Russian and Soviet history, as it does through the histories of many nations.

The initial outward thrust from Muscovy under Ivan the Great (reigned 1462–1505) was mainly northward and eastward. The rival principality of Novgorod was annexed and a domain secured which extended to the Arctic Ocean and eastward to the Ural Mountains. In this new realm the port of Archangel (now Arkhangelsk) was founded on the White Sea and served for more than a century as Russia's main seaport despite a harbor frozen for several months a year. It is still an important port today, mainly for timber shipments from the taiga. Tsar Ivan the Terrible (or Dread; reigned 1533–1584) added large new territories by conquering the Tatar khanates of Kazan and Astrakhan in 1552 and 1554, respectively, thus giving the Russian state control over the entire course of the Volga River. Later tsars pushed the frontiers of Russia westward toward Poland and southward toward the Black Sea. Peter the Great gained a secure foothold on the Baltic Sea by defeating the Swedes under Charles XII, and Catherine the Great secured a frontage on the Black Sea at the expense of Turkey. Peter founded St. Petersburg (now Leningrad) as Russia's main Baltic port, while Catherine expanded and modernized Odessa as the main port on the Black Sea. The two ports retain their leadership today.

Meanwhile the conquest of Siberia had proceeded rapidly. This wild and lonely territory, peopled by scattered aborigines, was already being penetrated by traders and Cossack military pioneers at the end of Ivan the Dread's reign. (The Cossacks were peasant-soldiers in the steppes, formed originally of runaway serfs and others fleeing from tsardom. Eventually they gained special privileges as military communities serving the tsars.) In 1639 a Cossack expedition reached the Pacific. Russian expansion toward the east did not stop at Bering Strait, but continued down the west coast of North America as far as northern California, where a Russian trading post (Fort Ross) existed between 1812 and 1841. In 1867, however, Alaska was sold to the United States and Russia withdrew from North America.

The movement across Siberia has been summarized as follows by a specialist on Russian expansion, Robert J. Kerner[3]:

> The earliest background of the eastward movement across the Urals is to be found in the fur-trading enterprise of Novgorod. Daring merchants and trappers from Novgorod exploited the lower reaches of the Ob from about the fourteenth century by portaging from the tributaries of the Pechora. They and the Muscovites, who carried out expeditions in 1465, 1483, and 1499, raided the inhabitants beyond the Urals for the purpose of obtaining tributes of furs, of which there was a diminishing supply in European Russia. The Russian raids were often followed by counterraids of Siberian natives, which endangered the safety of the Ural frontier. It was the latter which especially concerned Tsar Ivan the Dread. He received news that the Volga pirate Yermak, who was wanted for offenses against the laws of tsardom, had, in the employ of the Novgorodian family of the Stroganovs, raided beyond the Urals. The tsar, in fact, ordered the Stroganovs to bring him back for trial. He feared the Ural frontier would be overrun by the tribesmen of the Tatar khan, Kuchum. Yermak's success in capturing Sibir, the capital of Siberia, caused the tsar to change his intentions in regard to Yermak; instead of beheading him he gave Yermak his blessing and a real coat of armor. Incidentally, it was this heavy accoutrement that caused Yermak to lose his life by drowning.
>
> Moscow took over in 1538 and ended the practice of raids. It initiated a planned domination of rivers and portages through the building of blockhouses, called *ostrogs* in Russian. This was in line with centuries of Russian tradition in Europe. The original motive for the advance into Siberia was the acquisition of furs. Moscow sought to add to it the search for gold and silver. The conquest of the Tatar khanate gave security to the Ural frontier and created a base for further expansion. Thus from its origins to the present day, Russian rule in Asia was planned and regimented from Moscow.

The relentless quest for sable, sea otter, and other valuable furs brought great cultural changes, often devastating in character, to aboriginal peoples in Si-

beria and Pacific North America. The transformation has been epitomized as follows by a historical geographer, James R. Gibson[4]:

> In the process the intruders altered the numbers and the mores of the aborigines by introducing European diseases and spirits, Russian language, Orthodox faith, Slavic collectivism, capitalistic exploitation, authoritarian bureaucracy, and inegalitarian society. Although the native cultures were changed less drastically and less bloodily than in New Spain, New France, or New England, the natives and their lands were unmistakably Russified.

In the Amur River region near the Pacific, the Russians were not able to consolidate their hold for nearly two centuries, due to opposition by strong Manchu emperors who claimed this territory for China. Thus they were barred from the east Siberian area best suited to grow food for the fur-trading enterprise, and the one best provided with good harbors for maritime expansion. This situation changed in 1858–1860 following the defeat of China by European sea powers in the Opium Wars. Russia was able to add the Amur region to its earlier gains in the Ob, Yenisey, and Lena basins.

Most of the southern zone of high mountains and dry lands was conquered by Tsarist Russia very late. This took place in the 19th century and involved a long series of military actions in the Caucasus region and Turkestan (the arid or semiarid area east of the Caspian Sea, now containing the five republics of Soviet Middle Asia). The conquered territories, and Siberia as well, were administered as colonial areas by the tsars, and in more than a few aspects they are colonial today.

THE ROLE OF THE NATURAL ENVIRONMENT IN RUSSIAN AND SOVIET EXPANSION AND DEVELOPMENT

Interaction with a complex and demanding environment has been a major theme in Russian and Soviet expansion and development. The environment has posed great problems, but it also has provided large assets. We will now examine the rivers, topography, climates, plant life, soils, and minerals of the USSR in their relation to human purposes and needs.

---

[3]Robert J. Kerner, "The Russian Eastward Movement: Some Observations on Its Historical Significance." © 1948 by The Pacific Coast Branch, American Historical Association. Reprinted from *Pacific Historical Review*, 17 (1948), 136–137, by permission of the Branch.

[4]James R. Gibson, "Russian Expansion in Siberia and America," *Geographical Review*, 70 (1980), 128. Reprinted by permission.

THE ROLE OF RIVERS

Early Russian expansion in Eurasia followed the river lines. The Moscow region lies in a low upland from which a number of large rivers radiate like the spokes of a wheel. The longest rivers lead southward: the Volga to the landlocked Caspian Sea, the Dnieper (Dnepr) to the Black Sea, and the Don to the Sea of Azov, which is connected with the Black Sea through a narrow strait (see Fig. 10.1). Shorter rivers lead north and northwest to the Arctic Ocean and the Baltic Sea. These river systems are accessible to each other by easy portages. In the early history of Russia the rivers formed natural passageways for trade, conquest, and colonization. They were especially crucial in Siberia. The latter region, half again as large as the United States, is drained by some of the greatest rivers on earth: the Ob, Yenisey, Lena, and Kolyma, flowing to the Arctic Ocean, and the Amur, flowing to the Pacific. By following these rivers and their lateral tributaries, the Russians advanced from the Urals to the Pacific in less than a century.

It is not surprising that many of the largest Soviet cities are located on major rivers (see Fig. 10.1). In the western plains, Kiev and two other large cities lie on the Dnieper, and the seaport of Rostov is on the Don near the point where it flows into the Sea of Azov. Gorkiy, located at the junction of the Volga and the Oka, is the largest of many cities spaced along the Volga. Siberia's largest city, Novosibirsk, is on the Ob River at the point where the Trans-Siberian Railroad crosses it. Leningrad, the USSR's second largest city and its main Baltic seaport, is located on the Neva River where it empties into the Gulf of Finland. The Neva is relatively short, but it drains Lake Ladoga and hence it carries a large and regular flow of water. Another historic Baltic seaport, Riga, the capital and main city of the Latvian SSR, lies at the mouth of the Western Dvina River, which was a major link in the old Viking trade route between the Baltic and Black seas. Moscow is located on the small Moscow River, which is a headstream of the Oka and hence the Volga but is too shallow and strewn with rapids to be navigable to the Oka. Completion of the Moscow Canal in 1937 gave the city a good water connection to the upper Volga. The canal is used by large barges and river vessels, and it is also a major source of Moscow's water supply. Access to the Volga has made the city a ''port of five seas'' inasmuch as the Volga and various connecting canals and rivers give access to the Caspian, Azov, Black, Baltic, and White seas. A major link in the system is the Volga-Don Canal, opened in 1952 to tie together the two rivers where they approach each other in the vicinity of Volgograd.

Improvement of rivers for power, navigation, and irrigation has been an important and much-publicized aspect of economic development under the Soviets. Since World War II there has been very extensive development of hydroelectricity, which now provides about 13 percent of the USSR's electric power supply. Numerous power dams, some of which are very large, have been built along rivers in the mountains of Transcaucasia or Soviet Middle Asia, and still others are under construction or planned in those regions. But the main element in the program has been the construction of large dams, or chains of dams, along such major rivers as the Dnieper, Don, Volga, Kama, Irtysh, Ob, Yenisey, and Angara. In the process the middle and upper Volga has largely been converted to a series of lakes, with consequent improvement in navigation. This river is by far the most important internal waterway in the country, used by fairly large number of barges and log rafts as well as by a fleet of passenger vessels that carry tourists (Fig. 10.6). Nearly all cargo traffic on the Volga system, as on other internal waterways of

Figure 10.6 The passenger port at Kuybyshev on the Volga. The vessels are typical of the many tourist ships that ply the Volga and Don rivers and the Volga-Don Canal. Prominent tourist attractions on the Volga are Lenin's birthplace at Ulyanovsk and the great monument at Volgograd commemorating the Soviet triumph over Nazi Germany in the Battle of Stalingrad. Most visitors are Soviet citizens.

the Soviet Union, consists of bulk goods, especially unprocessed construction materials, timber (Fig. 10.7), and petroleum and its products.

The importance of inland waterway traffic should not be overstressed. In the Soviet Union not more than about one twentieth of domestic freight traffic, as measured in ton-miles, is carried by rivers and canals. The Soviet government has tried to foster internal waterways, and there has been a large increase in traffic during recent decades, but the importance of these waterways relative to other forms of transportation has declined. They are at a competitive disadvantage in several ways. River transportation is more circuitous and much slower; the channels are blocked by ice for several months a year; and facilities for transshipment of goods at river ports are often inadequate. The new reservoirs behind power dams stay frozen longer, and windstorms on these broadened sheets of water cause increased hazards and delays in the operation of river craft. The flow of goods in the Soviet Union is prevailingly east-west, whereas the river courses are oriented generally north and south. For many types of freight movement, transshipment to or from river craft is inconvenient and costly.[5]

THE ROLE OF TOPOGRAPHY

Most of the important rivers of the Soviet Union wind slowly for hundreds or thousands of miles across large plains. Early expansion was facilitated not only by the long, continuous highways provided by the rivers themselves but also by the easy overland connections between river systems. Only in the extreme south and east are the river basins separated by ranges of high mountains. Elsewhere the divides ordinarily lie in areas of low hills, or even in level plains where the gradients between headstreams are so gentle as to be scarcely perceptible.

Plains and low hills comprise nearly all the terrain from the Yenisey River to the western border of the country (see Fig. 10.2). The only mountains that rise in this vast lowland are the Urals, a low range (average elevation less than 2000 feet/610 m) located about midway between the western frontier of Russia and the Yenisey. The Urals trend due north and south but do not occupy the full width of the

Figure 10.7 *Sections of a log raft in a lock at the Kuybyshev Dam on the Volga River. Southward movement of huge timber rafts is a major type of traffic on the river. The rafts are disassembled for passage through locks at the various dams.*

lowland. A wide gap between the southern end of the mountains and the Caspian Sea permits uninterrupted east-west movement. Actually, the Urals themselves do not constitute a serious barrier to transportation, the main range being less than 100 miles (161 km) wide in most places and cut by river valleys offering easy passageways. The central Urals are especially low, and rail lines connecting Moscow with the important Ural city of Sverdlovsk cross this part of the range over a divide lying at only 1350 feet (a little more than 400 m) above sea level. The southern and central Urals are forested, but at the extreme north the mountains extend into the zone of arctic tundra.

The plains west of the Urals are mostly undulating or rolling rather than flat. North of Moscow they are liberally strewn with moraines and other glacially created irregularities surviving from the Ice Age. In places there are prominent limestone or chalk escarpments, often extending hundreds of miles. The Dnieper and Volga flow for long distances at the foot of eastward-facing escarpments, and such cities as Kiev and Volgograd were founded on the high western banks of their respective river valleys to take advantage of firmer ground and greater defensibility against the menace of raids by nomads to the east.

Between the Urals and the Yenisey River, the great lowland of western Siberia (West Siberian Plain) is one of the flattest areas on earth. Much of it is covered by immense swamps and marshes

---

[5]For more detailed discussion, see Robert N. Taaffe, "Volga River Transportation: Problems and Prospects," in Richard S. Thoman and Donald J. Patton, eds., *Focus on Geographic Activity: A Collection of Original Studies* (New York: McGraw-Hill Book Company, 1964), 185–193.

through which the Ob River and its tributaries slowly wend their way. This waterlogged country, underlain by a permanently frozen subsoil (*permafrost*) that blocks drainage, is a major barrier to land transport and is extremely uninviting to settlement. Large sections are almost devoid of people. In the spring, tremendous floods occur when the breakup of ice in the upper basin of the Ob releases great quantities of water while the river channels farther north are still frozen. Despite such conditions, the northern part of the West Siberian Plain has become a major center of oil and natural gas production. In the south the plain rises in elevation and becomes drier and far more thickly settled. Here it extends into the black-earth belt along the Trans-Siberian Railroad.

The area between the Yenisey and Lena rivers is occupied by hilly uplands (Central Siberian Uplands) lying at a general elevation of 1000 to 1500 feet (*c.* 300–450 m). East of the Lena River and Lake Baikal the landscape is dominated by mountains. Extreme northeastern Siberia is especially bleak and difficult mountain country, with a few peaks reaching elevations of 10,000 feet (*c.* 3000 m) or more.

High mountains border the Soviet Union on the south from the Black Sea to Lake Baikal, and lower mountains from Lake Baikal to the Pacific. Elevations of 15,000 feet (*c.* 4500 m) or more are reached in the Caucasus Mountains between the Black and Caspian seas and in the Pamir, Tien Shan, and Altai mountains east of the Caspian Sea. From the foot of the latter ranges and lower ranges between the Pamirs and Caspian Sea, arid or semiarid plains and low uplands extend northward and gradually merge with the West Siberian Plain and the broad plains and low hills lying west of the Urals.

THE ROLE OF THE CLIMATIC AND BIOTIC ENVIRONMENT

Russian and Soviet expansion and development have gone forward in a harsh climatic setting. This experience resembles that of Canada, except that Russian conditions are even more extreme. Severe winter cold, short growing seasons, desiccating summer winds that shrivel crops in the steppes, and a countrywide tendency toward drought are major handicaps. Associated with these rigorous conditions are large compensations in the form of tillable soils, forests, and natural pastures for livestock, plus a valuable wild fauna. But such resources often have a very marginal utility as a result of unfavorable climatic factors. The subsequent discussion examines

the origins and character of the climatic environment and its associated natural features, with frequent reference to Old Russian and Soviet experience.

General Nature of the Climate

Most parts of the Soviet Union have a continental climate characterized by long cold winters, warm to hot summers, and low to moderate precipitation. The severe winters are the result of a northerly continental location, coupled with mountain barriers on the south and east. Four fifths of the total area is farther north than any point in the conterminous United States (Fig. 10.8). The Soviet Union lies mainly in the higher middle and lower high latitudes of the Northern Hemisphere where land masses reach their greatest extent relative to the bordering oceans. In these latitudes the climatic influence of the land is paramount over the influence of the sea in continental interiors, and the most extreme continental climates of the world prevail. The effects of continental location are heightened by the Arctic Ocean, which is frozen most of the year and thus in a sense forms an extension of the land. Continuous chains of mountains and high plateaus occupying the southern and eastern margins of the country act as a screen against moderating influences from the Indian and Pacific oceans. In addition, most of the USSR is in a zone of high atmospheric pressure in winter, with outflowing winds. Westerly winds from the Atlantic moderate the winter temperatures somewhat, but their effects become steadily weaker toward the east. Thus Leningrad (60° North Latitude) on the Baltic Sea has an average January temperature of 18°F (−8°C), whereas Verkhoyansk (68°N) in eastern Siberia has a January average of −53°F (−47°C).

Aside from the interior of eastern Siberia, most parts of the USSR experience winter temperatures that do not differ greatly from the temperatures of places located at comparable latitudes and altitudes in the interior of North America. Indeed, due to westerly winds from the Atlantic, many places in the west and south of the Soviet Union are actually warmer in midwinter than comparable interior places in North America. For example, Odessa (46°N) is 14°F (8°C) warmer in January on the average than Montreal. Leningrad, located a thousand miles (*c.* 1600 km) north of Minneapolis, is several degrees warmer in January than the latter city.

Nevertheless, the Soviet Union as a whole is definitely handicapped by the winter climate. Huge areas are permanently frozen at a depth of a few feet. This condition, called *permafrost*, makes construction

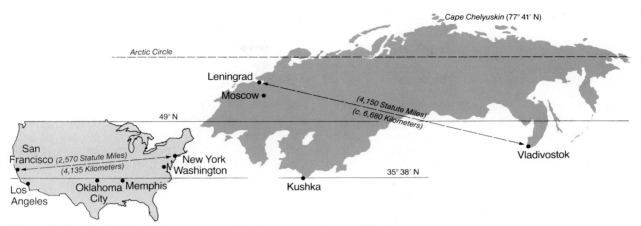

Figure 10.8 The Soviet Union compared in latitude and area with the conterminous United States.

difficult. Heat generated by buildings melts the upper layers of permafrost and causes foundations and walls to sink and tilt. Pipelines carrying crude oil, which is very hot when it comes from the ground, may likewise melt the permafrost and sink unless they are heavily insulated or built on elevated supports. The average frost-free season of 150 days or less in most parts of the Soviet Union except the extreme south and west is too short for a wide range of crops to mature.

Most places are relatively warm during the brief summer, and the southern steppes and deserts are hot. A factor partially offsetting the brevity of the summer is the length of summer days. Leningrad has 19 hours of daylight on June 22. Throughout the summer, long hours of sunlight facilitate the growth of plants and compensate somewhat for the short growing season.

From an agricultural standpoint, lack of moisture is probably an even greater handicap than low temperatures. The annual average precipitation is less than 20 inches (less than 50 cm) nearly everywhere except in the extreme west, along the eastern coast of the Black Sea and the Pacific littoral north of Vladivostok, and in some of the higher mountain areas. North of the Arctic Circle and in the southern desert zone, the average precipitation is less than 10 inches (less than 25 cm). Since most precipitation is derived from the Atlantic Ocean, the total amount tends to decrease from west to east. Precipitation also decreases toward the south, being 10 to 20 inches in the black-earth belt, but only 5 to 10 inches or even less in the dry steppes and deserts farther south. The country's most important grain-producing region, the black-earth belt (see Fig. 11.1), is subject to se-

vere droughts and to hot, desiccating winds (*sukhovey*) which may greatly damage or destroy a crop. Areas to the north of this belt in the western USSR have a somewhat greater and more dependable rainfall, coupled with a lower rate of evaporation resulting from lower temperatures. Thus they are better supplied with moisture than is the black-earth belt, but this advantage is offset by their poorer soils, cooler summers, shorter growing season, and larger proportion of poorly drained land.

Types of Climate

In the USSR five main east-west climatic belts—tundra, subarctic, humid continental, steppe, and desert—are customarily recognized in classifications of climate. These belts, each with its associated vegetation and soils, succeed each other from north to south in the order named (see Figs. 2.18, 2.19, 2.22, and climatic data in Table 10.1).

TUNDRA CLIMATE

The zone of *tundra climate* occupies a strip, 50 to several hundred miles in width, along the Arctic coast from the Norwegian frontier to Bering Strait. Tundra is both a climatic and a vegetational term. Climatically, it signifies a region which has at least 1 month averaging above 32°F (0°C) but no month averaging higher than 50°F (10°C). In most areas, 2 to 4 months average above freezing. Climatic conditions are too severe for trees, except for small localities where favorable conditions such as protection from wind allow stunted trees to grow. The typical vegetation consists of mosses, lichens, sedges, hardy grasses, and scrubby bushes. Some islands in

TABLE 10.1 CLIMATIC DATA FOR SELECTED SOVIET STATIONS

Station	Latitude and Longitude		Elevation Above Sea Level		Type of Climate	Average Temperature						Average Length of Frost-Free Season (days)	Average Precipitation	
						Annual		January		July				
Arkhangelsk	65°N	40°E	50′	15 m	Subarctic	34°F	1°C	10°F	−12°C	61°F	16°C	119	21″	54 cm
Igarka	67°N	86°E	115′	35 m	Subarctic	17°F	−8°C	−20°F	−29°C	59°F	15°C	80	16″	41 cm
Nordvik	74°N	111°E	102′	31 m	Tundra	7°F	−14°C	−21°F	−29°C	41°F	5°C	45	5″	13 cm
Verkhoyansk	68°N	133°E	400′	122 m	Subarctic	5°F	−15°C	−53°F	−47°C	61°F	16°C	69	6″	16 cm
Leningrad	60°N	30°E	30′	9 m	Humid continental	41°F	5°C	18°F	−8°C	64°F	18°C	159	22″	56 cm
Moscow	56°N	38°E	480′	146 m	Humid continental	39°F	4°C	14°F	−10°C	66°F	19°C	141	23″	58 cm
Sverdlovsk	57°N	61°E	925′	282 m	Humid continental	36°F	2°C	5°F	−15°C	64°F	18°C	110	18″	46 cm
Novosibirsk	55°N	83°E	436′	133 m	Humid continental	32°F	0°C	−3°F	−19°C	66°F	19°C	120	17″	43 cm
Irkutsk	52°N	104°E	1532′	467 m	Subarctic	30°F	−1°C	−6°F	−21°C	64°F	18°C	94	18″	46 cm
Vladivostok	43°N	132°E	95′	29 m	Humid continental	39°F	4°C	7°F	−14°C	68°F	20°C (Aug)	187	32″	82 cm
Odessa	46°N	31°E	210′	64 m	Steppe	50°F	10°C	28°F	−2°C	72°F	22°C	215	15″	39 cm
Yalta	45°N	34°E	135′	41 m	Mediterranean	56°F	13°C	39°F	4°C	75°F	24°C	247	22″	56 cm
Batumi	42°N	42°E	20′	6 m	Humid subtropical	57°F	14°C	43°F	6°C	73°F	23°C	304	99″	250 cm
Baku	40°N	50°E	0′	0 m	Steppe	57°F	14°C	39°F	4°C	77°F	25°C	293	9″	24 cm
Tselinograd	51°N	72°E	1148′	350 m	Steppe	35°F	2°C	1°F	−17°C	68°F	20°C	c.125	12″	30 cm
Tashkent	41°N	69°E	1568′	478 m	Steppe	55°F	13°C	32°F	0°C	81°F	27°C	204	16″	42 cm
Krasnovodsk	40°N	53°E	−56′	−17 m	Desert	57°F	14°C	36°F	2°C	82°F	28°C	278	4″	9 cm

the Arctic Ocean have temperatures averaging below freezing or only slightly above freezing even during the warmest month. Here much of the land is covered with glaciers, and an ice-cap climate similar to that of Antarctica or interior Greenland prevails.

SUBARCTIC CLIMATE

At the south, the tundra climate gradually merges with the zone of *subarctic climate*. Here 1 to 4 months average above 50°F. Thus the subarctic climate differs from the tundra in having a warmer summer. Winters, however, are extremely long and cold, most places averaging below freezing for 5 to 7 months. This climate occupies a wedge-shaped area extending from the Finnish border to the Pacific Ocean. It is narrowest at the west and broadens eastward to the neighborhood of Lake Baikal, where it occupies the full width of the country except for a narrow strip of tundra at the north.

The zone of subarctic climate is essentially coextensive with the *taiga*, or northern coniferous forest. The taiga of the Soviet Union is the largest continuous area of forest on earth. A similar belt covers much of Canada. The main trees are spruce, fir, larch, and pine. These needleleaf coniferous soft-

woods, although useful for pulpwood and firewood, are frequently too small, twisted, or knotty to make good lumber. Nevertheless, large reserves of timber suitable for lumber exist in parts of the taiga, and this part of the Soviet Union is one of the major areas of lumbering in the world. Intermixed with the conifers are stands of certain broadleaf species such as birch and aspen. They are ordinarily second-growth, replacing stands of conifers removed by fires or cutting, and have little commercial utility.

Some agricultural settlement occurs in the taiga, especially toward the south. However, farming here is marginal, being handicapped by (1) the short summers, (2) the winter cold, which makes it difficult or impossible for winter grains and fruit trees to survive, (3) unseasonable frosts that often occur in late spring or early fall, (4) the marshy or swampy character of much of the land, and (5) the prevalence of poor soils. The dominant soils of the taiga are the *podzols*. This term, derived from Russian words translated as "ashes underneath," refers to a group of soils which characteristically have a grayish, bleached appearance when plowed, are lacking in well-decomposed organic matter, are poorly structured, and are very low in natural fertility. The acidity of these soils is unfavorable for most crops and also for bacteria, earthworms, and other soil-improving organisms.

The boundary between the taiga and tundra is not sharp, although it appears so when generalized on small-scale maps. The changeover is so gradual that a discontinuous transitional area of "wooded tundra" (stands of stunted trees alternating with tundra) marks the contact in most places.

HUMID CONTINENTAL CLIMATE

South of the subarctic zone a triangular area of *humid continental climate* extends eastward from the western border of the USSR to the vicinity of Novosibirsk. At the west the triangle is more than 600 miles (more than 965 km) wide, but the Siberian portion is much narrower. The humid continental climate differs from the subarctic in having longer summers, milder winters, and more precipitation. However, these are only differences of degree. Even within the humid continental zone most places have an average frost-free period of less than 150 days except at the extreme west (Leningrad, 159 days; Kiev, 180 days). Winters are long and cold, with average cold-month temperatures ranging from $-3°F$ ($-19°C$) at Novosibirsk in the east to 24°F ($-4°C$) at Riga in the west. The average annual precipitation is relatively low, being only 18 to 25 inches (46–64 cm) at most places. However, the low rate of evaporation increases the effectiveness of the precipitation, so that the available moisture is generally adequate for the staple crops of the climate zone. The Soviet Union has the short-summer subtype of humid continental climate. A comparable climate is found in the Great Lakes region and in the northern Great Plains of the United States and adjoining parts of Canada. In this climate zone of the USSR mixed or broadleaf deciduous forest supplants the evergreen taiga forest. Oak, ash, maple, elm, and other broadleaf trees alternate with conifers in the vegetation cover. At the south is the "wooded steppe" or "forest steppe," transitional between forest and open steppe grassland. Here clumps of trees intermix with grassy areas.

Both climate and soil are more favorable for agriculture in the humid continental climate than in the subarctic. Soils developed under broadleaf deciduous or mixed forest are normally more fertile than podzols, although less so than grassland soils. The forest soils in the areas of humid continental climate are very important in Soviet agriculture, although areas with a steppe climate have a greater total output of agricultural products.

STEPPE CLIMATE

The grassy plains of the Soviet Union south of the forest are known as the *steppe,* or steppes. This term refers both to a type of climate and to the characteristic form of vegetation associated with it. On climatic maps the zone of steppe climate in the USSR is shown as an east-west band of varying width extending from the Romanian border to the Altai Mountains in Asia. Compared with the humid continental short-summer type of climate, the Soviet band of steppe climate is characterized by warmer summers, a somewhat longer frost-free season, and less precipitation. The average annual precipitation is 10 to 20 inches (*c.* 25–50 cm), an amount barely sufficient for growing crops without irrigation. Recurring periods of drought add to the hazards of farming. Nevertheless, this is the Soviet Union's most important area of crop and livestock production. The handicap of low and variable rainfall is partially offset by the fertility of the soils in the famous *black-earth belt.* This expanse of deep, black, exceedingly fertile soils extends for about 2500 miles (about 4000 km) in an east-west direction and 300 to 600 miles (roughly 500–1000 km) from north to south (see Fig. 11.1). Characteristic soils here are the *chernozems*—a term meaning "black earth." Chernozem soils are exceptionally thick, productive, and durable—being in fact among the best soils to be found anywhere. A similar belt occurs in the eastern Great Plains of North America. The great fertility of the chernozems is largely due to an abundance of the well-decomposed organic matter called *humus* in the topsoil, the presence of sufficient lime to neutralize excessive acidity, and less leaching of nutrients by rainfall. Strictly speaking, the term humus refers to organic remains, both plant and animal, which have been decomposed by bacteria and have become part of the soil. Complete decomposition of humus makes available to plants the nutrients it contains, and its presence in the soil is essential to the development of a friable (loosely compacted; crumbly) structure that permits easy cultivation and helps the soil retain moisture. The accumulated remains of annual grasses in mid-latitude prairies and steppes are a better source of humus than the leaves and twigs of a forest. The steppe zone of the USSR includes extensive areas of *chestnut soils* in the areas of lighter rainfall south of the chernozems (see soils map, Fig. 2.22). The chestnut soils are lighter in color than the chernozems and lack their superb fertility. Nevertheless, they are among the better soils of the world. In North America a belt of chestnut soils occurs west of the chernozems in the Great Plains.

Within the steppe climate a considerable range of natural vegetation types is found, varying from tall grass and scattered forest stands in the "wooded steppe," where moisture is more abundant, to a sparse cover of low grasses and shrubs in the drier areas. The most characteristic form of natural vegetation is short grass, forming a carpet which is continuous, or nearly so. The treeless steppe grasslands, stretching monotonously over a vast area between the forest and the southern mountains and deserts, were utilized as grazing lands from an early time by pastoral nomads. Today, however, much of the steppe is cultivated, with wheat being the main crop (see Fig. 3.5). In addition, the steppe is a major producer of sugar beets, sunflowers grown for vegetable oil, and various other crops.

DESERT CLIMATE

To the east and immediate north of the Caspian Sea, the steppe grades into desert. Here, rainfall is even more scanty and erratic. Widely spaced shrubs and occasional tufts of grass afford only the sparsest forage for livestock. Crop-growing is largely precluded by aridity, except for oases watered by the Syr Darya, Amu Darya, and other rivers that originate in high mountains bordering this region on the south and east. The oases produce most of the Soviet Union's cotton. The desert zone includes two extensive areas of sandy desert—a type much less common on a world basis than deserts floored by combinations of sand with gravel, rock fragments, or the bare bedrock.

SUBTROPICAL, MONSOON, AND HIGHLAND CLIMATES

A number of relatively small but distinctive climatic areas remain to be mentioned. The south coast of the Crimean Peninsula, sheltered by the Yaila Mountains, has the relatively mild temperatures and summer rainfall minimum associated with the *mediterranean* or *dry-summer subtropical climate.* This picturesque area, with its orchards, vineyards, and resorts, of which Yalta is the most famous, is sometimes called the "Russian Riviera." A subtropical climate also prevails in the coastal lowlands and valleys south of the high Caucasus Mountains. Mild winters and warm to hot summers are characteristic throughout the lowlands of Transcaucasia. However, the rainfall is very unevenly distributed. The lowlands bordering the Black Sea in western Transcaucasia receive the heaviest precipitation of the USSR—50 to 100 inches (*c.* 125–250 cm) a year—and are classed as *humid subtropical,* while the lowlands of eastern Transcaucasia, bordering the Caspian Sea, receive so little precipitation in most places that they are classed as steppe. It should be noted that the USSR's "subtropical" areas are rather marginally so, as they have quite a bit of freezing weather.

The coastal regions of the Soviet Far East, from Vladivostok northward to the mouth of the Amur River, have a humid continental climate with a distinct *monsoon* tendency. As in nearby parts of Korea and of China, most of the annual rainfall results from moist onshore winds of the summer monsoon. In contrast, the cold outflowing winds of the winter monsoon produce little precipitation. The average annual precipitation is 20 to 30 inches (*c.* 50–75 cm), of which three fourths or more falls from April through September.

The mountains of the USSR are characterized by climates varying according to altitude and exposure to wind and sun. However, the range of climates is smaller than in mountains located closer to the equator.

THE ROLE OF MINERALS

Until the latter 19th century, minerals played only a modest role in Russian expansion and development. Gold was sought with some success in Siberia, and iron working was developed in the Urals on a sizable scale during the reigns of Peter the Great and Catherine the Great. In fact, for a time 18th-century Russia was the largest producer of pig iron in the world and exported some of the product to Great Britain. Abundant forests in the Urals provided raw material to make the charcoal used to smelt many relatively small deposits of high-grade iron ore. A much larger development of iron smelting began in the Ukraine in the 1880s on the basis of the large deposits of coking coal in the Donets Basin and iron ore at Krivoy Rog. Meanwhile oil wells were being drilled at Baku on the Caspian Sea, which by 1900 had become for a few years the world's leading center of oil production. Other minerals, such as the copper of the Urals, were worked at various times and places, but on the whole the country's great mineral wealth was poorly developed prior to the 1917 Revolution. In fact, the real surge of mineral extraction and processing did not come until the era of five-year planning, which began in 1928. The emphasis on heavy industry and armaments in plan after plan required huge quantities of numerous minerals, most of which the Soviet Union was found to have in adequate quantities. Today, the country is remarkably self-sufficient in minerals, basing the heart of its economy on products from its own mines plus a few imports of modest size. The massive scale and rapid pace of Soviet development may eventually force large imports of some minerals, possibly including petroleum. But for the present the USSR meets most of its mineral needs and is a sizable exporter of mineral fuels, gold, diamonds, and some other minerals. Most of these exports go to European countries. In the case of some minerals the Soviet Union achieves self-sufficiency by mining low-grade deposits at a relatively high cost. This is often due to the fact that known deposits of higher grade were exhausted by the insatiable demands of heavy industry after 1928. World War II resulted in a heavy drain on high-grade ores, as was also the case in the United States.

The diversity of minerals in the USSR reflects the great range of geologic formations in this nation of continental size. Coal and lignite reserves associated with sedimentary formations are exceedingly widespread, but the known reserves of most other minerals tend to be rather heavily concentrated in a relatively few areas.

Major Mining Areas

The *Ukrainian SSR* is the most important producer of coal, iron ore, and the important ferroalloy manganese (see Fig. 11.4). It also has a large production of natural gas, salt, and sulfur, all of which form the basis for important chemical industries. The *Ural Mountains* are one of the world's great storehouses of diversified metals and chemical materials. Hundreds of mines are scattered through these mountains. In the western foothills of the Urals and along the Volga River, the *Volga-Urals oil fields* are one of the Soviet Union's two main areas of oil production. To the eastward, the *West Siberian oil and gas fields* have become the largest Soviet suppliers of these important fuels. Extraction is carried on under very difficult northerly conditions in the swampy West Siberian Plain. The *Kuznetsk Basin* near Novosibirsk in Siberia is the principal coal supplier in Soviet Asia, although there are many other coal fields. *Soviet Middle Asia*, with a diversified array of fuels, metals, and chemical materials extracted in many places, has become increasingly prominent in mineral production during recent times. Natural gas, copper, lead, and zinc are major specialties, as is the coal of Karaganda in the Kazakh SSR.

The bulk of Soviet mineral output comes from the areas just named, but many other areas make a contribution. Examples are the metals and phosphate of the *Kola Peninsula* in the ancient Fennoscandian Shield of the northwest, the iron ore present in huge quantities in the *Kursk Magnetic Anomaly* (KMA) between the Ukraine and Moscow, and the oil and gas of the *Caucasus-Caspian region*. Other mining areas and unmined reserves too numerous to mention here are scattered throughout the western Arctic and Siberia. The gold and diamonds of Siberia are prominent examples. Some of these mines and minerals are discussed in Chapter 11, which also provides greater detail concerning the mineralized areas.

THE SOVIET ECONOMIC SYSTEM AND ITS GEOGRAPHIC EFFECTS

No treatment of Soviet geography can be complete without considering the effects of the Communist economic system, which has played such a central role in Soviet life since the Bolshevik Revolution. In fact, this highly publicized system is so central that the Soviet state was founded in order to create it. The "state idea" espoused by the Bolshevik leaders and later stated formally in the Constitutions of 1936 and 1977 affirms that the Soviet Union is a country of workers and peasants (and "intelligentsia," according to the 1977 Constitution) who collectively own the means of production. All citizens are obligated to contribute to the economy "according to their abilities" and will be rewarded "according to their work." The state, guided by the Communist Party, operates or controls the economy and is obligated to ensure "democratic rights" and provide full employment and social welfare for all the people.

The socialized economy of the USSR represents an attempt to put into practice the economic and social ideas of the 19th-century German philosopher Karl Marx (who actually spent most of his adult life in England and is buried in London). According to Marx, the central theme of modern history is a struggle between two social classes: the capitalist class (*bourgeoisie*) and the industrial working class (*proletariat*). He forecast that ruthless exploitation of workers by greedy capitalists would eventually lead the workers to revolt, overthrow the capitalists, and turn over the ownership and management of the means of production to new workers' states. In the classless societies of these states there would be social harmony and justice, with little need for formal government.

Today, Marx's utopian society has not materialized, although his ideas have had a great impact on scholarly discussion, and they have stimulated revolutionary activity in many countries. The following comments appeared in a *Christian Science Monitor* editorial on the hundredth anniversary of Marx's death[6]:

> . . . His [Marx's] vision of a communist society, in which each person would contribute according to ability and receive according to need, captured the imagination of millions and inspired revolutionaries in many parts of the world.
>
> Yet, a hundred years after his death, it is baffling why Marxism still retains its ability to appeal. As an economic theory, it has long ago been discredited. It never did take hold in the industrialized countries as Marx expected but was adopted by underdeveloped nations in which the peasant class still predominated. In the industrialized nations, capitalism did not lead to pauperization of the working class, largely because of the emergence of trade unions. The supposed class struggle did not develop. Governments in the industrialized states, far from becoming instruments of repression, used their power to mitigate social injustice and the abuses of the capitalist system. Capitalism did not disappear.
>
> Wherever communism is practiced today, it has fallen far short of Marx's utopian ideal. Instead of the state withering away in such countries as the Soviet Union, China, Cuba, Vietnam, it has become the overriding presence in society. Instead of the old autocratic ruling class, communist countries have produced a new ruling class—the "New Class," Milovan Djilas called it—with its own privileged way of life.

MAJOR CHARACTERISTICS OF THE ECONOMIC SYSTEM

Despite the shortcomings of Marxism, the ideas of Marx as interpreted and implemented by Lenin ("Marxism-Leninism") remain the official philosophy of the Soviet state, and the system founded on this philosophy continues to operate—however inefficiently. Since it is dominant in a country covering more than one seventh of the earth's land surface, is copied to various degrees by many other nations, and, despite current modifications, seems likely to survive for a long period of time, we must consider the system a truly major feature of the world's economic geography. Space and the requirements of a geography text do not permit anything approaching a full analysis of this complex and controversial "command economy." Here we consider selected

major characteristics having broad import for the country's geography, including (1) centralized economic planning, (2) a large degree of national self-sufficiency, (3) strong regional specialization, (4) large dependence on gigantic projects, and (5) a multiplicity of legal and underground "economies."

Centralized Economic Planning

One of the characteristics for which the Soviet Union is best known is its tightly controlled, socialized type of planned economy. Since 1928 the economy has been operated under a series of Five-Year Plans touching every major sector of economic life. These plans prescribe the goals of production for the entire nation. They specify the types and quantities of minerals, manufactured goods, and agricultural commodities to be produced; the factories, transportation links, and dams to be built or improved; the locations of new residential areas to house industrial workers; and so on.

The Bolsheviks did not develop this state-controlled economic machine from scratch. A huge state economy had long existed under the tsars, gaining a strong impetus as far back as the reign of Peter the Great and continuing to develop up to World War I. Some of its major features are set out in the following selection by Bertram D. Wolfe[7]:

> The greatest industrializer of Russia prior to Lenin-Stalin was Peter the Great. His modernization, like that of his successors, was intended primarily for the purposes of war and the state. Industrialization was an aspect of militarization. It came from the top downward. It lifted, shoved, dragged, drove, booted the Russian people toward Western technique by fiat of the state. Overnight, Peter created alongside the cottage industry, a great textile and clothing industry to produce uniforms in millions and a great iron and steel industry to produce bayonets cannon and the munitions of war. The industries were created by state plan and *ukaz*. Sons of the nobility were ordered to study science and technology; capital was invested by royal command; a laboring class was created by assigning state serfs from the Crown estates. Thus, long before Marxian socialism was so much as dreamed of, the Russian state became the largest landowner, the largest factory owner, the largest employer of labor, the largest trader, the largest owner of capital, in Russia, or in the world. The needs of its huge armies made it the largest customer for private industry as well.

Further dominance by the state was achieved by stages following the Bolshevik Revolution. Essen-

---

[6]"Marx and His Legacy," *The Christian Science Monitor*, Wednesday, March 16, 1983, 24. Reprinted by permission from the Editorial page of *The Christian Science Monitor*. © 1983 The Christian Science Publishing Society. All rights reserved.

[7]Bertram D. Wolfe, *Three Who Made a Revolution* (New York: The Dial Press, 1948, 1960), 21. Reprinted by permission.

tially the aim of the revolutionary leaders was two-fold. They meant to abolish the old aristocratic and capitalist institutions of Tsarist Russia and to develop a strong socialist state able to stand on an equal footing with the major industrial nations of the West. For the first decade they were mainly occupied in consolidating their hold on the country and in putting a limited part of their program into effect. Large-scale industry, banking, and foreign trade were nationalized, but a certain amount of private trading, together with private ownership of small industries and agricultural land, was permitted under the New Economic Policy (NEP) announced in 1921. This compromise policy was the result of a near-breakdown in the newly instituted Communist economy during the difficult period of civil war and foreign intervention following the Revolution. However, it proved to be only a temporary expedient.

Lenin, the chief architect of the Revolution, died in 1924 and was subsequently interred in the mausoleum of red marble by the Kremlin Wall where millions view his remains each year (see Fig. 10.4). His memory, preserved in countless pictures, statues, and museums across the land, rallies the emotions of the Party faithful, and his writings, together with those of Marx, are the acknowledged foundation of Soviet ideology. But it was Lenin's successor, Joseph Stalin, who mainly forged the Soviet system as we know it. Abandoning the expediency of the New Economic Policy, his long and despotic regime embarked on full-scale Communism and swept aside real or suspected opposition by a policy of terror. A ruthless drive was launched for comprehensive planning, forced socialization of the economy, and massive industrialization. The motivation was partly ideological (Stalin's determination to "build socialism in one country") and partly defensive against a possible conflict with Western nations, especially Germany. Since Stalin's death in 1953, his successors have abolished or modified many rigidities and cruelties of the Stalin era and have been able to raise the Soviet standard of living somewhat. But such changes have not revolutionized the basic structure and ideology that developed under Lenin and Stalin, and worsening economic conditions led to official calls for fundamental reform in the mid-1980s. Revamping of the economic system (*perestroika*: "reconstruction" or "restructuring") and (*glasnost*: "publicity"; "openness") were announced as major priorities in the regime of Mikhail Gorbachev that began in 1985. In early 1989 it was not possible to forecast where these much-publicized prescriptions for change would lead. Much opposition to them existed among persons for whom change might mean loss of power or loss of economic security.

National plans for the Soviet Union are formulated by an agency in Moscow called Gosplan (Committee for State Planning). Gosplan is closely supervised by the Politburo of the Communist Party. Once the plans have been approved at the highest level, they are transmitted downward through several levels of a huge administrative bureaucracy (*apparat*) until they finally reach the operating level of individual factories, farms, or other enterprises. This unwieldy process generates inefficiency in a number of ways:

1. Fear of offending powerful superiors makes persons at lower levels reluctant to suggest ways to improve efficiency.

2. It is difficult for planners in Moscow to manage an area larger than North America as though it were one gigantic corporation. Failure to comprehend local conditions may lead to unworkable decisions, such as the attempts of Nikita Khruschev (in power 1953–1964) to grow corn for grain in areas not climatically suitable for it. Trying to coordinate such a huge and diverse body of enterprises, materials, labor, and consumer demand from one central point is too great a task.

3. In freer economies the market—the desires and abilities of consumers and businesses to buy things—largely determines what will be produced. The Soviet planning bureaucracy has had no free market to guide it; hence goods have often been produced that people would not buy, or have not been produced although people would have liked to buy them.

4. It is well known in the Soviet Union that greater freedom by enterprise managers to make production decisions would result in more efficient operation of plants and farms. Some progress has been made in granting such freedoms since Stalin's time, but attempts to do it on a thoroughgoing scale have been stymied by fears of fostering (or seeming to foster) free enterprise, or by the reluctance of powerful bureaucrats to surrender any of their decision-making powers.

5. Production targets have been stated by Gosplan in quantitative terms, and this has often resulted in shoddy workmanship and unsalable goods.

National Self-Sufficiency

In the beginning the Bolsheviks had to seek national self-sufficiency because they repudiated the debts of the Tsarist government to foreign investors, expropriated all foreign holdings, and hence found themselves with no credit in the outside world. Even today foreign investors are often reluctant to extend liberal credit, despite the fact that the Soviet government has had a good record in paying off its own debts. The country also suffers from the fact that its artificially controlled currency is not accepted in non-Communist countries; hence Soviet imports from the capitalist world have to be paid for in "hard" currencies of capitalist nations. In order to earn currency to pay for imports of grain and high-technology items from the West, the Soviets have pushed their exports of oil, gas, gold, and diamonds, thereby stimulating rapid development in the difficult northern areas from which most of these exports come. But there are other reasons for seeking self-sufficiency: desire to make the country less vulnerable in wartime; reluctance to become too dependent on the Western nations which the Soviet Union is trying to surpass; and pride of accomplishment in using the country's own resources to build socialism. In attempting to be economically self-sufficient and strategically invulnerable, the Soviet planners have often invested large sums of money to develop resources in difficult and out-of-the-way places. In some places good-sized cities have been developed from scratch, as was done in the 1970s at Bratsk, the site of a huge power dam in the Siberian taiga on the Angara River.

FOREIGN TRADE

Striving for a high degree of self-sufficiency does not mean that the Soviets have largely abandoned foreign trade. In fact, trade has increased manyfold since 1928, particularly since the 1960s. The USSR ranks in the bottom half of the world's ten largest traders—far behind the United States, West Germany, or Japan in total value of imports plus exports, but ranking in a general class with Italy or Canada. Soviet trade, however, is far less varied in composition than that of most other major traders. Fuels and raw materials provided more than half of all exports in 1987, while high-technology items and grain were among the most valuable imports. High-technology imports are products of industries (such as the computer industry) that are still seriously underdeveloped in the USSR.

Not surprisingly, much Soviet trade has political motivations—principally the desire to keep other Communist countries bound economically to the USSR. For example, the European satellites are very dependent on Soviet oil, natural gas, iron ore, and other minerals, and the USSR, in turn, profits from the inflow of machinery, vehicles, and miscellaneous consumer goods that satellites such as East Germany, Hungary, and Czechoslovakia provide.

Regional Specialization

The output of particular commodities is not evenly spread across the USSR, but the main production of each commodity tends to come from a few areas: textiles from the Moscow region; steel from the Ukraine, the Urals, and areas immediately south and north of Moscow; automobiles from the Volga and Moscow regions; raw cotton from Soviet Middle Asia; and so on. The manufacture of machinery is very widespread, but a given city will specialize in particular kinds of machinery, such as grain harvesters at Rostov-on-Don in the black-earth belt, mining and metallurgical machinery at Sverdlovsk in the Urals, or textile machinery in the Moscow region. In these respects the Soviet Union is not so very different from the United States, which also maintains a strongly regionalized economy. In both countries the present distribution of vital producing districts must be explained partly in terms of historical and political factors as well as economic factors of the present day. In both, the shipment of materials, energy, and products to and from the specialized districts involves a great deal of long-distance transportation, but the United States is much better equipped to handle it because of its more diversified transport net. Long-distance trucking is still in its infancy in the USSR; the internal water transport net is far less developed than that of the United States; and railroads, along with a growing net of oil and gas pipelines, are overwhelmingly the principal means for long-distance transport of goods. Mile for mile, Soviet railways are reputed to be the most heavily used in the world. Some of the strain on the railroads has been alleviated by programs to increase regional self-sufficiency through new mining and manufacturing enterprises to serve local needs in less-developed parts of the USSR. New development in Soviet Middle Asia is a major example.

Dependence on Gigantic Projects

Emphasis on grandiose economic projects has been a much publicized and important aspect of the Soviet economic system. Such enterprises have harnessed the energies and resources of the whole country to achieve specific objectives thought to be critically important at particular times. The Soviet people have been exhorted to make sacrifices for these projects in order to make the Soviet Union strong and provide a better life in the future. The term "Hero Project" has frequently been used to whip up enthusiasm. A few examples include (1) the construction of large tractor plants in the 1920s and 1930s at

Kharkov in the Ukraine, Stalingrad (now Volgograd) on the Volga, and Chelyabinsk in the Urals; (2) the "Virgin and Idle Lands" program of the 1950s to expand grain acreage east of the Volga; and (3) construction of the Baikal-Amur Mainline (BAM) Railroad during the 1970s and 1980s to provide an alternate link to the Pacific north of the Trans-Siberian Railroad in the far eastern USSR. The accomplishments have been impressive, but one can make too much of the "Hero Projects." They have not succeeded in giving the Soviet people a level of living approaching that of the Western democracies, and they certainly do not overshadow such projects of the American government as the Interstate Highway System or the Tennessee Valley Authority.

Multiple Economies

In the Soviet Union most of the value of production is accounted for by the planned and socialized national economy of manufacturing, mining, agriculture, forestry, fisheries, and services. It is with the impact of this mainstream economy on Soviet geography that we are mainly concerned. But side by side with it, several other legal, surreptitious, secret, or illegal economies exist:

1. *The legal private-enterprise economy.* This includes the personal plots allocated to farmers and many urban workers; the free markets in which surplus agricultural produce from the personal plots may be sold at unfixed prices; and many small-scale crafts, trades, and services. Products and services from the private sector are indispensable to the USSR at the present time.

2. *The military economy.* The Soviet military controls its own armaments production, which has first call on the nation's resources and talents. Although details are largely secret, production methods are reported to be technologically advanced, and there is better quality control than is generally true of Soviet industry. Many arms are marketed or donated abroad.

3. *The service economy for the elite.* A large network of stores and restaurants supplies the USSR's choicest food and consumer items (including many imports) at cut-rate prices to a privileged Party elite in government, the military, and other fields. The best housing, medical facilities, and other services are also available. Such facilities are not available to ordinary citizens. This economy of privilege extends into all corners of the Soviet Union, although the government does not officially admit its existence.

4. *The underground economy* ("counter-economy," "second economy," or *na levo*—"on the left"). Most of this traffic is illegal, often with severe penalties prescribed by law. But there is a general tendency for the authorities to overlook such transactions because they are so essential to the economy. The system has been described as follows by an experienced Moscow correspondent, Hedrick Smith, in a much-praised book on the Russians[8]:

Corruption and illegal private enterprise in Russia, "creeping capitalism," as some Russians playfully call it, grow out of the very nature of the Soviet economy and its inefficiencies—shortages, poor quality goods, terrible delays in service. They constitute more than a black market, as Westerners are accustomed to thinking of it. For parallel to the official economy, there exists an entire, thriving counter-economy which handles an enormous volume of hidden or semi-hidden trade that is indispensable for institutions as well as individuals. Practically any material or service can be arranged *nalevo*—from renting a holiday cottage in the country, buying a raincoat or a pair of good shoes in a state store, getting a smart dress made by a good seamstress, transporting a sofa across town, having the plumbing fixed or sound-proofing installed on your apartment door, being treated by a good dentist, sending your children to a private playschool, arranging home consultation with a top-flight surgeon, to erecting buildings and laying pipe in a collective farm.

But the heart of the Soviet economy in employment and value of product remains the nationwide net of socialized industries and farms directed from Moscow. Agriculture was dominant when five-year planning commenced, and taxation of farms largely paid for the initial expansion of industry. Today agriculture remains very important, but industry employs more people, and it exceeds agriculture many times over in value of product. We turn now to an analysis of these major economic sectors, commencing with agriculture.

THE USSR'S COLLECTIVIZED AGRICULTURE: PURPOSES AND RESULTS

Between 1929 and 1933 about two thirds of all peasant households in the Soviet Union were collectivized, and the class of more prosperous private farmers known as *kulaks* was liquidated. By 1940 collectivization was virtually complete. This program

---

[8]Hedrick Smith, *The Russians* (New York: Quadrangle/The New York Times Book Co., 1976), 86. Reprinted by permission.

Figure 10.9 *This view of a hamlet in steppes along the Don River illustrates the underdevelopment of rural Russia. The unpaved road connecting this little community to the outer world is often impassable in bad weather. The community does have electricity brought by the line at the right, but few other amenities such as indoor plumbing.*

was fiercely resisted by the land-hungry peasants, and drastic measures were employed to put the government decrees into effect. Millions of livestock were slaughtered by the peasants and nomads, and crops were burned to avoid turning them over to the "socialized sector." Wholesale imprisonments and executions followed, together with confiscation of food at gunpoint (often including the peasants' own food reserves and seed). Famines took millions of lives. It is estimated that at least 5 million people died as a result of Stalin's forced collectivization. But these costs were disregarded, and the countryside was reorganized into two main types of farm units:

the collective farm (*kolkhoz*) and the large factory-type state farm (*sovkhoz*).

A collective or state farm normally includes several villages, the majority of which are located on unpaved roads (Fig. 10.9). Machinery is kept in centralized places (Fig. 10.10). Unpainted log or frame houses (perhaps with painted window frames) are the rule; in the steppes many houses are made of adobe. Many villages consist of a single street, along which the houses are spaced. Behind each house extends a rectangular garden allotment ("private plot"), perhaps an acre (0.4 hectare) in size. Here the farm family is allowed to keep a few livestock

Figure 10.10 *Machinery park on a state farm in the black-earth belt near the Don River. The machines are relatively simple but sturdy. They are parked in the open on a dirt surface, as building materials cannot be spared for sheds and concrete.*

and grow crops for family use. Surplus products from the private plots can be sold for whatever they will bring. (Fig. 10.11). These plots provide a vital element in the nation's food supply. Around the villages stretch the large fields of the collectivized land that the villagers are obligated to work. On the state-operated farms (*sovkhozy*), workers receive cash wages in the same manner as industrial workers. Bonuses for extra performance are paid. Workers on the collective farms (*kolkhozy*) receive shares of the income after the obligations of the collective have been met. Payments are graduated according to the amount of time worked and the kinds of work done. As on the state farms, there is a system of bonuses for superior output.

As originally conceived, the system of collectivized agriculture was supposed to result in the following major advantages:

1. The old arrangement of small, fragmented individual holdings separated by uncultivated boundary strips would be replaced by a system of larger fields incorporating the boundary strips, thus increasing the amount of cultivated land and giving scope for mechanized farming.

2. Increased mechanization would release surplus farm labor for employment in factories and mines, thus facilitating industrialization and bringing into existence the large urban working class looked to as the principal support for the Communist system.

3. Mechanization, improved methods of farming, and reclamation of new land under state supervision would result in greater overall production.

4. Increased production, plus easier collection of surpluses from a greatly reduced number of farm units, would result in larger and more dependable food supplies for the growing urban populations and greater tax revenues to use in building industry.

5. Liquidation of individual peasant farming would remove the most important capitalist element still remaining in the USSR.

Early Handicaps of Soviet Farming

Soviet agriculture experienced many setbacks in the first quarter-century of collectivization, and overall production was slow to expand. Various factors have been cited as contributing to the stagnation of farm production, including (1) rapid draining of agricultural workers (including many of the more skilled workers) for employment in industry, without an adequate supply of farm machines being provided to

Figure 10.11   *Roadside scene in the Ukraine. The woman is selling apples from her family's personal plot on a collective farm. Any produce from the personal ("private") plot can be legally sold for whatever price it will bring.*

take their place; (2) peasant resistance to collectivization, including large-scale slaughter of draft animals and other livestock; (3) climatic handicaps; (4) rigid and inefficient management by the Soviet bureaucracy; (5) the devastation of many of the best farming districts in World War II; and (6) perhaps most of all, the government's policy of draining wealth from agriculture for the support of industry. Farm products were requisitioned from the collectives at low prices and were then sold in state-owned stores at inflated prices. By the time a collective had met its obligations for seed, fertilizer, other supplies, and services, only a meager surplus of cash and produce was left to be distributed among members. Under such conditions there was a natural tendency for collective farmers to shirk their obligations to the collectives in favor of work on their personal plots. Any surplus from these plots and from privately owned livestock could be sold on the collective-farm market in a nearby town or city, generally at much higher prices than those paid to the collectives by the state. Meanwhile, much of the country's farm land suffered from a lack of fertilizer. The production of chemical fertilizer was given a low priority by government planners and consequently it was not available for most crops. Animal manure for use as fertilizer was also in short supply, due to the slow recovery of livestock from the slaughter of the early collectivization era and the ravages of World War II.

Changes in Soviet Farming

After World War II, and particularly after Stalin's death in 1953, important changes were made in Soviet farming in the hope of increasing productivity. These changes were instituted over a considerable period and are still going on today. The system of agricultural procurement, pricing, and taxation, for example, was revamped to try to induce farmers to produce more on collectively farmed land. By merging adjoining units, the number of collective farms was sharply reduced. Some collectives were absorbed by or converted to state farms. In Tsarist Russia there had been about 25 million private farms, but by 1979 the country's agriculture was being carried on by 26,000 collective farms (*kolkhozy*) with an average sown area of about 9000 acres (3650 ha) and an average work force of about 525, and 20,800 state farms (*sovkhozy*) with an average sown area of

about 13,000 acres (5300 ha) and an average work force of 475. Beginning in 1958 the collectives were allowed to own and operate their own tractors and farm machinery instead of contracting for them from machine-tractor stations. Restrictions on the use of personal plots and privately owned livestock were lessened, as were restrictions on off-the-farm travel by collective farmers, and the latter were guaranteed a minimum income (payable monthly rather than annually). Prices of consumer goods, previously much higher in rural than in urban areas, were adjusted downward in rural areas as a means of raising living standards there. Attempts were made to give greater powers of decision-making to individual farm managers, but not with great success due to bureaucratic opposition. Various experiments have been in progress, such as the development of very specialized animal farms (for example, cattle feedlots, hog-feeding operations, and poultry hatchery and broiler units); the development of interfarm enterprises in which two or more collective and/or state farms collaborate in certain operations such as land improvement; and schemes under which a few farmers are assigned a tract of land and then are allowed to make most of the operating decisions to work it.

THE "VIRGIN AND IDLE LANDS" SCHEME

Another facet of the drive to increase the national supply of farm products and diversify the diet to include more animal products was a sizable enlarge-

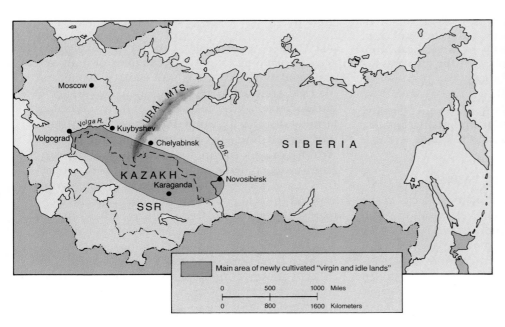

Figure 10.12   The heavy shading encloses most of the acreage of new lands brought under cultivation in the period 1954–1957. (After a map in N.N. Baransky, Economic Geography of the USSR, **Moscow: Foreign Languages Publishing House, 1956, p. 71.)**

Main area of newly cultivated "virgin and idle lands"

0   500   1000   Miles

0   800   1600   Kilometers

ment of the cultivated area. In 1954 a vast program was instituted to increase the amount of grain (mainly spring wheat and spring barley) by bringing tens of millions of acres of "virgin and idle lands" ("new lands") into production in the steppes of the northern Kazakh SSR in Soviet Middle Asia and adjoining sections of western Siberia and the Volga region (Fig. 10.12). Hundreds of new state farms were organized in the area and a network of narrow-gauge railroads was constructed to serve them. Between 1954 and 1960 the cultivated area of the USSR was enlarged by more than 90 million acres (36 million hectares), with most of the increase taking place in 1954–1956 in the new grain lands of Kazakhstan, Siberia, and the Volga region. In the agrarian history of Old Russia and the Soviet Union, expansion of cultivation into "new lands" has been a recurrent theme, but no expansion comparable in scale to that of 1954–1956 has ever occurred in such a short time. Despite adverse weather in some years, with consequent crop failures or low production, the "new lands" program has added large increments to Soviet grain output and has moved the center of gravity of grain farming eastward. This lessens the impact of bad weather in a given year, as a poor winter wheat crop in the western USSR may be offset by a good spring wheat crop in the east, and vice versa. Atmospheric influences in the USSR are such that there is seldom a drought year both west and east of the Volga. Most of the eastern production, coming from huge farms worked primarily with machines, is available to feed urban populations. However, there are important problems in maintaining satisfactory production on a long-term basis. Such lands are very marginal in precipitation, and they require extremely careful management to conserve moisture in the soil and prevent serious wind erosion. Much technical knowledge about these matters has accumulated in the United States and Canada as a result of disastrous experiences with "dust bowls" along the dry margins of agriculture in the Great Plains; and the Russians themselves have long encountered such problems in the drier steppes. The uncertainties of production in the east were highlighted in the late 1970s and early 1980s, when the Soviets had poor wheat harvests for several years in a row. This led to huge purchases of grain overseas.

The "new lands" expansion of cultivated acreage has accompanied an ambitious program to make the agriculture of older farm lands more productive and varied. These lands, located primarily west of the Volga, are superior to the "new lands" from the standpoint of annual moisture and temperature regimes. Their agricultural environment is distinctly

inferior, however, to that of the best lands in the United States—a fact that became very evident in the unsuccessful attempt of the Khruschev regime (1953–1964) to add large increments to Soviet grain supply by growing corn (maize). In the highly productive agriculture of the United States corn plays a key role, being in fact the single most important crop. It is grown primarily for livestock feed, and the Soviets attempted to foster its production on a large scale for the same purpose. On a visit to the United States, Khruschev was greatly impressed with the productivity of Midwestern corn and livestock farms. But climatic conditions in most farming areas of the Soviet Union are more comparable to those of Canada than they are to those of Iowa and Illinois, and the requirements of corn in regard to moisture, temperature, and length of growing season cannot be met satisfactorily. The program to grow corn for grain proved a costly failure, although there was greater success in growing corn for silage and green forage. Of the USSR's sown acreage in all crops, no more than 2 percent now grows grain corn, but an additional 8 percent is allocated to corn grown for harvesting in an immature stage. Actually, much grain corn comes from areas in the Moldavian and Georgian SSRs that are climatically suitable for corn and already were growing it before the Soviet era.

Despite their climatic limitations in regard to corn-for-grain (and for soybeans, a major American crop that is grown very little in the USSR), Soviet farm lands offer many possibilities for increased production of crops and livestock through greater use of chemical fertilizers and animal manures, liming of acid soils, use of improved varieties of crops and animals, better control of plant and animal diseases, increased irrigation, mechanization, and drainage, construction of better storage facilities, use of more scientific methods of cultivation and animal husbandry, and so on. Such measures have increased greatly since the Stalin era, and despite periodic reverses in poor crop years, there has been a marked upward trend in agricultural output. This reflects an increase in agricultural investment to the point that in the 1970s and 1980s investment in agriculture began to represent a larger percentage of overall Soviet investment than did agriculture's share of the gross national product. Today the average Soviet citizen has a diet comparable to that of the United States in caloric content and better balanced than that in Stalin's time. But the proportion of grain products and potatoes in the diet is still far larger than that in the American diet, and there are sizable deficiencies in meat, vegetables, and fruits. While output of diversified foods has been increasing in the socialized sector, it must be noted that personal plots and privately owned livestock are still vital in maintaining and improving the overall quality of the Soviet diet. Only about 3 percent of all tilled land is in the personal plots, and they contribute very little to the supply of grain, sugar, or vegetable oil, but in the production of many other foods their role is quite large. For example, in the early 1980s the "private sector" was producing about three fifths of the USSR's potatoes, a quarter of all vegetables, nearly one third of all meat and milk, and more than one third of all eggs. A large share of this produce is consumed by the farm families themselves, but the residue sold in urban markets is an indispensable element in the country's food supply.

In overall output of many farm products (for example, wheat, barley, oats, rye, potatoes, sugar beets, flax, sunflower seeds, milk, butter, mutton), the USSR is the world's leading nation. It does not achieve this rank, however, through high productivity of its agriculture per unit of land and labor. Compared with the United States, for example, the Soviet Union uses about one and one-half times as much crop acreage and about eight times as many farm workers to produce a volume of farm commodities that is probably smaller than that of the United States and must meet the needs of a larger population. In mechanization and the technical proficiency of its agriculture, the USSR lags. Although applications of chemical fertilizer have increased greatly in recent decades, a shortage of herbicides and pesticides makes weed and pest control a major problem. Such factors, along with climatic conditions and other handicaps, contribute to a yield per acre that is only 40 to 60 percent as great as that of the United States for many important crops.

A long-term problem of Soviet agriculture has been the stagnation of animal industries for long periods. Since the Stalin era the leadership has placed much stress on the need to increase the supply of livestock products and associated forage crops. This program has had ups and downs, but on the whole has achieved considerable success. For example, between 1953 and 1979 milk production more than tripled, and the milk supply of the USSR is reported to be reasonably adequate.

There are so many problems connected with Soviet agriculture that solutions will be slow in coming. Besides the difficulties previously noted, there is a strong tendency for farm machinery to stand idle because of improper maintenance (no *individual* owns the machine) and shortages of spare parts. Poor storage and transportation facilities, plus wholesale pilfering, cause an alarming degree of loss after the harvest. Younger people are deserting the farms in droves for urban work; often they continue to live with the family and commute to their city jobs. Hence women, children, and the elderly are having to do a very large share of the farm work. Thus agriculture limps along; in a sense responsibility for each collectivized field and livestock herd is everybody's business and nobody's.

INDUSTRIAL PROGRESS AND URBAN DEVELOPMENT UNDER THE FIVE-YEAR PLANS

The principal target of Soviet national planning since 1928 has been a large increase in industrial output, with emphasis on the production of heavy machinery and other capital goods, increased exploitation of minerals, increased development of electric power, improvement of transportation, and military strength. The drive to industrialize has had far more success than the planned expansion of agriculture. The Soviet Union was largely an agricultural country in 1928, but today it is probably the third industrial power of the world, after the United States and Japan. Masses of peasants have been converted into factory workers, new industrial centers have been created, and old ones have been enlarged. The increase in urban population has been phenomenal. In 1926 only 18 percent of the population lived in cities, but by the middle 1980s the figure had risen to an estimated 65 percent (world average, 45 percent; United States, 74 percent). In 1928, the iron and steel center of Magnitogorsk in the Urals did not exist, but today it has more than 400,000 people. Hundreds of other cities have been created or greatly enlarged under the Soviets. It should be noted, however, that in a recent year an estimated 35 percent of the Soviet population was still rural. The Soviet Union has far to go before it becomes a truly urban country on the order of the United States, Great Britain, West Germany, or Japan.

The prodigious drive to remake the USSR industrially has made the Soviet Union the world's largest producer of such items as steel, oil, gas, and cement. It has provided the industrial base for space flight, a huge military machine, atomic weapons, a large expansion of railways and ocean shipping, and the mass production of millions of new apartments. This has often been achieved at the expense of Soviet consumers, whose needs have been slighted in favor of heavy metallurgy and the manufacture of machinery, power-generating and transportation equipment, and industrial chemicals. The situation in regard to consumer goods has improved over time, despite relative stagnation in the 1980s. But long lines of consumers still wait in stores to purchase scarce items of clothing and everyday conveniences. Foreign-made goods are prized, although access to them tends to be reserved for the Soviet elite. Aside from the day-to-day shortages of particular foods, probably the most publicized consumer shortages are automobiles and housing. Automobiles are very expensive despite the large new production from a plant on the middle Volga built for the USSR by Italy's Fiat Company. To purchase an auto, one must often be on a waiting list for several years. There has been a chronic housing shortage in the cities ever since the freed serfs began leaving the countryside in the later 19th century. Access to housing is controlled by the state, and waiting lists for new apartments are very long.

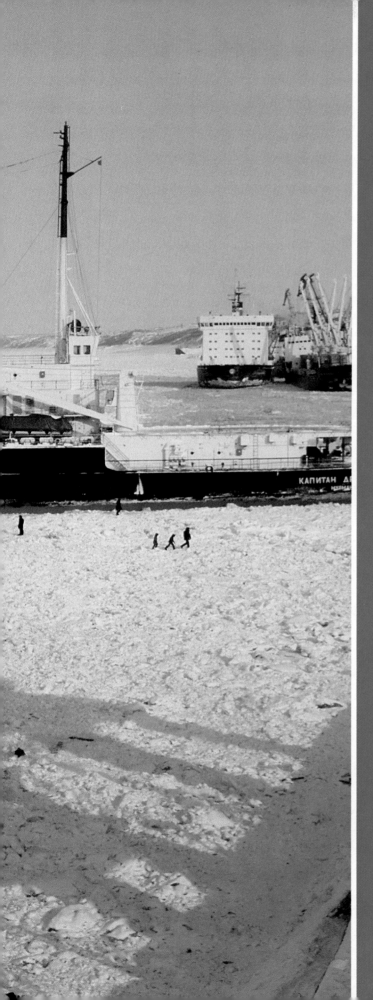

CHAPTER 11

Coreland and Outlying Regions in the USSR

This April view of the port of Dudinka in the Siberian Arctic symbolizes the difficult environment of most of the USSR. Ice-fishing on the frozen Yenisey River is in progress. The ship is an icebreaker of the Northern Sea Route.

L IKE other countries with extremely large areas, the Soviet Union has a very uneven spatial pattern of population and development. Nearly three fourths of the people and an even larger share of the cities, industries, and cultivated land are packed into a triangular coreland comprising roughly one fifth of the USSR's land (Fig. 11.1). Lying mainly west of the Urals but narrowing eastward into Soviet Asia, the triangle is often called the *Slavic Coreland, Fertile Triangle,* or *Agricultural Triangle.* The rest of the USSR lies mostly in Asia and may be termed the *Outlying Regions.* Here settlement is extremely spotty, being handicapped by environments that are essentially nonagricultural except in very limited areas. Most of the non-Slavic population of the USSR lives in the Outlying Regions, but many immigrant Slavs live there too, generally in cities. Great sections are thinly inhabited, often by peoples who are still tribal, but two areas—the Caucasus region and Soviet Middle Asia—contain districts that have relatively dense populations and civilizations older than the Slavic civilization of the Coreland. The Outlying Regions are a storehouse of minerals, timber, and waterpower, and a major share of investment there is poured into the development of these resources. The result tends to be a series of discrete production nodes widely separated from each other by taiga, tundra, mountains, deserts, swamps, and frozen seas.

THE SLAVIC CORELAND

Slavic peoples are the dominant population element in the USSR, both in numbers and political and economic power. The major groups are the Russians, Ukrainians, and Belorussians, constituting 52, 16, and 3 percent, respectively, of the national population (1983). These percentages have been dropping slowly for decades, due almost entirely to the much higher rates of natural increase among the Soviet Union's Muslims, the great majority of whom live outside the Slavic Coreland in Soviet Middle Asia or Transcaucasia.

Figure 11.1   *Major regional divisions of the Soviet Union as considered in this text.*

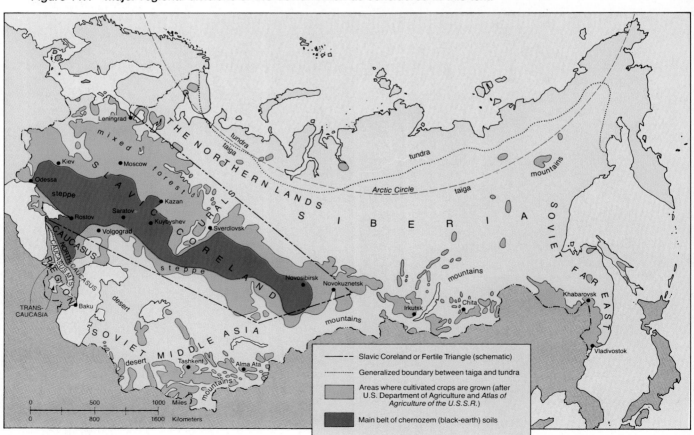

Most of the Slavic population lives in the triangular coreland extending from the Black and Baltic seas to the neighborhood of Novosibirsk in Siberia. The coreland is about half as large as the United States in area and has over four fifths as large a population. It is truly the "Soviet Union that matters," despite its minor share of the USSR's land. The Outlying Regions contribute various resources to it—most notably oil and natural gas—but the Soviet Union could still function as a superpower if the coreland stood alone.

Lowlands predominate in the coreland, the only mountains being the Urals, a small segment of the Carpathians, and a minor range in the Crimean Peninsula. The original vegetation was mixed coniferous and deciduous forest in the northern part and steppe grassland in the south. Moscow (12.4 million; city proper, 8.3 million) and Leningrad (5.6 million; city proper, 4.3 million) are by far the largest cities. (Note: In this chapter, city population figures are for metropolitan areas unless "city proper" is specified. All figures are rounded approximations for the mid-1980s.)

THE PREEMINENCE OF MOSCOW

Moscow, the capital of Russia before 1713 and of the Soviet Union since 1918, is today the most important manufacturing city, transportation hub, and cultural, educational, and scientific center. It is maintained as a showpiece city (see Fig. 10.4), and residence there is eagerly sought, although hard to achieve because of strict influx controls. Moscow has the highest priority of any city with regard to food and consumer goods, including automobiles. Large numbers of people who live outside the metropolis come there to shop or to sell produce from their private plots.

How Moscow's Cityscape Evolved[1]

Kiev was Russia's most important city, political capital, and religious center until it was destroyed by the Tatar invasion in the 13th century. The founding of Moscow predated the Tatars, but not until the latter were overthrown by the Russians in the late 14th and 15th centuries did the city become preeminent. The evolution of Moscow's urban layout and

architectural styles exemplifies that of many other Russian cities. Indeed, these cities have often taken their cues from developments in the metropolis.

Early Russian settlements that grew into cities were sited in good defensive positions, often on high riverbanks. Such citadels, fortified with walls, became known as kremlins. As population and business activity expanded outside the original wall, additional fortifications were thrown up. In Moscow the second wall protected the market square (Red Square) just outside the Kremlin, together with the nearby merchants' quarter. Eventually two other walls enclosed new spaces farther out. In later times all the walls except the Kremlin Wall were torn down and replaced by circular boulevards. Today the latter are intersected by roads that lead outward in a radial pattern from the city center. Far out at the edge of the present built-up area, a divided highway encircles the city, and beyond it lies a green belt of birch and pine woods, through which are scattered dormitory suburbs housing Moscow workers, and summer homes of the elite. Still farther out are Moscow's four airports. Then for many miles come satellite towns with industrial or other employment for their residents, and residential towns from which workers commute to Moscow by train or bus. The railways that reach Moscow in a radial pattern from all directions do not penetrate the innermost city but terminate at a ring of nine stations interconnected by a circular belt-line railway. The railway stations, bearing names such as Leningrad Station or Kiev Station, are stops on Moscow's famous subway system, the Metro, which again is comprised of radiating lines that connect the Kremlin vicinity with the edges of the built-up area and are themselves interconnected by a belt line.

Moscow's architecture mirrors a succession of changing styles surviving from different eras (Figs. 1.4 and 11.2). Medieval construction was mainly of wood and hence highly vulnerable to fire. Most buildings left over from this early time, frequently churches or monasteries with onion domes, are made of stone or brick. During Peter the Great's reign (1682–1725) architectural styles from western Europe were introduced into Russia for both public and private buildings. Foreign architects were brought in, primarily to St. Petersburg, where they produced buildings in a variety of Western styles. Russian architects then carried these styles to Moscow, where their work is reflected in numerous buildings of the inner city.

With the spurt in Moscow's population growth following the emancipation of the serfs and the first large development of railways in the 1860s, the Industrial Revolution came to the city. Teeming slums

---

[1]This section is based primarily on material in R. A. French, "The Changing Russian Urban Landscape," *Geography*, 68, Part 3 (June 1983), pp. 236–244; and Paul E. Lydolph, *Geography of the U.S.S.R.* (3d ed.; New York, John Wiley & Sons, 1977), pp. 21–28.

Figure 11.2 *View from a window of the huge Rossiya Hotel in central Moscow near Red Square and the Kremlin. The buildings exhibit a variety of styles dating from different historical periods. The oldest is the small Russian Orthodox church in the foreground, which dates from the late Middle Ages. Dominating the skyline are squarish high-rises built in the Soviet era. Other buildings are in European-derived styles of the 19th century. Most buildings in the view house offices of Moscow's enormous bureaucracy.*

soon crowded closely around the new factories, whose workers had to live within walking distance because of the inadequate public transportation. As industry expanded, wealth from industrial profits was often used to finance mansions and lavish civic buildings of various architectural styles. Following the 1917 Revolution, the deplorable state of workers' housing stimulated many idealistic plans for a more livable city, but such planning had few results because of Stalin's emphasis on investment in heavy industry. From the Revolution until Stalin's death in 1953, the limited amount of new construction in Moscow centered on flats for the elite, together with a few "wedding cake" skyscrapers and other showpiece structures such as the richly ornamented stations on the Metro. At Stalin's death, the available floor space per Moscow resident was less than it had been in 1928 at the beginning of the First Five-Year Plan. The intensive effort of Khruschev and his successors to catch up on housing construction then produced an overwhelming mass of high-rise blocks of flats, which are still accumulating today. Constructed mainly of greyish concrete slabs, these buildings emphasize prefabrication and standardization, and they are grouped into micro-regions designed to be close to places of work or to rapid transit and self-sufficient in services such as stores, education, clinics, and recreation. This kind of uniform urban landscape has now spread throughout the Soviet Union and into Communist Europe as well (see Fig. 11.13 and chapter-opening photo of Chapter 9). In Moscow the present structures at the city margin are rising much higher than the original five-story buildings, which were restricted in height because of the shortage of elevators. Hence the generally low skyline of inner Moscow is being walled in by higher construction in every direction.

THE UKRAINE

Russians are by far the largest ethnic group in the coreland, followed by Ukrainians. Most of the latter live in the Ukrainian SSR. Although closely related to the Russians in language and culture, the Ukrainians are a distinct national group. The name Ukraine is generally translated as "at the border" or "borderland." Here the Russian tsars fought for centuries against nomadic steppe peoples and also against Poles, Lithuanians, and Turks before the Ukraine was finally absorbed into the Russian Empire in the 18th century.

Today the Ukraine is one of the most densely populated and productive areas in the USSR. With about 51 million people (1988), it is only a little smaller in population than the United Kingdom or France. It is a major producer of coal, iron ore, manganese, natural gas, iron and steel, chemicals, machinery, ships, wheat, barley, livestock, vegetable oil (mainly from sunflower seeds), beet sugar, and many other products. The Ukraine lies partly in the forest zone and partly in the steppe. In the border between these vegetation realms is the historic city of Kiev (2.7 million), the capital of the SSR and a major industrial and transportation center (Fig. 11.3).

Figure 11.3 The heart of Kiev, third largest metropolis of the USSR, capital of the Ukrainian SSR, and mother city of the Ukrainian and Russian branches of Orthodox Christianity. This view features the St. Sophia Cathedral (foreground) and Bogdan Khmelnitsky Square, named for a 17th-century Ukrainian nationalist leader. Kiev was devastated by the military actions of World War II, and massive reconstruction of buildings took place after the war.

LESSER REPUBLICS OF THE CORELAND

The Belorussian (White Russian) SSR adjoins the Ukraine on the north. This republic is much smaller in area and population than the Ukrainian SSR (Table 11.1). In the past it developed rather slowly, due partly to a lack of minerals other than peat. Recently, however, major oil and gas pipelines from other parts of the USSR have provided raw material for new oil-refining and petrochemical industries. Some oil production has developed in Belorussia itself, and a large potash deposit is being exploited to make

TABLE 11.1   UNION REPUBLICS: AREA AND POPULATION DATA

Republic	Land Area		Population 1988 Estimates (millions)	Titular Nationality Percentage of Total, 1979
	thousand sq. mi.	thousand sq. km		
Russian SFSR	6593	17075	147.2	83
Ukrainian SSR	233	604	51.4	74
Belorussian SSR	80	208	10.2	79
Moldavian SSR	13	34	4.2	64
Lithuanian SSR	25	65	3.7	80
Latvian SSR	25	64	2.7	54
Estonian SSR	17	45	1.6	65
Azerbaijan SSR	33	87	7.0	78
Georgian SSR	27	70	5.3	69
Armenian SSR	12	30	3.5	90
Uzbek SSR	173	447	19.8	69
Kazakh SSR	1049	2717	16.7	36
Tadzhik SSR	55	143	5.0	59
Kirgiz SSR	77	199	4.3	48
Turkmen SSR	189	488	3.5	68
Totals	8600	22275	286.1	

fertilizer. Minsk (1.5 million), the capital and main industrial center, had to be almost completely rebuilt after World War II. Situated on a broad terminal moraine that stretches from Moscow to Poland, Minsk lay in the path of the main German thrust toward Moscow. The city saw fierce fighting as the Germans advanced in 1941 and again as they retreated toward Germany in 1944.

Belorussia is too cool, damp, and infertile to be prime agricultural country, although it has a substantial output of small grains, hay, potatoes, flax, and livestock products. Southern Belorussia incorporates the greater part of the Pripyat (Pripet) Marshes, large sections of which have been drained for agriculture.

The small Lithuanian, Latvian, and Estonian SSRs, often called the Baltic republics, lie between the Belorussian SSR and the Baltic Sea. They were part of the Russian Empire before World War I, but successfully asserted their independence following the Russian Revolution. In 1940, however, they were reabsorbed by the USSR as Union Republics. The Latvians and Lithuanians have borrowed culturally from their Slavic neighbors but are not themselves Slavs. The Estonians, related to the Finns, are also non-Slavic. In these republics dairy farms alternate with forested areas in a hilly, glaciated landscape. Few mineral resources of consequence exist, the most important being large deposits of oil-bearing shale in Estonia. The shale fuels power stations which transmit electricity to Leningrad and Riga as well as Estonia itself. Like the Belorussian SSR, the Baltic republics now benefit from natural gas and petroleum brought by pipelines from fields located hundreds or thousands of miles away. A large expansion of industry since World War II has helped make the Baltic region one of the more prosperous parts of the USSR. The largest city and manufacturing center is the seaport of Riga (950,000) in the Latvian SSR. Adjoining the Lithuanian SSR on the southwest is former German East Prussia, the northern half of which is now a part of the Russian Soviet Federated Socialist Republic (RSFSR). The German population was expelled to Germany, and Russians now comprise over three fourths of the population. There is also a sizable minority of Russians in Latvia and Estonia, many of whom have been attracted there by new industrial employment in the cities. The Estonians, Latvians, and Lithuanians do what they can to safeguard their respective national identities against Russian encroachment. Their small republics have recently achieved somewhat greater self-direction under the *glasnost* and *perestroika* policies of the Gorbachev regime.

Within the coreland at the extreme southwest is the small Moldavian SSR, largely made up of territory taken from Romania in 1940 (see chapter-opening photo of Chapter 10). The Moldavians, a people with many Slavic cultural characteristics, speak a dialect of the Romanian language. Mainly a fertile black-earth steppe upland, this SSR is primarily agricultural, although industry has shown marked gains since World War II. Corn, wheat, and other grains occupy the bulk of the crop acreage. Dairy cattle and hogs are the principal livestock in the north, but sheep and goats predominate in the drier south. An important specialty is the growing of vegetables and fruits, especially grapes. This republic is one of the major grape-growing and wine-producing areas of the USSR, its natural environment for vineyards being somewhat similar to that of the famous French district of Champagne.

MAJOR INDUSTRIAL CONCENTRATIONS

Although pre-Revolutionary Russia was largely an agricultural country, a slow development of modern industry, partially financed by foreign capital, took place before World War I. At the time of the Revolution industrial development was mainly confined to three areas: the Moscow region, Leningrad (then called Petrograd), and the Ukraine. Under the Soviets, these areas have continued to represent major concentrations of industry. In addition, industry has been greatly expanded in the Urals and in widely spaced cities along the Volga River, and an entirely new industrial concentration has been created in the Kuznetsk Basin. All these major industrialized areas are within the coreland, along with many smaller industrial centers (Fig. 11.4).

The Diversified Industries of the Moscow Region

The industrialized area surrounding Moscow is often referred to as the Central Industrial Region, Old Industrial Region, or Moscow-Tula-Gorkiy Region. These names indicate important characteristics of the region. It has a central location physically within the USSR's western plains and is functionally the major focus for the entire Soviet Union. It lies at the center of the Soviet rail and air networks and is connected by river and canal transportation with the Baltic, White, Azov, Black, and Caspian seas. From Moscow the region extends southward to the metallurgical and machine-building center of Tula (650,000) and eastward to the diversified industrial center of

Gorkiy (2 million) on the Volga (Fig. 11.5). To the north the region includes the important textile center of Ivanovo (city proper, 475,000) and Yaroslavl (city proper, 625,000), a very diversified center of oil refining, petrochemicals, and many other industries on the upper Volga. Textile milling, largely on the basis of imported American cotton and Russian flax, was the earliest form of large-scale manufacturing to be developed in the Moscow region. It gradually replaced the earlier handicraft industries in the 19th century. Although the area still has the USSR's main concentration of textile plants (cottons, woolens, linens, and synthetics), with Moscow and Ivanovo the leading centers, a great variety of other light and heavy manufactures have developed. Metal-fabricating industries, emphasizing types of construction requiring a relatively high degree of skill and precision, are the most important in value of product. In addition, all sorts of specialized chemicals such as pharmaceuticals are made here. This region ranks with the Ukraine as one of the two most important industrial regions in the USSR. It produces at least one fifth of the country's industrial output by value and a much higher proportion of high-technology goods.

The industrial eminence of the Moscow region has been achieved in spite of a notable lack of mineral resources other than lignite and peat—both burned in large thermal plants to produce electricity—phosphate deposits for fertilizer, and small iron-ore deposits (now practically exhausted) that provided the initial basis for an iron and steel industry at Tula. However, the well-developed railway connections of the capital, partly a product of political centralization, have provided good facilities for a constant flow of minerals, other materials, and foods and a return outflow of finished products to all parts of the USSR. Much smaller amounts of long-distance freight are handled by highway, water, or air transportation. A significant development in recent decades has been the construction of pipelines to bring petroleum and its products or natural gas to the Moscow region from fields in several areas. Piped oil and gas have become major fuels in the region and provide the basis for large petrochemical industries.

Leningrad's Pioneering Role in Soviet Industrialization

Leningrad does not form the center of an industrial region comparable in area and population to the Moscow region, the Ukraine, or the Urals. Nevertheless, the city and its immediate environs account

for possibly 5 percent of the USSR's manufacturing. As in the case of Moscow, this industrialization has not been aided much by local minerals. Peat, oil shale, and lignite from nearby deposits were once the main power sources, but they have largely been superseded by hydroelectricity from the region; by coal, oil, and gas brought from great distances; or by electricity from a large nuclear station near the city. Metals from outside the area provide material for the metal-fabricating industries that are the leaders in Leningrad's diversified industrial structure. This region has no large steel plant but is estimated to consume about one tenth of the USSR's steel output.

Supported by university and technological-institute research workers, Leningrad's highly skilled labor force played an extremely significant role in early Soviet industrialization, pioneering the development of many complex industrial products such as power-generating equipment and synthetic rubber, and supplying groups of experienced workers and technicians to establish new industries in various parts of the USSR. Industrial innovation now centers in Moscow, but Leningrad is still preeminent in some very specialized industries such as the making of huge turbines for hydroelectric stations all over the USSR.

The Ukrainian Industrial Region: Mineral Riches and Heavy Industries

From the beginning of the Five-Year Plans, the USSR has laid great stress on the heavier types of industry. Today heavy metallurgy and chemical manufacturing, supported by local mineral wealth, are mainly concentrated in two regions: the Ukraine and the Urals. A lesser, although important, concentration exists in the Kuznetsk Basin. Still other heavy-industrial plants and mining areas, some of which are very large producers, occupy scattered locations outside the major clusters. For example, a huge steel mill has been developed at Lipetsk (city proper, 450,000) in the black-earth belt to the south of the Moscow industrial region, and another big mill has been built at Cherepovets (300,000) to the north of the Moscow region. The first mill mainly supplies steel-using factories in and around Moscow, while the second provides much of the steel used at Leningrad.

Heavy industry in the Ukraine is based essentially on major deposits of five minerals: coal, iron, manganese, salt, and natural gas. The coal is found in the Donets Basin (Donbas) coal field. Mining began here under the tsars, and as late as 1913 the Donets

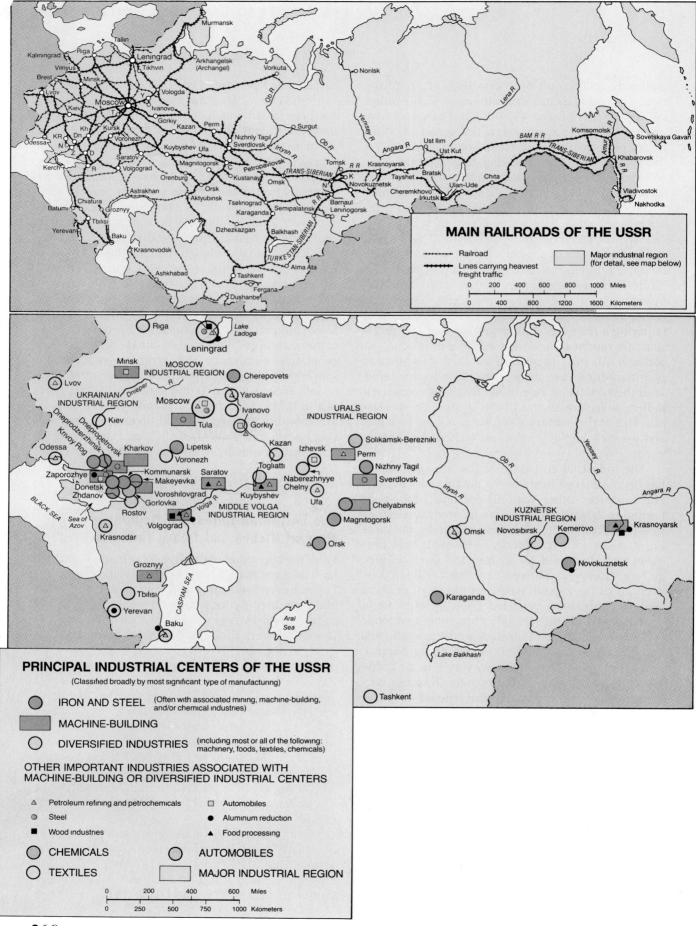

MAIN RAILROADS OF THE USSR

Murmansk
Kaliningrad
Riga
Tallin
Leningrad
Tikhvin
Arkhangelsk (Archangel)
Vorkuta
Vilnyus
Minsk
Vologda
Brest
Lvov
Moscow
Ivanovo
Kiev
Gorkiy
Kh
Kursk
Kazan
Perm
Nizhniy Tagil
Sverdlovsk
Irtysh R.
Surgut
Norilsk
Odessa
KR
Voronezh
N
Z
D
Saratov
Kuybyshev
Ufa
R
Magnitogorsk
Kerch
Volgograd
Petropavlovsk
Omsk
Tomsk
R R
Krasnoyarsk
Angara R.
Ust Ilim
Komsomolsk
Sovetskaya Gavan
Chiatura
Astrakhan
Orenburg
Orsk
Kustanay
Tayshet
Bratsk
Ust Kut
BAM R R
TRANS-SIBERIAN
Khabarovsk
Batumi
Grozhyy
Aktyubinsk
TRANS-SIBERIAN
K
Novokuznetsk
N
Cheremkhovo
Ulan-Ude
Chita
Amur R.
Tbilisi
Tselinograd
Semipalatinsk
Leninogorsk
Irkutsk
Vladivostok
Yerevan
Karaganda
Barnaul
Nakhodka
Baku
Krasnovodsk
Balkhash
Dzhezkazgan
TURKESTAN-SIBERIAN
R R
Ashkhabad
Tashkent
Alma Ata
Fergana
Dushanbe

MAIN RAILROADS OF THE USSR

········· Railroad
╋╋╋╋ Lines carrying heaviest freight traffic

☐ Major industrial region (for detail, see map below)

0   200   400   600   800   1000 Miles
0   400   800   1200   1600 Kilometers

Riga
Lake Ladoga
Leningrad
Minsk
MOSCOW INDUSTRIAL REGION
Lvov
Cherepovets
UKRAINIAN INDUSTRIAL REGION
Dnieper R.
Yaroslavl
Moscow
Ivanovo
Kiev
Tula
Gorkiy
URALS INDUSTRIAL REGION
Dneprodzerzhinsk
Kazan
Solikamsk-Bereznik
Dnepropetrovsk
Kharkov
Lipetsk
Izhevsk
Perm
Krivoy Rog
Voronezh
Togliatti
Nizhniy Tagil
Odessa
Zaporozhye
Kommunarsk
Saratov
Naberezhnyye Chelny
Sverdlovsk
Makeyevka
Donetsk
Voroshilovgrad
Kuybyshev
Ufa
Chelyabinsk
Zhdanov
Gorlovka
MIDDLE VOLGA INDUSTRIAL REGION
Ob R.
BLACK SEA
Rostov
Volga R.
Magnitogorsk
Sea of Azov
Volgograd
Yenisey R.
Krasnodar
Orsk
Irtysh R.
Angara R.
KUZNETSK INDUSTRIAL REGION
Grozhyy
Omsk
Novosibirsk
Kemerovo
Krasnoyarsk
Tbilisi
CASPIAN SEA
Novokuznetsk
Yerevan
Baku
Aral Sea
Karaganda
Lake Balkhash
Tashkent

PRINCIPAL INDUSTRIAL CENTERS OF THE USSR

(Classified broadly by most significant type of manufacturing)

⬤ IRON AND STEEL (Often with associated mining, machine-building, and/or chemical industries)

▬ MACHINE-BUILDING

◯ DIVERSIFIED INDUSTRIES (including most or all of the following: machinery, foods, textiles, chemicals)

OTHER IMPORTANT INDUSTRIES ASSOCIATED WITH MACHINE-BUILDING OR DIVERSIFIED INDUSTRIAL CENTERS

△ Petroleum refining and petrochemicals   ☐ Automobiles
● Steel   ● Aluminum reduction
■ Wood industries   ▲ Food processing

⬤ CHEMICALS   ◯ AUTOMOBILES

◯ TEXTILES   ▬ MAJOR INDUSTRIAL REGION

0   200   400   600 Miles
0   250   500   750   1000 Kilometers

Figure 11.4 Maps of major minerals, industrial areas, and railroads in the USSR. The symbols are based in part on qualitative data. Mineral symbols show producing deposits or areas. Symbols for industries are modified from Richard E. Lonsdale and John H. Thompson, "A Map of the USSR's Manufacturing," Economic Geography, Vol. 36, No. 1 (January 1960), pp. 36–52.

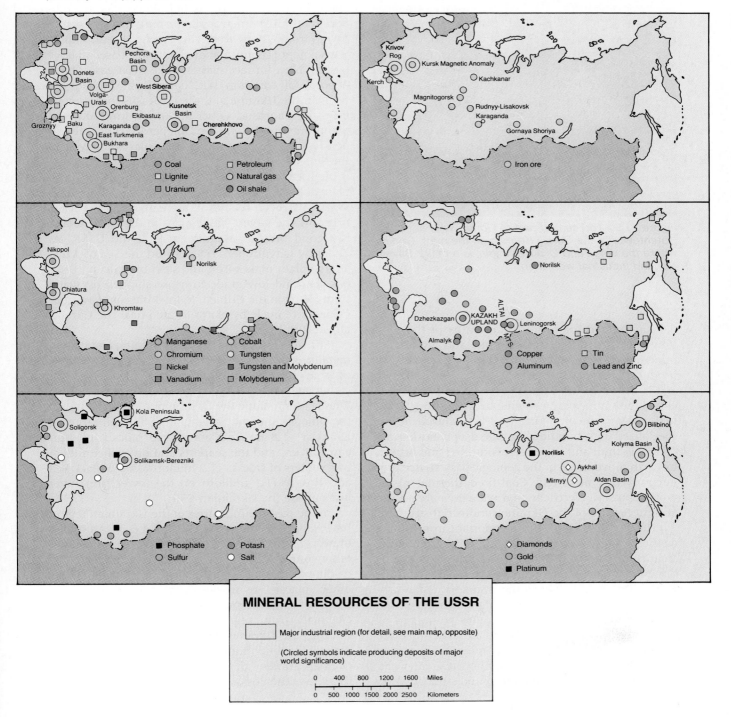

Figure 11.5   The central part of Gorkiy, with a watchtower of the Gorkiy Kremlin on the right. Gorkiy, where the Volga and Oka rivers join, is a major Soviet city and industrial workshop.

field accounted for nearly nine tenths of the total tonnage of hard coal (bituminous and anthracite) mined in Russia. It is still the country's largest coal producer; its annual output is many times that of 1913, but the field's relative importance has declined. The mines are valuable because they yield a high proportion of coking coal and anthracite for specialized uses, but the seams are deeply buried, are relatively thin, and have been contorted and broken by earth movements in the geologic past. Hence production costs are high. Coal-fired thermal plants which transmit electricity to a wide area are an important feature of the Ukraine's industrial structure.

The Soviet Union's most important iron-mining district lies in the western part of the Ukraine in the vicinity of Krivoy Rog (city proper, 675,000). Ore has been extracted here since tsarist days, and the Krivoy Rog field has been the leading Russian producer for many decades. The higher grades of ore are deep-mined, while lower grades are mined by large-scale open-pit methods and are then processed at concentrating plants prior to shipment. The output by each method is very large. By far the greater part of the production is used by iron and steel plants in

the Ukraine, but much ore is shipped to European Communist countries and some to Soviet regions outside the Ukraine. Manganese, the most important of the ferroalloys, is mined southeast of Krivoy Rog at Nikopol. The Soviet Union is the world's largest producer and has very large reserves.

The principal concentration of iron and steel plants in the Ukraine is located on the Donets Basin coal field, in the general vicinity of the coal-mining and heavy-industrial center of Donetsk (about 2 million in multi-city metropolitan area). Iron ore comes by rail from Krivoy Rog. This area is also an important center of chemical manufacturing, based in large part on blast furnace wastes, coke oven gases, huge deposits of common salt, natural gas piped from wells in the Ukraine and the Caucasus, or oil piped from the Caucasus. Other important iron and steel plants are found along the Dnieper River at Dnepropetrovsk (1.5 million) and other places, or west of the Dnieper at Krivoy Rog. This part of the Ukraine is often called the Dnieper Bend. An iron and steel plant at Krivoy Rog that was destroyed during World War II has been rebuilt on a greatly enlarged scale. It is now the largest single plant in the Ukraine and is second only to Magnitogorsk in the USSR. At Zhdanov (city proper, 525,000) on the Sea of Azov, a large plant makes iron and steel by utilizing Donbas coal and low-grade iron ore shipped by water from Kerch in the Crimea. Many kinds of heavy machinery are made by plants located near the Ukrainian steel mills.

Surrounding the inner core of mining and heavy-metallurgical districts in the Ukraine is an outer ring of large industrial cities which carry on metal-fabricating and various other types of manufacturing. These cities include the Ukrainian capital of Kiev on the Dnieper, the machine-building and rail center of Kharkov (1.8 million) about 250 miles (400 km) east of Kiev, and the seaports and diversified industrial centers of Odessa (1.2 million) on the Black Sea and Rostov (1.1 million) on the lower Don River just outside the Ukrainian SSR.

In the past the industries of the Ukrainian region have depended largely on Ukrainian coal as a source of power, and it is still a very important source, but other sources have risen rapidly in recent years, including natural gas and oil from the Ukraine and other places; hydroelectricity transmitted from chains of dams on the Dnieper and Volga rivers; and electricity from several nuclear stations scattered through the Ukraine. This whole region is so fundamentally important to the USSR that its loss to German armies early in World War II was a major calamity. The loss was partly made good by expanded development to the eastward, particularly in the Urals.

Mineral-Based Industrial Development in the Urals

The Ural Mountains contain an extraordinarily varied collection of useful minerals. Although deficient in coking coal, this highly mineralized area has valuable deposits of iron, copper, nickel, chromium, manganese, bauxite, asbestos, magnesium, potash, industrial salt, and many other minerals. Much low-grade bituminous coal plus lignite and some anthracite are mined, although relatively little of the coal is suitable for coking. The important Volga-Urals oil fields lie partly in the western foothills of the Urals, and a major gas field lies at the southern end of the mountains.

The Soviet regime has placed great emphasis on the development of the Urals as an industrial region well removed from the exposed western frontier of the Union. The major industrial activities are as follows:

1. Heavy metallurgy, including the manufacture of iron and steel and the smelting of nonferrous ores.

2. The manufacture of heavy chemicals based on some of the world's largest deposits of potassium and magnesium salts.

3. The manufacture of machinery and other metal-fabricating activities, carried on in the important industrial and transportation centers of Sverdlovsk (1.5 million), Chelyabinsk (1.3 million), Perm (1.1 million) on the Kama River, Ufa (1.1 million) in the Volga-Urals oil fields, and dozens of smaller places.

Old pre-Soviet metallurgical and machine-building plants in the Urals have been modernized and expanded, and a number of immense new plants have been constructed. The most famous and spectacular development has been the creation of an iron and steel center at Magnitogorsk (city proper, 425,000) in the southern Urals. This place, located near a reserve of exceptionally high-grade iron ore, was not even a village prior to 1931. In that year construction of a huge plant was begun and a city was built to house the workers. The iron and steel plant at Magnitogorsk has long been the largest in the Soviet Union. Other giant iron and steel mills are scattered through the Urals at Chelyabinsk, Nizhniy Tagil, and Orsk.

The years that saw the creation of Magnitogorsk also witnessed a large expansion of coal mining in the Kuznetsk Basin (Kuzbas), located in southern Siberia more than 1000 miles (c. 1600 km) to the east. A railway shuttle developed, with Kuznetsk coal and coke moving to Magnitogorsk and other industrial centers in the Urals, and Urals iron ore (primarily from Magnitogorsk) moving to a new iron and steel plant in the Kuznetsk Basin. Thus was created the famous Urals-Kuznetsk Combine (*Combinat* in Russian). Each end of the Combine, however, soon became partially independent of the other. A coal field was developed at Karaganda in Soviet Middle Asia to provide fuel for the Urals, while iron mining was developed south of the Kuznetsk Basin in foothills of the Altai Mountains. After 1937 the Combine, as such, was de-emphasized following a transportation crisis that led to a search for greater regional self-sufficiency. Nevertheless, due to qualitative shortcoming of Karaganda coal and Siberian iron ore, the Kuznetsk Basin continued to send large quantities of coal and coke to the Urals and received high-grade iron ore in return. In recent years the expanding iron and steel industries of the Urals and Kuznetsk Basin, as well as those elsewhere in the Soviet Union, have come to depend heavily on low-grade iron ore mined in open pits at a considerable number of sites. Concentrating plants upgrade the ore prior to marketing. Shortages of high-grade ore, especially ore minable at shallow depths, have led to the new reliance on lower-grade ore. For example, the famous deposit of high-grade ore at Magnitogorsk is nearly exhausted, and the plant there uses large amounts of ore from open-pit mines at sites in the Kazakh SSR.

Considerable amounts of iron ore now reach the Urals from mines that have been developed in a huge ore formation between the Moscow region and the Ukraine called the Kursk Magnetic Anomaly, or KMA (so-named because of its disturbing effect on compass needles). Technological difficulties delayed the exploitation of this resource until the 1960s, but it is now mined on a large and ever-increasing scale. The KMA deposits, containing both low-grade ores (mined in open pits) and high-grade ores (mined underground), represent one of the USSR's main reserves for the future. At present the primary markets are in iron and steel plants (such as the Lipetsk plant previously noted) within a 300-mile (480-km) radius of Moscow.

Sverdlovsk, located on the eastern flank of the mountains, is the largest city of the Urals and the preeminent economic, cultural, and transportation center. Direct trunk rail lines connect the city with Moscow and Leningrad. The second most important rail center is Chelyabinsk, 120 miles (193 km) south of Sverdlovsk. This grimy steel town had an infamous name in Old Russia as the point of departure for exiles being sent to Siberia. In western Siberia between the Urals and the large industrial and trading center of Omsk (1.1 million), rail lines from Sverdlovsk and Chelyabinsk join to form the Trans-Siberian Railroad, the main artery linking the Soviet Far East with the coreland (railroad map, Fig. 11.4). Omsk is a major metropolitan base for Siberian oil and gas development. It is Siberia's most important center of oil refining and petrochemical manufacturing.

The Kuznetsk Industrial Region: Siberia's Largest

From Omsk the Trans-Siberian leads eastward to the Kuznetsk industrial region, the most important concentration of manufacturing east of the Urals. Here the principal localizing factor for industry is an enormous deposit of coal, much of it suitable for coking. The reserves are estimated to be about four or five times those of the Donbas, and the seams are much thicker and easier to work. Production has expanded rapidly in recent years, primarily for use in the Urals but also in western Russia and in the industries of the Kuznetsk region itself.

The manufacture of iron and steel is a major industrial activity of the Kuznetsk region. The main center is Novokuznetsk (city proper, 575,000), where the first iron and steel plant was commenced in 1932, concurrent with the development of Magnitogorsk. A second large plant was constructed later. The industry draws its iron ore from various sources in Soviet Asia. The Kuznetsk Basin produces steel primarily for use by fabricating industries in Siberia, but sizable quantities move to factories west of the Urals. Chemical manufacturing, based in large measure on by-products of coke ovens, is also important.

The largest urban center of the Kuznetsk region is Novosibirsk, a diversified industrial, trading, and transportation center located on the Ob River at the junction of the Trans-Siberian and Turkestan-Siberian (Turk-Sib) railroads. Sometimes called the "Chicago of Siberia," Novosibirsk has developed from a town of a few thousand at the turn of the century to a metropolitan area of 1½ million people. Hundreds of factories produce mining, power-generating, and agricultural machinery, tractors, machine tools, and a wide range of other products. The city's much-publicized *Akademgorodok* ("Academy Town" or "Science Town"), Siberia's main center of scientific research, is an outlying suburb.

The Middle Volga Industrial Region: Oil, Hydropower, Petrochemicals, Automobiles

Numerous industrial cities outside of the five industrial concentrations previously discussed are scattered through the coreland. The most notable are a group of cities spaced along the Volga from Kazan southward. The largest are Kuybyshev (1.5 million), Volgograd, formerly Stalingrad (1.3 million), Kazan (1.1 million), and Saratov (1.1 million). These four cities in the middle Volga industrial region have diversified machinery, chemical, and food-processing plants.

In recent years the cities and towns along the Volga have seen a marked upsurge of industrial activity. One of the most publicized aspects has been the construction of a chain of dams and hydroelectric stations, not only along the Volga River itself but also along its large tributary, the Kama. The largest units are in the vicinity of Volgograd and Kuybyshev. (For a more extended discussion of waterway development, see Chapter 10.) A second major development has been the rapid rise of the Volga region to leadership in the automobile industry. In the past this industry was centered in Gorkiy and Moscow, and these places remain important. But more recently several cities along or near the Volga and Kama rivers have become major centers of automaking, especially Togliatti (city proper, 575,000) on the Volga, which makes more passenger automobiles than the rest of the Soviet Union put together. These come from one huge factory, the Volga Automobile Plant, which was built and equipped for the Soviet government by the Fiat Company of Italy. Another enormous factory, the Kama River Truck Plant at Naberezhnyye Chelny, was developed with technical aid and equipment purchased in several different foreign countries. This plant has become the main Soviet producer of the heaviest types of trucks. Such factories are facilitating a more rapid entry by the Soviet Union into the automobile age, although a vast amount of highway development will be necessary to make the USSR a truly modern motorized country.

THE VOLGA-URALS OIL FIELDS

Still another important economic thrust contributing to the industrial rise of the Volga cities has been the large-scale exploitation of petroleum in the nearby Volga-Urals fields. Prior to the rise of fields in western Siberia during the 1970s, the Soviet Union's most important area of oil production was the Volga-Urals fields, sometimes referred to as the "Second Baku." Stretching from the Volga River to the western foothills of the Ural Mountains, and containing many separate deposits of oil, these fields were producing somewhat over two thirds of the USSR's oil output in the middle 1960s, but by the early 1980s this figure had dropped to between one fourth and one third of the countrywide total. The relative decline was due to the opening of other oil fields, especially in western Siberia, plus a leveling off of production in the Volga-Urals area. These fields also produce some natural gas, although the USSR's main gas reserves and production are elsewhere. Pipelines carry crude oil or petroleum products from the Volga-Urals area to the Moscow region and many other areas, including several countries in Communist Europe. Important petrochemical industries based on oil and natural gas from these fields and other sources have developed rapidly in many parts of the USSR. They manufacture synthetic rubber, artificial fibers, nitrogenous fertilizers, plastics, and various other products. In the Volga-Urals region Kuybyshev and Ufa are especially well-known oil-refining and petrochemical centers, but most large cities along the Volga and Kama rivers, as well as some smaller cities in the oil fields, have a share of these industries.

MAJOR AGRICULTURAL ZONES IN THE CORELAND

The Fertile Triangle is the agricultural core of the Soviet Union. It is by far the leading area of production for all the truly major crops except cotton and for all the major types of livestock, including not only cattle, hogs, sheep, goats, and poultry, but also horses. Although tractors have largely replaced horses for plowing and harvesting, many horses are still used on farms for pulling wagons and as draft power for cultivation of the small personal plots (see chapter-opening photo of Chapter 10). Within the Triangle two main agricultural zones have long been recognized: (1) a black-soil zone in the southern steppes and (2) a nonblack-soil zone roughly corresponding to the region of mixed forest.

The Black-Soil Zone

The black-soil zone includes not only the chernozem soils proper, but also associated areas of chestnut and other grassland soils. Wheat does well on these soils, and this crop zone is one of the major wheat-growing areas of the world (Fig. 3.5). The USSR is the largest producer of wheat, which is ideally suited to the large-scale, mechanized agriculture stressed in Soviet planning. In general, winter wheat is grown west of the Dnieper, and spring wheat in the harsher climate east of the Volga; areas between the two rivers grow both winter and spring wheat. The black-soil zone is also the principal producing area for sugar beets, sunflowers, hemp (grown mainly for oil), barley, and corn. Diversions of water from the Dnieper, Don, Volga, and other rivers are being used to foster irrigation here.

The Nonblack-Soil Zone

In the nonblack-soil zone, with its cooler and more humid climate and poorer soils, rye has traditionally been the main grain crop for human food. However, increased liming and fertilization of soils, plus the development of frost-resistant and quick-ripening varieties of wheat, have made possible some gain in wheat acreage and production in recent times. Agriculture in the nonblack-soil zone has a strong emphasis on dairy farming and the growing of potatoes, oats, and flax. Sugar beets have been introduced in some areas, although the conditions are less suitable than in the black-soil zone. Clover and grasses are widely grown for hay. In recent years Soviet planners have allocated considerable investment to the nonblack-soil zone as a means of upgrading productivity and the level of living. Many villages have been in a very depressed state.

THE OUTLYING REGIONS

Beyond the coreland, and mainly in Asia, lies an immense periphery of mountainous, northern, and arid lands peopled originally by Asians but conquered and penetrated since the 15th century by immigrant Slavs. Four major outlying regions can be distinguished here: the Caucasus region, Soviet Middle Asia, the Soviet Far East, and the Northern Lands.

THE CAUCASUS REGION: INTRICATE DIVERSITY IN A MOUNTAINOUS ENVIRONMENT

The Caucasus region occupies the mountainous southern borderland of the USSR between the Black and Caspian seas. It includes the rugged Caucasus Mountains, a fringe of foothills and level steppes to the north, and the area to the south known as Transcaucasia.

The Greater Caucasus Range forms practically a solid wall from the Black Sea to the Caspian. It is similar in age and general character to the Alps but is considerably higher: Mt. Elbrus, the highest summit, stands at 18,510 feet (5642 m) compared with 15,771 feet (4807 m) for Mont Blanc, the highest peak of the Alps. Railroads to Transcaucasia follow narrow coastal lowlands at either end of the range. In the south of Transcaucasia is the Armenian Plateau, a mountainous, volcanic highland reaching over 13,000 feet (c. 4000 m) in elevation. Between the Greater Caucasus Range and Armenian Plateau are subtropical valleys and coastal plains where the majority of the people in Transcaucasia live. Russians and Ukrainians predominate in the North Caucasus, but non-Slavic groups form a large majority in Transcaucasia. The Caucasian isthmus between the Black and Caspian seas has been an important north-south passageway for thousands of years, and the population comprises many different peoples who have migrated into this region at various times. At least 25 or 30 nationalities can be distinguished, mostly small and confined to mountain areas that have served as refuges in past times. In addition to Russians and Ukrainians, most of whom live north of the Greater Caucasus, the nationalities of greatest importance are the Georgians (Fig. 11.6), Armenians, and Azerbaijanis, each represented by a separate Union Republic.

These nationalities have racial characteristics and cultural traditions that are primarily Asian or Mediterranean in origin. They differ in religion, with the Azerbaijanis (also known as Azeri Turks) being Muslims, while the Georgians belong to one of the Eastern Orthodox churches, and the Armenian Church is a very ancient, independent Christian body. There has been a history of animosity between the Armenians and Azeri Turks, growing out of the persecution of Armenians in the Ottoman Empire prior to and during World War I. Such feelings recently flared into massive violence between the two peoples over the question of Nagorno-Karabakh, a predominantly Armenian enclave governed by Azerbaijan but claimed by Armenia. Numerous lives were lost, and large numbers of Armenians fled from Azerbaijan to Armenia as refugees. Their troubles were then compounded by an earthquake that devastated much of Armenia in December 1988. The Georgians, Armenians, and Azerbaijanis have stubbornly maintained their cultures in the face of pressure by stronger intruders. The Georgian culture, for example, has survived an incredible number of conquests:

> The Georgians were conquered successively by Romans (first century), Persians (sixth century), Arabs (seventh century), Turks (eleventh century), Mongols (thirteenth century), Tamerlane's Tatars (fourteenth and fifteenth centuries, in a wave of eight separate invasions), and, alternately, Turks and Iranians (sixteenth and seventeenth centuries). Finally this Orthodox Christian outpost sought the protection of Orthodox Christian Russia and was annexed by Tsar Paul in 1801.[2]

Subtropical crops and minerals are important contributions of the Caucasus region to the economy of the USSR. The lowlands bordering the Black Sea, with the heaviest rainfall and warmest winters of the USSR, produce such specialty crops as tea (Fig. 11.7), tung oil, tobacco, silk, and wine, together with some citrus fruits that are grown on a rather marginal basis (there is too much freezing weather for the citrus industry to really flourish). Corn and rice are basic food crops. This area, the most densely populated part of Transcaucasia, is located in the Georgian SSR. The Caspian lowlands of eastern Transcaucasia, located in the Azerbaijan SSR, have colder winters and less rain; here irrigated cotton, grown in rotation with alfalfa (used for livestock feed), is the main crop except in the coastal lowlands at the south where warmer temperatures and greater precipitation permit the culture of some rice, tea, and

Figure 11.6 Ethnic Georgians selling produce in a free market at Tbilisi, the main city and capital of the Georgian SSR.

[2]Elizabeth Pond, *From the Yaroslavsky Station: Russia Perceived* (New York, Universe Books, 1981, 1984, 1988), p. 116. Reprinted by permission.

Figure 11.7 Harvesting tea leaves by machine on the large Ingirsky Tea-Growing State Farm in the Georgian SSR. Soviet agricultural planners have long tried to promote such mechanized factory-type farms. But a large share of the tea crop still comes from smaller collectives where the picking is done by hand. The Soviet Union is a major consumer of tea, most of which comes from the humid subtropical climatic area where the farm in the photo is located.

citrus. Temperate fruits such as apples and peaches are grown in many parts of Transcaucasia, as are chestnuts and walnuts, grapes for wine-making and other uses, and mulberry trees used to feed silkworms. Many livestock, principally sheep and cattle, are grown. Some are grazed on mountain pastures in summer and then are wintered in lowlands.

Oil and gas are the Caucasus region's most valuable minerals. Both were relatively more important in the past than they are at present. The main oil fields are along or underneath the Caspian Sea around Baku (1.9 million) or north of the mountains at Groznyy (city proper, 400,000) and other places. Wells recently developed in the Georgian SSR also produce some oil. Dry and windswept Baku, the largest city of the Caucasus region and capital of the Azerbaijan SSR, was the leading center of petroleum production in Old Russia (in fact, for a few years around the turn of the century it was the leading center in the world), and under the Soviets it continued its leadership within the USSR until it was decisively surpassed in the 1950s by the Volga-Urals fields and again in the 1970s by the West Siberian fields. However, Baku oil continues to be prized for its high quality, as it is low in sulfur and hence can be refined into high-octane gasoline and superior lubricating oils. For a time in the 1960s the Caucasus region was the USSR's leading producer of natural gas, mainly from fields in the North Caucasus, but its output is now far exceeded by fields in Siberia, Soviet Middle Asia, and the Ukraine. Manganese at Chiatura and miscellaneous other metals, together with some low-grade coal, are mined in Transcaucasia; there is a considerable output of both hydroelectric and thermal power; and a diversity of man-

ufacturing industries has developed on a relatively small scale. Most factories are located in or near the republic capitals: Baku; Tbilisi (1.3 million) in Georgia (Fig. 11.8); and Yerevan (1.2 million) in Ar-

Figure 11.8 Tbilisi, the capital and largest city of the Georgian SSR, extends for a long distance along the banks of Transcaucasia's longest river, the Kura. Distinctive towers of old cathedrals rise above the low skyline of the city.

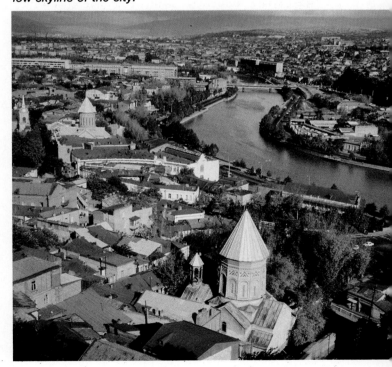

menia. Oil refining and petrochemical industries are significant in the areas that produce oil and gas.

The Caucasus region vies with the Crimea as a center of resort development. The main resorts are along the Black Sea coast and in the mountains of the Georgian SSR. They contribute further to the intricate diversity that is a major geographic characterisic of this exotic area.

SOVIET MIDDLE ASIA: MUSLIMS, DESERTS, OASES, MINERALS

Across the Caspian Sea from the Caucasus region lies a large expanse of deserts and dry grasslands, bounded on the south and east by high mountains. Like Transcaucasia this region, called Soviet Middle Asia in this text, has a preponderance of non-Slavic nationalities. Those with the largest populations are four Muslim peoples speaking closely related Turkic languages—the Uzbeks (Fig. 11.9), Kazakhs, Kirgiz, and Turkmen—and a fifth Muslim people of Iranian origins, the Tadzhiks. Each has its own Union Republic. By far the largest in area is the Kazakh Republic (Kazakhstan) (Table 11.1).

In a sense the "nationalities" of Soviet Middle Asia and their republics are artificial creations of the Soviet authorities. The peoples of this region are the heirs of local oasis civilizations that may go back 50 centuries or more and of the diverse cultures of nomadic peoples who captured oases from time to time or strongly influenced them. Life in earlier times tended to develop as a tissue of small-scale social and political units—clans, tribes, petty autocracies—that were associated with particular oases and might

or might not be controlled by a larger empire at any given time. The area lies on a very ancient route across Asia from China to the Mediterranean, and it repeatedly attracted the attention of conquerors: Alexander the Great, the Arabs who brought the Muslim faith in the 8th century, the Mongols, and the Turks. By the time of the Russian conquests in the 19th century, Turkestan (as it was then known) had become a cockpit of feuding, tradition-bound Muslim khanates having some political ties to China, Persia, Ottoman Turkey, or even British India. Cultural and political fragmentation continued up to the Russian Revolution and into the period of civil war. During World War I there was a revolt against the Tsarist government, which had begun to draft the Muslims for menial labor at the front. This set off a period of widespread disorder and bloodshed, which went on until the Bolshevik government finally gained control in the early 1920s and began to organize the area into "nationalities" and republics.

The 20th century has seen a large incursion of Russians and Ukrainians into Soviet Middle Asia, mainly into cities. They include administrative and managerial personnel, engineers, technicians, factory workers, and, in the north of Kazakhstan, a sizable number of farmers in the "new lands" wheat region. Most of the Slavic newcomers live apart from the local Muslims.

A visitor to Tashkent (2.2 million), the capital of the Uzbek SSR and the largest city in Soviet Middle Asia, has conveyed her impressions of Russian modernity grafted onto the older Islamic city[3]:

> Before I left Moscow, I had been told that Tashkent was a real metropolis, that its population of almost 2 million made it the fourth-largest city in the U.S.S.R. and that it was home to some of the Soviet Union's biggest factories and scientific institutions. Yet what struck me that day was not Tashkent's modernity but its essentially Asian character.
>
> As I looked out the car windows, what I saw was a series of long, dusty roads dotted with small adobe or wooden houses, all built in true Muslim style, without windows or doors looking out onto the street. The city center was a cluster of modern buildings—which in large part, however, already appeared to be crumbling. Old men on donkeys competed with buses and trolleys on the main thoroughfares. As we drove past the university, two young Uzbeks grazing their lambs in a park waved at me. All in all, Tashkent seemed a place where bits and pieces of today were stuck like tiles on the surface of another century.

Figure 11.9   Muslim men at prayer in Samarkand, Uzbek SSR. Women pray separately behind a screen.

[3]Nancy Lubin, "Mullah and Commissar," *Geo* 2, No. 6 (June 1980), p. 16. Reprinted by permission.

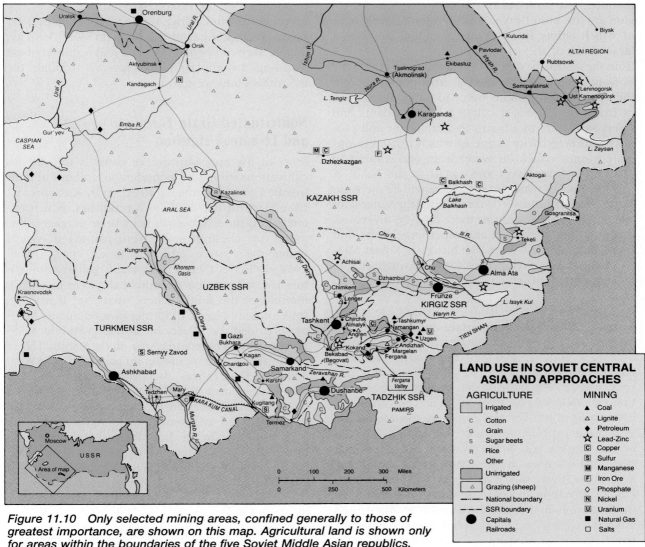

Figure 11.10   Only selected mining areas, confined generally to those of greatest importance, are shown on this map. Agricultural land is shown only for areas within the boundaries of the five Soviet Middle Asian republics. Modified and redrafted from a map by Robert N. Taaffe, "Transportation and Regional Specialization: the Example of Soviet Central Asia," Annals of the Association of American Geographers, *Vol. 52, No. 1 (March, 1962), p. 81. Used by permission of the author and the* Annals.

Physiography and Irrigated Agriculture

Soviet Middle Asia is predominantly composed of plains and low uplands except for the Tadzhik and Kirgiz SSRs, which are extremely mountainous. In these latter republics are found the highest summits of the USSR: Mount Communism (formerly Mt. Stalin), 24,590 feet (7495 m), in the Pamirs (Tadzhik SSR), and Pobeda (Victory) Peak, 24,406 feet (7439 m) in the Tien Shan (Kirgiz SSR) (Fig. 10.2). The Kazakh Upland in east central Kazakhstan is a hilly area with occasional ranges of low mountains. Soviet Middle Asia is almost entirely a region of interior

drainage. Only the waters of the Irtysh, a tributary of the Ob, reach the ocean; all the other streams either drain to enclosed lakes and seas or gradually dry up and disappear in the Middle Asian deserts.

The majority of the people in Soviet Middle Asia live in irrigated valleys at the base of the southern mountains (Fig. 11.10). Here most soils (often formed of loess) are fertile, the growing season is long, and rivers issuing from mountains provide large supplies of water for irrigation. The irrigated area has been substantially increased in size and productivity under the Soviets, and further increases are planned. The principal rivers in the heart of the region are the Amu Darya (Oxus) and Syr Darya, both of which empty into the enclosed Aral Sea. But a large share of the irrigation water is not obtained from these streams directly, but from their tributaries or from other rivers. Numerous irrigation canals tap the rivers and distribute water to oases. Expansion of irrigation is being facilitated by new dams and reservoirs in the mountains to store water and produce hydroelectricity, some of which is used to operate pumps which lift water from one conduit to another or from rivers and canals to the fields. A major source of water for the future will be the Amu Darya, which has two or three times the volume of the Syr Darya. In addition there may be diversions of water from the Irtysh and other large rivers that drain to the Arctic Ocean. The history of Soviet Middle Asia is replete with plans for large water-supply schemes, some of which have come to fruition and some not.

Most of the larger irrigated districts are in the Uzbek SSR. Especially important are the fertile Fergana Valley on the upper Syr Darya, almost enclosed by high mountains; and the oases around Tashkent, Samarkand (city proper, 525,000), and Bukhara (city proper, 200,000). These old cities arose and became powerful political centers on caravan routes connecting southwestern Asia and the Mediterranean basin with eastern Asia. Samarkand, an epitome of exotic places, is the successor to many earlier cities that occupied the site on loess hills beside the Zeravshan River. It was the capital of Tamerlane's empire which encompassed much of Asia and, like Bukhara, has many striking Muslim buildings and imposing ruins, some of which are being reconstructed. Streams fed by melting snow and ice in some of the world's highest mountains have sustained these places through their remarkably long histories.

Today the major crop of the oases is cotton. This area grows well over nine tenths of the USSR's output. Something like half of the national crop comes from the Uzbek republic alone. The Soviet Union is a major world producer and exporter of cotton. The bulk of the country's production is used to supply textile mills in the Moscow region, Tashkent, and many other Soviet cities in widespread locations. Alfalfa is rotated with cotton to replenish soil nitrogen, remove part of the salt that accumulates in irrigated soils, and provide winter feed for livestock.

Mulberry trees to feed silkworms are grown around the margins of many irrigated fields and along irrigation canals in the Middle Asian oases. Irrigated rice and other grains, sugar beets, vegetables, vineyards, and orchards of temperate fruits are important. Commercial orchard farming is especially prominent around Alma Ata ("father of apples"; 1.1 million), the capital of the Kazakh SSR. This city, largely Slavic in population, is a major military base guarding an important passageway through the mountains into western China.

Nonirrigated Grain Farming and Livestock Herding

Most of the area outside the oases is too dry for cultivation, although a considerable amount of nonirrigated grain farming is carried on by Russians and Ukrainians in the north of Kazakhstan, which extends into the black-earth belt. Huge acreages, primarily on chestnut soils, were planted in wheat under the "virgin and idle lands" program of 1954–1956 (see Chapter 10). Nonirrigated grain culture is also found on the rainier slopes in mountain foothills of the south, where there are often good soils formed of loess. In former times nomadic herding provided the livelihood of many Middle Asian peoples. It was based primarily on the natural forage of the steppes and on mountain pastures in the Altai and Tien Shan ranges. But over the centuries there was a slow drift away from nomadism, and this process was accelerated by the Soviet government. It collectivized the remaining nomads, often by harsh measures in the face of strong resistance, and settled them in permanent villages. Frequently such villages were made possible by the drilling of deep wells to provide a dependable year-round supply of water and by the planting of forage crops for supplemental feed. Livestock raising in Soviet Middle Asia has become a form of ranching, although herdsmen must often accompany the grazing animals for considerable distances as the latter are moved from one area of range to another. Sheep and cattle are the principal types of livestock, but goats, horses, donkeys, and camels are also raised (Fig. 11.11).

The Upsurge in Mining and Manufacturing

During recent decades Soviet Middle Asia has become increasingly important as a producer of minerals, particularly natural gas, coal, oil, iron ore, nonferrous metals, and ferroalloys. Large reserves of bituminous coal are mined at Karaganda in central Kazakhstan. The Karaganda coal basin has become a very important producer (although much less so than the Donets and Kuznetsk basins), and Karaganda itself has experienced spectacular growth from a tiny village in 1926 to a city of somewhat over 600,000 (city proper) today. Much of the coal mined

Figure 11.11 *Horse-drawn sledges at a village of a minor non-Slavic nationality in the foothills of the Altai Mountains just east of Soviet Middle Asia as defined in this text. The use of horses for draft power (and among some peoples for meat) is still very common in the USSR. This view is relatively near the southern border of the country, but the winters are still severe enough to yield considerable freezing weather and a fair amount of snow.*

here is shipped to metallurgical works in the Urals. Additional coal mining in large open pits has developed about 150 miles (*c.* 240 km) northeast of Karaganda. Some of this coal can be coked, although its ash content, like that of Karaganda coal, is high. Among other uses, it provides fuel for thermal-electric power production near the mines.

Coal from these and other locations in Soviet Middle Asia, as well as from the Kuznetsk Basin, provides fuel for large-scale metallurgy in the region. Of primary importance is the smelting of nonferrous metals, especially copper, lead, and zinc. The largest copper reserves and the most important copper mines of the USSR are found in central Kazakhstan, midway between Karaganda and the Aral Sea. Complex ores yielding zinc and lead are mined in the Altai Mountain foothills of eastern Kazakhstan. Impressive reserves of chrome and nickel in the Urals foothills of northern Kazakhstan, natural gas near the Amu Darya, petroleum in fields bordering the Caspian Sea, large iron-ore deposits in northern Kazakhstan, and deposits of these and many other minerals scattered through the region add further to the picture of Soviet Middle Asia as a rich and diversified mineralized area.

Manufacturing scarcely existed at the time of the 1917 revolution, aside from handicrafts. But the Soviets have developed textile milling (including silk) and agricultural-processing industries on a sizable scale, and these have been supplemented, especially since 1940, by machine-building and chemical industries, and by an integrated iron and steel plant at Temir Tau near Karaganda.

Since Soviet Middle Asia borders Iran, Afghanistan, and China, is so heavily Islamic in religion, and may have 100 million people in its five republics by the end of the century, its future is a major question mark in Soviet affairs. Undeniably, its Muslims are better off economically than most of those in the bordering countries. Thus far (early 1989), there is little sign that they share the fundamentalist fervor of revolutionary Iran or make common cause in any overt way with the peoples of Soviet-invaded Afghanistan. But many kinship relations do exist between Soviet Muslims and their neighbors on the other side of the international frontiers, and this circumstance, along with dissatisfaction over Russian and Communist attitudes and policies, may yet make Soviet Middle Asia a major political problem area.

THE SOVIET FAR EAST: RUSSIA'S PACIFIC WINDOW

The Soviet Far East is the mountainous Pacific edge of the USSR. Most of it is a thinly populated, underdeveloped wilderness in which the only settlements are fishing ports, lumber or mining camps, or the villages or encampments of aboriginal tribes. It is remote from the main centers of Soviet life, and its contribution to the national economy is relatively meager. Most of the Russians and Ukrainians who comprise the great majority of its population live in a narrow strip of lowland behind the coastal mountains in the southern part of the region. This lowland, drained by the Amur River and its tributary,

Figure 11.12 *Rail yard on the electrified Trans-Siberian Railroad at the busy port of Nakhodka on the Pacific. The view features passenger cars, timber for export, and a trainload of containers. Much container traffic moves between Pacific countries and European countries via the Trans-Siberian "bridge."*

the Ussuri, is the region's main axis of industry, agriculture, transportation, and urban development. Several small to medium-sized cities form a north-south line along two important arteries of transportation: the Trans-Siberian Railroad and the lower Amur River. At the south on the Sea of Japan is the port of Vladivostok (city proper, 600,000), now mainly a naval base of major importance. It is kept open throughout the winter by icebreakers. About 50 miles (80 km) east of the city, the main commercial seaport area of the Soviet Far East has been developed at Nakhodka (city proper, 150,000; Fig. 11.12) and nearby Vostochnyy (East Port). Both ports are nearly ice-free.

Not far to the north of Vladivostok lies a small district that is the most important center of the Soviet Far East's meager agriculture, producing cereals, soybeans, sugar beets, and milk for Far Eastern consumption. The Soviet Far East, however, is far from self-sufficient in food or, for that matter, in consumer goods of other types, and local production is supplemented by large shipments from the coreland or overseas sources. The diversified industrial and transportation center of Khabarovsk (city proper, 575,000; Fig. 11.13) is located at the confluence of the Amur and Ussuri rivers. Here the main line of the Trans-Siberian Railroad turns south to Vladivostok and Nakhodka, and the Amur River turns north to the Sea of Okhotsk. To the northeast in the Amur Valley is Komsomolsk (city proper, 300,000), which

has the only steel mill in the region. The mill has no blast furnaces and uses pig iron from the coreland together with scrap metal. A branch line of the Trans-Siberian reaches Komsomolsk and the port of Sovetskaya Gavan (Soviet Harbor) on the Gulf of Tatary (Fig. 11.4), and both of the latter places are on the new Baikal-Amur Mainline (BAM) Railroad, which has its Pacific terminus at Sovetskaya Gavan.

The Pacific littoral along the Sea of Okhotsk to the north of the Amur mouth is very mountainous and sparsely populated. The largest settlement is Magadan (city proper, 125,000), the supply port for an important gold-mining district in the basin of the Kolyma River.

Prior to World War II the Soviet Union and Japan held the northern and southern halves, respectively, of the large island of Sakhalin. But at the end of the war the USSR annexed southern Sakhalin and the Kuril Islands and repatriated the Japanese population. Sakhalin has some mineral wealth, including coal, petroleum, and natural gas, and important forest and fishing industries. The Kurils are a sparsely populated chain of small volcanic islands that screen the Sea of Okhotsk from the Pacific. Fishing is the main economic activity. At the north the islands approach the large peninsula of Kamchatka. Like the Kurils, Kamchatka is a region of active vulcanism, having many hot springs in addition to 23 active volcanoes, one of which reaches a height of 15,584 feet (4750 m).

Fisheries and Forest Industries

Aside from its important seaport functions, the Soviet Far East is of minor economic significance to the USSR. There is relatively little mineral output in the immediate coastal region, although some coal, oil, and a few other minerals are produced in various places. However, the region's ports will handle increasing shipments of Siberian minerals as deposits are developed that lie well inland in areas tributary to the new BAM Railroad. Fisheries and/or forest industries provide the main support for most Far Eastern communities.

Timber in the Soviet Far East consists primarily of softwoods but includes some oak and other hardwoods in the basin of the Ussuri River. Sawmills scattered along the seacoasts and rivers furnish sizable lumber exports, although much wood is shipped out of the region (mainly to the coreland or to Japan) in the form of logs (Fig. 11.12). Southern Sakhalin is a significant area of pulp and paper milling. Several mills developed by Japan were taken over by the USSR when it annexed the area in 1945. On the mainland a large complex of wood-chemical

Figure 11.13 Old and new in the Soviet Union, seen here from the Trans-Siberian Railroad at Khabarovsk in the Soviet Far East. Frame or log houses, often with painted window frames, are typical of the older rural landscape from the Baltic to the Pacific, just as standardized high-rise apartment buildings and electric-power lines are typical of the new urban and industrial landscape.

plants has been developed on the Amur River upstream from Komsomolsk. Large quantities of forest products from the Soviet Far East are sold in Pacific markets, particularly Japan.

THE NORTHERN LANDS: BLEAK STOREHOUSE OF RESOURCES

North and east of the Agricultural Triangle and west of, and partially including, the Pacific littoral, lie enormous stretches of coniferous forest (taiga) and tundra extending across the USSR from the Finnish and Norwegian borders to the Pacific. These outlying wilderness areas may for convenience be designated the Northern Lands, although parts of the Siberian taiga extend to the USSR's southern border. The Northern Lands include over half of the Soviet mainland and also island groups in the Arctic Ocean, among which the mountainous, fjorded islands of Novaya Zemlya are especially well known.

These difficult lands comprise one of the most sparsely populated large regions of the world. Ordinary types of agriculture are largely precluded by the climate, although hardy vegetables, potatoes, hay, and barley are grown in scattered localities, and some dairy farming is carried on. Most of the people are supported by a few primary activities, including lumbering, mining, reindeer herding, fishing, hunting, and trapping. Towns and cities of much size are limited to a small number of sawmilling, mining, transportation, and industrial centers, found in most cases along the Arctic coast to the west of the Urals, along the major rivers, or along the Trans-Siberian Railroad between the Agricultural Triangle and Khabarovsk.

The oldest city on the Arctic coast is Arkhangelsk (Archangel; city proper, 400,000), located on the Northern Dvina River a bit inland from the White Sea. It was founded by Tsar Ivan the Terrible in 1584 for the purpose of opening a seaborne trade with England. Today, the city is the most important sawmilling and lumber-shipping center of the USSR. Despite its restricted navigation season (late spring to late autumn), it is one of the country's more important seaports.

Another Arctic port, Murmansk (city proper, 420,000), located on a fjord along the north shore of the Kola Peninsula west of Arkhangelsk, is the headquarters for important trawler fleets that operate in the Barents Sea and North Atlantic. Murmansk is the largest center of the Soviet fishing industry, which has increased greatly since World War II. Fish processing is a major industry of the city. More importantly from a strategic standpoint, Murmansk is a major naval base, thought to be the most important home port for Soviet nuclear submarines. It is also a large cargo port, connected to the coreland by rail, and open to shipping all year. The harbor is normally ice-free, thanks to the warming influence of the North Atlantic Drift, and thin ice that occasionally forms can be broken with icebreakers. Murmansk and Arkhangelsk played a vital role in World War II, continuing to receive supplies by sea from the Soviet Union's Western allies after the other ports of the western USSR had been captured or closed off by the German invasion. The Soviet war effort was materially aided by the trucks, ammunition, and other supplies that reached the Arctic ports despite heavy losses in British and American convoys that had to run the gauntlet of attacks by German aircraft, submarines, and surface warships based in Norway.

Murmansk is a western terminus (along with Arkhangelsk) of the Northern Sea Route, a waterway developed by the Soviets to provide a connection with the Soviet Far East via the Arctic Ocean and to link the Arctic lands and seas more effectively with the rest of the country. The economic importance of the route is not very great. Navigation throughout the whole length is possible for only a few weeks in the year, despite the use of the world's most powerful icebreakers to lead convoys of ships, helicopters to seek out channels through the Arctic ice, and broadcasts concerning weather and ice conditions. Except for the timber of Igarka and the metals and ores of Norilsk, the areas along the route provide few cargoes. Most ships operating from Murmansk and Arkhangelsk go no farther east than the Yenisey River (see chapter-opening photo). Determined efforts, including the use of multiple icebreakers to escort convoys, have considerably lengthened the navigation season in this western section of the route during recent years. Supplies carried by coasting vessels to settlements along the Arctic shore of Siberia often are shipped northward by river transportation from cities along the Trans-Siberian Railroad.

The railroad that connects Murmansk with the coreland serves important mining districts in the interior of the Kola Peninsula and lumbering areas in Karelia, adjoining Finland. The Peninsula is one of the more important mineralized areas of the USSR, producing ores of nickel, copper, iron, aluminum, and other metals, as well as phosphate for fertilizer. The Kola Peninsula and Karelia are physically a prolongation of Fennoscandia, being located on the same ancient, glacially scoured, granitic shield which underlies most of the Scandinavian Peninsula and Finland. Karelia, with its thousands of lakes, short, swift streams, extensive softwood forests, timber industries, and hydroelectric power stations, bears a close resemblance to adjoining areas of Finland. In fact, some of it was annexed from Finland during World War II.

Important energy resources, including good coking coals, petroleum, and natural gas, are found in the basin of the Pechora River east of Arkhangelsk. The principal coal-mining center is Vorkuta (city proper, 100,000), infamous for the labor camp where great numbers of political prisoners died during the Stalin regime. Coal from the Pechora fields moves by rail to Leningrad and to a large iron and steel plant at Cherepovets north of Moscow, which processes iron-ore concentrates from the Kola Peninsula and ore from the Kursk Magnetic Anomaly to make steel that is mainly used in the industries of Leningrad.

East of the Urals the swampy plain of western Siberia north of the Trans-Siberian Railroad has been witnessing a surge of petroleum and natural gas production. The West Siberian oil and gas fields are extremely large, and they are now the largest Soviet producers. As is generally the case in the USSR, the oil and gas deposits do not coincide in

Figure 11.14 *Laying an oil pipeline in the Northern Lands. Immense quantities of steel have been required for thousands of miles of oil and gas pipes that have been laid in the Soviet Union since World War II.*

location; in this instance the main gas fields lie well to the north of the main oil fields. Extraction and shipment take place under frightful difficulties caused by severe winters, permafrost, and swampy terrain. Huge amounts of steel pipe (Fig. 11.14) and pumping equipment have been required to connect the remote wells with markets in the coreland, East Central Europe, and western Europe. One large pipeline, built in the 1980s with west European aid, supplies natural gas to West Germany and other west European countries.

Still farther east the town of Igarka, located about 425 miles (*c.* 680 km) inland on the deep Yenisey River, is an important sawmilling center accessible to ocean shipping during the summer and autumn navigation season. Beyond Igarka to the northnortheast lies Norilsk (city proper, 200,000), the most northerly mining center of its size on the globe. Rich ores yielding nickel, copper, platinum, and cobalt have justifed the large investment in this Arctic city, including the railway that connects the mines with a port area at Dudinka on the Yenisey downstream from Igarka (chapter-opening photo). Few settlements of special note occur north of the Arctic Circle between Norilsk and the Lena River, although east of the Lena a handful of scattered mining towns produce tin or gold.

In central and eastern Siberia the taiga forest extends southward beyond the Trans-Siberian Railroad. A vast but sparsely settled hinterland is served by a few cities spaced at wide intervals along the railroad: Krasnoyarsk (city proper, 875,000) on the Yenisey River, Irkutsk (city proper, 600,000) on the

Angara tributary of the Yenisey near Lake Baikal, and others. From east of Krasnoyarsk a branch line of the Trans-Siberian leads eastward to the Lena River. After several years of construction under very difficult conditions, this line has now been continued to the Pacific under the name Baikal-Amur Mainline, or BAM (Fig. 11.15). The chief purposes of the new railroad are to open important mineralized areas east of Lake Baikal and to lessen the strategic vulnerability caused by the close proximity of the Trans-Siberian to the Chinese frontier.

The railroad from the Trans-Siberian to the Lena River passes through Bratsk (city proper, 250,000), the site of a huge dam and hydroelectric power station on the Angara River. Lake Baikal, which the Angara drains, lies in a mountain-rimmed rift valley and is the deepest body of fresh water in the world, being more than a mile deep in places. Fed by hundreds of streams, it is said to contain one sixth of all the fresh water in the world, and it supplies an enormous and relatively constant volume of flow to the Angara. Some years ago, Lake Baikal became the center of a controversy because of pulp-mill wastes being discharged into the lake and possibly jeopardizing the unique plant and animal life. The lake contains hundreds of species not known to exist elsewhere on the globe. Like the United States and other industrialized countries, the Soviet Union has nationwide problems of environmental pollution caused by rapid economic development without due regard for environmental effects. As in the United States, numerous laws designed to protect the environment have been enacted, but progress in the environmental field has been spotty despite official

concern, a great deal of publicity, and widespread debate. In both countries environmental cleanup is hindered by the massive cost of repairing past damage and a reluctance to increase the future expense of producing goods (and hence the cost of goods to consumers) by stringent enforcement of pollution-abatement measures. Thus far, the Soviet Union appears to have made less progress than the United States in environmental protection and cleanup, although both nations have pressing problems in this regard.

In the development of Siberia, great emphasis is being placed on hydroelectric installations along the major rivers. The largest ones thus far are two on the Yenisey (at Krasnoyarsk and in the Sayan Mountains) and another two on the Yenisey's large tributary, the Angara (at Bratsk and Ust Ilim). Additional installations on these immense rivers are planned. The power stations at the Siberian dams are the biggest in the USSR and among the largest in the world. They supply great quantities of cheap electricity, much of which is consumed by large, highly mechanized industries that are voracious users of power. Prominent examples include cellulose plants that process Siberian timber and aluminum plants that make use of aluminum-bearing material (alumina) derived initially from various ores in Siberia, the Urals, and other areas. The aluminum and wood-industry complexes near the Siberian dams are among the most impressive of their kind. The Soviets claim that the Bratsk aluminum plant is the world's largest.

Important gold-mining centers are scattered throughout central and eastern Siberia. Although

Figure 11.15 *Linkup of tracklayers from east and west on the new Baikal-Amur (BAM) railroad, September 29, 1984. The two ends of the line met in the mountainous taiga between Lake Baikal and the USSR's Pacific margin. Opening of the entire line to traffic by the laying of a symbolic "golden rail" marked the completion of 10 years of work under difficult climatic and terrain conditions.*

production figures are not definitely known, this part of the Soviet Union is considered to be one of the major gold-mining areas of the world. Mining is mainly centered in the basins of the Kolyma and Aldan rivers and in other areas in northeastern Siberia. In Stalin's time, the Kolyma mines, located in one of the world's harshest climates, developed an evil reputation as death camps for large numbers of political prisoners who were forced to work there. Large coal deposits are mined along the Trans-Siberian Railroad at various points between Irkutsk and the Kuznetsk Basin. Some of this coal is used to process bauxite or nephelite into alumina, which is then transferred for final reduction to aluminum metal at the plants powered by electricity from the Angara and Yenisey dams. East of Irkutsk and Bratsk, varied minerals are secured in small to medium-sized quantities from widely scattered mining areas tributary to the railroad. The new BAM is expected to greatly increase this flow. Huge coal reserves in many areas between the Yenisey and the Pacific have scarcely begun to be exploited. In the 1950s, discoveries of diamonds were made in the eastern part of the Central Siberian Uplands. Soviet sources report that the deposits, found in kimberlite ''pipes'' similar to those of South Africa, are among the largest in the world. The USSR has become a major producer and exporter of both gem diamonds and industrial diamonds.

Aside from irregularly distributed centers of lumbering, mining, transportation, and industry peopled mostly by Slavs, the Northern Lands are mainly occupied by non-Slavic peoples who make a living by reindeer herding, trapping, and fishing and, in the more favored areas of the taiga, by cattle raising and precarious forms of cultivation. The domesticated reindeer is an especially valuable source of livelihood for the tundra peoples, providing meat, milk, hides for clothing and tents, and draft power. The Yakuts, a Mongoloid people inhabiting the basin of the Lena River, are among the most prominent of the non-Slavic ethnic groups in the Northern Lands. Their political unit, the Yakut ASSR (Yakutia) is larger in area than any Union Republic except the Russian SFSR, within which it is contained. However, it has only about 800,000 people inhabiting an area of 1.2 million square miles (3.1 million sq. km). Actually, the Yakuts do not form a majority in their republic, being outnumbered by immigrant Slavs. The republic's capital city of Yakutsk (city proper, 175,000) on

the middle Lena River has road connections with the BAM and Trans-Siberian railroads and with the Sea of Okhotsk. River boats on the broad and deep Lena provide a connection with the Northern Sea Route and the southern rail system during the warm season. Yakutia will figure prominently in future mineral development, as it has large reserves of natural gas, coking coal, and other minerals.

Although the efforts of the Soviet government to develop the Northern Lands have achieved successes in lumbering, hydroelectricity, fishing, transportation, and a limited amount of manufacturing, it seems probable that these lands will continue to be a thinly settled frontier for a long time to come. Considerable experimentation has been carried on with quick-growing and frost-resistant crops and with hothouse culture and forced plant growth, but it seems doubtful that these can form the basis of a significant agriculture except in very limited areas. At present there is no disposition on the part of Soviet planners to stimulate any large flow of population to these lands, although selected enterprises will continue to be developed. Even these enterprises, most of which will be located in Siberia, are apt to have problems in maintaining a stable labor supply. Living conditions in the east are generally less attractive, and living costs are higher, than in areas west of the Urals. The climate is harsher; housing, heating, and transportation are more expensive to provide and maintain; the food supply is less varied and more costly; and social services and amenities are less adequate. Incentives in the form of higher pay and fringe benefits have not been sufficient in many cases to counterbalance the disadvantages of Siberian life, although increased bonuses paid by the government during recent years have apparently had some success in slowing the tendency of workers to leave their jobs east of the Urals and seek employment in the more developed west. Even so, the Siberian Northern Lands will continue to have an extraordinarily scanty population—currently smaller than that of New York State, within an area larger than the United States.

There are various resemblances between the USSR's Northern Lands and those of Canada, and the experiences of the two nations in developing their sparsely inhabited northern territories will continue to afford interesting comparisons in the coming decades.

PART IV

THE MIDDLE EAST

CHAPTER 12

The Middle East as Culture Focus and Problem Area

Prevalence of the Muslim religion is a major Middle
Eastern characteristic. The photo shows Islamic
pilgrims on the Plain of Arafat at Mecca. Ceremonies
here are a central part of the annual haj (pilgrimage).

W HAT and where is the Middle East? This part of the world has generated so much controversy that its name is known everywhere, but how to define and bound it is something of a puzzle. The name has become identified with regional traits that overlap each other spatially in the heart of the region but do not correspond well at the margins. Hence, any boundary must be arbitrary, and the one used in this text involves a good deal of overlap with areas also assigned to other world regions (Fig. 1.3). Not all ''Middle Eastern'' traits apply to the whole extent of the region as defined here for a teaching purpose,

and some traits extend well beyond the region. This makes for a certain untidiness, but it does not invalidate the concept of a large and tolerably cohesive area that has become identified with a ''crossroads'' location, aridity, oil wealth, Islamic culture, Arabic language, early contributions to civilization, and a recent history of ferocious strife. In this chapter, we examine the Middle East as a region stretching for about 6000 miles (9700 km) across North Africa and southwestern Asia from Morocco and Western Sahara on the Atlantic to Pakistan in the Indian subcontinent, and reaching for 3000 miles north-south from Turkey, Iran, and Afghanistan to Sudan, Ethiopia, and Somalia (Figs. 12.1 and 12.2). Thus defined,

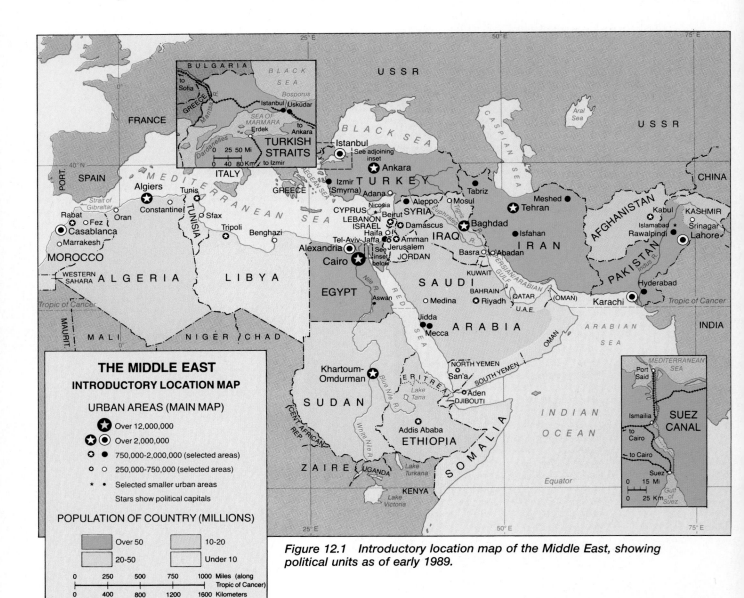

Figure 12.1 Introductory location map of the Middle East, showing political units as of early 1989.

the region incorporates some 28 countries (Table 12.1), occupying an area of 7 million square miles (18 million sq. km) and inhabited by about 492 million people in mid-1988. In the chapters dealing with the Middle East, the authors have elected not to discuss certain areas that lie along the edges of the region and have many Middle Eastern characteristics. The principal examples are Soviet Middle Asia, the region called Xinjiang (Sinkiang) in western China, and northern sections of western tropical Africa.

The title of this chapter asserts that the Middle East is a ''culture focus'' and a ''problem area.'' The term *culture focus* is meant to suggest a region that has played a central role as an innovator and diffuser of culture traits and that has also been active as a recipient of culture and as an interface of distinctive cultures existing side by side. Much of the world's cultural geography traces back ultimately to innovations developed in, and diffused from, certain areas in the heart of the region. Such places—for example, ancient Mesopotamia (now in Iraq)—are known to cultural geographers as *culture hearths*. The crossroads location of the Middle East at the junction of Europe, Asia, and Africa favored not only the dissemination of culture but also the receipt of culture, often brought in by invaders.

Frequent invasions of Middle Eastern areas by outsiders or by other Middle Easterners is a long-standing characteristic of the region that has continued in our own time. Recent major invasions of Af-ghanistan by the Soviet Union (1979), or Iran by Iraq (1980), and of Lebanon by Israel (1982) are cases in point. Such events contribute to the Middle East's worldwide reputation as a major problem area—that is, an area beset by problems that are very serious, recurrent, difficult to solve, and of great concern to outside nations. There are deep-seated environmental and economic problems throughout the region, and these will be touched on many times in the chapters dealing with the Middle East. But the problems that draw the most attention are the animosities and flareups that generate civil strife, terrorism, invasion, warfare, changes of government, shifts of territory, economic distress, and the flight of millions of refugees. One of the major purposes of Chapters 12 and 13 is to make these destructive events more intelligible by presenting the geographic context within which they occur.

PHYSICAL AND RELIGIOUS SETTING

The margins of the Middle East are mainly occupied by oceans, seas, high mountains, and deserts: to the west the Atlantic Ocean; to the south the Sahara Desert, the highlands of East Africa, and the Indian Ocean; to the north the Mediterranean, Black, and Caspian seas, together with mountains and deserts lining the southern land frontiers of the Soviet Union; to the east the Thar Desert of India and the great mountain knot of Inner Asia. The Middle East

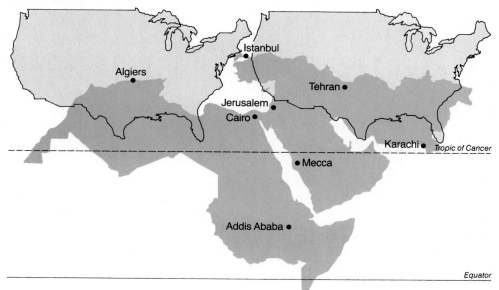

Figure 12.2   The Middle East compared in latitude and area with the conterminous United States.

TABLE 12.1 MIDDLE EAST: BASIC DATA

Political Unit[a]	Area (thousand sq. mi)	Area (thousand sq. km)	Estimated Population (millions, mid-1988)	Estimated Annual Rate of Natural Increase (% 1988)	Estimated Population Density (per sq. mi)	Estimated Population Density (per sq. km)	Infant Mortality Rate	Urban Population (%)	Cultivated Land (% of total area)	Per Capita GNP ($U.S.: 1986)
Arab League States										
Egypt	386.7	1001.4	53.3	2.8	138	53	93	45	2	760
Sudan	967.5	2505.8	24.0	2.8	25	10	112	20	5	320
Somalia	246.2	637.7	8.0	3.1	32	13	147	34	2	280
Djibouti	8.5	22.0	0.3	2.5	35	14	132	74	0	740 (1984)
Libya	679.4	1759.5	4.0	3.1	6	2	74	76	1	7500 (1985)
Tunisia	63.2	163.5	7.7	2.2	122	47	71	53	29	1140
Algeria	919.6	2381.7	24.2	3.2	26	10	81	43	3	2570
Morocco	172.4	446.4	25.0	2.6	145	56	90	43	19	590
Lebanon	4.0	10.4	3.3	2.0	825	317	51	80	29	1775 (1985)
Syria	71.5	185.1	11.3	3.8	158	61	59	49	31	1560
Jordan	37.7	97.7	3.8	3.6	101	39	56	59	4	1540
Iraq	167.9	434.8	17.6	3.5	105	40	86	68	13	2300 (1985)
Saudi Arabia	830.0	2149.7	14.2	3.3	17	7	85	72	1	6930
Kuwait	6.9	17.8	2.0	2.9	290	110	18	80	3	13890
Bahrain	0.3	0.7	0.5	2.8	1577	609	32	81	3	8530
Qatar	4.2	11.0	0.4	2.7	96	37	42	86	7	12520
United Arab Emirates	32.3	83.6	1.5	2.6	46	18	38	81	0.2	14410
North Yemen	75.3	194.9	6.7	3.3	89	34	175	15	7	550
South Yemen	128.6	333.0	2.4	3.3	19	7	116	40	1	480
Oman	82.0	212.5	1.4	3.3	17	7	117	9	0.3	4990
Totals	4884.2	12649.2	211.6	3.0	43	17	92	46	4	1915
Other Units										
Turkey	301.4	780.6	52.9	2.2	176	68	95	53	35	1110
Cyprus	3.6	9.2	0.7	1.1	203	79	12	62	47	4360
Iran	636.3	1648.0	51.9	3.2	82	32	113	51	9	3770 (1984)
Afghanistan	250.0	647.5	14.5	2.4	58	22	183	15	12	195 (1984)
Pakistan	310.4	803.9	107.5	2.9	346	134	121	28	25	350
Israel	8.0	20.8	4.4	1.6	566	217	11	89	21	6210
Ethiopia	471.8	1221.9	48.3	3.0	102	40	118	10	11	120
Western Sahara	103.0	266.8	0.2	2.5	2	1	–	–	0	–
Totals	2084.1	5398.7	280.4	2.8	135	52	115	34	16	1200
Grand Totals	6968.3	18047.9	492.0	2.9	71	27	105	39	8	1500

[a]Figures for Egypt, Syria, Jordan, and Israel are based on the extent of territory controlled by each of these countries prior to the 1967 Arab-Israeli War. Egypt was not an active member of the Arab League in mid-1988, having been suspended from membership following the Egyptian-Israeli (Camp David) Accords of 1979. Not shown in the table are five small enclaves still held by Spain along the Mediterranean coast of Morocco. Also not shown is the area called Kashmir, disputed among India, Pakistan, and China. Kashmir is marginally Middle Eastern in various ways. Its area is about 86,000 square miles (223,000 sq. km), with a population of perhaps 10 million. Figures for Pakistan do not include the Pakistani-occupied section of Kashmir. Figures for the politically

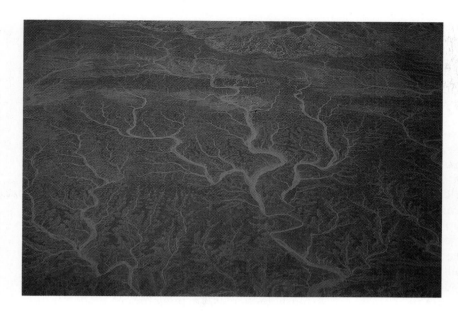

Figure 12.3 Africa and the Middle East contain the world's largest areas of dry land, much of which is so waterless that it supports little or no plant life. In past ages this land had greater rainfall, and branching river systems were formed that show up as fossil landscapes when viewed today from the air. The slide was taken from a plane flying across the Egyptian Sahara near the Nile Valley.

itself is composed mainly of arid or semiarid plains and plateaus, together with considerable areas of rugged mountains.

Although this region exhibits much variety from place to place, two important factors help to give it unity: (1) the dominance of dry climates (Figs. 2.18 and 12.3), and (2) the Islamic (Muslim or Moslem) religion (chapter-opening photo). The latter is the dominant religion by far in all the Middle Eastern countries except four: Israel, where Judaism prevails; Lebanon and Ethiopia, where ancient forms of Christianity are of major importance; and Cyprus, where the majority of the population adheres to the Greek Orthodox Church. Even these countries have many Muslims in their populations; indeed in Lebanon Muslims are a slight majority.

WORLD IMPORTANCE OF THE MIDDLE EAST

Throughout history the sparsely populated deserts and mountains of the Middle East, separating the humid lands of Europe, Africa, and Asia, have been a hindrance to overland travel between those regions. Yet circulation of people, goods, and ideas has taken place along certain favorable routes, and the scattered population centers of the region have had a history of vigorous interaction with the outside world and with each other. No areas of the world have been invaded so often, have seen so many empires rise and fall, or have been subjected to a greater variety of cultural influences and political pressures.

None have made more fundamental contributions to humankind. The earliest foundations of Western civilization were laid in the river valleys of ancient Egypt and Mesopotamia. In the Middle East the earliest cities arose, and the great monotheistic religions of Judaism, Christianity, and Islam were developed. Many of the plants and animals upon which the world's agriculture is based were first domesticated here: The list includes wheat, barley, apples, oxen, sheep, goats, and many others.

> We who eat roast beef [or lamb chops] . . . seldom wonder whom to thank for these gifts, other than the ultimate and divine Source of all bounty. It was the ancient hunters and earliest farmers of the Middle East who first rounded up these animals and tamed them for their use. Try to imagine yourself on foot, armed with a bow and arrow, a length of cordage, and a stone ax, either alone or accompanied by a dozen of your fellows, setting out to catch a wild bull in the forest or a wild sheep on the mountain crags.[1]

Most of the world's great empires have included portions of the Middle East. Some of these empires were indigenous, while others, such as the Roman Empire, were imposed from outside. Commencing in the 7th century a powerful Islamic empire, organized by Arabs, arose in the Middle East and evolved the most brilliant civilization of its day. Later it de-

---

[1]Excerpt from *Caravan: The Story of the Middle East* by (Carleton S. Coon, copyright 1951 and renewed 1979 by Holt, Rinehart and Winston, Inc., reprinted by permission of the publisher. Supplementary material has been inserted in brackets.

cayed. In the 16th century much of the territory it had controlled was conquered by the Ottoman Turks. Still later, when the Ottoman Empire was declining, Great Britain, France, and Italy seized control of sizable areas. For centuries, while the great powers of the Western world were industrializing and achieving high levels of living, most of the Middle East remained an underdeveloped area dominated by foreigners. But in the 20th century the peoples and countries of the region have reasserted their independence and importance. The Ottoman Empire was liquidated at the end of World War I, and the Italian Empire following World War II. Since World War I the Middle Eastern possessions of Great Britain and France have gained independence, as have the possessions of Spain along the Atlantic margin of the region except for some small enclaves still in Spanish hands on the Moroccan side of the Strait of Gibraltar.

As European colonialism has gradually vanished from the Middle East, many political trouble spots have developed. The region lacks political stability, a fact partly associated with the illiteracy and poverty of many of its people. Several international crises have been precipitated by Middle Eastern questions since World War II. These reflect the growing importance of the region in world affairs. One of the main factors is the rich oil deposits of the Persian/Arabian Gulf area and the Sahara Desert. France, Italy, West Germany, and other European countries import large quantities of oil from the Middle East, and various other countries in the world—for example, Japan—also rely heavily on Middle Eastern oil. The United States is much less dependent on oil from this source, although American companies pioneered oil development within much of the region and, like British, French, and Dutch companies, still have important interests there. The nationalization of the oil industry by Middle Eastern governments has by no means ended their technological and marketing dependence on outside companies. Even though a relatively minor share of American oil consumption now comes from the Middle East, the region's oil continues to be a vital strategic concern of the United States because that country's allies are so dependent on Middle Eastern oil sources.

A traditional route of sea transportation, the Mediterranean-Asiatic route, crosses the heart of the Middle East by way of the Suez Canal and Red Sea. It connects the nations of the Atlantic community with the Persian/Arabian Gulf oil fields, the eastern coasts of Africa, the Indian subcontinent, eastern Asia, and Oceania. Important air routes span the region, focusing on Cairo, Tel Aviv, Beirut, Algiers,

Karachi, Tehran, other major metropolises, and many smaller centers (Fig. 12.1). Long-distance rail and highway transportation are poorly developed, however. Today the importance of the region as a transit land lies principally in its traffic by sea and air, although in earlier times its caravan routes carried small but valuable amounts of intercontinental land traffic.

Although the age of European colonialism in the Middle East has closed, the great powers of the world continue to take a very active interest in the region, with its abundant oil, its strategic "crossroads" location, and its host of young nations that are modernizing economically and doing all they can to assert themselves politically. Jealous of their new liberties, and very conscious of their long histories as civilized lands, these countries are determined to exert increasing control over their own destinies and to push forward with programs of modernization which they themselves direct. In this enterprise, some nations are aided by large oil proceeds, but in general the region is handicapped by a natural-resource base that is seriously inadequate for the needs of modern societies. Some salient features of the Middle Eastern environment are discussed in the following section.

DOMINANT FEATURES OF THE ENVIRONMENT

On the whole, nature has not been kind to the Middle East. With a few conspicuous exceptions, this region is notably deficient in natural resources.

DOMINANCE OF DRY CLIMATES

For the Middle East as a whole, the most critical resource deficiency is lack of moisture. Most of this region is part of the Dry World—a vast belt of deserts and dry grasslands extending across Africa and Asia from the Atlantic Ocean nearly to the Pacific. At least three fourths of the Middle East has an average yearly precipitation of less than 10 inches (25 cm)—an amount too small for most types of nonirrigated agriculture under the prevailing temperature conditions. Some fair-sized areas bordering the Mediterranean Sea have 20 to 40 inches of precipitation, most of which falls during the cool season and thus is not available for growing the many crops that require the higher temperatures of the summer months. Rainfall sufficient for unirrigated summer cropping is concentrated in areas along the southern or northern margins of the region, most notably the highlands of Ethiopia and North Yemen; southern Sudan; strips of territory bordering the Black and Caspian seas; and limited areas in northern Pakistan.

Temperature Regimes

Middle Eastern climates exhibit the comparatively large seasonal and diurnal ranges of temperature which are characteristic of dry lands. Summers in the lowlands are very hot almost everywhere. Many places regularly experience daily maxima over 100°F (38°C) for weeks at a time. Shade temperatures of 130°F (54°C) or higher have been recorded in parts of the Sahara, Arabia, Iran, and Pakistan. Day after day a baking sun assails the parched land from a cloudless sky, and hot, dusty winds erode both the soil and human health. Only in the mountainous sections or in some places near the sea do higher elevations or sea breezes temper the intense heat of midsummer.

Lower temperatures of winter bring relief from the summer heat, and the more favored places receive enough precipitation to grow winter wheat or barley and a limited number of other cool-season crops. In general, Middle Eastern winters may be characterized as cool to mild. However, cold winters are experienced in the high interior basins and plateaus of Iran, Afghanistan, Pakistan, and Turkey. Only in the southernmost reaches of the region, such as the upper Nile Basin, do temperatures remain consistently high throughout the year.

The Middle Eastern deserts have exceptionally wide daily ranges of temperature—one of the most characteristic features of desert climates. Clear skies, the relatively low humidity of the air, and the lack of vegetation cover permit the sun's rays to heat the earth rapidly by day, but also promote a rapid escape of heat from the earth at night. Extreme cases are recorded of places in the Sahara that have witnessed a maximum temperature of over 90°F (32°C) and a minimum of below 32°F (0°C) within a 24-hour period.

Types of Climate

Maps showing climates classified by type reveal the Middle East to be predominantly an area of desert or steppe climate, with smaller fringing areas of mediterranean (dry-summer subtropical) climate, highland climate, and, in the extreme south, atypical tropical savanna climate. (See world climatic map, Fig. 2.18.)

The areas of *desert climate* include the great Sahara of northern Africa, the world's largest desert; the immense desert which covers most of the Arabian Peninsula and extends northward into Iraq, Jordan, and Syria; and smaller deserts in Iran and Afghanistan and along the lower Indus River plain in Pakistan. These areas have low and erratic rainfall averaging 5 to 10 inches (13–25 cm) a year or less. Occasional violent downpours alternate with rainless periods lasting for months or years at a time. Scattered areas of desert exhibit the familiar dunes of shifting sand, but rocky or gravelly surfaces are far more common. The characteristic vegetation is composed of widely spaced woody shrubs, with occasional tufts of grass.

Areas of *steppe climate*, while dry, are better supplied with moisture than the deserts because of greater rainfall, less evaporation, or a combination of these. Such areas are classed as semiarid rather than arid. An annual average of 10 to 20 inches (25–50 cm) of precipitation is typical, although deviations occur at both the upper and lower ends of this scale. The natural vegetation is more closely spaced than in the deserts and includes a larger proportion of grasses. Thus better forage is available for livestock, and the steppes are far superior to the deserts as grazing lands. However, the undependable nature of the rainfall renders crop-growing precarious except where irrigation water can be brought to the land.

The Middle Eastern areas of *mediterranean* or *dry-summer subtropical climate* are principally confined to the borderlands of the Mediterranean Sea in northwestern Africa and southwestern Asia. This climate zone encloses Lebanon and Cyprus, and the most productive farming areas of Morocco, Algeria, Tunisia, Turkey, Syria, Israel, and Jordan lie within it. Rainfall averaging 15 to 40 inches (40–100 cm) annually provides more moisture for crop growth than in the steppe climate, but the characteristic regime of rainy winters and dry summers precludes nonirrigated production of crops that require a combination of adequate moisture and high temperatures at the same season. Both the climate and associated agricultural activities conform generally to the pattern described for Southern Europe in Chapter 9.

Areas of *tropical savanna climate* are found near the equator in parts of Sudan, Ethiopia, and Somalia. This humid climatic type, occupying the outer margins of the rainy tropics, is not typical of the Middle East. It is characterized by high temperatures the year round, a complete absence of frost, an annual rainfall averaging 20 to 60 inches or more, and a dry season lasting most or all of the winter half-

TABLE 12.2 CLIMATIC DATA FOR SELECTED MIDDLE EASTERN STATIONS

Station	Latitude to Nearest Whole Degree	Elevation Above Sea Level (ft)	(m)	Type of Climate	Average Temperature Annual		January (or Coolest Month)		July (or Warmest Month)		Precipitation Annual Average (in.)	(cm)	% October–March
Tehran (Iran)	36°N	4002	1220	Steppe	61°F	16°C	39°F	4°C	86°F	30°C	8	21	75
Baghdad (Iraq)	33°N	110	34	Desert	73°F	23°C	50°F	10°C	95°F	35°C	6	16	83
Karachi (Pakistan)	25°N	13	4	Desert	79°F	26°C	64°F	18°C	88°F	31°C (June)	8	20	17
Cairo (Egypt)	30°N	67	20	Desert	72°F	22°C	57°F	14°C	84°F	29°C (July, Aug.)	1	2	91
Khartoum (Sudan)	16°N	1247	380	Desert	84°F	29°C	72°F	22°C	91°F	33°C (May, June)	6	16	2
Istanbul (Turkey)	41°N	164	50	Mediter-ranean	55°F	13°C	41°F	5°C	72°F	22°C	26	67	71
Algiers (Algeria)	37°N	194	59	Mediter-ranean	63°F	17°C	50°F	10°C	77°F	25°C (Aug.)	27	69	79
Jerusalem (Israel-Jordan)	32°N	2485	758	Mediter-ranean	63°F	17°C	48°F	9°C	75°F	24°C (Aug.)	19	49	94
Juba (Sudan)	5°N	1509	460	Tropical savanna	79°F	26°C	75°F 24°C (July, Aug.)		82°F	28°C (Feb., Mar., Apr.)	38	98	21
Addis Ababa (Ethiopia)	9°N	8038	2450	Tropical highland	63°F	17°C	59°F 15°C (July, Dec.)		68°F	20°C (May)	43	109	16

year. The tropical savanna climate is described more fully in Chapters 14 and 21.

Highland climates, varying in character according to altitude and distance from the equator, are found in the higher mountain areas of the Middle East.

Table 12.2 gives climatic data for some representative Middle East stations. Note the unbalanced precipitation regime at all stations listed.

ROLE OF MOUNTAIN WATER

Mountains play a vital role in the economy of the Middle East. Indeed, by furnishing the principal supplies of water for irrigation and household use, they make life possible for most of the inhabitants of this dry region.

The mountains of the Middle East are found in three principal areas (Fig. 12.4):

1. In northwestern Africa the Atlas Mountains of Morocco, Algeria, and Tunisia, reaching a maximum of over 13,000 feet (3962 m) in elevation, lie between the Mediterranean Sea and Atlantic Ocean and the Sahara.

2. A larger area of mountains occupies the northeastern quarter of the Middle East. It stretches across Turkey, Iran, Afghanistan, and Pakistan and includes the highest peaks in the region. The loftiest and best-known mountain ranges in Turkey and Iran are the Taurus, Anti-Taurus, Elburz, and Zagros mountains, which radiate outward from the rugged Armenian Knot in the tangled border country where Turkey, Iran, and the Soviet Union meet. The higher summits attain elevations of 10,000 to nearly 19,000 feet (c. 3000–6000 m). The mountains in this quarter of the Middle East culminate in the great Hindu Kush mountain

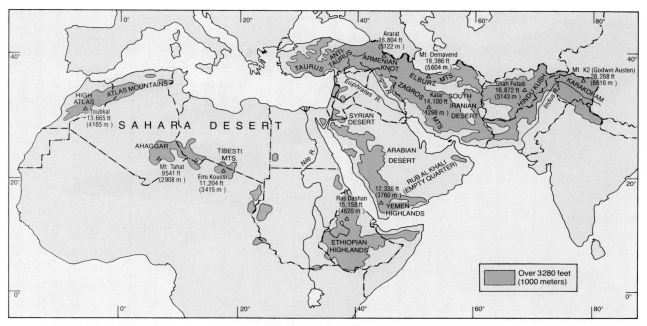

Figure 12.4   Principal highlands and deserts of the Middle East.

system of Afghanistan and Pakistan, which reaches over 25,000 feet (*c.* 7600 m), and the Karakoram Range, in the Pakistani-controlled part of Kashmir, where Mount K2 (Godwin Austen) rises to 28,268 feet (8616 m).

3. The third principal area of mountains is found in Ethiopia and North Yemen, bordering the southern end of the Red Sea. These mountains are generally lower than those just noted, although the extensive highland of Ethiopia has scattered summits higher than 13,000 feet (3962 m). Lower, discontinuous mountain ranges extend northward from Ethiopia and North Yemen along either side of the Red Sea.

Scattered mountain areas occur elsewhere in the Middle East, for example, the Tibesti Mountains in the central Sahara.

Water originating in mountain rainfall or snowfall often percolates for long distances underground and reaches the surface in springs or is drawn upon by wells or qanats in populated areas at the base of the mountains or beyond (Fig. 12.5). Water supplies of this kind are ordinarily sufficient for only a limited local development of irrigation farming. Most of the larger irrigated districts depend on water carried from the mountains by surface streams, a few of which are among the great rivers of the world. The Nile, Tigris-Euphrates, and Indus rivers provide water for many millions of farmers and urban dwellers in Egypt, Sudan, Iraq, and Pakistan. The earliest known civilizations arose in the valleys of these rivers, each of which has supported agriculture for several thousand years.

THE SCARCITY OF WOOD

Extensive forests existed in early times in the Middle East, but overcutting and overgrazing have almost wiped them out. Timber has been cut for fuel and construction faster than nature could grow it, and the young seedlings have been grazed off by sheep, goats, and camels, with the result that the forests have been unable to reproduce themselves. Lumber in commercial quantities can still be obtained from a few mountain areas, such as the Atlas region of Morocco and Algeria and the Elburz Mountains of Iran, but the total supply falls far short of the need.

SHORTAGES OF MINERALS

Many parts of the Middle East are further handicapped, especially with regard to industrialization, by a shortage of mineral resources. Good deposits of coal are rare—those in Turkey being the most important. The region is rich in petroleum and natural gas, but the largest deposits are confined to a few countries bordering the Persian/Arabian Gulf (Saudi Arabia, Kuwait, Iran, Iraq, United Arab Emirates, Qatar), plus Libya, Algeria, and Egypt in North Africa. Although scattered deposits of metals occur,

Figure 12.5 These shafts in Saudi Arabia mark the course of a tunnel used to carry irrigation and drinking water from an underground source by gravity flow. Such devices, called "qanats" (kanats), "karez," or "foggaras," are very common in some Middle Eastern countries, notably Iran, Afghanistan, Algeria, and Morocco. "The essential idea is that of a gently sloping tunnel, often along the radius of an alluvial fan, which extends upslope until the water table is tapped and emerges at the downslope end to supply an oasis. To give access to the tunnel, vertical shafts are dug at closely spaced intervals. The length of a qanat ranges from a few hundred yards to tens of miles, and the upper end may be several hundred feet below the surface. . . . By this means thousands of acres are irrigated and hundreds of villages receive their sole water supply." (George B. Cressey, "Qanats, Karez, and Foggaras," Geographical Review, Vol. 48, 1958, p. 27. Used by permission of the author and the American Geographical Society of New York. Students are urged to read the entire article, pp. 27–44.)

only a few are of much importance on a world scale. Good-sized salt deposits are fairly common, and phosphate rock, useful as a chemical and fertilizer material, is mined on a large scale in Morocco and on a lesser scale in Tunisia and other countries. Israel extracts potash, another chemical and fertilizer material, from the briny waters of the Dead Sea. But aside from oil and gas, which currently account for all but a tiny percentage of Middle Eastern mineral production by value, the outlook for mineral extraction seems rather poor. Vigorous prospecting and development may, of course, bring resources into play that are not now know or utilized. Currently, the most varied mineral production is found in Morocco and Turkey.

PEOPLES AND WAYS OF LIFE

The Middle East is principally inhabited by dark-haired peoples of the Caucasoid racial group. Especially important among the many languages spoken in the region are Arabic and other Semitic languages (including Hebrew), Berber and other Hamitic languages, Turkish, Persian, Pushtu (spoken in Afghanistan), and Punjabi and Urdu (spoken in Pakistan). Despite rapid urban growth in recent years, a majority of all Middle Easterners are still farmers. Nomadic herdsmen are a well-known element in the region, but they are dwindling in numbers and today represent a tiny minority of the population. The main areas of dense population are the irrigated val-

Figure 12.6 Characteristic traffic in an older section of Cairo, Egypt. Most Middle Eastern and other Third World cities have seen explosive growth in recent decades, and they are often choked with more traffic than they are equipped to handle. Cairo is infamous for its incessant din of blaring auto horns.

leys of the major rivers and the coastlands of the Mediterranean Sea. The few large cities are mostly political capitals or seaports; those with estimated metropolitan populations of 3 million or more (mid-1980s) include Cairo (Fig. 12.6), Tehran, Istanbul, Karachi, Alexandria, Ankara, Lahore, and Baghdad. The cities of the region, particularly the larger ones, have been increasing in population very rapidly due to declining death rates in the cities themselves and emigration from the countryside, where death rates also have been falling and the population has been growing much faster than the supply of arable land. For unemployed and poverty-stricken villagers the city has offered the possibility of employment, education, some medical care, and excitement and ferment to replace the monotony of village life and make poverty easier to bear. Unfortunately, the influx of villagers has generally outrun the ability of the city to provide jobs, housing, and services, and the result often has been unsightly and unsanitary shantytowns at the edge of the city.

Increased employment opportunities and a rising level of living in the Middle East will have to be provided primarily by manufacturing industries, despite the general shortage of raw materials and labor skills, and the governments of the region are making strenuous efforts to increase the number and output of modern factories. Most large factories are located in the larger cities, although many smaller factories and traditional handicraft industries are present in lesser cities and towns. Textiles, foods, and other relatively simple types of industries are the principal ones in most countries of the region. More intricate manufactures such as automobiles, typewriters, and electrical machinery must, with rare exceptions, be imported. Only in Israel and a handful of cities elsewhere has much of a start been made in the more sophisticated forms of industry.

The Middle East, like other world regions, is sufficiently varied in its ways of living that generalizations concerning the culture of the region as a whole nearly always contain a certain amount of distortion. Nevertheless, a surprising amount of similarity, at least on a superficial level, can often be discerned in the modes of life pursued in places as far apart as Morocco and Afghanistan. Because of Islam's powerful influence as a total way of life, it is possible to distinguish a "Middle Eastern culture" which is characteristic of the entire region in greater or less degree. This culture in its original form has been much affected by influences from Europe and America, particularly in the cities. However, much of the old culture still remains in rural areas and in the older sections of urban places.

The influence of Islam on Middle Eastern culture is so encompassing that any human geographic study of the Middle East must take it into account. It is necessary to understand not only the basic beliefs and practices of the dynamic Islamic faith but also the historical process of its origin and spread. Such knowledge is essential both for general understanding of the Middle East and for specific understanding of current world events. Middle Eastern conflicts now embroil the whole world directly or indirectly, and these situations generally have an important Islamic context. Only basic outlines can be explored in the brief discussion that follows, but further elaboration of particular situations can be found in Chapter 13.

ISLAM AND THE ISLAMIC WORLD[2]

Islam, Judaism, and Christianity are related monotheistic religions that arose in Palestine and adjacent Arabia. There is much overlap among the three in sacred figures, shrines, and precepts. Islam emerged as a separate religion in the 7th century of the Christian era and is based on the teachings of the Prophet Muhammad (Mohammed), an Arab trader. Today its believers number an estimated 860 million or more. The word *Islam* means "submission" to the will of Allah (God); a Muslim (Moslem) is "one who submits." Although this religion began in Arabia and is the main religion in the Arab world, the majority of today's Muslims live outside the Arab realm in such countries as Indonesia, Pakistan, Bangladesh, India, China, the USSR, Turkey, Iran, Afghanistan, and Nigeria. However, the holiest places for Muslims are the Arabian cities of Mecca and Medina, where the Prophet Muhammad lived. Some Islamic shrine cities lie outside Arabia, the principal one being Jerusalem.

Muslims believe that Allah began to reveal himself to humanity through Adam and continued through Abraham, Moses, and Jesus. All are praised in the Koran (Arabic: *Qu'ran*) the holy book of Islam. Muslims call Jesus the "prophet of mercy." Like the Jews, they regarded him as human and not divine, although they attribute miracles of healing to him. But for Muslims the last and greatest revelation came when the sayings now embodied in the Koran were transmitted from Allah to Muhammad by the Angel Gabriel in a series of religious experiences from 610–632 A.D. Some of the main features of the

Prophet's teachings and the Islamic faith are summarized in the following selection[3]:

> The Prophet was born in A.D. 570 at Mecca and died at Medina some sixty-two years later. His contribution was not so much a new religious concept as a new pattern of life. At the time of his birth, Arab society was virtually without moral precept or guidance. Muhammad conscientiously set out to correct what he thought were the worst features of that society. For example, he ended the practice of female infanticide, which before that time had been left to the judgment of the father. Through his teachings, the position of women greatly improved, as the number of wives was limited to four at a time and divorce was regularized. Muhammad laid down laws for what he considered the proper treatment of orphans, slaves, prisoners, and animals. He forbade bearing false witness, worshiping idols, or speaking ill of chaste women. Because the main drink of the Arabs then was a potent beverage made from the hearts of palms, which quickly reduced the user to a state of belligerent stupidity and led to many quarrels and sometimes to bloodshed, Muhammad forbade the use of intoxicating liquors. He also felt that the practice of gambling had weakened Arab society in his day and prohibited it too.
>
> In the eyes of true Muslims, the Koran is the word of God, or Allah, which was transmitted through the Angel Gabriel to Muhammad and then recited by him to the faithful. . . . The Koran is not only the Holy Book of Islam, . . . but it is the textbook used by most Muslims in learning to read; thus it standardizes written Arabic throughout the entire world. . . .
>
> The duties of a Muslim are built around the five Pillars of Islam. The first of these is the profession of faith: "There is but one God, and Muhammad is his apostle." The second is frequent daily prayer, which is largely supplication. . . . The most important prayer of the week occurs at noon on Friday, a public ceremony which must be observed by all Muslims. Muhammad emphasized the necessity for this gathering, and the need for the pilgrimage to Mecca, in order to bring Muslims together and weld them into a single community.
>
> The third Pillar of the Muslim faith is the giving of alms. This started as a voluntary act, became obligatory for a while, and in most parts of the Arab world today has again become voluntary. The fourth Pillar is fasting in the month of Ramadan, the month during which the Koran was first revealed to Muhammad. . . . During this month the true believer can eat no food and swallow no drink from dawn until sunset, unless he is ill or on a journey. . . .

---

[2]Much of the discussion under this heading is based on an exceptionally fine series of articles in *The Christian Science Monitor,* July 23–27, 1984, published under the general heading of "Islam behind the Veil."

[3]Richard H. Sanger, *The Arabian Peninsula* (Ithaca, N.Y., Cornell University Press, 1954), pp. 95–97. Used by permission of the publisher. "Mohammed" has been changed to "Muhammad" and "Moslem" to "Muslim" to conform to usage in this text.

Making the pilgrimage to Mecca is the fifth Pillar of Islam (see chapter-opening photo). Every true Muslim tries to do this once during his lifetime. . . . Special ceremonies peculiar to the pilgrimage include striking down the devil with stones in the Valley of Mina, sacrificing a sheep or camel there, entering into the prescribed holy places wearing a seamless garment, walking seven times around the Kaaba,[4] and devoting oneself to prayer.

In Islam's 3rd century (about the 9th century A.D.) scholars codified Muhammad's words as recorded by his friends and family into six recognized collections of tradition known as the Hadith. Together with accounts of Muhammad's deeds, they form the *Sunna* ("beaten path" or "custom"). A large majority of all Muslims accept them and therefore are known as Sunni or Sunnite Muslims. But there are various religious minorities within the Islamic faith, the largest being the Shiite or Shia Muslims. They live mainly in Iran, Iraq, Lebanon, Saudi Arabia, the smaller states along the Persian/Arabian Gulf, Turkey, and Pakistan. Shiites derive their name from the Arabic *Shi 'at Ali* ("follower of Ali") and claim that their leaders act in accordance with true Islamic authority which descended through Muhammad's son-in-law, Ali, and through Ali's son, Hussein. Hussein, facing greatly superior forces, was killed in battle in the year 680. His followers regard him as the ultimate martyr and commemorate his death with intense emotion.

Within the framework of obedience to the Koran, the followers of Islam vary markedly from one country to another. Principles are interpreted in the light of different traditions and cultures. Saudi Arabia is one of the countries where the most strict interpretations are used. The Koran does not say that women are required to wear veils, but it does urge them to be modest, and it portrays their roles as different from those of men. In Saudi Arabia, this is taken to require that a woman wear a floor-length, long-sleeved black robe and black veil in public (Fig. 13.7), that she not travel unaccompanied by a male member of her family, and that she not drive a car. In Egypt, by contrast, Muslim women are free to appear in public unveiled if they choose. However, in most Muslim countries conservative ideas about the role of women are still very strong: They should be modest, retiring, good mothers, keepers of the home. The Koran portrays women as equal to men in the sight of Allah, and in principle Islamic teachings guarantee the right of women to hold and inherit property. In practice women are usually subordinate to men and have fewer opportunities for education and for work outside the home, although this is gradually changing in the cities of more secularized states.

Recently a wave of Islamic fundamentalism has been sweeping the Islamic world. It rejects both the materialism of Western countries and the atheistic and materialistic doctrines of Communism. Its adherents advocate a restoration of Islamic purity under Koranic principles. This movement has affected some countries and governments much more than others, the most publicized case being Iran, where the Western-oriented government of the Shah was overturned in 1979 by a fundamentalist revolution headed by a politically powerful Shiite religious leader, the Ayatollah Khomeini.

THE EXPANSION OF ISLAM

After the death of the Prophet Muhammad in 632, his Arab followers spread the Islamic faith, mainly by conquest, through a vast area in Asia, Africa, and Europe. Before a century had passed, Arab armies had subjugated an empire reaching to Morocco and Spain on the west, to present-day Pakistan on the east and to Soviet Middle Asia and Transcaucasia on the north, and southward into the Sahara. Within the new empire a high civilization developed. Great architectural works were created, significant advances were made in agriculture, medicine, mathematics, and science; and a vigorous intellectual life flourished in such university centers as Cairo, Fez, Baghdad, Damascus, Cordova, Toledo, Seville, and Granada. At a time when western Europe was relatively stagnant, the Arab lands contained many of the foremost centers of culture and learning in the world.

But these lands did not enjoy an overall political unity for very long. The original empire, ruled from Medina, then from Damascus, and finally from Baghdad, commenced to disintegrate soon after it was established. Separate Islamic states arose in Spain, northwestern Africa, Egypt, and elsewhere. Beginning in the 11th century there were conquests of portions of the empire by Christian Crusaders and somewhat later by Turkic and Mongolian peoples. In the 15th century Christian armies of the Spanish kingdom, after centuries of intermittent warfare by Spaniards and Portuguese against Islam, completed the reconquest of the Iberian Peninsula. Subse-

---

[4]Editor's note: the Kaaba, a building in Mecca housing the venerated Black Stone, is the chief shrine of the Islamic world. It was a holy place long before Muhammad's time.

quently, most parts of the original Arab empire, weakened and disunited, were conquered by the Ottoman Turks. Still later, large areas came under control of European colonial powers. Since World War II these areas have achieved independence, and today they form a host of "developing" Islamic states.

CULTURAL GEOGRAPHY OF MIDDLE EASTERN CITIES[5]

In the Middle East religious life has focused in and been diffused from cities. Cities have been part of the Middle Eastern cultural landscape for thousands of years, first emerging in the southern Tigris-Euphrates Valley about 4000 B.C. With the spread of militant Islam in the 7th and 8th centuries A.D., a distinctive new landmark—the Friday-prayer mosque—was added to the center of the congested walled cities. Weekly prayer in the mosque became an observance bringing together all ranks of Islamic society in a common experience. Today the five daily calls by the muezzin for prayer by the faithful may be recorded and announced over a public address system, but they still provide a rhythm of life that has come down through the centuries.

The commercial life of the traditional Middle Eastern city centered in the twisting close-set lanes and merchants' stalls of the *bazaar* (central market). Goods and their sellers were arranged in a rough order from the center to the outer edge of the bazaar, with high-value goods and the most sought-after craftsmen having the more central locations. Around the mercantile core of the city were crowded residential quarters, each tending to represent a distinct group of people in terms of religion, ethnicity, occupation, or place of origin. Jews had separate quarters from Muslims, as did various groups of Christians. The separate neighborhoods, generally walled apart, tended to include a place of worship, a school, a public bath, a small market, and permanent workshops, especially for weaving. Houses faced inward from the narrow street to central courtyards of the walled family compounds that provided privacy and

security. In general, the elite lived near the city center for security and convenient access to the government, bazaar, and Friday-prayer mosque. Limits of space and unhealthful living conditions, together with restricted transportation for bringing produce from the hinterland, kept down the size of the traditional city.

Over the centuries Middle Eastern cities have tended to share a common historical experience, with occupation by various conquerors, a period of European colonial rule, and finally national independence after World War II. The last two periods have seen a great deal of Westernization. During the colonial age resident Europeans preferred to live in more spacious settings at the outer edges of the city, and, in time, this pattern was continued by the national elite. In recent times independent governments have followed Westernized building styles, with broad traffic arteries cutting through the old quarters, and large central squares near government buildings. This opening up of the cityscape has scattered commercial activity along the wide avenues, thus diluting the prime importance of the central bazaar as the focus of trade. However, the bazaar continues to be a major place to buy and sell, as well as to socialize, debate, and even conspire. For example, the Tehran bazaar played an important role in financing and coordinating the overthrow of the Shah of Iran in 1979.

As a consequence of increasing migration to the cities from the villages, governments have often built high-rise public housing and provided various social welfare services. In oil-rich countries with relatively small populations, an urban standard of living equaling that of affluent Western countries has been provided. But in more populous countries, the swelling rural-to-urban migration and declining mortality rates have overwhelmed resources, and deplorable shantytowns have sprouted. In Cairo many people have taken up residence in the tombs of a large cemetery called the City of the Dead.

CURRENTS OF CHANGE IN MIDDLE EASTERN VILLAGES

Historically, agricultural villages contained by far the majority of Middle Eastern populations. They produced food for the cities in return for goods and services. In this generally dry environment villages were placed near a reliable water source, with cultivable land near by. They tended to be composed of closely related family groups, with the land of the village often owned by an absentee landlord. Most often the villagers lived in closely spaced, flat-roofed

---

[5]The remainder of this chapter draws heavily on Paul Ward English, "Urbanites, Peasants, and Nomads: The Middle Eastern Ecological Trilogy," *Journal of Geography,* 66 (1957), pp. 55–59, also, English, "Geographical Perspectives on the Middle East: The Passing of the Ecological Trilogy," Chapter 7 in Marvin W. Mikesell, ed., *Geographers Abroad: Essays on the Problems and Prospects of Research in Foreign Areas* (University of Chicago, Department of Geography Research Paper No. 152, 1973), pp. 134–164; and Carleton S. Coon, "Point Four and the Middle East," *Annals of the American Academy of Political and Social Science,* 270 (1950), pp. 88–92.

houses made of adobe or mud brick. Production and consumption focused on a staple grain, with sheep and goats providing milk products and wool. Villagers shared common ties of blood and religion. They observed the same customs, and their work revolved in a timeless fashion around a repetitive cycle of agricultural activities.

However, villagers never lived in total isolation from urban areas. There were continuing trade relations. The landlord from the city dealt with the village sharecropper, often through his local overseer, and young villagers served in the military. One of the most cosmopolitan influences was the *haj,* the once-in-a-lifetime pilgrimage made by Muslims to the holy city of Mecca. This annual event widened the horizons of village pilgrims who mingled with people from afar. From the mid-18th century onward, village life was increasingly exposed to outside influences. Contacts with European colonialism brought significant economic changes. Cash crops were introduced and modern facilities were developed to ship them. Improved and expanded irrigation brought more land under cultivation. European capital financed much of this development. Recent agents of change in newly independent states have been government doctors, government teachers, and agents of land reform. Products of modern technology such as radios, bicycles, and sewing machines have modified old patterns of living. The young and more ambitious have been drawn to urban areas. Such changes have tended to undermine the traditional village values such as deep respect for religion and the elderly. Today urban influences are readily carried to villages by improved roads and communications, and village communities are becoming more integrated into national societies. One result has been a fundamentalist Islamic revival spread by modern communication techniques. Iran's Ayatollah Khomeini spread word of his fundamentalist revolution through the medium of cassette tapes. While in exile in France, he recorded messages for village religious leaders in Iran, and these helped pave the way for his triumphal return to Iran in 1979.

THE DECLINE OF NOMADISM

A romantic picture of the Middle Eastern nomad has been painted in such movies as "Lawrence of Arabia." More realistically, these tribesmen have been a classic instance of people adapting to life in a harsh environment. In a place of scarcity their lifestyle has accommodated to the meager resources at hand. For thousands of years they have followed the pattern of migratory grazing established by their ancestors: from waterhole to waterhole in deserts; from lowland winter range to highland summer range in mountains. The traditional life of nomads was extremely self-sufficient, with food, clothing, and shelter provided by their herds of sheep, goats, and camels. Wool and goat hair furnished material from which to weave tents, carpets, and clothing. The *jellaba*—the long white flowing robe worn by men— gave warmth in winter and during cold desert nights, as well as protection from the sun in summer. The woman's head-to-toe *chador* provided the modest dress required by the Koran. The nomads traded with villagers for needed food products such as wheat, barley, and dates, and with town-dwellers for tea, sugar, cloth, guns, and ammunition. In exchange they provided work animals, meat, milk, hides, and transport services. They were fiercely independent, disdainful of the sedentary life of settlements, and were redoubtable fighters who often were paid tribute by villagers fearful of nomad raiding parties.

Today only remnants of traditional nomadism can still be found. Modern forces of change have made it progressively less possible for this age-old mode of life to exist. Today trucks and airplanes cross the deserts and mountains, and in the face of this competition the camel as a vehicle for transport is no longer viable in many situations. Modern weapons in the hands of government soldiers have pacified raiding tribesmen, and governments have forced the nomads into fixed settlements where they can be taxed and kept under control. This pattern has become general throughout the Middle East.

CHAPTER 13

The Middle Eastern Countries: Common Threads, but National Distinctiveness

Water to irrigate dry land is vital to most Middle Eastern countries. The photo shows the Nile River near Aswan in Upper Egypt, with rocky desert and a grove of date palms in the background. Thousands of sailboats called feluccas *use the river.*

To comprehend such a large and complex area as the Middle East, it is necessary to evolve overall themes and generalizations such as those set out in the preceding chapter. Otherwise, comprehension is likely to be drowned in detail. But at the same time there is danger that too much homogeneity will be inferred for an area that in reality consists of a diverse array of countries, regions, and localities.

As a case in point, let us briefly compare Egypt and Libya, two countries that front the Mediterranean Sea and are located side by side in the North African sector of the Middle East. Libya is larger in area than Egypt, but both are among the larger Middle Eastern countries. Both are Muslim countries, although in Egypt a Christian minority comprises nearly one fifth of the population. In both countries Arabic is the national language. Both have national territories largely comprised of arid and nearly uninhabited plateaus in the Sahara, with over 95 percent of the people being crammed into not more than 5 percent of the national area. The two countries are basically subtropical in climate, both having extremely hot summers and mild winters, and both are nearly rainless except for a narrow strip near the Mediterranean where some rain falls in the cooler season of low sun. Libya benefits more from this winter rainfall than does Egypt, being closer to the Atlantic moisture source and being provided with some low mountains near the coast that generate orographic rain to supplement cyclonic rainfall from eastward-moving low-pressure systems off the Atlantic. Winter wheat and winter barley are grown in both Egypt and Libya, although not much is grown in Egypt without irrigation; and summer cropping in both countries depends totally on irrigation or drought-resistant trees and vines.

Thus far we have noted characteristics of Egypt and Libya that are rather typically Middle Eastern and would be shared in greater or less degree by many other countries and areas. But this is not to say that Egypt and Libya are practically two sides of a single coin or that either is closely similar to any other Middle Eastern country. Each has unmistakably a unique geographic personality that must be assessed if more than a superficial understanding is to be achieved. Egypt, although smaller in area than Libya, had 13 times as many people in 1988 (Table 12.1), which is in good part a reflection of the fact that Egypt has the Nile River to supply irrigation water, whereas Libya has no perennial surface streams. On the other hand, Libya had a per capita gross national product (GNP) that was ten times that of Egypt in a recent year, reflecting Libya's greater oil resources and smaller population. Egypt's historical experience has been infinitely more complex than that of Libya, which has few great monuments to ancient grandeur, little record of major cultural innovation, no great and historic metropolis like Cairo or Alexandria, no Suez Canal or Aswan High Dam. This is not to disparage Libya, but only to highlight the fact that Middle Eastern countries are very far from uniform, despite certain broad similarities.

As a further illustration, Iran and Saudi Arabia are quite extensive and arid Muslim countries, but they are greatly different from each other and from Libya or Egypt. All four countries have different international alignments and different outlooks on the world. One can easily become lost in such complexities, but it is possible in each country to single out key features and events that aid in understanding the country's essential character and its role in world affairs. This is attempted in the subsequent discussion, with special attention being paid to certain countries—such as Egypt, Iran, Saudi Arabia, Turkey, Lebanon, and Israel—that have an unusually momentous role because of some combination of such factors as large population size, oil riches, military strength, international alignments, and geographic position. To facilitate generalization, the countries of the Middle East will be grouped regionally under the headings of the Arab World and Israel; the Gulf Oil States; Turkey, Cyprus, and Afghanistan; and Africa's Eastern Horn. Pakistan, located at the eastern edge of the Middle East, is discussed with the Indian subcontinent in Chapter 15.

THE ARAB WORLD AND ISRAEL

The Arab world, stretching east-west from the Indian Ocean to Morocco and southward from the Mediterranean to the fringes of tropical Africa, is a major component of the Middle East. The peoples who inhabit it are racially and culturally diverse, but an overwhelming majority speak some form of Arabic and are thought of broadly as Arabs. Originally the Arabs were inhabitants of the Arabian Peninsula, but their language and associated Islamic culture were spread widely by conquest after their conversion to Islam.

Today most Arabs are in countries that belong to the League of Arab States, commonly known as the Arab League. This body, formed in 1945, is comprised (1989) of Algeria, Bahrain, Djibouti, Egypt, Iraq, Jordan, Kuwait, Lebanon, Libya, Mauritania, Morocco, Oman, Palestine Liberation Organization

(PLO), Qatar, Saudi Arabia, Somalia, Sudan, Syria, Tunisia, United Arab Emirates (UAE), North Yemen, and South Yemen. Quarrels and jealousies among these units have often hindered their functioning as a group. For example, Egypt was suspended from Arab League membership for a decade after it signed a peace treaty with Israel under the Camp David accords (1979). But such divisiveness is counteracted in some measure by a significant consciousness of overall Arab unity or solidarity, or at least some sense of community within a common Arab realm. Such feelings have several important roots. Partly they are the product of a common language, religion, and cultural heritage, coupled with pride in past achievements and a desire to regain past glories. Partly they are due to the realization of a need for united effort to improve social and economic conditions throughout the Arab lands. And partly they grow out of opposition to perceived economic and political domination by foreigners—a feeling most sharply focused on the state of Israel, which was founded under foreign auspices and is looked on as usurping territory that is rightfully Arab.

ISRAEL, JORDAN, LEBANON, AND SYRIA: UNEASY NEIGHBORS IN THE FERTILE CRESCENT

Since 1948, when Israel was formed from the British mandated territory of Palestine, antipathy to the new state has been a major rallying-point for the League of Arab States, which includes several "front-line" states that are Israel's immediate neighbors. To the east and north these neighbors include Jordan (formerly Transjordan), Lebanon, and Syria. All three, and Palestine, were held by Turkey before World War I. When the Ottoman Turkish Empire was liquidated following the war, Syria and Lebanon were mandated to France by the League of Nations, while Great Britain received a mandate for Palestine and Transjordan. Today Syria and Lebanon are republics, having secured independence by stages in the period 1943–1946. Transjordan received independence in 1946 as the Hashemite Kingdom of Jordan.

A Geographic Sketch of Israel

Palestine was divided among Israel, Jordan, and Egypt when Israel was constituted in 1948 as an independent homeland for the world's Jewish population. This partitioning resulted from warfare after British and United Nations' efforts at peaceful partitioning had failed. From 1948 to 1967 it gave Israel

some 77 percent of the territory of the Palestine mandate, including the coastal areas (except the Gaza Strip in the south, which was occupied by Egypt), the northern hill country of Galilee, the dry, thinly populated southern triangle called the Negev, and a portion of the city of Jerusalem in the Judean hills, plus a corridor leading to the city from the coastal plain (Fig. 13.1). Prior to the independence of the Israeli state, the Zionist movement had brought Jews back to Palestine since about 1880, although some had returned there even earlier. After independence the inflow greatly increased. A total of 687,000 Jewish immigrants, mostly from Europe or the Middle East, entered Israel between May 14, 1948, when the new state was officially proclaimed, and the end of 1951. This influx more than doubled the country's Jewish population. Since 1951 the annual immigration has been smaller and has fluctuated considerably from year to year. There has also been some migration out of the country; in fact, in 1986 outmigration exceeded inmigration. The population reached an estimated 4.4 million by 1988, exclusive of 1.6 million in territory conquered and occupied by Israeli forces in the Six-Day Arab-Israeli War of June 1967 and the subsequent "October War" in 1973. Somewhat over four fifths of Israel's present population within pre-1967 boundaries is Jewish, most of the remainder being comprised of Muslim or Christian Arabs.

Aided by large amounts of outside capital, primarily from the United States, the Israelis have developed their small republic as a modern Westernized state. Israel is still far from self-sufficient economically, despite strenuous efforts to lessen its dependence on foreign aid. It has great difficulty in financing both an extremely high military budget and extensive social programs. Its problems are compounded by levels of taxation that already are much higher than those of the United States and a shortage of most resources other than the skill and determination of the war-weary people.

One of the country's greatest achievements since independence has been the expansion and intensification of agriculture. Both cultivated land and irrigation in Israel proper have greatly increased. Israel produces the bulk of its food supply, and there is an export surplus of some agricultural commodities, primarily oranges and orange juice. Food production for domestic use concentrates on dairying, beef, poultry for eggs and meat, vegetables, and animal feedstuffs. The intensive, mechanized agriculture, oriented to nearby urban markets, resembles in many ways the agriculture of densely populated areas in western Europe. But in Israel a distinctive touch is added by the collectivized settlements called

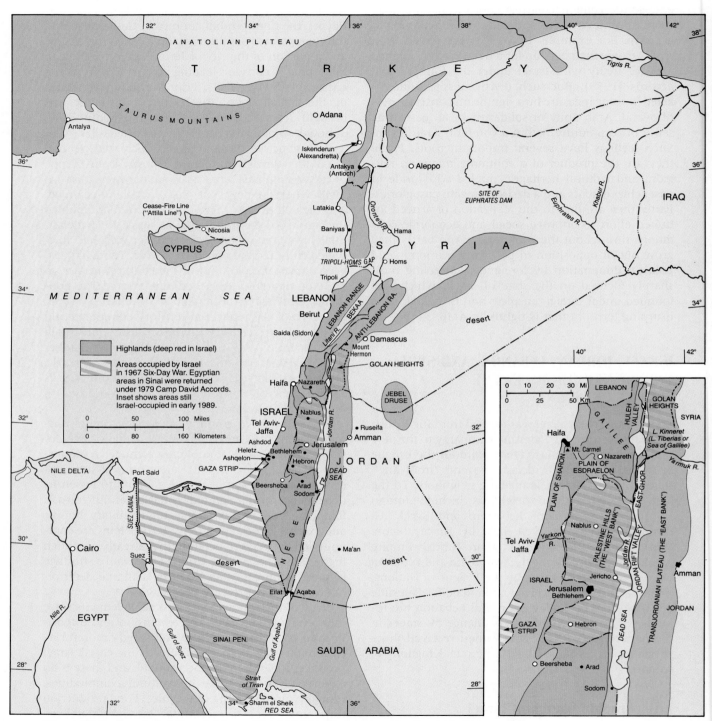

Figure 13.1 *Israel and its neighbors. Most highland areas are predominantly mountains except in Israel and Jordan, where hills predominate. International boundaries show the status prior to the 1967 Arab–Israeli War.*

kibbutzim (singular, *kibbutz*). Many of these lie near the frontiers and have defense as well as agricultural and industrial functions. Far more numerous and important in Israel's agricultural economy, however, are other types of villages. They include the small-holders' cooperatives called *moshavim* (singular, *moshav*) and villages of private farmers.

Existing agricultural development is concentrated in the northern half of the country, with its heavier rainfall (largely concentrated in the winter half-year) and ampler supplies of surface and underground water for irrigation. Annual precipitation averages 20 inches (50 cm) or more in most sections north of Tel Aviv and in the interior hills where Jerusalem lies, although it drops far below that figure in the deep rift valley of the Jordan River. But from Tel Aviv southward to Beersheba, the rainfall decreases to about 8 inches (20 cm), and in the central and southern Negev it drops to 4 inches or less. In the northern half of the country irrigation water from varied sources (streams, lakes, springs, wells, artificial reservoirs to catch runoff, and reclaimed sewage water) is used to intensify agriculture and overcome the summer drought. In the semiarid or arid south, on the other hand, surface water is scanty, and underground water, where found at all, is apt to be too saline for most crops. To bring agriculture to the Negev, Israel undertook the transfer of large quantities of water from the north by pipelines. The largest source is Lake Kinneret (Lake Tiberias or Sea of Galilee), which is fed by the upper Jordan River. Existing irrigation development in the Negev is most concentrated on fertile loess soils in the region around Beersheba. Local irrigation schemes exist in several parts of Israel, and these are coordinated by a national water planning agency. The Israelis have developed very advanced irrigation methods, such as the use of a technique called trickle or drip irrigation to conserve water and maximize yields of certain high-value crops by applying water individually to plants.

But Israel's main hope for continued support of its growing population lies in expanded industry, trade, and services, including tourism. Industry has increased greatly since independence, although not very much of Israel's industrialization can be based on its own mineral resources. Metal-bearing ores and mineral fuels are scanty, although potash, bromine, and other materials extracted from the Dead Sea have provided an important basis for expanded chemical manufacturing. Immigrants to Israel, and the country's own advanced educational system, have provided technical and scientific skills required for diverse industries, including many of a high-

technology character. The cutting and polishing of imported diamonds is one example of an industry that has capitalized on the skills of immigrants. Diamonds comprised 27 percent of all reported exports in 1987 and were the largest single item by value. Other prominent items in a rather diversified list of exports include machinery, electronics, and armaments. Under the stress of repeated warfare and strategic uncertainties, Israel has developed its own armaments industries, often of a very advanced character, and has become one of the world's more important sellers of arms to other nations. In manufacturing as in agriculture, the country is the most technically advanced in the Middle East.

Tel Aviv-Jaffa (1.4 million) on the Mediterranean coast is Israel's largest city and industrial center. The leading seaport and center of heavy industry (for example, steel milling and oil refining) is Haifa (425,000) in northern Israel. About three fifths of Israel's population (pre-1967 boundaries) is concentrated in Tel Aviv, Haifa, or the narrow coastal strip around and between them. The ancient town of Beersheba (city proper, 115,000) is the northern gateway, administrative center, and main industrial center of the Negev frontier region. To the south lies Israel's small Red Sea port of Eilat.

Arab-Israeli Disputes and Wars

Israel's relations with the Arab world continue to cloud the country's future. Central issues—all of which have much geographic significance—include (1) Israel's right to exist, (2) Israel's possession of occupied territories taken from Arab countries in warfare, (3) the rights of the Arab Palestinians who fled as refugees in wartime or were overrun by Israeli occupying forces, (4) the Jerusalem question, and (5) the Jordan River question.

FIVE WARS AND THEIR CONSEQUENCES

Bad relations between Israel and its Arab neighbors have exploded into full-scale war on five occasions since Israel's independence was declared on May 14, 1948. The geopolitics of the entire world, and the geography of the Middle East itself, have been profoundly affected for several decades by these conflicts. The character and results of each war are surveyed below, commencing with the 1948 war and continuing in sequence through the wars of 1956, 1967, 1973, and 1982.

The 1948 War. Immediately after the proclamation of Israel's independence in 1948, armed forces of Jordan, Egypt, Iraq, Syria, Lebanon, and Saudi Arabia invaded the new country, but they were defeated or stalemated by the Israeli army. The Arab invaders maintained that Israel had no right to exist as a state and was usurping territory that had to be recovered for the Arab population.

The 1956 War. This Israeli-Egyptian conflict, fought in Egypt's Sinai Peninsula, was a quick military victory for Israel, but the latter's occupation of Sinai was short-lived. Israeli forces invaded Egypt shortly before a British and French invasion of the Suez Canal Zone in the autumn of 1956. All three countries had bad relations with Egypt's President Nasser, who had nationalized the Suez Canal earlier in the year, thus antagonizing the British and French, and was regarded by the Israelis as their most powerful enemy in the Arab world. International pressure, led by the United States and the Soviet Union, quickly forced the invaders to withdraw from Egypt. This short war, often called the Suez War, climaxed a long buildup of hostility between Israel and Egypt. Israel emerged with enhanced military prestige and a reopening of the Gulf of Aqaba waterway from the Red Sea to Eilat. Egypt had previously closed both the Gulf and the Suez Canal to Israeli ships and cargoes, which continued to be barred from the Canal after the war.

The 1967 War. Often called the Six-Day War or June War, this struggle was precipitated in part by a further Egyptian closure of the Gulf of Aqaba. Israel fought on three fronts against Egypt, Jordan, and Syria. Again the Israeli triumph was rapid, but this time the outcome had a far more enduring significance. Israel occupied all of the Sinai Peninsula, the Gaza Strip, the West Bank area held by Jordan, and the Golan Heights section of Syria overlooking Israel's Huleh Valley in which the upper Jordan River flows. The war put the Suez Canal out of service, and it became a frontier between hostile armies. In 1979 Israel returned the Sinai to Egypt under the American-sponsored Camp David accords, but the Gaza Strip, West Bank, and Golan Heights have continued to be Israeli-held.

The 1973 War. This "October War" or "Yom Kippur War," fought by Israel against Egypt and Syria, was less clearly an Israeli victory than the previous wars, although the Israeli-occupied territory in the Golan Heights was somewhat expanded. In Sinai, the Egyptian army offered stouter resistance than before, and in a cease-fire negotiated under big-power auspices, came off with something resembling a draw. Israeli forces pulled back some miles from the Suez Canal at the end of the war, and the way was cleared for the Canal to be cleaned of wartime debris and reopened (in 1975). Then in 1979 Egypt's President Sadat and Israel's Prime Minister Begin, with the support of America's President Carter, ne-

gotiated the Camp David peace treaty under which the long-standing state of war between Israel and Egypt officially ended. Israel's rights as a sovereign state were recognized by Egypt, and Israel agreed to return occupied Sinai to Egypt, which was subsequently done. Disposition of the Jordanian and Syrian territories that had been occupied by Israel in 1967 was left to future negotiations in an ongoing peace process involving the entire question of Palestinian rights.

The 1982 War. This conflict was one facet of a larger war in Lebanon involving both civil strife and international intervention. Lebanon had become the main stronghold of the PLO, which used the country as a base for guerrilla operations against Israel. In September 1982 Israel launched an all-out invasion of Lebanon with the avowed intention of smashing the PLO and establishing security for the northern Israeli frontier. This move was hotly debated in Israel and was widely censured in the outside world. Syria, whose forces had previously occupied Lebanon's Bekaa Valley on an alleged peace-keeping mission for the Arab League, resisted the Israeli advance and gave support to the PLO. Both the Syrians and the PLO were routed, with heavy casualties. Syria suffered extremely heavy losses of aircraft and antiaircraft weapons that had been supplied by the USSR. The Israelis did not cross the border into Syria, and they refrained from attacking Damascus by air or long-range artillery. The Israeli push continued northward to the suburbs and vicinity of Beirut, and for a time Israel controlled the vital Beirut-Damascus highway through Lebanon's mountains. The Soviet Union soon replenished Syria's weaponry with shipments of more advanced types, accompanied by thousands of Soviet advisors and technicians, but further large-scale fighting between Israelis and Syrians did not develop. However, casualties continued to be inflicted on the Israelis by various combatants among the Syrians, the Lebanese, and what remained of the PLO. Weary of the sporadic sniping, firefights, and other resistance, and failing to achieve their political objective of a Christian-dominated Lebanese state linked to Israel by treaty, the Israeli forces largely withdrew in 1983, although Israel continued to exercise close surveillance over a "security zone" in southern Lebanon.

THE PALESTINIAN QUESTION

The creation of the state of Israel, the subsequent wars, and the Camp David accords left the future of the Arab Palestinians unresolved. Approximately 800,000 Arabs are Israeli citizens (1988), residing within the pre-1967 boundaries of Israel and enjoying much the same rights as other Israelis, but being very much a minority in a country that is 82 percent Jewish. Far larger numbers of Arabs deriving from the Palestine mandate and known collectively as Palestinians live outside of Israel proper, or even outside of the Israeli-occupied West Bank, Golan Heights, and Gaza. Prior to and during the fighting of 1948, over 500,000 Palestinian Arabs fled or otherwise departed from the new state of Israel to the neighboring Arab countries. Refugee camps for these displaced persons were established by the United Nations in Jordan, Egypt, Lebanon, and Syria. Subsequently, the number of refugees more than doubled through natural increase and illegal registration. Little was done to resettle them in permanent homes, and both the Arab governments and the refugees themselves continued to insist on the return of the refugees to Israel and the restoration of their properties there. New complexities were added as a result of the 1967 war, in which Israel took over the areas where most of the camps were located. Many persons in the camps again took flight, and an entirely new group of refugees was created by the departure of a sizable number of permanent residents from areas overrun by Israeli forces. Some of the latter subsequently returned to their homes, but many others either did not attempt to return or were denied permission by Israeli authorities. Further dislocations resulted from the October War of 1973.

Israel has been moving since 1967 to strengthen its grip on the occupied territories through security measures and the government-sponsored planting of new Jewish settlements, most of which are spread through the West Bank. For the most part these are not frontier farming settlements but are inhabited by relatively prosperous middle-class Jews who commute to jobs in Israel. Some of the motivation is ideologic, as the West Bank is composed of Judea and Samaria, regarded by many devout Jews as part of Israel's historic and inalienable homeland. The proliferation of new settlements has increased the bitterness between the Arab world and Israel, and it has driven a wedge into the Jewish population of Israel itself. Many Israelis are very opposed to further Israeli settlement in the occupied territories or to annexation of these territories to the Jewish state. If the latter were done, Israel would become a country approximately 60 percent Jewish, with an Arab minority having a higher birth rate than that of the Jews. Hence the Jews could eventually become the minority group. Commencing in 1987, the Israeli army used gunfire, beatings, and other harsh methods to quell Palestinian unrest in the West Bank and Gaza. These actions ordered by the Israeli government drew severe criticism in the outer world and roused much opposition in Israel itself.

As matters stand (early 1989), the scattered Palestinians have no territory where their political aspirations as a people can be realized. One solution that has been proposed is the creation of an independent Palestinian state composed of the West Bank and Gaza. Obstacles to this solution include (1) the presence of the large Jewish settlements on the West Bank and the furor within Israel that would accompany any attempt to remove them; (2) the insistence of Israel on rigorous guarantees of its own security; and (3) Israel's refusal to entertain the idea of having on its borders any independent Palestinian state headed by the PLO. In early 1989 the ultimate fate of the Palestinians remained unclear. The complex and vexing Palestinian question continued to be a major political issue for the entire world.

JERUSALEM AND JORDAN RIVER QUESTIONS

Since 1948 the status of Jerusalem (Fig. 13.2) has been intensely disputed between Israel and the Muslim world. The dispute originates in the fact that Jerusalem is a holy place for both Jews and Muslims (as well as for Christians). As a result of the first Arab-Israeli war, Jerusalem was partitioned between Israel and Jordan. Despite its perilous situation on the boundary with a hostile country, the Israeli-held section—incorporating most of the "New City"—was made the capital of Israel. Most of the sacred shrines, however, were in the "Old City" in the Jordanian sector. In the 1967 war, the Israelis took the Jordanian part of the city and since then have integrated it economically and politically with Israeli Jerusalem. However, no country has recognized Jerusalem as Israel's capital, and most of the foreign diplomatic community is quartered in Tel Aviv.

An American correspondent, David K. Shipler, imparts something of the city's timeless religious flavor in the following selection[1]:

Dawn comes gently to Jerusalem. Even before the sky begins to lighten, the thin wail of the muezzin's call to Moslem prayer sings across the city. It mingles with the mournful strains of Jewish prayer at the Wailing Wall. It reaches deeply into the darkened labyrinth of narrow, twisting alleys that make up the old walled city of Jerusalem; there, in the Church of the Holy Sepulchre, it mixes with the haunting chants of Armenian priests and acolytes at the Tomb of Christ.

· · · · ·

Jerusalem. Its name in Hebrew—Yerushalayim—means "City of Peace." In Arabic it is al-Quds, "The

---

[1]David K. Shipler, "In Search of Jerusalem," *The New York Times Magazine*, December 14, 1980, pp. 74, 77, 104. Copyright © 1980 by The New York Times Company. Reprinted by permission.

Figure 13.2 A portion of the walled Old City of Jerusalem, sacred to Jews, Christians, and Muslims. The prominent building at the right is the mosque called the Dome of the Rock, occupying the spot from which Muslims believe the Prophet Muhammad ascended to Heaven.

Holy." Since its first appearance in manuscripts as a Canaanite city-state in the Bronze Age nearly 4,000 years ago—and all through a bloody succession of conquerors and rulers from King David and King Solomon and the Kings of Judah, through the Babylonians, Macedonians, Egyptians, Seleucids, Greeks, Jewish Hasmoneans, Romans, Byzantines, Persians, Moslems, Crusaders, Mameluks, Ottoman Turks, British, Jordanians and now again the Jews—Jerusalem has been a city of angry piety, a city that has known no line between warfare and religion.

. . . . .

The three most sacred objects in Jerusalem are of ancient stone. The Church of the Holy Sepulchre is built partly around a rock believed by some denominations to have been the core of the hill called Calvary. On the Temple Mount, near Al Aksa Mosque and inside the ornate blue-tiled Mosque of Omar, or Dome of the Rock, is a massive outcropping from the earth, a stone that marks the intersection of Islam and Judaism. It is said to be the spot where Mohammed began his night ride to heaven. And part of it may also be the *Even Shetivah*, the rock that lay inside the Holy of Holies, the inner sanctum of the Jewish Temple. . . .

Still another dispute between Israel and its Arab neighbors concerns the use of water from the Jordan River. This famous (although comparatively small) stream flows southward in a deep rift valley and empties into the Dead Sea. After independence, Israel's government desired to use water from the river for irrigation and hydropower development, and expressed willingness to share the river with Jordan, across whose territory the lower river flows. However, Jordan, Syria, and Lebanon, which controlled the river's headstreams, objected strenuously to schemes for developing the Jordan which Israel, the United States, and the United Nations proposed. Israel nonetheless proceeded to implement its water diversion plans. In the 1967 war the Israeli army took over a section of Syria in which an important headstream of the Jordan rises, and it occupied the parts of northwestern Jordan lying west of the river (the "West Bank"). Jordan's most important irrigated district, known as the East Ghor, lies in the Jordan Valley east of the river and is watered by a canal from a Jordan River tributary, the Yarmuk.

Political and Economic Uncertainties in Jordan

Jordan, bordering Israel on the east, is mostly desert or semidesert. The main agricultural areas and settlement centers, including the main city and capital, Amman (800,000), are in the northwest. As it existed prior to the 1967 war with Israel, northwestern Jordan included two upland areas, the Palestine Hills on the west and the Transjordanian Plateau on the east, separated by the deep rift valley occupied by the Jordan River, Lake Kinneret, and the Dead Sea (Fig. 13.1). The valley is the deepest depression on the earth's land surface, being about 600 feet (183 m) below sea level at Lake Kinneret and nearly 1300 feet (400 m) below sea level at the Dead Sea (the bottom of the Dead Sea reaches −2600 feet). Although the annual precipitation increases northward to an average of more than 12 inches (30 cm) in the vicinity of Lake Kinneret, most of the valley bottom lies in deep rain shadow and receives less than 4 inches a year. But the bordering uplands, rising to 2000 feet (610 m) or more, and lying in the path of moisture-bearing winds from the Mediterranean, receive precipitation during the cool season that averages 20 inches or more annually within two north-south strips. It was within these belts, one in the Palestine Hills (the West Bank) and the other in the hilly Transjordanian Plateau (the East Bank), that most of Jordan's pre-1967 population was found. The largest percentage were villagers supporting themselves by cultivating winter wheat and barley, vegetables, and fruits, especially olives and grapes.

Prior to Israel's occupation of the West Bank in 1967, Jordan and Israel had roughly equal populations, but the present population under Jordanian control has shrunk to about 2.8 million, compared to 4.4 million in Israel proper. However, the rate of natural increase in Jordan—a country that is 93 percent Muslim and only 60 percent urban—is 3.6 percent a year (one of the world's highest rates) compared to Israel's 1.6 percent. The disparity between the two countries is even more striking in per capita GNP: $6210 in Israel and $1540 in Jordan in 1986.

Jordan's poverty is not apt to be significantly relieved except by a long, slow process of development. No oil or gas has been found, and other natural resources are scanty, although rock phosphate exists in sufficient quantity to provide a major export. The phosphate is shipped through Jordan's small Red Sea port of Aqaba, located on the Gulf of Aqaba adjoining Eilat in Israel. Some manufacturing is carried on, but most factories are small, uncomplicated enterprises such as flour mills, vegetable canneries, and soap factories, together with some textile, clothing, and shoe factories. A handful of larger and more complex enterprises exist, but the overall contrast with the large and technologically advanced industries of Israel is extreme.

Only about 4 percent of Jordan is cultivated, although much larger areas are used to graze sheep and goats. Rainfall is so scanty that only small areas can be cultivated without irrigation, and water available for irrigation is also very limited. Some food has to be imported, although irrigated vegetables (especially tomatoes), together with olives, citrus fruits, and bananas, provide substantial exports. Winter wheat occupies a greater acreage than any other crop, as it generally does in Middle Eastern countries.

Jordan's economic and political situation has always been precarious, but a variety of foreign nations have repeatedly given economic and military aid. Formerly the main source of aid was Great Britain; more recently large financial support has come from Saudi Arabia, and arms have been supplied by the United States. Jordan has been a front-line state in the Muslim world's conflict with Israel and hence has received funding from Saudi Arabia and other Middle Eastern oil states, but the country's central position vis-à-vis Syria, Iraq, Saudi Arabia, Egypt, and Israel has also given it strategic significance for Western nations interested in maintaining peace, opposing the spread of Soviet influence, and protecting the huge oil resources along the Persian/Arabian Gulf. Jordan has been very anti-Communist, and its relations with Syria and Iraq, both recipients of much Soviet aid, have been cool. Within the country King Hussein, supported by the native Bedouin, has withstood various crises related to the entry of great numbers of Palestinians, including the PLO. For a time Jordan was the chief operating base of the PLO, but in 1970–1971 the king expelled the armed Palestinian forces, who thereupon moved to Lebanon. Palestinians still comprise the majority of Jordan's population, not including those in the Israeli-controlled West Bank that was originally annexed by Jordan in 1950 and then was taken by Israel in the 1967 war.

Lebanon and Syria: A Vortex of Conflict

Israel's immediate neighbors on the north are Lebanon and Syria (Fig. 13.1). Since the beginning of the Lebanese civil war in 1975, world attention has been focused on the two countries to a degree seldom accorded areas so small. Lebanon is somewhat smaller than the state of Connecticut in area and has a population size (3.3 million) comparable to Connecticut's. Syria is larger, being very similar to North Dakota in area, and having about the same population size (somewhat over 11 million) as the state of Ohio. But most of Syria is very dry, and a large majority of its people are crowded into a more humid western strip about three or four times the size of Lebanon.

This mountainous tract bordering the eastern Mediterranean forms a vortex into which contending peoples and powers have been drawn for many centuries, but never with such a worldwide impact as in the past few years. Actually the idea of a geopolitical vortex could be expanded to the entire Fertile Crescent within which Lebanon and western and northern Syria lie. Northern Israel and northwestern Jordan also form part of it, as does the plain of the Tigris and Euphrates rivers (plain of Mesopotamia) in Iraq. The Crescent forms a famous cultivable strip between the Syrian Desert and the high mountains of southeastern Turkey and southwestern Iran. Through the centuries, from the beginning of recorded history, this semicircle of land, leading from the Persian/Arabian Gulf and the Iranian plateaus to the Mediterranean coast and southward toward the delta and valley of the Nile, was followed by the main caravan tracks from interior Asia to the Mediterranean and northern Africa. It also served as a pathway for marching armies—Assyrians, Hittites, Persians, Egyptians, Macedonians, Romans, Crusaders, Arabs, Mongols, Turks, and others came this way at various times. The countries of the Fertile Crescent have probably been overrun by foreign armies more times than any other group of countries in the world. This dismal record was recently continued in Lebanon (early 1980s) by large-scale fighting among armies of the PLO, Israel, and Syria. Added to fierce internal conflict among the Lebanese themselves, the warfare reduced tiny Lebanon to one of the more thoroughgoing shambles yet seen in the world's deplorable history of violence and destruction since World War II.

Israel's two northern neighbors exhibit a pattern of physical and climatic features broadly similar to that of Israel and Jordan. As in Israel, narrow coastal plains backed by highlands front the sea. But the highlands of Lebanon and Syria are loftier than those of Israel and Jordan; indeed the Lebanon, Anti-Lebanon, and Mount Hermon ranges are high mountains, reaching a maximum of over 10,000 feet (*c.* 5050 m) elevation in the Lebanon Range. The latter range, rising practically from sea level, is roughly as high as the Rocky Mountains when measured from base to summit. Thus to cyclonic precipitation brought by air masses off the Mediterranean is added a considerable fall of orographic rain and snow created by the mountain chains that parallel the coast and block the path of the prevailing westerly winds.

Lebanon and Syria conform to the mediterranean climatic pattern of winter rain and summer drought. The coastal plains and seaward-facing mountain slopes are settled by a rather dense population of villagers growing characteristic mediterranean crops such as winter grains, vegetables, and drought-resistant trees and vines. Much of Lebanon's agriculture is on terraces that climb the lower mountain slopes.

In Syria and Lebanon, as in Jordan, precipitation decreases toward the east and pastoralism increases. The Anti-Lebanon Range and Mount Hermon are drier than the Lebanon Range, which screens out moisture they would otherwise receive. These mountains, and the eastern slopes of the Lebanon Range as well, are used primarily for grazing. The Bekaa Valley, which has been a Syrian military stronghold since 1976, lies between the Lebanon and Anti-Lebanon ranges and is a northward continuation of the Jordan rift valley.

East of the mountains a semiarid zone covers central and northern Syria. In southeastern Syria it trends gradually into the Syrian Desert, most of which lies in Iraq and Jordan. The semiarid region, much of which has good soils, is Syria's main producer of cotton, wheat, and barley. Cotton is grown as an irrigated summer crop, while wheat and barley, primarily nonirrigated, are winter crops. The cultivated areas of interior Syria are not continuous but are separated by much land that is vacant or used for grazing. Cotton and wheat are grown in two main districts. One lies in the west along the Orontes River, the other along the northern edge of the country east of Aleppo. Irrigated districts growing cotton and grain are found along the valley of Syria's largest river, the Euphrates. A large earthen dam, built to supply irrigation water and hydroelectricity, impounds the Euphrates in north central Syria.

Syria is more industrialized than Lebanon, but neither compares with Israel industrially. Textile milling, agricultural processing, and other light industries predominate. Both countries are very deficient in minerals, although Syria has an oil field of moderate importance in the east.

Three of the five largest cities in the Fertile Crescent are located in Syria or Lebanon. All have been important for many centuries. Damascus (1.3 million), Syria's capital, is in a large irrigated district in the southwestern part of the country. The main city of northern Syria is Aleppo (city proper, 1.1 million), an ancient caravan center located in the "Syrian Saddle" between the Euphrates Valley and the Mediterranean. The old Phoenician city of Beirut (1.7 million), on the Mediterranean, is the capital, largest city, and main seaport of Lebanon. Equipped with modern harbor facilities and an important international airport, Beirut before 1975 was one of the busiest centers of transportation, commerce, finance, and tourism in the Middle East. Although devastated by warfare and terrorism, the city has proved remarkably resilient, and it may yet resume the international functions that made it a major Middle Eastern hub before the civil war.

Prior to 1975 Lebanon had a reputation for economic success despite meager natural resources. It never attained prosperity commensurate with that of the Gulf oil states or most Western nations, but compared to most Middle Eastern countries its economic situation was highly respectable. Lebanese prosperity was based on varied sources, principally (1) revenues from transit and entrepôt trade (including transit fees for oil shipped through Lebanese ports via pipelines from Saudi Arabia and Iraq), (2) profits from international financial and business transactions, (3) remittances to their families from the many Lebanese who have emigrated to the Americas and other overseas areas, and (4) profits from a considerable tourist industry. Tourism was based on the summer coolness and winter skiing of the Lebanon Range, plus Mediterranean beaches, the night life of Beirut, and historical monuments covering a span of thousands of years.

COMMUNAL STRIFE IN LEBANON

Commencing in 1975, Lebanon's prosperity was shattered and the country's political life fell apart under the murderous stresses of civil war and foreign intervention. The origins of this conflict reach deep into history and are rooted in Lebanon's extraordinary mixture of Arab religious communities which entered the area at various times in past centuries and became strongly localized in particular areas. Warfare among the different communities broke out from time to time, and antagonisms were heightened by foreign powers that gained influence with one or another group.

What the French put together into a new political unit in 1920 was a collection of small geographic areas, each dominated by one or a few minorities.

> . . . . Thus to Mount Lebanon, the traditional home of the Maronites and the Druse, were added the Bekaa Valley with a Shia majority and a large plurality of Greek Catholics, the coastal towns where the Sunnis and the Greek Orthodox predominated, the southern region inhabited by the Shiites and the northern region, where the Sunnis formed the majority.[2]

By 1946, when final independence was granted to the Republic of Lebanon, the French had managed to secure assent of most Lebanese to a National Pact under which a government was created which balanced the claims of different minorities by assigning government posts to particular groups.

> Thus Lebanon's President was always to be a Maronite, the Prime Minister a Sunni Muslim, and the President of the Chamber of Deputies a Shiite. Key ministries were also reserved for particular religious groups: the Foreign Minister was always to be a Christian and usually a Maronite, the Defense Minister a Druse, and so on. . . .[3]

In the 1950s, the "balancing act" in Lebanon broke down, and in 1958 a short civil war took place in which the governmental structure was saved by United States military intervention and the accession of a new Maronite President willing to give Muslims an increased share of power. Since 1946 an economic gulf had widened between the more prosperous Christians and the less prosperous Muslims, while at the same time the Muslims had increased their share of the population because of a higher birth rate.

For a time the Lebanese situation was patched together in an uneasy fashion, but then in 1975 the attempted assassination of a prominent Maronite leader precipitated a vicious civil war in which numerous factions were pitted against each other and the city of Beirut became a major battleground severed into hostile zones controlled by contending groups.

In 1976 Syrian forces came into Lebanon as "peace-keepers" sanctioned by the Arab League and operating from a base in the Bekaa Valley. Israel entered the fray in 1978 by invading southern Lebanon, then largely withdrew, but came back in force in 1982 with the declared objective of destroying the PLO and achieving security for its northern border region. Heavy fighting ensued, not only between the Israelis and the PLO, but also between Israelis and Syrians. After inflicting a devastating defeat on the PLO and heavy damage on the Syrians, the Israelis eventually withdrew, although they continued to patrol a "security zone" in the south. Meanwhile four Western countries—the United States, France, Britain, and Italy—sent token military contingents to shore up the almost powerless Lebanese central government and army, but this move did not bring peace and after some costly terrorist attacks that left hundreds of Western troops dead, the Western contingents left. Syria then asserted itself more boldly, a cease-fire was arranged, and the task of reconstructing Beirut began. But this effort was set back in 1988 and early 1989 when further factional warfare in Beirut caused large casualties and severe damage to parts of the city.

Syria, like Lebanon, is a country ridden by religious factionalism, except that in Syria the Muslim communities outnumber the Christians nine to one. In 1970 a coup overthrew the ruling Sunni Muslim establishment and brought to power Hafez al-Assad, a Shiite Muslim of peasant origins. Under his authoritarian rule as President, dominance of the government quickly passed to Assad's sect, the Alawites. Opposition within the other communities was crushed by the police and army. Assad then maneuvered for a widened Middle Eastern role. An impor-

---

[2]Nikola B. Schahgaldian, "Prospects for a Unified Lebanon," *Current History,* 83, No. 489 (January, 1984), p. 6. Used by permission of *Current History.*

[3]*Ibid.,* p. 7. Used by permission.

tant success was gained in Lebanon where, despite a heavy battering by the Israelis in 1982, Syrian troops established a continuing presence and Assad's influence on the Lebanese government became extremely strong. Syria has long nurtured ambitions for a territorially expanded and politically dominant "Greater Syria" in the heart of the Middle East.

EGYPT—THE "GIFT OF THE NILE"

At the southwest, Israel borders Egypt (officially the Arab Republic of Egypt), the most populous Arab country. This ancient land, strategically situated at Africa's northeastern corner between the Mediterranean and Red seas, is more utterly dependent on a single river than is true of any other nation in the world. Not only does the Nile (chapter-opening photo) supply the water that enables 53 million Egyptians to exist, but the Nile built the alluvial flood plain and triangular delta on which more than 95 percent of all Egyptians live (map, Fig. 13.3; photo, Fig. 2.23). Only 2 percent of Egypt is cultivated, and nearly all this land lies along and is watered by the great river. The conversion of the original marshes and swamps along the Nile to the thickly settled irrigated landscape of today is a dramatic story that has been unfolding for more than 50 centuries. Some of its essence is recounted in the following selection by Lord Kinross, a British diplomat and scholarly interpreter of the Middle East.[4]

The Nile and the Origins of Egypt

The river Nile, stretching across half of Africa, flows northward from the tropical mountains and forests of the equator to the temperate Mediterranean Sea.... Of Egypt, the land with which it is most closely associated and which the Nile fructifies for the last thousand miles of its course, Herodotus wrote that it is an acquired country, "the gift of the river." So it is. But the river itself is, in a sense, the gift of man. "Help yourself," runs an Egyptian proverb, "and the Nile will help you." The Nile as we see it today is the product of peoples who have been helping themselves for the past five thousand years....

. . . . .

During the millenniums preceding the dynastic history of Egypt, which begins around 3200 B.C., the ending of the Ice Age gradually dried up the grasslands which bordered the Nile, transforming the pastures

[4]Lord Kinross, "The Nile," *Horizon,* 8, No. 3 (Summer 1966), 80 ff. Reprinted by permission of the Pitus, Fraser & Dunlop Group, Ltd. One footnote has been added.

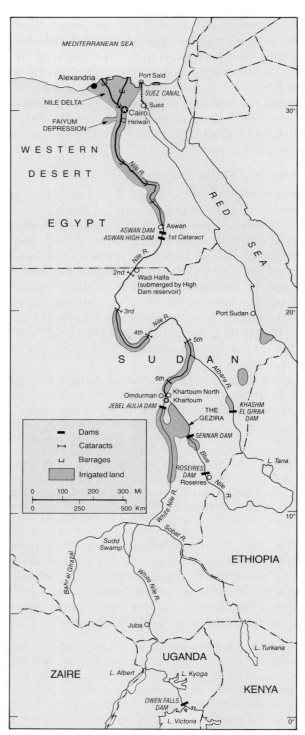

Figure 13.3   The Nile Basin. The width of the irrigated strips along the Nile is somewhat exaggerated. (Based partly on a map in Focus, *published by the American Geographical Society)*

of herdsmen and hunters into waterless desert. Yet the river itself remained, sprawling through this desert, overflowing its banks into jungle swamps and water-logged marshes, where hippopotamuses and crocodiles flourished and vegetation ran rife and unproductive.

. . . . .

The Nile valley was to develop into the Egyptian landscape so familiar to us now—a long ribbon of green cultivation which threads its way through the sunbleached desert, vividly coloring the deep-soiled plain of Thebes; continuing seaward above ancient Memphis (now merely a palm grove across the river from Cairo), it broadens into a delta crisscrossed by irrigation channels to cover the black earth with a thick plaid of ordered and perennial fertility. . . .

This transformation of the Nile valley and of the lives of its people was gradual; it spread over hundreds, indeed, thousands of years. Portrayed on the walls of tombs of the Old Kingdom are thickets of reeds and papyrus where men hunt the hippopotamus, the crocodile, and other creatures of the swamp no longer to be seen north of the First Cataract. But as time went on, murals of peasant activities came to predominate—ploughing fields with teams of oxen, sowing seed, drawing water with the *shaduf*, an apparatus of buckets on poles that is still in use today, reaping and threshing crops, treading grapes to make wine.

The process of domestication, slow as it may have been, was continuous from before the third millennium, when the two kingdoms of Upper and Lower Egypt[5] were drawn together into one united kingdom. This act of political union, so early in history, made possible the harnessing of the river and Egypt's subsequent economic development. It took a strong central authority to execute the major works of engineering required to control the great river (and incidentally to build the Pyramids as well), to mobilize labor and the other resources needed for the reclamation of land, and, above all, to maintain these works. . . .

Egypt's Irrigated Agriculture Today

Commencing in the 19th century, the construction of barrages and storage dams made it possible to gradually convert the traditional *basin irrigation* to *perennial irrigation* wherein water is fed to the land by canals. Six barrages are spotted along the Nile between Aswan in Upper Egypt and the Mediterranean (Fig. 13.3). These are low dams designed to raise the level of the river high enough that the water will flow by gravity into irrigation canals. The bar-

[5]Editor's note: Since ancient times, the Nile Delta has been referred to as Lower Egypt; the narrow Nile Valley between Cairo and Sudan is Upper Egypt.

rages permit a constant flow of water to irrigated land even in periods of low water. They are not designed for storage of large amounts of water, a function performed by two dams in Upper Egypt—the old Aswan Dam (completed in 1902) and the huge Aswan High Dam (completed in 1970)—and several others along the Nile or its tributaries in Sudan and Uganda. Perennial irrigation allows water to be released to the fields whenever it is needed by the crops, whereas the older basin irrigation depended on water brought to the land by the annual floods. In basin irrigation floodwaters were led into fields surrounded by low embankments; after the water had stood for several weeks the excess was drained back into the Nile and crops were sown in the muddy fields. For the period of growth they were dependent on the moisture stored in the soil. Only one crop a year could be harvested, whereas perennial irrigation has made it possible to harvest two or even three crops a year from the same field. Almost all of Egypt's cropping today is done by perennial irrigation. Water is transferred from irrigation canals to the fields primarily by motor-driven pumps, but older water-lifting devices such as the *shaduf* (a bucket on a counterweighted pole), the *saqia* (in which the water is scooped up by buckets attached to a rotary wheel, propelled by animal power), or the *Archimedean screw* are still used.

Long-staple Egyptian cotton has long been Egypt's main commercial export crop; food crops include corn (a staple in the Egyptian diet), wheat, barley, rice, millet, fruit, vegetables, and sugarcane. Berseem (clover), a legume, is grown for livestock feed and to enrich the soil with nitrogen. Egypt's population has grown to the point that the land can no longer feed it, and there are very large imports of food, especially wheat and flour.

Since 1960 the proportion of Egypt's labor force employed in agriculture has declined substantially, but the actual number of people living and working on the land has increased. Rural population densities already are among the highest in the world, and the crowding is growing worse despite a rapid migration to Egypt's main metropolis, Cairo (12 million), and other cities (Figs. 3.4 and 12.6).

Work, Housing, Diet, Sanitation, and Health

Until the 20th century, and, indeed, in many respects until our own day, the toilsome life of the Egyptian *fellah* (peasant) had known little basic change for many centuries: "his tools . . . those of his ancestors—the hoe, the wooden plow, the hand sickle, and the threshing board. His dwelling . . . a crude, fly-infested, two-room mud hut sheltering family, water buffalo, and chickens alike."[6] His diet was heavy in starches and low in protective foods—vegetables, fruits, eggs, meat, and milk. His life expectancy was very short. Even among the world's downtrodden peasants, the dire poverty of Egypt's fellahin was proverbial. Efforts of the Egyptian government to improve the rural situation during recent decades have had considerable success. Land reform has been carried out, cooperative farming fostered, water supply improved, high-yielding grain varieties introduced, and public health programs instituted. The average diet, at least in calories, is better than that of many other "developing" nations, and average life expectancy at birth has risen from 46 years in 1960 to 59 years in the 1980s. Birth rates, however, have continued to be high, and Egypt is confronted with rapid population increases that threaten to outrun the food supply and bring to naught all attempts to raise the average level of living. Alarmed by this prospect, Egypt's government has actively sponsored programs of birth control since the early 1960s. These have not been effective in checking the annual rate of natural population increase, which rose from 1.9 percent in 1947 to 2.8 percent in the middle 1980s—one of the world's higher rates.

Land Reform and Industrialization

In 1952 the government of King Farouk was overthrown by a revolution sponsored by elements of the Egyptian army. An Army lieutenant colonel, Gamal Abdel Nasser, soon rose to leadership in the new regime. In 1953 the revolutionary government proclaimed Egypt a republic. One of its earliest important acts was to institute a nationwide program of land reform. Much of the countryside had been dominated by large estates worked by tenants (generally sharecroppers) or wage laborers. Often such estates were absentee-owned, generally by urban merchants, professional men, bankers, or industrialists, and their operation was delegated to hired managers and overseers. Rents were often excessive. Rising land values and the dire poverty of most peasants made it exceedingly difficult for landless peasants to acquire land or for small owners to increase their holdings. To alleviate this situation, the government expropriated several hundred thousand acres from large landholders and redistributed the land among peasants. To enable them to pay for it, long-term government loans at low interest were extended.

The Egyptian government has also sponsored an increase in manufacturing, partly to draw surplus population off the farms. Many additional considerations have motivated this long-term program, among them the desire to overcome the economic and military weakness that made Egypt a pawn of foreign powers for many centuries and that recently led to humiliating defeats by Israel in several wars.

Actually Egypt has been industrializing in a modest way for a long period. Cotton textiles, food processing, clothing manufacture, and cement production are major industries, and there is a sizable output of chemicals, such as nitrate and phosphate fertilizers. The manufacture of automobiles and other consumer durables involves primarily the assembling of imported components. Egypt, in other words, has a fairly extensive range of industries but is far from being a highly developed industrial nation. A large share of its manufacturing plants, as well as many other business enterprises such as banks, are government-owned. Under the regimes headed by Nasser and his successors, Sadat and Mubarak, Egypt has become a socialistic country. Prices of basic dietary items such as bread are kept low by government subsidies, and rent controls keep down housing costs for poor families. The main industrial centers by an overwhelming margin are Cairo, the capital, and Alexandria (3 million), the main port, located, respectively, at the head and western seaward edge of the Nile Delta.

Cairo: Problems of an Overcrowded Metropolis

Both Alexandria and Cairo are very old and historic cities exhibiting survivals from the many empires that have ruled Egypt. Of Cairo it has been said[7]:

> A space traveler happening on such a city would probably react the same way as an American reporter. He would wonder at the age and variety and texture of life on earth.
> He might discover in Cairo a repository for one of almost everything from terrestrial history: over there a Pharaonic pyramid, a Roman temple, a Coptic cathedral, a Sunni mosque, a Napoleonic palace, a Brit-

---

[6]George H. T. Kimble and Dorothy L. Weitz, "Egypt," *Focus*, 2, no. 4 (December 15, 1961), p. 2. Used by permission of *Focus*, published by the American Geographical of New York.

[7]John Yemma, "Out of the haze of desert dust—Egypt's Cairo," *The Christian Science Monitor*, October 20, 1983, p. 25. Reprinted by permission from *The Christian Science Monitor* © 1983 The Christian Science Publishing Society. All rights reserved.

ish cricket field, a Marriott Hotel. There, near Mustafa Kemal square, is Groppi's ice-cream parlor, and there, in Maadi, is Kentucky Fried Chicken.

Cairo is not of the 20th century. It is of all the centuries. . . .

Of today's overcrowded Cairo, marked by massive disparities between great wealth and incredible poverty, another visitor wrote[8]:

> The lack of telephone service, and the fact that smog now blots out the view of the Pyramids from the windows of Cairo's high-rise hotels, may be things noticed mostly by foreign visitors to the city. But those are only two relatively superficial symptoms of the enormous infrastructural and environmental problems threatening to strangle the Egyptian capital. This is particularly true of those sections of the Old City in which, according to latest estimates, as many as 150,000 people are squeezed together within a single square kilometer, many of them without any sanitary facilities and living under the most miserable conditions. There are hours-long outages of the water and power supplies in these miserable slums and obsolete sewage pipes give way under their load, frequently turning the narrow lanes and alleyways into cloacae. In view of all this, however, it verges on the miraculous that Cairo's sanitation system does not collapse entirely, but continues to function to some extent. And yet—on what a basis! It is primarily youngsters, in some cases children under the age of 10, who begin in the early morning hours with the huge task of garbage collection. They operate in the most primitive manner, mostly by hand, sometimes with a small hand cart as vehicle or, at best, a donkey to carry the load.
>
> As regrettable as it is, this example provides some insight into the magnitude and complexity of the tasks facing the Egyptian government. The proper organization of the capital's sanitation system would not only involve the acquisition of trucks, for which there are no funds, it would also mean depriving the poor youngsters from the slums of the vitally important pittances which they earn through this foul labor.

Contribution of Minerals, the High Dam, and the Suez Canal

Egypt shares the customary handicaps of "developing" countries that attempt to industrialize: an impoverished home market, a shortage of investment capital and skilled managerial and technical personnel, and a dearth of skilled industrial labor. A rapid development of public education in recent years is

helping remedy deficiencies in technically trained personnel, although many children do not attend school because they must work to help support their desperately poor families. Mineral wealth to support industrialization is not outstanding, although some important minerals do exist in quantity. Chief among these is oil. Egypt's oil wells, located for the most part along the Gulf of Suez and including many offshore wells, provide the greater part of the country's exports; in 1985–1986 crude oil and oil products represented 65 percent of all exports by value. The country also produces natural gas (used primarily in industry), phosphate rock for fertilizer, and iron ore that supports a small output of iron and steel. Egypt's oil and other exports are far too small to pay for a great diversity of needed imports, and the country has had to depend heavily on foreign loans and grants to stay afloat. Income from tourism helps redress the unfavorable trade balance, as do transit fees from ships using the Suez Canal.

THE ASWAN HIGH DAM

For power with which to operate its industries, Egypt relies heavily on electricity produced at two hydroelectric stations on the Nile. The first was opened at the old Aswan Dam in 1960. The main source of hydroelectricity, however, is the generating station at the Aswan High Dam. This enormous dam, completed in 1970, is located about 5 miles (8 km) upstream from the older Aswan Dam. The reservoir (Lake Nasser), more than 300 miles (c. 480 km) long, reaches across the Egyptian border into Sudan. By storing water in years of high flood for use in years of low flood, the High Dam regularizes the flow of the river in Egypt to a much greater extent than has been possible heretofore. It provides a larger and more constant water supply for the pre-existing 6 million irrigated acres (2.4 million hectares), thus enhancing the productivity of the land. It also has made possible the conversion of practically all the remaining basin irrigation to perennial irrigation and the reclamation of considerable amounts of additional farm land. However, some negative aspects can be cited. For example, water loss from the reservoir due to evaporation, seepage, and upstream drought has been greater than was forecast; troublesome problems of waterlogging and salinization in irrigated land have developed; Egypt's sardine fisheries in the Mediterranean have been damaged by loss of nutrients formerly brought by the Nile floods; and the nation's bill for artificial fertilizers has greatly increased. Formerly the land's fertility was renewed during the annual flood by rich silt brought from volcanic highlands in Ethiopia, but most of this now

[8]Walter Günthardt, "Economic Survival in Egypt," *Swiss Review of World Affairs*, 31, No. 12 (March 1982), p. 27. Used by permission of the *Swiss Review of World Affairs*, Zürich.

settles in the High Dam reservoir. Nonetheless, the increased output of electricity, the regularized flow of irrigation water, and the psychological lift given Egypt by successful construction of the massive dam and power station are very positive assets.

THE SUEZ CANAL

In 1956 Egypt's High Dam project helped precipitate a major international crisis. In 1955 the United States and the United Kingdom had indicated their willingness to help Egypt finance the costly dam. But in July 1956 they withdrew their offers, and cancellation of a promised loan by the World Bank followed. President Nasser of Egypt then nationalized the Suez Canal, declaring that canal tolls would be used to build the High Dam. The canal had been constructed and operated by the Universal Suez Canal Company, a semiprivate Egyptian corporation in which the controlling financial interest was British but the actual management largely French. Great Britain and France protested Egypt's action and froze Egyptian financial assets in their countries. In October 1956, Israel invaded the Sinai Peninsula of Egypt for the avowed purpose of ending Egyptian border raids into Israeli territory. Great Britain and France promptly intervened, ostensibly to protect the canal. Port Said and Egyptian airfields were bombed, and troops were landed in the Canal Zone. These events, coupled with a revolt in Hungary against Soviet control, created a tense international situation. Diplomatic pressure by the United Nations, led by the United States and the USSR, plus protests by opposition parties in Great Britain and France, brought about a withdrawal of the invading forces. Commencing in 1950, Egypt barred Israeli ships from the Suez Canal and frequently interfered with ships of other nations carrying cargoes to or from Israel. The June War of 1967, however, put the Canal out of service, and it became a frontier between hostile armies. Not until 1975 was it finally cleared of sunken ships and returned to service.

The Canal connects the Mediterranean Sea with the Gulf of Suez and Red Sea across the low Isthmus of Suez (Fig. 12.1, inset). Unlike the Panama Canal, in which ships are raised or lowered to successive water levels by massive locks, the Suez Canal is a sea-level canal. It was completed in 1869 under the direction of a noted Frenchman, Ferdinand de Lesseps. The total length of the waterway, including dredged approach channels and the Bitter Lakes, is 107 miles (172 km). At opposite ends of the canal are the seaports of Port Said (375,000) on the Mediterranean and Suez (260,000) on the Gulf of Suez. Since 1975 the Canal has been widened, deepened, and equipped with communications gear to enable it to handle more ships. Petroleum and its products are the largest category of goods passing through the waterway, although the largest supertankers draw too much water to use it when fully loaded (most empty tankers can use it). These huge ships bound for Europe from the Persian/Arabian Gulf oil fields must still use the long route around Africa that the Canal was built to bypass.

SUDAN: A BRIDGE FROM THE MIDDLE EAST TO TROPICAL AFRICA

Egypt is bordered on the south by the vast, sparsely populated tropical republic of Sudan. Formerly an Anglo-Egyptian condominium, Sudan received independence in 1956. To Egypt its southern neighbor is vitally important, for it is from Sudan that the Nile River brings the irrigation and drinking water that sustains the Egyptian people. The Nile receives all its major tributaries in Sudan (Fig. 13.3), and storage reservoirs exist there on the Blue Nile, the White Nile, and the Atbara River, which benefit Egypt as well as Sudan itself.

The Nile's main branches, the White Nile and Blue Nile, originate, respectively, at the outlets of Lake Victoria (Uganda) and Lake Tana (Ethiopia). It is the Blue Nile, Atbara, and Sobat rivers, flowing from seasonally rainy highlands in Ethiopia, that are primarily responsible for the annual summer floods of the main river. In contrast to the Ethiopian rivers, however, the White Nile, fed by one of the world's largest lakes, maintains a fairly constant flow through the year. The Owen Falls Dam in Uganda increases the storage capacity of Lake Victoria and thus facilitates control of the White Nile. In southern Sudan the river passes through a region of swamps and marshes, the Sudd, where much water is lost by evaporation. An artificial channel to straighten the river's course and increase the velocity of its flow through the Sudd has been under construction for a considerable time. It is called the Jonglei Canal.

Approximately 5 percent of Sudan is cultivated, but only a minor share—perhaps 10 to 20 percent—of the cultivated land is irrigated. Sudan's largest block of irrigated land is found in the Gezira (Arabic, ''island'') between the Blue and White Niles. Water flows by gravity to the fields through canals from Sennar Dam and Roseires Dam on the Blue Nile (Fig. 13.3). Substantial irrigated areas exist in Sudan at many points along the main Nile and its tributaries.

The most important cash crop of Sudan is irrigated long-staple Egyptian-type cotton. In the Gezira project it is grown in a systematic rotation with *dura* (a general name for grain sorghums that form the basic food crop in most parts of Sudan) and *lubia* (a legume that yields fodder for cattle and beans for human or animal food). The Sudan Gezira Board, a government agency, administers the project. The land is farmed by tenants, who receive a share of the proceeds from the cotton crop and are allowed to retain crops other than cotton. Nonirrigated short-staple cotton is grown with the aid of summer rains in several parts of central and southern Sudan, but it accounts for only a minor part of the total cotton output. Cotton provided 44 percent of the country's exports by value in 1986.

In Sudan, as in most tropical countries, a variety of crops attain local importance (for example, peanuts, dates, wheat, barley, corn, tobacco, sugarcane, sesame, coffee, manioc, bananas, and rice). Some crops, especially sesame seeds, are exported in considerable quantities. An interesting export is gum arabic, used to make inks, adhesives, and a variety of other products. The bulk of the world supply is gathered in Sudan from resinous "tears" that form on certain acacia trees when incisions are made.

For the majority of Sudan's population, livestock raising (cattle, camels, sheep, goats), largely on a subsistence basis, is the principal means of support. The country's industries are relatively meager, and there are few minerals of consequence. Some oil deposits are being developed in southern Sudan, but their extent and potential significance are uncertain (early 1989).

Most of Sudan is an immense plain, broken frequently by hills and in a few places by mountains. Climatically and culturally, the area is transitional between the Middle East and Tropical Africa. A contrast exists between the arid Saharan North, peopled mainly by Arabic-speaking Muslims, and the seasonally rainy equatorial South, with its grassy savannas, papyrus swamps, and animistic or Christian peoples. About three fourths of the population is Muslim. There has been much political friction between the two sections, manifested by uprisings, guerrilla warfare, and a great deal of bloodshed in southern Sudan since independence. Among the reasons for this have been historical antagonisms and economic disparities between the most developed North and the less developed and poorer South, along with efforts by the Arab-dominated government to impose the Islamic faith and the Arabic language on the South and to exploit the South economically.

Sudan is overwhelmingly rural. The largest urban district, formed by the capital, Khartoum, and the adjoining cities of Omdurman and Khartoum North, is located toward the center of the country at the junction of the Blue and White Niles. The total population of the metropolitan area is perhaps 2 million or more. Transportation within Sudan as a whole is poorly developed. Through navigation on the Nile north of Khartoum is barred by the famous series of cataracts (stretches of rapids) (Fig. 13.3), although the White Nile is continuously navigable at all seasons from Khartoum to Juba in the deep south. Most of the country is devoid of railways or good highways, although there is a widespread air net and a good deal of auto traffic despite the lack of surfaced roads. Khartoum has rail connections with several widely separated sections (see Fig. 13.8). The main route for rail freight connects Khartoum and the Gezira with Port Sudan (over 200,000), a modern, well-equipped port which handles most of Sudan's seaborne trade.

LIBYA: DESERTS, OIL, ISLAMIC MILITANCE

Egypt's neighbor to the west, Libya, formerly an Italian colony, became an independent kingdom in 1951 and then in 1969 became a republic after a coup by army officers. Some 97 percent of the population is comprised of Muslim Arabs and Berbers. Three major areal divisions—Tripolitania in the northwest, Cyrenaica in the east, and the Fezzan in the southwest—are commonly recognized (map, Fig. 13.4). Libya's principal cities are the capital, Tripoli (city proper, 900,000), in Tripolitania, and Benghazi (city proper, 400,000) in Cyrenaica.

Most of Libya lies in the Sahara and thus is too dry to support cultivation except at scattered oases watered by wells or springs. The average density of population is only about 6 per square mile (2 per sq. km). This figure has little meaning, since most of Libya has, for practical purposes, no population at all. Libya's 4 million people are heavily concentrated in coastal lowlands and low highlands along the Mediterranean border, where winter rains produce an annual precipitation averaging 8 to 17 inches (20–43 cm) in coastal Tripolitania and 8 to 24 inches in the northern "hump" of Cyrenaica. A scanty agriculture emphasizes crops typically associated with areas around the Mediterranean Sea.

For many years the economy and government of Libya operated at a deficit made good by foreign subsidies. But in 1961, when the first Libyan oil was exported, the country's financial situation took a sharp turn for the better. Intensive prospecting by numerous Western oil companies revealed large deposits in widely separated areas; the main producing fields developed thus far are clustered in Cyrenaica. Pipelines connect the fields with shipping points on the coast (Fig. 13.4). The oil companies were na-

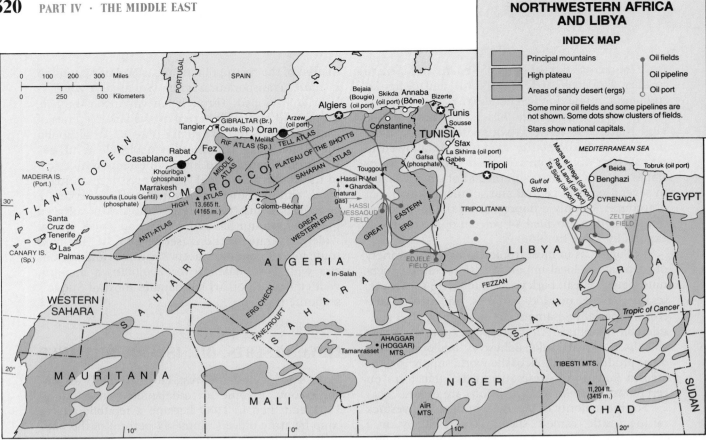

Figure 13.4   General reference map of Libya and the Maghrib countries.

tionalized in the early 1970s. Crude oil comprises practically all of Libya's exports, and returns from oil sales supply governmental revenues which increased enormously as a result of the steep jump in the price of crude oil in 1973 and after. Most oil moves to western Europe. Large amounts of oil revenue have been spent on public works, education, housing, aid to agriculture, and other projects for social and economic betterment, although vast sums have been spent for armaments.

Libya's revolutionary government headed by Colonel Muammar al-Qaddafi has instituted many socialistic programs, maintained a purist Muslim tone, and often adopted a bellicose stance vis-à-vis Israel, the Western nations, and Arab regimes not deemed sufficiently vigorous in their support of the Palestinian Arabs. In 1986, American warplanes bombed Tripoli and Benghazi in reprisal for alleged terrorist acts against U.S. citizens instigated by Libya.

NORTHWESTERN AFRICA— THE ARAB "MAGHRIB"

For many centuries Arabs have spoken of the northwestern fringes of Africa as the "Maghrib" ("West") of the Arab world. Despite its marginal location, this area is Middle Eastern in many essential characteristics. Today it includes three main units: Morocco, Algeria, and Tunisia. In the 19th and early 20th centuries France became the dominant power in the area. Although large sums were invested to develop mines, industries, irrigation works, power stations, railroads, highways, and port facilities, French rule was not popular. In 1956, after long agitation by local nationalists, Tunisia and Morocco secured independence. In Algeria, independence in 1962 came only after a civil war lasting for 8 years.

Muslim Arabs form the largest population element in all three countries of the Maghrib; Berbers, who antedate the Arabs in the area and were converted by Arabs to Islam, are numerous in Morocco and Algeria. Many Berbers now speak Arabic and many have adopted Arab customs. Nearly 2 million Europeans, primarily French but also of Spanish or Italian origin, formerly lived in the Maghrib. About two thirds were in Algeria. Most Europeans were concentrated in Algiers (estimated metropolitan population, 2.5 million), Oran (over 500,000), Casablanca (2.6 million), Tunis (well over 1 million), and other cities, where they formed a class of business and professional people, administrators, office workers, and skilled artisans. In Algeria possibly 25,000 Europeans were supported directly by agriculture.

Their farms, worked by Muslim laborers or tenants, were generally much larger and more mechanized than farms owned by Muslims, and they included a large share of Algeria's best agricultural land. Smaller numbers of European farmers also were found in Morocco and Tunisia. Independence for the Maghrib countries was followed by a large exodus of Europeans, mainly to France. Most European-held farm land was expropriated and redistributed to private Muslim farmers and cooperatives or placed in large units run by the various governments. There was also some redistribution of land held by large Muslim landowners. All three countries show a contrast between the small and generally poor farms of the "traditional sector" in which the bulk of the agricultural population still lives, and the larger and more mechanized and productive units in the "modern sector" inherited from the former European settlers.

Most people in the Maghrib live in the coastal belt of mediterranean climate (called in Algeria "The Tell"), with its cool-season rains and hot, rainless summers. Here, in a landscape of hills, low mountains, small plains, and scrubby mediterranean vegetation, are grown the familiar products associated with the mediterranean climate. Export production, largely from the "modern sector," concentrates mainly on grapes for wine, citrus fruits (largely oranges), olives, and vegetables. Such exports are relatively small, and each country imports a greater value of food than it exports. For each country France was still the most important trading partner in the late 1980s.

The majority of cultivators and livestock raisers in the Maghrib are in the "traditional sector"; they eke out a living from small farms or nomadic herding, or drift off to *bidonvilles* (shantytowns) in the cities. Most livestock (primarily sheep and goats) are raised in the traditional sector. About half of the population in the Maghrib still can be classed as rural. The average level of living is low (although higher than that of most African countries) and the rate of population increase is high. All three governments are faced with massive problems as they attempt to bring their populations up to acceptable levels of livelihood and welfare. In Algeria and Tunisia, oil revenues have been very helpful.

Inland from the coastal belt of the Maghrib, the Atlas Mountains extend in relatively continuous chains from southern Morocco to northwestern Tunisia. These mountains block the path of moisture-bearing winds from the Atlantic and the Mediterranean, and some mountain areas receive 40 to 50 inches (c. 100–130 cm) or more of precipitation annually. In places the precipitation nourishes good forests of cork oak or cedar. The highest peaks are in the High Atlas of Morocco, where one summit reaches 13,665 feet (4165 m). In Algeria the mountains form two east-west chains, the Tell Atlas nearer the coast and the Saharan Atlas (Fig. 13.4). In Tunisia the mountains are lower than in Morocco or Algeria, but they contain many deep valleys.

At the south all three countries reach the Sahara; indeed the greater part of Algeria is Saharan. A prominent line of oases fed by springs, wells, mountain streams, and *foggaras* (tunnels from mountain water sources) is found along or near the southern base of the Atlas Mountains. A sparse population of nomads roams the Sahara, and clusters of oases exist in a few places where mountains rise high enough to catch moisture from the passing winds. Several large areas of sandy desert (*ergs*) and a barren gravel plain, the *Tanezrouft*, occupy portions of the Algerian Sahara (Fig. 13.4).

The Maghrib economies benefit from a valuable endowment of minerals. Algeria's oil has the greatest value. The fields are located in the Sahara (Fig. 13.4), and the oil flows through pipelines to shipping points on the Mediterranean in Algeria and Tunisia. Algeria also extracts natural gas from the Sahara, some of which is exported in liquefied form. Petroleum products, crude oil, and gas comprise nearly all of Algeria's exports by value. Morocco's exports also include various minerals, although agricultural exports and clothing make up a sizable share of the total. Phosphate is by far the leading mineral export, followed by a diverse array of metals (Table 21.3).

Consumer-type industries, including plants that assemble foreign-made components, predominate in the modest industrial structures of the Maghrib countries, although Algeria's socialistic government has built a few state-owned heavy industries that refine oil, liquefy natural gas, make petrochemicals, or manufacture small quantities of iron and steel.

WESTERN SAHARA

In recent years Morocco and Algeria have been at odds over the future of Spain's former dependency known as Western Sahara. This sparsely populated coastal area of stark desert, which contains some rich phosphate deposits, adjoins Morocco and is claimed by that country. When Spain relinquished control in 1975, the territory was partitioned between Morocco and Mauritania, but in 1979 the latter country withdrew. Algeria objected to Moroccan control and, along with other outside parties, gave assistance to an armed resistance movement in Western Sahara called the Polisario Front. Meanwhile the Organization of African Unity (OAU), composed of all of Africa's sovereign states except South Africa, called for a referendum whereby the Western Saharans could determine their own future. At a 1984 summit meeting, the OAU gave a seat to a Western Saharan government-in-exile, the Saharan Arab Democratic Republic (SADR). Morocco then withdrew from the organization in protest. By 1989, Morocco and the Polisario Front had agreed to a cease-fire and a referendum on independence, but the status of Western Sahara remained far from resolved. Such uncertainties are very prevalent in present-day Africa and the Middle East.

THE GULF OIL REGION

Petroleum is vital to the contemporary world as a source of fuels, lubricants, and chemical raw materials. So essential is this greasy mixture of hydrocarbons that world production of crude oil grew from an average of 7 million barrels per day in 1945 to about 63 million barrels (2½ billion U.S. gallons) per day in the all-time peak year of 1979. A crucial fact of present world geography is the concentration of an estimated 53 percent of the world's proved petroleum reserves in a few states that ring the relatively shallow, desert-rimmed arm of the Indian Ocean commonly known as the Persian Gulf, but called by Arabs the Arabian or Arab Gulf. All but one of the oil-producing states that touch the Gulf belong to the Arab world; the remaining state, Iran, has had close historical relations with the Arab countries and shares their Islamic religion, but most of its people are not Arabs.

All but about 2 percent of the Gulf region's proved oil resources are held by five countries: Saudi Arabia, Kuwait, Iraq, Iran, and the United Arab Emirates (UAE). Saudi Arabia, by far the world leader in reserves, is considered to have about 24 percent of all the proved crude oil on the globe (1987; source: *Britannica Book of the Year*, 1988). Small Kuwait has an additional 13 percent, Iraq 6, Iran 5, and UAE 5. By comparison, the giant Soviet Union has 8 percent of world reserves, Venezuela 8, Mexico 8, and the United States only 4.

Most oil deposits in the Gulf countries lie along, underneath, or relatively near the Gulf (Figs. 13.5 and 13.6). These countries were producing about one fifth of the world's crude oil in 1985. Formerly they produced a considerably greater share, but a glutted world market for oil in the 1980s led to sharply reduced production within the Gulf countries in an effort to keep oil prices high. The five states named in the preceding paragraph are the region's leading producers, but two others—Oman and Qatar—are significant.

The oil produced in the Gulf region is largely marketed in the outside world, although the larger Gulf states require considerable amounts for their internal needs. Exports of goods from all of the eight states that touch the Gulf are comprised overwhelmingly of crude oil and/or refined products. Such exports, which are paid for in American dollars, have long yielded most of the vital foreign exchange used by the oil states in huge amounts to purchase equipment, materials, and technology for development and defense.

Oil from the Gulf is sold to many countries, but most of it is marketed in western Europe or Japan. The United States also imports some Gulf oil, but in recent years this has amounted to less than 5 percent of the national consumption. However, the Gulf region is very important to the United States because of such factors as the heavy dependence of close American allies on Gulf oil, the importance of the oil as a future reserve, the great involvement of American companies in oil operations and oil-financed development in the Gulf countries, and the large economic impact of Gulf "petrodollars" spent, banked, or invested in the United States. The American government has long proclaimed its intention to resist any threat to the continued flow of oil from the Gulf and has underscored this intent by providing arms to various Gulf countries and by military dispositions such as stationing naval forces in the Indian Ocean outside the Strait of Hormuz which connects the Gulf with the ocean. In 1987 American naval forces began convoying American-flagged vessels (mainly "reflagged" Kuwaiti tankers) through the Gulf as a protection against the hazards of the Iran-Iraq War. This led to various military incidents that claimed hundreds of lives.

Most oil produced in the Gulf region is shipped to market by tankers, generally the huge vessels called supertankers. These ships load oil at a handful of terminals spaced along the Gulf. From the Strait of Hormuz, most tankers sail around the southern tip of Africa to Europe or the Americas, or around Asia to Japan. A much smaller proportion of Gulf oil now moves to market via the Red Sea-Mediterranean route than formerly. From 1967 to 1975 the Suez Canal was out of service as a result of Arab-Israeli wars, and the Gulf oil industry became oriented to the supertanker route around Africa. Despite recent deepening of the Canal by Egypt, many loaded tankers have too deep a draft to use this waterway, and in any case the time required to transit the Canal, plus payment of tolls, may render the route less economical for a particular voyage than the longer trip around Africa. Still, enough loaded tankers do pass through the Canal to make crude petroleum the largest cargo item, and oil also is transferred from Red Sea to Mediterranean tankers by pipelines across Egypt and also across southern Israel. Some crude oil moves by pipelines from oil fields in certain Gulf countries to tanker terminals on the Mediterranean Sea or Red Sea (Figs. 13.5 and 13.6).

A growing proportion of extracted crude oil is refined in the Gulf states for local use or export. Petrochemical industries exist in most states, with the largest development by far in Saudi Arabia. Such industries use oil as a raw material, but they primarily use natural gas. Huge gas reserves are associated with oil deposits, and other major reserves exist apart from the oil fields in "nonassociated" deposits. Growing amounts of gas are exported in liquefied form; in fact, Saudi Arabia is the world's largest exporter.

The oil industry of the Gulf region was developed originally by foreign companies, principally British or American in ownership. These concerns made huge profits from their oil concessions, but with a few exceptions the companies eventually saw their properties nationalized by the Gulf governments. However, many outside companies still work closely with national oil companies in the Gulf states, providing the latter with technical and managerial expertise, skilled workers, equipment of many kinds, marketing channels, and miscellaneous services.

THE CHANGING FORTUNES OF OPEC

In addition to taking over the ownership and management of their oil industries, all of the Gulf states except Oman and Bahrain banded together with some other oil-exporting countries of the world to form the Organization of Petroleum Exporting Countries (OPEC), with the aim of joint action to maximize returns from oil for the participating countries. The organization was formed in 1960 and has a present membership of 13 (1989). In addition to Saudi Arabia, Iran, Kuwait, Iraq, UAE, and Qatar on the Gulf, the members include Algeria and Libya in North Africa, Nigeria and Gabon in low-latitude Africa, Venezuela and Ecuador in South America, and Indonesia in Southeast Asia. Many oil-exporting countries do not belong, some prominent examples being the Soviet Union, Mexico, the United Kingdom, Norway, and Oman.

In the early 1970s, OPEC began to play a prominent role in world economics and politics. The or-

ganization, which is a voluntary association of countries that actually compete with each other in world oil markets, fixes oil prices and production quotas to which the members promise to adhere. By concerted action it tries to keep oil prices high, and it has often received support from independent actions by non-OPEC exporters who are also concerned to maximize their oil incomes.

The organization was relatively obscure until 1973, when world oil prices took a jump as a result of the Arab-Israeli "October War." The hostilities caused frantic buying of oil by the developed world to build up reserve stocks, and the resulting price-rises were heightened when some Arab states, led by Saudi Arabia, embargoed oil shipments to the United States and the Netherlands to try to force those nations to modify their support of Israel. At this time the world economy was extremely dependent on the OPEC countries, which were supplying an overwhelmingly large share of the world's oil exports. Subsequently, OPEC coordinated a long series of price-rises, which peaked in 1981 when the organization's price for "Arab Light Crude" reached 34 U.S. dollars per barrel, compared with $2 per barrel in early 1973.

The successive oil price-rises had enormous repercussions on the world economy. Immense wealth was transferred from the developed world to the OPEC countries to pay for indispensable oil supplies. The skyrocketing cost of gasoline and other oil products helped cause serious inflation in the United States and many other countries. Desperately poor "developing" countries found that high oil prices not only hindered the development of their industries and transportation, but also hindered food production because of high prices for fertilizer made from oil and natural gas. For the developed nations, the transfer of wealth to OPEC had return benefits, as most of the money found its way back to the developed world in payments for goods and services or in the form of bank deposits and investments. Many billions of "recycled" petrodollars were subsequently lent by financial institutions in the developed world to help finance development in such emerging nations as Mexico and Brazil. Meanwhile such non-OPEC oil exporters as Mexico, Norway, Great Britain, and the USSR had a surge of profits, as did all segments of the complex oil business throughout the world.

In the Gulf countries the oil bonanza produced a huge wave of spending for military hardware, showy buildings, luxuries for the elite, and ambitious development projects of many kinds. Per capita benefits to the general populace were greatest in the oil states of the Arabian Peninsula, where small popu-

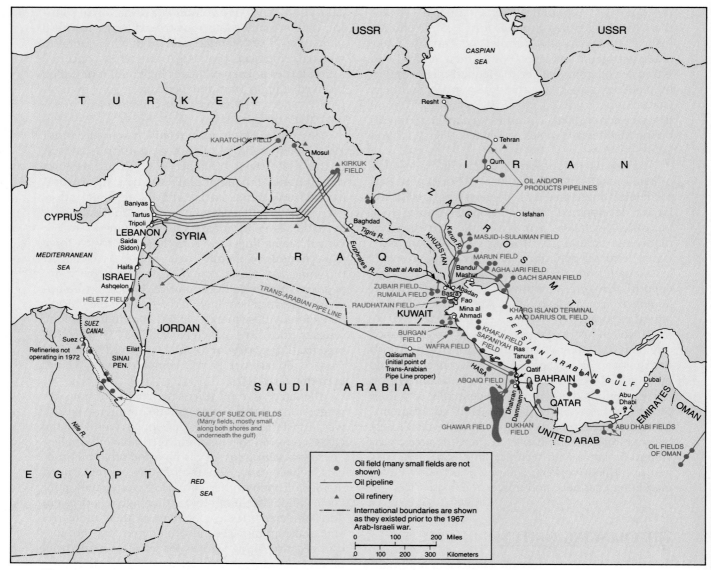

Figure 13.5 *Principal oil fields, pipelines, and refineries in the heart of the Middle East, as of 1972.*

lations and immense inflows of oil money made possible the abolition of taxes, the development of comprehensive social programs, and heavily subsidized amenities such as low-cost housing and utilities, including water distilled from the Gulf by desalting plants.

Then at the beginning of the 1980s, the era of continually expanding oil production, sales, and profits by the OPEC states came to a rather abrupt end. After 1973 the high price of oil stimulated oil development in countries outside of OPEC. Oil-conservation measures such as a shift to more fuel-efficient vehicles and furnaces were instituted. Substitution of cheaper fuels for oil increased. Coal, in

particular, replaced oil in many electric-generating stations. Oil refineries were converted at great cost to make gasoline from cheaper "heavy" oils rather than from the more expensive "light" oils previously relied on. Meanwhile the world entered a period of economic recession, which was due in part to high oil prices. Lessened business activity seriously decreased the demand for oil. By the early 1980s a worldwide oil glut was driving the world price of oil downward and forcing the OPEC states not only to lower their prices but also to greatly restrict their production. Profits of the world oil industry (and taxes paid to governments) were severely cut, large numbers of refineries had to close, and a good part

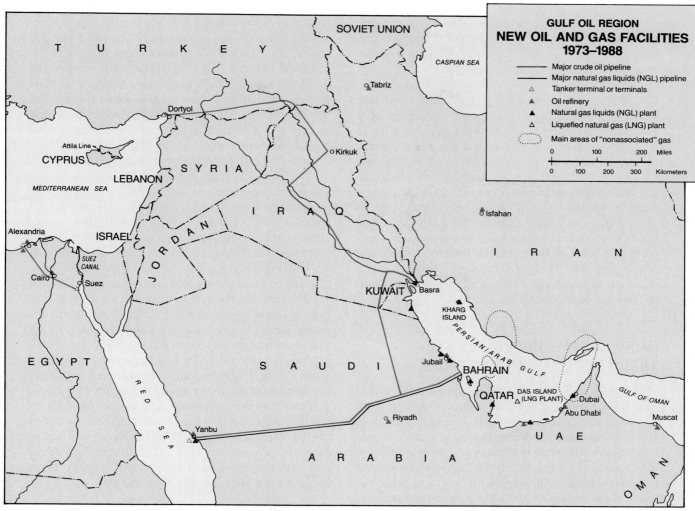

Figure 13.6 *Selected major facilities developed in the Gulf oil region after 1972. Natural gas pipelines, which are located primarily in Iran and Iraq, are not shown. Many relatively short gas pipelines are scattered through the region. Natural gas liquids are extracted from petroleum at many new plants shown on the map. The liquids are exported or used for petrochemical manufacturing at various places, especially Yanbu and Jubail in Saudi Arabia. Yanbu receives natural gas liquids via the major pipeline from eastern Saudi Arabia. Huge reserves of natural gas for the future are found in "nonassociated" deposits along or under the Gulf. One plant in the Gulf liquefies gas for export. Major new crude oil pipelines are shown schematically. Oil is exported from Iraq via pipelines to Dortyol, Turkey, to Yanbu, Saudi Arabia, to pre-1973 terminals in Syria and Lebanon, and to Iraq's own terminals (now being rebuilt after the Iran-Iraq War) near Basra. The pipeline from Suez to Alexandria, Egypt, enables crude oil to be offloaded at Suez (from tankers too large to transit the Suez Canal) and then reloaded on Mediterranean tankers at Alexandria.*

of the world tanker fleet was idled. In the meantime, OPEC's leverage on oil prices was being weakened by an increasing tendency of purchasers to buy their oil in the "spot market" (free market) on a day-to-day basis instead of contracting for longer-term deliveries at set prices.

By the late 1980s the Gulf oil states, together with the rest of OPEC, were in considerable disarray.

OPEC's member states were maintaining a shaky front in the face of pressures to break the line on prices and production quotas agreed upon by the organization. By 1987 the Gulf region was producing only about half as much oil as it had produced in 1979, and the revenue from oil was only half of what it had been in 1981. Ambitious development schemes were having to be cut back, and accumulated monetary reserves were being drawn upon to finance current budgets. But the immense oil and gas reserves still in the ground seemed to guarantee that this region would continue to have a major long-term impact on the world.

We turn now to discussions of the countries that comprise the Gulf region, commencing with the states of the Arabian Peninsula (including for convenience North and South Yemen), moving then to Iraq, and finally to Iran. Each of the three areas is distinctly different from the others physically and culturally, and each has its own proud history of imperial greatness in past times. All the countries are Muslim, but the range of Islamic beliefs and practices among them is broad, as is the range of social and political attitudes. Throughout the region modernizing trends are often in conflict with fundamentalist Muslim values. In Iran this conflict was a major cause of revolution in 1979; in all countries, it tends to foster political instability.

SAUDI ARABIA: MODERNITY AND TRADITION IN AN OIL COLOSSUS

The rags-to-riches saga of Saudi Arabia since World War II is so improbable that it might have come from *The Arabian Nights*. In 1945 this desert kingdom was still a chronically underdeveloped country, deeply bound by old Islamic traditions and peopled for the most part by roving bands of Bedouin herders or poor oasis farmers living in scattered mud-brick villages. Even as late as 1972 the modernized and urbanized Saudi Arabia of today would have seemed very far off. By that year the process of transformation was well begun, but crude oil was still selling at a price too low to portend the torrents of oil money that would revolutionize the kingdom from 1973 onward.

It was the escalating price-rises in world oil from 1973 to 1981 that loosed Saudi Arabia's oil genie. Annual oil revenues of the Saudi government rose from a value of $2.7 billion U.S. dollars in 1971 to $110 billion in 1981. Lavish spending created a new government-financed infrastructure, expansive social programs, and a modernized military establishment. Foreign goods, including great numbers of automobiles, poured in. A new consumerism replaced the spartan lifestyle of earlier times. Even today, however, the transition to a level of living commensurate with that of the industrialized Western nations is still far from complete. Despite the new affluence shown in dramatic landscape changes and a high per capita GNP, Saudi Arabia still bears various marks of a less-developed country. For example, in-

fant mortality (deaths during the first year of life) was recently estimated at a startlingly high 85 per 1000 live births (by comparison: Japan, 5; United States, 10; Israel, 11). Another mark of a "less-developed" country is the economic gulf between the moneyed elite of Saudi Arabia and the general mass of the population. Many upper and middle class Saudi Arabians have enriched themselves by acting as agents for foreign firms. But the sharing of oil wealth has been sufficiently widespread to bring undreamed-of prosperity to most of the population.

The new emphasis on material consumption in Saudi Arabia offers a strong contrast to the austere puritanism of the Wahhabi Muslims whose political leader, King Abdul Aziz al Saud (Ibn Saud: 1888–1953), consolidated many tribal and regional units into the Saudi kingdom before the advent of oil. However, many old-style Muslim traditions continue to be widespread, and such links with the past are fostered by the conservative Saudi family that rules this absolutist monarchy. The continued subordination of women is a very publicized facet of the Saudi attempt to preserve tradition while the tide of modernity rushes in (Fig. 13.7). As the keeper of Islam's holiest shrines, the ruling house feels a special responsibility to govern the country by Islamic precepts. Outsiders in Saudi Arabia, including thousands of Americans, must live discreetly so as not to openly transgress the kingdom's behavior codes. For example, all consumption of alcohol is forbidden, and women must not appear publicly in dress deemed immodest by official standards.

Physical Character and Population

With an area of roughly 830,000 square miles (2.1 million sq. km), Saudi Arabia occupies the greater part of the Arabian Peninsula, the homeland of Arab civilization and the Muslim religion. In past times this harsh land and the austere lives of its Bedouin camel people repeatedly caught the imagination and stimulated the pens of talented Western visitors such as Charles M. Doughty (*Travels in Arabia Deserta*, 1888, considered a classic of travel literature) and Wilfred Thesiger (*Arabian Sands*, 1959). Of the extreme deserts of southern Arabia, Thesiger wrote[9]:

> A cloud gathers, the rain falls, men live; the cloud disperses without rain and men and animals die. In the deserts of southern Arabia there is no rhythm of the seasons, no rise and fall of sap, but empty wastes

---

[9]Wilfred Thesiger, *Arabian Sands* (New York, E. P. Dutton and Company, Inc., 1959), p. 1. Used by permission of the publishers.

Figure 13.7 This photograph of Saudi Arabian women pictures law students leaving their classes in the city of Jidda (Jeddah) on the Red Sea. Although Saudi women have increasing opportunities for education and employment, their dress and deportment must still conform to conservative Muslim ideals espoused by the Saudi state.

where only the changing temperature marks the passage of the years. It is a bitter, desiccated land which knows nothing of gentleness or ease. Yet men have lived there since the earliest times. Passing generations have left fire-blackened stones at camping sites, a few faint tracks polished on the gravel plains. Elsewhere the winds wipe out their footprints. Men live there because it is the world into which they were born; the life they lead is the life their forefathers led before them; they accept hardships and privations; they know no other way.

Today the numbers of nomadic Bedouin are dwindling, and oil money has made their lives more secure. But much of the arid waste that has been their home since long before the Prophet Muhammad has a geography little changed by the age of oil. Some major features are described below.

The interior of Saudi Arabia is an arid plateau, fronted on the west by mountains that rise steeply from a narrow coastal plain bordering the Red Sea. Most of the plateau lies at an elevation of 2000 to 4000 feet (*c.* 600 to 1200 m), with the lowest elevations near the Gulf and in the eastern part of the desolate, sandy southern desert called the Empty Quarter (*Rub al Khali*). Mountains rise locally from the plateau, and the plateau surface is trenched by many *wadis* (stream valleys) that carry water whenever it happens to rain. Most interior sections average less than 4 inches (10 cm) of precipitation annually; often a year or more will pass with no precipitation at all. Such rainfall as there is comes generally in winter and is heaviest along the moun-

tainous western margin of the peninsula. The rainfall suffices for very little nonirrigated agriculture. It is estimated that 95 percent of Saudi Arabia's present crop production depends on irrigation. Cultivated land comprises under 1 percent of the country's area.

The estimated population of Saudi Arabia was about 14 million in mid-1988. This figure includes many foreign workers drawn to the kingdom by jobs ranging in character from purely menial to highly technical. Saudi Arab workers have not been available in sufficient numbers to fill these jobs. In recent years large numbers of foreign workers have been sent home as a result of financial stringency imposed by a world oil glut. Many would have returned home in any case due to the completion of major construction projects on which they had been employed.

Foreign workers have been selectively admitted to Saudi Arabia in order to exclude elements regarded as potential threats to the stability of the state. The Saudi monarchy, conscious of the vast treasure beneath its deserts and the vulnerability inherent in a relatively small national population, views uneasily the strident fundamentalism of Iran across the Gulf, the East-bloc ties of Marxist South Yemen which Saudi Arabia borders, and other potentially disruptive elements in the strife-torn Middle East. One potentially disturbing factor within the country is the presence of around 400,000 Shiite Muslims, mainly in the eastern oil-field region, within a national population that is largely Sunni Muslim. The Shiites share in Saudi affluence, but to a smaller degree than the Sunni majority. This underprivileged status is of long standing and seems to have originated in Sunni disdain for certain Shiite beliefs and rituals. But even within the Sunni majority there is much factionalism and tension associated with the division of Sunni society among many tribes and sectarian groups. This diversity requires much political balancing by the king and the senior princes who make the important government decisions.

The Oil Economy and Its Effects

Oil production in Saudi Arabia was developed by the Arabian American Oil Company (ARAMCO), which was itself originally owned by four United States companies but subsequently was nationalized (1976–1980). Despite the transfer of formal ownership, the American companies continue to work closely with ARAMCO. A host of other American business interests are deeply involved in Saudi Arabia, often in joint ventures with Saudi firms. The American government itself maintains a strong connection with the Saudi state.

The first oil discovery in commercial quantities was made at Dammam on the shore of the Gulf in 1938, but large-scale production was delayed until after World War II. The oil fields, the largest of which is the Ghawar field, are located for the most part in the eastern margins of the country. Much Saudi Arab oil is produced from deposits underlying the Gulf. The character of Saudi Arabia's oil fields has made it possible to secure an immense amount of oil from a remarkably small number of wells (780 producing wells in 1983). This is due to the great thickness and the high porosity of the oil-bearing strata, together with high reservoir pressures. These conditions make it possible for large amounts of oil to move quickly to the wells. Such advantages are widespread in the eight Gulf oil states. The great productivity of each well in the Gulf fields makes each barrel extremely cheap to extract, and makes it comparatively simple to increase or reduce production quickly in response to world market conditions.

Sizable amounts of Saudi Arabian oil are processed by refineries within the country or at a refinery in nearby Bahrain, but the bulk of the production still leaves Saudi Arabia in crude form via tankers (1989). Loading terminals exist on both the Gulf (at Ras Tanura) and on the Red Sea. In the future, the proportion of the national income derived from overseas sales of refined products and petrochemicals will increase. Huge new refining and petrochemical complexes, with associated industries of various types, have been developed at Jubail on the Gulf and Yanbu on the Red Sea (Fig. 13.6). Their petrochemical industries depend mainly on cheap natural gas associated with Saudi Arabia's oil fields. In the future large additional supplies can be tapped as needed from "nonassociated" reserves in Saudi Arabia, Qatar, and other Gulf states.

Formerly there was a large movement of crude oil through the Trans-Arabian Pipe Line to an export terminal on the Mediterranean coast in Lebanon, but supertankers loading directly in Saudi Arabia proved to be more cost-effective, and "Tapline" was discontinued in 1975 as a conduit for exports beyond Lebanon. The dangerous Lebanese war also contributed to this decision. However, the line continues to carry oil for internal needs in Jordan and Lebanon. A newer and larger oil pipeline paralleled by a pipeline carrying natural-gas liquids crosses central Saudi Arabia to the Red Sea coast. These lines bring hydrocarbons from eastern Saudi Arabia to the refining and petrochemical industries at Yanbu, and the oil pipeline carries additional crude oil to be loaded on tankers for export from Yanbu. Still more oil exports reach Yanbu via a new pipeline from Iraq.

An almost medieval land before the development of oil, Saudi Arabia has moved with remarkable speed into the complexities of the modern world. The state has spent petrodollars by the hundreds of billions to build infrastructure of all kinds, free Saudi citizens and companies from taxes, provide free education and medical care, subsidize low-cost food and housing, foster industrialization and scientific agriculture, and maintain an expensive military establishment. The push to modernize has created a mass of new air-conditioned high-rise office buildings, royal palaces, and supermarkets and other new shopping facilities, together with electric power lines, air-conditioned housing, and immeasurably improved communication and transport facilities, including more than a million new telephones and thousands of miles of surfaced highways and streets. Water supplies have been increased greatly by huge desalting plants that distill sea water and by drilling wells to tap underground water. Scientific agriculture and animal husbandry have been fostered by government subsidies, although imports still supply the greater part of all food. Irrigated wheat and vegetables, together with poultry, eggs, and milk, are major products from some of the highest-cost farms in the world. The remaining Bedouin nomads have been aided in various ways such as the delivery of subsidized water and feed to the herds by government-funded trucks. Government financing of agriculture and pastoralism has been in part a mechanism to help spread the benefits of oil wealth to the impoverished rural sector of the population, but it has also been a strategic measure to keep Saudi Arabia from becoming almost completely dependent on the outside world for food.

Industrial development has also been pushed, although Saudi Arabia is far from being a major industrial power. As of the middle 1980s there were reported to be about 3000 factories in the country, mainly producing or assembling consumer goods for Saudi Arabian consumption. The most spectacular industrial developments are the big diversified oil- and gas-based complexes at Jubail and Yanbu.

Saudi Arabia's best-known city is Mecca, the birthplace of the Prophet Muhammad and the principal holy city of the Islamic world. During recent years some 1½ to 2 million Muslims from many countries (including Saudi Arabia) have made the pilgrimage (*Haj*) to Mecca each year (see chapter-opening photo of Chapter 12). Handling the mass of worshippers in the few weeks of the pilgrimage season is an enormous logistic problem involving not only transportation, food, lodging, and medical care, but the provision of great numbers of animals (primarily sheep) for ritual sacrifices by the pilgrims. The interior city of Riyadh (1.5 million) is Saudi Arabia's capital. Here futuristic structures and traffic-clogged thruways have sprouted like magic in the dusty, bedraggled mud-brick town where King Ibn Saud established the government of the newly consolidated Saudi state in 1932. The port of Jidda (1.5 million) on the Red Sea some 40 miles (64 km) from Mecca is the main point of entry for foreign pilgrims. A large majority of the latter come by air, although some reach Jidda by ship.

As late as the early 1970s, Saudi Arabia was still overwhelmingly a rural country, but since then the urban population has increased with extraordinary speed. Recently the country's population was estimated to be 72 percent urban. Urbanization often has been due to the absorption of villages by an expanding city, but most of it has come about through migration from the countryside and the entry of foreign workers. Meanwhile, most of the Saudi Arabian landscape remains a rocky or sandy desolation, and some people in outlying areas still do not enjoy a life that would be deemed acceptable by the material standards of the industrialized Western world. But modernity will go on spreading, although the process has recently slowed because of sharply lessened sales of oil in an oversupplied world market.

LESSER COUNTRIES OF THE ARABIAN PENINSULA

Saudi Arabia is rimmed to the south and east by smaller Arab countries along the Persian/Arabian Gulf, the Indian Ocean, or the Red Sea. The small oil-producing countries of Kuwait, Bahrain, Qatar, United Arab Emirates, and Oman along the Gulf are linked with Saudi Arabia in the Gulf Cooperation Council (GCC). This organization tries to achieve better coordination among its members in regard to economic and security policies. Difficulties in this regard are created by a multiplicity of independent-minded tribes, sects, powerful families, and local rulers within the different countries.

At the southwestern corner of the Arabian Peninsula lie the mountainous countries of North Yemen and South Yemen. Their boundaries with Saudi Arabia in the Empty Quarter have never been demarcated, resulting in a kind of no man's land where a good deal of illegal cross-border movement of people and smuggling of goods has gone on. There also has been chronic border warfare in this physically difficult and little-known outlying sector of the political world. During the 1970s, full-scale war between the two Yemens erupted on two occasions.

The Wealthy Emirate of Kuwait

The little country of Kuwait, governed by an Emir who heads the ruling Sabah family, is about the size of New Jersey in area, but has only one fourth as many people (an estimated 2 million in 1988, including foreign workers). Kuwait has larger proved oil resources than any country except Saudi Arabia and very large gas reserves as well. Oil production centers in the Burgan field south of Kuwait City, although several other fields exist. Associated with Kuwait's oil is much natural gas, which is used as fuel within the country or is liquefied or made into petrochemicals for export. In addition to revenues from its own oil and gas production, Kuwait receives half the governmental revenue from large oil operations in the former Neutral Zone. The other half goes to Saudi Arabia. The Neutral Zone, created in 1922 to resolve a boundary dispute, was governed jointly by the two countries until 1969, when it was divided between them. The oil facilities are still owned by outside interests.

Oil revenues have changed Kuwait from an impoverished country of boatbuilders, sailors, small traders, and pearl fishermen to an extraordinarily prosperous welfare state. Generous benefits accrue to Kuwaiti citizens, although not, in general, to the many non-Kuwaitis who have been drawn to the emirate by employment opportunities.

Bahrain, Qatar, and UAE

The Bahrain sheikdom occupies a small archipelago in the Gulf between the Qatar peninsula and Saudi Arabia. The main island is connected to Saudi Arabia by a causeway. In addition to being an oil producer in a very minor way, Bahrain is an active center of oil refining, seaborne trade, ship repair, air transportation, business administration, international banking, and aluminum manufacture based on imported alumina and local natural gas. It became an oil-financed welfare state even before Kuwait.

East of Bahrain along the Gulf lie the small emirates of Qatar and United Arab Emirates (UAE). The latter country is a loose confederation of seven units, among which only three produce oil and only Abu Dhabi and Dubai produce it in major quantities. Qatar and UAE are extremely arid and prior to the oil era supported only a meager subsistence economy based on fishing, seafaring, herding, and a little agriculture, augmented by smuggling and piracy. Their large oil revenues have given them a world importance unknown to them previously and enormously accelerated internal development, including oil refineries and other industries.

The Southern Margins of Arabia

North Yemen and Oman occupy, respectively, the mountainous southwestern and southeastern corners of the Arabian Peninsula. These states receive the heaviest rainfall of the peninsula, particularly North Yemen. Elsewhere in the peninsula the scanty rainfall is confined mostly to the cool season and any unirrigated agriculture is built around fall-sown grains; but North Yemen has a summer rainfall maximum (derived from the same monsoonal wind system that brings summer rain to the Indian subcontinent), and its main subsistence crops are spring-sown millet and sorghums. Most of the country's people live in the rainier highlands rather than in the arid coastal zone along the Red Sea. Extensive areas in the highlands are terraced for cultivation.

Until the 1960s North Yemen was an autocratic, old-style Muslim monarchy whose rulers were frequently assassinated by rival claimants. The country was extremely isolated and little known, although historically it had some prominence as a trading center which prided itself on a higher culture than that of Arabia's roving Bedouin. In the middle 1960s a republican government supported by Egypt took control of large sections, although not without great bloodshed. Subsequently a military junta seized the government but has had trouble exerting its authority over the entire country. The north is peopled by conservative, fiercely independent, and heavily armed tribes who supported the monarchist cause; the south has been the scene of guerrilla warfare carried on by leftist rebels drawing support from Marxist South Yemen. North Yemen's government has long maintained a nonaligned stance in world affairs, accepting support from the Soviet Union, other East-bloc nations, Saudi Arabia, and the United States. Despite the aid (much of which has been military), North Yemen remains extremely poor and underdeveloped (Table 12.1).

San'a (city proper, 425,000), the capital of North Yemen, is a city of mud-brick houses and garden plots along dirt streets. Some modernity has come to the country and the pace of modernization may quicken if newly discovered oil can be exploited on a sufficient scale. In the meantime this remote and politically uneasy fastness, which is said to have two rifles for each of its 6.7 million people, remains a very poor relation of oil-rich peoples that it once outshone.

Most of the Sultanate of Oman lies along the Indian Ocean immediately west of the Strait of Hormuz entrance to the Persian/Arabian Gulf, although a small detached part of Oman borders the Strait or the Gulf. As in North Yemen, the coastal region is very arid (although verdant in some irrigated spots), and most of the population is found in interior highlands where mountains induce a fair amount of rain. The country once controlled a sizable empire and was a major center of trading in African slaves. Crude oil from fields in the interior is piped to the coast and comprises practically all of Oman's exports. The country's strategic position on the Strait of Hormuz has made it an object of recent concern by industrial nations, particularly the United States. The latter country has rights of access to base facilities in Oman in a military emergency threatening the continued flow of oil from the Gulf region.

South Yemen, until 1967 the British colony and protectorate of Aden, lies along the Indian Ocean between North Yemen and Oman. A Marxist state supported by the Soviet Union and other East-bloc nations, South Yemen has engaged in much contention with neighboring countries and the Western world since independence. The USSR has access to naval and air facilities in South Yemen (at Aden and on the island of Socotra) and is reported to have supplied the country with arms worth billions of dollars. South Yemen has given aid to various dissidents in North Yemen, Oman, and Saudi Arabia, and South Yemeni troops have fought against the armed forces of all these countries from time to time. Recently the country has followed less aggressive policies, but it continues to supply an important East-bloc foothold on the Arabian Peninsula near the strategic Straits of Bab el Mandeb connecting the Red Sea with the Indian Ocean. The main urban place, Aden city (perhaps 400,000), is located on a fine harbor near the straits. The city serves transshipment and entrepôt functions and has an oil refinery whose products comprise nearly all the country's exports. The refinery, which processes imported crude oil, was expropriated from the British Petroleum Company after independence. A South Yemeni offshore oil discovery in the Gulf of Aden has been reported. Any oil development will be extremely welcome, as South Yemen, despite the attention paid to it by the Communist world, is miserably poor.

IRAQ: GEOPOLITICAL COMPLEXITIES OF A RIVERINE OIL STATE

The Republic of Iraq at the eastern end of the Fertile Crescent occupies one of the world's most famous locales of early civilization and imperial power. Like the other countries of the Fertile Crescent, this riverine land has had a complex geopolitical history. It has been a seat of empire, a target of conquerors and, in the 20th century, a focus of oil development and political contention involving many other nations. Complex interaction with the outer world began here very early. The ancient Babylonians and Assyrians overcame neighboring peoples and built a succession of empires reaching from the Persian/Arabian Gulf to the Mediterranean; the earliest such empire dates to 2350 B.C. Irrigated land on the Tigris-Euphrates plain gave empire-builders a dependable food base, and the Fertile Crescent corridor between mountains to the north and deserts to the south was a natural passageway for military movement and trade. Much later, during the centuries of Arab ascendancy after the death of the Prophet Muhammad, Baghdad on the Tigris became for a time the capital of an imposing Arab and Muslim empire. Over thousands of years such periods of imperial triumph alternated with times of internal disorder and weakness when Iraq was subjugated by outsiders such as the Persians and the Romans. The area was devastated by Mongol invaders in the 13th century, and subsequently it became a part of the Ottoman Turkish Empire. Then in World War I British forces defeated the Turks in Iraq, and after the war Great Britain took over the country as a mandated territory under the League of Nations until independence was granted in 1932.

Iraq's government was a monarchy from 1923 to 1958, when a short revolution engineered by army officers made the country a republic. Over the next two decades, marked political instability was reflected in many coups or attempted coups as various factions struggled for power. By the middle 1970s Iraq had become an authoritarian one-party state ruled by a strongman President (Saddam Hussein) in the name of the Arab Baath Socialist Party. Aspiring to leadership in the Arab community, this regime stridently asserted its dedication to Arab solidarity, uncompromising opposition to Israel, freedom from Western imperialism, and strong internal development for a self-dependent, secularized, and socialistic Iraq. In 1972 the Iraqi government signed a treaty of friendship with the Soviet Union, and Iraq subsequently received large Soviet arms shipments and the equipment and technology for many new factories. After 1973 an increasing tide of oil money underwrote new social programs, large construction projects, and the creation of many new businesses under state, private, or joint ownership. New high-rise buildings, busy expressways, and an active commercial and night life gave metropolitan luster to ancient Baghdad. With oil revenues pouring in, Iraq seemed on the way to becoming a Middle Eastern showcase of planned socialist development.

Then in 1980 Iraq came to the forefront of world concern when President Hussein launched an invasion of neighboring (and much larger) Iran. This action, further described in the section on Iran, was preceded by an intermittent series of armed clashes along the boundary between the two countries during the 1960s and 1970s. A protracted war began that proved enormously costly to both sides. Although the actual fighting was mainly localized along the international border, a host of noncombatant nations became involved in various ways. For example, some nations gave Iraq large-scale financing or arms because they feared the destabilizing effects of an Iranian victory. No one could foresee what territories might be overrun or what populations subverted by a triumphant Iran filled with zeal to export its fundamentalist Shiite revolution. Huge monetary grants were made to Iraq by Saudi Arabia as well as Kuwait and the other small Gulf emirates, which were concerned about the vulnerability of their oil and gas fields to possible Iranian attack, and the vulnerability of their none-too-unified populace to possible subversion by the revolutionary Shiite ideology. The United States and the Soviet Union pursued their respective geopolitical interests in ways that generally benefited Iraq, although neither superpower declared formal support for either of the warring countries. Turkey, sharing a border with each belligerent, greatly expanded its trade with both; France sold Iraq sophisticated missiles and aircraft. All manner of other noncombatant nations were drawn in. Hence, an extraordinarily complex web of geopolitical relationships made the Iran-Iraq War (also called the Gulf War) more of a storm center in the political world than was the case with many other conflicts raging simultaneously in various parts of the world.

Iraq's Oil Industry

Iraq's oil was developed by the Iraq Petroleum Company, which was organized by a multiplicity of interests in several Western countries. After the revolution of 1958 the oil properties were taken over (mainly in 1972) by the state-owned Iraq National Oil Company. The largest oil field, first opened in 1927, is located in the vicinity of Kirkuk (500,000) in northern Iraq (Fig. 13.5). Many lesser fields exist, mainly in the north but including some in the far south near the Gulf. Natural gas is also present, largely in association with oil.

Prior to the war with Iran, some crude oil was exported by tanker from Gulf terminals, but the main export flow moved through pipelines to Mediterranean tanker terminals in Lebanon, Syria, and Turkey (Figs. 13.5 and 13.6). The flow via Iraq's Gulf terminals was cut off when these facilities were destroyed by Iranian attacks early in the Iran-Iraq War. Then in 1982 Syria cut off all Iraqi oil flow across Syrian territory as a means of aiding Iran in the war. This left only the pipeline through Turkey as a conduit for exports. Iraq then constructed two additional pipelines: a second line through Turkey and one across Saudi Arabia to the Red Sea oil port of Yanbu. Recently Syria has permitted Iraq to resume shipments of oil across Syrian territory to the Mediterranean.

Iraq's Internal Geography

Iraq, formerly known as Mesopotamia, occupies a broad plain drained by the Tigris and Euphrates rivers, together with fringing highlands (in the north) and deserts (in the west). In ancient times the country had an elaborate system of irrigation works on the Tigris-Euphrates plain, and its population at certain periods is thought to have been larger than it is today. But in the Middle Ages much of the irrigation system was destroyed or fell into disrepair during periods of political disorder and foreign invasion, and it has never been fully reconstructed. In recent decades Iraq's government has been expanding irrigation, partly through the construction of multipurpose dams (for flood control and hydroelectric development as well as irrigation storage) on the Euphrates and on the Tigris and its tributaries.

An estimated 68 percent of Iraq's 18 million people are urban dwellers—a proportion considerably higher than in most Middle Eastern countries. The rest of the population consists largely of villagers depending on agriculture for a living. Much of the agriculture today is carried on by Egyptians who have emigrated from overcrowded irrigated land along the Nile. Such immigrants can become Iraqi citizens with relative ease after a period of residence. Iraq prides itself on adherence to the concept of an Arab nation transcending present political boundaries. Small bands of nomadic Bedouin still raise camels, sheep, and goats in Iraq's western deserts. Some of the Kurdish tribesmen in the mountain foothills of the north are seminomadic graziers, although most Iraqi Kurds are sedentary farmers. The Kurds, a Muslim people of Aryan antecedents, are by far the largest of several non-Arab minorities in Iraq. They have revolted against the central authority on numerous occasions. The Kurdish question has international implications, as the area occupied by the Kurds extends into Iran, Turkey, and the USSR as well as Iraq. An independent state of Kurdistan was promised in treaties at the end of World War I, but the governments concerned never took steps to create it. Uprisings among the Kurds in Iran took place at the time of the 1978–1979 revolution; in Iraq, there were uprisings after the outbreak of the Iran-Iraq War in 1980. Turkish troops are reported to have crossed the border to help Iraq quell the risings. Within Turkey, the Kurds have long been under martial law.

Agriculture is dependent on irrigation in most parts of Iraq, although the northern highlands receive enough precipitation in the winter half-year to support nonirrigated fall-sown crops, principally wheat. Iraq's grain production, including a considerable amount of irrigated rice, is now far from sufficient for the country's needs, and large quantities of grain are imported. Smaller imports of many other foods also are required.

Iraq is not well developed industrially, and sizable quantities of machinery, transport equipment, and other industrial goods are imported. The country does have a considerable number of factories, most of which produce consumer goods for sale within Iraq. Many of the country's largest industrial and commercial enterprises, along with Iraq's banks and insurance companies, transportation lines, and utilities, are government-owned. Baghdad (4 million) on the Tigris River is the main industrial center as well as the political capital and largest city. Like other metropolises in "developing" nations, Baghdad has been receiving a huge influx of villagers seeking a better life or at least a more exciting one. The Iraqi preference for single-family homes finds expression in an endless spread of low-rise houses with garden plots. This mass of housing stretches outward for many miles from the cluster of office buildings and Western-style hotels in the downtown core of the city. In Baghdad both men and women generally dress in Western garb. Women are far more emancipated in secular Iraq than they are in the Gulf monarchies or revolutionary Iran. Basra (900,000) is Iraq's main seaport. It was blockaded by Iran early in the Iran-Iraq War and suffered extensive wartime destruction.

OIL WEALTH, ARIDITY, AND RELIGIOUS UPHEAVAL IN IRAN

Iran (Persia) was the earliest Middle Eastern country to produce oil in large quantities. The initial discovery dates from 1908, and commercial production commenced in 1912. The main oil fields form a line, oriented northwest-southeast, along the foothills of the Zagros Mountains in southwestern Iran (Figs. 13.5 and 13.8). The oil facilities were originally developed by the Anglo-Persian Oil Company, which was subsequently renamed the Anglo-Iranian Oil Company and is now the British Petroleum Company (BP). During World War I the British government gained a controlling interest in the company as a means of assuring a large oil source for the British fleet and merchant navy. The largest refinery in the world at that time was built on tidewater at Abadan, and pipelines were built to the inland oil fields. Abadan, which was heavily damaged by Iraqi shelling in the Iran-Iraq War, lies on the Shatt al Arab ("River of the Arabs": formed by the junction of the Tigris and Euphrates) about 50 miles (80 km) from the Gulf. It is only accessible to small tankers which can haul refined products economically but are not large enough for economical long-distance transportation of crude oil. Nearly all exports of crude oil from Iran pass through a newer terminal at Kharg Island which allows supertankers to load in deep water.

Revenues from the oil industry (fully nationalized in 1973) are crucially important to Iran, a country where a dry and rugged habitat makes it hard to provide a good living for the rapidly growing population of 52 million (1988). The greater part of the land is comprised of arid plateaus and basins, bordered by high, rugged mountains. Some large basins of interior drainage in southwestern Iran are practically uninhabited, being nearly rainless, often covered by sand dunes, and frequently underlain by treacherous salt-encrusted bogs where unwary people or animals can break through the crust and be swallowed by the deep muck underneath. Moisture-bearing air is blocked off from these basins by high encircling mountain ranges—a classic instance of

deserts caused by rain shadow. Only an estimated 9 percent of Iran is cultivated; the land actually under crops at any one time amounts to much less, since a great deal of unirrigated crop land is allowed to lie fallow at intervals to accumulate moisture and regain some of its fertility. Enough rain falls in the northwest during the winter half-year to permit unirrigated cropping, but rainfall sufficient for intensive agriculture is largely confined to a densely populated lowland strip between the Elburz Mountains and the Caspian Sea. The western half of this lowland receives more than 40 inches (c. 100 cm) annually, including both summer and winter rainfall, and produces a variety of subtropical crops such as wet rice, cotton, oranges, tobacco, silk, and tea. In the other regions of Iran, agriculture depends mainly or exclusively on irrigation. Many irrigated districts lie on alluvial fans at the foot of the Elburz and Zagros mountains; the capital, Tehran (Teheran: 8 million), is in one of these areas south of the Elburz. Qanats (tunnels) furnish the water supply for a large share of the country's irrigated acreage. However, they are expensive to build and maintain, and an increasing part of Iran's water supply during recent times has come from deep wells equipped with diesel pumps or from stream diversions. Wheat is the most widespread and important crop, as it generally is in Middle Eastern areas having winter precipitation and summer drought. Cattle are raised in areas where there is enough moisture to grow forage for them, but drier areas are characterized by many millions of sheep and goats. Many livestock are raised by seminomadic mountain peoples—the Qashqai (Kashgai), Baktiari, Lurs, Kurds, and others—whose independent ways have long been a source of friction between these tribesmen and the central government.

A little over half of all Iranians are reported to be urban dwellers, but many millions are still farmers or graziers. They have long been beset by the widespread Middle Eastern ills of drought, dusty winds, illiteracy, disease, poverty, and landlordism. Their access to modern conveniences is still scanty, although conditions have improved markedly in some areas during recent times. Traditional methods and tools, relatively unchanged for centuries, are still widely used.

Development under the Pahlavi Shahs

Prior to World War I, Iran was an exceedingly poor and undeveloped country, despite its ancient history of imperial grandeur when Persian kings ruled a great empire from Persepolis. In the 1920s a military officer of peasant origins, Reza Khan, seized control

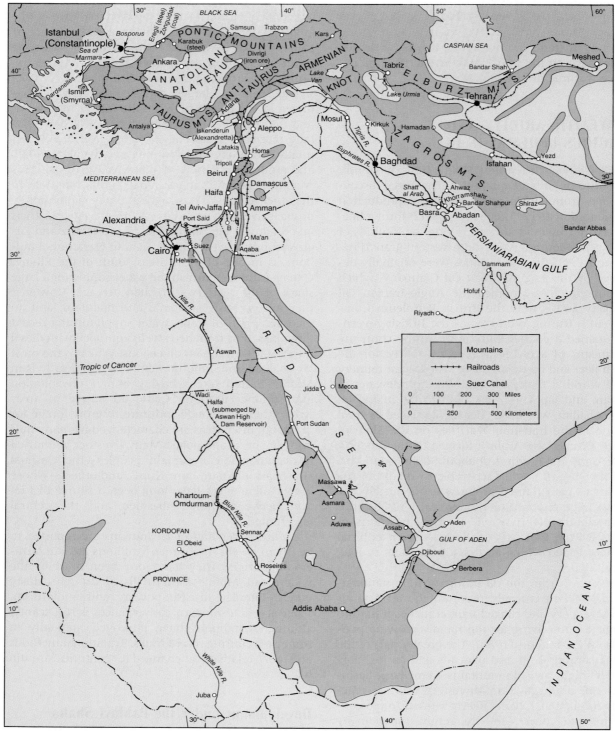

Figure 13.8 Mountains, railroads, and selected rivers and cities in the central part of the Middle East. The areas of mountains shown have been broadly generalized and often incorporate important agricultural valleys (as in western Turkey) or agricultural uplands (as in Ethiopia).

of the government and began a program to modernize Iran and free it of foreign domination. In 1925 he deposed the last shah of a family that had ruled Iran since 1794 and had himself crowned as Reza Shah Pahlavi, the founder of a new Pahlavi dynasty. Influenced by the modernizing efforts of Mustafa Kemal Ataturk in neighboring Turkey, the new Shah introduced social reforms, together with the beginnings of economic development on a broad front. To tie Iran together economically, the Trans-Iranian Railroad was constructed from the Persian Gulf across the Zagros and Elburz mountains to the Caspian Sea. Then during World War II the Shah displayed such pro-German leanings that his country was invaded by British and Russian troops (in 1941) for the purpose of securing this "back door" through which Allied supplies could reach the Russian front. (Subsequently, American forces played a large role in handling these supplies and improving both the Trans-Iranian Railroad and road connections across Iran.) As a result of the invasion, Reza Shah was forced to abdicate, and, in 1941, his young son took the throne as Mohammed Reza Shah Pahlavi.

Following the war, the young Shah continued and expanded the program of modernization and Westernization begun by his father. Progress was relatively slow until the decade of the 1970s, when Iran played a leading role in raising world oil prices. The mounting revenues that resulted underwrote explosive industrial, urban, and social development. New port facilities and surfaced highways were built, extraction of coal and a variety of metals and other minerals was stepped up, and hundreds of new factories were built. Foreign investors and corporations played a key role in this economic surge. The new industrial development centered in Tehran, but Isfahan (1.2 million) received a large steel mill, Tabriz (900,000) in the northwest became a center of machine-tool production to undergird general industrial advance, and a vast petrochemical plant on the Gulf was begun in association with Japanese interests. A nationwide attack on illiteracy was launched by a centrally administered Education Corps, and the literacy rate of the total population rose from a reported 16 percent in 1960 to 43 percent in 1980. Health services improved. Although such advances to date have fallen far short of Western or Japanese levels (Iran's infant mortality rate was recently reported to be over 20 times that of Japan), they seemed to indicate at the time that Iran was well on the road to modernization in the Western manner.

Meanwhile the countryside presented a mixed picture of outdated survivals and attempts at modernization. When the Shah took over in 1944, most rural villages and their land were owned either by the Crown, by large and very wealthy landowners

often called "the Thousand Families," or by Shiite priests (*mullahs*) who had been given land as a religious observance. Various attempts at land reform by the Shah's government resulted in the transfer of large amounts of Crown land to peasants, together with considerable land purchased from large private estates and religious holdings. Some 1½ million tenant farmers became small landowners. New owners were obliged to pay for the land over a period of years, but the annual charges were less than the rentals peasants had paid under traditional sharecropping arrangements.

Attempts were made by the Shah's government to upgrade agricultural productivity through the introduction of better methods, tools, seeds, and animal varieties, together with more fertilization of crops and better animal nutrition. Such measures had some success, although most farmers lacked the capital to take full advantage of them. A major effort was made to increase the supply of irrigation water and electricity by building large storage dams and hydropower stations along mountain rivers fed by melting snows. The largest river-control scheme involves the Karun River and smaller rivers in the southern oil-bearing region called Khuzistan, which adjoins Iraq. This part of Iran contains most of the country's Arab minority. Here a fringe of Iran is comprised of fertile alluvium like that of the adjoining Mesopotamian plain, and the climate permits an emphasis on such irrigated commercial crops as sugarcane, sugar beets, and citrus fruits. Organizationally, there were attempts to create larger and more efficient land units able to take advantage of mechanization and economies of scale. Large acreages were made available to giant agribusiness operations such as sugarcane and sugar beet plantations in Khuzistan. A few collective farms and a large number of cooperative farms were organized. Such measures created resentment among many landowners and farmers whose holdings were incorporated into larger units. Most farmers were little affected by the various reforms, and the reforms themselves were not sufficient to alter the generally depressed state of Iranian agriculture. A mounting exodus of poverty-stricken families from the countryside poured into Tehran and other cities. From this devoutly Muslim group of new urbanites came much of the support for the revolution that ousted the Shah in 1978–1979.

Religious Revolution and War with Iraq

The many kinds of development surveyed previously were funded largely by oil money, but the benefits to Iran's people were greatly lessened by the vast amounts spent on military forces and equipment by the Shah. Arms came primarily from the United States, which allied itself with Iran and looked to the Shah's army and air force to protect the Gulf oil region. Then in 1978–1979 these arrangements were shattered by a revolution that overthrew the Shah and abruptly took Iran out of the American orbit. The country became an Islamic republic governed by fundamentalist Shiite clerics. The latter include a handful of revered and powerful *ayatollahs* ("reflections of Allah") at the head of the religious establishment and an estimated 180,000 priests called *mullahs*. These religious leaders superintend all aspects of Iranian life and perform many functions allotted to civil servants in most countries. They base their authority on Shiite interpretations of the Islamic faith. The Shiite sect arose in Iran, and some 93 percent of all Iranians adhere to it. Iran is by far the largest Shiite country in the world.

The central figure among the revolutionaries who overcame the Shah was the Ayatollah Ruhollah Khomeini, an imposing elderly critic of the regime who was forced into exile by the Shah in 1964. Living in Iraq until 1978 (when he was expelled by President Hussein) and then in Paris, Khomeini sent repeated messages to Iran's Shiites that helped spark the revolution. During 1978 the country experienced a mounting series of riots and strikes, including strikes among oil-field workers that brought the oil industry to a halt. These disorders eventually caused the Shah to flee the country and abdicate the throne. Thereupon the Ayatollah Khomeini returned in triumph to Tehran, where he soon was able to take control of the government. Meanwhile some of his followers were allowed to hold 52 American diplomatic personnel as hostages in Tehran for 444 days—an enormously publicized event that helped defeat America's President Carter in the 1980 Presidential election. The revolution itself appears to have been the product of a rather diffuse but overwhelming buildup of opposition to the Shah and his regime among most elements of Iranian society. The opponents ranged from Communists through Westernized liberals to severely fundamentalist clerics and their followers. Grievances against the Shah were too numerous and complex to discuss here in detail, but a few major complaints can be cited. One was the widespread corruption within the ruling circles, whereby the Shah, his numerous relatives, members of the government, and court favorites enriched themselves by such means as accepting payoffs when government contracts were awarded to private firms, or demanding shares of firms seeking approval to do business in Iran. Further grievances included the huge sums spent on arms, the exaggerated pomp of the Shah's court, the heavy-handed control of dissent by the secret police, and the Shah's harassment of the influential merchants in the bazaars, who were denounced and often prosecuted for allegedly creating inflation by charging high prices for their goods. Landowners (including the mullahs) whose holdings had been diminished or amalgamated by the Shah's land reforms were resentful. But the real driving force of the revolution, as it now appears, was a furious tide of Shiite religious feeling directed against modernization, Westernization, Western imperialism, Communism, the exploitation of underprivileged people in Iran and elsewhere, and the very institution of monarchy, which was regarded by Shiite holy men as illegal under Shiite traditions of democracy and republicanism. What emerged was a theocratic state guided above all by edicts of the Ayatollah Khomeini, who undertook to create in Iran a society governed by strict Islamic principles and to spread his revolution to other lands. Nowhere was the new Shiite order in Iran more apparent than in the altered status of women, many of whom had become very emancipated under the Shah. In the new revolutionary Iran, women were obliged to veil themselves from head to foot in the black garment called the *chador;* many freedoms for women were revoked; and some mullahs even called for the restriction of women to their homes. All this was part of a thoroughgoing attempt to stamp out Westernization and restore Islamic purity. Within Iran, a merciless campaign to crush political and religious dissidence accompanied the accession of the Ayatollah to power. Many thousands were executed, including numerous Westernized middle-class intellectuals. Iran's Communists also were a major target.

Internal turmoil caused by the Khomeini repression soon was overshadowed by a murderous war with Iraq. This conflict was set off in 1980 when Iraqi President Saddam Hussein ordered an invasion of Iran by Iraqi troops, including armored and air forces. He seems to have assumed that revolutionary Iran would be too weak and chaotic to offer effective resistance and that several Iraqi purposes could be achieved with relative ease, including the overthrow of the Khomeini regime, regarded as a threat to Iraq, Saddam Hussein himself, and the Arab nation; the acquisition of territory in Iran that would give Iraq control of both sides of the Shatt al Arab waterway; and possible Iraqi acquisition of parts of Khuzistan containing rich oil resources and the bulk of Iran's Arab population. The Shatt al Arab, long a focus of acrimony between Iran and Iraq, was divided between them by a 1975 agreement fixing the boundary along the center of the navigable channel. This agreement was denounced by Saddam Hussein at the beginning of the war. As it turned out, Iran mounted a far better defense than had been anticipated, expending its greater manpower in bloody ''human wave'' assaults to counter Iraq's greater firepower from tanks, artillery, machine guns, and warplanes. A horrified world was told repeatedly of the slaughter of poorly armed or unarmed Iranian recruits—including thousands of children—who blew up minefields with their bodies to clear the way for Iranian tanks, and were killed in droves as they charged into Iraqi machine gun and cannon fire. Iran's Shiite mullahs actively recruited soldiers for the front as young as 12 to 15 years, promising them instant entry into heaven should they be killed. These youth brigades were adjuncts to Iran's regular army and the zealous followers of Khomeini called the Revolutionary Guards. The latter had been spearheads of the revolution (and the repression which followed it) and now they fought with equal zeal to save the revolution from the forces of Saddam Hussein.

Early in the war the Iraqis took Iran's main commercial port, Khorramshahr, but the Iranians retook it in fierce fighting and then launched offensive after offensive to drive the Iraqis out of Iran. The Iraqi air force bombed many Iranian cities, including Tehran; Basra was heavily shelled by the Iranians; and both Baghdad and Tehran were hit by long-range missiles. Both sides made attacks on Gulf oil installations and tankers, including many tankers belonging to noncombatant nations. By 1988 these attacks had brought about a sizable buildup of warships from the United States and several other outside nations to protect Gulf shipping sailing under their respective flags. In mid-1988 the total number of people killed in the war was estimated at about 500,000, representing a carnage greater than that of all the Arab-Israeli wars put together. By far the majority of the casualties were Iranian, although Iraqi casualties were heavy. The year 1988 saw the two sides holding military positions not greatly different from those at the beginning of the war. An armistice agreed to in midsummer, 1988, was still in force in June 1989, a month in which the Ayatollah Khomeini's death created new uncertainties in Iran.

TURKEY, CYPRUS, AND AFGHANISTAN

Turkey and Afghanistan contain some oil, although the deposits do not compare in size with those of the countries bordering the Persian/Arabian Gulf. In addition, Afghanistan has rather sizable reserves of natural gas, some of which is exported to the Soviet Union. These mountainous countries are Middle Eastern in many respects, but in other ways are transitional to other world regions. Like their joint neighbor, Iran, they lie outside the Arab-dominated part of the Middle East. But both are Muslim countries; indeed, Afghanistan has been a major stronghold of traditional Islam. The Mediterranean island of Cyprus off Turkey has a Christian (Greek Orthodox) majority in its population, but with a substantial Muslim (Turkish) minority. Ethnic differences on the island have been at the root of periodic flareups and fighting that have made Cyprus a focus of international tension and brought about a political partitioning of the island following a Turkish invasion in 1974. The island is described following the discussion of Turkey.

TURKEY: PAST AND PRESENT

Once a reactionary Muslim state at the center of a disintegrating empire, Tukey has been transformed in the 20th century into a relatively cohesive national unit with fewer ties to traditional Islam and a marked Western orientation. This rapidly growing nation of 53 million (1988) has recently been reaffirming its Islamic and Middle Eastern ties while continuing to function as a member of NATO and an associate member of the European Common Market.

Ataturk and the Transformation of Turkey

The Turks who organized the Ottoman Empire in the 14th and subsequent centuries were originally pastoral nomads from interior Asia, where major Turkic elements still exist in the Soviet Union and China. From the 16th to the 19th century, their empire was an important power. At its height in the 16th century its territories included North Africa as far west as Algeria and south to Sudan; most of southeastern Europe; the Fertile Crescent and parts of the Arabian Peninsula; and at the center of the empire the large Anatolian Peninsula between the Mediterranean, Aegean, and Black seas which forms most of the national territory of the present Republic of Turkey. That republic was created from the wreckage of the old empire after World War I. Its founder, Mustafa Kemal Ataturk, was determined to Westernize the country, raise its plane of living, and make it a strong and respected national state. He inaugurated social and political reforms designed to break the hold of traditional Islam and open the way for modernization and Turkish nationalism. Under the Ottoman Empire Islam had been the state religion, but Ataturk disestablished it. The Caliphate that had been held by the Ottoman Sultan was abolished. Wearing of the fez, an important symbolic act under the Ottoman caliphs, was prohibited, and the religious schools that had monopolized education were replaced by state-supported secular schools. To facilitate public education and remove further traces of Muslim dominance, the Arabic script of the Koran was supplanted by Latin characters. Polygamy was outlawed (not retroactively), along with slavery, and women were lifted from seclusion and given full citizenship. Legal codes based on those of Western nations replaced Muslim law, and the forms of democratic representative government were instituted, although Ataturk actually ruled as a dictator.

The transformation that Ataturk envisioned could not be accomplished overnight, although his reforms made remarkable changes in Turkish life. The peasantry, mostly illiterate and devoutly Muslim, submitted reluctantly to many elements in his program, although they supported his objective of creating a strong Turkey. Since Ataturk's death, the government has made concessions to them in religious matters. Due to a heterogeneous mixture of social groups and shades of political and religious opinion, Turkey has had trouble in establishing full democracy, and there have been frequent periods of military rule.

Characteristics of Turkish Agriculture

Although the value of manufactural output in Turkey is greater than that from agriculture, there is still a very strong rural and agricultural component in Turkish life. Some 47 percent of the population was still classed as rural in the middle 1980s. Most farmers are small landowners living in villages and owning their work animals. Cultivated land, including tree crops and vineyards, occupies perhaps one third of the country. Grains, primarily wheat and barley (wheat being the leader by far), are grown in all parts of the country; tree crops (principally olives, figs, nuts, citrus fruits) and vineyards are confined mainly to coastal sections. Sheep, goats, and cattle are an important component of agriculture in most areas. Turkey's farm production has greatly expanded under the republic, but yields are still low compared to those of western Europe. Expanded irrigation and greater use of commercial fertilizers are pressing needs, made even more imperative by the increasing use of new high-yielding grain varieties that require large applications of water and fertilizer. But irrigation is hindered by the difficulty of raising water to the level of the fields from streams that are often deeply intrenched; and farmers often are too poor to buy sufficient fertilizer. Relative poverty as compared to the great majority of European countries is a striking characteristic of present-day Turkey.

The Anatolian Plateau

From a physical standpoint Turkey is comprised of (1) the Anatolian Plateau and associated mountains, occupying the interior of the country (Fig. 13.8), and (2) the coastal regions of hills, mountains, valleys, and small plains bordering the Black, Aegean, and Mediterranean seas. The Anatolian Plateau is a country of wheat and barley fields and grazing lands. The annual precipitation, averaging only 8 to 16 inches (20–40 cm) and concentrated in the winter half-year, is barely sufficient for grain. The plateau lies at 2000 to 6000 feet (c. 600–1800 m); it is highest in the east where it adjoins the high mountains of the Armenian Knot. Its surface is rolling, treeless, windswept country, hot and dry in summer and cold and snowy in winter, with a natural vegetation of short steppe grasses and shrubs. The farming system combines cereals and the raising of livestock; traditional methods are still widespread, although tractors and other mechanized equipment are now fairly common. As in many developing countries, mechanization is held down by the high cost of machinery and fuel, as well as by the difficulty of servicing machinery in an undermechanized country.

Along its northern side, the Anatolian Plateau is bordered by the Pontic Mountains. These ranges, lying generally at 3000 to 6000 feet in the west but rising to a maximum of over 12,000 feet in the east, are clothed with Turkey's best forests. The trees, mainly hardwoods, are nourished by heavy precipitation throughout the year. The Taurus Mountains, bordering the plateau on the south, are in general somewhat higher than the Pontic Mountains. But their annual precipitation is smaller and their summers are dry; some good forests exist, but mediterranean woodland of little commercial value predominates.

Aegean, Black Sea, and Mediterranean Coastlands

Coastal plains and valleys along or near the Aegean Sea, Sea of Mamara, and the Black Sea are in general Turkey's most densely populated and productive areas. Here are grown such exports as hazelnuts, tobacco, grapes for sultana raisins, and figs. The Black Sea coastlands differ from the rest of Turkey in having summer as well as winter rain (although even here the maximum rainfall comes in the winter half-year) and in having a much greater total precipitation than other parts of the country. Growing of corn is a regional emphasis in this area. In the interior of Thrace ("European Turkey"), irrigated sugar beets are important, as they are in several other parts of the country.

The southern (Mediterranean) coast of Turkey is more sparsely populated than the Aegean or Black Sea coasts. Most of the coast is mountainous, and only a few sizable lowlands or valleys occur. The most important is the plain that centers on Adana (city proper, 600,000) in southeastern Turkey. This plain is Turkey's main area of cotton growing.

Industry, Mining, and Transportation

Except for handicrafts there was almost no industrial development in Turkey under the Ottoman Empire. During the latter part of the Ottoman period, European powers secured an increasing hold over Turkish finances by virtue of loans made to the Ottoman sultans. When the Turkish Republic was formed in 1923, its leaders were anxious to avoid further financial involvement with foreign nations, and they imposed severe restrictions on foreign investment in Turkey. This policy handicapped Turkish industry, because Turkey itself had insufficient capital to finance industrial development. Private capital was so lacking that the state, which was anxious to stimulate industry, was forced to play a large role in financing it. Many of the present industries are state-owned, although some have recently been sold to private owners in order to lessen state control and state protection of the economy as a means of achieving expanded production, greater exports, and lower inflation.

Turkey's predominant industries are those customarily found in developing nations: textiles, agricultural processing, cement manufacture, simple metal industries, assembly of vehicles from imported components, and so on. The country is very dependent on imports for much of its machinery as well as many other types of manufactured goods, although enough oil is refined in Turkey to supply domestic needs for refined products and permit some exports. An integrated iron and steel mill in the northern interior is the largest heavy-industrial establishment. Coke is manufactured here from coal supplied by a nearby coal basin. High-grade iron ore comes by rail from deposits 500 miles (800 km) away. In addition to coal and iron ore, Turkey has valuable deposits of chromium and a variety of other metallic and nonmetallic minerals. Transportation facilities are more adequate than those of most Middle Eastern countries, although not up to western European standards. In sum, Turkey is a kind of "in-between" country economically—well below the

Figure 13.9   The double-deck Karakoy (Galata) Bridge, which floats on huge pontoons, spans the Golden Horn inlet from the Bosporus in the heart of Istanbul, Turkey. In the background, minarets of historic mosques tower above the old walled city from which the Byzantine Empire and then the Ottoman Turkish Empire were ruled. Ferries such as the one in the foreground interconnect points along the Golden Horn and the nearby Bosporus. Road linkage between Istanbul and the main part of Turkey east of the Bosporus is provided by the high Bosporus Bridge, completed in 1973 and not shown in this view.

level of advanced Western industrial countries, but above most of the world's less-developed countries.

Istanbul and the Straits

Istanbul (5.5 million), formerly Constantinople or Byzantium, is Turkey's main metropolis, industrial center, and port (Fig. 13.9). One of the world's most historic and cosmopolitan cities, Istanbul was for many centuries the capital of the Eastern Roman (Byzantine) Empire. It became the capital of the Ottoman Empire when it fell to the Turks, after withstanding many sieges, in 1453. However, the capital of the Turkish Republic was established in 1923 at the more centrally located and more purely Turkish city of Ankara (3 million) on the Anatolian Plateau. Both Istanbul and Ankara have huge shantytowns that house migrants from impoverished rural Turkey. The same is true of Turkey's third largest city, the seaport of Izmir (1.5 million) on the Aegean Sea.

339

Istanbul is located at the southern entrance to the Bosporus, the northernmost of the three water passages (Dardanelles, Sea of Marmara, Bosporus) that connect the Mediterranean and Black seas and are known as the Turkish Straits (inset map, Fig. 12.1). The Straits have long been a focus of contention between Turkey on the one hand and Imperial Russia or the Soviet Union on the other. Today several of the main Soviet seaports are located on the Black Sea or Sea of Azov. The government to the north has often demanded a voice in the control of the Straits outlet to the Mediterranean. In recent years relations between Turkey and the Soviet Union have been relatively tranquil, although the Turks, who have fought many wars against the Russians, retain a wary posture. Within NATO, Turkey has more men under arms than any country except the United States. Some of Turkey's military preparedness has been dictated by fragile relations with its NATO neighbor, Greece. The main questions recently at issue have been the status of Cyprus (see the next section) and conflicting claims to oil and gas resources of the Aegean seabed.

ETHNIC SEPARATION IN CYPRUS

Most of the many islands that border the Aegean and Mediterranean coasts of Turkey are predominantly Greek in population and have long been part of Greece. But the large Mediterranean island of Cyprus, located near southeastern Turkey, came under British control in 1878 after centuries of Ottoman Turkish rule. Prior to its conquest by the Turks from the Venetians in the 16th century, Cyprus had formed a part of the Persian, Roman, and Byzantine empires. The island's strategic location in the eastern Mediterranean near Asia Minor, the Fertile Crescent, and Egypt made it an attractive forward base for imperial powers. Today its offshore situation in the area where Europe, Asia, and Africa meet continues to give it advantages as a rising center of international business.

Physical Character of Cyprus

The area of Cyprus—3572 square miles (9251 sq. km)—is about half that of the American state of Hawaii. Physically the island is a mosaic of small plains, hilly slopes, and low but steep mountains. The highest mountains (maximum elevations just over 6400 feet [c. 2000 m]) rise in the southern interior, while a lower range extends along the north coast. Picturesque whitewashed mountain villages dot both ranges. The mediterranean climate yields the cus-

tomary pattern of mild rainy winters and hot dry summers. Annual precipitation varies from a little under 10 inches (25 cm) in the driest lowlands to somewhat over 40 inches in the highest mountains. The heaviest precipitation nourishes oak and pine forests, but mediterranean scrub is the prevailing wild vegetation in most places. Rainfall sinks rapidly into the porous limestones that underlie most of the island, and there are few surface streams. Nearly half of Cyprus is under cultivation, with drier and mountainous districts being used for livestock herding.

Ethnic Conflict

An all-important circumstance in present-day Cyprus is the island's division between Greek Cypriots comprising about three fourths of the estimated population of 675,000 (1986) and Turkish Cypriots comprising about one fourth. Agitation by the Greek majority for union with Greece (*enosis*) was prominent after World War II and led in the 1950s to widespread terrorism and guerrilla warfare by Greek Cypriots against the occupying British. Violence also erupted between Greek advocates of *enosis* and the Turkish Cypriots, who greatly feared a transfer from British to Greek sovereignty.

An agreement to make Cyprus an independent republic was signed in 1959 by Britain, Greece, and Turkey. Approved by the leaders of both the Greek and Turkish Cypriots after lengthy negotiations, it guaranteed Britain's right to continued sovereignty over its bases on the island. Subsequent agreements between the new Republic of Cyprus (proclaimed in 1960) and Great Britain, Greece, and Turkey forbade both *enosis* and ethnic partitioning. Then in 1963 there was an upsurge of violence between the Greek and Turkish Cypriots after modifications in the agreements were proposed by Greek Orthodox Archbishop Makarios, who was serving as the young republic's first President. Troops from Britain, Greece, and Turkey were sent to restore order, and they were followed by a United Nations peacekeeping force which has since maintained a presence in the island. This force was unable to prevent further communal clashes later in the 1960s.

In 1974 a major national crisis erupted when a short-lived coup by Greek Cypriots temporarily overthrew President Makarios, who had generally followed a conciliatory policy toward the Turkish minority. The coup was fomented by the ruling military junta in Greece and was carried out by pro-*enosis* Cypriots (National Guard militia) mainly led by officers from Greece. Thereupon Turkey launched a military invasion which overran the northern part of the island. As a consequence, Cyprus became partitioned between the Turkish north and the Greek south. A buffer zone (the "Attila Line") sealed off the two sectors from each other, including sectors within the main city, Nicosia (perhaps 200,000). A separate government was established in the north, and in 1983 an independent "Turkish Republic of Northern Cyprus" was proclaimed. As of 1988, only Turkey had recognized this new state. Meanwhile the internationally recognized Republic of Cyprus has continued to function in the Greek-Cypriot sector, which comprises somewhat over three fifths of the island's land. Both republics have their capitals in Nicosia, which was the capital before partition.

Prior to the partitioning of Cyprus, the north had dominated the economy. But since partition the north has had severe economic difficulties, whereas Greek Cyprus has prospered. The Turkish sector was seriously weakened during the 1974 crisis by the flight of an estimated 200,000 Greek Cypriots to the south as refugees. There was a return flow of Turkish Cypriots entering the north, but this was far smaller and could not begin to compensate for the almost total loss of Greek entrepreneurs, farmers, skilled workers, and consumers. Most outside nations refused to trade directly with Turkish Cyprus after the invasion, and the trade and economic assistance proffered by an economically weak Turkey were insufficient to provide much momentum.

The Greek sector, by contrast, was able to make effective use of large amounts of economic aid from Greece, Britain, the United States, and the United Nations. A prolonged construction program provided new housing, business buildings, roads, and port facilities; tourism based on both beach resorts and mountain resorts was greatly expanded; and an efficient telecommunications system gave the south new links to all parts of the world. Many hundreds of new businesses came to the south, attracted there by such factors as favorable tax policies, modern facilities, dependable overseas communications, an educated and reasonably priced labor force, the relative security provided by the island location and a government friendly to foreign business, and the amenities provided by a Mediterranean tourist locale. Among the more prominent types of new or expanded enterprise are tourist-related businesses, international banking and insurance, shipping, transit trade, and miscellaneous light industries. Seaport and airport facilities in the Republic have been greatly expanded, and a free zone has been established that allows goods to be landed and transshipped without payment of customs duties.

On the whole the Greek sector has made a remarkable recovery from the chaos that accompanied the Turkish invasion in 1974. There is still consid-

erable dependence on foreign aid and imports of manufactures and food. But most refugees have been comfortably resettled and absorbed into the expanding economy, and the per capita GNP is probably three times that of the economically depressed north. The north, tied to the struggling economy of the Turkish mainland, is exceedingly dependent on aid from Turkey. Since 1974, many thousands of immigrants from Turkey have settled in the north.

RUGGED AFGHANISTAN—POOR, STRATEGIC, AND DEVASTATED

In the 20th century the highland country of Afghanistan (Fig. 13.10) has generally been remote from the main currents of world affairs, although the Soviet military intervention of 1979 and the ensuing devastation catapulted the country into world prominence. A landlocked nation largely occupied by high and rugged mountains, Afghanistan has limited resources, poor internal transportation, and relatively little foreign trade. Today it remains one of the world's poorest and most undeveloped countries. But the territory it occupies played a strategic

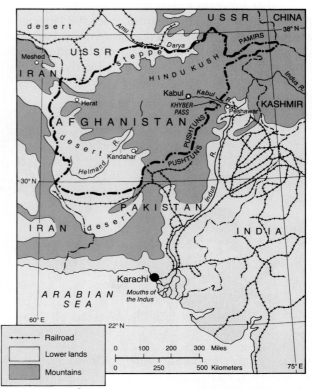

Figure 13.10 *Reference map of Afghanistan in its regional setting.*

role in empire building of the past, by virtue of important routes and passes leading across it from the steppes and oases of interior Asia and the plateaus of Iran to the plains of northern India that have been a goal of Asian conquerors for thousands of years.

The present Afghan state arose in the 18th century. Through much of the 19th century its independence was jeopardized by its position between the expanding British and Russian empires. But in 1907 Great Britain and Russia agreed to maintain it as an independent buffer state between their respective domains. British influence continued to be strong, however, until after World War I.

In 1973 the Afghan monarchy was overthrown by a military coup, and a Communist government oriented to the Soviet Union took power. Widespread rebellions ensued, and in 1979 the USSR was called on for assistance. The Soviets responded with sizable military forces, and a long guerrilla-type war began. The Soviet invasion roused an international furor, led by the United States. Widespread killing and maiming of civilians, destruction of villages, burning of crop fields, killing of livestock, and pollution or destruction of irrigation systems by Soviet ground and air forces caused several million Afghan refugees to flee into neighboring Pakistan (Fig. 13.11) or Iran. Meanwhile arms from various foreign sources filtered into the hands of rebel bands who kept up resistance in the face of heavy odds. The USSR kept general control over the cities and major routes, but large rural areas remained in rebel hands. Soviet troops were harassed by frequent raids and ambushes. Finally in 1988 and early 1989 the Soviet government withdrew its forces, but the wrecked country remained in the toils of civil war. What kind of regime would eventually emerge remained uncertain in early 1989.

Soviet military operations were hampered by Afghanistan's poor transport and communications network. Even today no rail line across Afghanistan connects Soviet Russia with the Indian subcontinent, although the Soviet and Pakistani rail systems extend to its borders (Fig. 13.10). In fact, within Afghanistan itself not a single mile of railway exists, and surfaced highways are very few. The limited foreign commerce of the country traditionally was routed through the port of Karachi in Pakistan. But trade is now primarily overland to the Soviet Union, which took about three fifths of Afghanistan's exports according to last reports in the early 1980s. Natural gas moving by pipeline to the USSR is the largest export.

Afghanistan's population before the Soviet invasion was estimated at about 15 million—only one third the size of neighboring Iran's. All but a small proportion of the people live in irrigated valleys around the fringes of a mass of high mountains occupying a large part of the country. The most heavily populated section is the southeast, particularly the fertile valley of the Kabul River in which is located the capital and largest city, Kabul (1.5 million; elevation 6200 feet [1890 m]). Most of the inhabitants of the southeast are Pushtuns, also known as Pashtuns or Pathans. The Pushtuns are the largest and most influential of the numerous ethnic groups that comprise the Afghan state. Their language, Pushtu,

Figure 13.11   This makeshift camp for Afghan refugees in the desert near Peshawar, Pakistan, symbolizes the thousands of camps for displaced persons that still exist over the world. Many millions of persons have fled or been forcibly consigned to such places in the 20th century as a result of warfare, repression, ethnic conflict, natural disasters, disease, and famine. Some camps with more permanent facilities than those shown here have existed for many years.

is related to Persian. The latter language is widely used in Afghanistan, being the main language of administration and commerce.

Afghanistan's second most populous area occupies a belt of foothills and steppes on the northern side of the central mountains. Most of its inhabitants live in a series of oases forming an east-west belt along the base of the mountains. Northern Afghanistan borders the Soviet Union, and millions of people on the Afghan side are related to peoples of Soviet Middle Asia.

Southwestern Afghanistan is comprised of a large desert basin, part of which extends into Iran. An area of interior drainage, it receives the waters of the Helmand River (Fig. 13.10) and several others originating in Afghanistan's central mountains. Extensive irrigation works supported a large population here in ancient times, but the water system was largely destroyed by Mongolian invaders in the 13th and 14th centuries, and the damage has never been fully repaired. The Afghan government, through its Helmand Valley irrigation and power project, has been reclaiming and resettling land by providing irrigation water from storage and flood-control reservoirs on the Helmand and other rivers.

Afghanistan is overwhelmingly a rural and agricultural or pastoral country. Recent times have seen a considerable development of small consumer-type industries and a few large industries, but most people gain their living from farming or livestock grazing. The country is so mountainous and arid that only an estimated 12 percent is cultivated. Enough rain falls in the main populated areas during the winter half-year to permit nonirrigated growing of winter grains. Livestock raising on a seminomadic or nomadic basis is widespread and important. In general, Afghanistan's agriculture bears many of the customary Middle Eastern earmarks: traditional methods, crude tools, inadequate fertilization, poor varieties of plants and animals, and low yields. The country exhibits a wide range of temperature conditions, corresponding in general to difference in altitude. A variety of cultivated crops, fruits, and nuts are locally important. A fair amount of mineral wealth exists, but the only mineral currently extracted in much quantity is the natural gas piped to the Soviet Union.

The Afghan-Pakistani Border Region

Present-day Afghanistan is an amalgam of many ethnic and tribal groups. Practically the whole population is Sunni Muslim (an estimated 87% in 1980) or Shiite Muslim (12%). Traditional modes of behavior based on family, tribal, or religious custom and allegiance are important influences, although a leaven of modernity was already bringing substantial social change before the Communist era. The most publicized region where tribalism remains very important lies along the eastern frontier with Pakistan. A variety of militant tribes and clans, mostly Pushtuns, inhabit this mountainous borderland on both the Afghanistan and Pakistan sides. Many migrate seasonally across the international boundary. These independent-minded people have always been slow to recognize the authority of central governments. In the days of Great Britain's Indian Empire, the area was proverbial for warfare among tribes, tribal raids on British-controlled areas, and British punitive expeditions against the tribes. In those days Peshawar on the Indian side of the Khyber Pass into Afghanistan became a noted garrison town for British and Indian troops. Since Britain's withdrawal from India in 1947, the border region has continued to be a source of friction between Afghanistan and the new state of Pakistan. Partially, the friction has been due to border incidents, but it has also grown out of proposals that the Pushtun-inhabited areas of Pakistan be incorporated in a separate state ("Pushtunistan"), either independent or affiliated with Afghanistan. Pakistan has firmly opposed such proposals. A new spotlight played on the region after the 1979 Soviet intervention in Afghanistan. Some 300 camps in Pakistan received an estimated 3 million Afghan refugees (Fig. 13.11). The numerous mountain passes scattered along the rugged border provided escape hatches for the refugees entering Pakistan. The passes also were supply routes for Afghan guerrilla forces fighting the Soviets.

ETHIOPIA, SOMALIA, AND DJIBOUTI: AFRICA'S EASTERN HORN

East of the Sudan section of the Nile Basin a great volcanic mountain mass rises steeply from the desert. This highland occupies the greater part of the "Eastern Horn" of Africa and is the heartland of the large country of Ethiopia (Fig. 13.8). Much of the area lies at elevations above 10,000 feet (3050 m) and one peak reaches 15,158 feet (4620 m). The highland receives its rainfall during the summer half of the year rather than the winter rain characteristic of mediterranean climatic areas. The Blue Nile, Atbara, and other Nile tributaries rise here (Fig. 13.3). Temperature conditions vary from tropical to temperate as altitude increases. Thus crops varying from bananas, coffee, and dates through oranges, figs, and temperate fruits to cereals can be produced without irrigation. Large expanses of upland pasture provide for a variety of livestock, primarily cattle and sheep.

Ethiopia is inhabited by perhaps 48 million people of very diverse racial and cultural origins. Although dark-skinned, most of the population is descended from the Hamitic or Semitic branches of the Caucasian race. Possibly one half, including the politically dominant Amhara peoples, adhere to the Coptic Christian faith, a very ancient branch of Christianity which penetrated Ethiopia from Egypt, where there is still a sizable Coptic minority. The rest of the people are divided between the Muslim faith, a variety of tribal religions, and some Protestant, Evangelical, or Roman Catholic Christians. The vast majority of the population is extremely poor and illiterate, and tribal forms of life remain important. A recent estimate placed the number of physicians in Ethiopia at around 450, or about 1 for every 90,000 people. Few parts of the world are so abysmally poor.

East of the mountain mass of Ethiopia, and partly included within the country, lower plateaus and coastal plains descend to the Red Sea, the Gulf of Aden, and the Indian Ocean. Extreme heat and aridity assert themselves at these lower levels, and nomadic or seminomadic Muslim tribesmen make a poor living from camels, goats, and sheep. Precarious forms of agriculture are carried on in scattered oases.

Ethiopia and its borderlands constitute a rather marginal part of the Middle East, especially the rainy tropical Ethiopian highlands. However, the arid lowland sections possess many typically Middle Eastern characteristics, and the entire area has had important cultural and historical links with the core of the Middle East in Egypt, the Fertile Crescent, and Arabia.

In 1974 the aging Emperor Haile Selassie, who had held Ethiopia's population together in a relatively loose political union, was deposed by a civilian and military revolt against the country's feudal order. Causes of the revolt included widespread drought and famine during the early 1970s, separatist dissension in some areas, and general discontent with the country's social and economic backwardness. A harshly repressive Marxist dictatorship strongly oriented to the Soviet Union emerged from the revolution. Thousands of middle-class intellectuals were executed or jailed; land ownership and industries were nationalized; and collectivization of agriculture was begun. American influence, which had been strong under Haile Selassie, was eliminated. Munitions and advisors from the Soviet Union and other East-bloc nations poured in to equip and train one of Africa's largest armies, and the Soviets gained access to base facilities in Ethiopian ports. The takeover of Ethiopia by Marxism, together with an earlier takeover of South Yemen, gave the USSR a foothold on both sides of the strategic southern entrance to the Red Sea. Attempts by outside powers to gain such footholds have a long history, as the following section will show.

EUROPEAN IMPERIALISM

European powers seized coastal strips of the Eastern Horn in the latter 19th century. Britain was first with British Somaliland in 1882, then France annexed French Somaliland in 1884, and lastly Italy asserted dominance over Italian Somaliland and Eritrea in 1889. The strategic importance of these holdings along the Suez-Red Sea route and adjacent to Ethiopia is obvious. Their economic importance has been very small.

Italy attempted to extend its domain from Eritrea over the much more attractive and potentially valuable land of Ethiopia in 1896, but the Italian forces were annihilated by the Ethiopian tribesmen at Aduwa. Forty years later, in 1936, a second attempt was successful. Hopes of developing the country's potential wealth and of using Ethiopia as an outlet for surplus Italian population were frustrated in World War II, when the Italian empire in eastern Africa was conquered by British Commonwealth forces (primarily South Africans). Ethiopia was restored to independence and Eritrea was federated with it as an autonomous unit in 1952. Subsequently Ethiopia incorporated Eritrea as a province (in 1962).

Following World War II, Italian Somaliland was returned to Italian control to be administered as the Trust Territory of Somalia under the United Nations, pending independence in 1960 as a republic. The present Somali Democratic Republic, or Somalia, includes not only the former Italian territory, but former British Somaliland as well. The British territory was granted independence at the same time and chose to unite with Somalia. The people of much smaller French Somaliland remained separate and eventually became independent as the Republic of Djibouti, with its capital in the seaport of the same name. Influential elements among the Somali peoples have advocated the formation of a "Greater Somalia" which would include the Somaliland units plus substantial parts of Ethiopia and possibly some of Kenya. Claims to Ethiopian areas rest primarily on seasonal use of these lands by Somali graziers who migrate with their livestock across the international boundary. The Ethiopian government opposes such proposals to annex parts of its territory and has advanced counterclaims to parts of Somalia. There has been a good deal of fighting between regular military forces of the two countries, and much guerrilla activity by anti-government rebels—within Somalia as well as within Ethiopia. It is estimated that over 1 million Somali refugees have fled from Ethiopia into Somalia as a result of warfare or famine. This entire struggle, centering in the dry lowland of eastern Ethiopia called the Ogaden, is perceived only dimly by the outer world and from that standpoint is representative of the incredibly numerous armed conflicts that have beset Africa in recent decades.

PROBLEMS OF ETHIOPIAN DEVELOPMENT

Ethiopia has substantial resources, possibly including a fair amount of mineral wealth. But its resource base remains largely undeveloped, although the government, with foreign assistance, has been able to develop a limited number of factories and a few power dams to harness some of the country's large hydroelectric potential. A poor transportation system—whose best parts often tend to be the 3000 miles (*c.* 4800 km) of good roads built by the Italians during their occupation in the 1930s—is a major factor in keeping the country isolated and underdeveloped. The French-built railroad from Djibouti to Addis Ababa (1½ million) lacks feeder lines and has never been as successful as hoped. Many routes into the Ethiopian heartland are mere trails, although better road connections have gradually been developing, and air services help somewhat in overcoming the deficiencies in ground transportation. But on the whole Ethiopia is badly handicapped by inaccessibility. A good share of its population lives in villages on uplands cut off from each other and from the outside world by precipitous chasms. These formidable valleys were cut by streams that carried huge amounts of fertile volcanic silt to the Nile and thus provided the main material of which the famous flood plains in Egypt are built.

Ethiopia has been further handicapped in recent times by political and military turmoil centering in Eritrea, the adjacent parts of northern Ethiopia, and the borderland between Ethiopia and Somalia. A secessionist movement has long existed in predominantly Muslim Eritrea. Both Muslims and Christians in that province have resented their political subjection to Ethiopia's Amhara majority. Terrorism and small-scale guerrilla actions on behalf of Eritrean self-determination escalated into civil war during re-

cent years. A major effort was made by the Ethiopian army to stamp out the revolts by two Eritrean "liberation fronts" (one Marxist and the other non-Marxist), but as of early 1989 resistance was continuing. Eritrea borders the Red Sea and contains the small ports of Massawa and Assab which handle the bulk of Ethiopia's foreign trade. Both have highway connections to Addis Ababa, and Massawa has a relatively short rail connection to adjacent highlands as well (Fig. 13.8). The railroad serves Eritrea's main city and manufacturing center, Asmara (275,000), located in highlands at an elevation of over 7000 feet (2130 m). Desire for greater autonomy has also led to warfare in Tigre Province adjoining Eritrea. In both Eritrea and Tigre, government forces have been holding the main towns and their vicinity, while rebels have controlled large sections of the countryside.

In addition to the devastation wrought by war, Ethiopia has been stricken since the late 1960s by droughts centering in the war-torn areas of the north. For years at a time, there has been little or no rain in areas inhabited by millions of people. Hunger has beset both people and livestock, to the point that desperate farmers have fed their families with grain needed for seed and have sold their oxen needed for plowing. The Ethiopian villagers were especially vulnerable to drought because of their heavy dependence on rainfed agriculture. The volcanic soils are fertile, but the streams are entrenched so far below the upland fields that little irrigation is feasible.

In the early 1980s drought in Ethiopia reached the point of widespread catastrophe. Great numbers of Ethiopians left their villages and gathered in makeshift camps in the hope of being fed. Lack of drinking water and outbreaks of disease compounded the basic problem of insufficient grain and other food. In the late summer of 1984, hundreds of thousands of penniless refugees fled to camps in neighboring Sudan, which was beset by droughts and famines of its own but at least was more accessible to outside relief agencies than was highland Ethiopia. Within Ethiopia, such agencies encountered great difficulties in getting food to the hungry due to such factors as the limited capacity of the available ports, competition for use of the ports by general export and import traffic and arms shipments, the insufficiency of trucks and servicing, the lack of road access to starving villages, and government bombing and strafing of relief trucks in rebellious areas.

Recent economic statistics for Ethiopia point to a country that is still desperately undeveloped. In 1986 the per capita GNP was calculated at $120 and Ethiopia was said to be possibly the least developed of the world's countries. Coffee, often picked from wild trees, made up nearly three fourths of the scanty exports in 1984–1985; there were practically no exports of manufactured goods or minerals. Manufacturing is largely confined to simple textiles or processed food. Addis Ababa and Eritrea account for the bulk of the manufacturing.

ETHNIC HOMOGENEITY AND ECONOMIC SIMPLICITY IN SOMALIA

Between Ethiopia and the Indian Ocean lies Somalia—a country only slightly less poor than Ethiopia. Some 99.8 percent of the people are Sunni Muslims and 95 percent are ethnic Somalis—surely one of the most homogeneous populations in the world.

Moisture is so scanty that only 2 percent of the country is cultivated. Livestock raising on a nomadic or seminomadic basis is the main support for Somalia's simple economy; in recent years live animals (goats, sheep, camels, cattle), shipped mainly to Saudi Arabia, have comprised three fourths of all exports. Industry is very minor and confined largely to food processing; mineral production is practically nonexistent. Somalia formerly received aid from the East bloc but became estranged when the bloc aided Ethiopia in the Ogaden war. The United States was then granted access to base facilities in the two main cities and ports: Mogadishu (500,000), the capital, in former Italian Somaliland, and the much smaller city of Berbera in former British Somaliland.

PART V

THE ORIENT

CHAPTER 14

Geographic Themes and Variations in the Orient

The packed housing of this Japanese town in a mountain valley on the island of Kyushu symbolizes the crowded Orient, where over half of the world's people occupy a relatively minor share of the Orient's total land.

HE Orient (East) is used in this book as a regional name for the countries occupying the southeastern quarter of Eurasia. In addition to mainland countries extending from India, Pakistan, and Bangladesh to China, Mongolia, and Korea, the Orient, as defined here, also includes an arc of island countries stretching for thousands of miles from Sri Lanka (Ceylon), through Indonesia and the Philippines, to Japan (Figs. 14.1 and 14.2). Extreme diversity marks the many countries grouped here as a major world region, yet certain common themes exist that provide vantage points from which to gain perspective on this important area as a whole. In this chapter these themes will be introduced, along with some supporting details. Subsequent chapters will then explore more localized aspects in countries grouped

Figure 14.1 Introductory location map of the Orient, showing political units as of January 1, 1989.

under the regional headings of the Indian subcontinent, Southeast Asia, the Chinese realm, and Japan and Korea. (For area, population, and other data, see Table 14.1.)

MAJOR ORIENTAL THEMES

Many of the geographic characteristics highlighted under the following themes are not unique to the Orient, but the total array of themes does set this area clearly apart from all other big sections of the world. The themes tend to mask profound differences from place to place within the Orient—as between India and China, for example—and a case can be made for distinguishing more than one "major world region" here. Still, the overall commonality does exist in more than a superficial way. The "major world regions" in this text are devices to further world comprehension, and introducing too many of them would be likely to complicate the learning process without adding much to broad-scale understanding. Using "the Orient," then, as a convenient

Figure 14.2 Major landforms and water features of the Orient.

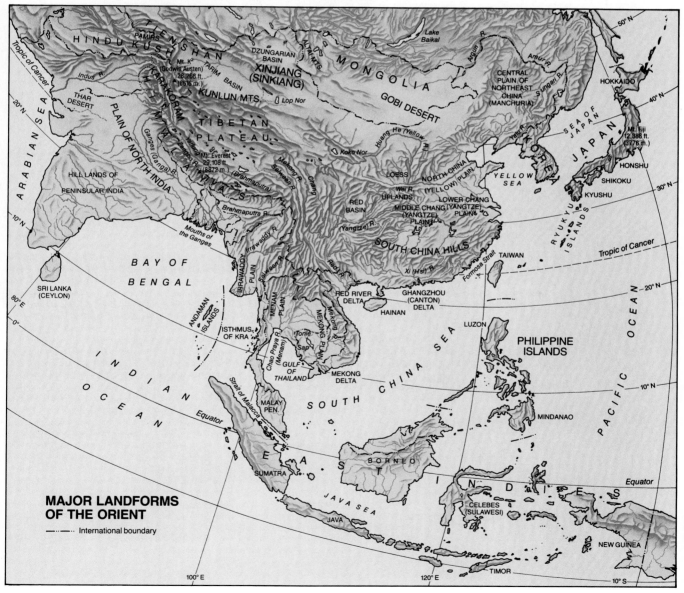

TABLE 14.1 ORIENT: BASIC DATA

Political Unit[a]	Area (thousand sq mi)	Area (thousand sq km)	Estimated Population (millions, mid-1988)	Estimated Annual Rate of Natural Increase (%, 1988)	Estimated Population Density (per sq mi)	Estimated Population Density (per sq km)	Infant Mortality Rate	Urban Population (%)	Cultivated Land (% of total area)	Per Capita GNP ($US: 1986)
Indian Subcontinent										
India	1266.6	3280.5	816.8	2.0	645	249	104	25	51	270
Pakistan	310.4	803.9	107.5	2.9	346	134	121	28	25	350
Bangladesh	55.6	144.0	109.5	2.7	1969	760	135	16	63	160
Nepal	54.4	140.8	18.3	2.5	336	130	112	7	16	160
Bhutan	18.1	47.0	1.5	2.0	75	29	139	5	2	160
Totals	1705.1	4416.2	1053.6	2.1	618	239	109	24	45	265
Southeast Asia										
Sri Lanka	25.3	65.6	16.6	1.8	656	253	31	22	34	400
Burma	261.2	676.3	41.1	2.1	157	61	103	24	15	200
Thailand	198.5	513.8	54.7	2.1	276	106	52	17	38	810
Vietnam	127.3	329.4	65.2	2.6	512	198	53	19	20	170 (1983)
Kampuchea	69.9	181.0	6.7	2.3	96	37	134	11	17	–
Laos	91.4	236.7	3.8	2.5	42	16	122	16	4	220 (1984)
Malaysia	127.3	329.6	17.0	2.4	134	52	30	35	13	1850
Singapore	.22	.58	2.6	1.0	11000	4250	9	100	10	7410
Brunei	2.2	5.8	0.3	2.7	111	52	11	64	1	15400
Indonesia	735.4	1903.8	177.4	1.7	241	93	88	22	11	500
East Timor	5.7	14.9	0.7	2.2	122	47	166	12	5	–
Philippines	115.8	299.9	63.2	2.8	546	211	51	41	38	570
Totals	1760.3	4559.0	449.3	2.3	255	99	72	25	17	650
Chinese Realm										
China (PRC)	3705.4	9598.3	1087.0	1.4	293	113	44	21	11	300
Hong Kong	.4	1.0	5.7	0.8	14000	5650	8	93	8	6720
Macao	.007	0.02	0.4	1.7	70000	27000	12	97	0	2270 (1983)
Taiwan	12.5	32.3	19.8	1.1	1584	613	7	67	24	3180 (1985)
Mongolia	604.2	1565.0	2.0	2.6	3	1	53	52	1	1000 (1984)
Totals	4322.5	11196.6	1114.9	1.4	258	100	43	21	8	390
Japan and Korea										
Japan	143.7	372.2	122.7	0.5	854	330	5	77	13	12850
North Korea	46.5	120.5	21.9	2.5	471	182	33	64	19	760 (1984)
South Korea	38.0	98.4	42.6	1.3	1121	433	30	65	22	2370
Totals	228.3	591.1	187.2	0.9	820	317	14	73	16	9100
Grand Totals	8016.2	20761.9	2805.0	1.8	350	135	71	26	18	970

aFigures for India include the Indian-held part of Kashmir; figures for Pakistan exclude the Pakistani-occupied part of Kashmir. Figures for China (PRC) do not include Taiwan or Chinese-occupied...

learning aid, we can begin to form some initial impressions from the following set of themes:

1. The Orient is a region where many geographic patterns were created or reshaped by Western colonialism.

2. During the 20th century the region has shown great political instability associated with decolonization and Communist expansion.

3. The natural setting shows extreme variety, but still exhibits a certain homogeneity in (a) the prevalence of land unsuited for agriculture because of roughness or other environmental handicaps, (b) the widespread monsoonal climates, and (c) the prevalence of tropical or subtropical temperatures in the areas where most Orientals live.

4. There is a strikingly uneven pattern of population distribution: Over half the world's people live here, but they are heavily concentrated on scattered pieces of land representing a relatively small part of the region's total area (see chapter-opening photo).

5. The complex racial and cultural pattern exhibits great differences from place to place, yet has some measure of unity in the widespread distribution of culture traits derived ultimately from India or China or both.

6. The region's people are very poor, aside from the affluent Japanese and relatively small upper and middle classes elsewhere; and the rapid growth of population in recent times has made the poverty hard to combat.

7. Relief of poverty and the attainment of power have been major motivations for Oriental countries to develop modern industries, but large-scale industrialization thus far has been scattered and not of outstanding world consequence except for the giant industrial economy of Japan.

8. Rural, agrarian, and village life prevails in most of the region, and urban dwellers comprise little more than one fourth of the population, but the number of people is so great that approximately 750 million—equal to the *total* population of Europe plus the United States of America—actually live in towns and cities.

9. Irrigated rice dominates the cropping pattern in areas where it can be grown, as it is dependable, high-yielding, and generally preferred to other grains in Oriental diets; but many other food crops are grown, along with limited numbers of animals kept for food and draft power, and some parts of the region have long been noted for specialty export crops such as natural rubber or tea.

10. The spread of new crop varieties by the "Green Revolution" has given an upsurge of food production in many areas and has increased hope (despite the heavy capital inputs required) that the Orient can provide an adequate food supply for its expanding population.

REGIONAL GROUPS OF COUNTRIES

For convenience in study in the chapters that follow, the countries of the Orient are divided into a number of regional groups on the basis of such factors as (1) proximity; (2) environmental, economic, and cultural similarities; (3) political relationships; and (4) historical ties.

1. The *countries of the Indian subcontinent* include India, Pakistan, Bangladesh, the disputed area called Kashmir, and the small Himalayan states of Nepal and Bhutan. Sri Lanka (formerly Ceylon) is often included with this group, although it is considered with Southeast Asia in this presentation.

2. The *countries of Southeast Asia* include (a) Burma; (b) Thailand; (c) the units that once comprised French Indochina—Vietnam, Kampuchea (Cambodia), and Laos; (d) Malaysia, comprised of the mainland unit called Malaya, together with Sarawak and Sabah occupying contiguous portions of the large island of Borneo; (e) the small but densely populated and commercially important island country of Singapore, essentially a single seaport; (f) the large island countries of Indonesia and the Philippines; (g) the smaller island country of Sri Lanka; and (h) the small but wealthy oil-exporting sultanate of Brunei on Borneo. East Timor is a former Portuguese colony held by Indonesia.

3. The *countries of the Chinese realm* as considered here include the People's Republic of China, Taiwan, the Mongolian People's Republic or Mongolia (placed in this group for convenience and because of its historical associations with China), and the European coastal possessions of Hong Kong (British) and Macao or Macau (Portuguese), which are largely Chinese in population and will be transferred to China before the end of the century.

4. Japan, for the sake of convenience, is discussed here with its former colony of Korea, now divided between North Korea (Communist) and South Korea (non-Communist).

THE COLONIAL BACKGROUND OF THE ORIENT

Modern European penetration of the Orient began at the end of the 15th century. The early comers established trading posts and gradually extended political control over limited areas near the coast. In the 18th and 19th centuries the pace of annexation quickened, and large areas came under European sway. By the end of the 19th century, Great Britain was supreme in India, Burma, Ceylon, Malaya, and northern Borneo; the Netherlands possessed most of the East Indies; and France had acquired Indochina and a number of small holdings around the coasts of India. Meanwhile Portugal, the supreme colonial power of the Orient in the 16th century, had been displaced from her early holdings with the exception of Macao, a part of the island of Timor, and Goa and two other small holdings on the west coast of India. China, although retaining a semblance of territorial integrity, was forced to yield possession of strategic Hong Kong to Britain in the mid-19th century and to grant special trading concessions and extraterritorial rights to various European nations and the United States. At the end of the century the Philippines, dominated by Spain after the mid-16th century, passed into American control. The only Oriental countries of any importance to escape domination by the Western powers during the colonial age were (1) Thailand, which formed a buffer between British and French colonial spheres in Southeast Asia; (2) Japan, which withdrew into almost complete seclusion in the 17th century, but emerged in the latter 19th century as the first modern, industrialized Oriental nation, and soon acquired a colonial empire of its own; and (3) Korea, which also followed a policy of isolation from foreign influences from the early 17th century until 1876, when a trade treaty was forced on it by Japan.

During the age of European expansion the Orient constituted an extraordinarily rich colonial area from which Western nations extracted vast quantities of such valuable commodities as rubber, sugar, tea, copra, palm oil, spices, and tin, and in which Western manufacturers found large markets for cheap textiles, metalwares, and other inexpensive types of goods. Westerners also found the Orient a fertile field for investment in plantations, factories, mines, transportation, communication, and electric power facilities.

END OF COLONIALISM IN THE TWENTIETH CENTURY

Western dominance of the Orient was ended in the 20th century by a complex chain of circumstances, including (1) the weakening effects of conflicts among the Western nations in the two world wars; (2) the rise of Japan to great-power status and its successful, although temporary, military challenge to the West in the early stages of World War II; and (3) the rise of anticolonial movements in areas subject to European control. World War II ended Western colonialism in China, and after the war all colonial possessions in the Orient, whether European, American, or Japanese, gained independence, except for Hong Kong (now scheduled by a British-Chinese agreement to be returned to China in 1997) and Macao (likewise slated for transfer from Portuguese to Chinese sovereignty, in 1999).

In the Orient the 20th century has been an age of revolution, war, and general turmoil. All Oriental nations have been affected by these stresses, and the spotlight of world attention has focused first on one area and then another. The trouble spots are too numerous and widespread to recount here. The major ones will be touched on in the chapters that follow.

THE VARIED PHYSIOGRAPHIC SETTING

The stage on which the Oriental drama is being enacted is a complex intermingling of many types of topography—high, rugged mountains, arid plateaus and basins, humid hill lands and river plains, and a vast number of offshore islands rising from the floor of shallow seas. Although the picture is extremely complicated in detail, a certain order appears if the surface features of the Orient are conceived of as three concentric arcs or crescents of land—an inner arc of high mountains, plateaus, and basins; a middle arc of lower mountains, hill lands, and river plains; and an outer arc of islands and seas (Fig. 14.2).

THE INNER HIGHLAND

The inner highland of the Orient is composed of the highest mountain ranges on earth, interspersed with plateaus and basins. At the south the great wall of the Himalaya, Karakoram, and Hindu Kush mountains overlooks the north of the Indian subcontinent (Fig. 15.4). At the north, the Altai, Tien Shan, Pamirs, and other mountains separate the Orient from the Soviet Union. Between these mountain walls lie the sparsely inhabited Tibetan Plateau, over 15,000 feet (*c.* 4500 m) in average elevation, and the dry, thinly populated basins and plateaus of Xinjiang (Sinkiang) and Mongolia.

RIVER PLAINS AND HILL LANDS

The area between the inner highland and the sea is principally occupied by river flood plains and deltas, bordered and separated by hills and relatively low mountains. Major components of the topography are (1) the immense alluvial plain of northern India, built up through countless ages by the Indus, Ganges, and Brahmaputra rivers; (2) the hill lands of peninsular India, geologically an ancient plateau, but largely hilly in aspect; (3) the plains of the Irrawaddy, Chao Praya (Menam), Mekong, and Red rivers in peninsular Southeast Asia, together with bordering hills and mountains; (4) the hill lands and small alluvial plains of southern China; (5) the broad alluvial plains along the middle and lower Chang Jiang (Yangtze River) in central China and the mountain-girt Red Basin on the upper Chang Jiang; (6) the large delta plain of the Huang He (Yellow River) and its tributaries in North China, backed by loess-covered hilly uplands; and (7) the broad central plain of Northeast China (Manchuria), almost completely enclosed by mountains with low to moderate elevations. The floor of the Red Basin and the central plain of Northeast China are structural rather than river-made plains and are rolling or hilly rather than flat.

OFFSHORE ISLANDS AND SEAS

Offshore, a fringe of thousands upon thousands of islands, mostly grouped in great archipelagoes, borders the mainland. On these islands high interior mountains with many volcanic peaks are flanked by broad or narrow coastal plains where most of the inhabitants live. Three major archipelagoes include most of the islands—the East Indies, the Philippines, and the Japanese Archipelago. Sri Lanka, Taiwan, and Hainan are large islands not included in an archipelago. Between the archipelagoes and the mainland lie the China Seas, and, to the north, the Sea of Japan. At the southwest the Indian peninsula projects southward between two immense arms of

the Indian Ocean—the Bay of Bengal and the Arabian Sea.

THE PATTERN OF CLIMATES AND VEGETATION

In detail the climatic, like the physiographic, pattern is one of almost endless variety. However, two unifying elements are present throughout most parts of the Orient inhabited by any considerable number of people. These are (1) the dominance of warm climates and (2) a characteristic monsoonal regime of precipitation.

TEMPERATURE AND PRECIPITATION

In the parts of the Orient where most of the population is found, temperatures are tropical or subtropical and the frost-free season ranges from around 200 to 365 days. The principal exceptions exist in northern sections of China, Korea, and Japan. Here summers are warm to hot in the lowlands, but the growing season is generally less than 200 days and winters are often severely cold. The arid, sparsely populated basins and plateaus of Xinjiang and Mongolia also have a continental type of climate with warm summers and cold winters. The higher mountain areas and Tibetan Plateau have highland climates varying with the altitude and latitude. Permanent snow fields and glaciers occur at the higher elevations.

Annual precipitation varies from near zero in parts of Xinjiang to an average of more than 400 inches (over 1000 cm) in parts of the Khasi Hills of northeastern India. A monsoon climate, or at least a climate with monsoon tendencies, prevails nearly everywhere in the populous middle arc of plains and hills, and in many parts of the islands as well. Technically a monsoon is not, as some people imagine, a violent downpour of rain with accompanying winds and lightning (although such are often its effects) but is simply a current of air blowing fairly steadily from a given direction for several weeks or months at a time. Although conditions vary from place to place, the Orient is broadly characterized by two monsoons: (1) a summer monsoon blowing from the sea to the land and bringing high humidity and rain, and (2) a winter monsoon blowing seaward and bringing little or no rain and cool or cold, clear weather. The characteristic features of the monsoonal type of climate, then, are (1) the seasonal reversal of wind direction, (2) the strong summer maximum of rainfall, and (3) the long dry season, typically lasting for most or all of the winter half-year.

TYPES OF CLIMATE

Seven main types of climate in the Orient are customarily recognized in climatic classifications: tropical rain forest, tropical savanna, humid subtropical, humid continental, steppe, desert, and highland (see world climatic map, Fig. 2.18).

Tropical Rain Forest

The rainy tropical climates of the Orient are found along or relatively near the equator. Consequently, high temperatures are experienced throughout the year in the lowlands. There is a complete absence of frost, and the year-round growing season offers the maximum possibilities for agriculture from the standpoint of temperature. However, the high temperatures and heavy rain promote rapid leaching of mineral nutrients and destruction of organic matter, with the result that most soils in the rainy tropics are relatively infertile despite the thick cover of deep-rooted trees and grasses they often support. Two main types of humid tropical climate, each associated with characteristic forms of vegetation, are recognized. These are (1) tropical rain forest climate and (2) tropical savanna climate.

The *tropical rain forest climate* is typically found in lowlands within 5 or 10 degrees of the equator. A rainfall of at least 30 to 40 inches (*c.* 70–100 cm), often 100 inches (*c.* 250 cm) or more, is spread throughout the year so that each month has considerable rain. Average temperatures vary only slightly from month to month: Singapore, for example, exhibits a difference of only 3°F between the warmest and coolest months. Monotonous heat prevails all year, although excessively high temperatures of 95° to 100°F (35°–38°C) or more are seldom or never experienced. Some relief is afforded by a drop of 10° to 25°F (6°–14°C) in the temperature at night, and conditions are more pleasant along coasts subject to periodic sea breezes.

The tropical rain forest produced by the climatic conditions just described is characteristically a thick forest of large broadleaf evergreen trees, mostly hardwoods, from 50 to 200 feet (*c.* 15–60 m) in height, and forming an almost continuous canopy of foliage. The trees are often entangled in a mass of vines, and dense undergrowth occurs wherever sufficient light can penetrate to the ground.

Tropical rain forest climate and vegetation are characteristic of most parts of the East Indies, the Philippines, and the Malay Peninsula. Rain forest vegetation is also found in certain other areas which experience a dry season, but in which the precipitation of the rainy season is sufficiently heavy to promote a thick growth of trees. Such areas exist (1) along the west coast of India south of Bombay and along the south coast of Sri Lanka; (2) along the coasts of Kampuchea, Thailand, and Burma, extending northward to the Ganges-Brahmaputra delta region in India and Bangladesh; (3) along part of the east coast of Vietnam; and (4) in parts of the northern Philippines.

Tropical Savanna

Although the *tropical savanna climate*, like the tropical rain forest climate, is characterized by high temperatures the year round, this climate type is customarily found in areas farther from the equator, and the average temperatures vary somewhat more from month to month. However, the most striking and important difference between the two climate types lies in the fact that the savanna climate has a well-defined dry season, lasting in some areas for as much as 6 or 8 months of the year. The annual precipitation is less, on the average, than in the rain forest climate, but it is the seasonal distribution of the rain rather than inadequate total precipitation which represents the principal handicap for agriculture. The main areas of tropical savanna climate in the Orient occur in southern and central India, the Indochinese Peninsula except for most coastal areas and some northerly areas, and eastern Java and smaller islands to the east. Much larger areas of this climate type occur in Africa and Latin America and a smaller area in northern Australia.

In the Orient the characteristic natural vegetation associated with the savanna climate is a deciduous forest of smaller trees than those found in the tropical rain forest. The forest growth deteriorates to scrub in drier areas. Tall, coarse tropical grasses, a very common vegetation form in the African and Latin American savannas, are found in only limited areas, and even there are thought to have been produced by repeated burning of forest growth during the dry season. In parts of Southeast Asia long-continued burning has fostered pure stands of certain tree species peculiarly resistant to extinction by fire. Among these are a number of economically valuable types, especially teak, which is exploited on a considerable scale in Burma and Thailand. In the more densely populated and long-settled parts of the Ori-

Figure 14.3   Towering dunes and irrigated oasis fields have resulted from natural and human processes at work in the arid region called Xinjiang, western China.

ental savanna lands the natural vegetation has been so modified by centuries of human occupation that the original conditions are difficult or impossible to determine.

Climates Outside the Humid Tropics

The *humid subtropical climate* occurs in southern China, the southern half of Japan, much of northern India, and a number of other Oriental countries. This climate type is characterized by warm to hot summers, mild or cool winters with some frost, and a frost-free season lasting 200 days or longer. The annual precipitation of 30 to 50 inches (*c.* 75–125 cm) or more is fairly well distributed throughout the year, although monsoonal tendencies produce a dry season in some areas. The natural vegetation, now largely removed over extensive areas (especially in China), is a mixture of evergreen hardwoods, deciduous hardwoods, and conifers. A generally comparable climate occurs in the southeastern part of the United States, although rainfall is more evenly distributed through the year in the latter area.

The northern part of Humid (eastern) China, most of Korea, and northern Japan have a *humid continental* type of climate marked by warm to hot summers, cold winters with considerable snow, a frost-free season of 100 to 200 days, and less precipitation than the humid subtropical areas. Most areas experience a definite dry season in winter. The predominant natural vegetation is a mixture of broadleaf deciduous trees and conifers, although prairie grasses are thought to have formed the original cover in parts of northern China. Many aspects of the humid continental climate of the Orient are duplicated in comparable latitudes of eastern North America, although the American areas lack the dry season in winter. The Orient has the long-summer subtype of humid continental climate except for Northeast China (Manchuria) and extreme northern Japan, which have the short-summer subtype.

Steppe and desert climates, whose characteristics have been previously described in Chapters 10 and 12, are found in Xinjiang (Fig. 14.3) and Mongolia and in parts of western India and Pakistan. A severe *highland climate* characterizes the Tibetan Highlands

TABLE 14.2   CLIMATIC DATA FOR SELECTED ORIENTAL STATIONS

Station	Country	Latitude to Nearest Whole Degree	Elevation Above Sea Level (ft)	Elevation Above Sea Level (m)	Type of Climate	Average Temperature Annual °F	Annual °C	January °F	January °C	July (or warmest month) °F	July (or warmest month) °C	Average Precipitation Annual (in)	Annual (cm)	Percent Occurring April–Sept.	Percent Occurring June–Sept.
Karachi	Pakistan	25°N	13'	4 m	Desert	78°F	26°C	66°F	19°C	87°F	30°C (June)	9"	23 cm	84%	83%
Bombay	India	19°N	37'	11 m	Tropical savanna	81°F	27°C	75°F	24°C	86°F	30°C (May)	82"	208 cm	95%	94%
Madras	India	13°N	51'	16 m	Tropical savanna	84°F	29°C	75°F	24°C	91°F	33°C (May–June)	49"	123 cm	37%	31%
Calcutta	India	24°N	21'	6 m	Tropical savanna	81°F	27°C	68°F	20°C	88°F	31°C (May)	62"	158 cm	83%	73%
Mandalay	Burma	22°N	252'	77 m	Tropical savanna	82°F	28°C	70°F	21°C	90°F	32°C (Apr.)	34"	87 cm	76%	54%
Singapore	Singapore	1°N	16'	5 m	Tropical rain forest	81°F	27°C	79°F	26°C	82°F	28°C (Apr.–June)	90"	228 cm	42%	29%
Jakarta	Indonesia	6°S	26'	8 m	Tropical rain forest	81°F	27°C	79°F	26°C	81°F	27°C	69"	176 cm	31%	16%
Delhi	India	29°N	718'	219 m	Humid subtropical	77°F	25°C	57°F	14°C	94°F	34°C (May, June)	28"	72 cm	86%	84%
Guangzhou (Canton)	China	23°N	29'	9 m	Humid subtropical	72°F	22°C	57°F	14°C	82°F	28°C (July, Aug.)	68"	172 cm	81%	55%
Shanghai	China	31°N	23'	7 m	Humid subtropical	59°F	15°C	37°F	3°C	81°F	27°C (July, Aug.)	45"	114 cm	69%	52%
Wuhan	China	31°N	121'	37 m	Humid subtropical	62°F	17°C	39°F	4°C	84°F	29°C	50"	127 cm	73%	47%
Chengdu	China	31°N	1611'	491 m	Humid subtropical	63°F	17°C	43°F	6°C	79°F	26°C (July, Aug.)	45"	115 cm	89%	76%
Beijing (Peking)	China	40°N	125'	38 m	Humid continental (long summer)	54°F	12°C	23°F	−5°C	79°F	26°C (July, Aug.)	24"	62 cm	92%	84%
Shenyang (Mukden)	China (Manchuria)	42°N	141'	43 m	Humid continental (long summer)	46°F	8°C	10°F	−12°C	79°F	26°C	28"	71 cm	84%	71%
Harbin	China (Manchuria)	46°N	526'	160 m	Humid continental (short summer)	37°F	3°C	4°F	−16°C	73°F	23°C	23"	58 cm	86%	75%
Tokyo	Japan	36°N	19'	6 m	Humid subtropical	59°F	15°C	39°F	4°C	79°F	26°C (Aug.)	62"	156 cm	61%	44%
Hakodate	Japan	42°N	13'	4 m	Humid continental (short summer)	47°F	8°C	25°F	−4°C	71°F	22°C (Aug.)	46"	117 cm	59%	45%
Kashgar	China (Xinjiang)	40°N	4296'	1309 m	Desert	55°F	13°C	28°F	−2°C	81°F	27°C	3"	8 cm	46%	30%
Ulan Bator	Mongolia	48°N	4347'	1325 m	Steppe	27°F	−3°C	−15°F	−26°C	61°F	16°C	8"	21 cm	92%	85%
Gyantse	China (Tibet)	29°N	13110'	4000 m	Highland		6°C		−4°C		14°C		20 cm		

358

(Fig. 17.4) and adjoining mountain areas. Some of the higher mountains in the East Indies are also best classified as having highland climates.

Climatic data for representative stations in the Orient are presented in Table 14.2.

POPULATION, SETTLEMENT, AND ECONOMY OF THE ORIENT

Somewhat over half of the world's people live in the Orient (see Table 3.1). They range from small tribal groups with local cultures to the Indians, Chinese, and other major culture groups. Mongoloid peoples form a majority in China, Japan, Korea, Burma, Thailand, Kampuchea, Vietnam, and Laos, but the majority of the people in India, although darker skinned than Europeans, are considered in many classification systems to belong to the Caucasian race; and brown-skinned Malays or kindred peoples form a majority among the native inhabitants of the Malay Peninsula, the East Indies, and the Philippines. This cursory survey does little justice to the tremendous variety of physical stocks found in the region as a whole.

The picture with respect to religion is also complicated. Hinduism is dominant in India, although many other religions are practiced, while the Islamic (Muslim) faith is dominant in Pakistan, Bangladesh, most parts of the East Indies, parts of the southern Philippines, parts of outer China, and among the native Malays of the Malay Peninsula. Various forms of Buddhism are dominant in Burma, Thailand (Fig. 14.4), Kampuchea, Laos, Tibet, and Mongolia. Sri Lanka and Nepal divide between Buddhism and Hinduism. The Chinese and Vietnamese are hard to categorize, even if effects of Communist rule are disregarded. Among them, Buddhism, Confucianism, and Taoism (Fig. 17.3) have all exerted an important influence, often in the same household. The same general situation has prevailed in Korea. In Japan religious affiliations also overlap, so that a reported 74 percent of the population adheres to Buddhism, and 93 percent adheres to the strongly nationalistic religion of Shintoism. The Philippines, with a large Roman Catholic majority, is the only Christian nation of the Orient, although Christian groups are found in various other areas such as South Korea, India, Indonesia, or (in a very minor way) Japan (Fig. 14.5). The religions just named are indigenous to the Orient except for Christianity and Islam, which originated in the Middle East. The indigenous faiths tend to seek converts less zealously than do Christians or Muslims. Veneration of ancestors is

Figure 14.4 *Buddhist monks in their saffron robes, Bangkok, Thailand. Buddhism is the main religion in several Southeast Asian countries.*

prominent, especially among Chinese, Vietnamese, and Koreans. Often an elaborate ritual for everyday living is followed, this being particularly characteristic of the Hindu religion. The more primitive hill tribes of the Orient are largely Animists (who believe that natural processes and objects have souls).

DISTRIBUTION OF POPULATION

The densest populations of the Orient are found, generally speaking, on river and coastal plains, although surprisingly high densities occur in some hilly or mountainous areas. The higher mountains and the steppes, deserts, and some areas of tropical rain forest are very sparsely inhabited.

Most countries in the Orient have experienced large increases in population during recent centuries, especially since the beginning of the 19th century. Most of the additional people have accumulated in areas that were already the most crowded. A rough calculation would suggest that about 2.8 billion peo-

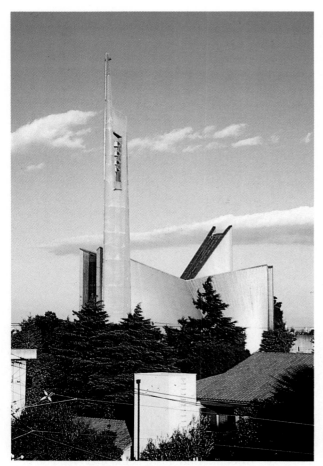

Figure 14.5   No major world region has a richer variety of religions than the Orient. The view shows a modernistic Roman Catholic cathedral in Tokyo, Japan. Oriental Christians do not compare in numbers with Buddhists, Hindus, or Muslims, but Christian congregations of many persuasions are widespread. Their origins are associated with Western activity in the Orient from the end of the 15th century onward, aside from a few Christian foundations that took place even earlier.

ple are now packed into a total area no larger than the 48 conterminous states of the United States. Hence the average density of the Oriental area is over 10 times that of the conterminous states. This makes the food situation precarious, although increased grain supplies from new high-yielding varieties and the availability of surplus grain from overseas countries (or from a few areas in the Orient) have been making it possible for most Oriental countries to maintain adequate levels of nutrition. How long this can continue in the face of rapid population increases remains to be seen. Japan has achieved a rate of natural population increase a little lower than that

of the United States and Canada; and China has considerably reduced its own rate by stringent birth-control measures; but most Oriental countries continue to have relatively high rates characteristic of the "developing" world.

DOMINANCE OF VILLAGE SETTLEMENT[1]

Although about 750 million Orientals live in cities and towns, the main settlement unit in the Orient is the village. Nearly three fourths of the people are residents of an estimated 1.9 million villages. Highly urbanized Japan, Singapore, Hong Kong, Macao, and Brunei do not fit this pattern, nor in lesser measure do Taiwan, South Korea, and Mongolia. But in all the other countries the typical inhabitant is a villager. It is true that scattered areas such as India's Himalayan foothills are characterized by dispersed farmsteads rather than compact villages. It is also true that small, isolated groups of people have seasonal, temporary, or shifting settlements associated with hunting, gathering, fishing, or shifting cultivation. But such exceptions are very minor when compared to approximately 2 billion Orientals for whom villages are home.

The Oriental village is essentially a grouping of farm homes, although some villages house other occupational groups such as miners or fishermen. Clusters of houses bunched tightly together are typical. Little space is normally allotted to adjoining gardens or yards. Defense may have been the original cause for these compact living arrangements.

As described by Spencer and Thomas, the main function of a village is to house people and their livestock, stored crops, and tools. Cheap and simple structures are characteristic. They are all that poor villagers can afford for privacy, shelter, and storage. Piped water and indoor plumbing are very exceptional in village homes, although a relatively small number now have electricity. Details of village life will vary according to culture; for instance, in Hindu villages there are segregated quarters for different castes.

The layout of villages varies, but some basic types can be distinguished:

1. Cluster Villages. These are very compact, some being circular, while others are shaped

[1]This section draws on material in J. E. Spencer and W. L. Thomas, *Asia, East by South: A Cultural Geography* (2d ed.; New York, John Wiley & Sons, Inc., 1971), pp 65–72; and J. E. Spencer, *Oriental Asia: Themes toward a Geography* (Englewood Cliffs, NJ, Prentice-Hall, Inc, 1973), pp. 41–43.

to fit a focal point such as a river fork or bridgehead.

2. **String Villages.** Their long and narrow layout conforms to some site feature, whether it be natural (*e.g.*, a riverbank or the seashore) or man-made (a dike or highway).

3. **Grid-Pattern or Rectangular Villages.** These might seem to be a modern development, but this type of village layout is actually very old. It appears to have originated in northwestern India, disappeared there, and later appeared in various parts of Japan, China, and India where space was available.

Whatever the layout, the original siting of villages was closely adapted to natural conditions. For example, in flood plains the villages will be slightly elevated—on natural levees, dikes, or raised mounds. Early villages in Indonesia were often built in defensible mountain sites, although the Dutch colonial administration gradually required the building of villages along main roads and trails in the lowlands. The latter siting made it easier for the government to exercise control, collect taxes, and draft soldiers or laborers for road work or other projects.

MEANS OF LIVELIHOOD

Japan was the first country of the Orient to develop modern types of manufacturing on a really large scale (see chapter-opening photo of Chapter 18). Since World War II, it has become one of the world's three most important industrial powers. But China and India also have important and expanding industries. These giant countries, by far the largest in the world in population, are much better supplied with mineral resources than Japan, and they have cheaper labor. But Japan's labor force is on the average more skilled, and Japan has shifted more and more to types of industry requiring skilled labor. Not only does it lead the Orient in manufacturing, but it is far and away the leader in services. The remaining countries of the Orient present a very mixed picture in their development of manufacturing and services. Hong Kong and Singapore are outstanding in proportion to their size. South Korea and Taiwan also have a relatively high development. At the other end of the scale are small and war-ravaged Kampuchea and Laos, where development is so pathetic as to place these countries among the world's very poorest.

For the Orient as a whole, agriculture remains the dominant source of livelihood. In most Oriental countries—Japan being a conspicuous exception—the majority of the population is supported directly by agricultural activities. A number of major types of agriculture may be distinguished. Two types, plantation agriculture and shifting cultivation, are discussed in some detail in Chapter 17 on Southeast Asia, the part of the Orient in which these forms of agriculture are the most prominent. In the steppes and deserts of outer China, the Mongolian People's Republic, and Pakistan, nomadic or seminomadic herding and oasis farming (Fig. 14.3) are practiced. Over large sections of the Orient, most farmers make a living by cultivation of small rainfed or irrigated plots worked by family labor (Fig. 14.6). In Com-

Figure 14.6 Despite huge growth in cities, industries, and services during the 20th century, agriculture *remains the leading source of employment in the Orient and in the world. All but a small proportion of the world's farmers still work with hoes, other simple hand tools, or crude animal-drawn plows. This scene is from northern India, in a section of the Indo-Gangetic Plain northwest of Delhi.*

Figure 14.7   Terraced fields of irrigated rice in a characteristic landscape of rural Japan. Such a system of fields may have numerous individual owners, necessitating cooperative effort to keep the terraces in repair and the channels open so that water can move slowly by gravity from one field to the next.

munist-held areas, collectivized agriculture has replaced many traditional family farms. Large amounts of very arduous hand labor are applied to the land, and production is often of a semisubsistence character. This type of agriculture, which is often referred to as intensive subsistence agriculture, is built around the growing of cereals, although other types of crops are raised. Where natural conditions are not suitable for irrigated rice, such grains as wheat, barley, millet, grain sorghums, or corn (maize) are raised. However, irrigated rice is the grain which yields the largest amount of food per unit of area where conditions are favorable for its growth, and this crop is the agricultural mainstay in the areas inhabited by a large majority of the Orient's people.

THE RICE ECONOMY AND THE "GREEN REVOLUTION"

For most people in the Occident, bread is the staff of life. But for many hundreds of millions in the Orient, rice is the basic source of carbohydrate. The rice is boiled and is often seasoned with the preparation called curry. Protein diet is most often provided by peas, beans, or fish instead of the meat, milk, and cheese of the Occident.

Irrigated rice ("wet rice") is grown in the tropics or subtropics of all the world regions discussed in this text, but nine tenths of the world's supply comes from the monsoonal lands of southern and eastern Asia. In the Orient it is still primarily a crop produced by hand labor or animal power, but mechanization has been taking hold in some areas, particularly Japan. Japanese rice farmers have largely converted to power-driven hand tractors, small power-operated threshing and hulling machines, and other labor-saving devices. Adequate water supplied from streams, wells, springs, or ponds is a crucial factor in the rice environment, as the grain is produced in fields which are submerged under several inches of slowly moving fresh water. It is customary to start the rice seedlings in small seed beds and then transplant them to the flooded paddy fields. Wet rice is grown on various types of soil and under a variety of environmental conditions. One condition, however, must always be present: an impervious subsoil which retards the downward seepage of water. Ideal conditions are often found on alluvial flood plains where fertile silt has been deposited on an impervious layer of clay.

In many parts of the Orient, level land has been created by terracing steep hillsides (Fig. 14.7). Smith and Phillips have briefly described this landscape and some of the traditional techniques in producing and using rice[2]:

> Lowland rice must be grown by irrigation, and the devices used in fitting and keeping the land for this service are among the greatest monuments of human diligence in the world. . . . In Ceylon [Sri Lanka], for example, the railway that goes from the seacoast to the highlands goes through an irrigated plain divided by low banks into ponds of small area—rice fields, each of which has by great labor been leveled so that the water may be of uniform and proper depth for rice growing. As the railroad climbs the slopes of the hills the rice patches continue, with smaller area and higher banks, turning at last into a giant flight of gentle water steps, one of the most beautiful landscapes that the world possesses. . . .
>
> The common treatment of . . . lowland rice is alternately to flood it and draw off the water during the early periods of its growth. It is kept under water dur-

ing a large part of its development, the water being entirely drawn off as it ripens. The water must not become stagnant, and to keep it in motion it is the common practice on the hillsides to lead a stream to the top terrace, and let the water pass from terrace to terrace down the slopes. . . .

     . . . . .

> When the Asiatic rice field is finally drained, the ripened grain is usually cut by hand, tied up in bundles, and allowed to dry. To accomplish this in moist places, it is often necessary to put the sheaves upon bamboo frames [Fig. 18.4]. It is usually threshed by hand with the aid of some very simple devices. One of these is a board with a slit in it. Drawing the rice through the slit pulls the grains from the heads and allows them to fall into a receptacle. The grain at this stage is called *paddy* because of a closefitting husk not unlike that which protects the oat kernel. As with oats, these husks cause the grain to keep much better than when the husk is removed and the final husking of rice for home use is usually deferred until the time of use approaches. Among the Oriental people the husking of the paddy to prepare it for food is a daily occurrence, commonly done by hand. One of the commonest sounds throughout the East . . . is the pounding of a heavy mallet or pestle as it falls into a vessel full of paddy in the process of pounding the grain and loosening the husk.

The Green Revolution[3]

Today the specter of widespread hunger hangs over large parts of the world. As a step toward eliminating hunger by using modern technology to increase rice yields, the International Rice Research Institute (IRRI) was founded in the Philippines in 1962. The institute is one facet of a worldwide research effort involving not only rice but also wheat and other grains. Notable success has been achieved in breeding new high-yielding varieties (HYV), and there has been a large upsurge of production in certain areas where the new strains have been widely introduced. The entire effort has come to be known as the Green Revolution. Whether the term is too optimistic remains to be seen, as the "revolution" is still in an early stage, but it does indicate an upsurge of hope that science has shown the way to a better food supply for impoverished people in the Orient and elsewhere.

A major aim of IRRI has been the breeding of new high-yielding rice varieties adapted to tropical

---

[2]J. Russell Smith and M. Ogden Phillips, *Industrial and Commercial Geography* (3rd ed.; New York, Holt, Rinehart and Winston, 1946), pp 432–434. One heading has been omitted and one word italicized in editing. Reprinted by permission.

[3]The discussion under this heading draws heavily on B. L. C. Johnson, "Recent Developments in Rice Breeding and Some Implications for Tropical Asia," *Geography*, 57, Pt. 4 (November 1972), pp. 307–320.

Asian conditions. This is, of course, a complex and difficult matter, and the varieties produced represent compromises among various objectives. In general, rice breeders have tried to evolve rice strains having the following characteristics:

1. *High tillering and heading capacity*—the production by each rice plant of numerous *tillers* (vegetative shoots) in addition to the main stem of the plant, with both main stem and tillers producing heads (*panicles*) bearing as many fully developed rice grains as possible.

2. *Superior responsiveness to fertilizers,* so that the rice plants will use nutrients efficiently to maximize their production of rice grains in relation to low-value stalks and leaves.

3. *Insensitivity to changes in length of day.* Many rice varieties are *photosensitive,* that is, they do not flower except in response to a decreasing length of day. Hence planting is limited to certain times of the year. Nonphotosensitive varieties, on the other hand, enable the grower to plant whenever other factors are favorable, and thus to get more rice crops from a given field.

4. *Early maturation,* to maximize the number of harvests from a field.

5. *Short, sturdy stalks and narrow, upright leaves,* to conserve the plants' vegetative energies for grain production and to reduce shading and the danger of *lodging.* The latter term refers to excessive bending and flattening of stalks and leaves due to wind, rain, or heavy rice heads. Lodging makes rice more difficult to harvest, and the tangle of vegetation increases shading and hence lessens the effectiveness of photosynthesis. The yield of food in the rice grains is closely related to the solar energy received by the leaves of the rice plant. Direct sunlight facilitates high yields, whereas the less intense light caused by cloudiness or shading promotes growth but produces lower yields.

6. *Resistance to shattering*—the spontaneous shedding of rice grains from the heads.

7. *Resistance to diseases* such as blight caused by fungi or bacteria, and viral diseases transmitted by plant hoppers and leaf hoppers.

8. *Proper milling qualities, texture, and taste* to appeal to the consuming market. Some consumers, for example, prefer rice to be dry and firm when cooked, whereas others are accustomed to softer, stickier rice.

To capitalize fully on the Green Revolution, it is necessary to overcome many economic, social, political, and environmental obstacles. For example, success requires inputs of capital that are often beyond the present means or inclination of farmers, landlords, and governments to provide. Such expenditures are needed for water-supply and water-control facilities (for example, the tubewells that have burgeoned by hundreds of thousands on the Indo-Gangetic Plain of the northern Indian subcontinent); chemical fertilizers, particularly nitrogen; and chemicals to control weeds, pests, and diseases. As agriculture becomes more mechanized, the costs of machinery and fuel have to be borne. Governments have to improve transportation facilities so that the heavy inputs of fertilizer required by the new varieties can be delivered to farmers in outlying areas. Grain-storage facilities, now subject to plundering by rats, have to be improved.

Overcoming the financial obstacles is rendered more difficult by the widespread system of share tenancy. If a farmer is a share tenant, he may not wish to assume any additional risk, even if credit on reasonable terms is available. He generally has no security of tenure, and thus he cannot be sure that money he invests in the Green Revolution will actually benefit him in the future. Even if he remains on his holding, the landlord may take up to half of the increased crop, while bearing little or none of the additional expense. Landlords, in their turn, may be content to collect their customary rents without expending additional capital, or they may endeavor to turn tenants off the land in order to create larger units which they themselves can farm with machinery and hired labor. In fact, large landowners often become the chief beneficiaries of the new technology in the countryside, with large numbers of farmers becoming a class of landless workers hired for low wages on a seasonal or migratory basis.

These comments by no means exhaust the range of problems posed by the Green Revolution. Others include damage to ecosystems by large infusions of agricultural chemicals, often poisonous, and economic dislocations when rice-importing countries become more self-sufficient in rice and thus cause hardships for rice exporters. Unfortunately, solutions to the almost innumerable problems may be thwarted by decisions taken on political grounds. In one instance, American agricultural technicians were virtually eliminated from India by such decisions, despite the fact that they were badly needed to maintain the highly scientific Green Revolution as a continuing process and "to find scientific answers to problems of environmental adjustment and ecological backlash as they crop up."[4]

---

[4]Richard Critchfield, "India: The Lost Years," *The New Republic,* 170, No. 224 (June 15, 1974), p. 18. Reprinted by permission of THE NEW REPUBLIC, © 1974, The New Republic, Inc.

What we call the green revolution is essentially the geographical transfer of new high-yielding seeds, irrigation, mechanization, and the massive application of chemical fertilizer, and, most important, the knowledge that goes with this. In countries like India in the late 1960s it came so fast that when the first spectacular results diminished, palpably absurd and trendy articles began appearing that the green revolution had "withered" or "failed" or whatever. But the green revolution is not an event but a process that will just go on, transforming for good and bad rural societies all over the earth.[5]

---

[5]*Ibid.* Reprinted by permission.

CHAPTER 15

Poverty and Progress in the Indian Subcontinent

The Indian subcontinent has made significant contributions to civilization in many fields, including mathematics and science. This photo shows an astronomical observatory at the Indian city of Jaipur, constructed in the 1700s by the astronomer-king of Jaipur, Jai Singh II.

THE northern reaches of the Indian Ocean are split into two enormous bays—the Bay of Bengal and the Arabian Sea—by a triangular peninsula which thrusts southward for a thousand miles (*c.* 1600 km) from the main mass of Asia. To the north the peninsula is bordered by the alluvial plain of the Indus and Ganges rivers, beyond which lie ranges of high mountains. The entire unit—peninsula, alluvial plain, and fringing mountains—is often called the Indian subcontinent. It contains the five countries of India, Bangladesh, Pakistan, Nepal, and Bhutan (Fig. 15.1), occupying an area a little over half the size of the 48 conterminous American states (Fig. 15.2). India far outweighs all the other countries in both area and population. Nepal and Bhutan, which lie on the southern flank of the Himalaya Mountains (Fig. 15.3), are relatively small, extremely rugged, and difficult of access. Their chief political significance lies in their location between India and China, which have engaged in bitter and sporadically violent border disputes since 1959.

The Indian subcontinent is enclosed on its landward borders by mountains. Its northern boundary lies in the highest ranges in the world, the Himalayas and the Karakoram. From each end of this massive wall, lower flanking ranges trend southward, to the Arabian Sea on the west and the Bay of Bengal on the east. Until 1947 it was customary to refer to the entire area except the small Himalayan states as "In-dia." It was for well over a century the most important unit in the British colonial empire. Then in 1947 it gained freedom but in the process became divided along religious lines into two sovereign nations, the predominantly Hindu nation of India and the Muslim nation of Pakistan. The latter was established in two parts, West Pakistan and East Pakistan, separated by Indian territory. In 1971 a revolt in the eastern part was successful due to the intervention of India's armies, and the former East Pakistan became the independent country of Bangladesh.

India is among the most important of the many countries that have gained independence since World War II. A major element in this is its sheer size, especially of population. With 817 million people (mid-1988 estimate), India contains nearly one sixth of the human race and is second in population to China, while its area of 1.3 million square miles (3.3 million sq. km) is exceeded by only six other countries. Military victory in the 1971 war with Pakistan, which created Bangladesh, established India as the leading power in its part of the world (although lacking the status of a major world power), and in 1974 the explosion of India's first atomic bomb made it the sixth country to join the "nuclear club." (Historically, India has made significant contributions to science and technology; see chapter-opening photo). Pakistan and Bangladesh are much smaller than India in area and population, but both are among the world's ten largest countries in population. Over one billion people, one of every five on earth, lived in the subcontinent in 1988.

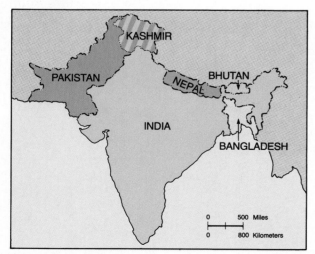

Figure 15.1   *Political units of the Indian subcontinent. The future status of Kashmir, disputed between India and Pakistan, with some parts occupied by China, remains unresolved at the time of writing.*

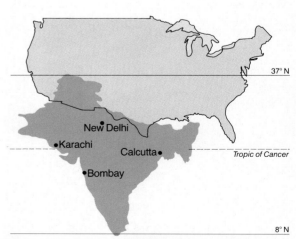

Figure 15.2   *The Indian subcontinent compared in latitude and area with the conterminous United States.*

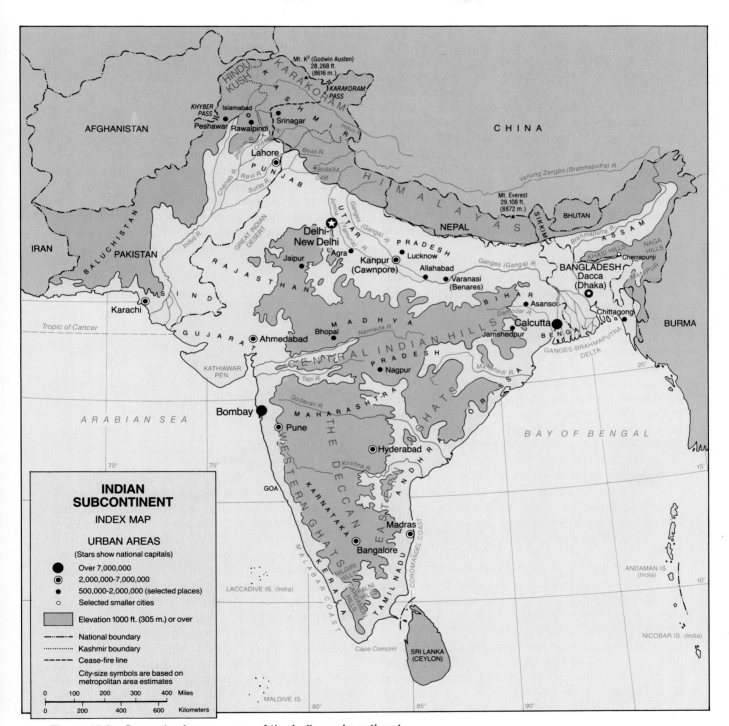

Figure 15.3 *General reference map of the Indian subcontinent.*

POVERTY AND POPULATION GROWTH

A major characteristic and core problem of the Indian subcontinent is the abysmal poverty of most of its people. All of the subcontinent's countries are among the very poorest in the world. Estimates of per capita gross national product (GNP) for 1986 (in U.S. dollars) assigned India a figure of $270, Pakistan $350, Bangladesh $160, Nepal $160, and Bhutan $160. A rounded average of these estimates yields $265 for the subcontinent as a whole. That is phenomenally low for such a huge population, although China's figure of $300 was not markedly better. The massive poverty of the subcontinent and of China, which together make up about two fifths of humanity, is a fundamental and deeply troubling fact of world geography and world relations in the late 20th century.

Measured by per capita GNP, the Indian subcontinent is clearly the poorest of the world's main areas. Outside this area, only about twenty widely scattered African and Asian countries, generally with small populations, had per capita GNPs lower than $265 in 1986. Africa, another major area beset by deep poverty, had an average figure ($620) well over twice that of the subcontinent, although quite a few individual African countries fell below $265. The exact figures will, of course, change from year to year, but most of the relative rankings will change much more slowly, and it is not to be expected that the Indian subcontinent, with a population over four times that of the United States, will soon lift itself from the low status in which it is now mired.

In measures of social welfare such as life expectancy and infant mortality, the subcontinent does not come off much better than it does in per capita GNP. Average life expectancy at birth was estimated in the middle 1980s at 54 years in India, 54 years in Pakistan, and 50 years in Bangladesh. These figures were about the same as the average for Africa (52 years), although many individual African countries fell well below this level. Such figures contrast remarkably with China's average life expectancy of 66 years, not to speak of the figure for the United States (75 years) or Japan (78 years). Data for infant mortality—the number of babies per thousand that do not survive to age one—again show the subcontinent in a poor light, although better than that of many African countries and scattered countries elsewhere. Recent figures assigned India an infant mortality of 104, Pakistan 121, and Bangladesh 135, compared to 110 for Africa, an incredibly high 175 for the West African country of Sierra Leone, 44 for

China, 10 for the United States, and a remarkable 5 for Japan.

It is true that the depressing averages for the subcontinent conceal some very wide variations among social groups and also among regions. In India, for example, a few people are very wealthy and there is a relatively prosperous middle class amounting to perhaps 50 or 60 million people (some estimates—depending on the definition of "middle class"—run as high as 70 to 100 million). But this only means that most people are even poorer than the averages indicate. The prosperous, relatively small minority is surrounded by a sea of overwhelming poverty.

There is much evidence that the subcontinent's poverty bears an unusually direct relationship to the presence of too many people on too little land. Some of this evidence is presented in the following table of population and arable land, which also gives comparative data for the United States:

	Bangladesh	India	Pakistan	United States
Population (millions, 1988)	110	817	108	246
Rural population (millions, 1988)	92	613	77	64
Arable land (%)	63	51	25	20
Arable land (sq. mi.: 000)	35	646	78	720
Arable land per person (acres):				
Total population	0.20	0.51	0.46	1.87
Rural population	0.24	0.67	0.65	7.20

Data: *1988 World Population Data Sheet* (Population Reference Bureau); *1988 Britannica Book of the Year.*

The figures in the table make apparent the extent to which the large, dense, and predominantly rural populations of the subcontinent lack adequate amounts of arable (crop-producing) land, with Bangladesh being considerably worse off than even India or Pakistan. The average rural dweller in Bangladesh is eking out a living from about one thirtieth the crop land that is connected with the support of the average person in rural United States. The crop land that these figures would assign an average Bangladesh farm family of five persons is less than the size of residential lots for which considerable parts of many American cities are zoned. Such crowding on the land was largely brought about by the spread of Western public health and medical techniques to the subcontinent. This process was begun by the British, who conquered the subcontinent in the 18th and early 19th centuries. It has accelerated rapidly since independence. Although health conditions are very unsatisfactory by Western standards, there is much better control of disease and hence death rates are lower than was once the case. But birth rates have remained relatively high, and population growth has been rapid. Some of the history of this growth, with projections to the year 2020, is summarized in the following table. The figures are for the subcontinent excluding Nepal and Bhutan:

Year	Subcontinent Population (millions)	Addition During Preceding Period (millions)	Percent Growth
1901 census	283	—	—
1941 census	386	103	36%
1981 census	855	469	121%
Projected:			
2020	1755	900	105%

It will be noticed that future growth is projected at a somewhat slower rate (percent) than was true during the explosive population surge of the middle 20th century. But growth is still expected to be quite rapid, and the base population is already so large that the projected addition to population in 39 years—900 million people—is massive. The projected slower rate of growth is based on the assumption that birth rates will fall with increasing urbanization and as a consequence of India's strenuous efforts to foster birth control among its people. The latter efforts may ultimately prove to be a disappointment, as they run counter to a traditional religious emphasis on fertility and also to the fact that loyal children usually represent the only social security system the peasant has. Of course, some catastrophe to the already hard-pressed food supply might change the projected picture drastically. A solution through industrialization has often been proposed and is being pursued, but a staggering increase in materials and energy would be required to raise the subcontinent to a Western standard of output and consumption. The long-term prospect remains fraught with uncertainty, and it promises at best a very slow advance for the subcontinent's poverty-stricken masses.

PHYSICAL REGIONS, RESOURCES, AND LIMITATIONS

Physical conditions are very important to such struggling societies as those of the Indian subcontinent. Since it would be impossible to feed the enormous population—or any substantial part of it—from outside sources, survival depends quite directly on the resources available to agriculture; and the general improvement of life in this desperately poor and overcrowded region will be very difficult or impossible unless the region can also provide a sizable share of the natural resources required for industrialization. The total resources available to the subcontinent are large, but they are hard to describe and assess because they are so varied in character and so uneven in spatial distribution. A multiplicity of physical areas can be distinguished, each with its own combination of terrain, rock types and minerals, moisture conditions, soils, and vegetation. But at the broadest scale it is possible to combine these into three major physical divisions: the outer mountain wall, the northern plain, and peninsular India.

THE OUTER MOUNTAIN WALL

This rampart forms an inverted "U" across the north of the subcontinent. In the west the mountains extend northeastward from the Arabian Sea to Kashmir (Fig. 15.3). From here the longest section of the wall trends southeast and then eastward to the subcontinent's northeast corner in the Indian state of Assam. Finally, the relatively short eastern leg of the "U" reaches southward along the border with Burma to the Bay of Bengal.

The western section of the mountain wall is traversed by Pakistan's borders with Iran and Afghanistan. Here the mountains become generally higher to the north, but almost everywhere they are quite rugged, and in most places they form a broad band extending well into the bordering countries. Desert and steppe climates reach upward to high elevations. Throughout history these mountains have provided traders and armies with only a few feasible crossing places, the most famous of which is the Khyber Pass on the border between Pakistan and Afghanistan. But such passes played a fateful role for thousands of years as gateways into India for conquerors from the outer world. In the second millennium B.C. came the Aryans, tribal pastoralists from the inner Asian steppes. They were the main ancestral bearers of many racial and cultural attributes that are dominant in the subcontinent today, notably the Caucasoid racial majority, the numerically dominant Aryan languages such as Hindi—most of which were derived from the Sanskrit used by the conquerors, and Hinduism, which developed from beliefs and customs of the Aryans. Then in medieval times the Islamic religion came into India from the west, originally brought in the 8th century to what is now Pakistan by Arabs entering along the narrow coastal corridor

Figure 15.4 *The jagged summit of Mount K2 in the Karakoram Range, a part of the Indian subcontinent's outer wall of high mountains. The view is from a climbers' camp on a glacial moraine at an elevation of nearly 15,000 feet (c. 4600 m).*

between the mountains and the Arabian Sea. This limited original lodgement of the subcontinent's Muslim faith was followed in later centuries by a far wider spread of Islam, brought about by waves of Muslim conquerors who came in from Afghanistan and central Asia through the mountain passes. The last great invasion, that of a Mongol-Turkish dynasty called the Moguls, began in the early 16th century, soon after the first Europeans came to India by sea in the Age of Discovery. At its height, the Mogul Empire ruled nearly all the subcontinent, but the power of the empire then decayed and control was gradually lost to the encroaching British. After the Soviet invasion of Afghanistan in 1979, great numbers of Afghan refugees fled through the mountains into adjacent Pakistan.

The northern and eastern segments of the subcontinent's outer mountain wall, formed by the world's highest mountains, have been much less passable. On the north, the Himalayas extend about 1500 miles (c. 2400 km) from Kashmir to the northeastern corner of India. Paralleling the Himalayas on the northwest, and separated from them by the deep

gorge of the upper Indus River, lies another towering range called the Karakoram. Beyond the Himalayas and the Karakoram is the rugged and very high Tibetan Plateau in western China. The two mighty ranges that wall off the north of the subcontinent contain 92 of about 100 world peaks with elevations of 24,000 feet (7315 m) or higher. Of these, 59 are in the Himalayas and 33 in the Karakoram, culminating in earth's two loftiest summits: Mt. Everest (29,108 feet/8872 m) in the Himalayas on the border between Nepal and China, and Mt. K2 or Godwin Austen (28,268 feet/8616 m) in the Karakoram within the Pakistani-controlled part of Kashmir (Fig. 15.4). Claims were recently made that K2 might actually be higher than Everest, but in 1987 an Italian expedition, using satellite signals as an aid to surveying, found Everest to be 840 feet higher. The expedition's figures are used here. Although most passes through the main ranges are higher than any peak in the Alps, a trickle of trade and cultural exchange has crossed these mountains for millennia. Modern China and India have fought some sharp military actions here over disputed boundaries. But neither commerce nor military forces have ever crossed the Himalayan-Karakoram barrier on any large scale.

The mountains along the subcontinent's border with Burma are much lower but are also nearly impenetrable. Not only are they rugged, but they are climatically difficult in being both tropical and one of the wettest areas on earth. Cherrapunji in the Khasi Hills has an average annual rainfall of 432 inches (1097 cm), of which over nine tenths falls in the summer half-year. Hence it ranks with a station on Kauai in the Hawaiian Islands as one of the two spots with the greatest annual rainfall yet recorded on the globe.

THE NORTHERN PLAIN

Just inside the outer mountain wall lies the subcontinent's northern plain (Fig. 15.5). This large alluvial expanse contains the core areas of the three major countries. In the west the plain is split between Pakistan and India. The part in Pakistan is crossed from north to south by the Indus River and its tributaries and distributaries. The climate here is desert and steppe, making irrigation water from the rivers essential to many millions of people. The sources and distribution of the water are matters of long-standing dispute, punctuated by some agreements, between India and Pakistan. The disputed mountain territory of Kashmir, claimed by both countries on ethnic and historical grounds, is traversed by the Indus and is the source area of some of its tributaries. The region called the Punjab ("Land of Five Streams": major tributaries of the Indus) is split between the two countries, with its rivers flowing through Indian territory before entering Pakistan. The division of these waters has been a major issue.

East of the Punjab lies the part of the northern plain traversed by the Ganges River and its tributaries, especially the Jumna. This is India's core region, containing perhaps 40 percent of the country's population as well as its capital, a string of other major cities, and various places of great religious significance to Hinduism. The area is relatively well-watered, especially toward the east, and an agriculture based primarily on irrigated rice supports population densities that are unusually high even for the subcontinent. At the relatively narrow western end of the Ganges-Jumna section of the plain, the old fortified capital of Delhi arose as a bastion against invaders from beyond the western mountains. The present (and adjoining) capital of New Delhi was founded later as the capital of British-controlled India. Today the metropolitan population of Delhi-New Delhi is well over 7 million, making it India's third largest metropolis (after Calcutta and Bombay).

The northeasternmost part of the northern plain, crossed by the Brahmaputra River, lies in the Indian state of Assam; and the southeastern part of the plain lies in Bengal, which is essentially the huge delta formed by the combined waters of the Ganges and the Brahmaputra. Assam has relatively good agricultural conditions and by Indian standards is relatively lightly populated—meaning that it only has around 700 people per square mile (*c.* 250 per sq. km). By contrast Bengal, which is divided between India and Bangladesh, is extremely densely populated. In the early 1980s thousands of illegal immigrants from Bangladesh were slaughtered in Assam by elements of the native population. But the total of such immigrants in Assam was already estimated at perhaps 4 million, and immigration has continued despite the massive violence. From the viewpoint of incredibly crowded Bangladesh, some parts of India such as Assam are lands of opportunity. Another magnet for migrants from Bangladesh (as well as from rural India) is the English-founded port of Calcutta in Indian Bengal. Despite a worldwide reputation for teeming slums and dire poverty (see chapter-opening photo of Chapter 3), Calcutta had reached an estimated metropolitan population of over 12 million by the late 1980s. The Ganges-Brahmaputra delta region could provide many millions of further immigrants, as it is one of the most overcrowded agricultural areas on earth.

PENINSULAR INDIA

The southern peninsular triangle of the subcontinent, located entirely within India, consists mainly of a huge plateau called the Deccan. It is relatively low, with much of its surface only 1000 to 2000 feet (*c.* 300–600 m) above sea level. As is characteristic of plateaus, the rivers of the Deccan tend to run in deep trenches cut far down into the plateau surface. This makes it impossible to use their water for irrigation of the upland surface without large inputs of

Figure 15.5 *A transportation scene that is characteristic of many "developing" areas in the world. The photo shows a section of the Indian state of Rajasthan on the subcontinent's northern plain. Use of camels for draft power is common in the drier western parts of the plain.*

capital and modern technology. Wells and the rain-catchment ponds called *tanks* are used, but water is in shorter supply and population density is less than in the better-watered parts of the northern plain. The edges of the Deccan are outlined by hills and mountains. In the north, belts of hills called collectively the Central Indian Hills separate the plateau from the northern plain. On both east and west the plateau is edged by ranges of low mountains called Ghats ("steps"), which overlook low-lying alluvial plains along the coasts of the Bay of Bengal and the Arabian Sea. The eastern mountains, quite low and discontinuous, are known as the Eastern Ghats, while the western mountains, which are higher and form a relatively continuous wall, are the Western Ghats. The two ranges merge near the southern tip of the peninsula in clusters of "hills" (Nilgiri, Palni, and Cardamon Hills) which are really rather high mountains, with some peaks rising above 8000 feet (*c.* 2400 m). The coastal plains between the Ghats and the sea are occupied by ribbons of extremely dense population, with the density reaching a peak in the far south along the Malabar Coast of western India and the corresponding Coromandel Coast in the east. The largest metropolis of the far south, Madras (5 million), is a seaport on the Coromandel Coast.

CLIMATE AND WATER SUPPLY

Climatic conditions in the subcontinent vary between remarkable extremes, from Himalayan ice and snow fields to the year-round tropical heat of peninsular India, and from some of the world's driest climates to some of the wettest. Some of the hottest monthly average temperatures on the globe occur in the plains of Pakistan and northwestern India. Delhi, for instance, averages 94°F (34°C) in both May and June (Table 14.2). The highest temperatures in the subcontinent tend to occur in late spring, before the moderating effects of summer rainfall commence after a long dry season (Table 14.2). Precipitation varies from near zero in the deserts of Pakistan and adjacent India, to rainfall heavy enough to create rain forest conditions on the seaward slopes of the Western Ghats and the coastal plain at their base, as well as in parts of the Ganges-Brahmaputra Delta and the eastern mountain wall near Burma. Climatic types in the subcontinent include highland climates in the northern mountains, desert and steppe in Pakistan and adjacent India, humid subtropical climate in the northern plains, tropical savanna in the peninsula, except for a patch of steppe in the rain

shadow of the Western Ghats, and the areas of rain forest mentioned previously.

Heat is a near-constant except in the mountains. The tropical peninsula is hot all year, while the subtropical north has stifling heat before the "break" of the wet monsoon, and warm conditions even in the winter. Such temperatures have consequences for agriculture. On the positive side they allow year-round crop production, if sufficient water is available, and they are favorable to a large variety of crops, including the wet rice which is the leading food of India and Bangladesh. On the negative side they lead to high evaporation, which dries out soils and desiccates plants rapidly when water is not being supplied. The high temperatures also accelerate chemical reactions in the soil, which in turn cause soils to be impoverished more quickly by leaching and the oxidation of organic material. Additionally, the lack of a freezing period allows plant and insect pests to flourish without check.

Another near-constant in the subcontinent's climatic conditions is the extreme seasonality of rainfall, offering a stark contrast between a short wet summer and desert or near-desert conditions the rest of the year. This striking and critical characteristic is apparent from Table 14.2. On the average, Bombay gets 77 inches (196 cm) of rain in the four-month period June to September and 5 inches (13 cm) through the other eight months; Calcutta 45 inches (114 cm) in four months and 17 inches (43 cm) through the other eight; Delhi 24 inches (61 cm) in four months and 4 inches (10 cm) in the other eight. Thus the "humid" parts of the subcontinent are deserts or near-deserts for most of the year, while floods are often a threat or a reality during the short rainy season, which is also a season of snow melt along upper river courses in the Himalayas. These conditions are caused by seasonally reversing surface winds called monsoons—a wet monsoon in the summer and a dry monsoon in the winter.

THE WET MONSOON

The southwest or wet monsoon is at its height during the months from June to September, and most parts of the subcontinent receive the bulk of their annual rainfall during those months. Two main arms of this monsoon can be discerned. One arm, approaching from the west off the Arabian Sea, strikes the Western Ghats and precipitates heavy rainfall on these mountains and the coastal plain between the mountains and the Arabian Sea. But the amount of rain diminishes sharply in the interior Deccan to the east of the mountains. Here the annual precipitation over a sizable area is barely sufficient for unirrigated ag-

Figure 15.6 Artificial ponds in which to store water are a device that helps communities in the Indian subcontinent survive the dry season. They are especially numerous in the Deccan plateau of Peninsular India. The one in this photo is in northwestern India. Some water buffalo used as draft animals can be seen bathing in the pond. This must be done frequently in the dry season, as the skins of these animals are vulnerable to dry heat.

riculture and in some years is so low as to result in serious crop failures.

The second major arm of the monsoon, approaching from the Bay of Bengal, brings moderate amounts of rain to the eastern coastal areas of the peninsula and heavy precipitation to the Bengal delta region (Ganges-Brahmaputra Delta) and the northeastern Indian state of Assam. Exceptionally heavy rainfall occurs on the forward slopes of low mountains that rise behind the delta. Part of the moving air in the Bay of Bengal arm of the wet monsoon passes up the Ganges Valley and precipitates moisture that diminishes with some regularity in total amount from east to west. Both major arms of the monsoon bring some rainfall to Pakistan, but the total is so small as to result in semiarid or desert conditions in most areas.

THE DRY MONSOON

The monsoon of the winter half-year is often called the northeast monsoon, although it generally blows from the west over most of the northern plain. The wind is often northeasterly, however, over the peninsula, the Bay of Bengal, and the Arabian Sea. To most parts of the subcontinent this monsoon, blowing from land to sea, brings dry weather, with occasional light rains in areas outside the tropics. An exception is found in the far south of the peninsula, where the heaviest rainfall of the year occurs along the eastern coast and in adjacent interior uplands during the four-month period from October through January. This precipitation is brought by retreating maritime air of the summer monsoon and by continental air of the winter monsoon which has accumulated moisture in its passage over the Bay of Bengal. In addition to widespread drought, the win-

ter monsoon brings cooler weather to the subcontinent, especially the north. But a period of extreme heat is experienced in the spring before the next onset of the rainy season.

THE STRUGGLE FOR WATER

With precipitation so concentrated into a short part of the year, and with the extreme pressure of too many people on too little land, it becomes vital to extend the season of good crop conditions by irrigation. Hence an enormous amount of labor is expended in getting additional water onto the land. In places huge modern dams impound rivers, and networks of canals distribute water from these reservoirs. The main locales in which such facilities have been developed are in the steppe and desert areas of the Punjab and the lower Indus Valley (Sind) in Pakistan. Here a system of dams and canals was begun by the British in the 19th century and is still being extended. Since the turn of the century rapid colonization has changed these dry lands from sparsely populated to rather densely populated areas. However, both the Indian and Pakistani parts of the Punjab still normally produce surpluses of wheat, and recent projects have made Sind a surplus producer of rice, much of which Pakistan exports.

Where large-scale modern irrigation systems do not exist, reliance for supplementary water must be placed on wells and "tanks." The latter are artificial ponds which fill in the wet season and serve as reservoirs in the dry season. There are hundreds of thousands of them in India (Fig. 15.6). Huge labor inputs are characteristic of irrigation from all sources because the lifting of water—from one canal or ditch to another, from canals to fields, from wells or tanks to canals or ditches—is done largely by hand labor using simple tools such as waterwheels, sweeps, leather sacks, pots, and scoops. A small minority of the more prosperous farmers now use motor-driven pumps, but more widespread use is hindered by the cost of fuel and electricity and the cheapness of labor.

Even if water is impounded in streams and tanks or lifted from wells which reach ground water, its availability depends ultimately on the rains of the summer monsoon. These are not entirely reliable from year to year. When they are below par, or even fail altogether in some areas, years may pass when rainfall is so scanty that tanks, wells, and some rivers dry up. Thus, the history of the subcontinent has always included frequent episodes of drought and attendant famine.

AGRICULTURAL CHARACTERISTICS, PROBLEMS, AND ACCOMPLISHMENTS

FOOD CROPS AND POPULATION DENSITIES

The major staple food crops of the subcontinent are distributed in relation to the average amount of water available, as are the associated population densities. Rice is the basis of life in the wetter areas and, with its high caloric yields per acre, is associated with the highest population densities. Its areas of dominance are the delta area of Bengal (in both India and Bangladesh), the adjacent lower Ganges Valley, and the coastal plains of the peninsula. Irrigated wheat is the staple crop and food in the drier upper Ganges Valley of India and the quite dry Punjab of India and Pakistan. Here the caloric yield and population densities are intermediate. Unirrigated millets and sorghums are dominant over most of the Deccan plateau, where the relatively low rainfall cannot be supplemented much by irrigation. Caloric yields are low, and so are population densities, although the "low" densities of this area may reach over 100 and often over 200 per square mile (c. 40; 80 per sq. km). Despite the prevalence of millets and sorghums, however, rice is a preferred food almost everywhere, and if enough water is available, patches of irrigated rice appear.

The staple grains are supplemented by a host of minor crops. Many kinds of fruits and vegetables are grown, as are peas and beans to supply protein, and oilseeds such as peanuts. Corn, barley, sugarcane, coconuts, bananas, and spices are very important in certain areas. The overall agricultural economy is so large that some "minor" crops represent important shares in world production. Sugarcane, for instance, has a subordinate role in Indian agriculture, but India is one of the two leading producers (with Brazil), accounting for about one fifth of world cane sugar output in the early 1980s. Almost all of this is grown for consumption within India.

INDUSTRIAL AND EXPORT CROPS

Cotton, jute, tea, and the rice of Pakistan's Sind region are the main crops grown in the subcontinent for industrial use or export. India ranks fourth in the world in total production of cotton. It is a small-farm crop, grown mainly in the interior Deccan on certain soils that hold moisture unusually well and produce a crop without irrigation. India's large textile industry absorbs most of the production. Small farmers growing irrigated cotton in the Punjab and lower Indus Valley make Pakistan the world's fifth largest producer, although its output is substantially smaller than India's. With a smaller textile industry, however, Pakistan is a large exporter of cotton.

The Ganges-Brahmaputra Delta is the world's greatest producing area for jute, the principal material for burlap. The canelike jute plant, grown in several feet of water, requires a long growing season with high temperatures. Its production also requires much hand labor. Thus, it is well suited to the delta, which is tropical, practically amphibious, and settled by a dense agricultural population. The plant became the commercial mainstay in the 19th century, after British entrepreneurs had developed the necessary technology. The delta lies in the former province of Bengal, which was partitioned between India and (East) Pakistan when British rule ended. Most of the jute-growing areas lay on the Pakistani side (now Bangladesh), while the jute-manufacturing plants, which had developed in the Calcutta area, went to India. Subsequently, India markedly increased raw jute production on its side of the border, and it now produces more than Bangladesh, from which it continues to import quantities of raw fiber. Meanwhile, Bangladesh has developed sizable jute-manufacturing industries of its own.

Of the three major commercial crops in the subcontinent, only tea is exported from the Republic of India on a sizable scale. In tea growing and exporting, India leads the world, accounting for around one third of world production and exports in recent years. Unlike cotton, jute, and most other crops of the subcontinent, tea is principally a plantation crop. The plantations were developed and are still largely owned by British interests. Production centers in the northeastern state of Assam, with a secondary center in the mountainous Nilgiri "Hills" and adjacent mountainous sections in the far south of peninsular India. Development of tea plantations in both Assam and south India was favored by the availability of lands which were lightly occupied, reasonably fertile and well drained, and supplied with ample moisture by a heavy rainfall.

LIVESTOCK

The situation with respect to livestock in the subcontinent is peculiar. Hinduism is a religion which encourages vegetarianism, and Muslims are forbidden to eat pork. Thus pigs, a major food resource in many developing countries, are almost absent. In contrast the subcontinent has about one fourth of the world's cattle. The great majority are in India, which has more cattle than any other country. These are mainly oxen and water buffalo whose primary function is to serve as draft animals, although they supply milk and also provide dung for fuel and fertilizer. The Hindu religion forbids them to be killed, and many are emaciated scavengers since the people are too poor to feed them. In Muslim areas their meat may be consumed, and their carcasses supply hides for considerable leather industries, in India as well as in Pakistan and Bangladesh. Sheep are valued for meat in Muslim areas and for wool in all three countries, and they are adapted to poor grazing conditions. The number raised is sufficient to rank both India and Pakistan fairly high among the world's countries, although even India, with over twice as many sheep as Pakistan, still has less than one third as many as the Soviet Union or Australia.

AGRICULTURAL PROGRESS

The basic nature of the relations between agriculture, population, and food supply in the subcontinent can be seen in the following summary of food production (caloric content) since 1921:

Period	Area	Change in Food Production	Change in Per Capita Food Production
1921–1941	Subcontinent	Little	−5%
1953–1969	India	+41%	−1%
	Pakistan	+64%	+3%
1976/1978–1986	India	+28%	+5%
	Pakistan	+33%	+1%
	Bangladesh	+26%	−3%

(Main data source: U.S. Department of Agriculture. Other sources and other periods yield somewhat different figures.)

Although there is still room for greater accomplishment, agricultural output has been increasing rapidly since independence. The droughts and near or actual famines which occur periodically receive media attention and can create a misperception of agricultural stagnation. This misperception can also arise from the fact that in many regions of the subcontinent expanding agricultural output has not been able to keep up with population growth, with the result that food supply has become progressively less adequate. The resulting hardship does not fall equally on all. Estimates for India in the early 1980s pointed to about 300 million people (two fifths of the population) who were inadequately fed, often to the point of chronic undernourishment. Hence India's population of underfed people was larger than the *total* population of any other country except China. The more fortunate majority ate regularly, or even luxuriously in the case of some of the 60 to 100 million people who might be considered middle or upper class.

The successes of agriculture in the subcontinent have been due mainly to increased use of artificial fertilizers, introduction of new varieties of wheat and rice associated with the "Green Revolution" (Chapter 14), application of more labor from the growing rural population, reclamation of new lands for cultivation, increased irrigation, the spread of education, and the development of government extension institutions to aid farmers. At the time of independence fertilizer use was very small. Artificial fertilizers were unavailable, unknown, and/or unaffordable. The dried manure of the world's largest cattle population was (and to a large extent still is) used as fuel for cooking. This was made necessary by lack of wood due to deforestation and by the lack of access to fossil fuels. Hence animal manure was not used in adequate quantities to replenish soils exhausted by centuries of cultivation. And the subcontinent's farmers, unlike the Chinese or Japanese, were culturally biased against fertilizing with human wastes. In this situation even small applications of mineral fertilizers could achieve dramatic results in increasing the tragically low yields. A large increase in fertilization began after the mid-1960s, especially in conjunction with the use of new strains of wheat and rice in the "Green Revolution."

The new developments took root first in the Punjab area of India and Pakistan and are continuing to spread from there. By the middle 1980s they had reached about one third of rural India and were making a considerable impact in Pakistan and Bangladesh as well. They have generally been associated with increasing irrigation, both from wells and from large and sometimes spectacular new reservoir and

Figure 15.7 Village children returning home from school in northwestern India (Punjab). Their parents, who are Sikhs, have better-than-average means, and they have elected to send the children to a private school. Public schools in poverty-stricken India generally are handicapped by a shortage of equipment, materials, and competent teachers. The cone-shaped structures in the village are bins for the dried cowdung cakes used as fuel for cooking. The bins shelter the cakes from the rain. The house at the village edge belongs to a wealthy farmer and is far superior to most housing in the village.

canal systems. Pumping by electric motors or by gasoline engines is essential to much of this new irrigation. Hence the whole development has required the spread of rural electrification and of fuel-distribution facilities.

India managed to increase literacy from a reported 16 percent in 1947 to 41 percent according to the 1981 census, but probably a much higher proportion of its rural village population remained illiterate. Nevertheless, education, including government extension, increasingly is reaching the villagers (Fig. 15.7), influencing their agricultural practices as well as their health and family planning. Even where fact-to-face efforts are not possible, one radio in a village represents a quantum jump in contact. Now an Indian communications satellite launched by the American space shuttle in 1983 is being used to extend educational television to villagers, although it will take a major effort to supply every village with a TV set. Such efforts and accomplishments, along with the possibility for much further improvement, indicate that it is not agricultural inadequacy or hopelessness which threatens disaster in the subcontinent, so much as population growth that keeps agriculture struggling to stay abreast of it.

INDUSTRY AND ITS PROBLEMS

In the partitioning of the Indian Empire, the Republic of India received almost all of the developed capacity in modern industry which existed in the subcontinent at that time. It also received the great bulk of the area's mineral and power potential. Pakistan, including the present Bangladesh, was left with a much smaller industrial plant and fewer natural resources. Even in India, however, handicraft workers still exceed factory employees by many millions. Handicraft industries supply many everyday needs of the subcontinent.

INDIA'S MAJOR INDUSTRIES

Cotton Milling

The chief branches of Indian factory industry are cotton textiles, jute, and iron and steel. Modern cotton mills are mainly concentrated in the Bombay (10 million) and Ahmedabad (2.5 million) metropolitan areas and neighboring areas along or near the west coast, and in smaller industrial cities located in or close to the mountains at the southern end of the peninsula. Of basic importance to cotton manufacturing are the large home market, abundant and cheap labor, hydroelectric power available from stations in the Western Ghats and the Nilgiri Hills, and the large domestic production of cotton. As a major world producer of cottons, India supplies cloth to enormous numbers of its impoverished citizens and is regularly a leading exporter of cotton textiles and clothing.

Jute Milling

India's jute industry is closely concentrated in and near the largest city of the subcontinent, Calcutta (over 12 million). The industry lost most of its raw material base, and consequently suffered a large loss of employment, when the jute-growing areas of East Pakistan were partitioned off from India. Supplies of raw jute subsequently imported from Pakistan were not always fully adequate or reliable. The two nations were highly competitive in the jute industry, with India attempting to increase the acreage in jute in West Bengal and Pakistan attempting to increase the number of jute-manufacturing plants in East Bengal. However, despite some success in both attempts, Calcutta continues to be the world's leading center of the jute-manufacturing industry. Jute manufactures are an Indian export, although one of less importance than several others. By contrast, the smaller jute-manufacturing industry of Bangladesh supplied 31 percent of that country's exports in 1985–1986, and raw jute another 15 percent.

Iron and Steel

India's iron and steel industry is concentrated in the northeastern corner of the peninsular uplands and in immediately adjacent parts of the Bengal lowland. Several iron and steel centers are scattered in a belt from near Asansol (1.2 million) in the north to the upper Mahanadi River on the south. The largest center is Jamshedpur (700,000), 150 miles (*c.* 240 km) due west of Calcutta. The most valuable collection of mineral resources in India—including iron ore, coal, manganese, chromium, and tungsten—is found in or near this belt. These resources have led to the rise of small to medium-sized industrial and mining cities in an area that in other respects is one of the least developed parts of the country, with much of the countryside occupied by relatively primitive tribal people. Steel production here was pioneered at Jamshedpur by the famous Tata family of Indian industrialists. Between World Wars I and II the Tata works at Jamshedpur became the single largest steel-producing unit in the British Empire.

In order to facilitate general industrialization, independent India has made expansion of the iron and steel industry a major objective. Most of the main steel plants now date from 1959 or later. Output has multiplied but is still under 2 percent of world output and is inadequate for India's needs. Except for coal of coking quality, which will probably have to be supplemented by imports, resources are adequate for much-expanded production in the future. The country has very large reserves of non-coking coal (very important in India's overall energy economy), iron ore, and manganese.

INDUSTRIAL DIVERSIFICATION: ENGINEERING AND CHEMICAL INDUSTRIES

As India's industrial expansion continues, its output of manufactured goods gradually becomes more diversified and sophisticated. Hence such smaller industries as engineering and chemicals are growing at a faster rate than the cotton-textile and jute industries. The machine-building industries are still quite inadequate for the country's needs, necessitating large imports, but they are growing rapidly. Between 1975 and 1981, for example, India's overall industrial output increased by 36 percent, compared with a worldwide growth of 24 percent. This relatively rapid growth has been continuing.

Figure 15.8 Motorcycles have become a popular means of family transportation in the Indian subcontinent. Often several members of a single family will be seen aboard a single vehicle. The riders of the motorcycle in the photo are Hindus; the locale is the Indian state of Punjab.

Today there are very few items among the great variety of machines used in a modern society which are not produced somewhere in India. There are several automobile and truck assembly plants in the country, motorcycles (Fig. 15.8) are manufactured, and bicycles, produced in many small factories, have become a considerable export item. Fairly recent additions to the array of manufactures include tractors, bulldozers, data-processing equipment, and silicon chips for electronics. Larger quantities and a greater variety of chemical products ranging from vitamins to fertilizers are steadily being achieved. High technology is represented by nuclear energy, atomic weaponry, rocketry, and communications satellites. But perspective on India's promising industrial range and growth must still be maintained. In the middle 1980s, with 16 percent of the world's people, India was estimated to have only about 1 percent of the world's total industrial output.

INDUSTRY IN PAKISTAN AND BANGLADESH

Pakistan and Bangladesh occupy areas which, compared to India, inherited little industrial development from Britain's Indian Empire. They are still far behind India in total industrial output. However, Pakistan has made rapid progress in some lines since independence. Its principal success has been the development of a cotton-textile industry. Cotton milling scarcely existed in Pakistan in 1947, but as early as 1954 the industry had reached the point where it supplied the bulk of domestic needs and even a small export. New jute mills in East Pakistan were developed, and these have been taken over by Bangladesh. Pakistan's main industrial centers are its two largest metropolitan areas: the seaport of Karachi (perhaps 6 million) at the western edge of the Indus River delta, and Lahore (perhaps 3.5 million) in the Punjab. In Bangladesh, industrial development centers in Dacca (Dhaka: 4.1 million), the capital and largest city, and Chittagong (1.6 million), the main seaport.

THE PROBLEM OF INADEQUATE POWER RESOURCES

Inadequate energy resources are apt to be a major problem confronting further industrialization in the subcontinent. The major domestic sources of commercial energy in the 1980s were India's coal deposits along the northeastern border of the Deccan and the hydropower generated in widespread areas. All countries in the subcontinent are heavily dependent on imported energy, especially oil, and would probably be more so if their level of development, and therefore their consumption, were higher. This will continue to be true despite the fact that India has developed a small but rapidly growing oil production from fields in Assam and offshore from Bombay. In all three major countries natural gas from domestic wells also plays a role in the energy picture. It is especially important in Pakistan.

PROBLEMS OF SOCIAL AND POLITICAL RELATIONS

On a physical map the Indian subcontinent looks like an obvious physical unit, marked off from the rest of the world by its mountain borders and seacoasts. Yet the social complexity of this area is so great that the apparent physical unity has never been paralleled by overall political unity except during a relatively brief period in the 19th century and the first half of the 20th century. During this time political unity was imposed from outside by Great Britain. Even so, however, a great variety of small to fairly large political units under native rulers retained varying degrees of autonomy, although all of these native states were ultimately under British control. As soon as British power was withdrawn, the divisive force of conflicting social groups again asserted itself and the subcontinent split into two nations. Within another quarter-century, divisiveness was again expressed—aided by geographic

separation—when Pakistan divided into two countries. But despite these successive divisions, India, Pakistan, and Bangladesh still have serious problems of national unity, as well as problems of relations with each other.

RELIGIOUS DIVISIONS AND CONFLICTS

The major social divisions within the subcontinent are in religion and language. Of these, religious divisions have generally appeared the more important and have received more attention. The most serious division has been between the two major religious groups, the Hindus and the Muslims. Hinduism is a religion native to the subcontinent and was the dominant religion at the time Islam made its appearance. It has continued to have the largest number of adherents—a reported 83 percent of the population of India, 12 percent of Bangladesh's, and 1 to 2 percent of Pakistan's.

Islam made its appearance in the subcontinent as a proselytizing and conquering religion in the 8th century A.D. Periodic later invasions penetrated the relatively weak northwestern frontier until at the peak of Islamic power in the 16th and 17th centuries the Mogul Empire dominated most of the subcontinent from its capital at Delhi (Fig. 15.9). British penetration of India was aided by the internal disintegration of this Muslim empire in the 18th century.

Seldom have two large groups with such differing beliefs lived in such close association with each other. Islam holds to an uncompromising monotheism and insists on uniformity in religious beliefs and practices, whereas Hinduism is monotheistic for some believers but polytheistic for others and holds the view that a variety of religious observances is consistent with the differing natures and social roles of humans. Islam is essentially intolerant of all other faiths, but Hinduism has an essentially tolerant attitude. Islam believes in its divine mission to convert all people to the true religion, but Hinduism holds that proselytizing is essentially useless and wrong. Islam's democratic belief in the essential equality of all believers is a total contrast to the inequalities of the caste system in Hinduism. Islam's use of the cow as food and for sacrifice is anathema to Hinduism, which considers the cow an especially sacred animal that must not be killed. The exuberant and noisy celebrations of the Hindu faith are a great contrast to the austere and silent ceremonials of Islam. Bloodshed has sometimes resulted when a Hindu parade with its jingling bells and firecrackers disturbed the solemnity of Muslims gathered for a particularly sacred religious rite.

Figure 15.9 The strong Muslim influence in the Indian subcontinent is evidenced by the architecture of mosques, palaces, and other structures. The photo shows a small section of the Taj Mahal at Agra, India, in the subcontinent's Northern Plain. The Taj Mahal was built by Mogul Emperor Shah Jahan as a mausoleum for his favorite wife, Mumtaz Mahal. The tall minaret at the right and the domes and pointed arches at the left are characteristic features of Muslim ceremonial architecture. Emperor and wife are buried in a vault in the building.

The antagonism to be expected between such differing groups was intensified when the formerly subordinate Hindus came, under the British occupation, to dominate most Indian business as well as the civil service. Many Muslims feared the results of being incorporated into a single state with the Hindu majority, and their demands for political separation led to the creation of two independent states from the Indian Empire rather than one. The creation of Pakistan in two widely separated parts was due to the distribution of the main areas of predominantly Muslim population at the time of partition. Immediately preceding and following partition violence broke out between the two peoples on a huge scale, and hundreds of thousands of lives were lost in wholesale massacres before the new governments could establish control. Mass migrations between the two countries involved some 12 million people. Then in 1971 the events which created Bangladesh—revolt, repression, and Indian intervention against Pakistan—again brought huge casualties and millions of refugees. Informal strife between adherents of the two religions, often involving riots and violence, remains common in areas where they are mixed.

Besides its Muslims, estimated at 11 percent of its population, India has many other religious minorities of some significance. There are about 20 million Christians, the bulk of whom live in the south of the peninsula. About 15 million Sikhs are concentrated mainly in the Punjab, where they are agitating to secure an autonomy approaching independence for this unusually developed and prosperous part of India. Among the numerous smaller religious minorities—including Parsees, Buddhists, Jains, and Jews—the Parsees, although very small in numbers, play an especially significant role. Numbering perhaps 200,000 and concentrated mainly in Bombay, this group has attained wealth and economic power in India far out of proportion to its size. The Parsees derive originally from Iran, although the group has been in India for more than a thousand years. Their religion is the ancient pre-Islamic Iranian (Persian) faith of Zoroastrianism.

TERRITORIAL DISPUTES BETWEEN INDIA AND PAKISTAN

The traditional religious conflict of the subcontinent between Muslims and Hindus has continued since independence in the form of frequent hostile relations between India and Pakistan. Partition created a number of problems, including territorial disputes, which have embittered subsequent relations. One dispute, involving the division of water resources in the Punjab, apparently was settled to the mutual satisfaction of the two nations by an agreement in 1960. The problem stemmed from the fact that all the major rivers—the Indus and the rivers of the Punjab—on which the life of Pakistan depends have their upper courses in Indian-controlled territory. How much of their water was India to be allowed to divert and how much could Pakistan count on? The agreement finally reached allocates the water of the three eastern rivers of the Punjab—the Beas, Sutlej, and Ravi—to India, which in return undertakes to allow unrestricted and undiminished flow of the Chenab, the Jhelum, and the Indus itself into Pakistan (Fig. 15.3). A system of new canals brings water from the western rivers to areas in Pakistan previously irrigated from the eastern streams.

More intractable and damaging, however, has been the dispute over Kashmir, in the mountainous northwestern corner of the subcontinent. A large majority (perhaps three fourths) of Kashmir's population of an estimated 10 million is Muslim, which is the basis of Pakistan's claim to the territory. Before independence the state was administered, within British India, by a Hindu ruler. Under the partition arrangements, each native ruler was to have the right to join either India or Pakistan, as he chose. The choice of India by Kashmir's ruler is the legal basis of India's claim, although this is somewhat clouded by the fact that India forcibly absorbed certain areas of Hindu population whose Muslim rulers had opted for Pakistan. Fighting after partition led to a cease-fire line leaving eastern Kashmir, with most of the state's population, in India. After 1959, Chinese claims on remote northern mountain areas of Kashmir adjoining China were pressed and some border territories controlled by India or Pakistan were occupied. Pakistan ceded the territory claimed by China to that country and established friendly relations with the Chinese, while India rejected Chinese claims. Recurrent small-scale military actions have not dislodged the Chinese from the section they occupy on the Indian side of the cease-fire line.

Two wars between India and Pakistan have effected little change in Kashmir. In 1965 conflict began in Kashmir, spread to the Punjab and escalated to a brief but indecisive full-scale war involving tanks, airborne forces, and widespread air raids. The United Nations achieved a cease-fire with no substantial change in the prewar situation. Renewed hostilities in Kashmir in 1971, as part of the war in which India supported the revolt of Bangladesh from Pakistan, again effected no substantial alteration in Kashmir. More recently (mid-1980s) the two countries have been making stronger efforts to stabilize peaceful relations with each other. This may be a reaction not only to the Chinese presence but also to the Soviet Union's intrusion into the general area with its 1979 invasion of neighboring Afghanistan.

PROBLEMS OF CASTE AND SECTARIANISM

In the Republic of India, serious internal problems of religious division still exist, not only between the Hindus and minority groups but also within the body of Hinduism itself. One of the fundamental features of Hinduism has been the division of its adherents into the most elaborate caste system ever known. Traditionally every Hindu is born into a par-

ticular caste, of which there are more than two thousand in India but many fewer in any particular location. Caste membership is determined by birth and cannot be changed. Castes form a hierarchy that determines a person's social rank, with the Brahman castes (somewhat over 5% of all Hindus) at the top and the untouchables (15%) at the bottom. Particular castes are associated with particular religious emphases, and their members must follow occupations that the caste is permitted. With certain exceptions, marriage outside the caste is forbidden, and meals may be taken only with fellow caste members.

Modernization of India comes inevitably into conflict with this rigid social system. Brahman privileges are being increasingly threatened and in some areas have been restricted by law. Untouchables, so-called because their touch has been held to defile a high-caste Hindu, must be drawn into a modern economic system in ways that undermine restrictions on their caste. Their cause has been championed for both moral and "practical" reasons by Indian leaders of higher caste, and untouchability is now officially outlawed in India, although with doubtful effectiveness. Agitation from below and reaction from above combine to create friction, and not infrequently violence, as the rigidity of the caste system loosens under the impact of modern conditions and needs. Especially in the cities the close intermingling of large numbers of people in factories, rooming districts, public eating places, and public transportation has been a major factor in hastening the disintegration of the caste structure, although it is still strong in the villages.

Pakistan also has internal problems of conflict centering on religious division. In its case, the major division is between the two main sects of Islam. About 70 percent of Pakistan's Muslims belong to the Sunni sect, while 30 percent are Shiites. In the 1980s Pakistan's military government conformed to a widespread fundamentalism in Islamic countries by moving to convert Pakistan into an officially "Islamic" state with laws derived from the Koran. This led to rioting by factions preferring a return to Western-style democracy and protests by women who objected to being officially branded as inferior and restricted in their activities. Beyond this, there was large-scale violence between Sunnis and Shiites in Karachi, as the Shiites protested the government's version of an Islamic state. These outbreaks were suppressed with vigor and violence, but the underlying antagonisms remain. Hence Pakistan will continue to share the general controversy roused by the new fundamentalist movement in the Islamic world, with its population split between adherents of an

Islamic state and supporters of a secular state, and with those of the first group being divided as to the form an Islamic state should take.

LANGUAGE DIVISIONS

Language supplements religion as a divisive factor in India and Pakistan. The languages of the subcontinent fall into two chief groups, the Aryan languages in the north (Fig. 15.10) and the Dravidian languages in roughly the southern half of peninsular India. The various languages within each of these groups are fairly closely related to one another. Languages of major importance, each with many millions of speakers, include four languages of the Dravidian group—Telugu, Tamil, Kanarese (Kannada), and Malayalam—and eight Aryan languages—Hindi, Bengali, Bihari, Marathi, Punjabi, Rajasthani, Gujarati, and Oriya. Dozens of other languages and dialects are spoken. Although bi- and trilingualism are common, the difficulties of communication created by so many languages are obvious. Such difficulties are increased by the low rates of literacy in the subcontinent's population: Bangladesh, 33 percent in 1985; Pakistan, 26 percent in 1981; and India, 41 percent in 1981. (The percentages represent literacy among all persons age 15 or older, or in India, 14 or older.)

English has been the *lingua franca* of the subcontinent, or at least of the educated classes. However, only a very small percentage of the population is literate in it. This fact plus nationalist feeling led to a desire to adopt some native tongue as an official language (the business of the governments is largely carried on in English) to be propagated as a common medium of instruction and communication. Disputes naturally arose in each country as to which language should be chosen. India decided on Hindi, which is spoken by the largest linguistic group, and made it the country's official language in 1965. But this decision on the part of the Indian government resulted in strong protests, especially from the Dravidian south, with the result that English was designated the "associate language." In Pakistan, Urdu, a language similar to Hindi but written with Perso-Arabic script, was chosen. English has been retained as the "official" language, and Urdu has been made the "national" language. In Bangladesh, Bengali is the predominant language. Bloody rioting between different language groups has occurred at times in areas of India where such groups are mixed, and demands by language groups have been major factors in re-

Figure 15.10 The linguistic diversity of the Indian subcontinent is reflected in this photo of an Indian school child learning to write the Punjabi language. This language is written in the script called Garmukhi. The young pupil, seated on the school ground, is using a whitewashed wooden tablet on which the letters are written with a homemade watercolor paint. When the makeshift slate is filled with letters, whitewash will be applied and the slate reused. The pupil's family is too poor to afford paper and pencils.

shaping the boundaries of a number of India's internal political units.

PERSPECTIVES

Given the problems outlined in this chapter, the subcontinent's accomplishments in recent decades may be regarded as remarkable. India has lowered its birth rate and achieved considerable growth in agriculture and industry. More uniquely among the world's poor countries, it has done so while maintaining, despite all its internal divisions and frictions, a Western-style representative democracy. Pakistan (at times) and Bangladesh have lapsed from disorderly democracy into military dictatorship, but Pakistan has achieved considerably better economic conditions than India. Even Bangladesh, the poorest of the three, has registered notable increases in food production and some advances in other lines. In all these countries the race between achievement and disaster continues.

CHAPTER 16

The Fragmented and Contentious World of Southeast Asia

The island city-state and seaport of Singapore has become a booming metropolis where old traditions and dynamic new businesses intermingle. Hence it is representative of the increasingly competitive centers of world business spotted along Asia's Pacific rim.

SOUTHEAST Asia, from Sri Lanka (Ceylon) on the west to the Philippines on the east, is a physically fragmented region of peninsulas, islands, and intervening seas. East of India and south of China, between the Bay of Bengal and the South China Sea, the large Indochinese Peninsula, occupied by Burma, Thailand, Kampuchea (Cambodia), Laos, and Vietnam, projects southward from the continental mass of Asia (Fig. 16.1). From it the long, narrow Malay Peninsula extends another 900 miles (*c.* 1450 km) toward the equator. Ringing the south and east of this continental projection are thousands of islands, among which Sumatra, Java, Borneo, Celebes (Sulawesi), Mindanao, and Luzon are outstanding in size. Another large island, New Guinea, east of Celebes, is culturally a part of the Melanesian archipelagoes of the Pacific World.

However, the western half of New Guinea, held by Indonesia, may be considered a marginal part of Southeast Asia in a political sense. To the west the island country of Sri Lanka, while merely a detached part of the Indian subcontinent in a geologic sense, may be considered the western outpost of the Southeast Asian region from the standpoint of economic development, culture, and climate.

Southeast Asia, thus conceived, is composed politically of eleven states: Sri Lanka, Burma, Thailand, Laos, Kampuchea, Vietnam, Malaysia, Singapore, Indonesia, Brunei, and the Philippines. Two of these states, Laos and Kampuchea, have been only nominally independent in recent years, being controlled by Vietnamese armies that occupied these countries and set up puppet governments there. With the exception of Thailand, which was never a colony, none of the states dates its present independence farther back than 1946. The last to become independent was

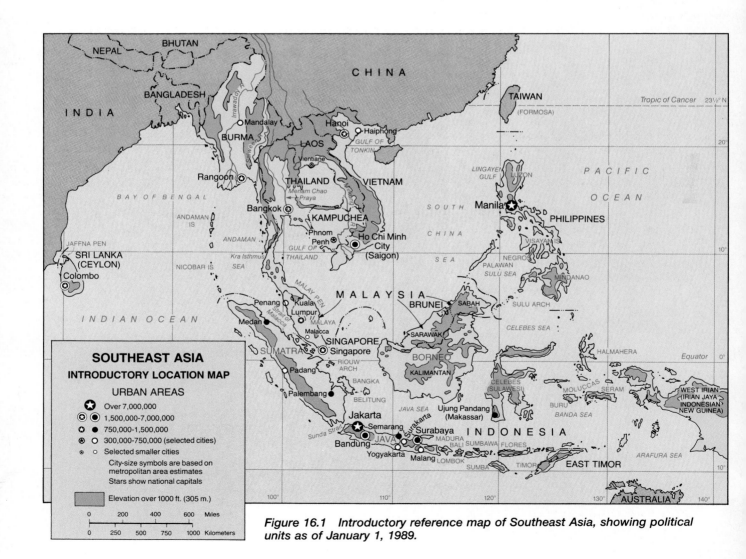

Figure 16.1   Introductory reference map of Southeast Asia, showing political units as of January 1, 1989.

Brunei in 1983. Since the 11 countries of Southeast Asia occupy a land area less than half that of China and are varied both in size and in form of government, it is apparent that political as well as physical fragmentation is a major Southeast Asian characteristic. As we shall see later, there is also a high degree of cultural fragmentation, which has often expressed itself in contention and bloody strife among differing cultural groups.

AREA, POPULATION, AND ENVIRONMENT

It is approximately 4000 miles (*c.* 6400 km) from Sri Lanka to central New Guinea and 2500 miles (*c.* 4000 km) from northern Burma to southern Indonesia (Figs. 16.1 and 16.2), but the total land area of Southeast Asia is only about 1.8 million square miles (4.6 million sq. km). The population was estimated at about 450 million in mid-1988, giving an overall density a little over 250 per square mile (*c.* 100 per sq. km)—an average which embraces very great extremes within the region (Table 14.1).

RELATIVE SPARSENESS OF POPULATION

The average population density of Southeast Asia is high compared to that of much of the world but low compared to most other areas on the seaward margins of the Orient (Table 14.1). For example, although Southeast Asia has nearly four times the overall density of the United States, its density is well under half that of the Indian subcontinent or China proper (the humid eastern part of China south of the

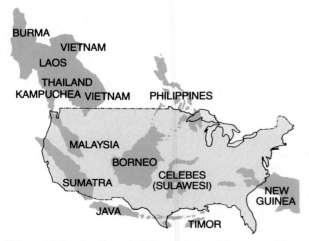

Figure 16.2 Southeast Asia compared in area with the conterminous United States.

Great Wall). Southeast Asia has apparently been less populous than India and China proper throughout history. It has been estimated that the whole region contained only about 10 million people around the year 1800. Despite the fact that the region was known in early times by Chinese, Japanese, and Indians, that trade has moved through it for many centuries, and that some early immigrants reached it by sea (particularly from India), widespread dense populations did not develop.

PROBLEMS OF A TROPICAL ENVIRONMENT

It seems likely that difficulties imposed by the natural environment of Southeast Asia had much to do with the slowness of population growth there. By land the region is relatively isolated, as the Indochinese Peninsula abuts in the north on high, rugged, and malaria-infested mountains. Over a period of many centuries the forebears of most of the present inhabitants entered the area as recurrent thin trickles of population, crossing the mountain barrier under pressure, usually as refugees driven from previous homelands to the north.

Environmental difficulties within the region itself are formidable. The most significant of them derive from the fact that Southeast Asia is truly tropical, with continuous heat in the lowlands, torrential rains, a prolific vegetation difficult to clear and keep cleared, soils which are generally leached and poor, and a high incidence of disease. The unfavorable agricultural effects of a 4- to 6-month dry season in the Indochinese Peninsula and scattered smaller areas in the islands are accentuated by a high rate of evaporation imposed by the tropic heat. The lush vegetation of Southeast Asia masks the infertility of most of its soils—an infertility that becomes apparent and generally increases rapidly when the land is cleared and cultivated. Heat and humidity speed bacterial and chemical action so that the humus in the topsoil is rapidly destroyed and the mineral plant foods become thoroughly depleted by leaching. In addition, much of Southeast Asia is mountainous, and erosion is so rapid that the rivers generally carry enormous volumes of mud and silt, eventually to be deposited on flood plains and deltas at or near the sea. More spectacular environmental difficulties include frequent volcanic eruptions and earthquakes in Indonesia and the Philippines, and the violent wind-and-rain storms known as typhoons which strike coastal Vietnam and the northern Philippines. Prescientific techniques were apparently unable to foster large populations under these conditions. Early civilizations did not expand widely or even maintain themselves, as is evidenced by the monumental ruins found in a few parts of the region. In other places, climate and vegetation have almost destroyed the last vestiges of early cities and states.

THE NEW RACE BETWEEN POPULATION GROWTH AND FOOD SUPPLY

Western scientific knowledge and technology have mitigated the effects of these environmental conditions somewhat and have permitted a tremendous population increase in the last century and a half. This increase, which accelerated after World War II, is now showing some signs of slowing but is still rapid. Between 1959 and 1973 the region's population grew from an estimated 212 million to 316 million, an increase of 104 million, or 49 percent, in 14 years. Between 1973 and mid-1988 it grew from 316 million to 449 million, a 14½-year increase of 133 million people, or 42 percent. Overpopulation is not yet the problem that it is in India or China, but growth must soon be slowed further if it is not to become so.

During recent times the developing race between population growth and food supply has been marked by significant progress in some countries but relative stagnation or regression in others. In the 1970s the growth in food production (caloric content) in Malaysia (108%), Thailand (83%), Indonesia (64%), and the Philippines (50%) was outstanding on a world comparison and easily outpaced the growth of these countries' populations. Burma and Sri Lanka also did reasonably well. Most of these increases were the result of better yields from the land due to new crop strains, more fertilizers, and better water control. Except in Burma, part of the growth was also due to a sizable expansion of the area planted to crops.

At the other end of the scale are Vietnam, Laos, and Kampuchea, where war, devastation, military forays, massacre, repression, and Communization have largely thwarted progress. For a time (1965–1975) the United States was heavily involved in some of these disasters through its participation in the Vietnam War, but since the American withdrawal from Vietnam in 1975, the evils have continued with little if any abatement. Additional millions have been slain, mainly in Kampuchea, by the various Communist regimes and factions in the three countries. A heavy dependence on Soviet aid has developed, but the aid has been primarily military rather than economic. As of 1980 the World Bank estimated that the food calories available in the three countries were inadequate to maintain moderate physical activity for all citizens: 97 percent adequate in Laos, 90 percent in Vietnam, and 88 percent in Kampuchea. This was in marked contrast to the situation before war and Communization, when Kampuchea and (South) Vietnam had been major exporters of rice. In estimated caloric adequacy the remaining Southeast Asian countries ranged in 1980 from 102 percent in Sri Lanka to 121 percent in Malaysia and a high 135 percent in the Singapore city-state. By the late 1980s there was no evidence that these divergent food trends and conditions within Southeast Asia were being reversed.

Despite progress in agriculture and food adequacy in most of its countries, Southeast Asia as a whole is one of the world's poorer regions. As of 1986 the per capita gross national product (GNP) in 7 of its 11 countries was below the average even of the world's "less-developed" countries. One country, Thailand, was somewhat above average; Malaysia was a relatively well-off "less-developed" country; while the ministates of Brunei and Singapore had per capita GNPs above the "less-developed" category. In fact, small Brunei, with sizable oil resources and relatively few people, is one of the world's wealthiest countries as measured by per capita product. The poorest countries in the region, by both per capita product figures and health measures such as infant mortality and life expectancy, are the three Communist countries and, by some measures, Burma and Indonesia (see Table 14.1).

THE ECONOMIC PATTERN

Despite rapid urbanization in recent years a large majority of the people in Southeast Asia are still farmers, and manufacturing is much less developed than in Japan, China, or India. Mining is relatively unimportant to the region as a whole, although it contributes significantly to the incomes of a few countries.

SHIFTING CULTIVATION

A permanent form of agriculture has proved extremely difficult to establish in most parts of this mountainous tropical realm. Much of the land is used only for a shifting, primitive subsistence form of cultivation in which fields are cleared and used for a few years and then allowed to lapse back into jungle while the cultivator moves on to clear a new patch. Clearing is ordinarily accomplished by fire, and the crops are often grown with little or no cultivation. Unirrigated rice is usually the principal crop, although corn, beans, and root crops (such as yams and cassava) are also grown. Agricultural activities are often supplemented by hunting and gathering in the forest. These migratory subsistence farmers ordinarily differ ethnically from adjacent settled populations, for they are the remnants of peoples driven to refuge in the back country by stronger invaders. Since it is estimated that 15 years are generally necessary for an abandoned clearing to regain its fertility, it is obvious that only a small proportion of the land can be cultivated at any one time. Therefore, populations living under such conditions must necessarily be sparse.

DENSELY POPULATED DISTRICTS OF SEDENTARY AGRICULTURE

Most of Southeast Asia's people live, often in extremely dense clusters, in scattered areas of permanent sedentary agriculture. Such areas form the core regions of the various countries and stand in striking contrast to the relatively empty spaces of the adjoining districts. A superior degree of soil fertility appears to have been the main locational factor in most instances. In Southeast Asia soils of better than average fertility have ordinarily resulted from one or more of the following factors:

1. The presence of lava and volcanic ash of proper chemical composition to weather into fertile soils.

2. The accumulation and periodic renewal of plant nutrients washed from upstream areas and deposited on river flood plains and deltas.

3. A slowing down or seasonal reversal of the leaching process in areas having a distinct dry season and relatively low rainfall.

4. Occurrence in some areas of uplifted coral platforms which weather into a superior soil.

A few areas of Southeast Asia exhibit fairly dense populations without any corresponding soil superiority. Such areas are ordinarily characterized by a plantation type of agriculture based on crops adapted to poor soils, such as rubber.

Side by side with the present areas of dense settlement, some of which are seriously overcrowded, are sizable areas still undeveloped despite the fact that they apparently are capable of supporting large populations. Their development had to wait until the need for their utilization became sufficiently great and the available resources sufficiently large to encourage large-scale clearing of forests, draining of swamps, and construction of irrigation systems. It also had to await the initiative and resourcefulness,

or the desperation, of a sufficient number of people willing to leave established homes and accept the risks of pioneering in areas often plagued by relatively low and undependable rainfall and/or by malaria. These conditions have now arrived in many areas, as witnessed by the large-scale expansion of arable land which has begun to occur in various countries of the region, but most markedly in Thailand, Indonesia, Sri Lanka, and Malaysia.

The typical inhabitant of the areas of permanent sedentary agriculture is a subsistence farmer whose main crop is wet rice, grown with the aid of natural flooding or by irrigation. In some drier areas unirrigated millet or corn, the latter especially in Indonesia and the Philippines, takes the place of rice. The major crop is supplemented, to a degree tending to vary positively with the pressure of population, by secondary crops which can be grown on land not suited for rice or other grains. Prominent secondary food crops include coconuts, yams, cassava, beans, and garden vegetables. If the year is a good one, a portion of the produce may be sold, and some farmers grow a secondary crop such as tobacco, coffee, or rubber primarily for sale. Fish are an important component of the food supply and a considerable source of income, not only along the coasts but also inland along streams and lakes. Many fish are harvested from artificial fish ponds or from flooded rice fields in which they are grown as a supplementary "crop."

THE ROLE OF COMMERCIAL AGRICULTURE

Subsistence production in Southeast Asia exists side by side with an important development of commercial agriculture, directed mainly toward supplying commodities for export. The region is one of the world's major supply areas for tropical plantation crops, and two of its countries—Thailand and Burma—are surplus rice producers and exporters. These commercial elements in Southeast Asian agriculture are principally an outgrowth of Western colonial enterprise in the region.

European imperialism in Southeast Asia began in the 16th century. Until the 19th century, however, the principal interest of the newcomers lay in the region's strategic location on the route to China, and they were generally content with effective control over scattered patches of land along the coasts and with gleaning for trade such surpluses of value as the native economies happened to offer. But during the 19th century increasing populations and an expanding technology in Europe and North America greatly enlarged the demand for products of tropical

Figure 16.3   This floating market in Bangkok, Thailand, symbolizes the amphibious character of much of Southeast Asia. Water transportation is vitally important in the river deltas and islands, where most Southeast Asians live.

agriculture. Consequently, effective European political control was forcibly extended over almost the whole of Southeast Asia, and European capital and knowledge were applied to bring about a rapid increase in production for export. Certain indigenous commodities, such as copra and spices, were given increased emphasis, and some new crops were introduced, one instance being the introduction in the 1870s of rubber seedlings grown in England's Kew Gardens herbarium from smuggled Brazilian seeds.

The usual method of introducing commercial production was to establish large estates or plantations managed by Europeans, but worked by indigenous labor or labor imported from other parts of the Orient. Development of these large commercial farming enterprises was assisted not only by the favorable climate but also by the availability of land in a relatively empty part of the Orient and by the amphibious nature of the topography, which made possible a close dependence on cheap, efficient, and easily established transportation by water (Fig. 16.3). The major difficulty to be overcome was in most cases the recruitment of an adequate labor force from a population already fully engaged in food production and not sufficiently pressed by poverty and hunger to be easily attracted from the communal life of the villages into labor for wages on the plantations. The eventual solution to the latter problem in many areas was large-scale importation of contract labor from India and China, which further complicated an already complex ethnic mixture.

Plantation activity came to have widespread repercussions on the economic life of the natives themselves, for many native farmers and plantation laborers learned by example and entered commercial production on their own account. Even shifting cultivators could plant a few rubber trees and return when the trees reached the producing state, 7 to 10 years later, to tap them for latex if the price warranted. Smallholders of the foregoing types now command an important share in the export production of most "plantation" crops.

Many different plantation-type commodities have been produced in Southeast Asia on a medium to large scale at one time or another. Wide fluctuations have occurred from time to time in the crops grown, the centers of production, the amounts exported, and the prosperity of the producers. These shifts have been occasioned by such factors as fluctuating world demands, intraregional and interregional competition, changing political conditions, and the occasional ravages of plant diseases. This general situation has prevailed not only in Southeast Asia but in plantation areas throughout the tropical world. The major plantation-type commodities now exported from Southeast Asia are:

1. *Rubber.* Over four fifths of the world's natural rubber is produced in Southeast Asia, primarily in Malaysia, Indonesia, and Thailand. Natural rubber, however, is much less crucial to the world than it was before the rise of large-scale synthetic rubber production during World War II. Synthetic rubber, most of which is made from petroleum, now supplies the bulk of the world's rubber consumption, although natural rubber still accounts for perhaps a third of all rubber used.

2. *Oil Palm and Coconut Palm Products.* These consist primarily of palm oil, coconut oil, and copra (dried coconut meat from which oil is pressed). Malaysia is by far the world leader in palm oil production, while the Philippines and Indonesia dominate the world output of coconut palm products.

3. *Tea.* This crop is important mainly in Sri Lanka, with Indonesia a fairly strong second. Sri Lanka is the world's third largest producer, and tea comprised 28 percent of its exports in 1986, but the country's output is far below that of India (which leads the world) or China.

A host of lesser crops are produced for export by various countries in scattered locations. These crops include coffee from Indonesia, cane sugar from the Philippines and Indonesia, pineapples—of which the Philippines and Thailand have become the world's leading producers and exporters—and many others. In general the countries of Southeast Asia are coming to depend less heavily on their agricultural exports, however, as mining and some manufacturing develop.

COMMERCIAL RICE FARMING

One very significant aspect of Western influence on the economy of Southeast Asia was the stimulation of commercial rice farming in certain areas that had formerly been unproductive. The development of plantation agriculture, mining, and trade provided a market for rice by bringing into existence a large class of people who worked for wages and had to buy their food. During the same period Western economic, medical, and sanitary innovations helped bring about an enormous increase in Oriental population generally and a growing demand for food. Western technology made possible the bulk processing and movement of rice, and aided in the development of drainage, irrigation, and flood-control facilities needed to produce it on a large scale in areas previously undeveloped. As a consequence, an Asian pioneer movement took place into certain areas capable of greatly expanded rice production. Among these latter areas three gained outstanding importance in commercial rice growing: the deltas of the Irrawaddy, Chao Praya, and Mekong rivers, located in Burma, Thailand, and (South) Vietnam and Cambodia (now Kampuchea), respectively.

Almost impenetrable swamps and uncontrolled floods had kept these deltas thinly settled previously, as other swampy areas in Southeast Asia are today, but adequate incentives and methods for pioneering have turned them into densely settled areas within the past century. Despite growing populations, the farms in these areas are larger than is common in the Orient, and the surplus of rice is such that three tenths of the world's total rice exports generally originated in the three deltas up to the 1960s. Long-continued warfare and economic dislocation then ended the ability of Vietnam and in the 1970s Kampuchea to generate surplus production. Indeed in Kampuchea, widespread starvation replaced surplus. Expansion in Thailand and Burma, which accounted for about one third and one twentieth, respectively, of total world rice exports in the period 1982–1984, has partially replaced the loss of the Mekong Delta as a surplus rice-producing area, and in Thailand corn (maize) has also become a major food export. Replacement of the grain supply within Southeast Asia is important, because most of the Southeast Asian countries are still major grain importers.

PRODUCTION AND RESERVES OF MINERALS

Petroleum is the most important mineral resource in Southeast Asia, although the region's reserves and production are not impressive on a world scale. Four countries—Indonesia, Malaysia, Brunei, and Burma—produce oil. Their combined output is generally no more than about 3 percent of world oil production. Among them only Indonesia, with approximately 2 percent of world oil production, at all resembles a major producer. Its oil fields are in Sumatra, Borneo (Indonesian Borneo is officially termed Kalimantan), and to a lesser extent Java and New Guinea. Malaysia's fields are in Sarawak on Borneo. There are offshore fields between Sumatra and Java, off northern Borneo, and out in the South China Sea. Possession of some of the latter is disputed between Malaysia and Vietnam.

Although it does not loom large on the world scene, Southeast Asian oil is very important to the countries that own and produce it. In 1986 oil and associated natural gas accounted for 97 percent of the exports of Brunei, and these exports make the little country one of the world's wealthiest nations on a per capita basis. Oil, gas, and small amounts of refined oil products supplied over half of Indonesia's exports in 1986, and crude or partly refined oil and liquefied natural gas provided about one fifth of Malaysia's exports. Singapore profits as a processor of oil, with 16 percent of its exports being refined products in 1987. The principal customer for Southeast Asian oil and natural gas (shipped in liquid form) is Japan, which is the leading trade partner of several Southeast Asian countries, with the three Communist countries (Vietnam, Laos, and Kampuchea) and the American-oriented Philippines being conspicuous exceptions.

Considerable mineral wealth other than oil and natural gas exists in Southeast Asia. A wide variety of metal-bearing ores are exploited to some degree in a great many locations, and it is thought that many unexploited reserves remain. Tin is the most important single metallic ore that is currently mined. About two fifths of the world output comes from this region (1985): one fifth from Malaysia and another fifth from Thailand and Indonesia combined. The Philippines produces the greatest variety of metals—notably copper, chromite, silver, and gold—in significant amounts. The most serious mineral deficiency in Southeast Asia is the near-absence of high-grade coal suitable for large-scale metallurgy. Other mineral resources, however, along with sizable waterpower potentials, seem adequate to support a considerable expansion of mining and manufacturing within Southeast Asia.

DISTINCTIVE QUALITIES OF THE INDIVIDUAL COUNTRIES

While the countries of Southeast Asia have many broad similarities, each also has its own distinctive qualities. Each exhibits a different combination of environmental features, native and immigrant peoples, economic activities, and culture traits. Many of the dominant characteristics and problems of these countries are the outcome of four centuries of European imperialism, but in no two countries have the influences and results of imperialism been the same.

The following brief portraits of the major countries in the region will perhaps be sufficient to give an idea of the distinctive characteristics of each.

SRI LANKA (CEYLON)

The island of Sri Lanka consists of a coastal plain surrounding a knot of mountains and hill lands in the south central part (Fig. 16.1). Most people live either in the wetter southwestern portion of the plain, in the hilly areas of the south center, in the drier Jaffna Peninsula of the north, or in a limited area on the east central coast. Coconuts and rice are the major crops of the low southwestern coast and Jaffna Peninsula, while tea and rubber plantations dominate the economy of the uplands. Colombo (1.8 million), in the southwest, is the capital, chief port, and only large city.

Centuries of recurrent invasion from India were followed by Portuguese domination in the early 16th century, Dutch in the 17th, and British from 1795 until 1948. In the latter year the island was granted independence as a dominion in the British Commonwealth. (In 1972 it became a republic but retained its Commonwealth membership.) This eventful history plus the island's long-standing commercial importance on the sea route around southern Asia has given Sri Lanka a polyglot population. The two major ethnic groups, distinguished from each other by language and religion, are the predominantly Buddhist Sinhalese, making up about 74 percent of the population, and the Hindu Tamils, constituting about 18 percent. The Tamils, whose main area of settlement is the Jaffna Peninsula and adjoining areas in the north, are descendants of early invaders and more recent imported laborers from southern India, and some of them have been involved in a bloody separatist movement employing terrorism and guerrilla warfare. In addition, there is a remnant of Arab population and the Burghers, descendants of Portuguese and Dutch settlers. About 8 percent of the people are Muslims and about 8 percent are Roman Catholics. In the interior a few thousand tribesmen known as Veddas practice shifting cultivation. Sporadic internal violence has punctuated the period of national independence, reflecting both antagonisms between ethnic groups and discontent with economic and political conditions, especially among the Tamils. Many hundreds (possibly thousands) of people have been killed and far greater numbers rendered homeless. Serious rioting and guerrilla warfare by Tamils in 1987 led to military actions by troops from India invited into the country to help the government restore order and preserve its authority.

Sri Lanka's economy is highly commercialized. Three export crops—tea, rubber, and coconuts—occupy most of the agricultural land and supply over two fifths of export earnings. During the 1970s clothing manufacture in the Colombo area grew into an export industry. Expansion of rice production during the same period was so successful that self-sufficiency replaced the need for major imports, a development helped by declining rates of population increase. But these notable accomplishments remain threatened by a population that already leads Southeast Asia (except for the Singapore city-state) in density, averaging over 650 per square mile (over 250 per sq. km) and still growing rapidly enough in the 1980s to double itself in just 38 years.

BURMA

Burma centers in the basin of the Irrawaddy River and includes surrounding uplands and mountains. Within the basin are two distinct areas of dense population: the Dry Zone, around and south of Mandalay (city proper, over 450,000), and the Delta, focusing on the capital and major seaport of Rangoon (city proper, 2.5 million) (Fig. 16.4). The Dry Zone has been the historical nucleus of Burma. The annual rainfall of this area (34 inches [87 cm] at Mandalay) is exceptionally low for Southeast Asia and there is a dry season of about 6 months (Table 14.2). The people are supported by mixed subsistence and commercial farming, with millet, rice, cotton, beans, peanuts, and sesame as major crops. During the past century the Dry Zone has been surpassed in population by the Delta, where a commercial rice-farming economy now supplies the country's largest export (41% of all exports in 1983–1984). The Irrawaddy River forms a major artery of transportation uniting these two core areas.

The native Burmese, most of whom live in the Irrawaddy Basin, number only an estimated 69 percent of the country's total population. Prior to World War II around 1 million Indians (nearly equally divided between Hindus and Muslims) lived in the Delta as laborers, merchants, moneylenders, and owners of farm land, much of it rented to Burmese tenants. By the late 1960s this Indian minority had been almost completely ejected through expropriations of property and other oppressive measures. Its migration back to India was associated with the thorough nationalization of Burma's economy, including land ownership, by the military government.

Figure 16.4 Index map of the Southeast Asian mainland. Stars show political capitals and the independent state of Singapore. Area symbols show river deltas, the Dry Zone of Burma, and the Tin and Rubber Belt of Malaya (West Malaysia).

The Shan Plateau to the east is inhabited by two distinct groups—the Shans to the northeast and the Karens to the southeast and in the delta of the Salween River. They account, respectively, for about 9 and 7 percent of the country's population. A variety of hill-dwelling tribes inhabit the Arakan Mountains of the west and the northern highlands. Burma is predominantly a Buddhist country, with an estimated 90 percent of its total population adhering to this religious faith.

Burma was conquered piecemeal by Great Britain in three wars between 1824 and 1885. In 1948 it abandoned all formal ties with Britain and became an independent republic. Japanese conquest and Allied reconquest in World War II have been succeeded by almost constant, although desultory, civil war since independence. At one time no less than eight different rebellions were in progress. Both Communism and ethnic separatism have supplied major motivations for rebel forces drawn principally from the Karens, the Shans, and a number of less numerous hill peoples. The economy has been badly damaged by the fighting, while at the same time being mismanaged under a form of rigid state control promoted as "The Burmese Way to Socialism." Hence Burma has become the poorest non-Communist country in Southeast Asia, with a black market estimated to account for one third of the national product. Japan and West Germany have given some foreign aid, and the establishment in the 1980s of Matsushita electronics and Mazda automobile plants at Rangoon may represent the beginnings of effective industrial growth.

THAILAND

Thailand, formerly known as Siam, centers in the delta of the Chao Praya River, often referred to as the Menam (a Thai word meaning river; the full name as used in Thailand is Menam Chao Praya). The annual floods of this river irrigate the rice that is the most important of several crops which provide most of the exports from this Southeast Asian land of relative agricultural abundance and progress. On the lower Chao Praya is Bangkok (6 million), the capital, main port, and only large city. Areas outside of the delta region are more sparsely populated, largely by sedentary farmers, although some shifting cultivators are present. These areas include mountainous territories in the west and north, inhabited mainly by Karens and a variety of mountain tribes, and the dry Korat Plateau to the east, populated by the Thai, related Laotian, and Kampuchean peoples. To the south in the Kra Isthmus, a part of the Malay Peninsula, live some 2 million Malays. An estimated 95 percent of the people of Thailand adhere to a branch of the Buddhist faith (Fig. 14.4), the only significant exception being the Malays, who are Muslims. Spectacular Buddhist temples are a prime attraction for one of Southeast Asia's largest tourist industries, as are ceremonials connected with religious or other historical traditions (Fig. 16.5).

The Thai monarchy, governed today as a military dictatorship modified by some democratic procedures, enjoys the distinction of being the only Southeast Asian country to preserve its independence throughout the period of Western colonialism. Its success appears to have been largely due to its position as a buffer between British and French colonial spheres. A number of border territories were

Figure 16.5 *Trained elephants have long played an important role in Southeast Asian ceremonies, traditional warfare, and everyday transportation and work. Those in the photo are participants in a ceremony held annually in eastern Thailand to commemorate the use of elephants in past wars.*

lost, however, and the government of the country has exhibited irredentist tendencies for many years. In return for Thailand's cooperation during World War II, the Japanese allowed the country to take over certain border territories from neighboring countries, but these areas were lost again at the end of the war. Recently, however, Thailand's attention has centered on combating insurgencies within its own outlying territories, rather than attempting to regain territories previously lost.

The major ethnic minority of Thailand is Chinese. An estimated 5 to 6 million Chinese are resident in the country and are claimed as citizens both by Thailand and by China. The Chinese control much of the country's business, a situation which has roused a considerable amount of ill feeling on the part of the Thais and makes the loyalty of the Chinese a matter of considerable importance. Up to the present time (early 1989) Thailand has been able to escape serious civil conflict, just as it escaped serious involvement in World War II. However, since the late 1960s guerrilla insurrections motivated both by Communist aims and by ethnic separatism have existed in northeastern and southern Thailand. Apparently these movements have been supported by outside powers objecting to Thailand's somewhat American-oriented political stance. In the late 1970s the country's situation was further complicated by a mass movement of refugees from adjoining Kampuchea and from 1979 on by clashes of the Thai military with Vietnamese troops pursuing fleeing Kampucheans into Thailand.

VIETNAM, KAMPUCHEA, AND LAOS

Vietnam, Kampuchea (formerly called Cambodia), and Laos are conveniently discussed together, as they are the three states which have emerged from the former French colony of Indochina. Vietnam is somewhat the largest in area of the three, has by far the largest population, and is geographically the most complex. Three distinct regions have long been recognized in Vietnam. Tonkin, the northern region, consists of the very densely populated delta of the Red River and a surrounding frame of sparsely populated mountains. Annam, the central region, includes the sparsely populated Annamite Cordillera and numerous small, densely populated pockets of lowland along its seaward edge. Cochin China, the southern region, lies mainly in the delta of the Mekong River and has a fairly high density of population.

Kampuchea and Laos occupy interior areas except for Kampuchea's short and little developed sea front on the Gulf of Thailand. But the territories they occupy are very different. Kampuchea is mostly plains country along and to the west of the lower Mekong River, although it has mountainous fringes to the northeast and southwest. Laos is an extremely mountainous, sparsely populated, and landlocked country which borders every other state on the Indochinese Peninsula and also China.

Each state is dominated by a particular people. At least nine tenths of Vietnam's 65 million people (1988) are Vietnamese. They are closely related to the Chinese in various cultural respects, including language and large religious elements of Confucianism, Buddhism and ancestor worship. A Catholic Christian minority of Vietnamese reflects the French occupation, and there are several minor native religions. Almost all the 6.7 million people of Kampuchea are Kampucheans, also known as Khmer people. Their language is Khmer and their predominant religion is a type of Buddhism. The dominant population group of Laos, the Lao, are linguistically and culturally related to the Thais. Their dominant religion is again Buddhism.

But the ethnic situation in the three states is complicated by many minority groups. Laotians form a minority in the mountains of northern Vietnam, and Kampucheans a small minority in southern Vietnam. In the 1980s Vietnam began to systematically settle hundreds of thousands of Vietnamese as a mi-

nority in the eastern part of conquered Kampuchea. Until the mid-1970s there was a large Chinese minority in the major cities of the three states, numbering perhaps 2 million altogether at that time and dominating much of the commerce of the region. Some probably remain, but how many is unknown since they have been heavily victimized by the repression which has created millions of Indochinese refugees since 1975. In the mid-1970s there were perhaps 5 million people belonging to a variety of relatively primitive mountain tribes in Vietnam and Laos. Some of these remain, but the number is uncertain, as many were allied with the United States during the Vietnam War and have been subjected to devastating attacks by Vietnam since they were abandoned by the Americans.

Between 1858 and 1907, France conquered by stages the area now occupied by these three states. The French first extinguished a Vietnamese Empire, covering approximately the territory of today's Vietnam, and defeated the Chinese whom the Vietnamese called upon for help. Later they took Laos (in 1893) and western Kampuchea (in 1907) from Thailand. France's administration of this colonial realm in Indochina, centered in Hanoi (now 3 million) and Saigon (now Ho Chi Minh City: 3.5 to 4 million), was relatively ineffective from an economic point of view, although it left a considerable cultural imprint. Its major success lay in opening the lower Mekong River area to commercial rice production, thus converting both southern Vietnam and Kampuchea into important surplus producers and exporters of rice. But neither plantation agriculture nor mining was developed on a scale comparable to that of other Western colonies in Southeast Asia.

French Indochina was overrun by Japan in 1941, and this began more than four decades of continuous warfare that Vietnam has now (late 1980s) been engaged in. Under Japanese occupation in World War II a Communist-led and nationalist-oriented resistance movement was formed, which carried on guerrilla warfare against the Japanese. When French forces attempted to reoccupy the area after the end of World War II, these forces fought to expel them. After 8 years of warfare the Vietnamese, with Chinese material support, destroyed a French army in 1954 in the mountains of Tonkin, and France withdrew.

Four states came into existence with France's departure. Laos and Kampuchea (then known as Cambodia) became independent non-Communist states. North Vietnam, where resistance to France had centered, emerged as an independent Communist state composed of Tonkin and northern Annam. South Vietnam, incorporating Cochin China and southern

Annam, gained independence as a non-Communist state which received many anti-Communist refugees from the north. Vietnamese Catholics were prominent among those who fled southward. This partition of Vietnam into two states, which deprived the Vietnamese insurgents of the full fruits of their victory, was largely the work of American power and diplomacy. From the outset South Vietnam was a client state of the United States, while Communist North Vietnam became a client first of Communist China and then of the Soviet Union.

Meanwhile the war in Vietnam continued, eventually to a different conclusion. Between 1954 and 1965 an insurrectionist Communist force, the Viet Cong, was built, grew, and achieved increasing successes in South Vietnam. It was supported by North Vietnam and indirectly at least by the Soviet Union. A similar situation existed in Laos. Increasing intervention by the United States in South Vietnam escalated from small-scale to large-scale in 1965 and eventually involved the full-scale military commitment of half a million men. North Vietnam responded by committing regular army forces against American and South Vietnamese troops. The United States never invaded North Vietnam, although American air strikes were devastating in both the North and the South. The restraint on actual invasion of the North was at least partly due to Chinese and Soviet support of the North. But in limiting the theater of ground warfare, the United States found itself unable to defeat a North Vietnamese army that was determined and tenacious, highly motivated, skillfully commanded, increasingly better equipped by its allies, ruthless, prodigal of manpower, and willing to bear the very heavy losses inflicted by the Americans. Under cover of a nominal peace agreement almost all American forces were withdrawn in 1973 from the costly war, which was extremely unpopular at home. Two years later, in 1975, North Vietnam completed the conquest of South Vietnam and the last Americans fled to waiting ships by helicopter from the roof of the American embassy in the South Vietnamese capital of Saigon—soon to be renamed Ho Chi Minh City after the original leader of Vietnam's Communists. Laos also came under Vietnamese control, enforced by an occupying army, although nominally the country remained an independent Communist state.

Kampuchea escaped these ills until relatively late, but then suffered even worse catastrophes. After 5 years of increasingly intense fighting in the country, an American-backed military government was overcome by a Communist insurgency in 1975. This Communist organization, known as the Khmer Rouge, ruled the country for 4 years with a savagery

almost unparalleled. In 1973 the population had been estimated at 7.5 million, but by 1979 it was estimated at only about 5 million—a decline on the order of one third which actually occurred in a 4-year period of mass murder between 1975 and 1979. Some of the intended victims managed to escape as refugees, principally to adjoining Thailand. Two million of those remaining were estimated to be starving in 1979. The Kampuchean Communists' stated policy was to build a new society, a "new kind of socialism," by eradicating education, money, cities, and the family. In their zeal against cities they forcibly expelled to the countryside the population of Phnom Penh, the capital and by far the largest city. The city's population dropped from over a million people in the metropolitan area in 1975 to an estimated 40 to 100 thousand in 1976 (by 1982 it had recovered to an estimated 600,000).

In 1979 the Communist/nationalist imperialism of Vietnam replaced the genocide being inflicted on Kampuchea by its native Communist government. There had been escalating border conflicts between the two countries, and in 1979 Vietnamese troops quickly conquered the greater part of Kampuchea and installed a puppet Communist Kampuchean government, just as earlier they had installed a Communist Laotian government in Laos. Thereupon the Khmer Rouge forces, supported by China (because Vietnam had become a close ally of the Soviet Union), began a guerrilla warfare resistance to the Vietnamese which has lasted to the present time (early 1989). Two non-Communist Kampuchean guerrilla forces also took the field. These three resistance forces have often quarreled among themselves, but all have maintained the conflict with the Vietnamese occupation army, which has periodically pursued them into Thailand and engaged Thai border units. At first the Vietnamese rejected international food aid for Kampuchea and followed a policy of systematically starving the country, apparently to weaken resistance against themselves. But more recently they have attempted to promote recovery under their puppet government. Some refugees have returned, although others have continued to flee, and the population has begun to rise again. In 1983, however, the United Nations estimated that 60 percent of the children in Kampuchea were malnourished.

After the United States abandoned Vietnam and the North conquered the South (1975), Vietnam entered a period of relative internal peace. However, this unusual condition has not precluded periodic military excursions, including border conflicts with Kampuchea, the 1979 conquest of Kampuchea, warfare against Kampuchean resistance groups, warfare against resistance groups in occupied Laos, attempted extermination of hill tribes which had been friendly to the United States, and a brief war against China. The latter consisted of a limited Chinese invasion of Vietnam's northern border regions in 1979 immediately after the latter country's conquest of Kampuchea, whose murderous Khmer Rouge was a Chinese ally. Like the Americans before them, the Chinese encountered sharp resistance in Vietnam, and, after devastating areas near the border, they soon withdrew.

Within its borders Vietnam unleashed, from 1975 onward, a repression against unwanted elements that was strong enough to cause a massive outpouring of refugees. This situation intensified, especially in regard to the country's ethnic Chinese, following the Chinese invasion of 1979. The desperation of many of the persecuted was made clear by the phenomenon of the "boat people"—refugees who fled onto the open ocean on large rafts, with chances of survival variously estimated at 40 to 70 percent. No one knows for sure, but it is thought by investigators that 300,000 to 400,000 had fled from southern Vietnam in this way by mid-1979. Somewhere near half of them were believed to have made their way to disease-ridden temporary refugee camps in various Asian countries, primarily Hong Kong, Malaysia, and Indonesia. Most of the others are believed to have died, principally by drowning. The total refugee flow from Vietnam, which was still continuing in the late 1980s, has been much larger. The United States has provided permanent homes for over half of those who have been resettled, with France, Canada, and Australia also receiving sizable numbers. Other Vietnamese and Laotian refugees in large numbers have been resettled in China near the border. They provide recruits for guerrilla resistance inside the new Vietnamese empire, especially in Laos.

Given these conditions of warfare, political turmoil, and oppression, with the added imposition of Communist economies, it is not surprising that the three Indochinese countries are, with Burma, the poorest in Southeast Asia. Up to the 1960s Kampuchea and Vietnam (the South Vietnamese part) were among the world's major rice exporters. But by the late 1980s all three countries were very short of food, despite much foreign aid from the Soviet bloc. Some signs of hope were appearing as strict socialization of economies was being relaxed somewhat in favor of small-scale free enterprise, and in Vietnam an illegal "underground" economy was providing perhaps two thirds or more of the increasing quantities of goods available. Rice harvests were showing a tendency to rise again. But conditions remained highly unsatisfactory even for such a poor region as Southeast Asia. There was some slowing in the numbers of people fleeing from these strife-torn countries, but no really pronounced upturn could be seen.

MALAYSIA AND SINGAPORE

The small island of Singapore, narrowly separated from the southern tip of the mountainous Malay Peninsula, lies at the eastern end of the Strait of Malacca, which is the major passageway through which sea traffic is funneled between the Indian Ocean and the China Seas. For centuries European sea powers contested with each other for control of this passageway. British control over the strait was continuous from 1824. It was exercised from the port of Singapore, founded on the southern side of the island 5 years earlier. Under the British, Singapore, with its relatively central position among the islands and peninsulas of Southeast Asia, developed not only into a major naval base but also into the region's major entrepôt.

From Singapore, the British gradually extended their political hold over the adjacent southern end of the Malay Peninsula. This expansion gave Britain control over the part of the peninsula now included, along with northwestern Borneo (except Brunei), in the independent country of Malaysia. This southern end of the peninsula is customarily referred to as Malaya, or Peninsular Malaysia, whereas the Borneo section is composed of the units (former colonies) of Sarawak and Sabah, which together are known as East Malaysia.

During the period of British administration Malaya developed a highly commercialized and, for Southeast Asia, a prosperous economy. Tin and rubber became the major commercial products, although other types of production such as oil-palm and coconut plantations were developed (Fig. 16.6). The tin-mining industry was pioneered in the later 19th century primarily by Chinese and British concerns. A line of rich tin deposits along the western foothills of the mountains was made accessible by the building of railroads. These same rail lines provided transportation for rubber when new electrical and automotive industries stimulated a greatly increased demand for that product in the world's industrial countries during the early decades of the 20th century. A densely populated belt of tin mines and rubber plantations developed in the foothills between Malacca and the hinterland of Penang (George Town: 500,000). Within this belt the inland

Figure 16.6 Harvesting coconuts with a long pole on a plantation in Malaya.

city of Kuala Lumpur (1.1 million) became the leading commercial center and capital (both before and after independence) of the country.

The development of a commercial economy gave Malaya an ethnically mixed population. The native Muslim Malays played only a minor role in this development, often preferring to remain subsistence rice farmers in small coastal deltas along both sides of the peninsula. Chinese and Indian immigrants and their descendants became the principal farmers, wage workers, and businessmen of the tin and rubber belt. So heavy was the immigration that 30 percent of Malaysia's population today is Chinese, and 8 percent is Indian. The growth of Singapore involved a proportionately heavier immigration from overseas, so that three fourths of Singapore's present population of 2.6 million is Chinese.

Ethnic antagonisms between Malays and Chinese have been fundamental in shaping the political geography of Malaysia and Singapore. Malaya accepted independence in 1957 only on condition that Singapore not be included. This was done to ensure that the Malays would have a majority in the new state. In 1963, however, the Federation of Malaysia was formed, composed of Malaya, Singapore, and the former British possessions of Sarawak and Sabah in sparsely populated northern Borneo. The non-Chinese majorities in Borneo were counted on to counterbalance the admission to the Federation of Singapore's Chinese. This experiment in union lasted until 1965, when Singapore was expelled from the Federation and left to go its own way as an independent state.

Malaysia and Singapore are among the more economically successful of the world's formerly colonial states that have received independence since World War II. The city-state of Singapore has become one of the outstanding centers of manufacturing, finance, and trade along the rim of Asia (chapter-opening photo). Its industries include oil refining, machine building, and many others. The tourists it attracts each year far outnumber its resident population. Singapore's economic success has been so great that the country has now reached a level of output and income higher than that of some of the poorer countries in Europe. Malaysia's dependence on its traditional exports of natural rubber and tin has greatly lessened, although it still leads the world as an exporter of both commodities. A more diversified export pattern now includes such items as crude oil (from northern Borneo), timber, palm oil, and electronic components. In 1987 rubber constituted only 3 percent of Singapore's exports by value, and tin even less. Malaysia's annual per capita product of $1850 in 1986 was far below Singapore's $7410, but was quite high for the "less-developed" world and exceeded that of any other state in Southeast Asia except Brunei ($15,400) and Singapore (see Table 14.1).

INDONESIA

Indonesia is by far the largest and most populous of the countries of Southeast Asia. Its several thousand islands comprise well over one third of the region's land area and contain about two fifths of its total population. The large population of Indonesia results from the enormous concentration of people which has developed on the island of Java. About 110 million people lived here in mid-1988, representing three fifths of Indonesia's population of 177 million and almost one fourth of the population of all Southeast Asia. The island's population density is well over 2000 per square mile (760 per sq. km). In contrast the remainder of Indonesia, which is about 13 times larger than Java in land area, has an average density of only about 100 per square mile (*c.* 40 per sq. km).

The extraordinary concentration of population on this one island can be partly accounted for by the superior average fertility of its soils, the best of which have been derived from materials poured out by its many volcanic peaks. However, it also is the result in good measure of the centering of Dutch colonial activities in Java. Other islands of Indonesia (with the notable exception of Borneo) have areas of volcanic soil, but such areas, while generally more densely populated than adjoining nonvolcanic areas, seldom attain the extremely high population densities found on Java. The economic development of some islands has undoubtedly been handicapped by the unfriendly nature of their coastlines, which present a front of coral reefs, steep cliffs, or extensive swamps, and thus create difficulties of access.

The Dutch East India Company, after long-continued hostilities with Portuguese and English rivals and with native states, secured effective control of most of Java in the 18th century. Large sections of the remaining islands, however, were not brought under control until the 19th century; and only in 1904 was the conquest of northern Sumatra completed. Strenuous efforts to develop and exploit the natural wealth of Java were undertaken by the government of the Netherlands with the introduction of the Culture System in 1830. Under this system forced contributions of land and labor were required of Javanese farmers for the intensive production of export crops under Dutch supervision. The system was undeniably harsh but, from the Dutch point of view, successful. Although the Culture System was abolished in 1870, native commercial agriculture continued to expand along with plantation production, and a means of support for increasing numbers of people was provided. Thus the introduction of the Culture System appears to have set off not only a great increase in the output of export commodities but also the enormous rise in Java's population, which is over 20 times larger than it was a little over a century and a half ago. Development of the other main islands began later and has been less intensive. Nevertheless, the eastern coastal plain of Sumatra, inland from the great fringing swamp, has now surpassed densely populated Java in agricultural exports.

Java's concentration of people and production, and Sumatra's importance in export production, are shown in the distribution of Indonesia's larger cities. Of 11 cities having estimated metropolitan populations of more than 500,000, 7 are on Java. These include four "million-cities," headed by the country's capital and main seaport, Jakarta. With 7.7 million people, Jakarta is the largest metropolis in Southeast Asia. Of the remaining four Indonesian cities of over half a million people, three are in Sumatra (including one "million city"), and one is in Celebes (Sulawesi). In all 11 cities, and many others as well, a minority of Chinese immigrants and their descendants form a commercial class that has frequently been the target of resentment and even bloodshed by other Indonesians.

The Indonesian state has had a turbulent and varied career. Increasing nationalism before and during the Japanese occupation of World War II led to a bitter struggle for independence from the Netherlands following the war. This struggle was eventually successful in 1949, and a further confrontation in 1962 brought western New Guinea (West Irian) from Dutch control into the Indonesian state. Being Papuan rather than Indonesian in culture, West Irian had not originally been included in Indonesia. However, Indonesian control of this territory was ratified by a United Nations plebiscite in 1969, and relatively friendly relations have been reestablished with the Netherlands.

Ethnic diversity, physical fragmentation and distances, and economic problems have made it difficult to attain peace, order, and unity in Indonesia. Although 87 percent of the population is at least nominally Muslim, great cultural diversity is reflected in the fact that over 200 languages and dialects are in use. Various groups in the outer islands have resented the political and economic dominance of the Javanese, and such animosities have escalated at times to armed insurrections. The government has been carrying on a long struggle against a guerrilla-type independence movement in East Timor, a former Portuguese possession which Indonesia occupied after the collapse of the Portuguese colonial empire in the mid-1970s. In 1965 an attempted Communist *coup d'état* in Indonesia was unsuccessful and resulted in the massacre of probably 300,000 Communists and Communist supporters by the Indonesian army and by Islamic and nationalist groups. It also brought to power a government in which army influence is predominant and which still controls the country (late 1980s). The main political opposition is comprised of Muslim fundamentalists who want to restructure state and society according to Islamic principles.

Indonesia is a state that has made some striking economic gains in recent years but still remains very poor. The gains have centered in the oil industry and in food production. Oil fields located mainly in Sumatra and Kalimantan produce 2 to 3 percent of world oil. Their production has been expanding. Crude oil, refined oil products, and liquefied natural gas accounted for more than half of Indonesia's exports in 1986 and have tied its economy very closely to that of Japan, the main market for these products. Indonesia is still a major source of various tropical agricultural products such as rubber and coconuts, and their production and export is still of major importance to many Indonesians, but they no longer dominate the country's commerce as they have done through nearly all its history. Along with the rising importance of oil, Indonesia has recently achieved some remarkable agricultural results, for instance, rice production in 1982 was 175 percent of output in the early 1960s; and total per capita food production increased 24 percent between the average for 1969–1971 and the average for 1981–1982. In the entire world, only Malaysia, Thailand, and two Latin American countries exceeded the latter record, although China matched it. But the per capita increase in food production would have been 64 percent if population had been stable, and oil revenues would have made more of a dent in Indonesia's poverty. Rapidly increasing population—51 million people, or 40 percent, in the 14 years from mid-1973 to mid-1988—continues to make a major contribution toward keeping Indonesia poor.

THE PHILIPPINES

The Philippine group includes over 7000 generally mountainous islands, but the two largest islands, Luzon and Mindanao, almost equal in size, account for two thirds of the total area. Most of the mountainous districts are inhabited by a sparse population of relatively primitive shifting cultivators, although in northern Luzon the Igorot tribes have developed a spectacular and world-famous system of wet rice cultivation on terraced mountainsides.

Most of the population is concentrated in three areas:

1. *The Visayan Islands in the center of the archipelago.* Here soils derived from volcanic materials and uplifted corals support an intensive subsistence agriculture based on rice and corn. Negros Island in this group is a major center of plantation sugar production.

2. *The plains extending from south of Manila to Lingayen Gulf and thence north along the west coast of Luzon.* Rice is the main food crop of these plains, and cane sugar, much of it produced on small native farms, is the main commercial product.

3. *The southeastern peninsula of Luzon.* Subsistence rice and commercial coconut production are basic to the economy here except in the extreme southern part, where large plantations utilize volcanic soils to produce a major share of the Philippines' output of abaca or Manila hemp.

The Philippines were a Spanish colonial possession from the later 16th century until 1898 and with the exception of the Japanese occupation of 1942–1944 were controlled by the United States from 1898 until 1946, when independence was granted. The Spanish legacy is still important in the country. Ever since its founding in 1571 the Spanish capital of Manila (7.5 million) has been the metropolis and only large city of the islands. The society created by Spain was composed of a relatively small upper class of Hispanicized Filipino landowners and a great mass of landless peasants. Problems created by the maldistribution of agricultural land have remained as a major source of difficulty for the new Republic of the Philippines. Discontented peasants gave much support to a Communist-led revolt after World War II. It was eventually suppressed, in large part through the granting of land (generally on sparsely populated Mindanao) to surrendered rebels, but flared up again during the 1960s and continues in the late 1980s. In the Philippines Spanish missionary activity succeeded in creating the only Christian nation in the Orient. Today 84 percent of the population is listed as Roman Catholic, while almost 10 percent adheres to other Christian denominations. However, a Muslim people, the Moros of eastern Mindanao and the Sulu Archipelago, successfully resisted Spanish control for 300 years and today carry on an intermittent secessionist guerrilla resistance against the Philippine state.

The American period, although it began with the suppression of a Philippine independence movement, eventually brought independence, along with economic development and tutelage in democracy. The growth of major Philippine export industries—coconut products, cane sugar, and abaca—was stimulated by preferential treatment in the American market. In the 1930s a large degree of self-government was granted and full independence was promised for 1946. At the same time education, in English, was furthered so that even now English is spoken by about 45 percent of the people and serves to some extent as a bridge between the diverse linguistic groups of the population. After independence one of the most widely used of the many native languages, Tagalog, was selected as the principal base for an official national language, Pilipino, which is now spoken by some 55 percent of the populace. Neither of these two main languages is native to most of those who speak them, however.

Since independence the Philippines have had considerable success in expanding their economy. Until the 1970s the most striking aspect of this expansion was the growth of exports from agriculture (coconut products; sugar; bananas; pineapples and other tropical fruits) and mining (copper, chrome, and other metals). More recently there has been outstanding growth in manufacturing and food production. Labor-intensive manufactures such as electronic devices and clothing now account for over a third of all exports by value. Total food production, mainly of rice and corn, increased 50 percent between the period 1969–1971 and 1981–1982. As in Indonesia, rapid population growth cut into this growth of food output, but even the 10 percent gain on a per capita basis was impressive. In the politically troubled 1980s, both total and per capita increases in production leveled off, with 1986 showing only small percentage gains over 1981 or by some figures an actual decline per capita. Japanese and American capital, together with American agricultural science and technology, have been very important in Philippine economic development. Over half of all Philippine exports are bought by the United States or Japan.

But in spite of encouraging progress in recent decades the Philippine republic remains poor and potentially explosive politically. Even the rapid economic expansion has been hard pressed to stay ahead of a population growth that reached a little over 50 percent (21 million people) in just 14 years between mid-1973 and mid-1988. And land and wealth are extremely unevenly distributed, as Philippine society has long been dominated by just a few hundred wealthy families. In the mid-1980s a repressive military regime, in power since 1972 and identified in many Philippine eyes as American-tolerated or even American-sponsored, was replaced by a democratic government. Since it took control, this government has been faced with extraordinarily difficult economic and political problems. Its struggle to resolve them is watched with deep concern by the United States, which has maintained friendly ties with its former colony since independence and continues to depend on important military bases there.

 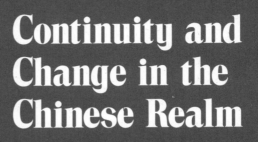

CHAPTER 17

Continuity and Change in the Chinese Realm

Battery production, seen here in Guangzhou (Canton), is very appropriate for an underdeveloped but industrializing country such as China, where power is needed for new machines and appliances.

Cʜɪɴᴀ is the world's largest country in population and is exceeded in area only by the Soviet Union and Canada. Its population (excluding Taiwan) was estimated in mid-1988 to be 1087 million, or a little over one fifth of all the people in the world. Based on the expected rate of natural increase—which stood at an estimated 1.4 percent a year in 1988—population experts project that China will have 1.2 billion people by the year 2000, or 1.4 billion by 2020. At present the net addition of over 15 million Chinese a year is a very serious matter for a developing country whose area of about 3.7 million square miles (9.6 million sq. km) is only about 2 percent larger than that of the United States (Fig. 17.1), but whose inhabitants outnumber the United States population 4½ to 1. This outsized mass of people lives in a state officially called the People's Republic of China (PRC), which plays a major role in world affairs but nonetheless is still poor and heavily agricultural despite notable industrial development under the Communist regime that came to power in 1949 as a result of civil war. In this chapter we examine the country in its recent Communist context, but also in the broader context of long-term cultural continuity, persistent regional differentiation, and the profound changes that resulted from China's modern encounter with the Western world. Briefer consideration is accorded Taiwan (formerly called Nationalist China), which operates independently but is claimed by the People's Republic; the European colonies of Hong Kong and Macao (Macau) on the coast of southern China; and Mongolia (the Mongolian People's Republic), which, un-like the others, is not Chinese in population, but occupies a territory controlled by China for long periods in past times. Great Britain is scheduled to return Hong Kong to Chinese sovereignty in 1997; Portugal will return Macao to China in 1999.

STRENGTH, WEAKNESS, AND CONTINUITY IN CHINA

Present-day China has important roots in the distant past, but many of its present characteristics are a more immediate product of civil unrest, disorder, and costly warfare beginning with the Opium War of 1839–1842 and ending with the Korean Armistice in 1953. During most of this period, China's unity and even its national existence were threatened by the combined and largely successful efforts of other nations to extract trading rights and other economic and political concessions. The humiliations of this dark era in Chinese history have been bitterly resented by most Chinese—a resentment intensified by China's long history as the dominant power in Asia and its great achievements in numerous fields. Although these achievements of the Chinese past and the country's highly developed culture were the envy of the West in the 18th century, China's preoccupation with the past inhibited its adoption and assimilation of the rapidly spiraling scientific and technological advances made in the Western world during the 19th century. Pride in culture, love of the past, and disinterest in the outside world contributed powerfully to the downfall of the old order in China, and violation of Chinese pride, as well as territorial encroachment by outside powers, helped foster a powerful nationalism during the 20th century.

The foregoing facts have gained added significance for the world at large with the emergence of mainland China as a military power under Communist rule. With the explosion of its first nuclear device in October 1964, and several successful nuclear tests since then, there can be no doubt that China has become an extremely important factor in world affairs. Even before their first nuclear explosion, the Chinese Communists had already demonstrated their military prowess in the Korean War and later in the short-lived border war with India in 1962. The haltingly successful efforts of Communist planners to build an industrial society and the continued development of an advanced weapons program have added to the country's power potential. The foreign policy of the People's Republic impinges directly on the many countries that border China (Figs. 14.1 and 17.2) and is a very important factor in world affairs generally.

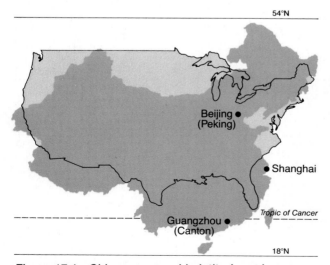

Figure 17.1   China compared in latitude and area with the conterminous United States.

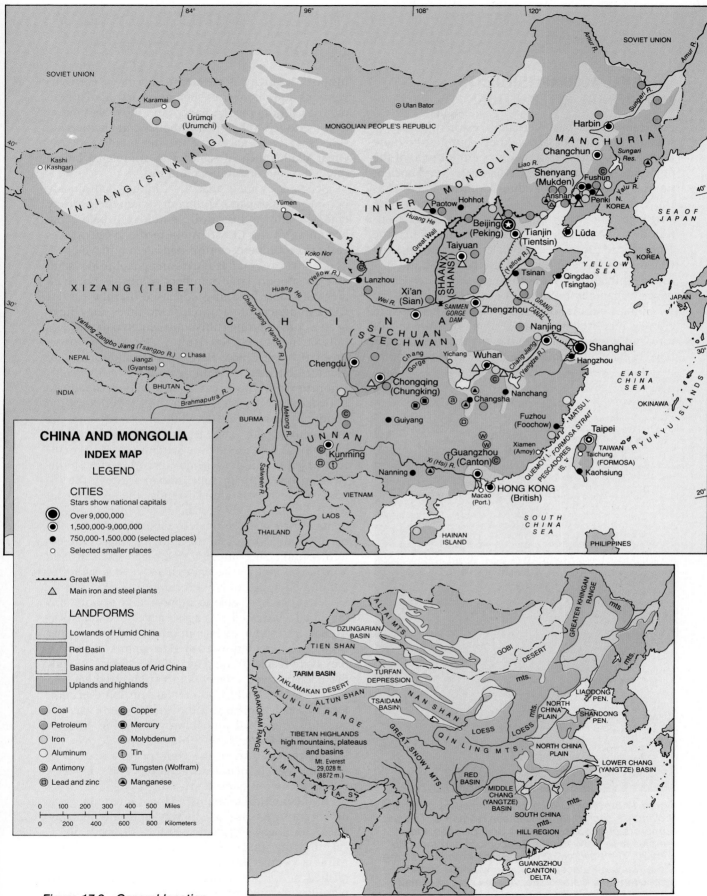

INDEX MAP

LEGEND

CITIES
Stars show national capitals

⊙ Over 9,000,000

⊙ 1,500,000-9,000,000

● 750,000-1,500,000 (selected places)

○ Selected smaller places

···· Great Wall

△ Main iron and steel plants

LANDFORMS

Lowlands of Humid China

Red Basin

Basins and plateaus of Arid China

Uplands and highlands

Ⓒ Coal   Ⓒ Copper

Ⓟ Petroleum   ▣ Mercury

○ Iron   Ⓐ Molybdenum

○ Aluminum   Ⓣ Tin

ⓐ Antimony   Ⓦ Tungsten (Wolfram)

▣ Lead and zinc   ▲ Manganese

0   100   200   300   400   500   Miles

0   200   400   600   800   Kilometers

Figure 17.2   General location map of the Chinese realm.

405

INTERNAL ACCOMPLISHMENTS
AND OUTSIDE INTERACTIONS

Politically China has alternated between periods of strength and periods of weakness, but despite these fluctuations it has continued to be the home of the dominant culture of East Asia since very ancient times (Fig. 17.3). Surrounding countries and regions show the long impact of Chinese culture on their languages and writing, religions, arts, crafts, institutions, ways of thinking, and histories. Since the 18th century, however, the balance has swung and China itself has come under the powerful and disruptive influence of Western culture, although China never was totally immune to outside influences.

China's cultural ascendency before the Western era was due in part to the creativity of Chinese culture, but it was also due to the fact that the Chinese have been an especially numerous people since pre-

Figure 17.3   Mount Wudang in northwestern Hubei Province is sacred to the Chinese religion of Taoism. Emperors of various dynasties built palaces and temples here between the 7th century and the 17th century. The building clinging to the precipitous crag is the Nanyan Palace, which is now under state protection as a historic site. A large part of Humid (eastern) China is hilly or mountainous country, often with very steep slopes such as those shown.

history. Creativity is attested to by many early inventions and discoveries such as a particular system of writing, gunpowder, paper, and movable type. It is also attested to by organizational and engineering skills that were necessary for the building of the Great Wall, much of which was completed before the end of the first millennium B.C.

Scholars estimate that the Chinese had already become a major part of humanity when their society emerged into its historical period, well before the beginning of the Christian era in the West. At this early time Chinese culture had already developed, as one of its essential features, an agricultural system able to support large numbers of people on small areas of arable land. It is supposed that the earliest Chinese agriculture in prehistoric times may have been a crude shifting cultivation centered in the loess hills of the Huang He (Hwang Ho: Yellow River) Basin of North China. Here the vegetation was light and the soil was fertile and easy for primitive cultivators to work. But when the mist of our inadequate knowledge begins to lift a bit, in the period of the Shang Dynasty (c. 1500–1000 B.C.), we find the great alluvial plain of the Huang He already occupied by Chinese peasants with a garden type of agriculture based on the hoe, irrigated crops, varied fertilizers, including night soil (human wastes), the intensive use of flat alluvial lands, the terracing of fertile hillsides to create flat land, a high degree of control over water, private ownership of land, the use of family labor, and the siting of villages on less fertile land. Little attempt was made to farm land which was not fertile enough to yield a living to families equipped only with hoes.

This system, so able to support large populations on small cultivated acreages, developed in North China. But some of its important elements came from elsewhere: Knowledge of irrigation techniques may have reached China by diffusion from the Middle Eastern river valleys, and certainly wheat and rice came originally from centers of domestication located elsewhere. But the overall system was apparently originated and improved by the Chinese themselves. From North China, it spread over the humid eastern part of the country by a process of diffusion which included migrations and conquests over a period of many centuries.

In the political realm, China was often almost as dominant in East Asia as it was culturally, but political dominance alternated historically with political disunity and weakness. Unification of the Chinese people into one empire was first accomplished in the 3rd century B.C., and after that time periods of strong central government, generally occurring soon after the beginning of a new dynasty, alternated with periods of weak central government and internal warfare. During periods of unity and strong central power, China generally extended its authority widely, both by direct conquest and by accepting the fealty and tribute of lesser powers around it. For example, under the Han Dynasty (206 B.C.–220 A.D.) Chinese armies pushed westward across what is now Soviet Middle Asia, and seem to have barely missed making contact with the Romans in the vicinity of the Caspian Sea. Under the Yuan (Mongol) Dynasty (1280–1367), China controlled Burma, Indochina, Korea, Manchuria, and Tibet; and under the Manchu Dynasty (1644–1911), the Chinese held large areas now in Soviet Middle Asia and Siberia. But strong dynasties tended to weaken with time, and then internal disorder, partition, and civil war would follow. During such periods of weakness, parts of China often were conquered by nomadic invaders (such as Mongols and Manchus) from the north. Sometimes these conquests were so extensive that the period of disorder ended and China reasserted its unity and strength under a new dynasty of foreign conquerors. Such was the case with both the Mongol and Manchu Dynasties.

Through all the rise and fall of dynasties, the distinctive Chinese civilization persisted and intensified.

The civilization that grew within the areas known as China was of a very high order. Learning was held in great regard, and officials were scholars, appointed to their posts under a democratic examination system that was fairly well developed as early as the sixth century. Literature, philosophy, poetry, and the arts flourished and rose to high levels of excellence. An ingenious handicraft system produced objects of great beauty and delicacy as well as sturdy objects for daily use.

Transportation and communications were well organized and amazingly swift. Post offices were set up at short intervals along the main roads and letters were carried by relays of riders on horseback so that the Emperor in Peking was informed within a very few days of events occurring on the outer edges of his empire. This stage post system began in the Han period before the Christian era. Public inns also served the main highways and were open freely to all who had occasion to travel, and there were stores of fodder for the horses. Some of the more important roads were paved with stones. In the Sui dynasty in the sixth century the Grand Canal was built to connect the north and the south. It was rebuilt in the thirteenth century by the great Khublai Khan to improve communications between the capital and the Yangtze Valley and to facilitate the administration of the empire. Great use was made of the rivers as means of internal communication. The Yangtze River [Pinyin: Chang Jiang] and its tributaries were then, as they are now, important factors in the life of the nation.

The Chinese felt no need of the outside world. Their civilization was superior to that of neighboring lands and was drawn upon and borrowed from by those lands. It reached also into the outside world and influenced distant cultures.[1]

CONTACTS WITH THE WEST

Despite the strong tendency of the Chinese to turn inward, there were always some contacts with the outside world. Caravans across inner Asia connected China with the Middle East and Europe, and there was some traffic across the Himalayas and by sea. Some Chinese became redoubtable navigators, making long voyages to India and even East Africa. Buddhism from India entered China and made converts, and both the Muslim faith and Christianity came in. During the Mongol Dynasty of the 13th and 14th centuries, China was one part of an extensive Mongol Empire with international interests. It was in this period that Europe's imagination was kindled by Marco Polo's account of the China of Khublai Khan.

Then in the 16th century European ships became frequent visitors to China, and the Europeans began to press for expanded trade. The Chinese had little desire for European goods in exchange for the cottons, silks, teas, and other high-quality Chinese products the Europeans wanted, and they also felt culturally superior to these intrusive "barbarians." The Chinese government strictly controlled trade and limited it to the single port of Guangzhou (Canton) in extreme southern China.

But as European science, technology, and industrialization advanced, the Westerners grew in power and became more insistent on large-scale trade with China. The weakening Manchu government was defeated by a British naval expedition in the Opium War of 1839–1842, forcing China to open its doors to opium from India and to other British goods. Once China's impotence in the face of European military technology and organization was demonstrated, a competitive race to extract concessions from China was entered rapidly by other European nations as well as by Russia and the United States. As a result, China became a kind of joint colony of many competitive foreign exploiters. A tottering Manchu Dynasty retained the throne until 1911 but was largely helpless against the foreigners, who got their way by occasional military actions and frequent

[1]*Geographical Foundations of National Power* (Army Service Forces Manual M103–3; Washington, D.C., Government Printing Office, 1944) pp. 1–2.

PINYIN SPELLINGS

The Chinese government has adopted a system of official place-name spellings known as "Pinyin," which is now widely used outside of China. It is superseding older spellings from the Wade-Giles transliteration system, so that "Peking," for example, has become "Beijing," "Canton" has become "Guang-zhou," and "Yangtze Kiang" has become "Chang Jiang." A few letters in the Pinyin scheme have sounds substantially different from the way these letters are customarily pronounced in English: "c" in Pinyin is pronounced as though it were "ts"; "q" is "ch"; "x" is "sh"; "z" is "dz"; and "zh" is "j." The system is based on the pronunciations of Chinese characters in the standard form of Chinese called Mandarin. Due to the widespread adoption of Pinyin spellings in Western news media, maps, and other published materials, it has been decided to cite the Chinese place names in this text in their Pinyin forms. But in each instance the familiar older spelling such as Peking or Canton is given in parentheses the first time or two that the word is used. After all, Pinyin was only adopted in 1958, and all but a very small fraction of the material in Western libraries uses the older spellings. In the case of some place names such as Shanghai, the newer and older spellings are the same. To make the reading less cumbersome, "Tibet" has been retained, although this region is called Xizang in Pinyin.

threats. China was opened to Western products and then to the establishment of Western-owned factories (especially textile plants) in its ports. Under the impact of these machine-produced goods, China's handicraft industries began to disintegrate. Foreigners in China were exempted from the control of Chinese law (an arrangement known as "extraterritoriality"), and the representatives of Western powers eventually gained control of China's taxation system as well as its trade policy.

The foreigners extorted territorial concessions also, although only a small part of China was ever reduced to an explicit colonial status. Britain secured Hong Kong; Portugal had been granted nearby Macao much earlier; and France and Germany obtained naval bases. Japan entered the contest late but took Taiwan and the Ryukyu Islands; and Russia conquered parts of Middle Asia, Mongolia, and Siberia. "Treaty Ports" were established, among which Shanghai and Tianjin (Tientsin) were especially prominent. Here the foreign powers were granted special concessions, including settlement areas under their own control. Chinese could not normally reside in such areas or in some cases even enter them. European and American naval vessels patrolled China's coasts and her great rivers to enforce foreign privileges.

The Chinese did not accept this situation passively. Disorder mounted within the country, but the submissive Manchu sovereigns were kept on the throne by Western power until they were finally overthrown by the Chinese Revolution of 1911. This revolution, led by Sun Yat-sen, established the Republic of China, but its authority was only nominal over much of the country. Local "warlords" controlled many areas until the army of the Nationalist Party (Kuomintang or KMT) under Chiang Kai-shek fought northward from Guangzhou (Canton) and managed to impose a higher degree of control on most of the country (1926–1928). The Nationalist regime, which established itself at Nanjing (Nanking), was able to abolish some foreign privileges, but its strength was sapped by almost continuous civil war against the Chinese Communists led by Mao Zedong (Mao Tse-tung) and then by war against Japan. Chiang finally drove the Communists from their original base areas in the hills of southeastern China in 1934, but in their famous "Long March" the Communist forces managed to survive and to reestablish themselves in the loessial hills of northwest China in 1935.

In 1931 Manchuria (known to Chinese as the Northeast) was invaded and taken over by Japan, and in 1937 the Japanese provoked a full-scale war with China. Although the resistance of the Chinese Nationalist armies proved tougher than the Japanese had anticipated, the Nationalist forces and the government were forced to retreat to the interior. Here, with its wartime capital of Chongqing (Chungking) protected by the mountains rimming the Sichuan (Szechwan or Red) Basin, the government was able to hold out, aided by a trickle of supplies from the outside, until final defeat of Japan by the Allies in 1945. Meanwhile, a kind of uneasy truce prevailed between the Nationalists and Communists. In the absence of pressure from the Nationalists, the Communists were able not only to engage in guerrilla warfare against the Japanese but at the same time to build and expand their bases in the countryside. Thus by 1945 they were relatively much stronger than before the war. Despite efforts at negotiation, full-fledged civil war once again commenced in 1946. Although aided by American equipment and supplies, the Nationalist government lost the support of large segments of the population, and the Nationalist armies gradually melted away. In 1949 the government, together with its remaining forces, fled to the large island of Taiwan about 100 miles (161 km) from the mainland. Since then, it has continued to hold Taiwan and some small islands, while the mainland has been ruled from Beijing (Peking) by the Communist Party and the People's Liberation Army.

MAJOR AREAL DIVISIONS

China is vast in size but like all extremely large countries contains a great deal of unproductive land—mainly in the western and northwestern parts of the country in the Tibetan Highlands, Xinjiang (Sinkiang), and Inner Mongolia (Fig. 17.2). These outlying areas are principally comprised of high mountains and plateaus, together with great expanses of arid or semiarid plains where rainfall, generally speaking, is insufficient for agriculture. This dry, sparsely settled country probably contains only 5 percent or so of China's population, consisting mostly of various peoples other than the Han Chinese.[2] This half of China is in marked contrast to the better-watered, densely settled eastern half of the country where about 95 percent of the population lives. Thus China is divided into a western half, or Arid China, and an eastern core region, or Humid China, the latter containing the bulk of China's population and most of its developed resources and productive capacity. A rough boundary between them would be a line drawn from the northeastern corner of India to the northern tip of Manchuria. This line corresponds in a general way to the line of 20 inches (c. 50 cm) average annual rainfall, with Arid China to the west of it and Humid China to the east.

---

[2]The term *Han Chinese* is commonly used to designate the dominant ethno-linguistic group in China, with the term *Han* derived from the first great dynasty of China. Some 50 non-Han Chinese ethno-linguistic groups, nearly all of them racially of Mongoloid stock, number perhaps 50 million and live primarily in the western and southwestern parts of the country.

ARID CHINA

The principal regions of Arid China are the Tibetan Highlands, Xinjiang (Sinkiang), and Inner Mongolia. All three contain much sparsely populated territory and are particularly identified historically with non-Han population elements, although Han Chinese now comprise the great majority of Inner Mongolia's population and may be a majority in Xinjiang.

TIBETAN HIGHLANDS

The very thinly inhabited Tibetan Highlands (Fig. 17.4) occupy about one fourth of China's area. Included are Tibet itself (Pinyin: Xizang) and fringes of adjoining provinces. Most of this vast area is a very high, barren, and mountainous plateau averaging nearly 3 miles (c. 5000 m) in elevation. To the northwest high basins of internal drainage are common, some containing large salt lakes. To the southeast the plateau is cut into ridge and canyon country by the upper courses of great rivers such as the Tsangpo (the Brahmaputra of India); the Mekong, which flows southward to the Indochinese Peninsula; and China's own Chang Jiang (Yangtze Kiang) and Huang He (Hwang Ho or Yellow River). Marked contrasts in elevation, precipitation, temperatures, and vegetation result from the great dissection of the plateau surface by these rivers and their tributaries. The combination of lower elevations, warmer temperatures, and greater precipitation in parts of the southeastern plateau results in some extensive grasslands and, in especially favored areas, stands of timber, mostly conifers. Around the edges of the Tibetan Highlands are huge mountain ranges: the Himalayas on the south; the Karakoram and other ranges to the northwest; the Great Snowy Range and others to the east; and several imposing ranges to the north.

The people of the Highlands, numbering perhaps 5 million (2 million in the political area called the Tibetan Autonomous Region), are very predominantly sedentary farmers. Agricultural land is extremely limited and found only at elevations lower than about 13,000 feet (c. 4300 m), in large part along the valleys of the Tsangpo and other major streams and their tributaries. A few hardy crops, especially barley and root crops, are basic to agriculture in this restrictive environment. In grasslands at higher elevations, a nomadic minority graze their flocks of yaks, sheep, and goats, and some of them combine herding with the growing of a little barley in the valleys.

Figure 17.4 A Tibetan village not far from Lhasa. This part of Tibet is low enough in elevation to permit a limited amount of crop-growing. Slopes in the background are used for grazing.

The Tibetan Highlands are under Chinese political control, but relatively few Han Chinese live there. The people are predominantly Tibetans, distinguished by their own language and by their adherence to Lamaism, the Tibetan variant of Buddhism. Prior to the Communist era China had sometimes exerted loose control over Tibet. The great distances from Chinese seats of power to southern Tibet and the lack of any significant agricultural land suitable for Chinese colonization contributed to the tenuous hold of China on the area. The last vestiges of traditional Chinese authority vanished with the overthrow of the Manchus, and from 1912 until the Chinese Communist military occupation of 1951 Tibet was in fact, if not legally, an independent state, with its capital and main religious center at Lhasa. After roads were completed to Lhasa and southern Tibet in late 1954, an increase in restrictive measures by the Chinese government contributed to the rise of guerrilla warfare. This culminated in the large-scale Tibetan revolt of 1959 and the flight of the Dalai Lama (the spiritual and political head of Lamaism) and many other refugees to India. The Chinese, freed from the necessity of dealing through the Dalai Lama and the Tibetan religious-political hierarchy, then moved more rapidly, driving most of the monks from their monasteries, expropriating the large monastic landholdings, completing their program of land reform, and making other institu-

tional changes similar to those designed to implement socialism in the rest of China. Continuing resentments against Chinese rule flared into serious rioting by Tibetans in 1987. This was quelled, and measures were taken by the government to limit access to the area by foreigners. Further rioting in March 1988 cost at least 18 lives.

XINJIANG (SINKIANG)

Xinjiang adjoins the Tibetan Highlands on the north. It has an area of roughly 635,000 square miles (c. 1.6 million sq. km) and a population somewhere between 13 and 25 million. In physical terms it consists of two great basins, the Tarim Basin to the south and the Dzungarian Basin to the north (Fig. 17.2). These basins are separated by the lofty Tien Shan range. The Tarim Basin is rimmed to the south by the mountains bordering Tibet, and the Dzungarian Basin is enclosed on the north by the Altai and other ranges along the southern border of the Soviet Union and Mongolia. Both basins are arid or semiarid. The Tarim Basin, particularly, is dry, being almost completely enclosed by high mountains which block off rain-bearing winds. The Tarim Basin is occupied by the Taklamakan Desert, perhaps the driest region in Asia. The basin varies in altitude from 2000 to 6000 feet (about 600–1800 m) above sea level; the smaller adjoining Turfan Depression drops to 928 feet (283 m) below sea level. The Dzungarian Basin averages about 1000 feet (305 m) in elevation, is more open than the Tarim Basin, and has somewhat more rain, although not enough for much agriculture.

The great majority of Xinjiang's population is concentrated in oases (Fig. 14.3), mainly located at points around the edges of the great basins where streams from the high mountains debouch on the basin floors. For many centuries these oases were stations on caravan routes crossing Central Asia from Humid China toward the Middle East and Europe. The most important routes followed the northern and southern piedmonts[3] of the Tien Shan. Under the People's Republic these ancient routes have been superseded by new transportation lines. A railroad connects Lanzhou (Lanchow: city proper, 1.5 million), a major industrial center and supply base in northwestern China proper, with Urumqi (Urumchi: city proper, 1 million), the capital of Xinjiang; several major roads link the important oasis centers; and several major airfields have been built in the region.

The transport links are key elements in a general drive to expand the economic significance of this remote part of China and bring it under firmer political control. Early in the Communist era, demobilized soldiers and urban youths from eastern China were organized into quasi-military production and construction divisions to spearhead expansion of irrigated land on state-operated farms. Cotton is a major crop, as it is in irrigated areas of adjacent Soviet Middle Asia. The region has deposits of coal, petroleum, iron ore, and other minerals. A considerable expansion of mining and manufacturing has occurred under the People's Republic, although regional development is handicapped by distance from the core area of China.

Development in Xinjiang serves Chinese political purposes by tying this outlying area more closely to China's core. The population was for centuries predominantly Muslim and Turkic and has often been restive under Chinese rule. In addition, some frontier regions have been contested between China and Russia, and Russian economic and political influence in the province was strong as recently as the 1930s and 1940s. However, Chinese control is probably tighter now than ever before, being strengthened by the new economic development which has brought so many Chinese immigrants into the area

that Xinjiang may now have a majority of Chinese in its rapidly growing population.

INNER MONGOLIA

North and northwest of the Great Wall, rolling uplands, barren mountains, and lifeless basins stretch into the arid interior of Asia. Here lie slightly more than a million square miles (c. 2.6 million sq. km) of dry terrain divided about equally between Inner Mongolia, the area nearest China, and the Mongolian People's Republic, often called Mongolia or Outer Mongolia. The latter unit, now a nominally independent Soviet satellite, is discussed in a separate section at the end of this chapter.

An overwhelming majority of Inner Mongolia's present population (an estimated 20 million in 1986) is Han Chinese, who are concentrated in irrigated areas along and near the great bend of the river Huang. By far the greater part of Inner Mongolia is still the habitat of a very sparse population of Mongol herdsmen. Although they have at their disposal some areas of grassy steppe, much of their territory is a notably barren, gravel-strewn desert—a harsh and desolate country of widely spaced springs and meager pasturage with less than 5 inches (13 cm) of precipitation per year.

The traditional picture of nomadic Mongol tribesmen herding their flocks of sheep and goats and using camels and horses for riding and pack purposes is fast disappearing. In part this is the result of a long history, which is still continuing, of incursions by Chinese settlers into the better-watered grasslands. For centuries Inner Mongolia, like other grassland areas along the dry margins of world agriculture, has been a zone of competition and conflict between grain farmers and herders. In addition, the comparatively few Mongol herdsmen who remain have been induced more and more by the Chinese to maintain fixed locations and use permanent pastures—an ancient device for keeping political control over the nomads and easing the transition to sedentary occupations.

The economic core of Inner Mongolia is comprised of the agricultural areas near the Huang. Here the irrigated areas, which have been expanded under the People's Republic, mainly produce oats and spring wheat, while unirrigated fields are used mainly for drought-resistant millet and kaoliang. The main city of the area is Baotou (Paotow: 1.2 million), an expanding industrial center on the Huang. Here one of China's major iron and steel works was opened with Soviet help in the 1950s, using nearby resources of coal and iron ore.

---

[3]A piedmont is a belt of country at an intermediate elevation along the base of a mountain range.

HUMID CHINA

Humid China, sometimes referred to as Eastern or Monsoon China, the core region of the country, includes the densely settled parts of China south of the Great Wall and also Manchuria. The ancient provinces of China south of the Wall are often referred to by outsiders as "China proper," although the Chinese themselves do not employ the term. China proper includes two major divisions, North China and South China, which differ from each other in various physical, economic, and cultural respects.

Although each area exhibits much variety from place to place, South China may be characterized in general as subtropical, humid, and hilly or mountainous, with irrigated rice as the main crop. North China, on the other hand, is continental and subhumid, has larger stretches of level land, and depends mainly on nonirrigated grain crops other than rice. Prior to collectivized agriculture, individual farms in the North averaged about twice the size of those in the South; in the latter region, however, double-cropping practices and the dominance of rice, with its higher yields per unit of land, compensated for the smaller size of farms. Oxen and other draft animals such as mules and donkeys are common in the North, but water buffaloes, with their ability to withstand heat and work in the muddy rice fields, predominate in the South. The North Chinese are typically taller, heavier, more purely Mongoloid, lighter complexioned, and mainly speak Mandarin (Chinese proper; also called North Chinese) or one of its variants. The South Chinese tend to be shorter, somewhat darker, and speak a variety of mutually unintelligible Chinese languages or dialects that are different from Mandarin. The use of Mandarin is spreading. Historically, the system of written characters that was common to all literate Chinese and the tradition of governing through an educated elite helped the Chinese to administer a large country with a population speaking a number of distinctive languages.

North China and South China are described at greater length in the quoted selection that follows. Although old, the description is extraordinarily concise and is not outdated.[4]

---

[4]*Geographical Foundations of National Power*, pp. 1 and 6–8. Supplementary material has been inserted in brackets. To make the reading less cumbersome, place names of the original have been altered to Pinyin spellings without brackets, and metric equivalents are supplied in parentheses rather than in brackets. Some unbracketed connectives are supplied without changing the sense of the original.

NORTH CHINA

North China is subhumid, average annual precipitation ranging from 17 inches (43 cm) in the Loess Upland to 21 inches (53 cm) in the North China Plain. But owing to the very high variability from year to year the area [has been] subject to famine from droughts and floods. The winters are cold, the summers hot. . . . The people depend on wheat, millet, and kaoliang (a sorghum) as their primary food. The great plain of the unruly Yellow River (Huang He) dominates the region and is densely populated. Chinese civilization originated and developed in this area. . . .

North China . . . comprises two very different subregions: (1) the uplands of the interior and (2) the great alluvial plain of the Huang or North China Plain, with the included uplands of Shandong (Shantung).

The Uplands

The Huang He, which rises in Tibet, drains the western portion of the uplands; thence it makes a great swing far to the north into the steppes of Mongolia, where it turns back south through the uplands, only to make a right angle bend to the east, 200 miles (c. 320 km) before flowing out into the plain. The upland area as a whole was the historic frontier zone between China proper and the dry lands of inner Asia. Much of the region is covered with deep deposits of loess, which the Chinese descriptively call "Yellow Earth," an exceedingly fine-grained but fertile soil, laid down by the wind [Fig. 17.5]. The scarce and unreliable rainfall renders agriculture precarious, however. In periods of drought famine [has occurred]. Conversely, when a series of years brings more than average rain, much of the steppe borderland can be tilled, and farmers push out into it. . . .

The North China Plain

The North China Plain (Yellow Plain) is an immense complex delta. The Huang shifts its course from time to time, reaching the sea first to the north and then to the south of the Shandong hills. Numerous [other streams] and ditches cross the Plain. These waterways are shallow and most of them are fordable and they have generally been diked in an effort to keep them within bounds during flood season. As the Huang deposited silt and debris, the dikes were built higher and higher to keep above the rising water level until the bed of the stream in many sections is above the level of the surrounding country. This situation placed the residents of the area in a precarious position. Few rivers are subject to such violent floods as the Huang. Hardly any of its waters, shallow and full of shifting sand bars, are navigable for large craft.

The country is so dry that it is a yellow and dusty land except during spring and early summer. There is

Figure 17.5   Flights of terraces for growing unirrigated crops climb the loess hills in this scene from the interior uplands of North China. Streams have cut miniature canyons into the loess, which has the property of standing in vertical walls when subjected to stream erosion. Many people in this part of China live in caves that have been excavated in the valley sides. Such dwellings are dry and are insulated by the loess from summer heat and winter cold.

generally sufficient rain, however, to bring crops to bearing without irrigation. . . .

SOUTH CHINA

South China is a different world from that of the North. Its abundant rainfall and warmer climate make it lush and green at times of the year when North China is parched with drought or withered with cold. Much of South China is hilly or mountainous. The principal lowlands are the three large basins drained by the Chang Jiang (Yangtze River) and the smaller Guangzhou (Canton) Delta.

The Chang Jiang (Yangtze) Basins

The three Chang basins are set off from each other by narrows in the river valley. The Lower Basin is a delta merging with [the North China Plain]. The Middle Basin is likewise largely built up of river sediments and is flat, low, and dotted with lakes. The soil of these two basins is not quite as fertile as that of the North China Plain, but the more abundant rainfall ensures a good crop every year and provides water for the irrigation of rice, which cannot be extensively grown farther north because of the long, cold winters. The third

or Red Basin lies farther upstream in the province of Sichuan (Szechwan); it is separated from the Middle Basin by a mountain wall through which the river has cut a deep gorge. The Red Basin stands several hundred feet above the Middle and Lower basins and is a land of hills, among which lies the highly productive Chengdu (Chengtu) plain.[5]

The three . . . basins are set off from North China by a combination of barriers. On the east the many rivers and lakes of the delta flats form a considerable obstacle. Farther west, first a range of hills and then the Qin Ling (Ch'in Ling; Tsinling) mountains, which rise to more than 10,000 feet (3048 m), mark the border zone. These barriers are crossed by minor trails in many places, but the main highways . . . are few and have remained little changed since the beginning of China's history. Even the construction of canals and railroads has not altered the pattern of the principal connecting routes. . . .

---

[5]The Basin of Sichuan or Red Basin contains one of the largest concentrations of population in China, currently estimated at around 100 million people. Nature has provided comparatively little level land in the basin, but rice and other crops are grown on enormous numbers of small fields in narrow ribbons of valley land and on terraced hillsides.

Throughout its course in China proper the Chang Jiang is broad, deep, and, in places, swift. It is an important element in the Chinese communication system. . . . It has always carried naval vessels as well as trading ships. Ships drawing ten feet of water [can reach Wuhan], the metropolis of the Middle Basin, and river craft drawing seven feet can ascend to Yichang (Ichang) at the foot of the gorge. Chongqing (Chungking), 1300 miles (c. 2100 km) from the river mouth, is accessible by small power vessels. The gorge between Chongqing and Yichang is hazardous because of rocks, whirlpools, and the swift current, and specially constructed, powerful steamers are required for its navigation. . . .

The Guangzhou (Canton) Delta

The principal river of China's far south is the Xi (Hsi), or West, River a shorter stream than either the Huang or the Chang Jiang [but second to the Chang in volume]. Its course is through uplands, except where its delta merges with those of lesser streams to form the Guangzhou Delta, a densely populated lowland, smaller in area than those farther north. This lowland, which is separated from the Chang Jiang by a wide belt of hills and low mountains, was not finally incorporated into China until the time of the Ch'in (Ts'in) and Han Dynasties, nearly 1000 years after the basins of the Huang and Middle and Lower Chang had been consolidated. . . .

The Hill Lands

The coastal zone between the deltas of the Chang and the Xi is so difficult to penetrate that it was not annexed to China until the third century A.D. The range of relief is between 1500 and 2500 feet (c. 460 and 760 m), although some peaks rise to 6000 feet (c. 1830 m). The rivers are short, swift, and unnavigable, and each basin constitutes a unit isolated except on the seaward side. The size of the towns at the river mouths is limited by the productivity of the basins they serve. Agriculture is restricted to small and scattered valley lands. The coast is dotted with fishing hamlets. This is the only section of China in which the people have taken much interest in seafaring.

Most of Yunnan, the southwesternmost province of China proper, is a deeply dissected plateau some 6000 feet high. In the long history of the southward advance of the Chinese, it was the area most recently penetrated. Many primitive groups still live there unassimilated to the Chinese way of life. . . .

CITIES OF CHINA

China's enormous rural population and the predominance of small rural villages and hamlets (estimated to number about 1 million) should not obscure the fact that the country has many sizable cities and a few very large ones. As of the mid-1980s, there were 12 Chinese cities with estimated metropolitan populations of more than 2 million and of course a great many smaller ones. In location the 12 largest cities fall into geographic groups which help to elucidate the spatial structure of China's economy. (1) *The Chang (Yangtze) Valley city-group* is a string of cities along and near the river. This group includes 5 of the 12 largest cities, a fact which reflects the importance of the Chang River basins as a major axis of the national economy. In the west, the group includes two cities in the Red Basin: Chengdu (Chengtu: 2.5 million), a node for transport lines connecting the Red Basin with Arid China; and Chongqing (2.8–7 million according to different sources), on the side of the Basin where the Chang leads downstream toward the rest of China proper. Downstream beyond the gorge lies Wuhan (3–6 million) in the river's Middle Basin. Relatively near the sea, in the Lower Basin, is Nanjing (Nanking: 2–5 million) and, near the mouth of the river, Shanghai (12 million). Shanghai, the main port for the Chang Jiang axis, is actually located on a navigable tributary of the main river. The city was originally built up from a 19-century fishing village by the trade-seeking foreign powers who forced their way into China. It is now China's largest city. In summation it may be seen that the Chang Valley city-group includes an inland commercial and industrial metropolis for each of the three Basins along the river, plus a seaport connecting the whole with overseas lands, and a far inland city providing connections with the high and arid interior farther west. (2) *The North China city-group* includes the capital city of Beijing (Peking: 9.5 million), the port of Tianjin (Tientsin: 8 million), and far to the southwest the very ancient city of Xi'an (Sian: 2.3 million). Each of these cities is located at an entryway to the North China Plain: Tianjin the entry by sea; Beijing near the end of a pass leading through the uplands to the north toward Mongolia; and Xi'an in the valley of the Wei River, a tributary of the Huang whose valley opens into the Plain downstream from Xi'an and affords a route upstream toward interior China. (3) *The Manchurian city-group* consists of three large cities: in the south, the port of Lüda (formerly Dairen-Port Arthur: 2 million); in the southern interior, Shenyang (also called Mukden: 4.2–5.3 million), near which several heavy-industrial centers cluster; and, to the north in central Manchuria, Harbin (2.6–3.8 million), a major focus of railways from surrounding regions in Manchuria, the Soviet Union, and Korea. (4) *South of the Chang Valley*, there is only 1 of China's 12 largest cities, the South China port of Guangzhou (3.2–7 million) in the Xi River delta. This fact indicates the relative retardation of South China's nonagricultural economy and the effectiveness of British-controlled Hong Kong (5.7 million) in dominating the region's trade and manufacturing.

MANCHURIA

Manchuria, or the Northeast as it is commonly termed by the Chinese, is the part of Humid China north and northeast of the Great Wall. Manchuria (three Chinese provinces) has an area of about 310,000 square miles (*c.* 800,000 sq. km) and an estimated population of approximately 94 million (1987). To the west are the Mongolian steppes and deserts; to the east Manchuria is separated from the Sea of Japan by Korea and by Soviet territory. In our century, this area has claimed an important share of world attention, chiefly through its role as a zone of rivalry and conflict among China, Russia, and Japan. Russian interest in Manchuria has been connected with the Chinese Eastern Railway, a shortcut across northern Manchuria between Vladivostok and the Trans-Siberian Railroad east of Lake Baikal, and with the naval and port facilities of Lüda (Dairen-Port Arthur), which the Russians have used extensively at various times. The Japanese were attracted to Manchuria as a source of supply for coal, iron, aluminum, timber, and other industrial resources, as a source of surplus foodstuffs, as an expanding market for Japanese manufacturers, and as a promising area for capital investment. After the Russo-Japanese War of 1904–1905, mainly fought in Manchuria, Japan was increasingly active in the area and took over actual control in 1931 with the establishment of the puppet state of Manchukuo. For the Chinese, Manchuria has served as an outlet for surplus population migrating from south of the Wall. Movement of Chinese colonists into Manchuria began centuries ago, but gained greatest momentum after 1900. Millions of Chinese farmers migrating to this area have made it one of the 20th century's major zones of pioneer agricultural settlement. However, of great significance to post-World War II China was the considerable development of heavy industry taken over from the Japanese, along with the rail network begun by the Russians and later expanded by Japan.

In physical terms Manchuria consists of a broad, rolling central plain surrounded by a frame of mountains which seldom rise higher than 6000 feet (a little over 1800 m). The mountains on the east and north are forested and contain much valuable timber, including oak and other hardwoods as well as conifers. The central plain, oriented northeast-southwest, is approximately 600 miles (966 km) long by 200 to 400 miles wide. The northern portions of the plain are drained by the Songhua (Sungari) River, a tributary of the Amur, and the southern portions by the Liao River, which flows to the Yellow Sea. The soils are relatively deep, dark-colored, and very fertile, and the summer rainfall is generally sufficient for the crops that are grown, although spring drought is often troublesome. Conditions are not favorable for irrigated rice, except in limited areas, and the winters are generally too severe for winter grains. However, soybeans, kaoliang, millet, corn, and spring-sown wheat do well during the relatively short but warm frost-free season of 150 to 180 days. Manchurian farms are larger, the density of the agricultural population is less, and much greater possibilities for future mechanization of agriculture are offered than in the parts of Humid China south of the Great Wall.

The industries of Manchuria are mainly centered in or relatively near the largest city, Shenyang, in south central Manchuria. The largest center of iron and steel production is Anshan (1.3 million), south of Shenyang. The industry is based on deposits of iron ore in a belt which crosses southern Manchuria. Substantial deposits of bituminous coal also occur in Manchuria, although the total estimated reserves are far smaller than those of China south of the Great Wall, and only a small fraction of Manchurian coal is suitable for coking.

CHINA'S POPULATION AND ECONOMY

AGRICULTURAL EMPHASES

Around 800 million Chinese in the People's Republic are directly engaged in and supported by agriculture. This enormous mass of peasant humanity must wring food for itself and for the remaining population from a country where roughly nine tenths of the land is not in cultivation because of steep slopes, dryness, short growing seasons, and/or technological inadequacies. China's arable area, concentrated in large measure on the plains and river valley lands of Humid China, is about half as large as that of the United States, but through multiple-cropping practices (growing successive crops on a given field throughout the year) the sown acreage may be larger. It provides, on the average, about one third of an acre of arable land per person to the agricultural population, and only one fourth of an acre per person to the total population (United States, 2 acres or 0.8 hectare per person of the total population). If the farming unit were the family farm—which for the most part it has not been under the People's Republic—the average farm for a family of five would be composed of about 2 acres of arable land. These are, of course, crude averages, and conditions vary—for better or worse—from one part of the country to another. Production per acre also varies markedly from region to region and from one type of land to another. Thus farmers in North China and Manchuria, where the arable acreage per person is above average, are not necessarily more productive or better off than farmers in South China, where the acreage per person is less but the productivity of each acre is often greater than in the north. Under the Communist regime, China has been able on the whole to increase agricultural output at a somewhat faster rate than the growth of population and to raise its levels of nutrition. But this has been, and remains, a difficult struggle. In this situation, food grains continue to be the main focus of agriculture.

MAJOR AGRICULTURAL REGIONS OF HUMID CHINA

On the basis of climate and related crops, Humid China falls into two major agricultural regions: a *subtropical region* essentially coextensive with South China as previously described and a *continental region* in North China and Manchuria. Rice is the dominant crop in the first region, but wheat is dominant in the second, with millets and sorghums also very important.

1. *The Subtropical (Rice) Region.* China's humid subtropics extend far enough north to include the Chang Jiang (Yangtze) basins. Precipitation is abundant (40–60 inches/100–150 cm per year), over half of the crop land is irrigated (Fig. 17.6), and long growing seasons allow double cropping (two successive crops a year from a given field) on most of this acreage. In the warmer, wetter southern part of the subtropical region, both these crops are likely to be rice, but in the northern part of the region the rice crop is more likely to be followed by winter wheat or an oilseed such as rapeseed. Secondary crops in the subtropical region include wheat, barley, sweet potatoes, corn, sugarcane, soybeans, peanuts, tobacco, tea, citrus fruits, jute, tung trees (whose fruits yield a drying oil with many industrial uses), and a large array of garden vegetables. In many areas interplanting (growing of two or more crops simultaneously on a field) is common (Fig. 17.7). In addition, the Chang Basins produce three important fiber crops—cotton, silk, and ramie.

2. *The Continental (Wheat) Region.* In this region agriculture must adapt to cold winters, hot summers, and much less moisture than in South China. North China, seaward from the interior steppes, is classed as humid continental in climate (Table 14.2), but it is barely humid. Beijing, for instance, averages only 24 inches (60 cm) of precipitation a year, which is hardly abundant considering the high evaporation and transpiration rates associated with summers in which the warmest month averages 79°F (26°C). Furthermore there is much less irrigated land than in the subtrop-

Figure 17.6 *The treadmill used here to pump water from an irrigation canal to the fields is one example of the arduous manual labor still applied in massive amounts to Chinese agriculture.*

Figure 17.7 Interplanting, seen here in South China, is one way to increase farm output. The darker crop is peanuts; the lighter is sugarcane. Peanuts, a legume, supply the soil with nitrogen, which helps fertilize the cane.

ical region. In these conditions winter wheat, a crop requiring only modest amounts of moisture, and summer millets and sorghums, which require even less, are the main bases of life in the North China Plain and in the loess uplands farther inland. The average precipitation is perilously near the minimum required for agriculture, and downward fluctuations have brought drought and famine here, especially in the drier uplands, many times in the past. Farther north, in Manchuria and Inner Mongolia, spring wheat replaces winter wheat in response to winters that are distinctly colder than those of the North China Plain. Secondary crops in the continental region are, as in the subtropics, quite numerous. In various sections corn, potatoes, sugar beets, soybeans, peanuts, and even a certain amount of rice are important secondary food crops. Oddly enough, considering the climate, cotton is the major industrial crop, while hemp, flax, and tobacco are also grown. As in South China, garden vegetables are ubiquitous.

In the crowded arable areas of Humid China, little land or production can be spared for feeding livestock, and relatively little meat is eaten by most of China's population. Hogs and poultry, both of which can be fed on kitchen scraps and waste, are common. China actually leads the world in pork production and ranks with the USSR and the United States as a top producer of poultry meat and eggs, but the output per person is far less impressive. In Arid China sheep, goats, and yaks grazed on natural forage are locally important for meat and milk. Sea, river, or pond fisheries make some contribution to the food supply in many areas. China is the third

nation in the world, after Japan and the USSR, in total fish catch, but as in the case of other foods the large total greatly shrinks when measured by the number of mouths to be fed.

ATTEMPTS TO INCREASE AGRICULTURAL PRODUCTION: ENDS AND MEANS

Since it took control of the country in 1949, the Chinese Communist government has attempted to increase China's agricultural production. Continued increases in farm output are regarded as necessary, not only to provide food for an increasing population but also to supply industrial raw materials, capital for industrialization, and agricultural exports to underwrite the costs of industrial equipment purchased in other countries.

The problem of increasing food production is many-sided, and the efforts and priorities given to it by Chinese planners have varied. The major lines of attack may be summarized under the headings of (1) water conservancy and increases in irrigation; (2) intensification of existing agricultural techniques, particularly an increase in the use of chemical fertilizers; (3) reclamation of land and increase in the cultivated acreage; (4) other noninstitutional measures, including mechanization, improved crop varieties, and elimination of insect pests; and (5) institutional and organizational measures. These approaches should be viewed in the overall context of Communist economic planning that initially relegated agriculture to a secondary role while giving priority in investment and resources to heavy industry. After the collapse, in 1960, of the program of forced economic development called the "Great Leap Forward," there was a reassessment of priorities and more recognition of the importance of agriculture and of industries supplying agriculture.

Water Conservancy and Expansion of Irrigation

Projects to control and conserve water and to expand the irrigated acreage have been given a major emphasis. Floods and droughts have always periodically afflicted Chinese crop production, especially in North China. In the past they have caused millions of deaths from famine. Thus a major objective of water control measures is to prevent water from running off in disastrous floods and to conserve it in reservoirs for times when it is needed. Three specific types of projects have been and are being carried out to achieve this objective:

1. *Construction of Small-scale Dams, Ponds, Canals, and Dikes by Local Communities.* These have been constructed in very large numbers. Usually such projects are undertaken during the agricultural off-season when large inputs of labor are available.

2. *Building of Large-scale Dams and Associated Structures to Control Larger Streams.* A number of large-scale, multipurpose projects were initiated during the early years of the Communist regime, and construction of more major dams continues today. Most construction of this type has been in North China, where drought and flood problems are the most frequent and serious.

3. *Afforestation and Reforestation.* These relate closely to water control. Only 14 percent of China is forested, and large areas are without significant vegetative cover. Most forest growth in China proper has long since been removed, primarily by peasants for use as fuel. The lack of vegetative cover on nonfarm lands causes a rapid runoff of much of the rain that falls, thus raising stream levels and increasing flood hazards as well as effecting serious erosion and the siltation of reservoirs and crop fields. The Chinese Communists have stressed large-scale afforestation and reforestation projects, plus shelter belts and other protective measures, and they have greatly increased the amount of standing timber, although much remains to be done.

A major objective of water conservancy measures has been to increase the amount of land under irrigation and thereby increase the output of crops. It appears that a considerable increase in the amount of land under irrigation has been achieved, many of the existing irrigation works have been improved, and use of power equipment has increased the efficiency of irrigation in some areas.

Intensification of Existing Agricultural Techniques

A very important emphasis in China's attack on the problem of agricultural production has been placed on further intensification of the existing agriculture in order to raise yields per unit of land. Irrigation and flood-control measures, where applied to land already farmed but not irrigated or protected from floods, may be seen as one phase of this effort. Another approach lies in the application of scientific knowledge and experimentation to develop better seed, equipment, and farming techniques, together with the development of administrative and propaganda machinery to spread their use. Still a third approach lies in increased fertilization. Chinese farmers have long fertilized their fields with animal and human manures, pond mud, plant residues and ashes, oilseed cakes, and green manure crops. Although the traditional fertilizers are still by far the most important source of plant nutrients, the regime has now begun to stress the use of chemical fertilizers. China's relatively small chemical-fertilizer industry has received a high priority and has been expanded rapidly by the purchase of whole plants from abroad. Despite sizable increases in the availability of chemical fertilizers, however, the amounts used are still low.

Land Reclamation

An obvious means for expanding the food output would appear to be an expansion of the cultivated acreage. One thing that has often impressed observers in China is the small use that is made of second-rate land even in places where population is excessively crowded on adjacent good land. For instance, in South China, narrow valleys often teem with an overcrowded and struggling population, but broader adjacent uplands are little used. The explanation seems to lie largely in the fact that the Chinese farmer and his family, unequipped with machinery, cannot till a sufficient amount of second-rate land by hand labor to make a living. Only by concentrating their labor on a smaller amount of good land can they feed themselves.

In the early 1950s, just after the Communists achieved control, their statements placed considerable stress upon reclamation and on the amount of potential land existing for future cultivation. Small gains in the amount of cultivated land were made through reclamation of poorly drained land, extension of irrigation into desert lands, and similar measures. Since the 1950s, however, there has actually been a slight decrease in the amount of land under cultivation. This suggests the difficulties of the land-reclamation approach, and in recent years the Chinese have scaled down their estimates of reclaimable land. Potential areas of significant size are claimed for Manchuria, Xinjiang, and scattered areas elsewhere, but these areas are agriculturally marginal. They suffer from short growing seasons, lack of water, poor drainage, and other deficiencies that make them uneconomic to bring into production under present conditions. Some potential is claimed for the hill lands of South China, but reclamation would require mechanized equipment and coordination with overall water and soil conservancy schemes so as not to create flood and silting problems for the intensively used and productive valley lands below.

Mechanization, Crop Breeding, and Control of Diseases and Pests

Other methods to increase farm output are not yet of first importance but may be crucial in the future. Mechanization has developed slowly and with uneven progress throughout the nation. The use of tractors is growing, but primarily in Manchuria and in the west where extensive farming techniques are feasible. In recent years some priority has been given to improved farming techniques, implements, and electric pumps. Thus there has been progress compared to the situation existing in 1949, but by and large the rural scene still is dominated by human and animal power and simple hand tools. Indeed, any rapid mechanization would release a probably ungovernable flood of surplus labor onto both the countryside and the cities.

Progress also has been reported in crop breeding and in the control of insects and plant diseases, but little definitive information is available concerning the overall success of such programs. The lack of trained agricultural scientists and technicians has been a hindering factor. Developmental work is being carried out with cotton, corn, rice, and other crops. There is evidence that insecticides and herbicides are coming into wider use. Pest control was greatly publicized during the early years of the People's Republic, with great campaigns to eradicate rats, mice, flies, sparrows, and other pests. Observers were amazed at the sight of Chinese peasants fanatically hunting down pests at the exhortation of the government. A major objective of these campaigns was to reduce food losses caused by pests, which have been credited with the destruction of many millions of tons of grain each year in the fields and in storage. Antipest campaigns against flies, mosquitoes, and rats have had, in addition, an obvious public health aspect. The evidence suggests that these programs have had at least a limited success and that progress may have been considerable.

Institutional and Organizational Measures[6]

Since the Communist Party gained control of China in 1949, it has mandated a series of reorganizations of Chinese agriculture in successive attempts to solve

the agricultural problem within the limits of Communist ideology and control. Thus far this effort has proceeded through four rather well-defined stages: initial collectivization; the strict commune system; the modified commune system; and the "household responsibility" system. They are summarized below:

1. *Initial collectivization* was carried out between 1949 and 1957, mostly in the earlier years. Land was transferred from individual owners to the state, and family farms were combined into collective farms on the Soviet model. Production was directed by the state, and farm products were sold to the state.

2. *The strict commune system* was in effect from 1958 into the early 1960s. Earlier economic progress had been substantial, but the government was not satisfied, especially with the performance of agriculture when measured against China's needs. So the commune system was conceived as the central institution in a much-publicized "Great Leap Forward" that engrossed China's energies from 1958 through 1960. The 750,000 collective farms were amalgamated into just 24,000 rural communes, each of which was a very large agricultural unit. These communes had combined responsibilities in agriculture, rural industrial development (including backyard iron smelters), commerce, education, militia affairs, and local administration. Their only market for products was the state, at prices set by the state. They were commanded by Party personnel deemed to have sufficient orthodoxy and enthusiasm (but not necessarily possessed of technical competence), with each commune being organized into subunits called production brigades, which in turn were subdivided into production teams. The incentives to produce consisted largely of threats, together with constant propaganda and indoctrination, rather than higher incomes for more or better work. With "politics in command," Mao Zedong (Mao Tse-tung) and the Communist Party held that all things were possible, including a doubling of China's food production within a single year. But this command economy failed. By 1960 output fell, the economy was in acute crisis, and the rural population was drained emotionally and physically. The leadership backtracked, and the period since 1960 has been marked by the progressive relaxation of the commune system.

3. From the early 1960s to the late 1970s a *modified commune system* was in effect. The communes became smaller and twice as numerous, and the production team, consisting generally of 20 to 30 families, became the most important local operating, management,

---

[6]The discussion of China in the remainder of this chapter draws substantially on Donald S. Zagoria, "China's Quiet Revolution," *Foreign Affairs*, 62, No. 4 (Spring 1984), pp. 879–904.

and decision-making unit in agricultural matters. Some economic incentive was allowed back into the system in that teams distributed income to their members on the basis of their work as measured in "labor-day" units; and small private plots and individual raising of pigs and poultry were reintroduced, along with rural free markets where surplus products from this small-scale private enterprise could be sold. But the system was still structured and regulated to strongly limit the scope of individual initiative and maintain a near-equality in income among the peasants.

4. Mao Zedong died in 1976 and, after a power struggle, a group of leaders with different ideas about economic development gained control of the Communist Party and hence the country. They saw the economy, including agriculture, as being in crisis, and the cure as the injection of a large measure of free enterprise to stimulate production through the profit motive. They continue to express themselves as convinced Communists, and they call the new economy "market socialism" rather than capitalism. In agriculture, this has resulted, since 1979, in the *"household responsibility"* system. Peasants and peasant families may now lease farm land for up to 15 years from the state, which still owns it. The peasant then contracts with local authorities to produce stated amounts of certain basic commodities for sale to the state, and production surpluses above the contracted amounts may be sold for profit in the free markets which are proliferating and thriving. Under only indirect state guidance, largely through manipulated prices as in Western agriculture, the peasant operates independently and profits from it if he is successful. Other features of the post-1979 system are expanded private plots for farm families, on which they can practice complete free enterprise; the end of restrictions on the feeding of grain to animals; and at least the beginnings of a farm credit system. Under this freer system the equality of income among peasants that was enforced under the previous Communist systems is rapidly eroding and a small class of wealthier peasants—"rich" by Chinese standards—is rising above the general mass.

Production results in the early years of this massive experiment with semicapitalism in agriculture were good. In the 5 years between 1978 and 1983 total grain production increased by about one third, and Western visitors had subjective impressions that much larger quantities of meat, fruit, and vegetables also became available. In the late 1970s and early 1980s China became one of the world's leading nations in the rapidity with which food production per

capita was increasing: For 1981 and 1982 it was estimated that about one fourth more food per person was being produced than had been the case in the years 1969–1971. By 1986 the output per capita had risen still further, but at a slower rate. Hence, questions remained concerning future food supply, but promising advances had been made under the freer system of production.

THE ROLE OF POPULATION POLICY

Besides increases in agricultural production, a drastic slowing of population growth was also responsible for bettering China's food situation from the late 1970s into the 1980s. Before this time, and after the initial recovery from war in the 1950s, there was scarcely any progress in *per capita* food output and availability. Increases in agricultural output were largely matched by rapid population growth, while the Mao Zedong regime made half-hearted and, later, no efforts to curb the birth rate. During the 1970s, however, attempts to limit births began again, and the post-Mao regime instituted probably the most stringent program of birth control in the world. This is the "one child campaign," which aims to limit the number of children per married couple to one. The government pours out propaganda in support of the campaign and maintains surveillance through local authorities. It dispenses free birth-control devices, free sterilization operations, free hospital care in delivery, free medical care for the child, free education for the child, an extra month's salary each year for the parents, and other favors and preferences to induce compliance with the one-child-per-family norm. Those who violate the norm are subjected to constant social and political pressure, denial of the privileges accorded one-child families, pay cuts, and fines. Women who become pregnant without permission are pressured to have a free state-supplied abortion, with a paid vacation provided.

Under such programs China's birth rate declined by the mid-1980s to the lowest in any major "less-developed" country—about two thirds that of India, for instance—and its rate of natural population increase dropped to about 60 percent of the average for other less-developed countries. But the situation remains desperate. China's population is so large that even the relatively low birth rate achieved by 1988 still gave a net increase of over 15 million people each year in a country whose death rate has also been drastically lowered by better food availability and better medicine and public health. The regime hopes to decrease childbearing to the extent that population in the 21st century will stabilize at 1.2 billion people, but some population experts are projecting a population of 1.4 billion by the year 2020. By the late 1980s, government officials were reporting widespread disregard of the "one couple, one child" policy, as peasants became wealthier and hence more ready to pay the fines imposed for violations of the national policy. The national birth rate was rising, but to what degree and to what probable effect was not clear.

THE PUSH TO INDUSTRIALIZE

The factory method of mass-producing industrial goods began to appear in China around the beginning of the 20th century. Previously, Imperial China had a system of handicraft industries which produced goods of such quality as to attract Western traders in spite of China's official disinterest in such trade. To this day handicraft and shop-scale industries continue to supply many of China's simpler consumer needs. The early factory industries were largely foreign-owned and were attracted to China by the huge market for cheap goods, by cheap labor, and sometimes by certain natural resources. As in so many countries, the first large-scale factory industry to develop was the manufacture of cotton textiles. Japanese and other foreign-owned firms, together with some native Chinese companies, developed a cotton industry that by the 1930s was exceeded in output only by the cotton-textile industries of the United States, the Soviet Union, Japan, and India. Along with other industries such as tobacco products and processed foods, textile milling developed in such coastal cities as Shanghai and Tianjin. Meanwhile the Japanese used the mineral resources of southern Manchuria to make that area the first center of modern heavy industry in China.

Under the People's Republic, strenuous Chinese efforts to industrialize have gone through four major phases: Communization and repair of wartime damage; heavy-industry development, Soviet style; near-isolation under late Maoism; and liberalization after Mao's death. Some major aspects are summarized below:

1. *Communization and Repair of Wartime Damage,* 1949–1952. The industrial structure, which had been badly hurt by 12 years of foreign invasion and civil war (1937–1949), was brought back to prewar production levels under the new system of state ownership and operation.

2. *Heavy-Industry Development on the Soviet Pattern,* 1952–1960. Aided by the Soviet Union,

Chinese planners gave priority to coal, oil, iron and steel, electricity, cement, certain machine-building industries, and armaments. China's resources for these industries are considerable and the country's centrally planned economy achieved major successes in developing them, as has been the case in many Communist countries. For instance, pre-Communist coal production had peaked at 66 million metric tons per year under Japanese military control in 1942, but by 1958 China's output was 270 million tons and ranked third in the world, after the USSR and the United States. Subsequently it increased to 880 million tons in 1986, making China the world's largest producer. Meanwhile a maximum pre-Communist steel production of 0.9 million tons per year was multiplied by a factor of 9, to 8 million tons, by 1958. By 1987 steel output had reached an estimated 56 million tons, placing China well ahead of West Germany for fourth position in the world, although well behind the USSR, Japan, and the United States.

This notable progress in heavy industry was begun with much Soviet material and technical aid during the 1950s and has also had the advantage of a very adequate resource base (Fig. 17.2). China has coal resources estimated at about one tenth of the world total. They are widely spread, but the major producing fields lie underneath the North China Plain and its bordering hills, or in southern Manchuria, while the largest coal reserves are in the loess hills that lie inland from the North China Plain. Iron-ore resources are less abundant, but workable deposits are found in a number of locations, with the largest deposits being in southern Manchuria, where they lie in convenient proximity to that region's coal fields. Southern Manchuria has been the country's leading heavy-industrial region since modern industries were developed there by the Japanese, but widespread coal and ore deposits have facilitated Chinese Communist development or expansion of iron and steel industries at other scattered centers in China proper, Inner Mongolia, and along the Chang River. A large new plant near Shanghai marks a departure from the previous pattern in that it relies on imported Australian ore. China's large production of alloys (tungsten, manganese, molybdenum, vanadium) and a number of other metals such as tin have further facilitated the development of basic heavy industry.

3. *Near-isolation under Late Maoism.* This third phase of Communist China's industrial development began with the country's rift with the Soviet Union in 1958–1960 and continued

until new economic policies began to be introduced, after the death of Mao Zedong, in the late 1970s and early 1980s. As the previously given figures for coal and steel output show, progress continued in heavy industry, which still received primary emphasis even without foreign aid. But the country's general industrial advance was probably slowed in pace and almost certainly in technical development, quality of output, and needed modernization by several factors: (a) the near-isolation of China from more advanced industrial countries, (b) the "Cultural Revolution" begun by Mao in 1966 and carried on until his death, and (c) the customary difficulties that non-market centrally planned economies on the Soviet model have had in developing consumer-oriented industries and a broad range of advancing technology. The "Cultural Revolution" was Mao's attempt to maintain an ongoing "permanent revolution" in China. Politically "pure" gangs of hoodlums known as "Red Guards," generally young, were loosed on the populace to harass those who could be accused of putting learning, skill, expertise, or personal matters above revolutionary enthusiasm or Marxist goals and principles. Government-sanctioned hooliganism plus government oppression in this campaign are thought to have led to a million assassinations, executions, or suicides, with millions more exiled to labor camps. Great numbers of responsible positions passed into the hands of persons demonstrating political orthodoxy and enthusiasm, but often very lacking in ability. The universities were decimated, and their contributions to productive efficiency or technical development were virtually wiped out. Many professors were sent off to be "re-educated" through hard labor and indoctrination.

4. *Post-Mao Liberalization.* The latest phase of China's industrialization was initiated by the post-Mao leadership in the late 1970s and early 1980s. The drastic liberalization in agriculture has already been described, and related measures were introduced into trade and industry, where emphasis shifted more toward light industry, consumer products, and profit-oriented enterprise (see chapter-opening photo). Much more private enterprise is being allowed in originating and operating small businesses (1988). State factories are being required to operate at a profit, part of which they may retain and invest in expansion. They are introducing pay incentives for superior performance and may now hire and discharge workers relatively freely rather than being obligated, as in the past, to provide lifetime jobs

regardless of performance. Capital investment by foreign companies is being welcomed and technology sought from Japan and the West, while many thousands of Chinese students are being sent abroad for higher education. Expanding foreign trade, rather than the relative isolation of the Mao era, is promoted. Whether these policies will secure the rapid development and modernization China seeks cannot yet be known, and whether they will even be long continued is uncertain. There is still widespread resistance to liberalization among ideologic purists and members of the 40-million-strong Communist Party who owe their positions to the Cultural Revolution and the old system.

A great variety of manufactured goods are produced to some degree in China, including some relatively sophisticated products such as those needed to maintain the country's nuclear arsenal. But China's trade pattern is still that of a country in the early stages of industrialization. Like other such countries, China emphasizes exports of relatively simple and labor-intensive manufactures, along with natural resources or resource-related products: such items as cotton textiles and clothing, crude oil, and oil products. There are scattered oil and gas deposits in various parts of the country, with the major producing fields being located in southern Manchuria and the North China Plain. These Chinese fields accounted for about 4 percent of world oil production in the mid-1980s, which was enough to supply China's low per capita usage and also furnish a major export. The country has signed agreements with foreign oil companies for offshore oil prospecting in the South China Sea, an area of interest to several other countries that border this sea and hope to initiate or expand oil development there. For China, a major expansion of oil revenues would be most useful in financing and fueling the whole modernization effort. The country's main trading partners and its main sources of foreign investment and technology are Japan, the United States, and Hong Kong, the latter acting partly as a middleman between China and the outside world.

IMPROVEMENT OF TRANSPORTATION

In order that mainland China's widespread resources and widely scattered producing and consuming centers may be integrated into a functioning whole, the Chinese government has placed great importance upon improvement in the transportation system. In spite of considerable progress, however, transportation facilities are still poorly developed and unevenly distributed throughout the country. Although trucks now play an important role, a good deal of China's internal commerce is still distributed by human porters, bicycles, pack animals, wheelbarrows, or carts drawn by animals or drawn and pushed by humans (Fig. 17.8). The country has a rudimentary but expanding highway system, now several times the length of the serviceable roads existing in 1949. However, much of it consists of earth roads subject to seasonal disruption. Important new roads have been constructed into western China, for example to Tibet, where feeder roads extend to the Sino-Indian border and to Nepal.

Water transportation was the main long-distance mover of Chinese goods in past times. The Chang Jiang and its many tributaries provided a massive, branching system reaching into many sections of the country. The Grand Canal linked the agriculturally rich lands along the Chang to the locus of political power in North China. But only one third of the present inland waterway network is located in the industrially developed North and Northeast, and not more than one fourth of the total length of waterways considered navigable can be utilized by cargo vessels other than relatively small, flat-bottomed junks and sampans. The internal waterway system, like the highway system, is being improved. But the major effort of the Communist government thus far has been directed to the improvement and extension of railways.

Prior to 1949, Manchuria had a fairly extensive rail system, developed originally by Tsarist Russia and later taken over and expanded by Japan. Manchuria and North China contained about 75 percent of China's rail mileage. Most of China south of the Chang River and all the western provinces were without rail lines. But, since 1949, the Chinese Communists, as a part of their large program of railway construction, have made South China accessible by building the first three bridges across the Chang, at Wuhan, Chongqing, and Nanjing. Meanwhile rail lines have been pushed not only into South China (including the Red Basin) but also into Xinjiang and across Mongolia to provide service between Beijing and Moscow via the Trans-Siberian Railroad.

CHINESE POTENTIALS AND UNCERTAINTIES

Although China's population remains predominantly rural (79% according to a 1989 Population Reference Bureau estimate), hundreds of millions are urban dwellers supported by nonfarm activities. The huge size of this urban sector gives some idea of the overall magnitude of China's nonagricultural development, even though per capita productivity is low. If China can check population growth, feed its people, and combine national order with a degree of freedom that taps the creative energies of the urban masses (Fig. 17.9), the country's potential is enormous. But China's potential has been extolled by outsiders throughout modern times, and the results have not met expectations thus far. Another swing of the pendulum, with a descent into chaos and tragedy, is always a possibility. Continuing uncertainty was manifest in June 1989 when massive student demonstrations in favor of greater democracy were savagely repressed.

Figure 17.8 Although larger tonnages are carried by rail, truck, or water, a great deal of China's internal commerce still moves by human porters, bicycles, wheelbarrows, or hand-drawn carts such as the melon cart seen here in Zhengzhou, Henan Province, North China.

Figure 17.9 View of a market in Shanghai. The signs in Chinese urge support for China's ''Four Modernizations'': the modernization of industry, agriculture, science and technology, and national defense. This drive to modernize is a major feature of present-day China.

HONG KONG AND MACAO (MACAU)

The two main remnants of European territorial encroachment on China are the British colony of Hong Kong and the Portuguese colony of Macao, or Macau, both located along the coast of the Guangzhou delta region of southern China. Hong Kong is by far the more important of the two. It consists of the island of Hong Kong and some smaller islands, located just offshore, and a much larger patch of the adjacent mainland known as the New Territories. Together these parts of the colony aggregate only a bit over 400 square miles (c. 1000 sq. km), but they had a 1988 population of 5.7 million, 98 percent of whom were Chinese and many of whom were refugees from China.

China ceded Hong Kong Island to Britain in 1842 as part of the settlement after the Opium War, and leased the mainland section of the colony to Britain in 1898 for 99 years. Hong Kong soon became a major naval base and commercial center dominating the trade of southern China and in recent decades has become a haven for refugee Chinese, an important financial center, a great tourist attraction, and an important East Asian manufacturing concentration. In the latter capacity, it exports a great variety of goods, especially textiles, clothing, and many kinds of machinery, among which timepieces and electronic equipment are prominent. The success of its capitalist economy in creating conditions very dif-

ferent from the norm in China is indicated by a gross national product of $6720 per capita for 1986—well over half that of Japan ($12,850)—while China's was $300.

But the British lease on the mainland portion expires in 1997, and a Chinese repudiation of Britain's ''perpetual'' lease on the island led in 1984 to a joint declaration of the two countries that the Hong Kong colony would be returned to Chinese sovereignty in 1997. China has asserted its intention not to damage the economy of Hong Kong, which is now extremely important to Chinese trade. Substantial autonomy for Hong Kong under a Chinese constitutional provision for establishing ''special administrative areas'' was agreed on by China and Britain after protracted negotiations leading to the joint declaration. The much smaller Portuguese colony of Macao will rejoin China in 1999. With around 450,000 people, nearly all of Chinese origin, Macao is all that is left of the worldwide empire which Portugal once possessed. The colony supports itself by transit trade with China, fishing, and tourism, including gambling.

TAIWAN

When it was driven from the mainland in 1949, the Chinese Nationalist (Kuomintang) government fled, with remnants of its armed forces and many of its more prominent adherents, to the island of Taiwan. Here, protected by American sea power, the government reestablished itself with its capital at Taipei (metropolitan population over 5 million by one estimate). It continues to hold the island, together with the Pescadores and other small neighboring islands, and claims still to be the legitimate government of all of China. Meanwhile the People's Republic claims Taiwan as an integral part of its own territory. For three decades the United States backed the Nationalist claim and supplied Taiwan with weapons and economic aid. But during the 1970s America and the Communist mainland developed closer relations, and as a result the United States withdrew official recognition from Taiwan. However, the United States opposes its annexation to China by force.

Taiwan, known by Westerners for centuries as Formosa (Portuguese: ''beautiful''), is separated from the coast of South China by the Formosa Strait, about 100 miles (161 km) wide. From 1895 to 1945 the island was governed by the Japanese, who developed its economy as an adjunct to that of Japan and called the island by its ancient Chinese name of Taiwan.

The island is about 12,500 square miles in area (c. 32,000 sq. km: a little smaller than Switzerland) and in 1988 had an estimated 19.8 million people, largely Chinese in origin. High mountains rise steeply from the sea on the eastern side of the island, but slope on the west to a broad coastal plain bordering the Formosa Strait. Taiwan lies on the northern margin of the tropics, and its temperatures are further moderated by the warm, northward-flowing Kuroshio or Japanese Current. Irrigated rice, the main food crop, is favored by the hot, humid summers of the lowlands and by an abundance of irrigation water provided by streams originating in the mountains. Monsoonal rainfall gives some of the higher mountain areas an annual precipitation averaging nearly 300 inches (762 cm).

The Chinese Nationalists, operating an authoritarian regime with some elements of political democracy, have been strikingly successful in fostering capitalist economic development on Taiwan. A major handicap has been lack of energy resources. Some coal and natural gas exist, and some hydropower has been harnessed, but the island depends heavily on oil imports from such countries as Saudi Arabia and Kuwait. However, there have been advantages also: Temperatures and water supply permit many irrigated rice fields to be double cropped and various tropical crops to be produced; the native rain forest supplies both soft and hard woods in export quantities; the Japanese had already done much to develop the island as a sugar-producing colony; and American aid, including educational aid and technical help, was large. On these foundations the Nationalists were able to unite cheap Taiwanese labor with foreign capital to build one of Asia's first urban-industrial countries. Its exports include clothing and textiles, electronic equipment and other machinery, sugar, shoes, toys, and sports equipment. Its per capita income is about ten times that of China (1985). Two thirds of its people are urban, and its infant mortality rate is lower than that of the United States. Although the United States and Japan together account for more than half of its foreign trade (1986), Taiwan trades to some degree with about 150 countries, most of which do not officially recognize its independent existence or the legitimacy of its government. It continues to resist China's overtures for peaceful reunion with the mainland, which include promises of autonomy and promises that Taiwan can keep its own system. But pressures on the island are growing. An unpredictable element in the situation lies in the fact that only about 15 percent of the people derive from the Nationalist refugees. Except for a relatively few aborigines, the rest are Chinese in culture but are possibly less China-oriented in political outlook. They have been discriminated against by the Nationalists, but their power is increasing, while outside pressures for reunification also increase. No reliable forecast of Taiwan's future seems possible at present.

MONGOLIA

The Mongolian People's Republic or Mongolia (referred to as Outer Mongolia in the past) was once part of the Chinese empire, but since 1924 it has been a separate country organized along Soviet lines and under strong Soviet influence. Mongolia includes an area of about 600,000 square miles and had an estimated population of about 2 million in 1988. About 90 percent of the population consists of various Mongol groups. The country contains large desert plains in the south and east, locally termed *gobis* (hence the "Gobi Desert"); mountain ranges in the west; and grassy valleys and wooded hills and mountains to the north. Vast herds of livestock, primarily sheep and goats, are grazed. Much of the population still derives its livelihood from animal husbandry, although about half is now urban. Collectivization of the nomadic population was accomplished (with some difficulties) in 1958–1959, the objective being to stabilize the population territorially and transform the rural economy to one based on crop tillage and dairying. Crops grown include wheat, barley, oats, and millet. Ulan Bator (city proper, 500,000) is the capital, major urban center, and focus of the major transportation routes.

CHAPTER 18

Japan and Korea: Economic Superpower; Geopolitical Arena

This Mazda automobile plant at Hiroshima gives evidence of Japan's prowess as a maker and exporter of automobiles. The cars will be driven on and off the specially designed ships through the large openings.

O N a cluster of islands off the eastern coast of Asia, the 123 million people of Japan operate an economy larger than that of any nation except the United States and probably the Soviet Union. In the mid-1980s, neighboring China, with more than one billion people, had an estimated output of goods and services (GNP) only about one fourth that of Japan. The economic prodigies of the Japanese have been achieved on mountainous, resource-poor, and crowded islands (see chapter-opening photo of Chapter 14) a little smaller in total area than the state of California, but containing half as many people as the entire United States (1988). Japan's dramatic rise from the ashes of World War II has drastically changed the economic and political geography of the world.

THE JAPANESE HOMELAND: RUGGED, HUMID, MEGALOPOLITAN

There are four main islands—Hokkaido, Honshu, Shikoku, and Kyushu—in the Japanese archipelago (Fig. 18.1). Hokkaido, the northernmost, is narrowly

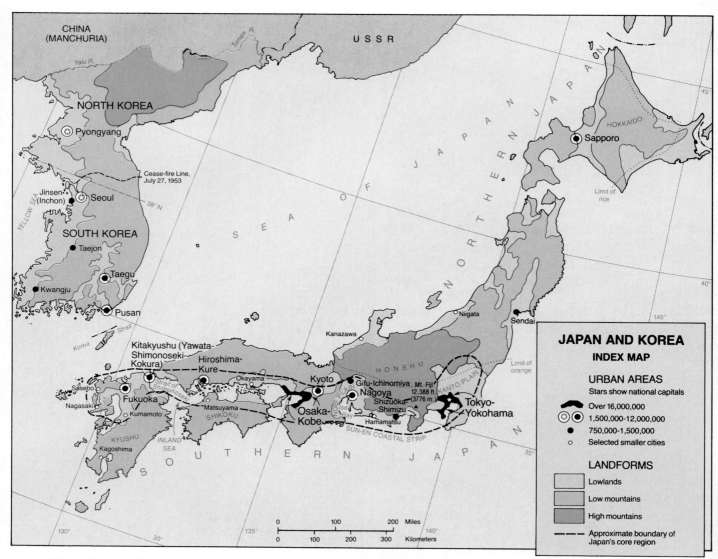

Figure 18.1 *Index map of Japan and Korea. Limits of rice and orange after Edward A. Ackerman,* Japan's Natural Resources and Their Relation to Japan's Economic Future *(Chicago, University of Chicago Press, 1953), p. 23. City-state symbols are based on metropolitan area estimates.*

separated from Soviet-controlled islands to its north. On the west the main islands are separated from the Soviet Union and Korea by the Sea of Japan and the Korea Strait. On the east, Japan faces the open Pacific. Honshu, the largest island, is separated from Shikoku and Kyushu by the scenic and busy Inland Sea. South of Kyushu the smaller Ryukyu Islands extend Japan southwestward almost to Taiwan. All the main islands consist largely of mountains, with the higher peaks generally between 5000 and 9000 feet (*c.* 1500–2750 m) in elevation, although the famous and majestic volcanic cone of Mount Fuji, west of Tokyo, exceeds 12,000 feet. The islands are volcanic (Fig. 18.2) and are subject to frequent and sometimes very severe earthquakes.

CLIMATES AND RESOURCES

Locations off the east side of the Eurasian land mass and latitudes comparable to those of the American east coast states from South Carolina to Maine (Fig.

18.3) strongly influence the climates of the main islands (see world climatic map, Fig. 2.18). Humid subtropical climate extends from southernmost Kyushu to areas north of Tokyo; humid continental long-summer climate characterizes northern Honshu; and Hokkaido has humid continental short-summer climate (Table 14.2). In lowland locations, where most of the people live, the average temperatures for the coldest month range from about 20°F (−7°C) in northern Hokkaido to about 40°F (4°C) in the vicinity of Tokyo and about 44°F (7°C) in southern Kyushu. Averages for the hottest month are only about 64°F (18°C) in northern Hokkaido, but rise southward to about 80°F (27°C) around Tokyo, and are above 80°F farther south. In the mountains conditions are cooler, with heavy snowfalls common in winter in Honshu and Hokkaido.

As the names of the climatic regions indicate, Japan is a humid country. Annual precipitation of over 60 inches (*c.* 150 cm) is common from central Honshu south, and 40 inches (*c.* 100 cm) or more in northern Honshu and Hokkaido. Location off the east coast of Eurasia puts the islands in the path of the inward-blowing summer monsoon from the Pacific, and summer is the main rainy season. But the islands are subject to the seaward-blowing winter

Figure 18.2   *The importance of vulcanism in the physical geography of Japan is symbolized in this photo of Mt. Sakurajima, an active volcano at Kagoshima in Japan's southern island of Kyushu.*

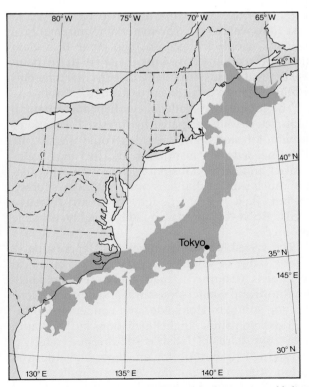

Figure 18.3   *Japan compared with the eastern United States in latitude and area. (After a map by E. O. Reischauer)*

monsoon after it has crossed the Sea of Japan, so that in Japan's case a good deal of precipitation also occurs in the winter half-year. Hence precipitation is much more evenly spread through the year than is common in Oriental areas with monsoonal climates (Table 14.2).

The Continuing Importance of Forests and Wood

Japan's natural-resource base was originally rather adequate for the nation's needs, but it has become increasingly less so with the passage of time. Up to the middle of the 19th century the country had a very high degree of self-sufficiency, although the level of living was low. Already a crowded land, Japan had a population now estimated at 30 million, but increases in population were kept in check by disease and starvation, and by the almost universal practices of abortion and infanticide. A rapid growth of population in the second half of the 19th century and after was made possible by industrialization, which has continued on an ever increasing scale except for the temporary debacle caused by Allied bombing in World War II and economic prostration for several years after the war. Population has continued to climb, although at a slowing rate—which still leaves Japan the seventh most populous nation of the world. Meanwhile affluence has grown to west European levels. The material requirements of so many people living on a high plane in a giant industrial economy have brought Japan to the point that it now imports the great bulk of its raw materials and fuels, plus a large share of its food as well.

The progressive inadequacy of Japanese natural resources is well illustrated by forests. The nation's heavily forested mountainsides could satisfy the wood needed for construction and fuel in a preindustrial and early industrial society with a population much smaller than it is now, but today's forests do not yield nearly enough wood for Japanese needs. However, forests are still a highly valuable resource.

The varied local environments of Japan support a wide range of valuable trees, including broadleaf evergreens in southern Japan, a mixture of broadleaf deciduous and conifers in central and northern Honshu and southern Hokkaido, and conifers in northern Hokkaido and at higher elevations farther south. About two thirds of Japan is still forested, which is a very high proportion for an important industrial nation. The mountainsides often are quilted with rectangular plots of different kinds of trees planted in rows. Most of Japan's first-growth timber is long since gone, but some virgin stands are still protected

in forest preserves. Although the charcoal formerly used for household cooking and heating has been supplanted by kerosene burned in inexpensive Japanese-made heaters, Japan still uses wood in endless ways. Single-family dwellings continue to be built largely of wood, which provides more safety from earthquake tremors than more rigid construction of other materials. Paper made from wood is used to cover the sliding partitions between rooms in traditional homes and apartments. Newer habitations, including apartments in larger earthquake- and fire-resistant buildings made of concrete and steel, may have a mixture of traditional rooms floored by mats, and Western-style rooms with durable floors and more furniture. Japan's large publishing industry requires a great deal of paper, and other industrial and handicraft uses of wood are so great that Japan has far outgrown its own timber supply and has become one of the world's largest importers of wood. Much of this comes in the form of sawlogs or pulpwood for processing in Japanese mills.

Minerals and Water

Early industrialization in Japan was supported by small deposits of many different minerals, notably coal, iron ore, sulfur, silver, zinc, copper, tungsten, and manganese. The supplies of all these minerals are grossly inadequate to Japanese needs today.

Japan's mountain streams are so swift, shallow, and rocky that they are of little use for navigation, but they provided much power for early industrialization, and hydropower still supplies 13 percent of the country's electricity. The bulk of the water used for irrigation in Japan comes from these streams, which emerge from the mountains, divide into distributaries, and then cross the plains in beds elevated above the general level of the crop land. The elevated beds make it possible to get much of the water to the land by gravity.

INTENSIVE FOOD PRODUCTION FROM SCANTY AGRICULTURAL LAND

Japan is not one of the world's major crop-growing nations, although it ranks no lower than eighth in output of rice, which is its basic food and most important crop. Helped by large government subsidies, the country's farmers actually achieve a net export of rice by intensive use of crop land amounting to only one eighth of the country's area. Such land is largely alluvium in scattered patches and small plains along the coasts and in the mountain basins. This level land, which grows most of Japan's crops, has been augmented by terracing of many slopes too steep for tillage in their natural state (Fig. 14.7). The flights of terraces often provide spectacular views in Japan's mountainous terrain, as they do in many other Oriental areas. A recent American visitor has described the agricultural scene as viewed on a trip to the mountains of Kyushu during the autumn rice harvest[1]:

> Early on a Sunday morning the main coastal road to the mountains was virtually empty as the Japanese, having to rise early six days a week to go to work or to school, enjoy sleeping late on Sunday. Even the farmers, usually to be found in the fields from sunrise to sunset during the harvest season, were late arriving. Upon entering the mountains, the road became one-laned and winding, usually following the steep course of a rapid stream. Wherever the stream valley broadened, wooden houses overlooked terraced levels of rice, vegetables, or tea. The rice was a golden yellow and on the steeper slopes and more inaccessible plots was being cut and gathered by hand, after which it was hung in sheaves on bamboo racks to dry [Fig. 18.4]. Levels reached by a ramp or narrow road were being harvested with mechanical cutters comparable in size to an American garden rototiller. The vegetables were largely cabbages and large Japanese radishes, with broad leaves that had grown to the point that the earth was no longer visible. Tea, planted in long rows closely resembling rounded boxwood hedges, was usually relegated to the most remote or steepest levels, as the leaves had to be picked by hand, and hence the plants could be placed on the land least suited to machines. As the road wound ever higher and the valley walls steepened, tea became the predominant crop.
>
> At last the final village before the climb to the pass came into view, its tiny fields clinging to the mountainsides on countless terraces. Some were on slopes so steep that only a single ''hedge'' of tea would fit before the next stone retaining wall began. As the route wound higher and higher, the views became more and more spectacular. Then they abruptly ended when the road, exceedingly narrow now, finally gave up its struggle to stay on the mountainside and burrowed into a tunnel.

Even on flat land, great numbers of Japan's fields are terraced in order to make possible the growing of irrigated rice. The country's mild winters and ample moisture permit double cropping—the growing of two crops a year on the same field—of most irrigated land from central Honshu south. Rice is grown in summer, and wheat, barley, or some other winter crop is then planted after the rice harvest. Overall, about one third of Japan's irrigated rice

Figure 18.4 *Hanging rice on racks to dry in the island of Kyushu, Japan.*

fields are sown to a second crop, and more than half of the unirrigated fields are double cropped. Intertillage—the growing of two or more crops simultaneously in alternate rows—is common (see Fig. 17.7).

Something like nine tenths of Japan's cultivated land grows food crops, of which rice is by far the most important. Irrigated rice is the basic crop nearly everywhere, including most parts of the northern island of Hokkaido. The intensive and widespread production of this grain is heavily subsidized by the Japanese government, which aims at rice self-sufficiency, with other foods being imported as needed. The country's latitudinal extent makes it possible for unirrigated fields to produce a great variety of crops: sugar beets in Hokkaido; all sorts of temperate fruits and garden vegetables; potatoes, peas, and beans, including soybeans; and in the south tea, citrus fruits, sugarcane, tobacco, peanuts, and mulberry trees to feed silkworms.

The surge in urban-industrial employment and general affluence in Japan during recent decades has been reflected strongly by changes in the agricultural sector. Rural labor has been siphoned off by the higher income available in towns, so that the number of agricultural workers has dropped dramatically (from about 25% of the national labor force in 1962 to 8% in 1987). Part-time farming has risen to the extent that only 13 percent of Japan's farming families were reported to be securing all their livelihood from their farms in a recent year. There has been a small decrease in total cultivated land due to the conversion of such land to nonfarm uses, but a huge

---

[1]Unpublished communication from H. Todd Stradford, Jr., used with his permission.

increase in farm mechanization has made it possible to till substantially the same amount of land with far less human labor. Millions of small tractors are now used to work Japan's tiny farms, together with small-scale power cultivators, sprayers, dusters, threshers, pickup trucks, and other mechanized equipment. Meanwhile some of the world's highest yields per unit of land continue to be maintained by modern agricultural technology, including heavy applications of fertilizer (chemical fertilizer and processed human wastes) to improved varieties of crops. The dominance of rice growing continues, upheld by high prices guaranteed by the government and supported by a powerful farm lobby. This raises food prices and creates large rice surpluses in a country where people are eating less rice and becoming more Westernized in their diet. Increased consumption of meat, milk, and eggs has been reflected in a growing number of cattle, pigs, and poultry on Japanese farms. Such production requires heavy imports of animal feeds, and the increasingly varied Japanese diet is requiring many kinds of imported human food. The United States is the largest supplier of imported feedstuffs and food.

Despite the new emphasis on varied animal products, sea fish remain the largest source of animal protein in the Japanese diet (Fig. 18.5). A variety of food fish are present in the waters surrounding Japan, and Japanese fishermen range widely through the world's major ocean fishing grounds. Japan and the Soviet Union currently vie for world leadership in total fish catch, both being far ahead of other nations.

JAPAN'S CORE AREA: THE KANTO-KYUSHU MEGALOPOLIS

Most of Japan's economic development and a large majority of its people are packed into a corridor about 700 miles (c. 1100 km) long from Tokyo-Yokohama and the surrounding Kanto Plain in the east to northern Kyushu in the west (Fig. 18.1). This highly urbanized core area—which contains more people than the Boston-to-Washington ''Megalopolis'' on the American east coast (see Chapter 28)—includes the southern coasts of Honshu and the northern parts of Shikoku and Kyushu. All of Japan's greatest cities, and many lesser ones, have developed here, on or near harbors along the Pacific or the Inland Sea. Typically they are sited on small, agriculturally productive alluvial plains between the mountains and the sea. The largest urban agglomeration in Japan and in the world, Tokyo-Yokohama and suburbs (30 million by one estimate; other figures differ according to how the metropolitan area is defined), has developed on the Kanto Plain, which is Japan's most extensive lowland. Tokyo grew originally as a political capital and Yokohama as the area's main seaport. This urban agglomeration functions as the national capital, one of the two main seaport areas (the other is Osaka-Kobe), the country's leading industrial center, and the commercial center for the Kanto Plain and all of Japan to the north. In the entire northern half of Japan there is only one major city, Sapporo (1.6 million), on Hokkaido. Around 200 miles (c. 320 km) west of Tokyo lies Japan's second largest urban agglomeration, in

Figure 18.5 Ocean fish are a major element in the Japanese diet. This photo shows an open-air fish market in Tokyo.

the Kinki District at the head of the Inland Sea. Here three cities—Osaka, Kobe, and Kyoto—together with their suburbs, form a metropolis of over 16 million people. The inland city of Kyoto was the country's capital from the 8th century A.D. until 1869 and is now preserved as a shrine city where modern industrial disfigurement is not permitted. Kyoto is the destination of millions of pilgrims and tourists annually. Osaka grew mainly as an industrial center, and Kobe was originally developed as the district's deepwater port, although each of the two cities now combines port and industrial functions. Still a third huge metropolis, Nagoya (5 million) is situated directly between Tokyo and Osaka, about 75 miles (120 km) east of the latter city. Located on the Nobi Plain at the head of a bay, Nagoya is both a major industrial center and a seaport.

West of the Kinki District smaller metropolitan cities are spotted through the coreland. The largest, all metropolises of over 1.5 million people, are Hiroshima on the Honshu side of the Inland Sea, Kitakyushu on the Strait of Shimonoseki between Honshu and Kyushu, and Fukuoka on Kyushu. The central city of the Hiroshima agglomeration was destroyed in 1945 when the United States, as a means of breaking Japanese resistance in World War II, dropped the first atomic bomb. To the westward on northern Kyushu are the Kitakyushu (a collective name for several cities) and Fukuoka agglomerations. They developed as the original centers of heavy industry in Japan, localized in true 19th-century fashion on or near the country's major coal deposits. The latter provided fuel for iron and steel mills, metal-using factories, and chemical factories. Kitakyushu and Fukuoka are still industrial centers, but the region's coal production has become very expensive and has declined. Industrial growth in the part of the coreland has lagged, at least by Japanese standards, in recent decades. Among several smaller cities of northern Kyushu is Nagasaki, which in 1945 became the second (and last) city to be atom bombed.

CYCLES OF EXPERIENCE IN JAPAN'S HISTORICAL BACKGROUND

Japan's eventful history has seen successive periods of racial and cultural immigration; nation building; withdrawal into seclusion; emergence as an industrial, commercial, and military power; military defeat and occupation; and reemergence as a sovereign state, shorn of colonies but with greatly expanded

industry and a thriving trade, and an overall position near the top of the world's economic powers. Some pertinent aspects of this history are summarized in the following sections.

EARLY JAPAN

The racial and ethnic origins of the Japanese are somewhat obscure. They are thought to be descended from a number of different peoples, primarily Mongoloids, who reached Japan from other parts of eastern Asia at various times in the distant past. An earlier people, the Ainu (probably non-Mongoloid), were driven into outlying areas where remnants still exist, principally in Hokkaido. Some of the Ainu may possibly have been assimilated by the invading peoples. The oldest surviving Japanese written records date from the 8th century A.D., but Japanese legendary traditions extend back to the reign of Jimmu, the first emperor, whose accession is ascribed to the year 660 B.C., although modern scholars have placed the date some centuries later. The acceptance of these traditions has been an important factor in Japanese psychology, since they ascribe a divine origin to the islands and the imperial family (Jimmu was supposedly descended from the Sun Goddess) and thus call attention to Japan as "the Land of the Gods."

The early emperors gradually extended control over all of their island realm. A society developed based largely on subsistence agriculture and organized into warring clans. By about the 12th century the emperors were being pushed into the background by powerful military leaders, or *shoguns*, who actually controlled the country while the emperors remained as figureheads. Meanwhile the provinces were ruled by nobles or *daimyo*, whose power rested on the military prowess of their retainers, the *samurai*. Thus evolved a governmental system resembling in some ways the European feudal system of medieval times.

EARLY CONTACTS WITH EUROPE

The adventurous and wide-ranging Europeans of the Age of Discovery reached Japan in the first part of the 16th century. In the 1540s came the Portuguese, and after them came representatives of other European nations. These early arrivals were mainly merchants and Roman Catholic missionaries. The merchants were allowed to set up trading establishments and open an active commerce, and the missionaries were allowed to preach freely. There were an estimated 300,000 Japanese Christians by the year 1600.

However, these early promising contacts were eventually nullified. Even while they were occurring, a series of strong military leaders were imposing internal unity and central authority on the disorderly feudal structure of Japanese society. By 1600 the Tokugawa family had acquired absolute power, which it retained until 1868. This period is generally known as the period of the Tokugawa Shogunate. In order to maintain power and stability the early Tokugawa shoguns desired to eliminate all disturbing social influences; the latter included the foreign traders and missionaries. In addition, the shoguns seem to have feared that the missionaries were the forerunners of an attempted conquest by Europeans, especially the Spanish, who held the Philippines. Consequently, the traders were driven out, and Christianity was almost completely eliminated in bloody persecutions during the early part of the 17th century. After 1641 a few Dutch traders were the only Occidentals allowed in Japan, and even they were segregated on the small island of Deshima in the harbor of Nagasaki. Japan settled under the Tokugawa into two centuries of isolation, peace, and stagnation.

WESTERNIZATION AND EXPANSION

When the foreigners again made a serious attempt to open Japan to trade two centuries later, the growing strength of the outside nations could not be thwarted by a Japan which had fallen far behind in the arts of both war and peace. This time the United States took the lead, and visits in 1853 and 1854 by American naval squadrons under Commodore Matthew Perry resulted in treaties opening Japan to trade with the United States. The major European powers were soon able to obtain similar privileges. Some of the great feudal authorities in southwestern Japan were strongly opposed to the new policy, but their opposition was quelled when coastal areas under their control were bombarded by American, British, French, and Dutch ships in 1863 and 1864.

These events so weakened the faltering power and prestige of the Tokugawa Shogunate that in 1868 the ruling shogun was overthrown by a rebellion. The revolutionary leaders restored the legitimate sovereignty of the emperor, who took the name Meiji or "Enlightened Rule." Thus the revolution of 1868 is generally termed the *Meiji Restoration*.

The men who came to power in 1868 were true revolutionaries in that they aimed at a complete transformation of Japan's society and economy. They saw that if Japan were not to fall under the control of the Western nations, its demonstrated military impotence would have to be remedied. They saw also that this would require a reconstruction of the Japanese economy and of many aspects of the social order. These tasks were approached with energy and intelligence. Feudalism was abolished, although not without the necessity of suppressing a bloody revolt in 1877. Thus was cemented the power of a strong central government which could be used as a vehicle for remodeling the country. Such a government was an institution that no other Oriental people possessed as an instrument of modernization and resistance to foreign encroachment. Under this government, which lasted until 1945, democracy was instituted but was strictly limited, so that Japan was generally ruled by relatively small groups of powerful men manipulating the machinery of government and the prestige of the emperor. Military leaders were generally very prominent in this power structure.

The new government set about the task of having its subjects learn and apply the knowledge and techniques which the Occidental countries had been accumulating during the centuries of Japan's isolation. Foreign scholars were brought to Japan, and Japanese students were sent abroad in large numbers. A constitution, modeled largely after that of imperial Germany, was promulgated and the legal system was remodeled to bring it into greater conformance with Occidental practices. On the economic side the government used its financial power, largely derived from oppressive land taxes, to foster industry in a variety of ways. Railroads, telegraph lines, and a modern merchant marine were constructed, and banks and other financial institutions were developed. Light industry was stimulated as a means of providing exports, and basic heavy industries were gradually developed. Wherever private interests could not achieve the desired economic development unaided, the government provided subsidies to private companies, or else built plants and operated them until they could be acquired by private concerns.

So spectacular were the results achieved that in 40 years Japan had become the first Oriental nation in modern times to attain the status of a world power. The Japanese were often spoken of in a derogatory fashion as mere imitators, but this seems unjust. It is obvious that rapid and excellent imitation was called for in the situation that confronted

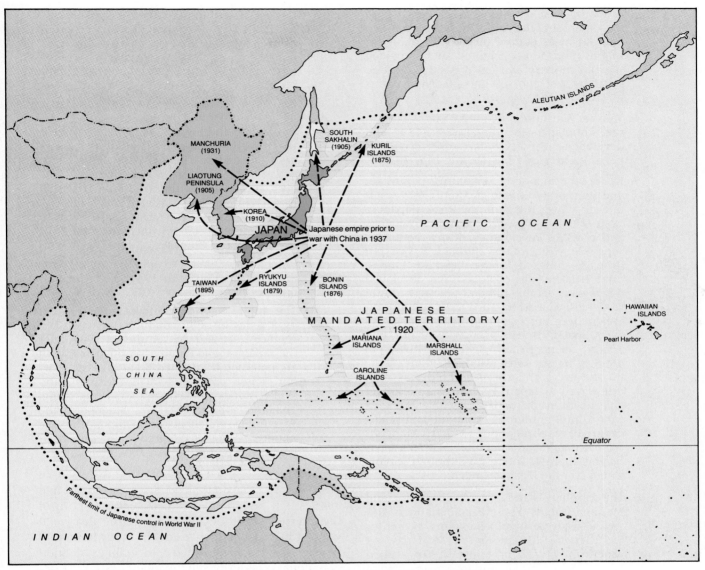

Figure 18.6 *Map showing overseas areas held by Japan prior to 1937 and the line of maximum Japanese advance in World War II.*

Japan, and the scope and success of that imitation represent a major accomplishment achieved in the face of great difficulties imposed by cultural inertia and the country's poverty in natural resources.

JAPAN'S PREWAR EMPIRE

In the political sphere, after Japan abandoned the stay-at-home policy of the Tokugawa period, it gradually emerged as an imperial power. Between the early 1870s and World War II the country pursued, with only brief interruptions, a consistently expansionist policy, and by 1941 Japan controlled one

of the world's most imposing empires (Fig. 18.6). The major components of the Japanese Empire, in the order of their acquisition, were as follows:

1. The Kuril Islands. Acquired from Russia in exchange for Japanese abandonment of claims on Sakhalin, 1875.

2. The Bonin Islands. Annexed without opposition, 1876.

3. The Ryukyu Islands. These had had a semi-independent status under the overlordship of both Japan and China. They were occupied, despite local and Chinese protests, in 1879, and Japan was confirmed in ownership by

China in 1895. The Ryukyus, together with the Bonins and the Kurils, were incorporated into Japan as part of the homeland. These islands, and others acquired later, provided sites for military and naval bases protecting Japan proper and serving as staging points for military expansion.

4. Taiwan (Formosa). Acquired from China in the Sino-Japanese War of 1894–1895 and developed to serve Japanese economic needs, particularly for rice and cane sugar.

5. Korea, the Liaodong (Liaotung) peninsula, and southern Sakhalin. Acquired as a result of victory over Russia in the Russo-Japanese War, 1904–1905. With Russian influence and pressure checkmated, Korea became a Japanese protectorate, although it was not formally annexed to the Japanese Empire until 1910. It was exploited and developed for its considerable mineral and agricultural resources. China remained the nominal sovereign over the Liaodong peninsula. However, actual control over Port Arthur and nearby territory had passed to Russia in 1898, when she forced China to grant a lease on the area, and this leasehold was now transferred to Japan. With Russia no longer able to intercede, Japan was able to extort from China numerous economic and political privileges throughout southern Manchuria, placing that area definitely within a Japanese "sphere of interest," although leaving China still the legal sovereign.

6. The Caroline Islands, Marshall Islands, and Mariana Islands. These Pacific island groups were former German possessions, except for the island of Guam in the Marianas, an American possession since 1898. When World War I broke out in Europe, Japan joined the Allied side and seized the German holdings in the Orient and the Pacific north of the equator. The peace settlements after the war confirmed Japanese occupancy of the islands under a mandate of the League of Nations. While the other major powers were embroiled in Europe, Japan was able, in addition, to exert pressure successfully on China for increased economic privileges, especially in Manchuria, Inner Mongolia, and northern China proper. Although Japan had occupied former German-controlled territory in China on the Shandong (Shantung) peninsula, she was induced by Chinese and international pressure to evacuate that area in 1922.

7. Manchuria. Conquered in the course of hostilities with China lasting from 1931 to 1933. Manchuria became the Japanese protecto-

rate of "Manchukuo," and the Japanese intensified their development of its economy as an adjunct to that of Japan itself. In addition, further Japanese privileges were extorted in Inner Mongolia and northern China proper.

8. General war with China in 1937 saw Japan by 1938 in control of most of North China, the middle and lower Chang Jiang (Yangtze River) Valley, and pockets along the southeast China coast. The war with China which Japan precipitated in 1937 continued, however, until 1945 without additional significant Japanese gains. Instead, Japan found it difficult to control the territory already conquered due to steadily increasing guerrilla warfare by the Chinese.

9. French Indochina. Occupied without resistance during 1940 and 1941, after the conquest of France by Germany.

10. The empire was brought to its greatest extent by a series of rapid conquests between December 1941 and the summer of 1942, after the Japanese had entered World War II with an attack on the United States at Pearl Harbor. These conquests included Guam and Wake islands, the Philippines, Hong Kong, Thailand, Burma, Malaya, Singapore, the East Indies, much of New Guinea, the Admiralty, Bismarck, Solomon, and Gilbert Islands, and part of the Aleutian chain.

After 1942, Japan was generally in retreat, and 1945 found her completely defeated and shorn of the overseas territories acquired during nearly 70 years of successful imperialism. The motives of Japanese expansionism were mixed, including a tradition of national superiority and "manifest destiny," a desire for security and great-power recognition, the desire of military leaders to aggrandize themselves and gain control of the Japanese government, and a desire on the part of various elements in Japan to gain assured markets for Japanese goods and assured sources of materials. Although the methods of gaining them were often illegitimate, the desire for materials and markets was solidly based on need. Expansion of industry and population on an inadequate base of domestic natural resources had made Japan dependent on sales of industrial products outside the homeland. Such sales provided, as they do now, the principal funds with which the country purchased the imported foods, fuels, and materials that it required for its large and growing population and its expanding industries. The trend toward militaristic solutions in prewar Japan was abetted by (1) the world depression that threatened the Japanese economy so perilously dependent on world trade, and (2) the growing political isolation of Japan caused by her continued expansion on the Asian mainland. These factors aided those who espoused the need for creation of an Asian realm controlled by Japan that, with the Western powers excluded, would insulate the Japanese economy from the vicissitudes of the world economy and, concurrently, would elevate the nation to the rank of a leading world power. In 1938 Japan proclaimed its "New Order in East Asia" and in 1940–1941 this widened into the "Greater East Asia Co-Prosperity Sphere"— a euphemism for Japanese political and economic control over China and the rich tropical lands of Southeast Asia. Since World War II, Japan, without benefit of colonies or spheres of influence, has become one of the world's leading industrial nations (chapter-opening photo) and has achieved an impressively high level of living. But the general circumstances of the world economy in the postwar era have been far more favorable for Japanese industrial and commercial expansion than they were in the 1920s and 1930s.

JAPAN'S PHENOMENAL DEVELOPMENT SINCE 1945: CHARACTERISTICS AND REASONS[2]

Japan's explosive economic growth after its defeat in World War II has been one of the most startling and significant developments of the later 20th century. The war showed that although the Japan of that time had developed industrially to a point where it could equip formidable armed forces on land, at sea, and in the air, it could still be overwhelmed by superior American productive capacity. At the end of the war, the country was an occupied and devastated shambles. Besides 1.8 million deaths in the armed services, it had suffered some 8 million casualties from American bombing of the home islands, and most of its major cities were over half destroyed. Tokyo, for instance, had fallen in population from nearly 7 million people to about 3 million, most of whom were living in shacks. In 1950, 5 years after the beginning of the occupation, national production still stood at only about one third of its 1931 level. The annual per capita income was $17 in 1946 and $32 in 1950. A study in this period showed the average urban worker to be subsisting on 1600 calories per day.

But by the 1970s Japan had become an economic superpower with a GNP far exceeding that of any country except the United States and the USSR. By 1986 her per capita GNP was higher than that of the great majority of European countries and her total production was nearly double that of Europe's leading economic power, West Germany. Although still substantially behind the United States in per capita GNP, the Japanese of the 1980s had already surpassed the United States in some important nonmaterial indices of standard of living. For example, infant mortality was 5 in Japan and 10 in the United States, while life expectancy at birth was 77 in Japan and 75 in the United States (data: Population Reference Bureau, 1988 *World Population Data Sheet*). Both America and Europe found themselves at a strong disadvantage against Japanese competitors in many industries, and both were busily trying to find ways to meet this competition or protect themselves from it without doing too much damage to their exporters (firms and workers), their consumers, and their overall trade relations.

THE AMERICAN OCCUPATION AND THE NEW POLITICAL FRAMEWORK

A great deal of the political framework within which Japan's postwar surge took place resulted from Japan's relations with the United States. The American Occupation governed Japan from 1945 until 1952 and made major changes there. The political system was democratized in a number of ways:

1. The divine status of the emperor was officially abolished.

2. Shinto, the strongly nationalistic state religion, was disestablished, although worship at Shinto shrines was allowed to continue (Fig. 18.7).

3. There was extensive land reform in the countryside to break up a near-feudal landlordism.

4. A new American-written constitution made Japan a constitutional monarchy with elective parliamentary government.

5. Some major Japanese companies were broken up as a means of lessening their monopolistic hold on the economy.

6. Women were enfranchised.

7. A democratic trade union movement was established, from which Communist leadership was subsequently expelled.

8. The formerly dominant military officer corps was purged.

9. Japan was forbidden to ever rearm, except for small "self-defense" forces.

---

[2]The remaining discussion of Japan draws in part on E. W. Spencer, "Japan: Stimulus or Scapegoat?" *Foreign Affairs*, 62, No. 1 (Fall 1983), pp. 123–137.

Figure 18.7 *The prominent red arch in this photo is called a* torii *and is the portal for a Shinto shrine in the Tokyo area. The shrine itself is out of sight amid the trees.*

When the Occupation ended in 1952, leaving American bases in Japan but returning control of Japanese affairs to the Japanese, Japan was placed under American military protection. One part of the country, the Ryukyu Islands, remained under American occupation, for military purposes, until 1972.

SOME REASONS FOR JAPAN'S ECONOMIC SUCCESS

In attempting to explain Japan's economic success, various types of experts emphasize various factors. One line of thought focuses on the American connection. Not only have many of the American-imposed reforms worked well (although often somewhat modified by the Japanese), but the relations between the two countries have continued to have beneficial aspects for Japan. Rapid recovery from the postwar depths began when Japanese production was called upon to support the United States forces in the Korean War of 1950–1953. By 1954 Japanese steel production was the largest in the country's history, and the textile and clothing industries supplied American military needs and then expanded exports

so rapidly that United States industries were asking for protection by 1956. Military procurement for American forces continued to supply a major market promoting Japanese recovery throughout the 1950s, and various types of American economic aid also were given. Many Japanese products were given relatively free access to the United States market for long periods, while Japan was allowed to continue the "protection" of its economy from imports of many American and other foreign products. This imbalance did not really begin to erode rapidly until the early 1980s. Japanese firms were allowed relatively free access to American technology, which they frequently improved on and then introduced the improvements into production ahead of American firms. And due to American military protection Japan was able to minimize its investment in defense and put more of its funds and efforts into economic growth. This became ironic as the Americans first forbade Japan to rearm and then, in the 1970s and 1980s, began to urge the Japanese to raise military expenditures (which had been kept below 1% of the GNP).

Another major line of thought on Japan's economic rise stresses the cheapness and quality of Japan's work force. During the 1950s and 1960s, when the Japanese economy had incredible growth rates averaging 10 percent a year, Japanese labor was very cheap compared to that of the United States and major European industrial countries. Not only were wages low, but the quality and productiveness of the work force were high. Japanese workers were relatively well educated, loyal to their companies, conscientious, meticulous, and hard-working. However, during the 1970s and early 1980s this advantage declined. Strikes by millions of Japanese workers won large wage increases. By the mid-1980s some analysts were pointing out that differential labor costs were no longer crucial in much of the competition between Japanese and Western firms.

At a higher level of the work force, certain peculiarities of Japanese management and technical personnel are sometimes singled out for emphasis. One is recruitment through a somewhat brutal educational system strongly emphasizing technical training. Admission to higher education is controlled by a testing system imposing such pressures that the suicide rate among failures is high. Within the universities science and technology have a relatively strong emphasis. For instance, Japan does not emphasize specific business training, and while the United States has about 20 times as many lawyers as Japan, the latter trains more electrical engineers. Executives and supervisory personnel are more likely to be trained in science and technology in Japan. Also cited as a Japanese advantage is the common style of management, featuring benevolence, fostering of group loyalty, and participation of workers in decision-making. Large Japanese companies are likely to be active in the provision of housing and recreational facilities for their employees. Compared to Western companies, they are said to pay more attention to employee suggestions and to stay in touch better with employees and give them more information about important matters affecting the company. A style of decision-making which involves more people "down the line," with attempts to get widespread agreement to decisions before implementing them, has often been noted. Some Japanese companies operating plants in the United States have apparently achieved remarkable results in using such methods with American workers. These management approaches may well be reinforced by the psychological returns from a spread between the pay of workers and management that tends to be much less wide in Japan than in the United States. Another characteristic of Japanese management often cited is a tendency toward higher research and development budgets and longer-range planning than is general

in American companies. This is said to be possible because Japanese companies can operate at lower rates of profit, due to management being less subject to threats from disgruntled stockholders or acquisitive outside interests when profits, dividends, and stock prices are low.

One very obvious and essential factor in Japan's postwar economic rise was a high level of investment in new and generally very efficient industrial plants. This was facilitated by some of the conditions already mentioned, such as a very low level of military expenditure and the ability of companies to operate at low profit margins. It has also been facilitated by government policies slighting expenditures on "social overhead capital" such as roads, antipollution measures, parks, housing, and, surprisingly, higher education. Thus Japanese industrial cities have grown explosively and have also become very highly polluted areas of dense and inadequate housing, with few public amenities (except for having very low crime rates) and snarled transportation. The money "saved" has been available for more direct investment in industrial growth.

Also crucial to the high level of investment has been a remarkably high rate of savings by the Japanese people. For instance, in a recent year the Japanese people saved 19 percent of their available income, compared to 8 percent saved by Americans. These savings increase the funds available for industrial investment. Such a high rate of savings is encouraged by several peculiarities of the Japanese system:

1. Very commonly workers receive only about two thirds of their earnings each year as regular wages or salary; the other third is in the form of one or two lump-sum payments during the year. Polls have indicated that Japanese workers prefer this because it automatically saves one third of their income, at least for a time.

2. Buying on time is relatively less common in Japan than in the United States and is frequently impossible. Essentially there is less lending to consumers, with more to industries, so that the Japanese consumer must save the money for a large-ticket item before a purchase is made.

3. Workers must save to take care of their children and themselves because Japanese higher education is expensive and much in demand and because the country's welfare system, pensions, and social security are quite inadequate. Most universities are private and depend on high tuition payments from parents' earnings and savings, although the best uni-

versities (*e.g.,* Tokyo and Kyoto) are public. Mandatory retirement, the end of the "lifetime" employment Japanese companies offer, is generally at 55, while social security payments do not begin until age 60. Thus people can often expect to have 5 years in which they must exist largely on savings. (Often they also go to work at lower wages for one of the country's multitude of small service and industrial enterprises.) Even with social security, retirement income is often quite inadequate without savings to supplement it. The pressure on older people is shown by the fact that a very high 45 percent of Japanese men over age 65 were still working in some capacity or looking for jobs in a recent year. But the savings this system makes so necessary—massive in the aggregate—are largely available through banks for investment in companies.

Elements of Japan's political culture and overall culture are also cited as economic advantages by some analysts. One political party, the Liberal Democratic Party, has been repeatedly reelected to power ever since Japan regained its sovereignty. It is conservative and strongly business-oriented and business-connected. It has promoted Japanese exports by financial policies tending to keep the yen (the Japanese unit of currency) and thus Japanese goods, cheap; and it has cooperated very closely with Japanese business in the development of new products and whole new industries. In the area of more general culture some analysts believe there is simply a relatively intense spirit of achievement and enterprise among the Japanese. For instance, although a number of the major Japanese companies are relatively old—Mitsubishi, for example, built the "Zero" fighter planes of World War II—some have originated and grown to prominence in the last few decades. Sony and Honda are prominent examples. Certainly there are an enormous number of small industrial firms, each attesting to a spirit of enterprise in one or more people. One small city north of Tokyo has hundreds of small textile "factories."

A TELESCOPED INDUSTRIAL PAST AND AN UNCERTAIN FUTURE

Japan's industrial evolution, proceeding under forced draft since 1868, has been finely adjusted at each stage to the country's limited and changing resources—both natural and human. The more gradual evolution that characterized Europe and America has been telescoped into little more than a century

in Japan. Development prior to World War II stressed textiles and coal-based heavy industry. Preindustrial Japan produced silks. This industry was expanded to export scale, and sales of silk abroad then provided some of the investment capital to establish other industries. A cotton-textile industry was established, which first supplemented and then overshadowed the silk industry. The competitiveness of both industries depended heavily on cheap Japanese labor. Before the Second World War Japan was already the world's leading exporter of cottons. Early iron and steel development was concentrated in northeast Kyushu, on and near the country's most significant coal resources. The product went largely into building a railway system and both commercial and naval shipping. Coal powered the country's factories, railways, and ships. Hydroelectricity was developed on many mountain streams and became an important element in the energy economy. Wood, also relatively abundant in this land of forested mountains, was used for building and home heating.

The phenomenal industrial rise since the Korean War period of the early 1950s has been marked by a series of overlapping booms, along with some declines, in various industries.

1. Cotton textiles and clothing, still based on cheap labor, led the way in the early 1950s. By 1956 the United States was already forcing "voluntary restraints" on Japan's exports of these products to the American market. Today Japanese labor is no longer cheap and these industries are declining in Japan, although they may be preserved there (and in other high-wage countries) by new robotized factories the Japanese are pioneering.

2. In the later 1950s boom periods in iron and steel, shipbuilding, and chemicals (especially fertilizers) set in. Japan's world lead in shipbuilding (40% of all new tonnage ordered in 1986) was built with a labor force which was not only comparatively cheap but also relatively well educated and skilled. And it involved Japanese pioneering in the design of increasingly larger ships. Thus the quality, and not just the cheapness, of Japanese labor became a major resource. Despite this fact, by the 1980s the Japanese shipbuilding industry was suffering from the competition of countries with cheaper, although less skilled, labor. The rise of Japan's iron and steel industry to a level of production in the 1980s where it has decisively surpassed American output for second place in the world (after the Soviet Union) has been based mainly on technological skill

Figure 18.8 A continuing search for expertise from the Western world has been a major theme in Japanese life for over a century. Here it is manifest in a recent photo of Japanese graduate students in science and engineering at the Massachusetts Institute of Technology (M.I.T.) in Cambridge, Massachusetts.

and the economies of operating at a very large scale. Kyushu's small coal resources have become quite inadequate to support this huge development, and coal mining is declining. Some of the world's largest steel plants, using the latest technology, have been built at coastal locations in the Japanese core area. They import their coal—much of it from the United States—and their iron ore, especially from Australia. Relatively low transport costs for these bulky materials are made possible by the oversized ocean carriers that Japan builds. The Japanese iron and steel industry is no longer expanding rapidly and may well suffer soon from new plants in countries with cheaper labor.

3. In the 1960s electronics, cameras and optical equipment, petrochemicals, synthetic fibers, and automobiles (chapter-opening photo) became boom industries. The cheapness of skilled labor still played a part, but increasingly these Japanese industries depend on the quality of labor and management and on technological innovation. The oil-based industries—petrochemicals and part of the synthetic fiber industry—reflect Japan's development into a country requiring gigantic oil imports, mainly from the Persian/Arabian Gulf. In the mid-1980s Japan, which is almost devoid of oil deposits, obtained one twentieth of its energy from hydropower, one tenth from nuclear power, one fifth from thermal electricity generated by coal, and the remaining two thirds from imported oil and gas.

4. In the 1970s and 1980s, Japan was becoming involved ever more deeply with the manufacture of computers and robots. In the computer industry, Japanese companies, supported by their government, were engaged in attempts to surpass the world lead of American companies. In robotization of industry, Japan already had a clear lead over all other countries by the early 1980s, although the making of robots was (and is) still in a very early stage of development. Thus an evolution from relatively simple industries based on domestic natural resources and cheap labor to highly sophisticated "futuristic" industries based on education and skills continues to run its course in Japan, even more vividly perhaps than in other industrialized countries (Fig. 18.8). Since Japan must import such a large part of its materials and food, it must literally continue to live by its wits. Edwin O. Reischauer, a noted authority on Japan, has written a fitting epitome of this versatile island nation[3]:

Japan's industrial progress . . . has been no steady, inevitable growth, based solidly on obvious economic advantages, as in the case of America's rise to industrial leadership. It has been more like the erratic progress of a broken-field runner, fighting his way

---

[3]Reprinted by permission of the publishers from Edwin O. Reischauer, *The United States and Japan.* (Cambridge, Mass.: Harvard University Press, Copyright 1950, 1957, 1965 by the President and Fellows of Harvard College), pp. 80–81.

against great odds by a quick getaway, brilliant improvisations, and daring reversals of the field. . . .With her industry-swollen population there can be no turning back. She cannot even stand still, for her less industrialized neighbors are always threatening to catch up with her and wipe out the technical advantages on which her people now live. . . .

But whatever may be the ultimate fate of Japanese industry, it has already laid a heavy imprint on Japan and the Japanese. It has made Japan into a land of huge cities and sprawling factories, of whirring machines and crowded commuters' trains.

DIVIDED KOREA: LAND BRIDGE AND GEOPOLITICAL FOCUS

Japan's closest approach to the Asian mainland is in the extreme west, where the peninsula of Korea lies only about 110 miles (177 km) away (Fig. 18.1). Korea's history during the 20th century, first as a colony of Japan and then as a divided land comprised of the separate states of North and South Korea, has tended to obscure its distinctive culture and contributions to the world. Although the Koreans have been strongly influenced by Chinese culture, and to a lesser extent by Japanese, they are racially and linguistically a separate people. Korea is grouped with Japan in this chapter for convenience and for certain comparisons between the two.

LIABILITIES OF KOREA'S GEOGRAPHIC POSITION

Korea is a country with an unfortunate geographic position. This relatively small country is surrounded by larger and more powerful neighbors who have frequently been at odds with one another and with the Koreans. Korea adjoins China along a land frontier which follows the Yalu and Tumen rivers. It faces Japan across the Korea Strait. In the extreme northeast Korea borders the USSR for a short distance.

For many centuries the Korean peninsula has served as a bridge between Japan and the Asian mainland. From an early time both China and Japan, the latter more intermittently, have been interested in controlling this bridge, and since Korea has generally been weaker than either, its history has usually been that of a subject or vassal state. However, from the late 7th century to the 20th century, Korea was a unified state, sometimes invaded, ravaged, and forced to pay tribute, but never destroyed

as a political entity. As noted previously, Chinese influence has been the most pronounced and continuous throughout Korean history, with Japanese contacts limited mainly to occasional pirate raids and the establishment of a few Japanese enclaves in the southeastern port cities. The decline of Chinese power in the 19th century and the concurrent rise of modern Japan greatly increased the influence of the latter, and at the same time Russia began to make itself strongly felt in eastern Asia. Korea became an object of contention among these powers. Japan emerged victorious in the struggles that followed, ousting the Chinese from Korea, as well as from certain other areas, in the Sino-Japanese War of 1894–1895, and ousting Russia from Korea and southern Manchuria in the Russo-Japanese War of 1904–1905. From 1905 until 1945 Korea was firmly under Japanese control, being formally annexed to the Japanese Empire in 1910.

Japan lost Korea, along with the rest of its empire, when it surrendered to the Allies, ending World War II, in 1945. In accordance with prearranged agreements, Russian forces occupied Korea north of the 38th parallel and American forces south of that line. Although it had been understood that Korea was to become a unified and independent country, no agreement could be reached between the occupying powers as to the establishment of a Korean government. Accordingly, separate governments were set up in north and south Korea under the aegis of the respective occupying powers: in the south, the Republic of South Korea, under the auspices of the United Nations; and, in the north, the Democratic People's Republic of Korea, which became a Communist satellite. The occupying powers withdrew the bulk of their forces in 1948 and 1949.

In 1950 North Korea attacked South Korea in force with great success, and United Nations units, mainly American, entered the peninsula to repel the aggression. In the latter part of that year, when the North Koreans had been driven back from the southern part of the peninsula almost to the Manchurian border, China entered the war and drove the United Nations forces south of the 38th parallel. Then the Chinese and North Koreans were in turn driven back slightly north of the parallel, where a stalemate developed until an armistice was arranged in the summer of 1953.

Few lands have ever been more devastated than Korea after several years of warfare covering the length and breadth of the peninsula. The tragedy was all the greater in that Korea is not a poor land by nature. During their period of control the Japanese developed transportation, agriculture, and industry, the latter from a sizable base of mineral and power resources. The industrial structure was to some extent integrated with that of southern Manchuria. The Korean people received few benefits from these developments, however, since the increased production was put mainly to Japanese uses. After 1945 the Korean economy was seriously disorganized and handicapped by the division between north and south and after 1950 by the enormous physical destruction. The physical scars of the Korean conflict, however, have been erased, and the economies of both the North and South Korean states rehabilitated and expanded with the help of considerable outside aid.

CONTRASTS BETWEEN THE TWO KOREAS

The Korean armistice line, which is still heavily garrisoned and subject to occasional violent incidents, divides one people into two very different countries (see Table 14.1). North Korea has an area of about 46,500 square miles (120,500 sq. km), inhabited by 22 million people in 1988. These figures indicate a dense population, averaging about 470 per square mile (c. 180 per sq. km). But in 1988 South Korea had 43 million people in only 38,000 square miles, thus averaging a little over 1100 per square mile (c. 435 per sq. km)—almost 2.5 times the density of the North. South Korea is a republic which has tended toward repressive military dictatorships and has a capitalist economy heavily dependent on relationships with the United States and Japan. North Korea is a rigid and very tightly controlled Communist state, governed by the same dictator since World War II and with a history of vacillation between ties with China and ties with the USSR.

There are important physical contrasts between the two countries. Although both are predominantly mountainous or hilly, North Korea is the more so. Relatively level lowland is found mainly along the western side of the Korean peninsula in both countries, but South Korea also has some fairly extensive areas of this type in the southeast. In both countries the mountains rise toward the eastern side of the peninsula and drop abruptly into the Sea of Japan with little or no coastal plain at their base. But they are highest and most rugged in northeastern North Korea adjoining Manchuria (Fig. 18.1).

North Korea is humid continental long-summer in climate, with hot summers but quite cold winters in lowland areas, while South Korea is mostly humid subtropical, with much shorter and milder winters. Both are strongly monsoonal, with precipitation concentrated in the summer. Given the Oriental preference for rice, and the need of dense populations for the high caloric output per acre supplied by irrigated rice, these conditions favor South Korea. Even in the North irrigated rice is the main crop, but it is increasingly less important northward. In the South the rice emphasis is more widely possible and, unlike the North, South Korea is able to double crop the irrigated rice fields by growing a dry-field winter crop such as barley after the rice has been harvested. In the North the main supplementary crop is corn, which must be grown in the summer and hence cannot occupy the same fields as rice. Of course these natural advantages of the South are very considerably offset by the lower population density of the North. Thus far both countries have been relatively successful in agriculture and neither has a problem of starvation. However, the population growth rates of 1988, if continued, would double the population of North Korea by the year 2016, and that of South Korea by 2040, possibly threatening the present relatively favorable situation.

Another physical contrast between the two Koreas is that most of the nonagricultural natural resources are in North Korea. Coal, iron ore, some lesser metallic ores, considerable hydropower potential, and forests are the peninsula's major resources. All are more abundant in the North than in the South. Accordingly, North Korea was originally the more industrialized state, featuring a typical Communist emphasis on mining and heavy industry, together with hydroelectric production and forest manufactures. However, the North has remained rather mired in this stage of development, while the South has taken over industrial leadership in the 1970s and 1980s by very rapidly developing a dynamic capitalist industrial economy incorporating a diversity of factories. In the middle 1980s South Korea's main export industries featured electrical equipment, passenger automobiles, clothing, textiles, shoes, iron and steel (from a huge plant on the southeast coast north of Pusan), and ships from a very major shipbuilding industry. Large investments from Japan and the United States, as well as markets in those countries, were fundamental to this explosive development. Also important were the skills of South Koreans receiving higher education abroad (and increasingly at home), cheap and increasingly skilled Korean labor, and vigorous backing of industrialization by strong and sometimes ruthless governments. Although still a relatively poor country, with a per capita GNP slightly above that of Panama in 1986, South Korea was growing rapidly in output and income during the mid-1980s, and com-

Figure 18.9 A shopping street in the dynamic metropolis of Seoul, South Korea.

petition from its products was causing concern to competing American and Japanese industries. Its burgeoning cities were very visible evidence of this growth, as the capital city of Seoul (Fig. 18.9) reached an estimated metropolitan population of over 11 million, the port of Pusan over 3.5 million, and the inland industrial center of Taegu over 2 million. By contrast, the only large city in North Korea was the capital of Pyongyang, with perhaps 2 million people in its metropolitan area.

THE PACIFIC WORLD

CHAPTER 19

Geographic Diversity in the Pacific Islands

High-rise buildings in Honolulu on the volcanic "high" island of Oahu, Hawaii, show a major aspect of the Pacific world. Although most land in this world region is deeply rural, a large majority of the region's people live in widely scattered metropolitan areas sustained by commerce, governmental activity, tourism, and limited amounts of manufacturing.

T HE Pacific world is mostly water. The Pacific itself, the largest of the oceans, is bigger than all the continents and islands of the world put together. But it has been a relative backwater of world commerce. Major shipping lanes reach mainly around the margins, and the products afforded by the multitudinous islands and the sparsely populated island continent of Australia do not bulk very large in world terms.

Prior to World War II the Western world had built up a legend about the Pacific and its islands as a kind of utopia. At an early time some of the islands did have an idyllic and appealing quality for visitors from Europe and America. However, on many islands the native peoples were reduced in numbers and their cultures were disrupted by a long period of unrestrained exploitation on the part of traders, whaling crews, labor agents ("blackbirders"), and other opportunists before the outside governments claiming jurisdiction began to impose tardy measures of regulation. On many islands, also, whatever idyllic quality remained was shattered by the military actions of World War II.

As here conceived, the term *Pacific world* is generally restricted to Australia, New Zealand, and the islands of the mid-Pacific lying mostly between the Tropics. The Pacific islands bordering the mainland of the Orient, the Soviet Union, and the Americas are excluded on the basis of their close ties with the adjoining continents, while large areas of the eastern and northern Pacific which contain few islands and are seldom visited are largely discounted. Because Australia and New Zealand are sufficiently different from the tropical island realms to the north, they could be considered a separate world region, but are included here in the Pacific world on the basis of their strong political and economic interest in the tropical islands, the ethnic affiliations of their original inhabitants with the peoples of those islands, and their insular character.

MELANESIA, MICRONESIA, POLYNESIA

The Pacific islands are commonly divided, mainly on an ethnic basis, into three principal regions: Melanesia, Micronesia, and Polynesia (Fig. 19.1). The islands of *Melanesia* (Greek: "black islands"), bordering Australia, are relatively large. New Guinea, the largest, is about 1500 miles (*c.* 2400 km) long and 400 miles (*c.* 650 km) across at the broadest point.

In general, these islands are hot, damp, mountainous, and overgrown with dense vegetation. *Micronesia* (Greek: "tiny islands") includes thousands of small and scattered islands in the central and western Pacific, mostly north of the equator. Prior to World War II, Micronesia was held by Japan under a mandate from the League of Nations, except for the Gilbert Islands, which were a British possession, and Guam, a United States possession. Since the war, the former Japanese islands have become the Trust Territory of the Pacific Islands, administered by the United States. *Polynesia* (Greek: "many islands") occupies a greater expanse of ocean than does either Melanesia or Micronesia. It is shaped like a rough triangle, with the corners at New Zealand, the Hawaiian Islands, and remote Easter Island.

Each of the three island regions contains a number of distinct island groups. The major groups, together with some noteworthy individual islands and the nations exercising political control, are indicated in Table 19.1.

THE NATIVE PEOPLES

It is commonly accepted among anthropologists that the original inhabitants reached the Pacific world from the mainland of Asia, the first migrants arriving in Melanesia and Australia thousands of years ago. Several successive physical stocks, including Australoid, Mongoloid, and Negroid, apparently settled in the islands as prehistoric migrants. Micronesia and Polynesia received a population of mixed Australoid and Mongoloid stock. Over a long period the Micronesians and the Polynesians became rather distinct from each other culturally and relatively homogeneous throughout their separate regions. The people of each region came to speak dialects of a single basic language and to share many features of a common culture. However, in the third region, Melanesia, physical types and cultural patterns are more complex. Here the Negroid physical stock is generally predominant but is often diluted by other elements. A variety of languages and dialects are found, and anthropologists have distinguished literally hundreds of fairly distinct cultures.

TYPES OF ISLANDS

The almost countless islands scattered across the tropical Pacific vary widely in size, type, and utility. A distinction is commonly made between "high" and "low" islands. Most of the "high" islands are

TABLE 19.1   SUMMARY OF MAIN PACIFIC ISLAND GROUPS

Island Realms	Main Island Groups	Well-Known Individual Islands or Atolls	Countries Exercising Political Control	Main Exports of Political Units (mainly 1986, 1985, or 1984 data)
Melanesia		New Guinea	Papua New Guinea	Copper 33%; gold 19%; coffee 13%
			Indonesia (Irian Jaya province)	
	Bismarck Archipelago	New Britain, New Ireland	Papua New Guinea	
	Solomon Islands	Guadalcanal	Solomon Islands	Fish 41%
		Bougainville	Papua New Guinea	
	Vanuatu (former New Hebrides)	Espiritu Santo	Vanuatu	Copra 29%; cocoa 13%; reexports 38%
	Fiji Islands	Viti Levu, Vanua Levu	Fiji	Sugar 55%; gold 16%
		New Caledonia	France (overseas territory; also includes the Loyalty Islands and other small islands)	Nickel 75%
Micronesia		Nauru	Nauru	Phosphate 100%
	Caroline Islands	Kusaie, Ponape (Palau, Truk, and Yap are noteworthy island groups)	United States (UN Trust Territory of the Pacific Islands)[a]	Coconut products 68%
	Mariana Islands	Guam	United States (self-governing territory)	Clothing 17%; miscellaneous other manufactures
		Saipan, Tinian	United States (UN Trust Territory of the Pacific Islands)[a]	
	Marshall Islands	Bikini, Eniwetok	United States (UN Trust Territory of the Pacific Islands)[a]	
	Kiribati	(Republic of Kiribati includes Tarawa and the rest of the Gilbert Islands, plus the Phoenix Islands, most of the Line Islands, and Ocean Island)	Kiribati	Fish 71%; copra 19%
Polynesia	Hawaiian Islands	Oahu, Hawaii, Maui, Kauai	United States (U.S. state)	Sugar; pineapples
	Tuvalu (former Ellice Islands)		Tuvalu	Copra 80%; developed cinema film 20%
	Tonga Islands		Tonga	Coconut products 58%
	Samoa Islands		Western Samoa	Coconut products 57%
			United States (American Samoa is a non-self-governing U.S. territory)	Canned tuna 93%
	Marquesas Islands	(The Marquesas, Tuamotus and Societies form French Polynesia, governed from Tahiti in the Society Islands)	France (overseas territory)	Reexports 69%; cultured pearls 25%
	Tuamotu Archipelago			
	Society Islands			
	Cook Islands	Rarotonga	New Zealand (self-governing territory)	Clothing 39%

[a]The Trust Territory of the Pacific Islands has devolved into (1) The Republic of the Marshall Islands and the Federated States of Micronesia (in the Carolines), where "compacts of free association" with the United States have replaced the original trusteeship agreements; (2) the Northern Marianas, a self-governing commonwealth forming a political union with the United States, with its people being U.S. citizens; and (3) Palau, where a compact of free association with the United States was not yet finalized in mid-1988 due to court challenges by Palauans favoring a nuclear-free status.
Source of data on exports: Mainly *Britannica Book of the Year,* 1988 or earlier years.

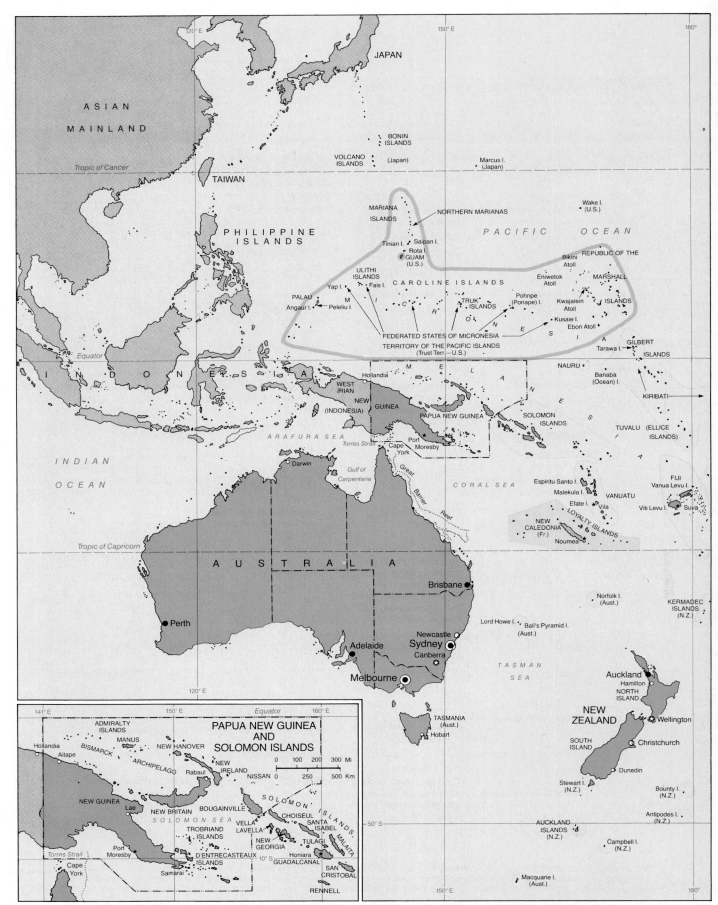

Figure 19.1   General reference map of the Pacific world, showing political units as of early 1989.

450

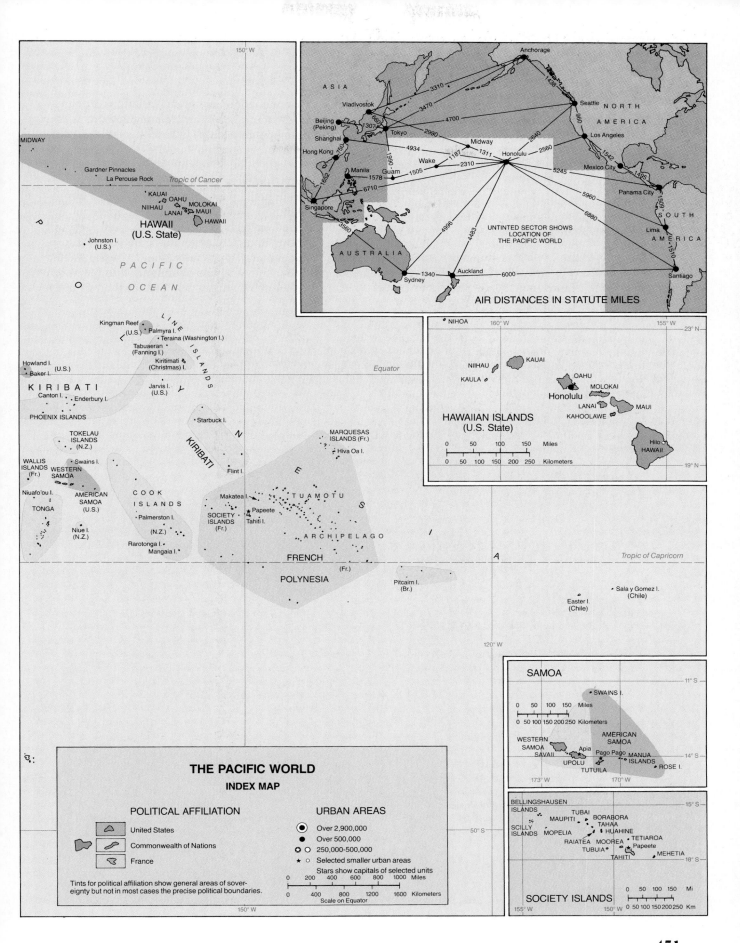

THE PACIFIC WORLD

INDEX MAP

POLITICAL AFFILIATION

United States

Commonwealth of Nations

France

Tints for political affiliation show general areas of sovereignty but not in most cases the precise political boundaries.

URBAN AREAS

Over 2,900,000

Over 500,000

250,000-500,000

Selected smaller urban areas

Stars show capitals of selected units

AIR DISTANCES IN STATUTE MILES

HAWAIIAN ISLANDS (U.S. State)

SAMOA

SOCIETY ISLANDS

the result of volcanic eruptions, but a relatively small number are so much larger and more complex in origin and character that they are called "continental" islands. The "low" islands are made of coral, a material comprised of the skeletons and living bodies of small marine organisms.

"HIGH" ISLANDS

The "high" islands of volcanic origin are often spectacularly scenic (Fig. 19.2). Steep slopes predominate, and some islands have peaks thousands of feet high. However, there are great variations in elevation, slope, soil, rainfall, and plant life from island to island or within a given island. Minerals of commercial value are scanty, but the soil is generally fertile. Among islands of this type, the Hawaiian, Samoan, and Society groups, all located in Polynesia, are especially well known. They are "weathered" islands on which the volcanic material has disintegrated enough to produce relatively deep soil that supports luxuriant vegetation. By contrast, certain "unweathered" volcanic islands in the Pacific have thinner soils and less abundant plant life, often consisting more of grasses and bushes than of trees.

Figure 19.2 A view of the volcanic "high" island of Hiva Oa in French Polynesia. The view shows the town of Atuona, the capital of the Marquesas Islands, in which Hiva Oa is located. The French colonial influence is very evident in the light-colored, red-roofed, and airy architecture.

Physical conditions in the volcanic islands are sufficiently varied to allow a diversity of tropical crops. The aboriginal peoples depended mainly on starchy foods such as breadfruit, plantain, taro, sweet potatoes, and yams. Pigs and sea fish supplied animal protein. Subsequently other crops and animals were introduced in various places, often by Europeans. Examples include arrowroot and cassava; bananas; tropical fruits such as mangoes, papayas, and citrus; coffee and cacao; sugarcane; and cattle, goats, and poultry. Of course, Westernization has brought packaged foods in great variety to various places, most notably the highly Americanized Hawaiian Islands, but the traditional staples continue to play a major role in most islands.

Sugar plantations were established by outsiders in some islands, most prominently in the Hawaiian Islands, Fiji Islands (Fig. 19.3), and Saipan in the Marianas. Labor was brought in from outside since the local peoples generally proved too few in number or too reluctant to work in the cane fields. A polyglot character was imparted to Hawaii by successive infusions of Chinese, Portuguese, Japanese, and Filipinos; while in Fiji the British plantation

owners imported Indian laborers (mainly Hindus), who multiplied so rapidly that they now outnumber the native Fijians. In Saipan the sugar plantations were developed by Japanese interests which brought in Japanese labor. The sugar industry did not survive World War II in Saipan (which suffered a devastating American invasion), but sugar is still important in Hawaii and Fiji.

Some high islands to the north and northeast of Australia are so large and have such complex rocks and varied environments that they are known as "continental" islands. This group includes New Guinea; New Britain and New Ireland in the Bismarcks; New Caledonia; Bougainville and smaller islands in the Solomons; the two main islands of Fiji; and a few others. Mountains over 16,000 feet (4877 m) high exist in New Guinea, and lower but still imposing mountains exist in various other islands. Valuable minerals are extracted in a few islands—an example being the nickel of the French-held island of New Caledonia. The island is one of the world's largest producers.

The main urban places and seaports in the tropical Pacific islands are in the volcanic "high" and the

Figure 19.3 Sugarcane is a major commercial crop in a few "high" islands of the Pacific world. The small locomotive in the view is hauling a load of cane on a plantation in the Fiji Islands.

Figure 19.4   A characteristic scene of the Pacific island world. The low coral islands in the view, located near New Guinea in the southwest Pacific, comprise an atoll. The islands and coral reefs surround a lagoon, visible beyond the large island in the center foreground. Coral reefs bordering the atoll are outlined by the white water. Countless numbers of atolls exist in the central and western Pacific.

"continental" islands. Honolulu on the island of Oahu in the Hawaiian group is incomparably the largest, having a metropolitan population of around 825,000 (1987) (see chapter-opening photo). No other city in the islands reaches 175,000, and very few reach 50,000 (late 1980s). (The New Zealand cities, while technically in Polynesia, are outside the tropics and are considered separately in Chapter 20.)

"LOW" ISLANDS

The "low" islands are generally smaller than the "high" islands and have fewer possibilities for supporting dense populations. Some are so dry and infertile that trees will not grow, and permanent settlement may be precluded by lack of drinking water. Typically the "low" islands are fringed by the waving coconut palms that are a mainstay of life and have become the trademark of the "South Sea Isles." The former Gilbert Islands (now in the independent republic of Kiribati) are typical of the picturesque, well-watered, and fertile islands idealized by Hollywood. Scattered around the Pacific are "raised coral" islands such as phosphate-rich Nauru, which have a little greater elevation because they have been produced by successive uplifts of coral reefs. Some major characteristics of the "low" islands are elaborated in the following selection by an American geographer, the late Otis W. Freeman[1]:

_____
[1]Otis W. Freeman, "The Pacific Island World," *Journal of Geography*, 44 (1945), pp. 25–28. Reprinted by permission of the *Journal of Geography*. Supplementary material has been inserted in brackets without compromising the sense of the original.

The "low" islands are made of coral and usually have an irregular ring shape around a lagoon. Such an island is called an atoll [Fig. 19.4]. Generally the coral ring is broken into many pieces, separated by channels leading into the lagoon, but the whole circular group is commonly considered one island, although quite often individual names are given to the larger islets. . . . The coral atolls are wholly within the tropics as the reef-forming organisms can only survive in warm waters.

[A number of] theories have been advanced to account for the coral atolls. That of Charles Darwin has three stages: (1) the coral builds a fringing reef around a volcanic island, (2) the island slowly sinks and the coral reef is built upward and forms a barrier reef separated by a lagoon from the shore, (3) the volcanic island has sunk out of sight and a lagoon occupies the former land area whose outline is reflected in the roughly oval form of the atoll. . . . Another theory is that of Sir James Murray who would have the coral forming a reef on the outer limits of a shallow submarine platform. Likely most such platforms would result from erosion of a volcanic island, so would have the approximately oval shape of a volcanic peak. [Probably no one theory explains all atolls.]

No matter how they may have been formed, the geographer finds that the atolls have similar characteristics wherever they occur, although in size they vary. Many are only a few miles in diameter, while a few huge affairs up to 50 to 100 miles [80–161 km] across occur. . . . Some coral islands lack the central lagoon, which has either been filled to become land or the island itself has been elevated enough for the feature to disappear. . . .

. . . . .

The soil of the low islands is broken from the coral by the waves and winds. It is thin and poor, and few plants except coconuts can grow on it or withstand the infiltration of the salt sea into the ground water supply. On the more barren islands, coconuts, the rather poor fruit of the pandanus, and fish are the only available [indigenous] foods. Sometimes the natives by much labor construct a garden of artificially fertile soil. Within a stone wall or pit . . . all available plant waste and other refuse is collected to supply needed humus to the sterile coral sand. . . .

The most significant animal life on the low islands consists of birds. . . . Birds have sometimes been slaughtered for their feathers, and both eggs and birds are eaten, although they have a fishy flavor. However, the most valuable product of the birds is guano and, indirectly, phosphate rock. On the low islands in the trade winds, rain is too scanty to wash away the bird droppings and other refuse which accumulate and change to the brownish guano that is a valuable fertilizer because it is high in nitrogen and phosphorus. On some islands, notably Nauru . . . just south of the equator, the phosphorus from the guano has interacted chemically with the coral limestone to form phosphate rock. This is quarried and exported [for fertilizer].

RELATIONS WITH THE WEST

Europeans began to visit the Pacific islands early in the Age of Discovery. Spanish and Portuguese voyagers were followed by Dutch, English, French, and, later, American and German. Many famous names are connected with Pacific exploration, including those of Magellan, Tasman, Bougainville, La Perouse, and Cook. By the end of the 18th century virtually all of the important islands were known.

For a long period the European governments exercised only nominal control over the islands, and the native peoples were subjected to the unrestrained abuses of whaling crews, sandalwood traders, indentured labor contractors and other adventurers. On island after island European penetration presented the dismal spectacle of decimation of the islanders and disruption of their cultures. The intruders introduced new diseases, alcohol, opium, forced labor, and firearms which greatly increased the slaughter in tribal wars. Although subsequent attempts by Westerners to give the islanders a "new deal" must be entered on the other side, the balance sheet for four and a half centuries of Western influence in the Pacific islands does not reflect much credit on the outsiders.

In recent times Western personnel and capital have been attracted to the islands by plantation agriculture, mining, fishing, military facilities and activities, and a large expansion of tourism. All these activities are rather widespread except for mining, which is largely confined to New Caledonia, Papua New Guinea, and Nauru.

VARIATIONS IN POLITICAL AND ECONOMIC DEVELOPMENT

A number of important conceptions about the current political and economic geography of the Pacific islands can be gained from Table 19.1. Politically this area is in midstream between colonialism and independence. Once entirely colonial, it now exhibits a mixture of colonial dependencies and states that have recently become independent. The remaining dependencies are in various stages of development toward independence.

Evolution of the Pacific islands away from colonial status began with the admission of the Hawaiian Islands to statehood within the United States in 1959 and the granting of independence to Western Samoa by New Zealand in 1962. As of the late 1980s nine independent states had emerged, with populations ranging from 3.6 million in Papua New Guinea—which is very exceptional in population and area for this region—to 8500 in tiny Nauru. Hence, these new states are some of the smallest of the world's nations. A "typical" Pacific island state has about 100,000 to 150,000 people in an area of 250 to 1000 square miles (c. 650–2600 sq. km). The country comprises a number of islands, is quite poor economically, and is an ex-colony of Britain, New Zealand, or Australia. The colonial dependencies vary from units resembling "typical" independent countries in area and population to units that are quite tiny. Independence has generally been delayed because such territories are viewed as too small and isolated for independence or because their possession confers military or economic advantages on the governing power. For example, Guam and American Samoa are useful to the United States for military purposes, French Polynesia is the locale for French atomic testing, and New Caledonia's mineral wealth holds a continuing attraction for France.[2]

Another characteristic of the Pacific islands is a general lack of industrial development, combined in many cases with population densities which are rather high for nonindustrial societies. This tends to spell poverty, with the major exception of Nauru, whose phosphate earnings provide a relatively high average income. Characteristic of less-developed countries is the fact that the major exports of the island units tend to be composed of a few primary products. Before the intrusion of outsiders the island

---

[2]Often there are disagreements among the islanders concerning the desirability of independence or the form an independent state should take. This is true of the American-held Trust Territory of the Pacific Islands, with the added complication that the United Nations must approve any final political settlement.

economies were heavily based on subsistence agriculture, gathering, and fishing, with coconuts a major element in the food supply. Many countries still rely heavily on modern commercialized versions of the coconut and fishing economy. But along with this commercial reliance goes a continuing dependence on coconuts and fish as major elements in a widespread pattern of subsistence activities. Indeed, the coconut is so important in these islands that they have sometimes been said to have a "coconut civilization." The latter is described in the following section by Douglas L. Oliver[3]:

> . . . The influence of the coconut . . . stretches from Truk to Tonga and from Hiva Oa to Hollandia, and directly or indirectly affects the life of nearly every islander in this vast area. . . .
>
> Coconuts require year-round warm temperature, a well-drained soil, and plenty of moisture and sunlight; they grow best of all in low altitudes near the coast. The palm grows out of the mature, fallen nut, and requires from eight to ten years to reach the bearing stage. After that it lives for nearly 80 years and bears nuts at the rate of about 50 a year for 60 or 70 years.
>
> The mature nut consists of a hollow kernel of oily white meat, one half inch thick, encased in a hard woody shell. Around this is a fibrous husk one half to two inches thick. The cavity of the unripe nut is filled with a thin "milk," a nutritious and refreshing beverage with a tangy taste. As the nut matures, this "milk" is absorbed into the coconut meat, which when dried and removed from the shell becomes the copra of commerce.
>
> Throughout Oceania the coconut leads a double life. In one way it is a source of food, shelter, and income which helps support native life nearly everywhere. In another way, however, it has been the instrument by which white men have done most to change native life.
>
> To islanders the milk of the unripe coconut is a prized beverage, the only one, in fact, on many islands lacking potable water or rain catchments. The meat is scraped from the shell and eaten either by itself or mixed into puddings of taro and yams and sago. Or, the oil is squeezed from the meat and used as a food, an unguent, or a cosmetic. The hollowed shell becomes a flask, a clean-scraped shell a cup or a spoon or a material for carved ornaments. Cord is manufactured from the fibrous husk; furniture, utensils, and building timbers from the tough trunk. Leaves are used to thatch huts or weave baskets; and even the pith of the palm is eaten when for some drastic reason the

palm is felled. On islands nearer to Asia's influence, the sugar-rich sap is drawn off by tapping the flower bud and is allowed to ferment into inebriating toddy. To deprive an islander of his coconut palms is to take away much of the basis of his living. Even the white man's world would be a much poorer place without them.[4]

Many of the island political units listed in Table 19.1 rely mainly on exports other than coconut products or fish. Such units fall into a number of major categories:

1. *Countries or colonies in which mineral extraction overshadows the commercialization of former subsistence economies.* Thus Papua New Guinea extracts copper from its island of Bougainville in the Solomons; New Caledonia profits from outstanding resources of nickel and other minerals; and tiny Nauru depends totally on exports of phosphates.

2. *Colonies where the economies tend to be supported and shaped by military activities of France or the United States.* The major examples are French Polynesia, which supplies some of the needs of the French atomic weapons program, and Guam, which is a major Pacific base of the United States.

3. *Fiji* was developed under British rule as a source of plantation sugar using labor largely imported from India. Sugar is still the largest export by far (Fig. 19.3). The modern state managed to function effectively after independence (1970) under a parliamentary system granting equal numbers of parliamentary seats to the native (Melanesian) Fijians and the Indian-descended majority. But in 1987 the continuing tensions between Fiji's ethnic groups gave rise to a military takeover of the government by native Fijians. The leader of the coup announced his intention to assure electoral supremacy for the native Fijians. The country was made a republic, and Britain's queen stepped down as head of state. As of early 1989, Fiji was not a member of the Commonwealth of Nations.

4. Only the *Cook Islands*, under New Zealand, show an early stage of evolution toward an industrial economy, with labor-intensive clothing manufacture supplying the leading export.

---

[3]Reprinted by permission of the publishers from Douglas L. Oliver, *The Pacific Islands* (Cambridge, Mass.: Harvard University Press, Copyright 1951 by the President and Fellows of Harvard College), pp. 135–136. One footnote added in editing.

[4]Coconut meat is used not only for pies, cakes, and candies, but pressed copra yields coconut oil for making such products as soap, margarine, and cooking oils. The residue from pressing, known as copra cake, is a nutritious stock feed. Most of the world's exports of copra and coconut oil come from Southeast Asia.

Thus the overall Pacific islands economic picture is one of nonindustrial economies (a) exporting products that were important in their precolonial subsistence activities, (b) exporting minerals desired by industrial nations, or (c) deriving income from activities connected with military needs of occupying powers. The most dynamic new element in the economic picture is a rapid and important growth of tourism as a basic industry, with America and Japan the main sources of visitors.

Environmental and Functional Relations of Australia and New Zealand

Widespread use of land to graze sheep and cattle is a major characteristic of Australia and New Zealand. This photo of sheep in interior New South Wales offers a classic view of middle-latitude steppe.

THE basic kinship of Australia and New Zealand is widely recognized. This kinship is derived from similarities in population, cultural heritage, political problems and orientations, type of economy, and location.

Both Australia and New Zealand are products of British colonization and are strongly British in ancestry and culture.[1] The Australians and New Zealanders speak English, live under parliamentary forms of government, acknowledge the British sovereign as their own, and attend schools patterned after those of Britain. Until recently the transplanted Anglican Church (Church of England) was considerably the largest religious denomination in each country. This is still true in New Zealand, but Australia has admitted so many immigrants from Catholic Europe since World War II that Catholics and Anglicans are now comparable in numbers. This is only one of many ways in which the immigration of recent decades has somewhat diluted Australia's Britishness.

Although they are fully independent nations, loyalty to Britain has been an outstanding characteristic of Australia and New Zealand in the past. This loyalty is still expressed in their membership in the Commonwealth of Nations and was plainly evident at the outbreak of both world wars, which the two countries entered immediately in support of Britain. However, in World War II, a threatened Japanese assault on their homelands was frustrated mainly by American forces, and since that time British power has declined sharply. Hence the two countries have also sought closer relations with the United States. Close international ties with strong and friendly outside powers, especially naval powers, have generally been seen by them as extremely important. This has been a logical outlook for units with such small populations and remote insular locations, far distant from other Western nations (Figs. 3.1, 19.1, and 20.1). Australia's 16.5 million people plus New Zealand's 3.3 million (1988) amount to a considerably smaller total than the population of California. But relative isolation from potential threats can also lead to feelings of security, and in the mid-1980s New Zealand felt secure enough to forbid nuclear-equipped naval vessels of its American ally from using its ports. The ensuing quarrel threatened the long-standing security pact between the two countries.

Since World War II, the strong historical ties of Australia and New Zealand to Britain have also diminished drastically in the field of trade. For most of their history the United Kingdom was far more important than any other country in their trade relations, and as recently as the early 1970s it accounted for about 30 percent of New Zealand's trade and 15 percent of Australia's. But by 1987 Britain was relatively insignificant in Australia's trade and accounted for only about a tenth of New Zealand's (Table 20.1). Britain's alignment toward Europe in the 1970s as a new member of the Common Market hastened this process. Today the minerals and agricultural products that comprise most of Australia's exports are sold largely to Japan and the United States, while Australia, Japan, and the United States have become New Zealand's leading partners in trade, and the agricultural exports on which New Zealand is heavily dependent go more to Japan and the United States than to Britain. Thus ties of history, tradition, sentiment, and culture between the two countries and Britain have become more important than more tangible ties of politics, defense, and trade. In trade and politics, Australia and New Zealand are no longer appendages of an Atlantic-oriented European power but have developed a new alignment with nations of the Pacific rim.

But in one tangible respect, besides their white populations and cultural antecedents, Australia and New Zealand are still "European" in that they are among the world's minority of prosperous nations. Australia's per capita gross national product (GNP) is well above that of Britain, although well below that of the United States or the most prosperous European countries. New Zealand fares less well in this respect. It has fewer resources and a less diversified economy than its larger neighbor, and it was more drastically affected by the loss of much of the British market when Britain entered the European Common Market. Its per capita GNP is only three

---

[1] Both Australia and New Zealand contain surviving minorities of prewhite native inhabitants. The primitive Australoid inhabitants of Australia were slaughtered or driven into outlying areas by the whites. Today the "aboriginal" population numbers about 200,000 (of whom a majority are of mixed blood), in contrast to an estimated 300,000 at the beginning of white settlement in 1788. Found mainly in the tropical north, this minority lives mostly on the fringes of white society, but partly independently in the "bush." In New Zealand the native inhabitants were the Maori, a Polynesian people concentrated (then and now) mainly on North Island. Not so primitive as the Australian aborigines, and more warlike, their hold on the land was broken only by a bloody war between 1860 and 1870. The Maori, who seemed destined for extinction about 1900, have rebounded to make up some 9 percent of New Zealand's population. While still something of a depressed group, their economic and social situation is much better than that of the Australian natives, and they are being increasingly integrated, both socially and biologically, with the rest of New Zealand's population. The cultural extinction threatened by this integration is being resisted by a Maori cultural and political movement.

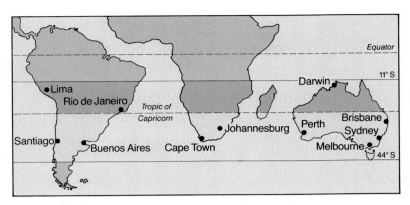

Figure 20.1 Australia compared in latitude and area with southern Africa and southern South America.

fifths that of Australia, is substantially below Britain's and is well under half that of Switzerland or the United States (1986). New Zealand's troubled economy was partly responsible for a sizable net emigration, particularly to Australia, which set in during the 1970s. But the country is still quite prosperous when compared to most nations of the world.

AUSTRALIA'S PHYSICAL AND CLIMATIC CHARACTERISTICS

Australia is not only a country but an island continent with an area, including the offshore island of Tasmania, of nearly 3 million square miles (7.8 million sq. km). It is exceeded in area by only five countries: the Soviet Union, Canada, China, the United States, and Brazil. However, most of the continent is very sparsely populated, and in terms of total pop-

ulation Australia is a relatively small country. The sparseness of Australia's population and its concentration into a comparatively small part of the total land area are closely related to the continent's physical characteristics, among which aridity (accentuated by high temperatures and rapid evaporation) and low average elevation are outstanding. On the basis of climate and relief, Australia may be divided into four major natural regions: (1) the humid eastern highlands, (2) the tropical savannas of northern Australia, (3) the "mediterranean" lands of southwestern and southern Australia, and (4) the dry interior (Figs. 20.2 and 20.3).

THE HUMID EASTERN HIGHLANDS

Australia's only major highlands extend along the east coast from Cape York to southern Tasmania in a belt 100 to 250 miles (161–402 km) wide. Al-

TABLE 20.1 PRINCIPAL TRADING PARTNERS OF AUSTRALIA AND NEW ZEALAND

	Australia (1986–1987)		New Zealand (1987)	
Percent of exports (based on total value) to	Japan	25	United States	16
	United States	12	Japan	15
	New Zealand	5	Australia	15
	China	4	United Kingdom	9
	South Korea	4	China	4
	All others	50	All others	41
Percent of imports (based on total value) from	United States	22	Japan	21
	Japan	21	Australia	18
	United Kingdom	7	United States	16
	West Germany	7	United Kingdom	10
	New Zealand	4	West Germany	6
	All others	39	All others	29

Source: *Britannica Book of the Year*, 1989.

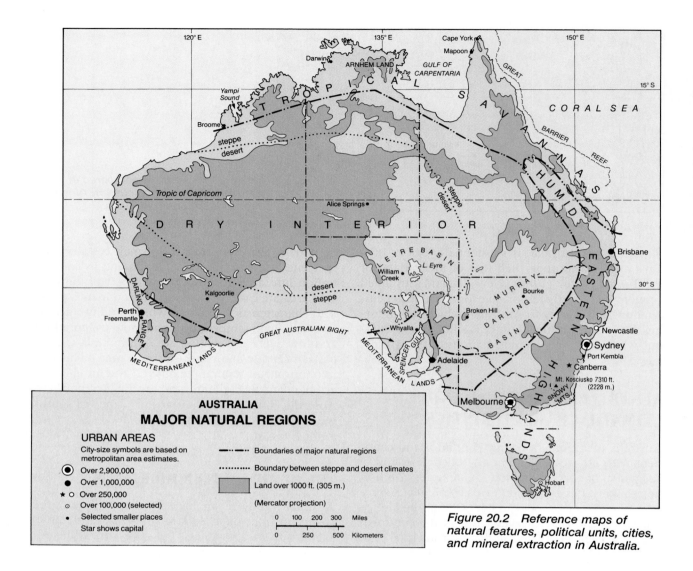

AUSTRALIA
MAJOR NATURAL REGIONS

URBAN AREAS
City-size symbols are based on
metropolitan area estimates.

◉ Over 2,900,000
● Over 1,000,000
★ ○ Over 250,000
○ Over 100,000 (selected)
• Selected smaller places
   Star shows capital

—·—·— Boundaries of major natural regions
·········· Boundary between steppe and desert climates
▨ Land over 1000 ft. (305 m.)
(Mercator projection)

0   100   200   300   Miles
0   250   500   Kilometers

Figure 20.2   Reference maps of natural features, political units, cities, and mineral extraction in Australia.

though complex in form and often rugged, these highlands seldom reach elevations of 3000 feet (914 m). Their highest summit and the highest point in Australia is Mount Kosciusko, which attains only 7310 feet (2228 m). The highlands and the narrow and fragmented coastal plains at their base constitute the only part of Australia which does not experience a considerable period of drought each year. However, although onshore winds from the Pacific bring appreciable rain each month, the strong relief reduces the amount of agricultural land in this most favored of Australia's climatic areas. South of Sydney and at higher elevations to the north the climate is commonly classified as marine west coast, despite the location. North of Sydney higher summer temperatures change the classification to humid subtropical, while still farther north, beyond approximately the parallel of 20°S, hotter temperatures and greater seasonality of rain cause essentially subhumid conditions.

TROPICAL SAVANNAS OF NORTHERN AUSTRALIA

Northern Australia, from near Broome on the Indian Ocean to the coast of the Coral Sea, receives heavy rainfall during a portion of the (Southern Hemisphere) summer season, but experiences almost complete drought during the winter 6 months, or more, of the year. This highly seasonal distribution of rainfall is essentially the result of monsoonal winds which blow onshore during the summer and offshore during the winter. The seasonality of the rainfall, combined with the tropical heat of the area, has produced a savanna vegetation of coarse grasses with scattered trees and patches of woodland. The effect of the long season of drought in reducing agricultural possibilities is compounded by the poverty of the soils and by a lack of highlands sufficient to nourish large perennial streams for irrigation. The alluvial and volcanic soils which support large pop-

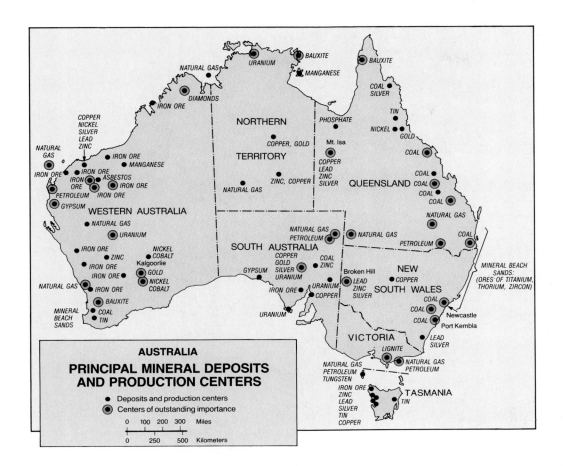

AUSTRALIA
PRINCIPAL MINERAL DEPOSITS AND PRODUCTION CENTERS
● Deposits and production centers
◉ Centers of outstanding importance
0   100   200   300   Miles
0   250   500   Kilometers

ulations in some tropical areas are almost completely absent in northern Australia.

"MEDITERRANEAN" LANDS OF SOUTHWESTERN AND SOUTHERN AUSTRALIA

The southwestern corner of Australia and the lands around Spencer Gulf have a mediterranean or dry-summer subtropical type of climate with subtropical temperatures, winter rain, and summer drought. In winter the Southern Hemisphere belt of the westerly winds shifts far enough north to affect these districts, while in summer this belt lies offshore to the south and the land is dry. Crops introduced from the mediterranean lands of Europe generally do well in these parts of Australia, but the agricultural possibilities of the Australian areas are limited by the lack of high highlands to catch moisture and supply irrigation water to the lowlands.

THE DRY INTERIOR

The huge interior of Australia is desert, surrounded by a broad fringe of semiarid grassland (steppe) which is transitional to the more humid areas

around the edges of the continent. Altogether, the interior desert and steppe cover more than half of the continent and extend to the coast in the northwest and along the Great Australian Bight in the south. This tremendous area of arid or semiarid land is too far south to get much rain from the summer monsoon, too far north to benefit from rainfall brought by the westerlies in winter, and is shielded from Pacific winds by the eastern highlands. Here again the lack of highlands is unfortunate from a climatic point of view. Approximately the western half of Australia is occupied by a vast plateau of ancient igneous, metamorphic, and hardened sedimentary rocks, but its general elevation is only 1000 to 1600 feet (c. 300–500 m) and its few isolated mountain ranges are too low to materially influence the climate or supply perennial streams for irrigation. To the east, between the plateau and the eastern highlands, the land is still lower in the great central lowland which stretches across the continent between the Gulf of Carpentaria and the Great Australian Bight. The effect of elevation on the climate is shown by the fact that the lowest part of this lowland of Australia, the Lake Eyre Basin, is the driest part of the continent (Fig. 20.4). Another part of the lowland, however, the Murray-Darling Basin, contains Australia's only major river system and has the

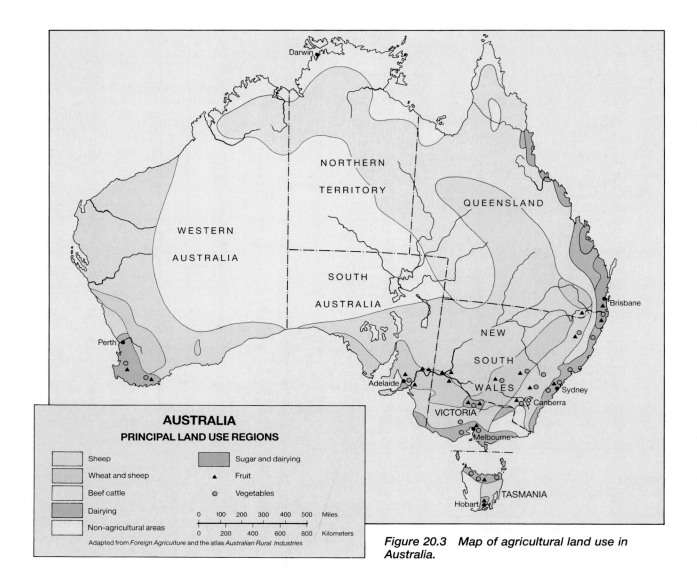

Figure 20.3 Map of agricultural land use in Australia.

most extensive development of irrigation works on the continent.[2]

Temperature and precipitation data for the four major natural regions of Australia are presented in Table 20.2.

---

[2]The irrigated areas scattered along these rivers have been expanded through the use of water supplied by Australia's Snowy Mountains scheme. This project is located in the Snowy Mountains part of the eastern highlands just south of the federal capital of Canberra. Streams flowing in their natural courses eastward to the Pacific are diverted into the headwaters of the westward-flowing Murray River and one of its tributaries, thus providing more water downstream on the dry side of the mountain divide. The engineering works of the Snowy Mountains scheme, begun in 1949, include numerous dams and reservoirs and nine hydroelectric stations. Hundreds of thousands of acres of newly irrigated land have been brought into production, much of it to grow rice for export.

THE SMALL PROPORTION OF ARABLE LAND

One very important result of the widespread seasonal or total aridity, and of the occupation of most of the only truly humid area by highlands, is that Australia offers very little good agricultural land relative to its total area. Although some estimates place the proportion of potentially cultivable land as high as 15 percent of the total, less than 6 percent of Australia is actually cultivated today. Fourteen percent of Australia's area is classed as forest and woodland, the greater part of which is of little value. Over half is classed as natural grazing land and about one fifth as complete wasteland. Despite the small proportion of its area which is arable, however, Australia has one of the most favorable ratios of crop-producing land to population in the world.

Figure 20.4 Australia's great grazing lands grade into stark desert in the continent's "dead heart." The frontier flavor of this arid backland is conveyed by the photo of the hotel at William Creek in South Australia. William Creek is a station on the railway that reaches northward from the seaport of Adelaide to a terminus at the oasis of Alice Springs in the Northern Territory. Other than the railroad station and the hotel, little exists at William Creek.

The distribution of population in Australia generally follows that of arable land. Thus, most of the country's small population is found in the humid eastern highlands and coastal plains, especially in the cooler south; in the areas of mediterranean climate; and in the more humid grasslands adjoining the southern part of the eastern highlands and the mediterranean areas.

AUSTRALIA'S NATURAL RESOURCE-BASED ECONOMY

AGRICULTURE AND RANCHING

The sectors of the Australian economy which regularly produce the country's leading exports are grazing, agriculture, and mining. Such a result seems a

---

TABLE 20.2 CLIMATIC DATA FOR SELECTED AUSTRALIAN STATIONS

Natural Region, Climate Type and Station	Latitude to Nearest Whole Degree	Elevation Above Sea Level (ft)	(m)	Annual °F	°C	January (or warmest month) °F	°C	July (or coolest month) °F	°C	Annual Average (in.)	(cm)	Months Averaging Less Than 1 in.
Humid eastern highlands												
Humid subtropical												
Brisbane	27°S	137'	42 m	68°F	20°C	77°F	25°C	59°F	15°C	43"	109 cm	0
Marine west coast												
Melbourne	38°S	115'	35 m	59°F	15°C	68°F	20°C	50°F	10°C	27"	69 cm	0
Tropical savannas												
Darwin	12°S	97'	30 m	82°F	28°C	84°F	29°C (Dec.)	77°F	25°C	62"	156 cm	5 (May–Sept.)
Mapoon (Cape York Peninsula)	12°S	20'	6 m	82°F	28°C	85°F	29°C	78°F	26°C	62"	156 cm	6 (May–Oct.)
Mediterranean areas												
Adelaide	35°S	140'	43 m	63°F	17°C	73°F	23°C	52°F	11°C	21"	52 cm	3 (Jan.–Mar.)
Perth	32°S	197'	60 m	64°F	18°C	75°F	24°C (Feb.)	55°F	13°C	35"	89 cm	5 (Nov.–Mar.)
Dry interior												
Bourke	30°S	361'	110 m	70°F	21°C	84°F	29°C	52°F	11°C	13"	34 cm	4 (July–Oct.)
William Creek	29°S	247'	75 m	68°F	20°C	82°F	28°C	48°F	9°C	5"	13 cm	12

logical outgrowth of a situation where 16.5 million people have at their command the natural resources of an entire continent, albeit a poor continent from the standpoint of agriculture. The existence of large and dependable markets for Australian primary products in Japan, the United States, Europe, and elsewhere—coupled with cheap and efficient transportation by sea to those markets, and the ability of the Australian people to supply or import the necessary skills and capital to exploit the continent on an extensive basis, utilizing a minimum of labor and a maximum of land and equipment—makes the emphasis on surplus production and export of primary commodities seem even more logical. Australian agricultural and pastoral output per person involved is very high compared to that of most of the world.

Sheep and Cattle Ranching

Until recently, sheep ranching was always the most important of Australia's rural industries (see chapter-opening photo). When the first settlement was planted at Sydney in 1788, the intention of the British government was merely to establish a penal colony, with the expectation that enough agriculture would be carried on to make the colony self-sufficient in food and perhaps provide a small surplus of some products for export. While self-sufficiency proved difficult to obtain in the early days due to the poverty of the leached soils around Sydney, it was soon discovered that sheep did well. At the time the market for wool was rapidly expanding with mechanization of the woolen textile industry in Britain. Thus, Australia found a major export staple in the early years of the 19th century. Sheep graziers rapidly penetrated the interior of the continent, and by 1850 Australia was already the largest supplier of wool on the world market, a position it has never lost.

Sheep raising has been tried practically everywhere in Australia that there seemed to be any hope of success. Much of the interior has been found to be too dry, parts of the eastern mountain belt too rugged and wet, and most of the north too hot and wet and with too coarse a forage in summer. Thus the sheep industry has become localized. It is concentrated mainly in a crescentic belt of territory which follows roughly the gentle western slope of the eastern highlands, from the northern border of New South Wales to the western border of Victoria (Fig. 20.3). These are the leading Australian states in sheep raising. Beyond the main belt of concentration the sheep industry spreads far northward on poorer pastures into Queensland and west into South Australia. A minor area of production rims the west coast in the vicinity of Perth. The sheep ranches or "stations" in Australia vary in size from a few hundred to many thousands of acres. They are generally larger on the poorer pasture lands, which lie outside the main belt of production.

The main product of the sheep stations is wool, a product which has been of utmost importance in Australia's development. Through most of the country's history, wool has been the principal export, amounting to as much as a quarter of all exports as late as the 1960s. Recently, however, it has declined rapidly in importance, due partly to the rise of other commodities in the Australian economy, and partly to wool's slowly losing battle with synthetic fabrics in the world economy. By 1985 wool still accounted for 8 percent of Australia's exports, but this represented only half the value of other kinds of agricultural and pastoral exports.

Lands in Australia which have not been found suitable for sheep or for a more intensive form of agriculture are generally devoted to cattle ranching. Such lands include particularly those of the north, with hot, humid summers and coarse forage. The main belt of cattle ranching extends east-west across the northern part of the country, with the greatest concentration in Queensland (Fig. 20.3). On many of the remote ranches, especially in the Northern Territory, the labor force is composed of remnants of the black aboriginal population of Australia.

The beef cattle industry is not as important in Australia as the sheep industry, but it is an industry which has expanded rapidly in recent decades. This expansion has been spurred by a growing demand for meat which accompanied the rising affluence of Japan, Europe, and the United States. Construction of roads for trucking Australian cattle to the coast also aided the expansion. By 1971 Australia had become the leading exporter of beef, and it has continued to be one of a small group of nations that dominate the world's beef exports.

Wheat Farming

Wheat is another long-time export of Australia, and still accounts for around a tenth of all exports (1985). As in the case of wool, wheat surpluses are largely a reflection of Australia's aridity and low density of population. In contrast to wool, however, wheat production did not begin as the result of a search for export staples. Wheat was the key crop in early attempts to make the settlement at Sydney self-sufficient. The soils among much of the coastal margin of Australia are leached and poor, although capable of much improvement through application of fertilizer. In the early years the colony almost starved as a result of wheat failures, and the problem of grain supply was not really solved until the development of the colony, now the state, of South Australia. There, fertile land and a favorable climate were found near the port settlement of Adelaide, so that wheat could be grown and shipped by sea to other coastal points.

In the decade of the 1860s, Australia passed from a deficit to a surplus position in grain production. Thereafter, wheat farming expanded rapidly. Among the factors making this expansion possible were (1) the development of the mechanical reaper and other types of machinery suited to extensive wheat farming, (2) the building of railroads inland from the ports, (3) scientific work in plant breeding and soil fertilization, (4) the enactment of legislation under which some of the large grazing estates were divided into smaller farms, and (5) an expanding market in Europe and subsequently in the world. The main belt of wheat production spread from the eastern coastal districts of South Australia into Victoria and New South Wales, generally following the semiarid and subhumid lands which lie inland from the crests of the eastern highlands. Thus the main wheat belt has come to occupy nearly the same position as the main belt of sheep production. In fact, the two types of production are very often combined in this area. A second, less important, wheat belt has developed in the southwestern corner of the continent inland from Perth. The combined exports from these two areas have normally been sufficient to give Australia high rank among the world's wheat exporters, exceeded in recent years by only the United States, Canada, and France. It should be emphasized that Australian wheat production is an extensive and highly mechanized form of agriculture, characterized by a small labor force, large acreages, and low yields per acre, although very high yields per worker.

Dairy Farming

Although far less important than grazing or wheat farming for the export trade, dairying is the leading type of agriculture in Australia in terms of employment. It has developed mainly in the humid coastal plains of Queensland, New South Wales, and Victoria (Fig. 20.3). Originally oriented entirely toward the local market, it has been mainly a response to the striking urbanization of the country and the necessity of a relatively high return per acre to defray the expense of clearing heavily forested land. Dairy-

ing has now become well established as the dominant type of farming in most of these coastal areas and regularly produces some export surplus beyond the requirements of Australia itself.

Fruit Growing

When Australia was settled, it was found not to have any native fruits of commercial value. Consequently, the fruit-growing industry which now exists is based entirely on plants transplanted to the Australian environment. The range of climates allows production of both tropical and mid-latitude fruits, ranging from apples, grapes, and oranges to bananas and pineapples. Much of this production is irrigated, with notable recent increases in the Murray Basin. The home market is abundantly supplied with fruit, and there is normally a fair-sized export.

Sugar Production

Australia is also an export sugar producer, although at a cost. Sugar production was begun on the coastal plain of Queensland in the middle of the 19th century to supply the Australian market. Laborers were imported from Melanesia to work in the cane fields on what often amounted practically to a slave basis. When the Commonwealth of Australia was formed in 1901 by union of the six states (New South Wales, Victoria, Queensland, South Australia, Western Australia, Tasmania), Queensland was required to repatriate the imported "Kanaka" laborers in the interest of preserving a "white Australia." However, since it was felt that production costs would be considerably raised by the use of exclusively white labor, the Commonwealth gave Queensland in return a high protective tariff on sugar. Operating behind this protective wall, the sugar industry has prospered in a tropical area where it was once argued that no white man could survive for very long at manual labor in the fields. Most of the original large estates have been divided into family farms worked by individual white farmers. This system of agriculture is very unusual for a commercial cane-sugar area. The Queensland sugar area is also unusual in the fact that it is probably the most thoroughly tropical area yet settled by Europeans doing hard manual work without benefit of native labor.

URBANIZATION AND INDUSTRIALIZATION

While mining, grazing, and agriculture produce most of Australia's exports, most Australians are city dwellers, and the high degree of urbanization is one

Figure 20.5 Sydney, Australia's largest city and port. The great automobile and railway bridge connects Sydney (left) with its large suburb of North Sydney. The modernistic building in the left foreground is Sydney's famous Opera House.

of the country's outstanding characteristics. In the mid 1980s the estimated metropolitan-area populations of the five largest cities were as follows: Sydney, 3.5 million (Fig. 20.5); Melbourne, 2.9 million; Brisbane, 1.2 million; Adelaide, 1 million; and Perth, 1 million. Thus nearly 40 percent, or two of every five people in the continent, were in the two largest metropolitan areas, and nearly 60 percent were in one of the five largest. When smaller cities and towns are added the Australian population may be counted as 86 percent urban.

It will be noted that all of the five largest cities are seaports and that each is the capital of one of the five mainland states of the Commonwealth. One of the reasons for the striking degree of urban development and its concentration in the port cities is the heavily commercial nature of the Australian economy. A large proportion of the production from the country's rural areas is destined for export by sea. In addition, much of Australia's internal trade moves from port to port by coastal steamer. Growth of the respective capital-city ports was further stimulated by the fact that each state originally built its own individual rail system focusing on its particular port. Since the different states used different gauges in building their rail lines, necessitating expensive break of bulk at or near state frontiers, each major port tended to effectively dominate the business of its own state. (Work has been under way for many years to change the rail system by substitution of a standard gauge throughout the Commonwealth.

Some long-distance links between major cities have been changed over, but other progress has been slow due to the costs involved.)

The high degree of urbanization in Australia is also indicative of a high degree of industrialization. A relatively small labor force is needed for the extensive forms of agriculture and grazing which are characteristic of Australia. As a means of supporting an increased population, the Australian government has encouraged industrial development. The latter has been viewed as highly desirable, not only as a means of giving Australia a population more in keeping with the size and resources of the continent, but also as a means of providing more adequate armaments for defense and of securing greater stability through a more diversified and self-contained economy.

Australia's Mineral Wealth

The rise of industry was facilitated by unification of the separate Australian colonies into the independent federal Commonwealth of Australia. This gave a unified internal market and tariff protection for developing industries. Industrial development also has been facilitated by mineral wealth. As in the case of agriculture, Australia's small population is able to draw upon the resources of a continent; and in mineral resources the continent is relatively rich. Consequently, besides supplying most of its own needs for minerals Australia has long been an important supplier to the rest of the world. In recent years large new discoveries have been expanding this long-time role in international trade. (See map of minerals, Fig. 20.2.)

Ample supplies of coal have been fundamental to Australian industrialization. Every Australian state contains coal, but the major reserves are near the coast in New South Wales and Queensland. The earliest major field to be developed outcrops on three sides of Sydney. Most of Australia's steel capacity is located near the coal mines at Newcastle to the north of Sydney and at Wollongong to the south. Ore is brought from distant deposits along or near other sections of the Australian coast. Australian coal not only meets the country's own needs but has also become a major export in recent times. Export shipments come mainly from fields in Queensland and New South Wales and go mainly to Japan.

Abundant iron ore located near the sea and thus conveniently accessible to cheap ocean transportation is another mineral resource of prime importance. Most of the ore for the steel plants near Sydney comes from deposits slightly inland from the port of Whyalla in South Australia. Whyalla itself is also a center of iron and steel production, making use of coal brought by ore freighters on the return trip from the east coast. Major new iron-ore discoveries and developments in recent decades have made Australia a leading world supplier, with over a tenth of the world's ore output. Ore in massive amounts is being shipped from enormous deposits in Western Australia. In the late 1960s and early 1970s, huge investments to develop these deposits created a wide scatter of modern mining towns and ore ports over some of the world's more desolate wastelands. This development was closely connected with the expanding ore market provided by the growth of the Japanese steel industry.

Major reserves of bauxite also have been found, mainly (1) near the north Queensland coast, (2) in the Darling Range inland from Perth, and (3) in coastal locations in Arnhem Land. By the 1980s Australia had become the world's leading bauxite producer, with an output over twice that of any other country in 1985. New aluminum plants are supplying the Australian market, and large quantities of the partially processed material called alumina are being exported for final processing into aluminum metal at overseas plants.

Mining has become increasingly important in the Australian economy during recent times as (1) the world's search for new sources of basic raw materials has extended more widely; (2) cheaper ocean transport has made Australia less remote; (3) Japanese industry, with its need for overseas materials, has

experienced explosive growth; and (4) science has provided improved methods of prospecting. But mining has long been a significant facet of Australian life. For example, gold played a large role in Australia's history, as it did in California's. Strikes in Victoria in the 1850s set off one of the world's major gold rushes and stimulated a threefold increase in the continent's population in 10 years. Today numerous mineral products are important besides those mentioned previously. They include copper, lead, titanium, zinc, tin, silver, manganese, nickel, tungsten, uranium, and diamonds. Most of these and some others have figured in new discoveries during recent decades. Such discoveries, for example, have uncovered individual deposits of diamonds (in Western Australia) and uranium (in South Australia) that are said to be the world's largest. But the most famous mining centers are three old settlements of the dry interior—Broken Hill in New South Wales, Kalgoorlie in Western Australia, and Mount Isa in Queensland—all of which are still very active. The Broken Hill Proprietary Company, which grew with the mines at that town, later branched into iron and steel production and other businesses and has become Australia's largest corporation.

Australia's general picture of mineral abundance has been marred in the past by a deficiency in petroleum. This began to be corrected in the late 1960s and early 1970s by exploration that located a number of relatively small oil fields. Most of these are on the mainland, but the main one is offshore on the continental shelf of Victoria. The new fields have alleviated the problems, and by 1985 Australia was producing nearly as much oil as it consumed. However, its proved reserves in 1986 were so small that they would be exhausted within a decade at current rates of production. New oil is being sought, and some oil is still imported. In the meantime, domestic oil production has substantially aided the country's economy, and the same is true of natural gas output from a series of relatively small but widespread fields.

Dependence on Overseas Manufactures

One might suppose that a wealthy, urbanized, and resource-rich country such as Australia would be a major exporter of finished industrial products. But over three fourths of the country's exports come directly, or with only early-stage processing, from mining, agriculture, and grazing, and Australia is a big importer of manufactures, primarily from Japan and the United States. Far from being an international industrial power of consequence, Australia can meet its own demands for manufactured goods in only a limited way, with heavy reliance on overseas manufacturers in many lines.

Much of the manufacturing that does exist in Australia is the result of government policies which shelter industries behind high tariff walls from more efficient overseas competitors and which have also promoted industrial development in other ways—for instance, by making foreign investment in Australia attractive. Internationally competitive efficiency in industry is not easy to attain in a country whose domestic market is so small and scattered, whose factories are so far distant from the rich markets of the major developed countries, and whose nearer markets in less-developed countries are so relatively close to Japanese competitors. Nevertheless, industrialization has been seen as necessary for political reasons, for national defense, and to provide jobs for a population growth which is seen as beneficial and has been fostered by policy. Between 1945 and 1971 net immigration into Australia was over 2 million people. Australian policy encouraged this immigration, favoring skilled, white, English-speaking immigrants. The majority were British, but many others, especially continental Europeans, arrived. Asian and black immigrants were excluded. Since then the rate of immigration has slowed, but immigration continues and for some years has been opened somewhat to skilled nonwhites, as Australia cultivates friendly relations with Asian neighbors. An expanding manufacturing sector has provided many of the jobs for new Australians, even if the factories have not been able to attain a high degree of international competitiveness. Meanwhile the country's very efficient agricultural and mining sectors provide the basic support for a relatively affluent population employed mostly in urban services and manufacturing.

PASTORALISM AND URBANISM IN NEW ZEALAND

New Zealand, over 1000 miles (c. 1600 km) southeast of Australia, consists of two large islands, North Island and South Island, separated by Cook Strait, and a number of smaller islands (Fig. 20.6). North Island is smaller than South Island but contains the majority of the population, which numbered about 3.3 million in mid-1988. Since the total area is approximately 104,000 square miles (269,000 sq. km), the population density averages just 32 per square mile (13 per sq. km). New Zealand is thus a sparsely populated country, although not so much so as is Australia.

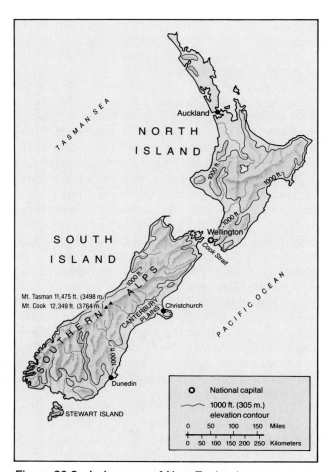

Figure 20.6   Index map of New Zealand.

PHYSICAL AND CLIMATIC CHARACTERISTICS

Rugged terrain is found throughout much of New Zealand. The topography of South Island is dominated by the Southern Alps, often cited as one of the world's most spectacular mountain ranges. These mountains present large areas above 5000 feet (c. 1500 m) and many glaciated summits above 10,000 feet. The mountains of North Island are less imposing and extensive, but many peaks exceed 5000 feet.

The highlands of New Zealand lie in the Southern Hemisphere belt of westerly winds and receive abundant precipitation. Lowlands generally receive over 30 inches (76 cm), fairly evenly distributed throughout the year, and highlands often receive over 130 inches (330 cm). Precipitation drops to less than 20 inches (51 cm) in small areas of rain shadow east of the Southern Alps. Temperatures are those of middle latitudes, moderated by the pervasive maritime influence. The result is a wet temperate climate com-

Figure 20.7   Dairy cattle on the lush pasture associated with New Zealand's cool and humid climate. The view is from a locality in the North Island not far from New Zealand's world-famous district of hot springs and geysers.

monly classified as marine west coast, with warm month temperatures generally averaging 60° to 70°F (16°–21°C) and cool month temperatures 40° to 50°F (4°–10°C). Highland temperatures are more severe, and a few glaciers are found on both islands.

IMPORTANCE OF PASTORAL INDUSTRIES

Rugged terrain and excessive precipitation in the Southern Alps and the mountainous core of North Island have resulted in almost a total absence of population from one third of New Zealand and have rendered about one fifth of the country completely unproductive, except for the notable attractions offered by the mountains to an expanding tourist industry. Well-populated areas are restricted to fringing lowlands around the periphery of North Island and along the drier east and south coasts of South Island.

The climate of New Zealand lowlands is ideal for growing grass and raising livestock, and over half of the country's total area is maintained in pastures and meadows which support a major sheep and cattle industry (Fig. 20.7). The outstanding importance and productivity of the pastoral industries in New Zealand are indicated by their dominance of the country's export trade. Meat, wool, and dairy products together account for nearly half of what New Zealand sells abroad (1987).

The earning power of its pastoral exports has been a major factor in New Zealand's relatively high standard of living. Until quite recently the United Kingdom was the main overseas market. But when the mother country joined the European Common Market in the 1970s, a new trading situation emerged, since the Common Market contains countries with surpluses competitive with those of New Zealand. This has put New Zealand products at a disadvantage in the British market. To counter this, the country has attempted to expand trade with other partners and to foster industrial growth. These efforts have not been entirely unsuccessful, but throughout much of the 1970s and into the 1980s the New Zealand economy suffered, the country slipped downward in the ranks of the world's affluent countries, and a sizable emigration took place, largely to Australia. Besides the economic troubles, a certain degree of boredom with an isolated, placid society in which tax policies make it difficult for almost anyone to make (or keep) really high income has also been cited as a factor behind the emigration.

INDUSTRIAL AND URBAN DEVELOPMENT

In New Zealand, as in Australia, a very basic dependence on the pastoral industries is accompanied by a high degree of urbanization and active attempts to develop manufacturing. The two countries are quite similar with respect to conditions and purposes fostering urban and industrial development. New Zealand's resources for manufacturing do not equal those of Australia, but coal, iron, and a number of other minerals are present in at least modest quantities, and there are very considerable potentials for hydropower development. Although somewhat depleted, New Zealand's magnificent natural forests enjoy excellent growing conditions, and production and exports of forest products have become increasingly important. To an even greater degree than Australia, New Zealand has serious problems deriving from a small internal market that militates against the mass production of a variety of manufactured goods at low cost.

New Zealand's degree of urbanization is somewhat less than Australia's and the scale of urbanization is much smaller. In the mid-1980s the country had only five urban areas with estimated metropolitan populations of more than 100,000: Auckland (800,000) and the capital of Wellington (350,000), at opposite ends of North Island; Hamilton (100,000), not far south of Auckland; and Christchurch (300,000) and Dunedin (100,000) on the drier east coast of South Island. Christchurch is the main urban center of the Canterbury Plains, which contain the largest concentration of cultivated land in New Zealand. About half of the country's population lives in these five modest-sized metropolises. But the largest, Auckland, has only 800,000 people, whereas the Sydney metropolitan area in Australia has about the same population as all of New Zealand.

Conditions and Problems of African Development

News from Africa tends to focus on disastrous events associated with political and military turmoil, drought, poverty, and exploitation. Less publicized is the strong family life of great numbers of Africans such as the Zairian couple seen here in front of their four-room home in the Kasai region.

RECENT decades have seen a worldwide surge of interest in the long-exploited, underdeveloped, and problem-ridden region of Africa. Although many facets of African life have drawn attention, interest has focused primarily on the area's mineral wealth, its new political independence from European colonialism, its warfare and strife accompanying and following independence, and its serious development problems. In trying to gain introductory perspective on this large and complex part of the world, which now contains nearly one third of all the independent nations on the globe, we shall focus on Africa's outsized dimensions, major population concentrations, European colonial heritage, major problems today, and diverse environments, peoples, and modes of life. Broad outlines will be sketched in this chapter, and Chapters 22 and 23 will examine selected aspects in more detail within the context of major African regions. Area and population data for African countries and regions are given in Table 21.1.

AREA, POPULATION, AND MAJOR AREAL DIVISIONS

Africa (including Madagascar and other islands) is the largest in land area of all the major world regions discussed in this book. Its 11.7 million square miles (30.3 million sq. km) make it more than three times the size of the United States. But much of it is sparsely populated (although in relation to resources and technology many areas are actually overpopulated). The region's population, increasing at the very high rate of 2.9 percent a year, was estimated at 623 million in mid-1988. This gives an average density that is substantially smaller than that of the United States (Table 3.1). However, this density gap will close rapidly, as Africa's current rate of natural population increase is four times that of the United States.

The relatively low population density of Africa as a whole obscures the fact that the majority of Africans are packed into a small number of densely populated areas which collectively occupy a very minor share of the region's total area. The principal ones are (1) the coastal belt bordering the Gulf of Guinea in West Africa, from the southern part of Africa's most populous country, Nigeria, westward to southern Ghana (for place locations, see Figs. 21.1 and 21.2); (2) the savanna lands in the northern third of Nigeria; (3) the Nile Valley and Delta in Egypt;

(4) the coastal fringes and some of the Atlas ranges of northwestern Africa in Algeria, Morocco, and Tunisia; (5) the highlands of Ethiopia; (6) the highland region surrounding Lake Victoria in Kenya, Tanzania, Uganda, Rwanda, and Burundi; and (7) the eastern coast and parts of the high interior plateau (High Veld) of the Republic of South Africa. Each of these population concentrations is a key area in the political and economic geography of Africa. Lesser concentrations are scattered irregularly through the sparsely inhabited deserts, steppes, and grassy or forested expanses of tropical "bush" that make up most of Africa.

Africa falls broadly into the two major divisions of North Africa, the predominantly Arab-Berber realm of the continent, and Africa south of the Sahara, often called sub-Saharan or Black Africa. North Africa is mostly comprised of desert or steppe with a fringe of mediterranean climate in northwestern coastal sections, while sub-Saharan Africa is mainly a region of tropical and subtropical grassland or deciduous forest, but with sizable areas of tropical rain forest and of dry lands, including some desert. Since North Africa has been considered in some detail in Chapters 12 and 13, major attention here will be focused on sub-Saharan Africa.

THE GEOGRAPHIC IMPACT OF EUROPEAN COLONIZATION

It now seems probable that the African continent was the original home of the human race, but by the dawn of history much of this immense area was out of touch with the outer world. While early civilizations were developing in Egypt, southwestern Asia, and Europe, the nature of Africa south of the Saharan desert barrier remained largely unknown. Egyptians, Romans, and Arabs developed contacts with the northern fringes of sub-Saharan Africa, and some trade filtered across the Sahara, but most of the "Dark Continent" was a self-contained, tribalized land of mystery. Even at the dawn of the 20th century the vast spaces of interior tropical Africa were still very poorly developed by European or American standards and were little known to Westerners.

The isolation of much of Africa from contact with more developed areas until comparatively recent times was due to a combination of circumstances. Among these were:

1. The great Sahara Desert barrier in northern Africa made it difficult to penetrate tropical Africa by land from the north and even hind-

ered early sailing expeditions due to the lack of water and fresh provisions along the shore.

2. The African shoreline was physically inhospitable to early visitors approaching from the sea. Africa has an exceptionally regular coast with comparatively few good harbors. In places the shore is lined by submerged rocks, reefs, or sandbars, and surf is often heavy enough to create difficulties for shipping. Thick tropical forests, including mangrove swamps, often border the shore in the rainy tropics, as do some of the world's driest deserts at the northern and southern ends of the continent.

3. The character of African rivers made it hard to reach the interior by water. Rivers are often shallow and full of sandbars near their mouths, generally fluctuate a great deal in water level from season to season, and are blocked inland at varying distances by falls and rapids.

4. Many coastal areas posed severe health hazards associated with a humid tropical environment, including a multitude of diseases such as malaria and tropical fevers.

5. Many African peoples were hostile to outsiders, due in considerable part to slave trading by Europeans, Americans, and Arabs.

6. The slave traders themselves resisted the penetration of the continent by other interests that might interfere with the supply of slaves.

7. Perhaps the most important reason was simply that the interior appeared to lack sources of wealth justifying the risks of penetration. At any rate, the earliest Portuguese voyages to sub-Saharan Africa in the European Age of Discovery were followed by some 400 years of Western activity that was largely confined to the immediate coasts.

Sub-Saharan Africa's isolation from the Western world began to end in the 15th century, when successive expeditions sent from Portugal by sea made their way southward along the Atlantic coast, until the continent's southern tip was finally rounded toward the end of the century and exploration of the Indian Ocean began. As these expeditions progressed, a fringe of trading posts and way stations to service passing ships was established around the Atlantic and Indian Ocean coasts of Africa. Other Europeans soon began to challenge the early Portuguese dominance, and a fierce competition developed for African slaves, ivory, gold, and other products. Slave trading, primarily by Arabs, was already extensively developed at the beginning of the European colonial age, but it was greatly intensified when the estab-

lishment of plantation agriculture in the Americas created a vast new market for slaves. An estimated 11 to 12 million Africans were shipped overseas for sale to slave plantations and mines, although many did not survive the voyage to the New World. This notorious traffic, from which many fortunes were made, provided the main early motivation for European commerce along the African coasts, and it inaugurated the long era when Africa was exploited for profit and political advantage by Europeans and their governments.

Not until the later 19th century did European enterprise begin to penetrate deeply into interior Africa. For the most past the early traders stayed on the coast and bought slaves and other goods from African middlemen. (Many slaves were captives in tribal wars—a process the traders often fostered by providing guns and ammunition to some tribes so they could prey on others.) In the early 19th century the European governments, led by Great Britain, made slaving illegal, but the profitable trade was hard to stamp out. One of the motives that led several governments to claim territory in the interior of tropical Africa during the last quarter of the 19th century was a desire to stop the slave traffic at its sources. Penetration of the interior was pioneered by a series of important journeys of exploration which began in the 1850s. These expeditions were undertaken by missionaries, traders, explorers motivated by scientific curiosity and often sponsored by geographic or other scientific associations, adventurers of various sorts, and in some cases government officials. Once the main outlines of inner African geography were revealed, a scramble for colonial territory in the interior took place among the European powers, beginning about 1881.

By 1900 Africa had become overwhelmingly a region of European dependencies—a status it retained for more than half a century. In these possessions Europeans were a privileged class, both socially and economically. Then after World War II (mainly in the 1960s and 1970s) the political map was remade as colony after colony gained independence—a peaceful process in many instances but accompanied by much bloodshed in places like Angola and Zaire. By the 1980s the political changeover was nearly complete (Fig. 21.3). But after five centuries of European exploitation, Africa remained so impoverished that economic dependence on European and other outside countries continued to be very strong after independence, and in South Africa, which gained independence as early as 1910, political control by the whites was firmly intrenched and the large black majority continued to form an underprivileged pool of low-wage labor.

(text continued on p. 482)

TABLE 21.1   AFRICA: AREA AND POPULATION DATA

Political Unit[a]	Area (thousand sq mi)	Area (thousand sq km)	Estimated Population (millions, mid-1988)	Estimated Annual Rate of Natural Increase (% 1988)	Estimated Population Density, 1988 (per sq mi)	Estimated Population Density, 1988 (per sq km)	Infant Mortality Rate	Urban Population (%)	Cultivated Land (% of total area)	Per Capital GNP ($US: 1986)
NORTHERN AFRICA										
Egypt	386.7	1001.4	53.3	2.8	138	53	93	45	2	760
Sudan	967.5	2505.8	24.0	2.8	25	10	112	20	5	320
Libya	679.4	1759.5	4.0	3.1	6	2	74	76	1	7500 (1985)
Tunisia	63.2	163.5	7.7	2.2	122	47	71	53	29	1140
Algeria	919.6	2381.7	24.2	3.2	26	10	81	43	3	2570
Morocco	172.4	446.4	25.0	2.6	145	56	90	43	19	590
Western Sahara	103.0	266.8	0.2	2.5	2	1	–	–	0	–
Ethiopia	471.8	1221.9	48.3	3.0	102	40	118	10	11	120
Somalia	246.2	637.7	8.0	3.1	32	13	147	34	2	280
Djibouti	8.5	22.0	0.3	2.5	35	14	132	74	0	740 (1984)
Totals	4018.3	10,407.4	195.0	2.9	49	19	101	34	5	880
WEST AFRICA										
Mauritania	397.9	1030.6	2.1	3.0	5	2	132	35	0	440
Senegal	75.7	196.1	7.0	2.6	92	36	137	36	27	420
Mali	478.8	1240.0	8.7	2.9	18	7	175	18	2	170
Niger	489.2	1267.0	7.2	2.9	15	6	141	16	3	260
Burkina Faso	105.9	274.1	8.5	2.8	80	31	145	8	10	150
Guinea	94.9	245.8	6.9	2.4	73	28	153	22	6	320 (1985)
Ivory Coast (Côte d'Ivoire)	124.5	322.3	11.2	3.1	90	35	105	43	12	740
Togo	21.9	56.8	3.3	3.3	151	58	117	22	25	250
Benin	43.5	112.6	4.5	3.0	103	40	115	39	16	270
Gambia	4.4	11.3	0.8	2.1	197	76	169	21	15	230
Sierra Leone	27.7	71.7	4.0	1.8	144	56	175	28	25	310
Ghana	92.1	238.4	14.4	3.1	156	60	72	31	12	390
Nigeria	356.7	923.8	111.9	2.9	314	121	122	28	34	640
Guinea-Bissau	13.9	36.1	0.9	2.4	67	26	132	27	8	170
Cape Verde	1.6	4.0	0.3	2.8	231	89	77	27	10	460
Liberia	43.0	111.3	2.5	3.1	58	23	127	42	3	450
Totals	2371.7	6142.7	194.2	2.9	82	32	124	28	10	520
EAST AFRICA										
Kenya	225.0	582.7	23.3	4.1	104	40	76	19	4	300
Tanzania	364.9	945.1	24.3	3.6	67	26	111	18	5	240
Uganda	91.1	235.9	16.4	3.4	180	70	108	10	28	230 (1984)
Totals	681.0	1763.7	64.0	3.7	94	36	98	16	8	260

EQUATORIAL AFRICA									
Cameroon	183.6	475.3	10.5	2.6	57	22	126	42	15	910
Gabon	103.3	267.6	1.3	1.6	13	5	112	41	2	3020
Congo	132.0	341.9	2.2	3.4	17	6	112	48	2	1040
Central African Republic	240.5	622.7	2.8	2.5	12	5	148	35	3	290
Chad	495.8	1284.1	4.8	2.0	10	4	143	27	2	70 (1984)
Zaire	905.6	2345.5	33.3	3.0	37	14	103	34	3	160
Rwanda	10.2	26.3	7.1	3.7	696	270	122	6	38	290
Burundi	10.7	27.8	5.2	2.9	486	187	119	5	47	240
Equatorial Guinea	10.8	28.0	0.3	1.9	31	12	130	60	8	197 (1983)
São Tomé and Príncipe	0.4	1.0	0.1	2.7	298	115	62	35	38	340
Totals	2092.9	5420.6	67.6	2.9	32	13	115	30	5	380
SOUTH CENTRAL AFRICA									
Zambia	290.6	752.6	7.5	3.7	26	10	87	43	7	300
Malawi	45.7	118.4	7.7	3.2	169	65	157	12	20	160
Zimbabwe	150.8	390.6	9.7	3.5	64	25	76	24	7	620
Angola	481.4	1246.7	8.2	2.6	17	7	143	25	3	1030 (1982)
Mozambique	309.5	801.6	15.1	2.6	49	19	147	19	4	210
Totals	1278.0	3310.0	48.2	3.1	38	15	124	24	7	440
SOUTHERN AFRICA									
South Africa	471.4	1220.9	35.1	2.3	75	29	66	56	11	1800
Namibia	318.3	824.3	1.7	2.8	5	2	111	51	1	1020
Botswana	231.8	600.4	1.3	3.4	6	2	67	22	2	840
Lesotho	11.7	30.3	1.6	2.6	137	53	106	17	10	410
Swaziland	6.7	17.4	0.7	3.1	105	40	124	26	8	600
Totals	1039.9	2693.3	40.4	2.4	39	15	70	53	6	1660
INDIAN OCEAN ISLANDS									
Madagascar	226.7	587.0	10.9	2.8	48	19	63	22	5	230
Mauritius	0.8	1.9	1.1	1.2	1336	516	27	42	58	1200
Comoros	0.7	1.8	0.4	3.4	607	234	96	23	43	280
Mayotte	0.14	0.4	0.1	4.9	537	207	—	53	—	—
Réunion	1.0	2.5	0.6	1.8	586	226	12	60	22	3940
Totals	229.3	593.9	13.1	2.7	57	22	58	26	5	480
Grand Totals	11,711.1	30,331.6	622.5	2.9	53	21	110	30	6	620

aSee notes to Table 12.1 on the Middle East. The Canary Islands, governed as part of Spain, and the Madeira Islands, governed as part of Portugal, lie in the Atlantic off northwestern Africa but are not included in the table. Figures for South Africa include that country's "independent" Homelands.

Figure 21.1   *Reference map of African countries, cities, and landforms, showing political units as of early 1989.*

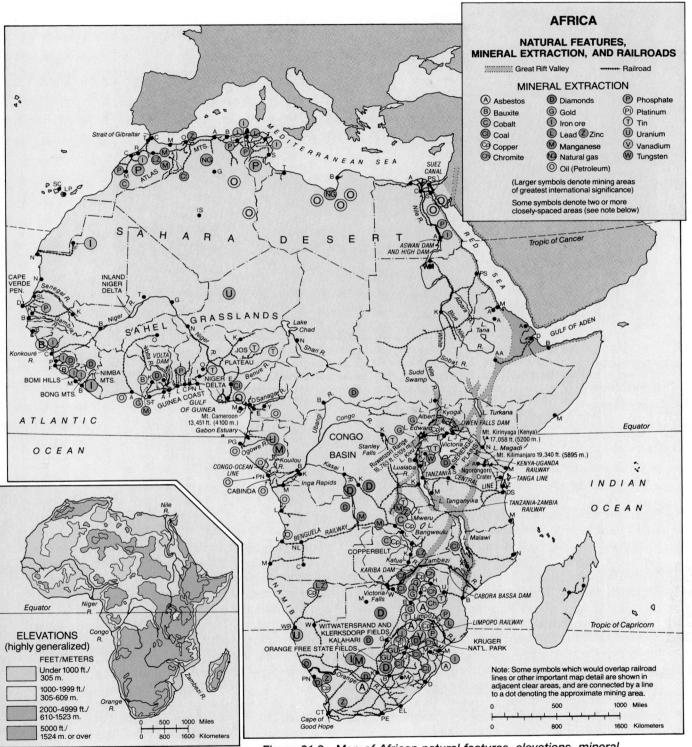

Figure 21.2 Map of African natural features, elevations, mineral extraction, and railroads. Not all minerals are shown, for example, Zimbabwe exports nickel. Large fluctuations in mineral output from year to year are characteristic of the world's mining areas, including the African areas shown by symbols on this map. In Africa, such fluctuations are generally caused by unstable markets, political and military changes and hazards, or the depletion of mineral deposits.

Figure 21.3   The new postcolonial Africa of independent states is symbolized here by the numerous flags at the Kenyatta Center in Kenya's capital, Nairobi. Named for Jomo Kenyatta, independent Kenya's first President, the Center hosts many international conferences and houses regional offices of various international organizations.

Of course, European colonial enterprise in Africa did not simply ransack the region for wealth, although a heavy-handed crudity and barbarity were often displayed in the earlier days. The colonies, and the independent nations which succeeded them, were the beneficiaries of new cities and transport links built by the colonizers (albeit with cheap African labor); new medical and educational facilities (often developed by Christian missions); new crops and better agricultural techniques; employment and income provided by new mines and modern industries in scattered places; new governmental institutions; and government-made maps that could be used as a basis for administration and planning. Such innovations were—and are—very helpful, but they were very unequally distributed from one colony to another, and in general they proved far from adequate for the needs of modern societies when independence came. African economies have provided wealth for overseas investors and prosperity for privileged elements in Africa itself but have left most of the region's people with a very precarious livelihood. When over 33 million people in large and mineral-rich Zaire have a per capita product as low as $160 a year (1986), it is apparent that

the Belgian colonizers left behind a country ill-equipped to participate in the world's "revolution of rising expectations."

AFRICAN PROBLEMS TODAY

Although most of Africa is still beset by deep poverty, recent decades have seen notable changes for the better in the conditions of African life. Improved standards of health and literacy in many areas have resulted from the work of government agencies and, in sub-Saharan Africa, of Christian missions. Better transportation facilities, particularly a wide extension of roads and airways, have made possible the marketing of farm products, including perishable items, from formerly inaccessible areas. Stores and markets in both rural and urban areas stock a variety of manufactured goods from overseas, and from African sources as well, although the supply is rather irregular and prices may often be unbelievably high in the many countries where inflation is rampant. Modern factories have been established in many urban centers, and improved agricultural techniques have been introduced in many areas. Today it is becoming very difficult to find peoples in Africa who are still essentially outside the range of modern influences.

Such changes have affected some peoples and areas much more than others, and the total impact of change represents only a beginning in lifting Africa from its current underdevelopment. One recent United Nations study placed 21 African countries among the 31 least developed in the world. Africa still shares the customary problems of such countries, including (1) a high incidence of illiteracy, poverty, hunger, and disease; (2) native customs, attitudes, and forms of social organization which tend to hinder the development of a modern economy; (3) inadequate overall facilities for transportation and communication (Fig. 21.4); and (4) a lack of domestic capital which could be used to foster increased agricultural and industrial production and thus to raise the standard of living. Other problems arise from a heavy dependence on outside markets. Africa's place in the commercial world is still basically that of a producer of foods and raw materials for sale outside the region. In the world markets where African goods are sold, there is a considerable fluctuation in demand and in prices over a period of years. In the political and social realm there are major problems of racial, tribal, and ideological friction. Their seriousness is attested to by the fact that many African countries have been ravaged by warfare and terrorism, often including large-scale massacres, at some time (often repeatedly) since the end of World War II. Many instances are cited in Chapters 13 and 22.

However, in the late 1980s it appeared that Africa's most pressing and vital problem was not to make its people prosperous or stop their widespread strife, but simply to feed them. Providing an adequate food supply has become increasingly difficult due to prolonged droughts; rapid population increases; unwise land-use practices brought on by population pressure; government policies favoring industry over agriculture and export crops over food crops; disruption of agriculture by warfare, inadequate transportation, and maintenance; and the high cost of imported oil, fertilizer, food, and other necessaries. These interrelated problems are surveyed in greater detail in Chapter 22.

THE DIFFICULT AFRICAN ENVIRONMENT

Many of the problems of Africa are due to a natural environment which is difficult to manage, particularly by peoples equipped with a rudimentary technology. This environment varies greatly in detail from place to place, but its broad outlines are relatively simple.

Figure 21.4 Women with characteristic headloads in the Shaba region of southeastern Zaire. Much of tropical Africa's produce moves to local markets in this manner. In the background at the left is an oil palm. The climate is tropical savanna.

SURFACE CONFIGURATION: A CONTINENT OF PLATEAUS

Most of Africa is a vast plateau or more precisely, a series of plateaus, at varying elevations (Fig. 21.1, inset). The plateau surfaces, predominantly level to rolling, are hollowed by shallow basins, often occupied by prominent river systems such as the Congo, Zambezi, and Orange. The principal lowland plains form a narrow band around the coasts, averaging 20 to 100 miles (c. 30–160 km) in width, although in some places much narrower, and in other places considerably wider. Inland from the coast, abrupt or ragged escarpments, often hilly or mountainous, mark the transition to the plateau, which customarily lies at an elevation of over 1000 feet (305 m). In southern and eastern Africa the general elevation rises to 2000 or 3000 feet, with considerable areas at 5000 feet or higher (Fig. 21.2, inset). The principal highlands above 5000 feet are found discontinuously within a broad zone extending from South Africa, Namibia, and Angola to Ethiopia. In this highland belt are located the highest peaks and largest lakes of the continent. The mountainous areas are found mostly in (1) Ethiopia, (2) the general region of the East African lakes, and (3) the eastern and southern parts of South Africa. The loftiest summits lie within a radius of 250 miles (c. 400 km) from Lake Victoria; they include mounts Kilimanjaro (19,340 feet/5895 m) and Kirinyaga (Kenya) (17,058 feet, 5200 m), which are volcanic cones, and the Ruwenzori Range (16,763 feet/5109 m), thought to be a nonvolcanic massif produced by faulting.

Lake Victoria, the largest lake in Africa, is surpassed in area among inland waters of the world only by the Caspian Sea and Lake Superior. Other large lakes in East Africa include lakes Tanganyika, Malawi, and several others. Although these lakes, particularly Lake Victoria, are used considerably for local water transportation and to some extent as connecting links in long-distance transportation routes, their total traffic is only a tiny fraction of that carried by the Great Lakes of North America.

Aside from the discontinuous mountain zone in eastern and southern Africa, the most extensive area of high mountains is the Atlas Mountains of northwestern Africa, described briefly in Chapters 12 and 13.

One of the most spectacular features of Africa's physical geography is the Great Rift Valley, a relatively broad, steep-walled trough, often with mountainous edges, which extends from Mozambique northward to the Red Sea and the valley of the Jordan River in southwestern Asia (Fig. 21.2). The Rift Valley has several branches, and much of it is occupied by lakes, rivers, seas, and gulfs. It contains most of the larger lakes of Africa, although Lake Victoria, located in a depression between two of its principal arms, is an exception. The valley floor, generally 20 to 60 miles (c. 30–100 km) in width, is far below sea level in some places such as the bottom of Lake Tanganyika, the second deepest lake in the world, but in other places is 5000 feet (1524 m) or more above the sea.

AFRICAN RIVERS

The physical structure of Africa has significantly influenced the character of African rivers. The main ones rise in interior uplands and descend by stages to the sea. At various points they descend abruptly, particularly at plateau escarpments, so that their courses are interrupted by rapids and waterfalls. These often block navigation a short distance inland from the sea. In addition, the navigation of many African rivers is hindered or prohibited by low water at certain seasons and by shallow and shifting delta channels. The maze of interlaced channels in some deltas made it difficult for early explorers to find the main stream. Among the important rivers which have built deltas are the Nile, Niger, Zambezi, Limpopo, and Orange. In contrast, the Congo River (called the Zaire in the Republic of Zaire) has scoured a deep estuary 6 to 10 miles (10–16 km) wide, which can be navigated by ocean vessels for a distance of about 85 miles (c. 135 km) to the seaport of Matadi in Zaire. The Congo is used more for transportation than is any other African river. A few miles above Matadi navigation is blocked by rapids, and goods must be transshipped by rail or truck a distance of about 230 miles (370 km) to Kinshasa. Above that city the Congo and its major tributaries are navigable for long distances by good-sized river craft, although the continuity of transportation is broken in places by falls and rapids which are bypassed by rail or road. The character of water transportation on the Congo system is generally representative of Africa's inland waterways, which are discontinuous and interconnected by rail or highway in a manner not duplicated in any other continent.

The frequent falls and rapids of African rivers have a more positive side. They represent a great potential reservoir of hydroelectric energy. Large power stations exist on the Nile River at the Aswan High Dam in Egypt; on the Zambezi River at the Cabora Bassa Dam in Mozambique and at the Kariba Dam which is shared by Zimbabwe and Zambia; at the Kainji Dam on the Niger River in Nigeria and at a large dam on the Volta River in Ghana. Many other stations of varying size are scattered over the continent. But only a very small fraction of the total hydroelectric potential has been developed thus far. Many of the best sites are remote from large markets for power. The largest single potential source of hydroelectricity in Africa, and possibly in the world, is the stretch of rapids on the Congo River between Kinshasa and Matadi. A sizable beginning has been made on Zaire's Inga Project to develop this resource.

CLIMATE AND VEGETATION

Africa is bisected by the equator. The distance from the equator to the northernmost point of the continent is approximately 2580 miles (*c.* 4150 km) and to the southernmost point, 2400 miles (*c.* 3860 km). Thus most of Africa lies within the low latitudes and is characterized by tropical climates. One of the most striking characteristics of Africa's climatic pattern is its symmetry or regularity. This is due mainly to the position of the continent athwart the equator, coupled with the generally level character of the surface and the comparative absence of extensive chains of high mountains.

Types of Climate

The broad pattern of climates in Africa may be outlined as follows. Areas of *tropical rain forest climate* in central and western Africa near the equator gradually merge into *tropical savanna climate* on the north, south, and east (Fig. 2.18). The savanna areas, in turn, trend into *steppe* and *desert* on the north and southwest. Along the northwestern and southwestern fringes of the continent are relatively small but important areas of *mediterranean climate,* while eastern coastal sections and adjoining interior areas of South Africa have a climate classified as *humid subtropical.* High elevations moderate the temperatures of extensive interior areas that lie within the realm of tropical savanna climate in the east and south of the continent.

TROPICAL RAIN FOREST CLIMATE

In Africa rainfall is heaviest, generally speaking, in areas near the equator in central and western Africa. Here abundant moisture and continuously high temperatures produce tropical rain forest as the characteristic type of climate and associated vegetation. The main area of rain forest forms a broad band along the equator from the Gulf of Guinea eastward to the highlands of East Africa. It includes most of southern Nigeria, plus Equatorial Guinea, most of Gabon, southern Cameroon, and large areas in the northern part of the Congo Basin. A much smaller area lies farther west, centering in Liberia. This second area has a 3- to 4-months' dry season but enough total rainfall to support rain forest vegetation. Still a third area, remote from the main body, rims the eastern, windward side of Madagascar. The tropical rain forests of Africa have been highly publicized, but they occupy only about one tenth of the continent.

TROPICAL SAVANNA CLIMATE

Areas of tropical savanna climate (Fig. 2.21) are far more extensive than areas of tropical rain forest climate in Africa. The most characteristic feature of this type of climate is the alternation of well-marked wet and dry seasons. The natural vegetation is most often dry forest or scrub, intermixed with expanses of tall, coarse, tropical grass, although one of the best known of the endless variations in savanna-type vegetation is "park savanna," comprised of grass and scattered flat-topped trees. Near the desert margins of the savanna climate, rainfall is lighter and there is characteristically a belt of shorter grass or shrubs often referred to as *tropical steppe.*

A broad belt of tropical steppe and savanna bordering the Sahara Desert on the south is known as the Sudan. Another name applied especially to the steppe portions is "Sahel." In the late 1960s and 1970s it came to world attention because of a devastating 7-year drought that wiped out many millions of livestock and is estimated to have brought death by starvation to tens of thousands of people. Seldom has a large area had its life so thoroughly disrupted by a natural calamity. The drought was an intensification of the precarious rainfall situation that characterizes this land along the southern margin of the Sahara. In the Sahel region, desert is creeping southward year by year at the expense of grazing land and crop land. This process, which is known as *desertification,* is widespread in Africa.

DESERT CLIMATE

Deserts border the tropical savannas and steppes on the north and southwest. The great Sahara, the world's largest desert, extends across the north of Africa from the Atlantic Ocean to the Red Sea. It is bordered on the north by a narrow belt of mid-latitude steppe climate. In South Africa and Namibia, a coastal desert, the Namib, borders the Atlantic. The "Kalahari Desert," which lies inland from the Namib, has a considerable growth of low grasses and shrubs, and it is better described as steppe or semidesert than as true desert.

MEDITERRANEAN AND HUMID
SUBTROPICAL CLIMATES

The coastal areas bordering the Mediterranean Sea in northwestern Africa and the southwestern tip of the continent around Cape Town have a mediterranean or dry-summer subtropical climate characterized by rainy winters and almost complete drought during the midsummer months. The southeastern coast of the continent in Natal Province of South Africa has rain at all seasons and is classed as humid subtropical. The high interior grasslands of South Africa, or High Veld, also have a subtropical climate, but one which resembles the tropical savanna climate in having a well-marked dry season during the period of low sun. Climatic data for selected African stations are given in Table 21.2.

WATER RESOURCES AND PROBLEMS

The total amount of precipitation received by the African continent is very large. Unfortunately, however, it is poorly distributed. Some parts of Africa receive too much rain, whereas other areas have scarcely any. Even in the rainier parts of the continent large areas have a dry season of considerable length, and wide fluctuations occur from year to year in the total amount of precipitation. In places considerable areas are inundated by waters which gather in swamps or marshes such as the Sudd Swamp along the Nile River in the country of Sudan. One of the major needs of Africa is better control over water, involving irrigation projects in some areas, drainage projects elsewhere, conversion of marshes and swamps to rice fields or pastures, and development of dams along both major streams and small tributaries to control floods, regularize the flow of water between seasons, and provide hydroelectric power.

In the typical village household water is carried laboriously by hand from a stream or lake or a shallow, polluted well—a task ordinarily performed by women. Especially in the seasonally rainy areas, more use of small dams is needed to provide water storage throughout the year, coupled with simple pipelines or flumes to carry water by gravity to households or the fields. Inhabitants of villages "can be taught to dig and line wells, and such a primitive device as an endless porous rope of fibers absorbing water and passing through two wooden rollers to squeeze out the supply is within the power of the least educated to make and maintain. Unfortunately, over much of Africa's underlying complex of ancient rocks the water table behaves irregularly, and a well is both difficult to dig and uncertain of its supply.

There is much to be said, in hilly or rolling country, for horizontal wells into the hillsides from which water would flow by gravity as an artificial spring."[1]

AFRICAN SOILS

The scientific study of African soils is still in an early stage. This represents a critical frontier of human endeavor in Africa, inasmuch as significant advances in the utilization of African lands for agriculture must be founded in large part on a better knowledge of the soils. Based on present knowledge, the following observations can be made.

Among the most productive soils of the continent are some alluvial soils found on river plains. Soils along the Nile River in Egypt, for example, are especially noted for their fertility and have supported agriculture for thousands of years (Fig. 2.23). Other especially fertile soils are found in scattered areas (particularly in parts of the East African highlands) where certain types of volcanic parent material occur. A third group of better than average soils is the grassland soils found in some areas of tropical steppe or tropical highland and in the mid-latitude grasslands of the High Veld in South Africa. These soils, however, do not appear to be entirely comparable to the chernozems, prairie soils, and chestnut soils of North America. They are more difficult to cultivate and seem to lose their fertility more quickly under continuous cropping.

Soils of the deserts and regions of mediterranean climate are generally thin and immature; over broad areas of desert, true soils are absent. In the mediterranean areas, fertile soils are found in some valleys where transported materials from adjoining slopes have accumulated.

In the tropical rain forests and savannas reddish tropical soils are generally dominant. There is a common idea that these soils are exceptionally fertile when cleared of their natural vegetation and used for agriculture. This idea is apparently based on the luxuriant plant growth which they often support in their natural state. However, the truth appears to be that they are rather poorly supplied as a rule with plant nutrients and tend to lose their fertility quickly when used for crops. They are usually low in humus and deficient in lime. These disadvantages of tropical soils are at least partly due to the rapid chemical action induced by the abundant heat and moisture

---

[1] L. Dudley Stamp and W. T. W. Morgan, *Africa: A Study in Tropical Development* (3rd ed.; New York, John Wiley & Sons, Inc., 1953, 1964, 1972), p. 82.

TABLE 21.2   CLIMATIC DATA FOR SELECTED AFRICAN STATIONS

Station	Political Unit	Latitude to Nearest Whole Degree	Elevation Above Sea Level (ft)	Elevation Above Sea Level (m)	Type of Climate	Average Temperature Annual °F	Average Temperature Annual °C	January (or coolest month) °F	January (or coolest month) °C	July (or warmest month) °F	July (or warmest month) °C	Average Precipitation Annual (in)	Average Precipitation Annual (cm)	Number of Months Averaging Less Than 1 in. of Rain
Casablanca	Morocco	33°N	190'	58 m	Mediterranean	64°F	18°C	54°F	12°C	73°F	23°C (Aug.)	17"	43 cm	5
Ghardaia	Algeria	32°N	1725'	526 m	Desert	70°F	21°C	51°F	11°C	93°F	34°C	3"	8 cm	12
Djibouti	Djibouti	12°N	23'	7 m	Desert	84°F	29°C	77°F	25°C	93°F	34°C	5"	13 cm	11
Dakar	Senegal	14°N	79'	24 m	Tropical steppe	75°F	24°C	68°F	20°C (Feb.)	82°F	28°C (Sept.)	23"	58 cm	8
Kano	Nigeria	12°N	1561'	475 m	Tropical savanna	79°F	26°C	70°F	21°C	88°F	31°C (Apr.)	34"	87 cm	7
Accra	Ghana	5°N	213'	65 m	Tropical savanna	79°F	26°C	75°F	24°C (Aug.)	82°F	28°C (Feb., Apr.)	31"	78 cm	2
Lagos	Nigeria	6°N	125'	38 m	Tropical rain forest	81°F	27°C	75°F	24°C (July)	82°F	28°C (Feb., Mar.)	64"	163 cm	0
Douala	Cameroon	4°N	43'	13 m	Tropical rain forest	79°F	26°C	77°F	25°C (July)	81°F	27°C (Dec., May)	162"	411 cm	0
Nairobi	Kenya	1°S	5897'	1797 m	Tropical savanna (upland)	64°F	18°C	59°F	15°C (July)	66°F	19°C (Mar., Apr., Oct.)	36"	93 cm	1
Lubumbashi	Zaire	11°S	4185'	1276 m	Tropical savanna (upland)	69°F	20°C	63°F	17°C (July)	73°F	23°C (Oct.)	48"	123 cm	5
Harare	Zimbabwe	18°S	4828'	1472 m	Tropical savanna (upland)	64°F	18°C	57°F	14°C (June, July)	70°F	21°C (Oct., Nov.)	34"	86 cm	5
Beira	Mozambique	20°S	26'	8 m	Tropical savanna	75°F	24°C	68°F	20°C (July)	82°F	28°C (Feb.)	56"	143 cm	0
Maun	Botswana	20°S	3100'	945 m	Tropical steppe (upland)	72°F	22°C	61°F	16°C (June, July)	81°F	27°C (Oct.)	19"	47 cm	6
Port Nolloth	South Africa	29°S	23'	7 m	Desert	57°F	14°C	53°F	12°C (Aug.)	60°F	16°C (Dec.)	2"	6 cm	12
Cape Town	South Africa	34°S	39'	12 m	Mediterranean	63°F	17°C	55°F	13°C (July, Aug.)	72°F	22°C (Jan., Feb.)	26"	65 cm	5
Durban	South Africa	30°S	26'	8 m	Humid subtropical	68°F	20°C	62°F	17°C (June)	75°F	24°C (Feb.)	41"	104 cm	0
Johannesburg	South Africa	26°S	5556'	1694 m	Humid subtropical (High Veld)	61°F	16°C	50°F	10°C (June, July)	66°F	19°C (Jan., Feb.)	30"	75 cm	5
Toamasina	Madagascar	18°S	20'	6 m	Tropical rain forest	75°F	24°C	69°F	21°C (July)	79°F	26°C (Feb.)	139"	353 cm	0

of tropical climates. Such conditions promote rapid leaching of nutrient materials by percolating waters and cause organic matter exposed to the air to combine with oxygen rather quickly and thus to be lost into the atmosphere as carbon dioxide rather than being converted to humus and plant food by soil bacteria. In places the highly leached subsoil forms a material called *laterite,* which hardens when exposed and makes cultivation useless.

DISEASES AND PARASITES

A high incidence of diseases and parasites affecting people, domestic animals, and cultivated crops has been one of the main hindrances to African development. Certain parts of tropical Africa, particularly the rainy lowlands of West Africa, long had a reputation in Western countries as a "white man's graveyard." Among native Africans diseases and parasites take a heavy toll of strength and energy, even when they are not fatal. Many of the major diseases are carried by insects. Those carried by mosquitoes include malaria, yellow fever, and dengue or "breakbone fever." Sleeping sickness is carried by the tsetse fly, which also transmits nagana, a destructive disease affecting cattle and horses. Plague is spread by fleas, and relapsing fever by ticks and lice. Large number of Africans are afflicted by digestive diseases and parasites, including dysentery, typhoid and paratyphoid fever, bilharziasis, and hookworm and other types of intestinal worms. Such afflictions may be traced in most instances to the use of contaminated water or to other unsanitary conditions. Other diseases common in Africa include tuberculosis, filariasis, nutritional deficiency diseases, pneumonia, yaws, leprosy, influenza, trachoma, venereal diseases, and many fungoid diseases of the skin. Recently AIDS (acquired immune deficiency syndrome) has spread so extensively in Africa as to amount to a new and major plague.

Despite the miserable conditions which still prevail in some areas, much progress has been made in recent decades toward the conquest of disease in Africa. Enough is known about the control of tropical diseases and parasites to greatly reduce their incidence if means were available for the technical knowledge to be fully applied. But in vast, poverty-stricken Africa, the need for medical assistance far outruns the money and personnel available for such assistance. Even where medical facilities are present, doctors and medical technicians are greatly handicapped by the ignorance and fears of their African clientele. Thus education is a vital sector in the fight against ill health and disease.

NATIVE ANIMAL LIFE

Africa's wild animals have been so well publicized that the region is often thought of as a sort of giant zoo. However, it is easy to overstate the abundance and significance of African animal life at the present time. While reptiles, monkeys, and birds are still numerous, the total numbers of most of the larger animals—the elephant, lion, giraffe, zebra, hippopotamus, rhinoceros, and so on—have greatly declined. This has been partly the result of unrestricted hunting in past times but has also been due to encroachment by an increasing human population and their livestock on the habitats of wild species. Whatever the cause, the numbers of some species have now diminished to the point that they are often protected by law against excessive hunting. Such laws are difficult to enforce, and poaching on a large scale is reported to be making serious inroads on animal populations. In some areas attempts are being made to conserve the remaining wildlife in large game reserves or national parks such as the Kruger National Park in the extreme northeast of South Africa or the Serengeti National Park in Tanzania.

The tropical grasslands and open forests of Africa have been the principal habitat of the larger herbivorous animals, such as the elephant, buffalo, antelope, zebra, and giraffe, and also of carnivorous and scavenging animals, such as the lion, leopard, and hyena. The tropical rain forests have been much more deficient in the larger species of animals than is commonly realized. In the rain forests the most abundant species have been birds, monkeys, and snakes, together with the hippopotamus, the crocodile, and a great variety of fish in the innumerable streams.

MINERAL RESOURCES AND PRODUCTION

Africa has become increasingly important in recent times as a producer of minerals (Fig. 21.2). Most of the production is exported, principally to Europe and the United States. In world terms, the continent is particularly important as a producer of (1) precious metals and precious stones, (2) ferroalloys, (3) copper, (4) phosphate, (5) uranium, (6) petroleum, and (7) high-grade iron ore. Numerous other minerals needed by industrialized nations are produced and exported. For about one third of Africa's countries, minerals are the most important category of exports.

Table 21.3 shows African production as a percentage of world production for a number of important minerals and indicates the principal African

TABLE 21.3 PRODUCTION DATA FOR SELECTED MINERALS IN AFRICA

Mineral	Percentage of World Output Mined in Africa (1985 estimates)	Leading African Producing Countries (figures represent percentage of total African production: 1985 estimates)
Diamonds		
Gem	66.5	Botswana 32.1; Zaire 30.4; South Africa 24.9; Namibia 4.9; Angola 2.1; Sierra Leone 1.3; Central African Republic 1.3
Industrial	71.9	Zaire 50.0; Botswana 25.2; South Africa 20.0; Ghana 2.1
Gold	46.7	South Africa 95.9; Zimbabwe 2.0; Ghana 1.3
Platinum	46.6	South Africa 99.8
Manganese ore	30.3	South Africa 53.9; Gabon 40.6; Ghana 4.6
Chromium	38.9	South Africa 86.5; Zimbabwe 12.9
Cobalt	68.2	Zaire 80.9; Zambia 18.7
Vanadium	41.5	South Africa 100
Copper	16.7	Zaire 41.2; Zambia 35.5; South Africa 14.9; Namibia 3.5
Tin	5.7	Zaire 26.6; South Africa 20.3; Zimbabwe 16.2; Nigeria 15.7; Rwanda 11.1; Namibia 9.1
Lead	7.7	Morocco 38.7; South Africa 37.7; Namibia 13.2; Zambia 6.1
Zinc	4.1	South Africa 35.8; Zaire 27.3; Zambia 13.3; Namibia 11.0
Asbestos	8.7	Zimbabwe 46.2; South Africa 46.2; Swaziland 7.3
Phosphate rock	20.7	Morocco 65.4; Tunisia 13.0; South Africa 8.8; Senegal 6.1; Algeria 3.8
Iron ore	6.8	South Africa 45.5; Liberia 27.8; Mauritania 14.5; Algeria 5.5
Antimony	15.7	South Africa 85.6; Morocco 11.5; Zimbabwe 2.9
Bauxite	16.5	Guinea 93.3; Sierra Leone 5.7
Coal (all grades)	3.9	South Africa 97.3; Zimbabwe 1.6
Petroleum (crude)	9.2	Nigeria 30.1; Libya 21.7; Egypt 18.2; Algeria 13.2
Uranium oxide (1984)	35.2	South Africa 42.1; Namibia 27.2; Niger 24.0; Gabon 6.7

Source: U.S. Bureau of Mines, *Minerals Yearbook*, 1985, preprints; data on uranium are for 1984, from *United Nations Statistical Yearbook*. Percentages are based on volume.

producing countries for each mineral. It is apparent from this table that four general areas in Africa are of primary importance. These are (1) South Africa and Namibia, (2) the Zaire-Zambia-Zimbabwe region, (3) northwestern Africa and Libya, and (4) West Africa, especially the areas near the Atlantic Ocean.

In *South Africa* gold is the leading mineral in value of production. Uranium secured as a by-product of gold mining makes South Africa the largest producer of this strategic mineral in Africa and one of the largest in the world. The principal gold-mining districts are found on the interior plateau in the southern Transvaal and northwestern Orange Free State; the most famous district is the Witwatersrand or Rand, centering on southern Africa's largest city, Johannesburg. South Africa is the world's leading gold-producing country and is a sizable producer of diamonds (Fig. 21.5), manganese, chromium, vanadium, platinum, copper, and several other important minerals. It has Africa's largest deposits of coal (in fact, they are the continent's only sizable deposits

of good coal except for those in neighboring Zimbabwe) and also has large deposits of iron ore. Coal and iron have provided the basis for the largest iron and steel industry in Africa. *Namibia,* which borders South Africa on the northwest, mines diamonds and various other minerals. The remarkable variety of minerals in South Africa and Namibia is associated with a highly diversified group of rock formations, ranging in age from very recent to extremely ancient rocks.

The *Zaire-Zambia-Zimbabwe region* is also an extremely rich mineralized area. In Zaire, mineral production is mainly centered in the Shaba region of the southeast, bordering Zambia. The country is a major producer of copper and cobalt from the Shaba region, and industrial diamonds from the Kasai region northwest of the Shaba. Copper is by far the most valuable mineral and export product. Zaire also mines relatively small amounts of various other metals. Often a particular ore body will yield one main metal and several others as by-products—a common situation in the world's mining districts. Zambia is

Figure 21.5   Workers in a South African diamond mine. Holes are being drilled for dynamite to blast the "blue ground" that yields diamonds.

even more dependent on copper exports than is Zaire. Its principal mining district (Copperbelt) borders the Shaba mining region of Zaire. Zimbabwe, separated from Zambia by the Zambezi River, has a more varied mineral production than that of Zambia; the leading minerals are gold, chromium, nickel, asbestos, and coal. These have fluctuated a good deal in relative importance over the years. Aside from coal, found in the west at Wankie, Zimbabwe's minerals come mainly from a belt in the central part of the country along or near the railroad connecting the country's two main cities: Bulawayo in the south and Harare (formerly Salisbury), the capital, in the north. At Que Que, midway between the two, a small iron and steel industry utilizes local iron ore and limestone and secures coking coal by rail from the large Wankie deposits about 300 rail miles (*c.* 480 km) to the west.

In *northwestern Africa and Libya* the leading minerals in value of production and exports are oil and gas from Libya and Algeria, and phosphate from Morocco and Tunisia. The region's phosphate deposits are among the largest in the world, and petroleum and natural gas reserves are also of major

size. Natural gas is produced in Algeria and Libya for domestic consumption, and liquefied gas is exported by tanker from Algeria. Many other minerals are produced and exported in varying quantities; of special note are the nonferrous metals and ferroalloys of Morocco.

In *West Africa,* where mineral production comes very largely from countries that touch the Atlantic, the principal items currently mined and exported are petroleum from Nigeria, iron ore from Liberia and Mauritania, bauxite from Guinea, diamonds from Ghana and Sierra Leone, and uranium from the inland country of Niger.

Some mineral production in Africa comes from countries other than those discussed here (for example, petroleum from Egypt), but the total value is very minor when compared with the aggregate production and exports of the four main regions.

Most mining in Africa is done by large corporations financed initially by investors in Europe or America. Mining has attracted far more investment capital to Africa than any other economic activity. Not only has money been invested in mines directly, but a great many of the transportation lines, port facilities, power stations, urban housing and business areas, manufacturing plants, and other elements in the continent's infrastructure have been developed primarily to serve the needs of the mining industry. Practically everywhere in Africa mining is carried on by modern large-scale methods. Managerial and supervisory personnel and skilled workers are often Europeans, although the proportion of non-Europeans in these posts is now quite large. Unskilled or semiskilled labor is done by Africans or Arabs. Great numbers of workers in the mines are temporary migrants from rural areas, often hundred of miles away. They are recruited for specified periods, ranging from several months to several years, and normally are given free transportation to and from the mines and free housing in mine "compounds." When their contracts expire, they return home. The use of migrant labor, not only in mines but also on farms and plantations or in other employment, is very widespread in Africa. Married men who work in mines generally are not accompanied by their families, although there are many exceptions to this, and some areas try to attract a more stable labor force by providing family housing. Thus the system of migrant labor tends to disrupt family life. On the other hand, although the wages paid are not high by the standards of America or Europe, they represent much larger amounts of money than it would be possible to earn in the home village. In addition, the mining companies generally provide free health care and an adequate diet (free or subsidized), and workers are apt to be in better physical condition on returning home than they were when they came to the mines. It may be noted that the recruitment of migrant workers has had important cultural effects. Many millions of Africans have worked for mining companies at one time or another, have come into contact with Western ideas, and have carried these with them on returning to their villages. Positive or negative aspects aside, it is a fact that the employers of migrant labor have been a powerful force in spreading Western influences throughout Africa. In addition, they have brought Africans from different localities into contact with each other and thus in some measure have leavened the parochialism of village life by making villagers more conscious of a wider African civilization. Needless to say, the radio—often bought with funds earned at a mine—also has had powerful effects in these directions. The impact of television is far smaller but is slowly growing (Fig. 22.1). In 1985 there were reported to be about 190 persons per television receiver in Nigeria, about 2500 in Zaire, and about 16,000 in Mali. South Africa, Egypt, and Algeria had a television receiver for every four to six persons, but in most of Africa this medium is either absent or available only to a privileged few.

AFRICAN PEOPLES AND THEIR AGRICULTURAL ECONOMY

Among the peoples who inhabit the African continent, except for a few million Europeans and somewhat over a million Asians (mainly Indians in South Africa), those living south of the Sahara exhibit Negroid physical characteristics in varying degree (see chapter-opening photo), while those north of the desert and in Egypt are basically Caucasoid and speak languages, principally dialects of Arabic or Berber, which are quite distinct from the languages of sub-Saharan Africa. In the east-west grassland belt known as the Sudan, in the Sahara, and in the East African highlands and coastal areas, there has been considerable mixing of Negroid and non-Negroid peoples over a long period. Some peoples in these latter areas speak Hamitic or Semitic languages, while others speak languages native to sub-Saharan Africa. Even the latter languages, however, often reveal Hamitic or Semitic influences.

In regard to language most of the peoples of sub-Saharan Africa belong to one of three broad groupings: (1) the Bantu peoples, most of whom live south of the equator; (2) the Guinea Coast peoples; and (3) the Sudanic peoples. The *Bantu peoples* speak languages belonging originally to a common language stock (Niger-Congo), although tribal variations are often different enough to be mutually unintelligible. The *Guinea Coast peoples* are linguistically related to the Bantu in that their many languages derive originally from the same Niger-Congo. The *Sudanic peoples*, on the other hand, speak languages unrelated to the first two groups. By one count, more than 800 languages are spoken in Africa.

Three minor population elements, the Pygmies, the San—also called Bushmen, now considered a demeaning term—and the Khoi, deserve mention. The *Pygmies*, averaging around 4 1/2 feet in height, are found principally in the tropical rain forests of the Congo Basin, although many are now farmers and herdsmen in nearby savannas. Traditionally, most Pygmies lived by hunting (using poisoned arrows); by gathering roots, nuts, fruits, and other food in the forest; and by trading with non-Pygmy groups. The *San*, found primarily in the Kalahari region of southwestern Africa, are also below average in height, although a little taller than the Pygmies. Traditionally they have been nomadic hunters, using poisoned arrows and spears and constructing crude temporary shelters as they moved about in search of game. They are expert trackers and stalkers of wild animals and are also skilled in finding water in the semidesert country where they live. Like the Pygmies they are gatherers as well as hunters; their diet has included such items as roots, honey, gum from acacia trees, ant eggs, ostrich eggs, lizards, and locusts. Today most of them herd cattle for other African peoples in their region.

The *Khoi*, assigned the name "Hottentots" centuries ago by South Africa's whites, have nearly been eliminated as a separate people. The remnants live mainly on tribal reservations in Namibia. They are primarily herdsmen, raising sheep, goats, and long-horned cattle. They live in semipermanent villages of oval huts built about a central open space. The village is surrounded by a thorn fence, and the animals are driven within the enclosure at night for protection. This type of village, known as a *kraal*, is often found among pastoral peoples of sub-Saharan Africa. Both the Khoi and the San (now often called collectively the Khoisan peoples) formerly ranged widely over southern Africa, but most of the lands they utilized were taken over by Bantu tribes pushing in from the north or by Europeans moving inland from the Cape of Good Hope.

AFRICAN AGRICULTURE

The peoples of the Guinea Coast live principally by tilling the soil and growing tree crops such as the cacao tree and the oil palm. Some tribes among the Sudanic peoples are also cultivators, some are herdsmen, and many are both. The same is true of the Bantu. Some of the chief characteristics of the native agricultural and pastoral economy of Africa have been described as follows by Stamp and Morgan[2]:

> Along the Mediterranean borders we find an agriculture established from very early times, with cattle, the plow, cereals, and Mediterranean fruits—especially barley, wheat, olives, figs, and the grape. Southward toward the desert margins nomadic pastoralism based on sheep replaces cultivation; in the deserts cultivation is limited to oases. . . .
>
> South of the deserts in the semiarid open lands of the Sudan are again pastoralists depending for their livelihood on cattle and sheep. Southward, with increasing rainfall and a modest rainy season, there is

again some cultivation, precarious, and dependent mainly on sorghum and millet. Cultivation is by the hoe; use of the plow remains unknown.

The area of intermediate rainfall of the savanna between the steppes of the desert margins which are too dry and the forests of the equatorial margins which are too wet is favorable to cultivation. Large villages are the rule; there have been in the past empires of considerable size. Dependence for food is on millet and corn (maize); latterly on peanuts on the uplands and rice along the rivers. Tobacco is grown, sometimes cotton; kola nuts are gathered. Often cattle are bred, sometimes sheep and horses, and among the cultivators or mixed farming communities are tribes almost exclusively pastoral.

Unfortunately, where the rainfall becomes more reliable and where mixed farming should reach its maximum intensity, we encounter the "fly belts" where the tsetse fly virtually eliminates domestic an-

Figure 21.6 A housewife in Zaire with a headload of manioc roots. The roots will be beaten to a starchy powder that is a major dietary staple in the African tropics. Cooked manioc has a high caloric value and is very filling but has little other food value. Hence it is generally eaten with more nutritious greens and sauces.

---

[2]L. Dudley Stamp and W. T. W. Morgan, *Africa: A Study in Tropical Development* (2nd ed.; New York, John Wiley & Sons, Inc., 1964), pp. 141–144. Used by permission. Some words added in brackets without altering the sense of the original.

imals. Here the basis of life becomes the yam, manioc [Fig. 21.6], banana, and palm oil . . ., with cocoa as a "cash crop" now developed to a great extent in [Ivory Coast], Ghana, and elsewhere.

. . . Madagascar differs somewhat from the continental mainland in that cattle breeding is there associated with rice cultivation.

. . . . .

Shifting Cultivation

The system of agriculture commonly though somewhat misleadingly known as "shifting cultivation" is practiced through a very large part of tropical Africa [Fig. 21.7]. As a system it is better described as "land rotation" or "bush fallowing." In those areas inhabited by sedentary farming peoples, each village or settlement has proper to it a tract of surrounding land. The tract is probably only loosely defined except where settlements are close together and population dense. In a given year the villagers working together as a community clear a part of this village land, cutting and burning the woodland or scrub and then planting the crops appropriate to the climate and soil of the area. In due course the crops are harvested communally and the land used for a second and perhaps a third year. It is then abandoned, and a fresh tract of the village land is cleared. The abandoned land quickly becomes covered with a second-growth woodland or scrub. In due course the clearings reach a full cycle, and if a given tract has been allowed to lie fallow for about fifteen years it may be regarded as fully rested. Bush clearing is largely man's work. There is also surrounding the village itself, often as a series of enclosed gardens or "compounds" attached to individual huts, the "women's land," cultivated regularly to afford a supply of vegetables for the pot. Often the kokoyams, peppers, beans, melons, bananas, etc., are scarcely grown at all in the open farmland. The women's land is enriched by house sweepings, ashes, refuse, and manure afforded by chickens, goats, and human beings.

The system has often been condemned as wasteful of natural forest, of land, and of labor. But it has many good points. The natural forest is probably second growth of little value anyway. The land cleared in small patches protected by surrounding woodland escapes the evils of soil erosion, and its nutrient status temporarily enhanced by the ashes of the burnt bush is maintained by the fallowing, the soil not being exposed to the atmosphere long enough for serious oxidation. Expenditure of labor is minimized by burning, and no attempt is made to remove large stumps. The cultivation is by hand—by hoeing—so the stumps do not constitute the obstacles they would be if the plow were used. We may accordingly agree with Lord Hailey when he says that shifting cultivation is "less a device of barbarism than a concession to the character of a soil which needs long periods for recovery and regeneration."

Figure 21.7   These remote villages on a ridgetop dirt road grow crops under a system of shifting cultivation. The light-colored patches on the slope at the lower right represent fields currently used. The locale is the Kasai region in southern Zaire. Few of the world's people are quite so poor, isolated, and bereft of conveniences as the African families in this view.

Cash Crops in Low-Latitude Africa

Perhaps two thirds of the people in Africa depend directly on agriculture or pastoralism for a livelihood. In many parts of the continent the proportion is much higher. In most areas a large share of the crops and livestock is grown for the use of the cultivators and their families or for a purely local sale. In other words, there is a strong subsistence component in agriculture. This is particularly true of low-latitude areas, although it applies also to various areas outside the tropics.

In much of Africa the land is deteriorating as population pressure grows. The traditional land-use method of shifting cultivation is still widespread, but fallow cycles have been shortened; hence erosion has increased, and the ability of the soil to restore itself during fallow periods has been lessened. Often land with slopes too steep to till without disastrous erosion has been brought into cultivation. Overgrazing and soil compaction have become major problems in areas where some of the traditional grazing land has been converted to tillage and the livestock now are grazed on a shrinking amount of land. Despite the increased use of chemical fertilizers, better crop varieties, and irrigation in Africa, unwise use is sapping the productive potential of the land and yields are tending to stagnate or actually fall.

In low-latitude Africa the proportion of crops grown for export to overseas destinations or for sale in African urban centers has risen significantly in recent times. This has been partly a result of taxation and other governmental pressures but has also been due to the desire of Africans for cash with which to purchase manufactured goods (hardware, utensils, bicycles, radios, clothing, etc.) or food. In many areas, farmers have become very dependent on the sale of cash crops. Governments tend to favor such crops as a means of gaining foreign exchange with which to buy such items as foreign technology, industrial equipment, arms, and consumption items for the elite.

Most export crops in low-latitude Africa are grown partly on small African farms and partly on plantations or estates,[3] but in the majority of cases the production comes primarily from small farms. Large plantations and estates have never become established to the same degree as in Latin America or Southeast Asia, and many of them have been forced out of business by political or economic pressures during the period of transition from European colonialism to independence. Some plantations have been nationalized by governments of new independent states, some have been obliged to sell their land to Africans, and others, for example in Zaire, have been unable to survive the civil warfare and other turmoil accompanying independence. Many of these units, such as the well-known Firestone rubber plantations in Liberia, continue to operate, and for some countries they provide a very important source of tax revenue and foreign exchange. But throughout tropical Africa as a whole there is a definite trend for export production to come increasingly from small farms. The export crops of greatest value in low-latitude Africa as a whole are coffee, cacao, cotton, peanuts, and oil palm products. Crops of lesser overall significance, but very important in some regions, include sisal, tobacco, tea (Fig. 22.2), rubber, pineapples, bananas, cloves, vanilla, cane sugar, and cashew nuts. Within Africa itself, as transportation improves and cities grow, there are increasing pos-sibilities for the sale of basic food crops such as corn, millets, manioc, rice, and fruit.[4]

Farming in Areas Outside the Tropics

In the parts of Africa that lie outside the tropics, much of the farming by Africans or Arabs is on a subsistence or near-subsistence basis. But these sections of the continent also have a considerable development of commercial agriculture. It is carried on by European growers of livestock, grain, fruit, and sugarcane in South Africa; Egyptian cotton growers in the Nile Delta and Valley; Arab or European fruit growers (citrus, grapes, olives) in Algeria, Morocco, and Tunisia; and a scattering of others. South Africa's farms are the most important producers and exporters in this group.

IMPORTANCE OF LIVESTOCK IN SUB-SAHARAN AFRICA

Many peoples in sub-Saharan Africa are pastoral. However, most of those who live by tilling the soil also keep some animals, even if only goats and poultry. The significance of livestock is portrayed in the following selection[5]:

> Many [African peoples] rely largely on cattle. This is especially true of the . . . Basuto, [Botswana], Hereros, Watusi, Masai, Gallas, Nilotic Negroes, and [Fulani], some of whom have become so dependent on cattle that if the latter were destroyed their whole economic and social order would be disrupted and even their religion affected.
>
> It is difficult to overestimate the importance of domestic animals in the lives of [Africans]. While wives represent to some extent real property or estates, cattle, sheep, and goats represent currency. In many tribes wives are secured only by the payment to the woman's family of a certain number of animals, say 30 goats or sheep or 10 cattle. . . . This is even the practice in the essentially crop-producing tribes of the

---

[3]These terms, which are more or less interchangeable, cover enterprises that vary widely in type of ownership, scale of operations, organization of production, and types of products. Some plantations and estates are corporate owned, some are government owned, and many are owned by individuals. Most of these units were established by Europeans, although there are significant exceptions such as the clove plantations developed by Arabs in the islands of Zanzibar and Pemba (now a part of Tanzania). They are commercial enterprises, generally producing one crop, or at most a very few crops, for export, and employing large numbers of African workers.

[4]It is an interesting fact that many of the most important food and export crops of Africa are not native to the continent. For example, corn (maize), manioc, peanuts, cacao, tobacco, and sweet potatoes were introduced from the New World. The complex story of plant introductions and dispersals in Africa (and other world regions as well) is still being pieced together by scientists and historians.

[5]H. L. Shantz, "Agricultural Regions of Africa. Part I-Basic Factors," *Economic Geography*, 16, No. 2 (April 1940), pp. 126–128. Used by permission of *Economic Geography*. Some words added in brackets without disturbing the sense of the original. One footnote has been added.

Figure 21.8 Cattle belonging to the Masai tribe in Kenya are shown here on the move to fresh pastures. The scrubby woodland is a common type of vegetation associated with the tropical savanna climate in Africa.

Bantu peoples.[6] Among the tribes such as the Watusi, Masai, and others which are dependent on cattle for their daily food, the whole social, economic, and religious pattern of their society is dependent on cattle. . . . These are extreme cases, but even in a crop-raising tribe such as the Kikuyu sheep or goats constitute the basis of the most important social and economic events in the life of the individual. . . .

Cattle, sheep, and goats occupy the greater part of the African continent. They are absent or scarce in the extreme deserts . . . . and the heavy tropical forest or coarse high grass savannas of the Congo and the Guinea Coast. With these exceptions, they are distributed over the remainder of the continent. . . .

The great grasslands between the tropical rain forests and the deserts and the mountain grasslands of central and southern Africa were occupied by cattle-raising peoples. They dominate the adjacent agricultural tribes and often hold within their organization a dependent group of soil tillers. The Masai of East Africa depend entirely on cattle which they milk and bleed for their daily food [Fig. 21.8]. So close is this relationship that they cannot migrate without their herds. Moreover, they regard these cattle as almost a divine gift and husband them as carefully as they would their own children. The same can be said of the Banyoro and the Watusi. Each animal has a name, the herds are often carefully sorted as to color, and the greeting of the nobles of the tribe asks first about the welfare of the cattle, then about the welfare of the wives and children.

One of the major needs of sub-Saharan Africa is the development of strains of grasses suited to African conditions which will be more nutritious for livestock than the native grasses now present. The latter are often coarse, tough, and not very nutritious, so that the carrying capacity of the native grasslands is relatively low. Considerable work with grasses has been done by government experiment stations in Africa, with results that seem promising. It is conceivable that Africa may eventually become a major exporter of livestock products, provided that better grasses can be widely introduced, the menace of the tsetse fly brought under control, and the tribesmen induced to take an interest in improving the quality of their herds, rather than thinking of them in terms of mere numbers. At present the largest exports of livestock products are from European farms and ranches in South Africa. Raising of chickens for family food and local sale is widespread in Africa, even in areas where the tsetse fly prevents or limits the raising of larger animals.

---

[6]In African tribal society wives are not ''bought,'' but it is a universal custom for the bridegroom to make a large gift to the bride's family in advance of the marriage. Payment of the ''bride price'' (or ''bride wealth,'' as it is more properly called) has customarily been made in cattle or other livestock but may also be made in cash or merchandise. The gift signifies that the groom's intentions are serious and that he has financial prospects for supporting the bride; in addition it compensates the bride's father for the loss of her labor in the household and the fields. Should the marriage break up on account of the wife's misdeeds, the husband is entitled to the return of the bride wealth.

CHAPTER 22

Regions of
Tropical Africa

Luanda, Angola, exhibits the common pattern of tropical African cities: A modest cluster of high-rise buildings in the city center is surrounded by a sea of small one-story houses, often on dirt streets and generally roofed with metal sheeting.

LMOST all of Africa's present states were European dependencies prior to World War II. At the outbreak of war in 1939, only three units—South Africa, Egypt, and Liberia—were independent. The rest were controlled by the United Kingdom, France, Belgium, Italy, Portugal, or Spain. But after the war, and particularly after 1955, a sustained drive for independence changed Africa from a colonial region to one comprising about one third of the world's independent states. Of these, all but a few at the extreme north or south lie wholly or predominantly in latitudes between the Tropics of Cancer and Capricorn (Fig. 21.1). Hence Africa has a heavy preponderance of tropical countries. Four such units—Sudan and the "Eastern Horn" countries of Ethiopia, Somalia, and Djibouti—already have been discussed under the Middle East. The remaining countries of Tropical Africa are surveyed here under the regional headings of West Africa, Equatorial Africa, East Africa, South Central Africa, and the Indian Ocean Islands. Statistical data are given in Table 21.1.

OVERVIEW OF SHARED TRAITS

The Tropical African countries are diverse in size and geographic personality, but they have many traits in common. Contrasts in size are enormous. Zaire in Equatorial Africa far exceeds the huge American state of Alaska in area, and large Texas is surpassed by the republics of Mali, Niger, Mauritania, and Nigeria in West Africa; Chad in Equatorial Africa; Tanzania in East Africa; and Angola, Mozambique, and Zambia in South Central Africa. But at the other end of the scale the West African countries of The Gambia and Cape Verde are smaller in area than the state of Connecticut, and some island countries in the Indian Ocean or the Gulf of Guinea are even tinier than the smallest American state, Rhode Island. Differences from country to country in population size and density are also huge. Africa's most populous country, Nigeria, is estimated to have two fifths to one half as many people as the entire United States, whereas a number of African countries are smaller in population than any American state. In overall population density, Mauritius in the Indian Ocean considerably exceeds any state, but arid Mauritania in West Africa has fewer people per unit of area than any state except Alaska. Such drastic variations in size and population density among Tropical African countries result from the play of history in colonial times, when African territories were acquired and organized by European powers in a piecemeal fashion over a period of several centuries. The resulting hodgepodge of large and small territories then was perpetuated in the sovereign nations of postcolonial Africa.

We turn now to a survey of some important characteristics shared by the individual countries of Tropical Africa. These traits are summarized here to provide an orderly overview of this jumble of nations and to avoid repetition in subsequent country-by-country analysis.

1. *Environmental variety within most countries is considerable.* In many countries it is sufficiently great that a wide range of agricultural, forest, mineral, and/or fishery products based on domestic resources is possible. These potentialities have been exploited very unequally from country to country. A glaring example of a country with inadequate exploitation is large Zaire, which has resources far surpassing those of many affluent industrial countries in the world but remains mired in abysmal underdevelopment and poverty.

2. *Drought is a persistent problem in all but a handful of countries.* Several countries include deserts where rain falls infrequently, but annual dry seasons are the rule even in most sections of Tropical Africa that are more humid. Although all droughts create problems, the most devastating effects in this heavily agricultural part of the world occur when the normal rainy season fails to come for a year or more. This condition was very widespread in Africa from the late 1960s to the 1980s. The years 1973–1974 and 1983–1984 were particularly disastrous. At least 24 African countries suffered from major food shortages associated with the drought of 1983–1984. One area of repeated droughts that has drawn much international news coverage is the east-west belt of dry grasslands north of the equator to which the name *Sahel* is commonly applied (it centers in the area shown as steppe on the world climatic map, Fig. 2.18). In such Sahelian countries as Mauritania, Mali, Burkina Faso, Niger, and Chad, great numbers of people have seen their ways of life disrupted when the brief rainy season has not come. Crops have failed, water sources have dried up, livestock have perished by the millions, stockpiles of food have run out, and human famine has been widespread. Actual death from starvation has been the fate of many. Such disasters have beset all countries whose populations live wholly or predominantly in the Sahel, but droughts of long duration also have

plagued many Tropical African areas that normally receive much greater rainfall than is ever available to the Sahel. For example, such droughts have intensified the misery of Mozambique, a country already devastated by civil war, and in 1984 the important East African country of Kenya suffered its worst drought in half a century.

3. *Great poverty is characteristic.* At present it shows little sign of abating, despite Tropical Africa's receipt of many billions of dollars in outside grants and loans during the post-colonial era. Measured by per capita GNP, many Tropical African countries are among the world's very poorest, and their dire poverty is underscored by other indices such as life expectancy and infant mortality (see Table 21.1). In the mid-1980s the average life expectancy at birth was estimated to be only 35 years in the West African country of Sierra Leone, and only 48 years within Tropical Africa as a whole (by comparison: United States, 75 years; Japan, 77). Infant mortality (the number of children failing to survive the first year of life) in Tropical Africa as a whole was about 120 per 1000 live births, rising to 175 in Sierra Leone (U.S., 10; Japan, 5). The brevity of life among both infants and adults in Tropical Africa results from such widespread afflictions as poor diet, disease (most often caused by lack of sanitation and the use of contaminated water), ignorance, and lack of medical care. Armed conflict and drought have added to the toll in particular areas. Serious international alarm has recently developed about the rapid spread of the lethal AIDS (acquired immune deficiency syndrome) virus in many African countries such as Zaire, Uganda, Rwanda, Burundi, Zambia, and Tanzania.

4. *Lack of education seriously hinders development.* In only a third of the countries surveyed in this chapter is more than half the population reported to be literate. In Mauritania, a Muslim country so set in tradition that slavery is still practiced (although officially banned), adult literacy was estimated at 17 percent in 1978; in Burkina Faso literacy in the total population was estimated in 1985 at 21 percent for males but an incredibly low 6 percent for females. Lack of schooling for females in much of Tropical Africa is a formidable barrier to improved family health and the creation of a more skilled labor force. All countries are short of skilled workers, particularly workers with administrative and managerial skills. College graduates were alarmingly few at the time of independence and the deficiency is still far from repaired. Television could be an important tool in education (Fig. 22.1), but relatively few Africans can afford it, and programming thus far has stressed government political telecasts and reruns of prime-time entertainment from Western networks rather than programs of educational benefit.

5. *There is a heavy preponderance of rural dwellers.* Individual countries range generally between an estimated 65 percent and 85 percent rural, with the most rural being Burundi (95%), Rwanda (94%), Burkina Faso (92%), and Malawi (88%). By far the most common rural home is a small hut, customarily made of sticks and mud, with a dirt floor, a thatched roof, and no electricity or plumbing. Life in

Figure 22.1   Education is a pressing need in Africa, but facilities and teachers are in very short supply in many areas. Educational television is a possible aid. This photo shows school children watching an experimental solar-powered television set in the country of Niger. Practically all the population of Niger lives in the Sahel region, where sunshine is abundant.

villages is the rule, although dispersed homes on individual farms are common in some areas. In southwestern Nigeria many farmers actually live in cities, from which they commute to their fields.

6. *A relatively unproductive agriculture is the main occupation in nearly all countries.* Though spots of better agriculture do occur, the great majority of Tropical African farmers operate largely on a subsistence basis, producing so little surplus for sale that they have little cash income and few savings. Women do a large share of the farm work in addition to household chores and the bearing and nurturing of numerous children. Mechanization is scanty, yields are lowered by a lack of affordable fertilizer, and storage facilities are often so poor that large quantities of harvested crops are eaten by rats and other pests or are ruined by weather. Protections against natural calamities are so lacking that a prolonged drought or an upsurge of pests such as locusts can be catastrophic, as can any kind of epidemic disease affecting crops, livestock, or humans.

7. *Per capita food output in most (perhaps all) countries has declined since independence.* The world's most rapid rates of population increase, government policies unfavorable to agriculture, and damage from warfare and drought have been among the factors that have thwarted efforts to maintain self-sufficiency in food. Large food imports now characterize the majority of countries. Tropical Africa desperately needs an all-out research

effort and better governmental support to lay the basis for a more productive agriculture. The much-publicized "Green Revolution" (see Chapter 14) has not proved directly transferable to African conditions, and governments have exacerbated the food problem by failing to provide incentives for farmers to produce more. In any event, the traditional emphasis on large families threatens to nullify any gains in food production that may be achieved. Children are desired for family labor on farms, and in both rural and urban areas they are desired as a mark of status, security for the parents in their old age or in sickness, and, in the case of girls, a source of the "bride wealth" paid to the family in a marriage settlement.

8. *The national economies of all countries are underindustrialized and dangerously dependent on the export of a few primary products.* Within the group of 38 sovereign countries considered here, recent figures showed coffee to be the leading export by value in 7 countries, cacao in 5, crude petroleum in 4, and petroleum products in 2. Cotton led in 3 countries, fish in 3, copper in 2, peanuts in 2, and tobacco in 2. A host of other commodities were the leading exports of single countries: iron ore, bauxite and alumina, diamonds, uranium, phosphate, palm products, sugar, and vanilla. Additional exports prominent in some countries included tea (Fig. 22.2), cobalt or other minerals, live animals, and wood. In most national units, one or two products supply over 50 percent of all exports. Hence each country is very vulnerable

Figure 22.2 *A large share of the farm output in Africa is used by the growers for their own subsistence. Areas of well-developed agriculture on a commercial basis form "islands" in the sea of subsistence farmers. Some of these islands are located on volcanic soils in the highlands of Kenya and several nearby countries. The view shows a Kenya tea plantation on the equator at an elevation of several thousand feet.*

to any international oversupply of an export on which it is vitally dependent. The economies of Zambia and Zaire, for example, were devastated by a long-term drop in copper demand and prices beginning in 1976, and Nigeria was similarly affected when world oil prices collapsed in the 1980s. Even Zimbabwe, which has the most diversified economy in Tropical Africa, is excessively dependent on the export of a few primary agricultural and mineral products. The value of imports generally exceeds that of exports in Tropical African countries, with imports consisting overwhelmingly of miscellaneous manufactures, oil products, and/or food.

9. *Almost all countries are heavily in debt to foreign lenders.* Loans amounting to billions of dollars have been made to some countries by the International Monetary Fund (IMF), the World Bank, private banks, or other lenders. The indebtedness was incurred primarily during a period in the 1970s when large amounts of loan capital became available to lending institutions in the form of "petrodollars" recycled from Middle Eastern oil states. Many costly development projects were undertaken with borrowed money. Western financiers, planners, and contractors gave optimistic assessments of the benefits to be expected from such projects, and African leaders were very ready to accept proffered loans as a way to reap quick benefits from newly won independence. Now, however, many countries are having great difficulty in even meeting the interest payments on their debts. Nigeria, Zaire, and Zambia are prominent examples of countries with outsized debts, but many other African countries could be cited. No country has indebtedness remotely approaching in total amount the huge debts incurred by Brazil and Mexico in Latin America, but the Tropical African countries have far smaller national economies and their debts are still formidable. Recent requests of African debtor nations that lenders reschedule interest payments and advance new loans have been resisted by the IMF and other lenders. As a condition of further support, there have been increasing demands that African debtors put their finances in better order by such measures as:

A. *Revaluation of national currencies downward to make exports more attractive.* African governments have kept the official exchange value of their national currency units (for example the *naira* in Nigeria or the *zaire* in Zaire) unrealistically high in relation to what the value of these units would be in the mar-

ketplace. This has decreased foreign demand for Tropical African exports but has helped African governments placate their urban supporters by lowering the price of imported food for the masses and luxury goods for the elite.

B. *Increasing the prices paid to farmers for their products* as a means of stimulating greater export production to earn foreign exchange and greater food production to lessen food imports and hence provide more money for repayment of loans and investment in economic development. It has been standard practice in postcolonial African governments to require farmers to market their crops through government marketing boards at prices set by the government. Prices paid to farmers have been kept low, enabling governments to accumulate operating funds (and venal officials to become wealthy) from sales of crops at higher prices to consumers. Some of the revenue skimmed off from agriculture has been passed along to urban consumers in the form of subsidized low prices for staple foods such as bread. This device has helped sustain a nonagricultural labor force to diversify the economy and has lessened political discontent in the cities, but it has lessened the incentive for farmers to produce much more than the food required for their own families.

C. *Assisting farmers to produce more* by making fertilizer available at affordable prices and by increasing the availability of better seeds, more farm credit, and information about better farming methods.

D. *Moving away from centrally planned and overregulated economies* in order to stimulate growth through free-market price incentives.

E. *Reducing governmental mismanagement, venality, extravagance, and waste* to free more capital for development and loan repayment. Central to this effort is the reduction of expenditures for arms, showy but unnecessary projects, and imported luxury items, as well as the scaling down of oversized bureaucracies. Also central, although very hard to implement, is the reduction of rakeoffs and payoffs that have enabled many government officials to enrich themselves at public expense. Some leaders of African states have accumulated immense fortunes, with most of the money being spirited out of their countries to private bank accounts and investments in Western countries. At lower levels a host of officials responsible for such public business as granting licenses and permits or awarding contracts have come to expect a private payoff for each transaction. At all levels illicit wealth has led to conspicuous consumption of goods and services from abroad, paid for with precious foreign exchange. A *Time* news story noted the addiction of the elite to expensive automobiles[1]:

> Even in the poorest African capitals, such as battle-scarred Ndjamena, Chad, government officials can be seen in convoys of Mercedes-Benz limousines, scattering cyclists and pedestrians as they pass. Owning a Mercedes is so potent an African status symbol that in East Africa a Swahili word was coined to describe the elite that drives them: *wabenzi*, literally, men of the Mercedes-Benz.

Although many African debtor nations have been attempting to conform to the above guidelines, progress is slowed by internal political factors. Loss of economic privileges angers urban elites on whom governments depend for support, and relaxation of price controls on food to stimulate production is apt to be met by rioting among city dwellers who have tight budgets already allocated for other costs of living. Nonetheless, substantial progress in meeting IMF guidelines has been achieved by some nations.

10. *Authoritarian governments have been the rule since independence.* Military governments are very common, and one-party states are nearly universal. The one-party state runs counter to Western ideas of multiparty democracy, but in Africa it does ward off a disuniting tribal politics wherein each tribe seeks to have its own party. Some postcolonial regimes have been charged with incredible atrocities, and reports of human rights violations by governments have been commonplace. One of many possible examples of oppression is provided by the government of the late Sekou Touré, who ruled Guinea from independence in 1958 until his death in 1984. Touré's disastrous regime, during which the economy of Guinea fell apart, was summarized by a *New Republic* commentator as follows[2]:

> For the Guineans, . . . his government brought the worst excesses of a one-party

state and personalized dictatorship: corruption, inefficiency, personality cult, ethnic polarization, and, above all, political intolerance and implacable suppression of dissent. Thousands of Guineans were imprisoned and killed during his rule. Victimization reached deep into the families of those fallen from grace. Every aspect of the country's life was under tight surveillance.

11. *There is serious political instability in many countries.* Often this is based on tribal rivalries and antagonisms of long standing. In country after country, political history since independence has been punctuated repeatedly by coups, thwarted coups, political murders, armed rebellions, and outright civil wars (Fig. 22.3). Orderly and peaceful transfers of power have been rare. Slaughter on the largest scale, often including massacres of unarmed civilians, has been seen in Nigeria, Uganda, Mozambique, Angola, Rwanda, Burundi, Zaire, and Zimbabwe, but various other countries have experienced serious civil disorder. One of the more publicized coups took place in the West African country of Liberia in 1980 when a military uprising overthrew the government drawn from the elite minority of Americo-Liberians. The latter group had ruled Liberia since the country's founding in 1847 as a haven for freed American slaves. Such recurrent conflicts and crises in Tropical African countries have drained national treasuries, discouraged foreign investment, and sidetracked progress toward nationhood. However, it must be noted that some African countries have been governed peacefully for many years by men who led independence movements and then took control of their countries when freedom came. Well-known names include Julius Nyerere in Tanzania (retired in 1985), Félix Houphouët-Boigny in Ivory Coast, Kenneth Kaunda in Zambia, Léopold Senghor in Senegal (retired in 1980), and the late Jomo Kenyatta in Kenya. Even these civilian leaders have governed in an authorization manner, with frequent jailings of political dissenters.

12. *A diverse medley of political, economic, and social ideas from the West, the East bloc, the Muslim world, and Tropical Africa itself has influenced governments since independence.* Western democratic ideas have rubbed shoulders with African ideas of chieftainship, East-bloc ideas of one-party dictatorship, and Muslim ideas of theocratic government. Capitalism has coexisted with socialism and individualism with collectivism. Different governments have adopted different combinations of the

---

[1]"A Continent Gone Wrong: After a Generation of Independence, Africa Faces Harsh Facts and Hard Choices," *Time,* January 16, 1984, p. 28. Copyright 1984 Time, Inc. Reprinted by permission.

[2]José Zalaquett, "Guinea's Worth: The Perils of Moving Toward Freedom," *The New Republic,* August 12 & 19, 1985, p. 13. Reprinted by permission of THE NEW REPUBLIC, © 1985, The New Republic, Inc.

Figure 22.3 *Warfare and drought in postcolonial Africa have driven millions of Africans from their homes as refugees. This harrowing scene shows a few of them on a road in Angola following the end of Portuguese control in 1975. Civil war between African political factions caused great numbers of Angolans to flee for safety to refugee camps.*

foregoing ideas. Various forms of socialism have been particularly widespread, resulting in such phenomena as centrally planned national economies, close state control over wages and prices, state ownership of major economic enterprises, and attempts at collectivization of agriculture and rural life. Such measures have produced disappointing economic results in one country after another, although some socialistic governments have been able to expand their social services. Poor economic performance has sparked a trend toward freer enterprise. But, at the same time, socialist rhetoric often continues to be used as a means of rallying diverse ethnic factions under a common banner. For example, private investment by Western capitalists has recently been encouraged in Congo, an Equatorial African country where the official philosophy is still Marxist and aid of various kinds is still received from the Communist world. Today, various African nations are exhibiting ''hybrid'' economies with various combinations of features derived from Communism, European-style liberal socialism, and Western capitalism.

13. *Although formal political colonialism has vanished, most countries still have important links with the colonial powers that formerly controlled them, and many foreign corporations that operated in colonial days still maintain an important presence.* A good example is Zaire, where all mineral deposits and production have been nationalized, but mining is still carried

on for the Zaire government by the same Belgian interests that monopolized the mining industry under the colonial regime.

14. *Poor transportation is a crippling bottleneck to development* (Fig. 22.4). Since independence the new nations have been able to build rela-

Figure 22.4 *Many African river crossings have no bridges and it is necessary to use ferries, although some streams can be forded during the dry season. This photo shows vehicles waiting for the ferry at a river crossing in the Shaba region of southeastern Zaire. Back-up of traffic at these bottlenecks may take hours or even days to clear.*

503

tively few surfaced roads or railroads, and those left over from colonial days have often deteriorated due to lack of maintenance or have been interrupted or even closed by guerrilla warfare in such countries as Angola and Mozambique. One estimate in the mid-1980s indicated that the total length of passable motor roads in Zaire had declined in the first quarter-century of independence from 85,000 miles (*c.* 137,000 km) to 12,000 miles (*c.* 19,000 km). Tropical Africa critically needs a good international transportation net, coupled with a lowering of trade barriers, in order to create market opportunities on a vastly enlarged scale. No semblance of America's Interstate Highway System or transcontinental railnet now exists, and creation of such vital underpinnings for development will doubtless be a distant dream for a long time. But surely this could be a highly rewarding long-term use for Western aid money and technological skills. However, as matters stand, overland transportation is both inadequate in extent and plagued by severe maintenance problems affecting both vehicles and routes. Some of the reasons include the tropical heat, high humidity, dust, the prevalence of unpaved roads, lack of lubricants and spare parts, low levels of technical skill, lack of a tradition of maintenance among the populace, and the weakness and venality of governments lacking a sense of disinterested public service.

WEST AFRICA

West Africa—here defined as extending from the Cape Verde Islands, Senegal, and Mauritania eastward to Nigeria and Niger, and southward to the Guinea Coast (Figs. 21.1 and 21.2)—was formerly a French or British colonial realm except for independent Liberia and two small Portuguese dependencies. Of the present countries, four—The Gambia, Ghana, Nigeria, and Sierra Leone—were British dependencies; nine—Benin (formerly Dahomey), Burkina Faso ("Land of the Upright Men"; formerly Upper Volta), Guinea, Ivory Coast (officially Côte d'Ivoire), Mali, Mauritania, Niger, Senegal, and Togo—were French; and Cape Verde and Guinea-Bissau were Portuguese. Cape Verde's dry volcanic islands lie well out in the Atlantic off Senegal, but in population, culture, economy, and historical relationships the islands are so akin to the adjacent mainland that they fit easily into a West African context.

In West Africa today the English, French, or Portuguese languages continue to be in widespread use, are taught in the schools, and are, in fact, official languages in the respective countries. European languages are a highly useful means of communication in this region where hundreds of indigenous languages and dialects, often unrelated, are spoken. Economic, political, and cultural relationships between West African countries and the respective metropolitan powers that formerly controlled them continue to be close. France, for instance, supplies economic and military aid to its former colonies in this and other parts of "Francophone" (French-speaking) Africa and further cultivates good relations with its former dependencies by inviting their leaders to summit conferences where matters of common interest are discussed. Similarly Britain maintains ties with "Anglophone" Africa through meetings of the Commonwealth of Nations to which all its former colonies in Tropical Africa belong. Both France and Britain were able to divest themselves peaceably of their West African possessions in the late 1950s and early 1960s. Portugal did not follow their path, and guerrilla warfare raged for several years in Guinea-Bissau before independence was finally granted in 1974. Subsequently Portugal improved relations with Guinea-Bissau and began to give it economic aid, although such aid is limited by Portugal's own poverty. In all but a few cases, the former colonial power remains the largest single trading partner for each of the West African states it once controlled. The former colonies have special trading privileges with the European Common Market to which Britain, France, and Portugal belong. These continuing economic connections with Europe are very important to the poor countries of West Africa.

ETHNIC AND POLITICAL COMPLEXITY AND PROBLEMS

As so often is the case in Africa, the governments of the region must deal with serious problems growing out of the ethnic complexity of national populations. West Africa is inhabited by innumerable peoples who speak an extraordinary variety of languages and dialects, hold varied religious beliefs, and often are characterized by a long history of suspicion and hostility toward one another. Many people hold traditional animistic beliefs, with endlessly varied customs and rituals. In the drier lands of the north, the Muslim faith is overwhelmingly dominant, and it has spread southward over a long period. To add further to the religious complexity, many West Africans were converted to Christianity by Protestant or Roman Catholic missionaries. As in other parts of Tropical Africa, Christian churches exist today in great variety, and their memberships are increasing rapidly. Many Christian symbols and practices are being altered to fit African traditions. For example, in some churches believers at the Communion table may eat consecrated bananas rather than the traditional bread and wine. In many countries Christian churches have long played an important role in education and medical care.

Only modest numbers of Europeans have ever lived in West Africa, a circumstance due to a variety of factors: the inhospitable climate, the numerous diseases, the historical opposition of slave-trading interests to other forms of European enterprise, the opposition of well-organized African kingdoms to encroachment by white settlers, and policies by European colonial powers that strictly limited the ownership of land by Europeans. Practically all Europeans in West Africa have been of a governmental, professional, managerial, commercial, or technocratic class, and the overwhelming majority have lived in the political capitals or in scattered other cities. Thus the presence of an entrenched European settler class owning large blocks of the best farm land has not been a circumstance contributing to internal political troubles of West African countries in the age of independence, as it did in Kenya, Zimbabwe, and various other countries outside West Africa. However, European governments did contribute initially to some present troubles by drawing arbitrary boundary lines around colonial units without proper concern for ethnicity. As a result, the typical West African country today, like most countries elsewhere in Tropical Africa, is a jumble of ethnic groups that often have little sense of identification with the national unit.

One country that has had unusually severe difficulties in maintaining political unity and stability is the Federal Republic of Nigeria. The country's largest ethnic elements—the Muslim Hausa and Fulani of the north, the Yoruba of the southwest, and the Ibo of the southeast—have frequently been at odds with one another. One group, the Ibo, heavily localized in the area east of the Niger River and south of the Benue, actually attempted secession in 1967 by establishing a separate country called Biafra. This action was resisted by the Nigerian central government, and a bloody civil war resulted. The Ibo received little support from other African governments or from non-African powers. After lives estimated at one to two million had been lost to fighting, starvation, and disease, Biafra collapsed in 1970 and

surrendered unconditionally. Contrary to many expectations that this would be followed by a murderous bloodbath, perhaps amounting to genocide, the victorious Nigerian government adopted a relatively generous and conciliatory policy toward the defeated Ibo, and reabsorbed them into the Nigerian federation.

It should be noted that West Africa was a relatively advanced part of Tropical Africa when the European Age of Discovery began. A series of strong pre-European kingdoms and empires developed here, both in the forest belt of the south and the grasslands of the north. Some states produced notable works of art, for example the bronze sculptures of the kingdom of Benin in what is now southern Nigeria. Empires in the grasslands, such as ancient Mali or ancient Ghana, were trading intermediaries between the Guinea Coast and Islamic North Africa in the days before this trade was largely taken over by Europeans approaching the Guinea Coast from the sea.

AREA, POPULATION, AND PHYSICAL ENVIRONMENT

Africa is a large continent, and West Africa alone is larger than one might realize from looking at the ordinary map unless careful attention is paid to scale. The 16 political units in the region aggregate over 2.3 million square miles (about 6 million sq. km), or nearly two thirds the area of the United States. Distances are correspondingly great. When using a map showing all of Africa, one may easily fail to realize that it is about 600 miles (somewhat under 1000 km) from the coast of Nigeria to the country's northern border, to say nothing of the fact that the distance along the savanna belt between Kano in northern Nigeria and Dakar in Senegal is about 1700 miles (c. 2700 km). The isolating effects of such distances are compounded by inadequate transportation. The region is less impressive in population totals. In 1988 the 16 countries had an estimated total population of 194 million, or some 52 million fewer than the United States. Eight countries had under 5 million people, and only three—Ivory Coast, with 11.2 million; Ghana, with 14.4 million; and Nigeria—exceeded 10 million. But the most striking single aspect of West Africa's population distribution is that well over half of the region's population is found within one country, Nigeria. Because of census irregularities the actual size of the Nigerian population cannot be determined accurately, but one estimate by the Population Reference Bureau suggests a figure of 112 million for 1988, or nearly one fifth of all the people in Africa (other estimates run as high as 120 million or more). Hence Nigeria is on an entirely different scale in population from the other West African countries. The unusual concentration of people apparently existed even before European contact, with its attendant commercialization and urbanization; it may have been related to the existence of powerful kingdoms which were able to impose a degree of internal order and security.

In much of West Africa population density is low. Large sections average less than 25 persons per square mile (10 per sq. km), and in the north the average density is less than 2 persons per square mile within a broad band of desert and semidesert stretching across sizable parts of Niger, Mali, and Mauritania. The greater part of Nigeria and Ghana and areas near the coast in a number of other countries are more densely populated than the average, and in a few places high densities occur. The principal areas in the latter category are the southern parts of Nigeria, Benin, Togo, and Ghana and an east-west belt of territory around Kano in northern Nigeria. In these areas the average density is over 100 persons per square mile (40 per sq. km) and in some places rises very much higher.

Environmental contrasts within West Africa are extreme. Climatically the region ranges from parts of the Sahara Desert in the north through belts of tropical steppe, dry savanna, and wetter savanna, to some areas of tropical rain forest along the southern and southwestern coasts (see world maps of climate and vegetation, Figs. 2.18 and 2.19). Most countries have their most populous areas in the zone of tropical savanna climate, although such areas may differ considerably with respect to the timing and intensity of the yearly wet season or seasons, as well as in the relative proportions of grass and deciduous woodland in the vegetative cover. A few countries deviate significantly from the general pattern. Mauritania, for example, lies far enough north that its more populated southern section is only wet enough to be classed as steppe, while Liberia lies in a climate that has a dry season but receives enough moisture to be classed as tropical rain forest. Due to the cutting of timber for export and the clearing of land by cultivators, large areas of rain forest vegetation in West Africa have undergone a change to scrub growth and coarse grasses more typical of the tropical savanna climate.

Throughout West Africa temperatures are tropical. Along the southern coast Fahrenheit temperatures in the "coolest" month of the year average in the high 70s and in the hottest month often average in the high 80s. Steaming heat is the rule in wetter areas and periods; blazing heat in drier areas and periods. Most uplands are not high enough to bring about significant reductions in temperatures. Innumerable diseases favored by tropic heat and unsanitary conditions still flourish, although the incidence is far less than when the Europeans first came. One example of a serious scourge of West African villagers in the savanna belt is the disease called river blindness. The eyesight of victims is destroyed by a parasitic worm carried by a fly that inhabits brush along streams. The disease and efforts by the international medical community to combat it are vividly described in the novel *Rivers of Darkness* (1974), by Ronald Hardy. Within West Africa river blindness has been especially destructive in northern Ghana and in Burkina Faso, where large numbers of adult villagers have lost their eyesight. However, recently there have been encouraging reports of progress in coping with this disease by spraying to control the flies and treatment of victims with a new drug.

West Africa exhibits much topographic as well as climatic variety (Figs. 21.1, 21.2, insets). In the dry north are vast reaches of low plateau, much of it with the appearance of plains. In places the monotonous landscape is broken by rougher terrain, including some areas of sand dunes, some sections dissected into hills, and a few areas of mountains. To the south lies an east-west belt of uplands, from interior Senegal to eastern Nigeria. It reaches almost to the coast in the south and extends some 400 to 600 miles (*c.* 650–965 km) northward into the interior. The detail of this latter area is very complex, including hills, low mountains, and plateaus in various stages of dissection. Along the immediate coast lies a narrow belt of coastal plain, widening in some sections such as the Niger Delta and tending to extend inland somewhat along the lower courses of major rivers. Parts of the belt are mangrove swamp, and its straight reaches of coast and barrier sandbars provide only a handful of good natural harbors.

ECONOMIC GEOGRAPHY

The West African countries exhibit the customary economic pattern of developing nations. They depend heavily on subsistence or near-subsistence crop-farming—or, in the drier north, livestock grazing—along with a few agricultural and/or mineral export specialties. In the wetter south subsistence agriculture relies mainly on root crops such as manioc and yams and on maize, the oil palm, and in some areas irrigated rice. In the drier, seasonally rainy grasslands of the north (as, for example, in northern Nigeria or northern Ghana), nomadic and seminomadic herding of cattle and goats is important, along with subsistence grain crops of millets and grain sorghums. In a number of grassland areas such as the flat "inland delta" of the Niger River in Mali (once an entrance to an extensive lake), irrigated rice has been introduced. Many livestock from the north are marketed in cities farther south.

The major export specialties of West African agriculture are cacao, coffee, and oil-palm products in the wetter, forested south, and peanuts and cotton in the drier north. Ivory Coast leads the world in cacao exports, followed by Brazil in South America and then Ghana and Nigeria in West Africa. Ghana formerly was the world leader, but its cacao production slipped badly during a long period of political turmoil and economic mismanagement after independence. In coffee exports, Ivory Coast again leads in West Africa and ranked third in the world in 1985 (after Brazil and Colombia). Nigeria was once the world leader in exports of palm oil and peanuts, but these exports vanished when Nigerian agriculture was allowed to deteriorate during the oil boom of recent decades. In Nigeria the main oil-palm belt is in the denser rain forest of the southeast, parts of which receive up to 100 inches (*c.* 250 cm) of rain per year; the main cacao belt is in the lighter forests of the southwest; and the peanut belt is in the open grasslands of the Muslim North. Various other export specialties such as the pineapples of Ivory Coast and the plantation rubber of Liberia are produced in scattered districts along or near the Guinea Coast. Liberia is Tropical Africa's largest exporter of natural rubber, which comes largely from corporate plantations of the American-owned Firestone Company. One peculiarity of West African commercial agriculture is the almost complete dominance of African smallholder production. With a few exceptions, plantations owned and managed by foreigners never achieved a notable foothold in the area, although foreign traders and companies have stimulated commercialization of small-farm agriculture. West African agriculture has had a long succession of difficulties in recent years, being beset by extended droughts, brushfires, plant and animal diseases and pests, low prices on world markets, ill-advised government policies, migration of younger workers to cities, and high expenditures for imported food and oil-derived fertilizer.

It is interesting that Ivory Coast, which is one of the West African countries most dependent on agricultural exports, has achieved a greater than average degree of economic success. In 1986 it had an estimated per capita GNP of $740, whereas no other country except oil-dependent Nigeria ($640) had a per capita figure over $500. The relative success of Ivory Coast can be attributed in large part to long-term stability of government under a capable French-educated President (Félix Houphouët-Boigny), a market economy allowed to operate without ideological hindrances, a relatively low level of taxation, and good relations with the industrialized West. Effective use has been made of French advisors, technicians, and administrators. Much of the manual work in Ivory Coast is done by impoverished temporary migrants from nearby countries, especially Burkina Faso.

The other main export industry of West Africa is mining (Fig. 21.2; Table 21.3). This region has not been neglected in the worldwide search for resources to supply expanding industrial needs, although its total mineral output to date has been rather modest on a world scale. For many West African countries, minerals have become very important as a source of export earnings, government finance, and funds for economic development. The most spectacular and significant mining development has been Nigeria's emergence as a producer and major exporter of oil. Commercial oil development began here in the 1950s. By 1983 crude oil constituted 96 percent of Nigeria's export by value, and this extreme dependence on the one product was still continuing in the late 1980s. Extraction thus far has mainly taken place in the mangrove-covered Niger Delta and offshore in the adjacent Gulf of Guinea. While most Nigerians remain very poor, the combination of a reasonably effective central government, a population estimated at over 100 million, and a relatively large income from oil has markedly increased Nigeria's international weight and influence, especially in Africa. But in the 1980s the country began to encounter massive difficulties due to a drop in international oil prices. Oil earnings fell from $26 billion in 1980 to $5 billion in 1986. Overcommitted to ambitious development projects, Nigeria was forced to cut back financially and to send home large numbers of workers who had migrated into the country illegally from Ghana and other West African countries. Most of Nigeria's people have benefited little from the country's large oil revenues.

Other prominent mineral products of West Africa include iron ore, bauxite, diamonds, phosphate, and uranium. Although the region supplies only a tiny fraction of the world's iron ore, such exports are very important to two countries, Liberia and Mauritania. The shipments consist largely of high-grade ore. The main bauxite resources found and developed thus far lie near the southwestern bulge of the West African coast. Guinea, by far the main producer and exporter, derives the bulk of its export revenue from bauxite and alumina (the second-stage product between bauxite and aluminum). Diamonds are important to Sierra Leone, and both Togo and Senegal export phosphate. Uranium is exported to France from the extremely poor and drought-prone country of Niger. Some offshore oil has been found in Ivory Coast.

URBANIZATION: RETARDED BUT GROWING

The scale and degree of urbanization in West Africa are low, although urban population is increasing very rapidly. Sizable cities in the world are mostly associated with trade, centralized administration, and large-scale manufacturing. As West Africa has little manufacturing of more than minor dimensions on a world scale, the region's principal cities have developed in its major areas of commercial agriculture, and the majority are seaports and/or national capitals. In nearly every country the present capital (early 1989) is far larger than any other city. Nigeria has by far the most impressive urban development. Of the ten largest metropolitan cities in West Africa, three are clustered in the Yoruba-dominated southwestern part of Nigeria, in or near the densely populated belt of commercial cacao production. Largest of the three—and by far the largest in West Africa—is *Lagos*, a teeming and untidy sprawl of shanties on dirt streets, surrounding a modest downtown knot of modern high-rise buildings and older low-rise structures with characteristic tropical architecture in the commercial, governmental, and residential core near the harbor. During the oil boom of recent decades, migrants from all parts of Nigeria (and many from outside the country) flooded into Lagos in such numbers as to defy current attempts to estimate the population. Conjectural estimates range as high as 5 million or even 7 million within the metropolitan area.[3] With its severe traffic congestion, rampant crime, inflated prices, and hopelessly inadequate public facilities, Lagos gives a strong impression of urbanism out of control. A large share of Nigeria's oil income has accumulated here in the hands of a moneyed elite that lives in comfort amid the general squalor. In slave-trading days Lagos had a notorious reputation as the largest West African port for the slave traffic to the New World. Today it is West Africa's leading general-cargo port and industrial center—a development that was greatly stimulated in the early 20th century when the British colonial ad-

ministration elected to make Lagos the ocean terminus for a trans-Nigerian railway linking the Gulf of Guinea with the north. Harbor facilities able to handle large modern ships had to be developed by deepening the original shallow harbor on a lagoon and providing a deeper channel through a barrier sandbar. The industries of Lagos are mainly of simpler types such as food processing, textiles, cement, and assembly of foreign-made automobiles. Factories are scattered through the metropolitan area but tend to cluster in or near the port area. Lagos is Nigeria's political capital, although a new capital has been under development at Abuja in the interior.

About 50 miles (80 km) north of Lagos on the railway lies Nigeria's second city, *Ibadan* (1 million and probably much larger), in the heart of the cacao belt, and another 50 miles northward is the third largest city, *Ogbomosho* (probably over 1 million). Ibadan is a city of traders and craftsmen which also includes a surprising number of farmers. Ibadan, in fact, has the character of both a city and a huge village—a circumstance also true of other sizable urban places in southwestern Nigeria. Farmers living in these cities may commute 10 to 20 miles or more to their fields outside the city. By contrast, Nigeria's main area of commercial oil-palm (and petroleum) production in the Ibo country of the southeast is remarkably lacking in large cities, although urban growth has been stimulated by the oil boom. The Ibo have always had a strong tradition of living in villages, as contrasted with city-dwelling traditions among the Yoruba.

Southern Ghana, another agricultural area with a relatively high degree of commercialization, contains two of the ten largest West African metropolitan cities. The larger of the two, *Accra* (over 1 million), is Ghana's national capital and until fairly recently was the main seaport despite the lack of deep-water harbor facilities and the consequent necessity of transferring cargo from ship to shore through the surf. A modern deep-water port has been developed at Tema, a short distance east of Accra, and that city is now the country's main port and industrial center. Tema has an oil refinery and also an aluminum mill that operates with electricity from a large dam on the Volta River. Well inland from Accra lies the country's second city, *Kumasi* (perhaps 600,000), located in the heart of Ghana's cacao belt, which was the world's greatest until it was recently surpassed by its westward continuation in Ivory Coast. The latter country contains yet another of West Africa's ten largest cities—*Abidjan* (perhaps 2 million). This clean and orderly city, which has a considerable Western veneer in its buildings and businesses, is Ivory Coast's main sea-

---

[3]City populations cited in this chapter are rounded approximations for metropolitan areas. Because of inadequate data, it is often not possible to give more than very crude estimates.

port. It is also the *de facto* capital, although in 1984 the national legislature formally approved the future transfer of the capital to inland Yamoussoukro, the birthplace of longtime President Félix Houphouët-Boigny (elected to a sixth consecutive 5-year term in 1985). Situated on a lagoon, Abidjan has deep-water facilities made accessible to ships by an artificial channel through a coastal sandbar. Still farther west are two more seaport capitals: *Freetown* (over 500,000) in Sierra Leone, on West Africa's finest natural harbor, and *Conakry* (over 700,000) in Guinea. With its enormous deep harbor in the estuary of the Sierra Leone River, Freetown became prominent during World War II as a haven for Allied convoys using the Cape of Good Hope route around Africa. On one celebrated occasion the world's two greatest passenger liners, the *Queen Mary* and the *Queen Elizabeth*—each capable of carrying an entire division of troops at a speed able to outdistance enemy submarines—were in the harbor at the same time. Like Monrovia, the capital of Liberia, Freetown was founded in the 19th century as a settlement for freed slaves. The town became an important early center of mission work and education along the Guinea Coast. Conakry is Africa's greatest shipping port for bauxite and alumina—commodities that make up most of Guinea's export trade.

Rounding out the ten largest metropolises are *Dakar* (over 1 million) in Senegal and *Kano* (well over 500,000) in northern Nigeria—both associated with commercial peanut production of the dry savanna belt. Dakar, Senegal's main port and capital, was formerly the capital of the immense group of eight colonial territories known as French West Africa. It is the commercial outlet for much of the peanut belt, is a considerable industrial center by African standards, and is a tourist and resort center with fine beaches and an exotic blend of French and Senegalese culture. Kano is the most important commercial and administrative center of Nigeria's North and has long been a major West African focus of Islamic culture.

TRANSPORTATION FACILITIES AND PROBLEMS

A common difficulty of developing areas such as West Africa is inadequate transportation. In this region a notable problem has been the scarcity of good natural harbors—a circumstance due largely to the frequency of offshore sandbars and the tendency of river mouths to be choked with silt. Often it has been necessary to transfer goods by lighter between ship and shore, and large expenses are incurred by these poor countries as they attempt to provide modern port conditions and facilities. Inland from the coast lies the formidable problem of adequate connection between the ports and the interior. Rapids, together with annual seasons of low water, limit the utility of rivers for transport, although a number of the latter carry some traffic. Most notable are the Niger and its major tributary the Benue, the Senegal River, and the Gambia River. In fact, the Gambia River is the main transport artery for The Gambia—a country consisting of a 20-mile-wide strip of savanna grassland and woodland extending inland for 300 miles (nearly 500 km) along either side of the river. On the whole, however, West Africa's rivers probably are more of an obstacle than a resource with respect to transportation, in view of the bridging problems along land routes. Railways are much more important than rivers in the present transport structure but are few and widely spaced. Instead of being a network, the rail pattern is composed basically of a series of individual fingers that extend inland from various ports but usually do not connect with other rail lines (Fig. 21.2). Examples are the railway linking Abidjan with Burkina Faso and the line linking Dakar with Mali.

The road net of the region is relatively dense in well-populated areas, but most roads are not paved or reliable in all weather. Much of the region's area and many of its people are still quite remote from any good highway. Nevertheless, truck traffic is sizable, although minuscule compared to such traffic in the United States or Europe. Not surprisingly, a very large part of the limited long-distance passenger travel within the region moves by air. But either air freight must become much more economical than it now is or road and rail facilities must be very greatly expanded, if much of West Africa is to become an effective part of the modern commercial world. Within the past two decades, inadequate transportation has been a major obstacle to Western relief efforts in the drought-ridden Sahel belt.

EQUATORIAL AFRICA

East and southeast of West Africa lies a varied group of republics to which the term *Equatorial Africa* often is applied. In colonial days four of the present units—Central African Republic, Congo, Chad, and Gabon—were held by France in the federation of colonies called French Equatorial Africa. Cameroon was composed of two trust territories administered by France and Britain, respectively. Belgium held Zaire, Rwanda, and Burundi; Spain possessed Equatorial Guinea; and Portugal held São Tomé and Príncipe. The equator crosses the heart of Equatorial Africa, and the climate is more truly "equatorial" than it is in African countries farther east (Uganda, Kenya, and Tanzania) that also lie on or near the equator but are, on the average, much higher in elevation. The latter are discussed separately in the section on East Africa. Three political units—Chad, Rwanda, and Burundi—diverge from the basic pattern of the area but for convenience are included with Equatorial Africa. Chad, the northernmost unit in the area, reaches far into the Sahara and is much drier and more dependent on livestock raising than are the other countries, and it has a larger proportion of Muslims in its population. Rwanda and Burundi, which are located in the highland margins of East Africa, are much too high in elevation, temperate in climate, and densely populated to be typical of Equatorial Africa.

Only scattered parts of Equatorial Africa, primarily along the eastern and western margins, can be described as truly mountainous. On the other hand, only a narrow band along the immediate coast is comprised of plains below 500 feet (c. 150 m) in elevation. As in West Africa, the coast is fairly straight and has few natural harbors; shallow lagoons behind coastal sandbars are common; and lagoons and delta channels often are fringed by mangrove swamps with their impenetrable tangle of stiltlike tree roots. Most of the region is comprised of plateaus that lie, in general, at elevations of 1000 to 3000 feet (c. 300–900 m) but in places rise considerably higher. Plateau surfaces usually are undulating, rolling, or hilly. They are capped in many places by remnants of older surfaces and are cut by innumerable stream valleys, often deeply incised. A conspicuous exception to the generally uneven terrain is a broad area of flat land in the inner Congo Basin, once the bed of an immense lake. The most prominent mountain ranges lie along the margins of the Great Rift Valley in Rwanda, Burundi, and the eastern part of Zaire, and in the west of Cameroon along or near the border with Nigeria. In both instances vulcanism has played an important role in mountain building. Strips of rough hilly country or low mountains often are found along the dissected edges of plateaus, particularly in the west where the plateaus drop away to the coastal plain. The greater part of Equatorial Africa lies within the drainage basin of the Congo River (called the Zaire River in Zaire). The heart of the region has a tropical rain forest climate, grading into tropical savanna to the north and south (Fig. 2.18). Most of Chad has a tropical steppe or desert climate; only in the extreme south does it reach into the zone of tropical savanna.

Some mountains in the east of Zaire rise high enough to have permanent snow.

Equatorial Africa is inhabited by hundreds of tribes speaking a multiplicity of languages and dialects. The majority are classed as Bantu, although Sudanic peoples predominate in the drier areas of grassland north of the rain forest. The tall Watusi of Rwanda and Burundi are thought to have affinities with Hamitic people of the upper Nile and the East African highlands. Further complexity is added by small bands of Pygmies in the deep rain forest of the Congo Basin. Within Equatorial Africa as a whole, population densities are low, and populated areas are sporadically distributed. Clusters of people tend to be isolated from each other by rough terrain, thick forests, swamps, or tsetse-infested areas. In general the savanna lands and highlands are more densely populated than the rain forests. Most countries have an overall density well below that of Africa as a whole (Table 21.1).

FORMER FRENCH EQUATORIAL AFRICA AND CAMEROON

Prior to independence, most of Equatorial Africa was under French or Belgian control. As previously noted, the units formerly held by France include Chad, the Central African Republic, the Republic of the Congo, Gabon, and the greater part of Cameroon. The latter republic was created by combining the United Nations trust territories of French Cameroons and part of British Cameroons (the rest of British Cameroons elected to join Nigeria). Both were held by Germany prior to World War I. The two parts of Cameroon have experienced some difficulty in functioning as a single country, due partly to disparities in size and economic strength (the former French part is by far the larger), partly to tensions between Muslim northerners and Christian southerners, and partly to previous development under very different colonial systems. Both French and English are official languages in Cameroon; French is official in all the other former French territories and in the former Belgian territories as well.

In most parts of the foregoing units the population is relatively sparse and is supported by agriculture or pastoralism, largely on a subsistence basis. Production of crops for export is comparatively small, being most significant in southern Cameroon. In that area coffee and cacao are the main export crops, supplemented by some exports of bananas, oil-palm products, natural rubber, and tea. Such products come partly from African small farms and partly from plantations; among the latter are plantations in western Cameroon that were developed in the German period before World War I and now are government-operated. They are found in an area of volcanic soil at the foot of Mt. Cameroon (13,451 feet/4100 m), a volcano which is still intermittently active. The mountain is notable for its extraordinarily heavy rainfall; a station on the lower slopes near the sea averages 392 inches (nearly 1000 cm) of precipitation a year, and the average on higher slopes is probably well over 400 inches. A chain of extinct volcanoes stretches northward from Mt. Cameroon, and the line of mountains is continued southward in the islands of Bioko (formerly Fernando Po), São Tomé, Príncipe, and Annobon, formerly held by Spain or Portugal. Most agricultural exports from former French Equatorial Africa and Cameroon come from areas of rain forest, although some cotton is exported from savanna areas. Limited exports of livestock products also come from the savannas and tropical steppes. Millions of cattle, their numbers fluctuating with the amount of rainfall, are raised in the grasslands by Muslim herdsmen, but they are kept primarily for wealth and prestige, ceremonial uses, or local food supply. The main subsistence crops in these countries are similar to those of West Africa: millets, sorghums, maize, and peanuts in the grasslands; manioc, yams, maize, rice, bananas, plantains, and oil palms in the rain forests.

Minerals, principally oil, have added a new dimension to the exports and financing of some countries, the most notable of which are Gabon, Congo, and Cameroon, where oil revenues have created per capital GNPs that are exceptionally high for Tropical Africa. Minerals—largely crude petroleum, but including some manganese—comprise about nine tenths of Gabon's exports, the remainder consisting almost entirely of wood in the form of logs or lumber. Congo, Gabon's immediate neighbor on the south and east is even more dependent on mineral exports, which consist overwhelmingly of crude oil. Oil is also the largest export by far from Cameroon. The three oil-producing states lie in the Atlantic oil belt from Nigeria to Zaire and Angola which has developed rapidly in recent decades. Both onshore and offshore fields are present along the narrow coastal plain and continental shelf. The oil industry is carried on by many Western companies.

Cities are few in these countries. The main ones are political capitals and/or seaports, two examples being the capital city of *Brazzaville* (well over 500,000) and the seaport of *Pointe Noire* (at least 300,000) in Congo. Brazzaville, located directly across the Congo River from the much larger capital

city of Kinshasa in Zaire, was once the administrative center of French Equatorial Africa. The 320-mile (515-km) Congo-Océan Railway connects the city to Pointe Noire. The Congo River and its large tributary the Ubangi provide a 740-mile (nearly 1200-km) navigable water connection from the Brazzaville railhead to the capital city of *Bangui* (500,000) at the head of navigation in the Central African Republic. The latter country, a landlocked state with poor highway connections to the sea, is heavily dependent on the waterway and the Congo-Océan rail line for transportation of its small foreign trade. In Cameroon two comparatively short railroads connect the main agricultural exporting areas with the port of *Douala* (perhaps 900,000). One of the lines (paralleled by a paved highway) reaches *Yaoundé* (600,000), the country's capital. Road transport reaching northward from Cameroon provides an important link to ocean transportation for the remote landlocked country of Chad. In Gabon a new Trans-Gabon Railway to tap minerals in the deep interior has been under construction for several years.

Aside from the handful of rail lines just mentioned, the five countries considered here are devoid of railways, and most areas are poorly served by roads. Isolation and poor transport facilities are major limiting factors for economic development, particularly in Chad and the Central African Republic. Manufacturing is little developed, aside from such establishments as cotton gins, oil presses, sawmills, and a limited number of factories that manufacture simple consumer items, process sawlogs into plywood and veneer, process imported alumina to make aluminum, or refine oil. Africa's first aluminum plant was established in 1957 on the Sanaga River in Cameroon, using hydroelectric power from a dam near by to process imported alumina. It is anticipated that future industrial development in the five countries will make extensive use of hydroelectricity for power.

Since independence the remote Muslim country of Chad has had serious internal political troubles, which have been accompanied by Libyan, French, and Zairian military intervention. Numerous tribally based factions have contended for more than two decades, with the key struggle being a contest between the Chadian government in the south and dissident northern elements supported militarily by Libya. Troops of the latter nation have been occupying a strip of Chad adjacent to Libya that is reported to contain uranium deposits. French military forces called to the government's aid have carried out operations against the Libyans on a number of occasions, and Zaire sent several thousand troops into Chad in 1983. In 1987 the government's forces—supported now by many of the northern rebels—routed Libyan forces in a number of small battles, but in 1989 Libya still held positions in Chad.

FORMER SPANISH AND PORTUGUESE POSSESSIONS

Until 1968, Spain's modest African empire included small areas in Equatorial Africa: Rio Muni on the mainland between Gabon and Cameroon, and Bioko and smaller islands in the Gulf of Guinea. These units received independence in 1968 as the Republic of Equatorial Guinea. Rio Muni, a densely forested area, mainly exports logs, together with some cacao and other tree crops. It is much less developed economically than Bioko, a volcanic island rising to a central peak some 9350 feet (2850 m) high. The tropic heat, abundant rainfall, high humidity, and fertile soils of the island provide fine conditions for cacao, the main export crop. Production developed mainly on European-owned plantations, but the plantation owners were expelled after the end of Spanish rule. The former Portuguese islands of São Tomé and Príncipe, southwest of Bioko, are smaller than the latter island but bear some similarity to it in environment and economy.

ZAIRE: A CASE STUDY IN COLONIAL POLICY AND ITS AFTERMATH

In 1960, Belgium's vast Congo colony in Equatorial Africa became the independent Republic of the Congo (later the name was changed to Democratic Republic of the Congo, and still later it took the present name, Zaire—which means "River"). The colony was ill-prepared to be a sovereign nation, as events swiftly proved and have continued to prove for over a quarter-century. But pressure for independence had built up rapidly in 1959, and Belgium had yielded to the demands of Congolese leaders. The new country was quickly faced with major crises that its government was unable to handle. Among these were (1) a large-scale mutiny within the Belgian-trained and -officered Congolese army; (2) regional separatist movements and rebellions, the most serious being a declaration of independence by the mineral-rich Katanga (now Shaba) Province in the southeast; and (3) ferocious outbreaks of tribal warfare, especially in Kasai Province, the country's main diamond-mining area. With the Congo disintegrating into chaos amid mayhem and confusion, Belgian troops entered to protect Belgium's nation-

als, large numbers of whom were evacuated to Belgium. Shortly afterward a United Nations emergency force, composed mainly of troops from African nations, was dispatched to restore order and assist the government in repairing the country's damaged political and economic fabric. After sharp fighting between United Nations and Katanga forces, the secession of Katanga ended in 1963. Withdrawal of UN military units in 1964 was followed by new rebellions and political crises, centering now in the remote east. Government forces eventually contained the rebels with the aid of white mercenaries, who themselves subsequently became disaffected and for a while exerted independent control over sizable areas in the east. This episode ended with the withdrawal of the mercenaries from the country under guarantees of safe passage. The authority of the central government was established over most of Zaire by 1966, and, by 1969, rebel activity was confined to isolated pockets. Internal order and some economic recovery characterized the early 1970s, with financial and technical assistance being received from Belgium, the United Nations, the United States, and other quarters. Many thousands of Belgians who had left the country during the worst crises returned. But the optimism of these years soon dissipated as Zaire's economy entered a disastrous era that was still continuing in the late 1980s. Among many contributory factors that have been cited are governmental corruption and mismanagement, squandering of funds on ill-advised projects, a decline in the world price of copper, and the cost of imported oil products in the era of high oil prices after 1973. We turn now to some geographic and historical circumstances of this big equatorial country that continues to shamble along as a land of large resources but unfulfilled promise.

Physical Variety of Zaire

Nearly all of Zaire lies within the drainage basin of the Congo River. This immense, shallow downwarp in the surface of the African plateau has a general elevation of 1000 to 2000 feet (c. 300–600 m) but is rimmed, particularly on the east and south, by land that lies considerably higher (Figs. 21.1 and 21.2 insets). Great thicknesses of sediments brought by streams from the bordering uplands have accumulated in the flat, swampy, and lake-strewn inner basin.

Zaire shows great variety in relief, climate, and vegetation. Most of the northern half, including nearly all the inner Congo Basin, is a huge expanse of lightly populated tropical rain forest. Southern Zaire has tropical savanna climate and a vegetation composed partly of woodland and partly of tropical grasses. Strips of trees ("gallery forests") are found along the streams. North of the rain forest and mostly outside Zaire lies a second savanna belt. The areas of tropical savanna climate in or near Zaire have a dry season of 2 to 6 months. In general the rainy season comes at the time of high sun; because the Congo drainage basin is bisected by the equator, the northern savanna is experiencing the high-sun rainy season at the time that the southern savanna is having the low-sun dry season, and vice versa. Since tributaries of the Congo River drain considerable areas both north and south of the equator, the main river tends to be much more constant in flow than such rivers as the Niger or the Zambezi, which are often very low at the height of the dry season in their respective hemispheres. In the extreme east of Zaire along the margins of the Great Rift Valley, and in neighboring parts of Rwanda, Burundi, and Uganda, there are highlands, frequently of volcanic origin, that rise above 5000 feet (*c.* 1500 m) over sizable areas. The highest summits are in the Ruwenzori Range of Zaire and Uganda, a nonvolcanic massif towering to 16,763 feet (5109 m) and capped by permanent snow and ice. In this eastern region of Zaire there is considerable vertical zonation of climate and vegetation according to altitude. Here in the western arm of the Great Rift Valley are found Lake Tanganyika and smaller lakes. In the vicinity of Lake Kivu there was some development of resort facilities and vacation homes during the period of Belgian rule. Among the attractions of this eastern area are the highland climate, cooler and more pleasant than the climate in most parts of Zaire, and some of the finest natural scenery in Africa. Besides the spectacular Ruwenzori, the mountains include a group of active volcanoes north of Lake Kivu. During the early and middle 1960s life in the east was severely disrupted and there was much killing, looting, and disorder as a consequence of forays and clashes of rebel bands, central-government troops, and white mercenaries. The southeastern part of Zaire in the Shaba region is an upland ranging in elevation from about 3000 to 6000 feet (*c.* 900–1800 m). Temperatures here are considerably moderated by altitude, although not as much as in the eastern lake region.

Colonial Background: From Frontier Lawlessness to Belgian Paternalism

Belgian penetration of the Congo Basin began in the last quarter of the 19th century, following explorations by an American, Henry M. Stanley. During this early period the colony was virtually a personal possession of King Léopold II, whose agents ransacked it ruthlessly for wild rubber, ivory, and other tropical products gathered by Africans. This lawless era inspired Joseph Conrad's famous novel *Heart of Darkness* (1902), in which the Congo trader Kurtz, at the point of death, evokes the ravaged Congo with the cry "The horror! The horror!" In the 1890s a railroad was built connecting the port of Matadi with the colonial capital of Léopoldville (now Kinshasa), thus avoiding a stretch of rapids on the Congo River and providing access to the interior. In 1908, the Belgian government, deluged by protest concerning the savage treatment of Africans, formally annexed the Congo.

From the beginning of administration by Belgium, economic life was dominated by large corporations given concessions to do business in certain lines and geographic areas. In many cases there was joint economic development by the government and private concerns. The corporations owned and operated plantations, factories, mines, concentration plants and smelters, stores, hotels, railroads, river boats, port facilities, and financial institutions. They collected, processed, and marketed the oil-palm products, cotton, coffee, cacao, and other commercial crops grown (often by compulsion) on African farms. These companies paid taxes to the Congo government which were used to meet the colony's expenses, including investment in projects designed to further economic and social development. No taxes went to Belgium. But the companies were mostly owned in Belgium, and dividends from company earnings went principally to Belgian shareholders. Many shares were owned by the Belgian government; some were owned in America or Britain.

The Congo colony was administered by a governor-general appointed by the government of Belgium and directly responsible to it. Subordinate officials, likewise, were appointed. Even the African chiefs were appointees. Neither whites nor Africans were allowed to vote, although both had recourse to courts of justice. Belgium took the position that the first requirement for the colony was a sound economic base and an acceptable standard of living for both whites and Africans, with political development to come later. Attempts were made to draw the Africans more fully into general economic life than was true of most African colonial possessions. Africans were not confined to unskilled labor, but through education and on-the-job training were brought into occupations requiring considerable skill. They learned to operate trains, river boats, steam shovels, bulldozers, and electric furnaces, to work as carpenters, masons, telegraphers, and typists, and to serve as postal clerks, nurses, elementary school teachers, and pastors of African Christian churches. However, almost none received training as doctors, lawyers, or engineers. In their school system the Belgians emphasized elementary education, and a larger proportion of the children attended elementary schools (operated in large measure by Roman Catholic or Protestant missions) than was true in most African colonies. But Belgian policies barred all but a tiny handful from receiving a college education. The lack of well-educated African leaders trained in management techniques proved disastrous when independence came.

During the period of Belgian rule, employers of African labor were required to conform to minimum wage scales set by the government. Workers received a share of their pay in the form of rations, clothing, housing, and services such as free medical care and education. Actual cash wages were low. However, the Belgians hoped to gradually stimulate among the Africans a desire for material possessions and thus to bring them more fully within a money economy. It was hoped, in other words, that the ideal of leisure could be replaced to some degree by the ideal of consumption, so that Africans would come of their own volition to do regular and steady work in order to accumulate money with which to buy goods. In this way a large internal market was to be developed and production of a variety of goods stimulated. Hence the colony would come in time to have a more balanced economy and be less dependent on exports of a few primary products. But Belgium's political control was terminated before these purposes could be achieved.

Economic and Urban Geography Today

Today, Zaire, like numerous other units in Tropical Africa, has a modest development of manufacturing plants, mainly producing simpler types of goods. Typical commodities are cotton textiles, shoes, bricks, cement, wood products, various types of processed foods, cigarettes, some chemicals, and simple metal products. Up to now, however, the country has been mainly important in the economic world as an exporter of minerals and tropical agricultural products. A large proportion of the minerals are smelted or concentrated in Zaire and then are further refined at plants in Belgium or other overseas industrial countries. Among the leading mineral exports, copper, cobalt, and other metals come principally from the Shaba region, and diamonds come from the adjoining Kasai region (Fig. 21.2). Copper, by far the country's leading export, is secured from both open-pit and underground mines. Ownership of the Shaba mining industry was taken over by the

Zaire government in 1967, and in 1969 a compensation agreement, providing for continued Belgian management, was reached with the Belgian company which had owned the mines. The Shaba region is the world's largest producer of cobalt, an alloy with vital high-technology uses (for example, it is used in jet aircraft engines to resist heat and abrasion). Diamonds, primarily industrial, but including some gem stones, are secured by open-pit mining in the Kasai region to the east and west of *Kananga* (over 300,000). Most nonmineral exports of Zaire, including palm products, rubber, some coffee, and wood, come from areas of tropical rain forest near the Congo River. Coffee is grown in many different parts of Zaire; the variety known as *arabica*, grown in the eastern highlands, brings a particularly high price. Corporate plantation agriculture is still important, but foreign ownership of agricultural land is prohibited.

Matadi is the only seaport of consequence. It handles a major share of the country's overseas trade. *Kinshasa* (over 3 million), located on the Congo River at the lower end of a stretch of navigable water about 1000 miles (*c.* 1600 km) long, is the capital, largest city, and principal manufacturing center. Another well-known city is *Lubumbashi* (over 550,000), formerly Elisabethville, in the copper-mining region of the southeast. One of Zaire's major arteries of transportation connects these two cities. It is comprised of a railway from Lubumbashi to Ilebo on the lower Kasai River (Fig. 21.2) and a navigable water connection from Ilebo to Kinshasa via the Kasai River and the Congo River. Zaire formerly used the Benguela Railroad across Angola to export copper through the Angolan port of Lobito, but this route to the sea has been largely useless for many years due to Angola's civil war. It was reported in 1985 that about half of Zaire's foreign trade was with South Africa or was moving through South Africa.

The Outlook for Zaire

Zaire is Africa's second largest country by area (after Sudan), but its estimated population in 1988 was only 33 million, or under one third that of the continent's most populous country, Nigeria. Nearly half of Zaire is covered by tropical rain forest, and here the population is still extremely thin. Two thirds of the national population is rural, composed of some 250 ethno-linguistic groups living typically in villages. By one count 75 different languages are spoken. Travel within the country is increasingly handicapped by the deterioration of roads. By 1986 Zaire was so deeply mired in poverty that its per capita GNP for that year was only $160. Only a handful of the world's countries were poorer.

In early 1989, there seemed little ground for optimism that Zaire's deplorable economic state would soon show any marked change for the better. Billions of dollars derived from mineral sales or foreign loans had been spent on grandiose ''white elephant'' projects or had been pocketed by political leaders and their families. A heavy foreign debt was having to be rescheduled because of Zaire's difficulty in even paying the interest. A great deal of hard and selfless work over many years by many Zairians and outsiders such as missionaries, Peace Corps volunteers, and the Agency for International Development (AID) had contributed to social welfare but had not been able to generate much economic progress. The following comments on Zaire's developmental follies are excerpted from a 1986 *New Republic* article by Steven Mufson, reporting from the seaport of Matadi[4]:

> Two vast legs of steel stand in the water of the mighty Congo River, and from them hangs Africa's longest suspension bridge. The new bridge, opened in 1984, boasts two decks, one for automobiles and one for trains. But there isn't any railroad here. The railroad deck runs into a solid rock cliff. A smoothly paved road crosses the river on the upper deck and runs past a granite inscription with the name ''President Mobutu Sese Seko Suspension Bridge.'' Then the road disappears around a bend and turns into gravel and broken blacktop. During the first six months the bridge was open, an average of 53 vehicles a day passed over it.
>
> When African nations gained independence . . . , they desperately needed sensible development to secure their economic future. Too often they received white elephant development projects like Zaire's bridge to nowhere, which have squandered both money and opportunity. There are hundreds of projects like these, ranging from a multimillion-dollar sugar plant that sank in a Sudanese swamp to a multibillion-dollar power line in Zaire. They have contributed to the continent's downward spiral toward even greater poverty, and have left sub-Saharan Africa [heavily] in debt with little to show for it.
>
> . . . . .
>
> The trouble with the bridge at Matadi is that it was designed to carry copper exports to markets that no longer exist through a port that was never built. . . .

In the early 1970s, European Community planners forecast that Zaire's copper exports would outgrow the river port [of Matadi]. They said the railroad delivering copper to Matadi should be extended over the river by bridge, and go all the way to the small oceanside town of Banana. There several European donor countries (and European companies) would build a deepwater superport and industrial zone.

Japanese companies built the bridge, commencing in 1979, and the Japanese government made Zaire a large loan to help cover the cost. The projected expense of building the railroad to Banana proved prohibitive, and this line was never begun. Nor have the superport and industrial zone been built.

Rwanda and Burundi

Adjacent to Zaire on the east lie the small but densely populated highland countries of *Rwanda* and *Burundi,* once a part of German East Africa. When Germany lost its colonies at the end of World War I, the two units were placed under Belgian administration by the League of Nations as the mandated territory of Ruanda-Urundi. Subsequently this area became a Belgian trust territory under the United Nations and then received independence in 1962 as two countries. Together the two units have over a third as many people as immensely larger Zaire which they adjoin, and which, in fact, drew on them extensively for labor during the Belgian era. The overall density of population for the two countries is nearly 600 per square mile (nearly 230 per sq. km) and becomes very much greater at the higher elevations where most of the people live. Only about 5 percent of the population is urban, and settlement on dispersed farms rather than in compact villages is the rule in the countryside. The deforestation, cultivation, and overgrazing of steep slopes in these mountainous countries have produced severe erosion, lessening the ability of the land to yield more than a bare subsistence to the extraordinarily dense and rapidly increasing population.

About 85 percent of the population in Burundi and 90 percent in Rwanda is comprised of the Hutu (Bahutu), Bantu-speaking farmers and herders who formerly lived in subjection to the tall Tutsi (Watusi) minority, a pastoral people of possible Hamitic antecedents. Ferocious violence between these two peoples has marked both countries intermittently. According to one estimate, some 200,000 of Burundi's Hutu were massacred in 1972 by the Tutsi after they were blamed for a failed coup. A decade earlier a successful Hutu rebellion in Rwanda had cost around 100,000 Tutsi lives. New violence erupted in northern Burundi in the summer of 1988. Several

---

[4]Steven Mufson, ''White Elephants in Black Africa: The Bridge to Nowhere and Other Developmental Follies,'' *The New Republic,* December 29, 1986, pp. 18–19. Reprinted by permission of THE NEW REPUBLIC, © 1986; The New Republic, Inc. Three words inserted in brackets without compromising the sense of the original.

thousand Hutu are reported to have been killed by the Tutsi-controlled Burundian army. Tens of thousands of refugees, the great majority of whom were Hutu, fled into neighboring countries. Reliable information about this latest bloodshed is scanty.

In 1986 coffee comprised 91 percent of Burundi's exports and 80 percent of Rwanda's. Rwanda exports some tin, and Burundi has good-sized nickel deposits, but low market prices and the high cost of transportation from the highlands to an ocean port minimize the importance of these minerals at the present time.

EAST AFRICA

In this chapter the term *East Africa* comprises three former British dependencies—Kenya, Tanzania, and Uganda—that lie between the Congo Basin and the Indian Ocean. Two of the countries—Kenya and Tanzania—front on the ocean, but Uganda is landlocked. Prior to independence from Britain, these units varied in political status. Uganda was a protectorate administered with African interests paramount, Kenya was a colony and protectorate in which a white settler class had great political influence, Tanganyika (formerly German East Africa) was a United Nations trust territory administered by Britain, and the islands of Zanzibar and Pemba, now a part of Tanzania, were a British-protected Arab sultanate. In 1964 a political union of Tanganyika and Zanzibar was formed, for which the name "United Republic of Tanzania" subsequently was chosen. Kenya and Uganda also are republics.

Maps of population density do not show most parts of East Africa to be thickly populated. The three countries aggregate some 681,000 square miles (1,764,000 sq. km)—substantially larger than France, Spain, Italy, the United Kingdom, and the Low Countries combined—but their total population in 1988 was only an estimated 64 million, or 3 million greater than that of West Germany. However, this comparison will be quickly outdated, as West Germany has a slightly negative rate of natural population increase, whereas the populations of the three East African countries were increasing in the middle 1980s at extreme rates: Kenya 4.1 percent natural increase per year (highest of any nation in the world); Tanzania, 3.6 percent; and Uganda, 3.4 percent. The estimated average density of population in 1988 for East Africa as a whole was 94 per square mile (36 per sq. km); as recently as 1973 the density had been only 55 per square mile (21 per sq. km).

Such averages have relatively little meaning, as the population tends to be heavily concentrated in a few areas of dense settlement that often support hundreds of people per square mile (Fig. 22.5). The two largest and most important areas with a greater-than-average density are found in a belt along the northern and eastern shores of Lake Victoria and in south central Kenya around and north of Nairobi. The population is still extraordinarily rural in location even for Africa: Kenya, 81 percent rural; Tanzania, 82; Uganda, 90. The heavy densities and explosive rates of population increase in these deeply rural and agricultural countries make their physical geographies crucially important.

PHYSIOGRAPHIC CHARACTERISTICS

Although a coastal lowland fringed by coral reefs and, in many places, mangrove swamps, occupies the eastern margins of Kenya and mainland Tanzania, the terrain of the three East African units is comprised mainly of plateaus. These lie generally at elevations of 3000 to 6000 feet (c. 900–1800 m), but rise higher in some places and in much of eastern and northern Kenya descend to relatively low plains. Individual plateau surfaces exhibit great variety not only in elevation but also in topography, climate, and vegetation. Often the land opens out broadly into undulating or rolling country, punctuated by remnant hills or mountains formed of unusually resistant rock. Volcanic cones are prominent in some areas. All three countries include sections of the East African rift valleys ("Great Rift Valley"). International frontiers follow the floor of the Western Rift Valley for long distances and divide lakes Tanganyika and Malawi and smaller lakes among different sovereignties. The more discontinuous Eastern Rift Valley crosses the heart of Tanzania and Kenya. Lake Victoria, lying in a shallow downwarp between the two major Rifts, is divided among the three East African countries. None of the three countries is predominantly mountainous, but all have mountains in some areas, most commonly in close proximity to the rift valleys. Often these more rugged areas are rift-valley escarpments, or plateau escarpments elsewhere, that have been carved by erosion into mountains or hills. Formation of the rift valleys was accompanied in many places by vulcanism, often on a massive scale. The two highest mountains in East Africa, Mount Kilimanjaro (19,340 feet/5895 m) and Mount Kirinyaga (formerly Mt. Kenya: 17,058 feet/ 5200 m), are extinct volcanoes. Both are majestic peaks crowned by permanent snow and visible for

Figure 22.5   Dense sedentary agricultural settlement in highland Kenya on the edge of a government forest preserve. Numerous huts of farmers are scattered in an irregular pattern through the landscape. Population density is far too great to permit shifting cultivation, but the favorable climate and soil of this volcanic equatorial highland are able to support the small farms on a continuing basis. This is primarily a district of subsistence cropping, but some livestock and commercial crops are raised.

great distances across the surrounding plains. In some places, most notably in southwestern Kenya, lava flows have been sculptured by erosion into a mass of hills or mountains. East Africa's most productive agricultural districts, including many that contribute greatly to the export trade, tend to be localized in these volcanic areas, which have unusually fertile soils.

CLIMATE AND VEGETATION

In most of East Africa the climate is classified broadly as tropical savanna—a term embracing a decided range of temperature and moisture conditions. Temperatures are, of course, much influenced by altitude, and some agricultural districts, even on the equator, are so tempered by elevation that mid-latitude crops such as wheat, apples, or strawberries do well. There are great variations from place to place in the amount, effectiveness, and dependability of precipitation and in the length and time of occurrence of the dry season (or seasons). Over much of East Africa moisture is too scanty for a truly productive nonirrigated agriculture. There is an unfor-

tunate tendency for long droughts to occur, even in what would normally be the rainy season. Relatively little of the present crop acreage is irrigated. A correspondence of adequate moisture and good soil is found only in scattered areas, and such lands support the bulk of East Africa's population. Often they attain remarkably high densities, in contrast to far larger expanses of dry, infertile, or tsetse-infested land with very sparse population.

East Africa's natural vegetation is composed largely of various types of deciduous woodland interspersed with grasslands. Forest growth of commercial value is scanty. Softwoods, in particular, are in short supply, and there has been some planting of such trees under government auspices. The most luxuriant forests are found on rainy mountain slopes. Park savanna, composed of grasses and scattered flat-topped trees, with belts of woodland along watercourses, stretches over broad areas and is the habitat for some of Africa's largest remaining communities of big game. Numerous parks and wildlife reserves have been established in the three countries, and these are visited by large numbers of tourists. Among the most famous are the Serengeti Plains and Ngorongoro Crater of Tanzania, and the Nairobi National Park, located just outside East Africa's largest city.

Over large sections of East Africa the tsetse fly is a menace; such areas are largely devoid of cattle and generally have a sparse human population as well. Of the three countries Tanzania has the largest area and proportion of tsetse-infested land. Tsetse-control measures include (1) the clearing of brush that harbors the fly; (2) the slaughter or fencing-off of wild animals that serve as intermediate hosts to the organisms (trypanosomes) transmitted by the fly that cause sleeping sickness in humans and nagana in cattle; (3) spraying; (4) the release of infertile flies to retard reproduction; and (5) the development and application of preventives and medicines. Such measures are applied in all three countries with some success, but the tsetse problem is still very far from solved.

The driest large area in East Africa is northern and northeastern Kenya. Here the land, with its sparse vegetation of low grasses and shrubs, wears a desert-like aspect except in the brief season of the rains.

ECONOMIC AND URBAN GEOGRAPHY TODAY

As previously noted, an overwhelming majority of East Africa's people are rural and agricultural. Most of them are cultivators, but a minority are pastoral. Practically the full range of subsistence and export crops typical of Tropical Africa is to be found in the region, although there are endless variations from one area to another in the specific crops grown. For the area as a whole the main subsistence crops are maize, millets and sorghums, sweet potatoes, plantains, beans, and manioc. Manioc is a particularly valuable plant in areas prone to crop failure, since the roots can be left in the ground for several years as a reserve food supply. Cultivators may depend exclusively on crops for subsistence, or they may keep animals; the latter tend to be both poor in quality and unscientifically husbanded. The East African governments have long exhorted farmers to improve the quality of their livestock (partly through crossbreeding of local cattle with imported strains), to sell excess stock where necessary to prevent overgrazing, to make more use of animals as draft power for breaking ground and cultivating crops (now done mainly with hoes), and to apply more animal manure to crop land. Progress in these directions tends to be slow, as it requires a considerable change in attitudes, values, and habits of long standing. For

East Africa as a whole and for each individual country, coffee is the leading export crop. Indeed in Uganda's shrunken economy, devastated by civil strife, coffee comprised nine tenths of all exports in the mid-1980s. Other export crops are varied, the most valuable being tea (Fig. 22.2). Prior to the new era of political independence, East Africa contained around 300,000 Asians (Indians, Pakistanis, and Goans), 100,000 Europeans, and 60,000 Arabs (the figures are estimates for 1958). Most Asians had commercial, industrial, or clerical occupations, although some owned estates producing export crops; they dominated retail trade and owned many small factories and some large ones. Many were clerical workers in government offices. Some were wealthy, and practically all were more prosperous than most Africans. Arabs were primarily a mercantile class. Europeans were a professional, managerial, and administrative class; in the "White Highlands" of southwestern Kenya about 4000 European families operated estates that were worked by African labor and produced export crops and cattle. Much smaller numbers of European estates were found in Tanzania; most of them were engaged in growing sisal, which was Tanganyika's main export until recently. Asians, Europeans, and Arabs were most numerous in Kenya and least so in Uganda. Since independence many Europeans have left and much farm land formerly in European estates is now occupied by Africans.

The most productive parts of East Africa are bound together by railways that form a connected system leading inland from the seaports of Mombasa in Kenya and Dar es Salaam and Tanga in Tanzania. *Mombasa* (450,000) is the most important seaport in East Africa and the second industrial center (after Nairobi). The city, situated on an island connected with the mainland by causeways, has two harbors: a picturesque old harbor still frequented by some Arab sailing vessels (*dhows*) and other small ships, and a deep-water harbor which has modern facilities for handling large ocean vessels. From Mombasa the main line of the Kenya-Uganda Railway leads inland to Kenya's capital, *Nairobi* (1.2 million), the largest city and most important industrial center in East Africa (Figs. 21.3 and 22.6). It is also the busiest crossroads of international air traffic and the main outfitting and departure point for safaris into East Africa's big-game country. Tourism is Kenya's second largest earner of foreign exchange (after agriculture). Nairobi was founded early in the 20th century as a construction camp on the Kenya-Uganda Railway, which continues westward to *Kampala* (500,000), the capital and largest city of Uganda.

Figure 22.6   One of the more prosperous-looking cities in Africa is Kenya's capital, Nairobi. The modern downtown section distracts attention from the slums that exist a few blocks away and, in fact, comprise a large part of the city.

Kampala, which has been greatly damaged by strife and neglect since independence, is located in southern Uganda near Lake Victoria. *Dar es Salaam* (perhaps 1.5 million or more) is the capital, main city, main port, and main industrial center of Tanzania. Its deep-water facilities can handle good-sized ships. From the city the Central Line of Tanzania's rail system extends westward to the railhead and lake port of Kigoma on Lake Tanganyika. From Tabora, once a major Arab slave-trading center, a branch line leads north to Lake Victoria and gives access to the country's main cotton-growing district. About midway between Tabora and the lake, the branch line serves a diamond mine which supplies most of the value of Tanzania's relatively meager mineral output. Well to the north of Dar es Salaam near the Kenya border, the port of *Tanga* (100,000) provides the principal outlet for Tanzania's sisal exports. Sisal

once supplied more than half of the country's exports by value, but the market price has declined in recent years and the crop has now been decisively surpassed by coffee and cotton as earners of foreign exchange. A railroad, the Tanga Line, leads inland through Tanzania's main area of sisal plantations to a coffee-growing region at the base of Mount Kilimanjaro, and, farther west, to Arusha, a tourist and coffee center.

In comparison with many other African countries, the East African nations are extremely lacking in known mineral reserves and in mineral production. Manufacturing is in an early stage, being confined primarily to agricultural processing, some textile milling, and the making of uncomplicated consumer items. Both Mombasa and Dar es Salaam have refineries that process imported crude oil. There is a large cotton-textile mill at Jinja, Uganda, that operates with hydroelectricity from a station at the nearby Owen Falls Dam on the Nile.

SOCIAL AND GEOPOLITICAL PROBLEMS AND PROSPECTS

The three East African countries are very representative of Tropical Africa's social and geopolitical dilemmas. None has had an easy road to internal peace, order, and prosperity, and none has had easy relations with its African neighbors. All have been beset by intense tribal rivalries and conflicts. Aside from comparatively small minorities of Asians, Europeans, and Arabs, the population of East Africa is comprised of a large number of African tribes—Tanzania alone having around 120 tribal groups. Among the better-known tribes are the Kikuyu and Luo of Kenya, the Baganda of Uganda, the Sukuma of Tanzania, and the pastoral Masai of Kenya and Tanzania. Most East African tribes speak Bantu languages, but Hamitic and Nilotic languages are present in some areas such as northern Uganda. Swahili, a Bantu language drawing heavily on Arabic for vocabulary, is a widespread *lingua franca*, as is English. In all three countries the population is divided among animist, Muslim, and Christian elements. The Muslim religion reflects in part the long history of penetration of the region by Arab traders, particularly slave traders, prior to the age of European control. Since independence the Asian minority has been driven from its commercially predominant position or, in the case of Uganda, abruptly expelled from the country as refugees. But in Uganda it is the Africans who have been subjected to the greatest violence. They have been massacred and their country pillaged on such a scale that Uganda was barely functioning in the late 1980s and faced an enormous task of reconstruction.

Such episodes of bloody violence have punctuated East Africa's history. In the days when mainland Tanzania was German-controlled, a great revolt in 1905–1907 against German authority was put down with an estimated loss of 75,000 African lives. Immediately prior to independence, Kenya experienced a bloody guerrilla rebellion centering on a secret society called the Mau Mau within the large Kikuyu tribe. Antagonism directed against British colonialism and European ways cost about 11,000 African and 95 white-settler lives over the period 1951–1956. But then this pre-independence turmoil was dwarfed in postcolonial Uganda by a savage orgy of murder, rape, torture, and looting during the notorious regimes of Idi Amin and Milton Obote.

Both of the latter were Ugandan northerners, Amin belonging to a relatively minor Muslim tribe, the Kakwa, and Obote being a member of the predominantly Christian Langi tribe. During the British era the Ugandan colonial army was recruited primarily from the Langi and two closely related northern tribes, the Acholi and the Teso. From independence in 1962 until 1971 Uganda was governed dictatorially by Obote as Prime Minister. His supporters initially were a coalition of northerners with the southern Baganda, Uganda's largest tribe and the one that had been favored by the British colonial government for recruitment of the civil service. But the Baganda did not support the Obote government for very long, and Obote was forced to rely on his northern kinsmen to maintain his power. In 1971 their allegiance proved insufficient to withstand a military coup by the army commander, Amin. At the outset Obote's northern soldiers were massacred in their barracks by soldiers that Amin had recruited from Sudan, Zaire, and his own Muslim tribe. Thereafter bands of soldiers terrorized the country until Amin was ousted in 1979.

> Amin's behavior recalled that of Mutesa, a nineteenth-century tribal monarch, whom the late Australian journalist Alan Moorehead described in his book *The White Nile* as "a savage and bloodthirsty monster" in the "vales of Paradise." During his eight-year rule, until his clumsy attempt to annex part of Tanzania resulted in an invasion that deposed him, Amin soaked this lush, sylvan country with the blood of several hundred thousand people. Several hundred thousand more were made homeless. The suffering was widespread, but not completely indiscriminate. All Langis and Acholis were in danger. So was any Bantu with a house, a car, or another possession that one of Amin's thugs might covet. Amin expelled the Asian business community in 1972 and this precipitated a steep economic decline from which Uganda has yet to recover. However, the basic machinery of state, including public services such as water and electricity, remained intact. To the outside world, Amin's bulk and buffoonery lent a comic-book quality to his atrocities. Yet, . . . even worse was to come after Amin, though there would no longer be a colorful madman to rivet attention on the slaughter.[5]

In 1981 Obote returned to power, and a new reign of terror by undisciplined soldiers began. It lasted until 1985, when Obote was exiled by an army coup. This wave of atrocities centered in the

---

[5]Robert D. Kaplan, "Uganda: Starting Over," *The Atlantic Monthly*, April 1987, p. 20. Reprinted by permission.

densely settled southern homeland of the Baganda. Finally in 1986 the ravaged country had something of a respite when the government was taken over by Yoweri Museveni, the leader of a southern guerrilla force that had developed in opposition to Obote. The formidable task of rebuilding began, with prospects very uncertain because of the collapsed economy and the continuing cultural and political fragmentation of Uganda among some 40 tribes, mostly unable to communicate with each other in their native tongues and bearing heightened suspicions and animosities from the recent disastrous decades. Sporadic internal violence and a serious border quarrel with Kenya have evidenced continuing political instability since 1986.

For a time an East African Community existed, within which Kenya, Tanzania, and Uganda cooperated in economic matters, but this broke apart in 1977. International borders within the area have been closed for long periods. Meanwhile the countries have followed divergent economic philosophies. Kenya has embraced capitalist ideas in the main and has attracted some Western investment. At the same time highly centralized control has been evident in certain aspects of the Kenyan economy such as agricultural marketing. Although much farm land formerly occupied by white owners has been redistributed among African smallholders (with compensation paid to the whites), there is still a body of several hundred whites whose large holdings, worked by African labor, make a vital contribution to Kenya's export trade in tea and other farm products. Kenya is regarded as a relatively successful Tropical African state, although a per capita GNP of only $300 in 1986 does not evince much prosperity. Still, Kenya is somewhat better off than Tanzania, where a long-time President, Julius Nyerere, instituted a socialist order that improved social services but did little for economic advance. In the countryside the forced resettlement of Tanzanian farmers into new communal villages roused resentment, and agricultural output lost ground. Industrial plants, which are state-owned or heavily state-influenced, were reported to be running at one third of capacity in 1984 due to lack of capital. Like many other Tropical African states, Tanzania has squandered money on various "white elephant" projects such as the multimillion-dollar paper mill that had to shut down because operating costs were so high that the paper was cheaper to import than to produce. In Uganda, government-inspired assaults on the largest and most productive tribe (the Baganda) dealt the economy heavy blows, as did the expulsion of the Asian business community.

Tanzania's islands of *Zanzibar* and *Pemba* deserve separate mention. Once a British-protected sultanate, the islands lie north of Dar es Salaam and some 25 to 40 miles (40–64 km) off the coast. At one time Zanzibar city was the major political center, slave-trading center, and entrepôt of a considerable Arab empire in East Africa. Today the islands rely mainly for support on millions of clove trees (primarily on Pemba) which provide the bulk of the world supply of cloves. Arabs formerly controlled political and economic life, but the majority of the population is African, and Africans have been in control since 1964 when a revolution overthrew the ruling sultan and considerable numbers of Arabs were killed.

SOUTH CENTRAL AFRICA

In the southern part of Tropical Africa are five countries—Angola, Mozambique, Zimbabwe, Zambia, and Malawi—that share the basin of the Zambezi River and long have had important functional relationships with each other. The first two are former Portuguese dependencies, while the latter three were formerly British. The five units are grouped here as the countries of South Central Africa.

ANGOLA AND MOZAMBIQUE

Portugal played an active role as an imperial power in South Central Africa from a very early time in the age of European colonial expansion. The Portuguese were the earliest colonial power to build an African empire, and Portugal was the last country whose African possessions survived intact in the new age of African independence. The epochal voyage of Vasco da Gama to India in 1497–1499 by the Cape route was the culmination of several decades of Portuguese exploration along the western coasts of Africa. During the 16th century Portugal controlled an extensive series of strongpoints and trading stations along both the Atlantic and Indian Ocean coasts of the continent. Later she was outdistanced by stronger powers but managed to retain footholds along both coasts. By 1964 her principal African possessions, Angola, or Portuguese West Africa, and Mozambique, or Portuguese East Africa, were the most populous European colonies still remaining in the world. In both units Portugal was harassed by guerrilla warfare for many years, with large military forces required to meet the threat, and there were insistent demands from newly independent African nations (supported by many states elsewhere) that the Portuguese be ousted from these territories by economic sanctions or military force. In 1974 Portugal's resistance came to a breaking point, and an army coup installed a regime in Lisbon that granted the two colonies independence.

Until after World War II the Portuguese did not do very much to develop their African territories. A poor country itself, Portugal did not have the resources to undertake development on a large scale, although it did receive aid from Great Britain, Belgium, and South Africa to build railways and harbor facilities servicing mineral-rich areas controlled by the latter countries. Following World War II Portugal did more to improve its territories, aided by income from Angolan mineral wealth: diamonds, iron ore, and, in the years immediately before independence, an increasing output of oil. Angola and Mozambique proved sufficiently attractive to white settlers that a policy of government-assisted immigration greatly increased the European population. By 1974 probably 450,000 Europeans, mainly Portuguese, lived in Angola and perhaps 200,000 more in Mozambique. They had almost a monopoly of the skilled occupations and controlled most of the wealth. During the disorders accompanying independence most of these settlers fled to Portugal or in many cases to South Africa or Zimbabwe (then Rhodesia). In 1988 both Angola and Mozambique were still so torn by guerrilla warfare (directed now against the new African governments) that the modest earlier gains had long since been wiped out and both the economies and the social services of these countries were in a catastrophic state.

Portugal's former west-coast possession of *Angola* is comprised physically of a narrow coastal lowland and expanses of plateau in the interior, including a fairly extensive area in the south central part of the country that averages over 1 mile (c. 1600 m) in elevation. The climate is tropical savanna, with a fringe of tropical steppe in the west. At the southwest the steppe grades into coastal desert. The vegetation associated with the tropical savanna climate consists of savanna grasses with scattered woody growth in some areas and open woodland or scrub in others.

Prior to independence, Angola had important functional linkages with Zaire and Zambia by virtue of a rail line, the Benguela Railway, linking Angola's seaport of Lobito with the Shaba (formerly Katanga) mining region of Zaire and the adjacent Copperbelt of Zambia. But during the many years of civil war in Angola, the railway has repeatedly been interdicted by guerrilla forces and rendered inoperative for long periods. Zaire now uses its own port of Matadi or South African ports for most overseas traffic, and Zambia mainly uses South African ports or the inefficient "Tazara" railway across Tanzania to the congested port of Dar es Salaam. The Tazara line, completed in 1976, was developed with Chinese aid, but the Chinese subsequently withdrew from further involvement and the line then lapsed into a state of chronic mismanagement and decrepitude. The Chinese-built locomotives provided to be underpowered; there have been severe shortages of equipment; maintenance has not been adequate; and the line has consistently operated at a deficit (recently $100 million a year). The blighted fortunes of the Tazara railway graphically illustrate Africa's enormous management and maintenance problems, as well as the precarious state of landlocked countries in a continent so beset by political upheavals.

When the Portuguese withdrew from Angola in 1975, a Marxist-Leninist liberation movement supported by the Soviet Union and Cuba took over the central government in the seaport capital and largest city, Luanda (well over 1 million: see chapter-opening photo). But its authority was not recognized by other liberation movements, one of which—the National Union for the Total Independence of Angola (UNITA) was still controlling large parts of the country in early 1989. This anti-Marxist movement headed by Jonas Savimbi (a member of Angola's largest tribe) had received aid from South Africa and the United States, while the government in Luanda was being supported by large quantities of Soviet arms, thousands of Cuban troops (estimated at 40,000 in 1986), and many Soviet and East German military advisors. Large Cuban contingents were being employed to guard vital oil fields, particularly in Angola's coastal exclave of *Cabinda* north of the Congo River mouth. Cabinda is the country's main oil source, although other fields are scattered along the Angolan coast between the Congo mouth and Lobito. In 1986 crude oil and relatively modest amounts of refined products made up 93 percent of Angola's exports. The oil industry is operated by Western companies, principally American. It is an oddity of world politics that the United States is the main customer for Angolan crude oil, despite American verbal and some monetary support for the UNITA forces who have been fighting Angola's Communist-oriented government. South Africa has sent military forces into Angola repeatedly, and there have been armed clashes between South African and Cuban troops. But the South Africans mainly have fought Angolan-based guerrilla forces supporting independence for South Africa's dependency of Namibia.

Besides oil and its products, Angola exports diamonds, coffee, and other items in relatively small amounts. The diamonds come from a northeastern area representing a southward continuation of the Kasai diamond fields in Zaire. Coffee was once Angola's main export, but its production fell from a reported 20 million bags in 1975 to 1 million in 1985 as a consequence of warfare in the countryside and the flight from Angola of the European farmers who grew most of the crop. Most Africans in Angola have been supporting themselves precariously in the midst of a wrecked economy by subsistence farming and cattle raising. Many thousands of civilian noncombatants have been killed or maimed by mines planted in fields or along paths and roads in the embattled countryside. In 1988, representatives of Angola, Cuba, and South Africa signed an American-mediated agreement under which Cuba and South Africa would withdraw militarily from Angola, and South Africa would grant independence to Namibia. In June 1989 a cease-fire was arranged in the Angolan civil war. But despite the hopeful initiatives concerning Angola, the fate of that ramshackle war-battered country remained very uncertain. The same was true of Portugal's other excolony, Mozambique, to which we now turn.

Like many other units in Tropical Africa, *Mozambique* consists of a coastal lowland rising to interior uplands and plateaus. Most interior areas are relatively low in elevation, although one peak in the northeast reaches 7992 feet (2436 m). The country has a tropical savanna climate, with a natural vegetation comprised primarily of grasslands with scattered trees; deciduous woodlands occur in places, but dense forests are infrequent. Marshy and mangrove vegetation is conspicuous along the coast. Mozambique is divided into two fairly equal parts by the lower Zambezi River. There are two main port cities: *Maputo* (900,000), the capital, in the extreme south, and the much smaller city of *Beira*, located about 125 miles (200 km) south of the Zambezi mouth. Under Portuguese rule, much of Mozambique's commercial economy was closely linked with the economies of nearby states. The two ports had rail connections with adjoining landlocked countries and handled much international transit trade as well as overseas trade for Mozambique itself. Maputo was an important port for South Africa's Transvaal, being closer than any South African port to the mining, industrial, urban, and agricultural districts which make the Transvaal southern Africa's most important generator of freight traffic. In addition to rail connections with South Africa and also with Swaziland, the port had a connection with Zimbabwe, which in turn provided a rail linkage to Zambia (Fig. 21.2). Beira was another outlet for Zim-

babwe and also Malawi, with rail connections to both countries. British capital played a major role in financing port facilities at Maputo and Beira, as well as rail lines connecting these ports with the interior.

This transportation net focusing on the Mozambican ports has suffered severely from many years of guerrilla warfare in Mozambique, which first involved African insurgencies against the Portuguese and then warfare by African dissidents against the Marxist government that took over from the Portuguese in Maputo. The rebel organization called RENAMO (Mozambique National Resistance) was still controlling large rural sections in early 1989. Meanwhile the ruling Mozambique Liberation Front (FRELIMO) in Maputo was trying to administer a country that had almost ceased to function in any organized way. Zimbabwe had committed several thousand troops to guard its rail, highway, and oil-pipeline connections across Mozambique to Beira.

Under the Portuguese the European population of Mozambique was mainly urban, although there were several settlement schemes for European farmers, and also a limited development of European-controlled plantation agriculture. Since independence Mozambique's African farmers, beset by civil war and years of extreme drought, have produced little more than a bare subsistence for themselves and their families. Actual starvation on a large scale accompanied the great drought of the early 1980s. In recent years there has been some export production of such crops as cotton, cashew nuts, tea, and sugar, but agriculture in general has been in a disastrous state and imports of farm products have far exceeded farm exports. There has been almost no recent mineral exploitation, though small amounts of coal are mined and a beginning in oil production was made under the Portuguese. The country is thought to contain large reserves of natural gas and other minerals such as titanium, bauxite, and gold. Meanwhile international initiatives have been under way to redevelop the damaged port of Beira and upgrade the transportation facilities in the "Beira Corridor" to Zimbabwe. These measures, which include extensive dredging in the silted harbor of Beira, have been pushed vigorously by Zimbabwean business interests within the frame of an international development corporation called the Beira Corridor Group.

Neighboring South Africa has exhibited a very mixed posture in regard to postcolonial Mozambique. The two countries cooperated in building the huge Cabora Bassa hydroelectric dam on the Zambezi River in Mozambique. But South Africa has given considerable aid to the RENAMO guerrilla fighters in Mozambique, initially justifying this action as retaliation for Mozambique's harboring of bases used by South Africa's anti-apartheid guerrilla movement, the African National Congress (ANC). In 1984 the two countries signed an accord wherein Mozambique promised to expel the ANC and South Africa promised to stop supplying munitions to RENAMO. But Mozambique's Marxist government continued its ties with the East bloc and also moved toward closer relations with Western nations strongly opposed to South Africa's racial policies. In the late 1980s there was evidence that South Africa was still giving some support to RENAMO.

ZIMBABWE, ZAMBIA, AND MALAWI

Prior to the post-World War II movement for African independence, the countries of Zimbabwe, Zambia, and Malawi (then known as Southern Rhodesia, Northern Rhodesia, and Nyasaland) were British dependencies. In 1953 the three became linked politically in the Federation of Rhodesia and Nyasaland, or Central African Federation. It had its own parliament and prime minister but did not achieve full independence. Serious strains within it were created by the racial policies and attitudes of Southern Rhodesia's white-controlled government, and these helped bring about dissolution of the Federation in 1963. Northern Rhodesia then achieved independence as the Republic of Zambia, and Nyasaland became the independent Republic of Malawi. Both countries have remained members of the Commonwealth of Nations. The government of Southern Rhodesia, the main area of European settlement in the Federation, was unable to come to an agreement with Britain concerning the future political status of the colony's African majority. In 1965 the government issued a Unilateral Declaration of Independence (UDI), and in 1970 it took a further step in separation from Britain by declaring Rhodesia a republic, no longer recognizing the symbolic sovereignty of the British Crown. Britain and the Commonwealth reacted with a trade embargo and there was a series of negotiations between Britain and the Rhodesian government, attempting to arrive at a formula to allow political rights and participation for the 95 percent of Rhodesia's population which was black and thus to achieve reconciliation and world recognition of Rhodesia's independence. Although the economic sanctions were relatively ineffective, pressure on the white government increased markedly in the mid-1970s due to guerrilla warfare by Africans against the government and the end of

white (Portuguese) control in adjacent Mozambique. Independence for Zimbabwe as a parliamentary state within the Commonwealth of Nations was achieved in 1980. The majority of the white population left the country, although by 1988 there were still perhaps 100,000 white residents.

Zimbabwe, Zambia, and Malawi occupy highlands with a tropical savanna type of climate. Over nine tenths of the annual rainfall comes in the 6 months from November to April. The natural vegetation is comprised primarily of open woodland, although tall-grass savanna predominates on the *High Veld* of Zimbabwe. The High Veld, lying generally at 4000 feet (*c.* 1200 m) or higher, forms a broad divide between the drainage basins of the Zambezi and Limpopo rivers. Composed of a band of territory about 50 miles (80 km) broad by 400 miles (*c.* 650 km) long, oriented southwest-northeast across the center of Zimbabwe, the High Veld has been the main area of European settlement in the three countries. Most of Zimbabwe's whites live on it. They include about 4500 farmers whose large holdings, worked by African labor, produce most of the country's agricultural surplus to feed urban populations and for export. Tobacco grown on these white farms was Zimbabwe's largest export in 1986, slightly exceeding gold. Other agricultural exports include cotton, cane sugar, and maize. The large commercial farms operated by whites raise beef and dairy cattle as well as crops and are such vital sources of urban food and foreign exchange that the socialist-leaning Mugabe government has made no move to redistribute this good land (about one fifth of all cultivated acreage) among Zimbabwe's impoverished black farmers. Some redistribution has taken place elsewhere in the country without notable results in the form of surplus production. Other parts of Zimbabwe are drier than the High Veld, and productive farming generally requires irrigation. The white-controlled government constructed a number of dams on rivers draining from the High Veld to the Limpopo in order to facilitate the growing of irrigated sugarcane and other crops in the "Low Veld" of the southeast near the Mozambique border. In all three countries maize occupies the largest acreage of any crop and is the most important subsistence crop for the African population.

The main axis of economic development in Zimbabwe lies along the railway that connects the largest city and capital, *Harare* (formerly Salisbury: 700,000), in the northeast, with the second largest city, Bulawayo (450,000), in the southwest. Along or near the axis are found not only the principal areas of European farming but also mines producing gold, chromium, asbestos, nickel, copper, and other minerals, primarily for export. Iron ore mined at Que Que, about midway between Harare and Bulawayo, is utilized at that place to produce steel in South Central Africa's only iron and steel plant. Coke comes to the plant by rail from the Wankie coal field in western Zimbabwe, the most important coal-producing area on the African continent outside of South Africa. Zimbabwe's manufacturing industries are more important, both in value of output and in diversity of production, than those of any other Tropical African nation except Nigeria. Factories are especially clustered in the two main cities, but a number of important plants are located in other places. Despite this production, the country's exports are dominated overwhelmingly by minerals, tobacco, food, and cotton, and many manufactures must still be imported.

Zimbabwe has internal political problems relating to the position of the Shona-speaking Bantu majority (71% of the population) vis-à-vis the Ndebele-speaking (Matabele) Bantu minority (16%), and the small but prosperous remnant of whites (2%). Dissidents in Matabeleland (the southwestern part of the country) carried on guerrilla warfare against the government after the accession to power of Prime Minister Robert Mugabe (a Shona) in 1981. Long-standing antagonisms between the Shona and Ndebele were exacerbated by the program of Mugabe, an avowed Marxist, to create a one-party state. Friction also resulted from heavy-handed actions by government forces in Matabeleland. Recent compromises have eased these tensions, at least for the present. Meanwhile the whites, who are currently indispensable to the struggling Zimbabwean economy, remain in a kind of precarious political limbo because of surging African nationalism determined to rid Africa of the last vestiges of European colonialism. Thus far (late 1980s) the whites have been treated well by the Mugabe government and have been participating meaningfully in Zimbabwean politics.

Zambia is separated from Zimbabwe by the middle course of the Zambezi River, on which the giant Kariba Dam is located. Hydropower from the dam flows to both countries. Zambia is far less developed economically than its neighbor to the south. Much of it is tsetse-infested wilderness with a sparse population. Swampy areas exist in several parts of the country, particularly around Lake Bangweulu in the north and in the upper basin of the Zambezi River in the west. As in Zimbabwe the African population lives mainly on a subsistence basis, but in Zambia African agriculture and cattle-keeping are even more tradition-bound, uncommercialized, and unproductive than in Zimbabwe. European farms worked by African labor have played a less important role in Zambia than in Zimbabwe, being confined almost entirely to a narrow strip of land along the railway connecting Zimbabwe with the Zambian Copperbelt.

Zambia's economy depends heavily on copper exported from several large mines developed by British and American capital. Production is mainly from underground mines, although there is some open-pit mining. An urban area has developed at each mine, and the mining area, known as the *Copperbelt*, consists of a series of separate population nodes strung close to the frontier with the adjacent copper-mining Shaba region of southeastern Zaire. Railroads and surfaced roads interconnect the different mining nuclei, which are separated from each other by open country. The Copperbelt, which has the majority of Zambia's manufacturing plants as well as its copper mines, is largely an urban region and contains a high proportion of the country's relatively small European population. Copper, most of which is shipped in refined form, comprises around nine tenths of Zambia's exports. Other mineral production includes cobalt, a by-product of copper mining; zinc and lead mined along the railway between the Copperbelt and Zambia's capital and largest city, *Lusaka* (over 800,000); and coal production from deposits in extreme southern Zambia.

Malawi is a long, narrow, densely populated country along the western and southern margins of Lake Malawi. Its population is almost entirely African and is even more rural and agricultural than that of most African states. Tobacco is the main item in Malawi's export trade, and tea and cane sugar comprise most of the rest. There is little mineral wealth, and manufacturing industries are exceptionally meager. The country has long supplied many migrant laborers to mines in Zimbabwe and South Africa, and their earnings are an important adjunct to the internal economy of this extremely poor unit.

Relations with South Africa are highly important to all three of the one-time partners in the Central African Federation. South African ports handle by far the greater part of Zimbabwe's overseas trade, and South Africa is the largest single source of imports, the largest market for exports, and the most important source of tourists for Zimbabwe. Much of Zimbabwe's mining and manufacturing is owned by South African interests, and it was recently estimated that nine tenths of Zimbabwe's export/import trade was dependent on South African transportation services. Malawi and Zambia also have major linkages with South Africa in trade and transportation.

Zimbabwe and Zambia have been strong critics of South Africa's white-supremacist (*apartheid*) policies, but their economic dependence on South Africa, coupled with the latter nation's far greater military strength, has caused the two countries to be wary of actions that might inflame relations with South Africa and bring about economic and/or military retaliation. Malawi, which is overwhelmingly dependent on South Africa in a variety of ways, is the only Tropical African country that has full diplomatic relations with South Africa. The latter nation aided Malawi in planning, financing, and building a new capital at *Lilongwe* (225,000). *Blantyre* (400,000) was formerly the capital.

INDIAN OCEAN ISLANDS

A number of islands or island groups in the Indian Ocean—Madagascar, the Comoro Islands, Réunion, Mauritius, and the Seychelles—are described briefly in this section. They exhibit African, Asian, Arab, and European ethnic and cultural influences in varying proportions. Madagascar and the smaller Comoro Islands are former dependencies of France, while Réunion remains an overseas department of France. Mauritius became independent of Britain in 1968 as a constitutional monarchy under the British crown, but the Seychelles, which gained independence in 1976, is a republic within the Commonwealth.

Madagascar (French, "La Grande Île") is one of the largest islands in the world. It is nearly 1000 miles (*c.* 1600 km) long and about 350 miles (*c.* 560 km) wide. It lies off the southeast coast of Africa and has geologic formations similar to those of the African mainland, although it has developed a distinctive flora and fauna of its own. The great majority of its native plant and animal species are unique to the island. Madagascar received independence in 1960 and is now officially the Democratic Republic of Madagascar. Some of the early inhabitants migrated from Indonesia, bringing with them the cultivation of irrigated rice and various other culture traits that may still be found in both areas. There was also an influx of Africans from the mainland, possibly consisting largely of wives and slaves acquired by Indonesian mariners along the mainland coast. Small numbers of French, Indians, Chinese, Arabs, Greeks, and immigrants from Réunion and Mauritius are resident on the island. The official language is Malagasy, a Malayo-Polynesian tongue that has acquired many French, Arabic, and Swahili words.

The east coast of Madagascar rises steeply from the Indian Ocean to heights of over 6000 feet (1829 m). Since the island lies in the path of trade winds blowing across the Indian Ocean, the east side receives the heaviest rain and has a natural vegetation of tropical rain forest. The remainder of the island has a vegetation consisting primarily of savanna grasses with scattered woody growth. An eastern coastal strip, low, flat, and sandy, is backed by hills and then by escarpments of the central highlands. Paddy rice is the principal food crop of the coastal zone, and coffee, vanilla, cloves, and sugar are grown for export. The most densely populated part of the republic is the central highlands. Here, the economy is principally a combination of rice growing in valleys and cattle raising on higher lands. *Antananarivo* (formerly Tananarive: 900,000), the capital and largest city, is located in the central highlands. It is connected by rail with *Toamasina* (formerly Tamatave: over 100,000), the main seaport, located on the east coast. Madagascar has experienced a much smaller degree of economic development than such large tropical islands as Java, Taiwan, or Sri Lanka. Deep economic troubles have marked its recent history under a socialist government. There is some mineral wealth in the form of chromite, bauxite, and graphite, although all mineral fuels must be imported.

The *Comoro Islands,* located about midway between northern Madagascar and northern Mozambique, are of volcanic origin. They exhibit a complex mixture of African, Arab, Malayan, and European influences. Aside from the island of Mayotte, the Comoro group forms the Federal Islamic Republic of the Comoros (short form: Comoros), in which Islam is the official religion and both Arabic and French are official languages. There are only three exports of consequence: vanilla (77% of all exports by value in 1986), cloves (12%), and ylang-ylang, an essential oil used in perfumes (9%). By its own choice Mayotte, which has a Christian majority, remains a dependency of France. *Réunion,* east of Madagascar, is a volcanic tropical island with a population mainly of French and African origin. The island was uninhabited prior to the coming of the French in the 17th century. Elevations reach 10,000 feet (3048 m). The soils are fertile, and tropical crops in great variety are grown. Cane sugar, however, mostly from plantations, supplied 75 percent of all exports by value in 1986, with rum supplying an additional 4 percent. Sugar is also a mainstay of the larger island of *Mauritius,* which lies nearby. Sugarcane, grown mainly on plantations, was reported in the 1960s to occupy 85 percent of all cultivated land. Mauritius has a larger proportion of lowland than does Réunion. Its population is comprised mainly of Indians but also includes descendants of 18th-century French planters and some blacks, Chinese, and racial mixtures. The island has developed a clothing industry which supplied more export value than sugar in 1985 (clothing 44.8% of all exports, sugar 39.2%). The coral-rimmed *Seychelles Islands,* far out in the Indian Ocean to the east of Kenya, have a small population of French, Mauritian, and African descent. Many food requirements have to be imported. There is a growing tourist industry. The country's principal exports by value consist of petroleum products, although the islands have no production of crude petroleum and no refining industry. The products are imported and then re-exported. Some fish and copra are exported.

CHAPTER 23

Geographic Variety and Racial Controversy in South Africa

Durban, South Africa's third largest city and principal Indian Ocean port and resort, presents a prosperous white-dominated face in this view. Resort hotels line miles of beaches. The harbor is in the background. The population of the city proper is about equally divided among whites, Indians, and Africans, who live in the separate areas prescribed by South Africa's apartheid laws.

THE far south of the African continent is occupied by a geographically varied state that has become one of the world's most controversial countries. Organized from four British-controlled units in 1910 as the Union of South Africa, it was for five decades a self-governing constitutional monarchy under the British crown. But in 1960 its white population voted by a narrow margin (52–48%) to make it a republic. The Republic of South Africa was proclaimed May 31, 1961, and on the same day it withdrew from the Commonwealth of Nations (more precisely, it decided not to apply for readmission upon assuming republican status). The country's official policy of "separate development of the races" (subsequently called "multinational development," and commonly known as *apartheid*) has drawn furious criticism in many other nations. As a result, South Africa has been subjected to sanctions of various kinds, such as denial of arms and strategic materials, and loss of diplomatic privileges. The decision to break away from Great Britain politically was in part a reaction against liberal racial policies of Britain. It was also occasioned by hostility to South Africa among African and Asian members of the Commonwealth of Nations.

Since 1961 the white-controlled government of the Republic has continued to follow its program of racial segregation, with limited regard for internal dissent or objections of the outer world. Recently some relaxation of racial laws has occurred in certain areas of South African life, but the fundamental structure of apartheid, designed to insure the racial integrity and political supremacy of the Republic's white minority, remains in place. In continuing its basic policies despite world opposition, South Africa's government is favored by such factors as sea protection by the Indian and Atlantic oceans, land frontiers with weak neighbors, relative economic self-sufficiency, a high degree of military strength for an African nation, and a reluctance on the part of other nations to interfere too much with South Africa's internal affairs for fear of damaging their own economies and strategic relationships. For example, the United Kingdom and other Western countries profit from large investments in South Africa, and many thousands of jobs in these countries are directly linked to trade with South Africa. In addition, South Africa is a major world supplier of many important minerals—for example, gold, platinum, chromium, manganese, and vanadium—some of which are vitally important for armaments and other industrial production in outside nations and could not easily be obtained in sufficient quantities elsewhere. One major mineral that South Africa does not supply is petroleum, but a vital strategic relationship exists even there, since South Africa flanks the major sea route around Africa used by many supertankers that carry vital oil supplies to Europe from Persian/Arabian Gulf fields.

This perplexing country, so African in many aspects yet reminiscent of Europe, America, or even Asia in others, is apt to rouse contradictory emotions in visitors. The stark racialism seen at a bus queue in Johannesburg, where every face is that of a black African awaiting transport to the huge all-black suburb called Soweto, may be briefly forgotten in feelings of admiration for the beauty of Cape Town on a bright subtropical afternoon, with the affluent city nestling on the slopes below the sheer cliff of Table Mountain (Fig. 23.1), and white sailboats dotting the nearby sea against the gaunt backdrop of the Cape Ranges. The search for understanding of what one sees is complicated, for South Africa has many facets imposed by nature or accumulated through a dramatic history. The introductory analysis which follows can only sketch a few highlights, but the student who wishes to explore further will find no lack of good writing on this controversial land. (See maps, Figs. 23.2, 21.1, 21.2).

DISTINCTIVE QUALITIES OF SOUTH AFRICA

With a white population of about 5.3 million (early 1989), South Africa is by far the main center of European settlement and advanced economic development in an underdeveloped continent that was dominated by Europeans until very recently but is overwhelmingly populated by non-Europeans. South Africa's Europeans form a highly prosperous class, but due to the poverty of the non-European majority, the country's overall per capita gross national product (GNP) is about like that of "developing" nations such as Mexico or Brazil and is far below that of most countries in Europe.

Lying almost entirely in middle latitudes, South Africa has mean annual temperatures ranging generally between 60° and 70°F (16°–21°C). Thus it is on the average a cooler land than Tropical Africa; indeed, most places in South Africa have some frost in winter. From the standpoint of temperature, as in many other respects, the country has proved well suited to European settlement, although it should not be forgotten that the temperatures have also been favorable for the black Africans and other nonwhites who form an overwhelming majority of the population.

Figure 23.1 Cape Town is South Africa's legislative capital and second largest city. Dominating the scene is the steep-faced plateau remnant called Table Mountain.

The Republic is today a land where visitors from western Europe or Anglo-America will find most institutions and facilities to which they are accustomed. Johannesburg and Cape Town, the largest urban places, are imposing cities; downtown Johannesburg has many towering office buildings that might have been transplanted from a large American city. Hundreds of smaller places manifest strong European or American influences in the arrangement of their business districts and white residential sections.

In economic diversity and importance, South Africa is in a class by itself among African countries. In the mid-1980s it accounted for about one fifth of the continent's international trade by value, two fifths of its manufactural output, and one half of its output of electricity. It had more than two fifths of all the registered motor vehicles and telephones in Africa, and its railways handled about two thirds of the continent's ton-miles of rail freight. The scale and diversity of its resources and production are im-

portant assets to South Africa as that nation carries forward its racial policies in the face of criticism from the outside world.

Whites (Europeans) in South Africa represent only 15 percent of the total population, being outnumbered about five to one by black Africans. (Statistics for "South Africa" in this chapter exclude the dependency of Namibia but include both the "independent" and nonindependent "homelands.") In addition, there are two other sizable racial groups, the Coloureds, of mixed origin, and the Asians, most of whom are Indians. Numbers and proportions of the four major population groups are given in Table 23.1.

THE AFRICANS AND THE RACIAL LAWS

The existence of the large African majority in South Africa, together with the denial of African rights by law, provides an obvious point of difference between South Africa and the United States. There is, of

TABLE 23.1 POPULATION ELEMENTS OF SOUTH AFRICA

Racial Group	Total Number (including "independent" homelands: 1987 estimates, millions)	Percent of Total Population
African	25.2	74
European	5.1	15
Coloured	3.0	9
Asian	0.95	3

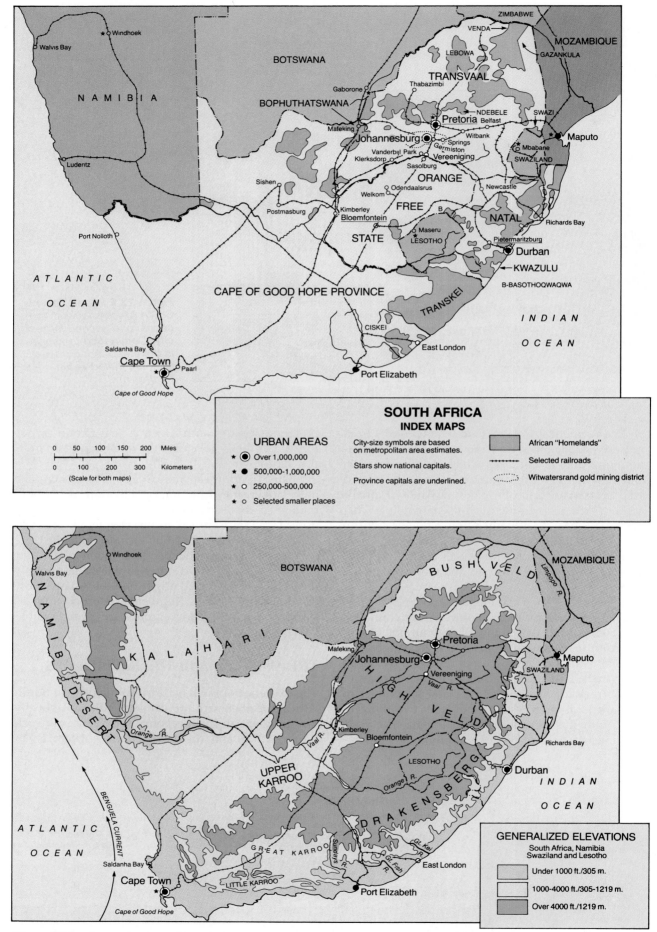

Figure 23.2 General maps of South Africa and nearby units. African "Homelands" include both "independent" and nonindependent homelands (see text). Arrows define general areas occupied by homelands named. Fragmented Bophuthatswana stretches from near Pretoria to near Kimberley, and fragmented Kwazulu covers much of Natal.

course, a sizable minority of African descent in the United States—a little larger than the entire non-white population of South Africa—but only in the South and some large metropolitan centers outside the South is it a large enough proportion of the population to be of major political, social, and economic significance. In South Africa, by contrast, numerous Africans are to be found in every province and important city. They make up roughly three quarters of the countrywide labor force employed by whites.

In contrast to the many laws of the United States that are designed to integrate all nonwhites into the social, political, and economic fabric of the nation, the present laws of South Africa are designed to systematize racial segregation. These laws restrict all racial groups, but they fall most heavily on black Africans. They rule out significant sharing of power by blacks in a unified South Africa. To justify its actions, the white government has declared that racial peace and justice can be found only if the races are separated territorially and socially so that each race can develop within its allotted niche. There is to be no such thing as a unified South African nationality, but the country is conceived of as a collection of "nationalities" living in segregated areas side by side.

South Africa's racial laws impose a startling array of restrictions and prohibitions on the population. The impact of these regulations has been changing the country's human geography drastically since the Nationalist—now the National—Party, heavily dominated by Afrikaners (whites, mainly of Dutch descent, who speak Afrikaans), came to power in 1948. The main restrictions on individual freedoms embodied in the laws (some of which have recently been modified) may be summarized as follows[1]:

1. *Mandatory classification and registration of all persons by race.* This procedure provides a uniform basis for application of the other racial laws. Not only must everyone be classified, but the classification by race is recorded on identity documents which are subject to scrutiny by the police.

2. *Prohibition of marriage or sex relations between whites and nonwhites.* These laws have recently been repealed. Offenders were liable for severe penalties, and they also risked social ostracism. The devastating effects of being "found out" in interracial sexual relationships are a central theme in such novels as Alan Paton's

[1]Much of the material in this section is based on *South Africa: Time Running Out,* The Report of the Study Commission on U.S. Policy toward Southern Africa (Berkeley and Los Angeles: University of California Press, 1981). Copyright 1981 by Foreign Policy Study Foundation, Inc.

Figure 23.3   The signs and the divergent paths tell the essential story of South African apartheid. The arrows point to roadside rest facilities in the Cape of Good Hope Province.

Too Late the Phalarope (1953) and *Ah! But Your Land Is Beautiful* (1981). The laws were never enforced against any great number of people, but their existence and periodic enforcement were among the aspects of apartheid most angrily criticized.

3. *Segregation of education by race.* There are separate educational systems at all levels for whites, Coloureds, Asians, and Africans. Funding per student is far greater for whites than for nonwhites, with Africans being the worst off. Segregation of higher education is not total, as white institutions permit some crossing of racial lines in specified fields of instruction. But only a tiny minority of nonwhites can afford the costs, and only limited funds are available for financial aid to disadvantaged students.

4. *Segregation of miscellaneous public facilities by race.* Until a selective loosening of "petty apartheid" began in recent years, nonwhites were comprehensively barred from hotels, restaurants, libraries, parks, hospitals, railway cars, buses, swimming pools, beaches, restrooms, and many other facilities reserved for whites. Signs designating facilities for "whites" or "nonwhites" were universal (Fig. 23.3). Selective desegregation of certain facilities (and events such as athletic contests) is now taking place, especially in larger cities, but South Africa is still a long way from racial integration even on a "petty" level, and the government continues to implement the territorial segregation of the races known as "grand apartheid."

5. *Denial of political rights to nonwhites.* South Africa's white government has proclaimed that members of each race and African tribe are to

hold citizenship within their own "nationalities." With limited exceptions, Africans are to be citizens of "homelands" defined on a tribal basis. As such, they may not vote or hold office in "white" South Africa, and they have no representation in the national parliament. The government has recently set criteria under which some Africans can be accorded citizenship in the Republic, but even they are denied the suffrage. Coloureds and Asians do have qualified parliamentary representation, as a result of a new constitution in 1984. A tricameral parliament with white, Coloured, and Asian chambers exists, but each chamber deals principally with legislative matters affecting its own "nationality," and continued overall control of the legislative process by the white majority party is assured by the governmental structure specified in the constitution. The keystone is a white executive president with very broad powers. The new constitution was adopted in 1983 by a 2 to 1 majority of the whites who voted (about three quarters of the eligible voters). Nonwhites were not allowed to vote. White opposition came primarily from Afrikaner conservatives who feared any dilution of apartheid; but the constitution also was opposed by white liberals who objected to the exclusion of blacks from all representation in the new parliament.

6. *Restriction of every domicile and business to a prescribed racial area.* South Africa has been divided into a mosaic of areas, each designated for a particular racial or tribal "nationality." The most publicized areas are the African "homelands," once called "Native Reserves," later called "Bantustans," and now officially "national states" (nonindependent) or "independent" Republics. But Africans living in cities of "white" South Africa long have been restricted to designated areas such as Soweto ("Southwest Township") outside Johannesburg. Coloureds and Asians do not have "homelands," but they do have their respective assigned areas of residence and business. The creation of the latter has involved much uprooting and transfer of nonwhites from "white" areas, along with a far more limited transfer of whites from "Coloured" or "Asian" areas. This kind of disruption, generally involving the forcible transfer of Africans to "homelands," has been occurring all over South Africa. Changes in the laws now allow limited local options under which some nonwhites will be permitted in the white areas, but very few nonwhites are likely to be affected.

7. *Restriction of movement through permits required for migration and travel.* The system of permits, created under various laws commonly known as the "pass laws," has applied mainly to Africans. Recently the system has loosened for Africans considered citizens of the Republic, but not for those in "homelands." A pass to leave a homeland generally requires a guarantee of employment in white South Africa, but many Africans leave illegally. In general, whites, Coloureds, and Asians may move about without passes.

8. *Job discrimination in favor of whites.* Nonwhites, especially blacks, have long been relegated by law and custom to lower-paying and less desirable jobs, and until recently were forbidden to participate in labor unions. These restrictions have been breaking down, partly as a result of pressure from foreign companies in South Africa and their home governments, but also because of an insufficient number of whites to fill all the better jobs in an economy that has greatly expanded during recent decades. The numbers of blacks and other nonwhites holding supervisory positions are still relatively small and are nearly nonexistent in cases where nonwhites would supervise whites. But the general situation of nonwhites has considerably improved, to the extent not only of better pay and entry into some skilled kinds of work but also desegregation of eating and washroom facilities in some workplaces. Black trade unionism in both all-black and mixed unions is now well established, although it is very far from universal.

9. *Repression of dissent.* South Africa's Minister of Justice and police forces have broad powers (intensified during declared states of emergency) to keep down dissent against the country's form of government and racial policies. Comprehensive laws prohibit a wide range of dissenting activities. Rights of accused persons are far more limited than in the United States. Under certain laws, persons deemed offensive or dangerous to the state may be held without trial or even formal charges for an indefinite period. Dissenters may be "banned" (closely confined to specified locales and denied a normal social life) without any reasons being stated. Censorship of media has not been routinely imposed, but editors must exercise caution. Books deemed by government censors to be offensive to the state or to public morals may be banned from sale or circulation—a fate that has befallen a long list of well-known novels about South Africa. The police exercise close surveillance of dissenting organizations, protest meetings, and demonstrations. Protest gatherings have often been broken up on the grounds that they did not have a proper permit or represented a threat to public order. Blood-

shed has frequently accompanied such events, as happened in 1960 at Sharpeville near Johannesburg, where 69 Africans protesting the "pass laws" were shot and killed by police. Again in 1976 many Africans in Soweto and other areas, including school children, were killed when they rioted against compulsory teaching of Afrikaans in the schools. Still others were killed in 1984 as a result of widespread labor protests and other unrest. Hundreds of South African organizations oppose the racial laws and have carried on peaceful protests of many kinds. Sporadic violent resistance is spearheaded by the outlawed African National Congress (ANC), which has claimed responsibility for a long series of terrorist bombings, including the destructive car-bombing of the South African Air Force headquarters in Pretoria in 1983.

In the period 1984–1986, black unrest became so widespread and violent that the government declared a state of emergency, which was still continuing in early 1989. Many blacks were killed by the police and the army, while others were killed by their fellows because they were perceived as collaborating with the whites. An intensive government crackdown resulted in wholesale banning of political activity by anti-apartheid groups, intensified restrictions on media, and the jailing of thousands of blacks perceived as protest leaders.

ECONOMIC EMPHASES IN SOUTH AFRICA

South Africa was extremely dependent on a single economic activity, mining (particularly gold mining) for a long time, although in recent decades it has lessened its dependence on the mining industry very markedly. This has been accomplished primarily by a steadily increasing development of diversified manufacturing, often carried on by South African branches of foreign firms. Today manufacturing employs more persons than do mining and agriculture combined, and it accounts for a greater value of production. The country's agriculture has been made more productive by the introduction of improved livestock and crop varieties (such as hybrid corn), increased irrigation, fertilization, and mechanization, better control of erosion, and other measures. Despite these changes, however, minerals continue to dominate the export trade and thus to provide foreign exchange with which to purchase the many kinds of manufactured goods and some raw materials and food that South Africa still must import. The mining industry also provides revenues (taxes

and royalties) used by the government to subsidize agriculture, manufacturing, and services.

SOUTH AFRICAN REGIONS

Most countries in the world are characterized by a certain degree of physical, cultural, and economic regionalism. In South Africa this characteristic is very pronounced. To gain much of an insight, the geography of the country must be considered not only as a whole but also in terms of regional units. The latter may be delimited in various ways according to the purpose in view. For an introductory overview the four political provinces and, for convenience, the unit called Namibia (South West Africa) will serve reasonably well as major units for study. They are discussed under the headings of (1) the Transvaal and Orange Free State, (2) Natal, (3) the Cape of Good Hope Province, and (4) Namibia. (See Fig. 23.2.) Some movement toward independence for Namibia was apparent in early 1989, but it was not possible to predict a date when South African control of this dependent territory would end.

TRANSVAAL AND ORANGE FREE STATE

The interior of South Africa is a plateau lying at a general elevation of 3000 to 6000 feet (*c.* 900–1800 m) above the sea. It represents the southernmost extension of the interior plateaus which occupy most of the African continent. In South Africa the plateau surface is highest at the east and tapers off gradually toward the west. Two river systems, the Orange (together with its large tributary the Vaal) and the Limpopo, drain most of the plateau area. At the extreme east of the plateau is the *High Veld,* originally an expanse of grassland somewhat similar to the original prairies of midwestern United States. The general elevation of the High Veld is 4500 to 6000 feet. It occupies most of the Orange Free State and the southern third of the Transvaal and extends into Lesotho. The northern two thirds of the Transvaal is primarily an area of woodlands and savanna grasses—the "Bush Veld."

Gold and Diamond Mining

South Africa's important gold-mining industry is almost entirely confined to the High Veld. The greater part of the gold mined thus far has come from the Witwatersrand (Rand), an area in the southern Transvaal about 115 miles (185 km) in an east-west

Figure 23.4   Johannesburg, the "City of Gold," is surrounded by huge dumps of waste material from former gold-mining operations. One of these dumps occupies the center of the view. Note the automobile expressway at the bottom of the view. In the distance lie residential areas reserved for whites.

direction by 5 to 30 miles (8–48 km) from north to south. Johannesburg, South Africa's largest city, is located in the central part of the Rand (Fig. 23.4). To the east and west are a number of smaller mining towns and cities. Johannesburg and its suburbs and satellites comprise a metropolitan area of 3.7 million people, and nearby Pretoria has an additional metropolitan population of about 1 million.

The Witwatersrand was the main area of gold mining from the original discovery of gold in 1886 until very recently, although production has now shifted in a major way to other South African fields. Immense amounts of capital, from British and other sources, are invested in the mines, which are operated by large corporations. In a few places mining operations have now reached depths of 9000 to over 11,000 feet (*c.* 2750–3350 m)—among the deepest mines in the world. The African workers employed in the mines are recruited from other African countries such as Lesotho and Malawi, or from South Africa's black "homelands."

The older Witwatersrand mines have passed their peak of production, and the major focus of gold-mining activity has shifted to newer mines located at the western end of the Witwatersrand district; in the southwestern Transvaal around Klerksdorp; and in the northern Orange Free State around Welkom. Besides the value of the gold itself, recovery of uranium from mine tailings and as a by-product of current gold mining has made South Africa a large uranium producer. The government has built a nuclear-electric plant near Cape Town and a ura-

nium enrichment plant near Pretoria to provide fuel. South Africa's potential capacity to make nuclear weapons has roused much apprehension in outside countries.

Since 1867 South Africa has been an important producer of diamonds. Formerly, the major production was from diamond-bearing kimberlite "pipes" (circular rock formations of volcanic origin) mined at Kimberley, Pretoria, and a number of other places (Fig. 21.5). Since 1927, however, production from alluvial deposits in the Transvaal, Cape Province, and Namibia has been of great importance. One of the major areas for this type of mining is found on closely guarded beaches along the coast of Namibia where diamonds are mixed with beach sands and gravels. Kimberley (150,000) in Cape Province is the administrative center for the diamond industry, although production of diamonds in the vicinity is now somewhat smaller than it was at the peak.

Coal and Steel

The mining industry in South Africa requires large quantities of electricity. Most of this is supplied by generating plants powered by bituminous coal, which is present in vast quantities and can be mined cheaply since much of it lies near the surface. The coal deposits are immensely important to South Africa, as the country has no petroleum or natural gas output (although active prospecting for these fuels is in progress) and has only a modest hydroelectric potential. Most of the important coal deposits are in the southern Transvaal and northwestern Natal, with some in the northern Orange Free State; they include various grades of bituminous and some anthracite coal. Much of the coal is of a rather poor grade, and some of it is difficult to mine because it lies underneath water-bearing rocks (which also create difficulties for some gold mines). But the deposits have been adequate for South African needs, and they support a large export trade.

Coking coal secured from fields around Witbank (Transvaal) and Newcastle (Natal) has facilitated development of Africa's leading iron and steel industry, which is concentrated at Pretoria and other places within a radius of about 40 miles (64 km) from Johannesburg. Iron ore comes by rail from large deposits in the Transvaal and from the northern Cape Province, which also supplies manganese from mines that are second in world importance only to those of the USSR.

Various districts in the Transvaal produce copper, asbestos, chromium, platinum, vanadium, antimony, phosphate, and other minerals. For all of those listed except copper and phosphate, South Africa ranks among the world's top two or three producers. The Transvaal and adjoining sections of the other three South African provinces comprise a remarkably rich and varied mining region in which the known reserves of a great many minerals are still very large. Only a handful of the world's largest nations possess a greater storehouse of mineral wealth.

Manufacturing

The government of South Africa, which is anxious to foster self-sufficiency in manufactured goods, has given assistance in various ways to the iron and steel industry (a large part of which is government-owned) and to other manufacturing industries as well. Foreign investments in manufacturing are important, but South Africa itself contributes a large share of the capital. The country's diversified industries include iron and steel, machinery (including much mining machinery), miscellaneous metal manufactures, automobiles, tractors, textiles, chemicals, processed foods, and others. The important chemical industry initially was developed to supply explosives for the mines; today, it also manufactures a wide range of industrial chemicals, fertilizers, paints, and numerous other products. At Sasolburg, south of Johannesburg, gasoline and petrochemicals are manufactured from coal. The production, which is being expanded elsewhere in the eastern Transvaal, supplies a substantial share of the country's needs for petroleum products. Imported crude oil

and its products are stockpiled in large quantities as a means of withstanding possible boycotts deriving from South Africa's racial policies. The country's main oil refineries, with associated petrochemical industries, are on the Indian Ocean coast a short distance south of the seaport of Durban and at Sasolburg. Crude oil moves to Sasolburg by pipeline from the Durban area via Richards Bay, a deepwater port north of Durban that is used primarily for exporting coal. Refined products also reach Sasolburg and the southern Transvaal by pipeline from Durban. South Africa is prospecting for oil and gas in offshore Atlantic waters, but only minor strikes have been reported.

Although the most important concentration of manufacturing industries in South Africa is located in the southern Transvaal and adjacent Orange Free State, sizable concentrations also are found in or near the seaports of Durban, Cape Town, Port Elizabeth, and East London, and there is a wide scatter of industries elsewhere in the country.

The Agriculture of the High Veld

The High Veld is not only the country's principal focus of mining and manufacturing but also the leading area of crop and livestock production. The main field crop in acreage and value of production is maize, although wheat and various other crops are grown. All but a minor fraction of the nation's maize acreage is located in the Orange Free State and Transvaal, mainly on the High Veld. Much of the production is fed to livestock, although in good years large amounts are exported. Maize is also a major item of diet for the Republic's African population. The principal farm animals of the High Veld are beef cattle, dairy cattle, and sheep. This is the only part of South Africa where commercial crop and livestock farming ("mixed farming") has developed on an extensive scale. Individual farms are large and farmsteads are widely spaced and include planted shade trees around the houses. Windmills are extensively used to pump water from wells for stock and household use. The Veld is rather short of surface water, but underground supplies are relatively abundant.

Some parts of the Transvaal, mostly outside of the High Veld, have a more specialized agriculture based mainly on tobacco, wheat, alfalfa, and fruit. These districts, which often depend on irrigation water from the rivers to supplement the summer rains, are generally at a lower elevation than the High Veld and have a longer frost-free season. The northern half of the Transvaal, in fact, is essentially an area without frost. Under these conditions exten-

TABLE 23.2 CLIMATIC DATA FOR SELECTED STATIONS IN SOUTH AFRICA AND NAMIBIA

Station	Climatic Area	Elevation (ft)	(m)	Average Temperature January °F	°C	July °F	°C	Average Precipitation Annual (in)	(cm)	Percent Oct.–Mar.	Percent Dec.–Feb.
Cape Town (Cape Province)	Mediterranean region	39′	12 m	72°F	22°C	55°F	13°C	26″	65 cm	19%	7%
Port Nolloth (Cape Province)	Namib desert	23′	7 m	60°F	16°C	53°F	12°C	2″	6 cm	30%	13%
Windhoek (Namibia)	Upland steppe	5669′	1728 m	73°F	23°C	55°F	13°C	15″	37 cm	89%	53%
Kimberley (Cape Province)	High veld	3927′	1197 m	75°F	24°C	50°F	10°C	16″	42 cm	76%	41%
Pretoria (Transvaal)	High veld	4491′	1369 m	70°F	21°C	50°F	10°C	29″	75 cm	85%	49%
Belfast (Transvaal)	High veld	6134′	1870 m	62°F	17°C	44°F	7°C	32″	82 cm	84%	47%
Pietermaritzburg (Natal)	Humid subtropical upland	2243′	684 m	71°F	22°C	56°F	13°C	37″	93 cm	80%	43%
Durban (Natal)	Humid subtropical coast	26′	8 m	73°F	23°C	62°F	17°C	41″	104 cm	69%	36%

sive production of citrus fruits has been possible, and the Transvaal is South Africa's leading province in citrus production.

The Moisture Problem

Although temperatures nearly everywhere in South Africa are favorable for agriculture, much of the country is too dry for nonirrigated farming. Most of the western half, and virtually all of Namibia, receives less than 15 inches (38 cm) of precipitation annually. Due to the high rate of evaporation, this amount is too small for most nonirrigated cropping. The High Veld gets about 20 to 30 inches on the average. But the total amount varies considerably from year to year, and there are frequent periods of drought during what would normally be the rainy season. Fortunately for agriculture, the High Veld receives most of its precipitation during the summer half-year, when it is most needed for crop production.

Precipitation in South Africa is mainly derived from the Indian Ocean. Moisture-bearing air masses blowing off the sea beat against the mountainous eastern escarpment of the plateau, are forced upward and cooled, and produce an annual rainfall of 40 inches (c. 100 cm) or more in most of Natal, northeastern Cape Province, and Lesotho. But moisture decreases sharply in the western part of South Africa. Western Cape Province and Namibia are mainly semidesert or desert.

Climatic data for some representative South African stations are given in Table 23.2.

The Orange River Project

As a means of conserving the country's none too plentiful water supplies and to provide hydroelectricity, many dams, most of which are relatively small, have been constructed on South Africa's rivers. In the early 1960s the country's most important water-control scheme was begun to augment preexisting dams on the Vaal tributary of the Orange River by construction of two dams on the main river, both of which are now in operation. The stored water is used partly to meet nearby urban and industrial needs, but a major share is allocated for irrigation of fields and orchards, not only in the drainage basin of the Orange River itself but also in districts to the south. Tunnels through mountain ranges transfer water to southward-flowing rivers that serve expanded irrigated areas in the Cape of Good Hope Province. Some of the water also goes to the Port Elizabeth urban-industrial area. The Orange River Project was not yet complete in the late 1980s and may not be completed before the end of the century.

NATAL

Natal occupies uneven and often hilly terrain between the Indian Ocean and the Drakensberg, a bold, mountainous escarpment marking the edge of the interior plateau. The highest elevations in South Africa are in the Drakensberg. Several peaks exceed 10,000 feet (c. 3000 m), and some in neighboring Lesotho exceed 11,000.

The leading commercial crop of Natal is sugarcane, most of which is grown in a narrow coastal belt lying mainly north of Durban but also extending for some distance to the south. Nearly all of South Africa's cane sugar comes from Natal. The production is sufficient for the country's needs and also supplies exports.

The Indians

Today, sugarcane is grown mainly on large white-owned farms worked by African labor. Formerly, however, the majority of workers in the cane fields were indentured laborers brought from India. The latter began coming to Natal in the 1860s. The majority chose to remain in South Africa when their terms of indenture ended, and the total number of Indians has steadily increased. In addition to indentured workers, some Indians came as free immigrants. No new immigration has been permitted since 1913, and internal migration of Indians from one province to another is restricted by law. About 85 percent of the total Indian population is in Natal, mainly in Durban or its vicinity. Today most Indians are employed in commercial, industrial, or service occupations—a great many having stores and shops. The presence of around 700,000 Indians—the main such concentration in Africa—is an important distinguishing characteristic of Natal. It may be noted that the Indians of South Africa, like the Europeans, Africans, and Coloureds, do not form a unified cultural group. They are divided among Hindu, Muslim, and Christian elements (well over two thirds are Hindu), and they speak a variety of Indian languages (Hindi, Tamil, etc.), although the use of English and Afrikaans is widespread.

Outside of Natal the main concentrations of Indians are in the Witwatersrand and in Cape Town. In these places they are employed very largely as tradesmen. Indians do not have equal political and economic rights with Europeans, although they rank higher than Africans. Their average level of living is better than that of Africans but well below that of whites. A good many Indians, however, are well off, and some are wealthy.

The Economy of Natal

Aside from sugarcane, the agriculture of Natal is built around maize and cattle, raised on a commercial basis on European farms and largely on a subsistence basis by Africans. There is some production of citrus fruits and vegetables, although in neither product does Natal compare with the Transvaal or the Cape Province. Recently there has been an upsurge of banana production on the south coast. Except for the coal fields of the northwest, mineral wealth is largely absent.

Durban (1.6 million), in southern Natal, is the province's largest city and main industrial center and port (chapter-opening photo). It is an important outlet for the High Veld as well as Natal itself, is a diversified manufacturing center, and has a thriving resort business centering in an impressive line of beachfront hotels.

CAPE PROVINCE

The Cape of Good Hope Province, like Natal, is distinct in various ways from the rest of South Africa. It is by far the largest in area of the four provinces, being half again as large as the other three provinces combined. It is also the most deficient in rainfall. Except for a mountainous, humid fringe along the southern and southeastern coasts, nearly all of the Cape Province is semidesert or desert. Parts of the province adjoining Natal share the humid subtropical climate of the latter area. The vicinity of the Cape of Good Hope has a mediterranean type of climate, being the only African area south of the Sahara where this climatic type occurs. Average temperatures at Cape Town are similar to those of Los Angeles, in the mediterranean climatic zone of southern California. The annual rainfall at Cape Town is greater, however, and the tendency to summer drought is less pronounced.

Cape Town (1.8 million) is the principal city of the Cape Province and the second largest city of South Africa (Fig. 23.1). Of the country's four main ports, it is the only one which fronts on the Atlantic. The others, including Durban in Natal and Port Elizabeth (725,000) and East London (325,000) in Cape Province, are Indian Ocean ports. North of Cape Town a deepwater harbor on the Atlantic at Saldanha Bay has been developed to export iron ore and other minerals brought by rail from the northern Cape Province.

The Cape Coloured

The Cape Province, like Natal, has a distinctive racial group—in this instance the Cape Coloured. Nearly nine tenths of the total Coloured population of South Africa is found in the province, primarily in or near Cape Town. This element had its origin in the early days of white settlement as a product of sexual relationships among Europeans (Dutch East India Company employees, settlers, or sailors) and non-Europeans, including slaves. Non-European ra-

Figure 23.5 *Home of a Coloured farm worker in the Cape of Good Hope Province. Most of South Africa's Coloureds are a depressed class economically, although not so badly off as the majority of Africans.*

cial stocks represented include Khoi (Hottentot), Malagasy, Bantu, West African, and various south Asian peoples. The term *Coloured* is reserved for these mixed-blood people, the Bantu peoples being referred to as *Africans.* The Coloureds vary in appearance from persons with pronounced African features to others who are physically indistinguishable from Europeans. About nine tenths speak Afrikaans as a customary language, and most of the remainder speak English; like great numbers of other South Africans, many are bilingual. Culturally they are much closer to the Europeans than to Africans, and many Coloureds with light skins have successfully "passed" into the European community. One intent of the national Population Registration Act, under which every South African must be officially registered as to race, was to prevent further "passing" by Coloureds.

Most of the Coloured inhabitants of South Africa work as domestic servants, factory workers, and farm laborers (Fig. 23.5) or perform other unskilled or semiskilled labor. Many have found employment in the fishing industry. There is a small but growing professional and white collar class. The Coloureds have always had a higher social standing and greater political and economic rights than the Africans, although ranking considerably below the Europeans in these respects, and far below them in level of living. It may be noted that among the Coloureds there is a social stratification, generally on the basis of lightness of skin. Here we note again an outstanding characteristic of South Africa: its population is not only divided among four major racial groups, but within each group there is further cultural or social splintering. It is a classic instance of a "plural society."

Distinctive Agriculture of Cape Province

Within the area of mediterranean climate, agriculture is adjusted to the characteristic regime of winter rain and summer drought. The early settlers established vineyards, for which the climate proved well adapted. Today, grape growing and wine making are a form of production helping to distinguish the Cape Province from the rest of the country. Although grapes are the most important fruit, the province also produces a variety of others, including citrus fruits (mostly oranges), deciduous fruits, and pineapples. These are grown principally in the area of mediterranean climate: on lowlands near the coast, or in valleys and basins among the Cape Ranges (Fig. 23.6). The various fruit-growing areas lie at different elevations and exhibit great diversity as to topography, moisture, temperature, sunshine, and winds. Each area specializes in fruits adapted to local conditions. Thus some valleys specialize in grapes for light, dry wines; others in grapes for heavy sweet wines; still others in grapes for raisins or for table use; and many areas specialize in deciduous tree fruits or in citrus fruits. Most vineyards and orchards are irrigated. Much of the production is marketed within South Africa, but there is also a sizable export trade. Since seasons in South Africa are the reverse of those in the Northern Hemisphere, fresh fruits can be marketed advantageously in western Europe when that area is having its off season for a particular fruit.

The principal field crop of the Cape Province is winter wheat, some of which is irrigated. Most of it is grown on small plains that lie between the main Cape Ranges and the sea within 100 miles (161 km) of Cape Town.

The Inland Districts

Most of the population of Cape Province resides in a band of relatively continuous settlement within 100 miles (161 km) of the southern and eastern coast, and extending from the vicinity of Cape Town to the border of Natal. This area, comprised mainly of low mountains, valleys, and small plains, receives most of the province's rainfall—winter rain in the west, but gradually changing to rainfall at all seasons in the east. Inland, the surface of the province rises by stages to the escarpment of the interior plateau, which is lower and less spectacular here than in the Drakensberg. Two well-known inland valleys located beyond the main belt of settlement are the Little Karroo and Great Karroo, respectively. Both are semidesert, with a vegetation dominated by scattered gray shrubs. This type of vegetation is known as "karroo" (a Khoi word meaning dry); it extends

Figure 23.6 *A characteristic rural landscape in the mediterranean climate of South Africa's Cape of Good Hope Province. The middle distance is dominated by fields of ripe winter wheat at the base of one of the Cape Ranges. Fruit and truck crops are visible nearer the camera. The time is late October.*

onto the western part of the interior plateau or Upper Karroo. Most of interior Cape Province is so dry that field agriculture is confined to scattered irrigated districts. However, sheep are able to forage on the sparse vegetation, and sheep ranching has become a very successful and major pursuit.

Aridity in the province increases toward the west, and northwestern Cape Province is desert. Most of the Atlantic shore is barren and sparsely settled. Aridity here is thought to be due primarily to high atmospheric pressures that block the entry of rain-bearing air masses from the Indian Ocean. It also results in some measure from the aridifying effects of the cool northward-flowing Benguela Current. Stable air masses moving inland from the cool offshore waters lower the temperatures of the coastal areas and produce much cloud and fog but yield very little rainfall.

Along this arid coast are found the most important fisheries in Africa. An upwelling of cool waters brings up nutrients from the ocean floor and provides a fertile habitat for plankton, the basic food of fish. These waters harbor many kinds of fish, with anchovies being by far the most abundant and caught in the largest quantities. Most of the catch is processed into fishmeal and fish oil for South African uses. In addition, this area is the locale for the rock lobster (crawfish) industry. South African rock lobster tails, canned or frozen, are exported in large quantities and have become a prized delicacy in the United States, western Europe, and Japan.

NAMIBIA (SOUTH WEST AFRICA)

The coastal desert of the western Cape Province extends northward into Namibia (population 1.7 million and perhaps substantially more). This former German colony was overrun by South African forces during World War I, and following the war it was mandated to South Africa by the League of Nations under the name of South West Africa. The terms of the mandate called for administration of the territory as an integral part of South Africa. After World War II the United Nations repeatedly called upon South Africa to place the territory under the trusteeship system, but the South African government refused to comply. In 1966 the General Assembly passed a resolution to end the mandate and subsequently appointed a governing council to administer South West Africa under a new name, "Namibia." South Africa declined to cooperate with the council and continued to integrate the territory into its own political and economic system. In 1988, South Africa agreed to grant independence to Namibia, provided that Cuban troops were withdrawn from neighboring Angola. A formal paper containing these provisions was signed by South Africa, Cuba, and Angola. South Africa has been supporting an Angolan guerrilla movement that is opposed to the Cuban-backed Marxist government; and South African military forces actively intervened in the Angolan struggle in the mid-1970s. Eventually they withdrew but have subsequently made attacks on Angolan bases of SWAPO (Southwest Africa People's Organization), a guerrilla group fighting for Namibian independence. In early 1989, the immediate political future of Namibia could not be reliably forecast, but there was good reason to believe that independence would eventually be granted.

The Namib Desert extends the full length of Namibia in the coastal areas. Inland is a broad belt of semiarid country, merging at the east with the Kalahari semidesert of Botswana. Furnishing a certain amount of sparse forage for stock, it is mainly peo-

541

pled by Bantu herders or by remnants of the Khoi population. Small and dwindling San (Bushman) remnants also survive there. Whites in Namibia, somewhat over 100,000 in number, are mostly found on an interior plateau. Windhoek (city proper, over 100,000), the capital, is located here. Export production is mainly confined to minerals (diamonds, uranium, and small quantities of others); livestock products, including Karakul lambskins; and products from fisheries.

SOUTH AFRICA'S ECONOMY AND THE RACIAL QUESTION

For unskilled and semiskilled labor with which to operate their mines, factories, farms, and services, the whites of South Africa rely mainly on low-paid black workers. The white economy, in its present form, could not operate without them, and the Africans, in turn, are extremely dependent on white payrolls. Thus the races are interlocked economically despite the political and social segregation imposed by innumerable laws and customs.

THE BRITISH-AFRIKANER DIVISION

There is every evidence that the great majority of white South Africans are in favor of continuing white supremacy, at least in political affairs. Or to put the matter another way, the whites are not prepared to yield political power to the extent that they might be outvoted by any combination of the other racial groups. In matters considered less fundamental, however, there is a division within the white population between the British South Africans and the Afrikaners, or Boers (Dutch: "farmers"). The latter speak Afrikaans, a derivative of Dutch, as a preferred language. They outnumber the British approximately three to two.

Early Boer Settlement at the Cape

The Afrikaners are the descendants of Dutch, French Huguenot, and German settlers who began coming to South Africa over three centuries ago. The earliest permanent settlement, at Cape Town, was established by the Dutch East India Company in 1652 as a way station to provide water, fresh vegetables, meat, and repairs for company vessels plying the Cape of Good Hope route to the Orient. Although it had not been the intention of the company to annex large areas of land, agricultural settlement slowly expanded in valleys near Cape Town, and a pastoral

frontier society developed in the back country. Most of the original Khoi and San inhabitants were killed or driven out, or, in the case of the Khoi, decimated by smallpox, with the survivors being put to work as servants or slaves. However, their numbers were not adequate for the labor requirements of the colony, and from the very beginning slaves were brought in from West Africa; later arrivals came from Madagascar, East Africa, Malaya, India, and Ceylon.

Particularly in the frontier districts, the early Boers became hardy individualists with an active dislike for central authority. During the Napoleonic Wars the Cape Colony was acquired by Great Britain. Occupation by the British on a permanent basis began in 1806, although a temporary occupation had taken place in 1795. Friction developed almost immediately between many Boers and the British authorities, who imposed tighter administrative and legal controls than the Boers were accustomed to; took an unfavorable view of the stern discipline, including use of the whip, applied by many Boer masters to their slaves and servants; and often took the side of the Africans when there were frontier clashes between Boer ranchers and African tribesmen. English was made the official language, despite the fact that most colonists did not speak it. Meanwhile an increasing shortage of range land developed as the population grew, and this added to Boer dissatisfaction. Then in 1833 slavery was officially abolished throughout the British Empire. The Boers were compensated for the loss of their slaves, but compensation was given in promissory notes redeemable only in London. Much of this paper was bought at a large discount by speculators, and the entire transaction further embittered relations between the Boers and the British governing officials.

The Great Trek

Boer discontent resulted in the Great Trek—a series of northward migrations through which groups of Boers, primarily from the eastern part of the Cape Colony, sought to find new grazing lands and establish new political units beyond the reach of the authorities at Cape Town. After some preliminary exploring expeditions in earlier years the main trek by horse and ox-wagon began in 1836. It resulted in the founding of Natal, the Orange Free State, and the Transvaal as Boer republics. Natal was annexed by Britain in 1845, but Boer sovereignty in the Transvaal and the Orange Free State was officially recognized in 1852 and 1854, respectively. The northward migration of the voortrekkers (pioneers) was a central event in the history of the Afrikaner nation and is a major focus of Afrikaner national cohesion. An estimated 12,000 Boers—men, women, and children—were involved in the movement during the decade following 1835. As a result of the Great Trek, roughly a fourth of the Boer population was withdrawn from the Cape Colony. At the time the interior plateau was comparatively empty due to depredations against other blacks by the formidable and disciplined Zulu tribe. The Boers clashed with remnants of tribes that had fled from the Zulus, and in Natal they encountered the main Zulu force. Although a good many Europeans were killed, the African spearmen, fighting on foot, eventually proved no match for mounted Europeans equipped with guns, and the Boers were able to take firm possession of the High Veld, although the majority withdrew from Natal. On the grasslands they grazed their animals and reestablished their traditional way of living, which was patriarchal in character and strongly affiliated with the Dutch Reformed Church. Feelings of nationhood were reinforced by memories of the Great Trek and the sense of being beleaguered within a *laager* (wagon circle). Today Afrikaner national consciousness is enshrined in the huge Voortrekker Monument overlooking the Afrikaner stronghold of Pretoria.

The Anglo-Boer War

While Boer disaffection with Britain built up, British settlers were coming to South Africa in increasing numbers. Port Elizabeth and East London were founded as British towns, and British influence also became dominant in Cape Town and Durban. It is conceivable that the British colonies and the Boer republics in South Africa might have developed peaceably side by side had not diamonds been discovered in the Orange Free State in 1867 and gold in the Transvaal in 1886. These discoveries set off a rush of prospectors and other fortune hunters and entrepreneurs from outside, and mining camps sprang up at Kimberley, Johannesburg, and other places. The Boers possessed little or no capital with which to work the deposits, but British capital soon poured in. The conservative Boer leaders did not welcome the influx of outsiders ("uitlanders"), the majority of whom were British. Ill feeling led to the Anglo-Boer War in 1899. It was an unequal contest since the British were able to draw support from the homeland and the Empire, whereas the Boers had to depend entirely on their own resources. After some early successes the Boer forces were decisively defeated, and the war ended in 1902. The Union of South Africa was established in 1910 as a self-governing dominion under the British crown. Dutch was recognized as an official language on a par with English (later this was altered to specify Afrikaans rather than Dutch), and Pretoria in the Transvaal was made the administrative capital as a concession to Boer sentiment. However, the national parliament meets at Cape Town, and the supreme court sits at Bloemfontein (250,000) in the Orange Free State.

Since 1910, and particularly since the Nationalist (now the National) party came to power in 1948, the Afrikaners have dominated the political life of South Africa by virtue of their greater numbers and cohesion. All of the country's prime ministers have been Afrikaners. Before 1948 the United Party, a coalition of Afrikaners and British elements, generally controlled the government, but since the triumph of the Nationalist Party the direction of affairs has been more purely in the hands of Afrikaners. Under this regime the policy of "separate development" or "multinational development" for different races has been enunciated in numerous laws and decrees, culminating in the scheme to establish a series of tribal states or "homelands," which one by one are being granted "independence."

THE "HOMELANDS" SCHEME

The South African government has been in the process of forming ten territorial units reserved for Africans and possessing elected African governments. These "homelands" (Fig. 23.2), organized essentially on a tribal basis, collectively incorporate hundreds of separate tracts of land that began to be set apart as African reserves soon after the Union of South Africa was formed. By 1988, four of the homelands—Transkei, Bophuthatswana, Venda, and Ciskei—had accepted "independence" as "republics" and were functioning under elected African governments, although no country except South Africa had given diplomatic recognition to any of them. Transkei, the largest of the four and the first to become "independent" (1976), is about the size of Tennessee in area and has an estimated 2.6 million residents—the population of the state of Mississippi. Nearly all of Transkei's population is classified as belonging to the large Xhosa tribe. The republic is located within the boundaries of the eastern Cape Province between the Natal border and East London. West of it lies the much smaller Ciskei homeland, which is also a Xhosa area. Northeast of Transkei is the territorially fragmented Zulu homeland called Kwazulu, which is embedded in Natal. This unit has refused to accept "independence" and thereby give up its right to share in the general development of South Africa. The fragmented republic of Bophuthatswana is located on the interior plateau north and west of Johannesburg; and Venda lies in the far northeast of the plateau near Zimbabwe. The remaining six homelands, which are in various stages of development toward "independence," are also located on the plateau, generally along or near a line between Bophuthatswana and Venda.

South Africa's laws permit the government to classify a black South African as a legal resident of one of the homelands, even though he may never have seen it and has no sense of affiliation with the tribe on which the homeland is based. By early 1989, several million Africans not wanted in "white" South Africa had been ejected from their homes there and transferred to homelands, and the government was planning to remove still others. Great numbers have been dumped into poverty-stricken and crowded areas unequipped to handle new influxes of people. The bulk of these unfortunates have been ejected either from urban areas or from white farms where they have provided farm labor under various legal or illegal arrangements. But some of the most harrowing examples of forced removal from "white" South Africa are the African families in hundreds of relatively small and scattered black enclaves known as "black spots." One instance occurred at a community called Dreifontein in the southeastern Transvaal. Here lived about 5000 Africans on land originally purchased in 1912 before the "Native Reserves" were established and it became illegal for Africans to own land outside them. But despite their long tenure in an established community, the Africans of Dreifontein were dispossessed of their land in 1983 and were trucked with their personal belongings to a bleak new existence in a crude homelands camp.

The degree of real independence that the new homelands will ultimately acquire is unclear. It is quite evident that millions of their African residents will continue to work in South Africa outside the homelands. These black millions are to be considered temporary visitors, who are to be permitted in "white" South Africa only in such numbers as needed by the economy of the whites. Thus will be perpetuated the system under which black breadwinners migrate for long periods to jobs outside their homelands while the rest of the family stays behind. The destruction of black family life that this system already has caused in South Africa is beyond calculation. It seems more than probable that many black families in "white" South Africa will be able to maintain at least a *de facto* permanence of residence there, but such matters are still very conjectural.

Meanwhile the black governments of the homelands are faced with problems that bear considerable

Figure 23.7 *African homes in warm, rainy hills of eastern South Africa near the Indian Ocean. A corn patch is at the right, and the grassy slope at the rear is used to pasture cattle. Living conditions among black farmers in South Africa are similar in many ways to those of African farmers in tropical countries to the north.*

similarity to those of many underdeveloped African states. They must attempt to administer areas that with some exceptions are very lacking in mineral resources, manufacturing plants, or even large towns. Agriculture in the homelands, generally built around subsistence production of livestock and maize, has long been substandard. The homelands embedded in the Cape Province and Natal are very hilly (Fig. 23.7), but they do have the advantage of plentiful rainfall and subtropical temperatures. The units on the interior plateau are flatter, but rainfall is much scantier and less dependable. To make these new units economically viable, it will be necessary to remove a large share of the population from agricultural occupations; to provide employment for them and for African newcomers from "white" South Africa in industrial and service occupations; and to build up the productivity of agriculture.

Prior to the time it commenced to grant the homelands "independence," South Africa had embarked on a scheme to foster "border industries," wherein manufacturing plants would be established just inside "white" territory but labor would commute from the homelands each day. Some industries of this kind are functioning, but they comprise an extremely minor share of South Africa's total industrial development.

By mid-1988 there had not been very much movement to establish new industries within the homelands themselves, but development initiatives of various kinds were under way, including holiday resorts in Bophuthatswana and Transkei that provide gambling casinos and other diversions patronized mainly by white South Africans (who lack these amenities in the strait-laced Republic). Bophuthatswana derives additional income from valuable minerals (especially platinum) within its borders.

The miserably poor Ciskei republic, governed by an authoritarian president-for-life, has undertaken an ambitious construction program that includes a large hospital built by an Israeli corporation, some showpiece buildings and mansions in the capital, and a projected seaport and airport. Some Israeli industrialists have expressed interest in establishing factories in Ciskei.

South Africa subsidizes the homelands in various ways, although by no means adequately, and the world waits to see what will happen next in this bizarre new set of "independent" countries that no nation except South Africa has recognized. In 1987 South African subsidies totaled a reported $1 billion (U.S.) for the four "independent" homelands and another $1 billion for the six non-"independent" homelands.

PROSPECTS FOR CHANGE

South Africa's future is one of the world's great question marks and will probably remain so for a long time. Will the country be able to continue its increasing role as an economic dynamo for all of southern Africa? The answer is probably yes, in view of the great dependence of neighboring countries (such as Zimbabwe, Mozambique, and Malawi) on South Africa as a source of imports, employment, technical expertise, transportation services, and other aids. Despite the criticism roused by its racial laws, South Africa has been able to make many accommodations with its black neighbors and may well continue to do so. Will there be continuing Western pressure on South Africa to change its racial laws? Undoubtedly, and South Africa seems likely to do what it can to keep good relations with the West. Concessions in the realm of "petty apartheid" will doubtless continue to be made, and advancement of nonwhites in health, education, economic status, and general welfare will continue. But segregation of the races territorially is still proceeding, and the determination of most whites to keep the upper hand politically seems unaltered. This determination is spearheaded by the Afrikaners, who are a strongly cohesive nation with a survival mentality forged originally in frontier warfare against blacks and then intensified by the struggle against the English in the Anglo-Boer War. Unlike the British South Africans, the Afrikaners have no emotional ties to a homeland outside South Africa. Hence

they cling to their privileged position with special intensity.

Various kinds of pressures, both external and internal, are being exerted on South Africa's government in opposition to the apartheid policy. Under the Anti-Apartheid Act of 1986, the United States Congress imposed limited sanctions. Over two dozen other countries have applied some form of sanctions in recent years. Great Britain has refused to do so on the grounds that sanctions are not effective and would be likely to harden South African resistance to change. In the United States there have been widespread attempts to induce or pressure institutions such as universities and union pension funds to divest themselves of investments in American corporations doing business in South Africa. As of 1987 over a hundred U.S. corporations, representing more than one third of such businesses in South Africa, had sold their assets and left the country. Stated reasons for these actions included unfavorable economic conditions in South Africa, shareholder pressure in the United States, and boycott threats in the United States. Actually, many American corporations in South Africa have set examples of fairness by integrating their workforces and providing amenities to nonwhite workers that are not available to such employees of indigenous firms.

Forces for change inside South Africa include the concerns and apprehensions of white business leaders, and opposition to apartheid by church groups, university students and faculties, and black labor unions. White businessmen have met outside the country with exiled leaders of the ANC to determine whether there are grounds for compromise in black demands for political participation. The outlawed ANC is still looked to by great numbers of blacks as the spearhead for their aspirations. In holding talks with the ANC, white business leaders have displayed not only their fears of racial warfare and internal chaos, but also their recognition that future growth of the economy must depend increasingly on the participation of skilled and educated blacks in technical jobs and responsible positions. Leaders in some church groups, both black and white, have worked for many years in support of nonviolent change in apartheid. For his role in this movement, Anglican Bishop Desmond Tutu of Johannesburg was given the Nobel Peace Prize in 1984. Subsequently this South African black prelate was designated Archbishop of Cape Town and head of the Anglican Church in southern Africa. Accompanying the protest movement in South African churches have been protests on university campuses. In fact, such complaints about the inequities of apartheid have been so widespread that the government has threatened cutbacks in education subsidies. Black labor unions in South Africa have been growing in membership and influence but probably not to the point of being able to force major concessions from the government concerning apartheid. The essential weakness of the black labor movement was revealed in 1987 when a major strike against mining companies not only resulted in refusal by management to meet workers' demands in regard to wage increases and upgrading of the kinds of jobs they were entitled to hold, but also caused the firing of a reported 45,000 workers who had walked out. The strike was settled in 3 weeks, and most of the discharged workers were subsequently rehired, but the limited power of black unions had been demonstrated.

An interesting and possibly significant move to break down apartheid took place in the province of Natal in 1986. Leaders from all racial groups held a conference and negotiated proposals for the future governance of the province. If carried out, these arrangements would essentially end apartheid in Natal. The national government, which has final authority in the matter, has done nothing to implement this initiative.

Although the activist role of some South African groups opposed to apartheid seems reminiscent of the drive for black civil rights in the United States during the 1960s, the situation in South Africa is far more resistant to change. In the United States blacks represent little more than a tenth of the total population, whereas in South Africa blacks form such a large majority as to raise white fears of being overwhelmed in any national election. Furthermore, in South Africa there are no such constitutional guarantees of full civil rights for blacks as exist in the United States. To the contrary, there is a large body of law meticulously crafted to ensure that black South Africans will be effectively excluded from participation in their country's political affairs. In May 1987 the white electorate of South Africa gave its response to limited international sanctions, internal black resistance, and the efforts of some whites within South Africa to find ground for accommodation. At the polls in a national parliamentary election, the voters returned a three-fourths majority of parliamentary seats for the National Party which is currently in control of the government. The second highest number of seats went to the very hard-line Conservative Party, whose policies call for even stricter apartheid than currently exists.

BOTSWANA, LESOTHO, AND SWAZILAND

Three units bordering South Africa—Botswana (formerly Bechuanaland), Lesotho (formerly Basutoland), and Swaziland—were protectorates ("High Commission Territories") of Britain until the 1960s, but Botswana and Lesotho gained independence in 1966 and Swaziland in 1968. Lesotho, a mountainous country containing the highest summits in the Drakensberg escarpment, is surrounded by South Africa, and Swaziland is nearly so, although it has a short frontier with Mozambique. Botswana, most of which is semidesert or, in the north, tsetse-infested swamp, is bordered on three sides by South Africa and Namibia and on the fourth side by Zimbabwe. Of the three, Swaziland has the most varied resources, diversified economy, and highest income per capita. Western Swaziland has mountains along the Drakensberg escarpment that receive heavy rainfall, and rivers that rise there and flow to the Indian Ocean are used for irrigation projects in the lower and drier east. Mountain slopes have been planted in conifers that form the basis for wood pulp and lumber industries. Mining is an important source of revenue, as is tourism which features gambling casinos. Most of the larger enterprises have been developed by Europeans, using Swazi labor, but racial relationships have been better than in many African countries and the economic benefits to Swaziland have been considerable. Although some of the population still lives outside of the money economy, increasingly the Swazi farmer is improving his methods and is marketing surpluses. Swaziland's main exports in 1985 were sugar, wood products, citrus fruit, and asbestos. Mining companies have been prospecting actively in Botswana and Lesotho; the principal result thus far has been a fair-sized production of diamonds, amounting in 1986 to 78 percent of the total value of exports from Botswana and 42 percent from Lesotho. The African populations of these two countries raise livestock and a little grain, largely on a subsistence basis; some Europeans own cattle ranches in eastern Botswana. Large numbers of African workers find employment in South Africa, primarily in mining. All three countries are very dependent on South Africa for trade and many kinds of services. Since independence, South Africa has established a customs union with the three.

LATIN AMERICA

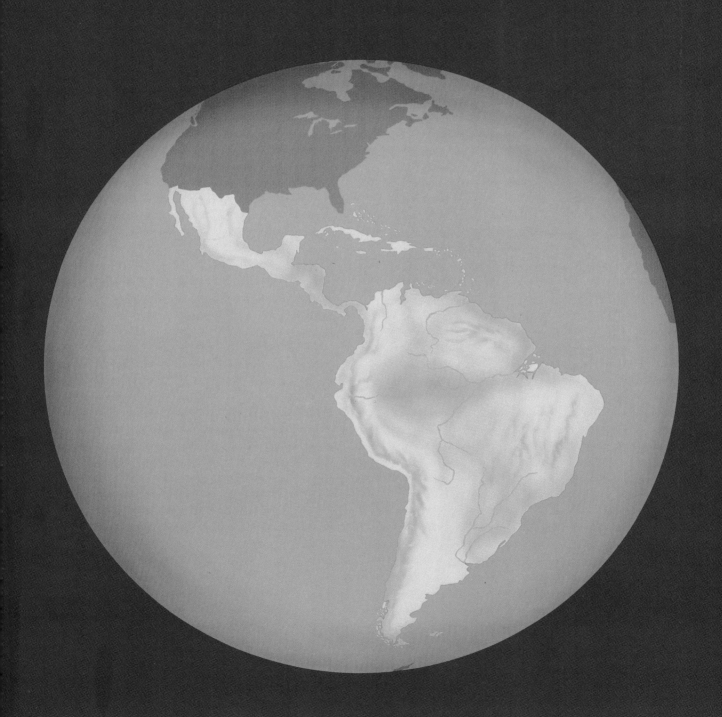

CHAPTER 24

The Many Faces of Latin America[1]

Political rally in São Paulo, Brazil. The large sign in Portuguese calls for freedom from repression, public ownership of public services, more public housing, land to the landless, and broadened political participation for the people.

T HE land portion of the Western Hemisphere to the south and southeast of the United States has come to be known as Latin America (Fig. 24.1). Both the name of this region and its ways of life reflect the importance of culture traits inherited from the Latin European colonizing nations of Spain, Portugal, and France. Spanish and Portuguese, for example, are overwhelmingly the major languages in Latin America as a whole, although Portuguese is the language only of the region's largest country, Brazil (chapter-opening photo). Dialects of French are spoken in the West Indian island republic of Haiti and in France's dependencies of French Guiana and the islands of Guadeloupe and Martinique. Some small states and dependencies in the West Indies or on the nearby mainland do not exhibit Latin culture except in minor ways. These units mainly speak English or Dutch, and some are still dependencies of the United Kingdom, the United States, or the Netherlands. Their presence in an area of original Spanish control reflects early colonial struggles among European nations for territories suitable for sugarcane plantations. Important precolonial influences in some Latin American countries are manifested by a multiplicity of native Indian languages and dialects. But Latin influences are predominant today in the region as a whole and are shown not only in Latin-derived (Romance) languages but also in the dominance of the Roman Catholicism that arrived in Latin America with the very first ships from Latin Europe (Fig. 24.2). Actually, language and religion are merely the most obvious cultural importations, which also included such major elements as legal systems, land tenure arrangements, governmental practices, social structures, and economic systems.

It should not be assumed that European cultures have been transplanted to Latin America without essential modification or that a uniform culture prevails today in even the most strongly Latinized parts of the region. Underneath a veneer of sameness, keen distinctions exist between regions, between countries, and within countries. Among the many factors responsible for this differentiation are diverse pre-European and colonial experiences, different resource bases, divergent political systems, and differential rates of development. The result is that Latin America presents many faces to the outer world: population groups of many kinds; extraordinary physical variety; economies that run the scale from technologically advanced to remarkably primitive; governmental philosophies and structures of diverse sorts; and countries that have distinctive personalities despite enough overall similarity and cohesion to fit together reasonably well in an identifiable major world region. This introductory chapter examines the physical dimensions of Latin America, the region's people and their diverse habitats, economic patterns in their geographic settings, and political arrangements and their origins. Chapter 25 then characterizes Latin America's countries by regional groups. Area, population, and other data for individual countries are given in Table 24.1, p. 554.

PHYSICAL DIMENSIONS

With a land area of slightly more than 7.9 million square miles (over 20.5 million sq. km), Latin America is surpassed in size by Africa, the Soviet Union, Anglo-America (including Greenland), and the Orient. However, its maximum latitudinal extent of more than 85 degrees or nearly 5900 statute miles (c. 9500 km) is greater than that of any other major world region, and its maximum east-west measurement, amounting to more than 82 degrees of longitude, is by no means unimpressive. Yet Latin America is not so large as these figures might suggest, for its two main parts are offset from each other. The northern part of the region, known as Caribbean or Middle America, trends sharply northwest from the north-south oriented continent of South America. The latter is thrust much farther into the Atlantic Ocean than is the Caribbean realm or Latin America's northern neighbor, Anglo-America. In fact, the meridian of 80°W, which intersects the west coast of South America in Ecuador and Peru, passes through Pittsburgh, Pennsylvania. Brazil lies less than 2000 statute miles (under 3200 km) west of Africa (Fig. 24.1, inset).

POPULATION GEOGRAPHY: UNEVEN SPACING, DIVERSE STOCKS, RAPID GROWTH

In mid-1988 the world was estimated to contain 5.1 billion people, of whom the Latin Americans represented some 429 million. Since Latin America is a large and diverse region with well over twice the land area of the United States, it is not surprising

---

[1]The authors wish to thank Brian K. Long for his valuable assistance in the revision of Chapters 24 and 25. These chapters contain some material written originally by Dr. Richard S. Thoman, but updated here and presented in a revised format.

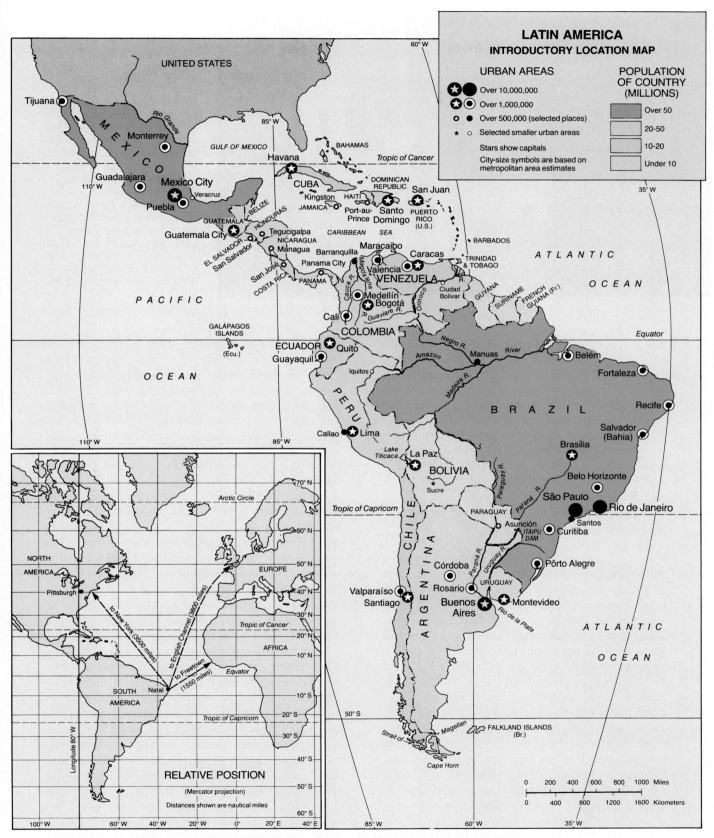

Figure 24.1 Introductory map of Latin America. For additional names of small island countries in the West Indies, see Figure 25.1.

Figure 24.2 Roman Catholicism is the leading religion in all Latin American countries except some relatively small units colonized by countries other than Spain or Portugal. The nearly universal cultural stamp of this religion is a very important distinguishing characteristic of Latin America. The old Cathedral of Lima, Peru, shown in the photo, is one of the more majestic among great numbers of Catholic buildings spread through Latin America from Mexico and Cuba to Argentina and Chile. The Bishop's Palace adjoins the cathedral to the left.

that its population is unevenly distributed and exhibits wide variations in density. However, there is some overall order in the fact that most of Latin America's people are packed into two major alignments, the larger of which forms a discontinuous ring around the margins of South America, while the other extends along a volcanic belt from central Mexico southward into Central America. The two alignments are outlined reasonably well by the patterns of population and crop land as shown in Figures 3.2 and 3.3.

MAJOR POPULATED AREAS

The South American "rim" of clustered population contains roughly two thirds of Latin America's people, with the majority living in and around cities located on or very near the coast. Two major segments of the "rim" can be discerned. One segment—much larger than the other in population and area—extends along the eastern margin of the continent from the mouth of the Amazon River in Brazil to the humid pampa (subtropical grassland) around Buenos Aires, Argentina. The second segment, located partly on the coast and partly in high valleys and plateaus of the adjacent Andes Mountains, stretches around the north end and down the west side of South America from the vicinity of Caracas, Venezuela, on the Atlantic to the vicinity of Santiago, Chile, on the Pacific. This second segment is more fragmented than the first, as it is broken in many places by steep Andean slopes or stretches of coastal desert. Between the two major segments on the Atlantic side lies a strip of hot rainy coast be-

tween the Amazon mouth and Caracas where the thinly populated interior of South America extends to the ocean. In the far south the population "rim" is again broken in the rugged, rainswept southern Andes and the dry lands of Argentina's Patagonia that lie in Andean rain shadow on the Atlantic side of the continent. These latter territories have only a scanty human population, although millions of sheep graze in Patagonia and adjacent Tierra del Fuego.

Before the European conquest, the populated "rim" of South America was not discernible except in the Andes, as the coasts were only lightly inhabited by migratory Indians. The high intermontane areas within the Andes already exhibited a pattern of settled Indian communities, ruled by the Inca Empire and distributed in a manner approximating that of the populated areas of today (Fig. 24.3). But when the Europeans began their quest for South American wealth in the 1490s, the coasts, lightly defended at best by the aborigines, lay open to whatever settlement pattern the newcomers might devise. Since the Europeans approached from the sea at many separate points, their first major task was to develop ports as bases for penetration of the interior. Many ports eventually grew into sizable cities, and a few became major metropolitan centers. In some instances—notably Lima, Caracas, and São Paulo—the main city developed a bit inland but retained a close connection with a smaller coastal city that was the actual port. Around the ports agricultural districts developed, spread, and shipped an increasing volume of surplus products overseas. The ships that took these goods away brought manufactures from Europe and carried passengers, including government officials, in both directions. Slave ships brought the Africans who provided the principal labor on European-owned plantations. Populations along or near the coasts multiplied and are still doing so today.

Meanwhile the lure of gold and silver stimulated the penetration of the Andes and the Brazilian Highlands. After looting the stores of precious metals that had been accumulated by the Indians, the newcomers opened mines or took over old Indian mines in many places, and urban centers arose to service the mines. Some highland cities gained an even wider importance as centers for new ranching or plantation areas. Such functions still endure, although great numbers of older mines have closed, and mining today is far more diversified and less dependent on gold and silver than was true in the early colonial age. A few highland cities—of which the largest is Bogotá in Colombia—have grown into sizable metropolises. These places are separated from the seaports by difficult terrain which required feats of engineering before satisfactory transportation links emerged. Some of the main seaports and highland cities became important centers of colonial government as well as economic nodes, and several are national capitals today. Rapid development of manufacturing industries and explosive population growth have characterized such cities in recent times.

The second major alignment of populated areas in Latin America lies on the mainland of Caribbean or Middle America to the northwest of the South American "rim." Composed of the majority of Middle America's people, it extends along an axis of volcanic land that dominates central Mexico and from there reaches southeastward through southern Mexico and along the Pacific side of Central America to Costa Rica. This belt is characterized by good soils, rainfall adequate for crops but not excessive, and enough elevation in most places to moderate the tropic heat but still permit many tropical crops such as coffee. The belt of high population density which extends across central Mexico from the Gulf of Mexico to the Pacific Ocean was already Mexico's center of population when the Spanish conquerors came. This was the principal domain of the large and technologically advanced Aztec Empire, which together with its vassal states comprised the bulk of Middle

Figure 24.3   The old city of Cuzco in the Peruvian Andes, shown here, was the capital of the Inca Empire when the Spanish conquerors came to Latin America. The city, still predominantly Indian in population, is located in a basin at an altitude of about 11,000 feet (c. 3400 m).

TABLE 24.1 LATIN AMERICA: BASIC DATA

Political Unit	Area (thousand sq. mi)	Area (thousand sq. km)	Estimated Population (millions, mid-1988)	Estimated Annual Rate of Natural Increase (% 1988)	Estimated Population Density, 1988 (per sq. mi)	Estimated Population Density, 1988 (per sq. km)	Infant Mortality Rate	Urban Population (%)	Cultivated Land (% of total area)	Per Capita GNP ($U.S.: 1986)
Caribbean America										
Mexico	761.6	1972.5	83.5	2.4	110	42	50	70	13	1850
Guatemala	42.0	108.9	8.7	3.2	210	80	65	33	17	930
Belize	8.9	23.0	0.2	3.0	20	8	36	50	2	1170
El Salvador	8.1	21.2	5.4	2.8	665	255	65	43	34	820
Honduras	43.3	112.1	4.8	3.1	110	42	69	40	16	740
Nicaragua	50.2	130.1	3.6	3.5	72	28	69	57	10	790
Costa Rica	19.6	50.7	2.9	2.9	148	57	19	45	13	1420
Panama	29.8	77.1	2.3	2.2	77	30	25	51	7	2330
Totals: Central America	201.9	523.0	27.8	3.0	138	53	58	42	13	1030
Guyana	83.0	215.0	0.8	2.0	10	4	36	32	2	500
Suriname	63.0	163.2	0.4	2.1	6	2	33	66	<1	2510
French Guiana	34.7	89.9	0.1	2.3	2	1	23	73	<1	2340 (1983)
Totals: Guianas	180.7	468.1	1.3	2.1	7	3	34	46	1	1300
Cuba	42.8	110.8	10.3	1.0	240	93	14	71	29	2690 (1984)
Dominican Republic	18.8	48.7	6.8	2.4	362	140	70	52	30	710
Haiti	10.7	27.7	6.3	2.8	589	227	117	25	33	330
Jamaica	4.2	11.0	2.5	1.7	595	227	20	54	24	880
Trinidad and Tobago	2.0	5.1	1.2	2.2	630	245	13	34	31	5120
Bahamas	5.4	13.9	0.2	1.9	46	18	26	75	1	7190
Barbados	0.2	0.4	0.3	0.8	1540	595	11	32	77	5140
Antigua and Barbuda	0.2	0.4	0.1	1.0	490	190	10	34	18	2380
Dominica	0.3	0.8	0.1	1.6	305	120	24	74	23	1210
Grenada	0.1	0.3	0.1	1.9	800	310	22	25 (1981)	41	1240
St. Christopher (Kitts) and Nevis	0.1	0.4	0.04	1.4	460	180	28	45	39	1700
St. Lucia	0.2	0.6	0.1	2.2	610	235	22	40	27	1320
St. Vincent and Grenadines	0.2	0.4	0.1	2.0	760	295	27	26 (1980)	50	960
Totals: Independent Island Countries	85.2	220.6	28.1	1.8	327	126	52	52	28	1500

Puerto Rico	3.5	9.1	3.3	1.2	964	372	15	67	15	5190
U.S. Virgin Islands	0.14	0.35	0.1	2.0	840	325	20	30 (1980)	7	9300 (1985)
British Virgin Islands	0.06	0.15	0.01	1.4	210	80	13	12 (1980)	—	7130 (1985 GDP)[a]
Cayman Islands	0.1	0.3	0.02	1.2	226	88	8	100 (1979)	—	12,100 (1985 GDP)
Montserrat	0.04	0.1	0.01	1.2	303	119	8	13 (1980)	47	2530 (1985)
Anguilla	0.04	0.1	0.01	1.7	195	75	14	—	—	—
Turks and Caicos Islands	0.2	0.5	0.01	2.4	55	21	10	—	—	3510 (1983 GDP)
Guadeloupe	0.7	1.8	0.3	1.3	494	191	17	46	23	3640 (1984)
Martinique	0.4	1.1	0.3	1.2	791	305	13	71	17	4260 (1983)
Netherlands Antilles	0.3	0.8	0.2	1.4	579	223	10	50	8	6110 (1985)
Totals: West Indian Overseas Affiliates	5.5	14.2	4.3	1.3	791	305	15	65	16	5220
Totals: Caribbean America	1234.8	3198.2	145.0	2.4	117	45	51	61	12	1720
Andean Countries										
Colombia	439.7	1138.9	30.6	2.1	68	26	48	65	5	1230
Venezuela	352.1	912.1	18.8	2.4	52	20	36	82	4	2930
Ecuador[b]	109.5	283.6	10.2	2.8	91	35	66	52	9	1160
Peru	496.2	1285.2	21.3	2.5	42	16	88	69	3	1130
Bolivia	424.2	1098.6	6.9	2.6	15	6	110	49	3	540
Totals:	1821.7	4718.4	87.8	2.4	47	18	62	67	4	1510
Brazil	3286.5	8512.0	144.4	2.0	44	17	63	71	9	1810
Southern Mid-Latitude Countries										
Argentina[c]	1068.3	2766.9	32.0	1.5	30	12	35	85	13	2350
Chile	292.3	757.0	12.6	1.6	43	17	20	83	7	1320
Uruguay	68.0	176.2	3.0	0.8	44	17	30	85	8	1860
Paraguay	157.0	406.7	4.4	2.9	28	11	45	43	5	880
Totals:	1585.6	4106.8	52.0	1.6	33	13	32	80	11	1950
Grand Totals:	7928.7	20,535.3	429.2	2.2	54	21	57	68	9	1720

[a]GDP (gross domestic product) is calculated on a different basis from GNP (gross national product), but either measure will serve as an approximation of national income.
[b]Figures for Ecuador include the Galapágos Islands, a dependency of Ecuador.
[c]The Falkland Islands, a British dependency in the South Atlantic located somewhat more than 400 miles (c. 650 km) east of Argentina and claimed by that country, are not included in the table.

Figure 24.4 A landscape in Central Mexico between Mexico City and Guadalajara. The corn fields and conifers betoken a relatively high altitude; the cinder cone of an extinct volcano is being quarried for road-building material.

America's population at that time. Then as today, the soils of this area's high volcanic plateaus and flat-floored basins provided the foundation for a productive agriculture based primarily on maize (Fig. 24.4). Spanish occupation did not alter the dominance of the area within Mexico, and roughly half of Mexico's 84 million people (1988) now live in, around, or between the region's two principal cities of Mexico City and Guadalajara.

In Pacific Central America also, the basic pattern of population distribution already was established when the Spanish took over. Today the majority of people live in highland environments, but considerable numbers live in lowlands. The ongoing vulcanism of the highlands periodically deposits ash into the highland basins, and this weathers into fertile soils that grow maize for local food and the coffee that is Central America's largest export. Between the highlands and the Pacific lies a coastal strip of rather dry lowlands where cattle and cotton are grown on an export basis. By contrast, the Atlantic side of Central America is lower, hotter, rainier, and far less populous than the Pacific side. Here on large corporate-owned plantations are grown most of the bananas that are Central America's agricultural trademark and second most valuable export.

VARIATIONS IN POPULATION DENSITY

Something like two thirds of Latin America's land lies outside the two major population alignments just described, and most of this land has very few people living on it. Hence figures on population density for the greater part of Latin America are extraordinarily low, averaging under two persons per square mile over approximately half of the entire region. This pulls down Latin America's average density, and that of most major Latin American countries, to levels well below the average density for the United States. Overall, Latin America had a density of 54 persons per square mile (21 per sq.

km) in 1988 and Brazil's density was 44 per square mile (17 per sq. km) compared to 68 per square mile (26 per sq. km) for the United States. Among the nine largest Latin American political units by area, only Mexico exceeded the United States in density (Table 24.1). This situation will change over time, as the population growth rate of the United States is far exceeded by that of Latin America as a whole and by nearly all the individual Latin American countries.

But the country-by-country pattern of population distribution will change far more slowly. Average population densities for entire countries conceal the fact that in nearly every mainland country there is a basic spatial configuration consisting of a well-defined population core (or cores) with an outlying sparsely populated hinterland. Densities within the cores are often greater than anything to be found within areas of comparable size in the United States. On the other hand, most mainland countries have a larger proportion of sparsely populated and relatively unorganized terrain than is true of the United States. Such outlying territories are comprised of tropical rain forests, tropical savannas, deserts, steppes, and/or rugged mountains. Long-term action to overcome environmental difficulties may well cause many of these areas to have a much greater population density in the future than they have today. But for the present they are largely beyond the reach of effective transportation or other services and hence are incompletely incorporated into the effective national territories of countries.

The pattern of core versus "outback" (to use an appropriate Australian term) is well exemplified in Brazil and Argentina. Most Brazilians live along or very near the eastern seaboard south of Recife, whereas huge areas in the interior are essentially without inhabitants other than scanty numbers of aborigines. Approximately three fourths of Argentina's population is clustered in Buenos Aires or the adjacent humid pampa—an area containing a little over one fifth of Argentina's total land. An extreme case is found in French Guiana, where the overall population density is under three persons per square mile and practically the entire population of about 90,000 lives in a very narrow coastal band containing the capital city of Cayenne (40,000).

Not every country displays the pattern of core versus outback. A very different picture of population density and distribution is evident in small El Salvador and on most West Indian islands. El Salvador, where an estimated population of 5.4 million (1988) clusters thickly on the volcanic land that comprises nearly all of the country, has an estimated overall density of over 660 per square mile (nearly 260 per sq. km). This figure puts El Salvador in a class by itself among the mainland countries. In fact, relatively few countries in the world have a greater overall density. But even El Salvador is surpassed within Latin America by several island countries (Table 24.1), most notably Barbados with over 1500 persons per square mile (nearly 600 per sq. km). Since the combined population of the West Indian islands is considerably less than that of central Mexico alone, these islands do not form a concentration of people comparable in scale to either of the two major Latin American concentrations discussed earlier. But they do stand out for exceptionally heavy densities, which have been created over a long period by the accumulation of population on territories of very restricted size. Often the crowding is aggravated by high rates of natural population increase and/or the prevalence of steep slopes that restrict the possibilities for further settlement. A case in point is the Republic of Haiti, which has 6.3 million people occupying an excessively mountainous territory a little larger in area than El Salvador. Haiti is both the most rural and the most poverty-stricken country in the Western Hemisphere, with a population that will double in the next quarter-century if the current rate of increase continues. Large numbers of Haitians live on tiny farms that cling to incredibly steep mountainsides. This situation is extreme in Haiti but is not unique to it, being found also in Jamaica and many other mountainous parts of Latin America. Serious erosion problems are often present in these areas where the pressure of numbers has forced people out of the valleys and up slopes too steep for normal tillage.

ETHNIC DIVERSITY

Thus far we have been looking at the spatial arrangement of population without regard for the ethnic diversity that is a major Latin American characteristic. Although most Latin Americans speak Spanish or Portuguese and embrace some form of Roman Catholicism, the region's inhabitants have highly varied ancestries—white European, black African, native Indian, Asian, or all manner of mixtures. The matter becomes even more complex when it is realized that the Europeans, Africans, and Asians came from various parts of their respective continents, while the native Indians represented a multiplicity of tribes. Here we can only sketch in broadest outline the major ethnic groups and how they are distributed in Latin America today.

Despite the overall dominance in Latin America of culture traits derived from Europe, only three nations—Argentina, Uruguay, and Costa Rica—have preserved white European racial strains on a large scale with little admixture by Indians or blacks. This is due in part to the fact that these nations received relatively large numbers of European immigrants during the postcolonial period. Scattered districts in other political units also are predominantly European.

Native Indians (Fig. 24.5) comprise an estimated 40 to 55 percent of the population in the highland nations of Guatemala, Bolivia, Ecuador, and Peru. They are also a major population element in Paraguay and southern Mexico. In outlying areas, especially in the basin of the Amazon River and in Panama, scattered lowland Indians maintain their aboriginal cultures. The outer world is pushing in on them despite some efforts on the part of governments to protect their ways of life. These Indians receive a good deal of publicity, but their numbers are too few to count for much in the total scale of Latin American affairs. Over the centuries after Columbus, the encounters of lowland Indians with Europeans generally had catastrophic results for the aborigines, who often resisted exploitation fiercely but were no match for the white man's guns, diseases, liquor, cupidity, and guile. This harsh phase of Latin American history is not entirely over in some remote parts of the region today. The highland Indians, being more numerous, more advanced, and better organized than the lowlanders, accommodated somewhat better to European encroachment and exploitation, but still suffered incredible population losses at European hands. The greater ability of the highlanders to maintain themselves in their mountainous redoubts was due partly to the fact that their lands were too cool and too removed from ocean transportation for plantation agriculture to flourish.

Black Latin Americans of relatively unmixed African descent are found in the greatest numbers on the West Indian islands and along hot, wet Atlantic coastal lowlands in Middle and South America.

Figure 24.5 Indian women at a market in Central Mexico. The scene is at Pátzcuaro, a small commercial focus and tourist center in a basin between Guadalajara and Mexico City.

Most political units in the islands have black majorities (often overwhelmingly so), although Puerto Rico, Cuba, the Dominican Republic, and Trinidad-Tobago are exceptions. In Haiti, for example, blacks are reported to comprise 90 percent of the population; in Barbados, 92 percent. On the mainland a major concentration of blacks is found along the lower Magdalena River Valley and on the Caribbean coast of Colombia. Other concentrations occupy districts in Panama, Venezuela, Guyana, Suriname, and French Guiana. The east coast of Brazil between Cape São Roque and Salvador has a sizable black population, as does Rio de Janeiro and its northern hinterland. Blacks are an important element in many Latin American cities.

Today the majority of Latin America's black population is still situated in the areas to which African slaves were brought during the colonial period, primarily as a source of labor for sugar plantations. An estimated 3 to 4 million blacks were sold in Brazil alone. Slavery was gradually abolished during the 19th century, although not until the 1880s in Brazil. By that time black servitude had generated large fortunes for many owners of plantations and slave ships, had provided a major basis for the permanent settlement of sizable areas in the New World, and had introduced African peoples who have made cultural contributions that are today a varied, colorful, and important part of Latin American civilization.

The ethnic diversity of Latin America is further enhanced by relatively small concentrations of Asians in scattered places. For example, about half of Guyana's population and two fifths of Trinidad-Tobago's is made up of people whose ancestors immigrated from India, and half of Suriname's population is Indian or Indonesian. These Asians came mainly as contract laborers for sugar plantations after the end of slavery, and many stayed after their contracts expired. Small areas in southern Brazil are inhabited predominantly by people of Japanese origin.

Much has been written about the degree of racial harmony which Latin America has achieved. Although such claims are often overstated, it does seem that the region has escaped many of the racial tensions that grip much of the world. This may be a result of the fact that the majority of Latin Americans have a mixed racial composition. Most of the region exhibits a primary mixture of Spanish and native Indians, resulting in a heterogeneous group known as *mestizos*. Mixed bloods of native Indian-black ancestry are usually termed *zambos*. European-black mixtures are called *mulattoes*.

While attempting to preserve treasured elements of their own heritages, non-Europeans in Latin America have been subject to strong European influences, and many have adopted a European lifestyle. For example, in Guatemala, where less than 3 percent of the population is classified as white, the term *Ladino* is applied to any person who speaks Spanish as a preferred language and espouses a way of life inherited from Spain. Most Ladinos are either mestizos or Caucasians, but full-blooded Indians may become Ladinos upon losing their native ways and adopting European modes of living. Despite widespread accommodation of this sort, however,

Latin American societies tend to put a premium on lightness of skin, and the upper classes often exhibit this to a marked degree. But skin color does not often become the basis for overt conflict to the same degree that some other parts of the world have witnessed during recent times.

PROBLEMS OF RAPID POPULATION GROWTH AND URBANIZATION

Irrespective of ethnic composition and type of culture, most Latin American countries today find themselves in the second stage of what has been called the *demographic transition*. This term refers to a model describing the changeover of societies from high birth and death rates to low birth and death rates. The model is divided roughly into three stages. The Latin American countries have passed beyond the first stage, in which there is relatively static natural increase because both birth and death rates are high. But there is a high potential for rapid increase if the death rate can be brought down before the birth rate drops. This occurs in the second stage, which now characterizes all but a very few Latin American countries. With the introduction of Western medical technology, improved diets, and better sanitation facilities, the death rate in Latin America has dropped dramatically during the 20th century, but the birth rate has remained high. The result has been several decades of tremendous population growth. The third stage of the demographic transition is a time when death rates are as low as they are likely to go, while birth rates have dropped to a relatively low level and may eventually result in negative population growth if the decline continues. This latter situation does not completely fit any Latin American country, but a handful now approach it fairly closely. Cuba and Uruguay are examples.

It is apt to be a long time before most of Latin America enters the final stage of low natural increase postulated in the "demographic transition" model. Birth rates for entire nations do not ordinarily drop swiftly and dramatically. Should Latin America's population continue to grow at its present rate, it would double in just three decades. Actually the rate is expected to slowly decline, but some population experts still project that the population will reach 711 million by the year 2020, with more than half the total in just two countries: Brazil (234 million) and Mexico (138 million). These are formidable numbers for two nations whose respective per capita GNPs were $1810 and $1850 in 1986 (U.S., $17,500). Patently there will have to be extraordinary exertions if their already depressed and debt-ridden economies are to keep the standard of living from slipping downhill as population accumulates.

The prospect is that Latin America will face increasing problems of population support unless human fertility rates come down sharply, and soon. An important aspect of the problem is the increasing proportion of younger ages in the population. In the early 1980s, an estimated 39 percent of the region's people were under the age of 15. Since 4 percent were 65 or older, the dependency ratio stood at 43 percent, meaning that a very large share of the population was essentially outside the labor force and dependent on others for support. Many individual countries had dependency ratios above 43 percent: Nicaragua and Honduras had ratios of 50 percent, and even Mexico, the region's second largest country in population, stood at 48 percent. The large proportion of Latin Americans under 15 years of age does not portend well for the future, as these individuals will soon move into their child-producing years.

Several reasons can be cited for the persistence of high fertility rates in Latin America:

1. In the past when infant mortality rates were much higher, it was necessary to have more than the desired number of children, for only a certain number could be expected to survive to adulthood. As prenatal and postnatal health care have improved, an increasing number of children are surviving, but fertility habits of long standing are slow to change. This is true even in the face of great poverty.

2. Motherhood is still widely perceived as the primary role to which Latin American women should aspire. This situation is changing, but a woman's status is still apt to be related to the number of children she bears.

3. In the societies of developing regions such as Latin America, large families are generally viewed as an economic resource. They provide the parents with a built-in labor pool as well as a form of old-age security. It is unlikely that population growth in Latin America will be moderated very much until large families are widely perceived as liabilities rather than assets.

Recognizing their demographic plight, some Latin American governments have taken steps to reduce population growth. In Mexico, for example, official policy favors contraception as a means of lessening poverty by limiting family size. Other countries are implementing family planning measures of their own. Yet some governments reject the idea that population growth needs to be controlled. Brazil's government, for example, tries to promote growth, which is felt to be necessary if the country is to attain a higher status in world affairs. Populating and developing the vast empty hinterland has long been a major national objective of Brazil. Throughout Latin America the artificial limitation of births is opposed by official dogmas of the Roman Catholic Church, although such teachings are followed with greater zeal by some Catholics than by others.

Explosive Urbanization: The Flight from Rural Slums to Urban Shantytowns

Startling rates of urban and metropolitan growth have been a major Latin American characteristic during recent times. Recent estimates give a figure of 68 percent urban for the population of the region as a whole, compared to a world average of 41 percent. A major element in the rapid growth of Latin American cities is a huge and continuing migration from countrysides that often are so overpopulated and poor as to be no better than rural slums. Even in rural areas that are better off, cities are viewed by the ambitious young as places to get ahead rapidly and enjoy exciting amenities while doing it. Deeply rural and tradition-bound villages offer little competition to the bright lights and swirling action of the metropolis. But a great many rural dwellers flee to the cities out of sheer desperation induced by drought, exhausted land, depressed farm prices, runaway inflation, chronic unemployment, guerrilla warfare, or other ills that beset the countrysides of many developing parts of the world in our time. In Latin America the urban explosion is most fully expressed in metropolitan Mexico City (Fig. 25.6) and São Paulo, Brazil, which have become two of the five largest metropolitan areas in the world (with Tokyo, Japan; New York City; and Osaka-Kobe, Japan). Buenos Aires, Argentina, and Rio de Janeiro, Brazil, provide other examples of skyrocketing growth. Caring for the huge influxes of people has strained the services of Latin American cities beyond their limits. As a result, every large city has big slums, which often take the form of ramshackle shantytowns on the urban periphery. Sometimes they are built on hillsides overlooking the city, as at Rio de Janeiro where some of the world's more unsightly housing commands one of the world's more imposing views. Such shantytowns, which are practically universal in the world's less-developed countries, are full of underemployed and ill-fed people, who still may prefer their present plight to the deplorable conditions of the back country from which they came.

PHYSICAL GEOGRAPHY: THE DIVERSITY OF HABITATS

TOPOGRAPHIC VARIATIONS

Latin America is characterized by pronounced differences in elevation and topography from one part of the region to another (Figs. 2.10, 25.1, and 25.2). A series of low-lying plains drained by the Orinoco, Amazon, and Paraná-Paraguay river systems dominates the central part of South America and separates older, lower highlands in the east from the rugged Andes of the west. In Mexico a high interior plateau broken into many basins lies between north-south trending arms of the mountains called the Sierra Madre. High mountains within Latin America—largely contained within the Sierra Madre and the Andes—form a nearly continuous line from northern Mexico to Tierra del Fuego at the southern tip of South America. Quite a few peaks in Mexico and Central America reach higher than 9000 feet (2743 m). But Latin America's highest peaks are in the Andes, where the highest crests continuously exceed 9000 feet for over 3500 miles (c. 5600 km) from northern Venezuela and Colombia to central Chile and Argentina. The maximum elevation in the Western Hemisphere is reached in Argentina's Mount Aconcagua (22,831 feet/6959 m). Most of the smaller islands of the West Indies are volcanic mountains that protrude above the level of the Caribbean, although some islands are made of limestone or coral only slightly upraised above the water. The largest islands have a more diverse topography, including low mountains.

HUMID LOWLAND CLIMATES OF LATIN AMERICA

Most Latin American lowlands are characterized by humid climates, which may be subdivided into "types" according to precipitation and temperature characteristics. Each type is associated with a dominant form of natural vegetation. The *tropical rain forest climate*, with its heavy year-round rainfall, monotonous heat, and superabundant vegetation dominated by large broadleaf evergreen trees, lies primarily along or very near the equator, although segments extend to the margin of the tropics in both the Northern and Southern hemispheres. The largest block lies in the basin of the Amazon River system (Figs. 2.18, 25.1, and 25.2). Additional areas are found along the southeastern coast of Brazil, on the Caribbean side of Central America and southern

Mexico, and along the eastern (windward) shores of some Caribbean islands.

On either side of the principal region of tropical rain forest climate, the *tropical savanna climate* extends to the vicinity of the Tropic of Capricorn in the Southern Hemisphere and, more discontinuously, to the Tropic of Cancer in the Northern Hemisphere. In this climate the average annual precipitation decreases and becomes more seasonal, the mean temperature decreases, and the broadleaf evergreen trees of the rain forest grade into tall savanna grasses or deciduous forests that lose their leaves in the dry season.

Still farther poleward in the eastern portion of South America lies a sizable area of *humid subtropical climate,* which has cool winters unknown to the tropical types. Its Northern Hemisphere counterpart is north of the Mexican border in the southeastern United States. In Latin America this climate is principally associated with a natural vegetation of prairie grasses in the humid pampa of Argentina and extreme southern Brazil. On the Pacific side of South America, a small strip of *mediterranean* or *dry-summer subtropical climate* in central Chile is reminiscent of southern California. To the south in Chile is a strip of *marine west coast climate,* which occupies the lower slopes of bleak, rainy, windswept, glaciated, and essentially uninhabited mountains at the southern end of the Andes.

DRY CLIMATES AND THE FACTORS THAT PRODUCE THEM

The humid climates of Latin America have a more or less orderly and repetitious spatial arrangement. One may expect to find generally similar climates in generally similar positions on all major land masses of the world. But the region's dry climates—*desert* and *steppe*—are often due, at least in part, to local circumstances such as the presence of high landforms. In northern Mexico (Fig. 2.7) such climates are partly associated with the global pattern of semipermanent zones of high pressure which create arid conditions in many areas of the world along the Tropics of Cancer and Capricorn. However, they are also due in part to the rain-shadow effect caused by the presence of high mountain ranges on either side of the Mexican plateau. In Argentina the extreme height and continuity of the Andean mountain wall are largely responsible for the aridity of large areas, particularly the southern region called Patagonia. Here the Andes block the path of the prevailing westerly winds, creating heavy orographic precipitation

on the Chilean side of the border, but leaving Patagonia in rain shadow. But in the west-coast tropics and subtropics of South America, the Atacama and Peruvian deserts cannot be explained so simply. Here shifting winds, cold offshore currents, and other complexities, as well as the Andes Mountains, are important conditions which combine to create one of the world's driest areas. However, the mountains serve to restrict this area of desert to the coastal strip.

ALTITUDINAL ZONATION: THE *TIERRA CALIENTE, TIERRA TEMPLADA,* AND *TIERRA FRÍA*

One of the most significant features of Latin America's climatic pattern with respect to population and development is the importance of a series of highland climates arranged into zones by altitude. This zonation results from the well-known fact that air temperature decreases with altitude, at a normal rate of approximately 3.6°F (1.7°C) per 1000 feet (304.8 m) of elevation. At least three major zones are commonly recognized in the higher lands of Latin America: the *tierra caliente* (hot country), the *tierra templada* (cool country), and the *tierra fría* (cold country).

The Hot, Wet *Tierra Caliente*

At the foot of the highlands, the *tierra caliente* is a zone embracing the tropical rain forest and tropical savanna climates discussed earlier. The zone reaches upward to approximately 3000 feet above sea level at or near the equator, and to slightly lower elevations in parts of Mexico and other areas near the margins of the tropics. In this hot, wet environment are grown such crops as rice, sugarcane, and cacao, often on a plantation basis. Many areas show heavy concentrations of blacks, zambos, and mulattoes.

The *Tierra Templada:* Zone of Coffee

The *tierra caliente* merges almost imperceptibly into the *tierra templada.* Although sugarcane, cacao, bananas, oranges, and other lowland products reach their respective uppermost limits at some point in this higher level, the *tierra templada* is most notably the zone of the coffee tree. Indeed, some scholars use the criterion of effective coffee culture as the dividing line from the *tierra caliente.* In the *tierra templada* coffee can be grown with relative ease (although by no means all the soils are suitable); at lower altitudes the crop encounters difficulties occasioned by excessive heat and/or moisture. The upper limits of this zone—approximately 6000 feet (somewhat over 1800 m) above sea level—tend also to be the upper limit of European-induced plantation agriculture in Latin America. In its distribution the *tierra templada* flanks the rugged western mountain cordilleras and, in addition, is the uppermost climate in the lower uplands and highlands to the east. Thickly inhabited sections occupy large areas in southeastern Brazil, Colombia, Central America, and Mexico. Although broadleaf evergreen trees characterize the moister, hotter parts of this zone, coniferous evergreens replace them to some degree toward the zone's poleward margins. In such places as the highlands of Brazil or Venezuela where there is less moisture, scrub forest or savanna grasses appear—the latter generally requiring the more water.

In brief, the *tierra templada* is a prominent zone of European-induced settlement and of commercial agriculture. Urban as well as rural settlement is very much in evidence: Of Latin America's metropolitan areas of 100,000 or over about a fourth are in or very near the *tierra templada*. Five metropolises exceeding 2 million—São Paulo, Belo Horizonte, Caracas, Medellín, and Guadalajara—are in this zone, while another—Mexico City—lies just above it. Others, like Rio de Janeiro, which are situated at lower elevations, have close ties with predominantly residential or resort towns in these cooler temperatures.

The *Tierra Fría:* Highland Indian Domain

The *tierra fría*, or cold country, may be distinguished from the other zones by two criteria. First and perhaps more important in a markedly agricultural region like Latin America, it is a zone where frost occurs. As one might expect, frosts are only occasional in the zone's lower reaches at approximately 6000 feet (*c.* 1800 m) above the sea but are much more frequent at higher elevations. The second criterion refers to type of economy: In contrast to the Europeanized *tierra templada*, the *tierra fría* is often the habitat of a native Indian economy with a strong subsistence component. Most extensive in Peru and Bolivia, this type of economy is also very evident in Ecuador, Colombia, Guatemala, and Mexico. The *tierra fría* is comprised of high plateaus, basins, valleys, and mountain slopes within the great mountain chain that extends from northern Mexico to Cape Horn. By far the largest areas are in the Andes, although areas in Mexico are sizable (see extent of the *tierra fría*, Figs. 25.1 and 25.2). The upper limit of the zone is generally placed at about 10,000 feet for locations near the equator and at lower elevations

toward either pole. This line is usually drawn on the basis of two criteria: (1) the upper limit of agriculture, as represented by such hardy crops as potatoes, barley, or the locally important cereal quínoa; and (2) the upper limit of natural tree growth. Above are the alpine meadows, sometimes called *paramos;* and still higher there may or may not exist barren rocks and permanent fields of snow and ice.

The *tierra fría* tends to be a last retreat and the major home of the native Indian (Fig. 24.5) and is characterized by small settlements and what Europeans and Americans might consider rather primitive ways of life. It is chiefly rural, containing less than one tenth of the metropolitan areas in Latin America that number 100,000 persons or more. However, nature has placed here certain valuable minerals like tin and copper which have attracted modern types of large-scale mining enterprise into the *tierra fría* of Bolivia and Peru as well as some other Latin American countries.

Climatic data for representative Latin America stations are given in Table 24.2. Further discussion of mineral resources has been reserved for the subsequent section on economic geography.

ECONOMIC GEOGRAPHY: PROBLEMS OF AN "EMERGING" REGION

Latin America is an "emerging" region that does not yet provide a good living for most of its people. It is not the worst off of the major world regions; in fact, its overall per capita GNP of $1720 in 1986 was well over two times that of Africa. But compared to Anglo-America, Europe, or the Soviet Union, the Latin American region as a whole is miserably poor. This tends to be masked by the glitter of the great metropolises with their forests of new skyscrapers, but even these places are rimmed by wretched shantytowns, and out beyond lie depressed countrysides. The financial ills of the region are rooted in underdeveloped and tradition-bound economies operated far too much in the interest of a wealthy few and excessively dependent on the sale of primary products to more industrialized nations overseas. In an attempt to overcome their economic deficiencies and keep down popular discontent, many Latin American governments have borrowed heavily from the international banking community, including leading banks in the United States. By the 1980s unpaid loans had reached staggering proportions, and the Latin American debtor nations were having great difficulty in mustering even the annual interest payments. Hence the economic woes of the lands "south

TABLE 24.2 CLIMATIC DATA FOR SELECTED LATIN AMERICAN STATIONS

Station	Latitude to Nearest Whole Degree	Elevation Above Sea Level (ft)	Elevation Above Sea Level (m)	Type of Climate	Annual °F	Annual °C	Coolest Month °F	Coolest Month °C	Warmest Month °F	Warmest Month °C	Annual Average (in.)	Annual Average (cm)	Months Averaging Less than 1 in.
Monterrey, Mexico	26°N	1752	534	Steppe	72	22	57	14	82	28	28	72	4
Bridgetown, Barbados	13°N	181	55	Tropical rain forest	79	26	77	25	81	27	50	127	0
Caracas, Venezuela	11°N	3418	1042	Tierra templada	72	22	68	20	73	23	34	85	3
Ciudad Bolivar, Venezuela	8°N	164	50	Tropical savanna	82	28	80	27	83	28	38	97	3
Manaus, Brazil	3°S	144	44	Tropical rain forest	81	27	79	26	82	28	83	210	0
Cuiabá, Brazil	16°S	541	165	Tropical savanna	79	26	73	23	81	27	54	137	3
São Paulo, Brazil	24°S	2608	795	Tierra templada	66	19	59	15	72	22	55	139	0
Rosario, Argentina	34°S	89	27	Humid subtropical	63	17	50	10	75	24	39	99	0
Mendoza, Argentina	33°S	2713	827	Desert	61	16	45	7	77	25	8	20	11
Santiago, Chile	33°S	1706	520	Mediterranean	59	15	46	8	70	21	13	32	7
Valdivia, Chile	39°S	43	13	Marine west coast	54	12	46	8	61	16	98	249	0
Lima, Peru	12°S	449	137	Desert	64	18	59	15	72	22	1	4	12
La Paz, Bolivia	17°S	11,910	3630	Tierra fría	50	10	45	8	53	12	22	56	5
Quito, Ecuador	0°	9243	2817	Tierra fría	55	13	55	13	55	13	49	123	2

of the border" impinge increasingly on American or other bank depositors whose money has been lent. In this section we examine the geographic context of Latin America's underproductive economy, commencing with the agriculture that is the largest source of employment in most Latin American countries, moving then to the mining industries that supply large revenues (but pose many problems) to a handful of nations, and concluding with the manufacturing industries and commerce that must expand and flourish if Latin America is to markedly improve its economic situation.

TWO SYSTEMS OF AGRICULTURE: *LATIFUNDIA* AND *MINIFUNDIA*

Although the total number of Latin Americans employed in agriculture has not declined very much in recent decades, there has been a sizable percentage decline in agriculture as a component of both employment and value of product in national economies. The actual percentages vary a great deal from country to country. According to World Bank figures, 77 percent of Haiti's labor force was employed in agriculture in 1965, and 70 percent was still so employed in 1980; whereas in Brazil 49 percent of all workers were in agriculture in 1965 but only 31 percent in 1980. In Mexico the percentage dropped from 50 in 1965 to 37 in 1980; in Argentina, from 18 percent in 1965 to 13 percent in 1980. In Mexico agricultural products represented 16 percent of the gross domestic product (GDP) in 1965, but only 13 percent in 1985.[2] Dependence on agricultural exports has also dropped as national economies have become more diversified, but in many countries more than half of all export revenue is still derived from products of agricultural origin (Table 24.3). An

_____

[2]Gross domestic product (GDP) is an internationally comparable measure of economic output devised as an alternative to gross national product (GNP). In calculating the GDP, statisticians state the value of a country's output in American dollars by using the purchasing power of the country's currency as a conversion factor. By contrast, the GNP is calculated by using the official exchange rate of the currency in converting to dollars. See World Bank, *World Development Report* (Washington, D.C., annual).

TABLE 24.3   LEADING EXPORT COMMODITIES OF THE INDEPENDENT COUNTRIES OF LATIN AMERICA

Country	Commodity and Percent of Total Exports by Value (1985 or specified year)
Mexico (1985)	Crude petroleum, 61; petroleum products, 6
(1986)	Crude petroleum, 35; petroleum products, 4; machinery and vehicles, 19
Guatemala (1984)	Coffee, 32
El Salvador	Coffee, 67
Honduras	Bananas, 35; coffee, 24
Nicaragua	Coffee, 39; cotton, 31
Costa Rica (1984)	Industrial products, 31; coffee, 27; bananas, 25
Panama (1984)	Bananas, 27; shrimp, 17; sugar, 12
Belize	Sugar, 38; garments, 24; citrus fruits, 19; reexports, 29
Cuba	Sugar, 74
Haiti (1985–1986)	Manufactured goods, 34; coffee, 27; assembled articles for reexport, 24
Dominican Republic (1984)	Sugar, 31; gold alloy, 15; ferronickel, 12; coffee, 11
Jamaica	Alumina, 37; bauxite, 14; sugar, 8
Trinidad and Tobago	Crude petroleum, 47; petroleum products, 33
Bahamas	Petroleum and products, 89
Barbados (1985)	Electrical components, 57; sugar, 13
Antigua and Barbuda (1983)	Reexports (largely manufactures), 40; miscellaneous domestic manufactures, 33; machinery, 8
Dominica (1984)	Bananas, 45; soap, 22
Grenada (1983)	Fresh fruit, 22; cocoa beans, 21; nutmeg, 17; bananas, 17
St. Christopher and Nevis (1983)	Sugar, 55
St. Lucia	Bananas, 58; cardboard boxes, 10
St. Vincent and the Grenadines (1986)	Eddoes, dasheens, yams, plantains, and tanias, 39; bananas, 28
Guyana (1984)	Bauxite, 44; sugar, 28
Suriname (1984)	Alumina, 55; bauxite, 13; aluminum, 12
Colombia (1986)	Coffee, 59; petroleum and products, 13
Venezuela (1981)	Crude petroleum, 79; petroleum products, 14
Ecuador (1985)	Crude petroleum, 63; petroleum products, 4
(1986)	Crude petroleum, 42; petroleum products, 3
Peru	Petroleum and products, 20; copper, 14; zinc, 11; lead, 7; silver, 7
Bolivia	Natural gas, 55; tin, 28
Brazil (1984)	Metals, 11; coffee, 10; animal feedstuffs, 6; a great diversity of others
Argentina	Cereals, 21; vegetables, fruits, and nuts, 15; vegetable oils, fats, waxes, 13; animal feedstuffs, 10; meat, 6
Chile	Copper, 43; other minerals, 11
Uruguay (1986)	Textiles and textile products, 29; live animals and products, 27; hides and skins, 16
Paraguay	Cotton fibers, 48; soybeans, 31

Source: *Britannica Book of the Year*, 1988.

extreme case is Cuba, where sales of cane sugar still generated three fourths of all income from exports in 1985. A much-publicized aspect of agriculture in some Latin American countries is the illicit production of cocaine and marijuana for clandestine sale, mainly in the United States. This traffic has reached such huge proportions that it constitutes an important international problem.

Farms in Latin America are often divided into two major classes by size and system of production. Large estates with a strong commercial orientation are *latifundia* (singular: *latifundio*); smaller holdings with a strong subsistence component are *minifundia* (singular: *minifundio*). The large estates, whether called haciendas, plantations, or by some other name, are owned by families or corporations. Some have been in the same family ownership for centuries. The desire to own land as a form of wealth and symbol of prestige and power has always been a strong characteristic of Latin American societies.

Figure 24.6 Sugarcane on a Jamaican plantation. Occupying a basin rimmed by hills, this family-owned plantation is one of the few remaining commercial sugarcane producers in Jamaica. It bears continuing witness to the early role of cane cultivation on slave plantations as a stimulus to colonial settlement of the West Indies and many areas of the Latin American mainland. The plantation has diversified into cattle raising and citrus production; the darker green in the photo at the base of the slope marks a grove of orange trees.

Huge tracts were granted by Spanish and Portuguese sovereigns to members of the military nobility (*conquistadors*) who led the way in exploration and conquest. Much of this land has been reallocated to small farmers by government action from time to time, but in many countries a very large share of the land is still in the hands of a small, wealthy landowning class.

Agricultural production on the large estates characteristically involves the employment of large numbers of landless, often illiterate, workers. In some areas, these workers are traditionally bound to the estate, and in others they are free to move at will. In actuality, most of them remain on the estate. The profits from these large landholdings benefit primarily the owners, while the workers are often underpaid and only seasonally employed. Today many owners have business or professional occupations in cities, leaving the actual direction of their estates to managers. In such cases an owner and his family may reside exclusively in the city, or they may live part of the time on the estate. There are great variations in the care and efficiency with which estates are managed and kept up. Some are flourishing enterprises, whereas others are drained of wealth by absentee owners and held primarily for the prestige they confer. Although one estate may produce several commodities commercially, there is a tendency toward specializing in only one or two. These large landholdings have been declining in number, but they still produce a sizable share of the coffee, sugar, cotton, livestock products, and other Latin American farm commodities that enter world markets.

A special variant of the large estate is the single-crop commercial plantation owned by a company or a syndicate. The capital to establish these enterprises has come principally from the United States or Europe. Most plantations grow sugarcane or bananas. Sugar plantations are especially prominent in the Caribbean islands (Fig. 24.6), while banana plantations are found principally in Central America or in Ecuador. A considerable amount of subsistence farming by individual families is carried on in plantation areas, often on plantation-owned land. Especially well known among the single-crop, corporate-owned plantations are the American-owned banana plantations found along hot, moist coastal lowlands in Honduras and some other Central American countries.

A considerable proportion of Latin America's agricultural population lives on a hand-to-mouth basis bordering on mere subsistence. These people have little contact with foreign commerce and exchange. Their small holdings are *minifundia*. Such farmers lack the capital to purchase large and fertile properties, and hence they are relegated to the farming of marginal plots, often on a sharecropping basis. Those individuals who do own land are frequently burdened by indebtedness and the fragmented nature of farms which are becoming smaller and smaller in size as they are subdivided through inheritance.

Minifundia produce food primarily for family use but also for the local market. The crops most commonly raised are maize, beans, and squash, although many other crops are locally important, especially in the different climatic zones of the highland regions. Although these small farmers make up the bulk of Latin America's agricultural labor force, food has to be imported into many areas and many of the people are poorly nourished. Productivity is low in part because of the marginal quality of the land, but it is also due to rather primitive agricultural techniques. There is little capital with which to buy machinery, fertilizers, and improved strains of seeds. Soil erosion and soil depletion are making serious inroads in many areas. Some small farmers, such as those in Costa Rica, are well above average in productivity and income; often they get a considerable share of their income from an export crop such as coffee that they grow in addition to food crops for home use or local sale.

The enormous economic and social gap between wealthy landowners on the one hand and the struggling masses of small farmers and landless workers on the other hand has led many Latin American governments to institute programs of land redistribution. Efforts of this kind in Mexico have been going on for a particularly long time and will be examined in Chapter 25. Although success in implementation has varied sharply from country to country, this movement is gaining momentum. Some attention is being given to the breakup of existing properties, while other efforts are exerted to bring more vacant land (whether owned by the public, by private individuals, or by the church) into cultivation, usually by small farmers. Some new farms have been structured as communal holdings reflecting Indian traditions or 20th-century revolutionary plans of agrarian reform. In Cuba the Communist government placed the land of expropriated estates in large farms owned and operated by the state. Workers on these farms are paid wages, as they are on comparable farms in the Soviet Union. Thus the actual details of land reform schemes vary greatly from place to place. Here again we see that Latin America is not monolithic but is a region of rich and complex variety.

MINERALS AND MINING: SPOTTY DISTRIBUTION AND UNEQUAL BENEFITS

Latin America is a large-scale producer of a small number of key minerals that are very significant to the outside nations where most of the production is sold. Only a handful of Latin American nations gain large revenues from such exports. In most countries mineral production is relatively minor. Even in the countries that do have a large value of mineral output—notably Mexico, Venezuela, and Brazil—much

of the profit has been dissipated in the form of showy buildings, corruption, ill-advised development schemes, and enrichment of the upper classes and foreign investors. Hence the benefits to the poverty-stricken masses have tended to be rather minimal. Nonetheless, many new highways, power stations, water systems, schools, hospitals, and employment opportunities have been made possible by mineral revenues. These have helped the general level of living, although their impact has been lessened by rapid population increases and the amount of money that has had to be diverted for interest payments on huge foreign debts.

Among Latin America's known mineral resources of greatest consequence are petroleum, iron ore, bauxite, copper, tin, silver, lead, zinc, and sulfur. However, local consumption of these minerals has been retarded by a lack of good coal, especially the coking coal that is very important in steel production.

All but a small proportion of Latin America's *petroleum* is extracted in the Caribbean Sea–Gulf of Mexico area, particularly the central and southern Gulf coast of Mexico and northern Venezuela. Among world producers, Mexico ranked sixth and Venezuela ninth in 1986. Other oil fields in Latin America are widely scattered, the principal ones being located (1) along or near the Atlantic coast in Brazil, Argentina (Patagonia), and Colombia; (2) in sedimentary lowlands along the east front of the Andes in Peru, Ecuador, Bolivia, and Argentina; or (3) on the island of Trinidad. Natural gas is extracted in many of the areas that produce oil, but Latin American production is not yet of major world consequence. Mexico and Venezuela are the largest producers.

Latin America is a major world area in the production of metal-bearing ores. Most of the ore extracted is shipped to overseas consumers in raw or concentrated form, although the region has scattered iron and steel plants, as well as smelters of nonferrous ores, which process metals for use in Latin American industries or for export. Deposits of high-grade *iron ore* in the eastern highlands of Brazil and Venezuela are the largest known in the Western Hemisphere and are among the largest in the world. The two countries are Latin America's main producers and exporters of ore. Brazil is the main producer of iron and steel, with Mexico second. Most of the region's production of *bauxite*—the major source for aluminum—comes from Brazil, Jamaica (Fig. 24.7), Suriname, or Guyana. The deposits in these countries are located relatively near the sea and are of critical importance to the industrial economies of the United States and Canada. Bauxite mining and associated alumina production have been in a slump during recent years as a result of glutted world markets. Alumina, or aluminum oxide, is an intermediate stage in bauxite processing, produced by the removal of impurities present in the ore. Final processing to secure metallic aluminum is done by electrolysis of the alumina. Much of Latin America's bauxite is processed into alumina in the country where the bauxite is mined, but a good share moves to alumina plants outside the region. Most of the final processing for aluminum metal takes place at a few large American or Canadian plants located near hydroelectric stations on the Columbia and Tennessee rivers, or on the St. Lawrence River and its tributaries. Within Latin America, relatively small amounts of aluminum are produced by plants in Venezuela, Brazil, and Argentina. Huge unexploited bauxite deposits in Venezuela are a major resource for the future.

Low-grade but comparatively abundant *copper* deposits occur in the Atacama Desert of northern Chile and the arid and semiarid sections of Mexico. Additional reserves of copper are found in the Andes, especially in Peru and Chile. Chile is overwhelmingly the largest copper producer in Latin America and was the world's leading producer in the mid-1980s, with the United States second. Most of the known reserves of *tin* in Latin America are in Bolivia or Brazil, and those countries produce most of the region's output. The *silver* of Mexico, Peru, and Bolivia, sought from the very beginning of Spanish colonization, is principally found in mountains or rough plateau country. Mexico and Peru are by far the largest Latin American producers of silver and of *lead* and *zinc* as well. Peru's deposits of all three minerals are in the Andes, and Mexico's are mainly in the dry northern and north central sections of the country. A characteristic feature of nonferrous and precious metal production (in Latin America or elsewhere) is the tendency for several metals to be associated in the same ore body. Hence an ore that yields silver as the most valuable product may also yield lead, zinc, copper, tin, gold, or other metals in varying proportions.

In recent decades Mexico has become a significant source of native *sulfur*. The main fields border the Gulf of Mexico, as do the fields of the world's largest producer, the United States. Sulfur is used primarily for the manufacture of sulfuric acid, which has many important industrial uses.

Altogether, Latin America makes a respectable contribution to world mineral supply but not a truly outstanding one, especially considering the many important minerals not produced in Latin America or produced in very minor quantities. Coal, natural

Figure 24.7 An alumina plant in Jamaica. Here bauxite is converted to the second-stage product called alumina, which will then be shipped overseas for final processing. Red dust from the bauxite ore carpets the plant.

gas, phosphate, potash, salt, gold, diamonds, and most ferroalloys are prominent examples. In these regards the region suffers by comparison with the Soviet Union, Anglo-America, or Africa. But its main mineral resources tend to be quite important to the countries producing them and to the major buyer, the United States.

Mining as a Source of Livelihood

The significance of mining to Latin America is not easily appraised. In terms of labor force it is comparatively unimportant, for the highly mechanized mining industry seldom employs more than 3 percent of the total labor force of a Latin American country. Since mining ventures are largely financed and usually managed by outside interests in order to get raw materials for the industrial economies of Anglo-America and Europe, few of the extracted products reach the Latin American people. However, revenues from mining—whether in the form of income, property, export tariff, or other taxes—are very important to some Latin American governments. In recent times there has been a considerable tendency for such governments to assume the formal ownership of the mineral enterprises in their countries, or at least to require majority control of these enterprises by their own nationals, but this has not greatly lessened the overall dependence on foreign financing, foreign markets, and foreign technical expertise and equipment. In this respect, as in so many others, Latin America is still very much a "developing" or "emerging" part of the world.

Latin America has a long history of both favorable and adverse effects from the exploitation of minerals. Mining tends to generate large foreign-exchange earnings in a relatively short period of time. While this does benefit countries in the short run, it often leads to a situation in which government spending and bureaucracy expand wildly to absorb the newly created revenues. In time, the countries' economies may become highly dependent on the mineral wealth, and the mining sector grows at the expense of others. This frequently leads to a situation in which depressed world prices for a single mineral can cause great economic distress in a Latin American country. Often, the capital required for extraction of a mineral is supplied by foreign investors who can make decisions that are in their own economic interest but are not necessarily in the interest of the producing country. Furthermore, the large sums of money pumped rapidly into the economy frequently produce serious inflation, which causes severe hardships for the people who do not directly participate in the profits from the mining operations. All of the foregoing characteristics of mining can increase the economic gulf between rich and poor in a Latin American country, although the country as a whole may be in better financial condition than it was before.

THE INCREASING IMPORTANCE OF MANUFACTURING

Although they are still very spotty in distribution and often are lacking in complexity and sophistication, manufacturing industries in Latin America are making an increasingly important contribution to national economies. Most parts of the region are still a long way from full industrialization, but certain districts have achieved an impressive array of factories, including some of the most modern types. Since the industrial revolution has come to Latin America very late, the region does not face large problems of industrial obsolescence and is able to bypass centuries of development in older industrial countries by importing modern technology and equipment. There is a regionwide push to attract industry from abroad with offers of tax exemptions, cheap labor, and other inducements. Although the efforts to industrialize are making headway, their overall impact on Latin America is slowed by many obstacles, some of which are summarized below.

Obstacles to Industrialization

1. Modernization and the urban-industrial way of living have come to Latin America from outside and not by gradual development from within. Deep-rooted habits of people who make their living from the land are slow to give way to the more restrictive habits demanded by regular factory hours and year-round work on assembly lines.

2. There are serious gaps in Latin America's resource base, particularly energy resources. Good coal is largely absent, and truly large petroleum and natural gas resources have been found only in Mexico and Venezuela (although interior Brazil may yet prove to have major resources).

3. There is an overall shortage of capital, skills, and infrastructure such as power stations. The complex support systems that undergird industry in Anglo-America, Europe, or Japan are still in an early stage of development in Latin America.

4. There is a reluctance among Latin American and foreign investors to risk capital in untried enterprises and in countries with unstable governments, and this is coupled with a strong expectation of large and rapid returns from investments. Such attitudes are inimical to the rational development of industries over a considerable period.

5. The ruling classes in Latin America tend to view innovation as a threat to their status, and in their eyes prestige is more associated with landholding than with business enterprise. Such attitudes are weakening, but they still play a demonstrable role in retarding industrial development.

Despite its generally retarded condition, Latin America does have a large number of factories. Most of these are household enterprises or small establishments which employ less than a dozen workers and sell their products mainly in home markets. The larger operations are almost always located in the larger cities and often are branch plants of overseas companies. The Latin American countries which have the largest output of factory goods are Brazil, Mexico, and Argentina. Brazil's industrial output is now sufficient to rank it eighth among the world's countries, although its per capita ranking would be much lower than that. Within Latin America the industrial scene is dominated by a handful of large metropolises, most notably São Paulo, Mexico City, and Buenos Aires.

COMMERCE AND SERVICES

The countries of Latin America are not among the world's main trading nations. In total value of foreign trade Brazil, the Latin American leader, ranks just below the world's top 20 countries and most other countries in the region are far below that. These low rankings are related to several other characteristics of Latin American commerce:

1. These countries are too poor to afford truly large and expensive imports of goods from overseas.

2. There is an excessive dependence on primary exports of relatively low value (mainly agricultural products and unmanufactured or semifinished materials) and a serious lack of high-value exports such as machinery, transportation equipment, and electronic equipment. In country after country there is an overwhelming dependence on a handful of primary commodities for export, and sometimes just one (Table 24.3). This reduces the options for developing new markets in other countries, and it makes the Latin American countries very vulnerable to the price fluctuations that afflict primary commodities on world markets. The prosperity of an entire country may depend heavily on the ability of the country to market its coffee or bauxite at a good price.

3. There is an excessive dependence on trade with the United States. The economies of many Latin American nations are so closely tied to the economy of the United States that a slight downturn in the latter can cause severe economic hardships south of the border. Mexico is an important example of a nation for which the United States is overwhelmingly the main trading partner. In 1987 the United States supplied 64.4 percent of Mexico's imports by value, and took 64.5 percent of Mexico's exports.

4. The Latin American countries do not trade very much with each other. Since they tend to market the same restricted range of primary products, their economies are essentially competitive rather than complementary. Strenuous competition by Brazil, Colombia, and numerous other countries for the United States coffee market is a case in point. In recent decades there have been attempts to foster more economic integration and interdependence through international organizations within the region, but these have had only modest success. It will be hard to achieve truly notable success as long as the countries remain so basically similar in the types of goods they produce for export.

The great importance of service occupations in the employment structures of many Latin American countries must be mentioned. The term *services,* as used here, embraces all occupations other than agriculture, forestry, hunting, fishing, mining, manufacturing, construction, and utilities. Since service-type work centers heavily in cities, it is not surprising that the most urbanized Latin American countries are also the ones that have the highest proportion of their respective labor forces in such occupations. The correspondence is by no means exact, but in a broad way urbanization and service employment go hand in hand. Argentina, for example, has a population reported to be 85 percent urban and an employment structure in which 53 percent of the labor force was in services in 1980. At the other end of the scale Haiti is 25 percent urban, and had only 22 percent of its labor force in services in 1980. But even Haiti has a higher percentage in services than is true of many countries in the developing parts of the world. In this respect Latin America tends to be more similar to the nations of the industrialized West than it does to India, China, or any number of other countries in Africa and Asia. The proportion of Latin Americans in service employment has risen sharply during the past quarter-century as people from the countryside have flooded into the cities. Expanding government and business bureaucracies, the increasing affluence of urban upper and middle classes, an increase in tourism, and a general increase in the facilities required for even a minimal servicing of hordes of new urban dwellers have created many millions of new service jobs. Even at general pay scales far below those of western Europe or Anglo-America, such jobs have given new opportunity to desperately poor migrants from overcrowded rural areas, and they have helped accommodate the explosive natural increase of population within the cities themselves. In many West Indian islands and scattered places on the mainland, tourism has become a major economic asset. An especially notable case is the Bahamas, where tourists in a recent year outnumbered the native Bahamians ten to one, and tourism supplied about two thirds of the employment. United States citizens are by far the largest national contingent of tourists in Latin America. The great majority spend their time in the West Indies (often traveling from island to island on cruise ships) or in Mexico, which receives more tourists than any other Latin American country. Brazil also has a very large tourist trade.

POLITICAL GEOGRAPHY: PROBLEMS OF FRAGMENTATION AND INSTABILITY

Latin America is organized into political units that vary tremendously in size and political status. All but about 1 percent of the population is concentrated in the region's independent states. Aside from some smaller countries (mainly Caribbean island states) that gained independence after World War II, Latin America broke away from European control in the 19th century. The combined population of Latin America's still-dependent units is only about 5 million (1988) and is found in units still affiliated politically with the United States, France, the Netherlands, or the United Kingdom. All of these except the United Kingdom's Falkland Islands colony are located in the Caribbean islands or, in the case of French Guiana, on the nearby mainland of South America. The different units vary in their relationships to the overseas nations with which they are linked. Most of them, however, have much latitude in handling their internal affairs.

In view of the fact that most of Latin America was once controlled politically by Spain, it seems curious that the present region is so politically fragmented. In the Caribbean region much of the fragmentation resulted from the conquest of different islands or island groups, along with some mainland territories, by outside nations other than Spain. But even in the part of Latin America that remained Spanish until independence, a total of 19 states eventually emerged. Much of the reason for this lies in the administrative practices of colonial Spain. The Spanish portion of Latin America was not administered from a single colonial capital, but eventually became divided into four viceroyalties with their capitals at Mexico City, Lima, Bogotá, and Buenos Aires. Lesser administrative divisions such as the Captaincy-General of Guatemala were established within the viceroyalties. It was Spanish policy that the four major units function independently of one another—to the extent that they were not even permitted to trade among themselves without first routing the goods through Spain. Under this imposed isolation, which was heightened by the transoceanic distance from Spain itself, the territory tributary to each administrative capital developed feelings of regional identity that subsequently deepened into national identity. Political loyalties became focused less on Madrid than on the local center that oversaw regional affairs and served as an economic focal point. As Spain's grip loosened, the various administrative units began to break away, forming separate countries. This did not happen in Brazil, which was administered as a single colony by Portugal and gained independence as a single nation.

Today the fragmented political order of Latin America stands in marked contrast to the two massive political units—the United States and Canada—that occupy a territory of comparable size in Anglo-America. One can visualize the complexities that would have resulted had the individual states of the Union and the provinces of Canada become separate nations divided from each other by international boundaries. Actually the Latin American situation is not very comparable due to the nature of international frontiers there. In the United States and the more populous parts of Canada, most state or provincial boundaries run through areas where settlement is continuous on both sides of the boundary and the boundary line is crisscrossed by numerous highways and railroads. People in great numbers and goods in great quantities move back and forth across these boundaries every day. But in Latin America there is no land communication at all among the many island units (Haiti and the Dominican Republic, which share one island, are an exception); and on the mainland international communication among countries by road or rail is very scanty or, in large areas, nonexistent. Most mainland boundaries run through outlying areas where there are few people, few roads, and almost no railroads. For example, the whole long boundary separating Brazil from Peru, Colombia, Venezuela, Guyana, Suriname, and French Guiana is not crossed by a single railroad, and it is crossed by highways (recently completed) in only two places. Farther south, only a handful of routes connect Brazil with Bolivia, Paraguay, Argentina, and Uruguay. All over the Latin American mainland, the core regions of the different countries tend to be well insulated from each other by distance, unpopulated wilderness, and a dearth of international land routes. From a functional standpoint, the national population nodes and their capitals might almost be in separate continents. Each country tends to go its own way, communicating with the outside world by air or sea, and maintaining its most crucial relationships with nations outside of Latin America. United effort to develop Latin America's resources and alleviate poverty through economic development remains rather minimal, despite increasing efforts in recent years to promote greater cooperation and integration in both the political and economic spheres.

Meanwhile the region continues to be troubled by the serious political instability that has long been a major Latin American characteristic. Much of this political unrest can be attributed to maldistribution of wealth, especially land (chapter-opening photo). Some of it is due to simple hunger for power, leading an aspiring strongman to try to overturn the ruling strongman and his government by force. Such power struggles have been going on in many countries since the end of European rule. The strongman may have ideologic motives, sometimes of the Right and sometimes of the Left. A prominent recent example of the latter is the long rule of Fidel Castro at the head of the Communist government of Cuba. Often a strongman will have powerful backing from moneyed upper classes and business interests in the country, including foreign corporations that have large investments to safeguard. Foreign governments may support either the ruling strongman or an insurgent strongman on the grounds of perceived national interest, although such perceptions may be cloaked in rhetoric alleging great benefits to the local populace. A general may assume the role of strongman, or a military *junta* may seize control of the government, generally with the announced intention of rooting out corruption, ending mismanagement and oppression, salvaging national pride, and returning the country to the rule of the people. But in many instances democratic rule has been very slow in coming. Latin America has had relatively little experience with representative electoral democracy, although some nations—such as Colombia or Costa Rica—have much better records than others. There was little in the colonial experience to pave the way for democracy, since each colony was ruled autocratically by officials appointed by the overseas colonial power. Considering this history, democratic ideals have often proved surprisingly strong, and some countries have made considerable progress toward democracy in the 20th century. However, dictatorships of various kinds are still common, seizures of government by coups still occur, and opposition to the government in power generates guerrilla warfare in some countries. For Latin America as a whole the picture is very mixed from country to country.

The overturning of a Latin American government by force handicaps the country concerned in various ways. The government that has just come to power will have an immediate interest in protecting its position and solidifying its power base. Hence it is very likely to purge the previous government's bureaucracy, thus disrupting the machinery of state. Moreover, the programs of the deposed government are generally scrapped in favor of new programs sponsored by the new government. Thus a succession of leadership changes tends to preclude farsighted long-term policies. Bolivia is a well-known example of a Latin American country that has been notoriously subject to frequent coups, countercoups, and constant political uncertainty. By one count, Bolivia has suffered more than 185 revolutions since it became independent in 1825. Such conditions create a serious economic handicap, because investors are wary of risking their capital in a country that offers so little guarantee of security.

The general conditions of political instability described in this section are not just a matter of history but are still very prevalent today. For example, Guatemala had two military coups in 18 months during the 1980s; serious guerrilla warfare was in progress in a number of countries during 1989; Grenada's Marxist government experienced a Left-oriented military coup in 1983 which was ended by a United States military invasion; and in early 1989 a military coup in Paraguay overthrew a strongman who had held the office of President, with dictatorial powers, for many years.

CHAPTER 25

The Diversity of Latin American Regions[1]

The Iguassu Falls, which Brazil shares with Argentina near the Paraguay border, symbolize the great energy potentials available to many Latin American countries in the form of falling water. Such potentials are vital to "emerging" nations that often lack adequate deposits of mineral fuels.

Gvariety. Mexico and Argentina, for ENERALIZATIONS about Latin America tend to conceal a remarkable degree of national and regional example, speak Spanish and espouse Roman Catholicism but in many respects are extremely different from each other. Native Indian influence is strong in Mexico but scanty in Argentina; Mexico's core region in volcanic tropical highlands around Mexico City has little in common with the expansive mid-latitude lowland plains of Argentina's core in the humid pampa around Buenos Aires. Not only do the countries of Latin America differ widely among themselves, but there generally is very strong internal diversity within each one. Even the many small countries in Central America and the West Indies exhibit far more internal variety than is commonly assumed.

In this chapter the strongly regionalized character of Latin America is surveyed under four regional groups of countries: (1) a northern realm known as *Caribbean America;* (2) the *Andean countries* of Colombia, Venezuela, Ecuador, Peru, and Bolivia; (3) giant *Brazil;* and (4) the *countries of the Southern Mid-Latitudes*—Argentina, Chile, Uruguay, and Paraguay. These blocks of countries will be assessed in the order named. (For area, population, and other statistical data, see Table 24.1; also see maps, Figs. 24.1, 25.1, 25.2, and 25.3.)

THE INTRICATE MOSAIC OF CARIBBEAN AMERICA

The northernmost of the four major Latin American realms is Caribbean America, also known as Middle America. As considered here, it includes (1) Mexico, by far the largest country in area and population; (2) the small Central American states of Guatemala, El Salvador, Belize, Honduras, Nicaragua, Costa Rica, and Panama; (3) the northern South American units

of Guyana, Suriname, and French Guiana; and (4) the numerous islands, large and small, in the Caribbean Sea or near it.[2] Of the many political units on islands, the largest are Cuba (which encompasses the largest island), Haiti and the Dominican Republic (which share the second largest), and Jamaica. The entire area presents an extreme patchwork of physical features, races, cultures, political systems, population densities, and pursuits. Most political units are independent, but some are still dependencies of outside nations. They vary greatly in size, shape, geographic structure, and functional organization. We begin with the most important unit, Mexico.

MEXICO: AN UNRECOGNIZED GIANT

The federal republic of Mexico, officially *Estados Unidos Mexicanos* (United Mexican States), is by far the largest, most complex, and most influential of the many countries in Caribbean America. By comparison with most of the world's countries, Mexico is a giant in area and population, but its large size tends to go unrecognized. Peninsular in shape and situated next to the compact bulk of the conterminous United States, Mexico looks comparatively small on a world map. But its area of 762,000 square miles (*c.* 2 million sq. km) is nearly eight times that of the United Kingdom, and its elongated national territory would stretch from Washington state to Florida. The country's huge population total (an estimated 84 million in 1988; expected to reach 105 million by the year 2000) makes Mexico by far the largest nation in which Spanish is the main language, and Mexicans comprise the largest national block of Roman Catholics except those of Brazil. The oversized political capital, Mexico City (metropolitan population 17 to 18 million by one estimate), ranks with Tokyo and New York as one of the three largest metropolises in the world.

Mexico is also a big country economically. It ranked 13th among the nations of the world in gross national product (GNP) in 1985 (exceeded in the hemisphere by the United States, Canada, and Brazil), and it stands quite high in annual output of varied commodities. For example, recent figures on Mexico's mineral output rank the country among the world's top five in antimony, petroleum, silver, and sulfur, and among the top ten in copper, lead, manganese, mercury, natural gas, salt, and zinc. To-

---

[1]The authors wish to thank Steven Fair for valuable help in updating the section on Caribbean America. Small parts of the present section were written originally by Dr. Richard S. Thoman. His wording is presented here in a revised and updated context. In reworking this chapter, the authors benefited from access to up-to-date information in Preston E. James and C. W. Minkel, *Latin America,* 5th ed. (New York, John Wiley & Sons, 1986), 578 pp. This admirable standard textbook on the geography of Latin America is strongly recommended to students and teachers. It contains a valuable list of references (books and articles) arranged by country. City populations cited in this chapter are rounded metropolitan-area approximations for the mid-1980s.

---

[2]Colombia and Venezuela touch the Caribbean Sea and could fit comfortably with Caribbean America, but they also have highly important Andean relationships.

Figure 25.1   General reference maps of Caribbean America. Curaçao and Bonaire comprise the Netherlands Antilles; Aruba is a separate entity (as of 1986). Antigua and Barbuda form one political unit, as do St. Christopher and Nevis, and St. Vincent and the Grenadines. Major highways in the transportation inset are supplemented and interconnected by other surfaced highways not shown. Places shown by lettered symbol in the inset but not named on the main map include Ciudad Obregón and Manizales on the western route, and Aguascalientes, Irapuato, and San Francisco del Rincón (not lettered) on the central route between Durango and Mexico City.

575

Figure 25.2 General reference map of natural areas and major cities in South America. Note the clearly defined arrangement of cities around the rim of the continent and the enormous empty expanses in the interior. See also Figure 2.10 on which the Brazilian Highlands are shown as "plateaus." Chuquicamata in Chile is that country's most famous copper-mining settlement.

576

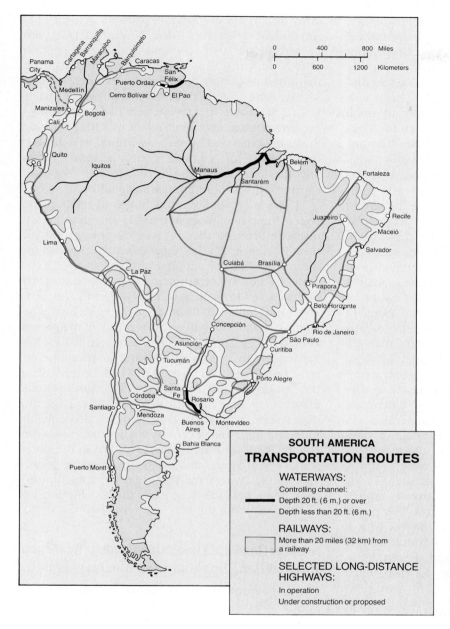

Figure 25.3 Only the basic skein of major highways in the Pan-American Highway System is shown, together with new long-distance roads reaching into the deep interior of Brazil.

tal production of a varied list of farm commodities is also impressive: "top five" rankings in cane sugar, citrus fruit, coffee, and corn (maize); "top ten" in bananas, cacao, cattle, coconuts, cotton, eggs, pigs, and tomatoes. Manufactural output is second only to Brazil's in Latin America.

Environmental Variety and Its Geographic Consequences

The remarkable diversity of Mexican mineral and agricultural production is made possible by an environment that encompasses great geologic, topographic, climatic, biotic, and soil variety. Metal-bear-

ing ores occur in many parts of the country, being mined particularly in mountainous terrain of northern and central Mexico where the range of metals is broadly comparable to that of the highlands in western United States. Iron ore from scattered locations supports a fairly sizable iron and steel industry, which can also draw on substantial coal deposits, some of which are suitable for coking. Mexico's coastal lowlands along the Gulf of Mexico are a southward continuation of the Gulf Coastal Plain of the United States, and their sedimentary rocks share the deposits of oil, natural gas, and sulfur that are such a marked feature of coastal Texas and Louisiana.

Mexico's agricultural variety is achieved despite an overall environment that is excessively mountainous and hence short of tillable land. Only about 13 percent of the country is cultivated. Especially fertile crop land mainly exists (1) on the floors and lower slopes of mountain basins with soils of volcanic origin, or (2) on scattered and generally small alluvial plains. Viewed as a whole, the country exhibits a complex mosaic of agricultural habitats, varying in character according to elevation, rock types, physiographic history, and exposure to atmospheric influences. Mountains too steep to cultivate without excessive erosion dominate the terrain in most areas, but they are flanked by plains, plateaus, basins, and foothills that offer agricultural possibilities if enough water is available. Areas actually used for crops lie at elevations ranging from below sea level north of the Gulf of California to over 8000 feet (*c.* 2400 m) in some highland basins. A little over half of Mexico lies north of the Tropic of Cancer and is an area dominated by desert or steppe climates with hot summers and winters that are warm to cool depending on the altitude. South of the Tropic, where about four fifths of all Mexicans live, temperatures at a given place vary less markedly from season to season, but altitudinal differences create conditions ranging from high temperatures in the *tierra caliente* of lowlands along the Gulf of Mexico, through more moderate heat in the *tierra templada* of low highlands, to cool temperatures in the *tierra fría* of high highlands. Nearly all areas inside the tropics are humid enough to support rainfed crops, although irrigation is often used in drier places to increase the range of crops and/or the intensity of agriculture. Large parts of Mexico's humid tropics have a dry season in the low-sun period ("winter"), when irrigation becomes a necessity for cropping.

Mexico's International Influence and Associated Problems

On the hemispheric and world stage Mexico speaks with a respected voice, although the country's poverty and its great economic dependence on the United States make its influence less potent than might otherwise be true of so large a nation. Commencing in the 1970s, huge new oil discoveries increased Mexico's weight in world affairs, but at the same time the country borrowed so much money abroad to finance oil and other development that it became a major debtor nation. By late 1986 unwise spending and a severe drop in world oil prices had put the economy in such a catastrophic state that a nationwide Mexican poll asking "when do you believe Mexico will emerge from the economic crisis?"

drew an answer of "Never!" from 54 percent of all respondents (*New York Times*, November 16, 1986).

Meanwhile the economic imbalance between Mexico and the United States remains far greater than their respective population sizes might suggest. Mexico has over a third as many people as the United States but in a recent year had a GNP only 4 percent as great. Economic relations with the United States are crucial to Mexico, as around two thirds of the country's foreign trade is with the United States, and the latter nation is overwhelmingly the main source of investment and loan capital in Mexico.

There is inevitably a touchiness in relations between these neighbors due to (1) the great present disparity between them in wealth, power, and cultural influence; (2) the historic fact that Mexico lost more than half its national territory to the United States in the 19th century; and (3) Mexican national pride derived from great cultural longevity. Mexico's aboriginal Indian civilizations were more highly developed than those within the territory of the present United States, and the court of the Spanish Viceroy in Mexico City was an opulent and far-reaching center of imperial power by the early 17th century when the first tiny English settlements of the future United States were struggling for survival at the edge of a vast wilderness. The University of Mexico dates its founding to 1536—a century earlier than the oldest United States university, Harvard. We look now at some key features of Mexico's eventful prehistory and history, with particular regard to the country's geographic heritage from early Indian empires, imperial Spain, and the Mexican republic.

Associations in Time: Geographic Residues from Indian, Spanish, and Mexican Eras

The place now known as Mexico is composed of a geographic system that has evolved slowly over time as part of a greater worldwide system and is still doing so. The system has been created by nature, which has provided habitats, and by people who have organized the habitats for group living. This long process has brought into being social, economic, and political arrangements and attendant landscape expressions that mesh into an ordered pattern susceptible of being mapped and geographically described and analyzed. The present geographic system contains discernible residues from many eras, and new components are constantly added while other components fade and may eventually disappear. A great many geographic features of the past have not survived in the present system, but one era after another has made enduring contributions to the phenomena of the present. These

Figure 25.4 The Pyramid of the Sun (shown in photo) and the Pyramid of the Moon are prominent pre-Aztec structures rising in the midst of corn and maguey fields near Mexico City. They are sited on a broad avenue floored by stone and lined with ruined stone temples and priests' quarters. The entire assemblage, known as Teotihuacan, is what remains of a city that may have housed 200,000 people prior to the city's destruction by fire some 14 centuries ago.

legacies must be recognized and appreciated if Mexico's present geography is to be understood. Here only the barest outline on the broadest scale is possible.

Brevity is especially necessary in regard to the long succession of geologic eras that have left residues in the diverse components of Mexico's physical geography. Here we note only that (1) the present landforms reflect most prominently the geologically recent era of Alpine mountain-building that created the mountains stretching along the western side of the Americas from Alaska to Cape Horn, and (2) Mexico was not subjected to continental glaciation and hence was not sculptured by ice sheets in the manner of large areas in Canada and the United States. But although the events of geologic time must be so nearly bypassed here for reasons of space, a more detailed comprehension of certain residues from the human past is indispensable to an introductory understanding of Mexico's present geography. Centrally important in the latter regard are (1) the country's rich Indian heritage, (2) the many features inherited from imperial Spain, and (3) the revolutionary changes that have resulted from internal Mexican struggles and external relations with the United States since the end of Spanish control. We therefore group our observations under three eras: the *Indian era* prior to the Spanish conquest of the early 16th century, the *Spanish colonial era* to the early 19th century, and the *Mexican era* since Mexican independence in 1821.

GEOGRAPHIC SURVIVALS FROM THE INDIAN ERA

At the time of the Spanish conquest, Mexico was the home of probably more than 1 million Indians speaking hundreds of languages or distinct dialects.

Artifacts from their varied cultures are carefully preserved in the world-famous National Museum of Anthropology at Mexico City. The invading Spaniards found Mexico a seat of Indian cultures more advanced than any others in the Americas except for those within the Inca Empire in the Andes of South America. Great builders in stone, the aboriginal peoples had constructed the mighty Pyramids near Mexico City (Fig. 25.4) and the huge temples and palaces of the Mayan culture in Mexico's peninsula of Yucatán. Today a rich variety of Indian cultures persists and the Indian racial component in the population is pronounced (Fig. 24.5), although Mexico's dominant cultural strain is Spanish, and the greater part of the country's people are mestizos with mixed Spanish and Indian blood. The present pattern of settlement has been powerfully influenced by the distribution of Indians at the time the Spanish came. The conquerors sought out concentrations of Indians as laborers and potential Christian converts. Hence the locales of Spanish-inspired settlement tended to conform to the population pattern already established. Maize, beans, and cacao are among the many pre-conquest Indian contributions to Mexico's present agriculture.

GEOGRAPHIC RESIDUES FROM THE SPANISH COLONIAL ERA

Mexico was part of Spain's colonial empire for nearly four centuries. Following the overthrow of the Aztec Empire by Hernando Cortez (Hernán Cortés) in 1521, the Spanish Crown established a Viceroyalty of New Spain, ruled from Mexico City and eventually encompassing not only the area now included in Mexico but also most of Central America and about a quarter of the territory now occupied by the 48 conterminous American states. Throughout this huge expanse a Hispanic pattern of life, still highly evident today, was established by Spanish administrators, fortune hunters, settlers, and Catholic priests. Racial, ethnic, and cultural residues of their activities are apparent in recent estimates that mestizos comprise 55 percent of Mexico's population, Caucasians (nearly all of Spanish ancestry) comprise 15 percent, Spanish is spoken as a first language by 93 percent of the populace, and religious affiliation is 93 percent Roman Catholic. Other strong Spanish influences are found in architectural styles and urban layouts. Towns and cities characteristically have a Spanish-instituted rectangular grid of streets surrounding a *plaza* at the town center. Cattle ranching begun by the Spanish endures as a way of life in large sections of Mexico, and the Spanish introduced not only cattle, but wheat, sugarcane, sheep, horses, donkeys, and mules. The consuming Spanish hunger for precious metals greatly expanded mining, particularly of silver. Many of Mexico's larger cities were founded as silver-mining camps or regional service centers for mining areas.

Under the absolutist government of imperial Spain a small Spanish aristocracy monopolized power, wealth, prestige, and education. Mexico was governed by a hierarchy of officials appointed by the Crown, and a large share of the best land passed into the hands of aristocratic Spanish owners of the large estates called *haciendas*. Haciendas dominated the countryside to the end of the Spanish colonial era and for about a century after Mexican independence. Meanwhile the Church, holding official status under devoutly Catholic Spain, spread its influence, structures, and extensive landholdings to every part of the imperial domain. Innumerable towns and cities of today originated essentially as Catholic missions in the midst of clustered Indian populations.

THE TURBULENT MEXICAN ERA: GEOGRAPHIC LEGACIES FROM MEXICO'S QUEST FOR NATIONAL COHESION, REFORM, SELF-DETERMINATION, AND ECONOMIC ADVANCEMENT

At the end of Spanish colonial control in the early 19th century, Mexico was overwhelmingly rural, illiterate, village-oriented, Church-oriented, and poor. Internal transportation was minimal. Huge landholdings belonging to a few thousand wealthy families or the Church enclosed nearly all of the best farm lands and a large share of the other land as well. In the capital an educated elite governed and exploited the country with little benefit to the downtrodden Indian and mestizo farmers in the countryside.

The bloody and chaotic Mexican Revolution of 1810–1821 ended Spanish control and instituted a short-lived Mexican Empire that expired in 1823 and was followed by a federal republic in 1824. The revolution against Spain essentially pitted the ruling class of Mexicans born in Spain (*peninsulaires*) against aspirants to greater status born in Mexico (*criollos*). Levies of villagers did most of the fighting in a civil war marked by savage slaughter and devastation, with a total loss of life of perhaps 600,000 out of a national population of 6 million. Then ensued a long period of political instability, with personalities of many persuasions contending for power. Government followed government in rapid succession. During its first three decades of independence Mexico had some 50 governments, none of which had resources enough to develop the country economically or power enough to extract more than limited tribute and token allegiance from powerful generals who governed local areas with the support of large landowners. The latter group greatly expanded their holdings by illegally appropriating communal Indian lands.

Meanwhile Mexico lost huge blocks of territory that had been governed from Mexico City by the Viceroy of New Spain. The Central American part of New Spain attained separate independence and soon broke into five nations, which endure today. Texas proclaimed its independence in 1835 and confirmed it with a military victory over the Mexicans at San Jacinto (near present Houston) in 1836. Subsequent annexation of Texas by the United States in 1846 was contested by Mexico, which then suffered military defeat in the Mexican War of 1846–1848 and lost not only Texas but also California and areas now comprising New Mexico, Arizona, and parts of some other states.

During the 1850s and 1860s a dynamic Indian reformist, Benito Juárez, attained power for a time, but the country then settled into a long period of despotic personal rule by President Porfirio Diaz that began in 1876 and did not end until 1910. During this time democratic processes were largely suspended and the rural population sank deeper into poverty, but heavy foreign investments were made in railroads, the oil industry, metal mining, coal mining, and manufacturing. Political discontent over national disunity and weakness, internal oppression and corruption, foreign exploitation, and poverty then led to the Mexican Revolution of 1910–1920, which ousted Diaz and subsequently turned into a chaotic and destructive struggle among personal armies of regional leaders. During the decade of the Revolution, Mexico's population is estimated to have declined by around one million. In 1917 a new Constitution was enacted, under which land reform was an urgent priority. Since 1910 huge acreages have been expropriated from haciendas by the government and redistributed or restored to landless farmers and farm workers. Large haciendas still exist, especially in the dry North, but they contain only a very small proportion of Mexico's cultivated land.

Although much of the best hacienda land has been transferred to smaller private landholdings, and some land has been restored to Indian groups whose communal holdings were illegally seized by hacienda owners in the mid-19th century, the best-known aspect of Mexican land reform has been the transfer of land to agricultural communities called *ejidos,* created under provisions of revolutionary legislation and the 1917 Constitution. Private holdings are far more important than ejidos in extent of land and total output of farm products, but the ejidos do comprise millions of farmers and are a major feature of the Mexican countryside. Their existence symbolizes the long-term importance of community ownership among Mexico's native Indians and the desire of Indian communities to have their own land to work. Peasant farmers, as ejido members, are assigned plots to cultivate as though the land were their own, but legal ownership remains with the ejido. Plots can be inherited, but they cannot be sold or mortgaged. Woodland and pasture are held in common by members of the ejido. Ejidos have been criticized for their relative lack of productivity caused by such factors as the small size of plots, peasant conservatism, inadequate technology and support services, and soil erosion. But these units do represent fulfillment of the revolutionary promise of "land for the landless" and thus have emotional support in a country where agricultural issues, particularly the ownership of land, have always loomed large.

Agriculture is still a highly important source of livelihood in Mexico, although recent statistics reported the country's population to be 70 percent urban. Corn is by far the most important crop in acreage and production, as it has always been. Mexico is almost certainly the area where corn was first domesticated in prehistoric times and from which it spread to become the premier crop of the New World. Although not a Mexican export of any importance, it is central to the Mexican diet. The crop is grown in all parts of the country, but the main producing area occupies the floors of numerous mountain basins of the core region around and between Mexico City and Guadalajara (Fig. 24.4). Beans, also an inheritance from the Indian past, are another universal Mexican crop and food. An extreme variety of other crops is grown in Mexico's diverse habitats, very largely for domestic consumption. The total value of output from Mexico's factories and mines is now several times that of output from farms; in 1986 the reported value of manufactured goods alone was nearly 3 times that of farm products. All of Mexico's economic production needs to be seen in a regional context, as the output of particular commodities tends to be rather sharply localized within particular areas. We turn now to a broad-scale view of Mexico's regional geography.

Regional Geography of Mexico

Mexico is strongly regionalized. Although innumerable areas have distinctive personalities, space restricts us here to regions of the broadest scale. We focus on (1) *Central Mexico*—the country's core region; (2) *Northern Mexico*—dry, mountainous, and increasingly intertwined with the United States; (3) the *Gulf Tropics*—comprised of hot wet coastal lowlands along the Gulf of Mexico, together with an inland fringe of *tierra templada* in low highlands; and (4) the *Pacific Tropics*—a poorly developed southern outlier of humid mountains and narrow coastal plains.

CENTRAL MEXICO: VOLCANIC HIGHLAND CORE REGION

It is very common in Latin America for a country to possess a sharply defined core region where the national life is centered. The core contains the main area of dense population and economic output, the largest city, and the political capital (which in nearly all cases is the largest city). The country's transportation routes converge on the core and connect it to the provincial districts and outlying areas that comprise by far the greater part of the national territory.

Mexico is a classic instance of this common Latin American pattern. Around half the total population is found within a contiguous series of highland basins clustered along an axis from the vicinity of Guadalajara (2.4 million) at the northwest to the vicinity of Puebla (1.1 million) at the southeast. Mexico City is located toward the eastern end of the axis. These basins comprise Mexico's core region, which is often referred to as *Central Mexico* or the *Central Plateau* (for typical scenes, see Figs. 24.4, 24.5, 25.5, and 25.6). The basin floors vary in elevation from about 5000 feet (1524 m) to as much as 9000 feet above sea level, with each basin being separated from its neighbors by a hilly or mountainous rim over which highways and railways climb via passes at 8000 to 10,000 feet or higher. The basins drain through canyons in their rims, generally to the

Figure 25.5 *A characteristic village in a highland basin of Central Mexico. Courtyards with trees and flowers lie inside the facade of adobe housing.*

Pacific Ocean, although the *Basin of Mexico* containing Mexico City is a conspicuous exception in having no natural outlet to the sea. Centers of basins contain stretches of flat land, which may be swampy. The swampiness has led populations to cluster on higher and more sloping land around basin margins, and population pressure has often induced cultivation of steep slopes, with much attendant soil erosion. Most parts of Central Mexico are too high in elevation for coffee or other crops of the *tierra templada,* although some areas are sufficiently low and warm for coffee or even sugarcane. The prevailing *tierra fría* environment generally dictates crops that are less subject to frost damage. Corn, grown for local subsistence and to feed livestock, is the leading crop in all basins, although the climate is too cool to be ideal, and corn yields are relatively low. Grain sorghums are common; wheat for sale in urban areas (where people generally prefer to eat white bread) is grown in some areas; potatoes are a staple crop; and temperate fruits and vegetables are grown. There is still a heavy reliance on human labor and animal draft power in agriculture. Cattle (both dairy and beef) and poultry are important. Very little of this produce is consumed outside Central Mexico; agriculture essentially supplies the needs of farmers and urban dwellers within the core. A productive agriculture has been maintained here for millennia, based on fertile and durable soils derived from lava and ash ejected by the many volcanoes.

Each basin of Central Mexico has a dominant urban center, which generally is the capital of a Mexican state (although state boundaries do not often coincide with basin boundaries). These cities are commercial and cultural foci for their basins, and they contain a diverse array of manufacturing industries. But with one towering exception their influence is basically regional rather than national. The exception is, of course, Mexico City (Fig. 25.6), which had nearly 9 million people in the city proper according to the 1980 census but is the core of a huge metropolitan area in the Basin of Mexico that incorporated an estimated 17 to 18 million people in the mid-1980s.

MEXICO CITY: RUNAWAY URBANISM IN A VENERABLE POLITICAL CAPITAL. Mexico City was a major urban and political focus long before its conquest by Cortez in 1521. Near the city are impressive pyramids and other monuments built by Indian precursors of the Aztecs (Fig. 25.4). But Indian political power reached its apex under the Aztec Empire in the centuries immediately before the Spanish came. The Aztecs, notorious for their insatiable practice of ritual human sacrifice, came to the Basin of Mexico from the north and built their capital city on an island in a shallow salt lake, with causeways to the shore. Most of the lake subsequently was drained over a period of centuries, and post-Aztec Mexico City arose on the drained land. The result was a gradual subsidence of buildings (still in progress today) as the lake sediments contracted. In recent times this problem has been aggravated by overpumping of ground water. But despite the unstable subsurface, including the fact that the city is underlain by a geologic fault, Mexico City now has an imposing collection of modern office towers with foundations especially designed to withstand both subsidence and earthquake shocks. Further environmental hazards are posed by severe pollution from the noxious emissions of automobile exhausts and great numbers of factory chimneys. Temperature inversions in the confined atmosphere of the basin keep pollutants from dissipating and regularly create some of the world's most notorious smogs. Disposal of household wastes from millions of shanties that have no indoor plumbing is another major problem. But despite such vexations and discomforts, there is a relentless flow of migrants to the Basin of Mexico. Unemployment is high, but many newcomers soon find low-wage service or industrial jobs, and others subsist with the aid of relatives until they can find work.

Explosive urbanism has outrun the efforts of Mexico City's planners to cope with it. One mayor of the city commented that administering this monstrous sprawl was like "repairing an airplane in flight" (Bart McDowell, "Mexico City: An Alarming Giant," *National Geographic,* August 1984, p. 144).

Figure 25.6   A view of downtown Mexico City. The building with the prominent archway (upper center) *is the Monument to the Revolution.*

Nonetheless, the city does have some spectacular planning achievements to its credit—not least is a modern subway system over 60 miles (100 km) long, and a massive pumping system to bring water to the city through the encircling mountain walls of the Basin of Mexico. Water available from sources within the Basin is now hopelessly inadequate and must be supplemented by large amounts pumped in. Meanwhile the provision of basic services to the expanding city becomes ever more difficult as the insurge of newcomers continues. In 1985 planning efforts received disastrous setbacks due to a destructive earthquake and a slump in the selling price of the oil exports on which Mexico's finances are vitally dependent.

In the face of these daunting events, Mexico City retains its dramatic image as the age-old focus of Mexican life. Towering above the crowded Basin of Mexico to the southeast of the city are the famous snow-crowned volcanoes Ixtacihuatl (17,343 ft/ 5286 m; last eruption, 1868) and Popocatépetl (17,887 ft/5452 m; last eruption, 1702). Northeast of the city the Indian past is perpetuated in the giant Pyramid of the Sun and Pyramid of the Moon built by the Toltecs who ruled in Central Mexico before the Aztecs. In the northern part of the metropolis is the shrine of Our Lady of Guadalupe, especially sacred to Mexico's Indians. The innermost city surrounds the central square (Plaza de la Constitución), which is connected by a broad tree-lined avenue, the Paseo de la Reforma, to Chapultepec Castle where Mexico's Presidents reside. It was at the castle that

the last defenders of the capital were overcome by American invaders in 1847.

Amid the many memorials to Mexico's eventful history, the modern city carries on its daily functions as a business and industrial center. It is estimated that approximately a third of all manufactural employees in Mexico work in thousands of factories in metropolitan Mexico City. Most factories are small, and manufacturing is devoted largely to miscellaneous consumer items, although some plants carry on heavier manufacturing such as metallurgy, oil refining, and the making of heavy chemicals. Mexico's manufacturing industries have come a long way in diversification and technological competence during the 20th century, but they do not yet come close to meeting the needs of the huge and rapidly growing population. Manufactured goods comprised around four fifths of all Mexican imports by value in the mid-1980s.

NORTHERN MEXICO: LOCALIZED DEVELOPMENT IN A DRY AND RUGGED OUTBACK

North of the Tropic of Cancer, Mexico is characterized by ruggedness, aridity, an extensive ranching economy supporting a generally sparse population, and some widely separated spots of more intense activity based principally on irrigated agriculture, mining, heavy industry, and diversified border industries adjacent to the United States. Most of Northern Mexico is above 2000 feet (610 m) in elevation, but there are lowland strips along the Gulf of Mexico, the Gulf of California, and the Pacific. Two ma-

583

jor north-south ranges of high mountains—the *Sierra Madre Occidental* in the west and the *Sierra Madre Oriental* in the east—tower above the general surface. Both are steep and rugged, with many precipitous canyons and few good passes. Surface transport across them is poorly developed. Between the two major ranges lie a great number of lower and shorter ranges, which are separated by valleys and basins of varying size. The climate is desert or steppe, with scanty xerophytic vegetation except where elevations are high enough to yield more rainfall. In the latter areas, some upland surfaces are carpeted with steppe grasses, and substantial mountain areas rising still higher have good coniferous forests. Most forest land is in the Sierra Madre Occidental. This is Mexico's main source of timber, and lumbering is an important economic activity in some places. Within Northern Mexico as a whole, ranching is the most widespread source of livelihood, although the total number of people supported by it is small. Many large landholdings still exist here. Scattered Indians in remoter areas still live on a semisubsistence basis.

In addition to being Mexico's main area of livestock ranching, Northern Mexico also provides the country's principal output of metals, coal, and products of irrigated agriculture. Iron ore, zinc, lead, copper, manganese, silver, and gold are the main metals produced. The ores, which often contain more than one metal, are smelted with coal mined in Northern Mexico. Some coal is made into coke and used for iron and steel manufacture at a number of places, particularly Monterrey (2.1 million), which is Mexico's leading center of heavy industries and Northern Mexico's most important business and transportation focus. Irrigated agriculture in the dry North has been greatly expanded since World War II by the building of many multipurpose dams to store the water of rivers originating in the Sierra Madre and emptying into the Gulf of California or the Rio Grande. New or expanded irrigated districts now are widespread, with the main clusters in the northwest and along or near the Rio Grande. Some large and important districts immediately along the American border have been in existence for a long time. They include (1) a district in the lower Rio Grande Valley directly opposite a larger district on the American side, and (2) the Mexican continuation of California's Imperial Valley north of the Gulf of California. Cotton is a major product of Northern Mexico's irrigated districts, and varied fruits and vegetables for shipment to American markets also are grown. In some irrigated areas the Mexican government has helped develop agricultural colonies of small landowning farmers who have migrated from more

crowded parts of the country. Throughout Northern Mexico, the new or expanded oases stand out as islands of agricultural productivity amid large areas that are either nonagricultural or used for ranching or nonirrigated grain farming yielding low returns per unit of land.

BORDER CITIES AND PROBLEMS: MEXICANS AND AMERICANS INTERMINGLED. Since World War II there has been a large upsurge of manufacturing in a series of widely separated Mexican cities immediately adjacent to larger American cities along the international boundary. The largest places are Tijuana (probably well over 1 million) adjacent to San Diego, California, and Ciudad Juárez (600,000 or more) across the Rio Grande from El Paso, Texas. In "free zones" within these and smaller places, American firms have established manufacturing plants that now employ large numbers of Mexican low-wage workers. The companies are allowed to move components into Mexico duty-free and ship finished products back to the United States, where duty is charged only on the value added to each item by labor cost and the cost of any components not shipped from the United States. More than 1000 such factories in the "free zones" are only one instance of the close intertwining of Mexicans and Americans along the border. Large numbers of Mexicans cross the boundary each day to work or shop in the United States, and there is a return flow of American shoppers and tourists. The Mexican border cities are collecting points for Mexican and Central American migrants who attempt to enter the United States illegally and often succeed despite the efforts of the American Border Patrol to apprehend them. A new American immigration law in 1986 gave amnesty to illegal migrants who could prove residence in the United States for a stated time, but the law prescribed tougher measures to try to halt the continuing illegal flow. It remained to be seen how well such measures could succeed along this bicultural border where desperate poverty lies closely adjacent to affluence and strong demand for low-wage Mexican labor exists on the American side.

THE GULF TROPICS: OIL AND GAS, PLANTATIONS, MAYAS, TOURISM

On the Gulf of Mexico side, Mexico has a sizable area of tropical lowlands that have a much lower population density than Central Mexico but are very important to the national economy. The most valuable product of these lowlands is oil, discovered in huge quantity in the mid-1970s. Just as in coastal Texas and Louisiana, some fields are onshore, while others lie underneath the Gulf. The oil industry is a government monopoly (carried on by a government corporation, *Petróleos Mexicanos* or PEMEX), and the United States is the largest foreign customer.

Actually, Mexico has seen two great oil booms. The earlier one took place in the first quarter of the 20th century and centered in fields farther north in the coastal lowlands than those discovered recently. Carried on by foreign companies (principally American), the oil industry for a time was a major world supplier. Then in 1938 the industry was nationalized as a gesture of resentment against what were deemed to be high-handed actions by foreigners in charge of a major Mexican resource. Following nationalization, Mexico's oil industry largely collapsed until it was reinvigorated by the bonanza strikes in the 1970s. These discoveries seemed so promising that the government borrowed heavily from foreign banks to finance both oil development and a wave of other economic and welfare schemes. Dependence on oil income grew to the point that crude oil represented 67 percent of all Mexican exports by value in 1983. Economic hardship then descended in 1985 when a worldwide oil glut caused prices to skid. In 1988 Mexico was continuing to have grave problems in regard to financial solvency.

Natural gas also is produced in the Gulf lowlands and is piped to customers in Mexico and the United States. Construction of a pipeline to the United States was widely criticized in Mexico due to sensitivity concerning the country's large and growing economic dependence on its large neighbor. There is much resentment over Mexico's role as a supplier of natural resources to feed the voracious American industrial economy. But Mexico's desperate need for foreign exchange has sidetracked questions of national sensibility in favor of exports to the United States not only of oil and gas but also of native sulfur from deposits in the Gulf Tropics and various minerals from the dry North.

The Gulf Tropics are Mexico's main producer of tropical plantation crops, which contribute in a relatively minor way to the export trade. Cacao, sugarcane, and rubber are among a considerable variety of crops in the lowlands, and coffee exports come from a strip of *tierra templada* along the border between the lowlands and Mexico's Central Plateau.

The greater part of Mexico's Gulf Tropics lies in the large Yucatán Peninsula. In pre-Hispanic times the Maya Indians developed a notable civilization here and in adjacent parts of Mexico and Guatemala. Its existence was unknown until ruined cities overgrown by jungle began to be found in the 19th century. Why their inhabitants abandoned them is still a mystery. Under the Spanish Empire, Yucatán was parceled into large haciendas where range cattle were raised, using forced Indian labor. After Mexican independence, henequen and sisal plantations were established and flourished for a time. But since the Spanish conquest this area has largely been a remote and poverty-stricken backwater, plagued by low rainfall and/or a long dry season over large areas, and further handicapped by water-supply problems caused by limestone bedrock into which rainfall sinks quickly without forming surface streams. Recently there has been an inflow of tourist income from a major new resort at Cancun on an island off the east coast. Attempts are being made to attract light industry to this poor corner of Mexico.

Cities of much size are few in the Gulf Tropics. The most important is Veracruz (400,000), which has been Mexico's leading seaport since the very beginning of Spanish colonization.

THE PACIFIC TROPICS: MOUNTAINOUS
SOUTHWESTERN OUTLIER

South of the Tropic of Cancer, a belt of rugged mountains with a humid climate lies between the densely populated highland basins of Central Mexico and the Pacific Ocean. Elevations rise to slightly over 12,000 feet (3658 m). This difficult terrain is far more thinly populated than Central Mexico, has a high incidence of relatively unmixed Indians living in traditional ways, and has relatively little economic production. Two very localized exceptions to the comparative underdevelopment are the famous coastal resort of Acapulco on a spectacular bay about 190 miles (*c.* 300 km) south of Mexico City and a major new iron and steel complex at a rivermouth location on the Rio Balsas about 150 miles up the coast from Acapulco. The latter industry is based on deposits of iron ore near by and is intended to undergird general economic growth within this rather retarded part of the country. Narrow lowlands along the Pacific shore are spotted with small fishing ports. The lowlands are of some importance for cotton production, and the highlands behind them grow a fair amount of coffee for export.

CENTRAL AMERICA: FRAGMENTED DEVELOPMENT IN AN ISTHMIAN BELT OF WEAK STATES

Between Mexico and South America the North American continent tapers southward through an isthmian belt of small and poor countries known collectively as Central America. Five of the countries—Guatemala, Honduras, Nicaragua, Costa Rica, and Panama—have seacoasts on both the Pacific

Ocean and the Atlantic (Caribbean Sea) (see map, Fig. 25.1). The other two countries have coasts on one ocean only: El Salvador on the Pacific and Belize on the Atlantic.

Geographic fragmentation is a major characteristic of Central America. A prime example is the fragmented pattern of political units that is so apparent on a map. Even if the seven countries formed one unit, they would still comprise an area only about one fourth that of Mexico (Table 24.1) and a population only one ninth that of the United States. But in actuality the individual countries have areas and populations comparable to those of medium-sized to small American states. Guatemala, the largest Central American country in population, had only an estimated 8.7 million people in 1988, analogous to the combined populations of Massachusetts and Connecticut. Small but much-publicized Nicaragua, with 3.6 million, was somewhat larger than Connecticut.

Five of the present Central American countries — Guatemala, El Salvador, Honduras, Nicaragua, and Costa Rica — originally were governed together as a part of New Spain in a unit called the Captaincy-General of Guatemala. Comprised of many small settlement nodes isolated from each other by mountains and empty backlands traversed by poor transportation, the unit never established strong geopolitical cohesion. Separate feelings of nationality emerged in various areas, and these became strong enough to fracture overall unity after the end of Spanish imperial rule in 1821. The old Captaincy-General was governed by newly independent Mexico for a few years, but in 1825 it broke away as a separate Central American Federation. Then in 1838–1839 this loose association of polities disintegrated into the five independent republics of today. Some major factors in the breakdown of the federation included fears of weaker states that they would be dominated by stronger states; controversies pitting more conservative states against more liberal ones; and the sheer physical isolation that made it difficult for any central governing body to exert effective control over such an elongated and poorly articulated domain. An ideal of overall unity persisted after the end of the federation, but various attempts to reestablish a political union did not succeed. In 1960 the five countries did manage to form a Central American Common Market (CACM) to facilitate trade and general economic advance, but even this limited association broke down when quarrels and conflicts developed among the members, including a short war between Honduras and El Salvador in 1969. Recent tensions within the five-country area have centered around (1) guerrilla warfare by American-aided "Contras" against the Soviet- and Cuban-backed Marxist government in Nicaragua; (2) the aiding of anti-government guerrillas in El Salvador by Nicaragua, with the Salvadorean government being aided in turn by the United States; (3) an American-aided military buildup in conservative Honduras, which has been embroiled in controversy with Marxist Nicaragua over various questions; and (4) long-continued anti-government guerrilla activity in Guatemala, with the government receiving American aid intermittently. In the 1970s and 1980s the entire Central American situation became a major question in the internal politics of the United States, which has long been deeply involved in Central American affairs for economic, strategic, and ideologic reasons.

The serious degree of political disunity within the part of Central America now occupied by Guatemala, El Salvador, Honduras, and Nicaragua reflects this area's fragmented social geography. In each of the four countries society is comprised of groups that differ sharply in wealth, social standing, ethnicity, culture, and opinion. The fragmentation is too diverse and complex to discuss here in detail, but a few general observations can be made. Wealthy owners of large estates have formed a social aristocracy since the first land grants were made in early colonial times by the Spanish Crown. Members of this group tend to be of relatively unmixed Spanish descent, although the great majority have some Indian blood. They have, in general, monopolized the national wealth, defended the status quo, and exercised great political power. Occupying a middle position in society is the class of small farmers, largely mestizos, who own and work their own land. At the bottom of the social and economic scale are millions of landless mestizo tenant farmers and hired workers, and the many Indians who live essentially outside of white and mestizo society. Most of Central America's Indians are in Guatemala, where they form somewhat over half the population. Hundreds of thousands work seasonally as laborers in commercial farming districts, where their services are essential at harvest time for such tasks as picking coffee or cotton. Aside from this they tend to lead very self-contained lives in their home villages. In the mid-1980s, urban dwellers formed a majority of the population in Nicaragua (57% urban) but were in the minority in the other three nations. In each country the national capital is overwhelmingly the largest city. Here are concentrated the social, business, and governmental elites, together with a growing mass of low-wage service and industrial workers, often newly arrived from the countryside or smaller towns and cities. There is much unemployment in the makeshift slums that ring the peripheries of all the capitals. In all countries, including even Marxist Nicaragua, the Roman Catholic Church is an important force, although the present Church is very divided among many shades of opinion regarding social and doctrinal questions.

Within the broad spectrum of Central American social groups, opinions on political, social, economic, and religious questions are so polarized that genuine democracy is hard to achieve. These countries have persistently been marked by authoritarian rule, uprisings and coups, the rise and fall of governments in rapid succession, and blatant use of governmental power to protect the interests of the privileged few. Military rule, generally rightist but sometimes leftist in philosophy, has been common. A recent history of great social unrest with accompanying violence and political instability is well exemplified by the overthrow of the repressive Somoza family dictatorship in Nicaragua in 1979 and the establishment of a Marxist-oriented government and social order there.

Costa Rica has been rather different in its historical development from the other four countries of the old Central American Federation. It was the area farthest away from the center of government in Guatemala and was left largely to its own devices in developing its economy and polity. Spanish settlers found few Indians there, and the ethnic pattern that developed was dominated overwhelmingly by whites of Spanish descent. By one estimate 87 percent of the present Costa Rican population is white, whereas the same source shows an overwhelming predominance of mestizos and/or Indians in El Salvador (94% mestizo), Honduras (90% mestizo), Nicaragua (69% mestizo, 14% white), and Guatemala (55% Mayan Indian; 42% mestizo). No precious metals of significance were found in Costa Rica, and no large landed aristocracy developed there. The basic population of small land-owning white farmers managed to create one of Latin America's most democratic societies despite some episodes of dictatorial rule. According to one source 93 percent of the present population over age 10 is literate—the highest proportion in Central America.

Panama and Belize were not part of the Central American Federation, and they achieved independence much later. Panama was governed from Bogotá as a part of Colombia until 1903, when it broke away as an independent republic. The United States was deeply involved in the political maneuvering that led to Panamanian independence and in 1904

was granted control over the Panama Canal Zone through which the famous transoceanic Canal was opened in 1914. In 1978 after long agitation in Panama for return of the Canal Zone and control over the Canal, the United States Senate ratified a treaty under which Panama would receive full control by stages ending in 1999.

Belize on the Caribbean side of Central America was held by Great Britain during the colonial age under the name of British Honduras. Remote, poor, undeveloped, and nearly uninhabited, the colony was used largely as a source of timber, with African slaves brought in to do the woodcutting. Not until 1981 did Belize become independent. It is now a parliamentary state governed by a prime minister and bicameral legislature; the British monarch, represented by a governor-general, is the chief of state. English is the official language. Great Britain maintains a small military garrison in Belize to guarantee the country's security. Neighboring Guatemala claims Belize and does not recognize it as an independent state.

Several Central American states are among the poorest in Latin America. Large areas in the region have little economic enterprise above a meager subsistence. Mineral wealth is generally lacking, although Guatemala has some oil production, and there is scattered mining of metals in all countries, including silver mining at a few sites that have been operational since early colonial times. Commercial agriculture producing coffee, bananas, cotton, sugarcane, beef, and a miscellany of other export commodities is very patchy in distribution and not very impressive in overall output, although it does provide the greater part of Central American exports (Table 24.3). All the countries except overcrowded El Salvador still have pioneer zones where new agricultural settlement is taking place. There is some manufacturing in the few large cities, but its products are limited largely to simple consumer goods such as textiles, clothing, shoes, wood products, and processed foods. Some assembly of foreign-made cars is carried on. But such bits and pieces of productive enterprise yield scanty returns in relation to the needs of the impoverished population, and the situation is worsened by the fact that a large share of the profit makes its way into very few pockets.

Within Central America as a whole, the distribution of settlement and economic life bears a strong relationship to three parallel physical belts: the *volcanic highlands,* the *Caribbean lowlands,* and the *Pacific lowlands.* These belts will be briefly assessed, with reference to individual activities and countries as appropriate.

Volcanic Highlands

All the Central American countries except Belize have a share of the volcanic highland belt that reaches southward from Mexico and generally lies inland a relatively short distance from the Pacific. Here are found the core regions, including the political capitals, of Guatemala, El Salvador, Honduras, Nicaragua, and Costa Rica. The only city with over a million people in its metropolitan area (mid-1980s) is Guatemala City (1.3 million); the other capital-city metropolises of San Salvador, Tegucigalpa, Managua, and San José range between 500,000 and 1 million. Dotted with majestic volcanic cones and scenic lakes, the highlands are the most densely settled large section of Central America. By far the greatest population density of any Central American country is found in El Salvador (Table 24.1), which is entirely a highland country except for a narrow fringe of Pacific lowlands and does not share the thinly settled Caribbean lowlands that reduce overall densities in the other countries.

Lower in average elevation than the highlands of Mexico, these Central American highlands are classed largely as *tierra templada* and here most of Central America's coffee is grown. Coffee culture was introduced at various times in the different countries, commencing with Costa Rica at the end of the 18th century. The crop is grown on farms of all sizes and most of it is exported, principally to the United States. Various other export crops are grown on a much smaller scale. Corn, rice, and beans are the major subsistence crops. Within the highlands increasing population pressure has resulted in deforestation to the point that only a minor fraction of the land is still forested. The remaining forests consist mainly of broadleaved tropical hardwoods such as mahogany and rosewood, although some stands of pine exist at higher elevations. Recently there has been a considerable development of hydroelectricity along highland rivers, and some geothermal power also is utilized to generate electricity. Such resources are important to these poor countries where a modest industrialization has begun and fossil fuels to power it are in short supply.

Caribbean Lowlands

Every Central American country except El Salvador includes a portion of the hot and rainy Caribbean lowlands that stretch unbroken from Colombia to Mexico. This long coastal strip of tropical rain forest still contains large wooded areas and is quite thinly and irregularly settled, although before the European conquest it supported in Guatemala and Belize the same advanced Mayan civilization that left such impressive stone structures in Mexico's Yucatán Peninsula. Throughout Spanish colonial times and far into the postcolonial era, economic activity in these disease-infested lowlands was confined to subsistence hunting, fishing, gathering, and shifting cultivation, plus an occasional bit of commercial activity such as woodcutting or the growing of sugarcane on scattered plantations. Some coastal settlements were established as bases for British pirate ships preying on Spanish commerce. A certain number of black slaves were brought in, and they provided the initial basis for Central America's relatively small black, mulatto, and zambo populations. In the latter 19th century some black workers migrated here from Jamaica to work on the banana plantations being developed by American fruit companies to serve the American market. At the turn of the century several companies became consolidated into the United Fruit Company (now part of United Brands), which subsequently became a symbol of American penetration into the Central American "banana republics." An up-and-down history of banana growing saw many plantations eventually shift to the Pacific side of Central America because of devastating plant diseases on the Caribbean side. Meanwhile the American banana interests were encountering various political and labor troubles, including some expropriation of their properties by governments. Some plantations provided living conditions for workers that were far superior to those generally available in Central America, but this was insufficient to overcome the resentments generated by a long history of intrigue and undeniable exploitation on the part of the banana companies. As recently as the mid-1980s, United Brands withdrew from banana growing in Costa Rica because of a long and costly strike. Today banana exports, grown in part on the Caribbean lowlands and in part on the Pacific lowlands, continue to be very important in the economies of Honduras, Panama, and Costa Rica (Table 24.3).

The Caribbean lowlands represent the principal area in Central America where agricultural settlement is expanding on a pioneer basis. The settlers generally come from overcrowded highlands and are often aided by government colonization schemes featuring grants of land and the building of roads to get products to market and bring in supplies. A major area of future settlement is northern Guatemala, where a large area of hot wet lowland has been nearly empty since the time when the Mayan civilization flourished there.

Pacific Lowlands

The relatively narrow Pacific lowlands, classed climatically as tropical savanna, are shared by all Central American countries except Belize. They were long the habitat of sparse and unproductive cattle ranching, but recently this pastoral pursuit has taken on a new dynamism emphasizing better breeds and pastures. Beef destined for market in the United States is loaded into refrigerated trailers, which then travel by all-weather highway to a Guatemalan port on the Caribbean. From here the trailers are transferred by seagoing ferry ships to Miami. This type of transportation offers numerous possibilities for the future shipment of perishable produce from Central America to United States markets. In addition to their beef output, the Pacific lowlands grow irrigated cotton for the Japanese market.

Various small fishing ports exist along the Pacific lowlands, a fact also true of the Caribbean lowlands. Shrimp are the most valuable product. There is no seaport of much importance on the Pacific side of Central America except Panama's capital, Panama City (650,000). It may be noted that Panama derives considerable income from the registry of vessels desiring to sail under the Panamanian flag in order to avoid the stricter maritime regulations of such countries as Great Britain and the United States.

The Panama Canal

The Panama Canal is one of the two major manmade waterways that have caused a gross reorientation of the world's ocean shipping lanes. Like its counterpart, the Suez Canal, its economic importance has lain chiefly in the intensity of ocean traffic that it has attracted by providing a passage through a narrow land bridge connecting two continents. The importance of the Canal to Panama could hardly be overstated. About half of the national population lives close to the Canal, and an overwhelmingly large share of the national income is generated in this central core area stretching for about 40 miles (64 km) across the narrowest part of the Isthmus of Panama from Panama City on the Pacific to Colón on the Atlantic. Although sovereignty over the Canal and the Canal Zone is being transferred to Panama, the American military presence is still very much in evidence. Because of increasingly heavy use of the waterway and the fact that many modern vessels are too large to transit the Canal, a need is growing for a new sea-level canal to be dug (possibly across Nicaragua) and/or the present canal deepened, widened, and otherwise improved. The presence of the Canal, combined with the security of United

States military protection, has stimulated many international banking and other business activities in Panama. It is fortunate that such opportunities exist, for Panama is not blessed with resources other than its geographic position. The eastern section of the country adjacent to Colombia contains a roadless stretch where the continuity of the Pan American Highway System is broken (Figs. 25.1 and 25.3). Thick rain forests, mountains, and swamps in this border area present severe engineering obstacles to future bridging of this last remaining gap in a surfaced highway linkage from Canada and the United States to Argentina, Chile, and southern Brazil.

THE HETEROGENEOUS CARIBBEAN ISLANDS

The islands of Caribbean America—commonly known as the West Indies—are very diverse. They encompass a wide range of sizes and physical types, great racial and cultural variations, and a notable variety of political arrangements and economic mainstays.

Physical Character: Rainy Tropical Lowlands, Steep Mountains, Volcanic Cones

Cuba, the largest West Indian island, is comprised primarily of lowlands with low to moderate relief—the principal exception being a mountainous area at the east end of the island. The islands that rank next in size—Hispaniola, Jamaica, and Puerto Rico—are steeply mountainous or hilly. Trinidad, a detached fragment of the South American continent, has a low mountain range in the north (a continuation of the Andes) and level to hilly areas elsewhere. The small islands of Aruba, Curaçao, and Bonaire, affiliated with the Netherlands, also lie near the mainland. They have low to moderate elevations and slopes. Most of the remaining islands, all comparatively small, fall into two broad physical types: (1) Low, flat limestone islands rimmed by coral reefs include the Bahamas (northeast of Cuba) and a few others; (2) most islands in the Virgin Islands, Leeward Islands, and Windward Islands, stretching along the eastern margin of the Caribbean from Puerto Rico toward Trinidad (see Fig. 25.1), are of volcanic origin. Each consists of one or more volcanic cones (most of which are extinct or inactive), with limited amounts of cultivable land on lower slopes and small plains. Some cones are thousands of feet high, while others have been greatly worn down by erosive processes. Intermixed with the volcanic islands are a small number of low limestone islands. Barbados, located east of the Windward Islands, is a limestone island with moderately elevated, rolling surfaces.

All the islands are warm throughout the year, although extremely hot weather is uncommon. The difference between the average temperatures of the warmest and coolest months in lowlands ranges from about 4°F (2.2°C) in Barbados to about 10°F (5.6°C) in northern Cuba. Precipitation, however, varies notably from island to island and from one section of mountainous islands to another. Windward slopes of mountains often receive very heavy precipitation, amounting in some instances to more than 200 inches (c. 500 cm) annually, whereas leeward slopes and very low islands may have rainfall so scanty as to create semiarid conditions. In some areas a moisture deficiency is created or abetted by porous limestone bedrock into which precipitation rapidly disappears. The natural vegetation of the islands varies from luxuriant forests in areas of abundant moisture to a sparse and scrubby woodland in very dry areas. Most areas experience two rainy seasons and two dry seasons a year. Hurricanes, approaching from the east and then curving northward along rather well-defined tracks, are a scourge of the northern islands in the late summer and autumn.

Racial, Cultural, and Political Complexity

The majority—in most islands, a very large majority—of the islanders are black or mulatto. People of relatively unmixed European descent comprise most of the remainder, aside from a considerable Asian element (mainly Hindu or Muslim Indians) in Trinidad. The cultural heritage of the different islands varies greatly and includes Spanish, British, French, Dutch, American, Danish, African, and (in Trinidad) Hindu and Muslim influences in practically endless combinations. Political arrangements also are diversified. Thirteen units are sovereign nations (five republics and eight parliamentary states); the others exhibit varying degrees of dependence on overseas nations and are tied to them under a variety of governmental forms. The republics are Cuba, Haiti, the Dominican Republic, Dominica, and Trinidad and Tobago; the independent parliamentary states are Jamaica, Barbados, the Bahamas, Grenada, St. Christopher (Kitts) and Nevis, St. Lucia, St. Vincent and the Grenadines, and Antigua and Barbuda. All the parliamentary states are former British colonies that continue to give formal allegiance to the British crown. Puerto Rico—a Commonwealth freely associated with the United States—and the French overseas departments of Guadeloupe and Martinique are the largest units that are not sovereign nations.

Economic Activities: Plantations, Small Farms, Tourism, Limited Mining and Manufacturing

For the islands as a whole and for most individual islands, agriculture is by far the leading source of livelihood. A large share of the product is for subsistence or sale in local markets. Prominent subsistence crops include corn, manioc, yams, rice, and bananas. But single-crop plantation agriculture is the economic activity for which these islands are best known, and it is still the commercial mainstay of some. Sugarcane, the crop around which the island economies were originally built, is still the most important export crop within the islands as a whole (Fig. 24.6). Other prominent commercial crops include coffee, bananas, tobacco, cacao, spices, citrus fruits, and coconuts. Although most of the value of commercial production for the islands as a whole has up to the present been accounted for by large plantations or estates worked by tenant farmers or hired laborers, some crops—such as the coffee of Haiti—are produced primarily by small landowning farmers. In Cuba agriculture was collectivized following the Communist revolution of 1959. On most islands mineral production is absent or unimportant, the most conspicuous exceptions being Jamaica (bauxite; Fig. 24.7), Trinidad (petroleum), and Cuba (metals). Manufacturing has made sizable gains in Puerto Rico and Cuba during recent times, but in other islands industrial progress has generally been slow or absent. Tourism is a major source of revenue in some islands, most notably the Bahamas. Most tourists come from the United States.

Emigration has long been an active option for this region's inhabitants, most of whom are economically underprivileged. Millions of West Indians now live in the United States, Great Britain, Canada, France, or the Netherlands—mainly in the cities. Money they send back to relatives has become a major source of island income.

The Greater Antilles

The four largest Caribbean islands are often called the *Greater Antilles;* the smaller islands are the *Lesser Antilles.* Of the five present political units in the Greater Antilles, Cuba, the Dominican Republic, and Puerto Rico were Spanish colonies, Haiti was French, and Jamaica was British. This diversity of colonial origins is only one of many ways in which these units have distinct geographic personalities.

CUBA: COMMUNIST DEVELOPMENT IN A HISTORIC SUGAR ISLAND

By any standard—size, resources, economic importance, or world influence—Cuba is the most important Caribbean island unit today. Centrality of location, large size, a favorable environment for growing sugarcane, and some useful mineral resources have been important factors contributing to Cuba's leading role in the Caribbean.

Cuba lies only 90 miles (145 km) across the Strait of Florida from the United States and about the same distance from Mexico's Yucatán Peninsula. The western end of the island, where Havana developed on a fine harbor as a major Spanish stronghold in the Caribbean, lies squarely between the two entrances to the Gulf of Mexico from the Caribbean Sea and the Atlantic. The long plow-shaped island would stretch from New York to Chicago if superimposed on the United States, and the island is not far from being as large in area as all the other Caribbean islands combined. Gentle relief, adequate rainfall, generally even distribution of rainfall, warm temperatures, and fertile soils create very favorable agricultural conditions. Agriculture has always been the core of the Cuban economy, and sugarcane has been the dominant crop. Grown on Spanish colonial plantations for centuries, the crop was expanded by American investments after Cuba gained independence in 1898. Prior to the Castro revolution in 1959, approximately one fifth of the world's sugar was produced each year, mostly for export. On the eve of the revolution, the United States was taking four fifths of Cuba's exports (mostly sugar) and was providing three fourths of its imports (mostly manufactured goods and food). The immediate aftermath of the revolution was marked by a period of experimentation in many aspects of Cuba's economy, but particularly in agriculture. By 1963, the output of sugar had fallen to less than half of the 1959 figure. Land formerly in cane was converted to varied subsistence and specialty crops such as rice, tomatoes, coffee, tobacco, and potatoes. Such a drastic conversion is not made easily, however, and crop yields in many instances were not satisfactory. Meanwhile, the world market price of sugar rose rapidly for a time, and Cuban planners decided to reinvest in sugar. Since then, Cuba has continued to emphasize sugar and also has continued growing in lesser amounts the diversified crops just mentioned. Because of the revolution and a 1961 trade embargo imposed by the United States, over four fifths of Cuba's foreign trade is now with the Soviet Union and other Communist countries. Cane sugar still accounted for about three fourths of Cuban exports by value in the mid-1980s. The 1959 revolution shifted the ownership of most Cuban land to the state. Today about 90 percent of all farm land is under state control, and practically all sugarcane is grown on large state farms.

Tobacco and metalliferous ores (primarily nickel and copper) account for most of Cuba's exports other than sugar and molasses. All mineral resources were nationalized in 1960. Mining is largely confined to two mountainous areas, one in the far south of the island and the other in the extreme northwest. Manufacturing centers on industries that process food and tobacco and that manufacture cotton and synthetic textiles, cement, chemical fertilizers, and automobile tires. Havana (Habana: 2 million) is Cuba's main industrial center, as well as its political capital, largest city, and main seaport.

Cuba's government, aided in a massive way by the Soviet Union, has had some success in raising the adequacy of food supply, medical care, housing, and education. As a part of Cuban foreign policy, civilian brigades of doctors, teachers, agronomists, construction workers, and others have been sent to various developing countries in Latin America and Africa. Cuban military intervention in war-torn African countries such as Angola and Ethiopia has drawn repeated censure by the United States. In 1963 a great international crisis erupted when the Soviet Union attempted to place nuclear missiles in Cuba, capable of striking Washington, New York, and other cities of the eastern United States. Although the missiles were withdrawn under pressure from the Kennedy administration, Cuba has continued to receive Soviet arms and has remained a storm center of political life in the Western Hemisphere.

HAITI AND THE DOMINICAN REPUBLIC: CONTRASTING AND UNEASY NEIGHBORS

The mountainous island of Hispaniola is shared uneasily by the French-speaking Republic of Haiti and the Spanish-speaking Dominican Republic. Black population forms an overwhelming majority in the Republic of Haiti, although a small mulatto elite (perhaps 1% of the population) holds a major share of the country's wealth. Population density is extreme, urbanization is scanty, and poverty is rampant. The present-day agricultural economy, largely of a subsistence character, reflects little influence from the former French colonial regime which was driven out by an extremely bloody revolt at the beginning of the 19th century. The republic, independent since 1804, has a very strong West African flavor in its landscape, crops, and other cultural expressions. Deforestation, erosion, and poor transportation are major problems. Aside from subsistence farming, Haiti's small economy concentrates on light manufacturing (aided by low-wage labor), coffee growing, and tourism. Many Haitians emigrate to other Caribbean islands or the United States (where 600,000 were estimated to reside in 1986) or filter across the border with the Dominican Republic.

The latter country, although containing some blacks and a large majority with some African blood, bears chiefly a Spanish cultural heritage. Occupying the east and center of the island, the Dominican Republic has an area nearly twice that of Haiti, but its population is somewhat smaller than that of its western neighbor. An independent country since 1844, the Dominican Republic has a history of internal war, foreign intervention (including periods of occupation by U.S. military forces), and misrule, though recent peaceful transfers of power point toward greater governmental stability. The economy is primarily agricultural with export sugar as the mainstay, but there is growing diversity based on tourism, mining, oil refining, and agricultural products such as meat, coffee, tobacco, and cacao. Urbanization is much more pronounced in the Dominican Republic than in Haiti; in each, the capital is the largest city, but metropolitan Santo Domingo (1.5 million) is considerably larger than Haiti's capital of Port au Prince (900,000).

THE PUSH FOR DEVELOPMENT IN PUERTO RICO

Puerto Rico, the smallest and easternmost of the Greater Antilles, is a dominantly mountainous island with coastal strips of lowland on all sides. The highest elevations—around 4000 feet (c. 1200 m)—are far lower than those of Hispaniola (9000–10,000 ft/c. 2700–3000 m). Puerto Rico was ceded to the United States by Spain after the latter nation's defeat in the Spanish-American War of 1898 and was administered by Congress as a Federal Territory until 1952. In that year, Puerto Rico became a self-governing Commonwealth voluntarily associated with the United States. The pre-World War II economy was agriculturally based, with a strong emphasis on sugar and accompanying wide disparities within the populace in income and living conditions. After the war a determined effort by Puerto Ricans to raise their living standard brought great changes. Hydroelectric resources were harnessed, a land-classification program provided a basis for sound agricultural planning, and a thriving development of manufac-

turing industries and tourism began, financed in large measure by capital attracted from mainland United States. Today, agriculture has been far outstripped in value of product by manufacturing. Petrochemicals, pharmaceuticals, electrical equipment, food products, clothing, and textiles are among the major lines. Sizable copper and nickel deposits have recently been found. Although Puerto Ricans have chosen thus far to remain a Commonwealth, there is strong pressure within the island for statehood. A small minority have expressed a desire for total independence.

JAMAICA: BAUXITE, TOURISM, DIVERSIFYING AGRICULTURE AND INDUSTRY

Jamaica, which has been independent from Great Britain since 1962, is located in the Caribbean Sea about 100 miles (c. 160 km) west of Haiti and 90 miles (c. 145 km) south of Cuba. The island is about equidistant from South America, Central America, and the Florida Peninsula. Its population of 2.5 million (1988) is largely descended from black African slaves transported there by the Spanish and (after 1655) by the British. When slavery was abolished in 1838, most blacks left the sugar plantations on the coastal lowlands and migrated to the mountainous interior, where the Blue Mountains reach 7000 ft. (c. 2100 m) in altitude.

Bauxite, tourism, and agriculture constitute Jamaica's economic base (Figs. 24.6 and 24.7). Since the production and export of aluminum-bearing bauxite ore and alumina began in 1952, Jamaica has been one of the world leaders in output and export. The island has also been an important focus of the West Indian tourist trade for many years, basing its tourism on distinctive cultural and historical attractions, a pleasant climate, scenic seaside and mountain landscapes, many miles of beaches, and extensive development of resorts. Agricultural production, long centered on sugarcane, has diversified into fruits, vegetables, rice, flowers, and cattle. Today only a handful of sugar plantations remain. Illegal production and export of marijuana from small mountain plots have been common. Labor-oriented light industries such as textiles, frozen foods, and electronics are being strongly pushed. The Jamaican government has tried to foster economic diversity in order to offset a falling world demand for bauxite and sugar. But unemployment and poverty remain chronic problems. The capital city of Kingston (550,000) had a notorious reputation in colonial times as a base for piracy. It lies on the southeastern coast, whereas the main tourist developments are in the north and northwest.

The Eastern Caribbean Islands: Common Threads but Differing Circumstances

Over 3 million people live on the eastern Caribbean islands scattered in a semicircular 1400-mile (2250-km) arc from the U.S. and British Virgin Islands east of Puerto Rico to the vicinity of Venezuela. This screen of small islands defining the eastern edge of the Caribbean Sea contains a variety of political entities. The Netherlands Antilles, still governed as a Dutch dependency, includes Curaçao and four other small areas (the island of Aruba recently separated from the unit in preparation for early independence). Guadeloupe and Martinique are French overseas departments, while Anguilla and Montserrat are still British dependencies. The British, however, have been withdrawing from Caribbean America in a massive way, starting with independence for Jamaica and Trinidad-Tobago in 1962. Since then, independence from British rule in the Lesser Antilles has been gained by Barbados (1966), Grenada (1974), Dominica (1978), St. Lucia (1979), St. Vincent and the Grenadines (1979), Antigua and Barbuda (1981), and St. Christopher and Nevis (1983). The British also granted independence to the Bahamas in 1973 and to the mainland unit of Belize in 1981 but still retain their colonies of the British Virgin Islands, Cayman Islands, and Turks and Caicos Islands. All their former colonies have retained membership in the Commonwealth of Nations.

The economies of the eastern Caribbean islands suffer from many serious problems, including unstable agricultural exports, limited natural resources, a constantly unfavorable trade balance, embryonic industries lacking economies of scale, and chronic unemployment. A shared trait of the islands is their history of European colonialism and plantation slavery. Such generalizations contribute to understanding but give little hint of the individuality these small units present when viewed in detail. A few examples are sketched below.

Barbados is the easternmost of the West Indies. Occupied continuously by the British since 1625, it was the first British-held island to experience the sugar revolution of the 17th century. Some 90 percent of all cultivated land is still used for sugarcane. Virtually all cultivable land is utilized—Barbados being the most densely populated country in the Americas. Yet political and economic stability are being maintained. At the time of independence, this most traditional and British of the former colonies already had a sense of nationhood due to such factors as relative isolation, compact territory, excellent communications, practically universal literacy, and strong social institutions such as the inherited system of parliamentary democracy. Barbados is noted within the Caribbean for its tolerant, well-informed, and educated population and its cosmopolitan atmosphere. These traits have aided the island in developing tourism as a counterpoise to a declining world market for sugar.

Of the island units in the eastern Caribbean, *Trinidad and Tobago* is the largest in area and population, the richest in natural resources, and the most heterogeneous in ethnic composition. Trinidad, by far the larger of the two islands, is a geologic extension of mainland Venezuela, and as such it benefits from offshore oil deposits in the south. Yet Trinidad refines more oil than it pumps. Imported crude oil from various sources is refined and then shipped out, mainly to the United States. Some 41 percent of the population is comprised of blacks, who live mainly in the urban areas and are often employed in the oil industry. Another 41 percent are the descendants of immigrants from the Indian subcontinent. This element lives primarily in the intensively cultivated countryside. About 16 percent are classified as "mixed," and there are small numbers of Chinese, Madeirans, Syrians, European Jews, Venezuelans, and others. Immigration from other West Indian islands is common. The ethnic complexity of Trinidad is reflected in religion, architecture, language, diet, social classes, dress, and politics. Tobago, some 22 miles (34 km) to the north, is more like the neighboring Windward Islands in character.

The islands of *Aruba* and *Curaçao* refine imported oil from Venezuela and the Middle East. Almost all food requirements are imported into these racially and culturally pluralistic islands. The local language is derived from a Spanish and Portuguese base, which is mixed with elements of Dutch, English, and African tongues. Tourism and manufacturing are being developed. Willemstad (130,000) on Curaçao has a distinctive blend of Dutch and Caribbean traits. Trinidad, Aruba and Curaçao, the United States Virgin Islands, and the Bahamas are the major centers of oil refining in the Caribbean islands.

Dominica illustrates the precarious situation of the smaller Caribbean islands. Soon after independence, hurricanes destroyed nearly all the island's banana plants and coconut trees, plus a large part of the fishing fleet. The coastal areas are the only settled parts of this extremely rugged island. The volcanic interior has elevations that reach almost 5000 feet (1524 m). About 2000 Carib Indians, descendants of the aboriginal inhabitants, live in a preserve—the only one of its kind in the Caribbean. Poor transportation hinders economic development. Settled late, this island is not strongly marked by a heritage of plantation agriculture and slavery. A French patois is widely spoken, although English is the official language.

The eastern Caribbean island of *Grenada* illustrates further complexities of the Caribbean microstates. This island's recent political history has been turbulent. Independence from British rule in 1974 was followed by a Marxist coup in 1979 that unseated a corrupt prime minister. Further internal disruption in 1983 resulted in a military invasion by the United States, with some American forces lingering in the island until 1985. Later in that year Queen Elizabeth II visited Grenada and publicly applauded its return to democracy from Marxist authoritarianism. These episodes, involving American, British, Soviet, and Cuban interests as well as the interests of the Grenadians themselves, reflect the long-term tendency of Caribbean islands to become involved in the geopolitical strategies of world powers.

The Tourist-Dependent Bahamas

The *Commonwealth of the Bahamas* stretches about 550 miles (885 km) from the island of Grand Bahama about 55 miles (c. 90 km) off the Florida coast, to Great Inagua, some 50 miles (80 km) off Cuba. The Bahamas are sometimes considered to be a separate group not included technically in the West Indies. Of the nearly 700 islands, only 22 are inhabited. The majority of the population (245,000 in 1987; 72% black, 13% white, 14% mixed) lives on New Providence Island where Nassau, the capital city (145,000), is located. After World War II, tourism grew so rapidly that today it is directly or indirectly responsible for four fifths of the GNP. Around 2 million tourists a year visit the Bahamas, mainly from the United States. Banking and business activities are a distant second to tourism as revenue producers. Small-scale farming and fishing help keep food imports low, and some income is derived from the refining of African and Middle Eastern crude oil. The refined products are exported to the United States.

GUYANA, SURINAME, AND FRENCH GUIANA

A kind of mainland outlier of the West Indies is formed by three political units—Guyana (former British Guiana), Suriname (former Dutch Guiana), and French Guiana—that front on the Atlantic between Venezuela and Brazil. Although not techni-

cally bordering the Caribbean Sea, they lie relatively close by and their geographies and histories fit easily into a Caribbean context. Here three non-Iberian European nations—Great Britain, the Netherlands, and France—established their only long-term colonies on the South American mainland. British Guiana gained independence in 1966 as Guyana, now a leftist republic and member of the Commonwealth of Nations. Dutch Guiana became independent as Suriname in 1975; it also is a republic, but its military government, brought to power by a coup in 1980, became estranged from the Netherlands and the latter country withdrew its financial support. French Guiana remains an overseas department of France. Guyana's population was about 800,000 in 1987, while Suriname had about 415,000 and French Guiana 90,000. About nine tenths of the people in the combined units live in a narrow strip of hot, rainy coastal lowlands where nearly all agriculture is carried on. The remaining tenth live among the heavily forested ranges of the rough Guiana Highlands in the interior or in lowland savannas of southern Guyana. Suriname has a very heavy export dependence on alumina (55% of all exports by value in 1984), aluminum (12%), and bauxite (13%), while the main exports of Guyana are divided between bauxite (44%) and sugar (28%). Sugar growing hugs the coast quite closely, whereas bauxite is a mined product from the interior. The three units share a heritage of plantation slavery and export sugar growing; in addition French Guiana, which has never had much economic output of any kind, was formerly a notorious French penal colony. There is considerable ethnic diversity, with prominent elements derived from West Africa, the Indian subcontinent, or Indonesia. No large cities exist. Guyana's capital of Georgetown, the largest urban place, has an estimated metropolitan population of perhaps 200,000.

THE ANDEAN COUNTRIES: HIGHLAND AND LOWLAND CONTRASTS IN SOUTH AMERICA'S INDIAN CORELAND

The South American countries of Colombia (estimated population 31 million in 1988), Venezuela (19 million), Ecuador (10 million), Peru (21 million), and Bolivia (7 million), all within the tropics and traversed by the Andes Mountains, are grouped in this text as the Andean countries. The five countries share many traits that are characteristic of Latin

America but often are present here with special intensity. A few outstanding traits are summarized below.

1. Environmental zonation, both vertical and horizontal, is very pronounced. Hence the natural setting of each country features innumerable local environments that offer extreme contrasts and often are very isolated from each other by natural barriers.

2. In conformity with environmental fragmentation, patterns of settlement and economic development also are fragmented—an especially notable characteristic of Peru.

3. Every country contains both populous highlands and an extensive pioneer fringe of lightly populated interior lowlands or low uplands awaiting development. Some of these interior areas already yield minerals (notably oil) on an export basis.

4. Every country except Bolivia has coastal lowlands with a dynamic pattern of present activity. Bolivia originally had a Pacific coastal zone but lost it to Chile during a 19th-century war, thus becoming a landlocked state—the only one in Latin America except adjacent Paraguay.

5. All countries had relatively large Indian populations in the mountains at the time of the 16th-century Spanish conquest, and the old Indian areas are still major zones of dense population and strong political influence today. In four countries—Colombia, Venezuela, Ecuador, and Bolivia—the Andean zones contain the national capitals.

6. Each country maintains strong Hispanic traditions but at the same time has an important heritage from Andean Indians. The Spanish language is official in all five countries (although the Quechua and Aymara Indian languages are also official in Bolivia, as is Quechua in Peru), and Roman Catholicism embraces over nine tenths of all religious adherents in every country. These badges of Iberian culture introduced by the Spanish conquerors were superimposed on advanced Indian cultures within the Inca Empire that the Spanish overcame. Today Bolivia, Peru, and Ecuador have Indian majorities in their populations, and cultural survivals from pre-Spanish times contribute to a keen sense of Indian identity in each country. In all the rest of Latin America, only Guatemala has an Indian majority, although Indian strains are very pronounced in many countries with important mestizo elements. The latter is true of Colombia and Venezuela, in each of which the Indian population is only 1 or 2 percent of the total but mestizos form the largest ethnic element (69% in Venezuela; 58% in Colombia).

7. All countries are relatively poor, although oil wealth makes Venezuela considerably better off than the others.

8. In every country except Ecuador the national capital towers above all other cities in size and importance.

The foregoing common themes recur constantly in the distinctive geographies of the individual countries sketched in the following sections. We commence with Colombia, which is considerably the largest of the five countries in population and is the most diversified in economic production.

COLOMBIA: LARGE RESOURCES AND PROBLEMS IN A DIFFICULT TERRAIN

Within its wide expanse of the Andes, Colombia includes a great many populous valleys and basins at varying elevations. Some are in the *tierra fría* and others in the *tierra templada*. The great majority of Colombia's mestizos and whites are scattered among these upland settlement clusters. The latter tend to be separated by stretches of sparsely settled mountain country with extremely difficult terrain. Lower valleys and basins, along with some coastal districts, developed a plantation agriculture employing slave labor on a sizable scale. Consequently, 14 percent of the country's present population is mulatto and 4 percent is black.

The Highland Metropolises: Bogotá, Medellín, Cali

Colombia's towns and cities have tended to grow as local centers of productive mountain basins, and there are now a number of mountain cities with over 100,000 people. The three largest—Bogotá, Medellín, and Cali—are not only outstanding in size, but diverse in geographic character. The largest Andean center in all of South America is *Bogotá*, Colombia's capital, which has over 4½ million people in its metropolitan area. The city is sited on an alluvial fan at the edge of a sizable mountain basin in the easternmost of three Andean ranges at this latitude in Colombia. The elevation of about 8700 feet (c. 2650 m: much higher than Denver, Colorado) places Bogotá in the *tierra fría* climatic zone. Here and in nearby basins, the Spanish conquerors subjugated a dense population of Chibcha Indians. The Chibchas were supporting themselves mainly by growing corn and potatoes. The Spaniards seized the land, put the Indians to work on Spanish estates, and improved agricultural productivity by introducing new crops and animals—notably wheat, barley, cattle, sheep, and horses. However, exploitation and disease reduced Indian numbers so drastically that they took a long time to recover. The Spaniards were not able to develop an export agriculture here. Bogotá's monthly average temperatures range from 57°F to 59°F (14°–15°C), with night coolness and frosts precluding such crops as sugarcane and coffee. In addition, early transport to the outside was extremely difficult, involving descent to the Magdalena River and difficult passage on and along that river to the north coast. Railroad connection with the Caribbean was only achieved in 1961. As a result, the Bogotá area was relatively isolated and self-sufficient, with easily transported emeralds and some gold its major exports to the outside world. Few blacks were introduced into the population. However, the growth of modern governmental functions and the provision of air, rail, and highway connections has led to explosive urban growth in recent decades. The basin has even found a large agricultural export in flowers grown for air shipment to the United States.

The other two large cities of the Colombian Andes illustrate conditions quite different from those of Bogotá. *Medellín* (2.2 million) lies in the *tierra templada* at about 5000 feet (1524 m) in a tributary valley above the Cauca River. The area lacked a dense Indian population to provide estate labor, was quite rugged and difficult to reach, and consequently remained for centuries a region of shifting subsistence cultivation emphasizing corn, beans, sugarcane, and bananas. The introduction of coffee and the provision of railroad connections changed this situation in the early 20th century. Excellent coffee could be produced under the cover of tall trees on the region's steep slopes, and the Medellín region became a major producer and exporter. Local enterprise and capital developed the city itself into a considerable textile-manufacturing center, and it grew from about 100,000 population in the 1920s to its present large size. Coffee is now Colombia's main official export, accounting for 59 percent of sales abroad in 1986. However, the Andean Indians traditionally chewed the leaves of the coca plant, and cocaine has become the leading, although unofficial, export of Colombia, Bolivia, and perhaps other Andean countries in the late 20th century.

Cali (1.5 million), located some 3000 feet (915 m) above sea level on a terrace overlooking the flood plain of the Cauca River, illustrates still a third set of conditions in the Colombian Andes. Compared to Bogotá and Medellín, this area of *tierra templada* had easier connections with the outside world via the Pacific port of Buenaventura. The Spanish developed sugarcane plantations in the Cali region and also produced tobacco, cacao, and beef cattle. Initial Indian labor was followed by that of African slaves. From the late 19th century onward, dams on mountain streams provided hydropower for industrial development. This asset, along with cheap labor, attracted foreign manufacturing firms, and Cali grew at a rate comparable to that of Medellín.

Coastal Colombia

The Caribbean and Pacific coastal areas of Colombia contrast sharply with each other in environment and development. The Pacific coastal strip has tropical rain forest climate and vegetation. One station in mountains somewhat inland from the shore reports over 400 inches (*c.* 1000 cm) of rainfall in the average year. In the north, mountains rise steeply from the sea, but there is a marked coastal plain farther south. The entire coastal area is thinly populated, with an unusually large proportion of blacks. The main settlement is Buenaventura, which is an important port but not a large city. Its principal advantage is that it can be reached from Cali via a low Andean pass at about 5000 feet.

The Caribbean coastal area is much more populous. Here three ports have long competed for the trade of the upland areas lying to the south on either side of the Magdalena and Cauca rivers. Barranquilla (975,000) and Cartagena (525,000) are the larger and more important in terms of legitimate trade. But the somewhat smaller city of Santa Marta, known as a beach resort as well as a port, may well be more important in trade today. It is reputed to be the control center for Colombia's huge illegal export trade in cocaine and marijuana, with many small unofficial airports on the peninsula to its east. Most of this Caribbean lowland section of Colombia is a large alluvial plain composed of materials washed down from the Andes. It is covered by an open-forest type of tropical savanna, with drier parts merging into steppe and wetter parts into rain forest. Agricultural development has been handicapped by the presence of large areas that are flooded either permanently or annually. Nevertheless this region has an economic history of ranching and banana and sugarcane production. The Colombian government is now fostering settlement by small farmers who grow corn and rice.

The Interior Colombian Lowlands

The very large part of Colombia comprised of interior lowlands east of the Andes forms a resource-rich but sparsely populated region that makes little contribution to the national economy as yet. It is a mixture of plains and hill country, with savanna grasslands in the north and rain forest in the south. The grasslands are used for ranching, and the forest for shifting cultivation. Considerable mineral wealth awaits exploitation.

Colombian Problems and Potentials

Colombia has the resources for notable economic development in the future. Energy resources include a large hydropower potential (second in Latin America only to that of Brazil); coal fields in the Andes, the Caribbean coastal area, and the interior lowlands; oil in the Magdalena Valley as well as in newly discovered and undeveloped fields in the interior; and large supplies of natural gas from the area near Lake Maracaibo. Iron ore from the Andes already supplies a steel mill near Bogotá, and large Andean reserves of nickel are under development. At least half of the country is forested, although most of this is made up of mixed hardwood species in the tropical rain forest. Due to the intermixture of many different kinds of trees, this type of growth has relatively low utility for forest industries other than the making of plywood. Of the foregoing resources only oil yet plays much of a part in the country's export economy. Coffee still comprises well over half the value of legal exports, while cocaine and marijuana are illegal agricultural products believed to represent a greater export value than all other exports combined.

There appear to be several major requirements for further economic development in Colombia. Solution of the country's transport difficulties is one such requirement. This is probably true of all developing countries, but the formidable Andes make Colombia's difficulties exceptionally great. Much has already been accomplished in developing air traffic and highways, but much remains to be done. Flight of capital from the country needs to be stemmed in favor of investment within Colombia. Again, this is a common problem in developing countries, but it is especially marked in a country whose largest generator of wealth is an illegal trade. Finally, internal violence has become a notable part of Colombia's national life, and the conflicts and inequities which breed it need to be resolved. For decades there has been guerrilla warfare aimed at overthrowing the country's government and revolutionizing its society. At its height between the late 1940s and the early 1960s this violence consumed many thousands of lives in the Cali-Cauca Valley region. It has since been less intense but has continued in various parts of the country. The guerrillas are Marxist groups of various persuasions, some inspired primarily by the USSR, and others by China or Cuba. In the face of these difficulties a reasonably democratic government has been maintained for over three decades, but it has not been able to substantially moderate the extreme social and economic inequities and the regional rivalries on which the conflict feeds.

VENEZUELA: SHIFTING AREAL FOCI OF HISTORICAL DEVELOPMENT

Venezuela's population is concentrated mainly in Andean highlands not far from the Caribbean coast, as it was when the Spanish arrived in the 1500s. The highlands contain the main city and capital, Caracas, which has a metropolitan population estimated at 3.3 million. Caracas fills a narrow valley 15 miles (24 km) long at about 3400 feet (c. 1000 m) elevation in the relatively low terminal branch of the Andes which parallels the country's northern coast. Elevation gives the city a *tierra templada* climate with monthly temperature averages ranging between 68°F and 73°F (c. 20°–23°C). Rainfall averages 34 inches (85 cm) annually, with a savanna (seasonally wet and dry) regime. The difference that elevation makes is shown by the climatic contrast with the city's port of La Guaira, located at sea level only 6 miles away. Here monthly average temperatures range between 78° and 83°F, and an average annual rainfall of only 11 inches (28 cm) results in desert conditions. The Spanish preferred the area around Caracas, where they found a relatively dense Indian population to draw on for forced labor, some gold to mine, and climatic conditions amenable to commercial production of sugarcane and later coffee. Labor needs associated with sugarcane led to the introduction of African slaves, as was the case in so much of Latin America, but Venezuelan plantation development was relatively small and only about a tenth of the present population is classified as black. Until the middle of the 20th century Caracas was a fairly small city, and its valley was still an important and highly productive area of farming. Now the city fills and overflows the valley, and Venezuelan agriculture is more associated with other highland valleys and basins, such as those around and near Valencia

(1.2 million) and Barquisimeto (725,000).

Oil resources of the coastal area around and under the large Caribbean inlet called Lake Maracaibo have made Venezuela a relatively fortunate Latin American country in recent decades. Here in a dry and hot savanna lowland the country produces about 4 percent of the world's oil, with crude oil and refined products supplying over nine tenths of Venezuelan exports. Maracaibo (1.3 million), located at the entrance to Lake Maracaibo, is the country's main lowland urban center. Oil revenues gave Venezuela a per capita income of $2930 in 1986. This was quite high for Latin America and well over twice the figure for any of the other Andean countries, although Venezuela is far from being a rich nation. The oil-based income has fueled rapid urbanization, attracted foreign immigration, and financed industrialization. It may well be largely responsible for Venezuela's recent political stability under democratic rule. However, in recent years the country's income has suffered because of the decline in oil prices. Such ills were one cause of serious riots in 1988.

Thus, the highlands have supported Venezuela in the past and the coastal lowland is crucial to the present. But the much larger share of the country that lies inland from the Andes is considered to be a major hope for the future. This sparsely inhabited section is mostly tropical savanna in the Guiana Highlands and in the lowlands (which are often swampy) along the Orinoco River and its tributaries. Tropical rain forest climate and vegetation mark the extreme south and areas near the Orinoco Delta. Some sections of the Guiana Highlands reach *tierra templada* elevations. For centuries remote interior Venezuela has been Indian country (in the highlands) or poor and very sparsely settled ranching country (in the lowlands). Now development is quickening. Government-aided agricultural settlement projects are beginning to occupy parts of the lowlands, and major oil and gas reserves are commencing to be exploited there. In the Guiana Highlands enormous iron-ore reserves south of Ciudad Bolívar and Ciudad Guayana began to be mined for export by American steel companies after World War II, while huge reserves of bauxite in the highlands await exploitation. Venezuela has nationalized its iron mining and has pushed the development of a major industrial center at Ciudad Guayana (city proper, perhaps 225,000). This has involved the development of (1) nearby hydroelectric supply from plants on an Orinoco tributary south of the city, (2) ocean freighter traffic on the Orinoco to export products and bring in foreign coal for industrial fuel, and (3) new transport connections from Ciudad

Guayana to other parts of Venezuela. Major steel, aluminum, oil-refining, and petrochemical industries are in operation, with further diversification planned.

ECUADOR: ANDEAN FASTNESS, DYNAMIC COAST, OIL-RICH AMAZON

In Ecuador the Andes form two roughly parallel north-south ranges. Crest elevations are very high, with some volcanic peaks reaching 15,000 to over 20,000 feet (*c.* 4600–6100 m). Between the two ranges are a series of intermountain basins floored with deep accumulations of volcanic ash. They lie mostly in the *tierra fría* at elevations between 7000 and 9500 feet (*c.* 2100–2900 m). Climatic conditions are represented reasonably well by Quito (1.1 million), Ecuador's capital and second largest city. Located at the equator, the city is high up on the rim of one of the northern basins. Here at an elevation of 9200 feet (2800 m) temperatures average 55°F (13°C) every month of the year, so that daily fluctuations between warm daytimes and cold nights are the principal temperature variations. Average annual rainfall is 49 inches (123 cm), which varies between two dry months and one which averages 7 inches (18 cm). Hence precipitation differentiates the seasons more than temperature does. The first Spanish invaders came from Peru in the 1530s. They found dense Indian populations in Ecuador's Andean basins. In their usual manner the conquerors appropriated both the land and the labor of the Indians and built Spanish colonial towns, often on the sites of previous Indian towns. But they found no great mineral wealth here, and both climate and isolation worked against the development of an export-oriented commercial agriculture. Even today these basins have an agriculture with a strong subsistence component and sales mainly in local markets. Potatoes, grains, dairy cattle, and sheep are the main crop and livestock emphases. The isolated Ecuadorian Andes attracted relatively few Spaniards, except to Quito, and did not develop a need for African slave labor. In consequence, the mountain population has remained predominantly pure Indian, although with some mestizo and white admixture.[3] Recently the isolation has decreased somewhat, and

---

[3]Another reputed reason for the small proportion of white population in the high Andes is that white women, lacking the genetic and physiologic adaptations of Indian women to the highland environment, suffered a devastating rate of miscarriages in such areas.

Quito and lesser Andean cities have seen the growth of a certain amount of industry.

The 20th century has been marked by a drastic shift of Ecuador's economy and population toward the coastal strip between the Pacific and the Andes. Through the 19th century something like nine tenths of the country's people still lived in the highlands. But during the 20th century, and especially since World War II, that proportion has been reduced to slightly under half. Topographically, the growth zone near the coast is partly plains and partly hills and low mountains. Its tropical lowland climate varies from rain forest in the north, through savanna along most of the lowland, to steppe in the south. The main focus of development has been Guayaquil (1.3 million), which began as a shallow-water port (using barges for ship-to-shore connections) but was provided with an artificial deep-water harbor in the 1960s. The port lies at the mouth of the Guayas River, which drains a large area of level land and alluvial soils north and east of the city. Bananas for export and rice for domestic consumption dominate the agriculture of this alluvial plain. Farther afield in the coastal zone, lesser agricultural areas also produce these crops, plus many others such as cacao and sugarcane. On adjoining slopes coffee is planted for export, and along the Pacific there is a considerable fishing industry. The population of this comparatively dynamic part of Ecuador is mainly mestizo and white.

In the 1960s major oil finds in the little-populated trans-Andean interior of Ecuador began to be exploited. The area lies under a tropical rain forest climate in the upper reaches of Amazon drainage. A pipeline across the Andes was constructed hundreds of miles to a small port on the Pacific, and Ecuador became an oil-exporting country. In 1986 oil and its products provided 45 percent of the country's (legal) foreign-exchange earnings. Other major mineral resources are not known to exist. Possible future mineral discoveries in the interior rain forest add urgency to a border dispute with Peru, which controls considerable territory of this type claimed by Ecuador. In 1981 a bout of armed conflict between the two countries occurred along their undefined boundary in the mountains of southeastern Ecuador.

PERU: FRAGMENTED DEVELOPMENT IN AN ENVIRONMENT OF EXTREMES

Peru is a poor nation with an extraordinarily fragmented distribution of population. The fragmentation has been induced by an environment that runs to extremes of ruggedness (in the Andes), dryness (along the coast) or wetness (in the trans-Andean Amazon lowland). The Peruvian Andes stretch across the country for approximately 1000 miles (c. 1600 km) between Ecuador and Bolivia. Over half of the population lives in scattered basins and valleys in the mountains. These highlanders are very predominantly pure Indian. Like the Andean Indians in other countries they are the descendants of an ancient civilization ruled in the last three centuries before Spanish times by the Inca Empire which originated around Cuzco (200,000) in Peru's southern Andes (Fig. 24.3). Indian predominance in Andean Peru gives the country a slight Indian majority. (A third of the population is classed as mestizo and an eighth as white.) But productivity in the Andean communities is so low that two thirds of Peru's national product is accounted for by metropolitan Lima (5.2 million), the coastal capital. Much of the rest comes from small oases and seaports scattered along the country's desert coast.

Andean Peru

The Peruvian Andes provide a notably poor environment for the development of surplus-producing agriculture. Here the ranges are considerably higher on the average than those of Ecuador or Colombia, with very large areas over 10,000 feet (c. 3050 m) and many peaks over 19,000 feet (c. 5800 m). Cultivable valleys and basins tend to be small and to have dense populations reflecting millennia of agricultural occupance. They are in the *tierra fría*, where cool temperatures have precluded the growing of such common export crops as coffee, sugarcane, or cacao and have largely limited the options of farmers to the raising of potatoes, grains, and livestock for home consumption and sale in local markets. Farmed areas in central and southern sections of the Peruvian Andes tend to be so dry despite their elevation that high productivity is limited to irrigated areas. Long before the Spanish invasion, Indian engineers were adept at constructing irrigation systems. The low productivity of present Indian communities in Peru's high Andes is reflected in the lack of large Andean cities. Surplus farm production is too scanty to give rise to sizable cities supporting themselves by agricultural marketing, processing, and services.

The Arid Coast

A narrow and discontinuous coastal lowland lies between the Peruvian Andes and the sea. Climatically it is a desert of remarkable aridity associated with cold ocean waters along the shore. Air moving landward becomes chilled over these waters, and over the land the air is stable. Onshore a relatively cool and heavy layer of surface air underlies warmer air. Hence the turbulence needed to generate precipitation is absent, although the surface air does produce heavy fogs and mists during the cooler part of the year, and it moderates temperatures somewhat. Lima, at 12°S, is representative of these conditions. Its monthly average temperatures range from 59°F (15°C) in the coolest month to only 72°F (22°C) in the warmest month, and only about 1 inch of precipitation is received in an average year.

Due to drier Andean conditions than occur farther north, plus the almost total aridity of the coastal strip, the great majority of rivers from the Andes carry so little water that they die out in the coastal zone before reaching the sea. But either at the coast or some way inland, irrigated agriculture has been practiced along these rivers, at least intermittently, for millennia. The Spanish founded Lima (1535) in an area of this type a few miles inland from a usable natural harbor. The city proved relatively well located for reaching the Indian communities and mineral deposits of the Andes, especially the Cerro de Pasco silver deposits which for a time led the world in silver production. Thus Lima was designated the colonial capital, and it became one of the main cities of Spanish America (Fig. 24.2). In the late 19th century, foreign companies became interested in the commercial agricultural possibilities of the Peruvian coastal oases. As a result, commercially oriented oasis enclaves now spot the coast like well-separated beads on a string. These irrigated areas produce varying combinations of cotton, rice, other grains, sugarcane, grapes, and olives. Small port cities and some inland centers have developed to serve the oases, whose further expansion is being pursued by the Peruvian government, with some foreign assistance. A major new oasis in the north is being developed by the Soviet Union. The small ports associated with the oases are generally fishing ports and also fish-processing centers, as the cold ocean along the coast is exceptionally rich in marine life. But this resource provides a very uncertain livelihood. For a few years in the 1960s and early 1970s Peru led the world in volume of fish caught (largely anchovies for processing into fishmeal and oil), and the port of Chimbote (city proper, 225,000) was the world's leading fishing port. But during the 1970s, a change in water temperature, plus overfishing, led to a drastic decline and widespread unemployment in the industry.

The Crucial Role of Minerals

Overall, the resources and performance of Peru's agriculture are so poor that the country must import a large part of its food. Consequently it remains dependent on mineral resources to supply the exports it needs to gain foreign exchange. Oil supplied a tenth of all export earnings in 1986. Some oil has been produced since 1863 from the northern coastal zone, but most of it now comes from a field developed during the 1970s in the remote Amazonian rain forest of northeastern Peru. Some is exported via pipeline across the Andes and some by river transportation across Brazil from the Amazon port of Iquitos (city proper, 200,000). However, the aggregate export value of metallic ores—mainly copper, zinc, lead, and silver—considerably exceeds that of oil. Metal-mining sites are divided between the coastal zone and the Andes, with the Andean sites being mainly in the central section near Lima. Iron ore from the southern coastal zone feeds Peruvian steel mills at Chimbote and Lima, and some ore is exported. There are also known reserves of gold, zinc, bismuth, coal, tungsten, phosphate, potash, and uranium, with some of these being mined in small amounts. But at their present state of development, Peruvian mineral resources can hardly pay the interest on Peru's large foreign debt, to say nothing of purchasing the imports of food and manufactures the country needs.

The Amazonian Pioneer Fringe

Over half of Peru is comprised of interior Amazonian plains with a tropical rain forest type of climate and vegetation. Fewer than 2 million people inhabit this large area, and its possible future importance to a country so pressed economically is obvious. Development will have to overcome year-round equatorial heat, extremely heavy rainfall, thick forest, leached soils, and isolation. For many years there has been a highway construction project to open the area to further settlement, but money has been short and progress slow. The main settlement in the area, Iquitos, originated in the natural-rubber boom of the Amazon Basin during the 19th century and has tended to be more oriented to outside markets via the Amazon than to Andean and Pacific Peru.

Political Handicaps to Peruvian Progress

Economic development in Peru has been seriously hampered by governmental instability. The country has gyrated wildly between civilian and military rule. Even the military governments recently in control have followed very divergent policies—sometimes leftist and sometimes rightist—with one government undoing the "reforms" of another. In 1980 a guerrilla insurrection began that purportedly took its inspiration from Communist China's Mao Zedong (Mao Tse-tung). By 1988 this challenge to national stability had resulted in over 5000 deaths. But despite the unrest, Peru succeeded in holding a democratic election in 1985, which brought about the first peaceful and democratic change of government in four decades.

LANDLOCKED BOLIVIA: ANDEAN CORE AND LOWLAND PIONEER FRINGE

Bolivia has extreme Andean conditions and is by far the poorest of the Andean countries. Like the other countries of this group it has a large and sparsely populated tropical lowland area east of the mountains. But unlike the other countries it has no coastal lowland to supplement or surpass the population and production of its mountain basins and valleys. Its original extension to the Pacific became northern Chile as the result of Bolivian defeat in a 19th-century war. Left largely to depend on the resources of its very high-altitude agricultural settlements and a few minerals, it has become one of the poorest states in Latin America, with a per capita production and income about half that of its poverty-stricken neighbor, Peru.

Agriculture still employs over two fifths of Bolivia's people and is limited by extreme *tierra fría* conditions combined with aridity. La Paz (1 million) is the main city and the *de facto* capital. Lying in a deep narrow valley at about 12,000 feet (*c.* 3700 m: more than twice the elevation of Denver, Colorado), La Paz averages 45°F (8°C) in its coolest month and only 53°F (12°C) in its warmest month. Aridity is a further problem. The city averages only 22 inches (56 cm) of precipitation per year, with 5 months essentially dry. A more rural cluster of people lives around the shores of Lake Titicaca on the Peruvian border. The surface of the lake is 12,507 feet (3812 m) above sea level, and the local climate is accordingly harsh. Smaller clusters of population are scattered through Andean Bolivia at similar elevations, mainly along the easternmost of the two main Andean ranges and the *Altiplano,* which is a very high depression between the two towering ranges. The next largest city in the highlands is Cochabamba, with only about 325,000 people. The small city of Sucre is the legal capital of the country, but the Supreme Court is the only branch of the government actually located there. Cochabamba and Sucre are both located in exceptionally populous areas of irrigated agriculture. At such high altitudes, however, export-oriented commercial agriculture has not been possible. Not only is the climate unfavorable, but products would have had to reach the sea via 13,000-foot (*c.* 4000-m) passes to the Pacific Coast or perhaps a long route to the Rio de la Plata. Ac-

cordingly, the infusion of Spanish blood was relatively light and the country has remained predominantly Indian. Mestizos are reported to comprise about a third of the population and whites about one seventh.

Bolivia was settled by the Spanish partly for the land and labor of its Indians but in good part for its minerals. In the late 16th century one Bolivian mine was producing about half of the world's silver output. Since the late 19th century tin has been the mineral most sought in the Bolivian Andes, and mining communities there now produce approximately one eighth of the world output. In 1952 the government in power sought to end a bloody history of recurrent armed conflict between exploited Indian miners and the mine owners (Bolivian and foreign) by nationalizing the mines, but conflict has continued. Meanwhile exports of tin have become far less valuable than exports of natural gas from the eastern lowlands. Since the 1920s a series of oil and gas discoveries has been made around the old frontier town of Santa Cruz at the edge of the savanna lowlands of southeastern Bolivia. Pipeline outlets to Argentina and Chile were built, a pipeline to Brazil was planned, and Santa Cruz (450,000) became a boom town. By the 1980s it was the second largest city in Bolivia. Besides Andean Indians, Europeans and Japanese were settling around it in growing agricultural settlements. Farther north, where the lowlands have a tropical rain forest climate and intermediate Andean slopes lie in the *tierra templada,* other pioneer agricultural settlements have been receiving immigrants from the highlands who come to pan for gold, produce cattle (meat is flown to La Paz), cut wood, and grow coffee, sugar, and the coca plant that yields cocaine. Lowland pioneering in the rain forest and savanna is offering a major opportunity for settlers from the crowded and poor Indian communities of the Andes.

BRAZIL: CYCLES OF DEVELOPMENT IN AN EMERGING TROPICAL POWER

Brazil is such a huge and important country that it is treated here as a major subdivision of Latin America. It is actually bigger in both area and population than any of the other three subdivisions used in this text, despite the fact that each of the latter contains four or more countries, some of which are sizable (Table 24.1). Even on a world scale Brazil is impressively large, and its general stature among the world's countries is growing. The latter is partly a function of Brazil's great size, but it is due primarily to a recent surge of dynamism that has enabled the country to shake off some of its traditional under-development and move toward fuller status as an industrialized tropical power.

AREA AND POPULATION COMPARISONS

With an area of about 3.3 million square miles (8.5 million sq. km) and a population estimated at 144 million in 1988, Brazil ranks fifth among the nations in area (after the Soviet Union, Canada, China, and the United States), and sixth in population (after China, India, the Soviet Union, the United States, and Indonesia). Although smaller in area than the United States by nearly 400,000 square miles, Brazil is larger than the 48 conterminous states (Fig. 25.7). Its maximum dimensions of 2688 miles (4326 km) east-west and 2683 miles north-south are a shade greater than the straight-line distance from New York to San Francisco (c. 2570 miles). This giant state had only about 60 percent as many people as the United States in 1988, but its annual rate of natural increase is so much higher (2.0% as opposed to 0.7%) that the gap between the two countries will narrow rapidly to the end of the century and beyond.

BRAZILIAN DYNAMISM AND POVERTY

Rapid population growth is an important facet of the dynamism that is giving Brazil an increasingly important role in hemispheric and world affairs. The dynamism is also dramatically expressed in such related phenomena as (1) explosive urbanism that has made metropolitan São Paulo (16 million) and Rio de Janeiro (10.3 million) two of the world's largest

Figure 25.7 Brazil compared in area with the conterminous United States.

cities, (2) the rise of manufacturing to the point that manufactural exports now exceed agricultural exports in value, and (3) the diversification of export agriculture to such a degree that the traditional overdependence on coffee has disappeared. In 1953, coffee provided 66 percent of *all* Brazilian exports by value, but by 1985 this figure had plummeted to 9.7 percent. No longer can Brazil be counted among the numerous emerging nations that have overwhelming dependence on one or two export products.

By providing an expanding internal market (body of consumers) and labor force, rapid population increase can make possible economies of scale and hence an accelerating economic development in Brazil. Certainly this happened in the United States at an earlier stage, when it, too, was an "emerging" nation with large frontier regions yet to be occupied with close settlement. But Brazil lacks many of the advantages that made the United States the world's greatest economic power, and today's burgeoning Brazilian population poses serious problems for the nation. Brazil is hard pressed to keep economic output ahead of population increase and to keep education and other social services at tolerable levels. Such problems are particularly acute in a country of so many millions where about two fifths of the population was under the age of 15 years in the early 1980s and only one tenth of all people over the age of 5 had received more than a primary-school education. Clearly Brazil has a massive educational problem if it is to find a place among the most advanced nations of the world. It also must find ways to further alleviate the poverty that gave the nation a per capita GNP of only $1810 in 1986. This figure was several times higher than that of the great majority of countries in Africa, and of many in Asia, and it was about average for Latin America. But it was still five to ten times lower than that of the United States, Canada, or nearly all countries in northwestern Europe. Brazil now has a sizable complement of millionaires, a substantial and growing middle class, and large numbers of technically skilled workers, but the benefits from development have been distributed so unevenly that many millions still live in gross poverty. And the country remains deeply mired in a huge foreign debt, owed by a government whose ambitions have outrun its finances in recent times. Some portents are favorable: Brazil does have abundant natural resources of many kinds, many friends among the world's advanced economic powers, and a remarkable degree of internal harmony among the diverse racial, ethnic, and cultural elements that make up the Brazilian population.

BRAZIL'S DIVERSE POPULATION ORIGINS

In many of the world's emerging nations, development is hindered by internal tensions and conflicts of a racial, ethnic, and cultural kind. Many African countries, for example, exhibit this problem to a serious degree. But Brazilians have managed to live together in relative concord, despite a multiplicity of potentially antagonistic elements in the population. Racially the population is classified in government statistics as 53 percent white (which may well be an overstatement), 22 percent mulatto, 11 percent black, and 12 percent mestizo, with smaller elements including the largest Japanese community outside of Japan, and hundreds of thousands of relatively unmixed Indians in the sparsely populated interior. The basically Portuguese nature of the culture is diluted not only by African elements, introduced during the slave era from the early 16th to the late 19th century, but by the diverse cultural contributions of millions of European immigrants who have come to Brazil from countries other than Portugal. For example, almost as many Italians as Portuguese have immigrated to Brazil. The numbers of Spanish and German settlers have been less imposing but still large, and considerable groups of Russians, other Slavs (principally Poles), and lesser elements have come. Brazil had a brutal early history in which Indians as well as blacks were enslaved, but the modern nation has gained an enviable reputation for racial and cultural tolerance. Subtle discriminations do remain, which may become blatant in regard to primitive Indians. Religious fragmentation is evident in the fact that the predominant Catholic Church is sharply split into several groups; and there are several Protestant denominations, as well as Spiritist and small Jewish and Buddhist elements. If Brazil can continue to realize its promise of development despite its potential for racial and cultural conflict, it will provide a worthy model for many other emerging countries with comparable situations.

ENVIRONMENTAL REGIONS AND THEIR UTILIZATION

Brazilian development must contend with many problems associated with humid tropical environments. The equator crosses Brazil near the country's northern border, and the Tropic of Capricorn lies just south of Rio de Janeiro and São Paulo. Hence only a relatively small southern projection extends into mid-latitude subtropics comparable to those of adjoining sections of Uruguay and Argentina. Within the Brazilian tropics the climate is classed as tropical rain forest or tropical savanna nearly everywhere, although relatively small areas of *tierra templada* exist in some highlands, and there is one section of tropical steppe climate near the Atlantic in the northeast. The overwhelming predominance of tropical rain forest and tropical savanna makes it hard to keep the soil fertile and in place, poses severe maintenance and disease problems, creates general health and sanitation problems in a country relatively devoid of air conditioning and flush toilets, and makes water supply a difficult matter in the huge areas that have a dry season. The latter problem is compounded by the fact that Brazilian rocks are apt to be old hard crystallines that do not harbor large accumulations of underground water.

On the opposite side of the coin, Brazil is not handicapped to any marked degree by excessively high or rugged terrain. Nowhere does the country reach the Andes. Mountains exist, but they are limited in extent and not generally very high. This leaves a country largely comprised of relatively low uplands or extensive lowlands, the latter found primarily along the Atlantic coast or in the drainage basin of the great Amazon River. We turn now to the large environmental regions of Brazil, outlining initially by landforms but looking also at associated natural features and patterns of human use.

The physiographic diversity of Brazil is very great when viewed in detail, but its major components can be outlined under three landform divisions: the Brazilian Highlands, the Atlantic Coastal Lowlands, and the Amazon Lowland. They are discussed here as environmental associations incorporating diverse climates, plant life, water features, soils, and minerals and are also considered as regions of human settlement and use.

Much the greater part of Brazil is upland, although high mountains are few. The northernmost fringe of the country incorporates a part of the *Guiana Highlands* that extends across the border from Venezuela. This region of hills and low mountains clothed partly in rain forest and partly in savanna is sparsely populated and has comparatively little importance in Brazil's present development.

The Brazilian Highlands

Of very large importance, however, is the area called the Brazilian Highlands (for its approximate extent, see the area shown as "plateaus" on Fig. 2.10). This is a huge triangular upland which extends from approximately 200 miles (*c.* 325 km) south of the lower Amazon River in the north to the border with Uruguay in the south. On the east it is generally separated from the Atlantic by a narrow coastal plain, but it reaches the sea in places. Westward the

Highlands stretch into parts of Paraguay and Bolivia. From here the northern edge runs raggedly northeastward toward the Atlantic, steadily approaching the Amazon River but never reaching it. The topography of the Highlands is mostly one of hills and river valleys, interspersed with tablelands and scattered ranges of low mountains. Elevations are highest toward the Atlantic. Here from Salvador to Pôrto Alegre the Highlands descend very steeply to the coastal plain or the sea. The descent in some places is a single formidable slope, in others a series of steps. Throughout its length this dropoff is called the Great Escarpment (Fig. 25.8). It has presented major obstacles to penetration of the interior from the coast. Although for the most part the Escarpment is only 2000 feet (610 m) or so in height, a central section from north of Rio de Janeiro nearly to Pôrto Alegre attains much higher elevations and forms the seaward edge of a belt of rather high and rugged mountains, with peaks often between 3000 and 9000 feet. This is the highest and most mountainous section of Brazil, and it forms a major drainage divide. Many short rivers come down the Escarpment directly toward the Atlantic, but all except one of the major streams drain down the long slope of the Highlands westward toward the Paraná. The one important exception is the São Francisco River, which flows northward from this elevated area and then turns east through the Great Escarpment to reach the sea north of Salvador (Fig. 25.2).

Most parts of the Brazilian Highlands have a tropical savanna type of climate, associated with vast expanses of grass and shrubs interrupted by scattered trees and patches or valley ribbons of forest. Cuiabá, near the Bolivian border, is a typical climatic station. Its monthly average temperatures range between 73°F and 81°F (23°–27°C), while its annual rainfall averages 54 inches (137 cm). This precipitation is very unevenly distributed through the year. Ordinarily the 3 winter months (June–August) are almost without rain, while January, in the height of summer, provides 10 of the year's 54 inches. As is typical of much of the savanna in the Highlands, the area around Cuiabá has long been sparsely occupied ranching country and is now being penetrated increasingly by some pioneer farmers.

However, some parts of the Highlands are atypical. The mountainous belt near the central eastern

Figure 25.8 *The Great Escarpment marking the edge of Brazil's mountainous eastern plateau, the Brazilian Highlands. A narrow coastal plain along the Atlantic shore of Brazil lies at the foot of the escarpment. The view shows the highway connecting the seaport of Santos with the inland metropolis of São Paulo.*

edge is high enough for temperatures to be significantly lowered, so that much of it is *tierra templada*, in which temperatures are actually subtropical rather than tropical. It is here that most of the core area of modern Brazil has developed during the 19th and 20th centuries, in the hinterland of São Paulo. Coffee production typical of Latin America's *tierra templada* was once overwhelmingly dominant but is now just one element in a relatively productive diversified agriculture. More importantly, this area has given rise to Brazil's greatest development of large-scale industry and urbanization. Climatic conditions in this elevated part of the Highlands are typified by São Paulo. Located at about 2600 feet (790 m) elevation near the Tropic of Capricorn, the city has pleasant temperatures averaging 59°F (15°C) in the coolest month and 72°F (22°C) in the warmest. The average annual rainfall is 55 inches (139 cm), and there is no dry season. Southward from São Paulo, temperatures tend to decrease with increasing latitude, and the type of climate becomes humid subtropical like that of southeastern United States. Occasional frosts in winter have restricted and thrown back attempts to expand coffee cultivation into this more southerly area. Nonetheless the humid subtropical area is a relatively productive and prosperous part of the country, with a more mid-latitude type of agriculture. For example, parts of it are important in growing soybeans. Brazil accounts for about one eighth of world soybean output and has a very significant share of the world market. One beneficial effect of the wet subtropical conditions in both the *tierra templada* and the southeastern mid-latitude extension of Brazil is that forest rather than savanna is the natural vegetation of most areas. This forest exhibits a mixture of deciduous and evergreen species, with deciduous trees generally dominant. However, in a large area south of São Paulo there is a strong admixture of needleleaf evergreens. This part of Brazil, especially the needleleaf evergreen section, has provided most of the timber for the urban-industrial growth of the core area.

Another rather exceptional section of the Brazilian Highlands occurs in the northeast. Here climatic maps show an interior area of semiarid steppe climate near Brazil's eastern tip. Actually the steppe area is only a shade drier than a larger surrounding area in which the tropical savanna climate is so dry as to be almost steppe. This larger area of low precipitation reaches the Atlantic along a considerable stretch on both sides of the city of Fortaleza and extends southward in a broad zone behind the Great Escarpment to the latitude of Salvador. The whole area, both climatic steppe and marginal tropical savanna around it, is sometimes called the *Caatinga*, a word referring to its natural vegetation of sparse and stunted xerophytic forest. Wet enough to attract settlement at an early date, it has been a zone of recurrent disaster because of wide annual fluctuations in rainfall. Years of lower than normal rainfall periodically witness agricultural failure, hunger, and mass emigration. The whole region called the Northeast, of which the Caatinga is a major part, is considered to be the poorest region of Brazil and the one most in need of government programs of support and development.

The Atlantic Coastal Lowlands

A narrow coastal plain lies between the edge of the Highlands and the Atlantic most of the way from northwest of Fortaleza to the border with Uruguay. Here and there, the plain disappears because of interruptions by seaward extensions of the Brazilian Highlands. The most important interruption occurs where the nearby Highlands are the most mountainous, from the vicinity of Vitória to that of Pôrto Alegre. Climatically, the northern part of the discontinuous ribbon of plain is tropical savanna, the central part is tropical rain forest, and the southern part is humid subtropical. A great deal of Brazil's history has centered along this coastal strip. It was the first part of the country to be settled by Europeans, and it has been in agricultural use for a changing variety of crops since the 1500s. Half of Brazil's metropolitan cities of a million or more population are seaports spotted along this coastal lowland: in the north, Fortaleza (1.6 million), Salvador (2.1 million), and Recife (2.4 million); in the south, Rio de Janeiro and Pôrto Alegre (2.3 million). Two of the three historic capitals are among them: Salvador (formerly Bahia), the colonial capital from the 16th century to 1763, and Rio de Janeiro (Fig. 25.9), which was first the colonial capital and then with independence (1822) the national capital from 1763 to 1960. In the latter year the capital was moved to its present location in the new inland city of Brasília. These shifts in the location of the capital were motivated by changing economic and political alignments within the country, although the shift to Brasília is more symbolic of the country's hopes for future inland development than it is of the locational realities of present development. Despite the fact that it is no longer the capital, Rio de Janeiro continues to be one of the world's greatest metropolises, with over 10 million people surrounding and spreading inland and along the coast from the capacious and beautiful mountain-girt harbor of Guanabara Bay.

Figure 25.9 *Rio de Janeiro occupies a spectacular setting of bays, peninsulas, islands, low mountains, and world-famous beaches serving a large international tourist clientele.*

The Amazon Lowland

The other major lowland area of Brazil is radically different from the coastal lowlands in development and size. This Amazon Lowland reaches the Atlantic along a coast which extends some distance on either side of the Amazon mouth. From this coastal foothold it extends inland between the Guiana Highlands and the northern edge of the Brazilian Highlands, gradually widening toward the west. Far in the interior, beyond Manaus, the Lowland widens abruptly and extends across the frontiers of Bolivia, Peru, and Colombia to the foot of the Andes. The Amazon Lowland is comprised generally of rolling or undulating plains country except for the broad flood plains of the Amazon and its many large tributaries. The flood plains are often many miles wide, lie below the general plains level and are nearly flat, but most of the Lowland lies slightly higher, is more uneven, and is not subject to annual flooding. Tropical rain forest climate and vegetation prevail nearly everywhere, although southern fringes of the Lowland are classed as tropical savanna. The central Amazon metropolis of Manaus (city proper, over 650,000) is climatically representative, with monthly average temperatures varying only between 79°F and 82°F (26°–28°C), 83 inches (210 cm) of annual precipitation, and no completely dry month. Manaus and the ocean port of Belém (city proper 1.1 million) are great exceptions to the rest of the Lowland in terms of development, as most of the

Amazon region is still very undeveloped and sparsely populated (Fig. 2.20). However, in recent decades the Brazilians have opened large areas by building trans-Amazon highways to augment river and air transportation (Fig. 25.10). By 1972 they

Figure 25.10 *Looking across the Amazon River from the north bank at Manaus. No bridge crosses the Amazon in Brazil, and road connections parallel to the river are very discontinuous. Hence auto traffic must utilize ferries such as the one in the foreground.*

had also accomplished the first complete mapping of the Amazon region as a foundation for development. The result has been a considerable influx of people into Amazonia—even to territories as remote as those on the Bolivian and Peruvian frontiers. These new settlers live, generally on a semisubsistence basis, by varying forms and combinations of ranching, agriculture, mining, lumbering, and fishing. Many ecologists fear that the push into the Amazon will eventually destroy the trees and the irreplaceable ecology of this largest of the world's remaining areas of tropical rain forest. The number of living species of plants and animals reaches its maximum here, and destruction of the forest would cause staggering losses in the world's gene pool.

CYCLES OF BRAZILIAN DEVELOPMENT

After the discovery by Columbus of land across the Atlantic from Europe, the Pope divided the world into Spanish and Portuguese spheres. In 1494 the two countries agreed on a line of demarcation along a meridian which, as it turned out, intersected the South American coast just south of the Amazon mouth. Hence the eastward bulge of the continent was allocated to Portugal. Later the Portuguese expanded beyond this line into the rest of what is now Brazil, with the international boundaries being drawn in remote and little-populated regions east of the Andes where Portuguese and Spanish penetration met.

The Sugarcane Era

The colonial foundations of Brazil were laid in the 16th century. A Portuguese fleet, well off-course on a voyage around Africa, discovered the Brazilian coast in 1500, and a sporadic trade with the local Indians for "Brazilwood" (a native source of dye) began. Settlement beyond the first small trading posts was promoted by Portugal in the 1530s in order to combat French incursions into the area. Only in 1567 was a French base on Guanabara Bay replaced by the Portuguese base which became Rio de Janeiro. Meanwhile, the coast became increasingly valuable as it rapidly developed exports of cane sugar to Europe and bought shiploads of slaves from Africa to labor in the cane fields. Slave traders in search of Indian slaves penetrated the interior to some degree, and Catholic missions also began to be founded among the Indians. One of these was established in 1554 at the site of present São Paulo. Unfortunately, the missions often brought uncontrollable disease epidemics and slaving expeditions in their wake.

Expansion of coastal sugar plantations and increasing penetration of the interior dominated 17th-century Brazil. The colony became the world's leading source of sugar. Although some sugarcane was grown as far south as the vicinity of the Tropic of Capricorn, the main plantation areas developed farther north in more thoroughly tropical areas that also lay closer to Europe. Recife (then Pernambuco) and Salvador (then Bahia) became the main ports and supply bases for the cane growers. In 1615 Belém was founded outside the sugar belt as a forward base against expansionist threats by the Dutch in the Guianas. This action proved unable to prevent a successful Dutch invasion in 1630 which took control over territory reaching southward to beyond Recife. Not until 1654 were the Dutch finally expelled. Meanwhile missionaries penetrated well up the Amazon, and a series of wide-ranging expeditions, largely based in São Paulo, took place in search of Indians to enslave or quick wealth through mineral discoveries. The expeditions often lasted for years, grazing cattle along the way and stopping periodically to raise food crops. In the late 1600s sizable gold finds were made in the Brazilian Highlands some hundreds of miles to the north and northwest of São Paulo in what is now Minas Gerais state.

The Gold and Diamond Era

During the 18th century, Brazil's center of gravity shifted southward, as symbolized by the transfer of the colonial capital from Salvador to Rio de Janeiro in 1763. The mineral discoveries in the Brazilian Highlands that began in the late 1600s continued into the next century. Gold and diamonds were mined at widely scattered locations in the highlands north and northwest of São Paulo and Rio de Janeiro and as far inland as the vicinity of the present cities of Goiás and Cuiabá. By the mid-1700s Brazil was producing nearly half the world's gold. In the latter part of the century this output declined, but by that time much of the highland around the present city of Belo Horizonte (founded only in 1896 as a state capital) had been settled, along with areas near São Paulo and spots in the far interior. Despite the difficulty of traversing the Great Escarpment between Rio de Janeiro's magnificent harbor and the developing interior, the city became the port and economic focus for the newly settled territories. Gold and diamonds were ideal high-value commodities to move by pack animals over poor roads and trails to the port.

Besides the attraction of mineral wealth, a factor contributing to the rise of the new areas was the economic distress of the sugar coast to the north.

This was a result of increasing development of sugar production in other New World colonies (notably in the West Indies) and hence increasing competition for European markets. One solution was emigration of owners with their slaves to the newly developing areas in Brazil. Another was emigration into the Caatinga area inland from the sugar coast. There a ranching economy began, which was augmented by cotton growing in the next century. These 18th-century events set a pattern for this Northeastern part of Brazil which has never been broken: (1) continued dependence on hard-pressed coastal sugar growing and inland cotton production; (2) poverty of unusual severity even for Latin America; (3) recurrent disasters brought by drought, and sometimes by flood, in the climatically unreliable interior; and (4) flows of emigration to other parts of Brazil. Such flows turn into strong currents in the worst times.

Rubber and Coffee Booms

Nineteenth-century Brazil was changed by rubber and coffee booms, by much-intensified settlement of the southeastern section, by the beginnings of modern industry, and by extension of the ranching frontier far into the Brazilian Highlands. It underwent these changes as a colony until 1822; as an independent empire, with a strongly federal structure, until 1889; as a military dictatorship until 1894; and then as a civilian-ruled republic. Its political history was turbulent. Not until 1888 did the country emancipate its large slave population.

Among the major economic developments of the 19th century, the one with the most ephemeral impact was a boom in wild-rubber gathering in the Amazon Basin. This followed the discovery of vulcanization, making rubber a valuable material, by the American Charles Goodyear in 1839. At that time Amazonia was the only home of the rubber tree (*Hevea brasiliensis*). A "rush" of sorts up the rivers followed, and thousands of widely scattered settlers began to tap the rubber trees and send latex downstream. Manaus grew as the interior hub of this traffic, and Belém as its port. But then some seeds of the rubber tree were smuggled to England and propagated in the herbarium at Kew Gardens in London. Seedlings transplanted to Southeast Asia became the basis for rubber plantations, and by 1920 the Amazon production, which depended on tapping scattered wild trees, was practically dead. Some of the stranded settlers remained to carry on a meager subsistence agriculture in Amazonia.

Colonial Brazil achieved only a very sparse pastoral occupation of the subtropical part of the country south of São Paulo. Parts of this were in dispute with Spain and subsequently with independent Argentina. The dispute with Argentina became intense, and there was also a threat of European intrusion, plus a rather fierce Indian resistance in some sections. The Brazilian answer was a deliberate policy of settling European immigrants in the area. Beginning in the 1820s German mercenary soldiers were brought in, and later thousands of German farmers. Still later, large contingents of Italians arrived, as did Russians and other Slavs. Descendants of the various European immigrant groups give this subtropical part of Brazil a distinctive ethno-cultural cast. The agriculture introduced by the immigrants was modified from that of their home countries, being based on mid-latitude grains and livestock, together with a large component of vineyards and wine production in Italian settlements inland from Pôrto Alegre.

However, coffee was the preeminent boom product of Brazil during the 19th century. It had long been produced in Brazil's Northeast in small quantities, and by the late 1700s was an important crop around and inland from Rio de Janeiro. During the 1800s it spread in the area around São Paulo and then increased explosively in planted area, labor requirements, and output during the last decades of the century. This growth was related to the growing world market for coffee, the emergence of expanded and faster ocean transport, nearly ideal climatic conditions and good soil conditions for coffee in this part of the Brazilian Highlands, and the arrival of railway transport to move the product. A spectacular and critical transport development was the completion by British interests in 1867 of a railway linking the coffee-collecting city of São Paulo with the ocean port of Santos at the foot of the Great Escarpment. The coffee frontier crossed the Paraná River to the west and spread widely to the north and south of São Paulo. Production doubled and redoubled. In the late 1800s over a million immigrants a year poured into Brazil, mostly into the region near São Paulo, while still more millions arrived in the São Paulo region from other parts of Brazil itself. The frontier of ranching and pioneer agriculture moved farther to the interior ahead of the coffee-planting frontier. São Paulo city began its rapid growth as the business center and the collecting and forwarding point for the coffee industry and also began a dynamic career as an industrial center. These developments were facilitated by very early hydroelectric production in the vicinity. In the 1880s the city's population was still on the order of 30,000 to 35,000 people. But a few years after the turn of the century São Paulo had more cotton-textile workers alone than it had had people 20 years before. (Cotton growing was a subordinate part of the agricultural expansion in the coffee region.) Eight decades later the city had an estimated 16 million people in its metropolitan area and was one of the largest metropolises of the world—just a century after being a relatively sleepy town of some 30,000.

Twentieth-Century Development to the End of World War II

Until after World War II, Brazil's development during the 20th century was not spectacular. The rubber and coffee booms of the 19th century both collapsed in the early years of the 20th century. Brazil became a minor factor in world rubber production. It continued to lead the world in coffee production and export, as it does today, but increasing international competition depressed coffee prices and the industry struggled economically. A large-scale attempt by the Ford Motor Company to produce rubber competitively on plantations in the Amazon began in 1927 but failed after a few years. The most positive developments of this period were (1) the continued growth of early-stage industrialization in the São Paulo area; (2) some continued expansion of the railway network; (3) the arrival, beginning in 1925, of Japanese immigrants who began to make important contributions to agriculture, especially truck farming; (4) successful government attacks on two widespread debilitating diseases—malaria and yellow fever—during the 1930s; and (5) a major expansion of irrigated rice production in the Paraíba Valley between Rio de Janeiro and São Paulo. Rice is a staple of the Brazilian diet, along with corn, beans, and manioc. It is widely grown, mostly as unirrigated "upland" rice, but the Paraíba Valley near the country's largest urban markets was found to be especially well suited to the far more productive irrigated varieties. Overall, however, the country struggled through this period without any major export boom. In 1930, at the beginning of the worldwide Great Depression, Brazil's economy practically collapsed, and in that year democratic government was replaced by a rightist dictatorship which lasted until 1945.

Quickened Development since World War II

Since 1945 the pace of development in Brazil has generally been very rapid. However, fluctuations have occurred from time to time, including a downturn in the 1980s. Growth took place under elected governments between 1945 and 1964 but occurred most rapidly between 1964 and 1985 under a military dictatorship with technocratic leanings and great ambitions. In 1985 the dictatorship relinquished power to a new elected government, which then had to face accumulated problems of development. Among these were (1) a crushing national debt that the generals had run up to finance expansion, and (2) widespread economic misery caused by governmental attempts to keep up with payments on the debt.

The provision of a far more adequate internal transportation system has been an essential foundation of Brazil's growth in the last few decades. This has involved primarily the building of a network of long-distance paved highways and the giving of primacy to trucks as movers of goods. Major urban centers have been connected, and additional highways have been pushed into little-populated and undeveloped areas. Highways now cross the interior of the country to the Bolivian and Peruvian borders from Brasília, Recife, and the north shore of the Amazon mouth. North-south connections in the interior are sparse, but some have been built. The new roads have attracted ribbons of settlement, have facilitated the flow of pioneer settlers to remote areas, and have furthered the economic development of such areas by connecting them to markets (for rubber, minerals, timber, fish, tourism, and some agricultural products) and sources of supply. But the single most spectacular instance of interior development has been the building of a completely new capital in a more interior and central location. Brasília was decided upon in 1957, built as a totally planned city on empty savanna in the Brazilian Highlands, and occupied by the government in 1960. Its growth, along with unwanted adjacent shantytowns (called *favellas* in Brazil) has been such that the new Federal District of about 2300 square miles (6000 sq. km) had over 1.6 million people by 1987. It is now a major highway focus, and the market it provides is stimulating agriculture in the surrounding region, especially by Japanese truck farmers.

AGRICULTURAL EXPANSIONS: SUGAR, CITRUS, SOYBEANS

Three major agricultural expansions in older-settled territories have contributed to the country's recent economic growth. A program of the military government to modernize and reinvigorate the old sugar industry had considerable success, and Brazil is again a major sugar exporter. In 1962 a severe freeze in Florida gave Brazil an entering wedge in the world market for orange concentrate. Orange growing had been gaining momentum for some years in the Rio de Janeiro-Paraíba Valley-São Paulo region, but only for the domestic market. By 1968 Brazilian exports of orange concentrate surpassed those of the United States, which had been the leading supplier, and by the 1980s Brazil was supplying about nine tenths of world exports. In the 1960s Brazil became a major soybean producer and exporter, with production centered in the subtropical southeast and in a new district in the tropical Brazilian Highlands. It continues to expand its share of the world market, in good part at the expense of American soybean growers. Although coffee was still the leading agricultural export in 1984, the three newer exports surpassed it in combined value.

BRAZIL'S NEW INDUSTRIAL AGE

But even more significant has been the rapid expansion of Brazilian manufacturing. In the 1980s the value of manufactures exported from Brazil began to exceed the combined value of other exports for the first time. The products involved today are not mainly the simple ones of early-stage industrialization, although textiles and shoes are still major items. Steel, machinery, automobiles and trucks, ships, chemicals, and plastics are all important exports. So are weapons, especially types developed by the Brazilian military during its armed suppression of urban guerrilla opposition during the 1960s and 1970s. Aircraft are also manufactured, although mainly for the extensive domestic market that one would expect to find in a country so large. An industrial core area has developed, and it accounts for the greater part of the foregoing manufactures. Its main cities are São Paulo and Rio de Janeiro, just over 200 miles (c. 320 km) apart, and Belo Horizonte (3 million) in a mineralized section of the Brazilian Highlands some 200 miles north of Rio de Janeiro. Other sizable industrial centers are scattered around and between these places.

Factors in Brazil's industrial rise have included (1) governments determined to accomplish it, (2) the country's large internal market, (3) cheap labor, (4) the growth of an educated class of technicians and managers, (5) a strong free-enterprise capitalist sector in the economy, (6) extreme hospitality to foreign capital and companies, and (7) borrowing on such a massive scale that the debt has become a national burden. But the industrial surge has also depended heavily on extraordinary natural-resource wealth. Waterpower has been a key element. A very large hydroelectric potential exists (1) along the short streams descending the Great Escarpment to the Atlantic, (2) along the entrenched course of the Rio São Francisco running northward parallel to the Escarpment, (3) at falls and rapids of the many rivers in the Brazilian Highlands, (4) along the large rivers that flow northward off the Highlands to the Ama-

zon, and (5) along the Paraná and Paraguay rivers. Although developed hydroelectric sites are widely scattered, half the developed capacity is on the Paraná River and its tributaries. Some of these plants are near São Paulo, but the world's largest hydroelectric installation, which began generating power in 1983, is located at Itaipu on the Paraguayan border just north of the border with Argentina. Over nine tenths of Brazil's electricity comes from falling water, and most of the potential is untouched (see chapter-opening photo). Metals represent the other main contribution of Brazil's resource wealth to Brazilian industry. The Brazilian Highlands are highly mineralized with a wide variety of ores yielding numerous metals, outstanding among which are iron, manganese, tin, and tungsten. Known reserves give Brazil about an eighth of the world's proven iron ore, and the country is a leading producer and exporter. The largest developed deposit is somewhat north of Belo Horizonte. The first modern steel mill in the country began production in the 1940s at Volta Redonda in the Paraíba Valley between Rio de Janeiro and São Paulo. Various others have been built in the industrial core zone since that time, and Brazil has become a considerable exporter of steel.

Brazil's greatest resource deficiency is in fossil fuels. Known supplies of coal are found only in the extreme southeast and are very inadequate for the country's needs, although they do provide some coking coal. Domestic petroleum supplies are known thus far only in the Northeast. Located offshore for the most part, they are the result of intense exploration since the drastic rise in world oil prices in 1973. But they supply only about half of Brazil's oil consumption (1987), and nearly half of all recent import costs have been for oil. This has been true despite a relatively successful program to develop and use gasohol, which includes a 20 to 25 percent alcohol content currently supplied by sugarcane grown in the Northeast and in the area around São Paulo. Unless the problem of oil supply can be solved, the country's continued development drive may be seriously hampered. Certainly the drastic slump in world oil prices in the mid-1980s was highly welcome in Brazil, at the same time that it was proving disastrous to oil-rich Mexico.

COUNTRIES OF THE SOUTHERN MID-LATITUDES

Four countries of southern South America—Argentina, Chile, Uruguay, and Paraguay—differ from all other Latin American countries in that they are es-

Figure 25.11 Large parts of mid-latitude South America are grasslands in which great numbers of cattle and sheep are grazed on the natural forage. The photo shows a corral and huts of herders on a large sheep ranch in the steppes of Argentina. Often such structures are used on a temporary basis due to shifting of the animals from one section of range to another.

sentially mid-latitude rather than low-latitude countries. Northern parts of all the countries except Uruguay do extend into tropical latitudes, but the core areas of these nations, where the great majority of their people live, lie south of the Tropic of Capricorn and are climatically subtropical. Argentina extends barely into the tropics in the extreme north, centers in subtropical plains around Buenos Aires, and reaches far southward into latitudes equivalent to those of southern Canada in North America. Chile also extends south of its core area to these latitudes. Uruguay's comparatively small territory is entirely mid-latitude and subtropical, and most Paraguayans live in the half of their country that lies south of the Tropic. The core areas of the three more eastern countries—Argentina, Paraguay, and Uruguay— have the same humid subtropical climatic classification as southeastern United States. Chile's core area is classed as having a mediterranean (dry-summer subtropical) climate like that of much of California.

Besides their atypical locations and climates these four Latin American countries share certain other regional characteristics. Agriculturally, for instance, they compete more with other mid-latitude countries (in both the Northern and Southern hemispheres) than with most other Latin American or other tropical countries. Argentina is an important export producer of such crops as wheat, corn, and soybeans, and thus competes with American farmers. Both Uruguay and Argentina are sizable exporters of meat and other animal products. Paraguay has recently come into the export market for soybeans as well as cotton. Chile imports more food than it exports, but it does export fruits, vegetables, and wine, as does California. Since these countries never developed labor-intensive and export-oriented plantation agricultures in colonial times, they have very few blacks in their populations. Two of them, Argentina and Uruguay, have been major magnets for European immigration and are exceptional in Latin America in having populations that are over 90 percent pure European in descent. Paraguay and Chile, in contrast, attracted fewer Europeans and are over 90 percent mestizo. Uruguay, Argentina, and Chile are among the most developed countries in Latin America, although their rate of development has been rather slow in recent years, possibly due in part to inadequate oil resources. They have low dependence on agricultural employment, with only 10 to 20 percent of their employed workers still in agriculture. Paraguay, with 43 percent of its labor force still on farms in the early 1980s, is less developed and poorer than the other three mid-latitude countries, at least partly because of its history of land-locked isolation from the outside world.

611

ARGENTINA: THE SHAPING OF GEOGRAPHY BY NATURE, IMMIGRATION, MARKETS, AND POLITICS

Argentina is a major Latin American country. Its area of nearly 1.1 million square miles (2.8 million sq. km), second in Latin America only to that of Brazil, is about a third the area of the 48 conterminous American states. Argentina's estimated population of 32 million in mid-1988 ranked it far behind Brazil and Mexico within Latin America, but a little ahead of Colombia (30.6 million) for third place.

The Agriculture of the Humid Pampa

Agricultural products which Argentina exports through its capital city of Buenos Aires (11 million), and to some extent through lesser ports, are sufficient in volume and value to make the country one of the world's principal sources of surplus food. The leading agricultural exports are wheat, corn, soybeans, and beef. Animal feedstuffs, wine, wool, apples, and apple juice are also significant. These characteristic mid-latitude products make Argentina an export competitor of such countries as the United States, Canada, France, and Australia.

Argentina's agriculture is concentrated heavily in the country's core region, known as the *humid pampa.* "Pampa" comes from an Indian word for plain. This is the Argentine part of the area identified on the map of natural regions (Fig. 25.2) as humid subtropical lowland prairie. The environment is outstanding for agriculture. Level to gently rolling plains lie under a climate similar to that of southeastern United States, featuring hot summers, cool to warm winters, and an annual precipitation averaging 20 to 50 inches (*c.* 50–130 cm). But the area is specially advantaged for agriculture by its soils, which are more fertile than might be expected from the humid subtropical climatic conditions. When the first Europeans arrived in the 16th century, they found the plains covered with tall-grass prairie. But planted trees do quite well there, so that there is considerable doubt whether the prairie was truly "natural" vegetation. It seems likely that this grassland resulted at least in part from the burning of originally forested areas by Indians in order to improve the grazing for game. (Grasses are more fire-resistant than trees.) Whatever their origin, the grasses of the prairie gave the soils an unusually high content of organic material (humus) just as they did in the prairie areas of North America. In addition, soil fertility has been enhanced by the soil material called *loess,* carried by the wind from drier areas to the west and deposited in a thick layer on the plains. The deposition of loess is still going on and providing new surface material.

Hence the soils have a loose structure of fine particles, affording both a high concentration of plant nutrients and the maximum opportunity for plant roots to be nourished by feeding on individual soil particles.

From the 16th century to the late 1800s the humid pampa was ranching country, and much of the southern part remained the domain of Indians until approximately 1880. Early Spanish settlers entered the pampa mainly from Paraguay, where still earlier settlers had found a more wooded environment to their liking. Cattle were introduced, and they flourished and multiplied remarkably on the open range. But markets were too small and distant to support any major commercial development. There were small exports of hides and skins, but the main commercial product was mules, bred on the prairie and then sold as beasts of burden to the Andean mining settlements. Population on the isolated ranches remained quite sparse.

In the late 19th and early 20th centuries Argentina experienced a rapid and dramatic change to commercial agriculture and urbanization. There was heavy European immigration, and population grew rapidly. Commercial contact with Great Britain contributed greatly to Argentina's new dynamism and was so close that Argentina was sometimes called an unofficial part of the British Empire. At the time, the British market for imported food was increasing rapidly. Refrigerated shipping came into use, allowing meat from distant sources such as Argentina to be sold in Britain. A British-financed rail network was built in Argentina to tap the countryside for trade. It focused on Buenos Aires, which rapidly became a major port and a large city. British cattle breeds, favored by the tastes of British consumers, were introduced onto Argentine ranches (*estancias*), along with alfalfa as a feed crop. Immigrant farmers assumed a major role in agriculture, generally as tenants on the large ranches which continued to dominate the pampa economically and socially. The immigrants came especially from Spain and Italy, but also from a number of other countries, giving present-day Argentina a complex mixture of Spanish-speaking citizens from various European backgrounds. Soon an export agriculture developed in which wheat, corn, and other crops supplemented and then surpassed the original beef exports.

The Less Productive Agriculture of Outlying Areas

Argentine areas outside the humid pampa are less environmentally favored, less productive, and less populous. The northwestern lowlands just inside the tropics have a tropical savanna climate and are used for cattle ranching and some cotton farming. Westward from the lower Paraná River and the southern part of the pampa the humid plains give way to steppe and then desert, with production and population declining accordingly. Population in these western areas is concentrated in oasis settlements along the eastern foot of the Andes, where mountain streams emerging onto the eastward-sloping plains have been impounded for irrigation. The production of the main oases has generated a number of sizable cities as service centers. Vineyards and, in the hotter north, sugarcane are special emphases in a varied irrigated agriculture. Córdoba (1.2 million) is the largest of these oasis cities and the second largest city in all of Argentina. It lies at the east flank of a range projecting eastward from the Andes, and its scenery and nearness to Buenos Aires make it a popular resort. To the south of the pampa lie thinly populated deserts and steppes in the hill country of Patagonia. This is a bleak region of cold winters, cool summers, and incessant strong winds. Indian country until the early 20th century, it now has millions of sheep on large ranches but little other agricultural production.

Urban and Industrial Argentina

Only about a tenth of Argentina's labor force is directly employed in agriculture and ranching—activities which produce the country's relatively abundant and cheap food and most of its exports. The great majority of Argentinians are city dwellers employed in an economy where service occupations predominate, and manufacturing employs about twice as many people as does agriculture. Metropolitan Buenos Aires has about 11 million people, a third of the national population. The next largest cities, Córdoba and Rosario, have only a little over 1 million people in their respective metropolitan areas. Buenos Aires was settled in the late 16th century as a port on the broad Paraná estuary called the Rio de la Plata and gradually gained some importance. But until Argentina became reoriented toward British and then world markets, the city remained a rather isolated outpost of a country centered on the north and the western oases, with relatively little trade by sea. In the 19th century Buenos Aires first became the national capital and then underwent explosive growth as a seaport, rail center, manufacturing city, and receiver of immigrants.

The growth of Argentine industry has conformed to a sequence often found in nations emerging from an underindustrialized state:

1. Industrialization began with small-scale industries supplying consumer items for sale within the country, together with plants processing agricultural exports (for example, meat-packing plants).

2. Some consumer industries such as textile manufacture and shoemaking were expanded with government support when the country was cut off from imported supplies during World War I and again during World War II.

3. After World War II a deliberate state-led policy of industrialization was instituted. There have been lapses in this policy and fluctuations in mechanisms of support, but considerable successes have been achieved in the quantitative growth of manufacturing and diversification of industrial output. Imported coal to supply power needs has been replaced by oil produced within Argentina (mainly from fields on the Atlantic coast of Patagonia), together with Argentine natural gas from Tierra del Fuego and the Andean piedmont, a considerable hydroelectric output, and some production of atomic power. Steel plants using imported ore and coal have been built near Rosario on the navigable Paraná River, and diversification has extended into the manufacture of chemicals, machinery, and automobiles.

However, the economic health of a relatively large urbanized service and industrial economy is open to considerable question in Argentina's case. Argentine manufacturing has never become competitive on the world market, and the country's trade pattern is still dominated by agricultural exports and industrial imports. Manufacturing contributes considerably less to national production than might be expected in a country of Argentina's size and economic status, and recent rapid growth of the small-enterprise service sector of the economy reflects in part a decline in some industries, with resulting unemployment. And despite sizable industrial gains, Argentina has not become a prosperous country. Its rate of growth in per capita income has not matched that of a number of other Latin American countries. In 1986 its per capita GNP of $2350 was only a few hundred dollars higher than that of Mexico ($1850), despite its superior agricultural resources and a much more superior relative position half a century ago.

The Disastrous Impact of Argentine Politics

Political turmoil and government mismanagement of the economy have taken a heavy toll. After initial disturbances following independence, Argentina became a constitutional democracy that was actually controlled by large and very wealthy landowners. This situation continued for many decades until

1930, when a military coup toppled the constitutional government. Between 1930 and 1983 the country was controlled by a shifting series of military juntas or by Juan Peron or his party, with brief intervals of semi-democratic government. Although there were variations, both the generals and the Peronists were ultranationalist and at least semi-fascist. They distrusted democracy, Communism, and free-enterprise capitalism, and believed in authoritarianism, violent repression of opposition (which itself was often violent), and state ownership and/or direction of major parts of the economy. They courted the support of a growing urban working class by providing it with jobs, pay, and benefits beyond what was justified by its productivity and output. And they paid for these things by inflating the currency and borrowing abroad. Inflation eventually reached hundreds of percent a year, lowering savings and leading to the flight of investment capital and an economy heavily focused on currency manipulation and speculation rather than on productive investment. The country's foreign debt became the heaviest in the world on a per capita basis by the early 1980s, requiring most of the proceeds from Argentina's exports to pay even the interest due foreign creditors. In 1982 the last military government started a war with Britain by invading the Falkland Islands, which had been claimed by Argentina during a century and a half of British occupation. When the war was lost and the government resigned, it left an economy in shambles, with raging inflation, massive debts, growing unemployment in an inefficient and declining industrial sector, and an industrial labor force organized in nondemocratic unions and accustomed to rewards politically given rather than economically justified. Revelation of the military's murder of thousands of people in the course of suppressing violent opposition during the 1970s further discredited military rule as an alternative to constitutional government. In 1982 an elected constitutional and democratic government took up the task of undoing the damage from long-term misrule.

CHILE: ASSETS AND LIABILITIES OF DIVERSE REGIONS

Chile's peculiar stringbean shape tends to conceal the country's sizable area, which is almost as great as that of the United Kingdom, West Germany, and Italy combined. From its border with Peru to its southern tip at Cape Horn, Chile stretches approximately 2600 miles (*c.* 4200 km), but in most places its width is only about 100 to 130 miles (*c.* 160–210 km). If superimposed on equivalent latitudes in the Northern Hemisphere, Chile would extend from Puerto Rico to Labrador, or along the Pacific Coast from a latitude just south of Mexico City to a latitude just north of Juneau in the Alaskan Panhandle. Given a favorable environment, Chile might not find its curious shape a serious obstacle to development, but the country has actually had to struggle with rugged terrain, extreme aridity, and cool wetness to such a degree that only one relatively small section in the center has any considerable population.

The country's long interior boundary lies mostly near the Andean crest, with the populous core areas of neighboring countries lying some distance away on the other side of the mountains. Although their elevations decline toward the south, the Chilean Andes are mostly very high, with some peaks over 20,000 feet (*c.* 6100 m), and even the passes near the core area of Chile in the vicinity of Santiago are over 10,000 feet high. Although the barrier has by no means been impassable or complete during Chile's four and a half centuries of colonial and independent existence, traffic across the Andes has been difficult and is still light.

Middle Chile: "Mediterranean" Core Region

Chile's populous central region of mediterranean (dry-summer subtropical) climate occupies lowlands between the Andes and the Pacific from about 31° to 37° South Latitude. Mild wet winters and hot dry summers characterize this area, as they do southern California at similar latitudes on the west coast of North America. Somewhat over 400 miles (644 km) long, this strip of territory has always been the heart of Chile's core area. Topographically, it consists of lower slopes in the Andes, a hilly central valley parallel to the foot of the Andes, and a strip of low coastal mountains. The area was the southern outpost of the Inca Empire and was occupied in the mid-1500s by a small Spanish army which approached from the north. Members of the army made use of land grants from the King of Spain to institute a rather isolated ranching economy. Grants varied in size by military rank of the grantees, and this fact, together with the conquered status of the Indians, created a society marked by great social and economic inequalities. At one end of the scale was the aristocracy, comprised of those holding enormous ranches. At the other end was a mass of landless cowboys and subsistence farmers. Men from all ranks of society married Indian women quite freely, sometimes several at a time. Thus was created the Chilean nation of today, some 92 percent of which is classified as mestizo.

The owners of great haciendas dominated the Chilean core area and the country itself well into the 20th century, and ranching remained the primary pursuit in Middle Chile. Agriculture, both irrigated and nonirrigated, was carried on but was secondary in importance and generally supplemental to ranching. Wheat, vineyards, and feed crops became agricultural specialties. A major part of the remuneration of ranch workers was paid in the form of permission to grow subsistence crops in assigned plots on the ranches. During the 1960s and early 1970s population pressure and resulting political demands led to large-scale land reform which somewhat reduced the role of large landowners. Small farmers have tended to specialize more on fruit and vegetable production than on livestock growing. As often happens, land reform has led to some drop in overall agricultural output, at least temporarily.

Growing population pressure in the core region has contributed to rapid urbanization in recent decades. The national capital of Santiago, founded at the foot of the Andes in the 16th century, now has a metropolitan population of over 4 million. Metropolitan Valparaíso, the core region's main port and coastal resort, has another million, and the port of Concepción has about 700,000. Altogether the area of mediterranean climate is the home of about 75 percent of Chile's 12.6 million people (1988). The population of the country was 83 percent urban by a recent estimate—exceeded in Latin America only by that of Uruguay (85%) and Argentina (85%).

To both north and south the core area contains transition zones where population density lessens until areas of very scanty population are reached. On the north, population declines irregularly across a narrow band of steppe to the arid and almost empty wastes of the Atacama. In the southern part of the core, the transition zone is much larger and more important economically. South of Concepción rainfall increases, average temperatures decline, and the mediterranean type of climate gives way to a notably wet version of the marine west coast climate, whose natural vegetation is impressively dense forests. The southernmost part of the mediterranean climatic area and the adjacent northern fringe of the marine west coast area as far south as the small seaport of Puerto Montt have become part of Chile's core only in the past century and still form a less populous transitional end of the core. About 15 percent of Chile's population lives here. The Araucanian Indians native to these forests fought off Inca expansion. Although a few Spanish outposts were established as early as the 16th century, the Araucanians, still aided by their forested habitat, were also able to check most Spanish expansion for some 300 years. Finally, they were assimilated and acculturated to Chilean society rather than truly conquered. Their full-blooded Indian descendants are still numerous in this part of Chile. Also distinctive is a less numerous German element descended from a few thousand immigrants who settled on this wild frontier in the mid-19th century. Today the German community is economically and culturally very important, even putting its stamp on architectural styles of the region. Most people in the southern transition zone, however, are wholly or partly descendants of mestizo settlers who came from the core of Chile in the past century as population pressure increased. Agriculture in this green landscape of woods and pastures is devoted largely to the raising of beef cattle, wheat, hay, and root crops. Thus it resembles the agriculture of many parts of northwestern Europe, but it lacks the intensive capitalization characteristic of the latter area.

The Atacama: Utter Drought and Mineral Wealth

North of the core, Chile is desert, except for areas of greater precipitation high in the Andes. This part of Chile, called the Atacama Desert, is one of the world's most utterly rainless areas. The central lowland and coastal mountains of the core continue northward into the Atacama, but even the coastal mountains (and lower parts of the Andes as well) are desert. Mild temperatures, high relative humidity, and numerous fogs are the result of almost constant winds from the cool ocean current offshore. At Iquique, near Chile's northern border, temperatures average in the 60s (F) every month of the year. Chile gained most of the Atacama by defeating Bolivia and Peru in the War of the Pacific (1879–1883). Previously Bolivia had included a corridor across the Atacama to the sea at the port of Antofagasta, while Peru owned the part of the desert farther north. The war was fought over control of mineral resources, primarily the Atacama's world monopoly on sodium nitrate, a material much in demand at that time for use in fertilizers and the manufacture of smokeless powder. Since then, other mineral resources of the region, principally copper, have eclipsed sodium nitrate in importance. In its few oases, small mineral-shipping ports, and mining settlements, the Atacama today is home to only about 300,000 people, but the region normally produces half or more of the value of Chile's exports.

The Difficult Environment and Scanty Development of Southern Chile

South of Puerto Montt the marine west coast section of Chile continues to the country's southern tip in Tierra del Fuego. In this little-populated strip the central valley and coastal mountains of areas to the north continue southward, but here the valley is submerged and the Pacific reaches the foot of the Andes along a rugged fjorded coastline, while the coastal mountains project above sea level only in higher parts which form a string of offshore islands. Extreme wetness, cool temperatures, violent storms, and dense mountain forests are characteristic. Little agriculture beyond some grazing of sheep has been possible, and population is very sparse. The largest settlement is Punta Arenas (c. 70,000) on the mainland side of the Strait of Magellan. Being in the lee of mountains, it is drier than most places in southern Chile. Punta Arenas is the central settlement of a region of oil and gas wells, some of which are on the mainland, others under the Strait, and still others beyond the Strait in the cool region of windswept sheep pastures and wooded mountains known as Tierra del Fuego.

Geography and the Chilean Economy

Chile's prosperity is highly dependent on the world prices of minerals that it exports from the Atacama and nearby Andes. Exports of natural sodium nitrate from the salt beds of dried lakes began in a small way in the early 19th century and then expanded greatly in the later part of the century. The War of the Pacific established a national monopoly in the product for Chile, with production financed largely by British capital. The monopoly was broken in the 1920s and 1930s when industrial nations built plants to extract nitrogen from the air. The resulting output reduced the mining of sodium nitrate in the Atacama to a relatively small export industry. But by then, vast reserves of copper had been discovered under the Andean slopes of northern Chile, and copper exports rose to decisively surpass nitrates. Important iron-ore discoveries and development occurred at about the same time, both in the Atacama and near Concepción in Middle Chile. Today minerals comprise over half of Chile's exports by value, with copper by far the leader (41% of all exports in 1987). Chile is the world's leading exporter of copper. A variety of other minerals have some importance.

A rather broad array of primary resources is allowing Chile to diversify its exports and could form the basis for major industrialization. In recent years forestry and fishing have expanded rapidly, and paper and fishmeal have become important exports. Chile generally ranks among the top five of the world's nations in total fish catch. Although its known oil and gas resources have thus far left it dependent on the world market for more than half of its hydrocarbon supplies, Chile actually has more than adequate power resources for major industrial-

ization in the future. These include coal fields, especially near Concepción where the country's iron and steel industry has developed; Andean waterpower potentials, little developed as yet; uranium deposits to support an atomic power industry which began in the early 1980s; and the possibility of using the high winds of the southern region to generate electricity in the future. Relatively little industrialization has been accomplished up to the present, although enough has been done to employ somewhat under 15 percent of the labor force in manufacturing. A somewhat larger proportion still is employed in agriculture, whose productivity remains so low that sizable imports of basic foods are required.

Like most Latin American countries, Chile must solve serious political problems if it is to achieve the development and prosperity its resources warrant. A period of democratic government eventuated in the election of a Marxist administration in 1970 by a minority vote against badly divided opposition. This government's drastic changes in the political and economic structure of the country threatened Chile's democracy and created chaos in its economy. A military coup in 1973 then fastened a repressive dictatorship of the Right on Chile. It undid many of the Marxist measures that had been instituted, and the country experienced considerable economic success in the new regime's early years. Hence the military dictatorship achieved a degree of popularity with much of Chile's political Right and center at the same time that it was violently repressing much of the Left. Then the economy was buffeted by world recession and depressed copper prices in the late 1970s and early 1980s, opposition to the regime expanded and intensified, and by 1989 Chile appeared increasingly ready for another major change in its political structure and economic policies.

URUGUAY: OVERDEPENDENCE ON AGRICULTURE IN AN URBANIZED WELFARE STATE

Uruguay, bordered only by Argentina and Brazil, is a part of the humid pampa with about the area of Missouri or Washington state. Its independence from Argentina and Brazil resulted historically from a buffer position between them. In colonial times there were repeated struggles between the Portuguese in Brazil and the Spanish to their south and west over possession of this territory flanking the mouth of the main river system in southern South America. After Argentine independence from Spain these struggles continued, at the same time that a movement for political separation grew in the territory north of the

Rio de la Plata. Eventually Great Britain intervened, fearing Portuguese expansion south of the river, and the two contenders signed a treaty in 1828 that recognized Uruguay as an independent buffer state between them.

Uruguay is very similar physically to the adjacent core region of Argentina, being largely composed of rolling plains, hilly in some sections, with a humid subtropical climate. The natural vegetation was tallgrass prairie, with ribbons of woodland in river valleys, when the Spaniards first entered Uruguay in the 16th century. Until the 19th century this was remote ranching country exporting only hides, some salted beef, and mules. Distant mines in Brazil were the main markets. Development then quickened, largely as a result of British initiatives and capital. During the 19th century sheep were introduced, along with techniques for breeding of better livestock. Barbed-wire fencing gave better control over pastures and breeding. The coming of refrigerated shipping aided the transportation of meat to overseas markets, and just after 1900 some relatively modern meat-packing plants were built. However, crop farming has not replaced pastoralism to the same degree as on the Argentine pampa, and only about 8 percent of Uruguay is used for crops. Ranching still provides the country's basic support to a far greater extent than in Argentina, accounting for about 38 percent of Uruguayan exports in 1987 compared to less than 15 percent for other farm products. A very wide range of crops is produced, with sugarcane, wheat, and rice being the most important. Government programs in the past few decades have had considerable success in expanding sugar and rice production in the warmer northern part of the country.

Uruguay has experienced the usual turbulent history of Latin American countries but is especially known for its attempt to become one of the world's first welfare states. This effort began during a period of democratic government in the early 20th century and achieved such initial success that Uruguay gained an international reputation as an enlightened Latin American state. But although much progress was made in spreading the benefits of a relatively prosperous economy in a socially equitable way, the economy proved unable to support the system during the 1960s and 1970s. This period was characterized by violent internal dissension, protracted guerrilla opposition, military dictatorship (1973–1985), and falling standards of living. Such conditions continued into the 1980s. Essentially, they resulted from the growth of a relatively inefficient urban, industrial, and governmental superstructure to such a size that the agriculturally dependent economy could not finance it.

Montevideo is the only large city in the country, but its metropolitan population of perhaps 1.4 million includes nearly half of all Uruguayans. Small cities and towns increase the total urban proportion in Uruguay to 85 percent. Montevideo has dominated the country almost from the city's inception in the 17th century as first a Brazilian (Portuguese) and then an Argentine (Spanish) fort. Its port facilities handle Uruguay's important waterborne trade. As the national capital Montevideo directs an economy which developed such a high degree of government ownership that the majority of workers came to be employed directly or indirectly by the government. The city is the main center of Uruguay's manufacturing industries, which employed 18 percent of the country's labor force in the mid-1980s but export little except textiles. Meanwhile Uruguayan agriculture, with 15 percent of the labor force, supplies the country's meat-rich diet and approximately half of all Uruguayan exports. This structure began to fail in the 1960s as costs outran production, and the structure was devastated by the high oil prices imposed by oil exporters in the 1970s. Aside from hydropower, which supplies nearly all electricity, Uruguay has almost no natural resources to support industrialization and is heavily dependent on imports of fuels, raw materials, and manufactures. These must be purchased with the proceeds from agricultural and textile exports. In the late 1980s Uruguay was attempting to cope with these problems and to preserve its recently restored democracy. Lower world oil prices were a factor easing the burdens of this conflict-racked country.

DEVELOPMENT PROBLEMS AND PROSPECTS IN LANDLOCKED PARAGUAY

The landlocked and underdeveloped country of Paraguay has long been isolated from the main currents of world affairs. The Spanish founded its capital city of Asunción in the 1530s at a site far enough north on the Paraguay River to promise security against the warlike Indians of the pampas to the south. In addition, high ground at the river bank gave security from floods. Asunción became the base from which surrounding areas were occupied. But when independence came, Paraguay had no way to reach the sea except through Argentina or Brazil. The main route that developed was down the Paraguay and Paraná rivers to Buenos Aires. But this route is much longer than it appears to be on maps, due to meandering of the rivers, and navigation was hindered by such difficulties as shallow and shifting channels, fluctuations in water depth between seasons, and the presence of sandbars. Transport costs were so high that Paraguay could not develop an export-oriented commercial economy. Thus the country remained locked into a very predominantly self-contained economy based on the resources of its own territory.

Today, after some boundary shifts due to past military conflicts, Paraguay has about the area of California. The Paraguay River flows southward across it and divides it into distinctly different eastern and western sections. The eastern section is hill country mixed with plains that are prone to flooding. The climate is humid subtropical, which here has produced a vegetation mainly of forest. Asunción has representative climatic conditions. Monthly average temperatures of the city range between 63°F and 84°F (17°C–29°C), and the average annual rainfall is 52 inches (132 cm). These conditions are similar to those of central Florida. It is in this eastern section, especially in the hill country east and southeast of Asunción, that the great majority of Paraguayans have always lived. Cassava, corn, sugar, bananas, citrus fruits, and livestock are important in their semisubsistence agriculture.

West of the Paraguay River is the region known as the *Chaco* ("hunting ground"), which extends into Bolivia, Argentina, and Brazil. This is a very dry area, increasingly so toward the west. Precipitation averages 20 to 40 inches (*c.* 50–100 cm) annually, but high temperatures cause rapid evapotranspiration and permeable sandy soils absorb surface moisture. These conditions produce an open xerophytic forest in the east, which thins westward and then gives place to coarse grasses on a vast plain of flat alluvium where wide areas are flooded during the wet season. Paraguayans in the entire western region number only in the neighborhood of 100,000. A striking element in this sparse population is provided by scattered Mennonite colonies that aggregate approximately 20,000 people.

Paraguay's comparatively low level of development manifests itself in a number of ways, one of which is total population. Only an estimated 4.4 million people occupied its sizable territory in 1988. This small population results partly from the fact that Paraguay has been a country of outmigration for most of its history and has never attracted any large immigration. The latter circumstance has given the country a racial mix containing few whites and a high proportion of Guaraní Indian blood in the 91 percent of the population that is classed officially as mestizo. Urbanization and industrialization have not gone far. Asunción (metropolitan population, over 700,000) is the only urban place with more than 100,000 people, while a high 43 percent of the labor force was still employed directly in agriculture in 1982. Per capita income is far below that of Argentina or Uruguay.

In addition to isolation, with consequent lack of commercial opportunities, two other factors need to be mentioned as causes for Paraguay's low level of development. One is an almost total lack of mineral resources that has deterred immigration of capital and people. The other is one of the world's more unfortunate military histories. A Paraguayan attempt to force an outlet to the sea resulted in a war between 1865 and 1870 in which Paraguay was pitted against the combined forces of Argentina, Brazil, and Uruguay. This conflict devastated the country and drastically checked its population growth. About four fifths of the entire population perished or fled. A fairly bloody and economically exhausting conflict, the Chaco War, then was fought against Bolivia in the years 1932–1935. Although Paraguay "won" this war by gaining ownership of the disputed Chaco, Bolivia kept the area around Santa Cruz which has become important in oil and natural gas production.

Recently the pace of Paraguay's development has quickened. Modern road and air links with the outside world, and between parts of Paraguay itself, have been forged. With this access to outside markets, exports of cotton and soybeans have grown rapidly and have come to dominate the country's trade. Ironically, recent immigration of Argentinians and Brazilians seeking opportunities in commercial agriculture has come to be seen as a problem. Paraguay has responded by promoting settlement of Paraguayans along the eastern and southern margins of its territory. The future may well see larger and more dramatic changes due to three huge new hydroelectric installations along the Paraná River on the southeastern border of the country. These were partly operative in early 1989, with construction work continuing. For a time, Paraguay will export electricity to pay its share of the costs of these projects, which it undertook jointly with Argentina and Brazil. The new hydroelectric output will give Paraguay a far more favorable energy situation in the future than this fuel-deficient country has had in the past.

ANGLO-AMERICA

CHAPTER 26

The Geography of National Success in Anglo-America

Corn fields and a dairy farm in the rolling Interior Plains section of southern Wisconsin convey a striking image of Anglo-American economic abundance.

THE colonizing activities of Great Britain have left a major imprint on the United States and Canada. Hence these countries are commonly distinguished, under the term "Anglo-America," from the rest of the Americas, which have received primarily the Latin imprint of Spain and Portugal. For some purposes the large Arctic island of Greenland (population, about 55,000) is conveniently included in the concept of Anglo-America. This procedure is justified by Greenland's proximity to Canada (Fig. 26.1) and its strategic importance to both of the Anglo-American countries, but not by the island's culture (a blend of native Eskimo and imported Danish elements) nor by its present political status as a self-governing part of the Kingdom of Denmark. The same general considerations apply to the small French island possession of Saint Pierre and Miquelon (population, about 6000), south of Newfoundland, except that the latter is French in culture and is administered as an overseas department of France.

TIES WITH BRITAIN

Effective British settlement of the present United States and of Newfoundland began in the early 17th century, at about the same time that France began to settle eastern Canada. Effective settlement of the Canadian mainland by Britain occurred in the later 18th century, after Canada had been wrested from France in the French and Indian Wars. Continued immigration from the British Isles and a high rate of natural increase in the newly colonized areas led to a rapid growth in the number of British settlers. Consequently, a basically British form of culture became firmly established and has remained dominant in Anglo-America to the present day, despite the entry of large numbers of settlers from continental Europe, Africa, Latin America, the Orient, and the Middle East. With a few notable exceptions these non-British elements in the Anglo-American population have been largely assimilated into the general culture of the United States and of Canada.

The most obvious and probably most important evidence of the cultural tie with Britain is the dominance of the English language in Anglo-America. Other evidences of the British heritage are found in the political and legal institutions of the Anglo-American countries. Present-day cultural and political affiliations with Great Britain are closer and more openly acknowledged in Canada than in the United States, despite the large French-Canadian element in Canada's population. Canada retains membership in the Commonwealth of Nations and gives formal allegiance to the British crown. The basic cultural bonds between the United States and Britain, on the other hand, were weakened by a long period of political antagonism beginning in colonial times and continuing for many years after the American Revolution, although the American and British peoples have been drawn closer together by the great world crises and military conflicts of the 20th century.

ANGLO-AMERICAN WEALTH AND POWER

Both the United States and Canada are wealthy and powerful nations, with Canada much the lesser of the two. Like the United States, Canada has achieved great national success in many fields, but its population is only one tenth that of the United States, and the resources available in its northerly environment have not provided the range of options available to its richer and more powerful neighbor. Favored by many circumstances such as those outlined in this chapter, the United States has been able to amass total wealth far exceeding that of any other nation. For example, its annual output of goods and services (GNP) in the mid-1980s was approximately double that of its nearest competitor, the Soviet Union. This output was about equal to that of Europe but much greater than that of any other major world region. Canada, a country smaller in population than the state of California, ranks quite high among the world's countries in productivity per capita, and its GNP in the mid-1980s amounted to about 2 percent of the world total. Anglo-America as a whole produced about one-quarter of the world's goods and services, a share that has been cut in half since the years immediately after World War II and is still slowly declining. When it comes to output of goods and services per capita, the United States was the world's second nation in 1986—slightly exceeded by Switzerland—but in other recent years it has been surpassed by various small countries with special resources or functions. Canada ranks somewhat farther down the scale, with over four fifths the United States output per person in recent years.

Although certain level-of-living indices such as infant mortality rate show the United States in a somewhat less favorable light (although still among the world's most favored countries), there can be no

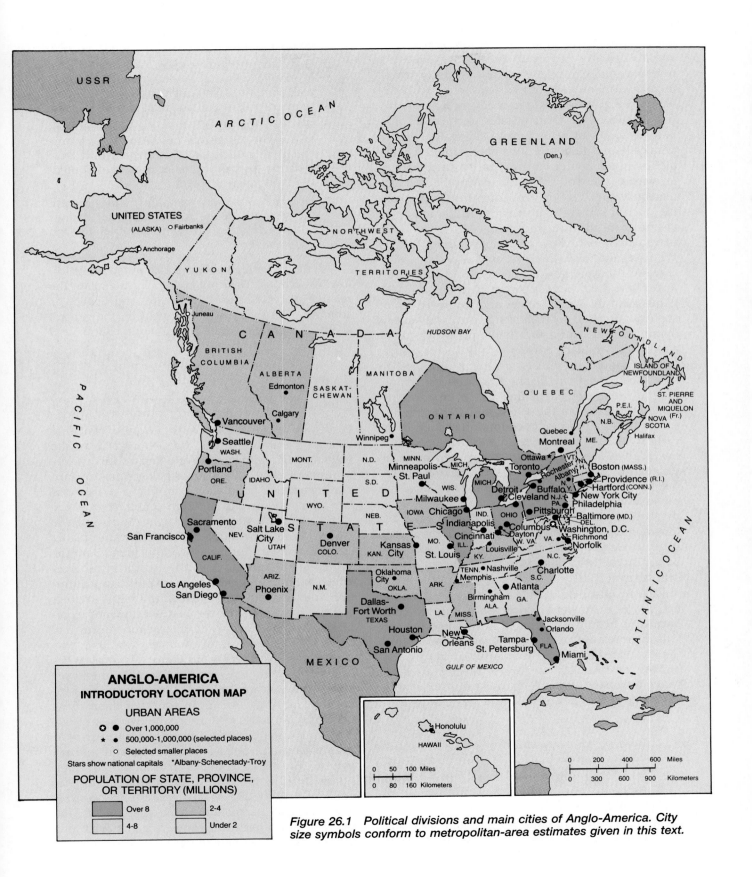

Figure 26.1   Political divisions and main cities of Anglo-America. City size symbols conform to metropolitan-area estimates given in this text.

doubt that the United States of the late 20th century possesses total wealth on such a scale that it is the world's leading economic power. Whether this translates into equivalent political power is difficult to gauge. Undoubtedly it translates into a great deal of military power in the form of advanced weaponry. It also allows the United States to maintain large military forces in being and to aid its friends economically and militarily. But purely military capabilities do not confer on America the same political leverage in the world that they did in the years immediately after World War II due to such factors as the nuclear standoff with the USSR, the increasing dispersion of advanced technology in the world, the fierce determination of "emerging" nations to resist Western (and Soviet) encroachment, and the reluctance of the American people to risk their military forces in strife-torn areas perceived as possible Vietnams. Both the United States and the Soviet Union are circumscribed in their actions by a climate of opinion in the world that looks with disfavor on attempts by military powers to influence events in weaker nations by threats or military actions. The two superpowers are clearly the strongest nations militarily, but such happenings as the American debacle in Vietnam in the 1960s and 1970s, the unsuccessful American intervention in Lebanon in the early 1980s, and the Soviet Union's inability to subdue Afghanistan after the 1979 Soviet invasion indicate that the world situation may make it unfeasible for "great powers" to impose their political will on lesser nations. The United States is still arguably the "world's greatest power" in a political sense, but the term has less force than it once did. There is even some evidence of strain on American economic power, in the form of massive national debt and recent trade deficits.

The development of such a high degree of wealth and power in Anglo-America is a matter too complicated to be fully explainable and is subject to various interpretations as to the primary reasons. But it is clear that the United States and Canada possess, or have possessed, a number of specific assets on which they have been able to capitalize. Some of the more important ones may be stated as follows:

1. Both countries are large in area.

2. Both possess comparatively effective internal unity, although Canada has had some difficult problems in this respect in recent decades.

3. Both are outstandingly rich in natural resources.

4. Their combined population is quite large (the United States alone being the fourth most populous nation of the world), yet neither country is overpopulated.

5. Both countries developed very mechanized economies rather early and under very favorable conditions, although this is considerably more true of the United States than of Canada.

6. The two countries occupy a geographic position in the world that has been, and to some extent still is, defensively strong due to relative isolation from other major centers of population and military power.

7. Despite occasional irritations and quarrels, the relations of the Anglo-American nations with each other are generally friendly and cooperative.

All the foregoing points are elaborated in the following sections. They not only help to explain the relative success of the United States and Canada as countries, but they also throw light on the general character of Anglo-America as a world region.

LARGE AREA OF THE ANGLO-AMERICAN COUNTRIES

Possession of a large area assures a country of neither wealth nor power. But it does afford at least the possibility of finding and developing a wider variety of resources and, other things being equal, of supporting a larger population than might be expected in a small country. In the 20th century two of the world's largest countries, the United States and the Soviet Union, have surpassed the older and smaller states of western Europe as world powers. Both the United States and the Soviet Union have resources which are much superior to those of the west European nations, but until the 20th century neither country had found the means to occupy and organize its national territory effectively and thus to take adequate advantage of its resources.

The fact that there is no direct relationship between area on the one hand and wealth and power on the other is amply attested by the comparative areas of the Anglo-American countries themselves. Canada is the larger country of the two in area but is much smaller than the United States in regard to total wealth and political, economic, and military power. The United States measures 3,619,000 square miles (9,373,000 sq. km), while the area of Canada is 3,850,000 square miles (9,971,000 sq. km). Canada's territorial extent is second in the world only to that of the Soviet Union, while the United States ranks behind these two countries and

China. Even the United States, however, has about twice the area of Europe outside the Soviet Union. France, the largest European country in area, is only one seventeenth the size of the United States.

INTERNAL UNITY OF THE UNITED STATES AND CANADA

The welding of their large national territories into effectively functioning units represents a major accomplishment by the Anglo-American countries and a major source of their wealth and power. They are not weakened by chronic political separatism in any of their parts to the same degree that is true of many of the world's countries. Their economic welfare and, indirectly, their political unity are promoted by constant and large-scale interchange of goods, people, and ideas between sections far removed from one another.

Political separatism in the world's countries is often based on ethnic differences. It is obvious that ethnic diversity not only exists but is a source of national pride in the United States and Canada. However, most ethnic minorities are small and/or geographically scattered and have been strongly Americanized or Canadianized. As a result, the degree of cultural unity in Anglo-America is strikingly in contrast to the large number of highly distinct ethnic groups, often with separatist feelings, that exist in many of the world's nations. The aboriginal population of American Indians was overwhelmed by the tide of European settlement and has now been reduced to a politically weak minority concentrated mainly in the western and northern parts of Anglo-America. The Indian population is gradually being absorbed into the main stream of American and Canadian life, although much of it still lives as a depressed group under conditions impossible to justify. Non-British immigrants have been gradually absorbed also, although not without making valuable contributions to (and modifications of) the Anglo-American society they have entered.

Each of the Anglo-American countries has one exceptionally large ethnic minority, and in each case this minority has been the focus of serious problems of national unity. About 27 percent of Canada's people belong to the French-Canadian group, which is differentiated from the rest of the Canadian population not only by language but also by religion, being overwhelmingly Roman Catholic while the rest of Canada's population is predominantly Protestant. The French-Canadians are concentrated, for the most part, in the lowlands along the St. Lawrence River in the province of Quebec, although they have also spread to some extent into adjoining areas. The ancestors of this group, some 60,000 to 70,000 in number, were left in British hands when France was expelled from Canada in 1763. They did not join the English-speaking colonies of the Atlantic seaboard in the American Revolution and by their refusal to do so laid the basis for the division of Anglo-America into two separate countries. Until the latter 20th century the French-Canadians increased rapidly in numbers, and they have clung tenaciously to their distinctive language and culture. At times major controversies arise between this group and other Canadians, and a provincial government with separatist aims was elected by Quebec's voters in the 1970s. However, the Quebec electorate then refused to approve its government's movement toward secession from Canada, and this danger receded during the early 1980s. The Canadian government has achieved and preserved national unity through the recognition of French as a second official language and the provision of legal protection for French-Canadian institutions, but the degree, details, and effectiveness of this bicultural policy are matters of frequent and sometimes acrimonious review. Although not assimilated in a cultural sense, French Canada has been incorporated into the nation to the extent that it has provided many Canadian leaders, including some prime ministers.

Black people of African descent, numbering almost one eighth of the American population, are the principal minority in the United States. Unlike the French-Canadians, they are not numerically dominant in any major political division or section of the country. Neither have black Americans as a group been an active force toward sectionalism in the United States. Nevertheless, a conflict of attitudes toward black slavery was among the important factors which nearly split the United States into two nations in the 19th century, and differing attitudes toward the status and aspirations of black people are still a major source of friction in American politics. During the 1960s the protest by blacks against discrimination, along with the white reaction, set off a period of racial violence, which subsequently abated. The violence centered in the big American cities where an increasing proportion of the black population has concentrated. Today (early 1989) several of the very largest cities, and a number of others, have black mayors.

Spanish-speaking people are a smaller ethnic minority in the United States that has been growing very rapidly in the 1970s and 1980s. In 1980 the census identified, through a self-classifying question-naire, over 14 million people of "Spanish origin." This was a little over 6 percent of the total population. Most of these people were of Mexican origin and strongly concentrated in California, Texas, New Mexico, Arizona, and Colorado. Some 2 million people of Puerto Rican origin were most heavily represented in New York; and about 800,000 of Cuban origin lived largely in Florida. Estimates varied as to how many million other Hispanic people, mainly Mexicans, were actually resident in the United States as illegal and unidentified immigrants. Like the blacks, American Hispanics are not numerically or politically dominant in any state. The closest approach to minority dominance of a state is in multi-ethnic Hawaii, where Asian-Americans have become very powerful politically (although without any hint of separatism). Among many Asian-American groups, Hawaii's Japanese-Americans are by far the most numerous, representing a little under 30 percent of the state's population in 1980.

Even without strong ethnic contrasts and antagonisms, however, regional conflicts of interest are bound to occur in countries of such large extent and varied physical and economic conditions. In this connection it is a striking and significant aspect of Anglo-America that both of its countries employ a federal structure of government. A multitude of powers and functions are assigned to the 50 states of the United States and the 10 provinces of Canada in an attempt to give latitude for governmental expression of regional differences and the solution of regional problems which might otherwise work toward the disruption of national unity. Of course, in both countries such federal structures themselves involve frequent disputes concerning divisions of power between the central governments and the state or provincial governments.

Another notable and important political characteristic of Anglo-America lies in the fact that this region represents an outstanding stronghold of democratic representative government. Governmental responsiveness to majority opinion, combined with safeguards for the minority and for individual freedoms, has tended to prevent revolutionary pressures from building up. The effectiveness of the political system is also strengthened by the fortunate economic circumstances of the region, which tend to allay political discontents.

The result is that Anglo-America has been characterized thus far by great stability of government. The American Civil War is the only large-scale violent civil conflict that has ever occurred in either nation. In neither country has the national government ever been overthrown by force. Thus the gov-

Figure 26.2 The Amtrak "Empire Builder" on the tracks of the Burlington Northern Railroad, en route from Chicago to Seattle. The train is stopped at Havre in the Great Plains of eastern Montana. Transcontinental railroads played a major role in shaping the human geography of the United States and Canada.

ernments of the United States and Canada have not been obliged to devote large energies merely to preserving the state's existence or its territorial integrity against internal stresses.

The governments of both countries have assiduously fostered national unity through the provision of adequate means of internal transportation. In both, the development of transportation networks has had to be accomplished not only over long distances but primarily against the "grain" of the land, since effective transport links between east and west have been the most imperative, while most of the mountains and valleys in Anglo-America have a north-south trend, thus opposing themselves as a series of obstacles to cross-country transport. Both the United States and Canada were tied together as effectively functioning units by heavily subsidized transcontinental railroads (Fig. 26.2) completed during the latter half of the 19th century. These fostered national unity not only directly, through the connections they afforded between sections, but also indirectly through their promotion of settlement and of general economic development. Later, automobile transportation on publicly constructed highway networks and, still later, the development of air transportation further solidified internal unity.

RESOURCE WEALTH OF ANGLO-AMERICA

The natural resources which the Anglo-American countries have within their boundaries are outstanding in both variety and abundance. In part, this fact is simply a reflection of the large territorial extent of each nation. Countries of such magnitude are almost bound to contain a sufficient diversity of physical conditions to yield resources that are out of the ordinary in size and utility. Nevertheless, it is probable that no area of the earth's surface of comparable extent contains so much natural wealth as does Anglo-America. The abundance of resources also reflects the inventiveness and skill of the Anglo-American peoples in the exploitation of nature, in the "making" of resources. It should not be forgotten, however, that the value of Anglo-American resources is partly due to scientific and technologic advances in other lands. Coal, for example, was made a major resource by discoveries in England long before the superior coal resources of Anglo-America were more than vaguely known. Certain important aspects of the world position of the Anglo-American countries with regard to resources and production are summarized in Table 26.1.

AGRICULTURAL AND FOREST RESOURCES

Large areas of Anglo-America are suited by topography and climate for cultivation (chapter-opening photo). The United States, although the fourth largest country in total area, is second only to the USSR in total amount of arable land and is probably first in the world in arable land of high quality. A much smaller proportion of Canada is arable, but even in that nation the total arable acreage is still greater than that of any country other than the Soviet Union, the United States, India, or China. Canada is the world's second richest country in arable land per capita, and the United States, despite its much larger population, ranks behind only Mongolia, Canada, Australia, Argentina, and the Soviet Union in this respect. Furthermore, the United States ranks third, after the USSR and Australia, in total area of meadow and pasture lands. Such abundance has helped Anglo-America become the main food-exporting region in the world.

Anglo-America is estimated to contain about 14 percent of the world's forested area. Canada and the United States rank third and fourth among the nations of the world in this regard, with the Soviet Union ranking first and Brazil second. Canadian forest resources are larger in extent than those of the United States but are not so varied in type (Figs. 26.3 and 27.5). For example, Canada has no forests comparable to the redwoods of California or the fast-growing pine forests of southeastern United States and is far behind the United States in both acreage and variety of deciduous hardwoods. Wood has been an abundant material in the development of both countries. Today the United States cuts more wood each year than any other country except the USSR. It both exports and imports large quantities, with the exports going largely across the Pacific (especially to Japan), and the imports coming overwhelmingly from Canada. Canada's timber industry is considerably smaller, although still one of the greatest in the world. But the country's small population compared to the size of its forest resources allows it to be the world's greatest exporter of wood.

MINERAL RESOURCES

In mineral resources also Anglo-America is outstandingly rich. Such resources played a major role in the development of the region's gigantic economy. Today the mineral output of the two countries is still huge. With its small population Canada is a major exporter of mineral products, largely to the United States. But the United States, while still a major producer, has also become a major importer. This situation reflects the partial depletion of some American resources, combined with growing foreign production, but it also reflects the enormous demands of the American economy.

Abundant energy resources have been basic to the rise of the Anglo-American economy and remain crucial to its operation and expansion (Fig. 3.6). *Coal* was the largest source of energy in the 19th-century industrialization of the United States. Today the country ranks with China and the Soviet Union as one of the three world leaders in coal reserves. In coal production China and the United States are the world leaders, each producing about a quarter of world output in recent years. But the United States is far more important as a coal exporter, being the world's largest by a wide margin. Canada is likewise an important coal exporter, with reserves and production which are small on a world scale but large compared to Canada's population and needs.

Petroleum fueled much of the development of the United States in the 20th century, and the country was for decades the world's greatest producer, with reserves so large that their long-term adequacy received little thought. In 1985 the United States was still the second largest producer of oil in the world

Item	Approximate Percent of World Total		
	United States	Canada	Anglo-America
Area	7	7	14
Population	5	0.5	5.5
Agricultural resources and production			
Arable land	13	3	16
Meadow and permanent pasture	8	1	9
Production of:			
Wheat	13	5	18
Corn (grain)	43	1	44
Soybeans	50	1	51
Cotton	15	—	15
Tobacco	10	1	11
Oranges	16	—	16
Milk	15	2	17
Number of:			
Cattle	10	1	11
Hogs	7	1	8
Forest resources and production			
Area of forests and woodlands	6	8	14
All wood cut	14	5	19
Sawn wood production	19	14	33
Wood pulp production	37	15	52
Minerals, mining, and manufacturing			
Estimated coal reserves	24	1	25
Coal production	24	1	25
Published proved oil reserves (excl. tar sands and oil shale)	5	1	6
Crude oil production	17	3	20
Crude oil refinery capacity	20	3	23
Natural gas production	26	4	30
Natural gas reserves	5	3	8
Potential iron ore reserves	5	6	11
Iron ore mined	6	5	11
Pig iron produced	8	2	10
Steel produced	10	2	12
Aluminum produced	25	8	33
Copper reserves	17	5	22
Copper mined	14	9	23
Lead reserves	22	13	35
Lead mined	13	8	21
Zinc reserves	13	15	28
Zinc mined	4	1	5
Nickel mined	2	20	22
Gold mined	4	6	10
Phosphate rock mined	34	0	34
Potash produced (K_2O equivalent)	5	23	28
Native sulfur produced	21	—	21
Asbestos mined	1	18	19
Uranium reserves	9	12	21
Uranium produced	15	29	44
Electricity produced	27	5	32
Potential waterpower	4	3	7
Production of hydroelectric power	17	15	32
Nuclear electricity produced	28	5	33
Sulfuric acid produced	26	3	29
Steel consumed industrially	21	2	23

Sources and comments: The table excludes Greenland. The notation (—) indicates none or no appreciable amount. Standard statistical sources, such as those cited under Part I in the References and Readings, were used to prepare the table. The percentages, drawn from the middle 1980s, show *volume* rather than *value* of production and reserves. Percentages fluctuate somewhat from year to year, but the total array gives a reasonably good impression of the comparative world importance of the United States and Canada in resources and production. Such figures are subject in varying degree to errors in estimation and compilation, as well as misinterpretations due to differences in quality or ease of exploitation of particular resources from one area to another.

In the late 1970s and the 1980s the world, and especially the United States, appeared to be retreating somewhat from earlier plans for expanding nuclear power generation based on *uranium*. Should uranium in quantity continue to be needed, there are huge reserves in Anglo-America.

Waterpower, harnessed by waterwheels, was the main source of energy for the earliest American industrialization, especially in New England. Today falling water continues to supply electricity by generating hydroelectricity, of which the United States and Canada (in that order) are the world's largest producers. However, the hydropower *potentials* of the two countries are far less impressive in a world comparison. As of 1985, the United States generated 11 percent of its electricity from waterpower, 15 percent from nuclear energy, and the remaining three quarters from fossil fuels. Canada, with a hydroelectric output almost as large as that of the United States, generated two thirds of its electricity from falling water and exported much electricity to the United States.

Non-fuel mineral resources in Anglo-America present a similar picture of abundance—too intricate to be detailed here. This abundance is qualified in the case of the United States by the fact that in instance after instance the country has large reserves and major production of a mineral but is also a major importer. Fairly long exploitation, at massive scales, has somewhat depleted the American reserves, especially of the ores easiest to mine and richest in content; while large mineral outputs within the United States are insufficient to supply the huge demands of the national economy. In many such cases Canada has been able to assume the role of principal foreign supplier. A major example is *iron ore*. The United States originally had huge deposits of high-grade ore, principally in Minnesota's Mesabi Range and lesser ore formations near western Lake Superior. By the end of World War II this ore had been seriously depleted due to the enormous scale of exploitation by the American iron and steel industry. New sources of high-grade ore then were developed in eastern Canada (Quebec and Labrador), principally to supply the continuing American demand for ore of this quality. Meanwhile the Lake Superior region somewhat revived its mining industry by extracting lower-grade ore which is present in vast quantities but must be processed in concentrating plants before it can be shipped at a profit to iron and steel plants.

Figure 26.3   The giant General Sherman sequoia tree in California's Sequoia National Park is a fitting symbol for the enormous timber resources of Anglo-America. The tree is the largest living thing on the planet and is estimated to be over 3500 years old. The person at the base provides a yardstick for this spectacular plant, which is 37 feet (11 m) in diameter at its widest and is 272 feet (83 m) high.

(after the USSR), with one sixth of world output. But this production was taking place from a base of about 4 percent of estimated world reserves, and the country was importing over a third of its needs. Canada's reserves and production were small in comparison but sufficient to be the basis of a long-continuing boom in the western province of Alberta.

Natural gas became a fuel of rapidly growing significance in the later 20th century. The United States has long been the world's largest consumer and today it is a major importer (from Mexico and Canada). In 1986 it ranked second in the world in output (behind the USSR), producing about a quarter of the world's gas from 5 percent of known world reserves. Canada, with smaller but rapidly increasing proven reserves, produced about 4 percent of world output and was a major exporter.

Among the many other important non-fuel minerals produced by the United States and/or Canada in outstanding quantities from large reserves are *copper*, *lead*, *zinc*, *nickel*, and *silver*; *asbestos*; and such chemical and fertilizer materials as *sulfur, salt, potash,* and *phosphate.* For several of these, either the United States or Canada is the leading world producer, and in each of the other cases one or both of the two nations ranks among the top three producers. *Gold* is still mined in fair quantity in both countries, although its relative importance is less than it was in earlier times when gold rushes were generators of development in California, Colorado, and various other parts of the Anglo-American West.

Such is the mineral wealth of Anglo-America that it is much easier to summarize gaps in the array than to describe the resources that are present. Some major minerals in which both countries appear to be truly deficient are chromite, tin, diamonds, high-grade bauxite, and high-grade manganese ore.

ADVANTAGES WITH RESPECT TO POPULATION

In mid-1988 the United States had an estimated population of 246 million, while Canada had an estimated 26 million inhabitants. Thus Anglo-America contained about 272 million people. Three of the world's countries—China, India, and the Soviet Union—surpass all of Anglo-America in population. Nevertheless, the United States ranks fourth among all countries, with a population total substantially larger than the combined totals of Europe's four most populous countries—West Germany, Italy, the United Kingdom, and France. Most of Anglo-America's people live in the southeastern quarter of the region, from the St. Lawrence Lowlands and Great Lakes south, and this part of Anglo-America is one of the world's four most extensive areas of dense population, along with the continent of Europe (in-

cluding the western Soviet Union), eastern China and nearby areas, and the Indian subcontinent.

From the beginning of settlement until a growth-rate decline in the 1960s and 1970s, Anglo-America was a region of rapidly growing population. By 1800, approximately two centuries after the first settlements were firmly established at Jamestown and at Quebec, there were over 5 million people in the United States and several hundred thousand more in Canada. Population growth since that time is summarized in Table 26.2. Continuing rapid—although slowing—growth is evident through the 1950s. A sharp downturn in the rate of growth came in the United States in the 1960s and in Canada in the 1970s.

The population of this region long increased at a rate more rapid than that of the world as a whole. Between 1900 and 1950 the increase in world population is estimated at around 50 percent, while that in Anglo-America was over 100 percent. Thus one element in the mounting world importance and power of Anglo-America, particularly of the United States, was its possession of an increasing share of the world's population. The older world powers of western Europe, especially, have been surpassed. In 1870, at the time of the Franco-Prussian War, the United States had almost exactly the same population as each of the two belligerents, Germany and France, but in mid-1988 it had 1.8 times as many people as France and the two Germanies combined. By 1988, however, the population growth rates of both the United States and Canada were well below those of most of the world, with important exceptions in the case of most European countries, Japan, and the Soviet Union. Together the Anglo-American countries account for a little over 5 percent of the estimated world population. This compares with

Year	U.S. Population (millions)	Canadian Population (millions)	Total (millions)	Total Increment (millions)	(%)
1850	23.3	2.4	25.7	—	—
1900[a]	76.1	5.4	81.5	55.8	217
1950[a]	151.1	14.0	165.1	83.6	103
1960[a]	179.3	18.2	197.5	32.4	20
1970[a]	203.2	21.6	224.8	27.3	14
1980[a]	226.5	24.3	250.8	26.0	12
1988	246.1	26.1	272.2	18.9	9

TABLE 26.2 POPULATION GROWTH IN THE UNITED STATES AND CANADA, 1850–1987

[a]Actual dates for Canada are 1901, 1951, 1961, 1971, and 1981. The decennial census is taken 1 year later in Canada than in the United States. The table uses census data except for 1988 estimates.

their possession of about 14 percent of the world's land area (excluding Antarctica). The estimated average population density of the two countries in 1988 was only 68 per square mile (26 per sq. km) for the United States and 7 per square mile (3 per sq. km) for Canada. The very low figure for Canada is due to its tremendous expanse of sparsely settled northern lands, the effectively occupied part of the country in the south being much more densely populated, although far smaller in areal extent.

It is evident from a population density figure of 89 per square mile (34 per sq. km) of land surface for the world as a whole that even the United States is far from overpopulated. This is even more true considering its exceptional resources and enormous production. And Canada, if anything, is underpopulated. In their development, both countries have been favored by abundant space and resources, and both are still rich in these respects by general world standards.

MECHANIZATION AND PRODUCTIVITY

The high productivity and high national incomes of the Anglo-American countries have essentially come about through the use of machines and mechanical energy on a lavish scale. Canada's achievements in these regards have been highly creditable for a nation with so few people, but they have been dwarfed by the massive outpouring of machines and machine-produced goods and services in the United States. From an early reliance on waterwheels to power simple machines, the United States moved on an increasing scale to exploit the power generated from coal by steam engines. Later the total amount of power used to drive machines enormously increased when internal-combustion engines and electric motors were developed. Thus did the United States eventually achieve mechanization on a scale unmatched by any other nation and a national reliance on machines that is one of the world's most all-embracing even today. In achieving such a mechanized economy and society in a relatively short span of time, the United States was able to take advantage of a unique set of circumstances. For one thing, the country had resources so abundant and varied that they attracted foreign capital and also made possible large domestic accumulations of capital through large-scale and often wasteful exploitation. There was also a labor shortage, which drew in millions of immigrants as temporarily low-paid workers but also encouraged higher wages and la-

bor-saving mechanization. A relatively free and fluid society encouraged strivings for advancement; large segments of the population subscribed with an almost religious fervor to ideals of hard work and economic success; and national energies did not have to be diverted excessively into defense and war.

Most of the foregoing American conditions were duplicated in Canada, and some of them in a number of other countries, but one additional American advantage was not. This was the existence in the United States of a large, unified, and growing internal market which allowed great economic organizations to specialize in the mechanized mass production of a few items, thus lowering the unit costs of production and in consequence further expanding the size of the market. This enormous American asset has been eroded, along with some others, in the latter 20th century. Transportation and communication have become so cheap and rapid that the world market rather than the domestic market has become the essential one for major industries. In a former time when world trade was much more restricted, an American company selling its goods in the huge United States market might have a considerable advantage over a Japanese or a European company operating in a much smaller and poorer home market. But today, so long as freedom of trade is fostered and protectionism is kept at bay, the American company and the Japanese or European company are likely to compete on a much more even basis in America, Japan, Europe, and the world. The powerful economic position of the United States is in good part a product of the country's earlier advantages, but today this position is under strain as a truly worldwide market and economy come into being.

DEFENSIVE ADVANTAGES OF RELATIVE ISOLATION

Anglo-America is essentially a huge island, bordered on the east by the Atlantic Ocean, on the north by the Arctic Ocean, and on the west by the Pacific Ocean. The only land connection with another world region is along the boundary between the United States and Mexico, which mostly runs through sparsely populated territory and is far removed from the core area of either country. There is practically no traffic by land from the United States across Mexico and Central America to South America. Anglo-America's contacts with the latter continent, as well as with other world areas, are made by sea and air.

The closest land approaches of Anglo-America to Eurasia are at the northeastern and northwestern corners of the region. A series of islands bridges the North Atlantic between Canada and Europe and includes Greenland, Iceland, the Faeroes, the Shetlands, and Svalbard. Only the Bering Strait separates Asia from Alaska. The first historic contact between Europe and North America, that of the Vikings in about the year 1000, was made via the island steppingstones of the North Atlantic. Contacts with Asia occurred earlier, since the North American Indians entered the continent in prehistoric times from Asia via the Bering Strait area. In the 18th century Russian fur traders entered Alaska, which passed from Russian to American control by purchase only in 1867.

But harsh climatic conditions in the latitudes of these land approaches have prevented them from serving as major avenues of movement between Anglo-America and Eurasia during historic times. The North Atlantic approach is dangerous to navigate and leads, on the Anglo-American side, only to the wasteland of tundra and coniferous forest which occupies northern Canada. Bering Strait is backed on both sides by similar stretches of wasteland. Between the two approaches, Eurasia and North America are separated by the ice-jammed Arctic Ocean, again backed by tundra and taiga on both sides. Consequently, the main connections between the two great land masses of the Northern Hemisphere have lain across the oceans farther south, although the routes generally followed do swing north to approximate great-circle courses and some air services now cross the Arctic. From the Atlantic seaboard ports of Anglo-America south of the taiga the sea distance to Britain is about 3000 miles (*c.* 4800 km; only about 2000 miles between the outer extremities of Newfoundland and Ireland), and from Puget Sound to Japan is about 4000 miles. If the Pacific is crossed via Hawaii to Japan, the distance becomes well over 5000 miles. These have been the real distances separating the states of Anglo-America from the world's other centers of political and military power—distances not overly long for peaceful commerce but constituting a major obstacle to military operations. Such distances (added to inherent strength) so reduced the danger of a successful military attack on Anglo-America that the region was able in the past to proceed with its domestic development while paying relatively little attention to considerations of defense and the entanglements of world politics.

However, three new factors have arisen in relatively recent times to lessen the value of Anglo-America's protected position. In the first place, the destruction of the balance of power in Eurasia has raised the possibility that while no individual Eurasian power could successfully overcome Anglo-America, a Eurasia united wholly or in large part under one dominion might well aspire to do so. A united Eurasia would change the position of Anglo-America from that of an isolated region protected by distance to that of an area semisurrounded by its possible antagonist. Second, the development of guided missiles has undermined the protective function of the oceans and the northern wastes with respect to attack from the air. Finally, the development of nuclear armaments has raised the possibility of devastating, or even decisive, effect by attack from the air.

The relatively isolated position of Anglo-America has allowed the region to develop in relative peace and security and remains a valuable asset and source of strength, since it is still harder to attack effectively over long than over short distances. But the protective value of this position has declined sharply in recent times, and the people of the region can no longer afford to view the rest of the world with the degree of detachment that was common in the past.

COOPERATIVE RELATIONS OF THE ANGLO-AMERICAN STATES

For many years after the American Revolution the political division of Anglo-America between the independent United States to the south and a group of British colonial possessions to the north was accompanied by serious friction between the peoples and governments on either side of the boundary. A heritage of antagonism was present due to the failure of the northern colonies to join the Revolution, their use as British bases during that war, the large element of their population which was composed of Tory stock driven from American homes during the Revolution, and uncertainty and rivalry as to ultimate control of the central and western reaches of the continent. The War of 1812 was largely fought as an American effort to conquer Canada. Even after the failure of this effort a series of border disputes occurred, and American ambitions to possess this remaining British territory in North America were openly expressed; suggestions and threats of annexation were made in official quarters throughout the 19th century and into the 20th century.

In fact, Canada as a unified nation is in good part a result of American pressure. After the American Civil War the military power of the United States took on a threatening aspect in Canadian and British eyes. Suggestions were made in some American quarters that Canadian territory would be a just recompense to the United States for British hostility toward the Union during the war. The British North America Act, passed by the British Parliament in 1867, brought an independent Canada into existence. Britain sought, and as the event has proved, successfully, to establish in Canada an independent nation capable of achieving unity, of morally deterring American conquest by its very independence of Britain, and of relieving Britain of some of the burden of defense in North America.

Hostility between the United States and its northern neighbor did not immediately cease with the establishment of an independent and unified Canada; a certain amount of friction over trade and policy differences has persisted. Nevertheless, relations between the two Anglo-American nations have improved gradually to the point where these countries are often cited as an outstanding example of international amity and cooperation, and the frontier between them has ceased to be a source of mutual insecurity and weakness. This frontier, stretching completely unfortified across a continent, has become more a symbol of friendship than of enmity.

The bases of Canadian-American friendship are cultural similarities, the material wealth of both nations, and the mutual need for and advantages of cooperation. A very large volume of trade moves across the frontier and strengthens both countries economically and militarily. Each country is a vital trading partner for the other, although Canada is much more dependent on the United States in this respect than is the United States on Canada. In 1986, for example, Canada supplied 18 percent of all United States imports by value and took 21 percent of all United States exports, making Canada second to Japan in total trade with the United States, while the United States supplied 69 percent of Canada's imports and took 77 percent of its exports. In addition, the United States supplies large quantities of capital which have been an important factor in the rapid economic development of Canada during recent decades but are also a frequent source of disquiet in Canada due to the amount of control over the Canadian economy they involve. However, Canadians, in turn, invest heavily in the United States. Except for Canadian exports of automobiles and auto parts to the United States, the main pattern of trade between the two countries is the exchange of Canadian raw and intermediate-state materials— primarily ores and metals, timber and newsprint, and oil and natural gas—for American manufactures. A free trade pact was signed in 1988.

In defense arrangements, the two nations work together so closely as to form almost a single unit.

Important aspects of their relationship include (1) a structure of joint command, (2) cooperative military maneuvers in far northern areas, (3) cooperative installations to warn against air attack, and (4) cooperation in production and supply of military equipment. These arrangements mean essentially that Canada is protected by its more powerful neighbor, which would certainly regard any attack on Canada as an attack on itself. But at the same time, of course, Canada's resources and strategic assets greatly strengthen that neighbor. Meanwhile, as an independent nation Canada continues to guard its sovereignty and cultural identity, to differ with American foreign policies at times, and to maintain its ties with the United Kingdom and the other states of the Commonwealth of Nations.

OPPORTUNITIES AND HANDICAPS IN DIVERSE ENVIRONMENTAL REGIONS

The foregoing discussion of Anglo-America as a whole largely bypasses the complex internal differentiation of this major world region. Here we commence to explore areal differentiation and its consequences through brief sketches of major environmental regions. Other kinds of regions based on human activity will be examined in Chapters 27 and 28.

Even in a broad-scale view which suppresses much detail, the natural environments of Anglo-America are remarkably diversified. This diversity reflects a number of major circumstances:

1. Due to its broad extent in both latitude and longitude the region contains a great many types of climate, each with its associated complement of natural vegetation, soils, and other climate-related features.

2. Due to a complex geologic and geomorphic history, landform types and features are also extremely varied.

3. Landform regions and climatic regions frequently do not correspond very well in location. Thus one climatic type often extends into several landform regions, and one landform region often embraces parts of several climatic regions. The result is a rather intricate, although reasonably orderly, mosaic of areas— large and small—wherein particular landform types occur under particular climatic conditions. Some areas have proved highly advantageous for human occupation and use; some have provided opportunities when used selectively; and some have been practically useless. In Anglo-America as a whole, the opportu-

nities have far outweighed the handicaps. In fact, nature endowed this region with an astonishing utility for a diversity of peoples moving through successive stages of development and change. Even today, when dependence on the outer world for certain resources has greatly increased, the inherent contributions of nature to human welfare in Anglo-America should not be underestimated. Widespread appreciation of these contributions has recently led to a new degree of concern for preservation and enhancement of the environment in the United States and Canada.

For clarity on an introductory level, climatic types and regions are considered separately from major landform divisions in the subsequent discussion, but interconnections of various natural elements with each other, and with economic development, are touched on at many points.

CLIMATIC TYPES AND REGIONS

The climatic pattern of Anglo-America exhibits both diversity and largeness of scale. Every major nontropical type of climate is found within the region, usually over broad expanses (Fig. 2.18). The United States includes a greater number of major climatic types within its boundaries than does any other country of the world. In addition to the country's middle- and high-latitude climates, the state of Hawaii and extreme southern Florida give it small areas of tropical climates. Even Canada is more diversified in climate than is commonly assumed. The variety of economic opportunities and possibilities afforded by the wide range of climates in Anglo-America is one of the basic factors underlying the economic strength of this world region. The region satisfies its own needs for most climate-related products of farm and forest or even produces large surpluses for export, although the lack of any large producing area for certain tropical specialty crops necessitates the import of such commodities as coffee, tea, cocoa, bananas, rubber, and tropical vegetable oils.

POLAR AND SUBARCTIC CLIMATES

Northern Anglo-America is handicapped by cold. A belt of *tundra climate,* with its associated vegetation of mosses, lichens, sedges, hardy grasses, and low bushes, rims the coast from Alaska to Labrador and extends into the Arctic islands of Canada and the coastal sections of Greenland (Fig. 26.4). In interior Greenland an *ice-cap climate* prevails, and the land is covered by an enormous continental glacier. In the ice-cap climate every month has an average temperature below 32°F (0°C). In the tundra climate from 1 to 4 months average above freezing, although no month averages above 50°F (10°C). South of the tundra a vast expanse of *subarctic climate* spreads across most of Alaska and occupies about half of Canada (Fig. 26.5). Like its counterpart in Eurasia, this belt extends completely across the continent between the Atlantic and Pacific oceans. The subarctic region of Anglo-America, with its long, cold winters and short, mild summers, has a natural vegetation of coniferous snow forest resembling the Russian taiga. Population is extremely sparse in the zones of tundra and subarctic climate and practically nonexistent in the ice-cap climate. Scattered groups of people engage in trapping, hunting, fishing, mining,

Figure 26.4   A classic view of tundra in the Keewatin District of Canada's Northwest Territories near the Eskimo village of Rankin Inlet on Hudson Bay. The ground is carpeted with lichens and grass, and not even the spindliest of trees is to be seen.

Figure 26.5 A scene in the Manitoba subarctic. Spindly conifers of little commercial value characterize this northern section of the subarctic. Marshes and swamps are common in the cool, glaciated environment. The road at the left is made of waste material from two operating levels of an underground nickel mine near by.

lumbering, and military activities. Many Eskimos and Indians in this region of limited economic opportunity live largely on welfare (Fig. 27.6).

HUMID CONTINENTAL AND HUMID SUBTROPICAL CLIMATES

Approximately the eastern half of Anglo-America south of the subarctic zone is an area of humid climate which accounts for most of Anglo-America's agricultural production and within which most of the people of this world region are found. Temperature variations associated with a spread in latitude of more than 25 degrees provide the basis for a division of this humid eastern area into three climatic regions, each taking the form of an east-west belt. At the north is a belt of *humid continental climate with short summers,* characterized by long and quite cold winters and short warm summers with few periods of really hot weather. Agriculturally this belt is characterized by a heavy emphasis on dairy farming, except in the extreme west, where spring wheat production is dominant. The central climatic belt of the three is one of *humid continental climate with long summers.* Here, winters are cold and summers are not only rather long but also hot. The leading forms of agricultural production are dairying in the eastern portion and a corn-soybeans-cattle-hogs combination in the midwestern (interior) portion. At the south is the *humid subtropical climate,* which characterizes the southeastern United States. Winters are short and cool, although with cold snaps, and summers are quite long and hot. Agriculture is rather diverse from place to place, with some major em-

phases in specialized areas being cattle (Fig. 28.14), poultry (Fig. 28.15), soybeans, tobacco, cotton (Fig. 26.6), rice, peanuts, and a wide range of fruits and vegetables. Within the three climatic regions the pattern of natural vegetation is complex, each region being associated with areas of coniferous evergreen softwoods, broadleaf deciduous hardwoods, mixed hardwoods and softwoods, and prairie grasses (Fig. 28.18).

Figure 26.6 A view in the humid subtropical climate zone of the Alabama Piedmont (near the town of Talladega), east of Birmingham. Fields of cotton and corn are bordered by mixed oak-pine woodland. The light-colored soil in which the cotton plants are growing is typical of the climate. Cotton, a clean-tilled row crop, contributed greatly to the American South's traditional problem of soil erosion, but relatively little of the crop is grown today on erosion-prone land.

Figure 26.7 *A dairy herd in the marine west coast climate of Oregon. One of the Coast Ranges is in the background. The herd is grazing on lush pasture in a stretch of alluvium between the mountains and the Pacific. The scene is in Tillamook County, one of the hundred leading dairy counties in the United States.*

MARINE WEST COAST AND MEDITERRANEAN CLIMATES

Due to the barrier effect of high mountains near the sea, ocean waters offshore which are warm in winter and cool in summer relative to the land, and winds prevailingly from the west throughout the year, a narrow coastal strip from southern Alaska to northern California has a distinctive *marine west coast climate*. The mild, moist conditions are strikingly similar to those of the corresponding climatic region in northwestern Europe. Most of the land near the coast is too rugged for much agriculture, and the mountains prevent the penetration of marine conditions for any great distance inland. Nevertheless, a certain economic resemblance to northwestern Europe is shown in a considerable development of dairying which has taken place in western Oregon and Washington and southwestern British Columbia (Fig. 26.7). A magnificent growth of giant conifers in this climate, including the famed redwoods of northern California and Douglas fir of Oregon and Washington, provides the basis for a great development of lumbering (Fig. 27.5).

In central and southern California, south of the area of marine climate, the rain comes in winter and temperatures are subtropical, giving the United States a small but very important area of *mediterranean* or *dry-summer subtropical climate*. Here irrigated production of cattle feeds, fruits, cotton, and a great range of other crops, together with associated livestock, dairy, and poultry production, makes California the leading American state in total agricultural output.

DRY CLIMATES AND HIGHLAND CLIMATES

In the immense region between the western littoral of the United States and the landward margins of the humid East, the dominant climatic characteristic is lack of moisture. The only large area of true *desert climate* occurs along the southern edge, in Arizona and adjoining parts of California, Nevada, Utah, New Mexico, and Texas. The rest is generally classified as having a semiarid *steppe climate*, although scattered areas of desert occur west of the Rocky Mountains as far north as southeastern Washington. The region of steppe climate extends northward into Canada both east and west of the Rockies. Temperatures in this vast region of dry climate range from continental in the north to subtropical in the south. The prevailing natural vegetation of short grass, bunch grass, shrubs, and stunted trees supplies forage varying greatly in utility from place to place but present in sufficient quantity for cattle raising to be the predominant form of agriculture. Areas that are less dry are often used for wheat, and both wheat and other crops are grown in scattered irrigated areas. High, rugged mountains rising in the midst of these dry lands have *highland climates* varying with latitude, altitude, and exposure to sun and to moisture-bearing winds. Mountain slopes having sufficient precipitation are forested up to the timberline with conifers, principally pine and fir (Fig. 26.8).

Climatic data for selected Anglo-American stations are given in Table 26.3.

Figure 26.8 *At timberline in the Colorado Rockies the "pointed conifers" that thrive in the cool high-altitude climate gradually give place to alpine tundra. The photo was taken at an elevation of about 11,000 feet (3350 m) from Monarch Pass west of Colorado Springs.*

TABLE 26.3 CLIMATIC DATA FOR SELECTED ANGLO-AMERICAN STATIONS

Station	Region	Elevation Above Sea Level (ft)	(m)	Type of Climate	Average Temperature: Annual °F	°C	January (or coolest month) °F	°C	July (or warmest month) °F	°C	Average Annual Precipitation (in.)	(cm)
Barrow, Alaska	Arctic Coastal Plains	22'	7 m	Tundra	10°F	−12°C	−18°F (Feb.)	−28°C	39°F	4°C	4"	11 cm
Fairbanks, Alaska	Yukon River Basin	436'	133 m	Subarctic	27°F	−3°C	−11°F	−24°C	59°F	15°C	11"	29 cm
Kapuskasing, Ontario	Canadian Shield	752'	229 m	Subarctic	34°F	1°C	1°F	−17°C	63°F	17°C	33"	85 cm
Regina, Saskatchewan	Interior Plains (Spring Wheat Belt)	1884'	574 m	Humid continental (short summer)	36°F	2°C	1°F	−17°C	66°F	19°C	16"	40 cm
Montreal, Quebec	St. Lawrence Lowlands	187'	57 m	Humid continental (short summer)	45°F	7°C	14°F	−10°C	70°F	21°C	38"	97 cm
Portland, Maine	Northern New England	61'	19 m	Humid continental (short summer)	46°F	8°C	22°F	−5°C	68°F	20°C	43"	108 cm
Des Moines, Iowa	Interior Plains (Corn Belt)	948'	289 m	Humid continental (long summer)	50°F	10°C	21°F	−6°C	77°F	25°C	30"	77 cm
Concordia, Kansas	Interior Plains (Winter Wheat Belt)	1375'	419 m	Humid continental (long summer)	53°F	12°C	31°F	−1°C	79°F	26°C	27"	67 cm
Lexington, Kentucky	Southeastern Interior Plains	979'	298 m	Humid subtropical	55°F	13°C	32°F	0°C	76°F	24°C	44"	111 cm
Norfolk, Virginia	Atlantic Coastal Plain	26'	8 m	Humid subtropical	60°F	15°C	41°F	5°C	79°F	26°C	45"	115 cm
Atlanta, Georgia	Southern Piedmont	975'	297 m	Humid subtropical	61°F	16°C	45°F	7°C	79°F	26°C	47"	120 cm
Houston, Texas	Gulf Coastal Plain	41'	13 m	Humid subtropical	70°F	21°C	54°F	12°C	84°F (Aug.)	29°C	46"	117 cm
Dallas, Texas	Gulf Coastal Plain	487'	148 m	Humid subtropical	66°F	19°C	45°F	7°C	85°F	29°C	34"	86 cm
Miami, Florida	Atlantic Coastal Plain	8'	2 m	Tropical savanna	75°F	24°C	66°F	19°C	82°F	28°C	60"	152 cm
Laredo, Texas	Lower Rio Grande Valley	500'	152 m	Steppe	74°F	23°C	58°F	14°C	88°F	31°C	19"	48 cm
Denver, Colorado	Great Plains	5292'	1613 m	Steppe	50°F	10°C	30°F	−1°C	73°F	23°C	16"	41 cm
Miles City, Montana	Great Plains	2629'	801 m	Steppe	46°F	8°C	15°F	−9°C	75°F	24°C	14"	35 cm
Calgary, Alberta	Great Plains	3540'	1079 m	Steppe	39°F	4°C	16°F	−9°C	62°F	17°C	16"	42 cm
Spokane, Washington	Columbia Plateau	2357'	718 m	Steppe	46°F	8°C	25°F	−4°C	70°F	21°C	17"	44 cm
Tucson, Arizona	Basin and Range Country	2558'	780 m	Desert	68°F	20°C	50°F	10°C	86°F	30°C	11"	29 cm
Los Angeles, California	Southern California Coast	312'	95 m	Mediterranean	64°F	18°C	55°F	13°C	73°F	23°C	14"	37 cm
Sacramento, California	California Central Valley	25'	8 m	Mediterranean	63°F	17°C	46°F	8°C	77°F	25°C	16"	41 cm
Seattle, Washington	Puget Sound Lowland	14'	4 m	Marine west coast	54°F	12°C	41°F	5°C	66°F	19°C	34"	86 cm
Juneau, Alaska	Southern Alaska Panhandle	15'	5 m	Marine west coast	39°F	4°C	23°F (Feb.)	−5°C	55°F	13°C	55"	139 cm
Alamosa, Colorado	Rocky Mountains	7536'	2297 m	Highland	41°F	5°C	17°F	−8°C	65°F	18°C	7"	18 cm

MAJOR LANDFORM DIVISIONS

The land surface of Anglo-America is extremely varied in form and can be classified into numerous landform types ranging from flat swampy plains to high mountains. The types, in turn, can be grouped regionally into a series of major landform divisions with commonly recognized names. This summation presents 11 divisions, as outlined in Figure 26.9. Five of these—the Canadian Shield, Arctic Coastal Plains, Gulf-Atlantic Coastal Plain, Piedmont, and Interior Plains—are basically areas of plains, although the Shield and the Piedmont differ from the other divisions in their greater proportion of rolling to hilly land and in their underlying rocks, which are largely igneous and metamorphic rather than the sedimentaries of the other plains. Plains areas surround two upland divisions—the Appalachian Highlands and Interior Highlands—that contain some areas of low mountains but are more generally comprised of hills. To the west of the Interior Plains, the physiography of Anglo-America is dominated by a spectacular display of high mountains, interspersed with lower mountains and hills, high plateaus, basins, and valleys. Most of the high mountains are in two major landform divisions—the Rocky Mountains and the Pacific Mountains and Valleys—while the plateaus and basins lie largely within the Intermountain Basins and Plateaus. Finally, remote Greenland may be recognized as an outlying major division. We now survey the landform divisions in a general progression from east to west.

THE CANADIAN SHIELD, ARCTIC COASTAL PLAINS, AND GREENLAND

The Canadian Shield, sometimes called the Laurentian Shield or Laurentian Upland or Plateau, occupies an immense area in northeastern Anglo-America. It extends from the Arctic Ocean to the line of the Great Lakes and the St. Lawrence Lowlands, and from the Atlantic to a line on the west traversing Great Bear Lake, Great Slave Lake, Lake Athabasca,

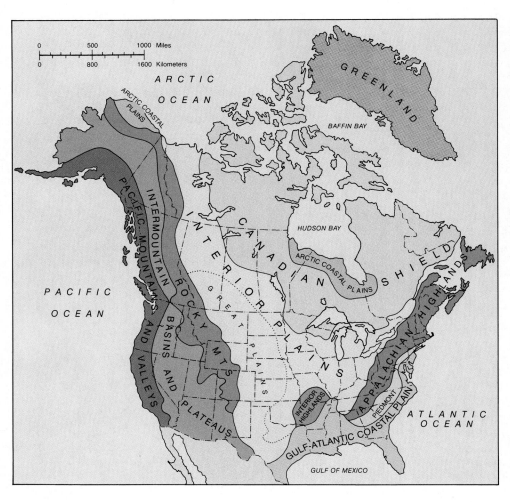

Figure 26.9 Major landform divisions of Anglo-America. Only certain portions of Canada's Arctic islands share the rock formations of the Canadian Shield. However, the overall physical conditions of the islands are sufficiently similar to conditions in the Shield that it is convenient for the purposes of this text to regard the entire island group as an extension of the Shield. (Physiographic boundaries on the map after Fenneman, Lobeck, and others.)

and Lake Winnipeg (Fig. 27.2). The Shield covers more than half of Canada and extends into the United States to the west and south of Lake Superior; this American section is known as the Superior Upland. The Adirondack Mountains in New York state are geologically an extension of the Shield, but for most purposes they are more conveniently discussed with the Appalachian Highlands. In general, the topography of the Shield is that of a rolling plain or an endless succession of low, rounded hills. Much of it has a rugged, rocky quality, but few sections are mountainous. Its surface was repeatedly scoured by the continental glaciers which spread from both sides of Hudson Bay in the Great Ice Age. Thus the ancient rocks composing the Shield are generally exposed at the surface or covered by a very thin layer of poor soil. Agriculture is extremely scanty, being deterred not only by lack of soil, but also by the harsh climate. Glaciation also disrupted the pre-existing drainage system, resulting in a plethora of lakes, swamps, and wandering streams with many rapids and waterfalls. Today, many parts of this tangled wilderness area often seem to consist more of water than of land. Population is very sparse despite major resources of wood, waterpower, and metallic minerals such as iron and nickel.

The Shield surrounds a fairly extensive area of Arctic Coastal Plains along the southern shores of Hudson Bay. A second segment of this landform division, far removed from the first, is in northern Alaska. The Alaskan coastal plain is an area of tundra, while that along Hudson Bay is largely subarctic in climate and forested, although grading into tundra in some sections. The latter area is extremely thinly populated and plays little part in the economic life of Anglo-America, although it has a few fur-trading posts. The Alaskan section, on the contrary, contains a major oil development that has come into being since 1968. Onshore and offshore oil fields in this "North Slope" part of Alaska are connected by the Alaska Pipeline to a tanker port at Valdez in southern Alaska. Aside from oil installations the Alaskan plain has little economic activity and a meager population.

Greenland is not considered a part of the Canadian Shield but may be thought of as an extreme phase of Shield conditions. It is especially comparable to Arctic islands of Canada that are included with the Shield. However, Greenland is far larger, being in fact the world's largest island. It also has higher and more extensive mountains, and the giant ice-cap that covers most of Greenland dwarfs the scattered areas of permanent ice in the Canadian islands. Despite this handicap, Greenland has some 55,000 permanent residents in coastal strips of tundra, mainly in the far southwest, and a few NATO military installations scattered over the island. The island's people are supported mainly by fishing. (For an additional discussion, see Chapter 9.)

THE GULF-ATLANTIC COASTAL PLAIN

An extensive coastal plain occupies the seaward margin of Anglo-America from New Jersey to the Rio Grande (Figs. 28.1–28.3, 28.10, and 28.11). North of New Jersey it appears in two disconnected sections, Long Island and Cape Cod, and south of the Rio Grande it continues into Mexico. On the landward side it is bordered by the higher ground of the Appalachian Highlands, the Piedmont, and the Interior Highlands, and it merges in certain sections with the Interior Plains. The Coastal Plain is narrow in the northeast but widens to the south and west. Beyond the Appalachians it extends northward along the line of the Mississippi River to the mouth of the Ohio before its margin again swings off southwestward toward the Rio Grande. Only three states—Florida, Mississippi, and Louisiana—lie entirely within the Coastal Plain, but the Plain touches 16 other states. This extensive portion of the United States includes a substantial part of both the South and the Northeast and reaches the margins of the Midwest and the West.

The Gulf-Atlantic Coastal Plain is generally low in elevation. Few places exceed 500 feet (c. 150 m) above sea level, and most lie considerably lower. In general, the surface is relatively level, although there are a few belts of rolling hills. Most of the soils are sandy and of low to medium fertility, and a large part of the land is in forest. Pine or mixed oak-pine forest is the prevailing natural vegetation. Stands of swamp hardwoods occur in poorly drained areas; and prairie grasslands were the original vegetation in certain coastal areas in Texas and southwestern Louisiana and in some inland areas underlain by limestone. The coastal prairies grade into steppe near the Rio Grande. Especially fertile soils are associated, for the most part, with limited areas of river alluvium or of chalky limestone bedrock. Agriculture involves many different specialties in islands of intensive development scattered through a sea of woods. The plain is crossed by numerous sluggish rivers and contains many swamps and marshes, especially along its seaward edge. In its southwestern section it contains great oil, gas, sulfur, and salt deposits which provide the basis for major oil-refining and chemical industries. The Florida section has some of the world's largest deposits of phosphate rock, a critically important fertilizer material.

THE PIEDMONT

The Piedmont lies in the United States between the Coastal Plain and the Appalachian Highlands (Figs. 28.1–28.3 and 28.10). Its elevation, varying generally between 400 and 1500 feet (c. 120–460 m), is distinctly higher than that of the Coastal Plain but lower than that of the Appalachians. Most of its surface is rolling to hilly. The northern end of the Piedmont is found in the vicinity of New York City, from which it extends generally southwestward, widening toward the south, until its southern end is reached in east central Alabama. Parts of 10 states lie within it. This landform division was originally an area of mixed forest, with hardwoods, especially oak, predominating. The Piedmont was once largely cleared for agriculture, but large acreages have now reverted to woodland, and there is a considerable production of hardwood lumber from some sections. Piedmont soils vary greatly in fertility, depending largely on the character of the underlying rock and the kind of treatment received since the beginning of settlement. Certain limestone-derived soils of southeastern Pennsylvania are among the most productive in the United States. These naturally fertile soils are farmed largely by the ''Pennsylvania Dutch'' (of German extraction), long known as some of America's most skillful and industrious farmers (Figs. 3.8 and 28.20). But most Piedmont soils are derived from metamorphic or igneous rocks, were never exceptionally fertile, and have been seriously depleted and eroded (especially in the central and southern Piedmont) as a result of clean-tilled row crops—such as tobacco, cotton (Fig. 26.6), and corn—in a hilly area with heavy rains.

The zone of contact between the Piedmont and the Coastal Plain is known as the *Fall Line*—a name deriving from the falls and rapids which mark the course of rivers as they descend from the ancient hard rocks of the Piedmont uplands to the lower, sedimentary Coastal Plain. These falls and rapids mark the head of navigation on many rivers and have long supplied waterpower. Consequently, many urban areas originated and developed into cities at sites where a river crosses the Fall Line, and a line of cities may now be traced along the latter. Among them are such large metropolises as Philadelphia, Baltimore, and Washington, D.C.

THE APPALACHIAN HIGHLANDS

The Appalachian Highlands extend from Newfoundland to northeastern Alabama (Figs. 27.2, 28.1–28.3, and 28.10). On the east they face the Atlantic and the Piedmont, on the west the Interior Plains. These Highlands touch 19 states and 5 Canadian provinces. The mountains which dominate the topography of this landform division are not very high (Fig. 2.12). The loftiest summits—in the Great Smoky Mountains of North Carolina, the White Mountains of New Hampshire, and the Adirondacks of northern New York—are below 7000 feet (2134 m), and most of the Appalachian area lies below 3000 feet (914 m). But the Highlands form a very extensive and complex system in which mountain ranges, ridges, isolated peaks, and rugged dissected plateau areas are interspersed with narrow valleys, lowland pockets, and rolling uplands. Climatic conditions range from subarctic in Newfoundland to humid subtropical in the south, and some sections are high enough to experience abnormal coolness and precipitation. The natural vegetation is forest—coniferous and mixed coniferous and deciduous as far south as Pennsylvania, predominantly deciduous (although with some mixture of coniferous stands) south from Pennsylvania.

The Appalachian Highlands exhibit some very great internal contrasts in population density and economic development, often within short distances, but as a whole they have a surprisingly large population for a highland area and their importance in the Anglo-American economy is very large. Many parts of the Highlands have a considerable rural population, but most of the land is in forest. The outstanding resource contributions of the region are enormous coal resources, waterpower, and wood. The coal is mostly in the section from Pennsylvania to Alabama, and the greatest waterpower development is on the Tennessee River and its tributaries.

THE INTERIOR HIGHLANDS

The Interior Highlands, often called the Ozark or Ozark-Ouachita Highlands, occupy most of southern Missouri, the northwestern half of Arkansas, and adjoining parts of eastern Oklahoma (Figs. 28.3, 28.10, 28.12, and 28.17). They constitute an island of hill country and low mountains in the midst of a sea of plains—the Interior Plains on the north and west and the Coastal Plain on the east (Mississippi Lowland) and south. The Interior Highlands are divided into two major segments by the east-west valley of the Arkansas River. The northern segment consists of the Boston Mountains, overlooking the Arkansas Valley, plus the more extensive Ozark Plateau of northern Arkansas, southern Missouri, and northeastern Oklahoma. South of the Arkansas Valley are the Ouachita Mountains, constituting the other prin-

Figure 26.10   A scene in the southwestern edge of the Missouri Ozarks. The water body lined with boathouses is an arm of one of the many artificial lakes in the Interior Highlands. The subdued hilly topography is characteristic of this physiographic area.

cipal segment of the Highlands. They consist of parallel ridges and valleys, oriented east-west. The peak elevations in the Interior Highlands lie generally between 1000 and 2000 feet (*c.* 300–600 m). No crest in the entire area reaches 3000 feet, although the highest summits of the Ouachitas approach that elevation. There are notable differences in local relief within the Highlands. A number of fairly level upland surfaces occur, but other areas are quite rugged. Most areas fall between these topographic extremes.

The Highlands were originally forested and most parts still are. Much land previously cleared for a semisubsistence agriculture on poor and stony soils has now reverted to forest, although some land in Missouri has been cleared recently for expanded cattle raising. Meanwhile there has been an expanding development of recreational and retirement facilities, based on low costs and a series of large artificial lakes (Fig. 26.10). The important mineral resources are lead, iron ore, and zinc—all from a district in eastern Missouri. The district is a major world center of lead mining.

THE INTERIOR PLAINS

The interior of Anglo-America, between the Rocky Mountains on the west and the Appalachian Highlands and Canadian Shield on the east, is essentially composed of a vast expanse of plains (Figs. 27.2, 28.1, 28.2, 28.3, 28.10, 28.11, 28.12, 28.17, 28.21, and 28.22). These Interior Plains lie mostly within the drainage basins of four important rivers: the Mississippi, the Mackenzie, the Saskatchewan, and the St. Lawrence. Most parts of the Interior Plains are many hundreds of miles from the sea; their location, in other words, is essentially continental rather than maritime. However, the Interior Plains reach the sea in the northwest and northeast of Anglo-America, respectively, along two narrow corridors—the Mac-

kenzie River Lowlands and the St. Lawrence Lowlands. In the south the Interior Plains extend into Mexico, merge with the Gulf-Atlantic Coastal Plain, or abut against the Interior Highlands and Appalachian Highlands. The Interior Plains include the whole of 6 states (the Dakotas, Nebraska, Kansas, Iowa, and Indiana), practically all of Illinois, and parts of 23 other states, provinces, and territories. Approximately the western third of the Interior Plains, from Mexico north to the southern part of Saskatchewan and Alberta, is a semiarid or subhumid area known as the Great Plains.

The Interior Plains are seldom flat, although considerable areas are very nearly so. Most of the land is gently rolling (see chapter-opening photo), but some areas are hilly and occasional areas are rather rugged. Most parts of the Interior Plains north of the Ohio and Missouri rivers were covered with a mantle of glacial debris during the Great Ice Age and the topography was somewhat smoothed thereby. Thus the more hilly sections tend to be outside the area of glaciation, although some unglaciated portions are also fairly level. The principal unglaciated hilly areas are in the southeast from southern Indiana across Kentucky into Tennessee and in an "island" of unglaciated terrain in southwestern Wisconsin. Other hilly areas of various types are scattered within the Plains, but their existence does not essentially damage the overall concept of a huge area with enormous expanses of land level enough for cultivation.

These lands, which lend themselves admirably to big-scale mechanized farming, are a major resource not only of Anglo-America but also of the world. South of the subarctic climatic area temperature and rainfall conditions allow much cultivation. Many of the soils, although not all, are fertile, and some are superb. The best soils developed under prairie or steppe grasslands (Fig. 28.18), but some very good soils in the east developed under deciduous forest

641

(Fig. 28.20). Favorable American and Canadian economic conditions have allowed these parts of the Interior Plains to become the leading surplus food-producing and -exporting section of the world. In this regard the United States Corn Belt—concentrating principally on corn, soybeans, cattle, and hogs—is outstanding. But the region also contains, in its semiarid western portions, the Spring Wheat Belt of the Canadian Interior Plains and adjacent United States, and the Winter Wheat Belt centering in Kansas. And much of the rest of the huge Plains region is also highly productive for agriculture—notably the dairy country along the northern edge of the Plains from Minnesota to the St. Lawrence Valley. Beyond its agricultural riches, the Interior Plains contains major resources of coal, oil, natural gas, and potash.

THE ROCKY MOUNTAINS

The Rocky Mountains, broadly defined, may be regarded as extending from northwestern Alaska to northern New Mexico (Figs. 27.2, 28.3, 28.21, 28.22, and 28.30). Within the mountain area are many distinct ranges. Some of the latter are known locally as the Rocky Mountains while others are not, but the whole may be viewed as a single enormous system. On the north the Rockies are bordered by the Arctic Coastal Plains, on the east by the Mackenzie River Lowlands and Great Plains sections of the Interior Plains, and on the west and south by various sections of the Intermountain Basins and Plateaus. In Canada, these mountains occupy the border zone between Yukon Territory and the Northwest Territories, and they extend south through eastern British Columbia and southwestern Alberta. In the United States, they occupy northeastern Washington, northern and eastern Idaho, western Montana, western Wyoming, northeastern Utah, central Colorado, and north central New Mexico.

The Rockies are lofty, rugged mountains with many snowcapped peaks. The highest elevations are found in Colorado, which contains a number of summits above 14,000 feet (4267 m). Mount Elbert, the highest peak in the entire system, has an elevation of 14,433 feet (4399 m). Throughout the Rockies, peak elevations of 10,000 feet (3050 m) or higher are very frequent, and the valleys and passes are generally above 5000 feet in elevation. Elevations decline toward the extreme north, where the Brooks Range of Alaska rises to general summit levels of 5000 to 6000 feet. The mountains generally form series of linear ranges trending roughly in conformance with the system as a whole. Valleys of varying size and shape are enclosed between and within the ranges. Only in the United States section north of Yellowstone Park does the linear outline break down. Here the mountains rise in jumbled masses. Easy passes through the Rockies are almost nonexistent, and the system has been a major barrier to transportation, but not an impassable one. Various major railroads and automobile thruways cross the mountains via selected passes.

In addition to spectacular scenery and other recreational assets, both the American and Canadian Rockies provide mineral resources of great variety and high value. Both metals and fuels are important. Some areas have timber industries based on mountain conifers, and livestock ranches utilizing mountain pastures are very widespread. The whole mountain area is a prime watershed, as many of the greatest American and Canadian rivers have their main headstreams there.

THE INTERMOUNTAIN BASINS AND PLATEAUS

The Rocky Mountains and the mountains near the western shore of Anglo-America are separated from each other, from Alaska to Mexico, by the Intermountain Basins and Plateaus (Figs. 28.3, 28.21, 28.22, and 28.30). This landform division occupies an immense part of western Anglo-America. In general, while lower than the bordering mountains, the Intermountain Basins and Plateaus lie at comparatively high elevations above the sea. Few sections lie below 3000 feet (914 m), and very few below 2000 feet. There is a great deal of variety in elevations, landforms, and local relief from place to place. Most of the land is composed of plateau surfaces in various stages of dissection so that rolling uplands and rugged hilly and mountainous sections are included, in addition to the large areas which are comparatively level or actually flat. In many places isolated mountain ranges project far above the general surface level. Over much of the area the river valleys are deeply incised, forming spectacular canyons and gorges. The most famous of these, the Grand Canyon of the Colorado River, is an enormous gash cut in the surface of the Colorado Plateau in northern Arizona.

Climatically, the Intermountain Basins and Plateaus are distinguished by low rainfall due to the position of this landform division between two shielding mountain systems. Most of the Intermountain area has a semiarid (steppe) climate, although there are a number of sizable desert areas, especially in the south. In the more northerly sections the effects of low precipitation are offset to some degree by the lessened evaporation attendant on lower temperatures. Temperature conditions range from subarctic in Alaska and northern Canada to subtropical along the Mexican border. Locally, there are great temperature as well as rainfall contrasts resulting from differences in elevation.

The Intermountain area may be divided into a number of major subsections, as follows: (1) the Basin and Range Country, (2) the Colorado Plateau, (3) the Columbia-Snake Plateau, (4) the Fraser-Nechako-Stikine Plateaus, and (5) the Yukon River Basin.

The *Basin and Range Country* extends from southern Oregon and southern Idaho to the Mexican border. It includes parts of Oregon, Idaho, California, Utah, Arizona, New Mexico, and Texas, as well as practically the entire state of Nevada. The section lying between the Wasatch Mountains of Utah and the Sierra Nevada of California is often referred to as the Great Basin, although the latter includes many smaller basins. The basins of the Basin and Range Country, many of which have no external drainage, are often separated from each other by blocklike mountain ridges rising high above the general level. Semiarid climatic conditions of northern sections and most eastern sections produce such vegetation forms as short grass, bunch grass, and sagebrush. In the southwestern part, true desert with extremely sparse vegetation prevails over a wide area in California, Nevada, and Arizona (Fig. 26.11), and extends eastward into southern New Mexico and southwestern Texas.

The *Colorado Plateau* occupies parts of Colorado, Utah, Arizona, and New Mexico. Rolling uplands lie at varying levels, often separated by steep escarpments. Mountain areas rise above the general surface in various places. Rivers, principally the Colorado and its tributaries, flow in deep canyons (see chapter-opening photo of Chapter 2 and Figs. 2.1 and 2.9). The latter have been a great hindrance to transportation and have kept many sections extremely isolated. The climate is generally semiarid, although some higher sections have sufficient precipitation to produce a forest growth.

The *Columbia-Snake Plateau* is found in eastern Oregon and Washington and southern and western Idaho. It is characterized by extensive areas of level land, the result of massive lava flows in the past which buried the previous topography. The soils formed from these volcanic materials are exceptionally fertile. Isolated mountains occur, and the streams often flow in canyons. The Columbia River, the master stream of the area, has cut many alternative channels in the past, which now form rugged ''scablands'' in parts of Washington. The Columbia's

Figure 26.11 *A view of desert and a major irrigation canal in Arizona's Basin and Range Country near Phoenix. The canal brings water from the Colorado River to central and southern Arizona; it is one link in the system of canals, pipelines, tunnels, and pumping stations called the Central Arizona Project. Note the spacing of the xerophytic vegetation to take advantage of all available moisture.*

giant tributary, the Snake, flows along the border between Oregon and Idaho in a spectacular canyon. The climate is semiarid.

The *Fraser-Nechako-Stikine Plateaus* of British Columbia form a considerably more narrow and constricted section of the Intermountain region than do their counterparts in the United States. The Fraser Plateau in the south is a deeply dissected and rugged area, but, in the north, the Intermountain section of British Columbia presents large areas of rolling upland, interrupted by several mountain ranges. The semiaridity of the area is moderated by lower temperatures toward the north, where the grasslands and parklike forests of the south trend into the subarctic (taiga) forest.

The *Yukon River Basin*, which occupies most of Canada's Yukon Territory and the greater part of Alaska, has a varied topography of rolling uplands, hill country, low mountains, intrenched streams, and comparatively small areas of flat alluvial plains. As the elevations decline toward the sea in the

Alaska section of the Basin, marshy areas become more prominent. The extreme subarctic climate of the Yukon Basin sets it apart from other areas in the Intermountain Basins and Plateaus. Cold rather than aridity is the dominant climatic factor, and the subarctic taiga forest is the prevailing vegetation type.

Resources as well as population are sparse in the Intermountain area. Subarctic cold in the north of the region, semiaridity over much of it, and desert conditions in the south limit agriculture largely to irrigated oases, although some dry farming is possible in certain areas, notably the Columbia Plateau wheat region of eastern Washington and Oregon. South of the subarctic, most land is used only for very low-capacity grazing, with agricultural production, population, and urban development concentrated in the irrigated areas. Mineral wealth is considerable, although widely scattered. Metals have been of major importance historically and some, especially copper, still are. Mining towns, both dead and alive, are common. In recent decades the extraction of oil, natural gas, coal, and uranium has become an important activity in the Colorado Plateau. Scenic spots such as Zion and Bryce canyons and an intermediate position between more densely populated areas to both the west and east provide the basis for a considerable travel and tourist industry in some American sections. Great quantities of hydroelectricity are produced and irrigation water is impounded at numerous dams along major rivers—the largest installations being Grand Coulee Dam on the Columbia River and Hoover and Glen Canyon dams on the Colorado River. Power from such dams is largely transmitted to points outside the Intermountain Basins and Plateaus (such as Los Angeles), although some electricity is diverted to local uses, including aluminum milling along the Columbia, gigantic neon signs for the gambling casinos in Las Vegas, Nevada, and air-conditioning for the huge retirement communities of the Phoenix metropolis in Arizona.

THE PACIFIC MOUNTAINS AND VALLEYS

The Pacific shore of Anglo-America is bordered by a series of mountain ranges extending from Mexico to the Aleutian Islands. In the United States, several large lowlands are included within this mountainous region. For purposes of introductory physical description, the Pacific Mountains and Valleys region may be conveniently discussed in terms of the following major subdivisions: (1) the Coast Ranges of California, Oregon, and Washington; (2) the Sierra Nevada; (3) the Central Valley of California; (4) the Klamath Mountains; (5) the Cascade Mountains; (6) the Willamette-Puget Sound Lowland; (7) the Coastal Ranges of British Columbia and southeastern Alaska; and (8) the Alaska Range and other mountains of southern Alaska north of the southeastern "Panhandle." (See maps, Figs. 28.21, 28.22, and 28.30.)

The coasts of California, Oregon, and Washington are fronted by the *Coast Ranges* (Figs. 26.7, 26.12, and 28.23). The section in northwestern Washington known as the Olympic Mountains has peaks reaching approximately 8000 feet (2440 m) and in the south some sections east of Los Angeles exceed 10,000 feet (3048 m). However, the peak elevations more commonly lie at 3000 to 5000 feet in California and at only 1000 to 3000 feet farther north. Along most of the western coast of the United States there is no coastal plain, or almost none, and even the lower parts of the Coast Ranges are often quite rugged. However, a few valleys are available for agriculture, especially in California; and from Los Angeles to the Mexican border the ranges lie a few miles inland and a lower, hilly district containing the greater part of California's population fronts the sea.

The *Sierra Nevada* (Fig. 2.3) forms the inland edge of the Pacific Mountains and Valleys in central California, merging with the Coast Ranges north of

Figure 26.12 *The Coast Ranges a short distance north of San Francisco. The climate here is classed as marine west coast. California's coast redwoods grow in a narrow band of lower slopes and valleys near the coast. The band of trees running diagonally across the center of the view is comprised of redwoods in the Muir Woods National Monument, located a few miles north of the Golden Gate Bridge. Despite the marine west coast designation of the climate, there is a strong tendency toward summer drought, but there is enough cloud and fog to maintain the moist, cool environment that redwoods require.*

Los Angeles and with the Klamath and Cascade ranges in northern California. The Sierra Nevada is an immense upraised, tilted, broken, and eroded block presenting a long and comparatively gentle slope to the west and a precipitous face eastward toward the Basin and Range Country. It is very high and rugged and constitutes a major barrier both climatically and with regard to transportation. Mount Whitney (14,494 feet/4418 m) in the southern Sierra Nevada is the highest peak in the conterminous United States.

The *Central Valley of California,* a level-floored alluvial trough some 500 miles (*c.* 800 km) long by 50 miles (*c.* 80 km) wide, occupies the center of California between the Coast Ranges (Fig. 28.25) and the Sierra Nevada. It is completely surrounded by mountains except where San Francisco Bay breaks the continuity of the Coast Ranges and brings the Central Valley into contact with the Pacific.

The *Klamath Mountains* form a link between the Coast Ranges and the Sierra Nevada and Cascade ranges and separate the northern end of the Central Valley of California from the southern end of the Willamette Valley of Oregon. The Klamath Mountains are physiographically a dissected plateau. The valleys are deeply incised, giving an extremely rugged aspect to the terrain. Summit levels are frequently at 6000 to 7000 feet (1829–2134 m) or higher.

The *Cascade Mountains* extend northward from the Sierra Nevada and the Klamath Mountains across Oregon and Washington and into British Columbia. Much of the area of the Cascades lies between 5000 and 9000 feet (1524–2743 m) in elevation, and the mountains are surmounted by a series of volcanic cones—such as Mt. Rainier and recently active Mt. St. Helens—reaching much higher elevations. The Cascades are broken into two sections at the Oregon-Washington boundary by a scenic gorge through which the Columbia River flows westward toward the Pacific.

The *Willamette-Puget Sound Lowland* lies between the Coast Ranges and the Cascades. Its Oregon section, from the Klamath Mountains in the south to the lower Columbia River in the north, is the valley of the Willamette River, while the northern section is commonly known as the Puget Sound Lowland. This latter section extends all the way across western Washington and north into British Columbia to include a small area along the lower course of the Fraser River.

In British Columbia and the southeastern "Panhandle" of Alaska, a northward extension of the Cascades lies along the coast and is known as the *Coastal Ranges.* These Ranges are generally higher and more rugged than the Coast Ranges in the United States. Many peaks reach 9000 to 10,000 feet (2743–3048 m). The mountains rise abruptly from the sea and are penetrated by fjords, resembling in this respect the coasts of Norway and southern Chile. West of the Coastal Ranges, a valley analogous in position to the Central Valley of California and the Willamette-Puget Sound Lowland has subsided below sea level and now forms the famous "Inside Passage" to Alaska. The mountains in British Columbia and southeastern Alaska which correspond in position to the Coast Ranges of the United States are partly submerged and form a string of rugged islands along the outer edge of the Inside Passage.

Most of southern Alaska is mountainous. Just north of the Panhandle, spectacular glaciers descend the mountains to the sea (Fig. 28.31). The highest mountains, however, are found farther to the north and west in the *Alaska Range,* where Mount McKinley in the Denali National Park reaches 20,320 feet (6194 m), the highest elevation in Anglo-America. In a more subdued form the mountains continue from the Alaska Range into the Alaska Peninsula and the Aleutian Islands. A fair-sized area of lower land is found south of the Alaska Range in the Susitna River Valley and along the western side of the Kenai Peninsula.

The economy of the Pacific Mountains and Valleys region has grown rapidly on the basis of a major collection of natural resources. Abundant forests are characteristic of the areas of marine west coast climate and adjacent high mountains from coastal Alaska southward. They include especially magnificent species in the redwoods of California and the Douglas fir of the Pacific Northwest. Unusually productive fisheries extend along the whole coast, and here and there are excellent natural harbors. Heavy orographic precipitation feeds numerous mountain streams, which support a large hydroelectric development, especially in California, and supply irrigation water for exceptional agricultural productivity in California's areas of dry-summer subtropical (mediterranean) climate (Figs. 1.1 and 1.2). Gold once drew settlers to California and British Columbia, and California has long been a major producer of oil and natural gas. Finally, exceptional scenery, beaches, mild temperatures, and sunshine in the California subtropics have drawn large numbers of both tourists and permanent residents. Natural assets of various kinds have had a strong attraction for film-making and aerospace industries in California, although much recent industrial development in the state (such as electronics) has been of a high-technology character not closely tied to natural conditions.

CHAPTER 27

National and Regional Characteristics of Canada

A portion of the harbor at Vancouver, British Columbia, Canada's busiest seaport. The container port is in the center of the photo, which was taken from the city's central business district.

CANADA is one of the world's more highly developed nations—an affluent industrialized, technologically advanced, and urbanized state. But in certain national characteristics it is curiously unlike most developed countries. These differences are closely related to Canada's internal geography and the country's location adjacent to the United States.

INDICATORS OF HIGH DEVELOPMENT IN CANADA

Canada stands appreciably below the United States in per capita output and income but still is more affluent than all but a comparative few of the world's other developed nations. Canadian affluence derives in considerable part from a diversified manufactural output sufficiently high in value to place the country among the world's top 10 or 12 manufacturing nations, although Canada is dwarfed in this regard by the very biggest industrial powers—the United States, the Soviet Union, Japan, and West Germany—and is considerably exceeded by a few other countries. In some manufactural lines, Canada ranks at or near the top, for example, newsprint, hydroelectricity, commercial motor vehicles, and aluminum metal. The country has a varied and advanced development of other economic activities, including the many kinds of services associated with highly developed states.

As would be expected from the foregoing traits, Canada is very urbanized, with 76 percent of its population classed as urban in a recent year (United States, 74%) and with well over one quarter of its people living in the three largest metropolitan areas: Toronto 3.4 million, Montreal 2.9 million, and Vancouver 1.4 million (1986). These attractive and dynamic metropolises convey a positive image of Canada as an advanced and prosperous country. Impressively good national statistics on health, housing, and education provide further evidence of high development.

DIFFERENCES FROM OTHER DEVELOPED STATES

Despite Canada's general conformity to expected patterns of developed countries, there are some striking departures. One difference lies in the country's unusual pattern of exports and imports. Most developed countries mainly export manufactured goods and import a diversified mixture of manufactures and primary products such as oil, ores, and food. But despite a fair-sized export of automobiles and parts, together with some other fabricated items, Canada is mainly an exporter of raw or semifinished materials, energy, and food and is mainly an importer of manufactures. In this respect as in some others, it rather resembles Australia. A second difference from other developed nations lies in Canada's overwhelming dependence on trade with a single partner, the United States. In 1986 the latter nation took 77 percent of Canada's exports by value and supplied 69 percent of Canada's imports. No other developed nation comes near these percentages of trade with just one outside country. Even in developing nations it is hard to find a parallel. Still another major divergence from other developed countries lies in the degree to which the Canadian economy is financed and controlled from outside the country, and again Canada has a dependence on the United States far greater than on all other countries combined. General Motors, Dow Chemical, IBM, Exxon, and many other huge American corporations own important shares of Canadian manufacturing, mining, and services. Sizable British, West German, and Japanese investment is present in Canada, but somewhat over three quarters of all foreign investment is American.

The extent of United States economic control has long been a sore point with many Canadians, who resent imputations that their country is an American economic "colony." Such irritations tend to be exacerbated by the degree to which American mass-produced culture has permeated Canada and the degree to which Canada is taken for granted in the United States. Few American schools, for example, offer entire courses on Canada, and tests have shown that a high percentage of American students cannot even name Canada's capital. American news media give relatively scanty coverage to events in Canada. There has been for some time an explicit movement in Canada to achieve a more predominant position for Canadians within their own economy, to strengthen Canada's economic relations with industrial countries other than the United States, and to heighten Canadian cultural self-determination. Not only do Canadians feel underprivileged when Americans own so much of their country and impact so much on their culture, but they also tend to feel rather helpless in being swept along by economic fluctuations in their huge neighbor. One result of the move to "Canadianize" is seen in the fact that Canadian capital invested in the country's manufacturing grew from 43 percent of total manufactural investment in 1974 to 51 percent in 1984. But however much the American connection may rankle, there seems no way for Canada to pull away drastically from the United States without risking its own prosperity. The country's economy is geared to close interchange with the gigantic economy next door, which provides, among other things, a largely irreplaceable market for Canadian exports. In 1988 the two nations signed a historic pact to abolish all tariffs on trade between the two and to remove or lower many other restraints on trade. Canada's population of 26 million—about the same size as California's—is affluent, but it is too small to provide an adequate market for the kinds of things Canada produces; and the difficult environment in most of Canada will continue to hamper the growth of population and settlement and hence expansion of the internal market.

CANADA'S REGIONAL STRUCTURE

Not only must Canada's government struggle with the difficult problems of maintaining a satisfactory relationship with the United States, but it also is confronted by related problems posed by Canada's regional structure. The basic regional division is between a thinly settled northern wilderness that occupies most of the country and a relatively narrow and discontinuous band of more populous regions stretching across Canada from ocean to ocean in the extreme south (Fig. 27.1). The latter regions are sharply different from each other geographically, and they often exhibit self-centered political attitudes that make it hard for the government at Ottawa to maintain national unity.

About four fifths of Canada suffers from the excessive cold of tundra and subarctic climates, more

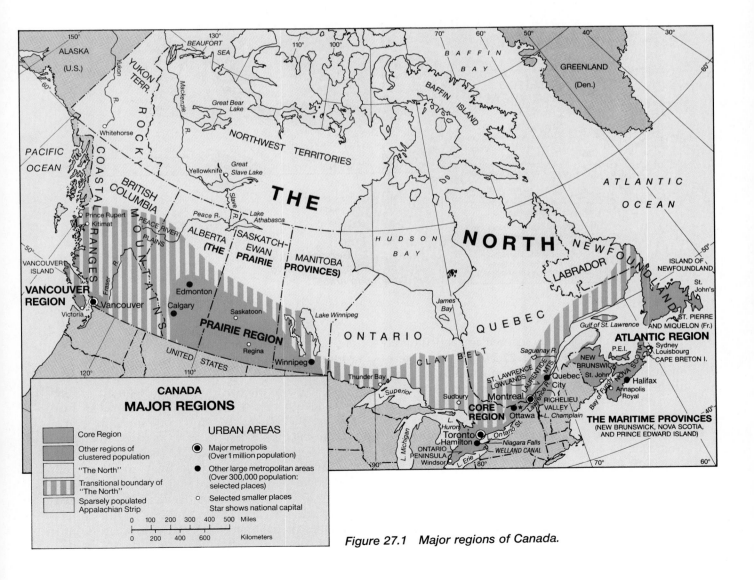

Figure 27.1 Major regions of Canada.

Figure 27.2 General reference map of the Canadian Shield and adjoining areas.

than half is in the glacially scoured and agriculturally sterile Canadian Shield (Fig. 27.2), and the western-most part near the Pacific is dominated by high and rugged mountains. Such conditions create the huge expanse of nearly empty land that Canadians call "The North." More than half of it lies in the North-west Territories and Yukon Territory administered from Ottawa, but large parts extend into 7 of the 10 provinces that comprise the major political subdivisions of Canada's federal system. The provinces are comparable in many ways to the states of the American federal union, although they tend to have much more power relative to their federal government than do American states. The provinces, for exam-

ple, have far more control over the natural resources within their boundaries than do the states.

The ten provinces, stretching across Canada from Nova Scotia on the east to British Columbia on the west, incorporate only 60 percent of Canada's area but have 99.7 percent of Canada's people. Even within the provinces—aside from tiny and more evenly populated Prince Edward Island—the greater part of the land is lightly inhabited, and nearly all the people are concentrated in limited areas of rela-tively close and continuous settlement. Separated from each other by wedges of thinly populated ter-rain, these population clusters are strung across Can-ada just north of the boundary with the 48 conterminous states. Even by a generous definition, Canada's effectively occupied districts comprise no more than 10 or 12 percent of the country's area. In few developed countries does the total area of close settlement occupy such a small part of the national domain. Australia is a fairly close parallel, not only in the relative smallness of its closely settled area but also in the attenuated and dispersed arrangement of its population clusters.

Within Canada's transcontinental band of population, all but a very small fraction of the people live within 300 miles (c. 500 km) of the American border, and an estimated 75 to 80 percent live within 100 miles of the border—closer to the United States than Minneapolis and Boston are to Canada. This positioning near a relatively open border facilitates close day-to-day interaction between the American and Canadian peoples and the development of strong ties between Canadian regions and adjoining American regions. Such bonds include family relationships between millions of people in Canada and their kinsmen who live on the American side of the border. Ties between families in French Canada and immigrant French-Canadians in New England are a striking example.

For the purposes of this introductory world-survey text, Canada's population clusters may be grouped into four main regions: (1) an *Atlantic region* of peninsulas and islands at Canada's eastern edge, (2) a culturally divided *core region* of maximum population and development along the lower Great Lakes and St. Lawrence River in Ontario and Quebec provinces, (3) a *Prairie region* in the Interior Plains between the Canadian Shield on the east and the Rocky Mountains on the west, and (4) a *Vancouver region* on or near the Pacific coast at Canada's southwestern corner. Each region is marked by a distinctive physical environment and cultural landscape; each has a different ethnic mix deriving from particular historical circumstances; and each has distinctive economic emphases associated with its particular location and environment. Each region plays a different role within Canada and has its own set of relationships with the United States and other countries. All the regions have important links with the North.

Here we portray Canada's four main regions of human settlement and activity in a sequence from east to west, making reference to functional relationships with the North and with other countries as appropriate for regional understanding. This chapter concludes with a discussion of some salient characteristics of the North considered as a whole. The North is very important to Canada and to individual provinces as a resource reservoir and development challenge, but it has so little population, overall cohesion, and self-determination that it forms a very different kind of area from the four main regions in Canada's south. (For statistical data on Canadian areas, see Table 21.1. For cartographic data, see Figs. 26.1, 26.9, 27.1, 27.2, 28.1, 28.2, 28.17, 28.22, and 28.30.)

THE ATLANTIC REGION: PHYSICALLY FRAGMENTED, ETHNICALLY MIXED, ECONOMICALLY STRUGGLING

The easternmost of Canada's four main populated areas is characterized by a series of strips or clumps of population, largely along or very near the seacoast, in the provinces of Newfoundland, New Brunswick, Nova Scotia, and Prince Edward Island. Deeply penetrated by arms of the ocean and fragmented into numerous islands and peninsulas, the four provinces are commonly called the Atlantic Provinces, and the last three, excluding Newfoundland, have long been known as the Maritime Provinces. The main population concentrations in the Atlantic region are separated from the far more populous Canadian core region in Quebec and Ontario by areas of mountainous wilderness in northern New Brunswick, eastern Quebec, and northern Maine.

Prince Edward Island, which is Canada's smallest but most densely populated province (Table 27.1), is a lowland largely occupied by a continuous mat of farms and small agglomerated settlements, but the main populated areas of New Brunswick, Nova Scotia, and the island of Newfoundland occupy valleys and coastal strips within a general frame of rough and lightly settled upland country. As a province, Newfoundland includes the dependent territory of Labrador on the mainland, but in common usage the name Newfoundland is restricted to the large island which lies opposite the mouth of the St. Lawrence. Newfoundland was the last province to join Canada, having done so as recently as 1949. Initially a British colony (Britain's oldest, founded in 1610), the island received parliamentary government in the mid-19th century and was formally constituted a self-governing Dominion in 1917 but then returned voluntarily to direct British rule in a time of extreme economic hardship during the world depression of the 1930s. Bleak and nearly uninhabited Labrador, situated on the Canadian Shield, is very much a part of Canada's North. The territory is important to the province for revenues and employment supplied by iron mining and hydroelectric production.

TABLE 27.1   CANADIAN PROVINCES AND TERRITORIES: BASIC DATA

Province or Territory	Land Area (thousand sq. mi)	Land Area (thousand sq. km)	Population (thousands: 1986 census)	Estimated Density, 1986 (per sq. mi of land)	Estimated Density, 1986 (per sq. km of land)	Economic Production (Value), 1982 — Percent of National Farm Receipts	Economic Production (Value), 1982 — Percent of National Mineral Output	Economic Production (Value), 1982 — Percent of National Manufacturing Shipments
Atlantic Provinces								
New Brunswick	27.8	72.1	710.4	26	10	1.0	1.6	2.0
Nova Scotia	20.4	52.8	873.1	43	17	1.2	0.9	2.0
Prince Edward Island	2.2	5.7	126.6	58	22	0.9	0.006	0.15
Newfoundland–Labrador	143.5	371.7	568.3	4	2	0.02	1.9	0.7
Totals	193.9	502.3	2278.5	12	5	3.1	4.4	4.9
Core Provinces								
Quebec	523.9	1356.8	6540.3	13	5	15.0	6.1	26.3
Ontario	344.1	891.2	9113.5	26	10	26.4	9.6	51.1
Totals	867.9	2248.0	15,653.8	18	7	41.4	15.7	77.4
Prairie Provinces								
Manitoba	211.7	548.4	1071.2	5	2	8.9	1.5	2.5
Saskatchewan	220.3	570.7	1010.2	5	2	21.2	6.6	0.5
Alberta	248.8	644.4	2375.3	10	4	20.2	60.9	6.6
Totals	680.9	1763.5	4456.7	7	3	50.3	69.0	9.6
Pacific Province								
British Columbia	359.0	929.7	2889.2	8	3	4.9	8.6	8.1
Territories								
Yukon	184.9	479.0	23.5	0.12	0.05	under 0.001	0.5	under 0.001
Northwest Territories	1271.4	3293.0	52.2	0.04	0.02	under 0.001	1.8	under 0.001
Totals	1456.4	3772.0	73.7	0.05	0.02	under 0.001	2.3	under 0.001
Grand Totals, Canada	3558.1	9215.4	25,354.1	7.1	2.8	100.0	100.0	100.0

A CENTURIES-OLD TRADITION OF FISHING

The fishing industry has always been of outstanding importance in the Atlantic region and actually initiated its development. Before the beginning of settlement in the early 17th century, and quite possibly before the discovery of America by Columbus in 1492, European fishing fleets began operating regularly in the waters along and near the Atlantic shores of Canada. These coasts are a part of North America that is relatively close to Europe, and their waters have always been exceptionally rich in fish. Elevated portions of the sea bottom known as ''banks'' are found off the coast of Anglo-America from near Cape Cod to the Grand Bank, largest of all, which lies just off the southeastern corner of the island of Newfoundland. The relative shallowness of the ocean and the mixing of waters from the cold Labrador Current and the warm Gulf Stream foster here a rich development of the tiny organisms called plankton, on which fish feed, and this part of the Atlantic has been for centuries one of the world's greatest centers of commercial fishing. Not only Newfoundland, but also the Maritime Provinces, Maine, Massachusetts, and transoceanic countries send fishing fleets here. Inshore fisheries in many parts of Atlantic Canada supplement the catch from the banks (Fig. 27.3). Although many kinds of fish and shellfish are caught today, the early fleets mainly fished for cod, which were cured on land before making the trip to European markets.

Both the British and the French established early fishing settlements on the Newfoundland coast, but struggles between Great Britain and France eventually saw the expulsion of the French, and Newfoundland became British in population. It has remained so, as its limited development has attracted very few other immigrants. In the Maritime Provinces French settlements combining fishing with subsistence farming began in the Annapolis-Cornwallis Valley of Nova Scotia at Port Royal (subsequently Annapolis Royal) in 1605. Tiny clusters of French gradually established themselves along or near the coasts elsewhere in the Maritimes.

AGRICULTURAL HARDSHIPS

In general the Atlantic margins of Canada provided hard environments for farming. Agriculture was usually the main basis of early settlement, although fishing often supplied an important supplement. The agricultural environments of the Atlantic region continue to present major problems today. The four provinces lie at the northern end of the Appalachian Highlands and have topographies dominated by hills and low mountains. Soils that are predominantly mediocre to poor have developed under a cover of mixed forest. They suffer from a wide range of deficiencies, being excessively thin or actually absent on many upland slopes, often stony or boggy, poorly supplied with plant food by their parent materials, and robbed of fertility by active leaching in the cool and rainy climate. Many soils that once were farmed wore out quickly and were abandoned. In the 20th century agriculture has become concentrated on relatively small patches of the best land in districts of very specialized farming.

Figure 27.3 Lobster traps line a small wharf along the Bay of Fundy in Nova Scotia. Due to the great tidal range of the Bay, the fishing boat will rest on the bottom at low tide.

Most of the first agricultural land used by European farmers in the Atlantic region did not result from forest clearing but from the reclamation of tidal flats along the Bay of Fundy in Nova Scotia. The funnel-shaped Bay is noted for its great range between low and high tide. Sizable areas of mud and marsh grass are exposed at low tide, and some of these were diked and drained for farms by French settlers. Although great labor was required for this reclamation, the marshland soils proved much more fertile and durable than soils that could have been secured by clearing adjacent forested land.

Not only do the soils of the Atlantic region leave much to be desired, but the same is true of the agricultural climate. A subarctic climate with strong marine modifications characterizes the island of Newfoundland, while the Maritime Provinces have a humid continental short-summer climate. The entire region is very humid and windy, with an average of 40 to 60 inches (c. 100–150 cm) of precipitation a year and much cloud and fog. Strong gales are frequent in the winter. Coupled with rather cool summers (excessively cool in Newfoundland), such conditions create difficulties for agriculture, although in parts of the Maritimes such crops as potatoes, hay, vegetables, small grains, and apples are successfully grown on a commercial basis. Colder conditions occur locally with slight increases in elevation, so that a good part of upland Newfoundland has a vegetation of tundra. Newfoundland has never developed a significant commercial agriculture, and most areas in the Maritimes have been agriculturally precarious and marginal throughout their history. Some of the more favored lowlands, notably Prince Edward Island, are exceptions, but they are small. Prince Edward Island, like some other lowlands such as the Annapolis-Cornwallis Valley of Nova Scotia, is underlain by sedimentary rocks and hence offers a contrast to the older, harder igneous and metamorphic rocks that underlie most of the Atlantic region. In the sedimentary areas soil fertility is generally better, but even in these areas most soils are not very good.

ETHNIC COMPLEXITY

Until France was expelled from Anglo-America in the mid-18th century (aside from the tiny islands of St. Pierre and Miquelon off Newfoundland that France still holds), the Maritime Provinces were the scene of frequent warfare and shifts of colonial territory between France and Great Britain. This strife has left its mark on the current population mix. The position of the Maritimes between more developed areas was the main reason the area was considered a prize justifying warfare. To the north the Gulf of St. Lawrence and the St. Lawrence River led inland to Quebec City (Fig. 27.4) and the heart of French settlement in North America; to the south lay England's possessions along the seaboard from New England to Georgia. Warships based in the Maritimes were in a good position to protect friendly vessels using North Atlantic routes or to capture or destroy vessels of hostile powers. Great Britain eventually triumphed in the Maritimes and subsequently expelled some 6000 to 10,000 "Acadian" French from their homes (Acadia was the French name for what became Nova Scotia). The Acadians were charged with disloyalty to the British crown because they refused to commit themselves to military service against France. In a series of deportations, principally in 1755, shiploads were sent to other British colonies or to Europe. Families were sometimes separated when men were sent to different destinations from their wives and children—one of many harrowing circumstances that inspired Longfellow's epic poem *Evangeline*. The Acadians were replaced by British and New England settlers. Then in the 18th and early 19th centuries came influxes of German settlers, American Loyalist refugees during and after the American Revolutionary War, and Scots and Irish from economically depressed areas of the British Isles. Some Acadian French remained or returned after the expulsion, and immigration from Quebec made northern New Brunswick largely French. Localized clusters of the various ethnic groups have persisted, making the Maritimes a very complex ethnic mosaic.

RECURRENT ECONOMIC DIFFICULTIES

During the first two thirds of the 19th century the Maritime Provinces became a relatively prosperous area with a preindustrial economy. Local resources supported this development: many harbors along indented coastlines; abundant fish for local consumption and export; timber—both softwood and hardwood—for export and for use in the building, sale, and operation of a great many wooden sailing ships; and land, albeit not very good, for the expansion of an agriculture partly subsistence in character and partly serving local markets. Mercantile and shipping interests carried on an extensive commerce with Great Britain and the West Indies.

Then followed an age—which is still continuing—of *relative* economic decline within an industrializing Canada. For more than a century the Atlantic region has struggled against odds, encountering hardships only occasionally inter-

Figure 27.4 *This photo shows Quebec City, once the capital of France's North American empire and now the capital of Quebec Province. At the right is the Lower Town along the St. Lawrence River; at the left, the Upper Town. Grain elevators symbolize the city's seaport function; the hotel called the Chateau Frontenac with its copper sheathing is a reminder of the city's heritage from Imperial France.*

rupted by better times. Despite the apparently advantageous position of the Maritime Provinces as a kind of Atlantic "wharf" between Canada's interior and Europe, no port in the region ever developed into a major commercial center serving Canada or the United States. The ports remained underdeveloped because they were too far from the developing interior of the continent, with the result that most shipping bypassed them in favor of the St. Lawrence ports or New York and other Atlantic ports of the United States. Halifax, Nova Scotia, and St. John, New Brunswick, are the main ports today. Both have fine natural harbors, and both became linked to the interior by rail in the late 19th and early 20th centuries. Today, each handles a considerable traffic, with a large emphasis on containers, but neither is in a class with Montreal or the main ports along the northeastern seaboard of the United States.

Meanwhile fishing, forestry, and agriculture suffered various disabilities. Fishing did not prove to be an adequate basis for a prosperous modern econ-

omy. The fisheries were handicapped by (1) tariff barriers imposed by the United States against Canadian fish, (2) competition from major world meat exporters and newly developed fishing areas in other parts of the world, (3) competition from foreign fleets on the banks, and (4) depletion of the fishery resource. Wood industries also suffered when ships began to be made of iron and steel, forests became depleted by overcutting, and competition set in from new areas of forest exploitation farther west. The position of agriculture was undermined when the building of railroads and the settlement of the Canadian Prairie region brought cheaper farm products into the Atlantic region. In general, agriculture has become quite unimportant in the region today except in Prince Edward Island, which has the highest proportion of tilled land of any Canadian province and derives a good income from a strong specialty on potato growing, coupled with livestock and milk production. Although quite a few people in the Maritime Provinces still farm, they do so largely on a small-scale, part-time, subsistence basis while earning their living mainly from other types of employment.

As economic challenges developed, the Atlantic region attempted to meet them (and is still doing so) by developing industry. Many small cotton-textile factories were built in the Maritime Provinces during the 19th century. But sales of their products in the United States were hindered by tariffs and by massive competition from mills in New England and the South; the mills were relatively remote from the main Canadian markets in southern Quebec and Ontario—which had competing mills of their own—and sales abroad were held down by international competition. The market in the Atlantic region was far too small to compensate for such disabilities, and these early industrial ventures largely failed.

Near Sydney on northern Cape Breton Island in Nova Scotia are sizable coal reserves, and iron ore was present on Newfoundland's Bell Island. On the basis of these resources a coal-mining industry and an iron-and-steel plant developed at Sydney and flourished for some time. The plant was built at the turn of the century, before Canada's transcontinental railroads had been completed, and it helped supply a large demand for steel rails. Later it prospered from the demand for steel to make war equipment during World Wars I and II. But eventually both the plant and the coal mines seriously declined, with consequent economic depression in the Sydney area. There is very little market in Atlantic Canada for Sydney's steel output, and the mill has proved unable to compete successfully at longer range except under wartime conditions. However, steel is still produced here on a much reduced scale, although the mines on Bell Island have closed and the mill operates with ore and scrap from various sources. The plant at Sydney was bought by the provincial government after World War II to forestall closure and is continued on a subsidized basis in order to alleviate the severe economic problems of the district. Some coal mines still operate in the Sydney district, although others have closed. The local coal industry has been handicapped by (1) the increasing cost of extracting the coal, which is deep-lying; (2) the shrinking market for coal in Atlantic Canada due to widespread replacement of coal by oil as a fuel; and (3) general inability to compete in major coal markets outside the region.

Isolation afflicts almost all major enterprises in the Atlantic region. Even the market within the region is small and is divided into many small clusters of people separated by little-populated land. In Newfoundland this fragmentation of a small population into small communities, mainly tiny fishing villages

and towns, is especially marked. The largest city in the Atlantic Provinces—Halifax, Nova Scotia—had a metropolitan population of slightly less than 300,000 in 1986; and only two other cities—St. John's, Newfoundland, and St. John, New Brunswick—surpassed 100,000.

The relatively meager population, not only of the main cities but also of the Atlantic region as a whole, indicates the comparative retardation of this part of Canada. The entire area has only 2.3 million people, or considerably less than the metropolitan area of Toronto or Montreal. A faltering economic development has resulted in a slow population growth. Incoming migrants to Canada have long bypassed this region, while its own people have migrated to the United States or to other parts of Canada in sufficient numbers to strongly retard growth. Incomes and employment have suffered correspondingly. This is the poorest of the four populated regions in per capita income. However, there are some positive factors and hopeful elements in this picture. After many years of being plundered by the large trawlers of foreign fishing fleets, Canada in 1977 extended control of its oceanic waters to 200 miles (322 km) from the coast. Fishing by foreigners is being licensed and regulated to protect Canada's own fisheries. Pockets of specialized commercial farming—notably the potato production of Prince Edward Island and the St. John River Valley of eastern New Brunswick—continue to prosper, and there is much land for subsistence agriculture to supplement other earnings by the region's unusually large rural nonfarm population. Pulp and paper production utilizing substandard forest resources has become an important sector in all the provinces except Prince Edward Island as cutting for lumber has declined. A few large metal-mining operations supply local employment and provincial taxes in the provinces of Newfoundland and New Brunswick. At its large Churchill Falls hydroelectric installation in Labrador, Newfoundland produces current that is sold for the most part to adjoining Quebec. That province, in turn, markets some of the current, plus surplus electricity of its own, in the United States. New Brunswick is also a sizable producer of hydroelectricity, much of which is marketed outside the province through an interlocking grid within the Atlantic region, Quebec, and the northeastern United States.

Canadian federal government assistance and a growing tourist industry are very helpful to the Atlantic Provinces, and further help may come from the development of offshore oil and natural gas. An unusually large share of overall income in the Atlantic Provinces comes from Canadian government payments. These take a variety of forms: (1) direct pension and welfare payments; (2) stationing of a large share of Canada's military forces in the region; (3) direct subsidies to provincial treasuries; (4) the location of many national administrative offices and personnel in the region; and (5) by no means least, federal schemes and subsidies for regional economic restructuring and redevelopment. Under the latter programs success has been achieved in attracting some industries, such as oil refining and petrochemicals on Cape Breton Island and Michelin tire plants in the vicinity of Halifax. Such developments brighten the economic picture somewhat, although they do not fundamentally alter the situation of Atlantic Canada as a region seriously underdeveloped in manufacturing. Expanding tourism in Atlantic Canada has been fostered by striking natural beauty, summer coolness attractive to visitors from sweltering American cities, quaint fishing villages and other landscapes unspoiled by industry or extensive agriculture, and reconstructions of history at such places as the restored French fortress of Louisbourg on Cape Breton Island. But tourism is handicapped by the distances separating these attractions from the major population centers which supply most of the tourists. In recent years oil and natural gas in quantity have been discovered by undersea drilling in a number of places off Atlantic Canada, particularly the Hibernia field about 200 miles (c. 320 km) off southeastern Newfoundland. There is a considerable possibility that oil and gas wealth may alleviate much of the Atlantic region's relative poverty at some time in the future.

CANADA'S CULTURALLY DIVIDED CORE REGION IN ONTARIO AND QUEBEC

The core region of Canada, which incorporates somewhat over half of the nation's population on 2 percent of the nation's area, has developed along lakes Erie and Ontario, and thence seaward along the St. Lawrence River. Here the Interior Plains of North America extend northeastward to meet the Atlantic Ocean. They are increasingly constricted seaward, on the north by the edge of the Canadian Shield and on the south by the Appalachian Highlands. The two provinces that divide the core region—Ontario and Quebec (French: *Québec*)—both incorporate large expanses of the Shield. In addition, Quebec has a strip of Appalachian country—the "Eastern Townships"—along its border with the United States. But the core region lies in lowlands bordering the Great Lakes or the St. Lawrence from

the vicinity of Detroit to Quebec City. The entire lowland area is often termed loosely the *St. Lawrence Lowlands*, although the Ontario Peninsula between lakes Huron, Erie, and Ontario is frequently recognized as a separate section. The Ontario Peninsula has a rather irregular surface bearing diverse forms of glacial deposition. The lowlands along the St. Lawrence River also have considerable local variety but consist basically of an old alluvial expanse where an ancient arm of the Atlantic was infilled.

CONTRASTS BETWEEN THE ONTARIO AND QUEBEC LOWLANDS

Agricultural landscapes predominate in the rural parts of the core region, but the Ontario and Quebec sections show marked differences in types of agriculture and cultural features. In the Ontario Peninsula the American Midwest is somewhat repeated on a small scale, with corn and livestock production, dairy farming, and specialty crops such as fruits, vegetables, potatoes, tobacco, and sugar beets. Here Canada protrudes farthest southward, and the climate is practically identical to that of adjacent southern Michigan or northern Ohio.

In the Quebec section of the St. Lawrence Lowlands, the French heritage is manifest in the "long-lot" pattern of strip-shaped landholdings. The long and narrow strips lie at right angles to the ribbons of comparatively dense settlement along roads where houses and farm buildings form elongated villages. In early times the settlements closely hugged the banks of the St. Lawrence and some of its tributaries, but as settlement expanded, the long-lot pattern was repeated away from the rivers in successive tiers along roads. Impressively large Catholic churches with towering spires are prominent landscape features. The Quebec lowlands lie farther north than the Ontario Peninsula and have a considerably harsher winter climate. Dairy farming is the predominant form of agriculture. Despite relatively small acreages of farm land compared with the Prairie region farther west, the provinces of Ontario and Quebec account for about two fifths of the value of products marketed from Canadian farms.

The Ontario-Quebec core area contains around 14 million people, or about 3 million less than the state of Texas. The Ontario section is basically English in language and culture but has many minority ethnic groups. By contrast, about four fifths of the people in the Quebec section are culturally French. This concentration of French-speaking people represents the biggest exception to the dominance of English culture in Anglo-America and is the only instance in which a national minority controls a provincial or state government in Canada or the United States. Within the coreland life focuses on Canada's two main cities: Toronto in Ontario and Montreal, the second largest French city in the world, in Quebec. There are also a number of smaller, but sizable, metropolitan areas, the largest being Ottawa (825,000), Quebec City (600,000), and Hamilton (560,000). Ottawa, the federal capital, is located just inside Ontario on the Ottawa River boundary between Ontario and Quebec provinces. Quebec City, the original center of French administration in Canada, is very heavily French in culture and is the capital of Quebec Province (Fig. 27.4). Both Ottawa and Quebec City have an important dependence on services, especially those connected with government, although both cities have a variety of industries. Hamilton, a Lake Ontario port not far from Toronto, is the main center of Canada's steel industry. Initially it supplied great quantities of steel rails for transcontinental railway building, and more recently it has profited from the market for steel afforded by the expanding automobile and other metal-fabricating industries of the coreland. A good part of the urban development in the core is associated with manufacturing. Three fourths of Canada's total manufactural development is here, with somewhat more in Ontario than in Quebec. But manufactural employment does not dominate the core; almost three times as many people are employed in services. The high proportion of service-type jobs reflects the coreland's status as the business and political center for a highly developed nation with an affluent population and an elaborate social welfare net.

HISTORICAL EVOLUTION

The coreland began as an entry to, and exit from, the interior of North America, and this continues to be one of its major functions and sources of wealth. The French founded a colonial outpost at Quebec City in 1608, at the point where the broad St. Lawrence Estuary leading to the Atlantic narrows sharply. Fortification of a bold eminence overlooking the river gave control over the seaward bottleneck of what rapidly became an extensive network of river and lake routes reaching far into the continent. French fur traders, missionaries, and soldier explorers used the St. Lawrence, its Ottawa River tributary, the Great Lakes, a few portages, and then the Ohio and Mississippi river systems to reach the Great Plains and the Gulf of Mexico. Montreal, founded some years after Quebec City on an island

in the St. Lawrence, became French Canada's forward post toward the interior Indian-dominated wilderness. Between Quebec City and Montreal a thin line of settlement evolved along the St. Lawrence and formed the agricultural and population base for the colony's far-flung fur-trading activities. Population grew relatively slowly in this northerly outpost where the winter climate was harsh and snowy and where none but French Catholics were welcome. In 1759, when a long series of wars with the British colonies to the south culminated in British conquest (formalized by treaty in 1763), there were only about 60,000 French-Canadians to come under British rule. Their ties were with monarchist France, and when the monarchist regime was overthrown by the French Revolution in 1789, the Quebec French were left with no emotional homeland to cling to except the St. Lawrence Valley.

A few British settlers came to Quebec after the conquest, and this immigration increased rapidly during and after the American Revolution, thus laying the foundations for the British segment of Canada's core. Many of the early English-speaking immigrants were Loyalist refugees from the newly independent United States, whose rebellion French Canada had refused to join. The Loyalists had sided with Britain in the American Revolution, either through conviction or through accidents and circumstances of the war. During and after the war many thousands fled from the United States, with the largest contingents going to Canada. They were soon joined by more English direct from Europe. These newcomers formed a sizable minority in predominantly French Quebec, including a British commercial elite in Montreal that persists to this day. But they also settled west of the French on the relatively good agricultural land of the Ontario Peninsula. Thus was laid the foundation for a Canadian core consisting of predominantly French Quebec along the St. Lawrence, and English-speaking Ontario in the adjacent lands north of the Great Lakes. By 1791 the British government already found it expedient to separate the two cultural areas into different political divisions—known at that time as Lower Canada (Quebec) and Upper Canada (Ontario).

THE RISE OF COMMERCE, INDUSTRY, AND URBANISM

Until Canada was formed as an independent country in 1867, Quebec and Ontario developed largely in a preindustrial fashion, with economies based primarily on agriculture and forest exploitation. With larger expanses of the better land and a somewhat milder climate (due to a more southerly location and more moderation by the Great Lakes), the Ontario Peninsula became more agriculturally productive and prosperous than Quebec. It specialized in surplus production and export of wheat. On a superior natural harbor the Lake Ontario port of Toronto developed and became the colonial (now the provincial) capital and Ontario's largest city. But the bulk of Canada's exports at this time moved eastward to meet ocean shipping, and this movement favored Montreal. Until about 1850 turbulence in the St. Lawrence made it hazardous for ocean ships in some places between Montreal and Quebec City, with the result that most ships from the Atlantic went no farther upstream than the latter port. But then improvements along the river enabled ocean ships to reliably penetrate as far inland as the major rapids (Lachine Rapids) at Montreal, and shipping rates were lowered for the developing Ontario Peninsula and areas to the westward. Quebec City was bypassed and began to grow relatively slowly, although it has continued to be an active ocean port. With deeper water than Montreal, Quebec City can handle much larger bulk carriers. But in the latter 19th century Montreal decisively surpassed its French-Canadian rival and became the leading port and largest city in Canada. Only recently has Montreal been exceeded in trade by the Pacific port of Vancouver and surpassed in metropolitan population by Toronto. Today oil imports, wheat exports from the Prairie region, and container shipments are major elements in Montreal's diverse port traffic. Imported crude oil is processed by a major group of refineries in the port area. Powerful icebreakers in the St. Lawrence now make Montreal accessible to ocean shipping throughout the year, whereas both the Seaway above Montreal and the Great Lakes are closed by ice from about mid-December to mid-April.

As Montreal's port traffic grew, the city's business interests facilitated it by pioneering a railway network focusing on Montreal. This network made use of lowland routes that radiate from Montreal: the St. Lawrence Valley leading east to the Atlantic and west to the Great Lakes; the Ottawa River Valley along which Canada's transcontinental railroads find a passage west and north into the Shield; and the Richelieu Valley-Champlain Lowland leading south toward New York City.

By the time of Confederation and national independence, initially arranged by Britain in 1867 through the joining together of Quebec, Ontario, New Brunswick, and Nova Scotia, the core area was already predominant within Canada in agricultural production and population. In the following century it was transformed into today's predominantly urban-industrial region. Small industries and generally small cities had previously appeared, and they often provided foundations for later development. But following Confederation the rapid growth of some industries and some cities made them dominant elements in the region and to a large extent in all of Canada. In its evolution to industrial dominance the coreland profited from a number of clear advantages over the rest of Canada: superior position with respect to transport and trade, the best access to key resources, close accessibility to the largest markets, and the most advantageous labor conditions.

Transportation Advantages

Very important among the coreland's advantages were lowland and water-transport connections between the Atlantic and the interior of North America. In all of eastern North America the only other connection of this sort through the Appalachian barrier is the lowland route connecting New York City with Lake Erie via the Hudson River and the Mohawk Valley. Rapids on the St. Lawrence River long barred ship traffic above Montreal, but the lowland along the river facilitated railway construction, and the building of some 19th-century canals bypassed the rapids and allowed small ocean vessels to enter the Great Lakes. Finally in the 1950s the river itself was tamed by a series of dams with locks in the St. Lawrence Seaway project, and since then good-sized ships have been able to reach the Great Lakes via the river. The Welland Canal, which bypasses Niagara Falls between Lakes Ontario and Erie by means of a stairsteps of locks, predated the Seaway. The canal admits shipping to the four Great Lakes above the falls, including the long and intricate shoreline of Ontario Province. Farther up the lakes, the "Soo" Locks and Canals enable ships to pass between Lake Huron and Lake Superior, although natural channels without locks interconnect lakes Erie, Huron, and Michigan.

Resource Availability

Industrialization in the coreland drew support from agricultural surpluses of the region, forests in and adjacent to the coreland, minerals, and waterpower. Today the region is no longer a major producer of surplus agricultural products due to the market for such products now provided by the coreland's relatively large population. But the coreland still accounts for about two fifths of Canada's farm output, and its agriculture made important contributions to the rise of industry. Before the Ontario Peninsula turned to specialization on dairy products, fruits, vegetables, and other production largely for the local market, the wheat and livestock surpluses of southern Ontario helped foster the growth of food-processing and farm equipment industries. Forest production also was important. Early lumber and timber exports were succeeded by the present massive development of pulp-and-paper mills, which sell their products largely to the adjacent United States. The latter industries, most of which are located in Quebec Province, draw their wood from the southern fringes of the forested wilderness in the adjacent Canadian Shield. The mills cluster along the St. Lawrence and its tributaries from the Ottawa River eastward. These streams provide transportation for logs, as well as the hydropower and water needed to process the wood. The Shield also is a storehouse of metallic minerals, of which Canada has long been a major world producer. These began to be tapped in the 19th century when railroads were built westward across the Shield from Montreal and Toronto. Among the many minerals are very large deposits of iron ore, some located in Canada and others on the United States side of the border around western Lake Superior. These deposits fed not only the United States steel industry but also the smaller Canadian industry that developed in Ontario, especially at Hamilton on Lake Ontario. Some of the ore deposits in the Canadian part of the Shield are in the Lake Superior area, while others, more recently discovered and developed, are in northeastern Quebec Province and adjacent Labrador.

Cheap water transportation of huge amounts of iron ore for much of the journey to the steel mills was an important aid to the development of both the American and the Canadian iron and steel industries. Another vital mineral available to the coreland was coal. This was provided to Ontario from the northern Appalachians of the United States and to Quebec from mines in Nova Scotia. Beginning in the 1890s this major power source was supplemented by the development of one of the world's larger concentrations of hydroelectric facilities. Many of these power stations are located near the edge of the Quebec and Ontario Shield where streams descend toward the lowlands along the Great Lakes and the St. Lawrence, but more recent development has focused on sites located deep in the northern wilderness. Hydroelectricity also comes from Niagara Falls and from installations along the St. Lawrence River that were developed in connection with the St. Lawrence Seaway. Electricity is relatively cheap in the coreland, and large quantities are exported to the United States. Oil is imported to the region via the St. Lawrence River, and both oil and natural gas come via pipelines from the Prairie Provinces to the west.

Market and Labor Advantages

Ever since large-scale industrialization in the coreland began, this region's population has represented the main cluster of Canadian consumers and available labor. The region also has had centrality within the rest of Canada's population, and it is adjacent to much of the United States market. American competitors selling in Canada might have stunted Canadian industrial growth, but a high-tariff policy to forestall this was adopted in Canada shortly after Confederation. The policy fell far short of shutting out the foreign (mainly American) industrial products which Canada imports in large quantities. But it seems to have favored the growth of a large variety of Canadian industrial enterprises, mainly in the core region and often including branch plants of American companies ''jumping'' the tariff wall. However, the policy has long aroused protest from other parts of Canada preferring to buy American goods that often are cheaper. It also has fostered many high-cost Canadian industries which lack international competitiveness. The exceptions to this latter generalization tend mainly to be industries whose products are desired in the United States or have been admitted freely to the United States for other reasons. Thus the Canadian pulp-and-paper and mining industries, with the American market available to their products, tend to be large and efficient. In 1965 the two countries made a free-trade agreement with respect to automobiles, and this sparked rapid expansion of the Canadian auto industry in southern Ontario. General Motors, Ford, and Chrysler all have plants there, some of which are in Detroit's Canadian satellite of Windsor (255,000), located immediately across the Detroit River. Metropolitan Toronto is the main Canadian center of the industry, however, and many other places have a share of it. Many factories are suppliers of auto parts, and there is a complex interchange of parts across the international boundary.

INDUSTRIAL DIFFERENCES BETWEEN ONTARIO AND QUEBEC

Canada's core region produces a very wide range of manufactures, and for most types of manufactures it accounts for more than three fourths of Canada's national output. Processed foods, paper, and primary metals are significant exceptions, but even in these instances the coreland produces over half of the total value. Within the core region there are marked divergencies between the types of manufactures emphasized in Ontario and those emphasized in Quebec. During the formative decades of industrialization, it was Ontario which had cheaper access to United States coal. This favored the development of iron-and-steel capacity and metal products industries there, and these types of industries are still more prominent in Ontario's economy than in that of Quebec. Major lines include automobiles and other transport equipment, machinery, and miscellaneous metal fabricating. In addition to making passenger cars, commercial motor vehicles, and short takeoff and landing aircraft, Ontario supplies parts and subassemblies to automobile and aerospace factories in the United States. In Quebec, metropolitan Montreal also has a considerable development of industries making transport equipment and other metal goods. But these are less typical of Quebec Province than are more labor-intensive industries such as apparel manufacturing. Such industries became established in the 19th century, partly on the basis of immigrant labor arriving at the Quebec ports but primarily on the basis of surplus labor within Quebec itself. The surplus was generated when unusually rapid increase of Quebec's Roman Catholic population created pressure on the available means of support within the province. This population surplus led to migration from the St. Lawrence Valley into other parts of Canada and the United States, but it also led to the establishment of manufacturing industries in Quebec that were especially labor-intensive and, hence, in need of notably cheap labor. The province's industrial structure came to incorporate large elements of such industries as textiles, shoes, and clothing. The differences between the two provinces still persist, and they result in lower average wages in Quebec. This accords with a marked difference in overall per capita incomes, with Ontario being more affluent by a considerable margin. Both provinces, however, are beset by overall contraction in industrial employment and slow growth or even decline in various aging industries and industrial areas. These problems are quite similar to those of adjacent northeastern and midwestern United States, being rooted in such factors as slowly growing markets, competition from Asian and Latin American imports, antiquated equipment, and lack of capital to modernize.

CULTURAL DIVERGENCE AND ITS CONSEQUENCES

The economic differences between the Quebec and Ontario parts of the Canadian core area are a much less serious matter than the cultural differences. In Quebec Province 82 percent of the people speak French as a preferred language and 11 percent speak English. This is the only province where French speakers predominate and control the provincial government, while Ontario, where 77 percent speak English as a first language and only 5 percent speak French, is the largest in population of the nine predominantly English-speaking provinces. Major political consequences for both Quebec and Canada have followed from this ethno-political situation.

For the better part of two centuries of British and then Canadian government in Canada these consequences did not seem serious. Quebec's government was French-controlled, while its economy was largely directed by an upper-class English minority. There was an unequal language situation: French speakers who wanted to rise economically might well find it necessary to become fluent in English, and French who migrated to other provinces would have to make adjustments to predominantly English-speaking environments; whereas the English-Canadians of Quebec could go about their lives with little or no knowledge of French. But Quebec's relatively large population gave it a large representation in Canada's Parliament, and the province produced some distinguished and powerful Canadian political leaders. At the same time the rapid natural increase of the Quebec French, largely rural and Catholic in a society heavily influenced by the Church, seemed to assure the continuation of the French language and French culture. For many years the language situation was relatively placid, although the placidity was broken occasionally by disturbances such as resistance to the draft in Quebec during both world wars.

Widespread and open French dissatisfaction arose after World War II. Its rise appears to have been related to marked demographic changes and probably also to sharply increased concern with the general situation of minorities in the Western world. Quebec's rate of population growth dropped below that of Canada as a whole in the two decades after 1941, and after 1961 the rate declined so sharply that by the 1970s it was only about half that of the country as a whole. Natural increase in Quebec slowed drastically as the province became more urbanized and secular. Furthermore, most new immigrants bypassed the province in favor of Ontario or provinces farther west, and immigrants who did settle in Quebec tended toward assimilation with the English-Canadian minority in the province rather than the French-Canadian majority. Quebec's population began to decline as a percentage of Canada's, and a perceived threat to the province's French culture and heritage intensified political and economic grievances of long standing.

The result was the rapid rise in the 1960s and 1970s of a separatist political party in Quebec. The *Parti Québecois* advocated the secession from Canada of an independent Quebec which could use its sovereign powers to protect its cultural heritage. The party achieved 24 percent of the Quebec vote in 1970, 30 percent in 1973, and 41 percent in 1976. The latter percentage was enough to bring it to power within the provincial government.

The Quebec separatists used their provincial authority to strengthen the position of French in the province through measures such as requiring the use of French rather than English in business and requiring all newcomers to the province to be educated in French. With respect to the major objective of independence from Canada, the party has moved cautiously. In 1980 the separatists held a provincial referendum asking the voters to authorize provincial negotiations with the federal government for independence, while holding out the hope of continued close ties with a diminished Canada. But Quebec's voters rejected their government's request for these negotiating powers. Only 42 percent of the voters gave approval, although probably 52 percent of the French-speaking voters did so. In 1981 the *Parti Québecois* was nevertheless returned to provincial office, thus keeping the separatist option alive. However, in 1985 the party was defeated at the polls, and a non-separatist government was reinstalled in Quebec City. Thus separatism receded, at least temporarily.

Realistically, however, the creation of a separate nation-state in Quebec now seems unlikely. The economy of the province has been having marked difficulty; Quebec's people already pay much higher taxes than the general Canadian average; and there are strong feelings that independence cannot be afforded. Especially among the large body of younger people who are well educated, technologically capable, and less traditional in their cultural attitudes, there is an increasing tendency to focus less on political separation than on the need for Quebec to become more competitive economically.

During the agitation for a separate nation in Quebec, the federal government's position—vigorously espoused by a federal Prime Minister (Pierre Trudeau) who was himself French-Canadian—was that all of Canada should be effectively bilingual. In a new constitution, achieved in 1982, the principle was laid down that English and French are both official languages of Canada, and either may be used in dealing with any branch of the federal government. French-Canadian minorities in provinces other than Quebec are entitled to have the public schools educate their children in French unless the children are too few in number to warrant financing such education. In Quebec, however, children of non-French-speaking parents may be educated in their home language only if their parents were educated in that language. Hence French-Canadians may not adopt English and then perpetuate it by having their children educated in it.

The end of this cultural-political struggle has probably not been reached, but it has already had important effects on Canada. Certainly it has sensitized the country to questions of minority rights and has frequently caused tempers to flare. It has probably weakened Quebec Province economically by causing investors to look elsewhere and some business to actually move out. It has contributed strongly to making Canada a federal state in which the political subdivisions are much stronger relative to the federal government than is true in the United States or in most federal states. Various concessions have been made to Quebec in respect to the handling of governmental matters, and these have weakened the position of the federal government in resisting demands from other provinces for more power over their own affairs. Such demands have come especially from the more western provinces, particularly mineral-rich Alberta. The new Canadian constitution strengthens the high degree of control over natural resources exercised by the provinces in which the resources lie and correspondingly weakens control over resources by the federal government at Ottawa.

THE PRAIRIE REGION: OIL AND WHEAT TRIANGLE

The provinces of Manitoba, Saskatchewan, and Alberta have long been identified as a regional group under the name "the Prairie Provinces." Relative isolation is one factor that has led to this grouping. The populations of these units are separated from other populous areas by hundreds of miles of very thinly inhabited territory on both the east and the west. To the east the Canadian Shield separates them from the populous parts of Ontario and Quebec, and on the west the Rocky Mountains and other highlands separate them from the populous Vancouver region in British Columbia. To the immediate north—including the northern parts of the three provinces—lies subarctic wilderness, and to the immediate south are lightly populated rural areas in Minnesota, North Dakota, and Montana.

The prairie environment which gives the provinces their regional name exists only in some southern sections of the three, and it is to these sections that we ascribe the name "Prairie region" in this text. At the time of settlement the region was characterized by grasslands occupying relatively small parts of southern Manitoba, all of southern Saskatchewan, and southeastern Alberta to about 300 miles (c. 500 km) north of the United States border. This expanse of open country is roughly triangular. It lies within the Interior Plains between the Shield and the Rockies, with the southern base of the triangle resting on the United States boundary. On all sides except the south the grasslands are bordered by vast reaches of forested country. Most of the southern part of the triangle represents a northward continuation of steppe grasslands from the Great Plains of the United States. However, toward the forest edges the soils are moister, and the original settlers found true prairies: taller grasses with scattered clumps and riverine strips of trees. It was to this prairie around the forested edges of the grassland triangle that settlers were most attracted, and this part of the Prairie region still forms a more densely populated, arc-shaped band. Drier short-grass steppe areas in the southern part of the Prairie triangle have been infilled with a sparser population.

The Prairie region was settled late, but quite rapidly, in a few decades after 1890. Although fur-trading posts and some small settlements had existed earlier, extensive settlement was delayed by the isolation of the region in Canada's deep interior and the availability of comparable land farther south in the United States. When prospective settlers began to arrive in large numbers, they found reasonably good land in a plains environment. A climate of long and harsh winters, short cool summers and rather marginal precipitation had to be endured, but hardy crops could be grown. Early subsistence farming was succeeded by a specialization on spring wheat for export, and the Prairie region became the Canadian part of the North American Spring Wheat Belt. The settlers who came in the late 19th and early 20th centuries were predominantly English-speaking people from eastern Canada and the United States. But they included notable minorities of French-Canadians, Germans, Ukrainians, and others of European extraction, and the ethnic composition of the region still reflects the prominence of these elements among the early settlers.

However, the economies of the three Prairie Provinces altered notably by the 1930s, and sharply after 1950. Agriculture is still basic in all of them. Together they produce around half the value of farm products in Canada and sell about a fifth of the world's wheat exports to importers around the world. Their agriculture has been diversified to include other hardy crops such as barley and rapeseed (for vegetable oil and fodder), together with significant numbers of livestock. But urbanization and nonagricultural industries have become increasingly important, especially in Alberta and Manitoba. The three main metropolises in the region are Edmonton (800,000) and Calgary (675,000) in Alberta, and Winnipeg (625,000) in Manitoba. Regina, the largest place in Saskatchewan, has only about 190,000 people in its metropolitan area. Thus, although agriculture is still a very important part of the economy in all three provinces, it is the most so in Saskatchewan. In that province 20 percent of all employment was still in farming in the early 1980s, while the figure was under 10 percent in the two more urbanized provinces. In this respect it has apparently been Saskatchewan's misfortune to be the central one of the three provinces, as the leading cities have all grown near the corners of the Prairie triangle where connections to the outside world are focused. Winnipeg developed where the rail lines through the Shield wilderness to the east crossed the Red River south of Lake Winnipeg and entered the Prairie Provinces; Calgary near a pass over the Rocky Mountains which gave a route toward Vancouver and the Pacific; and Edmonton near another Rocky Mountain pass leading to the Pacific. Both Winnipeg and Edmonton developed additional functions as metropolitan bases for huge sections of Canada's North. The cities of the Prairie region introduce an element of manufacturing into the regional economy, but such leading products as processed foods and farm machinery are an extension of the region's agriculture, and none of the three provinces has yet become a notable manufacturing center. Miscellaneous industries process local minerals and timber or make a variety of fabricated items.

Minerals, especially oil, are the foundation of recent advances in the Prairie region's economy, especially that of Alberta. Coal deposits underlie much of the plains section in Alberta, as well as the province's Rocky Mountain foothills, and also underlie western Saskatchewan. They have long been locally useful and in recent decades have begun to provide a major export. Nickel, copper, zinc, and other metal production from the Canadian Shield section of Manitoba is important to that province's economy, and mines in southern Saskatchewan produce major quantities of potash. The Shield of northern Saskatchewan produces uranium. But it is petroleum and associated natural gas that have had the most economic impact on the Prairie region.

The oil industry of the Prairies began near Calgary, Alberta, in the late 1940s. As production has multiplied and new fields have been brought in, the industry has remained very predominantly in Alberta, with a minor portion in Saskatchewan and a tiny share in southern Manitoba. On a world scale

Canada's total oil output is still small, but its growth in the Prairies has had some significant consequences for the region—including its relationships with the rest of Canada. One consequence has been explosive growth for Calgary and Edmonton, which are the oil industry's main business centers. The rapid population growth of the province of Alberta since 1950 has given it over half the population of the three provinces—2.4 million out of 4.5 million in 1987. There is an increasing tendency for Albertans to stress that their province is no longer just "one of the Prairie Provinces" but a distinct region in itself. In support of this position they cite not only oil, gas, and dynamic urbanism but also the different physical landscape of the spectacular Rocky Mountains that extend along Alberta's southwestern border. Another result of the oil and gas boom has been a running quarrel with the federal government at Ottawa, centering on resource control and especially on federal price control of oil piped from Alberta to eastern Canada. Alberta's economic dynamism and self-centered political stance seem likely to continue, as the province is much richer in potential energy than in present energy production. It is estimated that "tar sands" in northern Alberta may possibly contain two to three times the energy equivalent of all the oil now thought to exist in the rich Middle Eastern fields. Small amounts of oil products are being manufactured from the hard-to-process hydrocarbons in the sands, but major extraction must await improved technology and/or a time when greater scarcity of oil yields higher prices for oil products than exist today. An important product extracted from natural gas in Alberta is sulfur, of which Canada has become the world's largest exporter. The main markets are the United States and Japan—countries which use huge quantities of sulfuric acid in many industrial processes.

THE VANCOUVER REGION: CORE OF BRITISH COLUMBIA

Canada's narrow transcontinental belt of population-clusters is anchored at the Pacific end by a relatively small region centering on the important seaport of Vancouver at the extreme southwestern corner of British Columbia (chapter-opening photo). The region contains no more than 5 or 10 percent of British Columbia's area but has over half of the province's population of about 3 million. Metropolitan Vancouver has some 1.4 million, while another 260,000 reside in the metropolitan area of the province's capital city, Victoria. Vancouver is located on a superb natural harbor at the mouth of the large Fraser River, which rises in the Canadian Rockies and then flows southwestward to the Pacific. Its valley provides a natural route for railways and highways across the grain of rugged mountains and intermountain plateaus in Alberta and British Columbia. Victoria is located at the southern end of large Vancouver Island, the southernmost and by far the largest of a chain of islands along Canada's Pacific coast.

Victoria was the leading center of British Columbia before the rapid rise of Vancouver in the late 19th century. The latter development was set off by the completion of Canada's first transcontinental railroad, the Canadian Pacific, to Vancouver in 1886. The city quickly became the country's main Pacific port and has recently become the largest seaport of the entire country. The port has profited from a continuing buildup of economic development and foreign trade in British Columbia and the Prairie Provinces, as well as the increasing trade of Canada with Japan. Vancouver exports huge quantities of such bulk commodities as grain, wood in various forms, coal, sulfur from natural gas processing in the Prairie region, various metals, potash, and asbestos. It is also a very active container port. Sea trade is fundamental to Vancouver's economy, but the city is far more than just a seaport. Despite its peripheral location within British Columbia, Vancouver functions as the province's main industrial, financial, and corporate administrative center. It has numerous links with a diverse array of production nodes scattered through an expanse of sparsely populated and mountainous terrain larger than Texas and California combined. We turn now to an examination of the province for which Vancouver provides a metropolitan base, organizing center, and ocean outlet.

THE PROVINCE OF BRITISH COLUMBIA: CHARACTERISTICS AND CANADIAN COMPARISONS

The Pacific province of British Columbia combines an environment very different from that of the rest of Canada with a sparse population and resource-oriented economy that are very typical of Canada as a whole. High mountains trending southeast to northwest edge the province on two sides, the Coastal Ranges along the Pacific and the Rockies along the border with Alberta. The Pacific edge of the Coastal Ranges is extremely irregular and penetrated by fjords—long and narrow arms of the sea between mountain walls. The fjords, like those of Norway and southern Chile, resulted from the deepening of stream valleys by glaciers during the Ice Age. Tops of mountains beyond the mainland project as a multitude of islands, of which the largest is Vancouver Island in the southwest. They form a screen between the Pacific and the Inside Passage ship channel to Alaska. Many areas in the Coastal Ranges, and some in the Rockies, are capped by glacial ice and permanent snow fields. High plateau country lies between these massive fringing ranges. Much of it is also topographically rugged, with mountain ranges projecting above the general level and narrow valleys trenching the surface deeply. Only on its northeastern corner does British Columbia include some true plains territory—the Peace River section that forms an outlier of the Interior Plains.

British Columbia is also a distinctive Canadian area climatically. The islands and the Pacific slope of the Coastal Ranges have a marine west coast climate broadly similar to that of western Europe and of coastal Washington and Oregon to the south. Mild winters, cool summers, and frequent rains are characteristic in lower areas here, supporting a natural vegetation of dense coniferous evergreen forest. Climatic conditions in the southern interior of the province are quite varied, with both temperature and the amount of precipitation depending on the elevation and exposure of places. Some sheltered valleys are semiarid with a parklike vegetation of grass, shrubs, and woodlands, but most of the interior has enough moisture to support a forest growth somewhat sparser than that along the coast. In the northern half of the province subarctic temperatures and forests prevail except near the coast.

British Columbia is typical of Canada in its population distribution. Some 3 million people are scattered over its huge area, giving an average density of 8 per square mile (3 per sq. km). This figure is almost meaningless because over half the population is crowded into the small southwestern core area centering on Vancouver.

Like Canada as a whole, British Columbia has an economy based primarily on the export of products (''primary'' products) that are derived directly from natural resources. In the early and middle part of the 19th century, the first settlers were attracted to the area by furs, gold, coal, and fish. Subsequent development has multiplied this early population many times but has not yet moved fundamentally away from primary types of production. The latter are still developing at increasing scales and with spreading locations and increasingly varied products.

The province's forests (Fig. 27.5) and mines are its most important sources of support. Large sawmills and pulp-and-paper mills are scattered along the coast, with the largest concentration in the Vancouver area, to which logs are brought by sea from the north. In recent decades, however, active railway and road construction by the provincial government has made more of the interior accessible, and growing forest-products industries there have now surpassed the coastal mills in production. Mining of metals, financed for the most part by outside capital, has spread from early locations in the southern interior to newer sites farther north. Copper, molybdenum, silver, zinc, and lead are the major products. Coal has long been mined on a small scale in the Rocky Mountains of Alberta and British Columbia

Figure 27.5 Large-scale lumbering in progress on Vancouver Island, seen from a ship on the Inside Passage to Alaska. Logs in the water await shipment to a processing plant.

near the United States border, and in recent years production from that area has multiplied many times in response to market demand in Japan. Far to the north, new coal mining on a huge scale is under development in the Rockies of British Columbia west of the Peace River plains. Japanese capital has played a large role in financing the recent expansions of British Columbia coal production, and most of the coal is exported to Japan. Relatively small quantities of oil and natural gas from the Peace River plains add further to British Columbia's varied and valuable mineral output.

The sea and the mountains provide other important sources of support for the province. Spectacular scenery and other outdoor amenities are exploited by a considerable tourist industry. Commercial fishing, especially for salmon, has been an important export industry for a century. Mountain rivers, fed by glaciers and abundant precipitation, often flow in steep canyons, which are relatively easy to dam. The output of current from hydropower stations on these streams, plus a smaller output from coal-fired thermal stations elsewhere, provides enough electricity to supply the province's needs, with enough left over for a considerable export to the United States. Between 1950 and 1980, major hydroelectric plants were developed on the Columbia River near the American border, in the far northeastern Peace River country, and near the coastal town of Kitimat, where the power is used to produce aluminum by smelting imported alumina (the second-stage material from bauxite ore). Hydroelectricity is generated in this latter instance by falling water diverted from a river headstream by a tunnel through a coastal mountain range. Kitimat is on a deep harbor southeast of the seaport of Prince Rupert, which is tied to Canada's railnet by an arm of Canadian National Railways.

In contrast to British Columbia's abundant timber, mineral, and power resources are the relatively meager agricultural resources of the province. The rugged topography affords only limited areas that are level enough and warm enough to grow crops. Dairy products and fruit, especially apples, are marketed from small coastal acreages near Vancouver and from interior valleys; and the Peace River plains are a projection from the grain and livestock farming region in the Prairie Provinces. Considerable numbers of cattle are raised on large ranches in the interior of the province. But British Columbia is far from being an outstanding agricultural area.

Although deficient in agricultural resources and in advanced manufacturing, British Columbia has been an area of rapid growth for many decades. Continuing migration into the provinces comes mainly from the rest of Canada, although very little from French Canada. A small but distinctive population element has been provided over the years by immigration from the Orient, which has been mainly, although not exclusively, Chinese. As in the western United States, workers were imported from China in the 19th century as a labor force for railway construction, and today about 3 percent of the province's population is ethnically Chinese. The relatively high average income of British Columbians is viewed with reservation by some scholars who cite disadvantages in the high cost of living and the cultural poverty of so many scattered small towns isolated in the wilderness and dependent upon one major enterprise, whether sawmill, pulp mill, mine, smelter, or fish cannery. But cosmopolitan Vancouver and rather staid Victoria—the latter renowned for its ''Englishness'' in cultural and social traditions—have many amenities to offer, and their metropolitan areas have attracted sizable numbers of retirees from all over Canada.

THE CANADIAN NORTH: WILDERNESS AND RESOURCE FRONTIER

The North comprises most of Canada, but how far north this region begins is hard to say. Scholars have offered numerous interpretations. The North is certainly that major section of Canada's territory which is very sparsely populated because of its harsh physical environments, but population density fades gradually northward in such a manner that one is left with the problem of defining ''sparsely populated.'' How sparse is ''sparse''? Where should a line marking the southern edge of ''the North'' be drawn? Here we use the southern edge of the subarctic climatic zone as, for the most part, the southern edge of the North, but exempting the island of Newfoundland with its Atlantic associations and somewhat denser population. This definition places large northern sections of all the provinces except the Maritimes in the North. In addition, we may justify inclusion of parts of the Canadian Shield which project south of the subarctic climate in southeastern Manitoba, in Ontario, and in Quebec just north of the St. Lawrence Valley. These areas are so handicapped by poor soils as to be sparsely populated and appropriately categorized as basically Northern. The northern part of British Columbia, with a subarctic or severe highland climate and relatively few people, is clearly part of the North, but much of the southern part of the province outside the Vancouver region forms a broad zone of transition with too much development and too moderate a climate to be considered distinctively Northern. Within the provinces, only northern Quebec and part of Labrador reach the zone of arctic tundra. Most of Canada's tundra is in the Northwest Territories.

The southern fringes of the North, often called the "Near North," are spotted with widely scattered islands of population and development which are of considerable importance to Canada's economy and, through trade connections, to the economy of the United States. Mining settlements comprise many of these areas where population thickens. The Shield is endowed with a wide variety of metallic ores, although geologic processes have left it lacking in fossil fuels. The settlements which have been developed to exploit some of these mineral deposits range in size from mere hamlets through small towns to small metropolitan areas. The largest is Sudbury (metropolitan population: 150,000), located in Ontario north of Lake Huron. The city is actually the service center for a number of mining settlements producing major quantities of nickel and copper. Among many other metal-bearing ores produced in the North, iron ore, mainly from an area in Quebec and Labrador near their mutual border, and uranium, mined just north of Lake Huron and in northern Saskatchewan, are particularly noteworthy. In recent years, there have been many slowdowns, temporary suspensions, or even closures of metal-mining operations because of glutted markets and low prices for iron ore, nickel, uranium, and other metals.

Some Northern settlements carry on manufacturing, which consists mainly of the manufacture of pulp and paper or the smelting of metals. As in the case of mining settlements, a scatter of towns producing wood pulp and paper is spread across the southern fringes of the North. Most of these towns are small, but in two places clusters of mills combine with other functions to produce fairly sizable agglomerations near the southern edge of the Shield. One is a string of towns along the Saguenay River tributary of the St. Lawrence in Quebec, well downstream from Quebec City. Together these towns make up a metropolitan cluster of over 125,000 people, which is supported not only by pulp-and-paper plants but also by the manufacture of aluminum from raw materials of foreign origin brought in by ship. The other agglomeration is at Thunder Bay, Ontario (125,000), on the northwestern shore of Lake Superior. In addition to making pulp and paper, Thunder Bay derives support from its function

as the Canadian head of Great Lakes navigation. Its port links the Prairie Provinces with the Ontario-Quebec core area, the United States, and overseas points.

Water resources are of major importance in the southern fringe of the North. Hydroelectric developments support local mines and mills, but they also supply much electricity to more populous parts of Canada to the south, and even some exports of current to northeastern United States. Numerous rivers and lakes, which serve as natural storage reservoirs, are so characteristic of the Shield that much of it appears from the air as an amphibious landscape. Most hydroelectric developments are near the Shield's southern edge, with a particular cluster north of the St. Lawrence River, where tributaries from the north rush down through the Laurentide "Mountains"—the somewhat raised edge of the Shield in Quebec. But increasing demand, plus increasing ability to transmit electricity for long distances, has led since about 1960 to new and very major developments much farther north. These are in northern Quebec east of James Bay, in northern Ontario southwest of James Bay, in interior Labrador, and in Manitoba near Hudson Bay. Water resources also are fundamental to a recreational industry of increasing importance in the southern Shield.

Perhaps 2 million people now live in the North as defined in this text—a remarkably small number for such a vast area. With agricultural possibilities so limited, a large increase in population seems unlikely in the foreseeable future. Three small clusters of farming settlements account for a major share of the present agricultural development: (1) French-Canadian dairy farmers around the Saguenay River towns; (2) farmers utilizing scattered patches of arable land in an extensive area of Ontario and Quebec where sediments overlying the ancient rocks of the Shield have formed the Clay Belt; and (3) wheat and livestock growers in the Peace River plains of Alberta and British Columbia. These areas represent rare intrusions of agriculture into the subarctic climate, where winters are very cold, summers cool and quite short, and soils generally poor. For some decades before World War II the governments of Ontario and Quebec, as well as the Catholic Church of the latter province, promoted northern agricultural settlement, but for the last third of a century the settlements that were established, especially those of the Clay Belt, have been declining. Such agriculture as continues centers on the growing of hay and oats to feed dairy cattle and livestock, and much of it is carried on part-time by people with other jobs. In the Peace River section, patches of agricultural land, with attendant small towns, represent the northernmost point of the wheat belts of Anglo-America. Soils here are somewhat better than those of the Shield, and agriculture is more firmly established, although many handicaps had to be overcome during a pioneering period which lasted from about 1910 until recently.

"TERRITORIAL" CANADA: YUKON AND THE NORTHWEST TERRITORIES

North of 60 degrees lies the part of Canada not yet considered sufficiently developed and populous for provincial status. This bleak area—known as "Territorial Canada"—is under the direct administration of the federal government in Ottawa. Two administrative Territories have been established. In the northwest corner of the country, Yukon Territory is cut off from the Pacific by the Panhandle of Alaska, once claimed by Canada but acquired by the United States. The territory is made up of high and rugged plateau country over most of its extent but with glacier-capped mountains near the Pacific, a Rocky Mountain border on the east, and a bit of coastal plain on the Arctic Ocean. The larger Northwest Territories actually lie east of Yukon Territory. They include a fringe of the Rockies, a section of the Interior Plains along which the Mackenzie River flows northward to its delta on the Arctic Ocean, and a huge area of plains and hills in the Canadian Shield west of Hudson Bay and in the many islands of Canada's Arctic Archipelago. Two huge lakes, Great Bear Lake and Great Slave Lake, both drained by the Mackenzie River, mark the western edge of the Shield. Yukon Territory and the western part of the Northwest Territories have mainly a subarctic climate and taiga vegetation, with tundra near the Arctic Ocean and in highlands, but the center and east of the Northwest Territories, including the Arctic islands, are a treeless expanse of tundra (Fig. 26.4) except for a band of spindly taiga in the extreme south.

Together the two territories governed from Ottawa comprise over a third of Canada's area and almost half as large an area as the contiguous 48 states of the United States, but their combined population is only a little over 75,000 (1988). About a fifth of the inhabitants in each territory are native Indians, and in the Northwest Territories about a third of the people—some 15,000—are Eskimos. The rest are whites, aside from some who are racially mixed. Most whites are government, corporate, or Church employees: officials, clerks, teachers, social workers, medical personnel, policemen, military personnel, employees of mining companies or the Hudson's Bay Company, and Christian clergy of Anglican, Roman Catholic, or other persuasions. Some whites spend their careers in the Territories but most are relatively transient. White school teachers recruited from Canada's South, for example, may enter on their duties with enthusiasm but are eventually daunted by the isolation, the brevity of winter sunlight, and the incredibly low temperatures and wind chill that may persist for weeks on end.

Indians and Eskimos still carry on some of their traditional hunting, gathering, and fishing, but they are now far more concentrated in small fixed settlements than formerly (Fig. 27.6). Practically all Eskimos inhabit prefabricated government-built housing in widely separated villages along the Arctic Ocean or Hudson Bay. Unemployment among both Indians and Eskimos is very high, and a large proportion live essentially on welfare. A depressed economic status, high unemployment, and a heavy dependence on welfare also characterize the more numerous Indian minority in the provincial sections of the North to the south of the Territories. The Indians in the provinces have had their traditional way of life (and often the environment it depends on) severely disrupted by intrusions from the south. There are many complaints among these Indians that their treatment by the individual provinces (in regard to welfare payments, schooling, and so on) is inferior to that accorded the native inhabitants of the Territories—particularly the Eskimos—by Ottawa.

Europeans and white North Americans have been seeking resources from these far northern reaches since the 17th century, when England's Hudson's Bay Company first established fur-trading posts on the shore of the Bay. The chief 20th-century quest has been for minerals, and this has dramatically accelerated since World War II. The Klondike gold rush of 1898 first attracted outsiders into the Yukon on a considerable scale, and today the mining industry affords the principal commercial connection of both Territories to the outside world. Very recently the world's northernmost mine—the highly mechanized "Polaris" zinc and lead mine—was opened on an Arctic island just 80 miles (130 km) south of the North Magnetic Pole. The ore is concentrated in a plant at the mine and then stored to await the brief shipping season in the summer. The few other mines currently operating in the Territories also produce metals or, in one instance, asbestos. Meanwhile mineral-exploration parties are active, focusing especially on oil and gas known to exist in quantity in the Mackenzie Valley and in the Beaufort Sea section of the Arctic Ocean near the Mackenzie

Figure 27.6 *Hunting and trapping are relatively rare occupations in the present world, although small numbers of people still gain some income from them. This photo shows hides of arctic wolves curing in the sun at the village of Eskimo Point on Hudson Bay in Canada's Northwest Territories. The month is August, but the Eskimo woman is warmly dressed for this tundra climate. Today, Canada's Eskimos live in prefabricated houses like the one seen here. Fuel oil for household heating and cooking comes in oil drums brought by ship. The hides came from a hunt by the woman's husband during the preceding winter. He pursued a pack of wolves on his snowmobile and then shot them when they could run no more.*

Delta. Besides income from mining enterprises, the Territories also receive funding in the form of federal government payrolls in the small territorial capitals of Whitehorse in Yukon and Yellowknife in the Northwest Territories. Each Territory has an elected legislature, although the chief executive officer (Commissioner) is appointed by Ottawa, and the federal government controls territorial resources and governmental revenues. In the Northwest Territories, Indians and Eskimos comprise a majority of the legislators.

Energetic efforts to cope with the problems of the Territorial North are being made by government, business, and the region's inhabitants. Isolation has been mitigated by the airplane and by telecommunications via satellite that now bring not only telephone and radio communication but also television to many remote settlements. A network of well-provided schools and medical clinics blankets the Territorial North, which is an area where perhaps half of the present Indian and Eskimo population is under age 15. Air transportation enables the seriously ill to be flown to the few hospitals in the region or to hospitals in cities such as Edmonton or Winnipeg that serve as metropolitan bases for both the Provincial and the Territorial North. The small and scattered nature of settlements, plus the severe environment, makes the provision of services very costly, and the Territories consume far more federal tax dollars than would normally be required for such a small number of people. Development of minerals in the Territories may eventually ease the tax burden greatly, but for the present such development labors under severe handicaps. Costs of exploring for minerals and operating mines are exceedingly high in

this empty land of great distances, roadless terrain, frigid winters, permafrost, and icebound seas. Labor turnover is an acute problem, which mining companies cope with as best they can through high wages, the best of food (including fresh vegetables, fruit, and meat regularly flown in), snug housing, gymnasiums, and such other amenities and social life as are possible under such bleak conditions. One tactic in the more extreme situations is to establish a rotation whereby workers will spend a week on the job and then be flown out for a week in the South. Such expensive undertakings have been discouraged in recent years by generally depressed mineral demand and prices on world markets. Meanwhile the increasingly assertive Indian and Eskimo population is pressing land claims to mineralized areas and pipeline routes and demanding regulations to prevent environmental disruptions by mining companies. Under such circumstances, mineral development in the Territories seems likely to proceed slowly. In a less extreme way, these constraints also affect the Provincial North where all but a very small fraction of current Northern development is found.

The resource frontier in the north has always challenged Canadian ingenuity and resolve, just as Russians, Scandinavians, and Americans have been challenged by their own sectors of the arctic and subarctic lands. Motivated by national pride as well as by economic, social, and defense considerations, each people can be expected to pursue selective development of its Northern territories according to its own conceptions. The resulting variety of approaches and patterns will afford intriguing opportunities for comparative geographic study.

CHAPTER 28

Regional Geography of the United States

Boston, Massachusetts, has played a leading role in American life for nearly four centuries. Modern office buildings and hotels now tower above the subdued landscape of the older city. In the foreground, the Charles River.

E XTREME geographic variety is a major characteristic of the United States. Environmentally this huge country ranges from bleak Arctic tundra in northern Alaska to lush Pacific tropics in Hawaii; from temperate rain forests in Washington's superhumid Olympic Peninsula to nearly rainless deserts in the interior Southwest; from sweeping plains in the midcontinent to imposing peaks near the Pacific and in the Rocky Mountains. Economic variety also runs a wide scale, from Eskimo hunting and fishing in the Alaskan tundra to computer and robot manufacture in California's "Silicon Valley" near San Francisco. Prosperity ranges from extreme wealth in exclusive metropolitan neighborhoods such as Manhattan's Upper East Side or Chicago's Gold Coast, to abrasive poverty in inner-city ghettos and backwoods mining settlements. Great racial and cultural variety also pervades a national population created by immigration from all the world. The process of settling the country has generated cultural landscapes embodying diverse forms and layouts. Seen from above, these landscapes feature formal geometric patterns in some areas such as the Midwestern Corn Belt but are apt to be far more informal and irregular in other places such as the rural Southeast or rural New England.

In the 20th century much of the country's cultural variety has been muted by nationwide standardization of activities, structures, products, and habits. Yet the United States still is characterized geographically by marked differences from place to place that must be assessed if the country's complex geography is to be well understood. This chapter, with limited space for a very large subject, tries to foster understanding by analyzing broad-scale regions. Two systems of regional division—climatic regions and landform regions—have been introduced in Chapter 26. Students may want to review that material as preparation for studying this chapter, which employs a more comprehensive kind of regionalization. Here the United States is subdivided into just four major multi-state "general" regions—the Northeast, South, Midwest, and West—with further subdivision as needed for more detailed understanding. The states of Alaska and Hawaii are treated separately as distinctive outliers from the 48 conterminous states. State lines are used as boundaries of the major regions for convenience in using statistical data, although with full recognition that such boundaries are arbitrary separations of areas that merge with each other gradually in zones of transition. The four major regions, plus Alaska and Hawaii, are evaluated as to their location, extent, dis-

tinguishing features, overall coherence, national and world importance, internal differentiation and organization, sequential development, and major problems of a geographic character.

THE NORTHEAST: INTENSIVELY DEVELOPED AND COSMOPOLITAN CONTROL CENTER

Of the four major regions, the Northeast is the most intensively developed, densely populated, ethnically diverse, and culturally intricate. It incorporates the nation's main centers of political and economic control, and it is the area where relationships with foreign nations are the most elaborate. Strong traditions of intellectual, cultural, scientific, technological, political, and business leadership persist here. This was the area in which the country's political and economic systems had their main beginnings, and it has been historically the chief reception center for foreign immigrants, as well as the recipient of millions of immigrants from other parts of the United States. The long succession of newcomers has endowed the Northeast with a varied ethnicity and cosmopolitan character.

TWO SETS OF STATES: NEW ENGLAND AND MIDDLE ATLANTIC

As considered here, the Northeast consists of six New England states (Maine, New Hampshire, and Vermont in Northern New England; and Massachusetts, Connecticut, and Rhode Island in Southern New England), plus five Middle Atlantic states (New York, New Jersey, Pennsylvania, Delaware, and Maryland) and the District of Columbia. Delaware and Maryland have important historical links with the South, and they are reported as Southern states in federal statistics, but their present economic and urban character classifies them better as Middle Atlantic states within the Northeastern region.

THE NORTHEASTERN SEABOARD: A PINNACLE OF URBANISM

Within the Northeast lived an estimated 56 million people in 1986, or 23 percent of the national population, on 5 percent of the nation's area (Table 28.1). Hence the regional population density of about 310 per square mile (c. 120 per sq. km) was several times that of the South, Midwest, or West.

TABLE 28.1   UNITED STATES REGIONS: COMPARATIVE DATA

	Percent of National Totals			
	Northeast	South	Midwest	West
Total area	5.0	24.5	21.2	32.8
Total population	23.1	32.1	24.6	19.6
Land in farms	2.7	29.4	35.9	31.7
Crop land	4.1	25.7	54.5	15.7
Value of farm products sold	7.2	28.1	43.0	21.3
Value of livestock and products sold	9.1	30.1	43.6	17.3
Value of crops sold	5.2	25.9	42.4	25.8
Value added by manufacture	24.6	27.8	30.8	16.5
Value of mineral output	3.2	59.9	10.6	19.5
Value of retail sales	24.4	30.8	23.8	20.2
Total personal income	26.4	28.1	24.1	20.7
Total metropolitan population	26.7	28.5	22.8	21.5

Source of data: U.S. Bureau of the Census, *Statistical Abstract of the United States*, 1988. Most of the data are for 1986. National totals from which regional percentages are calculated include Alaska and Hawaii.

But such crowding is even more extreme within a rather narrow and highly urbanized belt that stretches for about 500 miles (*c.* 800 km) along the Atlantic margin from metropolitan Boston through metropolitan Washington, D.C. The belt is often called the Northeastern Seaboard, Northeast Corridor, Boston-to-Washington Axis, or "Megalopolis" (see Fig. 28.8). Here seven main metropolitan areas aggregated nearly 36 million people according to 1986 estimates: Boston 4.1 million, Providence 1.1, Hartford 1, New York 18, Philadelphia 5.8, Baltimore 2.3, and Washington 3.6.[1] Additional population within the belt amounted to roughly 6 million, giving a total of about 42 million for the belt, or three quarters of the people in the Northeast.

---

[1]City populations in this chapter are rounded approximations for MSAs, PMSAs, or CMSAs, with the most inclusive figure always being used. Figures are based on 1986 estimates in U.S. Bureau of the Census, *Statistical Abstract of the United States*, 1988 and Population Reference Bureau, *Metro U.S.A. Data Sheet*. In Census Bureau terminology, "MSA" (Metropolitan Statistical Area) refers to an urban aggregate comprised of a county, or two or more contiguous counties, meeting specified requirements as to metropolitan status. Within the MSA there must be at least one city ("central city") of 50,000 or more inhabitants, or a Census Bureau-defined urbanized area of at least 50,000 inhabitants within a total MSA population of at least 100,000 (75,000 in New England). Other contiguous counties may be included in the MSA on the basis of such criteria as commuting or population density. In New England, MSAs are defined by cities and towns rather than by counties. Designation of MSAs represents an attempt to give a realistic picture of functional urban aggregates, each of which often includes more than one incorporated urban place ("political city"), plus unincorporated urbanized areas and stretches of countryside where agriculture may be important. But even the countryside is closely bound to the services, markets, and employment afforded by the urban places. A PMSA (Primary Metropolitan Statistical Area) is, in general, a larger aggregate, and a CMSA (Consolidated Metropolitan Statistical Area) is the most inclusive unit of all. Figures cited in this chapter for the very largest cities, such as New York, Los Angeles, and a considerable list of others, are for CMSAs. Any of the three types of metropolitan units may include counties in more than one state.

Considered as a whole, the cities of this seaboard metropolitan belt comprise the country's leading center of political decision-making, business management, finance, trade, and services. Manufacturing is relatively less important here, but the belt still accounts for about one sixth of the nation's manufactural employment. Seen on a map as an elongated mosaic of metropolitan areas, this city-strip comprises an interlocking maze of central business districts (CBDs), residential neighborhoods and suburbs of many kinds, old factories and warehouses, new thruway commercial and industrial developments, governmental centers, transportation terminals, beaches and other recreational areas, and patches of intensive agriculture, all bound together by a dense net of surface and air transport lines.

Away from the seaboard belt, especially toward the north and in the interior, the population density of the Northeast fades, in some areas to very low levels. For example, in 1986 the state of Maine averaged only an estimated 35 persons per square mile (13 per sq. km) of total area, as opposed to 975 per square mile (376 per sq. km) for New Jersey, the nation's most densely populated state. Even within the seaboard belt there are many areas of relatively light population where woodlands are more prominent than cleared land.

THE NORTHEASTERN ENVIRONMENT: INTRICATE, SMALL-SCALE, RESOURCE-DEFICIENT

The Northeast is not a region in which topography opens out grandly into broad plains or rises to towering mountains. Instead, the region exhibits many small-scale and rather subdued habitats, intermingled in an intricate fashion. Most terrain is hilly, and some areas rise to low mountains, but valleys, bas-

673

ins, and small plains are plentiful. Not only is the topography relatively subdued, but natural resources tend to be so limited in size and variety, poor in quality, and spottily distributed that they are not very helpful to the present technologically advanced society of the region. Some resources such as furs, fish, varied timber, waterpower, small iron-ore deposits, and pockets of good soil met the limited needs of early settlers reasonably well, and later on the big-scale coal deposits of Pennsylvania fueled the nation's first great development of mechanized industries and transportation. But the sum total of the Northeast's original resource endowment would scarcely have foreshadowed the enormous economic development that has occurred in this region.

Today most materials required by Northeastern industries come largely from outside the region, although the Northeast's own coal deposits, water, and forests remain valuable industrial resources, and there is an abundance of stone, sand, gravel, and clay for construction. The varied physical environment provides little by way of oil, gas, and metal-bearing ores, but it is rather generously endowed with assets for tourism and recreation. Soils are generally poor, but they still provide enough of a base for a large production of milk and some other foods. Although the region's long-standing ocean fisheries are now greatly exceeded in output by the fisheries of states bordering the Gulf of Mexico and the Pacific, the Northeast continues to be an important producer of certain fish and shellfish such as flounder, cod, scallops, and lobsters. A major legacy from past resources is the Northeast's enormous financial power, which had important roots in the accumulation of wealth from resource exploitation and accompanying maritime trade in colonial times. Such wealth increased many times over when it was invested in the larger resources opened up as the nation expanded westward.

Most of the Northeast lies in the Appalachian Highlands, the exceptions being relatively small sections in the Atlantic Coastal Plain, the Piedmont, and the Interior Plains (Figs. 28.1–28.3). Some salient features of the various physical divisions and their human use are sketched in the following sections.

The Northeastern Coastal Plain

Atlantic Coastal Plain areas in the Northeast include (1) Cape Cod at the southeastern corner of Massachusetts; (2) Long Island in New York state; (3) New Jersey south of a line from New York City to Trenton; (4) the Delmarva Peninsula, which includes nearly all of Delaware plus the "Eastern Shore" of Maryland east of Chesapeake Bay; and (5) the "Western Shore" of Maryland inland to a line between Baltimore and Washington. The Coastal Plain is low and flat to gently rolling, with many sand dunes and marshes or swamps. Sandy beaches, often on low *barrier islands* just offshore, provide seaside recreation along much of the coast. Soils on the Plain are sandy and range from very low to mediocre in fertility. Many have been made very productive by intensive fertilization for the growing of high-value crops. The Plain is indented by many broad and deep *estuaries* ("drowned" lower portions of rivers), which have harbored shipping since the beginnings of European settlement. Today the largest Northeastern seaport cities are on estuaries: New York on bays and channels at the mouth of the Hudson River, Philadelphia on the lower Delaware, whose estuary broadens into Delaware Bay; and Baltimore on a tributary estuary (the Patapsco River) near the head of Chesapeake Bay, which is itself the oversized estuary of the Susquehanna River.

Northern Piedmont and Fall Line

Between the inner edge of the Coastal Plain and the Appalachians, the northern part of the *Piedmont* extends northward from the South as a narrowing belt across central Maryland, southeastern Pennsylvania, and central New Jersey to metropolitan New York where the Piedmont, the Coastal Plain, and the Appalachian Highlands converge. The Piedmont is generally higher in elevation than the Coastal Plain but lower than the Appalachians. Where the old erosion-resistant igneous and metamorphic rocks of the Piedmont meet the younger, softer sedimentary rocks of the Coastal Plain, many streams descend abruptly in rapids and/or waterfalls. Hence the boundary between the two physical regions is known as the *Fall Line*. Its location in the Northeast can be discerned from the line of cities which have developed along it: Washington, Baltimore, Wilmington, Philadelphia, Trenton, and New Jersey suburbs of New York (Fig. 28.1). Most of the Piedmont is distinctly rolling, with soils generally better than those of the Coastal Plain or Appalachians. Some Piedmont soils in southeastern Pennsylvania formed from an uncharacteristic sedimentary enclave of limestone are particularly fertile and support one of the most productive agricultural areas in the United States. Here in Lancaster and York counties lies the heart of the "Pennsylvania Dutch" (German) country with its distinctive pattern of superbly tended land on family farms, huge well-kept farmsteads, and close adherence to traditional values of

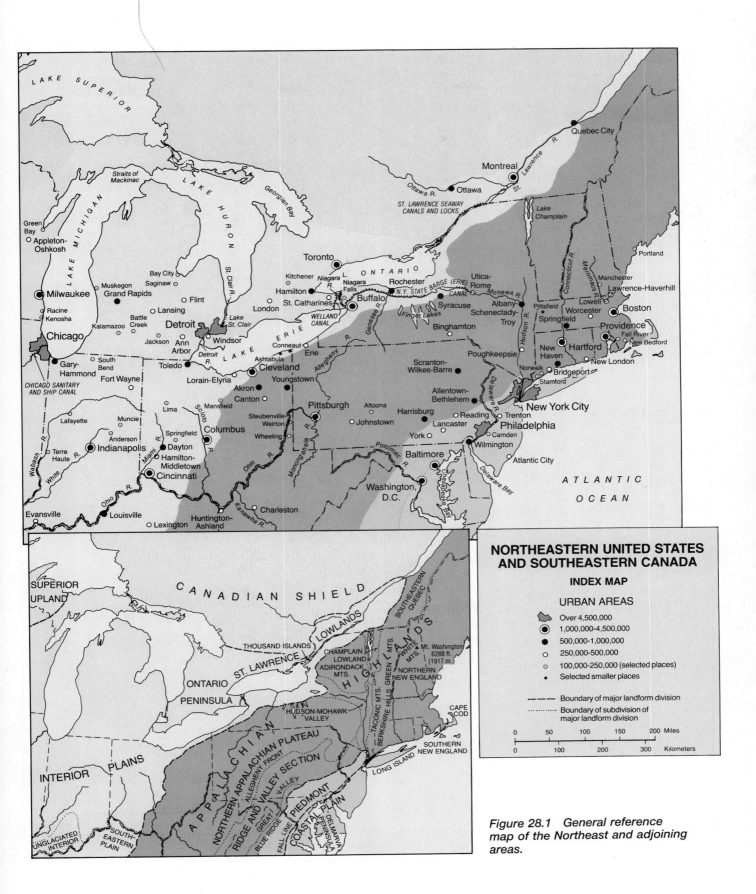

Figure 28.1 General reference map of the Northeast and adjoining areas.

NORTHEASTERN UNITED STATES AND SOUTHEASTERN CANADA

INDEX MAP

URBAN AREAS

Over 4,500,000

1,000,000–4,500,000

500,000–1,000,000

250,000–500,000

100,000–250,000 (selected places)

Selected smaller places

Boundary of major landform division

Boundary of subdivision of major landform division

0   50   100   150   200 Miles

0   100   200   300 Kilometers

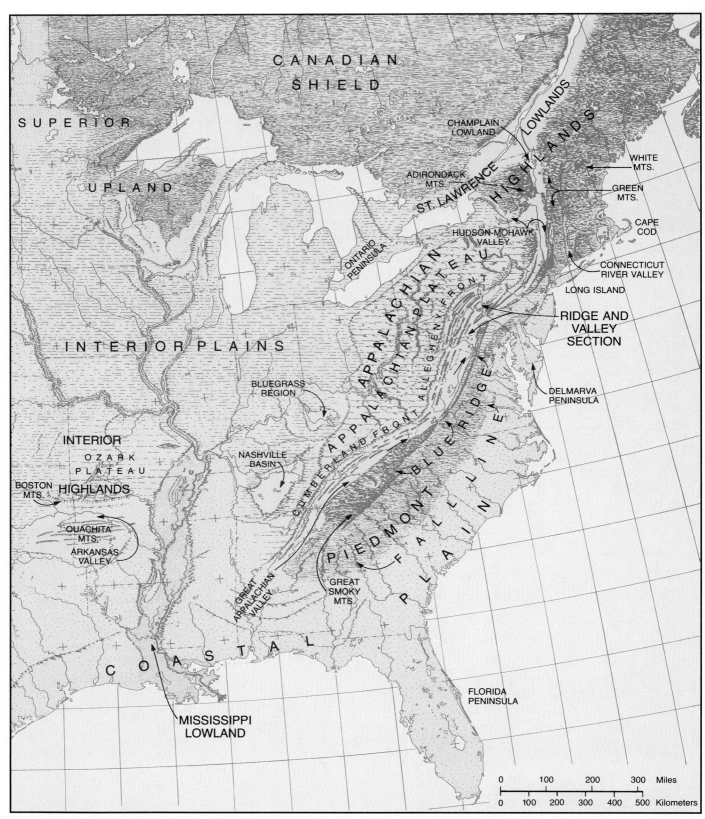

Figure 28.2   Major physiographic features of eastern United States and southeastern Canada. The base map is a portion of A. K. Lobeck's Physiographic Diagram of the United States, copyright, the Geographical Press, a division of C. S. Hammond & Company, Maplewood, New Jersey.

676

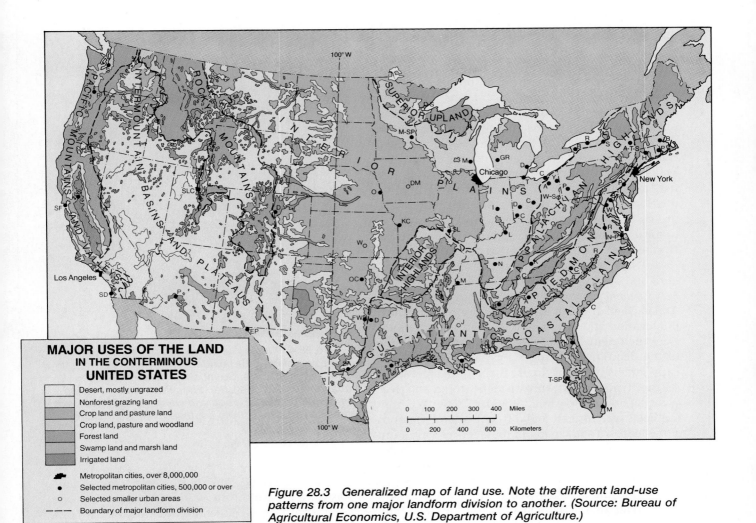

Figure 28.3 *Generalized map of land use. Note the different land-use patterns from one major landform division to another. (Source: Bureau of Agricultural Economics, U.S. Department of Agriculture.)*

the Amish religious sect (a branch of the Mennonites) (Figs. 3.8 and 28.20).

Physical Variety in the Appalachian Northeast

Aside from a narrow strip of the Interior Plains along the Great Lakes and St. Lawrence River (Figs. 28.1 and 28.2), the rest of the Northeast lies in the *Appalachian Highlands.* But in New York state a narrow lowland corridor, the *Hudson-Mohawk Trough,* breaks the Appalachians into two quite different subdivisions. This trough is composed of the Hudson Valley from New York City northward to the Albany area, and the Mohawk Valley westward from Albany to the Lake Ontario Plain. North and east of it the Appalachians are formed of old hard igneous and metamorphic rocks in all but a few places, but the section to the south and west has younger and generally softer sedimentary rocks except for the

ancient hard rocks of the narrow range called the Blue Ridge.

Northeast and north of the Hudson-Mohawk Trough, the mountains of New England and northern New York form several distinct ranges, characterized by rough terrain, cool summers, snowy winters, and poor soils. The *Adirondack Mountains* of northern New York rise as a roughly circular mass, surrounded by lowlands on all sides. Heavily forested and pocked by innumerable lakes, these glaciated mountains reach over 5000 feet (1524 m) in elevation. For a time in the mid-19th century the Adirondacks were the nation's leading area of logging, but today they are chiefly notable economically for their large resort industry.

The Adirondacks are bounded on the east by the *Champlain Lowland,* which contains Lake Champlain on the border between New York and Vermont. This Lowland reaches northward into Canada as a continuation of the Hudson Valley. East of it the *Green Mountains* occupy most of Vermont and reach over

4000 feet (1219 m). From here they extend southward, with declining elevations, and become the
Berkshire Hills (Berkshires) of western Massachusetts
and northwestern Connecticut. A southwesterly offshoot, the narrow *Taconic Range,* forms the eastern
edge of the Hudson Valley in New York state. East
of the Green Mountains across the narrow valley of
the upper Connecticut River, the *White Mountains*
occupy northern New Hampshire and extend into
Maine. Here Mount Washington in New Hampshire
rises to 6288 feet (1917 m), the highest elevation in
the Northeast. Like New York's Adirondacks, the entire assemblage of New England mountains is extremely important as a recreational outlet for the
Northeast's crowded urban areas.

In New England the areas between the mountains and the sea, sometimes termed collectively the
New England Upland, are considered to be part of the
Appalachian Highlands, but they are relatively low
in elevation and are more hilly than mountainous.
Here the original soils were notoriously stony (due
to the deposition of stones by glaciation), but the
early settlers were able to use them for a largely subsistence type of agriculture once the stones and trees
were laboriously cleared. Boulders removed from
the fields were used to build New England's famous
stone fences. Today many of these fences have been
torn down to facilitate modern farming or urban development. Some derelict fences may still be seen in
the many woodlands where farming has been abandoned and the land allowed to revert to trees and
brush (Fig. 28.4). The generally thin, leached, and
acidic soils of the New England Upland, lacking in
humus and often poorly drained, proved increasingly unable to support a stable long-term commercial agriculture once the pioneer subsistence days
were past. In the late 18th and 19th centuries there
was a large movement of New Englanders from unproductive farms to jobs in Northeastern cities, or to
more fertile agricultural areas in western New York
and the Midwest. Their migration westward eventually imparted a strong New England cultural flavor
to localities scattered all the way to the Pacific Coast.

Embedded in the old hard-rock New England
Upland are scattered and relatively small lowlands
with younger and softer sedimentary rocks and generally better soils. One of these is the *Aroostook Valley*
of extreme northeastern Maine, long notable as the
Northeast's most important area of potato growing.
In colonial times small sedimentary lowlands near
Boston and Providence were important agriculturally, although today this land has largely been
usurped by urban sprawl. In central Connecticut and
adjacent Massachusetts a colonial garden spot developed along the Connecticut River in the *Connect-*

Figure 28.4 *Originally, the farm land in New England
had to be cleared of boulders left behind by
continental glaciers. So much farm land has been
abandoned that one often finds stone walls winding
through patches of woodland where crops once were
grown. The view is from southern New Hampshire.*

icut Valley Lowland. This north-south strip of land
was once an arm of the sea and then the bed of a
late-glacial lake. Sedimentary deposits in the Lowland have given rise to some of New England's best
soils, and a productive agriculture has existed here
since the first settlements were planted in the 17th
century. Today small areas of surviving farm land
are interspersed with built-up land of metropolitan
Hartford and other industrial cities.

Elevations in the New England Upland decline
somewhat toward the sea, and a *Seaboard Lowland*
of varying width is sometimes recognized. But almost all of it is over 500 feet (152 m) in elevation,
and it is basically a continuation of the Upland rather
than a sharply different area like the Atlantic Coastal
Plain. The *New England Coast* is extremely irregular,
with many peninsulas, bays, and offshore islands.
Here lie a multitude of harbors that were important
to New England's early seafaring economy and now
give shelter to fishing boats and pleasure craft, although few harbors continue to handle seaborne
trade.

South of the Hudson-Mohawk Trough lies the
other major segment of the Northeastern Appalachians. It is comprised of three distinct sections:
the Blue Ridge in the east, the Ridge and Valley Sec-

tion in the center, and the Appalachian Plateau in the north and west. They lie roughly parallel to each other, with each trending northeast-southwest. The entire series extends far beyond the Northeast into the South.

The *Blue Ridge,* long and narrow, can be traced across New Jersey, Pennsylvania, Maryland, and on into the South. It has been given various local names, such as South Mountain in a Pennsylvania section near Reading. Often it forms the single ridge its name implies, but it broadens into many ridges in the South. The Blue Ridge tends to rise abruptly above the Piedmont to its east, although in the Northeast it is not very high. In fact, it often fails to reach 1000 feet (305 m) in elevation. Gaps through the Blue Ridge facilitated the building of railroads and highways from Northeastern port cities toward the interior.

West of the Blue Ridge, the *Ridge and Valley Section* of the Appalachians, characterized by folded sedimentary rocks, extends from southeastern New York into northern New Jersey, and thence follows a curving arc across eastern Pennsylvania, the Maryland panhandle, and on into the South. Here long, narrow, and roughly parallel ridges, often capped by erosion-resistant sandstone, trend generally north and south and are surrounded and separated by narrow valleys (Fig. 2.12). The ridges are commonly 3000 to 4000 feet (*c.* 900–1200 m) high, the valley floors 1000 to 2000 feet lower. On the east, immediately adjacent to the Blue Ridge, valley floors are wider and ridges are more scattered or absent. This valley-dominated eastern strip, extending almost continuously throughout the length of the Ridge and

Valley Section in the Northeast and the South, is essentially a single large valley that is known as the Great Appalachian Valley. It varies in width but is commonly about 20 miles (*c.* 30 km) wide. In various places it goes by local names such as Lehigh Valley in northeastern Pennsylvania or Shenandoah Valley in Virginia. The Great Valley has been a historic north-south passageway (followed by pioneers migrating south and west before the Civil War and by marching armies during that war), and its limestone floor has decomposed into some of the better soils of the Appalachians.

The *Appalachian Plateau* lies west and north of the Ridge and Valley Section. In the Northeast it occupies most of southern New York state, plus northern and western Pennsylvania and a small part of western Maryland. Beyond the Northeast it extends westward into Ohio and southward as a narrowing belt all the way to central Alabama. Although geologically a plateau (more precisely a series of plateaus) comprised of generally horizontal sedimentary strata, it is so deeply and thoroughly dissected in most places that it is actually a tangle of hills and low mountains separated by narrow, steep-sided, twisting stream valleys. Most summit elevations are between 2000 and 3500 feet (*c.* 600–1100 m). The northern part, including the section in the Northeast, is often called the *Allegheny Plateau,* whereas in parts of eastern Kentucky and farther south the plateau becomes known as the *Cumberland Plateau.* Notable east-facing escarpments, the *Allegheny Front* in the north (Fig. 28.5) and the *Cumberland Front* in the south, mark the eastern edge, from which elevations gradually decline toward the west.

Figure 28.5 The Allegheny Front, shown here a short distance west of Pennsylvania State University. The wooded escarpment marks the boundary between the Ridge and Valley Section and the Appalachian Plateau.

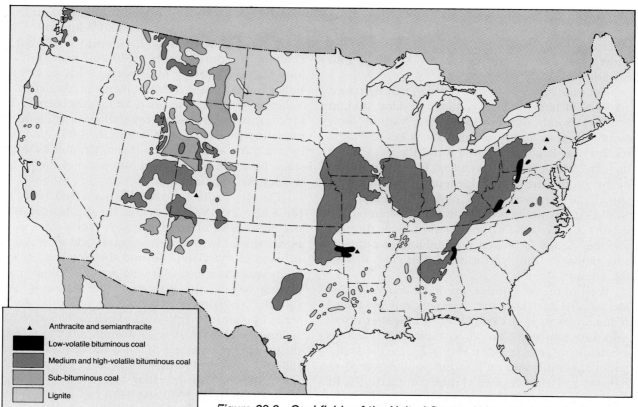

Figure 28.6   Coal fields of the United States. (After a map by W. S. and E. S. Woytinsky in World Population and Production: Trends and Outlook, *New York: Twentieth Century Fund, 1953)*

Legend:
▲ Anthracite and semianthracite
Low-volatile bituminous coal
Medium and high-volatile bituminous coal
Sub-bituminous coal
Lignite

Except in New York state the Appalachian Plateau is underlain by enormous and easily worked deposits of high-quality bituminous coal. Streams have sliced their valleys through the coal-bearing strata, making the coal seams accessible to mining along valley sides. From the mine mouths, often located far above the valley floors, horizontal mine tunnels burrow underground, often following a coal seam for miles. Where it is feasible to remove the overburden of earth and rock (often at or near the tops of hills and mountains) strip mining is carried on, scarring the landscape with huge cuts and producing unsightly accumulations of overburden called spoil banks. Often huge augers are used to extract coal by boring into seams laid bare by the power shovels employed in strip mining. The coal resources of the Plateau have been of fundamental importance to the development of the United States. Several Northeastern and Southern states share the deposits (Fig. 28.6), which fuel electric-generating plants, metallurgy, and other industries over a wide area in the Northeast, Midwest, and South and also provide large exports (Fig. 3.6).

The section of the Appalachian Plateau in New York state differs from the sections farther south in that it lacks coal deposits and in addition has been glaciated (together with small Plateau areas nearby in northern Pennsylvania and Ohio), whereas the rest of the Plateau is unglaciated. Hence the topography of the New York part is somewhat more subdued except along the eastern edge where the scenic low mountains called the *Catskills* rise boldly from the Hudson Valley and are the locale of a highly developed resort industry.

ORIGINS OF NORTHEASTERN URBANISM

The roots of the Northeast's extraordinary urban development go far back in time, at least by American standards. Boston was founded by English settlers and New York by the Dutch in the early 1600s, although Dutch New York was subsequently annexed by England in 1664. Philadelphia was founded by the English in the later 1600s, and Baltimore in the early 18th century. All four cities were outstanding urban centers by the American Revolution (1775–1783), although by today's standards they were quite small. Thus, their predominance in their region was already established before the existence of the United States or the settlement of its interior.

Figure 28.7 Office development shown here at Tysons Corner, Virginia, symbolizes the kind of sprawl that has revolutionized the peripheries of American cities in the age of expressways. Tysons Corner, which is not even an incorporated place, grew explosively in an area where several expressways intersect. By a recent estimate, this former crossroads in rural Virginia ranks 16th among the country's office-space complexes. A huge new shopping complex known as Tysons II is developing.

Washington, however, was not founded until after 1800 and only grew to be a large city as a consequence of the post-World War II expansion of the federal government. Today its metropolitan area encompasses both the District of Columbia and large suburban areas in Maryland and Virginia (Fig. 28.7). The city is not only a huge governmental center but is also a world financial center where the World Bank and the International Monetary Fund have their headquarters. When Washington's location on the Potomac was chosen as the site for a new and permanent national capital, the United States consisted of a narrow belt of states along the Atlantic margin. A location was picked on the border between the heavily agrarian Southern states with their large slave populations and the more diversified Northern states with fewer slaves and widespread abolitionist sentiments. The site also placed the capital on the Fall Line between seaboard and upcountry sectional interests. Many individual American states have their capitals similarly located between sections, for example, Albany in the border zone be-

tween "Upstate" New York and the southeastern section focused on New York City.

Early Development in the Seaboard Cities: The Crucial Role of Commerce

Except for Washington, the largest metropolises of the Northeastern Seaboard gained their initial impetus for growth as seaports and commercial centers. *New York* was founded at the southern tip of Manhattan Island, on the great natural harbor of the Upper Bay (Fig. 28.8). It shipped farm produce from lands now occupied by built-up areas of New York City and it suburbs and from estates up the Hudson River to the north. Using the Hudson waterway, its traders also made contact, at Albany, with Iroquois Indians who sold furs from the interior for export through New York.

Philadelpia (Fig. 28.9) was sited up the Delaware River where the latter stream is joined by a western tributary, the Schuylkill. Philadelphia's relatively large local hinterland—the Pennsylvania Piedmont

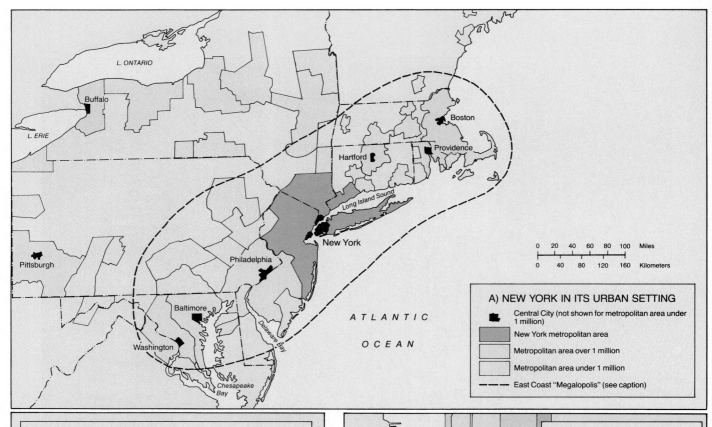

A) NEW YORK IN ITS URBAN SETTING

- **Central City** (not shown for metropolitan area under 1 million)
- New York metropolitan area
- Metropolitan area over 1 million
- Metropolitan area under 1 million
- – – – East Coast "Megalopolis" (see caption)

L. ONTARIO

Buffalo

L. ERIE

Boston

Providence

Hartford

Pittsburgh

Long Island Sound

New York

Philadelphia

Baltimore

Washington

ATLANTIC OCEAN

Delaware Bay

Chesapeake Bay

| 0 | 20 | 40 | 60 | 80 | 100 | Miles |
| 0 | 40 | 80 | 120 | 160 | | Kilometers |

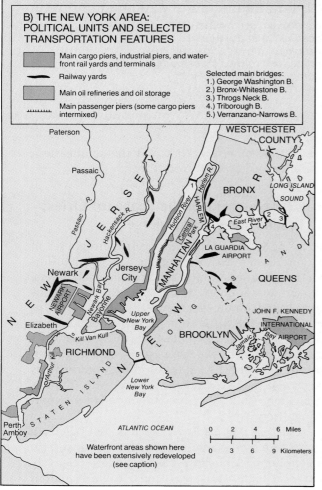

B) THE NEW YORK AREA: POLITICAL UNITS AND SELECTED TRANSPORTATION FEATURES

- Main cargo piers, industrial piers, and waterfront rail yards and terminals
- ◆ Railway yards
- Main oil refineries and oil storage
- Main passenger piers (some cargo piers intermixed)

Selected main bridges:
1.) George Washington B.
2.) Bronx-Whitestone B.
3.) Throgs Neck B.
4.) Triborough B.
5.) Verranzano-Narrows B.

Paterson

Passaic

WESTCHESTER COUNTY

NEW JERSEY

BRONX

LONG ISLAND SOUND

Passaic R.

Hackensack R.

Hudson River

Harlem R.

Central Park

MANHATTAN

LA GUARDIA AIRPORT

Newark

Jersey City

QUEENS

NEWARK AIRPORT

Newark Bay

Bayonne

Upper New York Bay

BROOKLYN

LONG ISLAND

Elizabeth

Kill Van Kull

JOHN F. KENNEDY INTERNATIONAL AIRPORT

Jamaica Bay

RICHMOND

Arthur Kill

STATEN ISLAND

Lower New York Bay

Perth Amboy

ATLANTIC OCEAN

| 0 | 2 | 4 | 6 | Miles |
| 0 | 3 | 6 | 9 | Kilometers |

Waterfront areas shown here have been extensively redeveloped (see caption)

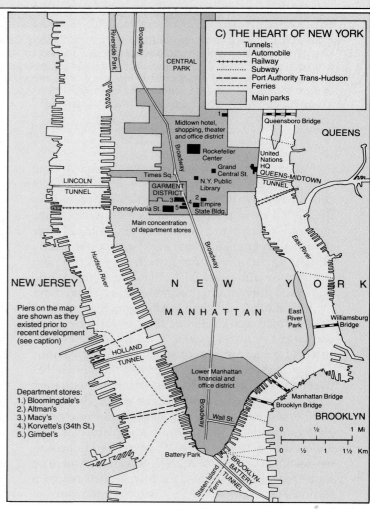

C) THE HEART OF NEW YORK

Tunnels:
- Automobile
- Railway
- Subway
- Port Authority Trans-Hudson
- – – Ferries
- Main parks

Riverside Park

CENTRAL PARK

Broadway

Midtown hotel, shopping, theater and office district

Rockefeller Center

Times Sq.

GARMENT DISTRICT

Grand Central St.

N.Y. Public Library

LINCOLN TUNNEL

Pennsylvania St.

Empire State Bldg.

Main concentration of department stores

Queensboro Bridge

QUEENS

United Nations HQ

QUEENS-MIDTOWN TUNNEL

NEW JERSEY

Piers on the map are shown as they existed prior to recent development (see caption)

Hudson River

N E W   Y O R K

M A N H A T T A N

Broadway

HOLLAND TUNNEL

East River

East River Park

Williamsburg Bridge

Lower Manhattan financial and office district

Department stores:
1.) Bloomingdale's
2.) Altman's
3.) Macy's
4.) Korvette's (34th St.)
5.) Gimbel's

Wall St.

Manhattan Bridge

Brooklyn Bridge

BROOKLYN

Battery Park

BROOKLYN-BATTERY TUNNEL

Staten Island Ferry

| 0 | ½ | 1 | Mi |
| 0 | ½ | 1 | 1½ | Km |

682

Figure 28.8 (left) *New York stands at the center of a strip of highly urbanized land that was christened "Megalopolis" by a French geographer, Jean Gottmann, in his book* Megalopolis: The Urbanized Northeastern Seaboard of the United States *(New York, Twentieth Century Fund, 1961). The dashed line is not intended to indicate a precise boundary, but shows broadly the area included by Gottmann in the Megalopolis concept. Many waterfronts and other areas shown on the two maps of the New York area are in a state of rapid change due to redevelopment for new commercial, residential, and recreational uses. This is especially true of the piers along both sides of the Hudson River. In the map, "The Heart of New York," the shading for the "Midtown Hotel, Shopping, Theatre, and Office District" covers the areas in which most of Manhattan's well-known hotels, restaurants, department stores, specialty shops, theatres, concert halls, and museums are found. The Midtown area also includes numerous office buildings, most notably the cluster of skyscrapers in Rockefeller Center. At the southern end of the area near the main department stores is found New York's famous Garment District with its huge work force and numerous workrooms and showrooms crowded into a remarkably small segment of Manhattan. The Lower Manhattan Financial and Office District includes the imposing cluster of office skyscrapers in the Wall Street area and, farther north, City Hall and a large assemblage of other government buildings occupied by city, county, state, and federal offices. Note the numerous bridges, tunnels, and ferries that tie Manhattan Island to the rest of metropolitan New York.*

and Great Valley, plus parts of southern New Jersey—was productive enough to make the city a major exporter of food, especially wheat, and helped it become ultimately the largest city in the Thirteen Colonies (population about 40,000 in 1776). Today it is the principal city of the nation's fourth most populous metropolitan area (after New York, Los Angeles, and Chicago), with a metropolitan territory encompassing much of southeastern Pennsylvania

and adjacent New Jersey, and reaching down the Delaware River to Wilmington, Delaware.

Baltimore was founded on a Chesapeake Bay harbor at the mouth of the Patapsco River and shipped tobacco and other farm products from its local hinterland in the Piedmont and the Coastal Plain.

Boston (chapter-opening photo) had a somewhat different early development in that most of its New England hinterland produced little agricultural sur-

Figure 28.9 *Philadelphia, seen from the New Jersey side of the Delaware River. In the center is the Walt Whitman Bridge. The port area in the foreground is devoted to container traffic. Note the pair of large cranes at the extreme left, used for transferring containers between ship and shore. In the distance is the skyline of downtown Philadelphia.*

plus for export. Instead, colonial New England placed its main commercial dependence on activities connected with the sea. Its forests yielded superb oak timbers with which to build sailing vessels and tall white pine trunks for masts. These forest products were exported in large quantities to Great Britain's shipbuilding industry, and the colonials themselves built ships for sale. Fishing vessels were a major item, and many were used by New Englanders to catch cod and other fish for export. Dried or salted fish from New England and Canada's Atlantic region became a staple food for slaves on West Indian plantations, and great quantities also were sold to Roman Catholic populations in Mediterranean Europe. Whaling, centered in New Bedford and nearby Nantucket Island in southeastern Massachusetts, was important for many decades in the 18th and 19th centuries. Meanwhile New Englanders were developing a wide-ranging merchant fleet, which, like the fishing and whaling fleets, continued to expand after independence from Britain. Their ships carried tobacco, rice, and later cotton to market from the plantation South, and even hauled ice, cut from New England ponds in winter and insulated in sawdust, for sale in hot climates as far away as India. Such activities came to characterize many New England ports, of which Boston was the largest.

The Race for Midwestern Trade

As the present Midwest of the United States was settled, primarily between 1800 and 1860, its agricultural surpluses and need for domestic and imported manufactures greatly expanded the trade of New York, Philadelphia, and Baltimore. So important was this commerce that the three ports engaged in a race to establish transport connections into the interior. In the competition New York surpassed its rivals, partly through superior enterprise but largely because of access to the Hudson-Mohawk Trough, the only continuous lowland passageway in the United States through the Appalachians. From New York City this route follows the Hudson Valley north to Albany and thence the Mohawk Valley westward between the northern edge of the Appalachian Plateau and the southern edge of the Adirondacks (Figs. 28.1 and 28.2). At its western end the route emerges onto the *Lake Ontario Plain,* a northeasterly projection of the Interior Plains. Westward the narrow *Lake Erie Plain* provided a corridor for land transport to the main body of the Interior Plains in the opening Midwest; and the port of Buffalo on Lake Erie gave access to transportation by lake boats.

New York won the race to the interior when the 364-mile (586 km) *Erie Canal* was completed in 1825. This new waterway from Lake Erie at Buffalo to the Hudson River near Albany provided the first all-water route from the Great Lakes to New York City. The canal required 87 locks to bridge the elevation difference of 571 feet (174 m) between the Hudson and Lake Erie, and the canal boats pulled by horses or mules on the banks were very slow, but the new means of transport was so superior to wagon haulage that transport costs between New York and the Midwest were reduced to a small fraction of their previous level. Hence rapid settlement of the Midwest was stimulated, and a flood of trade was directed from the new lands through the port of New York. In 1853, using the same route, New York interests became the first to connect a growing Chicago with an east-coast port by rail. By mid-century New York had become the leading United States port and city by a wide margin. Its chief rival for Midwestern trade at that time was not one of the other east-coast cities but New Orleans, located near the seaward end of the Mississippi River highway from the interior to the Gulf of Mexico.

Although greatly surpassed by New York, the competing ports of Baltimore and Philadelphia shared in Midwestern trade. Major railways were eventually pushed into the Midwest from both ports. But in both cases these lines to the interior had to wind through the Ridge and Valley Section, climb the steep Allegheny Front, and thread the twisting river valleys of the Appalachian Plateau westward to the Plains. The costs of development and operation per mile of track (and hence the costs to shippers) inevitably were higher than those borne by New York with its superior "water level" railway route via the Hudson-Mohawk Trough. Boston was never a real competitor in the commercial and transport race, although it has remained the largest seaport within New England. New York's connections to the interior were too good for Boston to overcome. Most shippers to and from New England have long found it convenient and economical to use the more extensive services and frequent sailings offered by New York.

Industrial Beginnings in the Northeast

Industrialization gained its first large impetus in the Northeast during the same period that the region established its connections to the interior. Favorable factors for industrial development included accumulated investment capital derived from commercial profits; access to industrial techniques already pioneered in Great Britain; skills derived from previous commercial operations and handicraft manufacturing; cheap labor supplied by immigrants; and the presence of certain energy and raw-material resources that were more important in earlier stages of industrialization than they are today. Also favorable was a United States high-tariff policy which restricted imports of foreign goods and secured much of the nation's internal market for its own emerging manufactures. The first American factory, a water-powered textile plant using pirated British technology, was established by Samuel Slater on the Blackstone River at Pawtucket, Rhode Island (near Providence) in 1790, and the subsequent growth of factory industry in the Northeast was rapid. Some of the necessary technology was imported, but much of it resulted from inventions by the region's own technically minded people. By 1815 America's industrial revolution was well under way in New England, and it spread from there to the rest of the country.

EARLY INDUSTRIAL EMPHASES ON THE NORTHEASTERN SEABOARD

The circumstances and results of industrialization were somewhat different in different parts of the Northeast. New England had superior waterpower, with many small and swift streams to turn early waterwheels, but the region lacked coal for use in metalworking. Its factories specialized particularly in woolen and cotton textiles, drawing early wool supplies from New England farms and receiving cotton by ship from the South. A New England inventor, Eli Whitney, originated the cotton gin (1793) to economically separate the fibers in the cotton boll from the cottonseeds, thus making cotton a major raw material. However, one New England state, Connecticut, took a different industrial path. It continued and expanded a colonial specialty in metal goods by becoming a major manufacturer of machinery, guns, and hardware, which it still is. The interior highlands of New England shared little in these developments, but mill towns sprang up across the Southern New England states and up the coast into southern Maine. In the Middle Atlantic states early industrialization included textile milling, but there was greater emphasis on metal smelting, metal fabrication, and the manufacture of clothing and chemicals. The emphasis on metals reflected local resources. Small deposits of iron ore were available, especially in southeastern Pennsylvania and in the Adirondacks of New York, and in the second quarter of the 19th century the largest deposits of anthracite coal in the United States began to be the locus of a booming mining industry and a major source of power for the Northeast. This coal is deeply buried under rough country in the Pennsylvania Ridge and Valley Section between Harrisburg (575,000) and Scranton–Wilkes-

Barre (725,000) but was made accessible despite the terrain by determined canal and railroad building as well as by large investment in the mines themselves. Mining here was expensive, not only because most of the coal lay far underground but also because the seams were contorted as a result of folding. In the 20th century the market for anthracite has plummeted and the mining area has suffered great economic distress. But for many decades the anthracite industry flourished. The coal was used extensively for household heating, and it was also very significant industrially. Usable in blast furnaces without the necessity for coking, Pennsylvania anthracite fueled large-scale iron- and steelmaking in such eastern Pennsylvania cities as Harrisburg and Bethlehem (660,000 with nearby Allentown). These early mills, which utilized small deposits of iron ore nearby, far antedate the construction of the large Sparrows Point iron and steel plant of the Bethlehem Steel Corporation at Baltimore and the post-World War II construction of the large Fairless Works of the United States Steel (now the USX) Corporation in eastern Pennsylvania on the Delaware River above Philadelphia. Iron and steel produced by the early plants in this eastern part of the Middle Atlantic region contributed greatly to the first industrial surge in the Northeast by furnishing raw material for the manufacture of machinery, steam engines, railway equipment, and other metal goods.

THE RISE OF INDUSTRY IN THE INTERIOR NORTHEAST

During the early and middle 19th century, two interior areas of the Northeast—the Hudson-Mohawk canal and railroad corridor, and western Pennsylvania—also began to develop industrially. Along the *Hudson-Mohawk corridor* across New York state a string of cities emerged: initially Albany, Troy, and Schenectady (tri-city metropolitan population, 845,000) on or near the Hudson; and Buffalo (1.2 million with adjacent Niagara Falls) at the Lake Erie end of the route. Subsequently, Rochester (980,000) and Syracuse (650,000) became important cities along the route. Buffalo, which became the largest city of the group, profited industrially from its role as the point of interchange between Great Lakes traffic and traffic along the Hudson-Mohawk lowland through the Appalachians. The other cities along this route tended to develop where the Erie Canal crossed or met some other significant feature. For example, major salt deposits provided raw material for a large chemical industry at Syracuse, and a waterfall on the Genesee River powered early industries at Rochester. As time went on, quite varied industries developed: flour milling at Buffalo and

Rochester, chemicals at Syracuse and Buffalo, the Eastman Kodak photographic company at Rochester, the General Electric Company at Schenectady, clothing manufacture at Troy, major iron and steel manufacturing at Buffalo, and the manufacture of miscellaneous metal goods and precision instruments at various locations.

A second area of early industrial development in the interior Northeast lay in the Appalachian Plateau of *western Pennsylvania*, especially at Pittsburgh (2.3 million). In the mid-18th century laborious land routes from Baltimore and Philadelphia were pioneered to the "Forks of the Ohio" where Pittsburgh developed. Here the Allegheny River from the north and the Monongahela from the south join to form the Ohio River. The Ohio was a broad highway on which pioneers could float into the developing Midwest. As traffic toward the interior increased, Pittsburgh became the main center for outfitting settlers and building river boats. Mercantile and banking firms of Atlantic Coast cities established branches there, and the town developed an ironworking industry based on small deposits of iron ore scattered over western Pennsylvania. The ore was smelted in small furnaces near the deposits by using charcoal made from the surrounding forests, and Pittsburgh blacksmiths then used the iron to make hardware (such as horseshoes, plows, chains, axeheads, wagon tires, hoes, pots, and nails) needed by settlers. Once the westward-moving settlers were established in new homes, they continued to draw on Pittsburgh for iron goods. As ironworking expanded to meet this demand, rolling mills to shape iron in larger quantities were built at Pittsburgh, and the town began to reach beyond western Pennsylvania for iron supplies. The metal came by steamboat from furnaces located in Ohio and Kentucky, or even as far away as Missouri.

Meanwhile the superb bituminous coal seams of western Pennsylvania began to be mined and were found to contain high-quality coking coal in large amounts. Shortly before the Civil War, a solution was found to the problem of bringing adequate supplies of iron ore to Pittsburgh blast furnaces fueled with local coal and coke. The ore came from huge high-grade deposits near Lake Superior in Wisconsin and Minnesota. Eventually the long and narrow ore-bearing formation in Minnesota called the Mesabi Range became the main source. Transportation of the ore to Pittsburgh involved the building of railways from the mines to loading points on Lake Superior, very cheap haulage by lake boat to the south shore of Lake Erie, and rail transportation to Pittsburgh. The main obstacle that had to be overcome was a stretch of rapids in the St. Marys River

connecting Lake Superior and Lake Huron. This bottleneck was bypassed by the Soo Canals at Sault Ste. Marie (Michigan and Ontario) in 1855, and the Pittsburgh area underwent a meteoric rise in iron smelting, including the manufacture of steel from the pig iron produced by the blast furnaces. For a considerable time the area was the world's leading producer of iron and steel, although it eventually lost its leadership to the Chicago area and more recently has been surpassed by various districts overseas. Meanwhile Appalachian coal mining spread into West Virginia, Kentucky, and Ohio, and the coal was shipped out in huge quantities to power industrial development in both the Northeast and Midwest. Much of the metal required to develop these regions, especially the Midwest, was supplied by the steel mills of Pittsburgh and smaller centers of steel production that developed in the Pittsburgh region.

During the latter 19th century and into the 20th century, the urban-industrial trends established in the Northeast during earlier times continued. The region experienced massive population growth and economic development. Meanwhile development in other parts of the United States gained momentum, so that the relative predominance of the Northeast gradually diminished. Quite recently the region has undergone economic stress, with a slowing or even reversal of growth in many industries and localities. But despite its altered economic complexion and the relative stagnation of its overall population growth, the Northeast continues to bear the stamp of its earlier urban-industrial surge in many ways, and not least in its social and cultural geography. Over a long period many millions of new workers entered its growing industrial cities from Europe, the American South, the Northeast's own countryside, and other parts of the world, swelling the populations of the major cities to truly large size and imparting a strongly "ethnic" character to the region as a whole.

THE PRIMACY OF NEW YORK CITY

America's greatest city is centrally located within the Northeastern Seaboard urban strip, being almost precisely midway between Boston and Washington. The enormous and protected harbor of New York, plus active business enterprise and a central location within the population and economy of the Thirteen Colonies and the young United States, had already allowed the city to become the country's leading seaport before the Hudson-Mohawk route through the Appalachians became of major importance. Superior access to the developing Midwest then proved a powerful asset as New York rose rapidly to unquestioned leadership among American cities in size, commerce, and economic impact. It reached this position of primacy well before the Civil War and has never relinquished it. With some 7.3 million people in 1986, New York City proper has well over twice the population of the next largest city, Los Angeles (city proper, 3.3 million); and metropolitan New York (18 million) is nearly one third larger than the second largest metropolis, which is again Los Angeles (13.1 million). The port of New York is still the nation's leading seaport in some respects such as volume of container traffic and total value of waterborne commerce handled. The three international airports of the metropolitan area (Kennedy, La Guardia, and Newark) make New York a major world center of air transportation. Although it lacks the towering importance industrially that it has in finance, management, and communications, the metropolitan area vies closely with the Los Angeles and Chicago metropolises as one of the country's three largest metropolitan concentrations of manufacturing. It is by far the largest center of wholesale trade, publishing, advertising, broadcasting, and performing arts, and is a major center of retail trade and services. The presence of the United Nations headquarters confers on the area great international prestige. But perhaps the most impressive aspect of New York's stature relative to other American metropolises lies in the city's leadership in finance and business management. The metropolitan area overshadows any other American center in funds controlled by banks and insurance companies, as well as in assets of industrial corporations headquartered there.

Spatial Structure of the New York Metropolis

New York City proper centers on Manhattan Island (Fig. 28.8). The island is narrowly separated by the Harlem River from the southern tip of the mainland in New York state. On the east Manhattan is separated from Long Island by the East River; on the west from New Jersey by the lower Hudson River. Manhattan is one of five boroughs comprising the city proper. Of the remaining boroughs, the Bronx lies on the mainland to the north, Brooklyn and Queens are on the western end of Long Island, and the borough of Staten Island is separated from Manhattan by Upper New York Bay. In addition to the boroughs of the city proper, the New York metropolis includes a large number of suburban and satellite cities and towns that are interlinked with the central city in various ways. The whole assemblage includes (1) much of the population and industry of New

Jersey (including Newark: city proper, 315,000); (2) communities on Long Island outside Brooklyn and Queens; (3) the southernmost mainland of New York state outside the Bronx; and (4) the southwestern corner of Connecticut, where Stamford has become a major national center of corporate headquarters. All parts of the metropolis are bound together by an elaborate net of expressways and mass-transit lines. These routes converge on a series of great bridges, tunnels, and ferries that link the five boroughs to each other and to New Jersey (Fig. 28.8). Port activity is heavily concentrated in New Jersey along the Upper Bay and Newark Bay, although some traffic is still handled in various New York sections of the waterfront.

INDUSTRIAL EMPHASES OF THE NORTHEAST TODAY

The huge increase of population in the Northeast that commenced in the mid-19th century went hand in hand with expansion of the region's manufacturing industries, for which the population provided all grades of labor, entrepreneurial experience, vast amounts of money to invest, and a very large market close at hand. The industrial surge expanded some types of industry founded earlier but included so many new lines that the present Northeast is especially marked by the astonishing diversity of its industrial structure. This variety is apparent in even the following sketchy summation of a few major specialties.

1. **Clothing manufacture** is carried on in many Northeastern locations but is especially concentrated in metropolitan New York. The nation's greatest organizing center in this industry is Manhattan's Garment District in the heart of the island (Fig. 28.8). Aside from some exclusive high-fashion firms that still do their manufacturing on the premises, the District has become primarily a center for management, design, and merchandising, with a major share of the actual manufacturing decentralized to other locations within the New York metropolis, the rest of the nation, or overseas locales such as Hong Kong. The clothing industry in Manhattan was originally an outgrowth of New York's role as an importer of European-made cloth and clothing. It was favored by a large immigration of skilled workers in the later 19th century, among them Jewish tailors fleeing from persecution in Tsarist Russia.

2. **Iron and steel manufacture** is still found at scattered places in the Northeast, although employment and output have been drastically reduced in

recent years because of competition from cheaper imported steel, large imports of products such as automobiles made from foreign rather than American steel, and the substitution of aluminum or plastics for steel in industrial processes. Baltimore, Pittsburgh, Buffalo, and eastern Pennsylvania remain the principal Northeastern centers of this old industry, which is burdened by much outmoded equipment and higher wage scales than those of competing industries in other countries.

3. Chemical manufacture is very diversified and widespread, with large production of both basic industrial chemicals in bulk and a multiplicity of consumer products. Metropolitan New York is the largest center, although the DuPont Company, which is the world's largest chemical-manufacturing concern, has both its headquarters and a major share of its manufacturing at Wilmington, Delaware.

4. Photographic equipment manufacture is dominated by the world's largest photographic company, Eastman Kodak, which has both its headquarters and its main plant at Rochester, New York. The Polaroid Corporation, based principally in the Boston area, is another prominent name in this industry.

5. Advanced electronics manufacture, centering on computers, is widespread in the Northeast, with International Business Machines (IBM) towering over a host of smaller companies. This industry is a major specialty in eastern Massachusetts, being especially concentrated in industrial parks along two beltways—Highway 128, the "Highway of Technology," and Interstate 495—that ring the Boston area. Graduates of the nearby Massachusetts Institute of Technology have been heavily involved in electronics development here. Some old textile towns such as Lowell and Lawrence on the Merrimack River have largely reconverted to electronics or other high-technology industries. They are examples of the widespread "recycling" of old mill towns in the Northeast. In recent years the New England electronics industry has benefited greatly from the award of federal defense contracts for the development of advanced communications equipment. Eastern Massachusetts was once the nation's largest center of textile milling, but most of this industry has changed its locale to newer plants in the South.

6. Electrical equipment manufacture, concentrated in Southern New England, New York state, and Pennsylvania, is dominated by the General Electric Company, headquartered at Stamford, Connecticut, in the New York metropolis. The company has plants at Schenectady and many other locations in the Northeast and elsewhere. Westinghouse Electric, headquartered in Pittsburgh, is another major company with plants in both the Northeast and other areas.

7. Manufacture of aircraft engines and helicopters is heavily concentrated at Hartford, Connecticut, where it is carried on by United Technologies Corporation, maker of the famous Pratt and Whitney engines and Sikorsky helicopters.

8. Manufacture of Trident nuclear submarines by the General Dynamics Corporation in the New London, Connecticut, area continues New England's long shipbuilding tradition, as well as Connecticut's long tradition of armaments manufacture. It was at a factory built near New Haven, Connecticut, in 1798 to supply muskets under a federal contract that Eli Whitney originated the use of interchangeable parts for the manufacture of firearms. This practice is central to modern mass-production of innumerable types of goods. Prior to Whitney, firearms had been laboriously made by individually shaping the parts for each gun.

9. Publishing and printing is a kind of manufacturing in which the Northeast decisively overshadows the rest of the country. Most of the industry is in the New York metropolis, although the Government Printing Office in Washington, D.C., is the world's largest publishing enterprise and printing plant. In New York many of the actual manufacturing phases of the publishing industry have been decentralized to other parts of the country, but editorial and management functions remain concentrated in or near Manhattan.

10. Food processing is widespread in the Northeast, which contains the headquarters of 3 of the nation's 14 largest food manufacturers (1988): Borden and CPC International in the New York metropolis, and the Campbell Soup Company at Camden, New Jersey, within the Philadelphia metropolis. They have worldwide operations, and only a minor share of their manufacturing is done in the Northeast. Many other companies have Northeastern plants. Specialized farms in the Northeast, including many that are owned by, or contracted to, food-processing companies, supply much of the needed produce. But a great deal of unprocessed food is brought in from the rest of the nation or overseas sources. We turn now to the spottily distributed but highly capitalized and productive agriculture of the Northeast, which is heavily dependent on markets within the region.

THE ROLE OF SPECIALIZED AGRICULTURE

The agriculture of colonial times provided a major basis for the Northeast's original urban growth. New England's agricultural environment was less promising than that of the Middle Atlantic colonies, but there were pockets of good land around Boston and Providence and in the lowland along the Connecticut River. Hard toil enabled farm families to subsist despite natural handicaps, and enough surplus was produced to feed considerable numbers of people engaged in fishing, trade, and handicraft manufacturing. Meanwhile, better agricultural hinterlands available to New York, Philadelphia, and Baltimore made it possible for those cities to flourish as exporters of surplus agricultural products. Such exports from New York and Philadelphia consisted largely of wheat and cured meat, while tobacco was the most valuable item shipped from Baltimore.

In the 19th century, competition from grain and meat producers in the agriculturally favored Midwest led Northeastern farmers to begin shifting toward specialties particularly adapted to the Northeast's natural conditions and markets. The main adaptation was a conversion from general grain and livestock farming to dairy farming. Distant agricultural areas with cheaper and/or better land could not ship fluid milk to Northeastern urban markets in competition with Northeastern farmers. Dairying was also logical for a region with so much mediocre or poor soil and sloping land more adaptable to pasture or hay than to tilled crops. In 1986 dairy products were the leading source of farm income in 5 of the region's 11 states, and were the second source in three others.

Other agricultural specialties developed in the Northeast as adaptations to the closeness of urban markets, the poor quality of much of the land, and the competition of products from other regions. The earliest was large-scale "truck farming," producing fresh summer vegetables and fruits to be hauled overnight to city markets. This type of agriculture developed mainly on the poor and sandy but easily tillable soils of the Coastal Plain, where high output value per acre justified high inputs of fertilizer. Orchards are still common in some Northeastern areas, and vegetable production is still widespread in the Coastal Plain, but newer specialties have recently been crowding truck farming out of its once-predominant place. One of these is chicken-broiler production, which has become the largest source of farm income in Maryland and Delaware. In three extraordinarily urbanized and densely populated states—Massachusetts, Rhode Island, and New Jersey—an extreme adaptation has taken place in that greenhouse products, largely from areas in or close beside the great metropolitan areas, have become the leading source of agricultural income.

Two general characteristics of agriculture in the present-day Northeast stand out.

1. *Agriculture is far overshadowed by other ways of life, occupations, and land uses.* In the Northeast as a whole, only a quarter of the land is in farms, and forest land far exceeds cleared farm land in extent. Since 1960 the acreage of land in farms has been cut in half. Although it has nearly a quarter of the country's population, the Northeast produced a little over 7 percent of the total value of farm products marketed in the United States in 1986.

2. *Northeastern agriculture is highly productive.* In fact, Delaware was the country's most productive state agriculturally in 1986, yielding more value of farm products per unit of total area than any other state. Maryland ranked fifth among the states in this respect, and New Jersey and Pennsylvania were also above average. The other states were distinctly lower, with New Hampshire and Maine very low. These outputs are produced mostly on relatively small farms, intensively operated, with large inputs of capital and labor resulting in large outputs per acre used. Such intensity is required by the high cost of land and is permitted by favorable market locations for such specialties as fresh milk, poultry and eggs, potatoes and other vegetables, fruits such as New York grapes and cherries and Massachusetts cranberries, or horticultural specialties grown in greenhouses.

"OLD METROPOLITAN BELT" VERSUS "SUN BELT": THE OUTLOOK FOR THE NORTHEAST

Recent statistics show that for the most part the total array of cities and industries in the United States now divides fairly evenly between two very different regions that we may term the *Old Metropolitan Belt* and the *Sun Belt.* Some places that are outlying or in-between do not fall clearly in either Belt, but they comprise only a minor fraction of the country's urban and industrial development.

What we are calling here the *Old Metropolitan Belt* contains the great majority of the present metropolitan areas that already had become prominent cities by the early 20th century. The Belt can be considered to stretch from the urbanized Northeastern Seaboard across the Midwest as far as Kansas City, Omaha, and Minneapolis-St. Paul, all of which emerged as important urban nodes when transcontinental railroads were built to the Pacific from them in the latter 19th century. On the north the Belt extends to the industrial metropolises of Southern New England, the Hudson-Mohawk corridor, and the Great Lakes; on the south it includes St. Louis, the Ohio River cities, and Washington, D.C. Nowadays the whole block of territory is often called loosely the Frost Belt, Snow Belt, or, at least in part, the Rust Belt. Many decades ago geographers began to call most of it the American Manufacturing Belt, a term often used to reach beyond the United States and include the industrial districts of Canada's core region along the Great Lakes and St. Lawrence River. Today this name is increasingly less appropriate, as manufacturing in other parts of the United States has grown to the point that the old Manufacturing Belt no longer enjoys the overwhelming preponderance that it once had.

Recently the Old Metropolitan Belt has been growing quite slowly in population and industrial employment, with some areas actually declining in one or both of these. Most of the country's old "smokestack industries"—a term applied particularly to the steel industry and other heavy metallurgy—are in this Belt, especially in the Middle Atlantic states and the eastern Midwest. Today these industries are often characterized by outmoded buildings and equipment, depressed sales and profits, high unemployment, and unattractive surroundings. It is primarily the smokestack industries that have inspired the derogatory name "Rust Belt."

By contrast, most of the *Sun Belt* in the South and West has been enjoying faster growth in population and jobs. *Sun Belt* is a very elastic term, but common usage and geographic logic both suggest an area encompassing most of the South, plus the West at least as far north as Denver, Salt Lake City, and San Francisco. The contrast between the Old Metropolitan Belt and the Sun Belt in manufactural employment can be illustrated by job gains and losses in the 7 years between 1977 and 1984. During this period, the United States economy underwent a net loss of 2.3 percent of its jobs in manufacturing. But large losses were concentrated in older industrial states of the Midwest and the Middle Atlantic region, whereas the Sun Belt states showed a sizable overall gain. In the Northeast, the New England states and the Middle Atlantic states differed sharply, with New England achieving modest gains in manufacturing jobs and most of the Middle Atlantic region experiencing sharp losses. These losses reflected such factors as (1) the disappearance of obsolete plants which went out of business, (2) the shifting of plants to more profitable locations in other American regions or overseas, (3) drastically reduced employment in plants that survived by re-equipping with labor-saving new machinery, and (4) the failure of new plants to locate in the Middle Atlantic region in sufficient numbers to offset the job losses in existing industries.

However, employment in services has been expanding quite rapidly in the Northeast, with the re-

sult that most of the region's states have tended during the 1980s to be above the national average in total expansion of employment. The major exception has been Pennsylvania, which was especially hard hit by the decline of older industries. But despite the overall expansion of employment, the Northeast has been growing in population relatively slowly, with little immigration to New England and heavy outmigration from the Middle Atlantic states. The main exceptions to the pattern of slow growth are some peripheries of metropolitan areas into which suburban and exurban development are expanding. Notable examples include expansion into New Hampshire from Boston; into Connecticut and New Jersey from New York City; and into Maryland from Washington, D.C. The state of Vermont, whose rate of population growth has recently been higher than that of any Northeastern state except New Hampshire, is not surburban, but much of this quiet and scenic state is exurban, attracting second homes, vacation homes, and the homes of some commuters, including commuters by air.

Redevelopment in the Emerging Northeast

Despite the weight of long-existent economic and urban problems, the Northeast has recently been the scene of spectacular new development in all the great metropolises, both in old decaying cores and around the peripheries. Old waterfronts, long notorious as slum areas, have been receiving special attention. Upgrading of such districts has replaced slum housing and rotting piers with attractive new parks and marketplaces, clusters of new office towers, and varied residential properties. The revitalizing process has given a decided new look to some of the shabbiest urban landscapes in the United States, although huge slum areas often remain nearby. Redevelopment has been particularly extensive in Lower Manhattan, focusing in and near the waterfront area called the Battery; but Boston's Quincy Market, Baltimore's Harborplace, and Philadelphia's Society Hill residential section with its neocolonial architecture have become showpieces of redevelopment. Attractive new residential properties have been stimulating a movement of some affluent suburban dwellers to the inner cities.[2]

Meanwhile, what is the outlook for the Northeast generally? By 1986 the estimated population of the entire region was only about 2½ million greater

---

[2]Upgrading of inner-city residential properties in a manner attractive to more affluent new occupants is termed *gentrification*. The process has drawn criticism because it often dispossesses poorer previous residents who then have difficulty in finding new lodgings that they can afford.

than it had been in 1967, representing an increase of 5 percent in two decades compared to a 22 percent increase for the entire United States, or 40 percent for the states outside the Northeast and Midwest. Total manufactural employment was below the level of 1982, and many older industries continued to be in deep trouble. On the other hand the unemployment rate within the total population was the lowest for any of the four major regions; in fact, in 1986 New Hampshire and Massachusetts had the lowest unemployment rates of any states in the Union. Thus the picture in the Northeast was very mixed, but it certainly could not be said that this historic region, with its great reserves of intrenched wealth, its formidable educational establishment, and its long tradition of getting things done, was simply stagnating. According to recent predictions nearly 85 percent of the nation's total population increase between 1985 and 2000 will take place in the South and West. But it is also predicted that Boston, Washington, Philadelphia, and New York will rank among the nation's 15 leading metropolitan areas in number of new jobs created. Most of these will be jobs in services or newer types of industry rather than the industries that have sustained the Northeast in the past. This changeover could be viewed as simply the latest episode in a long history of constructive regional responses to outmoded economic bases. Obviously, it is a situation far from unique to the Northeast, since the world's history is replete with instances of regions and even entire nations that have had to make similar adjustments to changing circumstances.

THE SOUTH: OLD TRADITIONS AND NEW DIRECTIONS

The South has long been recognized as a major region of the United States. By the middle of the 19th century its economic and social systems were so different from those of the North that 11 Southern states tried to secede and become a separate country, the Confederate States of America. They were held in the Union only by defeat in the desperately fought Civil War of 1861–1865. The regional distinctiveness of the South has lessened since then, especially in the past few decades. But "Southernness" remains well marked in many ways, and regional identity is underscored by the continued existence of such organizations as the Southern Baptist Convention and the United Daughters of the Confederacy.

As in the case of the other major American regions discussed in this chapter, various definitions of the South are possible, and none is entirely satisfac-

tory. Somewhat arbitrarily the region is discussed here as a block of 14 states: 5 along the Atlantic from Virginia to Florida; 4 along the Gulf of Mexico from Alabama to Texas; and the 5 interior states of West Virginia, Kentucky, Tennessee, Arkansas, and Oklahoma. Arguments could be made on various grounds for excluding some of these states. For example, West Virginia did not secede from the Union, although some parts of it were strongly secessionist. Instead, the state was part of Virginia until the middle of the Civil War, when it broke away from its parent state and was accepted into the Union as a separate state (1863). Kentucky also remained in the Union, although with divided allegiance. Oklahoma and Texas are rather typically Southern in their eastern parts, but their western sections lie in semiarid or (in extreme western Texas) arid environments more typical of the West. However, even their western parts tend to have a strong Southern flavor culturally. Missouri was a slave state and hence a northward extension of the South before the Civil War, which violently divided its people. But present economic and social characteristics make the state seem more typically Midwestern than Southern.

PHYSICAL ENVIRONMENTS AND THEIR UTILIZATION

Most of the South lies within the part of the United States classed as *humid subtropical* in climate. Summers are long and hot, with frost-free seasons of over 7 months almost everywhere, and over 8 months in much of the region. January temperatures average in the 30s (F) along the northern fringe, the 40s over the heart of the region, the 50s near the Gulf Coast, and the 60s in central and southern Florida and the southern tip of Texas. More than 40 inches (*c.* 100 cm) average annual precipitation is characteristic, rising to over 50 inches in large sections of the Gulf Coast and Florida, and over 80 inches in a small section of the Great Smoky Mountains. High humidity in both summer and winter is characteristic of the humid subtropical climate, intensifying heat discomfort in summer and chilliness in winter.

There are both advantages and disadvantages to these climatic conditions. They foster rapid and abundant forest growth, and forest-based industries are of major importance in the South. Agriculturally, they provide long growing seasons and plenty of heat and moisture for a wide variety of crops. But insects and pests also flourish if not controlled, while the heavy rainfall leaches out soil nutrients and causes severe erosion and flooding. The latter problems are increased by a tendency for the rain to occur in violent downpours.

Some parts of the South are exceptions to the general predominance of humid subtropical conditions. Higher parts of the Appalachians average below freezing in their coldest month and are classed as *humid continental*. In extreme southern Florida, winter temperatures are high enough to classify the climate as marginally tropical. With wet summers and dry winters, the area is climatically a northward projection from Latin America's large areas of *tropical savanna* climate. The parts of Texas west of approximately 100° W have a *subtropical steppe* type of climate, grading into *desert* in the extreme west. The climate of Oklahoma's small western panhandle could also be classed as subtropical steppe.

The Predominance of Plains

Most of the South is classified topographically as plains, but the plains form sections of three major landform divisions—the Gulf-Atlantic Coastal Plain, the Piedmont, and the Interior Plains (Figs. 28.2, 28.3, 28.10, 28.11, and 28.17).

The *Coastal Plain* occupies the seaward margin from Virginia to the southern tip of Texas. It is narrow in the northeast but widens to the south and west. The Plain encloses Florida, Mississippi, Louisiana, and portions of all the other Southern states except West Virginia. It is generally low in elevation, with only a few places (generally near the inner edge) exceeding 500 feet (*c.* 150 m) above sea level. Near the sea and along the Mississippi River large areas are flat, but from Georgia westward most inland sections have an irregular surface. Belts of low hills are found in some parts of the Plain, but nearly all the Plain is level enough to grow crops if other conditions permit. Much marshy and swampy land occurs near the sea and in parts of the northward-reaching section called the Mississippi Alluvial Plain or Mississippi Lowland. The Coastal Plain was originally forested (and the greater part still is), aside from isolated belts or patches of grassland such as the marsh grasses of the Everglades in southern Florida; the "coastal prairies" along the Gulf of Mexico in Louisiana and Texas; and prairie grasslands in some inland areas (such as the Black Waxy Prairie of Texas) that are underlain by limestone. Forest growth is dominated by pines or a mixture of pines and oaks except for gums, cypresses, and other "swamp hardwoods" in poorly drained areas. Soils in the Coastal Plain are generally sandy and easy to work, but they are often seriously lacking in natural plant nutrients and are heavily leached by a combination of abundant rain and warm temperatures. The Plain has a long history of soil depletion and abandonment of worn-out cropland. Many former

agricultural areas have reverted to forest, and the proportion of forested land within the region has increased considerably during the 20th century. The Plain's most fertile soils are associated particularly with river alluvium, chalky limestone bedrock, or organic material in drained swamps and marshes.

The *Piedmont* is a more extensive landform division in the South than in the Northeast, extending across central Virginia, the western Carolinas, most of northern Georgia, and into east-central Alabama. In position, elevation (generally between 400 and 1500 feet/*c.* 120–460 m), and topography it is intermediate between the Coastal Plain and the Appalachians. Its surface is mainly that of a rolling plain, with some areas distinctly hilly. There is a natural cover of mixed forest in which broadleaved hardwoods, especially oak, tend to predominate over pines. As in the Coastal Plain, large areas have reverted to forest from former agricultural use. Southern Piedmont soils tend to be poorer than those in the Northeastern Piedmont. Like the soils of the latter area they are mostly derived from underlying igneous or metamorphic rocks and were never first-rate. But Southern Piedmont soils have had an additional handicap in the form of serious damage through past cultivation of clean-tilled row crops (cotton, corn, and tobacco) on easily eroded slopes subject to frequent downpours. Between the Piedmont and the Coastal Plain is the *Fall Line* zone of transition where rivers flowing from the harder rocks of the higher division to the softer rocks of the lower division tend to have falls and/or rapids. Among the numerous cities that have developed at such sites are Richmond, Virginia; Raleigh, North Carolina; Columbia, South Carolina; Augusta, Macon, and Columbus, Georgia; and Montgomery, Alabama.

The *Interior Plains* section of the South occupies central and western Texas, plus all of Oklahoma except the easternmost part of the state. This plains area is the southwestern section of the Interior Plains of North America, including the southern end of the Great Plains (Fig. 26.9). In Oklahoma it is bounded on the east by the Interior Highlands and by a small section of the Coastal Plain. In Texas the *Balcones Escarpment* forms an abrupt boundary between this higher western Interior Plains country and the lower Coastal Plain to the east and south. A broad band of territory along the eastern edge of the higher land in Texas is rather hilly, especially toward the south in the *Edwards Plateau*. Such Texas cities as Fort Worth, Austin, and San Antonio lie on or near the Coastal Plain at the foot of this hilly margin of the Interior Plains. Westward from the belt of hills, the plains rise in elevation and become more level toward the west, where the *Llano Estacado* ("Staked Plain")

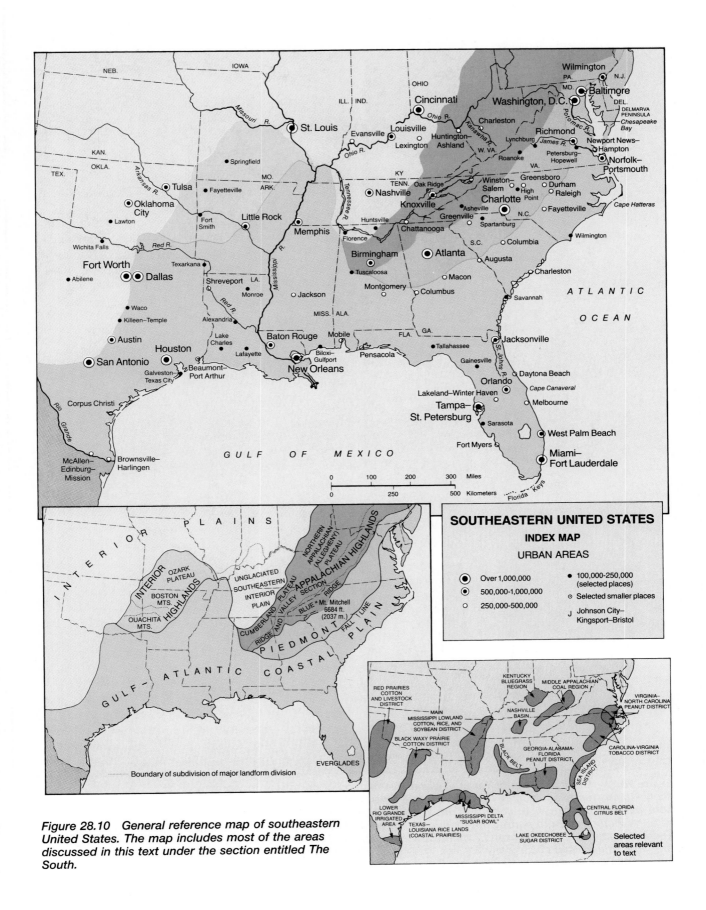

Figure 28.10 *General reference map of southeastern United States. The map includes most of the areas discussed in this text under the section entitled The South.*

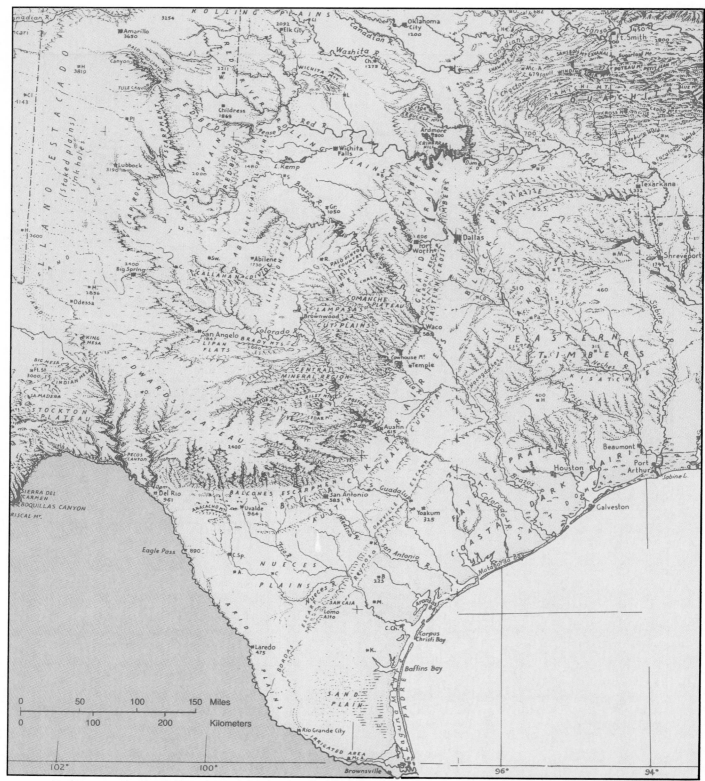

Figure 28.11   Landform map of Texas. A distinctive method of portraying topography is
shown in this representation of Texas by the late Erwin Raisz. Compare the method used
here with that used by A. K. Lobeck in the map of eastern United States (Fig. 28.2), and
with the method used to portray the topography of western United States in Figure 28.21.
The Texas map is a section from Erwin Raisz's "Landforms of the United States,"
prepared originally to accompany Wallace W. Atwood, The Physiographic Provinces of
North America (Boston, Mass.: Ginn and Company, 1940). Copyright by Erwin Raisz;
reproduced by permission of Ginn and Company.

along the Texas-New Mexico border presents a huge area of exceptional flatness.

The elevated plains of western Texas and Oklahoma are in many ways not typically Southern. Rather, they encompass the transition from the humid subtropical environment of the South to the steppe and desert environments typical of the West. Here forest gives way to grassland; irrigation becomes important, most notably in areas near Lubbock and Amarillo, Texas; ranching becomes prominent and also unirrigated farming of wheat and sorghums; and population thins. But one traditional "Southern" economic characteristic—an emphasis on cotton—remains strong.

Culturally the Interior Plains section of Texas and Oklahoma remains rather strongly Southern. Protestant Christians (principally Baptists, Methodists, Presbyterians, and Disciples of Christ) of white Anglo-Saxon origins predominate; in a large majority of counties the Southern Baptist Convention is the largest religious organization. Diverging from these generalities is a cluster of Texas counties in the "German Hill Country" near San Antonio where people of German extraction (both Catholics and Lutheran Protestants) are numerous. Catholic Hispanics predominate in the far south of the Interior Plains, but a large majority of the Hispanics in Texas (and black people as well) live in the Coastal Plain.

In the far southwest of Texas near El Paso, rainfall becomes so scanty that a band of territory along the Rio Grande and the New Mexico border is classified as desert. Physiographically, the band forms the easternmost section of the immense *Basin and Range Country*, a part of the *Intermountain Basins and Plateaus* (Figs. 28.21 and 28.22). Here scattered mountain ranges trend generally northwest-southeast and are separated and surrounded by arid basin floors. The area contains the highest peak in the Southern states (Guadalupe Peak: 8751 ft./2667 m near the New Mexico border), but even this mountain is very low by the standards of the American West. Physically and economically this far southwestern border of Texas is Western; culturally it is strongly Hispanicized. But even here there are important Southern cultural characteristics within the non-Hispanic population.

Hills, Mountains, and Small Plains of the Upland South

Although irregular or flat plains predominate in the South's topography, large parts of the region are hilly or mountainous. Areas of this kind in western and central Texas have already been noted, as have the scattered areas of low hills that occur farther east

in the Coastal Plain and the Piedmont. The remaining hills and mountains of the South lie in two large embayments of rough country that project into the interior South from the Northeast and Midwest. The larger of the two lies mainly in the Appalachian Highlands. But this embayment also includes some rough land immediately west of the Appalachians which is often considered to form the southeastern corner of the Interior Plains but is so dissected that much of it is better described as hill country than as plains. It is known as the Unglaciated Southeastern Interior Plain (Fig. 28.10). Farther west, the second large embayment is the southern part of the Interior Highlands. The three physical areas—Appalachians, Unglaciated Southeastern Interior Plain, and Interior Highlands—are often termed the Upland South.

SOUTHERN APPALACHIANS

Southern areas within the *Appalachian Highlands* include western Virginia, all of West Virginia, eastern Kentucky, westernmost North Carolina, a small tip of western South Carolina, eastern Tennessee, extreme northern Georgia, and much of northern Alabama. The Highlands include all of the major physical subdivisions already distinguished in the Middle Atlantic states, with the total complex narrowing southward (Figs. 28.2, 28.3, and 28.10).

The *Blue Ridge* extends southward from the Potomac across Virginia, comprises the Appalachian border zone between North Carolina and Tennessee, reaches into the western tip of South Carolina, and terminates in northern Georgia. From a single ridge in places toward the north, it becomes broader, more complex, and higher toward the south. Its highest section is south of Virginia and is called the Great Smoky Mountains. Here Mount Mitchell in western North Carolina is the highest point in the eastern United States at 6684 feet (2037 m).

The *Ridge and Valley Section* of the Southern Appalachians, including the *Great Appalachian Valley* in which ridges are less densely packed and valley floors more prominent, runs parallel to the Blue Ridge and inland from it. It occupies most of western Virginia beyond the Blue Ridge, easternmost West Virginia, a strip of Tennessee often called the Valley of East Tennessee, the northwestern corner of Georgia, and a section of Alabama extending southwestward to beyond Birmingham.

The Southern section of the *Appalachian Plateau*, known as the *Cumberland Plateau* in parts of Kentucky and areas to the south, is quite wide in the north where it occupies eastern Kentucky, most of West Virginia, and a small section of Virginia. It narrows rapidly across eastern Tennessee and widens somewhat to occupy much of northern Alabama before it terminates against the Coastal Plain. As in Pennsylvania the Plateau is edged on its east by an abrupt escarpment, is mostly dissected into hills and low mountains, and is underlain by very large deposits of high-quality coal. Some sections in Tennessee and Alabama form tablelands with fairly extensive upland surfaces.

UNGLACIATED SOUTHEASTERN INTERIOR PLAIN

Adjoining the Southern Appalachians on the west are sections of central Kentucky, central Tennessee, and northern Alabama that lie within the *Unglaciated Southeastern Interior Plain.* This part of the Interior Plains, bounded by the Appalachian Plateau on the east and the Tennessee River on the west and south, was not subject to the smoothing that glaciation accomplished on much of the Plains farther north. Consequently, a good part of this "plains" area is actually hill country, the result of dissection over a long period of time by streams fed by abundant rainfall and cutting downward through a somewhat elevated surface. True plains are present but are not extensive. Two of them—the *Kentucky Bluegrass Region* around Lexington and the hill-studded *Nashville Basin* around and south from Nashville, Tennessee—are underlain by limestone that has weathered into good soils. The two areas are famous for their agricultural quality in a region of poor to mediocre soils and difficult slopes.

SOUTHERN INTERIOR HIGHLANDS

In the South the *Interior Highlands* occupy the northwestern half of Arkansas and the eastern margins of Oklahoma except for a small strip of Coastal Plain along the Red River. From Arkansas they extend northward to cover most of the southern half of the Midwestern state of Missouri (Figs. 28.3, 28.10, and 28.12). Comprised of hill country, low mountains, and some areas smooth enough to be called plains, these highlands are split by the broad east-west *Arkansas Valley* followed by the Arkansas River. Immediately north of the valley, the *Boston Mountains* form the southern edge of the extensive *Ozark Plateau* of northern Arkansas, northeastern Oklahoma, and southern Missouri. South of the Arkansas Valley lie the *Ouachita Mountains* of west central Arkansas and adjacent Oklahoma.

The Interior Highlands display much internal variety of form and relief. Most of the Ozark Plateau has been dissected into hill country bearing some physical resemblance to the Appalachian Plateau, but lower in elevation, more subdued in relief, and lacking the coal resources of the Appalachian area. Some parts of the Ozark Plateau, especially the Boston Mountains, do attain the mountainous character

of higher parts of the Appalachian Plateau. The Ouachita Mountains consist of roughly parallel ridges and valleys resembling the Ridge and Valley Section of the Appalachians, except that in the Ouachitas the trend of the topography is east and west.

The Southern Interior Highlands have not been rich in resources. Most soils are quite poor, with many very stony. Most of the area is still forested, as it was when first settled, but it has been heavily cut over, and the timber today is generally rather poor. Deciduous hardwoods predominate north of the Arkansas Valley and coniferous softwoods in the Ouachitas. Mineral deposits are varied but often not of sufficient value to justify present exploitation. The main mineral wealth extracted in the Interior Highlands today, primarily lead, comes from the section outside the South in the state of Missouri.

HOW IMPRESSIVE A COUNTRY WOULD THE SOUTH MAKE?

Like each of the three other major American regions discussed in this text, the South is sufficiently large in area, populous, and productive to be an impressive nation in its own right. With an area of 886,000 square miles (2.3 million sq. km: 25 percent of the United States) and an estimated 1986 population of 77 million (32% of the nation), the 14 Southern states could constitute a country that would be exceeded in area only by the United States and ten other countries, and would also rank twelfth in the world in population. The South's area is over four times that of France, which is Europe's largest country by area, and the South's population exceeds by more than 16 million the population of West Germany, which is Europe's most populous country. Since the Great Depression of the 1930s, the South's level of economic development has increased greatly, and today the region is a major world producer of a long list of commodities such as wood products, coal, oil, natural gas, sulfur, phosphate, citrus fruits, soybeans, tobacco, electricity, textiles, clothing, aircraft, aluminum, chemicals, and electronic equipment. Such cities as Houston, Dallas-Fort Worth, Atlanta, and Miami would be impressive metropolises in any country.

COMPARISONS WITH OTHER MAJOR AMERICAN REGIONS

Although the South has become much more "standard American" in many ways since the Great Depression, it continues to have numerous distinc-

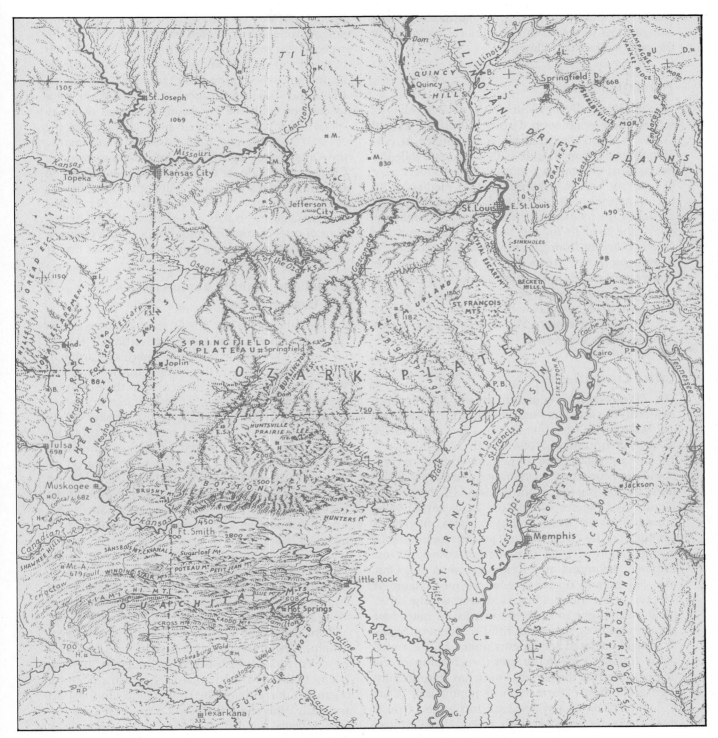

Figure 28.12   Map of the Interior Highlands and adjacent areas, from Erwin Raisz's "Landforms of the United States." On the original map the area in northern Missouri marked "till" was shown as "dissected loess-covered till prairies." Copyright by Erwin Raisz; reproduced by permission of Ginn and Company.

tive regional traits. For example, its patterns of population density, urbanism, ethnicity, and income are distinct from those of the other major regions. Some comparisons in regard to these indices are made in the paragraphs that follow.

Population Density

One significant distinguishing trait of the South is its intermediate population density within the United States. In 1986 the Southern states ranged in estimated density from Florida's 207 per square mile (80 per sq. km) of total area to only 39 per square mile (17 per sq. km) in Arkansas, with a median of 99 per square mile (38 per sq. km). The latter figure is far below the median for the Northeastern states, far above the median for the Western states, and not greatly different from that of the Midwestern states. It must be noted, however, that if one divides the Midwestern states into one group east of the Mississippi and another group west of the Mississippi, the Southern states emerge with a median density far below that of the eastern Midwest but far above that of the western Midwest.

Urban Comparisons

The difference in population density between the South on the one hand and the Northeast and eastern Midwest on the other reflects the South's lower level of urbanization. According to 1984 estimates, 18 American states had 75 percent or more of their population concentrated in metropolitan areas as defined by the Bureau of the Census. Of these, only Florida (91% metropolitan) and Texas (80%) were Southern states. Most Southern states, like most American states, were over 50 percent metropolitan in population in 1984 (although Mississippi, West Virginia, Arkansas, and Kentucky were not), but only Florida and Texas exceeded the mean figure of 76 percent for the nation as a whole. On the other hand, at the time of the 1980 census 16 states had populations over 40 percent rural (resident on farms or in agglomerated settlements under 2500), and 8 of the 16 states were in the South. Such comparisons do not mean that the present South has an unusually high proportion of farmers in its population. What it does indicate is a widespread incidence of rural nonfarm living, together with a very large number of small towns, small cities, and small metropolitan areas, and only a small number of the largest metropolitan areas. As of 1986 the South had far more metropolitan areas than any of the other three major regions but it had only 9 of the 37 metropolitan areas with populations over 1 million. The South's largest

metropolitan area, Dallas-Fort Worth, ranked only eighth in the country (Houston, almost identical in population, ranked ninth). However, many Southern metropolitan areas are growing very rapidly—especially as compared to metropolitan growth rates in the Northeast and Midwest—so that this regional distinction is becoming less marked as time passes. But in the foreseeable future the South will not have to contend with the enormous problems of urban sprawl and congestion that now characterize the coastal "Megalopolis" from Boston to Washington, the Great Lakes urban strip from Milwaukee and Chicago to Buffalo, or the Southern California urban cluster. Instead, the region will continue to have a great number of rather evenly spaced regional centers, manageable in size, and easily accessible to millions of rural commuters and shoppers via a highway net that has been greatly improved in recent times.

Racial and Ethnic Characteristics and Comparisons

The population of the South contains several distinctive racial and ethnic elements, although its population as a whole has less variety in this regard than that of the Northeast, the eastern Midwest, or the West. The great majority of people in the South today are white Protestants of British ancestry, black Protestants, or Mexican-American Catholics. There are relatively few people of continental European, Oriental, or Middle Eastern extraction, and aside from southern Florida's large Cuban-American community, relatively few are derived from parts of Latin America other than Mexico. A striking feature is the heavy preponderance of white Protestants with Anglo-Saxon origins. In the main, the original European settlers along the coasts were British Protestants—overwhelmingly English, but with some Scots, Welsh, and Scots-Irish. Small groups of Spanish Catholics settled in Florida and Texas, and some French Catholics settled in Louisiana and elsewhere along the Gulf Coast, but their numbers were insignificant compared to the tide of settlers from Great Britain. Western and interior areas of the South then were settled by British-descended migrants moving inland from the Atlantic Coast. The great streams of migration that entered the United States in the 19th and early 20th centuries from Catholic areas of Europe largely avoided the rural and economically underdeveloped South in favor of the more urbanized and industrializing North. Hence the South today has only minor numbers of people whose ancestors came from such Catholic countries as Ireland, Italy, or Poland. Nor did the South receive more than a trickle from the tide of German and Scandinavian landseekers who flooded into agricultural areas of the Midwest in the 19th century. The state of Florida's present white population is exceptional in that it contains not only a large body of British-descended Southern whites, but also a large and varied mix of Northern whites who have come to the state as retirees, business people, or winter residents.

THE SOUTH'S BLACK POPULATION

Another well-known way in which the South stands out socially from the other major regions lies in its high proportion of black population. This is largely a result of the importation of African slaves to the plantation economy of the pre-Civil War South. Consequently, high densities of black population today are especially marked in states that share the sections of the Coastal Plain where most plantations were located. The five American states that ranked highest in percentage of black population in 1980—Mississippi (35% black), Louisiana (29%), Alabama (20%), Georgia (27%), and South Carolina (30%)—form a Coastal Plain belt. Florida, which had few plantations, is outside the belt, although its figure of 14 percent black in 1980 still ranked it well up among the states. Most other Southern states also ranked well above the national figure of 11.6 percent black, although Oklahoma and Kentucky fell below this percentage and in West Virginia blacks comprised only 3 percent of the population. All of West Virginia and nearly all of Kentucky are in the Upland South where slave plantations were few; and Oklahoma was an American Indian domain with few whites or blacks until well after the Civil War.

Black migration out of the South has given such states as New York, California, Illinois, Michigan, Ohio, Pennsylvania, Maryland, and New Jersey black populations comparable in size to those of the majority of Southern states. In fact, there are more black people in New York or California than in any Southern state. But the *percentage* of black population in the eight states named is somewhat smaller than in all but a few states of the South. Of course, this is not true of large Northern or California cities where most black migrants have settled, including some cities where blacks are a majority in the city proper. Recently there has been a sizable reverse migration of blacks from North to South as economic and social conditions for blacks in the South have improved.

HISPANICS AND CAJUN FRENCH

Other distinctive racial and ethnic elements in the South's population are less widely distributed and tend to be concentrated in subregions of individual states. Especially numerous and notable are the Mexican-Americans, Cuban-Americans, and "Cajun" French.

The 1980 census reported that the four leading states in percentage of population "of Spanish origin" (known and reported) were New Mexico (37%), Texas (21%), California (19%), and Arizona (16%). The Texan minority of about 3 million Mexican-Americans was larger than that of any other state except California (4½ million). These people live mainly in a broad band of territory along the Gulf of Mexico and the Rio Grande. Here people of Mexican descent form a much higher proportion than the statewide 21 percent. The most distant origins of this Hispanic population are to be found in colonists who came from Mexico in the early 18th century when Texas was a northern outpost of Spain's large New World empire. San Antonio was dominant among a few small and scattered settlements. Natural increase and immigration have rapidly expanded the number of Mexican-Americans (in Texas and elsewhere) during recent decades. Concurrently, south Florida has received intermittent—although quite large—immigration of another "Spanish origin" group, the Cuban-Americans, and lesser numbers from other troubled Caribbean states. Almost a million refugees have fled from Communist Cuba since the Castro revolution of 1959, and the majority of them or their descendants now live in metropolitan Miami. Notable differences exist between the Mexican-Americans and the Cuban-Americans in regard to economic success and assimilation. Cuban-Americans, drawn largely from the middle and upper classes of pre-Communist Cuba, have shown rather rapid upward mobility economically. They also have been assimilating well into the general population through marriage and rapid adoption of American culture traits such as the English language. Meanwhile they have imprinted aspects of their own culture on south Florida. Their record of economic success and assimilation contrasts with that of the country's Mexican-American population, which largely continues to be an economically deprived group whose Hispanic identity remains strong.

The largest minority language group in the South which is not Hispanic is the French-descended "Cajun" group of southern Louisiana. Small French settlements, including New Orleans, existed along the Gulf Coast during the early 1700s. But it was the arrival of refugee "Acadians," expelled from Nova Scotia by the British in the middle of the 18th century, that implanted an enduring French-derived culture. Now strongly mixed with standard American elements, the Cajun culture characterizes a sizable part of southern Louisiana's swamp country.

Besides the Cajuns of French-Canadian colonial descent, Louisiana has a body of Creole French descended directly from France.

INCOME DIFFERENTIALS BETWEEN THE SOUTH AND THE NATION AS A WHOLE

A marked historical characteristic of the Southern states has been their relatively low average incomes compared to the nation in general. This distinction dates back to the days of slavery and is still generally valid, although the gap between the South and the rest of the nation has lessened markedly in recent decades. In 1984 Mississippi was the poorest state in the Union as measured by personal income per capita. Arkansas, West Virginia, Alabama, South Carolina, Kentucky, and Tennessee were also numbered among the 11 poorest states. The remaining Southern states were scattered upward along the income scale, but only Virginia, which ranked fifteenth, Texas (nineteenth), and Florida (twenty-first) were among the top 30 states. The South as a whole was the nation's poorest region, although not by a very wide margin over the West, which has several Mountain states that rank very low in income per capita. Southern poverty is especially concentrated in Coastal Plain states that were excessively dependent on cotton in slave-plantation and sharecropper days, or in states of the northern South that had large numbers of poor upland farms in slave times, and subsequently developed an excessive dependence on coal mining. These circumstances have been major aspects of the South's troubled history.

THE SOUTH'S GEOGRAPHIC HERITAGE FROM THE SLAVE-PLANTATION AND SHARECROPPER ERAS

For more than three centuries the South had an economy dominated by the production of cash crops for world markets. Cotton was by far the most valuable crop over the period as a whole, although tobacco and rice were important, and indigo (a dye-stuff) and sugarcane were locally grown. Production was concentrated primarily on large ownerships, but there was some production from a far greater number of small owner-operated farms. Before the Civil War the large ownerships were called plantations; after the Civil War many continued to exist in an altered form to which scholars have applied the name *dispersed plantations*. For two and a half centuries black slaves were the labor force for planta-

tions, and they also labored for small farmers able to afford them. However, the great majority of small farmers did not own slaves. After the emancipation of the slaves resulting from the Civil War, the reliance on cash crops continued, but with large numbers of landless families (black or white) working and living as tenants on plots of land into which the former plantations were subdivided. There were various tenancy arrangements, usually involving sharecropping. Thus developed the "dispersed plantation," still constituting a large landholding, but with its labor force living on small tenant farms instead of congregated in slave quarters near the owner's Big House. A sizable majority of the South's population from the beginning of settlement until the eve of World War II was made up of poor farmers or poor farm labor, although a small minority of large landowners, cotton ginners, merchants, bankers and other business or professional people prospered. Southern life was centered in the countryside, and both urbanization and industrialization were severely retarded.

Origins and Spread of Southern Cash-Cropping

Reliance on cash crops began at the dawn of settlement in the 17th century. Jamestown, Virginia, just upstream on the James River from present-day Norfolk, was founded in 1607 as the first enduring English settlement in what would become the United States. Land on the Coastal Plain was found to be suitable for growing tobacco, a New World plant that already had become known in Europe. The kind of tobacco grown by Virginia Indians was not marketable, but seeds of salable varieties from tropical America were brought to the colony from Europe. Tobacco cultivation is very labor-intensive, and enough workers could not be mustered from the small number of colonists or from the local Indians. The latter refused to work for the whites, and attempts to coerce them were not successful. But soon the labor shortage was solved by importation of African slaves, beginning in 1619. Expansion of tobacco growing through the 17th and 18th centuries made the Virginia "Tidewater" (the Coastal Plain adjacent to Chesapeake Bay) the first great region of slave-plantation agriculture in the South. Subsequently, production spread to adjacent Maryland and Delaware, and inland onto the Piedmont. So prominent did the plantation aristocracy of Virginia become that all but one of the first five Presidents of the United States were drawn from it.

As plantation agriculture spread over the Coastal Plain, many small farmers who lacked land and capital migrated to the Piedmont or the Appalachians, where they established a poor semisubsistence agriculture. They were joined by many new immigrants who pushed directly into the back country where land could be had at low cost. Thus originated a sturdy class of small landowners in the Piedmont and the Upland South who are often referred to as "yeoman farmers."

In the late 17th century and the 18th century a second outstanding plantation area developed in Coastal Plain swamps inland from the port of Charleston, South Carolina. In this hot wet environment, irrigated rice and indigo became the staple exports. Irrigation water for the rice fields was supplied primarily by tidal rivers. The fields were surrounded by embankments, with gates opening onto the rivers. Fresh river water ponded back by the insurge of sea water at high tide was led through the gates, which were then closed at the proper time. The fields could be drained as desired by opening the gates at low tide.

Then in the 1790s technological innovation made cotton the main basis of the plantation economy. Although long known in both the Old World and the New, cotton had been an expensive luxury due to the almost prohibitive amount of hand labor required to separate the cottonseeds from the clinging fibers of the cotton boll. A Connecticut inventor, Eli Whitney, solved this problem by inventing the mechanical cotton gin while on a visit to a Georgia plantation. Consequently, cotton became a cheap raw material for new mechanized textile factories that were already spearheading the Industrial Revolution in England and in the early 19th century began to do so in New England. A long boom in cotton growing ensued, with Southern production multiplying and remultiplying until the Civil War broke out in 1861.

The Westward Shift of Cotton and Tobacco

The cotton boom, plus continuing production of tobacco, caused a rapid westward expansion of the South and its slave-plantation economy. By the Civil War, cotton plantations extended westward as far as eastern Texas and tobacco plantations as far as central Missouri. For the most part, the slave plantations that grew cotton were a feature of the Coastal Plain in the more southerly parts of the South. Here, such plantations were very widespread, but they tended to cluster in certain favored areas: first the Sea Island District of coastal South Carolina and Georgia, then the Black Belt of Alabama and Mississippi, and finally the Mississippi Lowland (Mississippi Alluvial Plain) shared by Mississippi and several other states.

The first great center of cotton production was the offshore islands and adjacent mainland sections of Georgia and South Carolina. This *Sea Island District* had lost its primacy by the Civil War but continued to be an important producer of its superior long-staple (long-fibered) "Sea Island" type of cotton until the crop was devastated by the boll weevil in the 20th century. By 1860 the *Black Belt* of Alabama and Mississippi and the flood plain of the Mississippi River (*Mississippi Lowland*), growing short-staple cotton varieties, had become the principal centers of production, although the crop had spread farther west into Texas. The Black Belt, so-called for the color of its soils, is a crescent-shaped area extending east and west through central Alabama and then curving northward into northeastern Mississippi (Fig. 28.10). The virgin soils of this belt, developed on limestone bedrock, were exceptionally fertile but were worn out by excessive cropping in cotton, which, like tobacco, is notoriously hard on soils. Exhaustion of the soil plus the ravages of the boll weevil ruined cotton production in the Black Belt almost as completely as it was ruined in the Sea Island District. The Mississippi Lowland proved more durable as a cotton producer and remains an important area today. This alluvial plain, some 600 miles (966 km) long from the mouth of the Ohio to the Gulf of Mexico and often over 50 miles (80 km) in width between the bordering bluffs (Fig. 28.2), is a remarkable feature in the physical geography of Anglo-America—comparable in many ways to the great flood plain and delta areas which support dense populations in the Orient. The plain has deep and fertile soils, but many sections away from the river are swampy. The original cotton plantations clustered primarily on the *natural levees* of the main river and its tributaries. Natural levees, built up by deposition of sediments during flooding, are strips of land immediately alongside the streams. Higher and better drained than the rest of the flood plain, they exist because rivers in flood immediately drop most of their sediment load when they overflow their banks and lose their velocity. A natural levee slopes gently downward (often for many miles) from the river bank to lower land near the outer edge of the flood plain where flood waters and normal runoff collect in swamps.

North of the developing Cotton Belt, the tobacco economy spread westward into Kentucky, Tennessee, and Missouri. The crop was grown on slave plantations, but also on small farms. Hemp produced in these more northerly slave states was made into the coarse sacking used for baling cotton in areas farther south. Both Missouri and Kentucky bred mules for sale throughout the South.

Underdevelopment in the Plantation South

The dramatic expansion of the plantation economy that made the South the world's chief supplier of raw cotton was associated with a relative failure of other kinds of development. Most immigrants from Europe avoided the South, where they would be in competition with slave labor. Nothing rivaling the industrialization and urbanization of the Northeast occurred in the region. Investment went mainly into land and slaves, and few cities of much size developed. The largest was New Orleans, which ranked fifth in size among the nation's cities in 1810 (after New York, Philadelphia, Baltimore, and Boston). By the 1860 census it was still in fifth position, although after the Civil War it was surpassed by numerous other cities. Due to natural transportation afforded by the Mississippi-Ohio-Missouri river system, the port of New Orleans at the system's outlet was able to give strenuous competition to New York for the trade of the developing Midwest. But by the 1850s railways had connected the Middle Atlantic ports with the Midwest, and this foreshadowed the end of the river-boat era in which New Orleans had prospered. The traffic of the port went into a long period of relative decline, from which it only emerged in recent times with the revitalization of Mississippi River barge traffic.

The first census of the new United States in 1790 showed an almost even division of population between North and South, with most of the population of both regions located along or near the Atlantic Coast. But by 1860, on the eve of the Civil War, the South had only half as many people as the North and about a tenth as much industrial development. Meanwhile the nation had pushed westward through vast sparsely populated areas to the Pacific.

Adverse Effects of Postbellum Cash-Cropping

Although slavery was ended in the South by the Civil War, heavy dependence on cash crops was not. In the postbellum era there continued to be a basic dependence on cotton for cash sale, or on tobacco in the more northern areas. Both crops were very labor-intensive, and both generally failed to yield their cultivators a living above the poverty level. Cotton prices were adversely affected by increasing competition from foreign growers, and the South's own crop land gradually deteriorated as a result of continuous cropping in cotton, tobacco, and corn. These clean-tilled row crops not only depleted the soil of nutrients but also exposed it to severe erosion by heavy rains. In the case of cotton, such troubles

were augmented by entry of the devastating boll weevil from Mexico. The depths of distress were reached in the Great Depression of the 1930s, when the price of cotton sank to unprecedented lows and large numbers of cultivators and their families were forced to seek government relief in order to live.

ECONOMIC CHANGE SINCE THE GREAT DEPRESSION

Since the 1930s the South has made great progress in narrowing the economic gap that set the region apart from the rest of the nation for so long. Many factors have contributed to this economic rise. A number are general factors which have weakened the former dominance of the Northeast and the Great Lakes states over the South and West. But some factors have been more specifically Southern. Extremely important among these has been the movement away from dependence on labor-intensive cotton farming. Maps of United States agriculture as late as the 1920s show a Cotton Belt extending solidly across the South from an eastern extremity along the Atlantic Coast to a western extremity in the Texas Interior Plains. This represented the employment of millions of people producing a crop that required great amounts of hand labor during parts of the year, so that one worker produced relatively little and also found it difficult to combine cotton production with other remunerative types of agriculture. The crop had then to be sold at low prices on a highly competitive international market, with the result that the Cotton Belt was a poverty belt for most farmers.

The collapse of world trade in the Great Depression of the 1930s, with attendant record low prices for cotton, marked the beginning of the end of the Cotton Belt. Its demise was hastened by the introduction of a successful mechanical cotton-picker in the 1940s. This new machine ended the need for concentrated labor in the cotton fields. Being large and expensive, mechanical cotton-pickers can be used most advantageously by large farms on level land producing high yields. Hence their use favored a westward shift of cotton growing toward irrigated areas and away from the older sections of the Cotton Belt in the eastern South. Cotton has continued to be important on the broad and fertile Mississippi Alluvial Plain, but today is very minor east of there (Fig. 28.13). In 1986 nearly one third of the national cotton output by volume still came from the six states that share the Mississippi Plain, but even in this favored and traditional cotton area, soybeans and/or rice are now more valuable crops in the great majority of counties. Irrigated acreages in California and Arizona also produced a little under one third of the nation's cotton in 1986, and a little over one quarter came from Texas, which was the leading cotton state by a relatively small margin over California. Production in Texas comes primarily from the High Plains around Lubbock, but also from the Red Prairies, the Black Waxy Prairie, and the Coastal Prairies along the Gulf of Mexico. By far the greater part of Texas cotton is produced by irrigation, but some is grown by mechanized farming of huge rainfed acreages giving low and drastically fluctuating yields.

Disastrous unemployment might well have resulted from the mechanization of cotton production and the attendant abandonment of cotton over large areas. For many families such unemployment did create great hardship, at least for a time. Many displaced agricultural workers, both blacks and whites, joined a massive job-seeking migration to Northern industrial cities that already was well underway before the Great Depression. The large black component in this migration drastically altered the social geography of many Northern cities. But the bulk of the South's redundant cotton workers eventually were absorbed by an expanding job market in the South itself. The new jobs have been created in the accelerating nonagricultural sector of the Southern economy since the end of the Great Depression. Many factors have contributed to this industrial rise: labor cheaper than that in other parts of the country; federal military expenditures and programs of economic development; intensified development of Southern natural resources to meet regional and national needs; the introduction of air conditioning; and enormous improvements in transportation through the creation of the Interstate Highway System, expansion of airways, and improvements in waterways. New factories, offices, stores, and services now employ many workers from families that originally labored on poverty-stricken farms.

AGRICULTURAL GEOGRAPHY OF THE SOUTH TODAY

What are the main features of the South's agricultural geography after a half-century of revolutionary changes in the number of farmers, crop and livestock emphases, farm mechanization, and farm markets? In detail the picture is intricate, but some generalizations are possible:

1. *In appearance and land use, the greater part of the South today is not agricultural.* Over large areas less than half the land is in farms. Even where land is farmed, the amount of crop land is apt to be greatly exceeded by the acreage of woodland and/or pasture. In a few large parts of the South there are expanses of relatively continuous crop land, but from central Texas eastward the main element in the landscape generally is forest. In the drier climate of western Texas, pasture and range land dominate the landscape except where irrigation is strongly developed. The proportion of the entire South devoted to crops is much smaller than in the Midwest, about the same as in the Northeast, and much greater than in the West.

2. *The present South is not outstandingly agricultural compared to the country as a whole.* In 1980 only 2.5 percent of the national population still lived on farms. Only 6 of the 14 Southern states exceeded this proportion, whereas 11 of the 12 Midwestern states, several Western states, and one in the Northeast exceeded it. Within the South as a whole, 2.3 percent of the population lived on farms, compared to 4.9 percent in the Midwest, 0.8 percent in the Northeast, and 1.4 percent in the West. Not only is the farm population relatively small, but the South does not have an agriculture that is outstandingly productive per unit of total area. Only four Southern states rank reasonably high in this respect: North Carolina and Kentucky, where dollar output is boosted by tobacco sales; Florida, where citrus fruits perform the same function; and Arkansas, where inclusion of an unusually large segment of the highly productive Mississippi Alluvial Plain raises the state's output. Most Southern states are only moderately productive in proportion to their areas, and one, West Virginia, is among the least productive states in the Union. West Virginia's agricultural plight results from a location almost wholly within a particularly rough and infertile part of the Appalachians.

In the great majority of Southern counties the average production per farm is far below the national average, and this is true to a far greater extent than in any of the other major regions. Three Southern areas—central and south Florida, the Mississippi Alluvial Plain, and western Texas—are marked exceptions in having a relatively high average production value per farm. In central and south Florida the dominant sources of farm income are citrus fruits, beef cattle, vegetables, and sugarcane; in the Mississippi Alluvial Plain soybeans, cotton, and rice are dominant; and in western Texas (the part of the state west of the Coastal Plain) the leading farm income-producers are beef cattle, cotton, sorghums, and wheat. Scattered over the South are still other districts characterized by large-scale operations with high output value per farm. They lie mainly in the

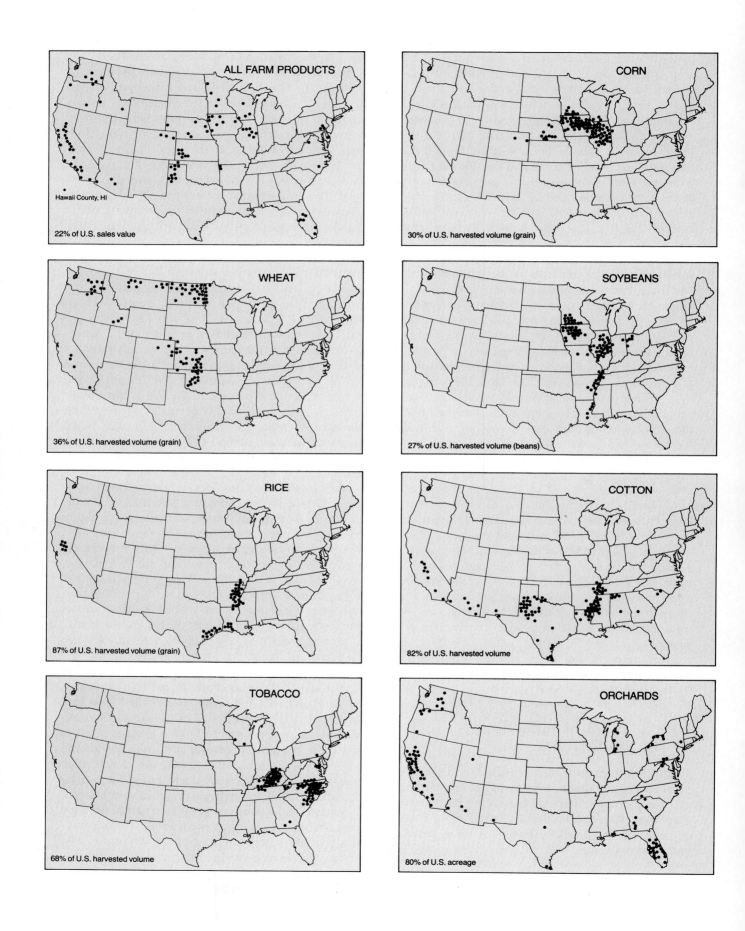

ALL FARM PRODUCTS

Hawaii County, HI

22% of U.S. sales value

CORN

30% of U.S. harvested volume (grain)

WHEAT

36% of U.S. harvested volume (grain)

SOYBEANS

27% of U.S. harvested volume (beans)

RICE

87% of U.S. harvested volume (grain)

COTTON

82% of U.S. harvested volume

TOBACCO

68% of U.S. harvested volume

ORCHARDS

80% of U.S. acreage

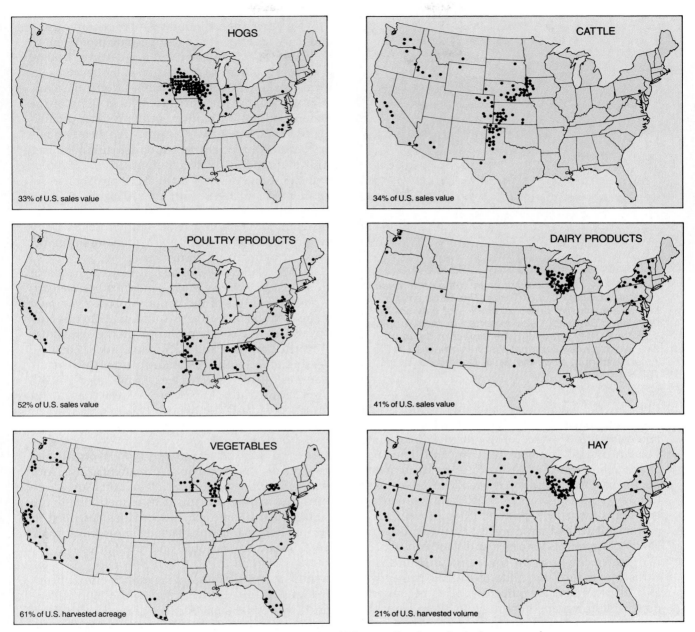

Figure 28.13 Panel of maps showing the 100 leading U.S. counties for selected crops and livestock products (50 counties in the case of rice). Percentages show the proportion of the national total represented by the 100 leading counties in each case. Data: U.S. Census of Agriculture, 1982. (Cartography by Joseph H. Astroth, Jr., and Jennifer Nichols.)

Coastal Plain from southeastern Virginia to southeastern Georgia and near the sea in the Coastal Plain of Louisiana and Texas.

3. *The South is outstandingly important nationally in reliance on small and part-time farms.* The region is characterized by a high incidence of both part-time farms and small full-time farms with low average production value. On the part-time farms, which themselves are generally small, more than half the total income normally comes from off-farm employ-

ment. Over large sections of the South, half or more of all farms are part-time. Within the South as a whole, such farms comprise a greater share of all farms than in the Northeast, Midwest, or West. As would be expected, the incidence of part-time farms is lowest in the exceptional Southern areas of large and highly productive farms noted previously.

4. *Within the South as a whole, animal products decisively surpass crops in total farm sales.* Beef cattle (Fig. 28.14), chicken broilers (Fig. 28.15), and dairy

Figure 28.14 Purebred Brahman cattle on a ranch in the Gulf Coastal Plain of Texas. The cattle shown here are expensive breeding stock, not only sold in the United States but also exported to many tropical countries. They represent one aspect of the impressive livestock development that has emerged in the American South since the Great Depression of the 1930s.

cattle are the main sources of animal products sold. Beef cattle have become the single most important and widespread source of Southern farm income. The farms involved range from tiny part-time operations to huge ranches. Rising cattle production during the past half-century has been fostered by many factors: (1) the presence of much land too poor for crops but amenable to pasture; (2) improvements in pasture grasses and cattle breeds; (3) increased availability of chemicals for pest and disease control; and (4) greater emphasis on soybeans, sorghums, and citrus by-products for cattle feed. Soybeans grown as a feedstuff and for other uses have become the most widespread and important Southern crop. The leguminous soybean plant adds nitrogen to the soil (through nodules on its roots containing bacteria that take nitrogen from the air and change it to a solid form); and the plant can be used for hay. But its greatest value lies in the beans that are pressed for soybean oil, with the residue (oilcake) being fed to cattle and poultry. Grain sorghums, which have low moisture requirements, have become a mainstay for cattle feeding in drier parts of Texas and Oklahoma. The greater part of Florida citrus fruits is now processed for frozen concentrate, and the residue is fed to cattle. The main market for the latter feed is a large ranch industry in southern Florida. Ranchers in this area also feed molasses derived from sugarcane grown and processed in south Florida.

Two other animal industries—broiler production and dairy farming—have also become prominent in Southern agriculture. In 1986, the five leading American states in the production of broiler chickens were Arkansas, Georgia, Alabama, North Carolina, and Mississippi. They produced three fifths of the national output, with other Southern states producing additional amounts. This American dietary staple has become a specialty of poorer agricultural areas, with outstanding concentrations in the Interior Highlands of Arkansas and the Appalachians of northern Georgia and northern Alabama. Considerable production also comes from Coastal Plain and Piedmont locations in several Southern states. Dairy farming is less prominent but is growing in importance as Southern urban markets grow.

5. *The historic Southern staples of tobacco and cotton are still major elements in the agriculture of some parts of the South.* Most of the nation's tobacco is grown in the eastern and northern South. In 1986 this crop was the leading source of income in the agriculture of North Carolina, South Carolina, and Kentucky. Nearly all the South's remaining cotton is grown farther west, being centered heavily on large mechanized farms in the Mississippi Alluvial Plain or in Texas, particularly the Interior Plains section.

6. *Many island-like districts of intensive agriculture produce a variety of specialties other than those previously emphasized* (Figs. 28.10 and 28.13). Such districts are scattered over the South in an irregular fashion, with the majority on the Coastal Plain. Often they represent the main producing areas of their kind in the United States. Some major specialties described in the following paragraphs include citrus fruits, truck crops, sugar, rice, peanuts, and race horses.

CITRUS FRUITS. Citrus growing in the South is heavily localized in the central part of the Florida Peninsula, with a far smaller concentration on irrigated land in the lower Rio Grande Valley of Texas. The danger of frost in areas farther north restricts commercial citrus growing to these extreme southern sections of the Coastal Plain, although damaging frosts occur in some winters even there. Most of Florida's citrus production comes from a belt in the center of the Florida Peninsula east of Tampa Bay. This part of the state has many lakes and low hills which give a certain amount of added protection against frost damage.

TRUCK CROPS. The Florida Peninsula is the outstanding Southern truck-farming area, although its proportion of the national output is far less than is true of citrus. With its warm winters and consequent

Figure 28.15 Raising and processing of chicken broilers has become a massive agricultural industry in the American South. The broilers in the photo will be processed at a plant in Arkansas operated by the food company ConAgra.

advantages for early harvesting and marketing, the state of Florida produces over one tenth of the nation's vegetables and small fruits by value. Tomatoes, sweet corn, and watermelons lead in acreage, but many other crops are grown. Much of the production comes from drained marshland, where the black organic soils comprised largely of plant remains are very fertile. Other truck-growing districts, all relatively small, are scattered along the Coastal Plain from the Delmarva Peninsula to the lower Rio Grande Valley.

SUGAR. No part of the conterminous United States is sufficiently tropical for really efficient production of cane sugar. However, two districts specializing in this product have developed in the Coastal Plain behind tariff protection. The larger and more important of the two is the so-called Sugar Bowl in the Louisiana part of the Mississippi Lowland; the other is near Lake Okeechobee in Florida. In the Louisiana area, where sugar is grown on alluvial land, pro-

duction began in French colonial times. The Florida area is a 20th-century development in a drained section of the northern Everglades. The combined production of the two sugarcane districts provides a substantially smaller percentage of American sugar consumption than that supplied by the beet-sugar industry, which is localized in other parts of the country. About half of the sugar marketed in the conterminous United States is cane sugar from tropical sources, most notably Hawaii, the Dominican Republic, and the Philippines. Federal market allocations and price supports enable American sugar producers to withstand competition from lower-cost sugar industries in other countries.

RICE. The United States supplies its own needs for rice and even exports enough to rank in the top three rice-exporting countries. Rice consumption in the United States is relatively small, and American rice is produced with extreme efficiency by machine methods. About half of the American crop comes from the Mississippi Lowland, where Arkansas is the outstanding producer. About three tenths of the crop comes from the prairies along the coasts of Louisiana and Texas. The remainder comes almost entirely from California. The total American production is quite small on a world scale, but rice growing is an important phase of the economy in the areas where it is centered.

PEANUTS. Over 95 percent of the peanut crop of the United States is grown on the Coastal Plain, mainly in southern Georgia and adjoining parts of Alabama and Florida, or in a district along the North Carolina-Virginia border. The crop is one of several major sources of vegetable oils used for making lard and butter substitutes, soap, and kindred products.

RACE HORSES. The Kentucky Bluegrass around Lexington is one of the world's most famous areas of thoroughbred horse-breeding. This landscape is a showpiece of conspicuous wealth, dominated by the mansions, large barns, and white board fences on the highly capitalized horse farms. The fences break the countryside into pastures and paddocks where some of the world's most expensive animals are kept. At nearby Louisville is the famous racecourse of Churchill Downs, scene of the annual Kentucky Derby.

7. *In sum, the agriculture of the modern South is increasingly diversified and productive, carried on by fewer and fewer people, and indicative of a regional economy that is steadily becoming less agricultural and more urban-industrial in character.*

INDUSTRIAL CHANGE IN THE SOUTH AND ITS GEOGRAPHIC OUTCOMES

The South emerged from the Civil War with a devastated economy that was overwhelmingly agrarian. Only slowly did rebuilding proceed in the period of federal military occupation known as Reconstruction. Conscious of the South's dire weakness in industry, Southern leaders began a drive to industrialize the region by marshalling internal investment and at the same time attracting firms and capital from outside. Although temporarily set back now and then by national economic downturns, this push has been gaining momentum ever since.

The Course of Southern Industrialization

The earliest large-scale industries to develop in the post-Civil War South were cotton-textile and apparel factories utilizing cheap labor drawn from economically depressed countrysides. In the late 19th and early 20th centuries this Southern development sapped the cotton-textile industry in New England and eventually gave national preeminence in the industry to the Southern Piedmont from Virginia to Alabama. Factories were built in a large number of small towns and cities—many of which offered inducements such as tax advantages and free land to attract the new mills. So great was this industrial surge that North Carolina had already displaced Massachusetts as the leading cotton-textile state by World War I. Much of the electricity to power the factories was supplied by the development of hydrostations on Piedmont streams flowing out of the rainy Blue Ridge. Piedmont mills sold large quantities of cotton cloth abroad, but they also marketed cloth to domestic apparel factories located in the Northeast. In the course of time the South developed a considerable apparel industry of its own, with factories at many locations in the Piedmont and Coastal Plain.

Then in the 1940s Southern industrialization accelerated, and this momentum has continued to the present. Not only did the number of factories and workers increase, but products became much more diversified. Low wages relative to the rest of the nation remained a powerful factor contributing to industrial success, but many other advantages developed. They included new hydroelectric facilities built by the federal government in the Tennessee River Basin and elsewhere, federal improvements in river transportation and harbors, and joint federal and state construction of the Interstate Highway System, which gave the South cheap and fast new connections with market concentrations elsewhere. Air conditioning made Southern workplaces and residences more comfortable and productive. When synthetic fabrics arose to challenge cottons and woolens, the South had abundant raw materials (wood and hydrocarbons) to make synthetic fibers. An extraordinarily valuable regional store of energy resources—oil, natural gas, coal, and falling water—powered new Southern industries and provided large energy exports to markets outside the region. Increasing technical skills became available through a combination of on-the-job experience in Southern factories, research and teaching by Southern universities, and importation of technology from outside the region to such places as the Oak Ridge atomic facility near Knoxville, Tennessee, and the space-flight facilities of the National Aeronautics and Space Administration (NASA) at Huntsville, Alabama; Houston, Texas; and Cape Canaveral, Florida. As population increased and incomes grew, Southern regional markets became large enough to attract plants that could profit from nearness to those markets.

Recent Gains in Manufactural Employment Vis-à-Vis the Northeast and Midwest

By the early 1980s the South was reaching a degree of industrialization generally commensurate with that of the entire nation, and its factories represented a very wide range of industrial specialties. Of the nation's total employment in manufacturing 29 percent was in the South in 1984, a percentage not far below the region's share of the national population (33%). At least two Southern states—Texas and North Carolina—are now major industrial states. Texas ranked fifth in the nation in total manufactural employment in 1984 and North Carolina ranked eighth. However, the total in Texas is more a function of that state's large area and population than of any exceptional degree of industrialization.

The degree of industrialization an area possesses is measurable in more than one way. One measure is the proportion of manufactural workers within the area's population. By this criterion the Southern states range from very highly to very little industrialized. The most highly industrialized are North Carolina and South Carolina, which ranked second and fifth in the country in 1984 and are comparable to the states of Southern New England. At the other end of the scale are Florida and Louisiana, which are among the lowest-ranking American states in degree of industrialization. Thus the highest degrees of industrialization in the South, as measured by manufactural employment, are in states associated with the old Piedmont textile belt where large-scale

Southern industrialization began. Another way to measure the industrialization of a state is to relate the number of manufactural employees to the state's area, giving an areal density of manufacturing. Under this procedure the most industrialized Southern states are again North Carolina and South Carolina, with Virginia and Georgia in the Piedmont textile belt and Tennessee to the immediate west also ranking high. All these states are well behind several Northeastern states by this measure, but they are comparable to the Great Lakes states in the Midwest. The lowest-ranking Southern states are Oklahoma, Texas, and Arkansas. As previously noted, Texas has more manufactural employees than any other Southern state, but the number does not bulk large when compared to the state's huge area.

The Major Southern Industrial Belts

The largest cluster of industrial development in the South lies along the Piedmont, with extensions into the adjacent Coastal Plain and Appalachians. Within this concentration, which may be termed the *Eastern South Industrial Belt,* about one tenth of the nation's manufactural labor force is employed. Textiles are the most prominent branch of manufacturing, but clothing, chemical products (including synthetic fibers), furniture, tobacco products, and machinery are important. The textile and clothing industries are very dependent on cheap labor, as they must meet strenuous competition from factories in low-wage countries such as South Korea and Taiwan. Chemical manufacture is especially important in Virginia, and some of this industry spreads into the nearby Appalachians of Tennessee and West Virginia. It is favored in this general area by the presence of Appalachian coal and wood to use as raw material for making such products as plastics and rayon, and also by relatively cheap electric power. Much of the latter is available from huge thermal stations of the Tennessee Valley Authority, as well as much smaller quantities from the many hydrostations at TVA dams along the Tennessee and its main tributaries. The TVA was founded in 1933 by the federal government during the Roosevelt New Deal era in order to promote economic rehabilitation of the deeply depressed Tennessee River Basin. The agency is a public corporation that undertook the construction of a great series of dams, locks, and hydrostations in order to control floods on the unruly Tennessee River, allow barge navigation as far upstream as Knoxville, Tennessee, and generate hydroelectricity. The demand for electricity soon outran the capacity of the hydrostations, and many thermal plants were built, fired by coal or using nuclear energy. Hydropower

now supplies under one tenth of the TVA's output of electricity. The success of the TVA in helping industrialize several Southern states is a well-publicized facet of the industrial surge that has spread ever more widely across the South in recent decades. In the Northeast and the Midwest, industrial development focused on the huge metropolises that arose in the railway age, but in the South most development came later and took advantage of long-distance power transmission and high-speed truck transportation to spread widely into small cities and towns.

The second main belt of industrial development in the South lies along and near the Gulf Coast from Corpus Christi, Texas, to New Orleans and Baton Rouge, Louisiana. The chemical industry, including oil refining, is the dominant industrial sector within this *West Gulf Coast Industrial Belt.* The industry is based on abundant deposits of oil, natural gas, sulfur, and salt, often occurring together in *salt domes* comprised of uparched sedimentary strata overlying huge masses of rock salt. The oil, gas, and sulfur occupy pockets in the domed strata. About 300 of these domes are scattered along the Gulf Coast, including many underneath the adjacent Gulf of Mexico. Often they are so deeply buried as not to be perceptible at the surface, but elsewhere they rise as low hills. The states of Texas and Louisiana produce over two fifths of the oil, two thirds of the natural gas, and practically all of the native sulfur output of the United States, with this Gulf Coast strip being the leading area of production for all three (1986 data). Much of the oil and gas output comes from offshore wells in the Gulf. The area is also an important producer of salt, an indispensable chemical raw material produced in quantity in several other parts of the United States. Machinery and equipment for oil and gas extraction, oil refining, and chemical manufacturing are also important products here. Major clusters of oil refineries and petrochemical plants based on oil and/or natural gas occur at several places, most notably the metropolitan areas of Houston, Texas; Beaumont, Texas; and Baton Rouge, Louisiana. Chemical manufacturing based on salt or sulfur is also very important.

Many other industries are present in the South, some of which are widespread while others are localized in a few places. Food-processing, machinery, and pulp-and-paper industries are widely distributed and important to many communities, but they do not form definable major concentrations. Food industries are especially important in Florida, with citrus concentrates the leading single item by value. Pulp-and-paper plants are particularly prominent near the Gulf Coast from eastern Texas to Florida and along the Atlantic Coast in Georgia. For raw material this industry depends largely on the pine trees that grow rapidly in the wet subtropical climate. Boats are built at many places along the Southern coasts, with major shipyards in the Norfolk, Virginia, metropolis and on the Mississippi coast. Aircraft and aerospace industries are prominent in the Atlanta and Dallas-Fort Worth areas, and both auto assembly and the manufacture of auto components are expanding, especially in Tennessee and Kentucky. Japanese and American companies have recently been investing large sums in entirely new auto plants in the South.

METROPOLITAN GROWTH IN THE SOUTH: RETARDED BUT ACCELERATING

The largest metropolitan areas of the South are not yet comparable in population to those of the Northeast, California, or the Midwest. In 1986 Dallas-Fort Worth (3,655,000) and Houston (including Galveston and other cities: 3,634,000) were the largest Southern metropolises. But they ranked only eighth and ninth in the United States, behind three metropolises in the Northeast, two on the Pacific Coast, and two in the Midwest. Nine Southern metropolitan areas had more than a million people each, but their combined population of about 20 million was only 2 million greater than that of metropolitan New York.

The retarded growth of large metropolitan areas in the South results from the dispersion of urban development among many small regional centers through centuries of history when the South's economy was overwhelmingly agrarian. The economic system was so decentralized, locally self-sufficient, and lacking in industrial development that masses of people did not congregate in centers of major size. Large-scale urbanization based in part on the South's trade and products did occur, but it took place in the North and in Great Britain. This situation persisted until World War II, but it has been changing rapidly since the war. Fast urban growth is characteristic of the new Southern economy oriented to the provision of manufactures and services, while slow growth is now widely characteristic of the cities of the North. The South is still very far from achieving a huge metropolis like New York, Los Angeles, or Chicago but instead is developing a large number of smaller but fast-growing metropolitan areas.

Most Southern metropolitan areas of 500,000 population or over are clustered into three major metropolitan zones: a *Texas-Oklahoma Metropolitan Zone*, a *Florida Metropolitan Zone* in the Florida Pen-

insula, and a *Southern Piedmont and Coastal Virginia Metropolitan Zone* stretching across Georgia, the Carolinas, and southern Virginia from Atlanta to Norfolk. Large areas of the eastern and southern Coastal Plain, the Appalachians, and the Interior Highlands have no urban clusters of this scale, although sizable metropolises do occur in scattered locations outside the major zones.

Texas-Oklahoma Metropolitan Zone

The Texas-Oklahoma Zone contains the two largest urban clusters in the South—Dallas-Fort Worth and Houston—plus four other urban concentrations of 500,000 or over: San Antonio (1.3 million) and Austin (725,000), Texas, and Oklahoma City (985,000) and Tulsa (735,000), Oklahoma. *Houston* is a seaport by virtue of the Houston Ship Channel, a federally financed waterway constructed in the late 19th and early 20th centuries. This access route for ocean shipping leads inland across the flat outer Coastal Plain to an artificial harbor at Houston. The city is well situated to serve Texas and the southern Great Plains, and it has capitalized on this location to become a major port. After the discovery of Gulf Coast oil in 1901 at Beaumont east of Houston, the growth of Houston speeded dramatically and the city became the principal control, supply, and processing center for the oil and gas fields and chemical industry of the Gulf Coast.

The other three Texas metropolitan areas with over 500,000 people—*Dallas-Fort Worth* (Fig. 28.16), *San Antonio,* and *Austin*—developed at the western edge of the Coastal Plain in a north-south strip of territory known as the Black Waxy Prairie. Here limestone-derived soils, unusually fertile for the South, supported the development of a productive agriculture and towns and cities to service it. Dallas-Fort Worth and San Antonio also extended their reach westward into the farming and ranching country of the Interior Plains. Like Kansas City and Minneapolis-St. Paul farther north, they serve large trading hinterlands stretching far to the west, within which there are no comparable competing centers. In the case of Dallas-Fort Worth the same is true in other directions except for the competition of Houston to the south. Aside from Houston, urban centers at all comparable in size and importance to Dallas-Fort Worth are Phoenix to the west, Denver to the northwest, Kansas City to the northeast, and Atlanta to the east (Fig. 26.1). Hence Dallas-Fort Worth is the major business center for a huge region. San Antonio is a major military base and retirement center, and Austin is the seat of the strongly funded University of Texas at Austin and a developing center of

Figure 28.16 Anglo-America's huge structure of business enterprise is largely directed from offices in impressive skyscraper structures such as the ones shown here in the central business district (CBD) of Dallas, Texas.

high-technology industries. All these cities are involved in many ways with the huge oil and gas industry of Texas. *Oklahoma City* and *Tulsa* are much smaller than Dallas-Fort Worth, but they combine on a smaller scale the same emphases on agricultural servicing, oil-based activities, and trade with an extensive commercial hinterland.

Florida Metropolitan Zone

A second area of exceptional metropolitan development in the South is the Florida Peninsula. In 1986 two Florida metropolitan areas had estimated populations of over a million—*Miami* (2.9 million) and *Tampa* (1.9 million)—and three more had over 500,000 each: *Jacksonville* (850,000), *Orlando* (900,000), and *West Palm Beach* (750,000). The West Palm Beach metropolis is just north of the Miami area. The two are not considered sufficiently integrated to form a single consolidated metropolitan area, but together they occupy a long narrow strip of highly urbanized territory aggregating over 3½ million people along Florida's southeast coast.

The distribution of Florida's main metropolitan areas reflects the importance to the state of amenities attractive to tourists and retirees. Proximity to the ocean is important, but winter warmth is even more crucial. For example, Miami's marginally tropical winters have undoubtedly helped that south Florida city grow faster than Jacksonville in the extreme northeast. In addition to the tendency for the greatest growth to occur in the south of the peninsula, there also is an obvious tendency for metropolitan development to hug the coasts. Florida's only inland metropolis of over half a million is Orlando, where

growth has been strongly spurred by the placement of Disney World there. If Florida's spectacular population surge continues, the pattern of southerly and seaside locations may be altered somewhat, as many smaller urban centers, some inland and some in the northern part of the state, are growing quite rapidly.

Florida's cities are by no means totally devoted to vacation and retirement functions. For example, Miami has become a major point of contact between the United States and Latin America in regard to air traffic, tourist cruises, banking, and the drug traffic. Tampa Bay affords a major harbor well placed for the trade of central Florida's citrus belt and important phosphate-mining area. Jacksonville is also a seaport. Both high-technology industries and regional corporate offices are increasingly attracted to Florida as the state's population grows and improved transportation and communication lessen the disadvantages of peripheral locations within the United States.

Southern Piedmont and Coastal Virginia Metropolitan Zone

The third major Southern concentration of metropolises over 500,000 lies along the Piedmont, Fall Line, and nearby Virginia Coastal Plain. Here most of the main metropolitan areas are Piedmont and Fall Line cities with 1986 estimated populations ranging from nearly 600,000 to a little over 1 million. Areas of this scale include *Richmond* (810,000), Virginia; *Charlotte* (1.1 million), *Greensboro* (900,000), and *Raleigh* (650,000), North Carolina; and *Greenville* (600,000), South Carolina. Metropolitan areas in this group have tended to form by the coalescing and

711

integration of cities located near each other. Hence there is a notable tendency toward multiple-centered metropolises, although each of the latter is identified here by the name of its leading central city. Such places, as well as some smaller ones along the Piedmont and Fall Line, represent the growth to metropolitan status of former mill towns which are now diversifying their economies into new types of manufacturing and more important commercial and service functions than those that existed previously.

The characteristics previously cited apply more to the heart of this Metropolitan Zone than to its southern and northern extremities, where the two largest metropolises are found. Both cities differ from the norm, not only in size but also in function. In the south *Atlanta* (2.6 million), the capital of Georgia, is by far the largest urban center of the Southern Piedmont and ranks fourth in metropolitan population within the entire South (after Dallas–Fort Worth, Houston, and Miami). Although its metropolitan area contains diverse industries, the city stands out within the South as a major inland transport center. Railroads and highways have generally avoided the Great Smoky Mountains, which lie just north of Atlanta. Instead, they skirt the southern end of the mountains and converge on the city. Superior location and transport facilities have been exploited to make Atlanta the largest Southeastern business and governmental center. Its ascendancy as a regional transport center today is perhaps most clearly seen in its airport, which is the second busiest airport in the United States (behind Chicago's O'Hare) as measured by volume of passenger traffic.

The other standout metropolis in this Metropolitan Zone is the seaport of *Norfolk* (1.3 million), Virginia, located in the Coastal Plain on the spacious Hampton Roads natural harbor at the mouth of the James River. The metropolitan area, which includes Norfolk city on the south side of the harbor and the smaller port of Newport News on the north side, is quite different in function from the typical Piedmont city. Its most significant commercial function is the shipment of coal from Appalachian fields by sea. But more important is its role as the main naval base for the Atlantic Fleet. Numerous functions connected with the Navy help support the area: shipbuilding, ship repair, administration, communications, and the servicing of Naval personnel and their families.

Metropolises of the Upland South

The Upland South (Appalachians, Unglaciated Southeastern Interior Plain, and Southern Interior Highlands) contains only four cities that had estimated metropolitan populations of over 500,000 in 1986. Widely separated from each other, these cities include Louisville, Kentucky; Nashville and Knoxville, Tennessee; and Birmingham, Alabama. *Louisville* (965,000), the largest, began as a river port at the low "Falls of the Ohio," where goods had to be transshipped around this break in transportation. It was also the closest point of contact on the Ohio for the state's first-settled and most productive agricultural area, the Kentucky Bluegrass Region of limestone soils immediately to the east of Louisville. By the time a canal had been built to bypass the falls (1830), Louisville was a leading place on the river, and it has continued to grow as a manufacturing and regional commercial center.

Two upland cities in Tennessee—Nashville and Knoxville—also grew in locations related to river transport and limestone soils which are relatively superior within the hill-country framework. *Nashville* (930,000) is sited where the Cumberland River, a tributary of the Ohio, extends farthest south into the hilly Nashville Basin of central Tennessee. *Knoxville* (590,000) is on the Tennessee River in an area of limestone valleys. The city is the regional economic capital for the Tennessee sector of the Appalachian Ridge and Valley Section. This area and the adjacent Blue Ridge (Great Smokies) contain numerous reservoirs behind dams constructed by the Tennessee Valley Authority on the Tennessee and its tributaries.

Farther south, *Birmingham* (910,000) is a very different type of city. Located near the southern end of the Ridge and Valley Section and adjacent Cumberland Plateau, metropolitan Birmingham grew as a steel-producing center at a place where low-cost production was facilitated by large deposits of coal and iron ore found close together. In its basic industrial pattern, Birmingham resembles the old heavy-industrial centers of the Northeast and Midwest, and it is the only large city of this type in the South. Today its plants rely on foreign ore, as the local ore is uneconomical to work, but coal for the furnaces still comes from the nearby Cumberland Plateau. By intensifying its role as Alabama's main center of trade and services, Birmingham has greatly lessened its dependence on a struggling steel industry, which is ridden by the same problems of obsolescence, high production costs, increasing foreign competition, and high unemployment that have recently beset the American steel industry generally.

Southern Metropolises on the Mississippi River

New Orleans and Baton Rouge, Louisiana; Memphis, Tennessee; and Little Rock, Arkansas, round out the list of Southern metropolitan areas with estimated populations of one half million or over in 1986. The first three are on the Mississippi River, and Little Rock lies on the Mississippi's large tributary, the Arkansas.

New Orleans (1.3 million) is located just upriver from the head of the "bird's-foot" delta where the Mississippi divides into separate channels to the Gulf known as "Passes." It is a very old city, founded by Imperial France in fur-trading days as a gatekeeper and port at the Gulf end of New France's great internal water system formed by the St. Lawrence River, the Great Lakes, and the Mississippi River and its tributaries. New Orleans became a major United States port in the river-boat era before the Civil War and for a time was New York's chief competitor in seaborne commerce. Its relative importance declined in the later 19th century when the consolidation of a national railnet directed most overseas commerce to the Northeastern ports on the Atlantic. But in recent decades New Orleans has risen once more to a position among the top American ports, being especially notable for huge shipments of grains and soybeans from the Midwest and Mississippi Alluvial Plain. Intensified barge traffic on the Mississippi and its tributaries has contributed greatly to the port's resurgence, as has a growing traffic in containers carried to and from the city by rail and interstate highway. Remnants of the French era, in the preserved "French Quarter," help make New Orleans a major tourist and convention center.

Somewhat under 100 air miles upriver from New Orleans is *Baton Rouge* (550,000), which has an unusual combination of functions in being not only the state capital and site of a major university (Louisiana State), but also an important center of heavy industry—in this case oil refining and associated petrochemical manufacturing. In addition to a huge refining and petrochemical complex at Baton Rouge itself, other plants are spotted along the Mississippi between the city and New Orleans. These industries, which are not associated with any oil or gas field in the immediate vicinity, are serviced by pipelines, barges, medium-draft seagoing tankers, railroads, and interstate highways. The entire array of plants is one of the country's largest refining and petrochemical clusters.

About 300 air miles to the north of Baton Rouge, *Memphis* (960,000) occupies a location where the Mississippi flows next to higher ground at one side of its broad flood plain. The city's present function as the main business center for the greater part of the Mississippi Alluvial Plain (and other areas in western Tennessee and northern Mississippi) dates back to antebellum days when a famous sector of the Cotton Kingdom developed in "The Delta" near Memphis. As used at that time, the term applied

particularly to the slave-plantation country on the Plain in northwestern Mississippi. Like many other regional economic capitals scattered over the South, present-day Memphis has a diversity of manufacturing industries, the great majority of which have been established since the Great Depression.

Little Rock (500,000) lies on the border between the Mississippi Alluvial Plain and the Interior Highlands and is an important business center for both. It is the capital of Arkansas, and its growth has been especially spurred by the great agricultural productivity of its service area in the Alluvial Plain.

THE MIDWEST: AGRICULTURAL AND INDUSTRIAL PRODIGY

The existence of a Midwest (also called Middle West) is widely recognized in the United States, but perceptions differ as to its extent. People in Nebraska, for example, might think of themselves as Midwestern but not consider people in Ohio to be, whereas Ohioans might be of the opposite opinion. In this chapter the term *Midwest* is applied to 12 states called in federal publications the *North Central states.* They are commonly subgrouped into the East North Central states of Ohio, Indiana, Illinois, Michigan, and Wisconsin, and the West North Central states of Minnesota, Iowa, North Dakota, South Dakota, Nebraska, Kansas, and Missouri. The two subgroups will be referred to here as the *eastern Midwest* and the *western Midwest.* The Mississippi River separates them except for the northern part of the Wisconsin-Minnesota boundary (Fig. 28.17). All states in the eastern Midwest have a frontage on one or more of the Great Lakes, but in the western Midwest this is true only of Minnesota, which borders Lake Superior in the north. In the eastern Midwest, the states of Ohio, Indiana, and Illinois are bounded on the south by the Ohio River, while in the western Midwest all states except Minnesota have long frontages on the Missouri River.

Use of the name Midwest for the 12 states just listed is arbitrary to some degree but can be justified by an impressive list of spatial (mappable) characteristics distinguishing these states as a group from those of the Northeast, South, and West. Among such characteristics are the Midwest's (1) interior "heartland" location; (2) distinctive plains environment; (3) critically important associations with major lakes and rivers; (4) areal differentiation in regard to various ethnic elements that have poured into the region during different eras; (5) outstandingly productive combination of large-scale agriculture, industry, and transportation; and (6) rural landscapes

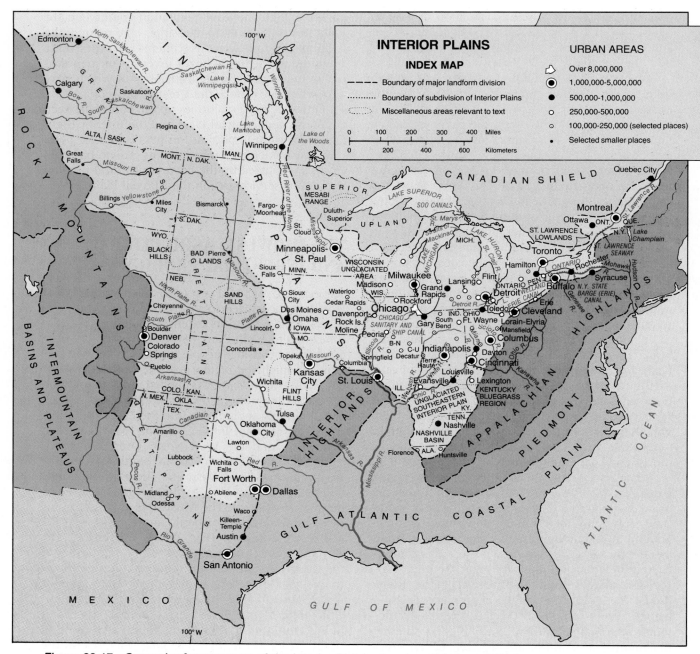

Figure 28.17 *General reference map of the Interior Plains, showing all but a relatively small part of the Midwest as discussed in this chapter. (See also Figures 27.2, 28.1, 28.2, and 28.10. For identification of cities shown by symbol but not by name, see Figure 28.1.) In Illinois B-N is Bloomington-Normal; C-U is Champaign-Urbana.*

famed the world over for their rectangularity, symmetry, and prosperous appearance.

Some question might be raised about the inclusion of the state of Missouri in the Midwest rather than the South. Before the Civil War Missouri was a slave state and it had some development of slave plantations. But this development was relatively weak, and during the war the state's people were so bitterly divided that effective support could not be given the Southern Confederacy. Today the state still has some demonstrably "Southern" characteristics culturally, but large elements in its population do not share these traits, and economically the state fits much better with the Midwest than with the South. The case of Missouri well illustrates the transitional character of the broad border zones that separate the major regions defined in this chapter on the basis of entire states.

THE MIDWEST IN A WORLD AND NATIONAL COMPARISON

Not only is the Midwest comparable to the other major American regions in being bounded by zones of transition, but it is also like them in having the size and production that would be expected of a major nation. Encompassing about 770,000 square miles (*c.* 2 million sq. km), the 12-state Midwest is surpassed in area only by the United States as a whole and 12 other countries. Its area is closely comparable to that of Mexico and is over five times that of Japan. The Midwest's population of about 60 million (1986 estimate) is exceeded only by the populations of the United States and 13 other countries. The region has about the same population as Europe's most populous country, West Germany.

In present output of certain kinds of goods—for example, corn, soybeans, pork, and farm machinery—the Midwest is not only the leader among the four major American regions surveyed in this chapter, but is unsurpassed by any nation other than the entire United States. Within the United States the Midwest is the largest regional producer of both agricultural commodities and manufactured goods, and it has an extraordinary web of transportation lines that is crucially important in binding the Northeast, South, and West together. This transport system focuses on the Midwest's greatest metropolis, Chicago. But perhaps the most notable geographic attribute of the region is its nearly solid occupance by intensive economic development. The continuously productive character of its broad plains contrasts sharply with the island-like character of intense development in the Northeast, South, and West.

THE WORLD'S MOST PRODUCTIVE PLAINS

By far the greater part of the Midwest is classified as plains, and this region forms a major segment of the huge Interior Plains of Anglo-America (Fig. 28.17). According to Edwin H. Hammond's standard landform classification, irregular surfaces predominate in the Midwestern plains, wherein 50 to 80 percent of the land is gently sloping or flat and the rest is steeper.[3] Most of this irregular terrain is level enough to be cropped, although erosion-control practices such as contour plowing and strip cropping are widely used to prevent loss of soil. Irregular plains are the most common type of terrain within the highly productive Midwestern agricultural areas west of the Mississippi River, and they also are dominant in western Illinois and some other parts of the eastern Midwest. But the Midwest also contains large areas classified as smooth plains or flat plains, where the surface is more nearly level and the local relief is less. Such areas of exceptional smoothness are widely characteristic of western Ohio, central and northern Indiana and Illinois, the Lower Peninsula of Michigan, western Minnesota, and the eastern part of the Dakotas. Expanses of relatively level land also are found in some other locations, including one very big area in the Great Plains of western Kansas. For the most part the smoother terrain in the Midwest has resulted from glacial deposition that buried more irregular previous terrain under a mantle of glacial debris (ground moraine or outwash). The very flattest areas often are *lacustrine plains*—the floors of former lakes where glacial meltwater accumulated and sediments washed in and settled. The largest such area is the floor of old Lake Agassiz in Minnesota, the Dakotas, and the Canadian province of Manitoba. Lake Winnipeg in Manitoba is a remnant of this gigantic glacial lake. Other prominent lacustrine plains border lakes Michigan and Erie, and represent overflow areas into which the two lakes expanded when glacial melting was at its height. The utter flatness of these old lake beds hindered natural drainage and resulted in many swamps. Artificial drainage has yielded some exceptionally productive farm land and also the sites on which the large cities of Chicago, Detroit, and Toledo have been built despite continuing problems of drainage and flash flooding.

Rougher topography is most common near the margins of the Midwest. Only two small areas are high and rough enough to be considered mountainous: the *Black Hills* of South Dakota (and Wyoming), and a narrow strip of the *Canadian Shield* (Superior Upland) immediately marginal to Lake Superior in Minnesota. Other notably rough areas along the margins of the region include the dissected *Ozark Plateau* of southern Missouri, the dissected *Appalachian Plateau* in eastern Ohio, the fantastically eroded South Dakota *Bad Lands* and other hilly parts of the western Dakotas, the extensive *Sand Hills* of northern Nebraska, the narrow band of *Flint Hills* extending north-south across east-central Kansas, and the ''*Knobs*'' of southern Indiana that form a narrow belt from just south of Indianapolis to the Ohio River. Away from the margins of the region, still other distinctly hilly areas occur in scattered places. One of these is the *Unglaciated Hill Land* of southeastern

---

[3]Edwin H. Hammond, *Classes of Land-Surface Form in the Forty Eight States, U.S.A.*, Map Supplement No. 4, *Annals*, Association of American Geographers, Vol. 54, No. 1, March, 1964. Scale: 1:5,000,000. An adaptation of the map can be found in U.S. Geological Survey, *The National Atlas of the United States of America* (Washington, D.C.: 1970). All serious students of United States geography should become acquainted with this outstanding portrayal of the nation's landforms.

Wisconsin, which seems to have escaped the most recent (Wisconsin) glaciation, although it bears some evidence of earlier glaciation. Other hilly areas are found in the sandy country of Michigan's northwestern Lower Peninsula and in southwestern Iowa and parts of northern Missouri. In the latter two areas glaciation occurred so early that streams have had time to dissect the glacial plains into relatively gentle hills. There also are many hilly strips, or actual cliffs, bordering the flood plains of Midwestern rivers. Here the land has been trenched with valleys by tributaries descending to the main rivers.

MIDWESTERN AGRICULTURE IN ITS ENVIRONMENTAL SETTING

These vast Midwestern plains produce a greater output of foods and feeds than any other area of comparable size in the world. The Midwest commonly accounts for over two fifths of all United States agricultural production by value. This output is unevenly spread among the 12 Midwestern states, with Iowa well ahead of the others. Within the nation Iowa is generally the second-ranking state in value of farm marketings, being exceeded only by much larger California with its extraordinarily productive irrigation-based agriculture, and by Texas in some years. In recent years Iowa has accounted for around 7 percent of the nation's farm output. The next five states in the national ranking—Illinois, Nebraska, Minnesota, Kansas, and Wisconsin—are all Midwestern. Individual state rankings shift from year to year, with each state producing 3 to 5 percent of all United States farm output. Total output falls off somewhat in the six remaining Midwestern states, although three of them—Indiana, Missouri, and Ohio—still rank in the nation's top 15 states. Indiana is the Midwest's smallest state by area and has quite a bit of rough land in its southern part. The southern half of Missouri lies in the agriculturally unrewarding Ozark Plateau, and the eastern half of Ohio is in the similarly handicapped Appalachian Plateau. The final three states—Michigan and the Dakotas—have even more pronounced environmental handicaps. Michigan has a great deal of infertile sandy land in its center and north, while North and South Dakota are handicapped by dryness and by rough topography toward the west.

The great agricultural productivity of the Midwestern plains has been made possible by a remarkable combination of natural conditions and resources allowing profitable cultivation of a very large proportion of the land. Gentle slopes are one element, but favorable climatic and soil conditions are also important. The *humid continental long-summer climate* predominates. Long, hot, and generally wet growing seasons are combined with winters that yield a good deal of additional moisture in the form of snow, together with enough freezing weather to retard leaching, erosion, and pests. In the north a broad strip of *humid continental short-summer climate* stretches across an extensive area in the Dakotas and the states of Minnesota, Wisconsin, and Michigan. Here the shorter and cooler summers are less favorable to cropping. In the western Dakotas and in western Nebraska and Kansas precipitation is low enough for the climate to be classified as *steppe*. This semiarid climate in the Great Plains permits a valuable output of crops such as wheat and grain sorghums, but yields are lower and crop possibilities are far less than in the more humid plains to the east.

Within most of the Midwest, soils are exceptionally good. The best soils developed under grasslands, which supplied abundant humus. Such outstandingly fertile soils are a normal accompaniment of steppe climates in the middle latitudes, but the fertility is counterbalanced by precipitation so low as to be marginal for agriculture. The remarkable feature of Midwestern soil development was that the soils of much of the humid continental climatic area (both long-summer and short-summer) developed under a tall-grass *prairie* (Fig. 28.18). These climatic conditions normally produce a natural vegetation of forest, and trees will grow well if they are planted. But in the early 19th century the first settlers found large expanses of prairie, or a prairie-forest mixture, as far east as northwestern Indiana. From here the

Figure 28.18   A remnant of undisturbed tall-grass prairie is seen here in the Flint Hills of eastern Kansas. The site, known as Konza Prairie, has a protected status. The view was taken in mid-April.

Figure 28.19   Banks of loess near the Missouri River. The photo was taken from Interstate 70 a few miles west of Columbia, Missouri.

prairie region extended west in a broadening triangle to the eastern parts of the Dakotas, Nebraska, and Kansas, and on into the South in Oklahoma and Texas. Of the Midwestern states, only Michigan and Ohio lacked considerable expanses of prairie. Thus a very large part of the Midwest had soils of a richness generally found only in semiarid areas but combined here with humid conditions more favorable to most crops. The existence of the anomalous triangle of prairie, bordered by forests both north and south, has been accounted for in two main ways. One explanation holds that prairie resulted from periodic drought conditions discouraging tree growth in this normally humid area. The other explanation suggests that Indians may have fired the forest to create grazing range for game. Possibly both factors were important, and prairie fires set by lightning may also have helped perpetuate the prairie once it was established (grass is less susceptible than tree seedlings to lasting damage by fire). Whatever the causes, the occurrence of grassland soils in a humid area where forests might normally be expected provided an extraordinarily favorable agricultural environment.

Another natural condition contributing to the unusual excellence of many Midwestern soils is the widespread occurrence of *loess* deposits made of dust particles (Fig. 28.19). These wind-laid soil materials are thought to have accumulated in glacial times when strong winds picked up particles from dry surfaces where the ice had melted but a protective mantle of vegetation had not yet become established. Since most of the nutrients that nourish plants are contained in the finer soil particles, loess is associated with soils of exceptional fertility in various parts of the world. In the Midwest, Iowa and Illinois are particularly characterized by such soils, but they are also found in some other areas. In the eastern Midwest, the soils formed under natural broadleaf forests

are not exceptionally fertile by nature, but neither are they poor (Fig. 28.20). The principal areas of poor soil in the Midwest lie in (1) the uplands of the Ozarks and the Appalachian Plateau; (2) sandy areas of needleleaf coniferous forest in the northern parts of Michigan, Minnesota, and Wisconsin; and

Figure 28.20   Amish farmer turning the soil with a horse-drawn plow in north central Ohio. Here a major Amish concentration centers in Wayne and Holmes counties near Akron and Massilon. While Lancaster County, Pennsylvania, remains the most famous area of Amish settlement, there are many other settlements in scattered parts of the United States. Deeply traditional cultural practices persist in such areas, just as they do in Lancaster County. The soil developed under broadleaf forest in the Ohio area is not first-rate, but it has been made highly productive by skillful management.

(3) rougher western parts of the four westernmost states.

After many decades of use, the present quality of Midwestern soils depends heavily on how they have been handled. Where properly fertilized and protected against depletion and erosion through the use of crop rotation, plowing on the contour, and other conservation measures, these soils have proved remarkably durable. Although unwise practices have caused much soil damage, huge expanses of Midwestern land remain highly productive. Aided by large inputs of modern agricultural science and technology, the soils of the Midwest as a whole are producing higher yields today than ever before.

The Legendary American Corn Belt

Corn (maize) is the single leading product of Midwestern farms, and the *Corn Belt* where this crop dominates the landscape has become an agricultural legend for its continuous display of high productivity. Cash receipts from sales of beef cattle in the 12-state Midwest are greater than receipts from corn, but the cattle are grown largely by feeding Midwestern corn, and the total value of corn produced on Midwestern farms exceeds that of cattle sold. Aside from an occasional subpar production year in some states, corn is the most valuable crop in every Midwestern state except Missouri (where soybeans lead, with corn second) and Kansas and North Dakota (where wheat leads overwhelmingly). In the mid-1980s, over four fifths of American corn production, and around two fifths of world production, was coming from the 12 Midwestern states. Highly favorable conditions for large-scale corn growing are provided by the Midwest's fertile soils, hot and wet summers, and huge areas of land level enough to permit mechanized operations on large farms. The corn plant itself and the methods of production have been greatly improved by extensive research. Today corn is primarily a cash crop, sold both in the United States and overseas for animal feed, industrial uses, or human food. But the crop also is fed to livestock on many of the producing farms.

LANDSCAPE GEOMETRY OF THE
CORN BELT AND ITS ORIGINS

The world area of most intensive corn production, long called the American Corn Belt, is limited on the east and south by rougher country, on the north by cooler summers and poorer soils, and on the west by semiaridity. This continuously productive belt extends over 800 miles (*c.* 1300 km) east-west and 300 to 600 miles (*c.* 500–1000 km) north-south. It reaches from western Ohio to eastern South Dakota,

Nebraska, and Kansas, including most of Indiana and Illinois, all of Iowa, and parts of Missouri, Minnesota, Michigan, and Wisconsin. The Corn Belt is famous for its rectangular landscape, created originally under federal laws requiring that federally owned land—the *public domain*—be surveyed into *townships* 6 miles square, *sections* 1 mile square, and *quarter-sections* one half mile square. The survey lines ran north-south and east-west, and most roads came to follow section boundaries. This system of rectangular survey, prescribed by the Confederation Congress under the Land Ordinance of 1785, imposed a uniformity on most of the United States that was absent in the earliest areas of American settlement, where various local survey systems were employed. As occupation of the land proceeded westward, the greater part of the public domain was disposed of by sale or gift to settlers, the most famous instance being the Homestead Act of 1862 under which each new settler family was granted a quarter-section (160 acres/65 ha) of free land. The new landholders then organized their plots into rectangular fields, most commonly 40 acres in size. A great deal of farm land in the Corn Belt already had been acquired by settlers before the Homestead Act was passed, but rectangularity was nearly universal in these earlier holdings and has persisted. Seen from the air, today's great sweep of Corn Belt rectangles, variegated in tone according to use and season, is an overpowering sight.

At ground level in areas of former prairie, Corn Belt farmsteads, with their planted shade trees and windbreaks, form visual "islands" spaced regularly through expanses of crop fields and pastures spreading to the horizon. In naturally forested areas of the eastern Corn Belt, rectangular woodlots are spotted through the landscape. Outside the Corn Belt, rectangularity is pronounced over vast sections of the United States settled after 1785, but such areas seldom duplicate the intricate intensity of the Corn Belt checkerboard. In many irrigated areas, including some in the western Midwest, the huge dots produced by center-pivot sprinkler irrigation have imposed a new kind of landscape geometry. Farmers in western Nebraska, for example, now feed underground water through sprinklers to thousands of these verdant circular fields.

THE INCREASING IMPORTANCE OF SOYBEANS

Soybeans, grown in the same areas as corn, are another major element in Corn Belt agriculture (Fig. 28.13). Long cultivated in the Orient (especially China), they have had a rapidly increasing impact on American and world agriculture during recent decades. The versatile soybean plant gives high yields, fixes nitrogen in the soil, and can be used in a great variety of ways. Both the plant and the beans provide livestock feed, and oil pressed from the beans is used in a wide and increasing range of foods and other products. The residue from oil extraction, known as soybean cake or soybean meal, is itself a nutritious feed. Growing well wherever corn grows well, soybeans have turned the Corn Belt into a region that would be more realistically named the Corn and Soybean Belt. About two thirds of the nation's soybeans are harvested in the Midwest, with the other third coming largely from the South. The United States is by far the leading world producer and exporter of soybeans, followed by Brazil.

Dairy Belt and Wheat Belts: Adaptations to Restrictive Environments

The northern and western margins of Midwestern agriculture do not share the corn-soybean-livestock emphasis of the Corn Belt. The agriculture of Wisconsin, most of Michigan, and much of Minnesota has adapted to cooler summers by specializing in dairy farming. This part of the Midwest lies in the western sector of what has long been called the *Dairy Belt*. The eastern sector of the Belt stretches across the northern part of the Northeast and adjacent southern Canada. Corn is widely grown in the Dairy Belt, but most of it is chopped in an immature state for silage rather than being harvested for grain. Much more of the land is kept in hay crops and pasture grasses to feed dairy cattle. Wisconsin, the country's leading dairy state, was producing around a sixth of the national output of fluid milk in the mid-1980s (see chapter-opening photo of Chapter 26 and Fig. 28.13). Its principal market is the huge Chicago metropolis.

To the west of the Corn and Dairy Belts, lessened precipitation focuses agriculture on wheat and ranching. Here central and western Kansas form the heart of the *Winter Wheat Belt*. It extends south into Oklahoma and Texas, and west into Colorado, but Kansas alone normally produces about a sixth of the country's wheat. In similar fashion, the *Spring Wheat Belt* centers in North Dakota, although it extends beyond that state in all directions. North Dakota is the second-ranking wheat state, with a little under one eighth of the national production. Wheat farming shares the two wheat belts with ranching, and cattle are a more important source of farm income than wheat in many localities. Feed crops adapted to minimal moisture, such as sorghums in Kansas and barley in North Dakota, support livestock production. Between the two premier wheat states, Nebraska and South Dakota grade from Corn Belt sections in the east to wheat and ranching in the west. In Nebraska the Corn Belt type of production extends particularly far west, due to large-scale irrigation from an abundant supply of underground water.

The Midwestern Emphasis on Livestock

In the Midwest as a whole, sales of animal products exceed sales of crops in value. In a recent year animal products represented 54 percent of all farm marketings from the 12 Midwestern states. Sales of beef cattle, including calves, are the largest source of income from animal products, with Nebraska, Kansas, and Iowa the leading Midwestern states. They are, in fact, the leading states in the nation except for Texas, which ranks first by a wide margin.

In marketing of hogs, the Midwest occupies an even more impressive national position. About four fifths of all marketings come from the 12 Midwestern states, with Iowa alone accounting for about one quarter of national sales. The Dairy Belt states of Wisconsin, Minnesota, and Michigan market a little over one quarter of the nation's milk.

Local Agricultural Specialties

Many local areas in the Midwest depend heavily on agricultural specialties that are less important in the total picture of Midwestern agriculture than are corn, soybeans, wheat, beef, pork, or milk. For example, a fruit belt is found in Michigan along the shore of Lake Michigan. It represents a response to climatic effects of the lake which make the beginning and end of the growing season more reliable by lessening the chance of untimely freezes. Cherries are the most important product. North Dakota has become the nation's leading state in growing sunflowers, cultivated for the oil pressed from their seeds. This production comes from rich flatlands near the Red River in the former bed of glacial Lake Agassiz. The sunflower district laps over into adjacent Minnesota, which is the second-ranking state in this crop. The bi-state district along the Red River is also prominent in sugar-beet and potato growing. Beans, sweet corn, berries, and asparagus are examples of still other crops grown commercially in local districts spotted through the Midwest.

The Small and Declining Farm Population

The Midwest's large agricultural output is produced by a small and decreasing minority of the region's people. According to the U.S. census of 1980, South Dakota had a higher proportion of farm population than any other state in the Union, but this only amounted to 16 percent of the state's people. North Dakota, Iowa, and Nebraska were the next three states in the national ranking. The other Midwestern states ranged downward to Michigan, where the farm population was only 1.9 percent of the total. But Michigan was the only Midwestern state that fell below the national figure of 2.5 percent in 1980, and the Midwest as a whole was well ahead of any other major region (Midwest 4.9%, South 2.3%, West 1.4%, Northeast 0.7%). The proportion of farm population was highest of all in the West North Central states (9.1%), not because of a dense farm population, but because of a relatively low degree of urban development. But the average Midwesterner is not a farmer, even in South Dakota, and many of those who are farmers also hold nonfarm jobs. The region's prodigious agricultural output comes overwhelmingly from large and medium-sized farms owned and/or operated by a remarkably small number of full-time farmers.

Of course, indirectly a much greater proportion of the Midwest's people contribute to the region's farm output. Very large sectors of the nonfarm economy supply the capital-intensive agriculture with equipment, chemicals, and services. Other sectors process and market farm products. Most Midwestern cities and towns began as nodes to service agriculture, and such activities still provide a large part of the business of many. Hence agriculture is a much larger part of the Midwestern economy than is indicated by the number of people on farms. This number has been decreasing for decades due to economic pressures and increasing opportunities for nonfarm employment.

The Risky Nature of Midwestern Agriculture

Despite its impressive resources and accomplishments, Midwestern agriculture is a risky enterprise, as is any capitalist business in a highly competitive market. Natural hazards such as drought and plant diseases increase the normal business risk, but the main hazards are sharp fluctuations in world supply and demand and therefore in market prices. Adverse natural and economic conditions in varying combinations have forced waves of farmers out of business at various times in the past. The latest crisis of this kind began in the early 1980s, but its roots are found in events of the previous half-century. Between 1930 and 1982 the number of farms in the Midwest decreased from slightly over 2 million to somewhat over 900,000. Most of the land of defunct farms was bought or rented by surviving farmers who expanded their acreage. Hence the amount of land in farms decreased during this half-century by only 6 percent, while the mean size of farms increased from about 180 acres (73 ha) to about 380 acres (154 ha). Increased mechanization allowed farmers to work larger acreages but also increased their capital costs and hence their vulnerability to low market prices for their products. This vulnerability became highly evident during the 1970s and 1980s, when a major upturn in farmers' fortunes was followed by an abrupt and drastic downturn. Between 1970 and 1980, strong foreign demand and vigorous promotion of agricultural exports by the American government led to a tenfold increase in United States corn exports and a fivefold increase in soybean exports. Farmers acquired more land and machinery by borrowing at high interest rates. Then in 1981 an economic slide began. A world economic recession led to lower demand in many countries; financial policies in the United States created a "strong" dollar, which made American products more expensive for foreign purchasers; and various overseas producers of grains and soybeans became increasingly effective competitors. Many farmers were in distress as they had to face lower sales and prices, with high-interest loans still to pay off. A notable feature of the downturn in the 1980s, as well as of previous such periods, was the inability of government subsidy programs to halt the process whereby some farms failed and other expanded. Such programs, costing many billions of dollars and brought into being by a mountain of legislation, have been a controversial facet of American economic and political life since the beginning of the Roosevelt New Deal in the 1930s. As American farms become fewer, larger, and more highly capitalized (in the Midwest and elsewhere), corporate forms of farm ownership and enterprise are increasingly appearing. Some of these holdings belong to diversified corporate interests, while others may be considered family farms in a new guise.

THE URBAN AND INDUSTRIAL MIDWEST

Across the Midwest the level of industrial development declines from east to west. Thus the East North Central states ("eastern Midwest") are highly industrialized and accounted for 22 percent of all United States manufacturing employment in a recent year, whereas the West North Central states ("western Midwest") accounted for only 7 percent. Concentration of most of the Midwest's manufacturing in the eastern Midwest reflects a number of facts:

1. The eastern Midwest was settled earlier, and industries were already established there

when the western Midwest was still open prairie. Once established, these industries tended to persist and expand in their places of origin.

2. Location of industry in the eastern Midwest was favored by that area's closer access to the markets of the Northeast and Europe.

3. Transport connections with a vast area reaching to the Pacific Coast converged on the Great Lakes cities (particularly Chicago) when the transcontinental railroads were built. This provided optimum conditions for industries in the eastern Midwest to reach markets and materials in developing areas to the westward.

4. The eastern Midwest benefited much more from access to water transportation (lake and river) than did the western Midwest.

5. The eastern Midwest had better access than the western Midwest to superior coal resources. In Ohio it contains part of the great mining region in the Appalachian Plateau, and it lies adjacent to other sections of this coal-bearing area in Pennsylvania, West Virginia, and Kentucky.

Agricultural Foundations of Midwestern Industry

Midwestern agriculture has been, and remains, one of the main foundations of Midwestern industry. Meat packing, grain milling, and other food-processing industries such as brewing and cheesemaking have always been of major importance. Cincinnati first developed as an early packing "Porkopolis" before the Civil War, and Carl Sandburg's poem "Chicago" celebrated that city's role as the "Hog Butcher for the World." The packing industry is still important in the Midwest, but it has moved westward with time. It now centers in western parts of the Midwest, notably in and near metropolitan Omaha in Nebraska and Iowa. The Chicago stockyards and slaughterhouses, once the largest in the world, have closed. The same westward migration has also been true of wheat milling, in which the Minneapolis-St. Paul metropolis of Minnesota is a leader.

In addition to processing the products of Midwestern agriculture, Midwestern industry also supplies many needs of the region's farms. A major item is farm machinery. Midwestern topography is unusually favorable to mechanized operations, and mechanization has grown steadily for more than a century. Many technical advances in farm equipment originated in the Midwest, a prominent example being the 19th-century invention by Cyrus Hall McCormick of the mechanical reaper for harvesting wheat. The world's largest farm-machinery corporations are headquartered in the region, and many Midwestern cities remain heavily dependent on this industry. The largest concentration of factories today is in Illinois and eastern Iowa. Famous corporations headquartered here include Navistar (formerly International Harvester) in Chicago, Caterpillar Tractor in Peoria, Illinois, and John Deere in Moline, Illinois. The manufacture of farm machinery in the Midwest helped foster the development of a broad range of other types of machinery.

The Midwestern Steel Industry

The emergence of over half the United States steel industry in the eastern Midwest was also based in part on Midwestern agriculture. In the last four decades of the 19th century and the first half of the 20th, enormous sums were invested in huge steel plants in Ohio, Illinois, Indiana, and Michigan. Much of the steel made in these mills was used to build and equip the transport net employed in hauling agricultural products. Additional steel was used by machinery industries serving agriculture and to construct the region's burgeoning cities, which were themselves heavily dependent on agricultural trade. As time went on, the markets for steel became more diversified. For example, the growth of the automobile industry in the Midwest created a major new demand for steel.

Besides a favorable market situation, the Midwestern steel industry had an excellent resource base in and near the region. High-grade coal was available from the Appalachian Plateau of Ohio, Pennsylvania, West Virginia, and Kentucky. Supplements of lower-grade coal came from the Eastern Interior coalfield of Illinois, southwestern Indiana, and western Kentucky. Vast deposits of high-grade iron ore were found near Lake Superior in Minnesota, Michigan, and Wisconsin, especially in the ore formation called the Mesabi Range, inland from Duluth, Minnesota. Until the end of World War II, these deposits provided the greater part of the ore used by the entire American steel industry. So heavy was the demand for ore, especially to support all-out wartime steel production, that the high-grade resources were largely exhausted by the end of the war. Then, in the 1950s, completion of the St. Lawrence Seaway made it feasible to bring ore from newly opened deposits in Labrador and Quebec, and beneficiation processes increased the utility of the remaining lower-grade ores in the fields near Lake Superior. Throughout the whole evolution of the Midwestern steel industry, the Great Lakes have provided large-scale water transportation for ore and coal. This has drastically reduced the transport costs of bringing together the enormous quantities of fuel and raw materials used by the large mills through their long history.

The steelmaking districts of the eastern Midwest have been a major source of the single most crucial material needed for American economic development. But by the 1980s they also had become major economic problem areas. The leading districts are located (1) within metropolitan Chicago along and near the shore of Lake Michigan in northwestern Indiana and south Chicago, (2) along and near the south shore of Lake Erie in Cleveland and other cities of northern Ohio, and (3) at Detroit, Michigan. Buffalo, New York, also belongs to this complex of steel districts fronting the Great Lakes. The main steel plants in these areas are enormous in scale, and they tend to be relatively old. Very large reinvestment is required to bring an old plant of this kind to the level of productivity and competitiveness represented by the new plants, using newer technology, that have been built in many parts of the world during recent times. If such investment is made, the work force must be drastically reduced, as the new technology is far more automated than the old. If the investment is not made, the mill becomes less able to compete and thus maintain production and employment. Because of these circumstances, the old steel districts have been saddled with major unemployment problems due to plant closings, production cutbacks, and massive layoffs of labor. The very high wage rates associated with the highly unionized steel industry tend to discourage the entry of new types of employment to replace the shrunken employment in steelmaking. An additional deterrent lies in the blighted landscapes and lack of amenities in these districts left over from an earlier age of industrialization.

The Enormous Automobile Industry

The availability of steel and a central location vis-à-vis markets were factors contributing to a heavy concentration of the United States automobile industry in Detroit and nearby Midwestern cities. When the industry began to develop during the early 20th century, production on a very large scale was necessary to achieve costs and prices low enough for mass sales. This required that production be centered in a few large plants and the vehicles marketed throughout the country. Steel was needed in huge quantities, plus a location from which distribution could reach the whole national market relatively cheaply. Detroit was nearly optimum from the latter point of view, as the national center of population at that time was in southern Indiana (it is now in eastern Missouri),

and buying power was higher toward the north and east. Detroit labor and management had a background in building lake boats and horse carriages, and experiments with internal-combustion motors had been carried on by various inventors, including the best-known Detroit inventor and entrepreneur, Henry Ford.

From southeastern Michigan the auto industry spread over much of the eastern Midwest: north to Flint and Lansing, east to Cleveland, south to Dayton and Indianapolis, and west to the Milwaukee area. The cities named, and still others within the auto-making region, became heavily involved with the new industry and very dependent on it. Their factories mainly supplied parts and subsystems to the assembly plants at Detroit. Assembly itself spread to St. Louis and Kansas City. But the American corporate giants of the industry—General Motors, Ford Motor, and Chrysler—still have their headquarters and main plants in metropolitan Detroit. As in the case of steelmaking, the motor-vehicle industry has provided much of the Midwest with a basic industry of great size, vital to employment, and paying high wages. But, again like the steel industry, it became a focus of employment problems in the 1980s because of sales by competing plants overseas, especially Japanese plants. Even before the challenge of overseas competition the industry was expanding at a faster rate in other regions of the United States than in the Midwest. Decentralization was a response to markets in California, the Northeast, and the South which became large enough to justify new assembly plants in those regions. Recently, Tennessee and Kentucky have been chosen for new American assembly plants of Japanese auto companies. Thus in automaking the Midwest has a second huge industry, historically basic to the region's industrial growth and prosperity, that now provides high wages to a shrinking labor force and is creating major unemployment problems in many communities.

Future Employment Uncertainties in the Industrial Midwest

The net regional result of such problems as those just outlined has been a sharp decline in overall manufactural employment in the more industrialized eastern Midwest. Between 1967 and 1982 (the latter a recession year) this five-state area lost 21 percent of its industrial jobs, which was only a slightly smaller percentage loss than the Middle Atlantic states experienced and was a greater loss than that of New England. Meanwhile the seven states of the western Midwest, containing only a quarter of all Midwestern industry, had a modest percentage gain in manufactural employment; and both the South and the West showed quite substantial gains. The severe net loss of manufactural jobs in the eastern Midwest indicates that this area is not attracting new high-growth (usually high-technology) industries in sufficient numbers to counterbalance the employment decline in older industries. The newer industries are growing primarily in parts of the country with lower average wages and/or more amenities such as seashores and warm winters. Within the Midwest this pattern is seen in the four westernmost states (the Dakotas, Nebraska, and Kansas) and Minnesota. The states in the western tier have lower wage rates than the other Midwestern states and are expanding their relatively small manufactural sectors, while Minnesota's lake-strewn environment has proved attractive to amenity-conscious high-technology industries. It remains to be seen whether Midwestern service industries can adequately fill the gap in employment and income left by declining industrial employment. Difficulties are created by the fact that so much of the Midwest suffers from a lack of climatic and other amenities attractive to basic services in an age when rapid communications and transport give such enterprises a large measure of freedom to locate where they will.

SPATIAL ASSOCIATIONS OF MIDWESTERN METROPOLISES

Despite the basic importance of agriculture in the Midwestern economy, about 70 percent of the region's population is metropolitan. In fact, in 1986 an estimated 30 million people, or 50 percent of all Midwesterners, lived in metropolitan areas of half a million or more people, and there are many smaller metropolitan areas. As in other parts of the United States, good roads and an increasing amount of commuting have been integrating more and more counties with the cities that form the centers of metropolitan areas. Hence most of the larger metropolitan areas are now sprawling multi-county districts. These are essentially urban in employment and function, even though outlying sections may not be highly urban in appearance. A typical instance of a multi-county metropolis is metropolitan St. Louis, which includes four counties in Missouri and five in Illinois, besides the city of St. Louis itself. As in most Midwestern metropolises, agricultural production is important and occupies much land within the metropolitan area defined by entire counties, but it employs only a tiny proportion of St. Louis's metropolitan population.

The larger Midwestern cities, around which the main metropolitan areas have formed, had their beginnings in the early 19th century, or occasionally earlier. Water transport was relatively more important then than it is now, and city locations tend to be related to lake and/or river features which conferred early importance in a developing pattern of water transportation. When railroads largely replaced water routes, the cities were already established and became rail foci. This new form of transportation was dominant for about a century after 1860. Then in the mid-20th century high-speed highways and airways added major new dimensions to transportation. But even today many of the major metropolitan areas still benefit from advantageous locations for freight movement by water. This is particularly true of the Great Lakes cities, to which we now turn.

Great Lakes Metropolises: Chicago, Detroit, Cleveland

The three largest Midwestern metropolitan concentrations center on Chicago, Illinois; Detroit, Michigan; and Cleveland, Ohio. Chicago is located at the southern end of Lake Michigan, Detroit is between lakes Huron and Erie, and Cleveland is on Lake Erie (Fig. 28.1). Much the largest is *Chicago* (8.1 million). The city began at the mouth of the small Chicago River, which afforded a harbor at the head of Great Lakes navigation where it projects farthest into the prime farm lands of the Midwest. A few miles up this river, a portage to tributaries of the Illinois River provided a connection via the Illinois to the Mississippi River in the vicinity of St. Louis. A canal built in 1848 along the portage route connected Lake Michigan with the Illinois and helped Chicago become the major shipping point for farm products moving eastward by lake transport from the rich farming areas in the prairies to the city's south and west. Then in the 1850s railways fanned out from Chicago into an ever broader hinterland and also connected the city with Atlantic ports. Chicago became the country's leading center of railway routes and traffic and has remained so. In the 20th century new harbors have been built south of the city center, and new large-scale canal connections have facilitated an increasing barge traffic using the Illinois River. At this major focus of transport and population growth a huge industrial structure rose and intensified. Several major industries developed very early: food processing, especially meat packing (which has now almost disappeared); wood products made from white pine and other timber cut in the forests to the north; clothing; agricultural equipment; steel; and a broad range of machinery of many kinds. Today the most important lines of manufacturing in the metropolitan area are machinery and steel, for both of which Chicago is the leading center in the country. The city's banks, corporate offices, commodity exchanges, and other economic institutions make it the overall economic capital of the Midwest. To the east its metropolitan area reaches into northwestern Indiana, and to the north its area is coalescing with that of *Milwaukee* (1.6 million), Wisconsin, which has grown around a natural harbor on the lake. Together with smaller metropolitan centers still farther north, Chicago and Milwaukee present a concentration of over 10 million people in a tri-state area stretching along and near the lakeshore from Green Bay, Wisconsin, past Gary, Indiana.

The second greatest concentration of people and production in the Midwest is metropolitan *Detroit*, together with the smaller adjoining area of Toledo, Ohio. The Detroit area extends all the way along the United States side of the water passages separating the Lower Peninsula of Michigan from the Ontario Peninsula in Canada and connecting Lake Huron with Lake Erie. From north to south, these passages include the St. Clair River, Lake St. Clair, and the Detroit River (Fig. 28.1). All are navigable for lake shipping without the necessity for locks, a fact also true of the Straits of Mackinac to the northward between Lake Huron and Lake Michigan. Detroit's estimated metropolitan population in 1986 was 4.6 million, and the adjacent Toledo metropolis contained 610,000 more. A closely connected Canadian population in metropolitan Windsor, Ontario, just across the Detroit River from Detroit, adds another 255,000 to the total metropolitan complex. Detroit is very old as Midwestern cities go, having been founded as a French fort and settlement along the strategic waterway between the lakes in 1701. It grew as a service center for a Michigan hinterland which was settled rapidly following completion of the Erie Canal in 1825. With water transportation provided all the way to the Northeastern Seaboard, settlers could move west and ship their surplus produce back to Northeastern markets and ports relatively cheaply. But Detroit remained modest in size until the automobile boom of the 20th century. Some of the circumstances and consequences of that development have already been discussed. *Toledo* was founded on a natural harbor at the mouth of the Maumee River at the extreme western end of Lake Erie. The Maumee flowed through what was originally an extensive swamp occupying the flat bed of an old glacial extension of Lake Erie. When this soggy plain was drained, it provided exceptionally

good farm land even for the Midwest, and Toledo profited as an agricultural service center. Natural gas and oil from early fields in western Ohio furnished fuel for a glass industry at Toledo which developed in the 19th century and now produces much of the glass for Detroit automobile production. The metropolitan area is heavily involved in the auto industry in still other ways.

The third-ranking concentration of population in the Midwest is in northeastern Ohio, where the Lake Erie port of *Cleveland* has a metropolitan population (including nearby Akron) of 2.8 million, and the adjacent inland metropolis of Youngstown, Ohio, adds 510,000 more. At metropolitan Cleveland the Appalachian Plateau, with its important coal fields, comes near the south shore of Lake Erie, while a narrow ribbon of lake plain separates the edge of the plateau from the lake, and is interrupted by natural harbors at the mouths of rivers entering the lake from the south. Around one of these river mouths, that of the Cuyahoga, the port city of Cleveland developed in the 19th century. The surrounding district was not outstanding agriculturally, and economic development centered on heavy industry. Today the district has the customary problems associated with that emphasis. Ore from the Lake Superior fields was landed at Cleveland, and subsequently at smaller ports near by, for shipment to Pittsburgh. A return flow of coal intended for many Great Lakes destinations moved by rail to the Lake Erie ports. The movement of ore one way and coal the other, as well as the mining of coal within the district itself, made it logical to develop the iron and steel industry at Cleveland, at adjacent Lake Erie ports, and at Youngstown on the route between Pittsburgh and the lake. Abundant coal supported the development of other industries such as machinery plants, and the Detroit-centered automobile complex eventually spread into parts of the district. A major facet of this latter development was the rise of Akron as the leading center of automobile tire manufacturing in the United States. Today northeastern Ohio is afflicted by the slow growth, contraction, or actual closure of many older industrial enterprises. Youngstown, for example, has been hard hit by a sharp decline in the steel industry, on which its employment depended to an unusual degree.

Metropolitan Areas along the Old National Road: Columbus, Dayton, Indianapolis

In the early 19th century the federal government built a pioneering national highway, the National Road, westward from Wheeling, West Virginia, to near St. Louis. This was the route subsequently followed in the 20th century by U.S. Highway 40 and then by Interstate 70. At certain strategic points along the Road the cities of Columbus (1.3 million), Ohio; Dayton-Springfield (935,000), Ohio; and Indianapolis (1.2 million), Indiana, developed. Columbus was already the state capital of Ohio and tied by a canal to the Ohio River and Cleveland when the National Road reached it. Dayton also was on a canal connecting Lake Erie with the Ohio (it ran from Toledo to Cincinnati). The National Road bypassed Dayton but ran through nearby Springfield. Indianapolis was the state capital of Indiana and on the most important land route between the Ohio River and Lake Michigan. The National Road, a rare phenomenon in those days of difficult overland travel, made major transport junctions of these places, all of which were already serving excellent agricultural areas. They have grown into commercial and industrial metropolises with a wide range of manufactures.

River Cities: St. Louis, Minneapolis-St. Paul, Cincinnati, Kansas City, Omaha

The remaining major metropolises of the Midwest originated as river cities—that is, cities which grew at strategic places on the major rivers which were early arteries of Midwestern traffic. These cities still benefit from their riverside locations, although their rail, road, and air connections have now relegated river traffic to a very subordinate position.

St. Louis, Missouri, is the center of a metropolitan population of 2.4 million people (1986) in Missouri and Illinois. It was founded by the French in the 18th century and remained the largest city in the Midwest until the later 19th century. Its position is that of a natural crossroads of river traffic using the Mississippi to the north and south, the Missouri to the west, the Illinois toward Lake Michigan, and (via a southward connection on the Mississippi) the Ohio toward the east. Early traffic had to adjust to the fact that the Mississippi downstream from the Missouri mouth was notably deeper than the upstream channel or the Missouri. Hence many river boats plying the Mississippi below St. Louis drew too much water to use the channels above the city, and St. Louis became a "break of bulk" point where goods were transferred between different kinds of craft. Coal from southern Illinois was available to support industrialization, and the metropolitan area eventually developed a highly varied industrial structure, including today such diverse manufactures as aircraft and aerospace equipment, automobile assembly, primary metals, machinery, electrical equipment, clothing, chemical products, shoes, meat, beer, grain

products, and printing. Three of the nation's 25 largest corporate exporters of manufactures have their headquarters and some of their manufacturing in St. Louis. They include McDonnell Douglas, exporting aircraft, missiles, and space and information systems; Monsanto (chemicals), and General Dynamics (tanks, aircraft, missiles, gun systems) (data on products: *Fortune*). McDonnell Douglas and General Dynamics are major suppliers of military equipment to the American government. Other major corporations with headquarters and plants in St. Louis include Emerson Electric (electronic and electromechanical components), Anheuser-Busch (beer), and Ralston Purina (food).

Minneapolis-St. Paul (2.3 million), Minnesota, developed at the head of navigation on the Mississippi River. The Falls of St. Anthony marking this point supplied waterpower for early industries such as the sawing of timber from the Minnesota forests into lumber for shipment to the timber-short prairie areas being settled to the south. At Minneapolis-St. Paul a major water-transport artery, later supplemented by a local railnet, met transcontinental rail lines which were built west from the Twin Cities to the Pacific. The spread of business influence westward along these lines made Minneapolis-St. Paul the economic capital of much of the northern interior United States. No competing center of comparable size emerged anywhere along the resource-rich strip of wheat-growing, ranching, mining, and lumbering country between Minneapolis-St. Paul and the seaports of Seattle and Portland on the Pacific. The hinterland of the Twin Cities lies in good part within the Spring Wheat Belt that stretches across the northern Interior Plains from Minnesota to the foot of the Montana Rockies, and Minneapolis-St. Paul has long been a major focus of the grain trade and of flour milling. Today a diversified industrial structure augments the long-standing transportation, commercial, and financial functions of this sprawling, lake-studded regional capital.

A much older river city than Minneapolis-St. Paul is *Cincinnati* (1.7 million) on the Ohio. Its present metropolitan area includes the southwest corner of Ohio and adjacent parts of Indiana and Kentucky. For a time its situation resembled the present one of Minneapolis-St. Paul in that it dominated a large hinterland to the west and north. The Ohio River turns sharply southward from the city, and early 19th-century settlers floating down the river from Pittsburgh often disembarked at Cincinnati to move overland into Indiana. The city quickly developed a major specialty in collecting and processing surplus produce, especially hogs, from Ohio and Indiana farms. Large quantities of salt pork were shipped downriver to markets in the South. This meat-packing function eventually was lost as the center of Midwestern agriculture migrated westward and competing packing centers arose. But coal mining developed in the nearby Appalachian Plateau, and Cincinnati became the economic capital for a considerable part of the mining region. It also profited from the fuel supply and its transport links in developing a diversified and relatively stable industrial base.

Kansas City, Missouri, with a 1986 metropolitan population of about 1.5 million in Missouri and Kansas, strongly resembles Minneapolis-St. Paul in its situation and development. The Kansas City area became established at what was essentially the head of navigation westward on the Missouri River. Although river transport was not impossible upstream, it was subject to various navigational hazards and, for a time, to danger from Indians, and the river led sharply northwest instead of directly westward. Consequently, two of the major overland trails to the West—the Santa Fe Trail and the Oregon Trail—had their main eastern termini in the Kansas City area. Here settlers and traders disembarked from river boats, outfitted themselves, and proceeded west by land. Later on, cattle drives came to Kansas City from range lands in the Great Plains, and railways reached westward from the city. Today Kansas City has a major competitor to the west in Denver, Colorado. Nonetheless, the hinterland of present Kansas City extends far to the west, and the city is the economic capital of the Winter Wheat Belt. It also has a growing industrial sector.

Omaha, Nebraska (615,000 metropolitan population in Nebraska and Iowa) is a smaller riverside city resembling both Kansas City and Minneapolis-St. Paul in its functions and westward reach. It lies a bit west of a line between the two larger cities. Like them, Omaha is in the border zone between the intensively farmed Corn Belt on the east and drier, less productive wheat and cattle country to the west.

THE WEST: ISLAND-LIKE DEVELOPMENT IN A DRY AND MOUNTAINOUS SETTING

Beyond the western borders of the South and the Midwest as defined in this text lie 11 states that make up the West (Figs. 26.1, 28.21, and 28.22). Eight of them are often termed the *Mountain states* or *Mountain West.* Of these, New Mexico, Colorado, Wyoming, and Montana comprise an eastern tier of four states lying partly in the Interior Plains (Great Plains), partly in the Rocky Mountains, and, in the

case of New Mexico and Colorado, partly in the Intermountain Basins and Plateaus. Idaho and Utah are partly in the Rockies and partly in the Intermountain Basins and Plateaus, while Arizona and Nevada lie wholly in the latter region except for western Nevada's small sector of the Sierra Nevada range in the Pacific Mountains and Valleys. Rounding out the West are the three *Pacific states* of California, Oregon, and Washington. They lie partly in the Pacific Mountains and Valleys and partly in the Intermountain Basins and Plateaus, with Washington and Oregon also including small sections of the Rocky Mountains. The last two states are often termed the *Pacific Northwest.* Federal publications include Alaska and Hawaii in statistics for the Western states, but these two states are so different in their locations, environments, economies, histories, and spatial associations that they are discussed here in a separate section and are not included in statistics and generalizations for the West.

Thus defined, the West comprises 33 percent of the United States by area, or 39 percent of the conterminous 48 states. The region is still lightly settled except for a huge cluster of people and development in central and southern California and a handful of smaller clusters in other widely dispersed locations. Especially prominent among the latter are clusters centering on the inland cities of Phoenix and Tucson, Arizona; Denver, Colorado; and Salt Lake City, Utah; and a seaboard cluster including the port cities of Seattle, Washington, and Portland, Oregon, in the Pacific Northwest. In 1986 the West had an estimated 48 million people (20% of the nation), of whom 55 percent lived in California. In that year, the West accounted for about one fifth of the nation's agricultural output, and over one sixth of its employment in manufacturing. California is the West's economic colossus, producing a little over two thirds of the region's manufactures and almost one half of its farm products by value. Even in California, however, there are large areas of sparse population and scanty development. The eight Mountain states contrast sharply with California in that they comprise 29 percent of the total area of the 48 conterminous states, but accounted for only 5½ percent of their population, 8 percent of their agricultural output, and 3 percent of their manufactural employment in the mid-1980s.

HANDICAPS TO DEVELOPMENT IN THE WEST

The generally low intensity of development in the West outside of central and southern California results principally from dry climates, rough topogra-

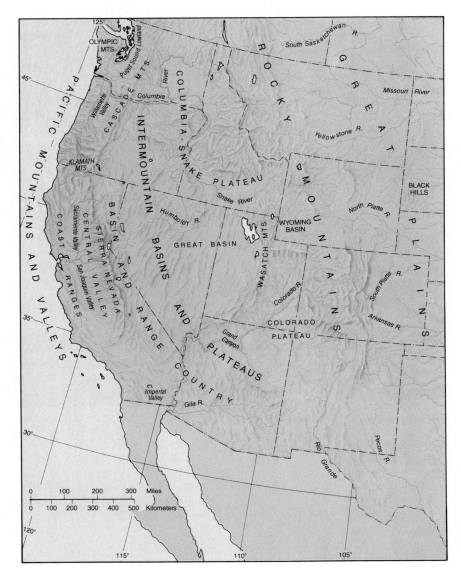

Figure 28.21 *Major physiographic features of western United States.*

phy, and late settlement. Absence of water transportation except along the Pacific Coast has been another handicap. Huge areas are very short of water, and most settlement is clustered in places where water in sufficient quantity can be had. Hence the population pattern is island-like, with many clusters being genuine *oases* in desert or near-desert climates. Local availability of water may be due to nature, as in the rainy seaward margins of Washington and Oregon, or it may be due to human investment in water collection and control. The settlement pattern runs a wide scale from major metropolitan areas to isolated ranches. Major areas of water availability and clustered population tend to be separated from each other by huge expanses of steppe, desert, and/or mountains—all with very few people.

The climates that exert so much effect on Western settlement are strongly related to topography. For example, precipitation is largely orographic, and dryness is largely due to rain shadow. Thus an understanding of topographic layout is needed to understand the climatic conditions that either handicap or facilitate settlement. The topography was summarized in Chapter 26 but will be elaborated here as appropriate.

WESTERN TOPOGRAPHY AND ITS EFFECTS ON CLIMATE

The Pacific shore is bordered by mountains along most of its length. The *Coast Ranges* front the sea, providing an often spectacular coastline, from the

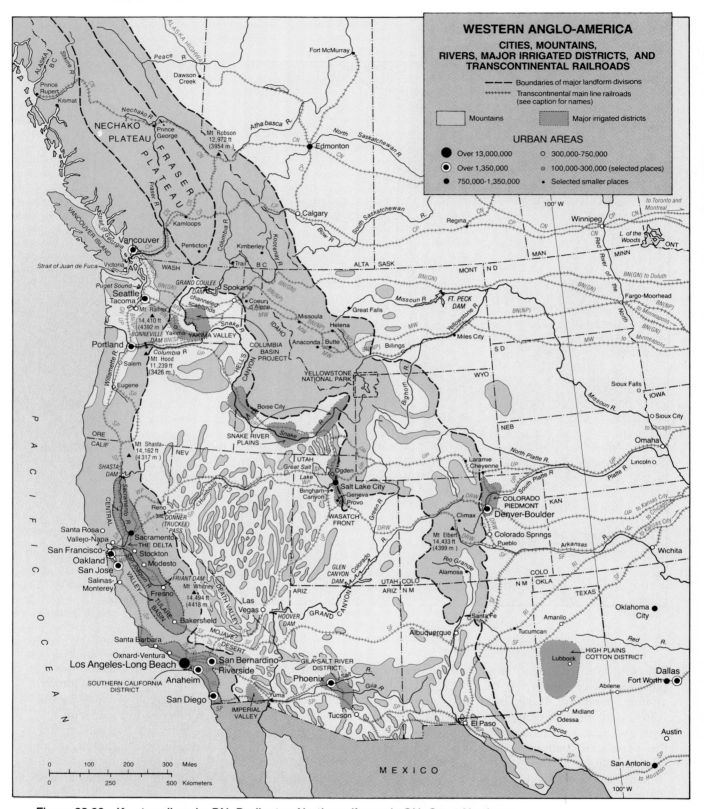

Figure 28.22   Key to railroads: *BN, Burlington Northern (formerly GN, Great Northern;
NP, Northern Pacific; or SPS, Spokane, Portland and Seattle); CN, Canadian National; CP,
Canadian Pacific; DRW, Denver and Rio Grande Western; MW, Milwaukee; RI, Rock Island;
SF, Santa Fe; SP, Southern Pacific; TP, Texas and Pacific; UP, Union Pacific; WP, Western
Pacific. For several decades United States railroads have been subject to reorganization
through mergers and the business arrangements of Amtrak and Conrail sponsored by the
federal government. On the map the population figure of over 13 million for Los Angeles–
Long Beach includes the populations of San Bernardino–Riverside, Anaheim, and Oxnard–
Ventura shown by separate symbols. Similarly, the metropolitan figure of 5.9 million cited
for San Francisco in the text includes Oakland, San Jose, Vallejo–Napa, and Santa Rosa.*

Olympic Mountains in northwestern Washington to just north of Los Angeles. The ranges are mostly low, although the Olympics reach nearly 8000 feet (*c.* 2400 m). North of Los Angeles the mountains swing inland a bit and then continue south to the Mexican border. In this southern section they tend to be higher, exceeding 11,000 feet in the San Bernardino Range east of Los Angeles. They are also different in structure, and physiographers do not generally include them with the Coast Ranges proper.

Inland, a line of truly high mountains parallels most of the length of the Coast Ranges. The *Cascade Mountains,* extending southward from British Columbia across Washington and Oregon to northern California, are surmounted by a series of volcanic cones, among which Mt. Hood near Portland reaches over 11,000 feet and Mt. Rainier near Seattle over 14,000 feet. Although long dormant, one of these volcanoes, Mt. St. Helens in southwestern Washington, recently came to renewed life with a disastrous eruption. At its southern end the Cascade Range joins the rugged *Klamath Mountains,* a deeply dissected plateau where the Cascades, Coast Ranges, and *Sierra Nevada* meet. The high and rugged Sierra increases in elevation southward to the "High Sierra," where Mt. Whitney at 14,494 feet (4418 m) is the highest peak in the 48 conterminous states.

Two large lowlands lie between the Coast Ranges and the inner line of Pacific mountains, and a smaller but highly important one lies outside the Coast Ranges in the extreme south. In Washington and Oregon the *Willamette-Puget Sound Lowland* separates the Cascade Range from the Coast Ranges. The Oregon sector, extending from the lower Columbia River to the Klamath Mountains, is the Willamette Valley. The Washington sector, north of the Columbia, is the Puget Sound Lowland, which is deeply penetrated by the deepwater inlet from the Pacific called Puget Sound. The lowland surface in both states is generally hilly. The second large lowland is the *Central Valley,* or Great Valley, of California, enclosed by the Klamath, Sierra Nevada and Coast Ranges. The Central Valley is a most unusual feature in the West—a large and practically flat alluvial plain, some 500 miles (*c.* 800 km) long by 50 wide. Its floor is made of detritus eroded from the bordering mountains. This material has filled to above sea level a trough (downfold) that was once an inlet of the Pacific connected to the open ocean by the gap in the Coast Ranges now occupied by San Francisco Bay. In California's far southwest, a smaller lowland lies between the coastal mountains and the shore. The wider northern part is the *Los Angeles Basin,* while a narrower coastal strip extends southward to San Diego and Mexico. The Los Angeles Basin is a plain dotted by isolated high hills and overlooked by mountains. The strip to the south is also hilly.

Precipitation is brought to this mountain and lowland terrain in the Pacific Mountains and Valleys by air masses moving eastward from the ocean. This occurs more frequently in the winter than in the summer, so that summer is much the drier season everywhere; and it occurs more frequently and to greater effect in the north than in the south. Hence total precipitation is greater in the north, while in the south summer conditions become desert-like, although winter still brings some rain and/or snow. The mountains receive the most precipitation, which is very heavy in the Coast Ranges from British Columbia to just north of San Francisco and all along the western (windward) slopes of the Cascades and the Sierra Nevada. The deepest snows of the United States accumulate in the higher areas. South from San Francisco even the Coast Ranges receive modest precipitation at best. Here the magnificent Douglas fir forests of the northern mountains, the overpowering coast redwoods of northwestern California, and the gigantic sequoias of the southern Sierra Nevada (Fig. 26.3) give place to sparse forests, grasses, and shrubs.

The lowlands of this Pacific strip vary from quite wet to arid. The Coast Ranges of Washington and Oregon are mostly low enough, and the incoming moisture is great enough, that the Willamette-Puget Sound Lowland is not severely affected by its position on the lee side of the mountains. Hence this lowland is wet and was naturally forested. Since its temperatures are moderated both winter and summer by incoming marine air, it is considered to have a *marine west coast climate* similar to that of western Europe. The Central Valley of California is much more drastically affected by rain shadow cast by the Coast Ranges. Although generally classed as having a *mediterranean climate,* with subtropical temperatures, wet winters, and dry summers, it is dry enough that its natural vegetation was steppe grass for the most part, grading into desert shrub in the arid Tulare Basin at the southern end of the Valley. The same observations apply to the Los Angeles Basin: The usual climatic classification is mediterranean subtropical, but the total precipitation, restricted to winter rains, is so low that a *steppe* or even a *desert* classification could be argued. (See climatic data in Table 26.3.)

But it is east of the high mountains—Cascades, Sierra Nevada, and southernmost ranges near the coast—that rain shadow becomes the most extreme. Here a vast area reaching eastward to the Rockies, and around their southern end in New Mexico, is predominantly desert or steppe in climate and natural vegetation. Little moisture is received from air masses that already have provided heavy precipitation (except in the extreme south) to the mountains near the Pacific. This region is known physiographically as the *Intermountain Basins and Plateaus,* for its location between the major ranges to its east and west. It includes the greater part of eastern Washington, the central and eastern parts of Oregon, northeastern and southeastern California, southern Idaho, practically all of Nevada, all of Utah except the northeast corner, westernmost Colorado, all of Arizona, and western and southern New Mexico.

This Intermountain area lies mostly at high elevations, with few sections below 3000 feet (914 m), but the area is lower than the high ranges which border it and deprive it of moisture. It is enormously varied in elevation, landforms, and local relief. Most of the land is composed of plateau surfaces in various stages of dissection, so that rolling uplands and rugged hilly and mountainous sections are included as well as large areas which are comparatively level. River valleys tend to be deeply incised, forming spectacular steep-sided gorges and canyons (see chapter-opening photo of Chapter 2 and Fig. 2.9). Only the highest parts of the Intermountain region (mountains or exceptionally high plateau surfaces) receive enough precipitation to be forested, and such forests often are sparse. The predominant vegetation today is desert shrub, with steppe grass in some areas, especially in the wetter and cooler north and on some high plateau surfaces.

Three major physical subdivisions of the Intermountain Basins and Plateaus are commonly recognized: (1) the Columbia-Snake Plateau, (2) the Colorado Plateau, and (3) the Basin and Range Country. The *Columbia-Snake Plateau* occupies eastern Washington between the Cascades and the Rocky Mountains, almost all of Oregon east of the Cascades, and Idaho south and west of the Rockies. It is characterized by (1) areas of relatively level terrain formed by massive lava flows in the past, (2) scattered hills and mountains, and (3) deep-cut river canyons. The vegetation is sagebrush steppe for the most part, with forests in the mountains. The *Colorado Plateau* occupies northern Arizona, northwestern New Mexico, most of eastern Utah, and most of extreme western Colorado. Here rolling uplands lie at varying levels, cut into isolated sections by deep river canyons, principally of the Colorado and its tributaries. Occasionally, mountains rise above the general level. Some higher elevations are forested, but steppe and desert vegetation predominates. The rest of the West's Intermountain area — in western Utah, Nevada, southern Arizona, southwestern and central New Mexico, and eastern California — lies in the *Basin and Range Country,* which continues beyond New Mexico into western Texas.

Part of this physical division, between the Wasatch Range of the Rockies on the east and the Sierra Nevada on the west, is not traversed by any river reaching the sea and is called the Great Basin. But the whole Basin and Range Country consists of small basins which tend to be separated from each other by blocklike mountain ranges. This is the driest part of the West, and desert shrub predominates except for steppe and/or forest on the higher mountain ranges. Extreme desert with very sparse vegetation prevails across the southern part of the Basin and Range Country from California into western Texas. In the northeast near the Wasatch Range is the Great Salt Lake, a remnant of the much larger Lake Bonneville which covered a good part of the Great Basin in late glacial times.

East of the Intermountain Basins and Plateaus, the *Rocky Mountains* extend southward out of Canada into parts of Idaho, Montana, Wyoming, Utah, Colorado, and New Mexico. They are mostly high and rugged, with frequent snowclad summits. The highest peaks are in Colorado, where a number rise above 14,000 feet (4267 m). Here Mt. Elbert, the highest mountain in the entire Rocky Mountain System, reaches 14,433 feet (4399 m). The Rockies form many separate ranges with individual names such as Front Range (in Colorado) or Wasatch Range (in Utah). Often the ranges are linear, with crestlines trending generally north-south or northwest-southeast, but many ranges depart from this pattern. Valleys and basins of many shapes, sizes, and orientations punctuate the ranges. The continuity of the system is almost broken in the Wyoming Basin, where high plains surfaces are extensive, with isolated hills and mountains rising above. This basin provides the easiest natural passageway through the Rockies and was the route of the first transcontinental railroad, the Union Pacific.

The Rockies are a formidable barrier to moisture-bearing air masses from either west or east. Hence they create rain shadow for both the Great Plains at their eastern foot and the Intermountain area at their western foot. The Rockies themselves rise high enough to receive much orographic precipitation, and much of their area is in forest. This tends to be denser in the north where precipitation is higher and temperatures and evaporation are lower. The highest elevations extend above the tree line to mountain tundra (Fig. 26.8), bare rock, and snow and ice fields.

The higher western part of the *Great Plains* at the western edge of the *Interior Plains* borders the Rockies in Montana, Wyoming, Colorado, and New Mexico. Here Denver is famed as the Mile High City. So elevated are these plains that the rivers which flow eastward across them from the Rockies have often been able to cut deep and wide valleys below the general surface. Thus the most common landscape is that of an irregular plain interrupted sharply at intervals by conspicuous valleys. Variations in this typical surface occur here and there: hilly areas, as in southeastern Montana; smooth plains, as in northeastern and extreme eastern Colorado and eastern New Mexico; and isolated mountain peaks or clumps, as in the Black Hills on the border between Wyoming and South Dakota.

These high plains are vegetated by a classical mid-latitude steppe grassland, marked by semiaridity and temperatures that are continental in the north and subtropical in the south. The plains are far removed from potential sources of moisture in the Pacific Ocean and the Gulf of Mexico, and they are on the lee side of the great assemblage of high mountains that separates them from the Pacific and blocks off most of the moisture that might otherwise reach them from the west. Enough precipitation does come from the Gulf or the Pacific to prevent actual desert conditions and support both steppe grasses useful for grazing and the growing of crops with low moisture requirements, principally wheat.

THE CRUCIAL ROLE OF WESTERN WATER

The West is so short of water that its availability is a critically important location factor for development. The greater part of the land is used only for grazing, if at all, and the carrying capacity is so small that only extremely low densities of animals and people are possible. Unirrigated farming of some steppe areas is important, notably wheat growing in the Great Plains and in the Columbia Basin to the east of the Cascade Range. But the intensity of production in unirrigated areas, while greater than that of grazing on natural range, is still too low to support dense populations or major cities. Wet areas in the mountains are too rugged and cool to be farmed. Thus intense agricultural production is concentrated in certain districts on valley and basin floors that are supplied with adequate water. Aside from the rainfed Willamette-Puget Sound Lowland, such areas require irrigation because of dryness, and they can properly be called oases. From the intensive cropping and related animal industries of a few large oases comes much the greater part of Western agricultural output. Except for unirrigated grains, the production of food crops for sale is concentrated in such areas. In addition, they grow large quantities of feed crops, particularly alfalfa hay, for dairy and beef cattle kept on feedlots in the oases. Such crops also are marketed to ranchers outside the oases as supplementary feed for livestock nourished principally on range grasses. In addition to the products just named, cotton is a major irrigated crop in California and Arizona.

THE MAJOR CLUSTERS OF WESTERN POPULATION AND ENTERPRISE

In general, the larger the irrigated acreage of a Western state, the greater is that state's agricultural output. And, in general, the greater the agricultural output, the larger is the state's population. But this does not mean that large numbers of people live on the West's farms and ranches. Numbers are particularly small on the cattle and sheep ranches that occupy huge areas of dry and often mountainous land. Such units are characteristically very big because the natural forage is generally so sparse that large acreages are required to sustain a profitable number of animals. Since the ranches are big and the number of people required to tend the grazing animals on each ranch is not large, the ranching population has a very low density and is widely dispersed. The same general circumstances apply to the population on unirrigated grain farms in the Great Plains and the Columbia Basin, where holdings must again be large because the scanty precipitation produces such a low yield of product per unit of cropped land. At the same time the use of large machines minimizes the need for labor. Hence the population of grain farmers is relatively small and thinly spread. But even in the West's intensively farmed oases, the number of people living on farmsteads in cropped areas is often remarkably small. While circumstances vary from one oasis district to another, a number of general reasons for the relatively sparse farm population can be cited:

1. Good highways have made it possible for many farm families to actually live in town and enjoy the benefit of conveniences and amenities there while at the same time commuting to their agricultural holdings as required.

2. The number of people needed to carry on farm operations has been sharply reduced in recent times through widespread adoption of labor-saving machinery.

3. Although large numbers of hired workers are still used at certain times of the year for such tasks as picking fruit or harvesting vegetables, this labor is temporary and generally migrant in character. Hence it does not form part of the farm population permanently settled on the land.

Figure 28.23   A relatively small but representative sector of the immense Los Angeles metropolis. Homes of the wealthy movie community perch on the chaparral-clad slopes below the "Hollywood" sign. At the lower right are buildings of one of the innumerable local business centers that are spread through the metropolitan area. Rising on the horizon are the San Gabriel Mountains, carpeted with a light winter snowfall.

4. The need for farm population in the countryside is lessened by the increasing employment of professional farm managers who make maximum use of machines. A single manager may operate the properties of several owners. Some holdings, especially those of corporations, are very large. Thus the population living permanently on the land and farming it may be remarkably thin despite the extremely high productivity of each acre.

The various circumstances just noted have given the 11-state West a farm population that amounted to only 605,000 in the census year of 1980 and has further declined since then. The decline has actually been in progress for a long time but has been particularly severe since World War II. As recently as 1960 the farm population was 1,520,000. Not only is the present farm population scanty, but the West also has a very low proportion of rural nonfarm people, most of whom actually live in towns or villages of under 2500 people and hence are counted as rural by the federal census. The immensely larger number of people counted as urban is sufficient to give the West a higher proportion of urban population than any of the other three major regions (West, 84% urban in 1980; Northeast, 79%; Midwest, 71%; South, 67%; U.S., 74%).

These Western urban dwellers are largely crowded into a handful of big metropolitan clusters located in or near major oases. The Pacific Northwest cluster centering on Seattle and Portland might be considered an exception, but even this rainfed area is a kind of oasis within the larger area of Western drylands. Today the West's major urban clusters have very diversified functions, among which the servicing of agriculture is generally the oldest and is still important. Individual cities characteristically gained early momentum as service nodes for small but productive agricultural areas and then gradually

added a multiplicity of other functions, some of which overshadow agricultural servicing today.

We turn now to an examination of the main Western clusters of population and intensive development, commencing with Southern California and continuing to San Francisco and the Central Valley of California; the Willamette-Puget Sound cluster in which Seattle and Portland are the main urban nodes; the Phoenix-Tucson cluster in Arizona; the Colorado Piedmont cluster centering on Denver; and Utah's Wasatch Front cluster centering on Salt Lake City (Fig. 28.22).

Southern California: The Evolution of a Pacific Megalopolis

The Southern California cluster is by far the West's largest. The name *Southern California* is commonly used for the Los Angeles Basin and the narrow extension of lowland southward to San Diego, along with their surrounding frame of mountains. The relatively dense population of Indian villagers who originally inhabited this area practiced no agriculture, but subsisted on wild foods such as seeds of the native wild grasses and acorns from the innumerable California oaks spaced through the grasslands. Then in the later part of the 18th century, small-scale agriculture utilizing alluvial lowlands, winter rainfall, and water from adjacent mountains was introduced by settlers from Mexico, together with cattle ranching based on the nutritious wild grasses. A series of small agricultural areas developed and eventually expanded and coalesced into an important agricultural district. Today much of the agriculture has been crowded out of the area by a sprawling and rapidly growing urbanization (Fig. 28.23) that in 1986 aggregated an estimated 15.3 million people in the combined metropolitan areas of Los Angeles (13.1 million) and San Diego (2.2 million). Packed into this evolving megalopolis in a small southwestern corner of the immense West are nearly three fifths of California's people, nearly one third of the population of the entire West, and about 1 in 16 of the people in the nation.

San Diego and Los Angeles were founded in late Spanish colonial times as remote footholds on the northern frontier of Spain's great empire in the Americas. The effort was spearheaded by Catholic priests from Mexico who founded a chain of missions in coastal locations as far north as San Francisco, including San Diego Mission (1769) and San Gabriel Mission (1781) in the Los Angeles area. San Diego was begun on a magnificent natural harbor that is now the main continental base for the United States Pacific Fleet. The San Gabriel Mission and the Pueblo (civil settlement) of Los Angeles were sited near the small Los Angeles River that crosses the Los Angeles Basin. When Mexico gained its independence from Spain in the early 19th century, the California settlements became part of the new country, and they subsequently came into possession of the United States in 1848 at the end of the Mexican War. Until the late 1800s, under all three regimes, San Diego and Los Angeles were little more than hamlets in one of the world's more isolated corners. Some 1500 miles (c. 2400 km) of tortuous overland routes through mountains and deserts separated them from the core of Mexico around Mexico City, and, in 1848, a comparable distance and terrain lay between them and the advancing frontier of continuous settlement in the eastern United States. The Orient lay an immense distance across the little-traveled Pacific Ocean, and Europe was also far away by sea. The tiny settlements of Southern California lived by ranching on the summer-dry steppe and by practicing a subsistence Mediterranean type of agriculture on small areas of alluvial land, using the limited rains of the mild winter rainy season and small-scale irrigation techniques introduced by the Spaniards to counteract the dry summers. Occasionally a ship came to trade for cattle hides. Some influx of American settlers occurred when sovereignty was transferred from Mexico, and this flow increased when gold mining in central California dropped off in the mid-1850s and some migrants originally attracted to California by the Gold Rush came south to Southern California.

The end of isolation and the onset of a nearly continuous boom that has lasted to the present time began with the arrival of railroad connections to the eastern United States. The first connection was indirect, via a line in 1876 to central California, which had already received a transcontinental link (1869). Then in the 1880s lines of the Southern Pacific and Santa Fe railroads tied Los Angeles directly to the East. They crossed the encircling mountains via the Cajon Pass at San Bernardino. To help settle the Los Angeles Basin and promote development that would supply traffic, the railroads advertised in the East for settlers and offered extremely low fares westward. Oranges shipped in refrigerated cars were the main local product that could stand the transportation cost to far-off eastern markets, and irrigated orchards spread rapidly. Los Angeles was the main transport and business center. Then in the 1890s a disastrous freeze in Florida's groves put California in the lead in American citrus production, and the state maintained this position until urban sprawl replaced many of the Southern California groves in the mid-20th century. The benign climate which favored citrus also attracted new population for reasons of health and comfort, as it still does.

The Southern California climate, plus irrigation water from streams descending the mountains, supported the early citrus boom; and these and other natural resources have continued to undergird a rush of settlement. Oil was discovered in the Los Angeles Basin in the 1890s, and the region is still an important producer and refiner. Fishing fleets on the Pacific (primarily tuna boats) soon became large and wide-ranging. The Pacific Fleet of the United States Navy was based on San Diego's harbor. In the early 1900s pioneering American movie-makers moved from the Northeast to Los Angeles to take advantage of Southern California's bright and warm weather and varied outdoor locales. Success of these ventures made the Los Angeles suburb of Hollywood synonymous with movies and led into production of programs for national television when that medium burgeoned after 1950. In the 1920s and 1930s early aircraft manufacturing was attracted to the area by the favorable climate, the space still available for capacious factories and runways, and the presence of a considerable labor force. This industry boomed to enormous size during World War II and afterward joined forces with the electronics industry in aerospace development. Today the aerospace industry is Southern California's largest manufacturing industry, being heavily supported by federal military expenditures. The Los Angeles metropolis receives more money from federal military contracts than does any other metropolitan area in the country. Beneath its diversity and glittery reputation, Los Angeles has become a major industrial city—an aerospace Detroit with palm trees.

The expanding population of Southern California brought other economic developments not so intimately related to the area's natural resources. A major natural harbor was not available north of San Diego, but large artificial harbors were built early in the 20th century at Los Angeles and adjacent Long

Beach, and today Los Angeles-Long Beach is one of the major seaport areas of the United States. The area's importance in this regard has risen with the increased importance of Pacific countries, especially Japan, in American foreign trade, and the same rise in importance of the Pacific has heightened San Diego's function as a naval base. Massed population made Southern California an increasingly attractive local market and led to the growth of a wide range of activities to serve that market. A major instance (among many) was the development of Los Angeles into the main Western center of auto assembly when that industry spread from the Midwest after World War II. Despite its extreme off-center location within the West, Southern California's large local population represented so much of the West's buying power that the area was the most economical single location from which to ship automobiles and various other products to the whole West. The region's casual subtropical lifestyle, its role as a media center, and its function as a home for many celebrities gave it the impetus to become a fashion center and this factor, along with cheap labor from Mexico, furthered its development as a major manufacturing location for the clothing industry.

All this development would have been impossible without the supply of vast and increasing amounts of water to this dry area. Streams from the surrounding mountains and wells tapping local groundwater resources became inadequate, and water has been brought by aqueduct from more distant sources at great expense: 242 miles (*c.* 390 km) from the lower Colorado River (the Colorado River Aqueduct from Parker Dam); 233 miles from the east foot of the Sierra Nevada (the Los Angeles Aqueduct from Owens Valley); and more recent supplies from the western slopes of the Sierra Nevada (Fig. 1.1) via the canal routes of the Central Valley (Fig. 1.2). Schemes have been proposed to acquire water from still more distant areas—the Klamath Mountains or perhaps the Pacific Northwest or western Canada. Economical desalinization of sea water would be one solution if technology permitted. Large-scale supply of water by this means already is present in oil states around the Persian/Arabian Gulf, but the cost is high and the needs of Southern California would be immensely greater.

The San Francisco Metropolis and the Central Valley

In 1986 another 5.9 million Californians lived in the San Francisco metropolis, which is the second largest metropolitan area in both California and the West. It centers on San Francisco city, located on a hilly peninsula between the Pacific and San Francisco Bay, but it includes urbanized counties on all sides of the Bay. The metropolitan area extends into scenic low mountains in the Coast Ranges both north and south of the Golden Gate entrance to the Bay, while eastward it extends into the flat plain of the Central Valley. The climate is unusual. Marine conditions are shown by San Francisco's very moderate temperatures, which range from a low-month average of 48°F (9°C) in December to a high-month average of only 64°F (18°C) in September. The area is in the border between the marine west coast and mediterranean climates of California, with the mediterranean influence being shown by the fact that about 80 percent of the rainfall comes in five wet months from November through March. San Francisco's mean annual precipitation is only 20 inches, but the moderate temperatures have enabled this amount of moisture to produce a naturally forested landscape. There are even groves of tall redwoods within a short distance both north and south of the Golden Gate (Fig. 26.12). In the part of the metropolitan area east of San Francisco Bay, somewhat warmer temperatures and lower precipitation have produced a vegetation of scrubby forest (chaparral or oakwoods) or natural grassland.

San Francisco began as a Spanish fort (*presidio*) and a mission on the south side of the Golden Gate and subsequently developed as a port for first the gold-mining and then the expanding agricultural economy of central California. The California Gold Rush of 1849 and subsequent years centered on the gold-bearing alluvial gravels of the many Sierran streams entering the Central Valley along a foothill belt extending both north and south of Sacramento. San Francisco, with its spacious and protected natural harbor, was the nearest port to receive immigrants and supplies for the gold camps. When the Gold Rush abated in the mid-1850s, many settlers turned to ranching and wheat farming in the grasslands of the Central Valley, and San Francisco continued to grow. Its trade area was enlarged by the extension of railroads into the Central Valley and across the conveniently located Donner (or Truckee) Pass of the Sierra Nevada leading toward Salt Lake City. Rail connection with the East was completed in 1869 at a point near Salt Lake City when San Francisco's rail line eastward (the Central Pacific) met the Union Pacific advancing westward.

THE SUPERPRODUCTIVE AGRICULTURE
OF THE CENTRAL VALLEY

Intensification and diversification of agriculture in the Central Valley has made California the leading American state in agricultural output (Fig. 28.13).

This result has been achieved mainly by progressive extension of irrigation. California now has more irrigated land by far than any other state, and the bulk of it is in the Central Valley. Small irrigation schemes which private enterprise began to develop over a century ago have now been overshadowed by massive works in the publicly funded Central Valley Project of the federal government and the California Water Project of the state. Dams in Sierran valleys store the water of streams carrying meltwater from the deep snows of the Sierra, and a huge canal system distributes the water. The latter is moved generally southward from the wetter northern part of the Central Valley (the Sacramento Valley, drained by the Sacramento River) and the adjacent Sierra to the markedly drier southern parts of the Valley (the San Joaquin Valley—drained by the San Joaquin River—and the Tulare Basin of the extreme south). Large supplies of irrigation water also come from wells that tap the deep infill of water-bearing sediments underneath the Central Valley.

In recent years, about one tenth of the value of all United States farm marketings has come from California. Largely produced on the irrigated lands of the Central Valley, these products represent an output of amazing variety. The state's two leading crops by value are grapes—retailed principally in the form of wine—and cotton. A large majority of the grapevines in the United States are in the southern and central parts of the Central Valley, or in a second major concentration in valleys of the Coast Ranges (the Napa Valley and others) slightly north of San Francisco. California's cotton acreage is concentrated in the southern half of the Central Valley. The state is a distant second to Texas in cotton acreage, but its yields average so much higher that it is a much closer second in total output, and actually ranked first in a recent year. Rice is another important crop of the state and the main crop of the Sacramento Valley. Fruits and vegetables are often the top-ranking crops in particular localities. Almost the complete range of subtropical and temperate fruits and vegetables is produced somewhere in California's orchard and truck-farming districts. These districts are largely in the Central Valley but also are found in the *Salinas Valley* and other broad valleys of the Coast Ranges south from San Francisco, as well as in the irrigated desert of the *Imperial Valley*, which lies below sea level in California's southeast (Fig. 28.24). The flat Imperial Valley, deeply infilled

Figure 28.24 *Infrared photography from space reveals here a solid block of highly productive irrigated land in California's Imperial Valley south of the Salton Sea. Extreme desert conditions make agriculture wholly dependent on irrigation water brought by a canal from the Colorado River.*

by sediments overlying a crustal downfold, receives its irrigation water by a canal from the Colorado River (the All-American Canal from Imperial Dam). From the Salton Sea at the north to the Mexican border, this valley is solidly in cultivation, and there is a smaller irrigated development on the Mexican side.

The production and sale of individual fruits and vegetables from California, and certain kinds of nuts as well, is often very large. California is the leading state, for example, in sales of grapes, peaches, pears, plums, lemons, figs, dates, asparagus, broccoli, cantaloupes, carrots, cauliflower, celery, lettuce, onions, tomatoes, strawberries, almonds, and English walnuts. The state's citrus industry, largely shifted now from Southern California to the San Joaquin Valley, is second to Florida's. California's large population, amounting to over one tenth of that of the nation, is a major market for the food crops just named, and huge additional quantities are sold in other parts of the country. But the two most valuable items sold from California farms are milk and beef cattle (Fig. 28.25) to supply markets within the state. California's population is overwhelmingly and increasingly urban and suburban, and the growing demand for basic foods tends to increase the proportion of farm land devoted to irrigated alfalfa and other feeds for beef and dairy cattle. Cattle also are fed on residues from the processing of the state's large production of sugar beets. A sizable output of poultry products adds further to California's rich agricultural variety.

San Francisco is the overall economic capital for the agriculture of the Central Valley and nearby Coast Range valleys. Like Los Angeles, it performs major financial and business functions for great areas in the West. But San Francisco's towering importance has not precluded the emergence of some fairly large cities as regional centers for parts of the Central Valley. *Sacramento* (1.3 million), the state capital, is the largest of these. It occupies a central location in the Valley, on the main rail and highway routes between San Francisco and Donner Pass over the Sierra. *Fresno* (590,000) is the dominant center for the San Joaquin Valley. Among still smaller cities *Bakersfield* (495,000) in the extreme south is outstanding as both a local agricultural service center and the operating base for oil fields in its area.

HIGH TECHNOLOGY IN "SILICON VALLEY"

San Francisco has added a sizable industrial sector to its original functions as a port and regional economic center. Manufacturing in the metropolitan area is highly varied, but in recent years the making of silicon chips in a strip of land near the south-

Figure 28.25 *A complex set of geographic relationships is reflected in this California scene of irrigated fruit-growing, livestock grazing on natural range, and transmission of electric power. The locale is the western edge of the Central Valley where it abuts on the Coast Ranges. Explanation of the scene would require analysis of numerous spatial, ecological, and regional factors and interrelations. Hence the view offers a stimulus to geographic thinking.*

western shore of San Francisco Bay has become the outstanding line. This area is now commonly referred to as "Silicon Valley." It is the main center in the United States for design and production of the chips which carry micro-circuits for computers and other products of modern electronics. The rise of Silicon Valley as a center of high technology is due in part to the attractiveness of the San Francisco area as a place to live and the nearby presence of two outstanding universities—Stanford University at Palo Alto in the valley (Fig. 28.26) and the University of California at Berkeley near the large industrial center of Oakland across the Bay from the city of San Francisco. However, by the mid-1980s the chip-based industrial boom of the valley was being seriously eroded by Japanese competition.

Figure 28.26   California's Hispanic heritage is evident in the buildings of the Quad at Stanford University. The view looks westward to one of the Coast Ranges.

Seattle and Portland in Their Pacific Northwest Setting

The Puget Sound Lowland of Washington and the Willamette Valley of Oregon to its south are not dry areas in need of irrigation. They have abundant precipitation and luxuriant natural forests, as do the adjoining Coast Ranges and Cascades. But these moist conditions end abruptly near the crest of the Cascades. Hence the lowlands and adjacent mountains may be viewed collectively as oasis-like in the sense of being a sharply limited area of abundant water and clustered population within the generally dry region of the American West. The clustering centers in the metropolitan areas of two port cities, each located on a navigable channel extending inland from the Pacific through the Coast Ranges. *Seattle* (2.3 million), located on the eastern shore of Puget Sound, has access to the interior via passes over the Cascades. *Portland* (1.4 million) is on the navigable lower portion of the Columbia River where the Willamette tributary enters it, and has a natural route to the interior via the Columbia Gorge through the Cascades. Besides the populations of these two metropolitan areas, the Willamette-Puget Sound Lowland contains somewhat over a million other people, most of whom live in a north-south line of small to medium-sized cities.

Fishing, lumbering, and sea trade were the main pursuits that built these cities in the latter 19th and early 20th centuries. But it was agricultural land in the Willamette Valley that drew the first large contingents of settlers from the East to the "Oregon Country" in the 1840s. Wheat, flour, butter, eggs, and vegetables from the area were marketed in California during the Gold Rush. Today agriculture is still of some importance in the Willamette-Puget Sound Lowland, although it is handicapped by hilly terrain and the prevalence of poor leached soils. The leading products are milk (also produced outside the Lowland; see Fig. 26.7) and truck crops for sale in nearly urban markets. A considerable amount of fishing (primarily for salmon) is still carried on, but the combined fish catch of Washington and Oregon is much smaller than that of either of the two leading states—Alaska and Louisiana. However, Oregon and Washington continue to be the leading states in production of lumber, and they still have about 30 percent of the sawtimber resources of the United States. The trees are mostly coniferous softwoods, and many are enormous in size. The famed Douglas fir has been the leading timber source from the beginning. The timber industry is still a basic element in the economies of both Oregon and Washington, although more so in Oregon's smaller and less diversified economy.

The trading hinterlands of Seattle and Portland extend across the Cascades and the Columbia-Snake Plateau to the Rocky Mountains. Along or near the eastern slope of the Cascades are scattered irrigated areas, including the Wenatchee and Yakima valleys that make Washington the leading state in apple production. Near the eastern foot of the Cascades in Washington the mighty Columbia River flows southward. Near the Oregon line the river is joined by its main tributary, the Snake, and then turns westward to the Pacific through the gorge it has cut into the Cascades. Beginning in the 1930s, the Columbia was harnessed by a series of federal dams, of which the largest is Grand Coulee Dam. These dams have supplied power to an area lacking in coal and oil, have improved navigation on the lower Columbia, and have attracted a major concentration of power-hungry aluminum plants to the Pacific Northwest. They also have supplied the water for a huge expansion of irrigation in recent decades in the Columbia Plateau of both Washington and Oregon.

In southeastern Washington the Columbia Plateau rises high enough to have a steppe rather than a desert climate and is mantled by a deep hilly covering of loess. This area, called the *Palouse,* grows unirrigated wheat, primarily winter wheat but with some spring wheat. Large-scale production is achieved despite an annual precipitation averaging as low as 9 inches (23 cm) in some places. Relatively low temperatures and water-retentive soils increase the effectiveness of the scanty moisture that falls. At the northeastern edge of the Palouse the small metropolis of *Spokane* (360,000), economically subordinate to the coastal ports, serves as the local capital of the "Inland Empire" in the northern Columbia Plateau and adjacent Rockies. Lead and silver are mined around Coeur d'Alene in the Idaho Rockies.

But the natural resources of the Pacific Northwest fail to account for its single leading industry: aircraft manufacturing by the Boeing Company. The economy of metropolitan Seattle is deeply affected by the fortunes of this huge concern. The company began here in the 1920s (at the outlying suburb of Everett north of Seattle), expanded vastly during World War II, and has since maintained its position as a world leader in aircraft manufacturing. It has done so despite a series of drastic production and employment fluctuations. The enormous Boeing 747s and various other commercial and military aircraft are marketed around the world (Fig. 28.27).

Population Clusters in the Mountain States

The eight inland states in the Mountain West are drier than the Pacific Mountains and Valleys, lack the abundant irrigation water of California, and lack the advantage of water transportation. These handicaps to development have been so constricting that the Mountain states had only an estimated 13 million people in 1986. About half of these people are concentrated into a handful of irrigated districts where extensive metropolitan growth has taken place. Outside of these clusters the common pattern

in one of the Mountain states is that of a few hundred thousand people scattered across one of the larger American states—on ranches and farms, in Indian reservations, or in small service centers and mining towns.

PHOENIX AND TUCSON

The largest population cluster in the Mountain West lies in the Basin and Range Country of southern Arizona. Here live about three quarters of Arizona's people, in the adjacent metropolitan areas of *Phoenix* (1.9 million) and *Tucson* (600,000). Water from the Salt and Colorado rivers, plus sizable resources of ground water, makes it possible to support this urban development despite the subtropical desert climate. Phoenix (Fig. 28.28) was originally the business center for an irrigated district along the valley of the Salt River. The district expanded after 1912, when the first federal irrigation dam in the West (Roosevelt Dam) was completed on the river. Cotton is the main crop of the valley, although hay to support dairy and beef cattle is important. Both Phoenix and Tucson accelerated in growth after the introduction of air conditioning made them attractive retirement centers. Each has a major state university, and Phoenix is the state capital. Both cities have mild winters, but summers are uncomfortably hot. Tourism and copper mining, the latter mainly in the southeastern part of the state, are important in Arizona. The state leads all others in mined copper, with a little over two thirds of the national output in recent years.

Figure 28.27 The Boeing plant at Everett, Washington. The photo shows a Boeing 747 under construction for export to Japan.

Figure 28.28   Phoenix, Arizona, has two relatively small business cores in the heart of the city, one of which is seen in this view. The Phoenix metropolis, which spreads through a desert landscape, has many suburban business centers serving residential subdivisions. These activities and land uses surround the barren mountains of the Basin and Range Country that are seen in the distance. The view is toward the west, but such mountains frame the city in all directions.

THE COLORADO PIEDMONT

Another important population cluster in the Mountain states stretches along a strip of Great Plains country at the immediate eastern foot of the Rocky Mountains. Within this area, called by geographers the Colorado Piedmont, metropolitan *Denver* contained an estimated 1.8 million people in 1986, and another 800,000 lived in metropolitan *Colorado Springs* (380,000) or smaller cities to the south and north of Denver.

These cities at the mountain foot developed on streams emerging from the Rockies onto the Great Plains. The stream valleys provided routes into the Rockies, where a series of mining booms began with a gold rush to Central City, upslope from Denver, in 1859. One mineral rush followed another through the later 19th century, with various minerals being exploited at scattered sites in the mountains. Denver and other Colorado Piedmont towns became supply and service bases for the mining communities, which often were short-lived. Irrigable land along Piedmont rivers was used to grow food for the mines, and agricultural servicing became an important function of the Piedmont urban nodes. In the course of time agriculture grew into a prosperous business with more stability than the mining industry in the mountains.

In the 20th century the Piedmont centers have grown at a rapid rate. This growth has been especially spurred by developments in agriculture, mining, tourism, and federal government activity. Irrigated acreage in Colorado has expanded, being aided the most spectacularly by the Colorado-Big

Thompson Project which transfers water by a tunnel from the west side of the Front Range to the Piedmont. This diversion of water from the upper Colorado River to the South Platte River helped foster a boom in irrigated feed to support some of the nation's largest cattle feedlots. Meanwhile Colorado's highly mineralized Rocky Mountains are an important producing area for molybdenum, vanadium, and tungsten. These comparatively rare alloys are much in demand for high-technology uses. The state has huge coal reserves, mainly in its Colorado Plateau section, and deposits of natural gas and oil exist in scattered locations. High energy prices in the 1970s led to expanded extraction of these fuels. Since World War II, tourism in the Colorado Rockies has profited greatly from increasing American affluence, the development of better road and air transportation, and the rise of skiing as a popular sport. Denver is the business capital for a series of major resort communities in the Rockies, as well as being the state capital, a major center for Western regional offices of the federal government, and an important transportation and general business focus for a huge area in the Rockies and the Great Plains. The present high development of tourism in Colorado was foreshadowed in the 19th and early 20th centuries by the arrival of many health-seekers who were attracted to the high plains and adjacent Rockies by the pure dry air and bracing but not excessively cold temperatures. Colorado Springs is an important focus of tourism and also profits economically from the nearby presence of several defense installations, including the United States Air Force Academy and the headquarters of the North American Air Defense Command, the latter located in a redoubt inside a mountain.

THE WASATCH FRONT (SALT LAKE) OASIS

Third in size among the major population clusters of the interior West is the Wasatch Front or Salt Lake Oasis of Utah, comprised of metropolitan *Salt Lake City-Ogden* (1 million) and *Provo* (240,000) to its south. The Oasis, which contains about three fourths of Utah's people, occupies an elongated north-south strip between the Wasatch Range of the Rockies and the Great Salt Lake, with a southward extension along the foot of the Wasatch. In all the rest of the huge state of Utah, encompassing sections of the Basin and Range Country, Colorado Plateau, and Rockies, there were only a little over 400,000 people in 1986.

The oasis along the Wasatch Front was developed as a haven for the Mormon religious sect when it was driven from the eastern United States in the 1840s. After a difficult but well-organized trek westward, the newcomers began irrigation development based on water from the Wasatch Range, and they built at Salt Lake City the Temple which is the physical and symbolic center of the Mormon religion and culture. From this center, Mormon settlers spread widely into all the surrounding states and California. United States expansionism soon overtook the Mormons, and Salt Lake City became the political capital of Utah Territory and then of the state of Utah (1896). In addition to its political significance the city remains a religious and cultural capital, and it has developed important economic functions as well. As the largest place on Interstate 80 between Sacramento and Chicago, and with no large competing metropolis within vast stretches in other directions, Salt Lake City is the business capital for a huge although sparsely populated area outside the Mormon core region along the Wasatch Front.

The agriculture on which the early Utah settlements depended has expanded through the years to supply a good part of local needs for farm products but is quite limited in scale. Needed supplements can be secured from much larger irrigated districts near at hand along the Snake River in southern Idaho, or can be brought by main-line transcontinental rail and highway links from California. Mining has long been a mainstay of the Utah economy. The leading products are oil and coal. They long included copper from one of the world's largest open-pit mines at Bingham Canyon near Salt Lake City but this mine closed in 1986. Besides coal, the Colorado Plateau section also has iron ore, and these materials are

used at Geneva, near Provo, by the only steel mill in the Intermountain area of the West. Many other minerals are produced at scattered locations in the state.

Less Populous States of the Interior West

The remaining five states of the interior West are still less populous, and their largest metropolitan areas are comparatively small, as may be seen in the following tabulation:

State	Estimated Population 1986	Largest Metropolitan Area	Estimated Population of Metropolitan Area, 1986
New Mexico	1,480,000	Albuquerque	475,000
Idaho	1,000,000	Boise City	195,000
Nevada	965,000	Las Vegas	570,000
Montana	820,000	Billings	120,000
Wyoming	510,000	Casper	70,000

The aggregate population of these five states is substantially smaller than that of metropolitan Philadelphia, but it is spread irregularly over an area somewhat larger than that occupied by more than 50 times as many people in the European nations of France, Italy, West Germany, the United Kingdom, and the Low Countries. The five states have all increased in population at a rapid rate during recent decades, but their aggregate increase in actual numbers has been less impressive: barely 1 million total between 1970 and 1980.

New Mexico's people are largely strung along a series of irrigated strips in the north-south valley of the upper Rio Grande, although less than 2 percent of the state's population is now comprised of farmers. Similarly, a substantial majority of Idaho's people inhabit irrigated districts in the southern Idaho section of the volcanic Columbia-Snake Plateau called the *Snake River Plains*. This area is the nation's largest producer of potatoes and also has a substantial production of sugar beets and of beef cattle fattened in feedlots. Idaho has a larger proportion of farm population than any other Western state except Montana, but in each of these two states the number of farm people amounted to only about 7 percent of the total population in 1980. About four fifths of Nevada's people live in metropolitan Las Vegas or the smaller cluster of metropolitan Reno (225,000)—cities in which the main sources of support are tourism, gambling, and retirement. Las Vegas is located in stark desert and could not be sus-

Figure 28.29 *A roaming band of Rocky Mountain goats at Glacier National Park in Montana symbolizes the treasured wildlife resources of Federal lands in the American West. Such animals are a major asset to Western tourism.*

tained on its present scale without the water supply impounded by nearby Hoover Dam on the Colorado River. This fact highlights the vital importance of Colorado River water to all the states that share the river's basin. The river itself, fed by melting snows and rainfall in the Rocky Mountains, carries a volume of water far smaller than that of the Columbia or the Mississippi, but this flow is crucial to so many communities and agricultural areas that numerous court battles have been fought over water rights. The most important controversies have erupted between the state of California, which appropriates large quantities of water for use in Southern California and the Imperial Valley, and the state of Arizona, which has pressed for water to sustain explosive urbanism in Phoenix and Tucson and a growing output of irrigated cotton. Arizona's search for new water has recently been rewarded by the Central Arizona Project, an expensive federally funded series of dams and canals that now feeds Colorado River water to metropolitan Phoenix, with an extension of Tucson under construction (Fig. 26.11).

Competition for scarce water is only one of many resource conflicts in the West. Another prominent controversy in recent years has centered in Montana and Wyoming, where huge deposits of coal underlying range land are being exploited by strip mining. The coal has a low proportion of sulfur and hence is desirable fuel for thermal-electric plants, since its smoke pollutes the air less than smoke from coal with a higher sulfur content. In this area the mining companies that hold rights to the coal are often in conflict with ranchers who own the grazing land on the surface. Much litigation has resulted from the ranchers' insistence that the coal companies go to the expense of reconverting the land to livestock range once the coal has been removed. Such matters are important to Montana and Wyoming, which

are very dependent on livestock raising and mining. Both states also derive considerable income from tourism (Fig. 28.29); and rainfed wheat-growing is important in Montana and small sections of Wyoming.

ALASKA AND HAWAII

In the middle of the 19th century the United States extended its national territory to the Pacific and subsequently began to acquire a widespread array of Pacific dependencies. Two of these, Alaska and Hawaii, were admitted to the Union as states in 1959. The United States gained Alaska by purchase from Russia in 1867, while Hawaii was annexed in 1898 following a turbulent period of American penetration. Acquisition of a Pacific empire was motivated by defense considerations to a considerable degree, and these have continued to play an important role in the development of both Alaska and Hawaii. Large and highly important federal military installations exist in both states, and federal defense expenditures have long been a major element in the economies of both.

ALASKA'S DIFFICULT ENVIRONMENT

Alaska (from *Alyeska,* an Aleut word meaning "the great land") comprises about one sixth of the United States by area, but its territory is so rugged, cold, and remote that its population only amounted to an estimated 530,000 in 1986. The state's difficult environment can be conveniently assessed under four major physical areas: the Arctic Coastal Plain, the Brooks Range, the Yukon River Basin, and the Pa-

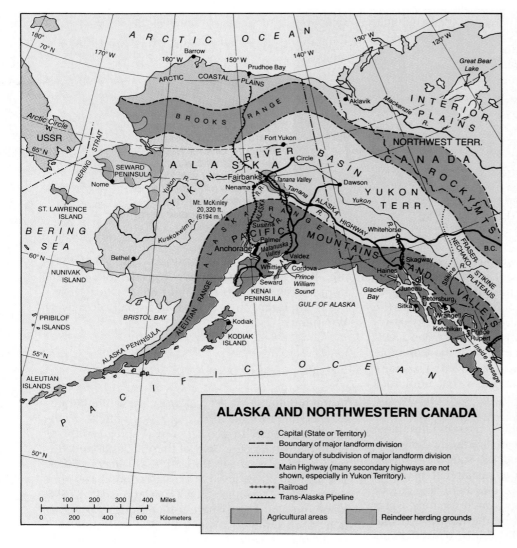

ALASKA AND NORTHWESTERN CANADA

- ⊙   Capital (State or Territory)
- – – –   Boundary of major landform division
- ··········   Boundary of subdivision of major landform division
- ——   Main Highway (many secondary highways are not shown, especially in Yukon Territory).
- +++++   Railroad
- ·······   Trans-Alaska Pipeline
- ▨   Agricultural areas   ▨   Reindeer herding grounds

Figure 28.30   Agricultural and reindeer-herding areas of Alaska are adapted from a map in Karl E. Francis, "Outpost Agriculture: The Case of Alaska," Geographical Review, 57 (1967), 498. Copyrighted by the American Geographical Society of New York; by permission of the author and the Review.

cific Mountains and Valleys. Each represents the northernmost sector of one of the major landform divisions of Anglo-America discussed earlier in Chapter 26 (Figs. 26.9 and 28.30).

1. Alaska's *Arctic Coastal Plain* is an area of tundra climate and vegetation along the Arctic Ocean north of the Brooks Range. Known as "the North Slope," this area contains very large deposits of oil and natural gas, some of which are located underneath the mainland, while others lie offshore.

2. The barren *Brooks Range* is considered to be the northwestern end of the Rocky Mountains. Summit elevations range from about 4000 to 9239 feet (*c.* 1200–2816 m).

3. The *Yukon River Basin* lies between the Brooks Range on the north and the Alaska Range on the south. This is the Alaskan portion of the Intermoun-

tain Basins and Plateaus. The Yukon River which drains it is navigable by river boats for some 1700 miles (*c.* 2700 km) from the Bering Sea to Whitehorse in Canada's Yukon Territory. The Yukon Basin in Alaska is an area of varied topography, generally rolling or hilly, with some sizable areas of flat alluvium that often are swampy. The extreme subarctic climate is associated with a taiga vegetation of coniferous forest that generally is thin and composed of relatively small trees. *Fairbanks* (city proper, 25,000) in the Basin is interior Alaska's main town and focus of transport routes. The *Alaska Highway* connects Fairbanks with the "Lower 48" conterminous states via western Canada, and both paved highways and the *Alaska Railroad* lead southward to Anchorage and other ports on the Gulf of Alaska. Fairbanks is a major service center for the *Trans-Alaska Pipeline* from the North Slope oil area to the port of Valdez in the south. The city is located on the Tanana River tributary of the Yukon and is associated with a small agricultural area in the Tanana Valley.

4. The *Pacific Mountains and Valleys* occupy southern Alaska, including the southeastern *Panhandle.* The mainland ranges of the Panhandle are a northward extension of the Cascade Range and the British Columbia Coastal Ranges, while the mountainous offshore islands are an extension of the Coast Ranges of the Pacific Northwest and the islands of British Columbia. On the mainland many peaks rise above 8000 feet (2438 m) and some in the extreme northwest exceed 15,000 feet (4572 m). Fjorded valleys extend long arms of the sea inland. Level land is rare. A very cool and wet marine west coast climate prevails on both the mainland and the islands, fostering coniferous forests that are a major resource exploited in a sizable way for the Japanese market. Heavy snows have generated huge accumulations of ice at higher elevations. From Juneau northwestward, one of the world's most majestic displays of glaciers, descending from interior ice fields, punctuates a long stretch of coast reaching west to Anchorage and beyond. Many of the ice fields extend across the international border into Canada. Glacier Bay National Monument in the northern Panhandle is a famous attraction for cruise ships using the island-screened *Inside Passage* from Vancouver and Seattle (Fig. 28.31). However, the highest mountains of Alaska do not lie along the immediate coast but are well inland in the great arc of the *Alaska Range,* where Mt. McKinley rises to 20,320 feet (6194 m) about 100 miles north of Anchorage. This is the highest peak in either the United States or Canada. At the southwest the Alaska Range merges into the lower chain of mountains that occupies the long and narrow *Alaska Peninsula* and the rough, almost treeless *Aleutian Islands.* Immediately south of Anchorage the *Kenai Peninsula* lies between *Cook Inlet* and the *Gulf of Alaska,* and still farther to the southwest is Alaska's largest island, *Kodiak.* These features mark the western end of the horseshoe of coastal mountains that curves across the north of the Gulf of Alaska from the coast and islands of the southeastern Panhandle. The climate of lower areas in this western coastal section is harsher than in the Panhandle, being classified as a relatively mild subarctic.

Southern Alaska is very much a part of the "Pacific ring of fire," with many volcanoes in the Aleutians and Alaska Peninsula, some 36 of which are reported to have erupted since the first Russian explorations in the mid-18th century. Frequent earthquakes give further evidence of crustal instability. Anchorage was devastated by a giant quake in 1964.

Figure 28.31 *The spectacular river of ice is the Margerie Glacier at Glacier Bay, Alaska. In the foreground is the cruise ship* Noordam *of the Holland America Line.*

FROM FUR TRADE TO OIL AGE: CYCLES OF ALASKAN DEVELOPMENT

After approximately two centuries of occupation by Russia and then the United States, the population of Alaska was still only 73,000 in 1940. Eighteenth-century Russian fur traders from Siberia spread very thinly along the southern and southeastern coasts. Evidence of this era persists in the many Russian names of places and features in Alaska. In 1812 the Russians established an agricultural colony on the California coast at Fort Ross north of San Francisco Bay. This effort to more adequately provision the Alaskan posts was abandoned in 1841, and then in 1867 remote and unprofitable Alaska was sold to the United States. By that time relentless hunting had nearly wiped out the sea otters whose glossy and expensive pelts had provided the main incentive for Russia's Alaskan fur trade. Only recently under American and Canadian governmental protection has a substantial beginning been made at replenishing these mammals that once were incredibly numerous from California north. Under the Americans there was a burst of settlement connected with gold rushes in the late 1890s and for a few years thereafter, followed by a very slow increase connected largely with fishing and military installations.

Beginning with World War II, during which Japanese troops invaded the Aleutian Islands, growth became more rapid. By 1986 an estimated 530,000 people lived in Alaska: more than 7 times the population of 1940 but still an extremely small number for such a large territory. About 235,000 of these lived in metropolitan *Anchorage* at the head of Cook Inlet. The rest were erratically distributed over the state, but living mainly in or near Fairbanks, the state capital of *Juneau* (somewhat over 20,000 with suburbs) in the Panhandle, or other small and widely spaced towns and villages. Of the total population about one sixth is comprised of "Alaska Natives" (Eskimos, American Indians, or Aleuts). Long the victims of exploitation and neglect by whites, these peoples have recently received some redress in large land grants and cash payments under the Alaska Native Claims Settlement Act passed by the federal Congress in 1971.

The burst of growth since 1940 has been principally due to (1) increased military and governmental employment; (2) the rise of the fishing industry to major importance, with salmon by far the main product; (3) revenues accruing from the development of Arctic oil (discovered in 1965), including the building of the pipeline to Valdez; (4) the rise of air transport as a major carrier of both people and freight; and (5) visits by increasing numbers of tourists, who are principally attracted to the state by the magnificent scenery. Agriculture consists largely of high-cost dairy farming in a few restricted spots. Both food and other consumer goods come overwhelmingly from the "Lower 48" states. The result is the highest cost of living among the states, which largely negates a per capita personal income that is also the country's highest.

HAWAII: AMERICANIZATION IN A POLYGLOT SOUTH SEAS SETTING

The state of Hawaii consists of eight main tropical islands (Fig. 19.1), lying just south of the Tropic of Cancer and somewhat over 2000 miles (*c.* 3200 km) from California. The islands are volcanic and largely mountainous, with a combined land area of about 6400 square miles (*c.* 17,000 sq. km), which is somewhat greater than the combined areas of Connecticut and Rhode Island. The island of Hawaii is by far the largest of the group and contains the only active volcanic craters. The state lies in the path of trade winds which produce extreme precipitation contrasts within short distances. Windward slopes receive heavy precipitation throughout the year (rising to an annual average of 460 inches/1168 cm on Mt. Waialeale in the island of Kauai—the wettest spot

on the globe), while nearby leeward and low-lying areas may receive as little as 10 to 20 inches (25–50 cm) annually. Some areas are so dry that crop-growing requires irrigation. Temperatures are tropical except for colder spots on some mountaintops. Honolulu averages 81°F (27°C) in August and 73°F (23°C) in January; the average annual precipitation at the city is 23 inches (58 cm), with a relatively dry high-sun period ("summer") and a wetter low-sun period ("winter").

Hawaii was more populous than Alaska and was the site of a Polynesian monarchy when it was acquired by the United States. By 1986 its estimated population was a little over 1 million, of whom about 820,000 were concentrated in metropolitan Honolulu (chapter-opening photo of Chapter 19). The metropolis occupies the island of Oahu, with the city proper being located near the southeastern corner of the island. Immigration from the Pacific Basin has given the state's population a unique mix by ethnic origin: one third white Caucasian, one quarter Japanese, somewhat under a fifth native Polynesian, 14 percent Filipino, 6 percent Chinese, and with many other elements such as Koreans and Samoans. Much racial and ethnic mixing has taken place. In general, the population is thoroughly Americanized.

The economy which sustains these people is based overwhelmingly on military expenditures, tourism, and commercial agriculture. The main defense installations, including the Pearl Harbor naval base, are on Oahu. The main agricultural products and exports—cane sugar and pineapples—have long been grown on large estates owned and administered by Caucasian families and corporate interests, but worked mainly by non-Caucasians. Work on such estates has often been the first employment of new immigrants. Markets are mainly in the conterminous United States. Tourism has been the great growth industry of recent decades, propelled by the development of air transportation and the growing affluence of much of the mainland American population. Visitors from Japan and other Pacific rim countries are common. The main attractions for tourists are the tropical warmth, beaches, surf, spectacular scenery, exotic cultural mix, and highly developed tourist facilities. Some of the "South Seas" glamour of the islands has been eroded by the remorseless spread of commercial and residential development.

References and Readings

Geography and related subjects provide an overwhelming number and variety of references and readings appropriate to users of this text. Only a careful sampling is attempted here, heavily weighted toward quite recent material. Most citations are dated later than 1975. Most frequently cited are books and articles by professional geographers, but especially useful publications by scholars in other fields are included, as well as some by journalists and novelists. The citations are grouped under the nine part titles of the book, with appropriate subgroupings by topic and geographic area.

PART I: THINKING ABOUT GEOGRAPHY

The citations under this heading pertain to the entire text or to Chapters 1–3.

General Bibliography *Current Geographical Publications* (University of Wisconsin—Milwaukee Library for the American Geographical Society; 10 issues a year) provides the most useful running list of books, articles, and maps. A huge compilation of such citations is contained in the multi-volume *Research Catalog of the American Geographical Society* (1961) and *Supplements* (1972, 1974, 1978), published by G. K. Hall. *Bibliographie Géographique Internationale* (Paris: Centre National de la Recherche Scientifique; quarterly) is another major source, as is *Geo Abstracts* (Norwich, England, published in 7 bimonthly series). Citations in the latter two sources are annotated. Broad bibliographic guidance is offered by S. GODDARD, ed., *A Guide to Information Sources in the Geographical Sciences* (Barnes & Noble, 1983); this source cites bibliographies and also lists thousands of book and article titles arranged by selected topical and regional fields. See also C. D. HARRIS, *Bibliography of Geography*, Part I, *Introduction to General Aids* (University of Chicago, Department of Geography Research Paper 179, 1976). A massive annotated list of 2903 books, atlases, and serials is contained in C. D. HARRIS, editor in chief, *A Geographical Bibliography for American Libraries* (Washington, D.C.: Association of American Geographers/National Geographic Society, 1985). Valuable bibliographic discussions and reference footnotes on many parts of the world will be found in M. W. MIKESELL, ed., *Geographers Abroad: Essays on the Problems and Prospects of Research in Foreign Areas* (University of Chicago, Department of Geography Research Paper 152, 1973).

Encyclopedias, Gazetteers, Dictionaries Geography students and teachers should not overlook the utility of standard encyclopedias, which often contain geographic articles written by professional geographers. The one-volume *Columbia Encyclopedia* (Columbia[1], 1975), a marvel of succinct writing, contains so much essential information from all fields of knowledge that it is practically a library in itself. Among other things, it functions as a world gazetteer, containing entries for thousands of places. A far larger gazetteer is L. E. SELTZER, ed., *The Columbia-Lippincott Gazetteer of the World, With 1961 Supplement* (Columbia, 1962). *Webster's New Geographical Dictionary* (Merriam-Webster, 1984) is a world gazetteer describing 47,000 places. See also W. G. MOORE, *The Penguin Encyclopedia of Places* (Penguin Books, 1978); D. MUNRO *et al.*, eds., *Chambers World Gazetteer: An A–Z of Geographical Information* (Cambridge and W. R. Chambers, 1988); and A. ROOM, comp., *Place–Name Changes since 1900: A World Gazetteer* (Scarecrow, 1979). Dictionaries of geographic terms, concepts, and quotations include A. N. CLARK, *Longman Dictionary of Geography: Human and Physical* (Longman, 1985); B. GOODALL, *The Facts on File Dictionary of Human Geography* (Facts on File, 1987); T. P. HUBER, R. P. LARKIN and G. L. PETERS, *Dictionary of Concepts in Human Geography* (Greenwood, 1988); R. J. JOHNSTON ed., *The Dictionary of Human Geography*, 2d ed. (Basil Blackwell, 1986); F. J. MONKHOUSE, *A Dictionary of Geography* (2d ed.; Edward Arnold, 1970); J. SMALL and M. WITHERICK, *A Modern Dictionary of Geography* (Edward Arnold, 1986); L. D. STAMP and A. N. CLARK, eds., *A Glossary of Geographical Terms* (3d ed.; Longman, 1979); U.S. CENTRAL INTELLIGENCE AGENCY, *The World Factbook* (annual); J. O. WHEELER and F. M. SIBLEY, *Dictionary of Quotations in Geography* (Greenwood, 1986).

Atlases and World Histories *Goode's World Atlas* (17th ed.; Rand McNally, 1986) is a versatile and relatively inexpensive all-purpose student atlas incorporating not only topical and regional maps, but many kinds of statistical data and a pronouncing index. Larger and more elaborate reference atlases include the *Times Atlas of the World* (7th Comprehensive ed.; Times Books, 1985); *New Interna-*

---

[1]"Columbia" denotes Columbia University Press. Subsequent citations to university presses follow this model.

tional Atlas (Rand McNally, 1984); *Prentice-Hall American World Atlas* (Prentice-Hall, 1985); *Rand McNally Commercial Atlas* (annual); *National Geographic Atlas of the World* (5th ed.; National Geographic Society, 1981); *Hammond Ambassador World Atlas* (Hammond, 1987). G. BARRACLOUGH, ed., *The Times Atlas of World History* (rev. ed.; Times Books/Hammond, 1984) is a magnificently produced combination of multicolored maps with textual summaries. See also *The Harper Atlas of World History* (Harper & Row, 1987). These atlases can be highly useful in providing historical context for all parts of the present text, as can the historical summaries and maps in such books as J. A. GARRATY and P. GAY, eds., *The Columbia History of the World* (Harper & Row, 1972); W. H. McNEILL, *Rise of the West: A History of the Human Community* (Chicago, 1970), also, *A World History* (3d ed.; Oxford, 1979); and *The Human Condition: An Ecological and Historical View* (Princeton, 1980); G. PARKER, *The World: An Illustrated History* (Harper & Row, 1986); H. THOMAS, *A History of the World* (Harper & Row, 1979). See also D. LOWENTHAL, *The Past Is a Foreign Country* (Cambridge, 1985).

Sources of Statistics The annual statistical publications of the United Nations are basic, especially the *Statistical Yearbook, Demographic Yearbook, Yearbook of International Trade Statistics*, and *FAO Production Yearbook*. The annual *Population Data Sheet* published by the Population Reference Bureau, Washington, D.C., offers an extraordinarily handy tabulation of social and economic data by country. Annual statistical profiles of countries and tables of comparative international data occupy many pages in the *Britannica Book of the Year;* the *Statesman's Year-Book* also includes much statistical data, as do other yearbooks and almanacs. The annual *Geographical Digest* (George Philip) is both a convenient source of world statistical data and a record of significant geographic changes. The annual *Statistical Abstract of the United States* (U.S. Bureau of the Census) is a vast summation of all kinds of statistics, including some international comparisons by country. Comprehensive data on minerals are reported annually by country, U.S. state, and mineral in the *Minerals Yearbook* (U.S. Bureau of Mines). *Agricultural Statistics* (annual; U.S. Department of Agriculture) reports agricultural data for both the United States and foreign countries. V. SHOWERS, *World Facts and Figures*, 3d ed. (Wiley, 1989) is a general compilation of data concerning natural features, climate, cities, population, and other subjects. Other statistical sources are noted under appropriate topical and regional headings below.

Periodicals and Regional Book Series *The Geographical Magazine*, written by professional geographers for a general readership, is especially valuable as an adjunct to this text. Illustrated with maps and photos in color, it carries large numbers of short and timely articles on geographic

areas and topics. Often the articles comprise a long series in successive issues. Smaller numbers of articles of the same general kind are available in *Focus* (American Geographical Society: quarterly). The familiar *National Geographic* (National Geographic Society: monthly) is invaluable for maps, and its lavishly illustrated articles provide up-to-date information for general readers. Within the more strictly professional geographic journals in the English language (most of which are quarterlies), articles directly pertinent to this text are most likely to be found in *Geography, GeoJournal, Geographical Review, Journal of Geography, Annals of the Association of American Geographers* (cited below as "*Annals AAG*"), *Geographical Journal*, Institute of British Geographers *Transactions* (cited as "*Transactions IBG*"), *Geoforum*, and *Landscape*. Names of still other journals in geography and many other fields will become apparent in the lists of citations for specific areas and topics.

In our rapidly changing world, it is difficult to keep abreast of day-to-day developments. Regular use of reliable geographically oriented newspapers and magazines such as the *Christian Science Monitor, New York Times, Wall Street Journal*, and *Economist* can be very helpful. *Foreign Affairs* (quarterly) provides serious analyses of important international areas and situations; *Current History* (9 issues a year) regularly devotes entire issues to articles on major world areas such as Latin America, the Soviet Union, or Africa; and the *Swiss Review of World Affairs* (monthly; published in English) provides fine journalistic coverage of world areas in articles that are exceptionally concise. Many well-written articles of geographic import appear in such mass-circulation periodicals as *Technology Review, The Scientific American, Natural History, Fortune, The New Yorker, The Atlantic Monthly, The New Republic*, and the weekly newsmagazines. The total flow of geographically related material in such publications for a general readership is great, and the space available here will permit only a minute sampling of truly outstanding articles.

Two series of volumes on selected countries are TIME–LIFE BOOKS, *Library of Nations* (1984–1987) and U.S. DEPARTMENT OF THE ARMY, *Country Studies* (Government Printing Office, from 1980).

Geography as a Discipline The nature of geography as a field of study and action has been explored by professional geographers in an overwhelming flood of articles, scholarly addresses, papers, and books over a period of many years. A good summation of the field's history is P. E. JAMES and G. J. MARTIN, *All Possible Worlds: A History of Geographical Ideas* (2d ed.; Wiley, 1981). Also basic, although not easy reading for undergraduates, are R. HARTSHORNE, *The Nature of Geography: A Critical Survey of Current Thought in the Light of the Past, Annals AAG*, 29 (1939), 171–658, subsequently reprinted in book form by the AAG; and *Perspective on the Nature of Geography*

(AAG, 1959). R. J. CHORLEY and P. HAGGETT, eds., *Frontiers in Geographical Teaching* (2d ed.; Methuen, 1970) is a valuable collection of essays on many aspects of geography. Other useful books and articles include ASSOCIATION OF AMERICAN GEOGRAPHERS, *Geography and International Knowledge* (1982); T. J. BAERWALD, "Thirteen Tips for Teaching Geography in Any Setting," *Journal of Geography*, 86 (1987), 165–170; J. O. M. BROEK *et al.*, *The Study and Teaching of Geography* (Merrill, 1980); M. CHISHOLM, *Human Geography: Evolution or Revolution?* (Penguin, 1975); C. A. FISHER, "Whither Regional Geography?" *Geography*, 55 (1970), 373–389; P. GOULD, *The Geographer at Work* (Routledge & Kegan Paul, 1985); L. GUELKE, *Historical Understanding in Geography* (Cambridge, 1982); P. HAGGETT, *Geography: A Modern Synthesis* (3d ed.; Harper & Row, 1983; a major introductory text); R. A. HARPER, "Geography in General Education: The Need to Focus on the Geography of the Field," *Journal of Geography*, 81 (1982), 122–139; J. F. HART, "The Highest Form of the Geographer's Art," *Annals AAG*, 72 (1982), 1–29; R. KING, ed., *Geographical Futures* (Sheffield, England: Geographical Association, 1985); J. T. KOSTBADE, "A Brief for Regional Geography," *Journal of Geography*, 64 (1965), 362–366, also, "The Regional Concept and Geographic Education," *Journal of Geography*, 67 (1968), 6–12; W. E. MALLORY and R. SIMPSON-HOUSLEY, eds., *Geography and Literature: A Meeting of the Disciplines* (Syracuse, 1987); D. W. MEINIG, ed., *On Geography: Selected Writings of Preston E. James* (Syracuse, 1971); R. MURPHEY, *The Scope of Geography* (3d ed.; Methuen, 1982); NATIONAL COUNCIL FOR GEOGRAPHIC EDUCATION and ASSOCIATION OF AMERICAN GEOGRAPHERS, *Guidelines for Geographic Education: Elementary and Secondary Schools* (1984); NATIONAL GEOGRAPHIC SOCIETY, *Maps, the Landscape, and Fundamental Themes in Geography* (1986); S. J. NATOLI and A. R. BOND, *Geography in Internationalizing the Undergraduate Curriculum* (AAG, 1985); W. D. PATTISON, "The Four Traditions of Geography," *Journal of Geography*, 63 (1964), 211–216; D. C. D. POCOCK, "Geography and Literature," *Progress in Human Geography*, 12 (1988), 87–102; J. PORRITT, "Education for Life on Earth," *Geography*, 73 (1988), 1–8; C. L. SALTER, *et al.*, eds., *Essentials of Geography* (Random House, 1989), also, with W. J. LLOYD, *Landscape in Literature* (Association of American Geographers, Resource Papers for College Geography, no. 76–3, 1977); D. R. STODDART, "To Claim the High Ground: Geography for the End of the Century," *Transactions IBG*, 12 (1987), 327–336; G. F. WHITE, "Geographers in a Perilously Changing World," *Annals AAG*, 75 (1985), 10–15; T. F. WOOD, "Thinking in Geography," *Geography*, 72 (1987), 289–299.

Maps and Mapping A few key references include L. A. BROWN, *The Story of Maps* (Little, Brown, 1949); P. GOULD and R. WHITE, *Mental Maps*, 2d ed. (Allen & Unwin, 1986); D. GREENHOOD, *Mapping*, rev. ed. (Chicago, 1964); P. D. A. HARVEY, *The History of Topographical Maps: Symbols, Pictures and Surveys* (Norton, 1980); D. HOWSE, *Greenwich Time and the Discovery of the Longitude* (Oxford, 1980); J. S. KEATE, *Understanding Maps* (Longman, 1982); P. LEWIS, *Maps and Statistics* (Wiley, 1977); P. C. MUEHRCKE, *Map Use: Reading, Analysis, and Interpretation*, 2d ed., (J. P. Publications, 1986); A. H. ROBINSON *et al.*, *Elements of Cartography*, 5th ed. (Wiley, 1984); M. M. THOMPSON, *Maps for America: Cartographic Products of the U.S. Geological Survey and Others* (Government Printing Office, 1979); N. J. W. THROWER, *Maps and Man: An Examination of Cartography in Relation to Culture and Civilization* (Prentice-Hall, 1972); E. R. TUFTE, *The Visual Display of Quantitative Information* (Cheshire, CT: Graphics Press, 1983); J. TYNER, *The World of Maps and Mapping* (McGraw-Hill, 1973); J. N. WILFORD, JR., *The Mapmakers* (Knopf, 1981; a history of cartography for the general reader).

General Physical Geography Multi-volume series include TIME–LIFE BOOKS, *Planet Earth* (1982–1984; 11 lavishly illustrated volumes for the general reader) and D. W. GOODALL, ed., *Ecosystems of the World* (Elsevier, 1977–: to include 29 vols.; very technical). Some individual books and articles for general readers are D. ATTENBOROUGH, *The Living Planet: A Portrait of the Earth* (Collins/BBC, 1984); D. BRUNSDEN and J. DOORNKAMP, eds., *The Unquiet Landscape: Series from the Geographical Magazine* (Wiley, 1978); D. J. CRUMP, ed., *Our Awesome Earth: Its Mysteries and Its Splendors* (National Geographic Society, 1986); "Dynamic Earth," *Geographical Magazine*, vols. 15–17 (1983–1985; 15 articles by various authors); F. W. LANE, *The Violent Earth* (Croom Helm, 1986), a book about destructive natural phenomena; NATIONAL GEOGRAPHIC SOCIETY, *Earth '88: Changing Geographic Perspectives* (1988); S. P. Parker, editor in chief, *McGraw-Hill Encyclopedia of Environmental Science* (McGraw-Hill, 1980).

Dictionaries of physical geography and environment include A. S. GOUDIE, ed., *The Encyclopaedic Dictionary of Physical Geography* (Basil Blackwell, 1985); R. W. DURRENBERGER, *Dictionary of the Environmental Sciences* (National Press Books, 1973); S. E. STIEGELER, ed., *A Dictionary of Earth Sciences* (Barnes & Noble, 1983); J. B. WHITTOW, ed., *The Penguin Dictionary of Physical Geography* (Allen Lane/Penguin, 1984). Numerous references to the environment are contained in K. A. HAMMOND, G. MACINKO, and W. B. FAIRCHILD, eds., *Sourcebook on the Environment: A Guide to the Literature* (Chicago, 1978).

Some general articles on physical geography include J. C. J. DOOGE, "Waters of the Earth," *GeoJounal*, 8 (1984), 325–340; P. KAKELA and R. W. CHRISTOPHERSON, "Life Geosystems: Or New Life to Physical Geography," *Journal of Geography,* 71 (1972), 140–146; G. MANNERS, "Our Planet's Resources," *Geographical Journal,* 147 (1981), 1–22; A. N. STRAHLER, "The Life Layer," *Journal of Geography,* 69 (1970), 70–76; K. WALTON, "The Unity of the Physical Environment," *Scottish Geographical Magazine,* 84 (1968), 5–14.

Among the many general textbooks in physical geography are R. E. GABLER *et al., Essentials of Physical Geography,* 3d ed. (Saunders College, 1987); P. GERSMEHL *et al., Physical Geography* (Saunders College, 1980); A. GOUDIE, *The Nature of the Environment: An Advanced Physical Geography* (Basil Blackwell, 1984); M. P. McINTYRE, *Physical Geography,* 4th ed. (Wiley, 1985); T. L. McKNIGHT, *Physical Geography: A Landscape Appreciation,* 2d ed. (Prentice-Hall, 1987); W. M. MARSH and J. DOZIER, *Landscape: An Introduction to Physical Geography* (Addison-Wesley, 1981); J. G. NAVARRA, *Contemporary Physical Geography* (Saunders College, 1981); A. N. STRAHLER and A. H. STRAHLER, *Modern Physical Geography,* 3d ed. (Wiley, 1987); also, *Elements of Physical Geography,* 3d ed. (Wiley, 1984). See also C. F. BENNETT, *Man and Earth's Ecosystem* (Wiley, 1975); and M. J. SELBY, *Earth's Changing Surface* (Oxford, 1986).

Landforms G. H. DURY, *The Face of the Earth,* 5th ed. (Allen & Unwin, 1986) is a concise introduction. A. K. LOBECK, *Geomorphology: An Introduction to the Study of Landforms* (McGraw-Hill, 1939) is a classic book of timeless value, well illustrated with diagrams and photographs. R. E. MURPHY, "Landforms of the World," is a large map in color which appeared as Map Supplement 9, *Annals AAG,* 58, no. 1 (March 1968). See also R. E. SNEAD, *World Atlas of Geomorphic Features* (Robert E. Krieger/Van Nostrand Reinhold, 1980); F. M. BULLARD, *Volcanoes of the Earth,* 2d rev. ed. (Texas, 1984); R. DECKER and B. DECKER, *Volcanoes* (Freeman, 1981); physiographic diagrams by A. K. LOBECK (available from Hammond, Inc.); and landform maps by E. RAISZ (available from Erwin Raisz, 130 Charles Street, Boston, MA, 02114). For spectacular photos from space, see O. W. NICKS, ed., *This Island Earth* (Washington, D.C.: NASA, 1970); N. M. SHORT, *Mission to Earth: LANDSAT Views the World* (NASA, 1976); N. M. SHORT and R. W. BLAIR, JR., *Geomorphology from Space: A Global Overview of Regional Landforms* (NASA, 1986).

Climate, Oceans, Biogeography, and Soils: General Some general books on climate include R. A. BRYSON and T. J. MURRAY, *Climates of Hunger: Mankind and the World's Changing Weather* (Wisconsin, 1977); L. S. ELLIOTT, *Climate and Man* (McGraw-Hill, 1969); J. R. GRIBBIN, ed., *Climatic Change* (Cambridge, 1978); J. T. HOUTON, ed., *The Global Climate* (Cambridge, 1984); H. H. LAMB, *Climate, History, and the Modern World* (Methuen,

1982); J. G. LOCKWOOD, *World Climatology: An Environmental Approach* (Edward Arnold, 1974); P. E. LYDOLPH, *The Climate of the Earth* (Rowman & Allanheld, 1985); J. E. OLIVER and R. W. FAIRBRIDGE, eds., *The Encyclopedia of Climatology* (Van Nostrand Reinhold, 1987); W. O. ROBERTS and H. LANSFORD, *The Climate Mandate* (Freeman, 1979); M. J. TOOLEY and G. M. SHEAIL, eds., *The Climatic Scene* (Allen & Unwin, 1985); G. T. TREWARTHA, *The Earth's Problem Climates,* 2d ed. (Wisconsin, 1981), also, with L. H. HORN, *An Introduction to Climate,* 5th ed. (McGraw-Hill, 1980). See also C. F. COOPER, "What Might Man-Induced Climate Change Mean?" *Foreign Affairs,* 56 (1978), 500–520; A. RAMIREZM, "A Warming World: What It Will Mean," *Fortune,* July 4, 1988, 102–107.

A massive technical discussion of the world's climates is contained in H. E. LANDSBERG, editor in chief, *World Survey of Climatology* (Elsevier, 1969–1984; 15 vols. by various authors). See also A. D. COUPER, ed., *The Times Atlas of the Oceans* (Times Books, 1983); J. G. HARVEY, *Atmosphere and Ocean: Our Fluid Environment* (Horsham, Sussex, England: Artemis Press, 1976); S. P. PARKER, editor in chief, *McGraw-Hill Encyclopedia of Ocean and Atmospheric Sciences* (McGraw-Hill, 1980).

On biogeography, soils, and diseases, see E. M. BRIDGES, *World Soils,* 2d ed. (Cambridge, 1978); also, "Soil: The Vital Skin of the Earth," *Geography,* 63 (1978), 354–361; A. S. COLLINSON, *Introduction to World Vegetation* (Allen & Unwin, 1977); R. T. COUPLAND, ed., *Grassland Ecosystems of the World* (Cambridge, 1979); S. R. EYRE, ed., *Vegetation and Soils: A World Picture,* new ed. (Edward Arnold, 1975); G. HARRISON, *Mosquitoes, Malaria and Man: A History of the Hostilities since 1880* (Dutton, 1978); G. M. HOWE, ed., *A World Geography of Human Diseases* (Academic, 1977); A. S. MATHER, "Global Trends in Forest Resources," *Geography,* 314 (1987), 1–15; M. MEADE, J. FLORIN, and W. GESLER, *Medical Geography* (Guilford, 1988); I. G. SIMMONS, *Biogeography: Natural and Cultural* (Edward Arnold, 1979); D. STEILA, *The Geography of Soils: Formation, Distribution, and Management* (Prentice-Hall, 1976); J. TIVY, *Biogeography: A Study of Plants in the Ecosphere* (2d ed.; Longman, 1982).

The Tropics: Geography and Development For general views, see M. J. EDEN and J. T. PARRY, eds., *Remote Sensing and Tropical Land Management* (Wiley, 1986); J. HATT, *The Tropical Traveler: An Essential Guide to Travel in Hot Climates* (Pan Books, 1982); M. HINTZ, *Living in the Tropics: A Cultural Geography* (Franklin Watts, 1987); B. W. HODDER, *Economic Development in the Tropics,* 3d ed. (Methuen, 1980); I. J. JACKSON, *Climate, Water and Agriculture in the Tropics* (Longman, 1977); J. G. LOCKWOOD, *The Physical Geography of the Tropics: An Introduction* (Oxford, 1976); J. OLIVER, "A Study of Geographical Imprecision: The Tropics," *Australian Geographical Studies*, 17 (1979), 3–17.

Books and articles on tropical rain forests and savannas include A. C. CHADWICK and S. L. SUTTON, "The Preservation of Tropical Moist Forest to Safeguard Mankind," *International Relations*, 7 (1983), 2304–2322; J. R. FLENLEY, *The Equatorial Rain Forest: A Geological History* (Butterworth, 1979); B. N. FLOYD, "The Rain Forest and the Farmer: Observations and Recommendations," *Geo-Journal*, 6 (1982), 433–442; N. GUPPY, "Tropical Deforestation: A Global View," *Foreign Affairs*, 62 (1984), 928–965; D. R. HARRIS, ed., *Human Ecology in Savanna Environments* (Academic, 1980); D. C. MONEY, *Tropical Rainforests* (Evans Brothers, 1980); N. MYERS, *The Primary Source: Tropical Forests and Our Future* (Norton, 1984); S. NORTCLIFF, "The Clearance of Tropical Rainforest," *Teaching Geography*, 12 (1987), 110–113; J. PROCTOR, "Tropical Rain Forest: Ecology and Physiology," *Progress in Physical Geography*, 9 (1985), 402–413, also, "Tropical Rain Forest: Structure and Function," *ibid.*, 10 (1986), 383–400; P. W. RICHARDS, *The Tropical Rain Forest* (Cambridge, 1952, 1964), also, "The Tropical Rain Forest," *Scientific American*, 229, no. 6 (December 1973), 58–67; A. SASSON, "Development of Forest Resources in Tropical Regions," *Impact of Science on Society*, 30 (1980), 211–216; P. A. STOTT, "Tropical Rain Forest in Recent Ecological Thought: The Reassessment of a Non-renewable Resource," *Progress in Physical Geography*, 2 (1978), 80–98.

Arid Lands: Geography and Development A small selection of titles includes Y. GRADUS, ed., *Desert Development: Man and Technology in Sparselands* (Reidel, 1985); A. T. GROVE, "Desertification," *Progress in Physical Geography*, 1 (1977), 296–310; R. L. HEATHCOTE, *The Arid Lands: Their Use and Abuse* (Longman, 1983); D. L. JOHNSON, ed., "The Human Face of Desertification," *Economic Geography*, 53 (1977), 317–432; R. P. KANE and N. B. TRIVEDI, "Are Droughts Predictable?" *Climatic Change*, 8 (1986), 209–223; A. WARREN, "Shifting Margins of the Desert," *Geographical Magazine*, 16 (1984), 457–462.

Polar and Mountain Environments N. J. R. ALLEN, ed., *Human Impact of Mountains* (Barnes & Noble, 1987); T. ARMSTRONG *et al.*, *The Circumpolar North* (Methuen, 1978); R. G. BARRY, *Mountain Weather and Climate* (Methuen, 1981); F. BRUEMMER *et al.*, *The Arctic World* (Sierra Club, 1985); D. K. HAGLUND and M. H. HERMANSON, "Changing Resource Use in the Arctic," *Focus*, 33, no. 5 (May–June 1983), 1–16; J. D. IVES and R. G. BARRY, eds., *Arctic and Alpine Environments* (Methuen, 1974); B. S. JOHN, *The World of Ice: The Natural History of the Frozen Regions* (Orbis, 1979); L. W. PRICE, *Mountains and Man: A Study of Process and Environment* (California, 1981); D. E. SUGDEN, *Arctic and Antarctic: A Modern Geo-*

graphical Synthesis (Barnes & Noble, 1982); J. H. ZUMBERGE, "Mineral Resources and Geopolitics in Antarctica," *American Scientist*, 67 (1979), 68–77.

Environmental Deterioration; Natural Hazards K. N. BAIDYA, "Firewood Shortage: Ecoclimatic Disasters in the Third World," *International Journal of Environmental Studies*, 22 (1984), 255–272; I. BURTON, R. W. KATES, and G. F. WHITE, *The Environment as Hazard* (Oxford, 1978); J. E. BUTLER, *Natural Disasters* (Heinemann, 1976); J. CORNELL, *The Great International Disaster Book* (Scribner, 1976); M. T. FARVAR and J. P. MILTON, eds., *The Careless Technology: Ecology and International Development* (Natural History Press, 1972); A. GOUDIE, *The Human Impact on the Natural Environment* (MIT, 1987); "Planet of the Year: Endangered Earth" (a special issue), *Time*, January 2, 1989; J. L. SIMON and H. KAHN, eds., *The Resourceful Earth: A Response to Global 2000* (Basil Blackwell, 1984); R. WARD, *Floods: A Geographical Perspective* (Macmillan, 1978); A. WIJKMAN and L. TIMBERLAKE, *Natural Disasters: Acts of God or Acts of Man?* (Earthscan, 1984); J. WHITTOW, *Disasters: The Anatomy of Environmental Hazards* (Georgia, 1979).

General Economic Geography, Economic Development, and Resources R. J. BARNET, "The World's Resources," a three-part article, *The New Yorker*, March 24, March 31, April 7, 1980; B. J. L. BERRY, E. C. CONKLING and D. M. RAY, *Economic Geography* (Prentice-Hall, 1987); L. R. BROWN *et al.*, "State of the Earth, 1985," *Natural History*, 94, no. 4 (April 1985), 51–86; J. H. BUTLER, *Economic Geography: Spatial and Environmental Aspects of Economic Activity* (Wiley, 1980); M. CHISHOLM, *Modern World Development: A Geographical Perspective* (Barnes & Noble, 1982); I. M. CLARKE, *The Spatial Organisation of Multinational Corporations* (Croom Helm, 1985); J. P. COLE, *The Development Gap: A Spatial Analysis of World Poverty and Inequality* (Wiley, 1981); J. P. DICKENSON *et al.*, *A Geography of the Third World* (Methuen, 1983); P. F. DRUCKER, "The Changed World Economy," *Foreign Affairs*, 64 (1986), 768–791; R. A. EASTERLIN, "Why Isn't the Whole World Developed?" *Journal of Economic History*, 41 (1981), 1–17; "Economic Development" (special issue), *Scientific American*, 243, no. 3 (September 1980); D. K. FORBES, *The Geography of Underdevelopment: A Critical Survey* (Croom Helm, 1984); N. GINSBURG, J. OSBORN, and G. BLANK, *Geographic Perspectives on the Wealth of Nations* (University of Chicago, Department of Geography Research Paper 220, 1986); A. GROTEWOLD, "Nations as Economic Regions," *GeoJournal*, 15 (1987), 91–96; K. R. HOPE, "Urbanization and Economic Development in the Third World," *Cities*, 3 (1986), 41–57; B. S. HOYLE and D. HILLING, eds., *Seaport Systems and Spatial Change* (Wiley, 1984); P. P. KARAN *et al.*, "Technological Hazards in the Third World," *Geographical Re-*

view, 76 (1986), 195–208; M. KIDRON and R. SEGAL, *New State of the World Atlas* 2d ed.; (Simon & Schuster, 1984); J. MATTILA, "Innovation: A Neglected Locational Factor," *GeoJournal,* 9 (1984), 187–197; R. M. PROTHERO and M. CHAPMAN, eds., *Circulation in Third World Countries* (Routledge & Kegan Paul, 1985); "Recession and the Third World," *Geographical Magazine,* 55–56 (1983–1984; 6 articles by various authors); J. B. RIDDELL, "Geography and the Study of Third World Underdevelopment," *Progress in Human Geography,* 11 (1987), 264–274; B. ROBERTS, *Cities of Peasants: The Political Economy of Third World Urbanization* (Edward Arnold, 1978); D. A. RONDINELLI, *Secondary Cities in Developing Countries: Policies for Diffusing Urbanization* (Sage, 1983); E. S. SIMPSON, *The Developing World: An Introduction* (Wiley, 1987); "Toward the 21st Century" (a special issue), *Current History,* 88, no. 534 (January 1989); D. WHEELER, *Human Resource Policies, Economic Growth, and Demographic Change in Developing Countries* (Oxford, 1984); WORLD BANK, *World Development Report* (Oxford, annual); WORLD RESOURCES INSTITUTE & INTERNATIONAL INSTITUTE FOR ENVIRONMENT AND DEVELOPMENT, *World Resources* (annual; Basic Books, from 1986); WORLDWATCH INSTITUTE, *Worldwatch Papers* (Washington, D.C.; a continuing series of paperbound reports on economic, social, political, and economic subjects), also, *State of the World* (Norton, annual).

Agriculture and Rural Life B. ANDREAE, *Farming, Development and Space: A World Agricultural Geography,* translated from the German by H. F. Gregor (Walter de Gruyter, 1981); H. BLUME, *Geography of Sugar Cane* (Berlin: Verlag Dr. Albert Bartens, 1985); P. P. COURTENAY, *Plantation Agriculture,* rev. ed. (Bell & Hyman, 1980); R. CRITCHFIELD, *Villages* (Anchor Press/Doubleday, 1981); B. CURREY and G. HUGO, eds., *Famine as a Geographical Phenomenon* (Reidel, 1984); H. J. DE BLIJ, *Wine Regions of the Southern Hemisphere* (Rowman & Allanheld, 1985); L. T. EVANS, "The Natural History of Crop Yields," *American Scientist,* 68 (1980), 388–397; "Food and Agriculture" (special issue), *Scientific American,* 235, no. 3 (September 1976); A. W. GILG, *An Introduction to Rural Geography* (Edward Arnold, 1985); R. J. A. GOODLAND *et al., Environmental Management in Tropical Agriculture* (Westview, 1984); E. GRAHAM and I. FLOERING, eds., *The Modern Plantation in the Third World* (St. Martin's, 1984); H. F. GREGOR, *Geography of Agriculture: Themes in Research* (Prentice-Hall, 1970); D. GRIGG, *An Introduction to Agricultural Geography* (Hutchinson, 1984), also, *The World Food Problem, 1950–1980* (Basil Blackwell, 1985); and "World Patterns of Agricultural Output," *Geography,* 71 (1986), 240–245; J. G. HAWKES, *The Diversity of Crop Plants* (Harvard, 1983); B. W. ILBERY, *Agricultural Geography: A Social and Economic Analysis* (Oxford, 1985);

D. Q. INNIS, "The Future of Traditional Agriculture," *Focus,* 30, no. 3 (January–February 1980), 1–16; E. J. KAHN, JR., *The Staffs of Life* (Little, Brown, 1985; a survey of major food crops, written for the general reader); C. L. A. LEAKEY and J. B. WILLS, eds., *Food Crops of the Lowland Tropics* (Oxford, 1977); W. MANSHARD, *Tropical Agriculture: A Geographical Introduction and Appraisal* (Longman, 1974); W. B. MORGAN, *Agriculture in the Third World: A Spatial Analysis* (Westview, 1978); M. PACIONE, ed., *Progress in Agricultural Geography* (Croom Helm, 1986); C. O. SAUER, *Agricultural Origins and Dispersals: The Domestication of Animals and Foodstuffs,* 2d ed. (MIT, 1969); J. L. SIMON, "World Food Supplies," *Atlantic Monthly,* July 1981, 72–76; B. L. TURNER II and S. B. BRUSH, eds., *Comparative Farming Systems* (Guilford, 1987).

Industrial and Transportation Geography S. COX, "Ancient Trade in a Modern Form" [world shipping], *Geographical Magazine,* 56 (1984), 521–527; C. J. DIXON, *Atlas of Economic Mineral Deposits* (Chapman & Hall, 1979); F. D. F. EARNEY, *Petroleum and Hard Minerals from the Sea* (Wiley, 1980); J. GEVER *et al., Beyond Oil: The Threat to Fuel and Food in the Coming Decades* (Ballinger, 1986); J. GRUNWALD and K. FLAMM, *The Global Factory: Foreign Assembly in International Trade* (Brookings Institution, 1985); P. JAMES, *The Future of Coal,* 2d ed. (Macmillan, 1984); D. T. JONES and J. P. WOMACK, "Developing Countries and the Future of the Automobile Industry," *World Development,* 13 (1985), 393–407; Y. KARMON, *Ports Around the World* (Crown, 1980); P. G. MASEFIELD, "The Challenge of Change in Transport," *Geography,* 61 (1976), 206–220; M. PACIONE, ed., *Progress in Industrial Geography* (Croom Helm, 1985); J. P. RIVA, JR., *World Petroleum Resources and Reserves* (Westview, 1983); W. R. SIDDALL, "Transportation and the Experience of Travel," *Geographical Review,* 77 (1987), 309–317; E. N. TIRATSOO, *Oilfields of the World,* 3d ed. (Beaconsfield, England: Scientific Press, 1984); J. E. VANCE, JR., *Capturing the Horizon: The Historical Geography of Transportation since the Transportation Revolution of the Sixteenth Century* (Harper & Row, 1986); K. WARREN, "World Steel: Change and Crisis," *Geography,* 70 (1985), 106–117.

Urban Geography J. AGNEW, J. MERCER, and D. SOPHER, eds., *The City in Cultural Context* (Allen & Unwin, 1984); J. BIRD, *Centrality and Cities* (Routledge & Kegan Paul, 1977); S. D. BRUNN *et al., Cities of the World: World Regional Urban Development* (Harper & Row, 1983); D. CLARK, *Urban Geography: An Introductory Guide* (Croom Helm, 1982); D. DRAKAKIS-SMITH, *The Third World City* (Methuen, 1987); D. J. DWYER, "Urban Geography and the Urban Future," *Geography,* 64 (1979), 86–95; P. HALL, *The World Cities,* 2d ed. (McGraw-Hill,

1979); D. T. HERBERT and C. J. THOMAS, *Urban Geography: A First Approach* (Wiley, 1982); S. LOWDER, *The Geography of Third World Cities* (Barnes & Noble, 1986); H. M. MAYER et al., *A Modern City: Its Geography* (National Council for Geographic Education, 1970; articles reprinted from the 1969 *Journal of Geography*); L. MUMFORD, *The City in History: Its Origins, Its Transformations, and Its Prospects* (Harcourt, Brace & World, 1961); R. J. ROSS and G. J. TELKAMP, eds., *Colonial Cities* (Martinus Nijhoff, 1985); J. R. SHORT, *An Introduction to Urban Geography* (Routledge & Kegan Paul, 1984); TIME–LIFE BOOKS, *The World Cities* (1976–1978; volumes on individual cities); M. P. TODARO, ''Urbanization in Developing Nations: Trends, Prospects, and Policies,'' *Journal of Geography*, 79 (1980), 164–174; J. E. VANCE, JR., *This Scene of Man: The Role and Structure of the City in the Geography of Western Civilization* (Harper's College, 1977).

Population Geography D. J. M. HOOSON, ''The Distribution of Population as the Essential Geographical Expression,'' *Canadian Geographer*, 17 (1960), 10–20; H. R. JONES, *A Population Geography* (Harper & Row, 1981); N. KLIOT, ''The Era of Homeless Man,'' *Geography*, 72 (1987), 109–121 (world survey of refugees); G. J. LEWIS, *Human Migration: A Geographical Perspective* (St. Martin's, 1982); *Population Bulletin* (6 issues a year), *Population Today* (11 issues a year), and other publications of the Population Reference Bureau, Washington, D.C. (issues of *Population Today* include useful one-page summaries of individual countries); R. M. PROTHERO, ''The People Problem'' [world population], *Geographical Magazine*, 58 (1986), 69–73; F. D. SCOTT, ed., *World Migration in Modern Times* (Prentice-Hall, 1968).

Cultural Geography A. GETIS, J. GETIS, and J. FELLMANN, *Human Geography: Culture and Environment* (Macmillan, 1985); J. F. HART, *The Look of the Land* (Prentice-Hall, 1975); J. B. JACKSON, *Landscapes: Selected Writings of J. B. Jackson* (Massachusetts, 1970); W. A. DOUGLAS JACKSON, *The Shaping of Our World: A Human and Cultural Geography* (Wiley, 1985); T. G. JORDAN and L. ROWNTREE, *The Human Mosaic: A Thematic Introduction to Cultural Geography*, 3d ed. (Harper & Row, 1986); J. LEIGHLY, ed., *Land and Life: A Selection from the Writings of Carl Ortwin Sauer* (California, 1963, reprinted, 1974); G. J. LEVINE, ''On the Geography of Religion,'' *Transactions IBG*, n.s., 11 (1986), 428–440; D. W. MEINIG, ed., *The Interpretation of Ordinary Landscapes: Geographical Essays* (Oxford, 1979); ''New Christendom: The Southern Cross,'' *The Economist*, December 24, 1988, 61–66 (a survey of world Christianity); D. E. SOPHER, *Geography of Religions* (Prentice-Hall, 1968); J. E. SPENCER and W. L. THOMAS, JR., *Introducing Cultural Geography*, 2d ed. (Wiley, 1978); C. L. SALTER, *The Cultural Landscape* (Dux-

bury, 1971); W. L. THOMAS, JR., ed., *Man's Role in Changing the Face of the Earth* (Chicago, 1956; a massive, impressive survey, with contributions by many notable authorities); P. L. WAGNER and M. W. MIKESELL, eds., *Readings in Cultural Geography* (Chicago, 1962).

Political Geography; Ethnicity D. G. BENNETT, ed., *Tension Areas of the World: A Problem-Oriented World Regional Geography* (Park Press, 1982); R. F. BETTS, *Uncertain Dimensions: Western Overseas Empires in the Twentieth Century* (Minnesota, 1985); P. BRASS, ed., *Ethnic Groups and the State* (Croom Helm, 1985); M. A. BUSTEED, ed., *Developments in Political Geography* (Academic, 1983); J. CHAY and T. E. ROSS, eds., *Buffer States in World Politics* (Westview, 1986); C. CLARK et al., ed., *Geography and Ethnic Pluralism* (Allen & Unwin, 1984); K. R. COX, *Location and Public Problems: A Political Geography of the Contemporary World* (Methuen, 1979); E. S. EASTERLY, III, ''Global Patterns of Legal Systems: Notes toward a New Geojurisprudence,'' *Geographical Review*, 67 (1977), 209–220; E. FISCHER, *Minorities and Minority Problems* (Vantage, 1980); S. GOODENOUGH, *War Maps: World War II* (St. Martin's, 1982); J. GOTTMANN, ed., *Center and Periphery: Spatial Variation in Politics* (Sage, 1980); R. HARTSHORNE, ''The Functional Approach in Political Geography,'' *Annals AAG*, 40 (1950), 95–130; D. L. HOROWITZ, *Ethnic Groups in Conflict* (California, 1985); R. J. JOHNSTON and P. J. TAYLOR, eds., *A World of Crisis? Geographical Perspectives* (Basil Blackwell, 1986); J. KEEGAN and A. WHEATCROFT, *Zones of Conflict: An Atlas of Future Wars* (Jonathan Cape, 1986); P. KENNEDY, *The Rise and Fall of the Great Powers: Economic Change and Military Conflict from 1500 to 2000* (Random House, 1987); M. W. MIKESELL, ''The Myth of the Nation State,'' *Journal of Geography*, 82 (1983), 257–260; R. MUIR and R. PADDISON, *Politics, Geography and Behaviour* (Methuen, 1981); R. NATKIEL, *Atlas of World War II* (Military Press, 1985); R. E. NORRIS and L. L. HARING, *Political Geography* (Merrill, 1980); J. O'LAUGHLIN, ''Superpower Competition and the Militarization of the Third World,'' *Journal of Geography*, 86 (1987), 269–275; P. O'SULLIVAN and J. W. MILLER, JR., *The Geography of Warfare* (St. Martin's, 1983); G. PARKER, *Western Geopolitical Thought in the Twentieth Century* (Croom Helm, 1985); W. PFAFF, ''On Nationalism,'' *The New Yorker*, May 25, 1987, 44–56; J. R. ROGGE, ed., *Refugees: A Third World Dilemma* (Barnes & Noble, 1987); J. R. SHORT, *An Introduction to Political Geography* (Routledge & Kegan Paul, 1982); R. SOLLEN, ''A World at War,'' *The Nation*, January 9/16, 1989, 46–47; J. L. STOKESBURY, *A Short History of World War II* (Morrow, 1980); P. J. TAYLOR, *Political Geography: World-Economy, Nation-State and Locality* (Longman, 1985); P. L. WHITE, ''What Is a Nationality?'' *Canadian Review of Studies in Nationalism*, 12 (1985), 1–23.

PART II: EUROPE

General Geographic Surveys and Analyses M. ANDERSON, ed., *Frontier Regions in Western Europe* (Frank Cass, 1983); C. Bertram, "Europe's Security Dilemmas," *Foreign Affairs*, 65 (1987), 942–957; L. BARZINI, *The Europeans* (Simon & Schuster, 1983); L. BERG *et al.*, *Urban Europe* (Oxford, 1982); J. H. BIRD and E. E. POLLOCK, "The Future of Seaports in the European Communities," *Geographical Journal*, 144 (1978), 23–48; M. BLACKSELL, *Post-war Europe: A Political Geography*, 2d ed. (Hutchinson, 1981); F. BRAUDEL, *Civilization and Capitalism 15th-18th Century*, 3 vols. (Collins/Harper & Row, 1981); D. BURTENSHAW, *et al.*, *The City in West Europe* (Wiley, 1981); W. J. CAHNMAN, "Frontiers between East and West in Europe," *Geographical Review*, 39 (1949), 605–624; H. D. CLOUT, *Regional Variations in the European Community* (Cambridge, 1986), also, ed., *Regional Development in Western Europe*, 2d ed. (Wiley, 1981), and, *et al.*, *Western Europe: Geographical Perspectives*, 2d ed. (Wiley, 1989); J. CUISENER, ed., *Europe as a Culture Area* (Mouton, 1979); G. DEMKO, ed., *Regional Development: Problems and Policies in Eastern and Western Europe* (St. Martin's, 1984); A. DIEM, *Western Europe: A Geographical Analysis* (Wiley, 1979); *Europe: Magazine of the European Community* (10 issues per year); "Europe's Internal Market," *The Economist*, July 9, 1988, 1–44; A. FONTAINE, "The Real Divisions of Europe," *Foreign Affairs*, 49 (1971), 302–314; C. R. FOSTER, ed., *Nations without a State: Ethnic Minorities in Western Europe* (Praeger, 1980); J. GOTTMANN, *A Geography of Europe*, 4th ed. (Holt, Rinehart & Winston, 1969; dated, but still very valuable for regional description and historical detail); P. HALL and D. HAY, *Growth Centres in the European Urban System* (Califonia, 1980); G. W. HOFFMAN, ed., *A Geography of Europe: Problems and Prospects*, 5th ed. (Wiley, 1983; a standard, authoritative textbook); R. HUDSON *et al.*, *An Atlas of EEC Affairs* (Methuen, 1984); B. W. ILBERY, *Western Europe: A Systematic Human Geography*, 2d ed. (Oxford, 1986), also, "Core-Periphery Contrasts in European Social Well-Being," *Geography*, 69 (1984), 289–302; T. G. JORDAN, *The European Culture Area: A Systematic Geography* (Harper & Row, 1973); P. L. KNOX, *The Geography of Western Europe: A Socio-Economic Survey* (Barnes & Noble, 1984); F. LEWIS, *Europe: A Tapestry of Nations* (Unwin Hyman, 1987); V. H. MALMSTRÖM, *Geography of Europe: A Regional Analysis* (Prentice-Hall, 1971); R. E. H. MELLOR and E. A. SMITH, *Europe: A Geographical Survey of the Continent* (Macmillan/Columbia, 1979); Y. MÉNY and V. WRIGHT, eds., *Centre-Periphery Relations in Western Europe* (Allen & Unwin, 1985); G. PARKER, *The Countries of Community Europe: A Geographical Survey of Contemporary Issues* (St. Martin's, 1979); W. H. PARKER, "Europe: How Far?" *Geographical Journal*, 126 (1960), 278–297; G. E. PEARCY, "Geographical Terminology of Europe," *U.S. Department of State Bulletin*, 48, no. 1236 (March 4, 1963), 330–338; D. M.

PEPPER, "Geographical Dimensions of NATO's Evolving Military Strategies," *Progress in Human Geography*, 12, no. 2 (June 1988), 157–188; D. A. PINDER, "Crisis and Survival in Western European Oil Refining," *Journal of Geography*, 85 (1986), 12–20; D. I. SCARGILL, gen. ed., *Problem Regions of Europe* (Oxford, 1973–1976; short paperbound vols.); C. TUGENDHAT, *Making Sense of Europe* (Columbia, 1988); S. VAN VALKENBURG and C. C. HELD, *Europe*, 2d ed. (Wiley, 1952; a classic textbook, now outdated, but sections can still be read with profit); P. WHITE, *The West European City: A Social Geography* (Longman, 1984); A. M. WILLIAMS, *The Western European Economy: A Geography of Post-War Development* (Barnes & Noble, 1988).

Europe: Historical Circumstances and Historical Geography From an overwhelming number of possibilities, a few selections are: J. BAECHLER, *et al.*, eds., *Europe and the Rise of Capitalism* (Basil Blackwell, 1988); J. P. V. D. BALSDON, *Romans and Aliens* (North Carolina, 1979); D. J. BOORSTIN, *The Discoverers* (Random House, 1983); M. E. CHAMBERLAIN, *Decolonisation: The Fall of the European Empires* (Basil Blackwell, 1985); W. G. EAST, *An Historical Geography of Europe*, 5th ed. (Dutton, 1966); D. K. FIELDHOUSE, *The Colonial Empires* (Dell, 1966); E. A. GUTKIND, *International History of City Development*, 8 vols. (Free Press, 1964–1972); A. G. HARGREAVES, "European Identity and the Colonial Frontier," *Journal of European Studies*, 12 (1982), 166–179; R. C. HARRIS, "The Simplification of Europe Overseas," *Annals AAG*, 67 (1977), 469–483; E. L. JONES, *The European Miracle: Environments, Economies and Geopolitics in the History of Europe and Asia*, 2d ed. (Cambridge, 1986); W. R. MEAD, "The Discovery of Europe," *Geography*, 67 (1982), 193–202; H. M. MUNRO, *The Nationalities of Europe and the Growth of National Ideologies* (Cambridge, 1945); J. H. PARRY, *The Age of Reconnaissance: Discovery, Exploration, and Settlement 1450–1650* (California, 1981), also, *The Discovery of the Sea* (California, 1981); N. J. G. POUNDS, *An Historical Geography of Europe 450 B.C. to 1330 A.D.* (Cambridge, 1973), also, *An Historical Geography of Europe 1500 to 1840* (Cambridge, 1979), also, *An Historical Geography of Europe 1800–1914* (Cambridge, 1985), and, with S. S. BALL, "Core-Areas and the Development of the European States System," *Annals AAG*, 54 (1964), 24–40; W. P. WEBB, *The Great Frontier* (Houghton Mifflin, 1952); E. R. WOLF, *Europe and the People Without History* (California, 1982); W. WOODRUFF, *Impact of Western Man: A Study of Europe's Role in the World Economy* (Macmillan, 1966); D. S. WHITTLESEY, *Environmental Foundations of European History* (Appleton-Century-Crofts, 1949).

British Isles: General Background and General Geography *Atlas of Britain and Northern Ireland* (Oxford: Clarendon Press, 1963; an extraordinary array of large, detailed topical and regional maps in color; perhaps the finest national atlas in the world); R. BECKINSALE and M. BECKINSALE, *The English Heartland* (Duckworth, 1980); P. COONES and J. PATTEN, *The Penguin Guide to the Landscape of England and Wales* (Penguin, 1986); H. C. DARBY, "The Regional Geography of Thomas Hardy's Wessex," *Geographical Review*, 38 (1948), 426–443; "Devolution and the UK," *Geographical Magazine*, 50 (1978; 7 articles by various authors); A. E. GREEN, "The North-South Divide in Great Britain: An Examination of the Evidence," *Transactions IBG*, n.s., 13 (1988), 179–198; J. FERGUSON-LEES and B. CAMPBELL, *Mountains and Moorlands* (Hodder & Stoughton, 1978); A. V. HARDY, *The British Isles*, new ed. (Cambridge, 1981; a geography); "Human Geography of Contemporary Britain," *Geographical Magazine*, 53–54 (1981–1982; 10 articles by various authors); R. J. JOHNSTON and J. C. DOORNKAMP, eds., *The Changing Geography of the United Kingdom* (Methuen, 1978); R. I. KIRKLAND, JR., "The Great Rebound: Britain Is Back," *Fortune*, May 9, 1988, 114–123; D. LOWENTHAL and H. C. PRINCE, "The English Landscape," *Geographical Review*, 54 (1964), 309–346, also, "English Landscape Tastes," *ibid.* 55 (1965), 186–222; P. C. LUCAS, *Britain: An Aerial Close-up* (Crescent Books, 1984; spectacular color photos in an oversized format); G. MANNERS *et al.*, *Regional Development in Britain*, 2d ed. (Wiley, 1980); "Map of Cherished Land," *Geographical Magazine*, 46, no. 1 (October 1973; a folded, multicolored separate map of Britain's scenic and recreational resources); J. B. MITCHELL, ed., *Great Britain: Geographical Essays* (Cambridge, 1962; a distinguished volume of regional studies by major authorities); J. J. NORWICH, ed., *Britain's Heritage* (Continuum, 1983; an oversized, lavishly illustrated survey touching many major topics); ORDNANCE SURVEY, *The Ordnance Survey National Atlas* (Hamlyn Publishing, 1986); D. C. D. POCOCK, "The Novelist's Image of the North," *Transactions IBG*, n. s., 4 (1979), 62–76; G. PRICE, *The Languages of Britain* (Edward Arnold, 1984); L. D. STAMP and S. H. BEAVER, *The British Isles*, 6th ed. (St. Martin's, 1972; a well-known text); J. A. STEERS, ed., *Field Studies in the British Isles* (Thomas Nelson, 1964); R. STIRLING, *The Weather of Britain* (Faber, 1982); J. W. WATSON and J. B. SISSONS, eds., *The British Isles: A Systematic Geography* (Thomas Nelson, 1964); "Transport in Britain," *Geographical Magazine*, 50 (1978; 6 articles by various authors); S. R. J. WOODELL, ed., *The English Landscape: Past, Present, and Future* (Oxford, 1985).

British Isles: Historical Background and Historical Geography

M. ASHLEY, *The People of England: A Short Social and Economic History* (Louisiana State, 1982); A. BRIGGS, *A Social History of England* (Viking, 1983); I. CAMERON, *To the Farthest Ends of the Earth: 150 Years of World Exploration by the Royal Geographical Society* (Dutton, 1980); A. J. CHRISTOPHER, *The British Empire at Its Zenith* (Croom Helm, 1988); H. C. DARBY, ed., *A New Historical Geography of England* (Cambridge, 1973); J. DARWIN, "British Decolonization since 1945: A Pattern or a Puzzle?" *Journal of Imperial and Commonwealth History*, 11 (1984), 187–209; R. F. DELDERFIELD, *God is an Englishman* (Simon & Schuster, 1970; an extraordinarily geographic novel offering a panorama of mid-19th century England); R. FREETHY, *The Making of the British Countryside* (David & Charles, 1981); "The Historical Geography of England from Earliest Settlement to the Victorian City," *Geographical Magazine*, 52–53 (1970–1971; 16 articles by various authors); K. HUDSON, *Industrial History from the Air* (Cambridge, 1985); C. H. KNOWLES, *Landscape History* (London: Historical Association, General Series, no. 107, 1983); J. LANGTON, "The Industrial Revolution and the Regional Geography of England," *Transactions IBG*, n.s., 9 (1984), 145–167; B. LAPPING, *End of Empire* (St. Martin's, 1985); T. O. LLOYD, *The British Empire, 1558–1983* (Oxford, 1984): R. McCRUM, W. CRAN, and R. MacNEIL, *The Story of English* (Viking, 1986); O. RACKHAM, *The History of the Countryside* (Dent, 1986); I. G. SIMMONS and M. J. TOOLEY, eds., *The Environment in British Prehistory* (Duckworth, 1981); P. WARWICK, "Did Britain Change? An Inquiry into the Causes of National Decline," *Journal of Contemporary History*, 20 (1985), 99–133; J. G. WILLIAMSON, "Why Was Britain's Growth So Slow during the Industrial Revolution? *Journal of Economic History*, 44 (1984), 687–712; S. WINCHESTER, *Outposts* [British Empire] (Hodder & Stoughton, 1985); E. A. WRIGLEY, *Continuity, Chance and Change: The Character of the Industrial Revolution in England* (Cambridge, 1988); E. M. YATES, "The Evolution of the English Village," *Geographical Journal*, 148 (1982), 182–206.

British Isles: Climate, Agriculture, and Rural Geography

R. BLYTHE, *Akenfield: Portrait of an English Village* (Pantheon, 1969); J. K. BOWERS and P. CHESHIRE, *Agriculture, the Countryside and Land Use: An Economic Critique* (Methuen, 1983); T. J. CHANDLER and S. GREGORY, eds., *The Climate of the British Isles* (Longman, 1976); B. A. HOLDERNESS, *British Agriculture since 1945* (Manchester, 1985); B. ILBERY, "A Future for the Farms?" *Geographical Magazine*, 59 (1987), 249–254; H. H. LAMB, *The English Climate* (English Universities Press, 1964); G. MANLEY, *Climate and the British Scene* (Collins, 1952); M. J. MOSELEY, "Changing Values in the British Countryside," *Geographical Magazine*, 53 (1981), 581–586; G. MOSS, *Britain's Wasting Acres: Land Use in a Changing Society* (Architectural Press, 1981); D. PHILLIPS and A. WILLIAMS, *Rural Britain: A Social Geography* (Basil Blackwell, 1984); J. A. TAYLOR, "The British Upland Environment and Its Management," *Geography*, 63 (1978), 338–353.

British Isles: Mining, Manufacturing, Transport, and Cities

F. ATKINSON and S. HALL, *Oil in the British Econ-*

omy (Croom Helm/St. Martin's, 1983); J. BIRD, *The Major Seaports of the United Kingdom* (Hutchinson, 1963); P. BORSAY, "Culture, Status, and the English Urban Landscape," *History*, 67, no. 219 (1982), 1–12; M. CHISHOLM, ed., *Resources for Britain's Future: A Series from the Geographical Magazine* (David & Charles, 1972); M. CHISHOLM and R. A. GIBB, "The Impact of the Channel Tunnel," *Geographical Journal*, 152 (1986), 314–353; J. P. COLE and K. DeBRES, "A New Channel Crossing: Light at the End of the Tunnel?" *Focus*, 38, no. 4 (Winter 1988), 1–9, 28; D. A. FARNIE, *The Manchester Ship Canal and the Rise of the Port of Manchester, 1894–1975* (Manchester, 1980); T. W. FREEMAN, *The Conurbations of Great Britain* (Manchester, 1959); J. B. GODDARD, "British Cities in Transition," *Geographical Magazine*, 53 (1981), 523–530, also, with A. G. CHAMPION, eds., *The Urban and Regional Transformation of Britain* (Methuen, 1983); A. G. HOARE, *The Location of Industry in Britain* (Cambridge, 1983); P. N. JONES and J. NORTH, "Unit Loads Through Britain's Ports: A Further Revolution?" *Geography*, 67 (1982), 29–40; E. H. FRANCIS, "British Coalfields," *Science Progress*, 66 (1979), 1–23; G. MANNERS, *Coal in Britain: An Uncertain Future* (Allen & Unwin, 1981); I. R. MANNERS, *North Sea Oil and Environment Planning: The United Kingdom Experience* (Texas, 1982); J. PATON, "Navigation in the English Channel and the Southern North Sea," *Journal of Navigation*, 31 (1978), 62–81; D. SADLER, "Works Closure at British Steel and the Nature of the State," *Political Geography Quarterly*, 3 (1984), 297–311; G. SHANKLAND, "The Liverpool Experience," *Geographical Magazine*, 53 (1981), 562–567; D. WATTS, "Thatcher's Britain—A Manufacturing Economy in Decline? The North Could Rise Again," *Focus*, 38, no. 3 (Fall 1988), 1–5; M. J. WISE, "The Birmingham-Black Country Conurbation in Its Regional Setting," *Geography*, 57 (1972), 89–104.

British Isles: London M. C. BORDER, *The City of London: A History* (Constable, 1977); A. CHURCH and J. HALL, "Discovery of Docklands" [redevelopment of London's Docks], *Geographical Magazine*, 58 (1986), 632–639; R. CLAYTON, ed., *The Geography of Greater London* (George Philip, 1964); H. CLOUT, ed., *Changing London* (University Tutorial Press, 1978; 14 contributions by various authors), also, with P. WOOD, *London: Problems of Change* (Longman, 1986); P. DOLPHIN *et al.*, *The London Region: An Annotated Geographical Bibliography* (Mansell, 1981); C. HAMNETT, "Life in the Cocktail Belt" [Britain's Home Counties], *Geographical Magazine*, 56 (1984), 534–538; R. W. HORNER, "The Thames Barrier Project," *Geographical Journal*, 145 (1979), 242–253; R. C. JARVIS, "The Metamorphosis of the Port of London," *London Journal*, 3 (1977), 55–72; B. LENON, "The Geography of the 'Big Bang': London's Office Building Boom," *Geographical Magazine*, 72 (1987), 56–59; "London Docklands," *The*

Economist, February 13, 1988, 67–69; R. J. C. MUNTON, "London's Green Belt: Containment in Practice" (Allen & Unwin, 1983), also, "Green Belts: The End of an Era?" *Geography*, 71 (1986), 206–214; F. J. OSBORN and A. WHITTICK, *New Towns: Their Origins, Achievements and Progress*, 3d ed. (Leonard Hill, 1977); S. PAGE, "The London Docklands: Redevelopment Schemes in the 1980s," *Geographical Magazine*, 72 (1987), 59–63; A. SAMPSON, "City of London: Bridge to the Past," *Geo*, 1, no. 7 (November 1979), 38–56; D. P. SHAW, "A Barrier to Tame the Thames," *Geographical Magazine* 55 (1983), 129–131; J. SHEPHERD *et al.*, "Londoners Are Moving Out," *Geographical Magazine*, 54 (1982), 669–673; A. E. SMAILES, "Greater London—The Structure of a Metropolis," *Geographische Zeitschrift*, 52 (1964), 163–189; D. THOMAS, "London's Green Belt: The Evolution of an Idea," *Geographical Journal*, 129 (1963), 14–24, also, *London's Green Belt* (Faber, 1970).

British Isles: Scotland I. H. ADAMS, "One Hundred Years Later: Does Scotland Exist?" *Scottish Geographical Magazine*, 100 (1984), 113–122; J. A. AGNEW, "Place and Political Behaviour: The Geography of Scottish Nationalism," *Political Geography Quarterly*, 3 (1984), 191–206, also, "Political Regionalism and Scottish Nationalism in Gaelic Scotland," *Canadian Review of Studies in Nationalism*, 8 (1981), 115–129; "Land of the Scots," *Geographical Magazine*, 51–52 (1979; 6 articles by various authors); K. J. LEA, "Greater Glasgow," *Scottish Geographical Magazine*, 96 (1980), 4–17; D. McCRONE, "Explaining Nationalism: The Scottish Experience," *Ethnic and Racial Studies*, 7 (1984), 129–137; A. MATHER, "Trees Over Scotland," *Geographical Magazine*, 59 (1987), 335–339; K. PANTON, "Scots Who Speak Gaelic," *Geographical Magazine*, 56 (1984), 168–170; M. L. PARRY and T. R. SLATER, eds., *The Making of the Scottish Countryside* (Croom Helm, 1980); W. RITCHIE, "Giant Oil Terminal on Shetland Farmland," *Geographical Magazine*, 53 (1981), 489–496; *Scottish Geographical Magazine* (3 issues a year; numerous articles and comprehensive bibliography on Scottish geography); D. J. SINCLAIR, "Scottish Identity," *Geographical Magazine*, 50 (1978), 803–807; H. D. SMITH, "The Role of the Sea in the Political Geography of Scotland," *Scottish Geographical Magazine*, 100 (1984), 138–150; H. D. SMITH *et al.*, "Scotland and Offshore Oil: The Developing Impact," *Scottish Geographical Magazine*, 92 (1976), 75–91; E. A. TIRYAKIAN, "Quebec, Wales, and Scotland: Three Nations in Search of a State," *International Journal of Comparative Sociology*, 21 (1980), 1–13; D. TURNOCK, *The Historical Geography of Scotland since 1707: Geographical Aspects of Modernisation* (Cambridge, 1982); G. WHITTINGTON and I. D. WHYTE, eds., *An Historical Geography of Scotland* (Academic, 1983); I. D. WHYTE, "Scottish Historical Geography: A Review," *Scottish Geographical Magazine*, 94 (1978), 4–23; C. W. J. WITHERS,

" 'The Image of the Land': Scotland's Geography Through Her Language and Literature," *Scottish Geographical Magazine*, 100 (1984), 81–95.

British Isles: Wales D. BALSON *et al.*, "The Political Consequences of Welsh Identity," *Ethnic and Racial Studies* 7 (1984), 160–181; C. S. DAVIES, "Wales: Industrial Fallibility and Spirit of Place," *Journal of Cultural Geography*, 4 (1983), 72–86; F. V. EMERY, *Wales* (Longman, 1969; "The World's Landscapes"); J. MORRIS, *The Matter of Wales: Epic Views of a Small Country* (Oxford, 1984); D. THOMAS, ed., *Wales: A New Study* (David & Charles, 1977); E. THOMAS, *Wales* (Oxford, 1983); G. WILLIAMS, "What Is Wales? The Discourse of Devolution," *Ethnic and Racial Studies*, 7 (1984), 138–159; J. ZARING, "The Romantic Face of Wales," *Annals AAG*, 67 (1977), 397–418.

British Isles: Ireland F. H. A. AALEN, *Man and the Landscape in Ireland* (Academic, 1978); F. W. BOAL and J. N. H. DOUGLAS, eds., *Integration and Division: Perspectives on the Northern Ireland Problem* (Academic, 1982); K. S. BOTTIGHEIMER, *Ireland and the Irish: A Short History* (Columbia, 1982); E. E. EVANS, *The Personality of Ireland: Habitat, Heritage and History* (Cambridge, 1973); *Irish Geography* (annual); J. H. JOHNSON, "The Political Distinctiveness of Northern Ireland," *Geographical Review*, 52 (1962), 78–91; IRISH NATIONAL COMMITTEE FOR GEOGRAPHY, *Atlas of Ireland* (Dublin: Royal Irish Academy, 1979); K. A. MILLER, *Emigrants and Exiles: Ireland and the Irish Exodus to North America* (Oxford, 1985); F. MITCHELL, *The Irish Landscape* (Collins, 1976); J. K. MITCHELL, "Social Violence in Northern Ireland," *Geographical Review*, 69 (1979), 179–201; "Northern Ireland: Brotherly Hate," *The Economist*, June 25, 1988, 19–22; A. R. ORME, *Ireland* (Aldine, 1970; "The World's Landscapes"); "Poorest of the Rich: A Survey of the Republic of Ireland," *The Economist*, January 16, 1988, 1–26; D. G. PRINGLE, *One Island, Two Nations? A Political Geographical Analysis of the National Conflict in Ireland* (Wiley, 1985); R. N. SALAMAN, *The History and Social Influence of the Potato* (Cambridge, 1949; discusses the causes of the 19th-century Irish emigration); L. STEPHENS and M. A. KRAMER, "Irish Landscapes," *Landscape*, 27, no. 3 (1983), 18–25; C. WOODHAM-SMITH, *The Great Hunger: Ireland 1845–1849* (Harper, 1963; a graphic account of the Irish famine of the 1840s, in the perspective of general relationships between Ireland and England).

West Central Europe: France J. BEAUJEU-GARNIER, *France* (Longman, 1975; "The World's Landscapes"); H. D. CLOUT, "A New France?" *Geography*, 67 (1982), 244–250, also, ed., *Themes in the Historical Geography of France* (Academic, 1977); I. EVANS, "Powerhouse in Decline" [Lorraine], *Geographical Magazine*, 59 (1987), 78–83; N. EVERSON, *Paris: A Century of Change, 1878–1978*

(Yale, 1979); L. GALLOIS, "The Origin and Growth of Paris," *Geographical Review*, 13 (1923), 345–367; R. HARTSHORNE, "The Franco-German Boundary of 1871," *World Politics*, 2 (1950), 209–250; J. W. HOUSE, *France: An Applied Geography* (Methuen, 1978); P. L. KNOX and A. SCARTH, "The Quality of Life in France," *Geography*, 62 (1977), 9–16; P. PINCHEMEL *et al.*, *France: A Geographical, Social and Economic Survey* (Cambridge, 1986); I. SCARGILL, *Urban France* (Croom Helm, 1983); A. SIEGFRIED, *France: A Study in Nationality* (Yale, 1930); J. TUPPEN, *The Economic Geography of France* (Croom Helm, 1983).

West Central Europe: Germany W. H. BERENTSEN, "Regional Change in the German Democratic Republic," *Annals AAG*, 71 (1981), 50–66; T. H. ELKINS with B. HOFMEISTER, *Berlin: The Spatial Structure of a Divided City* (Methuen, 1988); K. FILIPP, "Facing the Political Map of Germany," *Political Geography Quarterly*, 3 (1984), 251–258; J. G. HAJDU, "Phases in the Post-war German Urban Experience," *Town Planning Review*, 50 (1979), 267–286; J. M. HALL, "Berlin and the Wall," *Geographical Magazine*, 53 (1981), 817–821; A. HULL and S. KENNY, "Ruhr Economy in Decline," *Geographical Magazine*, 55 (1983), 516–521; M. R. LERSIUS, "The Nation and Nationalism in Germany," *Social Research*, 52 (1985), 43–64; R. E. H. MELLOR, *The Two Germanies: A Modern Geography* (Harper & Row/Barnes & Noble, 1978); G. MOHS *et al.*, "The Regional Differentiation of the German Democratic Republic: Structure, Dynamics, Development," *GeoJournal*, 8 (1984), 7–22; R. H. PLATT, "Berlin: Island of Détente," *Journal of Geography*, 86 (1987), 263–268; H. A. TURNER, JR., *The Two Germanies since 1945* (Yale, 1987); M. T. WILD, *West Germany: A Geography of Its People* (William Dawson, 1979/Barnes and Noble, 1980); A. ZIMM and J. BRÄUNIGER, "The Agglomeration of the GDR Capital, Berlin: A Survey of Its Economic Geography," *GeoJournal*, 8 (1984), 23–31.

West Central Europe: Benelux A. K. DUTT and F. J. COSTA, eds., *Public Planning in the Netherlands* (Oxford, 1985); A. K. DUTT and S. HEAL, "Delta Project Planning and Implementation in the Netherlands," *Journal of Geography*, 78 (1979), 131–141; *I.D.G. Bulletin*, issued (from 1973) by the Information and Documentation Centre for the Geography of the Netherlands, Utrecht, Netherlands; A. M. LAMBERT, *The Making of the Dutch Landscape: A Historical Geography of the Netherlands*, 2d ed. (Academic, 1985); A. B. MURPHY, *The Regional Dynamics of Language Differentiation in Belgium* (University of Chicago, Department of Geography Research Paper 227, 1988); E. SCHRIJVER, "How the Dutch Made Holland," *Geographical Magazine*, 51 (1979), 567–572; G. STEPHENSON, "Cultural Regionalism and the Unitary State Idea in Belgium," *Geographical Review*, 62 (1972), 501–523; [The

Netherlands]: special issue, *Tijdschrift voor Economische en Sociale Geografie,* 63, no. 3. (May–June 1972), 124–235; P. WAGRET, *Polderlands* (Barnes & Noble, 1972); G. G. WEIGEND, "Stages in the Development of the Ports of Rotterdam and Antwerp," *Geoforum,* no. 13 (1973), 5–15; H. L. WESSELING, "Post-Imperial Holland," *Journal of Contemporary History,* 15 (1980), 125–142.

West Central Europe: Switzerland and Austria
E. BONJOUR, *Swiss Neutrality: Its History and Meaning,* 2d ed. (Allen & Unwin, 1952); E. A. BRUGGER *et al.,* eds., *The Transformation of Swiss Mountain Regions* (Verlag Paul Haupt Bern und Stuttgart, 1984); A. DIEM, "The Alps," *Geographical Magazine,* 56 (1984), 414–420; E. EGLI, *Switzerland: A Survey of Its Land and People* (Bern: Paul Haupt/London: Macdonald & Evans, 1978); A. GILG, "Land Use Planning in Switzerland," *Town Planning Review,* 56 (1985), 315–338; K. HALTIMER and R. MEYER, "Aspects of the Relationship between Military and Society in Switzerland," *Armed Forces and Society,* 6 (1979), 49–81; G. W. HOFFMAN, "The Survival of an Independent Austria," *Geographical Review,* 41 (1951), 606–621; R. MAGGI and P. K. HAENI, "Spatial Concentration, Location and Competitiveness: The Case of Switzerland," *Regional Studies,* 20 (1986), 141–149; R. M. NETTING, *Balancing on an Alp: Ecological Change and Continuity in a Swiss Mountain Community* (Cambridge, 1981); C. STADEL, "The Alps: Mountains in Transformation," *Focus,* 32 no. 3 (January–February 1982), 1–16; J. STEINBERG, *Why Switzerland?* (Cambridge, 1976); B. WALLACH, "Geneva," *Focus,* 36, no. 3 (Fall 1986), 10–13.

Northern Europe P. M. COLE and D. M. HART, eds., *Northern Europe: Security Issues for the 1990s* (Westview, 1986); D. M. EPSTEIN and A. VALMARI, "Reindeer Herding and Ecology in Finnish Lapland," *GeoJournal,* 8 (1984), 159–169; C. G. GUSTAVSON, *The Small Giant: Sweden Enters the Industrial Era* (Ohio, 1986); F. HALE, "Norway's Rendezvous with Modernity," *Current History,* 83 (1984), 173–177, 184; J. C. HANSEN, "Regional Disparities in Present-Day Norway," *Norsk Geografisk Tidsskrift,* 39 (1985), 109–124; J. C. HANSEN, "Regional Policy in an Oil Economy: The Case of Norway," *Geoforum,* 14 (1983), 353–361; R. K. HELLE, "Reindeer Husbandry in Finland," *Geographical Journal,* 145 (1979), 254–264; J. J. HOLST, "The Northern Region: Key to Europe?" *Atlantic Community Quarterly,* 19, no. 1 (Spring 1981), 28–36; A. HOLT-JENSEN, "Norway and the Sea: The Shifting Importance of Marine Resources Through Norwegian History," *GeoJournal,* 10 (1985); A. HOLT-JENSEN, "The Norwegian Oil Economy," *GeoJournal* (Supplementary Issue), no. 3 (1981), 81–92; B. S. JOHN, *Scandinavia: A New Geography* (Longman, 1984); R. D. LIEBOWITZ, "Finlandization: An Analysis of the Soviet Union's 'Domination' of Finland," *Political Geography*

Quarterly, 2 (1983), 275–287; D. C. MacCASKILL, "Norway's Strategic Importance," *Marine Corps Gazette,* 65, no. 2 (February 1981), 28–33; W. R. MEAD, *An Historical Geography of Scandinavia* (Academic, 1981), also, "Problems of Norden," *Geographical Journal,* 151 (1985), 1–10, also, "Recent Developments in Human Geography in Finland," *Progress in Human Geography,* 1 (1977), 361–375, and "Sweden in Retrospect," *Geography,* 70 (1985), 36–44; M. REUTER, "Swedish in Finland: Minority Language and Regional Variety," *Word,* 30 (1979), 171–185; S. SARIOLA, "Finland and Finlandization," *History Today,* 32 (March 1982), 20–26; S. STRAND, "Roadless Norway," *Scottish Geographical Magazine,* 100 (1984), 49–59; K. SCHERMAN, *Daughter of Fire; A Portrait of Iceland* (Little, Brown, 1976); A. SÖMME, ed., *A Geography of Norden: Denmark, Finland, Iceland, Norway, Sweden* (Wiley, 1962; a major work by Scandinavian and Finnish geographers); S. THORANINSSON, "Living on a Volcano" [Iceland], *Polar Geography and Geology,* 8 (1984), 89–112; U. VARJO and W. TIETZE, eds., *Norden: Man and Environment* (Berlin and Stuttgart: Gebrüder Borntraeger, 1987).

Southern Europe M. BECKINSALE and R. BECKINSALE, *Southern Europe: The Mediterranean and Alpine Lands* (London, 1975); H. BRÜCKNER, "Man's Impact on the Evolution of the Physical Environment in the Mediterranean Region in Historical Times," *GeoJournal,* 13 (1986), 7–17; K. W. BUTZER *et al.,* "Irrigation Agrosystems in Eastern Spain: Roman or Islamic Origins?" *Annals AAG,* 75 (1985), 479–509; R. COLLINS, *The Basques* (Basil Blackwell, 1987); R. COMMON, "The Study of Regional Geography in Southern Europe," *Scottish Geographical Magazine,* 98 (1982), 180–188; P. P. COURTENAY, "Madrid: The Circumstances of Its Growth," *Geography,* 44 (1959), 22–34; R. HUDSON and J. LEWIS, eds., *Uneven Development in Southern Europe* (Methuen, 1985); R. KING, gen. ed., "Italy," *Geographical Magazine,* 59, 60 (1987, 1988: articles by various authors in separate issues, from no. 4, April, 1987); R. KING, *The Industrial Geography of Italy* (St. Martin's, 1985), also, "Southern Europe: Dependency or Development?" *Geography,* 67 (1982), 221–234; A. MORRIS and G. DICKINSON, "Tourist Development in Spain: Growth versus Conservation on the Costa Brava," 72 (1987), 16–25; A. B. MOUNTJOY, *The Mezzogiorno,* 2d ed. (Oxford, 1982; "Problem Regions of Europe"); J. NAYLON, "Industry on the Brink" [Spain], *Geographical Magazine,* 57 (1985), 436–440, also, "Modern Spain's Bitter Legacy," *Geographical Magazine,* 56 (1984), 570–574; E. C. SEMPLE, *The Geography of the Mediterranean World: Its Relation to Ancient History* (Holt, 1931); "Settlement and Conflict in the Mediterranean World" (special issue), *Transactions IBG,* n.s., 3 (1978), 255–380; C. D. SMITH, *Western Mediterranean Europe: A Historical Geography of Italy, Spain and Southern France since the Neolithic* (Academic, 1979); D. STANISLAWSKI, *The*

Individuality of Portugal: A Study in Historical-Political Geography (Texas, 1959).

East Central Europe M. ALDER, "A State Apart" [Albania], *Geographical Magazine*, 58 (1986), 132–137; H. C. DARBY et al., *A Short History of Yugoslavia from Early Times to 1966* (Cambridge, 1966; many useful maps); W. PFAFF, "Where the Wars Came From," *The New Yorker*, December 26, 1988, 83–90; J. J. PUTNAM, "Different Communism: Hungary's New Way," *National Geographic*, 163 (1983), 225–261; P. RAMET, *Yugoslavia in the 1980s* (Westview, 1985); D. S. RUGG, *Eastern Europe* (Wiley, 1986); D. TURNOCK, *Eastern Europe* (Westview, 1978), also, "Postwar Studies on the Human Geography of Eastern Europe," *Progress in Human Geography* 8 (1984), 315–346, and, "The Danube-Black Sea Canal and Its Impact on Southern Romania," *GeoJournal*, 12 (1986), 65–79.

PART III: THE SOVIET UNION

General S. BIALER, "Inside Glasnost," *Atlantic Monthly*, February, 1988, 64–72; A. BROWN, J. FENNELL, M. KASER, and H. T. WILLIAMS, gen. eds., *Cambridge Encyclopedia of Russia and the Soviet Union* (Cambridge, 1982); J. P. COLE, "China and the Soviet Union: Worlds Apart?" *Soviet Geography*, 28 (1987), 459–484; G. J. DEMKO and R. J. FUCHS, eds., *Geographical Studies on the Soviet Union: Essays in Honor of Chauncy D. Harris* (University of Chicago, Department of Geography Research Paper 211, 1984); J. C. DEWDNEY, *USSR in Maps* (Holmes & Meier, 1982); M. EDWARDS, "Chernobyl—One Year After," *National Geographic*, 171 (1987), 632–653; G. FEIFER, "The New God Will Fail—Moscow: Skeptical Voices on *Perestroika*," *Harper's Magazine*, October 1988, 43–49; R. G. KAISER, "The U.S.S.R. in Decline," *Foreign Affairs*, 67, no. 2 (Winter 1988/89), 97–113; R. I. KIRKLAND, JR., "The Death of Socialism," *Fortune*, January 4, 1988, 64–72; J. LUKACS, "The Soviet State at 65," *Foreign Affairs*, 65 (1986), 21–36; M. MATTHEWS, *Poverty in the Soviet Union: The Life-Styles of the Underprivileged in Recent Years* (Cambridge, 1986); J. PALLOT, "Recent Approaches in the Geography of the Soviet Union," *Progress in Human Geography*, 7 (1983), 519–542; W. PFAFF, "The Question Not Asked" [Leninism], *The New Yorker*, August 1, 1988, 60–65; E. POND, *From the Yaroslavsky Station: Russia Perceived*, 3d ed. (Universe, 1987; distinguished journalism); D. K. SHIPLER, *Russia: Broken Idols, Solemn Dreams* (Times Books, 1983; by a journalist); H. SMITH, *The Russians* (Quadrangle/The New York Times Book Co., 1976; a widely read account by a noted foreign correspondent); *Soviet Geography* (bimonthly; has both specially written articles and translations from the Russian); "The Soviet Union, 1988," *Current History*, October 1988, entire issue.

Geography Textbooks J. P. COLE, *Geography of the Soviet Union* (Butterworth, 1984); P. E. LYDOLPH, *Geography of the U.S.S.R.*, 3d ed. (Wiley 1977; a detailed text organized by economic regions), also, *Geography of the USSR: Topical Analysis* (Elkhart Lake, WI: Misty Valley Publishing, 1979); R. E. H. MELLOR, *The Soviet Union and Its Geographical Problems* (Macmillan, 1982); W. H. PARKER, *The Soviet Union*, 2d ed. (Longman, 1983; "The World's Landscapes"); L. SYMONS, ed., *The Soviet Union: A Systematic Geography* (Barnes & Noble, 1983).

Historical Background D. N. COLLINS, "Russia's Conquest of Siberia: Evolving Russian and Soviet Historical Interpretations," *European Studies Review*, 12 (1982), 17–44; E. CRANKSHAW, *The Shadow of the Winter Palace: Russia's Drift to Revolution, 1825–1917* (Viking, 1976); W. G. EAST, "The New Frontiers of the Soviet Union," *Foreign Affairs*, 29 (1951), 591–607; J. ELLIS, *The Russian Orthodox Church: A Contemporary History* (Indiana, 1986); H. J. ELLISON, "Economic Modernization in Imperial Russia: Purposes and Achievements," *Journal of Economic History*, 25 (1965), 523–540; J. R. GIBSON, "Russian Expansion in Siberia and America," *Geographical Review*, 70 (1980), 127–136; C. J. HALPERIN, *Russia and the Golden Horde: the Mongol Impact on Medieval Russian History* (Indiana, 1985); "Holy Russia's Millennium: The Church Is Risen Indeed," *The Economist*, April 2, 1988, 17–19; S. J. LINZ, ed., *The Impact of World War II on the Soviet Union* (Rowman & Allanheld, 1985); S. MASSIE, *Land of the Firebird: The Beauty of Old Russia* (Simon & Schuster, 1980); E. E. MEAD, ed., *Makers of Modern Strategy: Military Thought from Machiavelli to Hitler* (Princeton, 1943), chap. 14, "Lenin, Trotsky, Stalin: Soviet Concepts of War," pp. 322–364; A. S. MORRIS, "The Medieval Emergence of the Volga-Oka Region, *Annals AAG*, 61 (1971), 697–710; P. N. PAVLOV, "Fur Trade in the Economy of Siberia in the 17th Century," *Soviet Geography*, 27 (1986), 43–82; H. E. SALISBURY, *Black Night, White Snow: Russia's Revolutions 1905–1917* (Doubleday, 1978); D. W. TREADGOLD, *Twentieth Century Russia*, 6th ed. (Westview, 1987); B. D. WOLFE, *Three Who Made a Revolution: A Biographical History*, 2d ed. (Dial, 1960).

Ethnicity R. S. CLEM, "Russians and Others: Ethnic Tensions in the Soviet Union," *Focus*, 31, no. 1 (September–October 1980), 1–16; P. DOSTAL and H. KNIPPENBERG, "The 'Russification' of Ethnic Minorities in the USSR," *Soviet Geography*, 20 (1979), 197–219; M. EDWARDS, "Ukraine," *National Geographic*, 171 (1987), 595–631; R. KARKLINS, *Ethnic Relations in the U.S.S.R.: The Perspective from Below* (Allen & Unwin, 1986); H. KRAMER, "Soviet Islam: Adjusting to an Atheistic State," *Swiss Review of World Affairs*, 29, no. 8 (November 1979), 23–27; G. W. LAPIDUS, "Ethnonationalism and Political Stability: The Soviet Case," *World Politics*, 36 (1984),

555–580; U. MEISTER, "Islam in Azerbaijan," *Swiss Review of World Affairs*, 37, no. 1 (April 1987), 10–11; M. B. OLCOTT, "Soviet Islam and World Revolution," *World Politics*, 34 (1982), 487–504; T. SHABAD, "Ethnic Results of the 1979 Soviet Census," *Soviet Geography*, 21 (1980), 440–488; G. WHEELER, "Islam and the Soviet Union," *Asian Affairs*, 15 (1979), 245–251; S. E. WIMBUSH, ed., *Soviet Nationalities in Strategic Perspective* (Croom Helm, 1985); R. WIXMAN, *The Peoples of the USSR: An Ethnographic Handbook* (M. E. Sharpe, 1984).

Physical Environment *Fiziko-Geograficheskiy Atlas Mira* [Physico-Geographical Atlas of the World] (Moscow: Akademiya Nauk SSSR i Glavnoye Upravleniye Geodezii i Kartografii GGK SSSR; a magnificent reference atlas with text in Russian; an English translation of legends is available in *Soviet Geography: Review and Translation*, 6, nos. 5–6, May–June 1965, entire issue); P. M. KELLY *et al.*, "Large-Scale Water Transfers in the U.S.S.R.," *GeoJournal*, 7 (1983), 201–214; P. E. LYDOLPH, "The Russian Sukhovey," *Annals AAG*, 54 (1964), 291–309; P. P. MICKLIN, "The Status of the Soviet Union's North-South Water Transfer Projects before Their Abandonment in 1985–86," *Soviet Geography*, 27 (1986), 287–329; P. R. PRYDE, "The Future Environmental Agenda of the U.S.S.R.," *Soviet Geography*, 29 (1988), 555–567; D. TOLMAZIN, "Recent Changes in Soviet Water Management: Turnabout of the 'Project of the Century,' " *GeoJournal*, 15 (1987), 243–258; V. V. VOROB'YEV, "Problems of Protecting the Environment in Siberia," *Geoforum*, 15 (1984), 105–111.

Economy and Resources B. M. BARR, "Soviet Timber: Regional Supply and Demand, 1970–1990," *Arctic*, 32 (1979), 308–328, also, with K. BRADEN, *The Disappearing Russian Forest* (Barnes & Noble, 1988); L. DIENES, "Modernization and Energy Development in the Soviet Union," *Soviet Geography*, 21 (1980), 121–158, also, "Soviet-Japanese Economic Relations: Are They Beginning to Fade?" *Soviet Geography*, 26 (1985), 509–525, and, with T. SHABAD, *The Soviet Energy System: Resource Use and Policies* (Wiley, 1979); F. J. M. FELDBRUGGE, "Government and Shadow Economy in the Soviet Union," *Soviet Studies*, 36 (1984), 528–543; MARSHALL GOLDMAN, "Why Not Sell Technology to the Russians?" *Technology Review*, 87, no. 2 (February–March 1984), 70–80; R. HEILBRONER, "The Triumph of Capitalism," *The New Yorker*, January 23, 1989, 98–109; R. A. HELIN, "Soviet Fishing in the Barents Sea and the North Atlantic," *Geographical Review*, 54 (1964), 386–408; R. G. JENSEN, T. SHABAD, and A. W. WRIGHT, eds. *Soviet Natural Resources in the World Economy* (Chicago, 1983; massive, detailed, authoritative, with many maps); P. KENNEDY, "What Gorbachev Is Up Against: The Problem Isn't *in* the System—It *Is* the System," *Atlantic Monthly*, June, 1987, 29–43; T. SHABAD,

"Geographic Aspects of the New Soviet Five-Year Plan, 1986–90," *Soviet Geography*, 27 (1986), 1–16; S. F. STARR, "Technology and Freedom in the Soviet Union," *Technology Review*, 87, no. 4 (May–June 1984), 38–47; "The Soviet Economy: Russian Roulette," *The Economist*, April 9, 1988, 1–18; J. WINIECKI, "Are Soviet-Type Economies Entering an Era of Long-Term Decline?" *Soviet Studies*, 38 (1986), 325–348; C. S. WREN, "Breaking Out" [reform in the U.S.S.R. and China], *New York Times Magazine*, August 14, 1988, 22–25+.

Agriculture and Rural Life L. R. BROWN, *U. S. and Soviet Agriculture: The Shifting Balance of Power*, Worldwatch Paper 51 (October 1982); R. CONQUEST, *Harvest of Sorrow: Soviet Collectivization and the Terror-Famine* (Oxford, 1986); N. C. FIELD, "Environmental Quality and Land Productivity: A Comparison of the Agricultural Land Base of the USSR and North America," *Canadian Geographer*, 12 (1968), 1–14; S. HEDLUND, *Crisis in Soviet Agriculture* (Croom Helm, 1984); Z. A. MEDVEDEV, *Soviet Agriculture* [inefficiencies as seen by an exiled scientist] (Norton, 1988); I. STEBELSKY, "Agriculture of the Soviet Union and Eastern Europe," *Journal of Geography*, 84 (1985), 264–272, also, "Milk Production and Consumption in the Soviet Union," *Soviet Geography*, 29 (1988), 459–475.

Cities, Industries, and Transportation A. R. BOND, "Urban Planning and Design in the Soviet North: The Noril'sk Experience," *Soviet Geography*, 25 (1984), 145–165; also, "Economic Development at Noril'sk," *Soviet Geography*, 25 (1984), 354–368; and, "Transport Development at Noril'sk," *Soviet Geography*, 25 (1984), 103–121; J. ERONEN, "Routes of Soviet Timber to World Markets," *Geoforum*, 14 (1983), 205–210; R. A. FRENCH, "The Changing Russian Urban Landscape," *Geography*, 68 (1983), 236–244; P. E. GARBUTT, "The Trans-Siberian Railway," *Journal of Transport History*, 6 (1954), 238–249; M. E. HAMM, ed., *The City in Russian History* (Kentucky, 1976); C. D. HARRIS, "City and Region in the Soviet Union," in R. P. Beckinsale and J. M. Houston, eds., *Urbanization and Its Problems* (Barnes & Noble, 1968), pp. 277–296, also, *Cities of the Soviet Union: Studies in Their Functions, Size, Density, and Growth* (AAG Monograph 5, 2d printing; Washington, D.C.: Association of American Geographers, 1972); G. HAUSLADEN, "Containing the Growth of Moscow: Comparisons with London," *Cities*, 2 (1985), 55–69; D. KOENKER, "Urbanization and Deurbanization in the Russian Revolution and Civil War," *Journal of Modern History*, 57 (1985), 424–450; O. A. KONSTANTINOV, "Types of Urbanization in the USSR," *Soviet Geography*, 18 (1977), 715–728; G. M. LAPPO, "The City of Moscow and the Moscow Agglomeration," *GeoJournal*, Supplementary Issue, 1 (1980), 45–52; E. B. MILLER, "The Trans-Siberian Landbridge, A New Trade

Route between Japan and Europe: Issues and Prospects," *Soviet Geography,* 19 (1978), 223–243; H. W. MORTON and R. C. STUART, eds., *The Contemporary Soviet City* (M. E. Sharpe, 1984); V. L. MOTE, "A Visit to the Baikal-Amur Mainline and the New Amur-Yakutsk Rail Project," *Soviet Geography,* 26 (1985), 691-719, also, "Containerization and the Trans-Siberian Land Bridge, *Geographical Review,* 74 (1984), 304–314; W. H. PARKER, "The Soviet Motor Industry," *Soviet Studies,* 32 (1980), 515–541; R. H. ROWLAND, "Changes in the Metropolitan and Large City Populations of the USSR: 1979–85," *Soviet Geography,* 27 (1986) 638–658; T. SHABAD, "New Major Soviet Cities," *Soviet Geography,* 27 (1986), 59–65; D. J. B. SHAW, "Planning Leningrad," *Geographical Review,* 68 (1978), 183–200; D. A. STAPLE, "The BAM: Labor, Migration and Prospects for Settlement," *Soviet Geography,* 27 (1986), 716–740; C. THOMAS, "Moscow's Mobile Millions," *Geography,* 73 (1988), 216–225; K. WARREN, Industrial Complexes in the Development of Siberia," *Geography,* 63 (1978), 167–178; N. WEIN, "Bratsk—Pioneering City in the Taiga," *Soviet Geography,* 28 (1987), 171–194; C. ZumBRUNNEN and J. P. OSLEEB, *The Soviet Iron and Steel Industry* (Rowman & Allanheld, 1986).

Siberia and the North V. CONOLLY, "The Soviet Far East and Eastern Siberia: Impressions of a Recent Visit," *Asian Affairs,* 12 (1981), 35–48; P. DE SOUZA, "The Nature of the Manpower Problem in the Development of Siberia," *Soviet Geography,* 27 (1986), 689–715; L. DIENES, "Regional Planning and the Development of Soviet Asia," *Soviet Geography,* 28 (1987), 287–314, also, "The Development of Siberian Regions: Economic Profiles, Income Flows and Strategies for Growth," *Soviet Geography,* 23 (1982), 205–244; JOHN D. HARBRON, "Modern Icebreakers," *Scientific American,* 249, no. 6 (December 1983), 49–55; P. ROSTANKOWSKI, "The Decline of Agriculture and the Rise of Extractive Industry in the Soviet North," *Polar Geography and Geology,* 7 (1983), 289–298; J. SALLNOW, "Forbidden Kolyma," *Geographical Magazine,* 53 (1981), 261–267, also, "The Future in the Bear's Back Yard" [Soviet Far East], *Geographical Magazine,* 60, no. 6 (June 1988), 12–16; PETER SEIDLITZ, "Gas from Siberia," *Swiss Review of World Affairs,* 33, no. 3 (June 1983), 14–19; K. SOMERVILLE, "Western Siberia and Future Prospects for Soviet Oil," *International Relations,* 7 (1983), 2351–2362; A. WOOD, ed., *Siberia: Problems and Prospects for Regional Development* (Barnes & Noble, 1987).

PART IV: THE MIDDLE EAST

Middle East as a Whole I. R. AL FARUQI and L. L. AL FARUQI, *The Cultural Atlas of Islam* (Macmillan, 1986); S. R. ALI, *Oil, Turmoil, and Islam in the Middle East* (Praeger, 1986); *Aramco World* (bimonthly; a richly illustrated magazine for general readers published by Aramco Corporation; wide-ranging subject matter on the Arab and Islamic worlds); F. BARNABY, "The Nuclear Arsenal in the Middle East," *Technology Review,* 90, no. 4 (May–June 1987), 27–34; P. BEAUMONT *et al., The Middle East: A Geographical Study,* 2d ed. (Wiley, 1988); J. S. BIRKS, "The Mecca Pilgrimage by West African Pastoral Nomads," *Journal of Modern African Studies,* 15 (1977), 47–58; G. BLAKE, J. DEWDNEY, and J. MITCHELL, *The Cambridge Atlas of the Middle East and North Africa* (Cambridge, 1987); G. H. BLAKE and R. I. LAWLESS, eds., *The Changing Middle Eastern City* (Barnes & Noble, 1980); M. BRAUER, *Atlas of the Middle East* (Macmillan, 1988); J. I. CLARKE and H. BOWEN-JONES, eds., *Change and Development in the Middle East: Essays in Honour of W. B. Fisher* (Methuen, 1981); J. L. CLOUDSLEY-THOMPSON, ed., *Sahara Desert* (Oxford, 1984); F. J. COSTA and A. G. NOBLE, "Planning Arabic Towns," *Geographical Review,* 76 (1986), 160–172; V. F. COSTELLO, *Urbanization in the Middle East* (Cambridge, 1977); G. B. CRESSEY, *Crossroads: Land and Life in Southwest Asia* (Lippincott, 1960; a standard geography; dated but still useful for detail), also, "Qanats, Karez and Foggaras," *Geographical Review,* 48 (1958), 27–44, and, "Water in the Desert," *Annals AAG,* 47 (1957), 105–124; F. M. DONNER, *The Early Islamic Conquests* (Princeton, 1981); A. DRYSDALE and G. H. BLAKE, *The Middle East and North Africa: A Political Geography* (Oxford, 1985); J. L. ESPOSITO, ed., *Islam in Asia: Religion, Politics, and Society* (Oxford, 1987); W. B. FISHER, *The Middle East: A Physical, Social, and Regional Geography,* 7th ed. (Methuen, 1978), also, "Progress in the Middle East," *Progress in Human Geography* 5 (1981), 432–438, and, "Middle East," *Progress in Human Geography,* 7 (1983), 429–435; *Horizon,* 13, no. 3 (Summer 1971), articles as follows: J. MORRIS, "What Is an Arab?" 4–17; S. DE GRAMONT, "Mohammed: The Prophet Armed," 18–23; and D. DAICHES, "What Is a Jew?" 24–35; C. ISSAWI, *An Economic History of the Middle East and North Africa* (Methuen, 1982); P. JABBER, "Forces of Change in the Middle East," *Middle East Journal,* 42 (1988), 7–15; D. L. JOHNSON, *The Nature of Nomadism: A Comparative Study of Pastoral Migrations in Southwestern Asia and Northern Africa* (University of Chicago, Department of Geography Research Paper 118, 1969); P. P. KARAN and W. A. BLADEN, "Arabic Cities," *Focus,* 33, no. 3 (January–February 1983), 1–8; R. KING, "The Pilgrimage to Mecca: Some Geographical and Historical Aspects," *Erdkunde,* 26 (1972), 61–73; D. LAMB, *The Arabs: Journeys beyond the Mirage* (Random House, 1987; by a journalist); *The Middle East and North Africa,* (Europa Publications, annual, from 1948); *Middle East Journal* (quarterly); T. NAFF and R. C. MATSON, eds., *Water in the Middle East: Conflict or Cooperation?* (Westview, 1984); H. MUNSON, JR., *Islam and Revolution in the Middle East* (Yale, 1988); D. PIPES, "Fundamentalist Muslims between America and Russia," *Foreign Affairs*, 64 (1986), 939–959; I. A. RAGAB, "Islam and Development," *World Development*, 8 (1980), 513–521; H. ROBERTS, *An Urban Profile of the Middle East* (St. Martin's, 1979); F. ROBINSON, *Atlas of the Islamic World since 1500* (Facts on File, 1982); R. PATAI, "The Middle East as a Culture Area," *Middle East Journal*, 6 (1952), 1–21; F. J. SIMOONS, *Eat Not This Flesh: Food Avoidances in the Old World* (Wisconsin, 1961), especially chap. 3, "Pigs and Pork," 13–43; C. G. SMITH, "Water Resources and Irrigation Development in the Middle East," *Geography*, 55 (1970), 407–425; "The Middle East, 1989," *Current History*, February 1989, entire issue; H. VON WISSMAN *et al.*, "On the Role of Nature and Man in Changing the Face of the Dry Belt of Asia," in W. L. Thomas, Jr., ed., *Man's Role in Changing the Face of the Earth* (Chicago, 1956), 278–303; J. M. WAGSTAFF, *The Evolution of Middle Eastern Landscapes: An Outline to* A.D. *1840* (Barnes & Noble, 1985); J. WELLARD, *Samarkand and Beyond: A History of Desert Caravans* (Constable, 1977).

Middle East: African Sector J. A. ALLAN, "High Aswan Dam Is a Success Story," *Geographical Magazine*, 53 (1981), 393–396; H. ANSARI, *Egypt: The Stalled Society* (SUNY, 1986); K. M. BARBOUR, "The Sudan since Independence," *Journal of Modern African Studies*, 18 (1980), 73–97; R. E. BENEDICK, "The High Dam and the Transformation of the Nile," *Middle East Journal*, 33 (1979), 119–144; C. CLAPHAM, *Transformation and Continuity in Revolutionary Ethiopia* (Cambridge, 1988); S. EL-SHAKHS, "National Factors in the Development of Cairo," *Town Planning Review*, 42 (1971), 233–249; H. M. FAHIM, *Dams, People and Development: The Aswan High Dam Case* (Oxford, 1981); ALLAN M. FINDLAY and ANNE M. FINDLAY, "Regional Disparities and Population Change in Morocco," *Scottish Geographical Magazine*, 102 (1986), 29–41; A. T. GROVE, "Egypt Has Too Much Water," *Geographical Magazine*, 54 (1982), 437–441; G. HAILE, "The Unity and Territorial Integrity of Ethiopia," *Journal of Modern African Studies*, 24 (1986), 465–487; G. HANCOCK, *Ethiopia: The Challenge of Hunger* (Gollancz, 1983); A. HERACLIDES, "Janus or Sisyphus? The Southern Problem of the Sudan," *Journal of Modern African Studies*, 25 (1987), 213–231; H. E. HURST, *The Nile: A General Account of the River and the Utilization of Its Waters*, rev. ed. (Constable, 1957); M. JENNER, "Cairo in Peril" [historic preservation], *Geographical Magazine*, 57 (1985), 474–480; R. D. KAPLAN, "The Loneliest War" [Ethiopia], *Atlantic Monthly*, July 1988, 58–65; A-A I. KASHEF, "Technical and Ecological Impacts of the Aswan High Dam," *Journal of Hydrology*, 53 (1981), 73–84; G. KEBBEDE and M. J. JACOB, "Drought, Famine and the Political Economy of Environmental Degradation in Ethiopia," *Geography*, 73 (1988), 65–70; R. I. LAWLESS, "The Concept of *tell* and *sahara* in the Maghreb: A Reappraisal," *Transactions IBG*, 57 (November 1972), 125–137; J. MADELEY,

"Ethiopia's New Villagers," *Geographical Magazine*, 58 (1986), 246–249; N. J. MIDDLETON, "Dust Storms in the Middle East," *Journal of Arid Environments*, 10 (1986), 83–96; M. OTTOWAY, "Drought and Development in Ethiopia," *Current History*, 85 (1986), 217–220, 224; J. D. PENNINGTON, "The Copts in Modern Egypt," *Middle Eastern Studies*, 18 (1982), 158–179; N. POLLARD, "The Gezira Scheme: A Study in Failure," *The Ecologist*, 11 (1981), 21–31; S. RADWAN and E. LEE, *Agrarian Change in Egypt: An Anatomy of Rural Poverty* (Croom Helm, 1986); A. RICHARDS, "The Agricultural Crisis in Egypt," *Journal of Development Studies*, 16 (1980), 303–321; J. R. ROGGE, *Too Many, Too Long: Sudan's Twenty-Year Refugee Dilemma* (Rowman & Allanheld, 1985); R. SAID, "The Future Use of the Waters of the Nile," *Die Erde*, 117 (1986), 165–177; B. WALLACH, "The Nile Valley," *Focus*, 36, no. 1 (Spring 1986), 16–19, also, "The Sudan Gezira," *Focus*, 35, no. 4 (October 1985), 10–13; J. WATERBURY, *Hydropolitics of the Nile Valley* (Syracuse, 1979); C. WEIL and K. M. KVALE, "Current Research on Geographical Aspects of Schistosomiasis," *Geographical Review*, 75 (1985), 186–216; U. WIKAN, "Living Conditions among Cairo's Poor: A View from Below," *Middle East Journal*, 39 (1985), 7–26; C. WILLIAMS, "Islamic Cairo: Endangered Legacy," *Middle East Journal*, 39 (1985), 231–246.

Israel, Jordan, Lebanon, Syria F. AJAMI, "Lebanon and Its Inheritors," *Foreign Affairs*, 63 (1985), 778–799; D. H. K. AMIRAN, "Geographical Aspects of National Planning in Israel: The Management of Limited Resources," *Transactions IBG*, 3 (1978), 115–128; *Atlas of Israel*, 3d ed. (Macmillan, 1985); Y. BEN-ARIEH, "Urban Development in the Holy Land," in J. PATTEN, ed., *The Expanding City: Essays in Honour of Professor Jean Gottmann* (Academic, 1983), 1–37; H. CATTAN, *The Palestine Question* (Croom Helm, 1988); A. CATTANI, "Israel: The Struggle for Existence," *Swiss Review of World Affairs*, 38, no. 3 (June 1988), 8–11; H. COBBAN, *The Making of Modern Lebanon* (Hutchinson, 1985); P. COSSALI *et. al*, "Whose Promised Land? Israel and the Palestinians," *Geographical Magazine*, 60 (1988), 2–19; R. COCKBURN, "A State in Devastation" [Lebanon], *Geographical Magazine*, 55 (1983), 561–568; A. DRYSDALE, "Political Conflict and Jordanian Access to the Sea," *Geographical Review*, 77 (1987), 86–102; E. EFRAT, "Israel's Map of Inequality in Spatial Development," *GeoJournal*, 13 (1986), 401–411; A. ELON, "Letter from Israel," *The New Yorker*, February 13, 1989, 74–80; M. GILBERT, *The Arab-Israeli Conflict: Its History in Maps*, 3d ed. (Weidenfeld & Nicolson, 1979); C. L. HALLOWELL, "The Glory That Was Jerusalem," *Natural History*, 82 (1973), 39–49; B. HALPERN, *The Idea of the Jewish State* (Harvard, 1961); L. HALPRIN, "Israel, the Man-Made Landscape," *Landscape*, 9, no. 2 (Winter 1959–1960), 19–23; J. E. HAZLETON, "Land Reform in Jordan: The East Ghor Canal Project,"; *Middle East Studies*,

15 (1979), 239–257; A. KELLERMAN, "Tel-Aviv," *Cities,* 2 (1985), 98–105; S. KHALAF, *Lebanon's Predicament* (Columbia, 1987); R. G. KHOURI, *The Jordan Valley: Life and Society below Sea Level* (Longman, 1981); N. KLIOT, "Lebanon—A Geography of Hostages," *Political Geography Quarterly,* 5 (1986), 199–220; A. D. MILLER, "The Arab-Israeli Conflict, 1967–1987: A Retrospective," *Middle East Journal,* 41 (1987), 349–360; J. MUIR, "Lebanon: Arena of Conflict, Crucible of Peace," *Middle East Journal,* 38 (1984), 204–219; E. ORNI and E. EFRAT, *Geography of Israel,* 4th ed. (Jerusalem: Israeli Universities Press, 1980); D. PERETZ, *The West Bank: History, Politics, Society, and Economy* (Westview, 1986), also, "Intifadeh: The Palestinian Uprising," *Foreign Affairs,* 66 (1988), 964–980; F. E. PETERS, *Jerusalem and Mecca: The Typology of the Holy City in the Near East* (NYU, 1987); B. REICH, *Israel: Land of Tradition and Conflict* (Westview, 1985); B. REICH and G. R. KIEVAL, eds., *Israel Faces the Future* (Praeger, 1986); G. ROWLEY, "Divisions in a Holy City," *Geographical Magazine,* 56 (1984), 196–202; E. A. SALEM, "Lebanon's Political Maze: The Search for Peace in a Turbulent Land," *Middle East Journal,* 33 (1979), 444–463; S. SHERMAN, "Gaza: A History of Conflict," *Geographical Magazine,* 60, no. 12 (December 1988), 18–23; D. K. SHIPLER, *Arab and Jew: Wounded Spirits in a Promised Land* (Times Books, 1986; by a journalist); C. G. SMITH, "The Disputed Waters of the Jordan," *Transactions IBG,* 40 (December 1966), 111–128; E. STERN, Y. HAYUTH, and Y. GRADUS, "The Negev Continental Bridge: A Chain in an Intermodal Transport System," *Geoforum,* 14 (1983), 461–469; E. STERN and Y. GRADUS, "The Med-Dead Sea Project: A Vision or Reality?" *Geoforum,* 12 (1981), 265–272; G. SZPIRO, "Israel and the Arabs in the Occupied Territories," *Swiss Review of World Affairs,* 37, no. 1 (April 1987), 12–14; M. VIORST, "The Christian Enclave" [Lebanon], *The New Yorker,* October 3, 1988, 40–71; S. WATERMAN, "Ideology and Events in Israeli Human Landscapes," *Geography,* 64 (1979), 171–181; D. WEINTRAUB *et al., Moshava, Kibbutz, and Moshav: Patterns of Jewish Rural Settlement and Development in Palestine* (Cornell, 1969); M. ZAMIR, *The Formation of Modern Lebanon* (Croom Helm, 1985).

Arabian Peninsula and Persian/Arabian Gulf

P. ADAMS, "Yemen," *Geographical Magazine,* 60, no. 7 (July 1988), 26–33; F. AL-FARSY, *Saudi Arabia* (Routledge & Kegan Paul, 1986); M. ALIREZA, "Women of Saudi Arabia," *National Geographic,* 172 (1987), 422–453; H. K. BARTH and F. QUIEL, "Riyadh and Its Development," *GeoJournal,* 15 (1987), 39–46; P. BEAUMONT, "Water and Development in Saudi Arabia," *Geographical Journal,* 143 (1977), 42–60; R. BIDWELL, *The Two Yemens* (Westview, 1983); M. BLACKSELL, "A Crucial State" [Kuwait], *Geographical Magazine,* 58 (1986), 282–287; H. BOWEN-JONES, "Ancient Oman Is Transformed," *Geographical Magazine,* 52 (1980), 286–293; C. Drake, "Oman: Traditional and Modern Adaptations to the Environment," *Focus,* 38, no. 2 (Summer 1988), 15–20; S. A. EL-ARIFI, "The Nature of Urbanization in the Gulf Countries," *GeoJournal,* 13 (1986), 223–235; W. B. FISHER, "The Good Life in Modern Saudi Arabia," *Geographical Magazine,* 51 (1979), 762–768; S. T. HUNTER, "The Gulf Economic Crisis and Its Social and Political Consequences," *Middle East Journal,* 40 (1986), 593–613; G. LENCZOWSKI, "The Soviet Union and the Persian Gulf: An Encircling Strategy," *International Journal,* 37 (1982), 307–327; T. R. McHALE, "A Prospect of Saudi Arabia," *International Affairs,* 56 (1980), 622–647; A. MELAMID, "Dhofar," *Geographical Review,* 74 (1984), 106–109, also, "New Oil from Arabia," *Geographical Review,* 78 (1988), 76–79, and "Qatar," *Geographical Review,* 77 (1987), 103–105; I. I. NAWWAB *et al.,* eds., *Aramco and Its World: Arabia and the Middle East* (Washington, D.C.: Arabian American Oil Company, 1980); I. R. NETTON, ed., *Arabia and the Gulf: From Traditional Society to Modern States* (Barnes & Noble, 1987); "OPEC 2000: The Cartel that Fell Out of the Driver's Seat," *The Economist,* February 4, 1989, 17–19; R. K. RAMAZANI, *The Persian Gulf and the Strait of Hormuz* (Sijthoff & Noordhoff, 1979); S. SEARIGHT, "Farmers of the Desert" [Saudi Arabia], *Geographical Magazine,* 58 (1986), 127–131; W. D. SWEARINGEN, "Sources of Conflict over Oil in the Persian/Arabian Gulf," *Middle East Journal,* 35 (1981), 315–330; W. THESIGER, *Arabian Sands* (Dutton, 1959).

Turkey, Cyprus, Iraq, Iran, Afghanistan, Pakistan

F. AJAMI, "Iran: The Impossible Revolution," *Foreign Affairs,* 67, no. 2 (Winter 1988/89), 135–155; N. J. R. ALLAN, "Afghanistan: The End of a Buffer State," *Focus,* 36, no. 3 (Fall 1986), 2–9; S. A. ARJOMAND, "Iran's Islamic Revolution in Comparative Perspective," *World Politics,* 38 (1986), 383–414; B. W. BEELEY, "The Greek-Turkish Boundary Conflict at the Interface," *Transactions IBG,* 3 (1978), 351–366; G. BLAKE, "Turkish Guard on Russian Waters" [Turkish Straits], *Geographical Magazine,* 53 (1981), 950–955; G. D. CAMP, "Greek-Turkish Conflict over Cyprus," *Political Science Quarterly,* 95 (1980), 43–70; R. COCKBURN, "Two Sides of Cyprus," *Geographical Magazine,* 57 (1985), 144–149; W. B. FISHER, ed., *The Land of Iran* (The Cambridge History of Iran, vol. 1; Cambridge, 1968); J. F. HANSMAN, "The Mesopotamian Delta in the First Millennium, B.C.," *Geographical Journal,* 144 (1978), 49–61; J. F. JARRIGE and R. H. MEADOW, "The Antecedents of Civilization in the Indus Valley," *Scientific American,* 243, no. 2 (August 1980), 122–133; R. D. KAPLAN, "Afghanistan: Driven Toward God," *Atlantic Monthly,* September 1988, 16–20; Z. KHALILZAD, "The War in Afghanistan," *International Journal,* 41 (1986), 271–299; D. LESSING, "The Catastrophe" [Soviets in Afghanistan], *The New Yorker,* March 16, 1987,

74–93; P. G. LEWIS, "Iranian Cities," *Focus*, 33, no. 3 (January–February 1983), 12–16; J. NAGEL, "The Conditions of Ethnic Separatism: The Kurds in Turkey, Iran, and Iraq," *Ethnicity*, 7 (1980), 279–297; J. H. SIGLER, "The Iran–Iraq Conflict: The Tragedy of Limited Conventional War," *International Journal*, 41 (1986), 424–456; M. STERNER, "The Iran—Iraq War," *Foreign Affairs*, 63 (1984), 128–143; "Turkey: Half Inside, Half Out," *The Economist*, June 18, 1988, 1–30; R. WRIGHT, "Teheran Summer," *The New Yorker*, September 5, 1988, 32–72; H. E. WULFF, "The Qanats of Iran," *Scientific American*, 218, no. 4 (April 1968), 94–105.

PART V: THE ORIENT

Orient: General I. ADAMS, "Rice Cultivation in Asia," *American Anthropologist*, 50 (1948), 256–278; "America, Asia and Europe," *The Economist*, December 24, 1988, 29–41; R. BARKER *et al.*, *The Rice Economy of Asia* (Resources for the Future, 1985); L. W. BOWMAN and I. CLARK, eds., *The Indian Ocean in Global Politics* (Westview, 1981); G. B. CRESSEY, *Asia's Lands and Peoples*, 3d ed. (McGraw-Hill, 1963; quite dated but still a standard reference); "East Asian Security," *The Economist*, December 26, 1987, 35–40; J. FALLOWS, "When East Meets West: Oxford in Asia Paperbacks," *Atlantic Monthly*, 262, no. 3 (September 1988), 91–93; J. S. FEIN and P. L. STEPHENS, eds., *Monsoons* (Wiley, 1987); N. S. GINSBURG, ed., *The Pattern of Asia* (Prentice-Hall, 1958; a major geography text, still very valuable despite its date); R. E. HUKE, "The Green Revolution," *Journal of Geography*, 84 (1985), 248–254; L. KRAAR, "The New Powers of Asia," *Fortune*, March 28, 1988, 127–132; D. F. LACH, *Asia in the Making of Europe*, 2 vols. (Chicago; vol. 1, 1965; vol. 2, 1970, 1978); R. MURPHEY, *The Outsiders: The Western Experience in India and China* (Michigan, 1977); C. W. PANNELL, ed., *East Asia: Geographical and Historical Approaches to Foreign Area Studies*, (Kendall/Hunt, 1983); E. O. REISCHAUER and J. K. FAIRBANK, *A History of East Asian Civilization*, 2 vols. (Houghton Mifflin, vol. 1, *East Asia: The Great Tradition*, 1958, 1960; vol. 2, *East Asia: The Modern Transformation*, 1965); M. SMITH *et al.*, *Asia's New Industrial World* (Methuen, 1985); J. E. SPENCER, *Oriental Asia: Themes Toward a Geography* (Prentice-Hall, 1973), also, with W. L. THOMAS, *Asia, East by South: A Cultural Geography*, 2d ed. (Wiley, 1971); G. T. TREWARTHA, "Monsoons: With a Focus on South Asia and Tropical East Africa," *Journal of Geography*, 81 (1982), 4–11; C. E. TWEDDELL and L. A. KIMBALL, *Introduction to the Peoples and Cultures of Asia* (Prentice-Hall, 1985); P. J. VESILIND, "Monsoons," *National Geographic*, 166 (1984), 712–747; T. C. WHITMORE, *Tropical Rain Forests of the Far East*, 2d ed. (Oxford, 1984); R. O. WHYTE, *The Industrial Potential of Rural Asia* (University of Hong Kong, 1982); Y. YEUNG, "Controlling Metropolitan Growth in Eastern Asia," *Geographical Review*, 76 (1986), 125–137.

Indian Subcontinent R. AKHTAR and A. LEARMONTH, "Malaria Returns to India," *Geographical Magazine*, 54 (1982), 135–139; C. P. BAKER, "Observatories of the Raja," *Pacific Discovery*, 40, no. 4 (October–December 1987), 4–15; A. L. BASHAM, *The Wonder That Was India*, 3d ed. (Sidgwick & Jackson, 1967); C. BAXTER, *Bangladesh: A New Nation in an Old Setting* (Westview, 1984); T. P. BAYLISS-SMITH and S. WANMALI, *Understanding Green Revolutions: Agrarian Change and Development Planning In South Asia; Essays in Honour of B. H. Farmer* (Cambridge, 1984); S. M. BHARDWAJ, *Hindu Places of Pilgrimage in India: A Study in Cultural Geography* (California, 1973); *Cambridge Economic History of India*, 2 vols. (Cambridge, 1982); J. CHANG, "The Indian Summer Monsoon," *Geographical Review*, 37 (1967), 373–396; B. N. CHATURVEDI, "The Origin and Development of Tank Irrigation in Peninsular India," *Deccan Geographer*, 6 (1968), 57–86; B. DOGRA, "Traditional Agriculture in India: High Yields and No Waste," *The Ecologist*, 13 (1983), 84–87; J. J. DOUGLAS, "Traditional Fuel Usage and the Rural Poor in Bangladesh," *World Development*, 10 (1982), 669–676; A. K. DUTT and M. M. GELB, *Atlas of South Asia, Fully Annotated* (Westview, 1987); A. K. DUTT *et al.*, "Spatial Pattern of Languages in India: A Culture-Historical Analysis," *GeoJournal*, 10 (1985), 51–74, also, DUTT, "Bengal: A Search for Regional Identity," *Focus*, 34, no. 5 (May–June 1984, 1–12; S. DUTT, "India and the Overseas Indians," *India Quarterly*, 36 (1980), 307–335; B. H. FARMER, *An Introduction to South Asia* (Methuen, 1983), also, "Perspectives on the 'Green Revolution' in South Asia," *Modern Asian Studies*, 20 (1986), 175–199; J. FRYER, "Benefits from the Indian Sacred Cow," *Geographical Magazine*, 53 (1981), 617–624; A. GEDDES, *Man and Land in South Asia* (New Delhi: Concept Publishing Company, 1982); B. GOPAL, "Holy Mother Ganges," *Geographical Magazine*, 60, no. 5 (May 1988), 38–43; J. H. HUTTON, *Caste in India: Its Nature, Function and Origins*, 4th ed. (Oxford, 1963); B. L. C. JOHNSON, *Development in South Asia* (Penguin, 1983), also, *India: Resources and Development*, 2d ed. (Heinemann, 1983), and, *Pakistan* (Heinemann Educational, 1979); P. P. KARAN and S. IIJIMA, "Environmental Stress in the Himalaya," *Geographical Review*, 75 (1985), 71–92; M. KOSAMBI and J. E. BRUSH, "Three Colonial Port Cities in India" [Bombay, Calcutta, Madras], *Geographical Review*, 78 (1988), 32–47; R. KRISHNA, "The Economic Development of India," *Scientific American*, 243, no. 3 (September 1980), 166–178; D. O. LODRICK, *Sacred Cows, Sacred Places: Origins and Survivals of Animal Homes in India* (California, 1981); G. MOORHOUSE, *India Britannica* (Harvill Press, 1983); R. MURPHEY, "The City in the Swamp: Aspects of the Site and Early Growth of Calcutta," *Geographical*

Journal, 130 (1964), 241–256; V. S. NAIPAUL, *India: A Wounded Civilization* (Knopf, 1977); A. G. NOBLE and A. K. DUTT, *India: Cultural Patterns and Processes* (Westview, 1982); C. C. O'BRIEN, "Holy War Against India" [Sikh separatism], *Atlantic Monthly,* August 1988, 54–64; J. E. SCHWARTZBERG, ed., *An Historical Atlas of South Asia* (Chicago, 1978); F. J. SIMOONS, *Eat Not This Flesh: Food Avoidances in the Old World* (Wisconsin, 1961), especially chap. 4, "Beef," 45–63; D. E. SOPHER, "Place and Landscape in Indian Tradition," *Landscape,* 29 (1986), no. 2, 1–9; T. S. RANDHAWA, "When the Monsoon Fails," *Geographical Magazine,* 56 (1984), 93–97; A. H. SIDDIQI, "Society and Economy of the Tribal Belt in Pakistan," *Geoforum,* 18, (1987), 65–79; O. H. K. SPATE and A. T. A. LEARMONTH, *India and Pakistan: A General and Regional Geography,* 3d ed. (Methuen, 1967), with a chapter on Ceylon (now Sri Lanka) by B. H. Farmer—a huge, authoritative textbook that is still indispensable, although aging; I. STONE, *Canal Irrigation in British India* (Cambridge, 1948); A. R. TAINSH, "Hunger in India: The Human Factor," *International Relations,* 7 (1981), 1053–1062; R. TURNER, ed., *India's Urban Future* (California, 1962), especially chap. 3, "The Morphology of Indian Cities," by J. E. Brush, 57–70; R. G. VINES, "Rainfall Patterns in India," *Journal of Climatology,* 6 (1986), 135–148; J. R. WOOD, "Separatism in India and Canada: A Comparative Overview," *Indian Journal of Political Science,* 42, no. 3 (July–September 1981), 1–34.

Southeast Asia: General A. K. DUTT, ed., *Southeast Asia: Realm of Contrasts,* 3d rev. ed. (Westview, 1985); C. A. FISHER, *South-East Asia: A Social, Economic, and Political Geography,* 2d ed. (Dutton, 1964; a detailed and authoritative survey that is still very valuable despite its age); D. W. FRYER, *Emerging Southeast Asia: A Study in Growth and Stagnation* (George Philip, 1979); L. M. HANKS, *Rice and Man: Agricultural Ecology in Southeast Asia* (Aldine, 1972); L. KONG, "The Malay World in Colonial Fiction," *Singapore Journal of Tropical Geography,* 7 (June 1986), 40–52; Y. L. LEE, "Race, Language, and National Cohesion in Southeast Asia," *Journal of Southeast Asian Studies,* 11 (1980), 122–138; V. R. SAVAGE, *Western Impressions of Nature and Landscape in Southeast Asia* (Singapore, 1984); J. E. SPENCER, "Southeast Asia," *Progress in Human Geography,* 8 (1984), 284–288; P. A. STOTT, ed., *Nature and Man in South East Asia* (University of London, 1978); R. ULACK and G. PAUER, *Atlas of Southeast Asia* (Macmillan, 1988); C. F. YOUNG, ed., "Ethnic Change in Southeast Asia" (special issue), *Journal of Southeast Asian Studies,* 112 (1981); L. YOUNG-LENG, *Southeast Asia: Essays in Political Geography* (Singapore, 1982).

Southeast Asia: Sri Lanka B. AXELSEN, "How Dry Is Sri Lanka's Dry Zone? Some Comments on Agricultural Potential, Perception, and Planning," *Norsk Geografisk Tidsskrift,* 37 (1983), 197–209; K. M. DE SILVA, *Sri Lanka: A Survey* (Hawaii, 1977); S. K. HENNAYAKE and J. S. DUNCAN, "A Disputed Homeland: Sri Lanka's Civil War," *Focus,* 37, no. 1 (Spring 1987), 20–27; T. GOONERATNE, "The Present Situation in Sri Lanka," *Asian Affairs,* 17 (1986), 33–45; B. L. C. JOHNSON and M. SCRIVENOR, *Sri Lanka: Land, People and Economy* (Heinemann, 1981); R. N. KEARNEY, "Ethnic Conflict and the Tamil Separatist Movement in Sri Lanka," *Asian Survey,* 25 (1985), 898–917; R. SENANAYAKE, "The Ecological, Energetic, and Agronomic Systems of Ancient and Modern Sri Lanka," *The Ecologist,* 13 (1983), 136–140; S. J. TAMBIAH, *Sri Lanka: Ethnic Fratricide and the Dismantling of Democracy* (Chicago, 1986).

Southeast Asia: Burma, Thailand, Kampuchea, Laos, Vietnam S. ARNOLD, "South East Asian Stalwart" [Thailand], *Geographical Magazine,* 58 (1986), 578–583; D. P. CHANDLER, "The Tragedy of Cambodian History," *Pacific Affairs,* 52 (1979), 410–419; R. A. CROOKER, "Forces of Change in the Thailand Opium Zone," *Geographical Review,* 78 (1988), 241–256; T. DE RUBEMPRÉ, "Buddha's Own Country" [Burma], *Geographical Magazine,* 59 (1987), 183–188; B. B. FALL, "Two Thousand Years of War in Viet-Nam," *Horizon,* 9, no. 2 (Spring 1967), 4–22; J. FALLOWS, "The Philippines: The Bases Dilemma," *Atlantic Monthly,* February 1988, 18–30; C. A. FISHER, "The Vietnamese Problem in Its Geographical Context," *Geographical Journal,* 131 (1965), 502–515; B. K. GORDON, "The Third Indochina Conflict," *Foreign Affairs,* 65, (1986) 66–85; B. HODGSON, "Time and Again in Burma," *National Geographic,* 166 (1984), 90–121; J. HOSKIN, "Life Returns to Kampuchea," *Geographical Magazine,* 60, no. 1 (July 1988), 22–26; C. LEINSTER, "Vietnam Revisited: Turn to the Right?" *Fortune,* August 1, 1988, 84–102; J. D. RIGG, "The Role of the Environment in Limiting the Adoption of New Rice Technology in Northeastern Thailand," *Transactions IBG,* n.s., 10 (1985), 481–494; J. SILVERSTEIN, "Burma Through the Prism of Western Novels," *Journal of Southeast Asian Studies,* 16 (1985), 129–140; N. THRIFT, "Vietnam: Geography of a Socialist Siege Economy," *Geography,* 72 (1987), 340–344.

Southeast Asia: Malaysia, Singapore, Brunei C. CHOO, "Accelerated Industrialization and Employment Opportunities in Malaysia," *Geoforum,* 13 (1982), 11–18; P. P. COURTENAY, "The New Malaysian Sugar Plantation Industry," *Geography,* 69 (1984), 317–326, also, "The Plantation in Malaysian Economic Development," *Journal of Southeast Asian Studies,* 12 (1981), 329–348; J. DRYSDALE, *Singapore: Struggle for Success* (Singapore: Times Books International, 1984); B. FIELD and J. SMITH, "Singapore," *Cities,* 3 (1986), 186–199; J. GOODRIDGE, "A New Life for Tin," *Geographical Magazine,* 57 (1985), 424–429; K. GRICE and D. DRAKAKIS-SMITH, "The

Role of the State in Shaping Development: Two Decades of Growth in Singapore," *Transactions IBG*, n.s., 10 (1985), 347–359; I. HAMID, "Islamic Influences on Malay Culture in Malaysia and Indonesia," *Islamic Culture*, 56 (1982), 275–282; R. D. HILL, "Singapore: An Asian City-State," *GeoJournal*, 4 (1980), 5–12; B. HODGSON, "Singapore: Mini-Size Superstate," *National Geographic*, 159 (1981), 540–561; W. NEVILLE, "Economy and Employment in Brunei," *Geographical Review*, 75 (1985), 451–461.

Southeast Asia: Indonesia, Philippines R. BONNER, "The New Order" [Indonesia], I, II, *The New Yorker*, June 6/13, 1988; T. M. BURLEY, "The Philippines: An Economic and Social Geography (G. Bell, 1972); C. DRAKE, "The Spatial Pattern of National Integration in Indonesia," *Transactions IBG*, n. s., 6 (1981), 471–490; C. A. FISHER, "Indonesia—A Giant Astir," *Geographical Journal*, 138 (1972), 154–165; D. W. FRYER and J. C. JACKSON, *Indonesia* (Ernest Benn, 1977); A. GORDON, "Indonesia, Plantations and the 'Post-colonial' Mode of Production," *Journal of Contemporary Asia*, 12 (1982), 168–187; P. A. KRINKS, "Rural Changes in Java: An End to Involution?" *Geography*, 63 (1978), 31–36; H. LEITNER and E. S. SHEPPARD, "Indonesia: Internal Conditions, the Global Economy, and Regional Development," *Journal of Geography*, 86 (1987), 282–291; J. MADELEY, "People Transplanted; Forests Uprooted" [Indonesia] *Geographical Magazine*, 60, no. 7 (July 1988), 22–26; L. G. NOBLE, "Muslim Separatism in the Philippines, 1972–1981: The Making of a Stalemate," *Asian Survey*, 21 (1981), 1097–1114; J. PEACOCK, "The Impact of Islam" [Indonesia], *Wilson Quarterly*, 5 (1982), 138–144; *Philippine Geographical Journal* (quarterly); M. C. RICKLEFS, "Cultural Encounter: Islam in Java," *History Today*, 34 (1984), 23–29; F. L. WERNSTEDT and J. E. SPENCER, *The Philippine Island World: A Physical, Cultural, and Regional Geography* (California, 1967; comprehensive, authoritative; aging, but still valuable); W. A. WITHINGTON, "Indonesia's Significance in the Study of Regional Geography," *Journal of Geography*, 68 (1969), 227–237, also, "The Major Geographic Regions of Sumatra, Indonesia," *Annals AAG*, 57 (1967), 534–549; W. B. WOOD, "Intermediate Cities on a Resource Frontier" [Indonesia], *Geographical Review*, 76 (1986), 149–159.

China: General A. D. BARNETT, "Ten Years after Mao," *Foreign Affairs*, 65 (1986), 37–65; R. BAUM, "Science and Culture in Contemporary China: The Roots of Retarded Modernization," *Asian Survey*, 22 (1982), 1166–1186; C. BLUNDEN and M. ELVIN, *Cultural Atlas of China* (Facts on file, 1983); K. BUCHANAN *et al.*, *China: The Land and People* (Crown, 1981); F. BUNGE and R. S. SHINN, eds., *China, A Country Study*, 3d ed. (Washington, DC: Government Printing Office, 1981; U.S. Department of the Army,

Area Handbook Series); T. CANNON and A. JENKINS, "China: A Country So Changed," *Geography*, 71 (1986), 343–349; *China Geographer* (Westview, from 1975); P. J. M. GEELAN and D. C. TWITCHETT, ed., *The Times Atlas of China* (Van Nostrand, 1984); N. GINSBURG and B. LALOR, eds., *China: The 80's Era* (Westview, 1984); B. HOOK, ed., *The Cambridge Encyclopedia of China* (Cambridge, 1982; a massive, indispensable work); R. ISRAELI, "The Muslim Minority in the People's Republic of China," *Asian Survey*, 21 (1981), 901–919; P. JENKINS, *Across China* (Morrow, 1986); A. J. JOWETT, "The Growth of China's Population, 1949–1982," *Geographical Journal*, 150 (1984), 157–170; N. KEYFITZ, "The Population of China," *Scientific American*, 250, no. 2 (February 1984), 38–47; R. G. KNAPP, *China's Traditional Rural Architecture: A Cultural Geography of the Common House* (Hawaii, 1986), also, ed., *China's Island Frontier* (Hawaii, 1980); "Modern China," *Geographical Magazine*, 58–59 (1986–1987; articles in separate issues by various authors); I. MABBETT, *Modern China: The Mirage of Modernity* (Croom Helm, 1985); L. W. PYE, "Reassessing the Cultural Revolution," *China Quarterly*, 108 (1986), 597–612; C. L. SALTER, "The New Localism in China's Cultural Landscape," *Landscape*, 25, no. 3 (1981), 10–14, also, "Windows on a Changing China," *Focus*, 35, no. 1 (January 1985), 12–21; T. R. TREGEAR, *China: A Geographical Survey* (Wiley, 1980); D. S. ZAGORIA, "China's Quiet Revolution," *Foreign Affairs*, 62 (1984), 879–904.

China: Historical B. CATCHPOLE, *A Map History of Modern China* (Heinemann, 1976); K. CHAO, *Man and Land in Chinese History: An Economic Analysis* (Stanford, 1986); C. DIETRICH, *People's China: A Brief History* (Oxford, 1986); J. K. FAIRBANK, *The Great Chinese Revolution: 1800–1985* (Harper & Row, 1986); J. GERNET, *A History of Chinese Civilization*, trans. J. R. Foster (Cambridge, 1982); A. HERRMANN, *An Historical Atlas of China*; gen. ed., N. Ginsburg, prefatory essay, P. Wheatley (Aldine, 1966); K. S. LATOURETTE, *The Chinese: Their History and Culture* (Macmillan, 1964); F. MICHAEL, *China Through the Ages: History of a Civilization* (Westview, 1986); A. N. WALDRON, "The Problem of the Great Wall of China," *Harvard Journal of Asiatic Studies*, 43 (1983), 643–663.

China: Environment A. K. BISWAS, "Long Distance Water Transfer: The Chinese Plans," *GeoJournal*, 6 (1982), 481–487; L. CHANGMING and L. JIYANG, "Huang He: The Yellow River" ("Mighty Rivers of the World"), *Geographical Magazine*. 60, no. 2 (February 1988), 41–45; P. HO, "The Loess and the Origin of Chinese Agriculture," *American Historical Review*, 75 (1969), 1–36; L. J. C. MA and A. G. NOBLE, eds., *The Environment: Chinese and American Views* (Methuen, 1981); R. MURPHEY, "Man and Nature in China," *Modern Asian Studies* 1 (1967); V. SMIL, "China's Environment," *Current History*, 79, no.

458 (September 1980), 14–18, also, "Controlling the Yellow River," *Geographical Review,* 69 (1979), 253–272, and, *The Bad Earth: Environmental Degradation in China* (Sharpe, 1984); Z. SONGQIAO, *Physical Geography of China* (Wiley, 1986); A. S. WALKER, "Deserts of China," *American Scientist,* 70 (1982), 366–376; D. E. WALLING, "Yellow River Which Never Runs Clear," *Geographical Magazine,* 53 (1981), 568–575; "Water in China" (special issue), *GeoJournal,* 10, no. 2 (1985); C. XUEMIN, "Water Resources Planning on the Yellow River," *Natural Resources Forum,* 4 (1980), 315–324; Z. ZHANG, "Loess in China," *GeoJournal,* 4 (1980), 525–540.

China: Economy and Development M. C. BERGERE, "On the Historical Origins of Chinese Underdevelopment," *Theory and Society,* 13 (1984), 327–337; D. CHEN, "The Economic Development of China," *Scientific American,* 243, no. 3 (September 1980), 152–165; S. S. HARRISON, *China, Oil, and Asia: Conflict Ahead?* (Columbia, 1977); L. KRAAR, "The China Bubble Bursts," *Fortune,* July 6, 1987, 86–89; R. MURPHEY, *The Fading of the Maoist Vision: City and Country in China's Development* (Methuen, 1980); C. W. PANNELL, "Recent Chinese Agriculture," *Geographical Review,* 75 (1985), 170–185, also, "Less Land for Chinese Farmers," *Geographical Magazine,* 54 (1982), 324–329, and, with J. C. L. MA, *China: The Geography of Development and Modernization* (Wiley, 1983); D. R. PHILLIPS, "Oil in Chinese Waters," *Geographical Magazine,* 56 (1984), 444–445; O. SCHELL, "The Wind of Wanting to Go It Alone" [new economic pragmatism in China], *The New Yorker,* January 23, 1984, 43–85; V. SMIL, "Food Production and Quality of Diet in China," *Population and Development Review,* 12 (1986), 25–45; C. D. STOLTENBERG, "China's Special Economic Zones: Their Development and Prospects," *Asian Survey,* 24 (1984), 637–654; F. S. WORTHY, "Why There's Still Promise in China," *Fortune,* February 27, 1989, 95–101.

China: Urbanization D. BRONGER, "Metropolitanization in China?" *GeoJournal,* 8 (1984), 137–146; S. CHANG, "Distribution of China's City Population, 1982," *Urban Geography,* 7 (1986), 370–384, also, "Modernization and China's Urban Development," *Annals AAG,* 71 (1981), 202–219, and, "Peking: The Growing Metropolis of Communist China," *Geographical Review,* 55 (1965), 313–327; D. J. DWYER, "Chengdu, Sichuan: The Modernisation of a Chinese City," *Geography,* 71 (1986), 215–227; D. W. EDGINGTON, "Tianjin," *Cities,* 3 (1986), 117–124; R. KIRKBY, "China Goes to Town" [urban growth], *Geographical Magazine,* 58 (1986), 508–511, also, *Urbanization in China: Town and Country in a Developing Economy, 1949–2000 A.D.* (Columbia, 1985); L. J. C. MA and E. W. HANTEN, *Urban Development in Modern China* (Westview, 1981); K. C. TAN, "Small Towns in Chinese Urbanization," *Geographical Review,* 76 (1986), 265–275;

M. K. WHYTE, "Urbanism as a Chinese Way of Life," *International Journal of Comparative Sociology,* 24 (1983), 61–85.

Arid China, Hong Kong, Macao, Taiwan, Mongolia P. ALLEN, "Inside Chinese Tibet," *Geographical Magazine,* 52 (1980), 830–837; T. B. ALLEN, "Time Catches up with Mongolia," *National Geographic,* 167 (1984), 242–269; C. DUNCAN, "Macau," *Cities,* 3 (1986), 2–11; D. J. DWYER, "The Future of Hong Kong," *Geographical Journal,* 150 (1984), 1–10; N. GROVE, "Taiwan," *National Geographic,* 161, no. 1 (January 1982), 92–119; D. R. HALL, "Economic and Urban Development in Mongolia," *Geography,* 72 (1987), 73–76; I. KELLY, *Hong Kong: A Political-Geographic Analysis* (Hawaii, 1986); M. LeVASSEUR, "Destination: Tibet," *Focus,* 37, no. 1 (Spring 1987), 34–35; R. H. MYERS, "The Economic Transformation of the Republic of China on Taiwan," *China Quarterly,* 99 (1984), 500–528; W. PETERS, "The Unresolved Problem of Tibet," *Asian Affairs,* 19 (1988), 140–153; S. WONG, "Modernization and Chinese Culture in Hong Kong," *China Quarterly,* 106 (1986), 306–325.

Japan E. A. ACKERMAN, *Japan's Natural Resources and Their Relation to Japan's Economic Future* (Chicago, 1953; a massive survey, dated in many ways but still valuable); T. AKAHA, *Japan in Global Ocean Politics* (Hawaii, 1985); J. D. ALDEN, "Metropolitan Planning in Japan," *Town Planning Review,* 55 (1984), 55–74; ASSOCIATION OF JAPANESE GEOGRAPHERS, ed., *Geography of Japan* (Tokyo: Teikoku-Shoin, 1980); W. G. BEASLEY, "Tradition and Modernity in Post-war Japan," *Asian Affairs,* 11 (1980), 261–277; W. BURGESS, "Hokkaido: Japan's New Frontier," *Geography,* 67 (1982), 64–68; R. BUCKLEY, *Japan Today* (Cambridge, 1985); R. CYBRIWSKY, "Shibuya Center, Tokyo," *Geographical Review,* 78 (1988), 48–61; P. N. DALE, *The Myth of Japanese Uniqueness* (Croom Helm, 1986); P. F. DRUCKER, "Japan's Choices," *Foreign Affairs,* 65 (1987), 923–941; B. DUMAINE, "Japan's Next Push into U.S. Markets," *Fortune,* September 26, 1988, 135–142; J. D. EYRE, "Water Controls in a Japanese Irrigation System," *Geographical Review,* 45 (1955), 197–216; C. A. FISHER, "The Expansion of Japan: A Study in Oriental Geopolitics: Part I, Continental and Maritime Components in Japanese Expansion; Part II, the Greater East Asia Co-Prosperity Sphere," *Geographical Journal,* 115 (1950), 1–19, 179–193; C. HABERMAN, "The Presumed Uniqueness of Japan," *New York Times Magazine,* August 28, 1988, 38–43+; C. B. HARRIS, "The Urban and Industrial Transformation of Japan," *Geographical Review,* 72 (1982), 50–89; D. HENWOOD, "Behind Japan's Miracle: Playing the *Zaiteku* Game," *The Nation,* October 3, 1988, 266–271; *Japan: A Regional Geography of an Island Nation* (Tokyo: Teikoku-Shoin, 1985); "Japanese Property: A Glittering Sprawl," *The Economist,* October 3, 1987, 25–28; "Japan's Troubled Future: Special Report," *Fortune*, March 30, 1987, 21–53; J. D. KATZ and T. C. FRIEDMAN-LICHTSCHEIN, eds., *Japan's New World Role* (Westview, 1985); C. KOKUDO, *The National Atlas of Japan* (Tokyo: Japan Map Center, 1977); K. J. S. KO-KUSAI, *Atlas of Japan: Physical, Economic, and Social*, 2d ed. (Tokyo: International Society for Educational Information, 1974); D. KORNHAUSER, *Japan: Geographical Background to Urban-Industrial Development*, 2d ed. (Wiley, 1982); S. LOHR, "The Japanese Challenge," *New York Times Magazine*, July 8, 1984, 18–23+; L. H. LYNN, "Japanese Technology at a Turning Point," *Current History* 84 (1985), 418–421, 432; D. MacDONALD, *A Geography of Modern Japan* (Paul Norbury, 1985); G. M. MACKLIN, "Time and History in Japan," *American Historical Review*, 85 (1980), 557–571; I. J. McMULLEN, "How Confucian Is Modern Japan?" *Asian Affairs*, 11 (1980), 276–283; K. MURATA and I. OTA, eds., *An Industrial Geography of Japan* (St. Martin's, 1980); S. MATSUI *et al.*, "Farming at the Pace of Industry" [Japan], *Geographical Magazine*, 52 (1980), 740–746; M. NISHI, "Regional Variations in Japanese Farmhouses," *Annals AAG*, 57 (1967), 239–266; B. O'REILLY, "Will Japan Gain Too Much Power?" *Fortune*, September 12, 1988, 150–153; K. K. OSHIRO, "Mechanization of Rice Production in Japan," *Economic Geography*, 61 (1985), 323–331; J. PEZEU-MASSABUAU, *The Japanese Islands: A Physical and Social Geography* (translated from the French by P. C. Blum; Tuttle, 1978); H. C. REED, "The Ascent of Tokyo as an International Financial Center," *Journal of International Business Studies*, 11 (1980), 19–35; R. B. REICH, "The Rise of Techno-Nationalism" [U.S. and Japanese technology], *Atlantic Monthly*, May 1987, 63–69; E. O. REISCHAUER, *The Japanese Today: Change and Continuity* (Harvard, 1988); J. SARGENT, "Industrial Location in Japan with Special Reference to the Semiconductor Industry," *Geographical Journal*, 153 (1987), 72–85; E. SEIDENSTICKER, *Low City, High City: Tokyo from Edo to the Earthquake* (Knopf, 1983); H. SHITARA, "It Is Cold, It Is Hot, But It Is Seldom Dry" [Japanese climate], *Geographical Magazine*, 52 (1980), 547–554; T. C. SMITH, *The Agrarian Origins of Modern Japan* (Stanford, 1959); P. SPRY-LEVERTON and P. KORNICKI, *Japan* (Facts on File, 1987); H. TANAKA, "Landscape Expression of Buddhism in Japan," *Canadian Geographer*, 28 (1984), 240–257; S. TATSUNO, *The Technopolis Strategy: Japan, High Technology, and the Control of the Twenty-First Century* (Prentice-Hall, 1986); A. TAYLOR III, "Japan's Carmakers Take On the World," *Fortune*, June 20, 1988, 66–76; G. T. TREWARTHA, *Japan: A Geography* (Wisconsin, 1965; a standard, authoritative work, now somewhat outdated); E. F. VOGEL, "Pax Nipponica?" *Foreign Affairs*, 64 (1986), 752–767; J. WORONOFF, *Japan's Commercial Empire* (Sharpe, 1984).

Korea "A Survey of South Korea: Foursquare to the Future," *The Economist*, May 21, 1988, 1–22; J. FALLOWS, "Trade: Korea Is Not Japan," *Atlantic Monthly*, October 1988, 22–33; C. A. FISHER, "The Role of Korea in the Far East," *Geographical Journal*, 120 (1954), 282–298; P. IYER, "The Yin and Yang of Paradoxical, Prosperous Korea," *Smithsonian*, 19, no. 5 (August 1988), 46–59; N. JACOBS, *The Korean Road to Modernization and Development* (Illinois, 1985); L. KRAAR, "Korea: Tomorrow's Powerhouse," *Fortune*, August 15, 1988, 74–81; S. McCUNE, *Korea's Heritage: A Regional and Social Geography* (Charles E. Tuttle, 1964); "South Korea's Miracle," *The Economist*, March 4, 1989, 77–78.

PART VI: THE PACIFIC WORLD

Pacific World: General; Pacific Islands G. BARCLAY, *A History of the Pacific from the Stone Age to the Present Day* (Sidgwick & Jackson, 1978); J. C. BEAGLEHOLE, *The Exploration of the Pacific*, 3d ed. (Stanford, 1966); P. BELL-WOOD, *Man's Conquest of the Pacific* (Oxford, 1979), also, "The Peopling of the Pacific," *Scientific American*, 243, no. 5 (November 1980), 174–185; G. BERTRAM, "Sustainable Development in Pacific Micro-economies," *World Development*, 14 (1986), 809–822; S. G. BRITTON, "The Evolution of a Colonial Space-Economy: The Case of Fiji," *Journal of Historical Geography*, 6 (1980), 251–274; H. C. BROOKFIELD and D. HART, *Melanesia: A Geographical Interpretation of an Island World* (Barnes & Noble, 1971); K. BROWER, *Micronesia: The Land, the People, and the Sea* (Louisiana State, 1981); H. CLEVELAND, "The Future of the Pacific Basin," *Pacific Viewpoint*, 25 (1984), 1–13; J. CONNELL, "Independence, Dependence and Fragmentation in the South Pacific," *GeoJournal*, 5 (1981), 583–588, also, "New Caledonia: The Transformation of Kanaky Nationalism," *Australian Outlook*, 41 (1987), 37–44; K. B. CUMBERLAND, *Southwest Pacific: A Geography of Australia, New Zealand, and Their Pacific Island Neighbors*, rev. ed. (Praeger, 1968); C. DARWIN, *The Structure and Distribution of Coral Reefs* (California, 1962; first published in 1889); B. R. FINNEY, "Anomalous Westerlies, El Niño, and the Colonization of Polynesia," *American Anthropologist*, 87 (1985), 9–26; O. W. FREEMAN, ed., *Geography of the Pacific* (Wiley, 1951; a standard reference work, out of date in many respects but still useful); H. R. FRIIS, ed., *The Pacific Basin: A History of Its Geographical Exploration* (American Geographical Society, 1967); D. W. FRYER, "Changing Trade Patterns of the West Pacific," *GeoJournal*, 4 (1980), 479–484; E. LEACH, "Ocean of Opportunity" [the Pacific], *Pacific Viewpoint*, 24 (1983), 99–111; D. LEWIS, *We, The Navigators: The Ancient Art of Landfinding in the Pacific* (Hawaii, 1972); G. R. LEWTH-WAITE, "Man and Land in Early Tahiti: Polynesian Agriculture through European Eyes, *Pacific Viewpoint*, 5 (1964), 11–34, also, "Man and the Sea in Early Tahiti," *ibid.*, 7 (1966), 28–53; W. MENARD, "Coconut," *Pacific Discovery*, 21, no. 2 (March–April 1968), 19–24; A. MOOREHEAD, *The Fatal Impact: An Account of the Inva-*

sion of the South Pacific, 1767–1840 (Harper & Row, 1966); R. W. MURPHY, "American Micronesia: A Supplementary Chapter in the Regional Geography of the United States," Journal of Geography, 79 (1980), 181–186, also, "'High' and 'Low' Islands in the Eastern Carolines," Geographical Review, 39 (1949), 425–439; D. L. OLIVER, The Pacific Islands (Harvard, 1951; reprinted, Doubleday Anchor Books, 1962; a much-cited book by an anthropologist); "Pacific Island States" (special issue), GeoJournal, 16, no. 2 (March 1988); Pacific Islands Year Book (Sydney, Australia: Pacific Publications, annual); Pacific Viewpoint (semiannual; a geographical journal of high quality); A. G. PRICE, The Western Invasion of the Pacific and Its Continents: A Study of Moving Frontiers and Changing Landscapes, 1513–1958 (Oxford, 1963); R. J. SAGER, "The Pacific Islands: A New Geography," Focus, 38, no. 2 (Summer 1988), 10–14; M. D. SAHLINS, Islands of History [Oceania] (Chicago, 1985); B. SMITH, European Vision and the South Pacific, 2d ed. (Yale, 1985); O. H. K. SPATE, The Spanish Lake (Minnesota, 1979; vol. 1 in Spate, The Pacific since Magellan); D. R. STODDART, "Catastrophic Human Interference with Coral Island Ecosystems," Geography, 53 (1968), 25–40; S. D. THOMAS, "The Puzzle of Micronesian Navigation," Pacific Discovery, 35, no. 6 (November–December 1982), 1–12; R. TRUMBULL, Tin Roofs and Palm Trees: A Report on the New South Seas (Washington, 1977); H. J. WIENS, Atoll Environment and Ecology (Yale, 1962).

Australia B. ANSELM, "Australians Exploit Their Iron Mountains," Geographical Magazine, 53 (1981), 712–718; Atlas of Australian Resources (3d series; issued serially from 1980 by the Australian Division of National Mapping, Canberra); Australian Geographer (semiannual); Australian Geographical Studies (semiannual); J. BIRD, Seaport Gateways of Australia (Oxford, 1968); G. BLAINEY, The Tyranny of Distance: How Distance Shaped Australia's History (St. Martin's, 1968); R. BRITTON, "Australia's Resources Boom," Focus, 32, no. 4 (March–April 1982), 1–8; D. J. CARR and S. G. M. CARR, eds., Plants and Man in Australia (Sydney: Academic Press, 1981); P. P. COURTENAY, Northern Australia: Patterns and Problems of Tropical Development in an Advanced Country (Melbourne: Longman Cheshire, 1982); B. D. DAVIDSON, European Farming in Australia: An Economic History of Australian Farming (Elsevier, 1981); "Environment and Development in Australia" (a special issue), Australian Geographer, 19, no. 1 (May 1988), 1–220; R. FREESTONE, "Urban Australia: Postwar to Postindustrial," Focus, 32, no. 4 (March–April 1982), 9–14; R. HALL, "An Australian Country Townscape," Landscape, 24, no. 2 (1980), 41–48; R. L. HEATHCOTE and B. G. THOM, eds., Natural Hazards in Australia (Canberra: Australian Academy of Science, 1979), J. H. HOLMES, ed., "Queensland: A Geographical Interpretation," Queensland Geographical Journal, 4th Series, 1 (1986), 343 pp.; R. HUGHES, The Fatal Shore: The Epic of Australia's Founding (Knopf, 1986); D. N. JEANS, ed., Australia: A Geography (St. Martin's, 1978; 24 contributions by various authors); T. LANGFORD-SMITH, "New Perspectives on the Australian Deserts," Australian Geographer, 15 (1983), 269–284; E. LINACRE and J. HOBBS, The Australian Climatic Environment (Wiley, 1977); M. J. LOEFFLER, "Australian-American Interbasin Water Transfer," Annals AAG, 60 (1970), 493–516; R. E. LONSDALE and J. H. HOLMES, eds., Settlement Systems in Sparsely Populated Regions: The United States and Australia (Oxford, 1981); J. W. McCARTY and C. B. SCHEDVIN, eds., Australian Capital Cities: Historical Essays (Sydney, 1978); T. L. McKNIGHT, Australia's Corner of the World: A Geographical Summation (Prentice-Hall, 1970); D. J. MULVANEY, "The Prehistory of the Australian Aborigine," Scientific American, 214, no. 3 (March 1966), 84–93; D. PARKES, ed., Northern Australia: The Arenas of Life and Ecosystems on Half a Continent (Academic, 1984); J. M. POWELL, An Historical Geography of Modern Australia: The Restive Fringe (Cambridge, 1988), also, with M. WILLIAMS, eds., Australian Space, Australian Time: Geographical Perspectives (Oxford, 1975); Reader's Digest Atlas of Australia (Sydney, Australia: Reader's Digest Services, 1978); W. H. RICHMOND and P. C. SHARMA, eds., Mining and Australia (Queensland, 1983); M. SANT, guest ed., "Equity in Australia" (special issue), Geoforum, 19, no. 3 (1988); G. SIVIOUR, "A New Railway Age in Australia?" Geography, 68 (1983), 331–335; O. H. K. SPATE, Australia (Praeger, 1968); M. J. TAYLOR and N. THRIFT, "Large Corporations and Concentrations of Capital in Australia: A Geographical Analysis," Economic Geography 56 (1980), 261–280.

New Zealand K. B. CUMBERLAND and J. S. WHITELAW, New Zealand (Aldine, 1970; "The World's Landscapes"); E. M. K. DOUGLAS, "The Maori," Pacific Viewpoint, 20 (1979), 103–109; G. R. HAWKE, The Making of New Zealand: An Economic History (Cambridge, 1985); R. P. JORDAN, "New Zealand: The Last Utopia?" National Geographic, 171 (1987), 654–681; J. I. KELLY, "A Population Cartogram of New Zealand," New Zealand Journal of Geography, (October 1985), 7–11; New Zealand Geographer (semiannual); R. P. WILLIS, "Farming in New Zealand and the E. E. C.—The Case of the Dairy Industry," New Zealand Geographer, 40 (1984), 3–11.

PART VII: AFRICA

Africa: General, Historical, Cultural, Political Africa News (biweekly); Africa Report (bimonthly); J. F. AJAYI and M. CROWDER, Historical Atlas of Africa (Cambridge, 1985); J. B. BOYD, JR., "African Boundary Conflict: An Empirical Study," African Studies Review, 22, no. 3 (December 1979), 1–14; P. CALVOCORESSI, Independent Africa and the World (Longman, 1985); G. M. CARTER and P. O'MEARA, eds., African Independence: The First Twenty-

Five Years (Indiana, 1985); A. J. CHRISTOPHER, *Colonial Africa* (Barnes & Noble, 1984), also, "Continuity and Change of African Capitals," *Geographical Review*, 75 (1985), 44–57; M. CROWDER, ed., *The Cambridge History of Africa*, vol. 8: *c. 1940 to c. 1975* (Cambridge, 1984), also, "Whose Dream Was It Anyway? Twenty-five Years of African Independence," *African Affairs*, 86 (1987), 7–24; S. DENYER, *African Traditional Architecture: An Historical and Geographical Perspective* (Holmes & Meier, 1978); J. D. FAGE, *An Atlas of African History*, 2d ed. (Africana/Edward Arnold, 1978); D. K. FIELDHOUSE, *Black Africa, 1945–80: Economic Decolonization and Arrested Development* (Allen and Unwin, 1986); T. FORREST, "Brazil and Africa: Geopolitics, Trade, and Technology in the South Atlantic," *African Affairs*, 81 (1982), 3–20; N. GORDIMER, *The Essential Gesture: Writing, Politics and Places* (Knopf, 1988); I. L. GRIFFITHS, *An Atlas of African Affairs* (Methuen, 1984), also, "Famine and War in Africa," *Geography*, 73 (1988), 59–61; A. T. GROVE, *Africa*, 3d ed. (Oxford, 1978; a standard geography), also, with F. M. G. KLEIN, *Rural Africa* (Cambridge 1979); T. A. HALE, "Africa and the West: Close Encounters of the Literary Kind," *Comparative Literature Studies*, 20 (1983), 261–275; R. HALL, *Stanley: An Adventurer Explored* (Collins, 1974); J. D. HARGREAVES, *Decolonization in Africa* (Longman, 1988); C. HIBBERT, *Africa Explored: Europeans in the Dark Continent 1769–1889* (Allen Lane, 1982); R. H. JACKSON and C. G. ROSBERG, "Why Africa's Weak States Persist: The Empirical and the Juridical in Statehood," *World Politics*, 35 (1982), 1–24; G. W. JOHNSON, ed., *Double Impact: France and Africa in the Age of Imperialism* (Greenwood, 1985); A. M. JOSEPHY, ed., *The Horizon History of Africa*, 2 vols. (American Heritage Publishing Co., 1971); D. LAMB, *The Africans* (Random House, 1983; a wide-ranging report by a journalist); A. A. MAZRUI, *The Africans: A Triple Heritage* (London: BBC Publications, 1986); C. McEvedy, *The Penguin Atlas of African History* (Allen Lane, 1980); D. MOUNTFIELD, *A History of African Exploration* (Hamlyn, 1976); J. MURRAY, ed., *Cultural Atlas of Africa* (Facts on File, 1982); R. OLIVER and J. D. FAGE, *A Short History of Africa*, rev. ed. (NYU, 1963); R. OLIVER and M. CROWDER, eds., *The Cambridge Encyclopedia of Africa* (Cambridge, 1981; a major one-volume reference); T. RANGER, "White Presence and Power in Africa," *Journal of African History*, 20 (1979), 463–469; P. RICHARDS, "Spatial Organization as a Theme in African Studies," *Progress in Human Geography*, 8 (1984), 551–561; L. SANNEH, "The Domestication of Islam and Christianity in African Societies," *Journal of Religion in Africa*, 11 (1980), 1–12; T. SEVERIN, *The African Adventure: Four Hundred Years of Exploration in the 'Dangerous Continent'* (Dutton, 1973); T. C. SMOUT, ed., *Africa and Europe: From Partition to Interdependence or Dependence?* (Croom Helm, 1986); J. S. TRIMINGHAM, *The Influence of Islam upon Africa*, 2d ed. (Longman, 1980); S. J. UNGAR, *Africa: The People and Politics of an Emerging Continent* (Simon & Schuster, 1985); R. VAN CHI-BONNARDEL, Director, *The Atlas of Africa* (Paris: Jeune Afrique/New York: Hippocrene, 1973); J. S. WHITAKER, *How Can Africa Survive?* (Harper & Row, 1988); M. C. YOUNG, "Nationalism, Ethnicity, and Class in Africa: A Retrospective," *Cahiers d'Études Africaines*, 26 (1986), 421–495.

Africa: Environment, Population, Economy, Development W. M. ADAMS and F. M. R. HUGHES, "The Environmental Effects of Dam Construction in Tropical Africa: Impacts and Planning Procedures" [features Kenya and Nigeria], *Geoforum*, 17 (1986), 403–410; D. R. ALTSCHUL, "Transportation in African Development," *Journal of Geography*, 79 (1980), 44–56; A. AKINBODE, "Population Explosion in Africa and Its Implications for Economic Development," *Journal of Geography*, 76 (1977), 28–36; D. ANDERSON, "Declining Tree Stocks in African Countries," *World Development*, 14 (1986), 853–863; K. R. M. ANTHONY et al., *Agricultural Change in Tropical Africa* (Cornell, 1979); J. BAKER, "Oil and African Development," *Journal of Modern African Studies*, 15 (1977), 175–212; R. BAKER, "The African Experience: Drought and Famine in the Dry Zone," *Great Plains Quarterly*, 6 (1986), 238–245; T. BASS, "Feeding a Continent" [Africa], *Geographical Magazine*, 60, no. 5 (May 1988), 32–37; M. BELL, *Contemporary Africa: Development, Culture and the State* (Wiley, 1986); R. J. BERG and J. S. WHITAKER, eds., *Strategies for African Development* (California, 1986); D. BOURN, "Cattle, Rainfall and Tsetse in Africa" [features Ethiopia], *Journal of Arid Environments*, 1 (1978), 49–61; J. I. CLARKE and L. A. KOSINSKI, *Redistribution of Population in Africa* (Heinemann, 1982); R. E. CLUTE, "The Role of Agriculture in African Development," *African Studies Review*, 25 (1982), 1–20; C. K. EICHER, "Facing Up to Africa's Food Crisis," *Foreign Affairs*, 61 (1982), 151–174; W. S. ELLIS, "Africa's Sahel: The Stricken Land," *National Geographic*, 172 (1987), 140–179; M. GRIFFIN, "Dinka and Their Cattle Defy Time," *Geographical Magazine*, 53 (1981), 760–765; A. HANSEN and D. E. McMILLAN, eds., *Food in sub-Saharan Africa* (Lynne Rienner Publishers, 1986); J. J. HIDORE, "Population Explosion in Africa: Further Implications," *Journal of Geography*, 77 (1978), 214–220; H. KLOOS and K. THOMPSON, "Schistosomiasis in Africa: An Ecological Perspective," *Journal of Tropical Geography*, 48 (1979), 31–46; J. LEVI and M. HAVINDEN, *Economics of African Agriculture* (Longman, 1982); J. M. MANN et al., "The International Epidemiology of AIDS," *Scientific American*, 259, no. 4 (October 1988), 82–89; A. MOUNTJOY and D. HILLING, *Africa: Geography and Development* (Rowman & Littlefield/Barnes & Noble, 1987); S. E. NICHOLSON, "Climatic Variations in the Sahel and Other African Regions during the Past Five Centuries," *Journal of Arid Environments*, 1 (1987), 3–24; A. O'CONNOR, *The African*

City (Hutchinson, 1983); P. RICHARDS, "Ecological Change and the Politics of African Land Use," *African Studies Review*, 26, no. 2 (June 1983), 1–72, also, "The Environmental Factor in African Studies," *Progress in Human Geography*, 4 (1980), 589–600; J. B. RIDDELL, "The Geography of Modernization in Africa: A Re-examination," *Canadian Geographer*, 25 (1981), 290–299; R. SANDBROOK and J. BARKER, *Politics of Africa's Economic Stagnation* (Cambridge, 1985), also, "The State and Economic Stagnation in Tropical Africa," *World Development*, 14 (1986), 319–332; R. STEWART, "Prospects for Livestock Production in Tsetse-Infected Africa," *Impact of Science on Society*, 142 (1986), 117–125; S. WATTS, "The Crippling Worm" [Guinea worm disease], *Geographical Magazine*, 58 (1986), 54–55; D. WESTERN and V. FINCH, "Cattle and Pastoralism: Survival and Production in Arid Lands," *Human Ecology*, 14 (1986), 77–94; C. WINTERS, "Urban Morphogenesis in Francophone Black Africa," *Geographical Review*, 72 (1982), 139–154.

West Africa A. AIREY, "The Role of Feeder Roads in Promoting Rural Change in Eastern Sierra Leone," *Tijdschrift voor Economische en Sociale Geografie*, 76 (1985), 192–201; A. ARECCHI, "Dakar," *Cities*, 2 (1985), 186–197; A. ARMSTRONG, "Ivory Coast: Another New Capital for Africa," *Geography*, 70 (1985), 72–74; M. BARBOUR *et al.*, eds., *Nigeria in Maps* (Holmes & Meier, 1982); T. J. BASSETT, "Fulani Herd Movements," *Geographical Review*, 76 (1986), 233–248; J. A. BINNS, "Agricultural Change in Sierra Leone," *Geography*, 67 (1982), 113–125, also, "People of the Six Seasons" [Nigeria's Fulani], *Geographical Magazine*, 56 (1984), 640–644; B. A. CHOKOR, "Ibadan," *Cities*, 3 (1986), 106–116; R. J. H. CHURCH, *West Africa: A Study of the Environment and Man's Use of It*, 8th ed. (Longman, 1980; a major geographical text); J. DERRICK, "The Great West African Drought, 1972–1974," *African Affairs*, 76 (1977), 537–586; BUCHI EMECHETA, *The Bride Price* (1976) and *The Slave Girl* (1977) (Braziller; short novels of Nigerian life by a Nigerian novelist); E. O. ETEJERE and R. B. BHAT, "Traditional Preparation and Uses of Cassava in Nigeria," *Economic Botany*, 39 (1985), 157–164; C. FYFE, "The Cape Verde Islands," *History Today*, 31 (May 1981), 5–9; T. S. GALE, "Lagos, The History of British Colonial Neglect of Traditional African Cities," *African Urban Studies*, n.s., 17, no. 5 (Fall 1979), 11–24; A. T. GROVE, "Geographical Introduction to the Sahel," 144 (1978), 407–415, also, ed., *The Niger and Its Neighbours: Environmental History and Hydrobiology, Human Use and Health Hazards of the Major West African Rivers* (Balkema, 1985); J. D. HARGREAVES, "The Berlin West Africa Conference: A Timely Centenary?" *History Today*, 34 (November 1984), 16–22; D. HILLING, "The Evolution of a Port System: The Case of Ghana," *Geography*, 62 (1977), 97–105; A. M. HOWARD, "The Relevance of Spatial Analysis for African Economic History: The Sierra Leone-Guinea System," *Journal of African History*, 17 (1976), 365–388; A. JONES and M. JOHNSON, "Slaves from the Windward Coast" [West Africa], *Journal of African History*, 21 (1980), 17–34; R. LAW, "Trade and Politics Behind the Slave Coast: The Lagoon Traffic and the Rise of Lagos, 1500–1800," *Journal of African History*, 24 (1983), 321–348; J. G. LOCKWOOD, "The Causes of Drought with Particular Reference to the Sahel," *Progress in Physical Geography*, 10 (1986), 111–119; J. C. NWAFOR, "The Relocation of Nigeria's Federal Capital: A Device for Greater Territorial Integration and National Unity," *GeoJournal*, 4 (1980), 359–366; J. K. ONOH, *The Nigerian Oil Economy: From Prosperity to Glut* (Croom Helm, 1983); J. O. C. ONYEMELUKWE and M. O. FILANI, *Economic Geography of West Africa* (Longman, 1983); J. J. PARSONS, "The Canary Islands Search for Stability," *Focus*, 35, no. 2 (April 1985), 22–29; J. D. Y. PEEL, ed., "Rice and Yams in West Africa" (special issue), *Africa*, 51, no. 2 (1981), 553–705; R. POULTON, "Cooperation Against the Drought" [the Sahel], *Geographical Magazine*, 55 (1983), 524–531; P. RICHARDS, "Farming Systems and Agrarian Change in West Africa," *Progress in Human Geography*, 7 (1983), 1–39, also, *Indigenous Agricultural Revolution: Ecology and Food Production in West Africa* (Westview, 1987); R. D. STERN *et al.*, "The Start of the Rains in West Africa," *Journal of Climatology*, 1 (1981), 59–68; D. E. VERMEER, "Collision of Climate, Cattle, and Culture in Mauritania during the 1970s," *Geographical Review*, 71 (1981), 281–297; M. J. WATTS and T. J. BASSETT, "Politics, the State, and Agrarian Development: A Comparative Study of Nigeria and the Ivory Coast," *Political Geography Quarterly*, 5 (1986), 103–125; R. WELCOMME, "The Niger Falters," *Geographical Magazine*, 60, no. 7 (July 1988), 40–45; G. WENDLER and F. EATON, "On the Desertification of the Sahel Zone," *Climatic Change*, 5 (1983), 365–380.

Equatorial Africa B. AMMANN, "Zaire: 'Everything Is For Sale,' " *Swiss Review of World Affairs*, 38, no. 5 (August 1988), 25–29; B. EMERSON, *Leopold II of the Belgians: King of Colonialism* (Weidenfeld & Nicolson, 1979); P. FORBATH, *The River Congo* (Secher & Warburg, 1978); R. W. HARMS, *River of Wealth, River of Sorrow: The Central Zaire Basin in the Era of the Slave and Ivory Trade, 1500–1891* (Yale, 1981); T. B. HART and J. A. HART, "The Ecological Basis of Hunter-Gatherer Subsistence in African Rain Forests: The Mbuti of Eastern Zaire," *Human Ecology*, 14 (1986), 29–55; G. C. KABWIT, "Zaire: The Roots of Continuing Crisis," *Journal of Modern African Studies*, 17 (1979), 381–407; J. MacGAFFEY, "How to Survive and Become Rich Amidst Devastation: The Second Economy in Zaire," *African Affairs*, 82 (1983), 351–366; J. MADELEY, "Cameroon Grows Its Own," *Geographical Magazine*, 59 (1987), 296–300; V. S. NAIPAUL, *A Bend in the*

River (Knopf, 1979; a short, classic novel); ALEX SHOU-MATOFF, *In Southern Light: Trekking through Zaire and the Amazon* (Simon & Schuster, 1986); T. O'TOOLE, *The Central African Republic: The Continent's Hidden Heart* (Westview, 1986); H. WINTERNITZ, *East Along the Equator* [Zaire] (*Atlantic Monthly*, 1987); C. YOUNG and T. TURNER, *The Rise and Decline of the Zairian State* (Wisconsin, 1985).

East Africa R. G. ABRAHAMS, *Villagers, Villages, and the State in Modern Tanzania* (Cambridge, England: African Studies Centre, 1985); R. H. BATES, "The Agrarian Origins of Mau Mau," *Agricultural History*, 61, no. 1 (Winter 1987), 1–28; J. DOHERTY, "Tanzania: Twenty Years of African Socialism 1967–1987," *Geography*, 72 (1987), 344–348; N. HARMAN, "East Africa: Turning the Corner," *The Economist*, June 20, 1987, 1–18; B. S. HOYLE, *Seaports and Development: The Experience of Kenya and Tanzania* (Gordon & Breach, 1983); J. J. JORGENSEN, *Uganda: A Modern History* (St. Martin's, 1981); J. D. KESBY, *The Cultural Regions of East Africa* (Academic, 1977); C. LEO, *Land and Class in Kenya* (Toronto, 1984); R. B. MABELE *et al.*, "The Economic Development of Tanzania," *Scientific American*, 243, no. 3 (September 1980), 182–190; S. H. OMINDE, ed., *Population and Development in Kenya* (Heinemann, 1984); I. SINDIGA, "Sleeping Sickness in Kenya," *Erdkunde*, 41 (1987), 133–146.

South Central Africa J. BEST and L. M. ZINYAMA, "The Evolution of the National Boundary of Zimbabwe," *Journal of Historical Geography* 11 (1985), 419–432; D. DRA-KAKIS-SMITH, "Zimbabwe: The Slow Struggle Towards Socialism," *Geography*, 72 (1987), 348–352; P. S. FALK, "Cuba in Africa," *Foreign Affairs*, 65 (1987), 1077–1096; B. FETTER, "Malawi: Everybody's Hinterland," *African Studies Review*, 25 (1982), 79–116; R. HARDY, *Rivers of Darkness* (Putnam, 1979; a gripping novel of tropical medicine and guerrilla warfare in Mozambique); G. KAY, "Zimbabwe's Independence: Geographical Problems and Prospects," *Geographical Journal*, 147 (1981), 179–187; R. PELISSIER, "Africa without the Portuguese," *Geographical Magazine*, 52 (1980), 793–802; E. P. SCOTT, "Development Through Self-Reliance in Zambia," *Journal of Geography*, 84 (1985), 282–290; J. SHEPHERD, "Zimbabwe: Poised on the Brink," *Atlantic Monthly*, July 1987, 26–31; D. WEINER *et al.*, "Land Use and Agricultural Productivity in Zimbabwe," *Journal of Modern African Studies*, 23 (1985), 251–285; G. WILLIAMS, "Pulse of the Zambezi," *Geographical Magazine*, 59 (1987), 608–613; A. WOOD, "When the Bottom Goes Out of Copper" [Zambia], *Geographical Magazine*, 56 (1984), 16–20; L. ZINYAMA and R. WHITLOW, "Changing Patterns of Population Distribution in Zimbabwe," *GeoJournal*, 13 (1986), 365–384;

"Zambia the Slow, Malawi the Poor," *The Economist*, February 18, 1989, 88–89.

Southern Africa H. ADAM, "Variations of Ethnicity: Afrikaner and Black Nationalism in South Africa," *Journal of Asian and African Studies*, 20 (1985), 169–180; K. S. O. BEAVON, "Black Townships in South Africa: Terra Incognita for Urban Geographers," *South African Geographical Journal*, 64 (1982), 3–20, also, "Trekking On: Recent Trends in the Human Geography of Southern Africa," *Progress in Human Geography*, 5 (1981), 159–189; A. BRINK, *Writing in a State of Siege* [South Africa] (Simon & Schuster, 1983); A. J. CHRISTOPHER, "Partition and Population in South Africa," *Geographical Review*, 72 (1982), 127–238, also, *South Africa* (Longman, 1982; "The World's Landscapes"), also, "Southern Africa and the United States: A Comparison of Pastoral Frontiers," *Journal of the West*, 20 (1981), 52–59, also, *South Africa: The Impact of Past Geographies* (Cape Town: Juta & Co., 1984), and, "African Land: European Enterprise," *South African Geographer*, 13 (1985), 151–162; B. A. COX and C. M. ROGERSON, "The Corporate Power Elite in South Africa: Interlocking Directorships among Large Enterprises," *Political Geography Quarterly*, 4 (1985), 219–234; D. CRUSH and J. CRUSH, "A State of Dependence" [Lesotho], *Geographical Magazine*, 55 (1983), 24–29; J. CRUSH and P. WELLINGS, "The Southern African Pleasure Periphery, 1966–83," *Journal of Modern African Studies*, 21 (1983), 673–698; B. DAVIES, "Erratic Orange" [Orange River], *Geographical Magazine*, 60, no. 10 (October 1988), 35–39; R. J. DAVIES, "When! Reform and Change in the South African City," *South African Geographical Journal*, 68 (1986), 3–17; D. DEWAR, *et al.*, "Industrial Decentralization Policy in South Africa: Rhetoric and Practice," *Urban Studies*, 23 (1986), 363–376; W. R. DUGGAN, *An Economic Analysis of Southern Africa's Agriculture* (Praeger, 1986); A. DU TOIT, "No Chosen People: The Myth of the Calvinist Origins of Afrikaner Nationalism and Racial Ideology," *American Historical Review*, 88 (1983), 920–952; T. J. D. FAIR, *South Africa: Spatial Frameworks for Development* (Cape Town, South Africa: Juta & Company, 1982); B. FREUND, "Development and Underdevelopment in Southern Africa: An Historical Overview," *Geoforum*, 17 (1986), 133–140; H. GILIOMEE, "Constructing Afrikaner Nationalism," *Journal of Asian and African Studies*, 18 (1983), 83–98; K. W. GRUNDY, "The Social Costs of Armed Struggle in Southern Africa," *Armed Forces and Society*, 7 (1981), 445–466; D. HART, "A Literary Geography of Soweto," *GeoJournal*, 12 (1986), 191–195, also, with G. H. PIRIE, "The Sight and Soul of Sophiatown," *Geographical Review*, 74 (1984), 38–47; R. F. HASWELL, "South African Towns on European Plans," *Geographical Magazine*, 51 (1979), 686–694; L. HOLZNER, "Thriving Downtown amid Sprawl: Johannesburg," *Urbanism Past and Present*, 8, no. 2 (Summer/

Fall 1983), 23–30; H. LAMAR and L. THOMPSON, eds., *The Frontier in History: North America and Southern Africa Compared* (Yale, 1981); J. LELYVELD, *Move Your Shadow: South Africa Black and White* (Times Books, 1985); T. LODGE, "The Destruction of Sophiatown," *Journal of Modern African Studies*, 19 (1981), 107–132; M. LOEB, "What the U.S. Must Do in South Africa," *Fortune*, July 18, 1988, 88–90; P. LOPEZ, "Landscapes Open and Closed: A Journey through Southern Africa," *Harper's Magazine*, July 1987, 51–58; P. MAYLAM, *A History of the African People of South Africa: From the Early Iron Age to the 1970s* (St. Martin's, 1986); M. E. MEADOWS, *Biogeography and Ecosystems of South Africa* (Juta & Co., 1985); C. MERRETT, "The Significance of the Political Boundary in the Apartheid State, with Particular Reference to Transkei," *South African Geographical Journal*, 66 (1984), 79–93; JAMES A. MICHENER, *The Covenant* (Random House, 1980; a vast panoramic novel of South African history; a tremendous store of other fine fiction about South Africa has been created by such able writers as Peter Abrahams, André Brink, J. M. Coetzee, Nadine Gordimer, Alex La Guma, Doris Lessing, Alan Paton, and many others); G. H. PIRIE, "The Decivilizing Rails: Railways and Underdevelopment in Southern Africa," *Tijdschrift Voor Economische en Sociale Geografie*, 73 (1982), 221–228; P. POOVALINGAN, "The Indians of South Africa: A Century on the Defensive," *Optima*, 28 (1979), 66–91; *Reader's Digest Atlas of Southern Africa* (Cape Town: Reader's Digest, 1984); C. M. ROGERSON, "Patterns of Indigenous and Foreign Control of South African Manufacturing," *South African Geographer*, 10 (1982), 123–134; R. I. ROTBERG et al., *South Africa and Its Neighbors: Regional Security and Self-Interest* (Lexington, 1985); A. SAMPSON, *Black and Gold* (Pantheon, 1987; a British journalist's view of South Africa); J. S. SAUL and S. GELB, *The Crisis in South Africa*, rev. ed. (Monthly Review Press, 1986); G. B. SILBERBAUER, *Hunter and Habitat in the Central Kalahari Desert* [Botswana] (Cambridge, 1981); D. M. SMITH, *Apartheid in South Africa* (Cambridge, 1985), also, "Conflict in South African Cities," *Geography*, 72 (1987), 153–158, and, ed., *Living under Apartheid: Aspects of Urbanization and Social Change in South Africa* (Allen & Unwin, 1982); W. SMITH, "Cape Coloureds Who Survive in Limbo," *Geographical Magazine*, 53 (1980), 188–195; *South African Geographical Journal* (annual); "Southern Africa," *Geographical Magazine*, 58–59 (1986–1987; 13 articles by various authors); [Southern Africa: a supplementary issue], *GeoJournal*, 2 (1981); E. STERN, "Competition and Location in the Gaming Industry: The 'Casino States' of Southern Africa," *Geography*, 72 (1987), 140–150; STUDY COMMISION ON U.S. POLICY TOWARD SOUTHERN AFRICA, *South Africa: Time Running Out* (California, 1981); M. O. SUTCLIFFE, "The Crisis in South Africa: Material Conditions and the Reformist Response," *Geoforum*, 17 (1986), 141–159; G. G. WEIGEND, "Economic Activity Patterns in White Namibia," *Geographical Review*, 75 (1985), 462–481, also, "German Settlement Patterns in Namibia," *ibid.*, 75 (1985), 156–169; P. WELLINGS and A. BLACK, "Industrial Decentralization under Apartheid: The Relocation of Industry to the South African Periphery," *World Development*, 14 (1986), 1–38; B. WIESE, *Seaports and Port Cities of Southern Africa* (Wiesbaden: Franz Steiner Verlag GMBH, 1981); J. WESTERN, *Outcast Cape Town* (Minnesota, 1981; Allen & Unwin, 1982), also, "Undoing the Colonial City" [Cape Town and Tianjin], *Geographical Review*, 75 (1985), 335–357, and, "South African Cities: A Social Geography," *Journal of Geography*, 85 (1986), 249–255.

Indian Ocean Islands B. E. DAVIS, "Quality of Life in Small Island Nations in the Indian Ocean," *Human Ecology*, 14 (1986), 453–471; D. W. GADE, "Madagascar and Nondevelopment Culture," *Focus*, 35, no. 4 (October 1985), 14–21; M. GRIFFIN, "The Perfumed Isles" [Comoros], *Geographical Magazine*, 58 (1986), 524–527; A. JOLLY, *A World Like Our Own: Man and Nature in Madagascar* (Yale, 1980); P. LENOIR, "An Extreme Example of Pluralism: Mauritius," *Cultures*, 6 (1979), 63–82; R. TURLEY, "Madagascar," *Geographical Magazine*, 56 (1984), 28–33.

PART VIII: LATIN AMERICA

Latin America: General J. P. AUGELLI, "Food, Population, and Dislocation in Latin America," *Journal of Geography*, 84 (1985), 274–281; R. G. BOEHM and S. VISSER, eds., *Latin America: Case Studies* (Kendall/Hunt, 1984); H. BLAKEMORE and C. T. SMITH, eds., *Latin America: Geographical Perspectives*, 2d ed. (Methuen, 1983); R. D. F. BROMLEY and R. BROMLEY, *South American Development: A Geographical Introduction* (Cambridge, 1982); D. BUTTERWORTH and J. K. CHANCE, *Latin American Urbanization* (Cambridge, 1981); C. N. CAVIEDES, "Natural Resource Exploitation in Latin America: Espoiliation or Tool for Development?" *GeoJournal*, 11 (1985), 111–119, also, *The Southern Cone: Realities of the Authoritarian State in South America* (Rowman & Allanheld, 1984); J. CHILD, "Geopolitical Thinking in Latin America," *Latin American Research Review*, 14 (1979), 89–111, also, *Geopolitics and Conflict in South America: Quarrels among Neighbors* (Praeger, 1985); J. A. CROW, *The Epic of Latin America*, 3d ed. (California, 1980); C. GIBSON, "Conquest and So-Called Conquest in Spain and Spanish America," *Terrae Incognitae*, 19 (1980), 1–19; E. GRIFFIN and L. FORD, "A Model of Latin American City Structure," *Geographical Review*, 70 (1980), 397–422; A. HENNESSY, *The Frontier in Latin American History* (Edward Arnold, 1978); L. A. HOBERMAN and S. M. SOCOLOW, eds., *Cities and Society in Colonial Latin America* (New Mexico, 1986); P. E. JAMES and C. W. MINKEL, *Latin America*, 5th ed. (Wiley,

1986; a major textbook); H. S. KLEIN, *African Slavery in Latin America and the Caribbean* (Oxford, 1986); J. O. MAOS, *The Spatial Organization of New Land Settlement in Latin America* (Westview, 1984); D. R. MEYER, "The World System of Cities: Relations between International Financial Metropolises and South American Cities," *Social Forces,* 64 (1986), 553–581; A. S. MORRIS, *Latin America: Economic and Regional Differentiation* (Hutchinson, 1981); also, *South America,* 3rd ed. (Barnes & Noble, 1987; a geography); J. H. PARRY, *The Discovery of South America* (Taplinger, 1979); J. PEREZ, "The Hispanic Element in Latin America," *Cultures,* 7 (1980), 45–61; R. S. PLATT, *Latin America: Countrysides and United Regions* (McGraw-Hill, 1942; a classic book based on extended field research); D. PRESTON, ed., *Latin American Development: Geographical Perspectives* (Wiley, 1987); C. E. REBORATTI, "Human Geography in Latin America," *Progress in Human Geography,* 6 (1982), 397–407; I. ROXBOROUGH, "Unity and Diversity in Latin American History," *Journal of Latin American Studies,* 16 (1984), 1–26; "Spain Rediscovers the New World," *The Economist,* July 30, 1988, 17–20.

Caribbean America: General T. D. ANDERSON, *Geopolitics of the Caribbean: Ministates in a Wider World* (Praeger, 1984); C. BROCK, ed., *The Caribbean in Europe* (Frank Cass, 1986); M. E. CRAHAN, ed., *Africa and the Caribbean: The Legacies of a Link* (Johns Hopkins, 1979); M. W. HELMS, *Middle America: A Culture History of Heartland and Frontier* (Prentice-Hall, 1975); K. R. HOPE, *Economic Development in the Caribbean* (Praeger, 1986); P. HULME, *Colonial Encounters: Europe and the Native Caribbean, 1492–1797* (Methuen, 1986); J. D. MOMSEN, "Migration and Rural Development in the Caribbean," *Tijdschrift voor Economische en Sociale Geografie,* 77 (1986), 50–58; J. MacPHERSON, *Caribbean Lands* (Longman, 1980); L. PEAKE, "Guyana: A Country in Crisis," *Geography,* 72 (1987), 356–360; C. O. SAUER, *The Early Spanish Main* (California, 1969); P. SUTTON, ed., *Dual Legacies in the Contemporary Caribbean: Continuing Aspects of British and French Domination* (Frank Cass, 1986); R. C. WEST and J. P. AUGELLI, *Middle America: Its Lands and Peoples,* 2d ed. (Prentice-Hall, 1976).

Mexico D. D. ARREOLA, "Nineteenth-Century Townscapes of Eastern Mexico," *Geographical Review,* 72 (1982), 1–19; M. BARKE, "Mérida, Yucatán: A Core within the Periphery," *Scottish Geographical Magazine,* 100 (1984), 160–170; D. M. BRAND, *Mexico: Land of Sunshine and Shadow* (Van Nostrand, 1966); L. B. CASAGRANDE, "The Five Nations of Mexico," *Focus,* 37, no. 1 (Spring 1987), 2–9; P. G. CASANOVA, "The Economic Development of Mexico," *Scientific America,* 243, no. 3 (September 1980), 192–204; J. H. COATSWORTH, "Indispensable Railroads in a Backward Economy: The Case of Mexico," *Journal of Economic History,* 39 (1979), 939–960; W. E. DOOLITTLE,

"Aboriginal Agricultural Development in the Valley of Sonora, Mexico," *Geographical Review,* 70 (1980), 328–342; also, "Agricultural Expansion in a Marginal Area of Mexico, *Geographical Review,* 73 (1983), 301–313; J. S. HENDERSON, *The World of the Ancient Maya* (Cornell, 1981); G. W. JAMESON, "Defining Critical Landscape on Mexican Ejidos," *Landscape Planning,* 11 (1984), 109–123; B. McDOWELL, "Mexico City: An Alarming Giant," *National Geographic,* 166 (1984), 138–172; "Mexico's Population: A Profile," *Interchange* (Population Reference Bureau), 16, no. 2 (May 1987); J. D. NATIONS, "The Rainforest Farmers" [Mexico], *Pacific Discovery,* 34, no. 1 (January–February 1981), 1–9; B. O'REILLY, "Doing Business on Mexico's Volcano," *Fortune,* August 29, 1988, 72–74; A. RIDING, *Distant Neighbors: A Portrait of the Mexicans* (Knopf, 1985); R. A. SANCHEZ, "Oil Boom, A Blessing for Mexico?" *GeoJournal,* 7 (1983), 229–245; T. G. SANDERS, "Mexico: After the Revolution That Failed," *Focus,* 29, no. 5 (May–June 1979), 1–12; S. SANDERSON, *Land Reform in Mexico: 1910–1980* (Academic, 1984); L. SUAREZ-VILLA, "The Manufacturing Process Cycle and the Industrialization of the United States–Mexico Borderlands," *Annals of Regional Science,* 18 (1984), 1–23.

Central America J. P. AUGELLI, "Costa Rica's Frontier Legacy," *Geographical Review,* 77 (1987), 1–16, also, "The Panama Canal Area," *Focus,* 36, no. 1 (Spring 1986), 20–29; W. CROWLEY and E. C. GRIFFIN, "Political Upheaval in Central America," *Focus,* 34, no. 1 (September–October 1983), 1–15; J. M. CURRY-ROPER, "Nicaragua: Land of Conflict," *Focus,* 38, no. 3 (Fall 1988), 12–17; S. L. DRIEVER, "Insurgency in Guatemala: Centuries-Old Conflicts over Land and Social Inequality Spawn Guerrilla Movements and Hope for Democratic Change," *Focus,* 35, no. 3 (July 1985), 2–9; R. EVANS, "Central America's Slide to Ruin," *Geographical Magazine,* 59 (1987), 582–591; C. HALL, *Costa Rica: A Geographical Interpretation in Historical Perspective* (Westview, 1985); H. MARROQUIN, "Guatemala City," *Cities,* 4 (1987), 203–206; S. E. PLACE, "Export Beef Production and Development Contradictions in Costa Rica," *Tijdschrift voor Economische en Sociale Geografie,* 76 (1985), 288–297; R. L. WOODWARD, JR., *Central America: A Nation Divided,* 2d ed. (Oxford, 1985).

West Indies J. P. AUGELLI, "Nationalization of Dominican Borderlands," *Geographical Review,* 70 (1980), 19–35; D. J. AUSTIN, "Culture and Ideology in the English-Speaking Caribbean: A View from Jamaica," *American Ethnologist,* 10 (1983), 223–240; *Caribbean Geography* (semiannual); R. CHARDON, "Sugar Plantations in the Dominican Republic," *Geographical Review,* 74 (1984), 441–454; C. E. COBB, JR., "Jamaica: Hard Times, High Hopes," *National Geographic,* 167, no. 1 (January

1985), 114–140, also, "Haiti—Against All Odds," *ibid.*, 172 (1987), 644–670; J. I. DOMINGUEZ, "Cuba in the 1980s," *Foreign Affairs*, 65 (1986), 118–135; L. A. EYRE, "The Ghettoization of an Island Paradise" [Jamaica], *Journal of Geography*, 82 (1983), 236–239, also, "Political Violence and Urban Geography in Kingston, Jamaica," *Geographical Review*, 74 (1984), 24–37; B. FLOYD, "Agricultural Reform in Castro's Cuba," *Geographical Magazine*, 50 (1978), 808–815, also, *Jamaica: An Island Microcosm* (St. Martin's, 1979); B. W. HIGMAN, "The Spatial Economy of Jamaican Sugar Plantations: Cartographic Evidence from the Eighteenth and Nineteenth Centuries," *Journal of Historical Geography*, 13, (1987), 17–39; D. LOWENTHAL, *West Indian Societies* (Oxford, 1972); S. B. MacDONALD and F. B. JOSEPH, "The Caribbean Sugar Crisis: Consequences and Challenges," *Journal of Interamerican Studies and World Affairs*, 28 (1986), 35–58; R. B. POTTER, "Tourism and Development: The Case of Barbados, West Indies," *Geography*, 68 (1983), 46–50; F. REDMILL, "The Independent Way for St. Kitts," *Geographical Magazine*, 55 (1983), 636–640; B. RICHARDS, "The Uncertain State of Puerto Rico," *National Geographic*, 163 (1983), 516–543; R. J. TATA, *Haiti: Land of Poverty* (University Press of America, 1982); "The Caribbean: Columbus's Islands," *The Economist*, August 6, 1988, 3–18; J. H. WILLIAMS, "Cuba: Havana's Military Machine," *Atlantic Monthly*, 262, no. 2 (August 1988), 18–23.

Andean Countries C. J. ALLEN, "To Be Quechua: The Symbolism of Coca Chewing in Highland Peru," *American Ethnologist*, 8 (1981), 157–171; R. BROMLEY, "The Colonization of Humid Tropical Areas in Ecuador," *Singapore Journal of Tropical Geography*, 2 (1981), 15–26; A. CARDICH, "Native Agriculture in the Highlands of the Peruvian Andes," *National Geographic Research*, 3 (1987), 22–39; C. N. CAVIEDES, "El Niño, 1982–83," *Geographical Review*, 74 (1984), 267–290; C. CLAPPERTON, "Fire and Water in the Andes" [northern Andes volcanoes], *Geographical Magazine*, 58 (1986), 74–79; W. M. DENEVAN, "Peru's Agricultural Legacy: Ancient Methods May Be Useful in Reviving Today's Food Production," *Focus*, 35, no. 2 (April 1985), 16–21; D. A. EASTWOOD and H. J. POLLARD, "Lowland Colonization and Coca Control: Bolivia's Irreconcilable Policies," *Singapore Journal of Tropical Geography*, 8 (1987), 15–25; D. W. GADE, "Inca and Colonial Settlement, Coca Cultivation and Endemic Disease in the Tropical Forest," *Journal of Historical Geography*, 5 (1979), 263–279; M. HIRAOKA and S. YAMAMOTO, "Agricultural Development in the Upper Amazon of Ecuador," *Geographical Review*, 70 (1980), 423–445; L. McINTYRE, "The High Andes: South America's Islands in the Sky," *National Geographic*, 171 (1987), 422–459; M. MÖRNER, *The Andean Past: Land, Societies and Conflict* (Columbia, 1985); J. S. OTTO and N. E. ANDERSON, "Cattle Ranching in the Venezuelan Llanos and the Florida Flatwoods: A Problem in Comparative History," *Comparative Studies in Society and History*, 28 (1986), 672–683; J. J. PARSONS, "Geography as Exploration and Discovery" [Colombia], *Annals AAG*, 67 (1977), 1–16; R. B. SOUTH, "Coca in Bolivia," *Geographical Review*, 67 (1977), 22–33; "The Cocaine Economies," *The Economist*, October 8, 1988, 21–24; C. WEIL, "Migration among Landholdings by Bolivian Campesinos," *Geographical Review*, 73 (1983), 182–197.

Brazil J. L. ALEXANDER, "South of São Paulo," *Geographical Magazine*, 57 (1985), 598–600; B. K. BECKER, "The State and the Land Question of the Frontier: A Geopolitical Perspective" [Brazil], *GeoJournal*, 11 (1985), 7–14; A. BOTELHO, "Brazil's Independent Computer Strategy," *Technology Review*, 90, no. 4 (May–June 1987) 36–45; R. H. BROOKS, "The Adversity of Brazilian Drought," *GeoJournal*, 6 (1982), 121–128; J. P. DICKENSON, *Brazil* (Longman, 1982; "The World's Landscapes"); P. M. FEARNSIDE, *Human Carrying Capacity of the Brazilian Rainforest* (Columbia, 1986); P. FORESTER, *Capital of Dreams* [Brasília], *Geographical Magazine*, 58 (1986), 462–467; J. GREENWOOD, "Riches and Debt in Venezuela," *Geographical Magazine*, 56 (1984), 463–469; P. GRENIER, "The Alcohol Plan and the Development of Northeast Brazil," *GeoJournal*, 11 (1985), 61–68; R. HARVEY, "Clumsy Giant: A Survey of Brazil," *The Economist*, April 25, 1987, 1–26; F. I. JOHNSON, "Sugar in Brazil: Policy and Production," *Journal of Developing Areas*, 17 (1983), 243–256; J. MAIN, "Brazil's Tomorrow Is Finally in Sight," *Fortune*, September 15, 1986, 72–86; S. MITCHELL, ed., *The Logic of Poverty: The Case of the Brazilian Northeast* (Routledge & Kegan Paul, 1981); J. W. ROSSI, "Income Distribution in Brazil: A Regional Approach," *Journal of Development Studies*, 17 (1981), 226–234; A. J. R. RUSSELL-WOOD, *The Black Man in Slavery and Freedom in Colonial Brazil* (St. Martin's, 1982); R. F. SKILLINGS, "Economic Development of the Brazilian Amazon: Opportunities and Constraints (1984), 48–54; M. Y. UNE, "An Analysis of the Effects of Frost on the Principal Coffee Areas of Brazil," *GeoJournal*, 6 (1982), 129–140; P. J. VESILIND, "Brazil: Moment of Promise and Pain," *National Geographic*, 171 (1987), 348–385; L. N. WILLMORE, "The Comparative Performance of Foreign and Domestic Firms in Brazil," *World Development*, 14 (1986), 489–502.

Southern Mid-Latitude Countries J. KING, "Civilisation and Barbarism: The Impact of Europe on Argentina," *History Today*, 34 (1984), 16–21; P. McGRATH, "Paraguayan Powerhouse," *Geographical Magazine*, 55 (1983), 192–197; A. MELAMID, "The Future of the Falkland Islands?" *Geographical Review*, 73 (1983), 211–213; W. WEISCHET, "Climatic Constraints for the Development of the Far South of Latin America," *GeoJournal*, 11 (1985), 79–87.

Amazonia D. L. CLAWSON, "Obstacles to Successful Highlander Colonization of the Amazon and Orinoco Basins," *American Journal of Economics and Sociology*, 41 (1982), 351–362; R. E. DICKINSON, ed., *The Geophysiology of Amazonia: Vegetation and Climate Interactions* (Wiley, 1987); P. M. FEARNSIDE, *Human Carrying Capacity of the Brazilian Rainforest* (Columbia, 1986); R. B. HAMES and W. T. VICKERS, eds., *Adaptive Responses of Native Amazonians* (Academic, 1983); J. HEMMING, ed., *Change in the Amazon Basin*, vol. 1, *Man's Impact on Forests and Rivers*, vol. 2, *The Frontier after a Decade of Colonization* (Manchester, 1985); M. HIRAOKA, "Settlement and Development of the Upper Amazon: The East Bolivian Example," *Journal of Developing Areas*, 14 (1980), 327–347, also, "The Development of Amazonia," *Geographical Review*, 72 (1982), 94–98; C. F. JORDAN, "Amazon Rain Forests," *American Scientist*, 70 (1982), 394–401; B. KELLY and M. LONDON, *Amazon* (Harcourt Brace Jovanovich, 1983); J. KIRBY, "Agricultural Land-Use and the Settlement of Amazonia," *Pacific Viewpoint*, 17 (1976), 105–132; E. F. MORAN, *Developing the Amazon: The Social and Ecological Consequences of Government-Directed Colonization along Brazil's Transamazonian Highway* (Indiana, 1981), also, ed., *The Dilemma of Amazonian Development* (Westview, 1983); M. SCHMINK, "Land Conflicts in Amazonia," *American Ethnologist*, 9 (1982), 341–357, also, with C. H. WOOD, eds., *Frontier Expansion in Amazonia* (Florida, 1984); N. J. H. SMITH, *Rainforest Corridor: The Transamazon Colonization Scheme* (California, 1982); B. WALLACH, "Manaus," *Focus*, 35, no. 1 (January 1985), 8–11; C. WEIL, "Amazon Update: Developments since 1970," *Focus*, 33, no. 4 (March–April, 1983), 1–12.

PART IX: ANGLO-AMERICA

Anglo-America: General C. AGOCS, "Ethnic Groups in the Ecology of North American Cities," *Canadian Ethnic Studies*, 11, no. 2 (1979), 1–18; J. AXTELL, *The Invasion Within: The Contest of Cultures in Colonial North America* (Oxford, 1985); R. G. BAILEY et al., "Ecological Regionalization in Canada and the United States," *Geoforum*, 16 (1985), 265–275; B. BAILYN, *The Peopling of British North America: An Introduction* (Knopf, 1986); S. S. BIRDSALL and J. W. FLORIN, *Regional Landscapes of the United States and Canada* (Wiley, 1985; a geography text); C. F. DORAN, "The United States and Canada; Intervulnerability and Interdependence," *International Journal*, 38 (1982–1983), 128–146; S. J. FIEDEL, *Prehistory of the Americas* (Cambridge, 1987); O. J. FURUSETH et al., "Farmland Preservation in North America" (special issue), *GeoJournal*, 6 (1982), 498–588; J. R. GIBSON, *Imperial Russia in Frontier America: The Changing Geography of Supply of Russian America, 1784–1867* (Oxford, 1976), also, ed., *European Settlement and Development in North America: Essays on Geographical Change in Honour and Memory of An-*

drew Hill Clark (Toronto, 1978); J. GARREAU, *The Nine Nations of North America* (Houghton Mifflin, 1981; a book for general readers); W. E. GARRET, ed., *Atlas of North America: Space Age Portrait of a Continent* (National Geographic Society, 1985); P. GUINNESS and M. BRADSHAW, *North America: A Human Geography* (Barnes & Noble, 1985); C. B. HUNT, *Natural Regions of the United States and Canada* (Freeman, 1974); P. E. LYDOLPH and T. B. WILLIAMS, "The North American Sukhovey," *Annals AAG*, 72 (1982), 224–236; R. D. MITCHELL and P. A. GROVES, eds., *North America: The Historical Geography of a Changing Continent* (Rowman & Littlefield, 1987); C. MULVEY, *Anglo-American Landscapes* (Cambridge, 1983); J. E. H. MUSTOE, *An Atlas of Renewable Energy Resources in the United Kingdom and North America* (Wiley, 1984); A. G. NOBLE, *Wood, Brick and Stone: The North American Settlement Landscape* (Massachusetts, 1984); J. H. PATTERSON, *North America: A Geography of the United States and Canada*, 8th ed. (Oxford, 1989); K. B. RAITZ, "Ethnic Maps of North America," *Geographical Review*, 68 (1978), 335–350; J. R. ROONEY, JR., W. ZELINSKY, and D. R. LOUDER, eds., *This Remarkable Continent: An Atlas of United States and Canadian Society and Culture* (Texas A&M, 1982); C. F. RUNGE, *The Future of the North American Granary: Politics, Economics, and Resource Constraints in North American Agriculture* (Iowa State, 1986); V. E. SHELFORD, *The Ecology of North America* (Illinois, 1963); M. J. TROUGHTON, "Industrialization of U.S. and Canadian Agriculture," *Journal of Geography*, 84 (1985), 255–263; T. R. VALE, *Plants and People: Vegetation Change in North America* (Association of American Geographers, 1982); J. L. VANKAT, *The Natural Vegetation of North America: An Introduction* (Wiley, 1979); C. L. WHITE, E. J. FOSCUE, and T. L. McKNIGHT, *Regional Geography of Anglo-America*, 6th ed. (Prentice-Hall, 1985); M. YEATES, *North American Urban Patterns* (Wiley, 1980); W. ZELINSKY, "North America's Vernacular Regions," *Annals AAG*, 70 (1980), 1–16.

Canada: General "A Survey of Canada," *The Economist*, October 8, 1988, 1–18; B. M. BARR, "Canadian Geography in a Multilingual World: The Implosion of Relevance?" *Canadian Geographer*, 30 (1986), 290–301; R. BOULTON, *Canada Coast to Coast* (Oxford 1983; "Contemporary Problems in Geography"); "Canada: A Special Issue" (articles and bibliography by numerous authors), *Journal of Geography*, 83, no. 5 (September–October 1984); *Canadian Geographer* (quarterly; a scholarly journal of high quality); *Canadian Geographic* (bimonthly; a well-written magazine for general readers); P. G. GOHEEN, "Canadian Communications Circa 1845," *Geographical Review*, 77 (1987), 35–51; L. E. HAMELIN, *Canadian Nordicity: It's Your North, Too* (Montreal: Harvest House, 1979); C. HARRIS, "Presidential Address: The Pattern of Early Canada," *Canadian Geographer*, 31 (1987), 290–298, also, ed., *Historical Atlas of Canada: From the Beginning to 1800* (Toronto, 1988), and, with J. WARKENTIN, *Canada before Confederation: A Study in Historical Geography* (Oxford, 1974); W. C. HEINE, "Canada from Space: Eye-in-the-Sky Satellites Produce Vivid Views of Earth's Surface," *Canadian Geographic*, 106, no. 6 (December 1986–January 1987), 42–55; W. W. JOYCE, ed., *Canada in the Classroom: Content and Strategies for the Social Studies* (Washington, DC: National Council for the Social Studies, Bulletin 76, 1985); L. D. McCANN, ed., *Heartland and Hinterland: A Geography of Canada* (Prentice-Hall of Canada, 1982); P. C. NEWMAN, "Three Centuries of the Hudson's Bay Company: Canada's Fur-Trading Empire," *National Geographic*, 172 (1987), 192–229; J. L. ROBINSON, *Concepts and Themes in the Regional Geography of Canada* (Vancouver: Talonbooks, 1983); M. C. STORRIE and C. I. JACKSON, "Canadian Environments," *Geographical Review*, 62 (1972), 309–332; J. WARKENTIN, ed., *Canada: A Geographical Interpretation* (Methuen, 1968; essays, mainly historical, by 23 geographers).

Canada: Ethnicity; Political Geography; French Canada I. M. BARLOW, "Political Geography and Canada's National Unity Problem," *Journal of Geography*, 79 (1980), 259–263; E. R. BLACK, *Divided Loyalties: Canadian Concepts of Federalism* (McGill-Queen's, 1975); R. BRETON *et al.*, *Cultural Boundaries and the Cohesion of Canada* (Montreal: Institute for Research on Public Policy, 1980); G. CALDWELL, "Discovering and Developing English-Canadian Nationalism in Quebec," *Canadian Review of Studies in Nationalism*, 11 (1984), 245–256; D. CARTWRIGHT, "Changes in the Patterns of Contact between Anglophones and Francophones in Quebec," *GeoJournal*, 8 (1984), 109–122; W. D. COLEMAN, *The Independence Movement in Quebec, 1945–1980* (Toronto, 1984); K. J. CROWE, "Why the New Names for Eskimos and Indians?" *Canadian Geographic*, 99, no. 1 (August–September 1979), 68–71; J. L. ELLIOTT, ed., *Two Nations, Many Cultures: Ethnic Groups in Canada* (Prentice-Hall of Canada, 1979); D. FULLERTON, "Whither the Capital? Ottawa's Symbolic Role as a Unifying Force in the Nation Is Becoming More Important Than Ever," *Canadian Geographic*, 107, no. 6 (December 1987), 7–19; J. C. GOYDER, "Ethnicity and Class Identity: The Case of French- and English-Speaking Canadians," *Ethnic and Racial Studies*, 6 (1983), 72–89; J. D. HARBRON, "Canada and Big Power Politics," *Canadian Geographic*, 101, no. 3 (June–July 1981), 14–22; R. C. HARRIS, *The Seigneurial System in Early Canada: A Geographical Study* (Wisconsin, 1966), also, "Brief Interlude with New France," *Geographical Magazine*, 52 (1980), 274–280; V. KONRAD, "Recurrent Symbols of Nationalism in Canada," *Canadian Geographer*, 30 (1986), 176–180; J. A. LAPONCE, "The French Language in Quebec: Tensions between Geography and Politics," *Political Geography Quarterly*, 3 (1984), 91–104; C. M. Mac-MILLAN, "Language Issues and Nationalism in Quebec," *Canadian Review of Studies in Nationalism*, 14 (1987), 229–245; G. S. MAHLER, *New Dimensions of Canadian Federalism: Canada in a Comparative Perspective* (Fairleigh Dickinson, 1987); A. H. MALCOLM, *The Canadians* (Times Books, 1985); K. McROBERTS, "The Sources of Neo-Nationalism in Quebec, *Ethnic and Racial Studies*, 7 (1984), 55–85; J. MEEKER, "Canada: Path to Constitution," *Focus*, 33, no. 2 (November–December 1982), 1–16; A. L. SANGUIN, "The Quebec Question and the Political Geography of Canada," *GeoJournal*, 8 (1984), 99–107; G. WYNN, "Ethnic Migrations and Atlantic Canada: Geographical Perspectives," *Canadian Ethnic Studies*, 18 (1986), 1–15.

Canada: Economy; Cities M. BARLOW and B. SLACK, "International Cities: Some Geographical Considerations and a Case Study of Montreal," *Geoforum*, 16 (1985), 333–345; A. BLACKBOURN and R. G. PUTNAM, *The Industrial Geography of Canada* (Croom Helm, 1984); R. BOULDING, "Forestry in Canada," *American Forests*, 85, no. 7 (July 1979), 38–43; M. F. FOX, "Regional Changes in Canadian Agriculture," *Geography*, 71 (1986), 67–70; M. A. GOLDBERG and J. MERCER, *The Myth of the North American City: Continentalism Challenged* (British Columbia, 1986); D. G. HAGLUND, "Canadian Strategic Minerals and United States Military Potential," *Journal of Canadian Studies*, 19, no. 3 (Fall 1984), 5–31; H. HARDIN, *A Nation Unaware: The Canadian Economic Culture* (Vancouver: J. J. Douglas, Ltd., 1974); H. F. HEALD, "A Hundred Years of Ever Better Harvests" [Canada], *Canadian Geographic*, 106, no. 3 (June–July 1986), 34–41; J. U. MARSHALL, "Industrial Diversification in the Canadian Urban System," *Canadian Geographer*, 24 (1981), 316–332; G. A. NADER, *Cities of Canada*, 2 vols. (Macmillan of Canada: vol. I, 1975; vol. II, 1976); G. A. STELTER and A. F. J. ARTIBISE, eds., *Power and Place: Canadian Urban Development in the North American Context* (British Columbia, 1986); D. F. WALKER, *Canada's Industrial Space-Economy* (Wiley, 1980).

Canada: Provinces and Regions F. BRUEMMER, "Churchill: Polar Bear Capital of the World," *Canadian Geographic*, 103, no. 6 (December 1983–January 1984), 20–27; W. BURGESS, "Recent Mining Developments in the Canadian Arctic," *Geography*, 68 (1983), 50–53; J. B. CANNON, "Explaining Regional Development in Atlantic Canada: A Review Essay," *Journal of Canadian Studies*, 19 (1984), 65–86; J. COULL, "Canada's Atlantic Fisheries: New Opportunities and Problems," *Geography*, 69 (1984), 353–356; G. DACKS, "The Politics of Development in Canada's North," *Current History*, 83 (1984), 220–224, 226; D. L. ELLIOTT-FISK, "The Stability of the Northern Canadian Tree Limit," *Annals AAG*, 73 (1983), 560–576; R. J. FLETCHER, "Settlement Sites along the Northwest Passage," *Geographical Review*, 68 (1978), 80–93; G. FRIE-

SEN, *The Canadian Prairies: A History* (Nebraska, 1984); R. GEORGE, "More Energy for B.C. in Peace River Coal," *Canadian Geographic,* 100, no. 1 (February–March 1980), 26–33; D. HUMPHRIES, "What Saskatchewan Heavy Oil May Do for Us," *Canadian Geographic,* 100, no. 6 (December 1980–January 1981), 46–51; S. KRECH, III, ed., *The Subarctic Fur Trade: Native Social and Economic Adaptations* (British Columbia, 1984); G. W. LEAHY, "Quebec: Our City of World Heritage Renown," *Canadian Geographic,* 106, no. 6 (December 1986–January 1987), 8–21; P. MARCHAK, *Green Gold: The Forest Industry in British Columbia* (British Columbia, 1983); H. MILLWARD, "The Development, Decline, and Revival of Mining on the Sydney Coalfield," *Canadian Geographer,* 28 (1984), 180–185; J. L. ROBINSON, "Sorting Out All the Mountains in British Columbia," *Canadian Geographic,* 107, no. 1 (February–March 1987), 42–53; J. R. ROGGE, ed., *Developing the Subarctic,* Manitoba Geographical Studies, 1 (1973); J. SCHREINER, "The Port of Vancouver," *Canadian Geographic,* 107, no. 4 (August–September 1987), 10–21; T. G. SMITH, "How Inuit Trapper-Hunters Make Ends Meet," *Canadian Geographic,* 99, no. 3 (December 1979–January 1980), 56–61; E. STRUZIK, "Sister Cities North of Sixty—Yellowknife and Whitehorse: Twin Capitols, but Cast in Quite Separate Moulds," *Canadian Geographic,* 106 (1986), 66–71; "The Canada Series" (McGraw-Hill Ryerson: short books on Canada's provinces, 1979); L. TROTIER, gen. ed., *Studies in Canadian Geography* (Toronto, 1972: vol. 1: A. MACPHERSON, ed., *The Atlantic Provinces;* vol. 2: F. GRENIER, ed., *Quebec;* vol. 3: R. L. GENTILCORE, ed., *Ontario;* vol. 4: P. J. SMITH, ed., *The Prairie Provinces;* vol. 5: J. L. ROBINSON, ed., *British Columbia;* vol. 6: W. C. Wonders, ed., *The North*); B. G. VANDERHILL, "The Passing of the Pioneer Fringe in Western Canada," *Geographical Review,* 72 (1982), 200–217; B. WALLACH, "Vancouver Island," *Focus,* 36 (1986), 18–21; G. WYNN, "A Province Too Much Dependent on New England" [Nova Scotia], *Canadian Geographer,* 31 (1987), 98–113.

United States: General J. AGNEW, *The United States in the World Economy: A Regional Geography* (Cambridge, 1987); J. GLASSBOROW et al., eds., *Atlas of the United States: A Thematic and Comparative Approach* (Macmillan, 1986); C. D. HARRIS, *Bibliography of Geography, Part II: Regional,* vol. 1, *The United States of America* (University of Chicago, Department of Geography Research Paper 206, 1984); J. F. HART, ed., *Regions of the United States* (Harper & Row, 1972, published originally as a special issue, *Annals AAG,* 62, no. 2, June 1972; regional essays by geographers, of which the following are especially pertinent to this text: R. W. Durrenberger, "The Colorado Plateau"; J. F. Hart, "The Middle West"; E. C. Mather, "The American Great Plains"; D. W. Meinig, "American Wests: Preface to a Geographical Introduction"; J. E. Vance, Jr.,

"California and the Search for the Ideal"); P. L. KNOX et al., *The United States: A Contemporary Human Geography* (Wiley, 1988); S. I. SCHWARTZ and R. E. EHRENBURG, *The Mapping of America* (Abrams, 1980); U.S. BUREAU OF THE CENSUS, *County and City Data Boook* (Washington, DC: Government Printing Office, 1983 and periodically); U. S. GEOLOGICAL SURVEY, *Maps for America: Cartographic Products of the U.S.G.S. and Others,* 2d ed. (Washington, DC: Government Printing Office, 1981), also, *The National Atlas of the United States of America* (Washington, DC: 1970); J. H. WHEELER, "U.S.A.," chap. 8 (pp. 131–178) in S. GODDARD, ed., *A Guide to Information Sources in the Geographical Sciences* (Croom Helm/Barnes & Noble, 1983; chap. 8 is an extensive bibliography of books and articles on the United States through 1981).

United States: Historical Geography T. H. BREEN, "An Empire of Goods: The Anglicization of Colonial America, 1690–1776," *Journal of British Studies,* 25 (1986), 467–499; R. H. BROWN, *Historical Geography of the United States* (Harcourt, Brace, 1948; still a major work); M. P. CONZEN, "The Woodland Clearances [United States]," *Geographical Magazine,* 52 (1980), 483–491; C. V. EARLE, "The First English Towns of North America," *Geographical Review,* 67 (1977), 34–50; "Fashioning the American Landscape," *Geographical Magazine,* 52–53 (1979–1980; 12 articles by various authors); J. T. LEMON, "Early Americans and Their Social Environment," *Journal of Historical Geography,* 6 (1980), 115–131; D. W. MEINIG, *The Shaping of America: A Geographical Perspective on 500 Years of American History,* vol. 1, *Atlantic America, 1492–1800* (Yale, 1986; a major work of historical geography); NATIONAL GEOGRAPHIC SOCIETY, *Historical Atlas of the United States* (1988); U.S. BUREAU OF THE CENSUS, *Historical Statistics of the United States: Colonial Times to 1970* (Washington, DC: Government Printing Office, 1975); D. WARD, ed., *Geographic Perspectives on America's Past: Readings on the Historical Geography of the United States* (Oxford, 1979).

United States: Natural Environment W. W. ATWOOD, *The Physiographic Provinces of North America* (Ginn, 1940); R. G. BAILEY, *Ecoregions of the United States* (map, 1:7,500,000, U.S. Forest Service, 1976), also, comp., *Description of the Ecoregions of the United States* (U. S. Forest Service, 1978); C. F. BENNETT, *Conservation and Management of Natural Resources in the United States* (Wiley, 1983); T. R. COX et al., *This Well-Wooded Land: Americans and Their Forests from Colonial Times to the Present* (Nebraska, 1985); N. M. FENNEMAN, *Physiography of Western United States* (McGraw-Hill, 1931), also, *Physiography of Eastern United States* (McGraw-Hill, 1938); C. B. HUNT, *Natural Regions of the United States and Canada,* rev. ed. (Freeman, 1947); T. R. KARL and A. J. KOSCIELNY, "Drought in the United States: 1895–1981," *Journal of Climatology*, 2 (1982), 313–329; D. M. LUDLUM, *The Weather Factor* (Houghton Mifflin, 1984; climatic events in American history); G. D. ROBINSON and A. M. SPIEKER, eds., *"Nature to be Commanded . . .": Earth-Science Maps Applied to Land and Water Management* (Washington, DC: Government Printing Office, 1978); W. D. THORNBURY, *Regional Geomorphology of the United States* (Wiley, 1965); U. S. DEPARTMENT OF AGRICULTURE, *Climate and Man: Yearbook of Agriculture, 1941* (Washington, DC: 1941; especially Part Two, "Climate and Agricultural Settlement"); J. L. VANKAT, *The Natural Vegetation of North America: An Introduction* (Wiley, 1979).

United States: Cultural Geography (General)
B. BIGELOW, "Roots and Regions: A Summary Definition of the Cultural Geography of America," *Journal of Geography*, 79 (1980), 218–229; J. CONRON, ed., *The American Landscape: A Critical Anthology of Prose and Poetry* (Oxford, 1974); M. CONZEN, "What Makes the American Landscape," *Geographical Magazine*, 53 (1980), 36–41; R. D. GASTIL, *Cultural Regions of the United States* (Washington, 1975); J. F. HART, "The Bypass Strip as an Ideal Landscape," *Geographical Review*, 72 (1982), 218–223; P. LEWIS, "Learning from Looking: Geographic and Other Writing about the American Cultural Landscape," *American Quarterly*, 35 (1983), 242–261; D. B. LUTEN, *Progress against Growth: On the American Landscape* (Guilford Press, 1986); J. R. SHORTRIDGE, "Changing Usage of Four American Regional Labels," *Annals AAG*, 77 (1987), 325–336; G. R. STEWART, *Names on the Land: A Historical Account of Placenaming in the United States* (4th ed., Lexikos, 1982), also, *A Concise Dictionary of American Place Names* (Oxford, 1986); W. ZELINSKY, "The Changing Face of Nationalism in the American Landscape," *Canadian Geographer*, 30 (1986), 171–175, also, *The Cultural Geography of the United States* (Prentice-Hall, 1973).

United States: Ethnicity
D. D. ARREOLA, "Urban Mexican Americans," *Focus*, 34, no. 3 (January–February 1984), 7–11; T. D. BOSWELL and T. C. JONES, "A Regionalization of Mexican Americans in the United States," *Geographical Review*, 70 (1980), 88–98, also, "Mexican American Housescapes," *Geographical Review*, 78 (1988), 299–315; T. D. BOSWELL and M. RIVERS, "Cubans in America: A Minority Group Comes of Age," *Focus*, 35, no. 2 (April 1985), 2–9; W. K. CROWLEY, "Old Order Amish Settlement: Diffusion and Growth," *Annals AAG*, 68 (1978), 249–264; D. K. FELLOWS, *A Mosaic of America's Ethnic Minorities* (Wiley, 1972); L. H. GANN and P. J. DUIGNAN, *The Hispanic in the United States: A History* (Westview, 1986); J. F. JAKUBS, "Recent Racial Segregation in U.S. SMSAs," *Urban Geography*, 7 (1986), 146–163; K. E. McHUGH, "Black Migration Reversal in the United States," *Geographical Review*, 77 (1987),

171–182; K. B. RAITZ, "Themes in the Cultural Geography of European Ethnic Groups in the United States," *Geographical Review*, 69 (1979), 45–54; J. R. SHORTRIDGE, "Patterns of Religion in the United States," *Geographical Review*, 66 (1976), 420–434; I. SUTTON, "Sovereign States and the Changing Definition of the Indian Reservation," *Geographical Review*, 66 (1976), 281–295; R. J. VECOLI, "Return to the Melting Pot: Ethnicity in the United States in the Eighties," *Journal of American Ethnic History*, 5 (1985), 7–20; D. WARD, "The Ethnic Ghetto in the United States: Past and Present," *Transactions IBG*, n.s., 7 (1982), 257–275.

United States: Political and Governmental
S. D. BRUNN, *Geography and Politics in America* (Harper & Row, 1974); M. CLAWSON, *The Federal Lands Revisited* (Washington, DC: Resources for the Future, 1983); J. W. HOUSE, ed., *United States Public Policy: A Geographical View* (Oxford, 1983); R. J. JOHNSTON, "The Changing Geography of Voting in the United States, 1946–1980," *Transactions IBG*, n.s., 7 (1982), 187–204; L. SMITH, "Redrawing the Line on Defense" (U.S. global commitments), *Fortune*, April 25, 1988, 267–280.

United States: General Economic Geography; Industrial Geography
M. BATEMAN and R. RILEY, eds., *The Geography of Defence* (Barnes & Noble, 1987); J. R. BORCHERT, "Major Control Points in American Economic Geography," *Annals AAG*, 68 (1978), 214–232; S. D. BRUNN, "Sunbelt USA," *Focus*, 36 (1986), 34–35; P. P. CHRISTENSEN, "Land Abundance and Cheap Horsepower in the Mechanization of the Antebellum United States Economy," *Explorations in Economic History*, 18 (1981), 309–329; D. CLARK, *Post-Industrial America: A Geographical Perspective* (Methuen, 1985); S. S. COHEN and J. ZYSMAN, "The Myth of a Post-Industrial Economy," *Technology Review*, 90, no. 2 (February–March 1987), 55–62; J. F. HART, "Small Towns and Manufacturing," *Geographical Review*, 78 (1988), 272–287; S. P. HUNTINGTON, "The U.S.—Decline or Renewal?" *Foreign Affairs*, 67, no. 2 (Winter 1988/89), 76–96; R. J. JOHNSTON, *The Geography of Federal Spending in the United States of America* (Wiley, 1980); L. KRAAR, "Japan's Gung-Ho U.S. Car Plants," *Fortune*, January 30, 1989, 98–108; R. E. LONSDALE and H. L. SEYLER, eds., *Nonmetropolitan Industrialization* (V. H. Winston, 1979); J. D. LORD, "Banking across State Lines," *Focus*, 37, no. 1 (Spring 1987), 10–15; R. B. McKENZIE, "Myths of Sunbelt and Frostbelt," *Policy Review*, 20 (1982), 103–114; M. MAGNET, "The Resurrection of the Rust Belt," *Fortune*, August 15, 1988, 40–46; A. MARKUSEN *et al.*, *High Tech America* (Allen & Unwin, 1986); D. R. MEYER, "Emergence of the American Manufacturing Belt: An Interpretation," *Journal of Historical Geography*, 9 (1983), 145–174; S. NASAR, "America's Competitive Revival," *Fortune*, January 4,

1988, 44–52; R. PEET, "Relations of Production and the Relocation of United States Manufacturing Industry since 1960," *Economic Geography*, 59 (1983), 112–143; J. M. RUBENSTEIN, "Changing Distribution of the American Automobile Industry," *Geographical Review*, 76 (1986), 288–300, also, "Further Changes in the American Automobile Industry," *ibid.*, 77 (1987), 359–362, also, "Changing Distribution of American Motor-Vehicle-Parts Suppliers," *ibid.*, 78 (1988), 288–298, and, "The Changing Distribution of U.S. Automobile Plants," *Focus*, 38, no. 3 (Fall 1988), 12–17; L. SAWERS and W. K. TABB, eds., *Sunbelt/Snowbelt: Urban Development and Regional Restructuring* (Oxford, 1984); R. SCHOENBARGER, "Foreign Manufacturing Investment in the United States: Competitive Strategies and International Location," *Economic Geography*, 61 (1985), 241–259; S. H. SCHURR *et al.*, *Energy in America's Future: The Choices before Us* (Johns Hopkins, 1979); J. SZEKELY, "Can Advanced Technology Save the U.S. Steel Industry?" *Scientific American*, 257 (1987), 34–41; "The Rhine and the Ohio: A Tale of Two Rivers," *The Economist*, May 21, 1988, 21–24; B. L. WEINSTEIN and R. E. FIRESTINE, *Regional Growth and Decline in the United States: The Rise of the Sunbelt and the Decline of the Northeast* (Praeger, 1978); C. H. WHITEHURST, JR., *The U.S. Shipbuilding Industry: Past, Present, and Future* (Naval Institute Press, 1986).

United States: Agriculture and Land Use S. S. BATIE, *Soil Erosion: Crisis in America's Croplands?* (Conservation Foundation, 1983); J. N. BELDEN, *Dirt Rich, Dirt Poor: America's Food and Farm Crisis* (Methuen, 1986); L. R. BROWN, "The Growing Grain Gap" [decline in world grain reserves], *World-Watch*, no. 5 (September–October 1988), 10–18; W. W. COCHRANE, *The Development of American Agriculture: A Historical Analysis* (Minnesota, 1979); C. EARLE and R. HOFFMAN, "The Foundation of the Modern Economy: Agriculture and the Costs of Labor in the United States and England, 1800–60," *American Historical Review*, 85 (1980), 1055–1094; W. EBELING, *The Fruited Plain: The Story of American Agriculture* (California, 1979); B. GIBBONS, "Do We Treat Our Soil Like Dirt?" *National Geographic*, 166 (1984), 350–389; H. F. GREGOR, *Industrialization of U.S. Agriculture: An Interpretive Atlas* (Westview, 1982); J. F. HART, "The Persistence of Family Farming Areas," *Journal of Geography*, 86 (1987), 198–203; R. G. HEALY and J. S. ROSENBERG, *Land Use and the States*, 2d ed. (Johns Hopkins, 1979); R. D. HURT, *The Dust Bowl: An Agricultural and Social History* (Nelson-Hall, 1981); R. H. JACKSON, *Land Use in America* (V.H. Winston, 1981); E. G. SMITH, JR., "America's Richest Farms and Ranches," *Annals AAG*, 70 (1980), 528–541; I. VOGELER, *The Myth of the Family Farm: Agribusiness Dominance of U.S. Agriculture* (Westview, 1981); M. D. WINSBERG, "Agricultural Specialization in the United States since World War II," *Agricultural History*, 56 (1982), 692–701.

United States: Urbanization C. ABBOTT, *The New Urban America: Growth and Politics in Sunbelt Cities* (North Carolina, 1987); R. F. ABLER, J. S. ADAMS, and K. S. LEE, eds., *A Comparative Atlas of America's Great Cities: Twenty Metropolitan Regions* (Minnesota, 1976); J. S. ADAMS, ed., *Contemporary Metropolitan America*, 4 vols.: 1, *Cities of the Nation's Historic Metropolitan Core*; 2, *Nineteenth Century Ports*; 3, *Nineteenth Century Inland Cities and Ports*; 4, *Twentieth Century Cities* (Ballinger, 1976); J. BORCHERT, "American Metropolitan Evolution," *Geographical Review*, 57 (1967), 301–332, a classic article of enduring value, also, "Instability in American Metropolitan Growth," *Geographical Review*, 73 (1983), 127–149; S. D. BRUNN and J. O. WHEELER, eds., *The American Metropolitan System: Present and Future* (Wiley, 1980); E. K. BURNS, "The Enduring Affluent Suburb," *Landscape*, 24, no. 1 (1980), 33–41; E. S. DUNN, JR., *The Development of the U.S. Urban System* (Johns Hopkins; vol. 1: 1980; vol. 2: 1983); J. GOTTMANN, "The Mutation of the American City: A Review of the Comparative Metropolitan Analysis Project," *Geographical Review*, 68 (1978), 201–208; J. JACOBS, *Cities and the Wealth of Nations: Principles of Economic Life* (Random House, 1984); R. J. JOHNSTON, *The American Urban System: A Geographical Perspective* (Longman, 1982); R. PALM, *The Geography of American Cities* (Oxford, 1981); U.S. BUREAU OF THE CENSUS, *State and Metropolitan Area Data Book, 1986: A Statistical Abstract Supplement* (Washington, DC: Government Printing Office, 1986); J. E. VANCE, JR., "Cities in the Shaping of the American Nation," *Journal of Geography*, 75 (1976); J. O. WHEELER, "Similarities in the Corporate Structure of American Cities," *Growth and Change*, 17 (1986), 13–21; M. YEATES and B. J. GARNER, *The North American City*, 3d ed. (Harper & Row, 1980).

U.S. Northeast G. W. CAREY, "New York: World Economy, Feudal Politics," *Focus*, 31, no. 4 (March–April 1981), 1–16; W. CRONON, *Changes in the Land: Indians, Colonists, and the Ecology of New England* (Hill & Wang, 1983); R. A. CYBRIWSKY and T. A. REINER, "Philadelphia in Transition," *Focus*, 33, no. 1 (September–October 1982), 1–16; J. E. DILISIO, *Maryland: A Geography* (Westview, 1983); C. V. EARLE, *The Evolution of a Tidewater Settlement System: All Hallow's Parish, Maryland, 1650–1783* (University of Chicago, Department of Geography Research Paper 170, 1975); E. P. ERICKSEN *et al.*, "The Cultivation of the Soil as a Moral Directive: Population Growth, Family Ties, and the Maintenance of Community among the Old Order Amish" [Lancaster County, Pennsylvania], *Rural Sociology*, 45 (1980), 49–68; F. FEGLEY, "Plain Pennsylvanians Who Keep Their Faith" [Pennsylvania Germans], *Geographical Magazine*, 53 (1981), 968–975; J. V. FIFER, "Washington, D.C.: The Political Geography of a Federal Capital," *Journal of American Studies*, 15 (1981), 5–26; R. T. T. FORMAN, ed., *Pine Barrens:*

Ecosystem and Landscape (Academic, 1979); J. GOTT-MANN, *Megalopolis: The Urbanized Northeastern Seaboard of the United States* (Twentieth Century Fund, 1961); C. L. HEYRMAN, *Commerce and Culture: The Maritime Communities of Colonial Massachusetts, 1690–1750* (Norton, 1984); E. C. HIGBEE, "The Three Earths of New England," *Geographical Review,* 42 (1952), 425–438; P. JACKSON, "Neighborhood Change in New York: The Loft Conversion Process," *Tijdschrift voor Economische en Sociale Geografie,* 76 (1985), 202–215; J. T. LEMON, *The Best Poor Man's Country: A Geographical Study of Early Southeastern Pennsylvania* (Johns Hopkins, 1972); T. R. LEWIS and J. E. HARMON, *Connecticut: A Geography* (Westview, 1986); D. R. McMANIS, *Colonial New England: A Historical Geography* (Oxford, 1975); B. MARSH, "Continuity and Decline in the Anthracite Towns of Pennsylvania," *Annals AAG,* 77 (1987), 337–352; S. E. MORISON, *The Maritime History of Massachusetts, 1783–1860* (Houghton Mifflin, 1921); H. S. RUSSELL, *A Long, Deep Furrow: Three Centuries of Farming in New England* (University Press of New England, 1976), also, *Indian New England before the Mayflower* (New England, 1980); G. SHANKLAND, "Boston—The Unlikely City," *Geographical Magazine,* 53 (1981), 323–327; G. STERNLIEB and J. W. HUGHES, *Revitalizing the Northeast: Prelude to an Agenda* (Center for Urban Policy Research, Rutgers University, 1978); R. L. SWANSON *et al.,* "Is the East River, New York, a River, or Long Island an Island?" *International Hydrographic Review,* 60 (1983), 127–157; J. H. THOMPSON, ed., *Geography of New York State,* 2d ed. (Syracuse, 1977); B. WARF, "Japanese Investments in the New York Metropolitan Region," *Geographical Review,* 78 (1988), 257–271; W. ZELINSKY, "The Pennsylvania Town: An Overdue Geographical Account," *Geographical Review,* 67 (1977), 127–147.

U. S. Midwest T. J. BAERWALD, "The Twin Cities," *Focus,* 36, no. 1 (Spring 1986), 10–15; B. H. BALTENSBERGER, *Nebraska: A Geography* (Westview, 1985); B. W. BLOUET and F. C. LUEBKE, eds., *The Great Plains: Environment and Culture* (Nebraska, 1979); B. W. BLOUET and M. P. LAWSON, eds., *Images of the Plains: The Role of Human Nature in Settlement* (Nebraska, 1975); M. B. BOGUE, "The Lake and the Fruit: The Making of Three Farm-Type Areas," *Agricultural History,* 59 (1985), 492–522; J. BORCHERT, *America's Northern Heartland* (Minnesota, 1987); C. E. COBB, JR., "The Great Lakes' Troubled Waters," *National Geographic,* 172, no. (July 1987), 2–31; I. CUTLER, *Chicago: Metropolis of the Mid-Continent,* 3d ed. (Kendall/Hunt, 1982); S. L. FLADER, ed., *The Great Lakes Forest: An Environmental and Social History* (Minnesota, 1983); I. FRAZIER, "Great Plains," I, II, III, a 3-part article, *The New Yorker,* February 20, February 27, March 6, 1989; R. L. GERLACH, *Immigrants in the Ozarks: A Study in Ethnic Geography* (Missouri, 1976), also, *Settlement Patterns in Missouri: A Study of Population Origins, with a Wall Map* (Missouri, 1986); F. HAPGOOD, "The Prodigious

Soybean," *National Geographic,* 172, no. 1 (July 1987), 66–91; J. F. HART, "Change in the Corn Belt," *Geographical Review,* 76 (1986), 51–72, also, "Two Corn Belt Farms" [Iowa], *Journal of Cultural Geography,* 5 (1984), 7–17; A. D. HORSLEY, *Illinois: A Geography* (Westview, 1986); J. C. HUDSON, "Great Plains, U.S.A.: The 1980s," in J. ROGGE, ed., *The Prairies and Plains: Prospects for the 80s,* Manitoba Geographical Studies, 7 (1981); H. B. JOHNSON, *Order Upon the Land: The U.S. Rectangular Land Survey and the Upper Mississippi Country* (Oxford, 1976); M. P. LAWSON and M. E. BAKER, eds., *The Great Plains: Perspectives and Prospects* (Nebraska, 1981); T. L. McKNIGHT, "Great Circles on the Great Plains: The Changing Geometry of American Agriculture," *Erdkunde,* 33 (1979), 70–79; P. F. MATTINGLY and G. APSBURY, "Devolution of Cotton from the Piedmont," *Tijdschrift voor Economische en Sociale Geografie,* 77 (1986), 197–204; H. M. MAYER and R. C. WADE, *Chicago: Growth of a Metropolis* (Chicago, 1973); "The Midwest" (special issue), *Journal of Geography,* 85, no. 5 (September–October 1986; seven articles, with extensive reference lists); M. D. RAFFERTY, *Missouri: A Geography* (Westview, 1983), also, *The Ozarks: Land and Life* (Oklahoma, 1980); N. J. ROSENBERG, "Climate of the Great Plains Region of the United States," *Great Plains Quarterly,* 7 (1987), 22–32; W. A. SCHROEDER, *The Eastern Ozarks: A Geographic Interpretation of the Rolla 1:250,000 Topographic Map* (National Council for Geographic Education, 1967); J. R. SHORTRIDGE, "Cowboy, Yeoman, Pawn and Hick" [U.S. Great Plains], *Focus,* 35 (1985), 22–27, also, "The Emergence of 'Middle West' as an American Regional Label," *Annals AAG,* 74 (1984), 209–220, and, "The Vernacular Middle West," *Annals AAG,* 75 (1985), 48–57; N. SMITH and W. DENNIS, "The Restructuring of Geographic Scale: Coalescence and Fragmentation of the Northern Core Region," *Economic Geography,* 63 (1987), 160–182; L. M. SOMMERS, *Michigan: A Geography* (Westview, 1984); I. VOGELER, *Wisconsin: A Geography* (Westview, 1986); B. WALLACH, "The Return of the Prairie," *Landscape,* 28, no. 3 (1985), 1–5; W. P. WEBB, *The Great Plains* (Ginn, 1931); D. WORSTER, *Dust Bowl: The Southern Plains in the 1930's* (Oxford, 1979).

U.S. South: General J. F. HART, *The South,* 2d ed. (Van Nostrand Reinhold, 1976); E. PESSEN, "How Different from Each Other Were the Antebellum North and South?" *American Historical Review,* 85 (1980), 1119–1149; M. C. PRUNTY, "Two American Souths: The Past and the Future," *Southeastern Geographer,* 17 (1977), 51–59.

U.S. South: Agriculture and Rural Life C. S. AIKEN, "New Settlement Pattern of Rural Blacks in the American South," *Geographical Review,* 75 (1985), 383–404; T. D. CLARK, *The Greening of the South: The Recovery of Land and Forest* (Kentucky, 1984); J. M. CLIFTON, "Jehossee Island: The Antebellum South's Largest Rice Plantation,"

Agricultural History, 59 (1985), 56–65; P. A. COCLANIS, "Bitter Harvest: The South Carolina Low Country in Historical Perspective," *Journal of Economic History*, 45 (1985), 251–259; A. E. COWDREY, *This Land, This South: An Environmental History* (Kentucky, 1983); P. DANIEL, "The Crossroads of Change: Cotton, Tobacco and Rice Cultures in the Twentieth-Century South," *Journal of Southern History*, 50 (1984), 429–456, also, "The Transformation of the Rural South: 1930 to the Present," *Agricultural History*, 55 (1981), 231–248; C. V. EARLE, "A Staple Interpretation of Slavery and Free Labor," *Geographical Review*, 68 (1978), 51–65; J. F. HART, "Cropland Concentrations in the South," *Annals AAG*, 68 (1978), 505–517, also, "Land Use Change in a Piedmont County," *Annals AAG*, 70 (1980), 492–527, and, "The Demise of King Cotton," *Annals AAG*, 67 (1977), 307–322; R. G. HEALY, *Competition for Land in the American South: Agriculture, Human Settlement, and the Environment* (Conservation Foundation, 1985); T. G. JORDAN, *Trails to Texas: Southern Roots of Western Cattle Ranching* (Nebraska, 1981); J. T. KIRBY, "The Transformation of Southern Plantations c. 1920–1960," *Agricultural History*, 57 (1983), 257–276; C. F. KOVACIK and R. E. MASON, "Changes in the South Carolina Sea Island Cotton Industry," *Southeastern Geographer*, 25 (1985), 77–104; F. McDONALD and G. McWHINEY, "The South from Self-Sufficiency to Peonage: An Interpretation," *American Historical Review*, 85 (1980), 1095–1118; I. R. MANNERS, "The Persistent Problem of the Boll Weevil: Pest Control in Principle and in Practice," *Geographical Review*, 69 (1979), 25–42; J. J. MOLNAR, *Agricultural Change: Consequences for Southern Farms and Rural Communities* (Westview, 1986); D. H. REMBERT, JR., "The Indigo of Commerce in Colonial North America," *Economic Botany*, 33 (1979), 128–134.

U.S. South: Cities and Industries B. A. BROWNWELL and D. R. GOLDFIELD, eds., *The City in Southern History: The Growth of Urban Civilization in the South* (Kennikat, 1977); J. R. FEAGIN, "The Global Context of Metropolitan Growth: Houston and the Oil Industry," *American Journal of Sociology*, 90 (1985), 1204–1230; D. R. GOLDFIELD, *Cotton Fields and Skyscrapers: Southern City and Region, 1607–1980* (Louisiana State, 1982), also, "The Urban South: A Regional Framework," *American Historical Review*, 86 (1981), 1009–1034; M. L. JOHNSON, "Postwar Industrial Development in the Southeast and the Pioneer Role of Labor-Intensive Industry," *Economic Geography*, 61 (1985), 46–65; L. H. LARSEN, *The Rise of the Urban South* (Kentucky, 1985); P. J. WOOD, *Southern Capitalism: The Political Economy of North Carolina, 1880–1980* (Duke, 1986).

U. S. South: People and Culture C. S. AIKEN, "Faulkner's Yoknapatawpha County: A Place in the American South," *Geographical Review*, 69 (1979), 331–348, also, "Faulkner's Yoknapatawpha County: Geographical Fact into Fiction," *Geographical Review*, 67 (1977), 1–21; D. P. ARREOLA, "The Mexican American Cultural Capital," *Geographical Review*, 77 (1987), 17–34; R. ARSENAULT, "The End of the Long Hot Summer: The Air Conditioner and Southern Culture," *Journal of Southern History*, 50 (1984), 597–628; R. BERTHOFF, "Celtic Mist over the South," *Journal of Southern History*, 52 (1986), 523–550; W. J. CASH, *The Mind of the South* (Knopf, 1941); L. E. ESTAVILLE, JR., "Mapping the Louisiana French," *Southeastern Geographer*, 26 (1986), 90–113; D. W. MEINIG, *Imperial Texas: An Interpretive Essay in Cultural Geography* (Texas, 1969); R. A. MOHL, "An Ethnic 'Boiling Pot': Cubans and Haitians in Miami," *Journal of Ethnic Studies* 13 (1985), 51–74; L. W. NEWTON, *The Americanization of French Louisiana: A Study of the Process of Adjustment between the French and the Anglo-American Populations of Louisiana, 1803–1860* (Arno, 1980); J. S. REED, *One South: An Ethnic Approach to Regional Culture* (Louisiana State, 1982); D. RIEFF, "The Second Havana" [Miami], *The New Yorker*, May 18, 1987, 65–83; S. A. SMITH, "The Old South Myth as a Contemporary Southern Commodity," *Journal of Popular Culture*, 16, no. 3 (Winter 1982), 22–29.

U. S. South: States and Regions M. J. BRADSHAW, "Public Policy in Appalachia: The Application of a Neglected Geographical Factor?", *Transactions IBG*, n.s., 10 (1985), 385–400, also, "TVA at Fifty," *Geography*, 69 (1984), 209–220; A. B. CRUICKSHANK, "Development of the Deep South: A Reappraisal," *Scottish Geographical Magazine*, 96 (1980), 91–104; J. F. HART, "Land Rotation in Appalachia," *Geographical Review*, 67 (1977), 148–166; C. M. HEAD and R. B. MARCUS, *The Face of Florida* (Kendall/Hunt, 1984); J. W. HOUSE, *Frontier on the Rio Grande: A Political Geography of Development and Social Deprivation* (Oxford, 1982); T. G. JORDAN et al., *Texas: A Geography* (Westview, 1984), also, JORDAN, "The Imprint of the Upper and Lower South on Mid-Nineteenth-Century Texas," *Annals AAG*, 57 (1967); P. P. KARAN, ed., *Kentucky: A Regional Geography* (Kendall/Hunt, 1973); C. F. KOVACIK and J. J. WINBERRY, *South Carolina: A Geography* (Westview, 1987); J. MASLOW, "Trade: Blues in the Gulf" [U.S. Gulf Coast seaports], *Atlantic Monthly*, May 1988, 25–31; K. B. RAITZ et al., *Appalachia: A Regional Geography: Land, People, and Development* (Westview, 1984); R. REINHOLD, "Texas in a Tailspin" [effects of oil-price decline], *New York Times Magazine*, July 20, 1986, 22–25+; D. E. WHISNANT, *Modernizing the Mountaineer: People, Power, and Planning in Appalachia* (Boone, NC: Appalachian Consortium Press, 1980); M. R. WOLFE, "Changing the Face of Southern Appalachia," *American Planning Association*, 47 (1981), 252–265.

U. S. West: General V. CARSTENSEN, "Making Use of the Frontier and the American West," *Western Historical*

Quarterly, 13 (1982), 5–16; B. DeVOTO, *The Course of Empire* (Houghton Mifflin, 1952); P. HORGAN, *The Heroic Triad* (Heinemann, 1974); W. H. GOETZMAN, *New Lands, New Men: America and the Second Great Age of Discovery* (Viking Penguin, 1986); R. D. LAMM and M. McCARTHY, *The Angry West: A Vulnerable Land and Its Future* (Houghton Mifflin, 1982); M. P. LAWSON and C. W. STOCKTON, "Desert Myth and Climatic Reality," *Annals AAG*, 71 (1981), 527–535; G. D. NASH, *The American West Transformed: The Impact of the Second World War* (Indiana, 1985); "Privatising America's West," *The Economist*, October 22, 1988, 21–24; M. REISNER, *Cadillac Desert: The American West and Its Disappearing Water* (Viking, 1986); E. G. SMITH, JR., "Changing Population Patterns in the American West," *Association of Pacific Coast Geographers Yearbook*, 45 (1983), 71–84; D. WORSTER, "New West, True West: Interpreting the Region's History," *Western Historical Quarterly*, 18 (1987), 141–156.

U. S. West: California E. S. BAKKER, *An Island Called California: An Ecological Introduction to Its Natural Communities* (California, 1984); S. L. BOTTLES, *Los Angeles and the Automobile: The Making of the Modern City* (California, 1987); L. M. DILSAVER, "After the Gold Rush," *Geographical Review*, 75 (1985), 1–18; M. W. DONLEY et al., *Atlas of California* (Portland, OR: Professional Book Center, 1979); R. W. DURRENBERGER and R. B. JOHNSON, *California: Patterns on the Land*, 5th ed. (Palo Alto, CA: Mayfield, 1976); R. L. GENTILCORE, "Missions and Mission Lands of Alta California," *Annals AAG* 51 (1961), 46–72; O. GRANGER, "Increasing Variability in California Precipitation," *Annals AAG*, 69 (1979), 533–543; D. HORNBECK, *California Patterns: A Geographical and Historical Atlas* (Palo Alto, CA: Mayfield, 1983); W. L. KAHRL, ed., *The California Water Atlas* (Sacramento, CA: The Governor's Office of Planning and Research, distributed by William Kaufmann, 1979; a magnificent oversized atlas in color); D. W. LANTIS et al., *California: Land of Contrast*, 3d ed. (Kendall/Hunt, 1981); E. LIEBMAN, *California Farmland: A History of Large Agricultural Landholdings* (Rowman & Allanheld, 1983); C. LOCKWOOD and C. B. LEINBERGER, "Los Angeles Comes of Age," *Atlantic Monthly*, January 1988, 31–57; T. L. McKNIGHT, "Center Pivot Irrigation in California," *Geographical Review*, 73 (1983), 1–14; J. McPHEE, "The Control of Nature: Los Angeles against the Mountains," I, II, *The New Yorker*, September 26, 1988, 45–78; October 3, 1988, 72–90; C. S. MILLER and R. S. HYSLOP, *California: The Geography of Diversity* (Palo Alto, CA: Mayfield, 1983); W. H. MILLER, "Where Is Southern California?" *California Geographer*, 22 (1982), 67–95; M. S. MUSOKE and A. L. OLMSTEAD, "The Rise of the Cotton Industry in California: A Comparative Perspective," *Journal of Economic History*, 42 (1982), 385–412; H. J. NELSON, *The Los Angeles Metropolis* (Kendall/Hunt, 1982); J. J. PARSONS, "A Geographer

Looks at the San Joaquin Valley," *Geographical Review*, 76 (1986), 371–389; T. L. PETERS, "Trends in California Viticulture," *Geographical Review*, 74 (1984), 455–467; J. N. RUTGER and D. M. BRANDON, "California Rice Culture," *Scientific American*, 244, no. 2 (February 1981), 42–51; A. J. SCOTT, "High Technology Industry and Territorial Development: The Rise of the Orange County Complex, 1955–1984," *Urban Geography*, 7 (1986), 3–45; M. SCOTT, *The San Francisco Bay Area: A Metropolis in Perspective* (2d ed.; California, 1985); R. STEINER, "Large Private Landholdings in California," *Geographical Review*, 72 (1982), 315–326; W. L. THOMAS, JR., ed., "Man, Time and Space in Southern California," *Annals AAG*, 49, Supplement (September, 1959; entire issue); B. WALLACH, "The West Side Oil Fields of California," *Geographical Review*, 70 (1980), 50–59; C. WILVERT, "San Diego/San Francisco—Sansan," *Geographical Magazine* 53 (1981), 268–272.

U. S. Interior West R. H. BROWN, *Wyoming: A Geography* (Westview, 1980); D. B. COLE and J. L. DIETZ, "The Changing Rocky Mountain Region," *Focus*, 34, no. 4 (March–April 1984), 1–11; I. G. CLARK, *Water in New Mexico: A History of Its Management and Use* (New Mexico, 1987); M. L. COMEAUX, *Arizona: A Geography* (Westview, 1981); J. DAVIS, "King Coal in Cattle Country," *Geographical Magazine* (1984), 368–373; P. L. FRADKIN, *A River No More: The Colorado River and the West* (Knopf, 1981); P. GOBER, "The Retirement Community as a Geographical Phenomenon: The Case of Sun City, Arizona," *Journal of Geography*, 84 (1985), 189–198; W. L. GRAF, "An American Stream" [the Colorado], *Geographical Magazine*, 59 (1987), 504–509; N. HUNDLEY, *Water and the West: The Colorado River Compact and the Politics of Water in the American West* (California, 1975); R. H. JACKSON, "Mormon Perception and Settlement," *Annals AAG*, 68 (1978), 317–334; B. LUCKINGHAM, "The American Southwest: An Urban View," *Western Historical Quarterly*, 15 (1984), 261–280; D. W. MEINIG, *Southwest: Three Peoples in Geographical Change, 1600–1970* (Oxford, 1971), also, "The Mormon Culture Region: Strategies and Patterns in the Geography of the American West, 1847–1964," *Annals AAG*, 55 (1965), 191–220; S. P. McLAUGHLIN, "Economic Prospects for New Crops in the Southwestern United States," *Economic Botany*, 39 (1985), 473–481; J. A. McPHEE, *Basin and Range* (Farrar, Straus, Giroux, 1981); T. MILLER, ed., *Arizona: The Land and the People* (Arizona, 1986); R. L. NOSTRAND, "The Hispano Homeland in 1900," *Annals AAG*, 70 (1980), 382–396; D. F. PETERSON and A. B. CRAWFORD, *Values and Choices in the Development of the Colorado River Basin* (Arizona, 1978); R. REDFERN, *Corridors of Time: 1,700,000,000 Years of Earth at Grand Canyon* (Times Books, 1980; a spectacular oversized volume in color); R. SYMANSKI, *Wild Horses and Sacred Cows* (Northland

Press, 1985); B. WALLACH, "Sheep Ranching in the Dry Corner of Wyoming," *Geographical Review,* 71 (1981), 51–63.

U.S. Pacific Northwest W. A. BOWEN, *The Willamette Valley: Migration and Settlement on the Oregon Frontier* (Washington, 1978); J. R. GIBSON, *Farming the Frontier: The Agricultural Opening of the Oregon Country, 1786–1846* (Washington, 1985); I. HAMILTON, "From Roses to Microchips" [Portland, Oregon], *Geographical Magazine,* 58 (1986), 356–361; A. J. KIMERLING and P. L. JACKSON, eds., *Atlas of the Pacific Northwest* (7th ed.; Oregon State, 1985); W. G. LOY and S. ALLAN, *Atlas of Oregon* (Oregon, 1976); G. MACINKO, "The Ebb and Flow of Wheat Farming in the Big Bend, Washington," *Agricultural History,* 59 (1985), 215–228; A. C. McGREGOR, *Counting Sheep: From Open Range to Agribusiness on the Columbia Plateau* (Washington, 1982); D. W. MEINIG, *The Great Columbia Plain: A Historical Geography, 1805–1910* (Washington, 1968); J. W. SCOTT and R. L. DELORME, *Historical Atlas of Washington* (Oklahoma, 1988); G. SWAN, "The Beautiful and Dammed" [Columbia River], *Geographical Magazine,* 60, no. 8 (August 1988), 35–40.

Alaska D. T. KRESGE, T. MOREHOUSE, and G. W. ROGERS, *Issues in Alaska Development* (Washington, 1977); D. F. LYNCH *et al.,* "Alaska: Land and Resource Issues," *Focus,* 31, no. 3 (January–February, 1981), 1–16; W. MATELL, ed., "Alaska's Forest Resources," *Alaska Geo-graphic,* 12, no. 2 (1985), 1–199; J. McPHEE, *Coming into the Country* [Alaska] (Farrar-Strauss-Giroux, 1977); T. A. MOREHOUSE, ed., *Alaskan Resources Development: Issues of the 1980s* (Westview, 1984); T. A. MOREHOUSE and L. LEASK, "Alaska's North Slope Borough: Oil, Money and Eskimo Self-Government," *Polar Record,* 20, no. 124 (January 1980), 19–29; J. B. RAY, "Selection of the Marine Terminal for the Trans-Alaska Pipeline," *Journal of Geography,* 78 (1979), 147–151; J. REARDEN, ed., "Alaska's Salmon Fisheries," *Alaska Geographic,* 10, no. 3 (1983), 1–123; B. RICHARDS, "Alaska's Southeast: A Place Apart," *National Geographic,* 165, no. 1 (January 1984), 50–87; J. R. SHORTRIDGE, "The Collapse of Frontier Farming in Alaska," 66 (1976), 583–604.

Hawaii J. R. MORGAN *et al., Hawaii: A Geography* (Westview, 1983); E. C. NORDYKE, *The Peopling of Hawaii* (Hawaii, 1977); T. J. OSBORNE, "Trade or War? America's Annexation of Hawaii Reconsidered," *Pacific Historical Review,* 50 (1981), 285–307; UNIVERSITY OF HAWAII AT MANOA, DEPARTMENT OF GEOGRAPHY, *Atlas of Hawaii,* 2d ed. (Hawaii, 1983; a handsome atlas in color).

Greenland A. J. TAAGHOLD, "Greenland and the Future," *Environmental Conservation,* 7 (1980), 295–300, also, "Greenland's Future Development: A Historical and Political Perspective," *Polar Record,* 21, no. 130 (January 1982), 23–32.

Photo Credits

Photographs not listed were taken by the senior author, Jesse H. Wheeler, Jr.

Chapter 1
Chapter Opening Colour Library Books, Ltd.

Chapter 2
Chapter Opening RON SANFORD/Black Star Figure 2.6 GORDON WILTSIE Figure 2.7 F. GOHIER/Photo Researchers Figure 2.8 M. BOULTON/Photo Researchers Figure 2.11 B. BACHMAN/Photo Researchers Figure 2.20 GIULIO J. BARBERO

Chapter 3
Chapter Opening GEORGE SHELLEY Figure 3.4 P. TURNLEY/Black Star Figure 3.5 *Tass* from Sovfoto Figure 3.6 NICKELSBURG/Gamma Liaison Agency Figure 3.7 KATRINKA EBBE Figure 3.8 U.S.D.A.

Chapter 4
Chapter Opening P. GUERRINI/Gamma Liaison Agency Figure 4.2 N. KOMINE/Photo Researchers

Chapter 5
Chapter Opening P. KOCH/Photo Researchers Figure 5.6 K. MURRAY/Photo Researchers Figure 5.8 J. SUTTON/Gamma Liaison Agency

Chapter 6
Chapter Opening Colour Library Books, Ltd. Figure 6.10 J. ALLEN CASH/*Time Magazine* Figure 6.11 P. JORDAN/Gamma Liaison Agency

Chapter 7
Chapter Opening PERRIN/Gamma Liaison Agency Figure 7.5 P. KOCH/Photo Researchers Figure 7.8 B. HAYES/Photo Researchers

Chapter 8
Chapter Opening AeroCamera Hoffmeister Figure 8.2 R. NORMAN MATHENY/*Christian Science Monitor* Figure 8.3 LEE FOSTER Figure 8.5 Austrian Information Service Figure 8.6 Swiss National Tourist Office

Chapter 9
Chapter Opening ARDELLA REED Figure 9.2 P. KOCH/Photo Researchers Figure 9.3 Norwegian Information Service Figure 9.5 P. JORDAN/Gamma Liaison Agency Figure 9.6 LEN KAUFMAN Figure 9.8 JAY FREIS Figure 9.9 TOSHI Figure 9.10 MIGUEL/The Image Bank Figure 9.13 H. TODD STRADFORD, JR. Figure 9.14 H. TODD STRADFORD, JR.

Chapter 10
Chapter Opening *Tass* from Sovfoto Figure 10.3 *Tass* from Sovfoto Figure 10.4 *Tass* from Sovfoto Figure 10.5 *Soviet Life* Figure 10.6 *Tass* from Sovfoto Figure 10.11 KATRINKA EBBE

Chapter 11
Chapter Opening *Tass* from Sovfoto Figure 11.3 *Tass* from Sovfoto Figure 11.5 *Tass* from Sovfoto Figure 11.6 KATRINKA EBBE Figure 11.7 *Soviet Life* Figure 11.8 *Tass* from Sovfoto Figure 11.9 GERD LUDWIG/Woodfin Camp & Associates Figure 11.11 *Tass* from Sovfoto Figure 11.12 *Tass* from Sovfoto Figure 11.13 H. TODD STRADFORD, JR. Figure 11.14 *Tass* from Sovfoto Figure 11.15 Novosti Press Agency

Chapter 12
Chapter Opening R. AZZI/Woodfin Camp & Associates Figure 12.5 Exxon Figure 12.6 IVERSON/Gamma Liaison Agency

Chapter 13
Chapter Opening Porterfield/Chickering Figure 13.2 J. RAJS/The Image Bank Figure 13.7 A. NOGUÉS/Sygma Figure 13.9 R. & S. MICHAUD/Woodfin Camp & Associates Figure 13.11 GAAL/Gamma Liaison Agency

Chapter 14
Chapter Opening H. TODD STRADFORD, JR. Figure 14.3 H. TODD STRADFORD, JR. Figure 14.4 LOVINA EBBE Figure 14.5 H. TODD STRADFORD, JR. Figure 14.6 JOSEPH H. ASTROTH, JR. Figure 14.7 B. BRAKE/Photo Researchers

Chapter 15
Chapter Opening ROBERT HOLMES Figure 15.4 GALEN ROWELL/Mountain Light Figure 15.5 FOTO

Index

Note: Page numbers in *italics* refer to figures. Page numbers followed by a "t" refer to tables.